HOLE'S ESSENTIALS OF HUMAN ANATOMY & PHYSIOLOGY

Fourteenth Edition

Charles J. Welsh
DUQUESNE UNIVERSITY

CONTRIBUTOR

Cynthia Prentice-Craver
CHEMEKETA COMMUNITY COLLEGE

DIGITAL AUTHORS

Leslie Day
NORTHEASTERN UNIVERSITY

Julie Pilcher
UNIVERSITY OF SOUTHERN INDIANA

PREVIOUS EDITION AUTHORS

David Shier
WASHTENAW COMMUNITY COLLEGE

Jackie Butler
GRAYSON COLLEGE

Ricki Lewis
ALBANY MEDICAL COLLEGE

HOLE'S ESSENTIALS OF HUMAN ANATOMY & PHYSIOLOGY, FOURTEENTH EDITION

Some ancillaries, including electronic and print components, may not be available to customers outside the United States.

This book is printed on acid-free paper.

1 2 3 4 5 6 7 8 9 LWI 24 23 22 21 20

ISBN 978-1-260-25134-0 (bound edition)
MHID 1-260-25134-9 (bound edition)
ISBN 978-1-260-42595-6 (loose-leaf edition)
MHID 1-260-42595-9 (loose-leaf edition)

Portfolio Manager: *Matt Garcia*
Product Developer: *Krystal Faust*
Marketing Manager: *Valerie L. Kramer*
Content Project Managers: *Ann Courtney/Brent dela Cruz*
Buyer: *Sandra Ludovissy*
Designer: *David W. Hash*
Content Licensing Specialist: *Beth Cray*
Cover Image: *©LightField Studios/Shutterstock*
Compositor: *MPS Limited*

Library of Congress Cataloging-in-Publication Data

Welsh, Charles J., author. | Shier, David. Hole's Essentials of Human Anatomy & Physiology.
Hole's Essentials of Human Anatomy & Physiology / Charles J. Welsh, Duquesne University;
contributor, Cynthia Prentice-Craver, Chemeketa Community College.
Hole's essentials of human anatomy and physiology
Fourteenth edition. | Dubuque : McGraw-Hill Education, 2021. |
Revised edition of: Hole's Essentials of Human Anatomy & Physiology /
David Shier, Jackie Butler, Ricki Lewis. Thirteenth edition. 2018.
LCCN 2019039476 | ISBN 9781260251340 (hardcover) | ISBN 9781260425895 (ebook)
LCSH: Human physiology. | Human anatomy.
LCC QP34.5 .S49 2020 | DDC 612—dc23
LC record available at https://lccn.loc.gov/2019039476

The Internet addresses listed in the text were accurate at the time of publication. The inclusion of a website does not indicate an endorsement by the authors or McGraw-Hill Education, and McGraw-Hill Education does not guarantee the accuracy of the information presented at these sites.

mheducation.com/highered

BRIEF CONTENTS

UNIT 1
LEVELS OF ORGANIZATION

UNIT 2
SUPPORT AND MOVEMENT

UNIT 3
INTEGRATION AND COORDINATION

UNIT 4
TRANSPORT

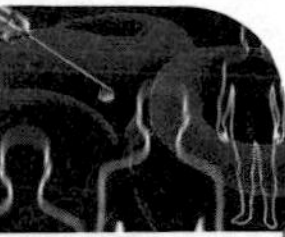

UNIT 5
ABSORPTION AND EXCRETION

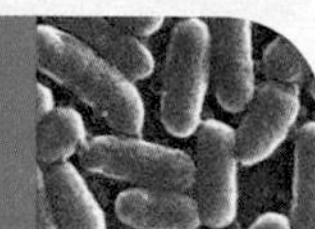

UNIT 6
THE HUMAN LIFE CYCLE

ABOUT THE AUTHORS

Courtesy of Leeanna Smith

CHARLES J. WELSH began his Anatomy & Physiology teaching career upon graduating with a B.S. in Biology from the University of Pittsburgh in 1989. He entered graduate school in 1992 and continued teaching night classes. He accepted his first full-time teaching position at Clarion University of Pennsylvania in 1996. In 1997, he completed his Ph.D. in Comparative Anatomy, Evolutionary Biology, and Ornithology at the University of Pittsburgh. Teaching primarily in nursing and other allied health programs, he now brings his thirty years of classroom experience to the fourteenth edition of *Hole's Essentials of Human Anatomy & Physiology.* Since 2009, he has been teaching at Duquesne University in Pittsburgh, Pennsylvania. During this time, he has received several teaching awards, as well as the Mentor of the Year Award for training graduate students to teach Anatomy & Physiology. Chuck and his wife, Lori, have three children and three grandchildren. They live in the historic town of Harmony, thirty miles north of Pittsburgh, with their youngest son, where they raise chickens and have a huge garden.

CONTRIBUTOR

Cindy Prentice-Craver

CYNTHIA PRENTICE-CRAVER has been teaching human anatomy and physiology for over twenty years at Chemeketa Community College and is a member of the Human Anatomy and Physiology Society (HAPS). Cynthia's teaching experience both in grades 6–12 and in college, her passion for curriculum development, and her appetite for learning fuel her desire to write. Her M.S. in Curriculum and Instruction, B.S. in Exercise Science, and extended graduate course-work in biological sciences have been instrumental in achieving effective results in the online and on campus courses she teaches. Cynthia co-authored the Martin *Laboratory Manual for Human Anatomy & Physiology,* 4e. Beyond her professional pursuits, Cynthia's passions include reading and listening to books, attending exercise classes, walking outdoors, attending concerts, traveling, and spending time with her family.

DIGITAL AUTHORS

Leslie Day

LESLIE DAY earned her B.S. in Exercise Physiology from UMass Lowell, an M.S. in Applied Anatomy & Physiology from Boston University, and a Ph.D. in Biology from Northeastern University. She currently works for Texas A&M University in the College of Medicine teaching Anatomy and Neuroanatomy to dual major medical and engineering students. Leslie has won several university and national awards for her teaching, including the ADInstruments Sam Drogo Technology in the Classroom Award from the Human Anatomy and Physiology Society (HAPS). Her current research focuses on the effectiveness of technology and pedagogical approaches in an anatomy-based curriculum. She brings her love for anatomy and willingness to try new technology in the classroom, both in person and online, to make for a dynamic evidence-based teaching style that is inclusive for all students. She is excited to bring this approach to the digital content for this book.

Courtesy of Gary Pilcher

JULIE PILCHER began teaching during her graduate training in Biomedical Sciences at Wright State University, Dayton, Ohio. She found, to her surprise, that working as a teaching assistant held her interest more than her research. Upon completion of her Ph.D. in 1986, she embarked on her teaching career, working for many years as an adjunct in a variety of schools as she raised her four children. In 1998, she began teaching full-time at the University of Southern Indiana, Evansville. Her work with McGraw-Hill began with doing reviews of textbook chapters and lab manuals and in content development for LearnSmart. In her A&P course at USI, she used Connect and enjoyed the challenge of writing some of her own assignments. She later accepted the opportunity to be more involved in the authoring of digital content for McGraw-Hill, understanding the importance of such content to both the instructors and the students.

DEDICATION

To my wife, Lori, our three children, Leeanna, Timothy, and Brady, and our three grandchildren, Milla, Holden, and Carolina, for the love and joy they bring me.

To the memory of my parents, Margaret Susan and Herman Joseph Welsh, for their love and support of all my passions.

Also to the memory of Dr. Robert J. Raikow, my mentor in graduate school. He saw the educator and scientist in me long before most. A true gentleman and scholar, his wisdom abounds throughout the pages of this book.

ACKNOWLEDGMENTS

I am honored and privileged to have directed the revision of this book that is based upon the hard work, efforts, and expertise of the previous authors: David Shier, Jackie Butler, Ricki Lewis, and John Hole, the original author of this classic work. I especially thank David Shier for his time and consultation during the revision. A project of this magnitude also requires the recognition of a large, dedicated, and talented team. I would like to thank the editorial team of Matt Garcia, Krystal Faust, and Michael Koot for their unwavering support and belief in my ability; marketing team Jim Connelly, Valerie Kramer, and Krissy Rellihan; and the production team of Ann Courtney, Sandy Ludovissy, David Hash, Beth Cray, and Brent dela Cruz. A thank you also goes to copyeditor Mike McGee and proofreaders Jennifer Grubba and Sharon O'Donnell for helping improve this work, and much thanks to Cindy Prentice-Craver for contributing considerably to the book with her keen eye and sharp pen. Provost Dr. David Dausey and the Dean, Dr. Philip Reeder of Duquesne University are thanked for their vision and support in granting me a leave during the preparation of the manuscript. Mark Cheskey and Glenn Sauer, my best friends since my formative years, are thanked for many hours of playing guitars and discussing music and sports when I needed such distractions. Most importantly, I thank my wife, Lori, for her love, patience, and more than thirty years of support for all of my academic endeavors.

REVIEWERS AND CONTRIBUTORS

A special thanks for the valuable contributions of all the professors, and their students, who have provided detailed recommendations for improving chapter content and illustrations, as well as suggestions regarding the development of ancillary resources for this new edition. They have played a vital role in building a solid foundation for *Hole's Essentials of Human Anatomy & Physiology.*

Reviewers

Michelle Barr
California State University, Fullerton

Janet Brodsky
Ivy Tech Community College

Nickolas Butkevich
Schoolcraft College

Jonathon P. Cohen
Kankakee Community College

Steven B. Hammer
Indian River State College

Mark Jaffe
Nova Southeastern University

Dean V. Lauritzen
City College of San Francisco

Marybeth Linse
Harper College

Mindy M. Murray
San Jacinto College–Central Campus

Rosser Panggat
Hartnell College

Amber D Ruskell-Lamer
Southeastern Community College

Amber M. Samuel
Lansing Community College & Macomb Community College

Peter C. Sayles
North Country Community College

Robyn York
California State University, Fullerton

Delon Washo-Krupps
Arizona State University

A NOTE FROM THE AUTHOR

To the Student

Welcome! As you read this (with your eyes) and understand it (with your brain), perhaps turning to the next page (with muscle actions of your fingers, hand, forearm, and arm), you are using your body to do so. Indeed, some of you may be using your fingers, hand, forearm, and arm to read through the eBook on your computer, tablet, or smartphone. The structure and function of the human body can be complex, and comprehending the material might not always seem easy. But what could be more fascinating than learning about your own body? To assist your learning, the fourteenth edition of *Hole's Essentials of Human Anatomy & Physiology* continues the tradition of presenting material in a conversational, accessible style.

Many of you are on a path toward a career in health care, athletics, science, or education. If you have not yet committed to a particular area of study, be sure to check out the Career Corner in every chapter for ideas and inspiration. They present interesting options for future careers. Balancing family, work, and academics is challenging, but try to look at this course not as a hurdle along your way but as a stepping stone. The book has been written to help you succeed in your coursework and to help prepare you in your journey to a successful and rewarding career.

To the Teacher

Written for ease of readability and organized for classroom use, this text serves the student as well as the instructor. This fourteenth edition of *Hole's Essentials of Anatomy & Physiology* continues the Learn, Practice, Assess approach that has substantially contributed to instructional efficiency and student success.

Each chapter opens with Learning Outcomes, contains many opportunities to Practice throughout, and closes with Assessments that are closely tied to the Learning Outcomes. Instructors can assign these, and students can use these features not only to focus their study efforts, but also to take an active role in monitoring their own progress toward mastering the material. All of these resources are described in more detail in the Chapter Preview / Foundations for Success beginning on page 1. In addition, thanks to the expertise of Leslie Day and Julie Pilcher, the LearnSmart and Connect digital platforms continue to enhance the printed content and the Learn, Practice, Assess approach. We are proud to have developed and to offer the latest and most efficient technologies to support teaching and learning.

Chuck Welsh

NEW TO THIS EDITION

Global Changes

Chapter openers: Featuring new and contemporary chapter-opening vignettes that are relevant to today's students.

Chapter introductions: Revised for ease of readability and to capture student attention.

Chapter Assess Integrative Assessments/Critical Thinking: Questions added to every chapter.

Chapter structure: Streamlined section structures.

Facts of Life boxes: Are now titled "Of Interest."

Boxed material: Small boxes have been integrated into the text for better flow or have been transformed into Clinical Application boxes.

Art: Revised colors, and placement of colors, to create better contrast.

Learning and Practice Outcomes: Feature updated numbering structures to help streamline outcomes within the section.

Specific Chapter Changes

Chapter 1	Revised opening vignette.
	Figure 1.2 revised to show sense of hearing.
	Figure 1.3 revised to show levels of organization using the cardiovascular system.
	Figure 1.6 changed for ease of understanding.
Chapter 2	New opening vignette about gluten.
	Reorganized section on bonding.
	Figure 2.3 revised for better understanding.
	Figure 2.7 revised for better understanding.
	Reorganized and revised section on acids and bases.
	Rewrote section on pH.
Chapter 3	Figure 3.3 changed to better show various membrane proteins.
	Reorganized discussion of organelles.
	Reorganized and revised sections on membrane transport for clarity: diffusion, facilitated diffusion, osmosis, tonicity, and active transport.
	Figure 3.19 revised to show both pinocytosis and phagocytosis.
	Reorganized section on cell division.
Chapter 4	Figure 4.1 now gives a better explanation of anabolic versus catabolic reactions.
	Figures 4.2, 4.4, and 4.5 all now have hydrolysis and dehydration synthesis labels.
	Figure 4.9 shows the coupling of the anabolic and catabolic components of cellular respiration.
	Figure 4.10 edited for clarity.
Chapter 5	Reorganized Epithelial Tissues section.
	Rewrote section on glands.
	Figure 5.14 changed for clarity.
	Reorganized Connective Tissues section.
	Reorganized Types of Membranes section.
	Reorganized Muscle Tissues section.
	Rewrote Nervous Tissue section.
Chapter 6	New opening vignette about tattoos.
	Reorganized section on epidermis.
	Reorganized section on skin functions.
Chapter 7	New opening vignette about forensic skeletal analysis.
	Reorganized Bone Development, Growth, and Repair section.
	Reorganized sections on axial and appendicular skeleton.
Chapter 8	Rewrote section on muscle fatigue.
	Added discussion of fast versus slow muscle fibers.

continued next page–

Specific Chapter Changes—*Continued*

	Added discussion of isotonic versus isometric contraction.
	Revised Figure 8.13 for clarity and better color contrast.
	Reorganized Skeletal Muscle Actions section.
	Revised Figures 8.14 and 8.15 for clarity.
Chapter 9	New opening vignette about CTE.
	Figure 9.1 new to show nervous system input, integration, and output.
	Reorganized much of chapter for better flow and use in classroom.
	Reorganized section on neurons and neuroglia.
	Rewrote section on neuron structure.
	Added discussion of satellite cells.
	Figure 9.3 revised to show neurilemma.
	Rewrote sections on membrane potential and action potential for brevity and clarity.
	Figure 9.16 new to show graphs for excitation and inhibition.
Chapter 10	New opening vignette about cochlear implants.
	Added discussion of labeled line principle.
	Added section on proprioception and baroreceptors.
	Figure 10.2 new to show muscle spindles.
	Combined old Figure 10.2 with Figure 10.3 for clarity.
	Figure 10.24 new to show and explain nearsightedness and farsightedness.
Chapter 11	Reorganized chapter.
	Switched Figures 11.1 and 11.2 for better flow.
	Changed Figure 11.17 for clarity on the actions of insulin and glucagon.
	Made separate sections for the pineal and thymus glands.
Chapter 12	New opening vignette about blood doping.
	Revised sections on formed elements and red blood cell production.
	Figure 12.4 revised for clarity.
	Added new table for types of anemia.
	Figure 12.5 revised for clarity.
	Figure 12.7 revised for clarity.
	Revised section on white blood cell function.
	Revised sections on platelets and coagulation to include clinical relevance.
	Revised Figures 12.8, 12.9, and 12.20 for better coloring and clarity.
	Table 12.5 updated.
Chapter 13	Figure 13.2 new to add more detail.
	Section on blood flow through the heart, lungs, and tissues revised for clarity.
	Figure 13.6 now includes schematic of blood flow.
	Sections on heart actions reorganized and revised for clarity.
	Figure 13.23 changed to show more detail.
	Figure 13.24 expanded to show more detail.
Chapter 14	Reorganized chapter.
	Revised discussion on the complement system.
	Revised discussion of fever.
Chapter 15	Revised discussion of the teeth.
	Revised discussion of the pharynx and esophagus.
	Figure 15.12 now shows action of HCl on pepsinogen.
	Revised discussion of the liver and gallbladder.

Specific Chapter Changes—*Continued*

	Figure 15.17 now has hepatic triad labeled.
	Figure 15.18 now has schematic showing the blood flow into and out of a liver lobule.
Chapter 16	New opening vignette about the effects of cigarette smoke, and electronic cigarettes and vaping.
	Figure 16.1 edited for more detail.
	Revised discussion of pleural cavity.
	Figure 16.12 changed for more detail and clarity.
	Revised discussion of spirometry.
	Revised discussion of factors affecting breathing.
	Figure 16.17 changed to show neurological control over breathing.
	Figure 16.22 and 16.23 consolidated and revised to better show relationship between the bicarbonate buffering system, plasma, and red blood cells.
Chapter 17	Figure 17.1 revised to show transverse section.
	Reorganized and rewrote section on kidney structure for clarity.
	Section on renal blood supply revised for clarity.
	Figure 17.3 revised for clarity and to show two nephron types.
	Figure added to show renal blood flow.
	Section on nephron structure revised, along with associated Figure 17.7.
	Rewrote section on urine formation for clarity.
	Revised sections on glomerular filtration, filtration rate, and tubular reabsorption for clarity.
	Revised sections on the urethra and micturition and added discussion of incontinence.
Chapter 18	Revised discussion of body fluid composition.
	Edited label on Figure 18.1 for accuracy.
	Edited label on Figure 18.4.
	Revised introduction to electrolyte balance.
	Reorganized acid-base balance section.
Chapter 19	Revised Figure 19.2 for clarity and more detail.
	Reorganized section on organs of male reproductive system.
	Created new section on spermatogenesis.
	Created new section on oogenesis and the ovarian cycle.
	Figure 19.10 new for clarity and more detail.
Chapter 20	Created new section Aging: The Human Life Span.
	Reorganized Genetics section.
	Revised discussion of dominant and recessive inheritance.
	Figure 20.19 new to better show dominant and recessive inheritance.
	Created section on extensions to Mendelian inheritance to include discussions of pleiotropy, co-dominance, incomplete dominance, and multiple alleles.
	Expanded discussions of chromosomal inheritance and polygenic inheritance.
	Added figures to show inheritance of sex and trisomy 21.

DYNAMIC ART PROGRAM

Art is vibrant, three-dimensional, and instructional. The authors examined every piece to ensure it was engaging and accurate. The fourteenth edition's art program will help students understand the key concepts of anatomy and physiology.

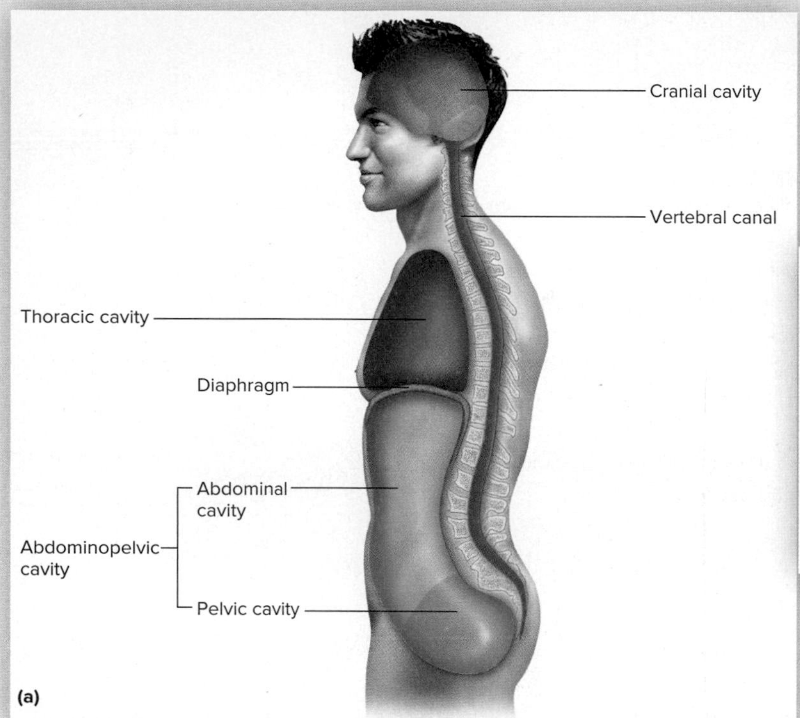

Realistic, three-dimensional figures provide depth and orientation.

Line art for micrographs is three-dimensional to help students visualize more than just the flat microscopic sample.

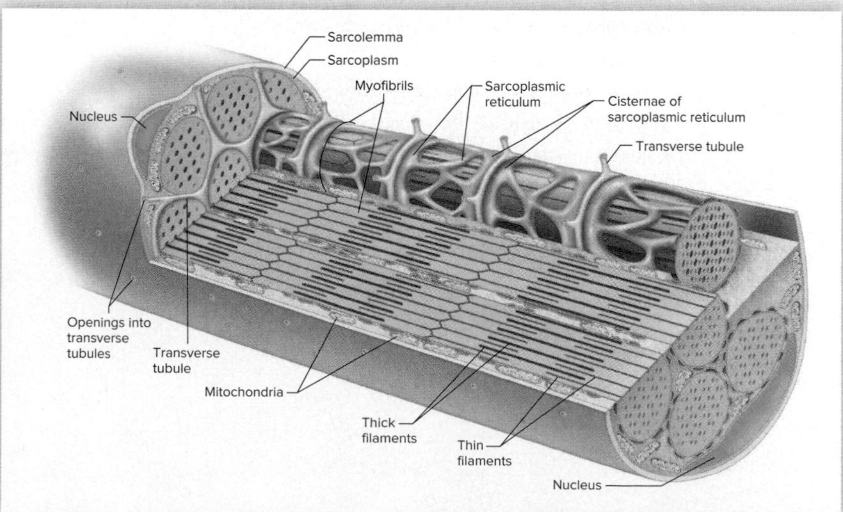

A longitudinal section shows the interior structures of a muscle fiber and reveals detail of the myofibrils, thick and thin filaments.

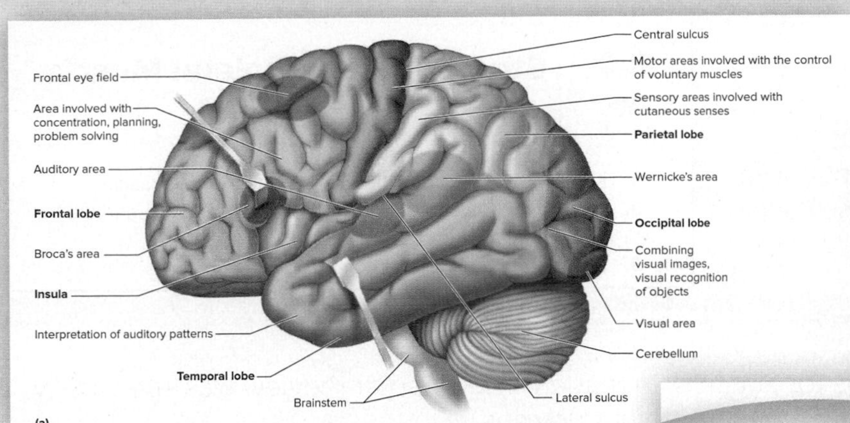

Colors readily distinguish functional areas.

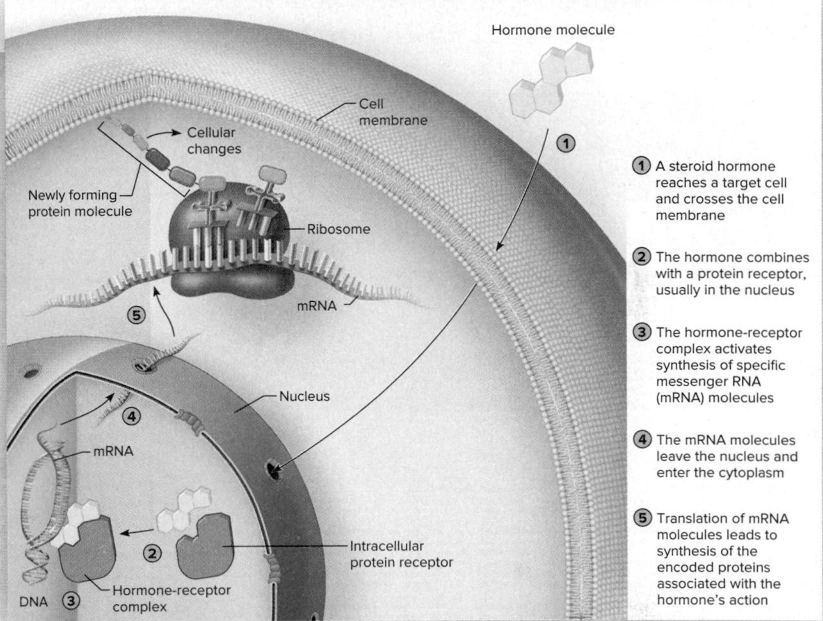

The explanation is part of the figure, not lost in the legend.

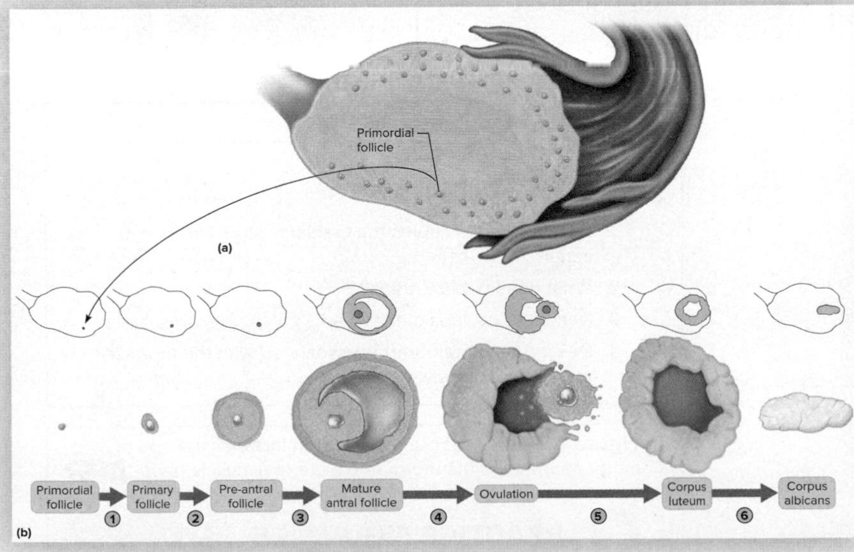

Locator icons help portray the process more accurately.

Learn, Practice, Assess!

Learn 

Learning Outcomes have been moved! They now follow the appropriate heading within the chapter. They continue to be closely linked to Chapter Assessments and Integrative Assessments/Critical Thinking questions found at the end of each chapter.

8.2 | Structure of a Skeletal Muscle

 **LEARN**

1. Identify the structures that make up a skeletal muscle.
2. Identify the major parts of a skeletal muscle fiber, and the function of each.
3. Discuss nervous stimulation of a skeletal muscle.

Learning tools to help you succeed . . .

Check out the Chapter Preview, *Foundations for Success,* on page 1. The Chapter Preview was specifically designed to help you **LEARN** how to study. It provides helpful study tips.

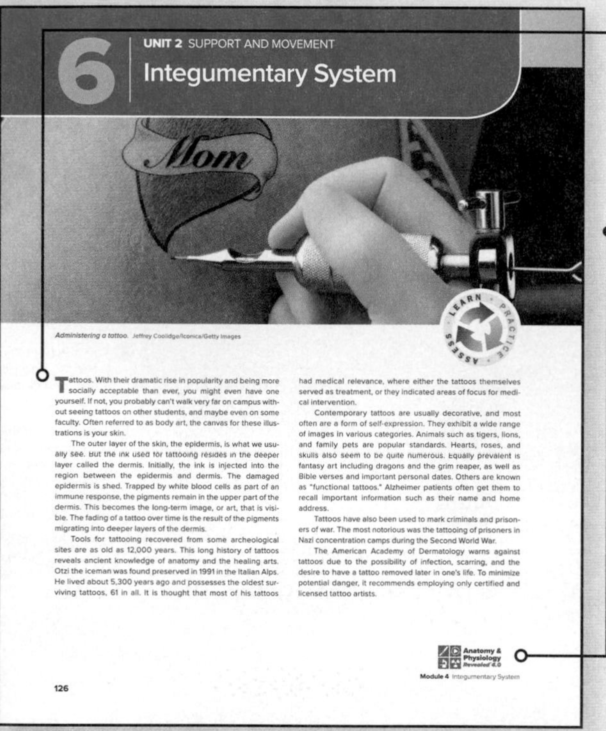

6 UNIT 2 SUPPORT AND MOVEMENT

Integumentary System

Administering a tattoo. Jeffrey Coolidge/Iconica/Getty Images

Tattoos. With their dramatic rise in popularity and being more socially acceptable than ever, you might even have one yourself. If not, you probably can't walk very far on campus without seeing tattoos on other students, and maybe even on some faculty. Often referred to as body art, the canvas for these illustrations is your skin.

The outer layer of the skin, the epidermis, is what we usually see. But the ink used for tattooing resides in the deeper layer called the dermis. Initially, the ink is injected into the region between the epidermis and dermis. The damaged epidermis is shed. Trapped by white blood cells as part of an immune response, the pigments remain in the upper part of the dermis. This becomes the long-term image, or art, that is visible. The fading of a tattoo over time is the result of the pigments migrating into deeper layers of the dermis.

Tools for tattooing recovered from some archeological sites are as old as 12,000 years. This long history of tattoos reveals ancient knowledge of anatomy and the healing arts. Otzi the iceman was found preserved in 1991 in the Italian Alps. He lived about 5,300 years ago and possesses the oldest surviving tattoos, 61 in all. It is thought that most of his tattoos had medical relevance, where either the tattoos themselves served as treatment, or they indicated areas of focus for medical intervention.

Contemporary tattoos are usually decorative, and most often are a form of self-expression. They exhibit a wide range of images in various categories. Animals such as tigers, lions, and family pets are popular standards. Hearts, roses, and skulls also seem to be quite numerous. Equally prevalent is fantasy art including dragons and the grim reaper, as well as Bible verses and important personal dates. Others are known as "functional tattoos." Alzheimer patients often get them to recall important information such as their name and home address.

Tattoos have also been used to mark criminals and prisoners of war. The most notorious was the tattooing of prisoners in Nazi concentration camps during the Second World War.

The American Academy of Dermatology warns against tattoos due to the possibility of infection, scarring, and the desire to have a tattoo removed later in one's life. To minimize potential danger, it recommends employing only certified and licensed tattoo artists.

Anatomy & Physiology Revealed 4.0

Module 4 Integumentary System

126

Vignettes lead into chapter content. They connect you to many areas of health care, including technology, physiology, medical conditions, historical perspectives, and careers.

Anatomy & Physiology REVEALED® (APR) icon at the beginning of each chapter tells you which system in APR applies to this chapter.

Aids to Understanding Words examines root words, stems, prefixes, suffixes, and pronunciations to help you build a solid anatomy and physiology vocabulary.

Reference Plates offer vibrant detail of body structures.

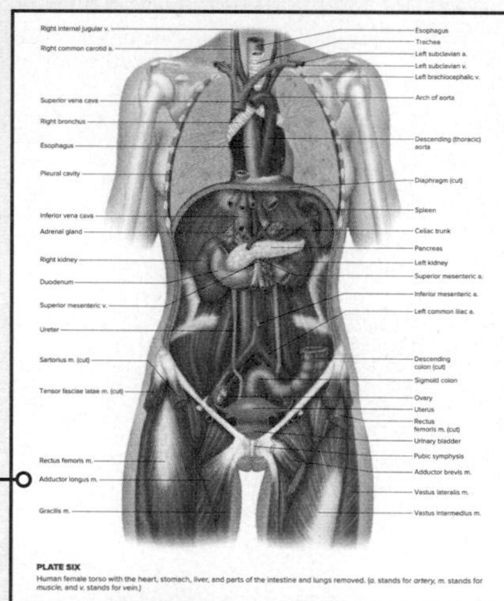

PLATE SIX

Human female torso with the heart, stomach, liver, and parts of the intestine and lungs removed. (*a.* stands for *artery*, *m.* stands for *muscle*, and *v.* stands for *vein*.)

Practice 

Practice with a question or series of questions after major sections. They will test your understanding of the material.

Interesting applications help you practice and apply knowledge . . .

PRACTICE 1.6

1. Which organ occupies the cranial cavity? The vertebral canal?
2. What does *viscera* mean?
3. Name the cavities of the head.
4. Describe the membranes associated with the thoracic and abdominopelvic cavities.

Figure Questions allow an additional assessment. They are found on key figures throughout the chapter.

Figure 8.5 A neuromuscular junction includes the end of a motor neuron and the motor end plate of a muscle fiber. APR

 **PRACTICE FIGURE 8.5**

How does acetylcholine released into the synaptic cleft reach the muscle fiber membrane?

Answer can be found in Appendix E.

"Of Interest" boxes provide interesting bits of anatomy and physiology information, adding a touch of wonder to chapter topics.

OF INTEREST The skeleton of an average 160-pound body weighs about 29 pounds.

Clinical Application sections present disorders, physiological responses to environmental factors, and other topics of general interest and applies them to clinical situations.

CLINICAL APPLICATION **18.1**

Water Balance Disorders

Dehydration, water intoxication, and edema are among the more common disorders that involve a water imbalance in body fluids.

Dehydration

In *dehydration*, water output exceeds water intake. Dehydration may develop following excessive sweating or as a result of prolonged water deprivation accompanied by continued water output. The extracellular fluid becomes more concentrated, and water leaves cells by osmosis (fig. 18A). Dehydration may also accompany prolonged vomiting or diarrhea that depletes body fluids.

During dehydration, the skin and mucous membranes of the mouth feel dry, and body weight drops. mechanism decreases with age, and physical disabilities may make it difficult for them to obtain adequate fluids.

The treatment for dehydration is to replace the lost water and electrolytes. If only water is replaced, the extracellular fluid will become more dilute than normal, causing cells to swell (fig. 18B). This may produce a condition called water intoxication.

Water Intoxication

Until recently, runners were advised to drink as much fluid as they could, particularly in long events. But the death of a young woman running in the Boston marathon, following brain swelling, from low blood sodium (*hyponatremia*) due to excessive water intake lead-

Assess

Tools to help you make the connection and master Anatomy and Physiology!

Chapter Assessments check your understanding of the chapter's learning outcomes.

Integrative Assessments/Critical Thinking questions allow you to connect and apply information from previous chapters, as well as information within the current chapter.

Chapter Summary Outlines help you review the chapter's main ideas.

ASSESS

CHAPTER ASSESSMENTS

8.1 Introduction

1. The three types of muscle tissue are ______, ______, and ______.

8.2 Structure of a Skeletal Muscle

2. Describe the difference between a tendon and an aponeurosis.
3. Describe how connective tissue associates with skeletal muscle.
4. List the major parts of a skeletal muscle fiber, and describe the function of each part.
5. Describe a neuromuscular junction.
6. A neurotransmitter ______.
   a. binds actin filaments, causing them to slide

14. Explain the causes of skeletal muscle hypertrophy and atrophy.

8.4 Muscular Responses

15. Define *threshold stimulus.*
16. Sketch a myogram of a single muscular twitch, and identify the latent period, period of contraction, and period of relaxation.
17. Define *motor unit.*
18. Which of the following describes the addition of muscle fibers to take part in a contraction?
    a. summation
    b. recruitment
    c. tetany
    d. twitch
19. Explain how skeletal muscle stimulation produces a

ASSESS

INTEGRATIVE ASSESSMENTS/CRITICAL THINKING

Outcomes 4.4, 8.3

1. As lactate and other substances accumulate in an active muscle, they stimulate pain receptors and the muscle may feel sore. How might the application of heat or substances that dilate blood vessels relieve such soreness?

Outcomes 5.3, 8.2

2. Discuss how connective tissue is part of the muscular system.

Outcomes 5.5, 8.2

3. What purpose is served by skeletal muscle cells being multinucleated?

Outcomes 8.2, 8.8

Outcomes 8.3, 8.4

5. A woman takes her daughter to a sports medicine specialist and asks the specialist to determine the percentage of fast- and slow-twitch fibers in the girl's leg muscles. The parent wants to know if the healthy girl should try out for soccer or cross-country running. Do you think this is a valid reason to test muscle tissue? Why or why not?
6. Following an injury to a nerve, the muscle it innervates may become paralyzed. How would you explain to a patient the importance of moving the disabled muscles passively or contracting them using electrical stimulation?

Outcomes 8.3, 8.6

7. Make an argument as to why cardiac muscle is

Chapter Summary

10.1 Introduction

Sensory receptors sense changes in their surroundings.

10.2 Receptors, Sensations, and Perception

1. Types of receptors
   a. Each type of receptor is most sensitive to a distinct type of stimulus.
   b. The major types of receptors are **chemoreceptors, pain receptors, thermoreceptors, mechanoreceptors,** and **photoreceptors.**
2. Sensations
   a. A **sensation** is the awareness of sensory stimulation.
   b. A particular part of the cerebral cortex interprets every impulse reaching it in a specific way.
   c. The cerebral cortex projects a sensation back to the region of stimulation.
3. **Sensory adaptation** may involve receptors becoming unresponsive or inhibition along the CNS pathways leading to the sensory regions of the cerebral cortex.

10.3 General Senses

General senses are associated with receptors in the skin, muscles, joints, and viscera.

1. Touch and pressure senses
   a. Free ends of sensory nerve fibers are receptors for the sensation of itching.
   b. **Tactile corpuscles** are receptors for the sensation of light touch.
   c. **Lamellated corpuscles** are receptors for the sensation of heavy pressure.
2. Temperature senses
   Temperature receptors include two sets of free nerve endings that are warm and cold receptors.
3. Body position, movement, and stretch receptors
4. Sense of pain
   a. Pain receptors are free nerve endings that tissue damage stimulates.
   b. Visceral pain
      (1) Pain receptors are the only receptors in viscera that provide sensations.

McGraw-Hill Connect® empowers students to learn and succeed in the Anatomy and Physiology course with user-friendly digital solutions.

SMARTBOOK®

SmartBook 2.0 provides personalized learning to individual student needs, continually adapting to pinpoint knowledge gaps and focus learning on concepts requiring additional study. **The result? Students are highly engaged in the content and better prepared for lecture.**

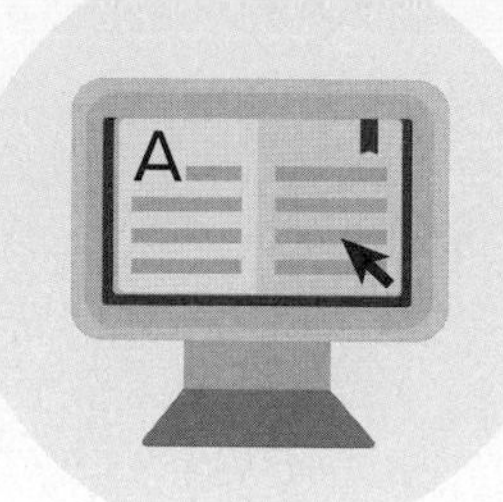

LEARNSMART PREP®

LearnSmart Prep helps students thrive in college-level A&P by helping solidify knowledge in the key areas of cell biology, chemistry, study skills, and math. **The result? Students are better prepared for the A&P course.**

Stop the Drop!

50% of the country's students are unable to pass the A&P course*

Practice Atlas for A&P is an interactive tool that pairs images of common anatomical models with stunning cadaver photography, allowing students to practice naming structures on both models and human bodies, anytime, anywhere. **The result? Students are better prepared, engaged, and move beyond basic memorization.**

Anatomy & Physiology Revealed® (APR) 4.0 is an interactive cadaver dissection tool to enhance lecture and lab that students can use anytime, anywhere. **The result? Students are prepared for lab, engaged in the material, and utilize critical thinking.**

Ph.I.L.S. 4.0 (Physiology Interactive Lab Simulations) software is the perfect way to reinforce key physiology concepts with powerful lab experiments. **The result? Students gain critical thinking skills and are better prepared for lab.**

Concept Overview Interactives are groundbreaking interactive animations that encourage students to explore key physiological processes and difficult concepts. **The result? Students are engaged and able to apply what they've learned while tackling difficult A&P concepts.**

*Statistic courtesy of *The New England Journal of Higher Education*

DIGITAL & LAB EXPERIENCE

In this edition of *Hole's Essentials of Human Anatomy & Physiology,* the digital author team, Leslie Day and Julie Pilcher, worked hand-in-hand with the print author team to deliver a seamless experience for instructors and students.

The digital authors make sure there is a variety of questions with different Bloom's Taxonomy levels. In this edition, we have increased the number of questions that are higher-level Bloom's to about 30 percent.

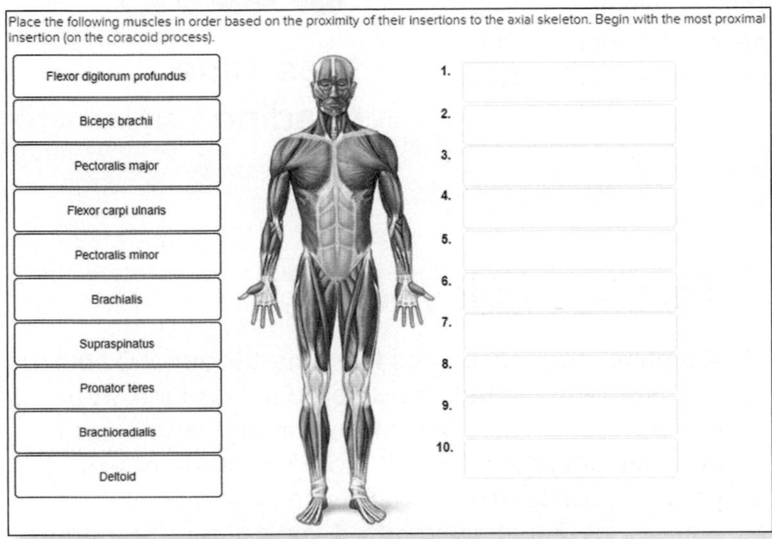

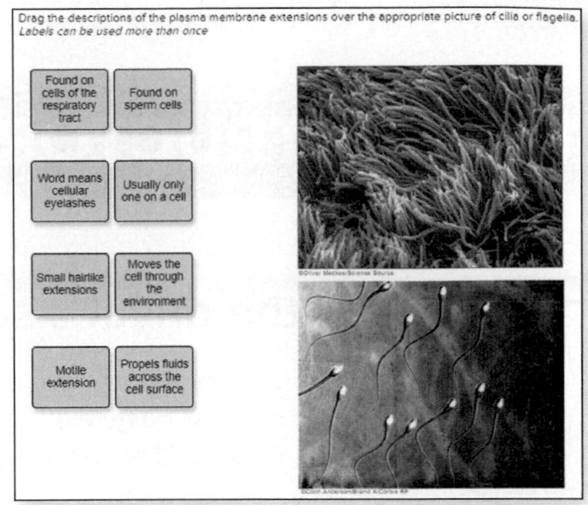

Leslie and Julie ensure that there is an appropriate number of questions for each learning outcome in the chapter. They tagged questions to textbook learning outcomes and to the Human Anatomy & Physiology Society (HAPS) learning outcomes. This makes it easy for instructors to find questions to assign in their course.

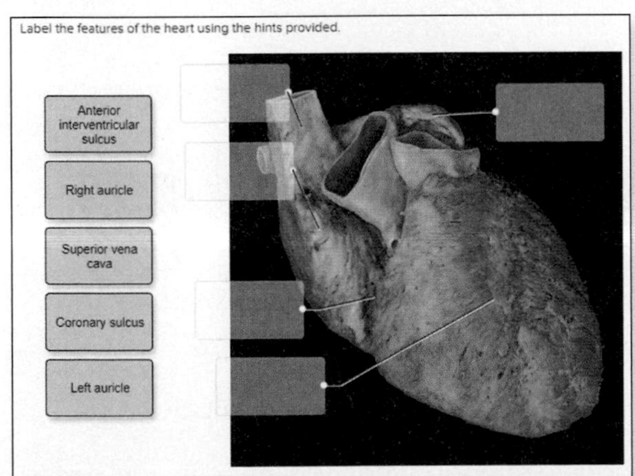

McGraw-Hill Connect® gives the instructor access to additional course-wide material for A&P. Instructors can access questions for Anatomy & Physiology REVEALED®, a variety of animations, diagnostic exam for LearnSmart Prep®, concept application questions, and supplemental laboratory questions.

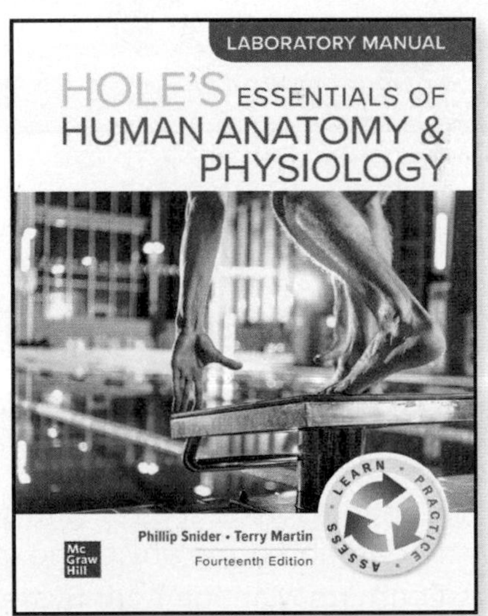

Laboratory Manual

Laboratory Manual for Hole's Essentials of Human Anatomy & Physiology, Fourteenth Edition, by Phillip Snider, Gadsden State Community College, and Terry R. Martin, Kishwaukee College, is designed to accompany the fourteenth edition of *Hole's Essentials of Human Anatomy & Physiology.*

FOR INSTRUCTORS

You're in the driver's seat.

Want to build your own course? No problem. Prefer to use our turnkey, prebuilt course? Easy. Want to make changes throughout the semester? Sure. And you'll save time with Connect's auto-grading too.

65%

Less Time Grading

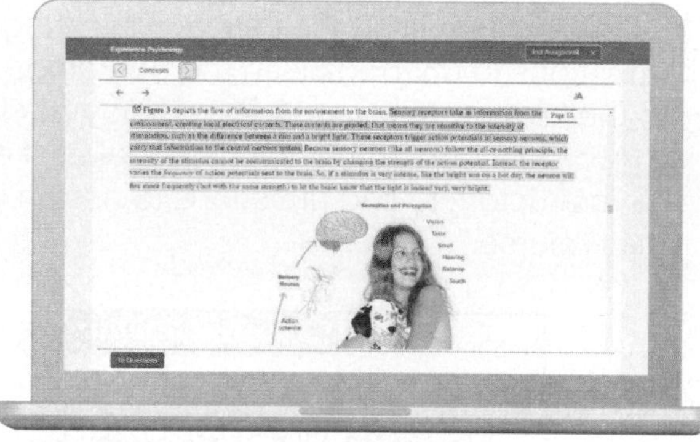

Laptop: McGraw-Hill; Woman/dog: George Doyle/Getty Images

They'll thank you for it.

Adaptive study resources like SmartBook® 2.0 help your students be better prepared in less time. You can transform your class time from dull definitions to dynamic debates. Find out more about the powerful personalized learning experience available in SmartBook 2.0 at **www.mheducation.com/highered/connect/smartbook**

Make it simple, make it affordable.

Connect makes it easy with seamless integration using any of the major Learning Management Systems—Blackboard®, Canvas, and D2L, among others—to let you organize your course in one convenient location. Give your students access to digital materials at a discount with our inclusive access program. Ask your McGraw-Hill representative for more information.

Padlock: Jobalou/Getty Images

Solutions for your challenges.

A product isn't a solution. Real solutions are affordable, reliable, and come with training and ongoing support when you need it and how you want it. Our Customer Experience Group can also help you troubleshoot tech problems—although Connect's 99% uptime means you might not need to call them. See for yourself at **status.mheducation.com**

Checkmark: Jobalou/Getty Images

FOR STUDENTS

Effective, efficient studying.

Connect helps you be more productive with your study time and get better grades using tools like SmartBook 2.0, which highlights key concepts and creates a personalized study plan. Connect sets you up for success, so you walk into class with confidence and walk out with better grades.

Study anytime, anywhere.

Download the free ReadAnywhere app and access your online eBook or SmartBook 2.0 assignments when it's convenient, even if you're offline. And since the app automatically syncs with your eBook and SmartBook 2.0 assignments in Connect, all of your work is available every time you open it. Find out more at **www.mheducation.com/readanywhere**

"I really liked this app—it made it easy to study when you don't have your textbook in front of you."

- Jordan Cunningham, Eastern Washington University

Calendar: owattaphotos/Getty Images

No surprises.

The Connect Calendar and Reports tools keep you on track with the work you need to get done and your assignment scores. Life gets busy; Connect tools help you keep learning through it all.

Learning for everyone.

McGraw-Hill works directly with Accessibility Services Departments and faculty to meet the learning needs of all students. Please contact your Accessibility Services office and ask them to email accessibility@mheducation.com, or visit **www.mheducation.com/about/accessibility** for more information.

Top: Jenner Images/Getty Images, Left: Hero Images/Getty Images, Right: Hero Images/Getty Images

CONTENTS

Comstock Images/Getty Images

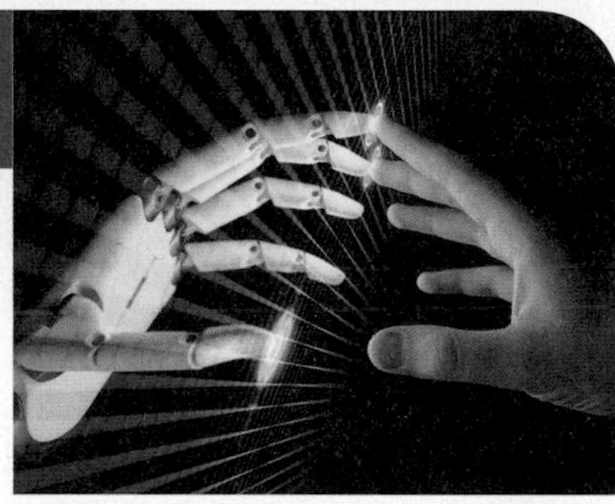

©Shutterstock/maxus

UNIT 2
SUPPORT AND MOVEMENT

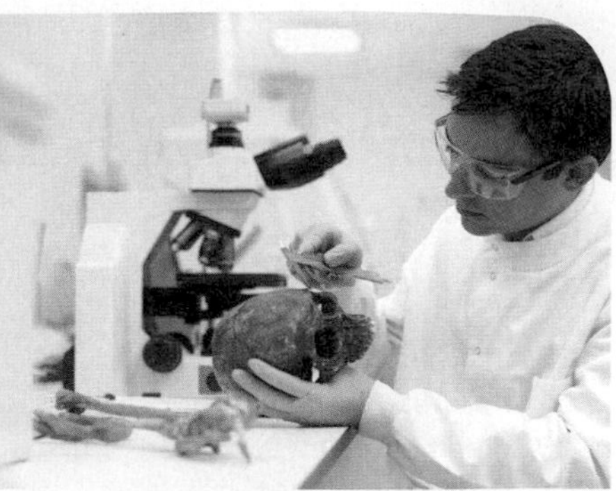
Adam Gault/Getty Images

UNIT 3
INTEGRATION AND COORDINATION

JoeSAPhotos/Shutterstock

UNIT 4 TRANSPORT

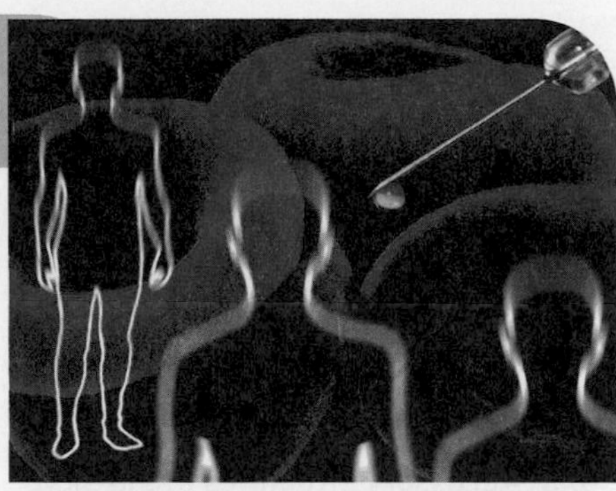

Steve Allen/Getty Images

UNIT 5 ABSORPTION AND EXCRETION

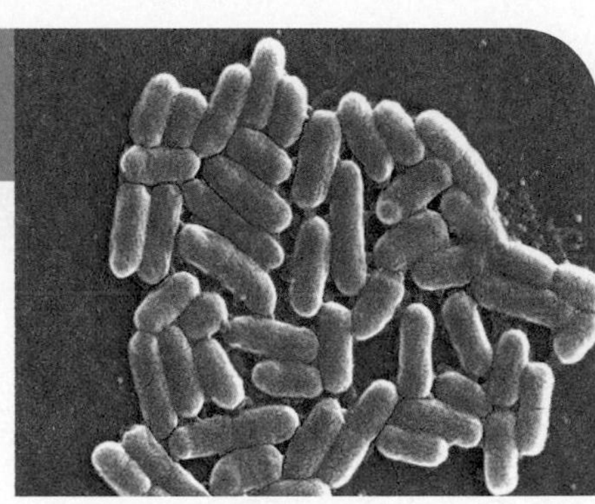

Source: CDC/Janice Haney Carr

UNIT 6
THE HUMAN LIFE CYCLE

Profs. P.M. Motta & J. Van Blerkom/Science Source

PREVIEW
Foundations for Success

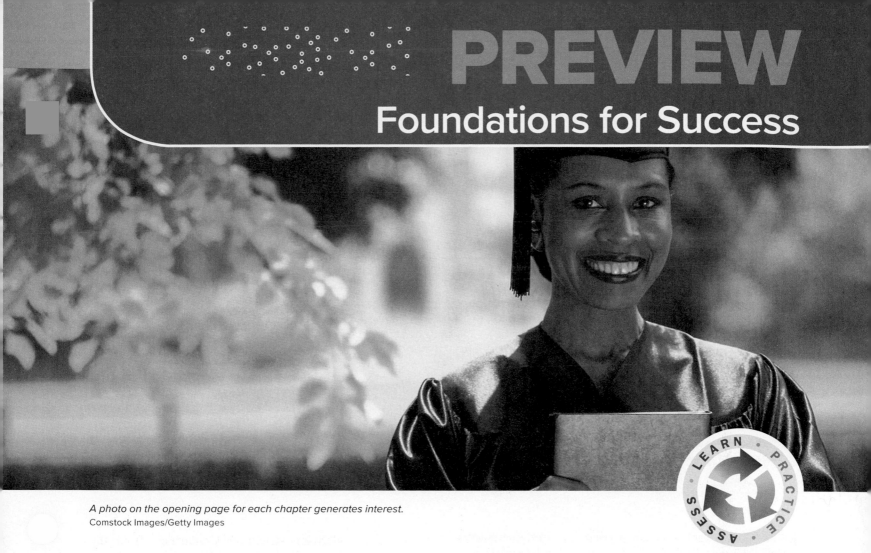

A photo on the opening page for each chapter generates interest.
Comstock Images/Getty Images

AN OPENING VIGNETTE discusses current events or research news relating to the subject matter in the chapter. These vignettes apply the concepts learned in the study of anatomy and physiology.

Pay attention. It is a beautiful day. You can't help but stare wistfully out the window, the scent of spring blooms and sound of birds making it impossible to concentrate on what the instructor is saying. Gradually the lecture fades as you become aware of your own breathing, the beating of your heart, and the sweat that breaks out on your forehead in response to the radiant heat from the glorious day. Suddenly your reverie is cut short—a classmate has dropped a human anatomy and physiology textbook on the floor. You jump. Your heart hammers and a flash of fear grips your chest, but you soon realize what has happened and recover.

The message is clear: Pay attention. So you do, tuning out the great outdoors and focusing on the class. In this course, you will learn all about the events you have just experienced, including your response to the sudden stimulation. This is a good reason to stay focused.

This chapter Preview not only provides great study tips to offer a foundation for success, but it also offers tips on how to utilize this particular text. Those tips will be found in boxes just like this.

This digital tool, as indicated below and with the APR icons within the chapters, allows you to explore the human body in depth through simulated dissection of cadavers and histology preparations. It also offers animations on chapter concepts.

LEARNING OUTLINE

After studying this chapter, you should be able to complete the "Learning Outcomes" that follow the major headings throughout the chapter.

Each chapter has a learning outline introducing what will be discussed in the chapter.

The section below introduces building blocks of words that your instructor may assign. Learning them is a good investment of your time, because they can be used over and over and apply to many of the terms you will use in your career. Appendix A has a comprehensive list of these prefixes, suffixes, and root words.

AIDS TO UNDERSTANDING WORDS

ana- [up] *ana*tomy: the study of breaking up the body into its parts.

multi- [many] *multi*tasking: performing several tasks simultaneously.

physio- [relationship to nature] *physio*logy: the study of how body parts function.

(Appendix A has a complete list of Aids to Understanding Words.)

Major divisions within a chapter are called "A-heads." They are numbered sequentially and set in very large, blue type to designate major content areas.

P.1 | Introduction

After each A-head is a list of learning outcomes indicating the knowledge you should gain as you work through the section. (Note the blue learn arrow preceding the "LEARN" heading.) These outcomes are intended to help you master the similar outcomes set by your instructor. The outcomes are tied directly to assessments of knowledge gained.

LEARN

1. Explain the importance of an individualized approach to learning.

Studying the human body can be overwhelming at times. The new terminology, used to describe body parts and how they work, can make it seem as if you are studying a foreign language. Learning all the parts of the body, along with the composition of each part, and how each part fits with the other parts to make the whole requires memorization. Understanding the way each body part works individually, as well as body parts working together, requires a higher level of knowledge, comprehension, and application. Identifying underlying structural similarities, from the macroscopic to the microscopic levels of body organization, taps more subtle critical thinking skills. This chapter will catalyze success in this active process of learning. (Remember that although the skills and tips discussed in this chapter relate to learning anatomy and physiology, they can be applied to other subjects.)

Learning occurs in different ways or modes. Most students use several modes (multimodal), but are more comfortable with and use more effectively one or two, often referred to as learning styles. Some students prefer to read the written word to remember it and the concept it describes, or to actually write the words; others learn best by looking at visual representations, such as photographs and drawings. Still others learn most effectively by hearing the information or explaining it to someone else. For some learners, true understanding remains elusive until a principle is revealed in a laboratory or clinical setting that provides a memorable context and engages all of the senses. This text accommodates the range of learning styles. Read-write learners will appreciate the lists, definitions (glossary), and tables. Visual learners will discover many diagrams, flow charts, and figures, all with consistent and purposeful use of color. For example, a particular bone is always the same color in figures where bones are color-coded. Auditory learners will find pronunciations for new scientific terms to help sound them out, and kinesthetic learners can relate real-life examples and applications to their own activities.

After each major section, a question or series of questions tests your understanding of the material and enables you to practice using the new information. (Note the green practice arrow preceding the "PRACTICE" heading.) If you cannot answer the question(s), you should reread that section, being on the lookout for the answer(s).

PRACTICE P.1

1. List some difficulties a student may experience when studying the human body.
2. Describe the ways people learn.

P.2 | Strategies for Your Success

LEARN

1. Summarize what you should do before attending class.
2. Identify student activities that enhance classroom experience.
3. List and describe several study techniques that can facilitate learning new material.

Many strategies for academic success are common sense, but it might help to review them. You may encounter new and helpful methods of learning.

The major divisions are subdivided into "B-heads," which are identified by large, reddish-orange type. These will help you organize the concepts upon which the major divisions are built.

Of Interest provides bits of anatomy and physiology information that add wonder and awe to some of the chapter concepts.

OF INTEREST The skeleton of an average 160-pound person contributes about 29 pounds total body weight.

Before Class

Before attending class, prepare by reading and outlining or taking notes on the assigned pages of the text. If outlining, leave adequate space between entries to allow room for note-taking during lectures. Or fold each page of notes taken before class in half so that class notes can be written on the blank side of the paper across from the reading notes on the same topic. This strategy introduces the topics of the next class discussion, as well as new terms. Some students team a vocabulary list with each chapter's notes. Take the notes from the reading to class and expand them. At a minimum, the student should at least skim the text, reading the A-heads and B-heads and the summary outline to become acquainted with the topics and vocabulary before class.

Many students who use this book and take various other courses in the health sciences are preparing for careers in health care. Some students may be undecided as to a specific area or specialty. The Career Corner feature presents a description of a particular career choice with each chapter. If it doesn't describe a career that you seek, perhaps it will give you a better sense of what some of your coworkers and colleagues do!

CAREER CORNER
Massage Therapist

The woman feels something give way in her left knee as she lands from a jump in her dance class. She limps away between her classmates, in great pain. At home, she uses "RICE"—rest, ice, compression, elevation—then has a friend take her to an urgent care clinic, where a physician diagnoses patellar tendinitis, or "jumper's knee." Frequent jumping followed by lateral movements caused the injury.

Three days later, at her weekly appointment with a massage therapist for stress relief, the woman mentions the injury. Over the next few weeks, the massage therapist applies light pressure to the injured area to stimulate circulation, and applies friction in a transverse pattern to break up scar tissue and relax the muscles. She also massages the muscles to improve flexibility.

A massage therapist manipulates soft tissues, using combinations of stroking, kneading, compressing, and vibrating, to relieve pain and reduce stress. Training includes 300 to 1,000 hours of class time, hands-on practice, and continuing education. Specialties include pediatrics, sports medicine, and even animal massage.

As you read, you may feel the need for a "study break" or to "chill out." At other times, you may just need to shift gears. Try the following: Look for the Clinical Application boxes throughout the book that present sidelights to the main focus of the text. Some of these may cover topics that your instructor chooses to highlight. Read them! They are interesting, informative, and a change of pace.

CLINICAL APPLICATION **P.1**

Factors Affecting Synaptic Transmission

Many chemicals affect synaptic transmission. A drug called Dilantin (diphenylhydantoin) treats seizure disorders by blocking gated sodium channels, thereby limiting the frequency of action potentials reaching the axon terminal. Caffeine in coffee, tea, cola, and energy drinks stimulates nervous system activity by lowering the thresholds at synapses. As a result, postsynaptic neurons are more easily excited. Antidepressants called "selective serotonin reuptake inhibitors" keep the neurotransmitter serotonin in synapses longer, compensating for a still little-understood decreased serotonin release that presumably causes depression.

Remember when you were very young and were presented with a substantial book for the first time? You were likely intimidated by its length but were reassured that it contained "a lot of pictures." This book has many "pictures" (figures) too, all designed to help you master the material. Some of the figure legends are followed by a question pertaining to that figure, intended to reinforce a concept or usage of terminology.

Photographs and Line Art

Photographs provide a realistic view of anatomy.

J and J Photography

Line art can present different positions, layers, or perspectives.

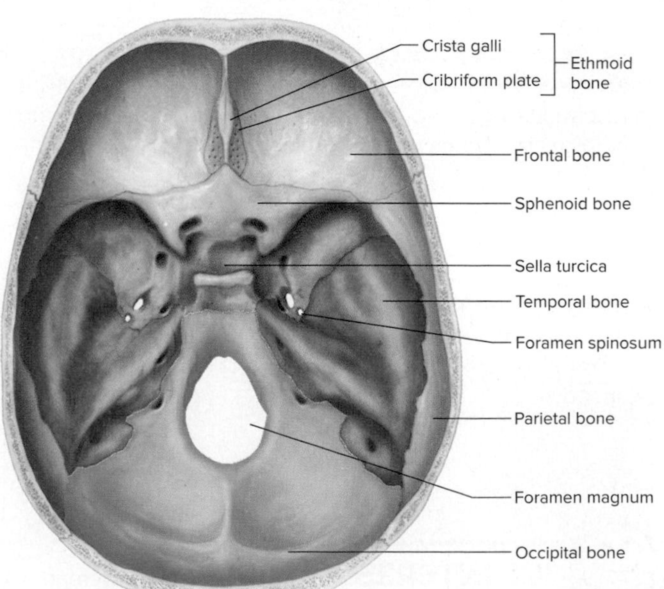

Figure questions encourage you to think about what you are seeing and "PRACTICE" making connections between the visual representation and the words in the text.

PRACTICE FIGURE P.2

What is the most posterior bone on the skull?

Answer can be found in Appendix E.

Macroscopic to Microscopic

Many figures show anatomical structures in a manner that is macroscopic to microscopic (or vice versa).

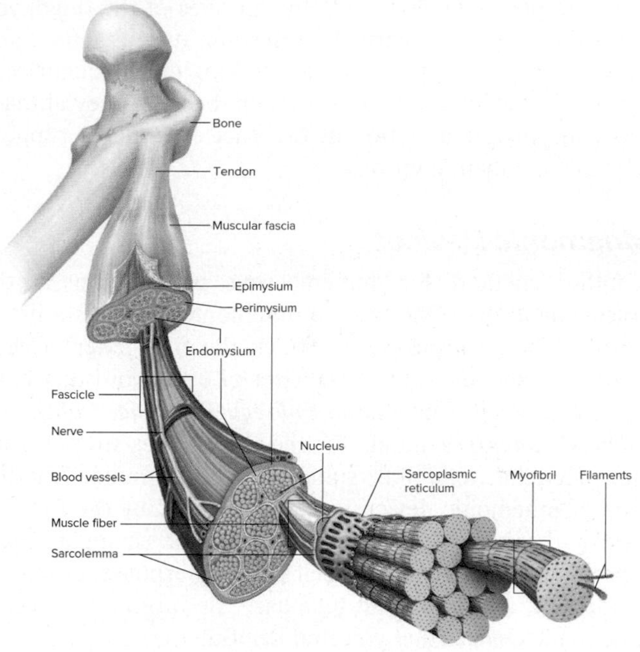

Flow Charts

Flow charts depict sequences of related events, steps of pathways, and complex concepts, easing comprehension. Other figures may show physiological processes.

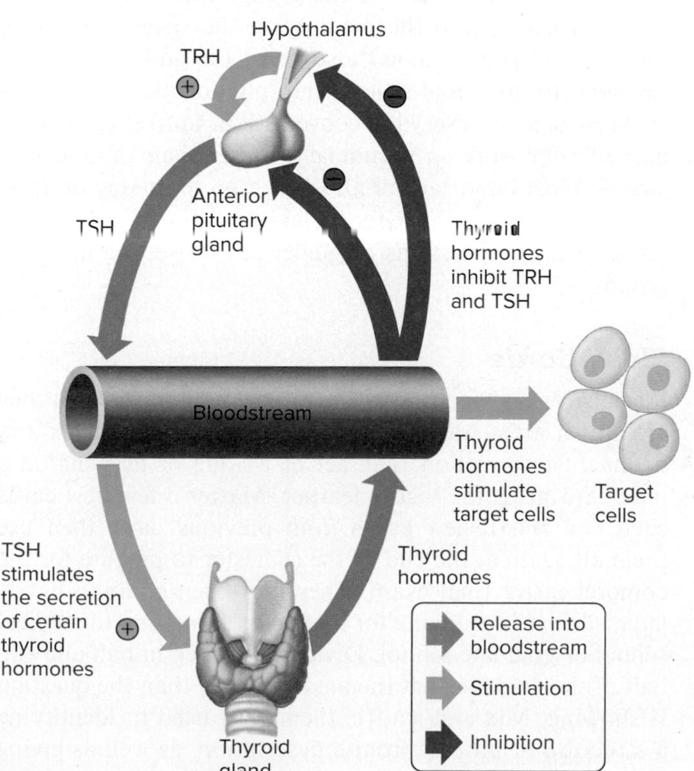

Anatomical Structures

Some figures illustrate the locations of anatomical structures.

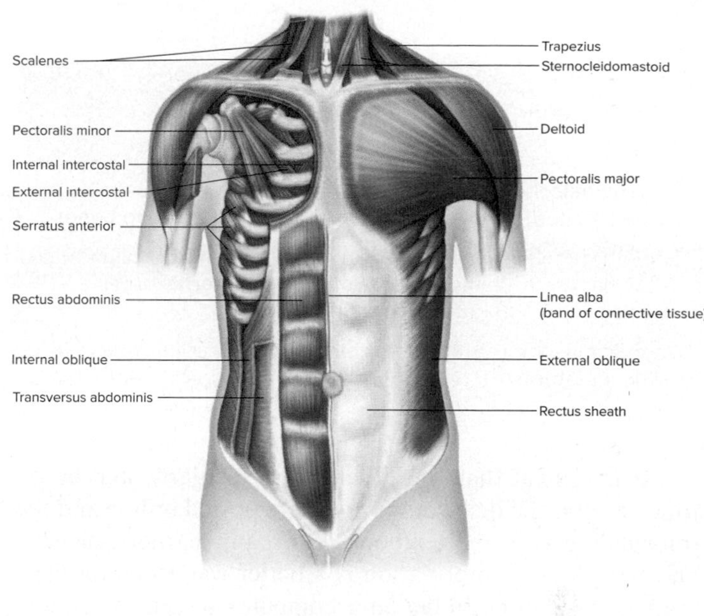

Other figures illustrate the functional relationships of anatomical structures.

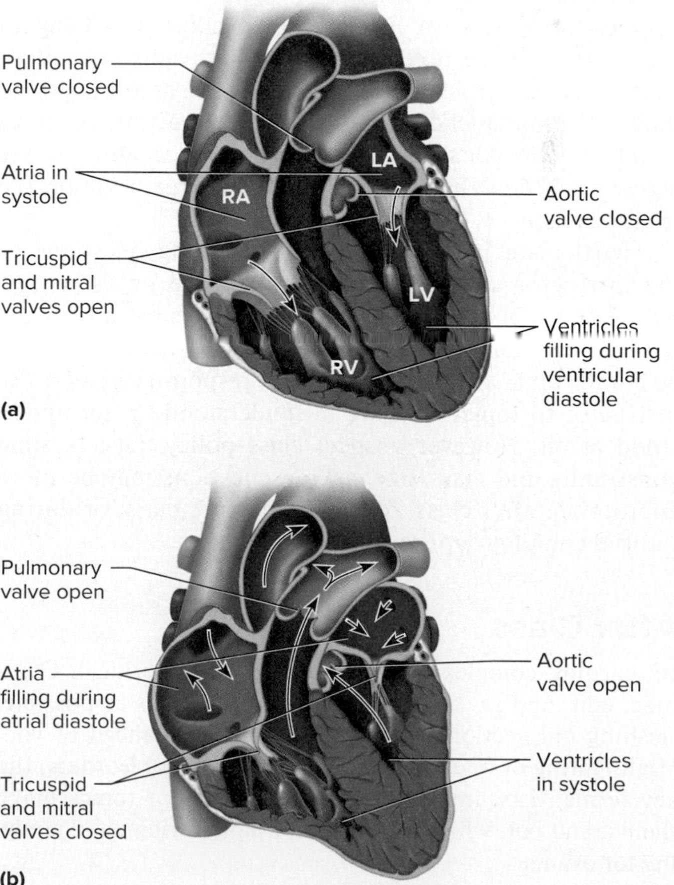

Organizational Tables

Organizational tables can help "put it all together," but are not a substitute for reading the text or having good notes.

TABLE 5.6 Muscle Tissues		
Type	**Function**	**Location**
Skeletal muscle tissue (striated)	Voluntary movements of skeletal parts	Muscles usually attached to bones
Smooth muscle tissue (lacks striations)	Involuntary movements of internal organs	Walls of hollow internal organs
Cardiac muscle tissue (striated)	Heart movements	Heart muscle

It is critical that you attend class regularly, and be on time—even if the instructor's notes are posted online and the information is in the textbook. For many learners, hearing and writing new information is a better way to retain facts than just scanning notes on a computer screen. Attending lectures and discussion sections also provides more detailed and applied analysis of the subject matter, as well as a chance to ask questions.

During Class

Be alert and attentive in class. Take notes by adding to your outline or the notes you took while reading. Auditory learners benefit from recording the lectures and listening to them while doing chores that do not require your cognitive attention. This is called **multitasking**—doing more than one activity at a time; however, with mental focus being on the lecture content.

Participate in class discussions, asking questions of the instructor and answering questions he or she poses. All of the students are in the class to learn, and many will be glad someone asked a question others would not be comfortable asking. Such student response can alert the instructor to topics that are misunderstood or not understood at all. However, respect class policy. Due to time constraints and class size, asking questions may be more appropriate after class, for a large lecture class, or during tutorial (small group) sessions.

After Class

In learning complex material, expediency is critical. Organize, edit, and review notes as soon after class as possible, fleshing out sections where the lecturer got ahead of you. Highlighting or underlining (in color, for visual learners) the key terms, lists, important points, and major topics make them stand out, which is helpful for daily reviews and studying for exams.

Lists

Organizing information into lists or categories can minimize information overload, breaking it into manageable chunks. For example, when you study the muscles of the thigh, you will find it easier to learn the insertion, origin, action, and nerve supply of the four muscles making up the quadriceps femoris if you study them as a group, because they all have the same insertion, action at the knee, and nerve supply—they differ only in their origins.

Mnemonic Devices

Another method for remembering information is the **mnemonic device.** One type of mnemonic device is a list of words, forming a phrase, in which the first letter of each word corresponds to the first letter of each word that must be remembered. For example, ***Frequent parades often test soldiers' endurance*** stands for the skull bones **f**rontal, **p**arietal, **o**ccipital, **t**emporal, **s**phenoid, and **e**thmoid. Another type of mnemonic device is a word formed by the first letters of the items to be remembered. For example, ***ipmat*** represents the stages in the cell cycle: **i**nterphase, **p**rophase, **m**etaphase, **a**naphase, and **t**elophase. Be inventive! Develop mnemonic devices that you find helpful!

Study Groups

Forming small study groups helps some students. Together the students review course material and compare notes. Working as a team and alternating leaders allows students to verbalize the information. Individual students can study and master one part of the assigned material, and then explain it to the others in the group, which incorporates the information into the memory of the speaker. Hearing the material spoken aloud also helps the auditory learner. Be sure to use anatomical and physiological terms, in explanations and everyday conversation, until they become part of your working vocabulary, rather than intimidating jargon. Most important of all—the group must stay on task, and not become a vehicle for social interaction. Your instructor may have suggestions or guidelines for setting up study groups.

Flash Cards

Flash cards may seem archaic in this computer age, but they are still a great way to organize and master complex and abundant information. The act of writing or drawing on a note card helps the tactile learner. Master a few new cards each day and review cards from previous days, then use them all again at the end of the semester to prepare for the comprehensive final exam. They may even come in handy later, such as in studying for exams for admission to medical school or graduate school. Divide your deck in half and flip half of the cards so that the answer rather than the question is showing. Mix and shuffle them. Get used to identifying a structure or process from a description, as well as giving a description when provided with the name of a process or structure. This is more like what will be expected of you in the real world of the health-care professional.

Manage Your Time

For each hour in the classroom, most students will spend at least three hours outside of class studying. Many of you have important obligations outside of class, such as jobs and family responsibilities. As important as these are, you still need to master this material on your path to becoming a health-care professional. Good time-management skills are therefore essential in your study of human anatomy and physiology. In addition to class, lab, and study time, multitask. When you are waiting for a ride or sitting in a doctor's waiting room, use your time by reviewing notes or reading the text.

Daily repetition is helpful, so you should schedule several short study periods each day instead of an end-of-semester crunch to cram for an exam. This does not take the place of time spent to prepare for the next class. If you follow these suggestions for learning now, you can maximize your study time throughout the semester and will give yourself your best prospects for academic success. A working knowledge of the structure and function of the human body provides the foundation for all careers in the health sciences.

PRACTICE P.2

1. Why is it important to prepare before attending class?
2. Name two ways to participate in class discussions.
3. List several aids for remembering information.

ASSESS

CHAPTER ASSESSMENTS

Chapter assessments that are tied directly to the learning outcomes allow you to assess your mastery of the material. (Note the purple assess arrow.)

P.1 Introduction

1. Explain why the study of the human body can be overwhelming.

P.2 Strategies for Your Success

2. Methods to prepare for class include ______________.
   a. reading the chapter
   b. outlining the chapter
   c. making a vocabulary list
   d. all of the above
3. Describe how you can participate in class discussions.
4. Forming the phrase "I passed my anatomy test" to remember the cell cycle (interphase, prophase, metaphase, anaphase, telophase) is a ______________ device.
5. Name a benefit and a drawback of small study groups.
6. Give an example of effective time management used in preparation for success in the classroom.

ASSESS

INTEGRATIVE ASSESSMENTS/CRITICAL THINKING

A textbook is inherently linear. This text begins with Chapter 1 and ends with Chapter 20. Understanding physiology and the significance of anatomy, however, requires you to be able to recall previous concepts. Critical thinking is all about linking previous concepts with current concepts under novel circumstances, in new ways. Toward this end, we have included in the Integrative Assessments/Critical Thinking exercises referencing sections from earlier chapters. Making connections is what it is all about!

Outcomes P.1, P.2

1. Which study methods are most successful for you?

Outcome P.2

2. Design a personalized study schedule.

Chapter Summary

A summary of the chapter provides an outline to review major ideas and is a tool for organizing thoughts.

P.1 Introduction

Try a variety of methods to study the human body.

P.2 Strategies for Your Success

Although strategies for academic success seem to be common sense, you might benefit from reminders of study methods.

1. Before class
   Read the assigned text material prior to the corresponding class meeting.
   a. Photographs give a realistic view, and line art shows different perspectives.
   b. Figures depicting macroscopic to microscopic show an increase in detail.
   c. Flow charts depict sequences and steps.
   d. Figures of anatomical structures show locations.
   e. Organizational charts/tables summarize text.
2. During class
   Take notes and participate in class discussions.
3. After class
   a. Organize, edit, and review class notes.
   b. Mnemonic devices aid learning.
      (1) The first letters of the words you want to remember begin words of an easily recalled phrase.
      (2) The first letters of the items to be remembered form a word.
   c. Small study groups reviewing and vocalizing material can divide and conquer the learning task.
   d. Making flash cards helps the tactile learner.
   e. Time management skills encourage scheduled studying, including daily repetition instead of cramming for exams.

Check out McGraw-Hill online resources that can help you practice and assess your learning.

Connect Interactive and Integrated Activity Questions Reinforce your knowledge and practice your understanding.

Classify each muscle that moves the foot based on its major (primary) action.

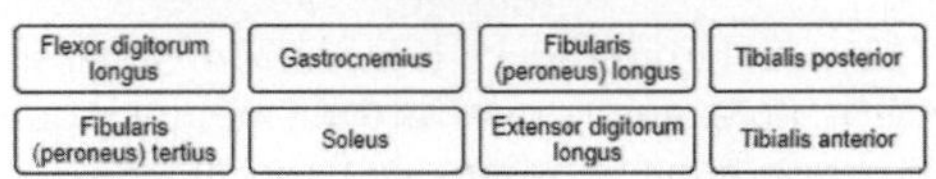

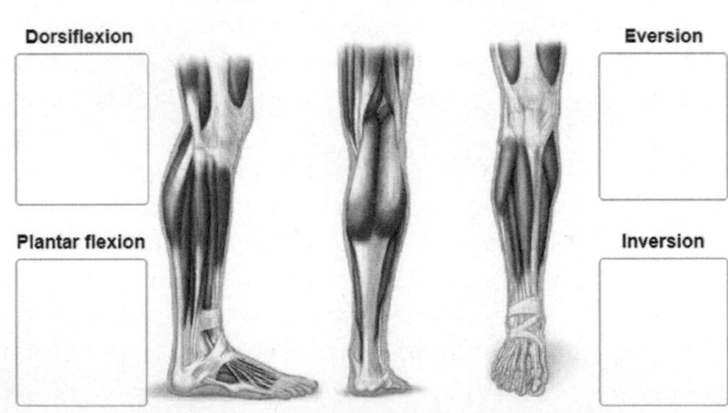

SmartBook continually adapts to the individual learner's needs, creating a more productive learning experience.

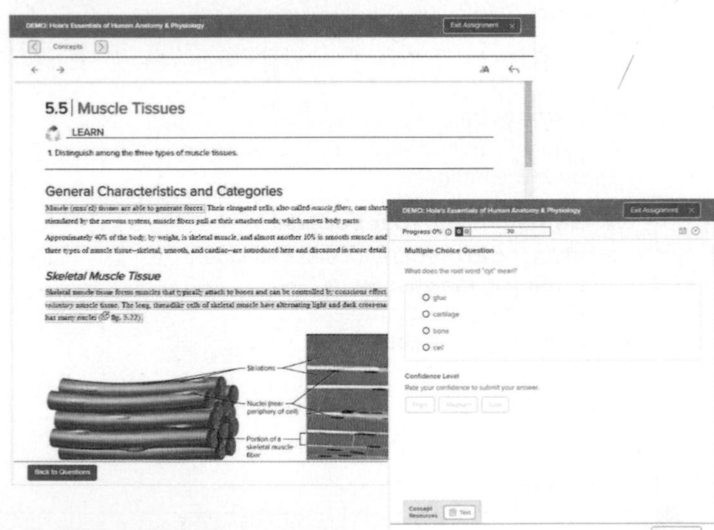

Anatomy & Physiology Revealed Go more in depth using virtual dissection of a cadaver.

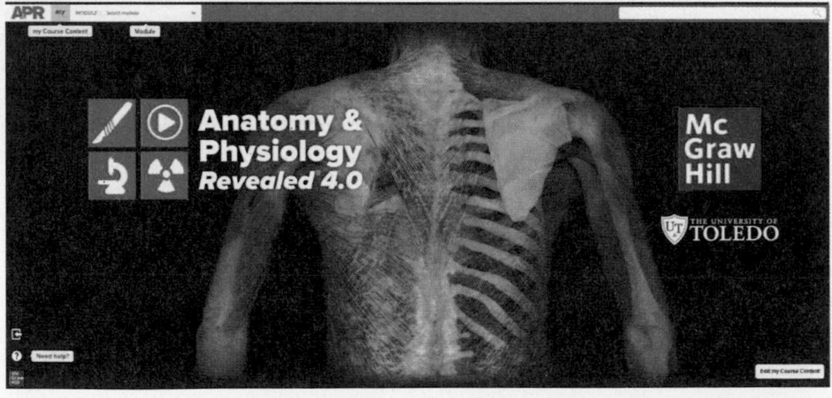

UNIT 1 LEVELS OF ORGANIZATION

1 Introduction to Human Anatomy and Physiology

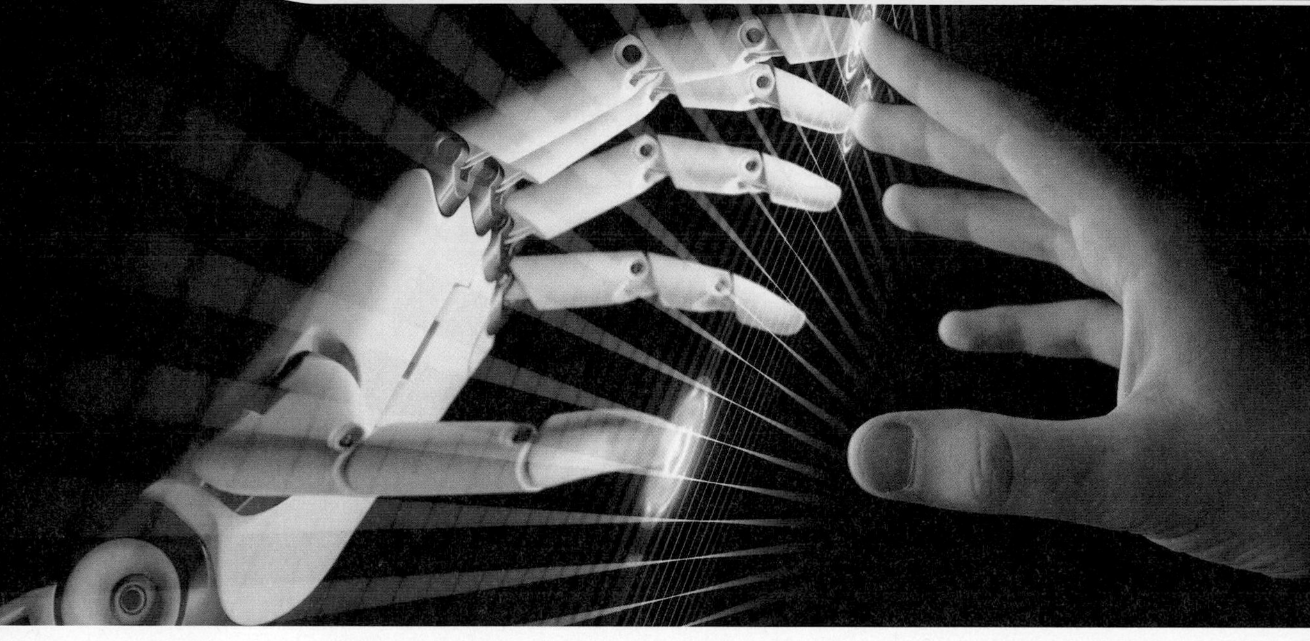

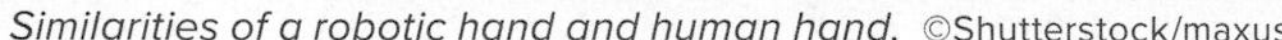
Similarities of a robotic hand and human hand. ©Shutterstock/maxus

Noticing the metal peeking from her uncle's pant leg, the inquisitive little girl's eyes widened in awe. She didn't hesitate with her questions, asking "Why do you have it?" and "How does it work?," as he took off the prosthetic limb to show her and explain.

The design of a prosthetic is intended to replace the anatomical structure and function of the original body part, providing a wholeness to the person. Artificial toes composed of wood and leather were the earliest prosthetics, discovered in ancient Egypt as part of the mummified remains of an Egyptian noblewoman. The artificial toes allowed for the distribution of body weight and forward propulsion, and looked anatomically similar to real toes. Enormous casualties in World War I resulted in a demand for artificial limbs for veterans, so they could return to work after the war. Over the years, prosthetic limbs have evolved from the more rudimentary prosthetics to those that can be controlled by using the mind.

Materials used in the modern-day construction of prosthetic body parts include strong and lightweight materials like carbon fiber. Harnessing these new materials and the use of 3D printing has improved function and aesthetics. Cutting-edge technology, called targeted muscle reinnervation, which surgically reassigns nerves that once controlled the arm or hand, has recently been developed by engineers at the Applied Physics Laboratory at Johns Hopkins University. Imagine the excitement of receiving a robotic arm that allows individual and simultaneous finger control, two degrees of movement at the wrist, and multiple grips, enabling a person to control the prosthetic device by just thinking of the action. To avoid the need for a harness that can be uncomfortable, modern devices use osteointegration, a surgical procedure that fixes a titanium implant into the marrow space of bone, and which eventually becomes part of the bone. A few weeks after surgery, a titanium extension is brought out through the skin so that the prosthetic can be attached. These technologies can improve the lives of people who were born without their limbs, or who have lost them due to infection, cancer, trauma, or combat.

Anatomy & Physiology *Revealed 4.0*

Module 1 Body Orientation

LEARNING OUTLINE

After studying this chapter, you should be able to complete the "Learning Outcomes" that follow the major headings throughout the chapter.

AIDS TO UNDERSTANDING WORDS

append- [to hang something] *append*icular: pertaining to the limbs.

cardi- [heart] peri*cardi*um: membrane that surrounds the heart.

cran- [helmet] *cran*ial: pertaining to the portion of the skull that surrounds the brain.

dors- [back] *dors*al: position toward the back.

homeo- [same] *homeo*stasis: maintenance of a stable internal environment.

-logy [study of] physio*logy*: study of body functions.

meta- [change] *meta*bolism: chemical changes in the body.

pariet- [wall] *pariet*al membrane: membrane that lines the wall of a cavity.

pelv- [basin] *pelv*ic cavity: basin-shaped cavity enclosed by the pelvic bones.

peri- [around] *peri*cardial membrane: membrane that surrounds the heart.

pleur- [rib] *pleur*al membrane: membrane that encloses the lungs and lines the thoracic cavity.

-stasis [standing still] homeo*stasis*: maintenance of a stable internal environment.

-tomy [cutting] ana*tomy*: study of structure, which often involves cutting or removing body parts.

(Appendix A has a complete list of Aids to Understanding Words.)

1.1 | Introduction

LEARN

1. Identify some of the early discoveries that led to our understanding of the body.

Medicine and the study of the human body have always been intimately tied together. Serious thoughts of the structure and function of the human body more than likely arose from observations of a malfunction. That is, little attention was needed until something went wrong, such as a broken bone or even a general condition of not "feeling well." Healers, the first "physicians," originally relied on superstition and a certain sense of magic to provide health care. This eventually evolved into a system where they would seek causes for effects, what we now call symptoms, and then provide remedies. The discovery that certain herbs and other naturally occurring chemicals could alleviate ailments such as coughs, headaches, and fevers was the beginning of medicine and pharmacology.

Much of what we know about the human body is based on the scientific method, an approach to investigating the natural world. The scientific method consists of testing a hypothesis and then rejecting or accepting it, based on the results of experiments or observations. This method is described in greater detail in Appendix B, Scientific Method, but it is likely that aspects of its application are already familiar to you. Imagine buying a used car. The dealer insists that the car is in fine shape, but you discover that the engine doesn't start. That's an experiment! It tests the hypothesis: If this car is in good shape, then it will start. When the car doesn't start, the wary consumer rejects the hypothesis and doesn't buy the car.

As techniques for making accurate observations and performing careful experiments evolved, knowledge of the human body expanded rapidly (fig. 1.1). At the same time, early medical providers coined many new terms to name body parts, describe the locations of the parts, and explain their functions and interactions. These terms, most of which originated from Greek and Latin words, formed the basis for the language of anatomy and physiology that persists today. (The names of some modern medical and applied sciences are listed in section 1.7, Anatomical Terminology.)

Rather than giving us all the answers, science eliminates wrong explanations. Our knowledge of the workings of the human body reflects centuries of asking questions, and testing, rejecting, and sometimes accepting hypotheses. New technologies provide new views of anatomy and physiology, so that knowledge is always growing. One day you may be the one to discover something previously unknown about the human body!

PRACTICE 1.1

Answers to the Practice questions can be found in the eBook.

1. What factors probably stimulated an early interest in the human body?
2. What kinds of activities helped promote the development of modern medical science?

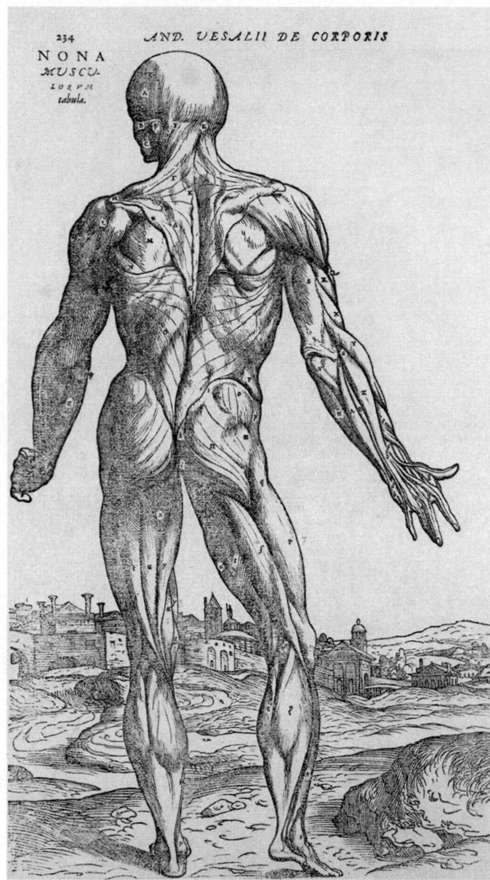

Figure 1.1 The study of the human body has a long history, as evidenced by this illustration from the second book of *De Humani Corporis Fabrica* by Andreas Vesalius, issued in 1543. (Note the similarity to the anatomical position, described later in this chapter.)

1.2 | Anatomy and Physiology

LEARN

1. Explain how anatomy and physiology are related.

Anatomy (ah-nat′o-mē) is the study of the structure of body parts—their forms and how they are organized. **Physiology** (fiz″ē-ol′o-jē) is the study of the functions of body parts—what they do and how they do it.

The topics of anatomy and physiology are difficult to separate because the structures of body parts are so closely associated with their functions. Body parts form a well-organized unit—the human organism—and each part functions in the unit's operation. A particular body part's function depends on the way the part is constructed—that is, how its subparts are organized. For example, the organization of the parts in the human hand with its long, jointed fingers makes it easy to grasp objects; the hollow chambers of the heart are adapted to pump blood through tubular blood vessels; the shape of the mouth enables it to receive food; and the eyes and ears are built to receive light waves and sound waves, respectively (fig. 1.2).

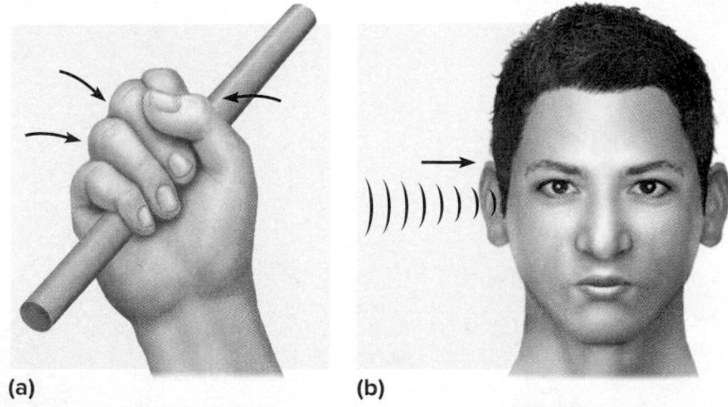

Figure 1.2 The structures of body parts make possible their functions: **(a)** The hand is adapted for grasping, **(b)** the ear is built to receive sound waves. (Arrows indicate the movements and direction of sound waves associated with these functions.)

As ancient as the fields of anatomy and physiology are, employing the scientific method, we are always learning more. Recent research using imaging technology has identified a previously unrecognized part of the brain, the planum temporale, which enables people to locate sounds in space. Many discoveries today begin with investigations at the microscopic level—molecules and cells. In this way, researchers discovered that certain cells in the small

CAREER CORNER
Emergency Medical Technician

The driver turns a corner and suddenly swerves as a cat dashes into the road. She slams on the brakes but hits a parked car, banging her head against the steering wheel. Onlookers call 911, and within minutes an ambulance arrives.

The driver of the ambulance and another emergency medical technician (EMT) leap out and run over to the accident scene. They open the driver-side door and quickly assess the woman's condition by evaluating her breathing and taking her blood pressure and pulse. She is bleeding from a laceration on her forehead, and is conscious but confused.

The EMTs carefully place a restraint at the back of the woman's neck and move her onto a board, then slide her into the ambulance. While one EMT drives, the other rides in the back with the patient and applies pressure to the cut. At the hospital, the EMTs document the care provided and clean and restock the ambulance.

EMTs care for ill or injured people in emergency situations and transport patients, such as from a hospital to a nursing home. The work is outdoors and indoors and requires quick thinking as well as strength. Requirements vary by state, but all EMTs must be licensed. Basic EMTs take 120–150 hours of training; paramedic EMTs take 1,200–1,800 hours of training. Paramedics may give injections, set up intravenous lines, and give more medications than can basic EMTs.

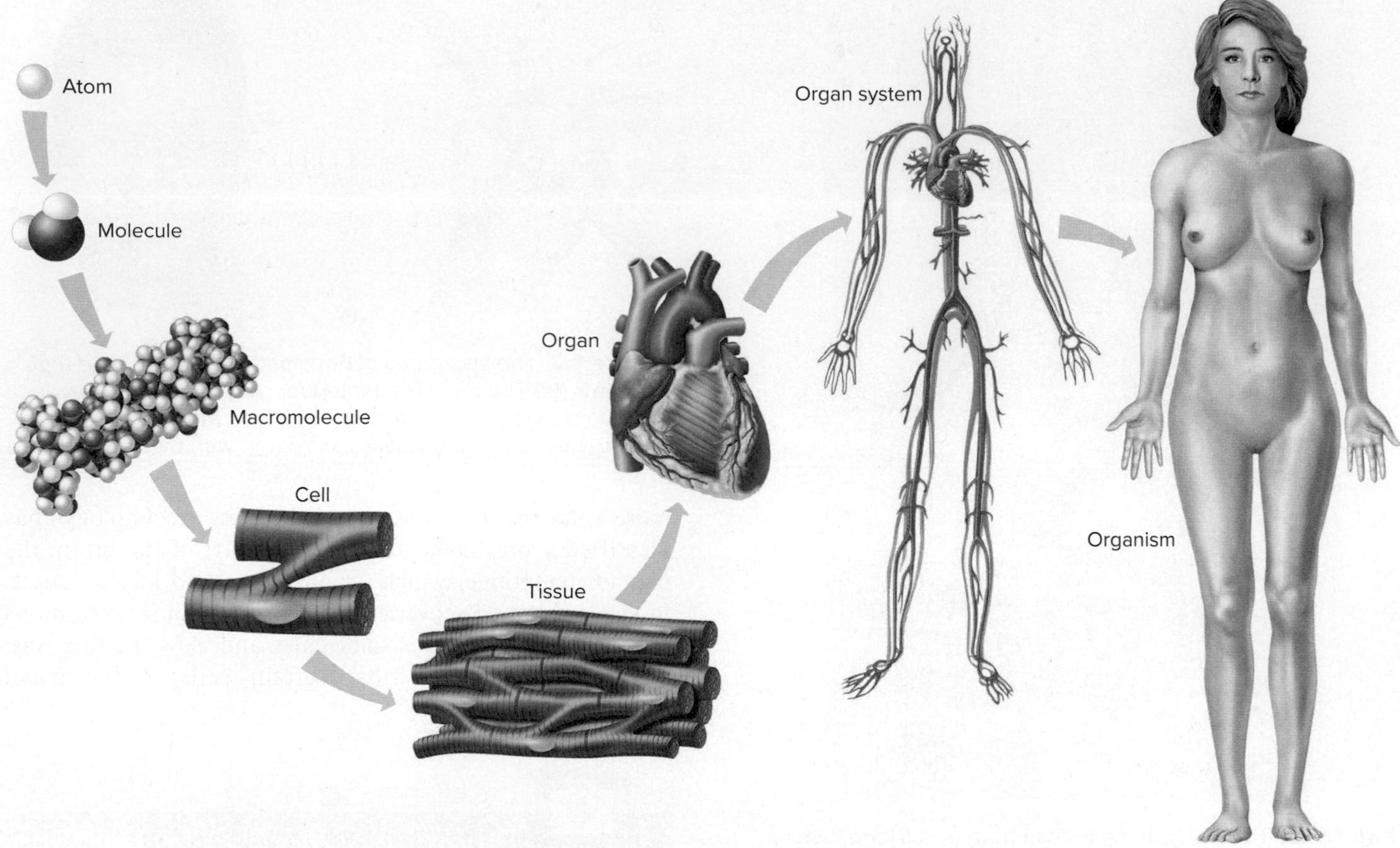

Figure 1.3 A human body is composed of parts made up of other parts, with increasing complexity.

intestine bear the same types of taste receptor proteins as found on the tongue. At both locations, the receptors detect molecules of sugar. The cells in the tongue provide taste sensations, whereas the cells in the intestines help regulate the digestion of sugar. The discovery of the planum temporale is anatomical; the discovery of sugar receptors in the intestine is physiological.

PRACTICE 1.2

1. Why is it difficult to separate the topics of anatomy and physiology?
2. List examples that illustrate how the structure of a body part makes possible its function.

1.3 | Levels of Organization

LEARN

1. List the levels of organization in the human body and the characteristics of each.

Until the invention of magnifying lenses and microscopes about 400 years ago, anatomists were limited in their studies to what they could see with the unaided eye—large parts. But with these new tools, investigators discovered that larger body structures are made up of smaller parts, which in turn are composed of even smaller ones.

Figure 1.3 shows the levels of organization that modern-day scientists recognize, increasing in complexity and size from atoms to organisms. All materials, including those that make up the human body, are composed of chemicals. Chemicals consist of microscopic particles called **atoms,** which join to form **molecules.** Small molecules can combine in complex ways to form larger **macromolecules.**

In all organisms, including humans, the smallest unit that exhibits all of the five characteristics of living systems is the **cell** (table 1.1). Cells are composed of the macromolecules proteins, carbohydrates, lipids, and nucleic acids. These macromolecules are the cornerstone of the physiology we will examine that occurs at the cellular level. Our bodies are built from trillions of cells that are organized into **tissues** to perform specific functions. Groups of different tissues that interact form **organs**—complex structures with specialized functions—and groups of organs that function closely together compose **organ systems.** Organ systems make up an **organism** (or'gah-nizm), which in this case is a multicellular living thing, a human. Figure 1.3 reveals this relationship using cardiac muscle cells and cardiac muscle tissue, the heart, and the entire cardiovascular system.

TABLE 1.1	Characteristics of Life
Process	**Description**
Growth	Increase in cell number and size and increase in body size
Reproduction	Production of new cells and organisms
Responsiveness	Reaction to a change inside or outside of the body
Movement	Change in body position or location; motion of internal organs
Metabolism	The sum of all chemical reactions in a living system: Energy production and nutrient cycling
	• **Respiration:** Making energy. Most organisms do it by taking in oxygen and giving off carbon dioxide • **Digestion:** Breaking down food into usable nutrients for absorption into the blood • **Circulation:** Moving chemicals and cells through the body fluids • **Excretion:** Removing waste products

PRACTICE 1.3

1. How does the human body illustrate levels of organization?
2. What is an organism?
3. How do body parts at different levels of organization vary in complexity?

1.4 | Characteristics of Life

LEARN

1. List and describe the major characteristics of life.
2. Give examples of metabolism.

All living organisms share five common qualities (table 1.1). We will use humans as our example. **Growth** is the increase in size from birth until adulthood. It involves both the increase in the size of cells and number of cells. **Reproduction** is the process of producing more of the same organisms. That is, having babies. **Responsiveness** is the ability to sense and react to changes inside or outside of a body such as eating when you are hungry or feeling pain when injured. **Movement** can be external, such as the change in position and location of a body in the environment; for example, when you walk or talk. Internal movements include the flow of blood and the beating of the heart. **Metabolism** is the sum of all chemical reactions in a body at any given time. It can also be described as *nutrient cycling*: taking in food, building and breaking down chemicals to make energy, and then giving off waste. This includes respiration, digestion, circulation, and excretion. All of these will be studied in great detail in coming chapters.

PRACTICE 1.4

1. What are the characteristics of life?
2. How are the characteristics of life dependent on metabolism?

1.5 | Maintenance of Life

LEARN

1. List and describe the major requirements of organisms.
2. Explain the importance of homeostasis to survival.
3. Describe the parts of a homeostatic mechanism and explain how they function together.

The structures and functions of almost all body parts help maintain life. Even an organism's reproductive structures, whose primary function is to ensure that the organism's species will continue into the future, may contribute to survival. For example, sex hormones help to strengthen bones.

Environmental Requirements to Maintain Life

Being alive requires certain environmental factors, including the following:

1. **Chemicals: Water** is the most abundant chemical in all living systems. It is required for many metabolic processes and provides the environment in which most of them take place. Water also carries substances within the organism and is important in regulating body temperature. Water inside the cells, along with substances dissolved in it, constitutes the *intracellular fluid.* Similarly, outside of the cells, including the tissue fluid and the liquid portion of the blood (plasma), is the *extracellular fluid* (fig. 1.4). Other chemicals such as carbon dioxide and oxygen are readily exchanged between living systems and the environment. Food, often called nutrients, is brought in and waste chemicals are eliminated.
2. **Heat** is a form of energy. It is a product of metabolic reactions, and the degree of heat present partly determines the rate at which these reactions occur. Generally, the more heat, the more rapidly chemical reactions take place. (*Temperature* is a measure of the degree of heat.)
3. **Pressure** is an application of force to something. For example, the force on the outside of the body due to the weight of air above it is called **atmospheric pressure.** In humans, this pressure is important in breathing. Similarly, organisms living underwater are subjected to **hydrostatic pressure**—a pressure a liquid exerts—due to the weight of water above them. In humans, heart action produces blood pressure (another form of hydrostatic pressure), which forces blood to flow through blood vessels.

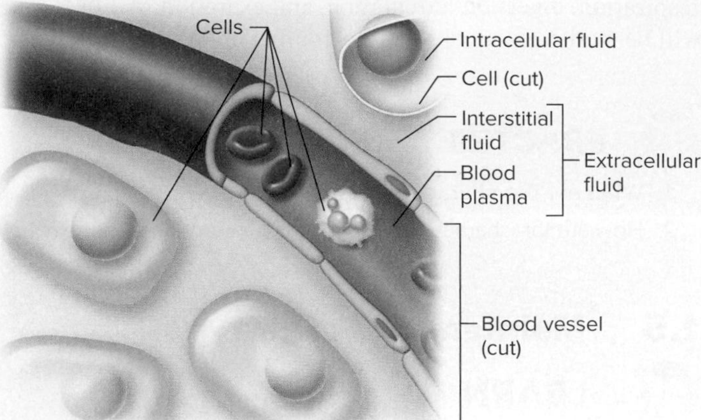

Figure 1.4 Intracellular and extracellular fluids. The extracellular fluid constitutes the internal environment of the body.

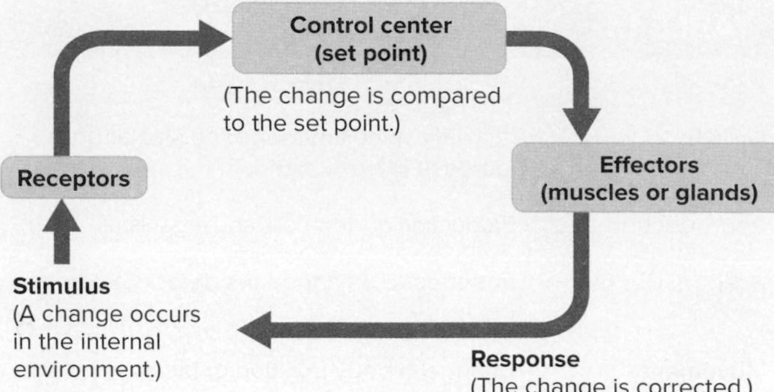

Figure 1.5 A homeostatic mechanism monitors a particular aspect of the internal environment and corrects any changes back to the value indicated by the set point.

Just as all animals do, humans require water, food, oxygen, heat, and pressure, but these alone are not enough to ensure survival. Both the quantities and the qualities of such factors are also important. For example, the volume of water entering and leaving an organism must be regulated, as must the concentration of oxygen in body fluids. Similarly, survival depends on the quality as well as the quantity of food available—that is, food must supply the correct nutrients in adequate amounts.

Homeostasis

Factors in the outside world, the external environment, may change. If an organism is to survive, however, conditions within the fluid surrounding its body cells, which compose its **internal environment,** must remain relatively stable. In other words, body parts function only when the concentrations of water, nutrients, and oxygen and the conditions of heat and pressure remain within certain narrow limits. This condition of a stable internal environment is called **homeostasis** (hō″mē-ō-stā′sis).

The body maintains homeostasis through a number of self-regulating control systems, called **homeostatic mechanisms,** that share the following three components (fig. 1.5):

- **Receptors** provide information about specific conditions (stimuli) in the internal environment.
- A **set point** tells what a particular value should be, such as body temperature at 37°C (Celsius) or 98.6°F (Fahrenheit). More about metric equivalents can be found in Appendix C; metric units are used throughout this text.
- **Effectors** bring about responses that alter conditions in the internal environment.

A homeostatic mechanism generally works as follows. If the receptors measure deviations from the set point, effectors are activated that can return conditions toward normal. As conditions return toward normal, the deviation from the set point progressively lessens and the effectors are gradually shut down. Such a response is called a **negative feedback** mechanism, both because the deviation from the set point is corrected (moves in the opposite or negative direction) and because the correction reduces the action of the effectors. This latter aspect is important because it prevents a correction from going too far.

To better understand the idea of negative feedback, imagine a room equipped with a furnace and an air conditioner (fig. 1.6*a*). If the room temperature is to remain near 20°C (68°F), the thermostat is adjusted to an operating level, or set point, of 20°C. A thermostat, which senses temperature changes, signals the furnace to start and the air conditioner to stop whenever the room temperature drops below the set point. If the temperature rises above the set point, the thermostat stops the furnace and starts the air conditioner. As a result, the room maintains a relatively constant temperature.

Body temperature is regulated by a homeostatic mechanism that is similar to control of room temperature (fig.1.6*b*). Temperature receptors are scattered throughout the body. The "thermostat" is a temperature-sensitive region in a temperature control center of the brain. In healthy people, the set point of the brain's thermostat is at or near 37°C (98.6°F).

If a person is exposed to cold and body temperature begins to drop, the temperature receptors sense this change and the temperature control center triggers heat-generating and heat-conserving activities. For example, small groups of muscles are stimulated to contract involuntarily, an action called *shivering.* Such muscular contractions produce heat, which helps warm the body. At the same time, blood vessels in the skin are signaled to constrict so that less warm blood flows through them. In this way, deeper tissues retain heat that might otherwise be lost.

If a person is becoming overheated, the brain's temperature control center triggers a series of changes that promote loss of body heat. Sweat glands in the skin secrete perspiration, and as this fluid evaporates from

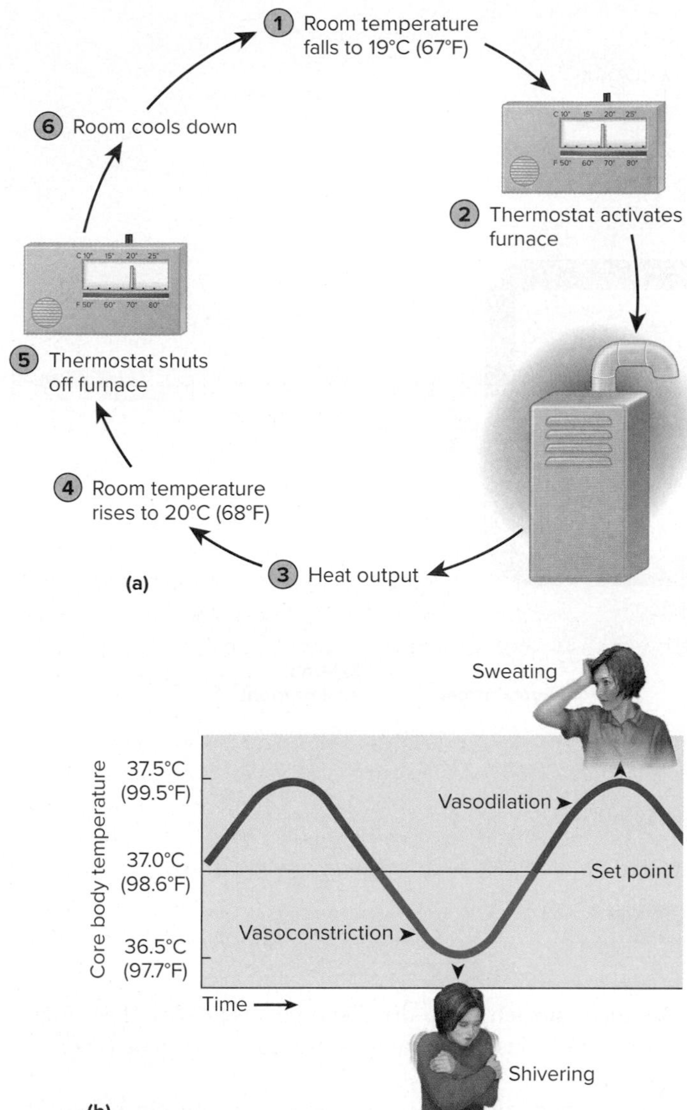

Figure 1.6 Negative feedback. **(a)** A furnace uses a negative feedback loop to maintain room temperature. **(b)** A homeostatic mechanism regulates body temperature.

PRACTICE FIGURE 1.6

What would happen to room temperature if the set point were turned up?

Answer can be found in Appendix E.

the surface, heat is carried away and the skin is cooled. At the same time, the brain center dilates blood vessels in the skin. This action allows more blood carrying heat from deeper tissues to reach the surface, where the heat is lost to the outside (fig. 1.6*b*). The brain stimulates an increase in heart rate, which sends a greater volume of blood into surface vessels, and an increase in breathing rate, which allows the lungs to expel more heat-carrying air. Body temperature regulation is discussed further in section 6.4, Skin Functions.

Another homeostatic mechanism regulates the blood pressure in the blood vessels (arteries) leading away from the heart. Pressure-sensitive receptors in the walls of these vessels sense changes in blood pressure and signal a pressure control center in the brain. If blood pressure is above the set point, the brain signals the heart chambers to contract more slowly and with less force. This decreased heart action sends less blood into the blood vessels, decreasing the pressure inside them. If blood pressure falls below the set point, the brain center signals the heart to contract more rapidly and with greater force. As a result, the pressure in the vessels increases. Section 13.5, Blood Pressure, discusses regulation of blood pressure in more detail.

Human physiology offers many other examples of homeostatic mechanisms. All work by the same general process as the two preceding examples. Just as anatomical terms are used repeatedly throughout this book, so can the basic principles of a homeostatic mechanism be applied to the different organ systems. Homeostatic mechanisms maintain a relatively constant internal environment, yet physiological values may vary slightly in a person from time to time or from one individual to another. Therefore, both normal values for an individual and the *normal range* for the general population are clinically important.

Organ systems contribute to homeostasis in different ways. Resources brought in by the digestive and respiratory systems are delivered to all body cells by the cardiovascular system. The same blood that brings needed nutrients to cells carries away waste products, which are removed by the respiratory and urinary systems (fig. 1.7).

Most feedback mechanisms in the body are negative. However, sometimes change stimulates further change. A process that moves conditions away from the normal state is called a *positive feedback mechanism.* In blood clotting (blood coagulation), for example, the chemicals that carry out clotting stimulate more clotting, minimizing bleeding (see section 12.4, Hemostasis). Another positive feedback mechanism increases the strength of uterine contractions during childbirth, helping to bring the new individual into the world (see section 20.3, Pregnancy and the Prenatal Period).

Positive feedback mechanisms usually produce unstable conditions, which might seem incompatible with homeostasis. However, the examples of positive feedback associated with normal function have very specific roles and are short-lived.

PRACTICE 1.5

1. Which requirements of organisms does the external environment provide?
2. Why is homeostasis important to survival?
3. Describe two homeostatic mechanisms.

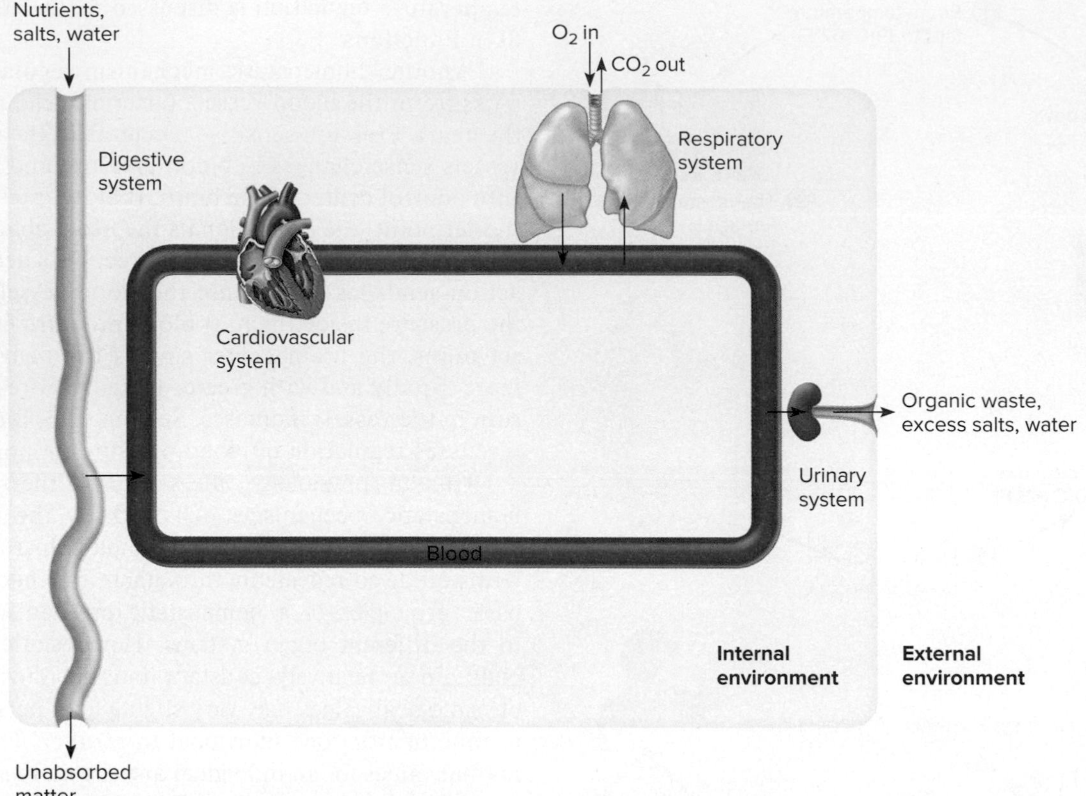

Figure 1.7 Examples of how organ systems contribute in different ways to maintenance of the internal environment.

1.6 | Organization of the Human Body

LEARN

1. Identify the locations of the major body cavities.
2. List the organs located in each major body cavity.
3. Name and identify the locations of the membranes associated with the thoracic and abdominopelvic cavities.
4. Name the major organ systems, and list the organs associated with each.
5. Describe the general functions of each organ system.

The human organism is a complex structure composed of many parts. Its major features include several body cavities, layers of membranes within these cavities, and a variety of organ systems.

Body Cavities

The human organism can be divided into an **axial** (ak'sē-al) portion, which includes the head, neck, and trunk, and an **appendicular** (ap"en-dik'ū-lar) portion, which includes the upper and lower limbs. Within the axial portion are the **cranial cavity,** which houses the brain; the **vertebral canal,** which contains the spinal cord within the sections of the backbone (vertebrae); the **thoracic** (tho-ras'ik) **cavity;** and the **abdominopelvic** (ab-dom"ĭ-no-pel'vik) **cavity.** The organs within these last two cavities are called **viscera** (vis'er-ah) (fig. 1.8*a*).

A broad, thin skeletal (voluntary) muscle called the **diaphragm** separates the thoracic cavity from the abdominopelvic cavity. The thoracic cavity wall is composed of skin, skeletal muscles, and various bones.

A compartment called the **mediastinum** (me"de-as-ti'num) forms a boundary between the right and left sides of the thoracic cavity. The mediastinum contains most of the thoracic cavity viscera (including the heart, esophagus, trachea, and thymus) except for the lungs. The right and left lungs are on either side of the mediastinum (fig. 1.8*b*).

The abdominopelvic cavity, which includes an upper abdominal portion and a lower pelvic portion, extends from the diaphragm to the floor of the pelvis. Its wall consists primarily of skin, skeletal muscles, and bones. The viscera within the **abdominal cavity** include the stomach, liver, spleen, gallbladder, kidneys, and most of the small and large intestines.

The **pelvic cavity** is the portion of the abdominopelvic cavity enclosed by the hip bones (see section 7.11, Pelvic Girdle). It contains the terminal portion of the large intestine, the urinary bladder, and the internal reproductive organs.

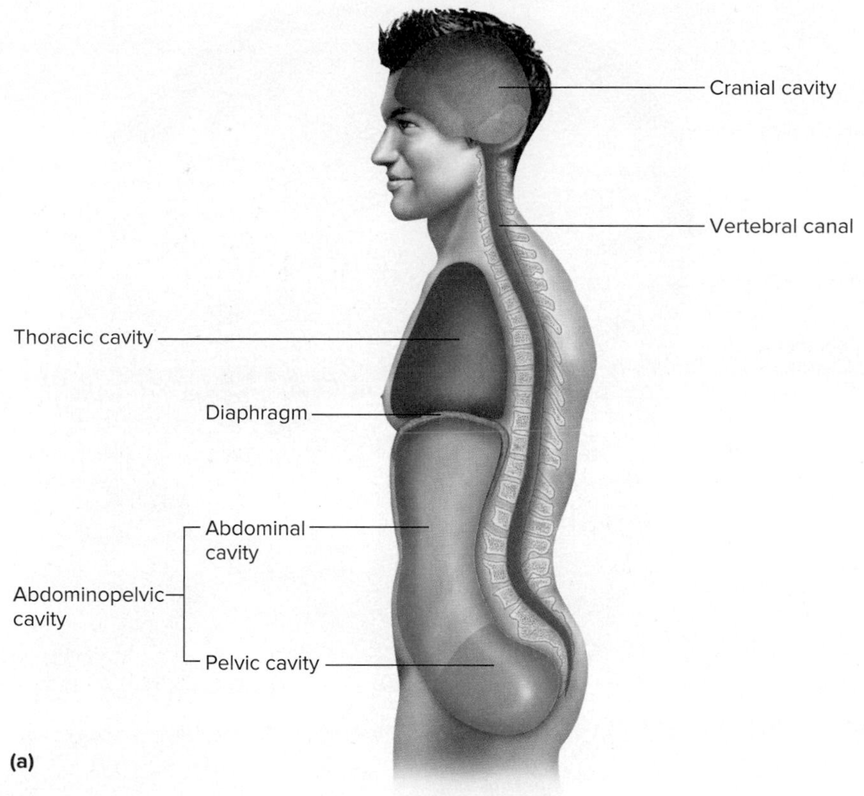

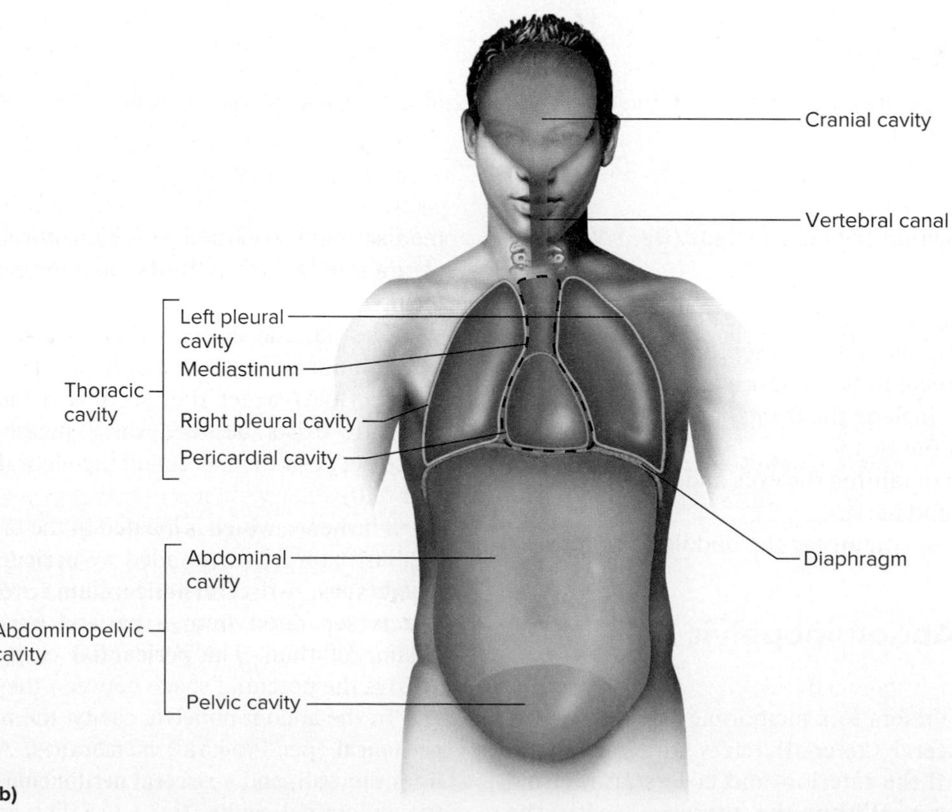

Figure 1.8 Major body cavities. **(a)** Lateral view. **(b)** Anterior view. APR

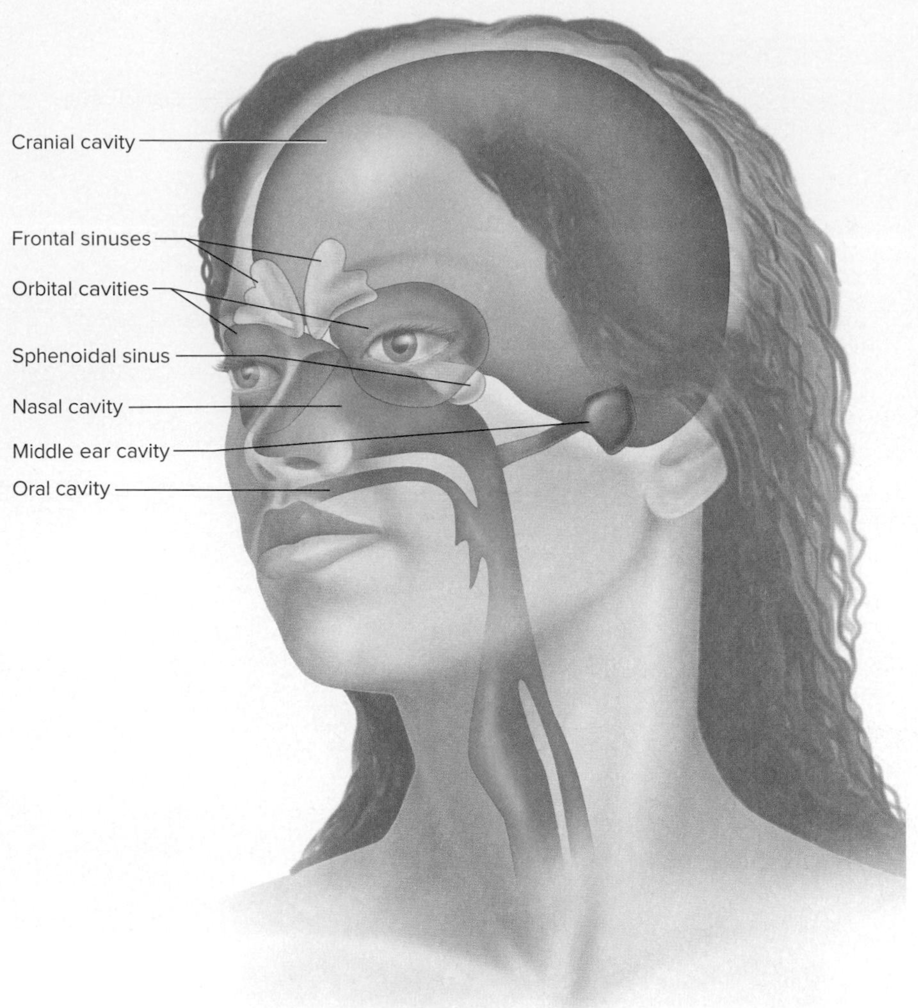

Figure 1.9 The cavities within the head include the cranial, oral, nasal, orbital, and middle ear cavities, as well as several sinuses. (Not all of the sinuses are shown.)

Smaller cavities within the head include (fig. 1.9):

1. **Oral cavity,** containing the teeth and tongue.
2. **Nasal cavity,** located within the nose and divided into right and left portions by a nasal septum. Several air-filled *sinuses* connect to the nasal cavity (see section 7.6, Skull). These include the frontal and sphenoidal sinuses shown in figure 1.9.
3. **Orbital cavities,** containing the eyes and associated skeletal muscles and nerves.
4. **Middle ear cavities,** containing the middle ear bones.

Thoracic and Abdominopelvic Membranes

Parietal (pah-rī′ĕ-tal) refers to a membrane attached to the wall of a cavity; **visceral** (vis′er-al) refers to a membrane that is deeper—toward the interior—and covers an internal organ, such as a lung. Within the thoracic cavity, the compartments that contain the lungs, on either side of the mediastinum, are lined with a membrane called the **parietal pleura** (fig. 1.10). A similar membrane, called the **visceral pleura,** covers each lung.

The parietal and visceral **pleural membranes** (ploo′ral mem′brānz) are separated only by a thin film of watery fluid (serous fluid), which they secrete. Although no actual space normally exists between these membranes, the potential space between them is called the **pleural cavity** (see figs. 1.8*b* and 1.10).

The heart, which is located in the broadest portion of the mediastinum, is surrounded by **pericardial** (per″ĭ-kar′dē-al) **membranes.** A **visceral pericardium** covers the heart's surface and is separated from a **parietal pericardium** by a small volume of fluid. The **pericardial cavity** (see figs. 1.8*b* and 1.10) is the potential space between these membranes.

In the abdominopelvic cavity, the membranes are called **peritoneal** (per″ĭ-to-nē′al) **membranes.** A **parietal peritoneum** lines the wall, and a **visceral peritoneum** covers each organ in the abdominal cavity (fig. 1.11). The **peritoneal cavity** is the potential space between these membranes.

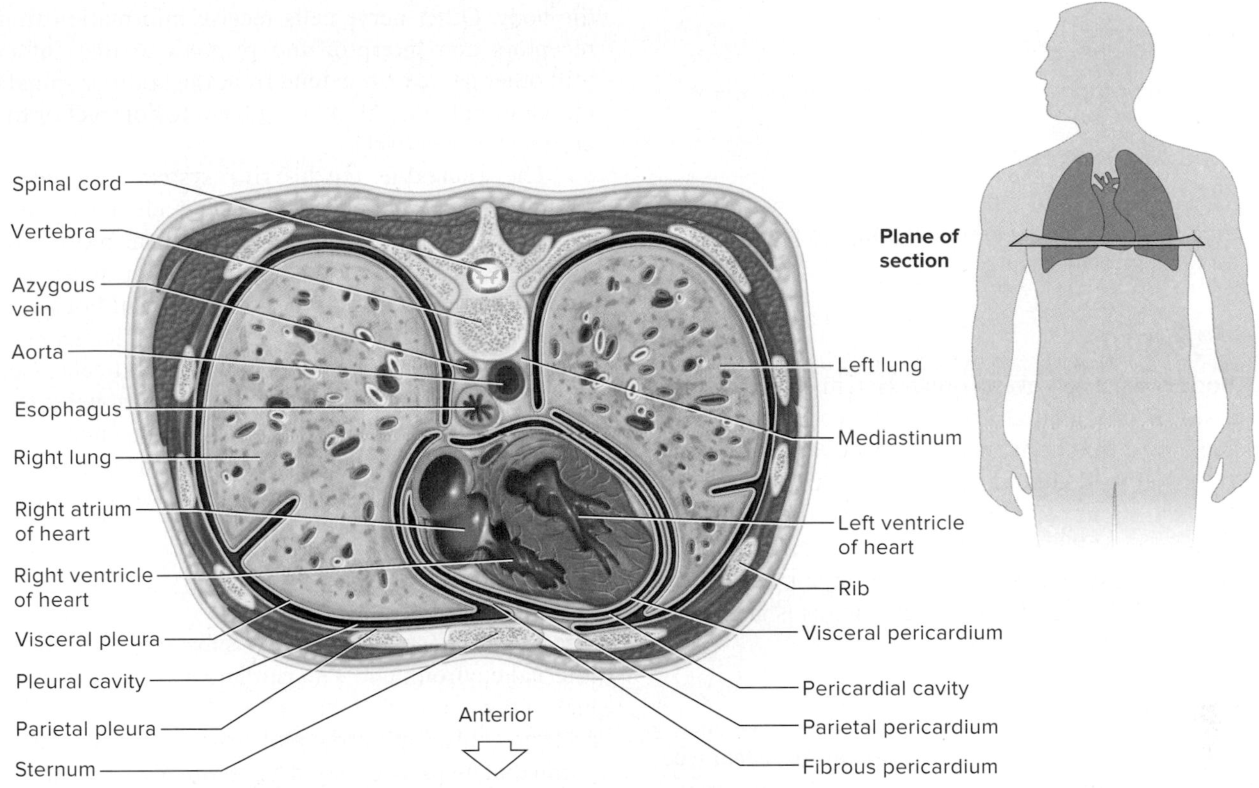

Figure 1.10 A transverse section through the thorax reveals the serous membranes associated with the heart and lungs (superior view). APR

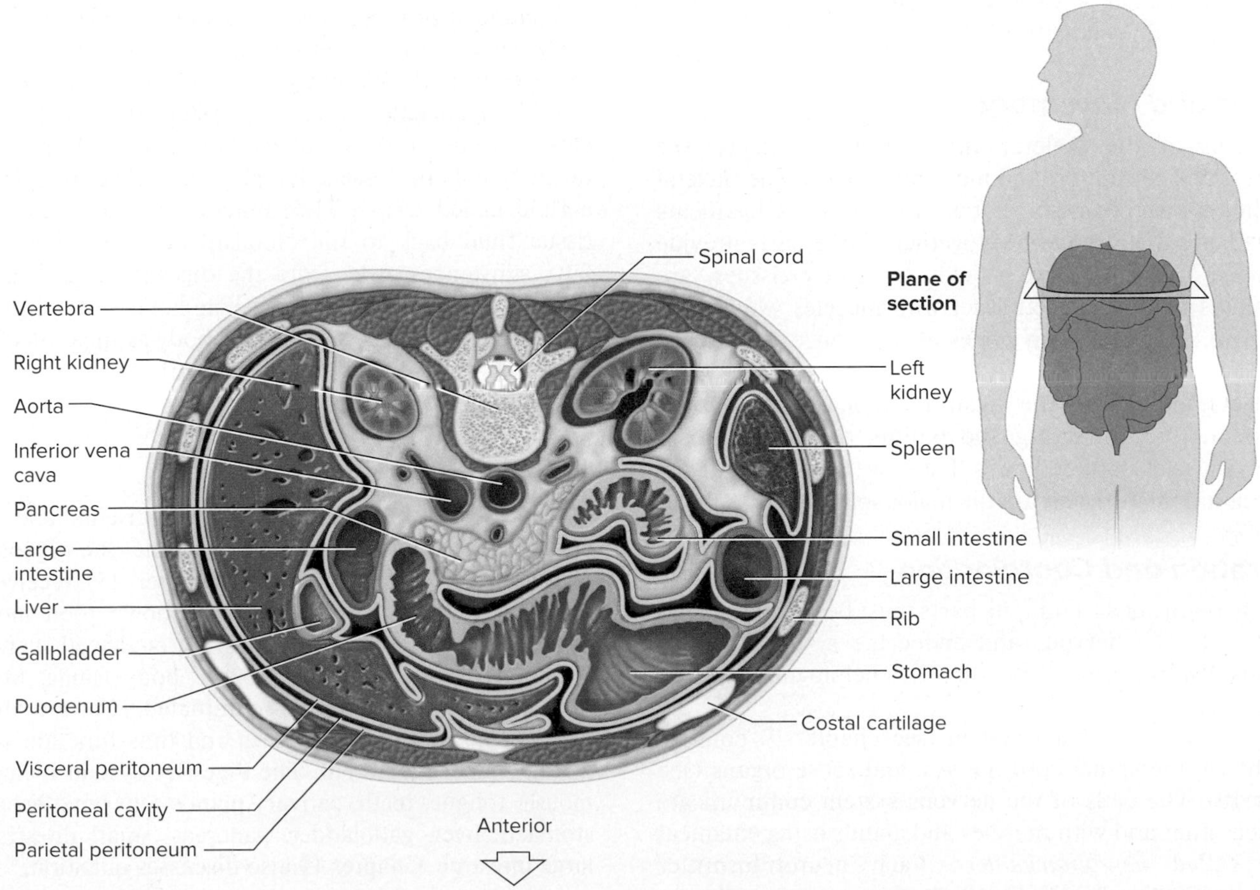

Figure 1.11 Transverse section through the abdomen (superior view). APR

PRACTICE 1.6

1. Which organ occupies the cranial cavity? The vertebral canal?
2. What does *viscera* mean?
3. Name the cavities of the head.
4. Describe the membranes associated with the thoracic and abdominopelvic cavities.

Organ Systems

A human body consists of several organ systems. Each system includes a set of interrelated organs that work together to provide specialized functions that contribute to homeostasis (fig. 1.12). As you read about each system, you may want to consult the illustrations of the human torso in the Reference Plates (see Reference Plates–The Human Organism, at the end of chapter 1) and locate some of the organs described. The following introduction to the organ systems is presented according to overall functions.

Body Covering

Organs of the **integumentary** (in-teg-ū-men′tar-ē) **system** (see chapter 6) include the skin and various accessory organs, such as the hair, nails, sweat glands, and sebaceous glands. These parts protect underlying tissues, help regulate body temperature, house a variety of sensory receptors, and synthesize certain products.

Support and Movement

The organs of the skeletal and muscular systems (see chapters 7 and 8) support and move body parts. The **skeletal** (skel′ĕ-tal) **system** consists of bones, as well as ligaments and cartilages that bind bones together. These parts provide frameworks and protective shields for softer tissues, are attachments for muscles, and act with muscles when body parts move. Tissues within bones also produce blood cells and store inorganic salts.

Skeletal muscles are the organs of the **muscular** (mus′ku-lar) **system.** By contracting and pulling their ends closer together, muscles provide forces that move body parts. They also maintain posture and are the major source of body heat.

Integration and Coordination

For the body to act as a unit, its parts must be integrated and coordinated. The nervous and endocrine systems control and adjust various organ functions, thus helping to maintain homeostasis.

The **nervous** (ner′vus) **system** (see chapter 9) consists of the brain, the spinal cord, nerves, and sense organs (see chapter 10). The cells of the nervous system communicate with each other and with muscles and glands using chemical signals called *neurotransmitters.* Each neurotransmitter exerts a relatively short-term effect. Some nerve cells are specialized receptors that detect changes inside and outside the body. Other nerve cells receive information from these receptors and interpret and respond to that information. Still other nerve cells extend from the brain or spinal cord to muscles or glands, stimulating them to contract or to secrete products, respectively.

The **endocrine** (en′do-krin) **system** (see chapter 11) includes all the glands that secrete chemical messengers called *hormones.* The hormones, in turn, move away from the glands in body fluids, such as blood or tissue fluid (fluid from the spaces within tissues). A particular hormone affects only a particular group of cells, called its *target cells.* A hormone alters the metabolism of its target cells. Compared to the action of a neurotransmitter, hormonal effects start less rapidly and last longer. Organs of the endocrine system include the hypothalamus of the brain; the pituitary, thyroid, parathyroid, and adrenal glands; and the pancreas, ovaries, testes, pineal gland, and thymus.

Transport

Two organ systems transport substances throughout the internal environment. The **cardiovascular** (kahr″dē-ō-vas′ku-lur) **system** (see chapters 12 and 13) includes the heart, arteries, veins, capillaries, and blood. The heart is a muscular pump that helps force blood through the blood vessels. Blood carries gases, nutrients, hormones, and wastes. It transports oxygen from the lungs and nutrients from the digestive organs to all body cells, where these biochemicals are used in metabolic processes. Blood also transports wastes from body cells to the excretory organs, where the wastes are removed from the blood and released to the outside.

The **lymphatic** (lim-fat′ik) **system** (see chapter 14) is closely related to the cardiovascular system. It is composed of the lymphatic vessels, lymph nodes, thymus, spleen, and a fluid called *lymph.* This system transports some of the tissue fluid back to the bloodstream and carries certain fatty substances away from the digestive organs and into the bloodstream. Cells of the lymphatic system are called lymphocytes, and they defend the body against infections by removing disease-causing microorganisms and viruses from tissue fluid.

Absorption and Excretion

Organs in several systems absorb nutrients and oxygen and excrete various wastes. For example, the organs of the **digestive** (dī-jest′iv) **system** (see chapter 15) receive foods from the outside. Then they break down food molecules into simpler forms that can pass through cell membranes and thereby be absorbed into the body fluids. Materials that are not absorbed are eliminated. Certain digestive organs also produce hormones and thus function as parts of the endocrine system. The digestive system includes the mouth, tongue, teeth, salivary glands, pharynx, esophagus, stomach, liver, gallbladder, pancreas, small intestine, and large intestine. Chapter 15 also discusses nutrition.

The organs of the **respiratory** (re-spi′rah-to″rē) **system** (see chapter 16) move air in and out of the lungs and

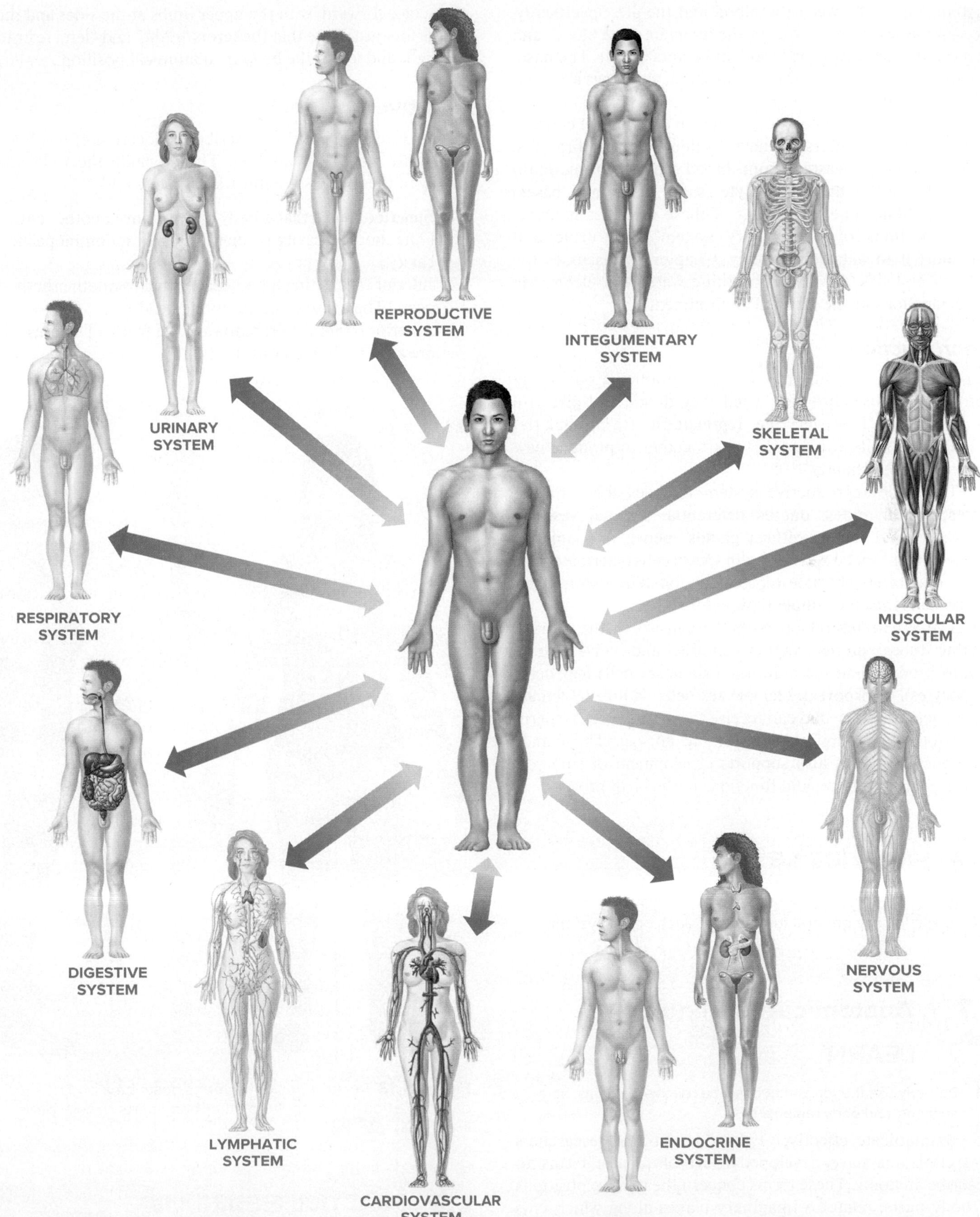

Figure 1.12 The organ systems in humans interact, maintaining homeostasis. APR

exchange gases between the blood and the air. Specifically, oxygen passes from air within the lungs into the blood, and carbon dioxide leaves the blood and enters the air. The nasal cavity, pharynx, larynx, trachea, bronchi, and lungs are parts of this system.

The **urinary** (ū′rĭ-ner″ē) **system** (see chapter 17) consists of the kidneys, ureters, urinary bladder, and urethra. The kidneys remove wastes from blood and help maintain the body's water and electrolyte (salts, acids, and bases) concentrations. The by-product of these activities is urine. Other portions of the urinary system store urine and transport it to outside the body. Chapter 18 discusses the urinary system's role in maintaining water and electrolyte concentrations of the internal environment.

Reproduction

Reproduction is the process of producing offspring (progeny). Cells reproduce when they divide and give rise to new cells. However, the **reproductive** (rē″pro-duk′tiv) **systems** of a male and female work together to produce new organisms (see chapter 19).

The male reproductive system includes the scrotum, testes, epididymides, ductus deferentia, seminal vesicles, prostate gland, bulbourethral glands, penis, and urethra. These parts produce and maintain sperm cells (spermatozoa). Components of the male reproductive system also transfer sperm cells into the female reproductive tract.

The female reproductive system consists of the ovaries, uterine tubes, uterus, vagina, clitoris, and vulva. These organs produce and maintain the female sex cells (egg cells, or oocytes), transport the female sex cells within the female reproductive system, and can receive the male sex cells (sperm cells) for the possibility of fertilizing an egg. The female reproductive system also supports development of embryos, carries fetuses to term, and functions in the birth process.

PRACTICE 1.6

5. Name and list the organs of the major organ systems.
6. Describe the general functions of each organ system.

1.7 | Anatomical Terminology

LEARN

1. Properly use the terms that describe relative positions, body sections, and body regions.

To communicate effectively with one another, researchers and clinicians have developed a set of precise terms to describe anatomy. These terms concern the relative positions of body parts, relate to imaginary planes along which cuts may be made, and describe body regions.

Use of such terms assumes that the body is in the **anatomical position.** This means that the body is standing erect, face forward, with the upper limbs at the sides and the palms forward. Note that the terms "right" and "left" refer to the right and left of the body in anatomical position.

Relative Positions

Terms of relative position describe the location of one body part with respect to another. They include the following (many of these terms are illustrated in fig. 1.13):

1. **Superior** means that a body part is above another part. (The thoracic cavity is superior to the abdominopelvic cavity.)
2. **Inferior** means that a body part is below another body part. (The neck is inferior to the head.)
3. **Anterior** (*ventral*) means toward the front. (The eyes are anterior to the brain.)

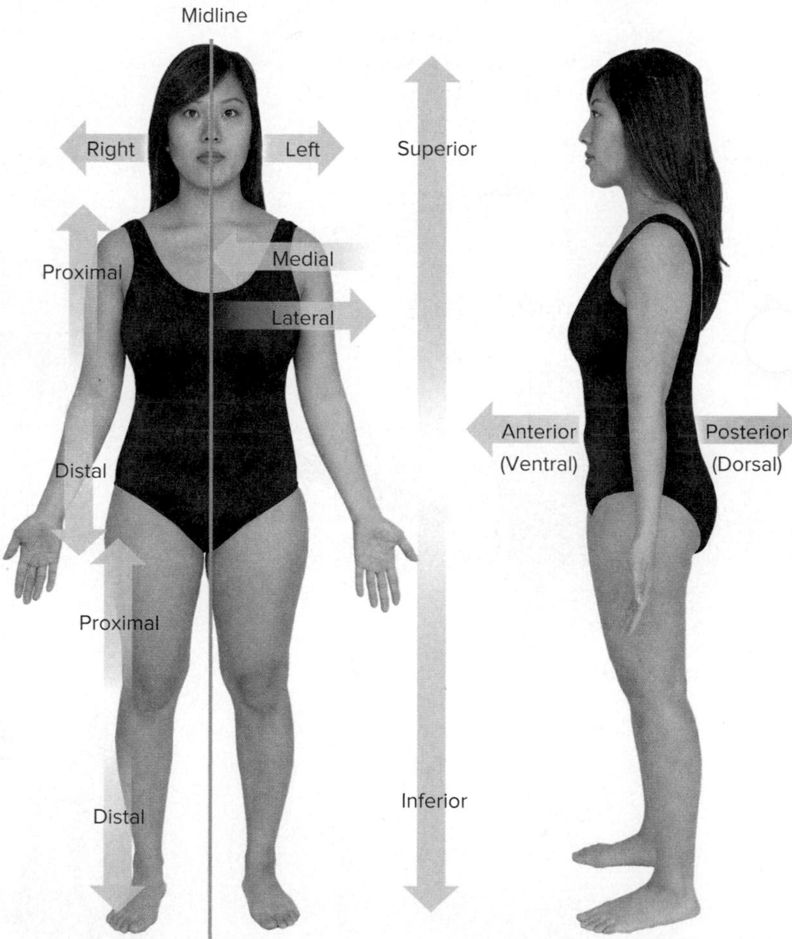

Figure 1.13 Relative positional terms describe a body part's location with respect to other body parts. ©Aaron Roeth Photography APR

PRACTICE FIGURE 1.13

Which is more lateral, the hand or the hip?

Answer can be found in Appendix E.

4. **Posterior** (*dorsal*) means toward the back. (The pharynx is posterior to the oral cavity.)
5. **Medial** refers to an imaginary midline dividing the body into equal right and left halves. A body part is medial if it is closer to midline than another part. (The nose is medial to the eyes.)
6. **Lateral** means toward the side, away from midline. (The ears are lateral to the eyes.)
7. **Bilateral** refers to paired structures, one of which is on each side of midline. (The lungs are bilateral.)
8. **Ipsilateral** refers to structures on the same side. (The right lung and the right kidney are ipsilateral.)
9. **Contralateral** refers to structures on the opposite side. (A patient with a fractured bone in the right leg would have to bear weight on the contralateral–in this case, left–lower limb.)
10. **Proximal** describes a body part that is closer to a point of attachment to the trunk than another body part is. (The elbow is proximal to the wrist.) *Proximal* may also refer to another reference point, such as the proximal tubules, which are closer to the filtering structures in the kidney.
11. **Distal** is the opposite of proximal. It means that a particular body part is farther from a point of attachment to the trunk than another body part is. (The fingers are distal to the wrist.) Distal may also refer to another reference point, such as decreased blood flow distal to blockage of a coronary artery.
12. **Superficial** means situated near the surface. (The epidermis is the superficial layer of the skin.) *Peripheral* also means outward or near the surface. It describes the location of certain blood vessels and nerves. (The nerves that branch from the brain and spinal cord are peripheral nerves.)
13. **Deep** describes parts that are more internal than superficial parts. (The dermis is the deep layer of the skin.)

Body Sections

Observing the relative locations and organization of internal body parts requires cutting or sectioning the body along various planes (fig. 1.14). The following terms describe such planes and the sections that result:

1. **Sagittal** refers to a lengthwise plane that divides the body into right and left portions. If a sagittal plane passes along the midline and thus divides the body into equal parts, it is called *median* (midsagittal). A sagittal section lateral to midline is called *parasagittal.*

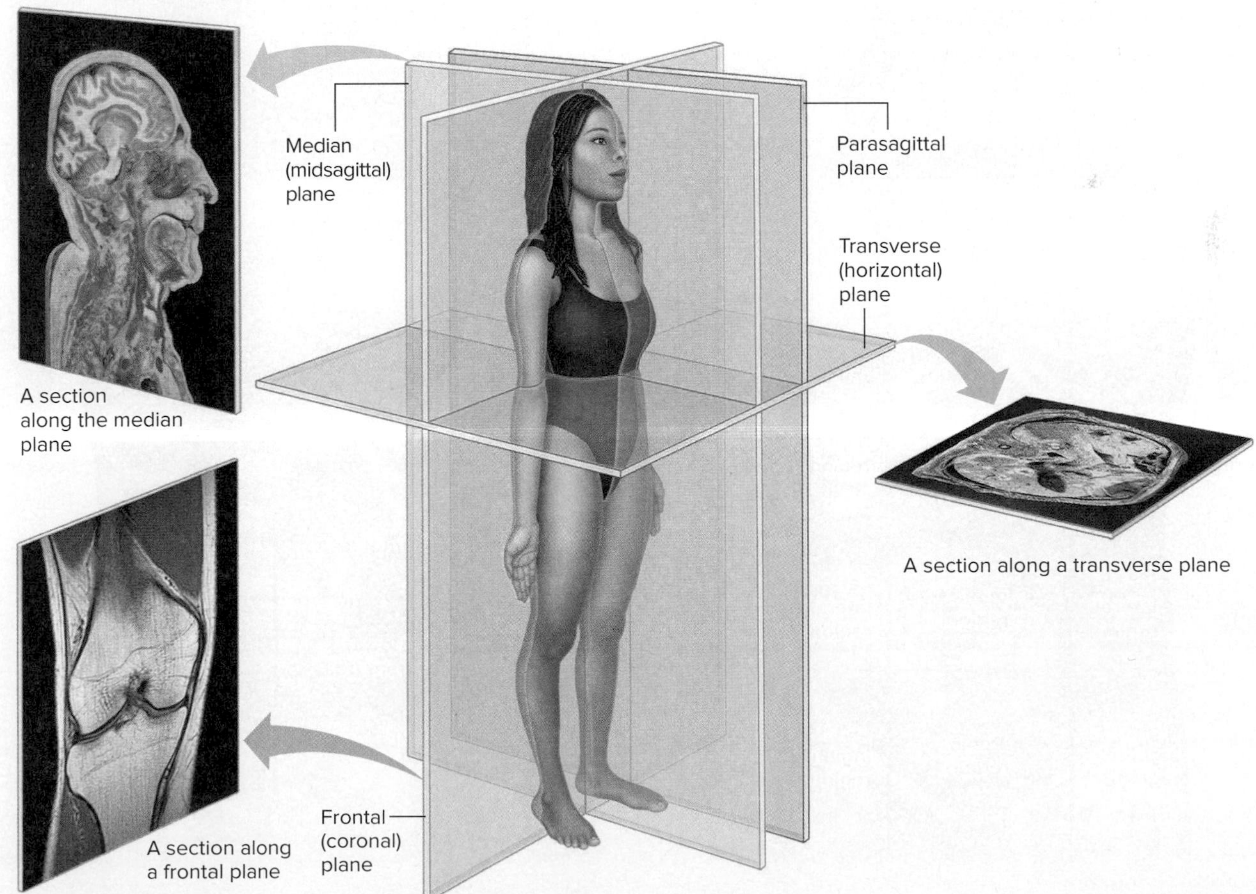

Figure 1.14 Observation of internal parts requires sectioning the body along various planes. (top left, right): ©McGraw-Hill Education/Karl Rubin, photographer; (bottom left): ©Living Art Enterprises/Science Source; (center): ©McGraw-Hill Education/Joe DeGrandis, photographer APR

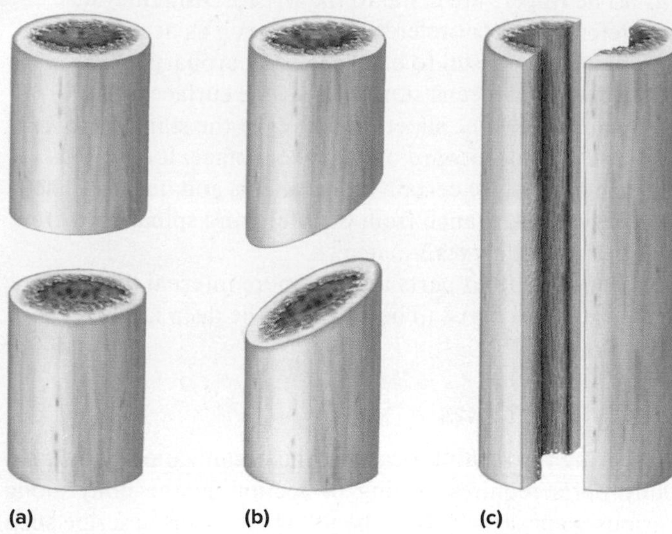

Figure 1.15 Cylindrical parts may be cut in **(a)** cross section, **(b)** oblique section, or **(c)** longitudinal section.

2. **Transverse** (*horizontal*) refers to a plane that divides the body into superior and inferior portions.
3. **Frontal** (*coronal*) refers to a plane that divides the body into anterior and posterior portions.

Sometimes, a cylindrical organ such as a long bone is sectioned. In this case, a cut across the structure is called a *cross section,* an angular cut is an *oblique section,* and a lengthwise cut is a *longitudinal section* (fig. 1.15). Clinical Application 1.1 discusses using computerized tomography to view body sections.

Body Regions

A number of terms designate body regions. The abdominal area, for example, is subdivided into the following nine regions, as figure 1.16*a* shows:

1. The **epigastric region** is the upper middle portion.
2. The **right** and **left hypochondriac regions** lie on each side of the epigastric region.

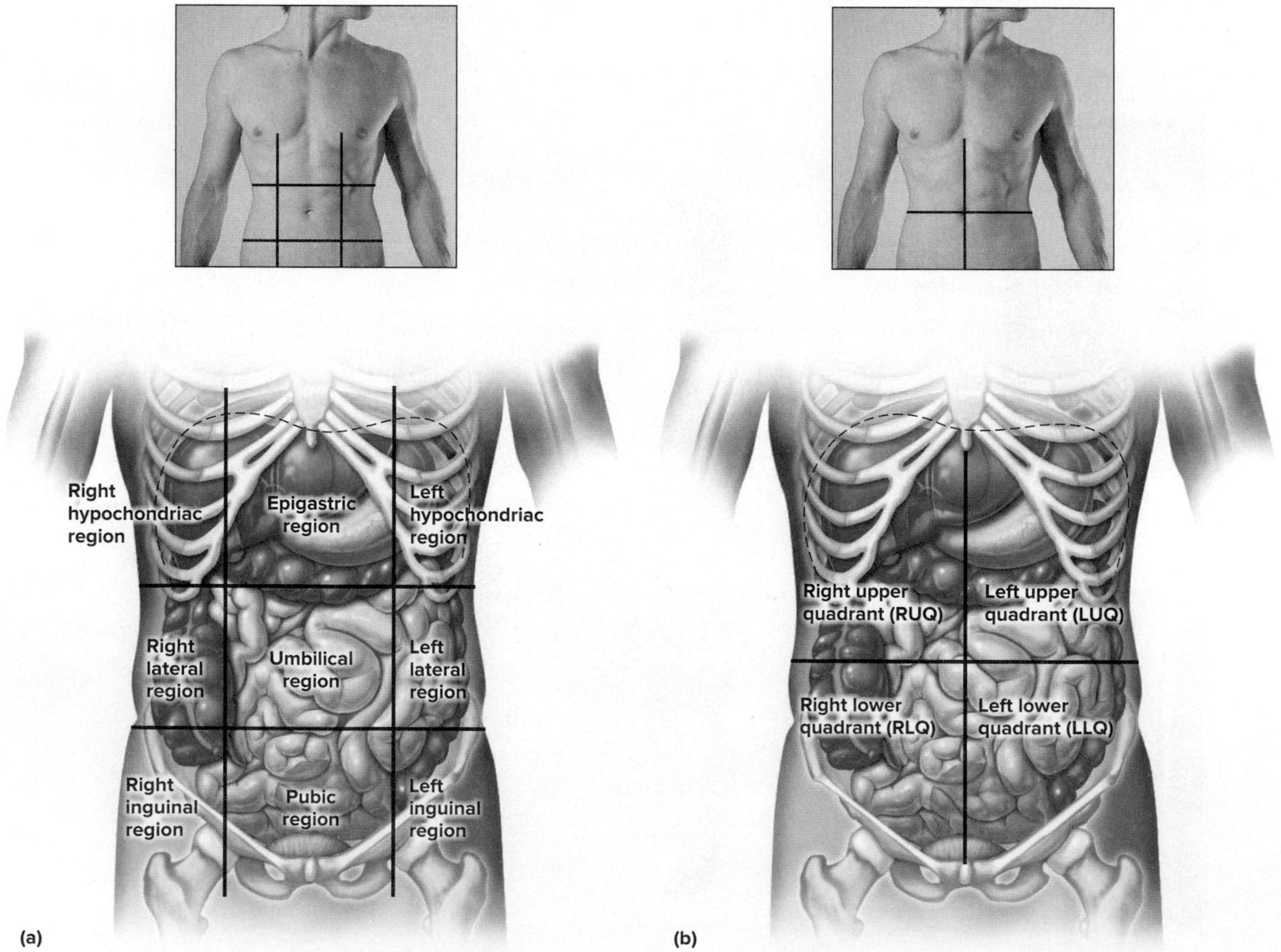

Figure 1.16 The abdominal area is commonly subdivided in two ways: **(a)** into nine regions and **(b)** into four quadrants.
Photos: ©Juice Images/Alamy RF APR

3. The **umbilical region** is the middle portion.
4. The **right** and **left lateral (lumbar) regions** lie on each side of the umbilical region.
5. The **pubic (hypogastric) region** is the lower middle portion.
6. The **right** and **left inguinal (iliac) regions** lie on each side of the pubic region.

The abdominal area is also often subdivided into four quadrants, as figure 1.16*b* shows.

The following adjectives are commonly used to refer to various body regions, some of which are illustrated in figure 1.17:

abdominal (ab-dom′ĭ-nal) The region between the thorax and pelvis.
acromial (ah-kro′me-al) The point of the shoulder.
antebrachial (an″tē-brā′ke-al) The forearm.
antecubital (an″te-ku′bĭ-tal) The space in front of the elbow.
axillary (ak′sĭ-ler″ē) The armpit.
brachial (brā′kē-al) The arm.
buccal (buk′al) The cheek.
calcaneal (kal-kā′-nee-al) The heel.
carpal (kar′pal) The wrist.
celiac (sē′lē-ak) The abdomen.
cephalic (sĕ-fal′ik) The head.
cervical (ser′vĭ-kal) The neck.
costal (kos′tal) The ribs.
coxal (kok′sal) The hip.
crural (kroor′al) The leg.

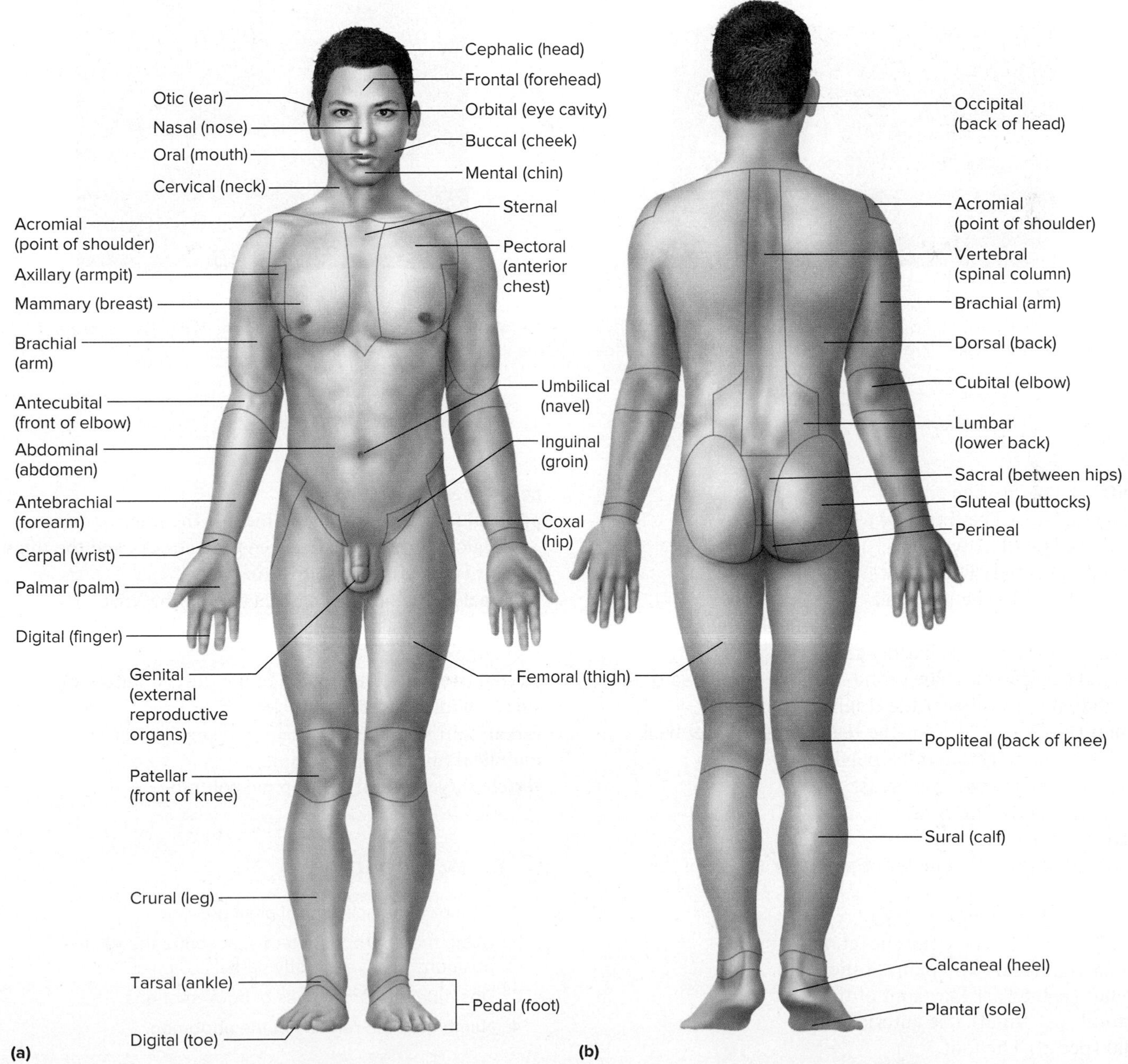

Figure 1.17 Some terms used to describe body regions. **(a)** Anterior regions. **(b)** Posterior regions. APR

CLINICAL APPLICATION 1.1
Computerized Tomography

Radiologists use a procedure called *computerized tomography,* or CT scanning, to visualize internal organ sections (fig. 1A). In this procedure, an X-ray-emitting device moves around the body region being examined or it may image the entire body. At the same time, an X-ray detector moves in the opposite direction on the other side. As the devices move, an X-ray beam passes through the body from hundreds of different angles. Because tissues and organs of varying composition within the body absorb X rays differently, the amount of X ray reaching the detector varies from position to position. A computer records the measurements from the X-ray detector, and combines them mathematically to create a sectional image of the internal body parts that can be viewed on a monitor.

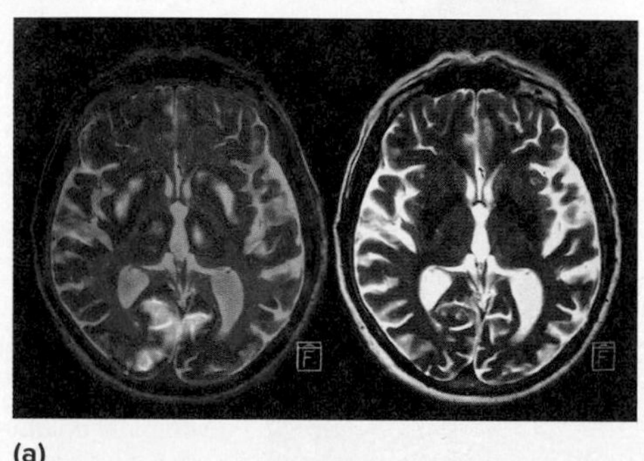

(a)

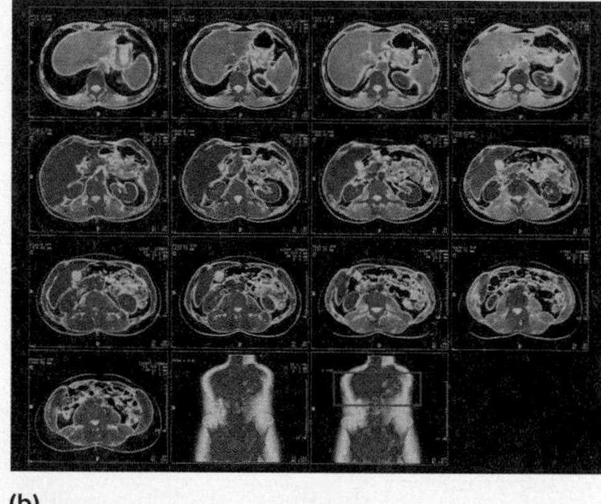

(b)

Figure 1A Falsely colored CT (computerized tomography) scans of **(a)** the head and **(b)** the abdomen. *Note: These are not shown in correct relative size.* (a): ©Ohio Nuclear Corporation/Science Source; (b): ©CNRI/Science Source

cubital (ku'bĭ-tal) The elbow.
digital (dij'ĭ-tal) The finger or toe.
dorsal (dor'sal) The back.
femoral (fem'or-al) The thigh.
frontal (frun'tal) The forehead.
genital (jen'ĭ-tal) The external reproductive organs.
gluteal (gloo'tē-al) The buttocks.
inguinal (ing'gwĭ-nal) The groin—the depressed area of the abdominal wall near the thigh.
lumbar (lum'bar) The loin—the region of the lower back between the ribs and the pelvis.
mammary (mam'er-ē) The breast.
mental (men'tal) The chin.
nasal (nā'zal) The nose.
occipital (ok-sip'ĭ-tal) The lower posterior region of the head.
oral (o'ral) The mouth.
orbital (or'bi-tal) The bony socket of the eye.
palmar (pahl'mar) The palm of the hand.
patellar (pah-tel'ar) The front of the knee.
pectoral (pek'tor-al) The anterior chest.
pedal (ped'al) The foot.
pelvic (pel'vik) The pelvis.
perineal (per''ĭ-nē'al) The perineum—the inferior-most region of the trunk between the buttocks and the thighs.
plantar (plan'tar) The sole of the foot.
popliteal (pop''lĭ-tē'al) The area behind the knee.
sacral (sa'kral) The posterior region between the hip bones.
sternal (ster'nal) The middle of the thorax, anteriorly.
sural (su'ral) The calf of the leg.
tarsal (tahr'sal) The ankle.
umbilical (um-bil'ĭ-kal) The navel.
vertebral (ver'te-bral) The spinal column.

PRACTICE 1.7

1. Describe the anatomical position.
2. Using the appropriate terms, describe the relative positions of several body parts.
3. Describe the three types of body sections.
4. Name the nine regions of the abdomen.

SOME MEDICAL AND APPLIED SCIENCES

cardiology (kar″dē-ol′o-jē) Branch of medical science dealing with the heart and heart diseases.

cytology (sī-tol′o-jē) Study of the structure, function, and abnormalities of cells. Cytology and histology are subdivisions of microscopic anatomy.

dermatology (der″mah-tol′o-jē) Study of the skin and its diseases.

endocrinology (en″dō-krĭ-nol′o-jē) Study of hormones, hormone-secreting glands, and their diseases.

epidemiology (ep″ĭ-dē″mē-ol′o-jē) Study of the factors determining the distribution and frequency of health-related conditions in a defined human population.

gastroenterology (gas″trō-en″ter-ol′o-jē) Study of the stomach and intestines and their diseases.

geriatrics (jer″ē-at′riks) Branch of medicine dealing with older individuals and their medical problems.

gerontology (jer″on-tol′o-jē) Study of the aging process.

gynecology (gī″nĕ-kol′o-jē) Study of the female reproductive system and its diseases.

hematology (hēm″ah-tol′o-jē) Study of the blood and blood diseases.

histology (his-tol′o-jē) Study of the structure and function of tissues. Histology and cytology are subdivisions of microscopic anatomy.

immunology (im″ū-nol′o-jē) Study of the body's resistance to infectious disease.

neonatology (nē″o-nā-tol′o-jē) Study of newborns and the treatment of their disorders.

nephrology (nĕ-frol′o-jē) Study of the structure, function, and diseases of the kidneys.

neurology (nu-rol′o-jē) Study of the nervous system and its disorders.

obstetrics (ob-stet′riks) Branch of medicine dealing with pregnancy and childbirth.

oncology (on-kol′o-jē) Study of cancers.

ophthalmology (of″thal-mol′o-jē) Study of the eye and eye diseases.

orthopedics (or″tho-pē′diks) Branch of medicine dealing with the muscular and skeletal systems and their problems.

otolaryngology (o″to-lar″in-gol′o-jē) Study of the ear, throat, and larynx, and their diseases.

pathology (pah-thol′o-jē) Study of structural and functional changes that disease causes.

pediatrics (pē″dē-at′riks) Branch of medicine dealing with children and their diseases.

pharmacology (fahr″mah-kol′o-jē) Study of drugs and their uses in the treatment of disease.

podiatry (pō-dī′ah-trē) Study of the care and treatment of feet.

psychiatry (sī-kī′ah-trē) Branch of medicine dealing with the mind and its disorders.

radiology (rā″dē-ol′o-jē) Study of X rays and radioactive substances and their uses in the diagnosis and treatment of diseases.

toxicology (tok″sĭ-kol′o-jē) Study of poisonous substances and their effects upon body parts.

urology (ū-rol′o-jē) Branch of medicine dealing with the urinary system, apart from the kidneys (nephrology) and the male reproductive system, and their diseases.

ASSESS

CHAPTER ASSESSMENTS

1.1 Introduction

1. Briefly describe the early discoveries that led to our understanding of the human body.

1.2 Anatomy and Physiology

2. Explain the difference between anatomy and physiology.
3. Identify relationships between the form and the function of body parts.

1.3 Levels of Organization

4. List the levels of organization within the human body, and describe the characteristics of each.

1.4 Characteristics of Life

5. List and describe the characteristics of life.
6. Define *metabolism,* and give examples.

1.5 Maintenance of Life

7. List and describe the environmental requirements to maintain life.
8. Define *homeostasis,* and explain its importance.
9. Identify the parts of a homeostatic mechanism and explain how they work together.
10. Explain the control of body temperature.
11. Describe a homeostatic mechanism that helps regulate blood pressure.

1.6 Organization of the Human Body

12. Explain the difference between the axial and the appendicular portions of the body.
13. Identify the cavities within the axial portion of the body.
14. Define *viscera.*
15. Describe the mediastinum and its contents.
16. List the cavities of the head and the contents of each cavity.
17. Distinguish between a parietal and a visceral membrane.

18. Name the major organ systems, and describe the general functions of each.
19. List the major organs that compose each organ system.

1.7 Anatomical Terminology

20. Write complete sentences using each of the following terms to correctly describe the relative positions of specific body parts:
    a. Superior b. Inferior c. Anterior
    d. Posterior e. Medial f. Lateral
    g. Proximal h. Distal i. Superficial
    j. Peripheral k. Deep
21. Sketch the outline of a human body, and use lines to indicate an example of each of the following sections:
    a. Sagittal b. Transverse c. Frontal
22. Sketch the abdominal area, and indicate the locations of the following regions:
    a. Epigastric b. Umbilical c. Pubic
    d. Hypochondriac e. Lateral f. Inguinal
23. Provide the common name for the region to which each of the following terms refers:
    a. Acromial b. Antebrachial c. Axillary
    d. Buccal e. Celiac f. Coxal
    g. Crural h. Femoral i. Genital
    j. Gluteal k. Inguinal l. Mental
    m. Occipital n. Orbital o. Otic
    p. Palmar q. Pectoral r. Pedal
    s. Plantar t. Popliteal u. Sacral
    v. Tarsal w. Umbilical x. Vertebral

ASSESS

INTEGRATIVE ASSESSMENTS/CRITICAL THINKING

Outcomes 1.2, 1.3, 1.4, 1.5

1. Which characteristics of life does a computer have? Why is a computer not alive?

Outcomes 1.2, 1.3, 1.4, 1.5, 1.6

2. Put the following in order, from smallest and simplest to largest and most complex, and describe their individual roles in homeostasis: organ, molecule, organelle, atom, organ system, tissue, organism, cell, macromolecule.

Outcomes 1.4, 1.5

3. What environmental characteristics would be necessary for a human to survive on another planet?

Outcomes 1.5, 1.6

4. In health, body parts interact to maintain homeostasis. Illness can threaten the maintenance of homeostasis, requiring treatment. What treatments might be used to help control a patient's (a) body temperature, (b) blood oxygen concentration, and (c) water content?

Outcomes 1.5, 1.6, 1.7

5. How might health-care professionals provide the basic requirements of life to an unconscious patient? Describe the body parts involved in the treatment, using correct directional and regional terms.

Outcome 1.6

6. Suppose two individuals develop benign (noncancerous) tumors that produce symptoms because they occupy space and crowd adjacent organs. If one of these persons has the tumor in the thoracic cavity and the other has the tumor in the abdominopelvic cavity, which person would be likely to develop symptoms first? Why? Which might be more immediately serious? Why?

Outcomes 1.6, 1.7

7. If a patient complained of a "stomachache" and pointed to the umbilical region as the site of discomfort, which organs located in this region might be the source of the pain?

8. If a radiologist wants an image of a patient that shows both kidneys, which body section(s) would accommodate her request?

Chapter Summary

1.1 Introduction

1. Early interest in the human body probably developed as people became concerned about injuries and illnesses.
2. Primitive doctors began to learn how certain herbs and potions affected body functions.
3. The belief that humans could understand forces that caused natural events led to the development of modern science.
4. A set of terms originating from Greek and Latin words is the basis for the language of anatomy and physiology.
5. Used to investigate the human body, the scientific method tests a hypothesis, then rejects or accepts it based on the results of experiments or observations.

1.2 Anatomy and Physiology

1. **Anatomy** describes the form and organization of body parts.
2. **Physiology** considers the functions of anatomical parts.
3. The function of a body part depends on the way it is constructed.

1.3 Levels of Organization

The body is composed of parts with different levels of complexity.

1. Matter is composed of **atoms.**
2. Atoms join to form **molecules.**
3. Organelles are built of groups of large molecules **(macromolecules).**
4. **Cells,** which contain **organelles,** are the basic units of structure and function that form the body.
5. Cells are organized into **tissues.**
6. Tissues are organized into **organs.**
7. Organs that function closely together compose **organ systems.**
8. Organ systems constitute the **organism.**
9. Beginning at the atomic level, these levels of organization differ in complexity from one level to the next.

1.4 Characteristics of Life

Characteristics of life are traits all organisms share.

1. These characteristics include:
   a. Growth—increase in cell number and body size.
   b. Reproduction—production of new cells and organisms.
   c. Responsiveness—reaction to internal and external changes.
   d. Movement—change in body position, or the motion of internal organs.
   e. Metabolism—all the chemical reactions in a living system.

1.5 Maintenance of Life

The structures and functions of body parts maintain the life of the organism.

1. Environmental requirements to maintain life
   a. **Chemicals: Water** is used in many metabolic processes, provides the environment for metabolic reactions, and carries substances. Other chemicals include oxygen, carbon dioxide, and nutrients.
   b. **Heat** is a product of metabolic reactions and helps govern the rates of these reactions.
   c. **Pressure** is an application of force to something. In humans, atmospheric and hydrostatic pressures help breathing and blood movements, respectively.
2. Homeostasis
   a. If an organism is to survive, the conditions within its body fluids, its **internal environment,** must remain relatively stable.
   b. Maintenance of a stable internal environment is called **homeostasis.**
   c. **Homeostatic mechanisms** help regulate body temperature and blood pressure.
   d. Homeostatic mechanisms act through **negative feedback.**

1.6 Organization of the Human Body

1. Body cavities
   a. The **axial** portion of the body includes the **cranial cavity,** the **vertebral canal,** the **thoracic cavity,** and the **abdominopelvic cavity.** The **appendicular** portion includes the upper and lower limbs.
   b. The organs in a body cavity are called **viscera.**
   c. The **diaphragm** separates the thoracic and abdominopelvic cavities.
   d. The **mediastinum** forms a boundary between the right and left sides of the thoracic cavity.
   e. Body cavities in the head include the **oral, nasal, orbital,** and **middle ear** cavities.
2. Thoracic and abdominopelvic membranes
   a. Thoracic membranes
      (1) **Pleural membranes** line the thoracic cavity **(parietal pleura)** and cover each lung **(visceral pleura).**
      (2) **Pericardial membranes** surround the heart **(parietal pericardium)** and cover its surface **(visceral pericardium).**
      (3) The **pleural and pericardial cavities** are the potential spaces between the respective parietal and visceral membranes.
   b. Abdominopelvic membranes
      (1) **Peritoneal membranes** line the abdominopelvic cavity **(parietal peritoneum)** and cover the organs inside **(visceral peritoneum).**
      (2) The **peritoneal cavity** is the potential space between the parietal and visceral peritoneal membranes.
3. Organ systems

The human organism consists of several organ systems. Each system includes a set of interrelated organs.

   a. Body covering
      (1) The **integumentary system** includes the skin, hair, nails, sweat glands, and sebaceous glands.
      (2) It protects underlying tissues, regulates body temperature, houses sensory receptors, and synthesizes various substances.
   b. Support and movement
      (1) **Skeletal system**
         (a) The skeletal system is composed of bones, as well as cartilages and ligaments that bind bones together.
         (b) It provides a framework, protective shields, and attachments for muscles. It also produces blood cells and stores inorganic salts.
      (2) **Muscular system**
         (a) The muscular system includes the muscles of the body.
         (b) It moves body parts, maintains posture, and is the major source of body heat.
   c. Integration and coordination
      (1) **Nervous system**
         (a) The nervous system consists of the brain, spinal cord, nerves, and sense organs.
         (b) It receives information from sensory receptors, interprets the information, and stimulates muscles or glands to respond.
      (2) **Endocrine system**
         (a) The endocrine system consists of glands that secrete hormones.
         (b) Hormones help regulate metabolism.
         (c) This system includes the hypothalamus of the brain and the pituitary, thyroid, parathyroid, and adrenal glands, as well as the pancreas, ovaries, testes, pineal gland, and thymus.
   d. Transport
      (1) **Cardiovascular system**
         (a) The cardiovascular system includes the heart, which pumps blood, and the blood vessels, which carry blood to and from body parts.
         (b) Blood carries oxygen, nutrients, hormones, and wastes.
      (2) **Lymphatic system**
         (a) The lymphatic system is composed of lymphatic vessels, lymph fluid, lymph nodes, the thymus, and the spleen.
         (b) It transports lymph fluid from tissues to the bloodstream, carries certain fatty substances away from the digestive organs, and aids in defending the body against disease-causing agents.
   e. Absorption and excretion
      (1) **Digestive system**
         (a) The digestive system receives foods, breaks down food molecules into nutrients that can pass through cell membranes, and eliminates materials that are not absorbed.

(b) It includes the mouth, tongue, teeth, salivary glands, pharynx, esophagus, stomach, liver, gallbladder, pancreas, small intestine, and large intestine.
(c) Some digestive organs produce hormones.

(2) **Respiratory system**
(a) The respiratory system takes in and sends out air and exchanges gases between the air and blood.
(b) It includes the nasal cavity, pharynx, larynx, trachea, bronchi, and lungs.

(3) **Urinary system**
(a) The urinary system includes the kidneys, ureters, urinary bladder, and urethra.
(b) It filters wastes from the blood and helps maintain water and electrolyte concentrations and the acidity of the internal environment.

f. Reproduction
(1) The **reproductive systems** produce new organisms.
(2) The male reproductive system includes the scrotum, testes, epididymides, ductus deferentia, seminal vesicles, prostate gland, bulbourethral glands, urethra, and penis, which produce, maintain, and transport male sex cells (sperm cells).
(3) The female reproductive system includes the ovaries, uterine tubes, uterus, vagina, clitoris, and vulva, which produce, maintain, and transport female sex cells (oocytes).

1.7 Anatomical Terminology

Terms with precise meanings help investigators and clinicians communicate effectively.

1. Relative positions

These terms describe the location of one part with respect to another part.

2. Body sections

Body sections are planes along which the body may be cut to observe the relative locations and organization of internal parts.

3. Body regions

Special terms designate various body regions.

REFERENCE PLATES

THE HUMAN ORGANISM

The series of illustrations that follows shows the major parts of the human torso. The first plate illustrates the anterior surface and reveals the superficial muscles on one side. Each subsequent plate exposes deeper organs, including those in the thoracic, abdominal, and pelvic cavities.

Chapters 6–19 of this textbook describe the organs and organ systems of the human organism in detail. As you read them, refer to these plates to visualize the locations of various organs and the three-dimensional relationships among them.

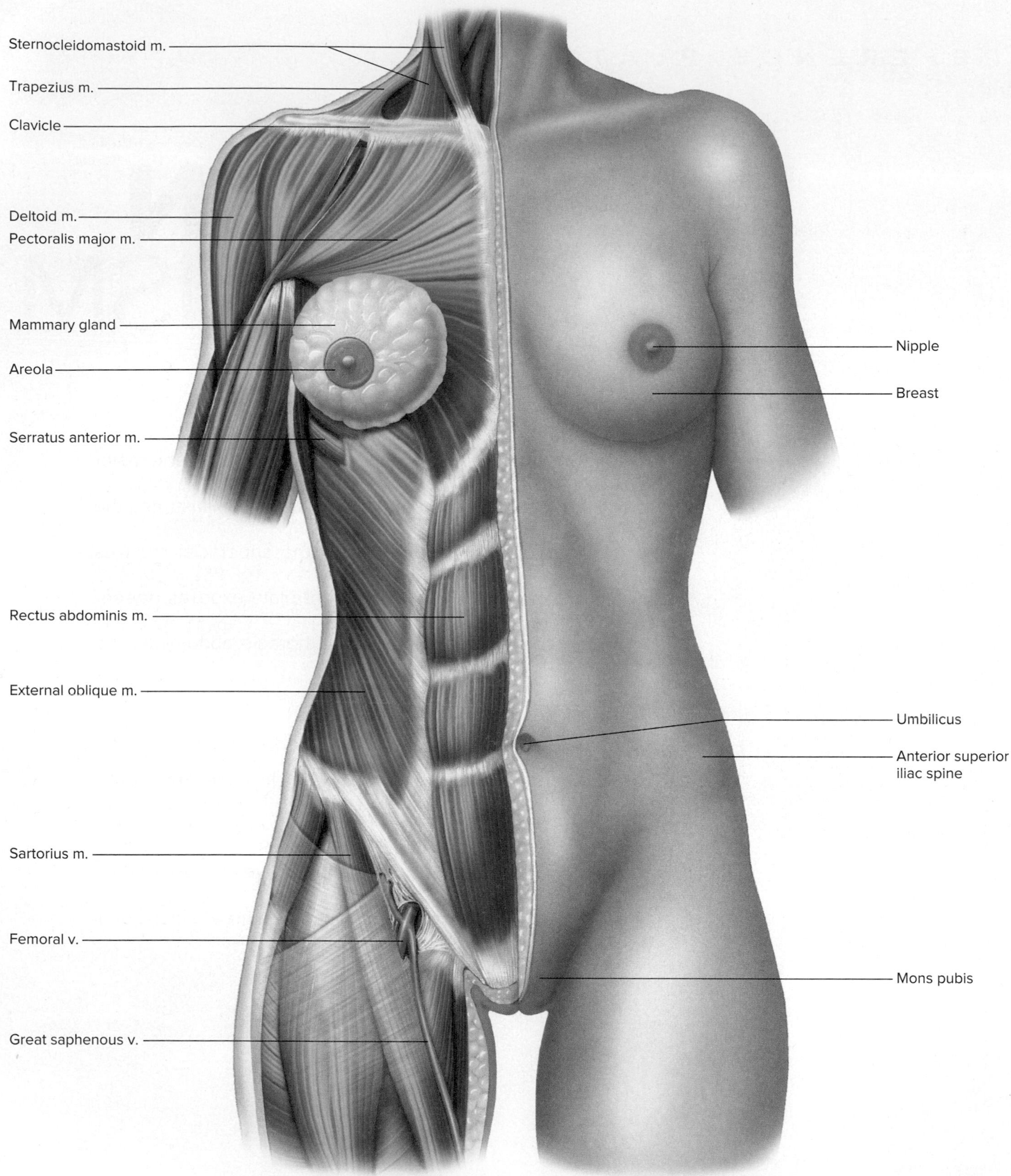

PLATE ONE

Human female torso showing the anterior surface on one side and the superficial muscles exposed on the other side. (*m.* stands for *muscle,* and *v.* stands for *vein.*)

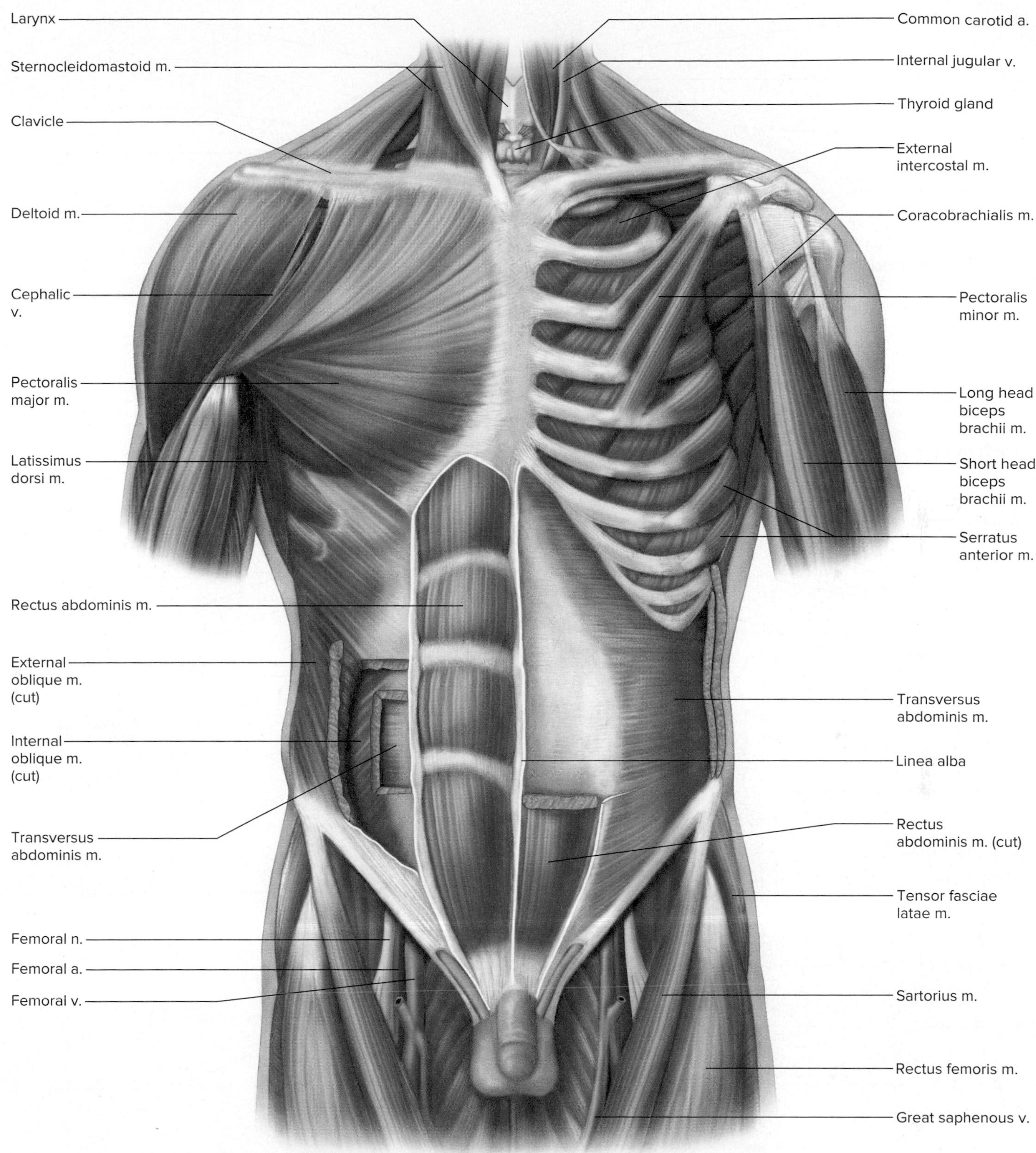

PLATE TWO

Human male torso with the deeper muscle layers exposed. (*n.* stands for *nerve*, *a.* stands for *artery*, *m.* stands for *muscle*, and *v.* stands for *vein*.)

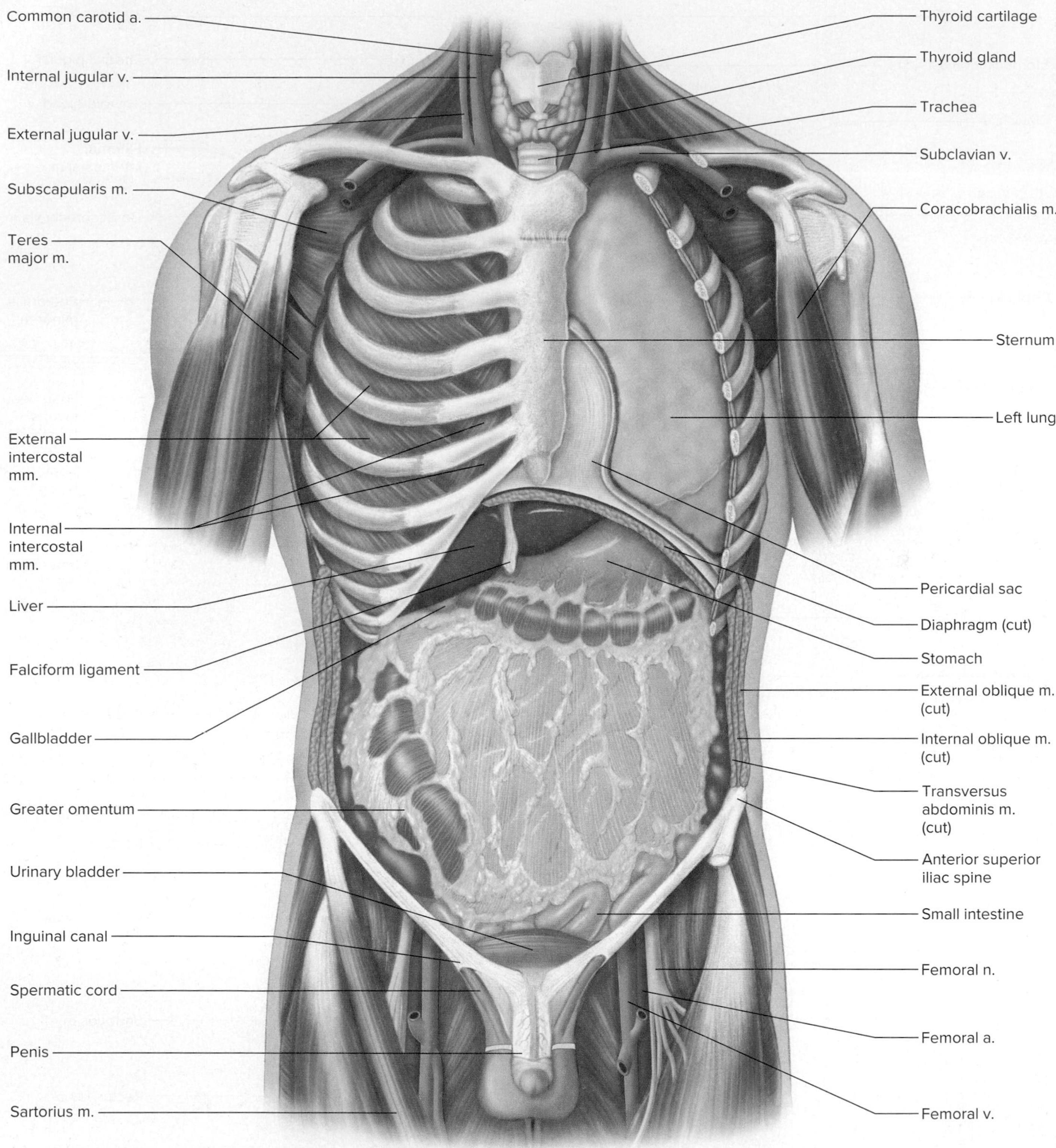

PLATE THREE

Human male torso with the deep muscles removed and the abdominal viscera exposed. (*n.* stands for *nerve, a.* stands for *artery, m.* stands for *muscle, mm.* stands for muscles, and *v.* stands for *vein.*)

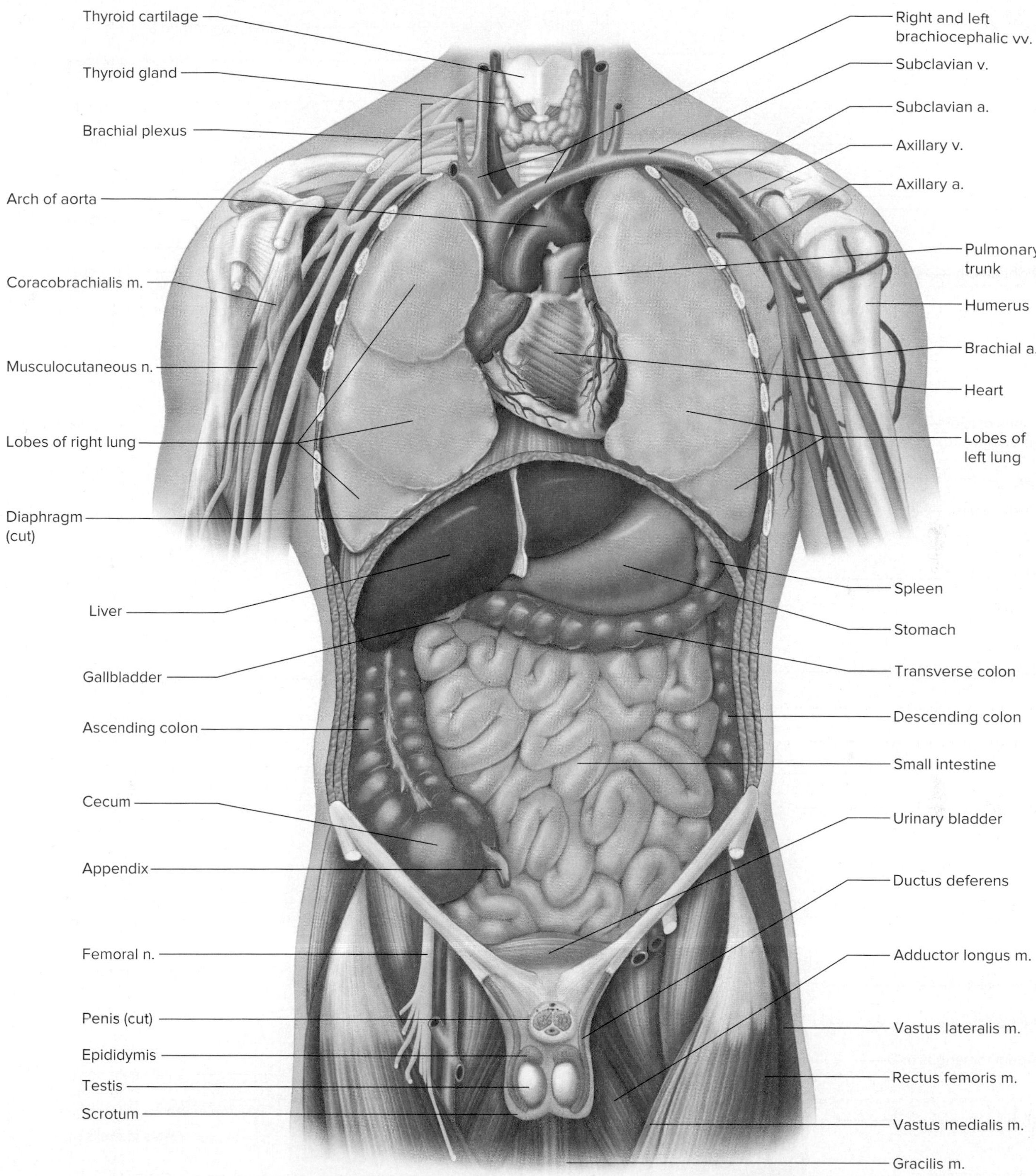

PLATE FOUR

Human male torso with the thoracic and abdominal viscera exposed. (*n.* stands for *nerve, a.* stands for *artery, m.* stands for *muscle, v.* stands for *vein,* and *vv.* stands for *veins.*)

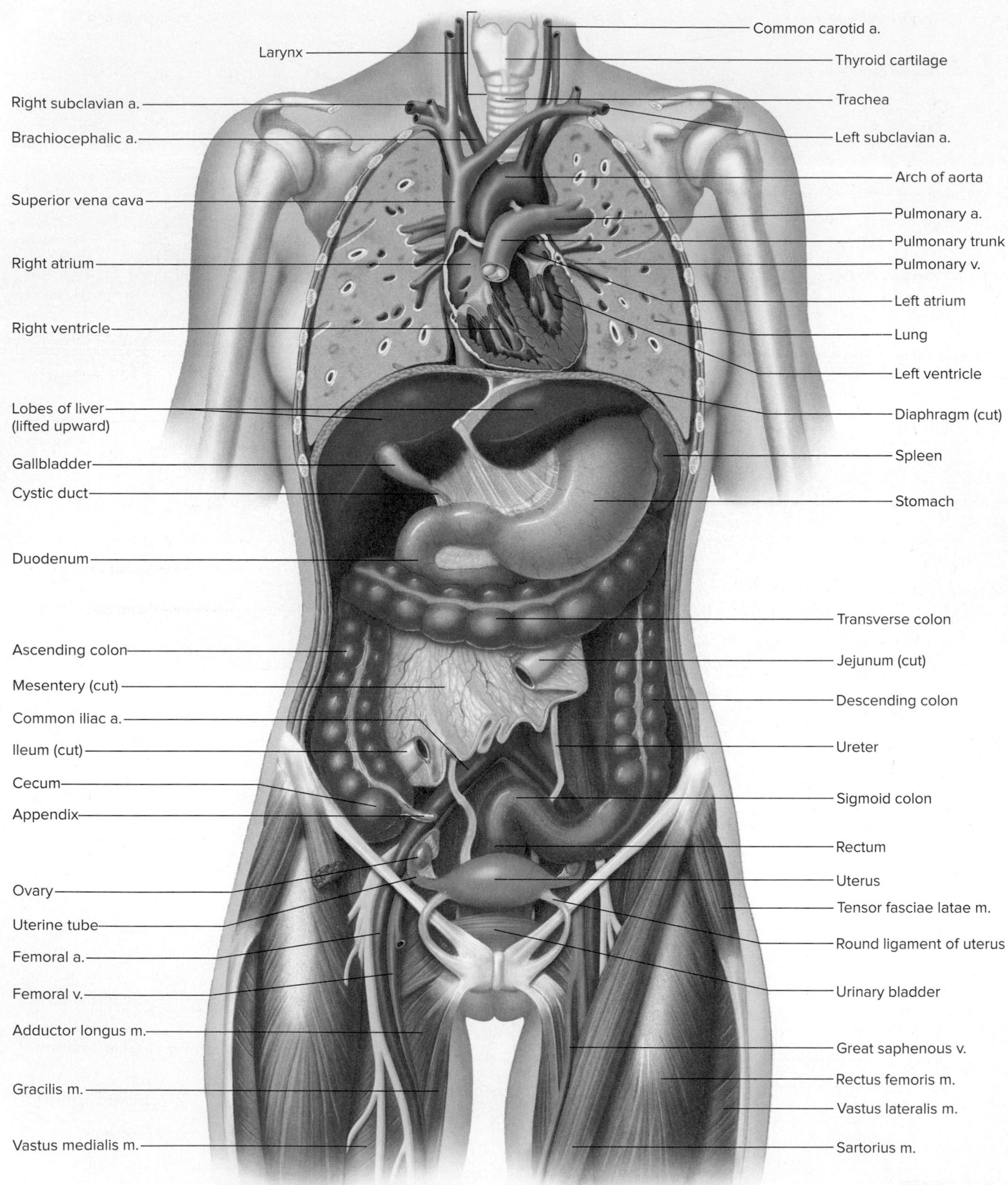

PLATE FIVE

Human female torso with the lungs, heart, and small intestine sectioned and the liver reflected (lifted back). (*a.* stands for *artery,* *m.* stands for *muscle,* and *v.* stands for *vein.*)

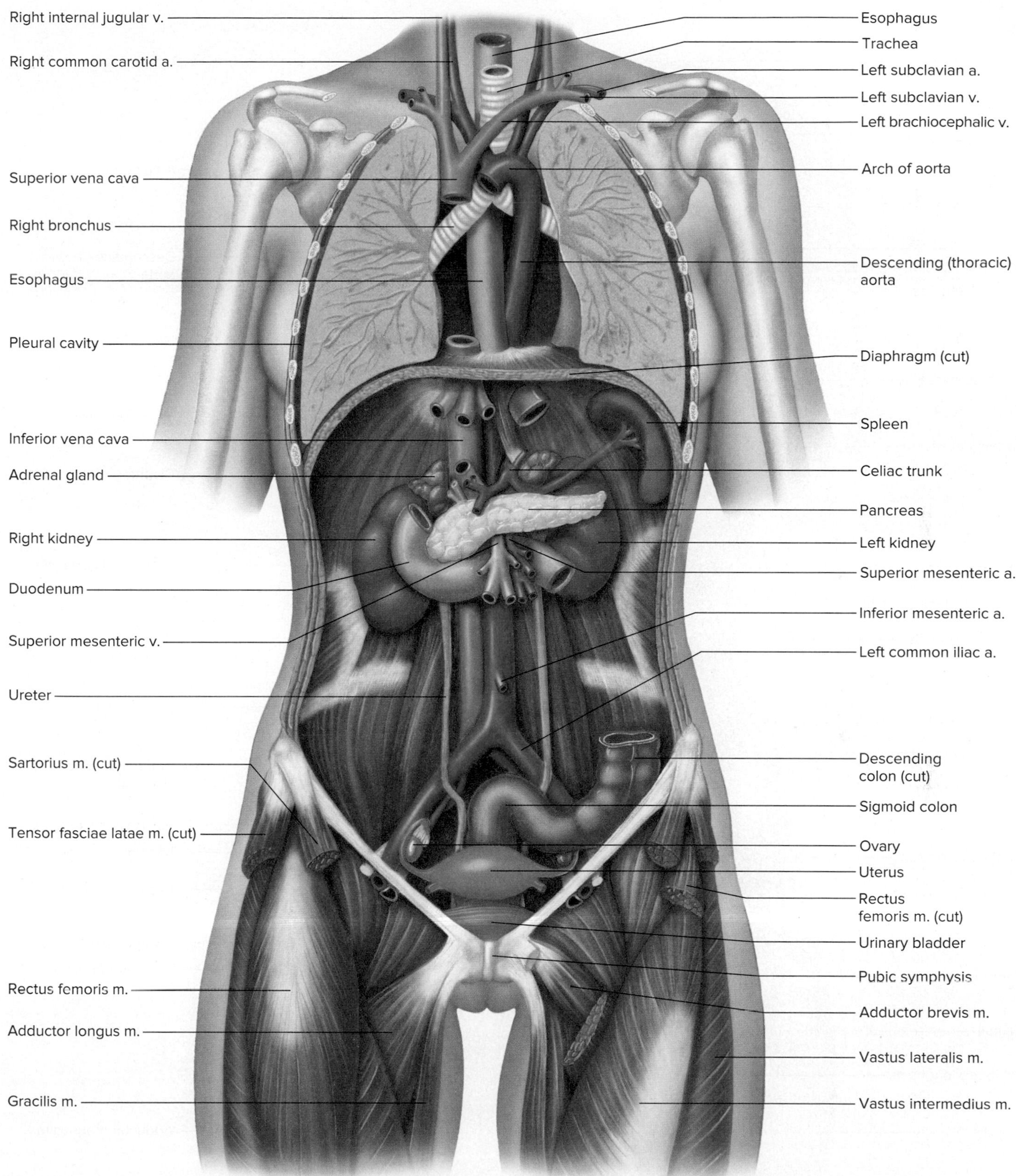

PLATE SIX

Human female torso with the heart, stomach, liver, and parts of the intestine and lungs removed. (*a.* stands for *artery, m.* stands for *muscle,* and *v.* stands for *vein.*)

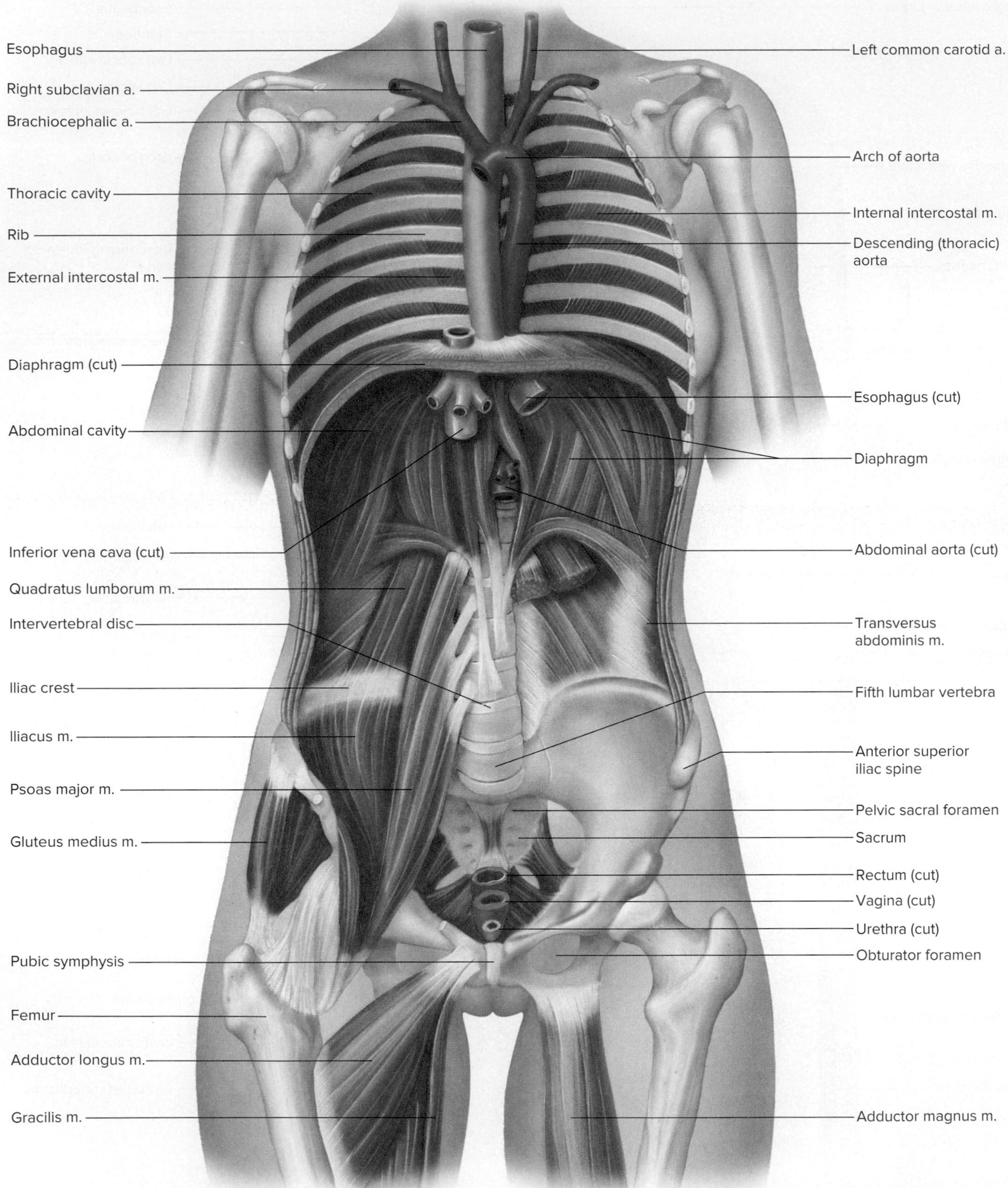

PLATE SEVEN

Human female torso with the thoracic, abdominal, and pelvic viscera removed. (*a.* stands for *artery,* and *m.* stands for *muscle.*)

2 Chemical Basis of Life

Many artisan breads, such as these, may contain gluten as a natural ingredient. ©Jman78/iStockphoto/GettyImages

Walking down the aisle of a grocery store, the variety of gluten-free options are plentiful. Gluten-free bread, pasta, pizza dough, and other food products are often available in restaurants. Just a few years ago, not many people had ever heard of gluten. You may be wondering what all the hype is about, and asking yourself what gluten is and if you should be on a gluten-free diet.

Gluten is a natural protein composite found in many grains such as wheat, barley, and rye. Some individuals develop an immune response to gluten where their body's own immune system attacks and gradually damages the villi of the small intestine where nutrient absorption normally occurs. Consequently, nutrients that are not absorbed draw water toward them as they are eliminated from the body, resulting in diarrhea, among other symptoms. Affecting about 3 million Americans, this hereditary condition is known as *celiac disease*, and a strict gluten-free diet is essential. Some people are *gluten sensitive,* where they experience symptoms similar to celiac disease but do not test positive for celiac disease. A wheat allergy is a hypersensitive response to one or more proteins found in wheat and may be the cause of the symptoms, which an allergy test can confirm. There is some evidence that certain short-chained carbohydrates may be the cause of gastrointestinal distress, and these carbohydrates happen to be found alongside gluten in some foods.

For the majority of the population, there is no health need to go on a gluten-free diet. In fact, gluten-free foods often have less nutritious value, fewer fortified vitamins and minerals, and often have more sugar and fat. Gluten is considered a prebiotic, upon which "good" bacteria in the gut feed, protecting our colon from associated gastrointestinal disease, such as colorectal cancer. Whole grains, in which gluten is a natural ingredient, are linked to improved health outcomes, like a lowered risk of both heart disease and type 2 diabetes mellitus. Gluten-free foods tend to be more expensive. There may be an added charge for a gluten-free option on a restaurant's menu. Some dieters have claimed that a gluten-free diet has helped them lose weight. However, it is more likely that their weight loss resulted from cutting out unhealthier foods like baked goods and processed snacks. One should not assume that digestive symptoms are caused by gluten, because eliminating gluten from the diet may cause the body more harm than good. Rather, one should consult a physician for persistent digestive issues.

Module 2 Cells & Chemistry

LEARNING OUTLINE

After studying this chapter, you should be able to complete the "Learning Outcomes" that follow the major headings throughout the chapter.

AIDS TO UNDERSTANDING WORDS

di- [two] *di*saccharide: compound whose molecules are composed of two bonded simple sugar units.

glyc- [sweet] *glyc*ogen: complex carbohydrate composed of many glucose molecules bonded in a particular way.

lip- [fat] *lip*ids: group of organic compounds that includes fats.

-lyt [dissolvable] electro*lyte:* substance that breaks down and releases ions in water.

mono- [one] *mono*saccharide: compound whose molecule consists of a single simple sugar unit.

poly- [many] *poly*unsaturated: molecule that has two or more double bonds between its carbon atoms.

sacchar- [sugar] mono*sacchar*ide: molecule consisting of a single simple sugar unit.

syn- [together] *syn*thesis: process by which substances join to form new types of substances.

(Appendix A has a complete list of Aids to Understanding Words.)

2.1 | Introduction

LEARN

1. Give examples of how the study of living materials requires an understanding of chemistry.

Chemicals are all around us. Most of us benefit daily from the chemicals found in soap, toothpaste, shampoo, or deodorant. The food and liquids we eat and drink and medications we take are also chemicals. Therefore, it is easy to appreciate that the human body is composed of chemicals, including salts, water, proteins, carbohydrates, lipids, and nucleic acids.

As we saw in chapter 1, cells are built from complex chemicals. The study of anatomy and physiology, therefore, begins with the study of chemistry. We first examine the fundamentals and basic structure of all matter.

2.2 | Fundamentals of Chemistry

LEARN

1. Describe the relationships among matter, atoms, and molecules.
2. Describe the general structure of an atom.

Everything nonliving and living in the universe is made up of **matter** and occupies space. Mass refers to the amount of matter present. Weight is a measure of the gravitational pull on mass. If your weight on earth is 150 pounds, on the moon it would be only about 25 pounds, but your mass would be the same in both places. That is, your composition is the same on the moon as it is on Earth, but you weigh less on the moon because the force of gravity is lower there. Because we are dealing with life on Earth, and constant gravity, we can consider mass and weight as roughly equivalent.

We can now formally define **chemistry** as the branch of science that studies the composition, properties, and interactions of matter.

Elements and Atoms

The simplest examples of matter with specific chemical properties are called **elements** (el′ĕ-mentz). Common examples include iron, copper, silver, gold, aluminum, carbon, hydrogen, and oxygen. A few elements exist in a pure form, but most combine with other elements.

Living organisms require about twenty elements. Of these, oxygen, carbon, hydrogen, and nitrogen make up more than 95% (by weight) of the human body (table 2.1). As the table shows, each element is represented by a one- or two-letter symbol.

Elements are composed of particles called **atoms** (at′omz), which are the smallest complete units of elements. Atoms of an element are chemically the same, but they differ from the atoms that make up other elements. Atoms of different elements vary in size, weight, and the ways they interact with other atoms. Some atoms can interact or even combine with atoms like themselves or of other elements by forming attractions called **chemical bonds,** whereas other atoms cannot form such bonds.

Atomic Structure

An atom consists of a central portion, called the **nucleus,** and one or more **electrons** (e-lek′tronz) that constantly move around the nucleus. The nucleus contains one or more relatively large particles called **protons** (pro′tonz). The nucleus also usually contains one or more **neutrons** (nu′tronz), which are similar in size to protons (fig. 2.1).

TABLE 2.1 Elements in the Body

Major Element	Symbol	Approximate Percentage of the Human Body (by weight)
Oxygen	O	65.0
Carbon	C	18.5
Hydrogen	H	9.5
Nitrogen	N	3.2
Calcium	Ca	1.5
Phosphorus	P	1.0
Potassium	K	0.4
Sulfur	S	0.3
Chlorine	Cl	0.2
Sodium	Na	0.2
Magnesium	Mg	0.1
		Total 99.9%
Trace Elements		
Chromium	Cr	Together less than 0.1%
Cobalt	Co	
Copper	Cu	
Fluorine	F	
Iodine	I	
Iron	Fe	
Manganese	Mn	
Zinc	Zn	

Electrons, which are extremely small, each carry a single, negative electrical charge (e^-), whereas protons each carry a single, positive electrical charge (p^+). Neutrons are uncharged and thus are electrically neutral (n^0) (fig. 2.1).

Because the nucleus contains the protons, it is always positively charged. However, the number of electrons outside the nucleus equals the number of protons in the nucleus, such that the charges cancel. Therefore, an atom is always electrically uncharged, or neutral. If an atom gains or loses electrons, it becomes charged and is called an *ion*. Ions are discussed in a later section.

Atomic Number

The atoms of different elements have different numbers of protons. The number of protons in the atoms of a particular element is called the element's **atomic number.** Hydrogen, for example, whose atoms each have one proton, has the atomic number 1; carbon, whose atoms each have six protons, has the atomic number 6.

Atomic Weight

The **atomic weight** of an atom of an element approximately equals the number of protons and neutrons in its nucleus (electrons have very little weight). Thus, the atomic weight of hydrogen, with one proton and no neutrons, is 1, whereas the

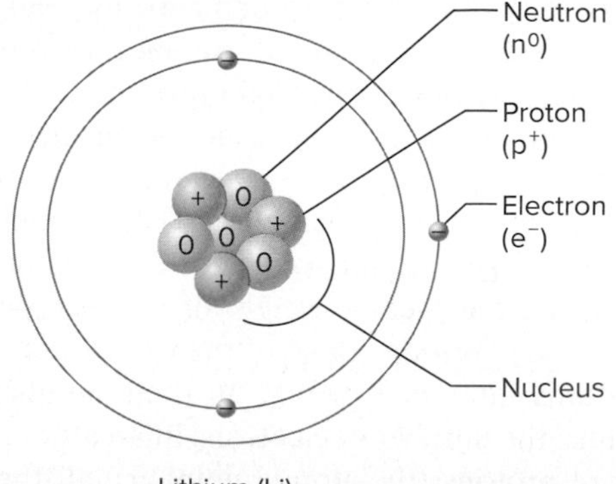

Figure 2.1 An atom consists of subatomic particles. In an atom of the element lithium, three electrons move around a nucleus that consists of three protons and four neutrons. APR

atomic weight of carbon, with six protons and six neutrons, is 12. In other words, an atom of carbon weighs twelve times more than an atom of hydrogen.

Isotopes

All the atoms of a particular element have the same atomic number because they have the same number of protons. However, the atoms of most elements vary in their number of neutrons; thus, they vary in atomic weight. Atoms of an

CAREER CORNER

Pharmacy Technician

The flu season is in full force and the supermarket pharmacy line snakes all the way to the bakery. The pharmacy technician speedily yet carefully updates customers' records and processes insurance information, accepts payment, and hands customers their prescriptions. He or she can answer practical questions, such as when to take a medication, but asks the pharmacist to address health-related concerns.

In addition to these customer service skills, the pharmacy technician gathers information from health-care professionals or from patients about particular prescriptions, checks drug inventories, prepares ointments, counts pills, measures liquid medications, and packages and labels drug containers. The pharmacist verifies that the prescription has been prepared and labeled properly. The technician may also assist at special events, such as vaccination clinics and education sessions.

Pharmacy technicians work in stand-alone pharmacies, in supermarkets and big-box stores, in hospitals and skilled nursing facilities, and at mail-order dispensaries. A high school diploma and in some states a training program and certification are required to be a pharmacy technician. The job requires stamina, attention to detail to avoid errors, and a friendly approach to serving customers.

element with different numbers of neutrons, and thus different atomic weights are called **isotopes** (ī'sō-tōps). For example, all oxygen atoms have eight protons in their nuclei, but these atoms may have eight, nine, or ten neutrons, corresponding to, respectively, atomic weights of 16, 17, and 18.

Because a sample of an element is likely to include more than one isotope, the atomic weight of the element is often considered to be the average weight of the isotopes present (see Appendix D, Periodic Table of the Elements).

How atoms interact depends on their number of electrons. Because the number of electrons in an atom is equal to its number of protons (the atomic number), all the isotopes of a particular element have the same number of electrons and react chemically in the same manner. Therefore, any of the isotopes of oxygen can have the same function in an organism's metabolic reactions.

Isotopes may be stable, or they may have unstable atomic nuclei that decompose, releasing energy or pieces of themselves. Unstable isotopes are called **radioactive,** and the energy or atomic fragments they give off are called *radiation.*

Clinical Application 2.1 discusses some practical applications of radioactive isotopes.

PRACTICE 2.2

Answers to the Practice questions can be found in the eBook.

1. What are elements?
2. Which elements are most common in the human body?
3. Where are electrons, protons, and neutrons located in an atom?
4. What is the difference between atomic number and atomic weight?

2.3 | Bonding of Atoms

LEARN

1. Describe how atomic structure determines how atoms interact.
2. Identify the three types of bonding.
3. Describe the difference between ionic and covalent bonding.

Atoms can attach to other atoms by forming **chemical bonds.** When atoms form chemical bonds, they do so by gaining, losing, or sharing electrons.

The electrons of an atom occupy one or more areas of space, called *shells,* around the nucleus. For the elements up to atomic number 18, the maximum number of electrons that each of the first three inner shells can hold is as follows:

First shell (closest to the nucleus)	2 electrons
Second shell	8 electrons
Third shell	8 electrons

More-complex atoms may have as many as eighteen electrons in the third shell. Simplified diagrams, such as those in **figure 2.2**, depict electron locations within the shells of atoms.

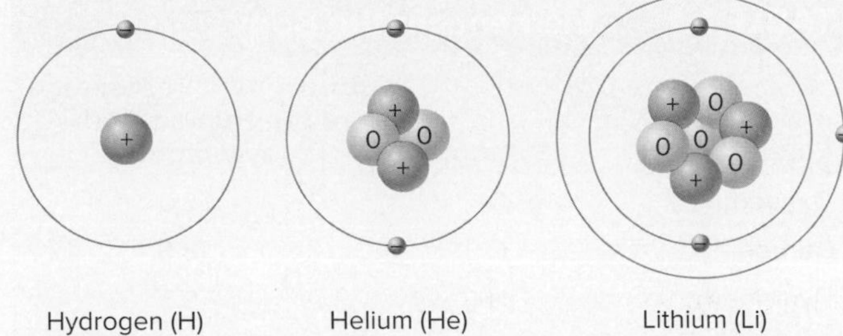

Figure 2.2 Electrons orbit the atomic nucleus. The single electron of a hydrogen atom is located in its first, yet outermost shell. The two electrons of a helium atom fill its outermost shell. Two of the three electrons of a lithium atom fill the first shell, and one is in the outermost shell.

The electrons in the outermost shell of an atom determine its chemical behavior. Atoms such as helium, whose outermost electron shells are filled, have stable structures and are chemically inactive, or **inert.** Atoms such as hydrogen or lithium, whose outermost electron shells are incompletely filled, tend to gain, lose, or share electrons in ways that empty or fill their outer shells. This enables the atoms to achieve energetic stability.

Ionic Bonding

Atoms that gain or lose electrons become electrically charged and are called **ions** (īonz). An atom of sodium, for example, has eleven electrons: two in the first shell, eight in the second shell, and one in the third shell (**fig. 2.3**). This atom tends to lose the electron from its outer shell, which leaves the second (now the outermost) shell filled and the new form stable (**fig. 2.4*a***). In the process, sodium is left with eleven protons (11^+) in its nucleus and only ten electrons (10^-). As a result, the atom develops a net electrical charge of +1 and is called a sodium ion, symbolized Na^+.

A chlorine atom has seventeen electrons, with two in the first shell, eight in the second shell, and seven in the third, outermost shell. This atom tends to accept a single electron, filling its outer shell and achieving stability (fig. 2.4*a*). In the

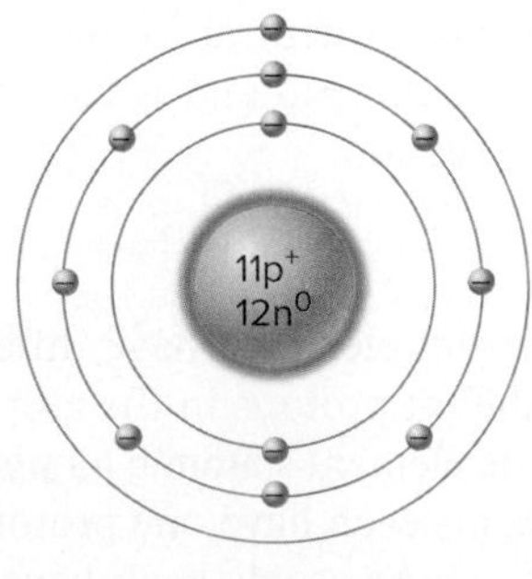

Sodium atom contains
11 electrons (e^-)
11 protons (p^+)
12 neutrons (n^0)

Atomic number = 11
Atomic weight = 23

Figure 2.3 A sodium atom.

process, the chlorine atom with seventeen protons (17^+) in its nucleus now has eighteen electrons (18^-). The atom develops a net electrical charge of -1 and is called a chloride ion, symbolized Cl^-. (Some ions have an electrical charge greater than 1—for example, calcium ion, Ca^{+2}, also written as Ca^{++}.)

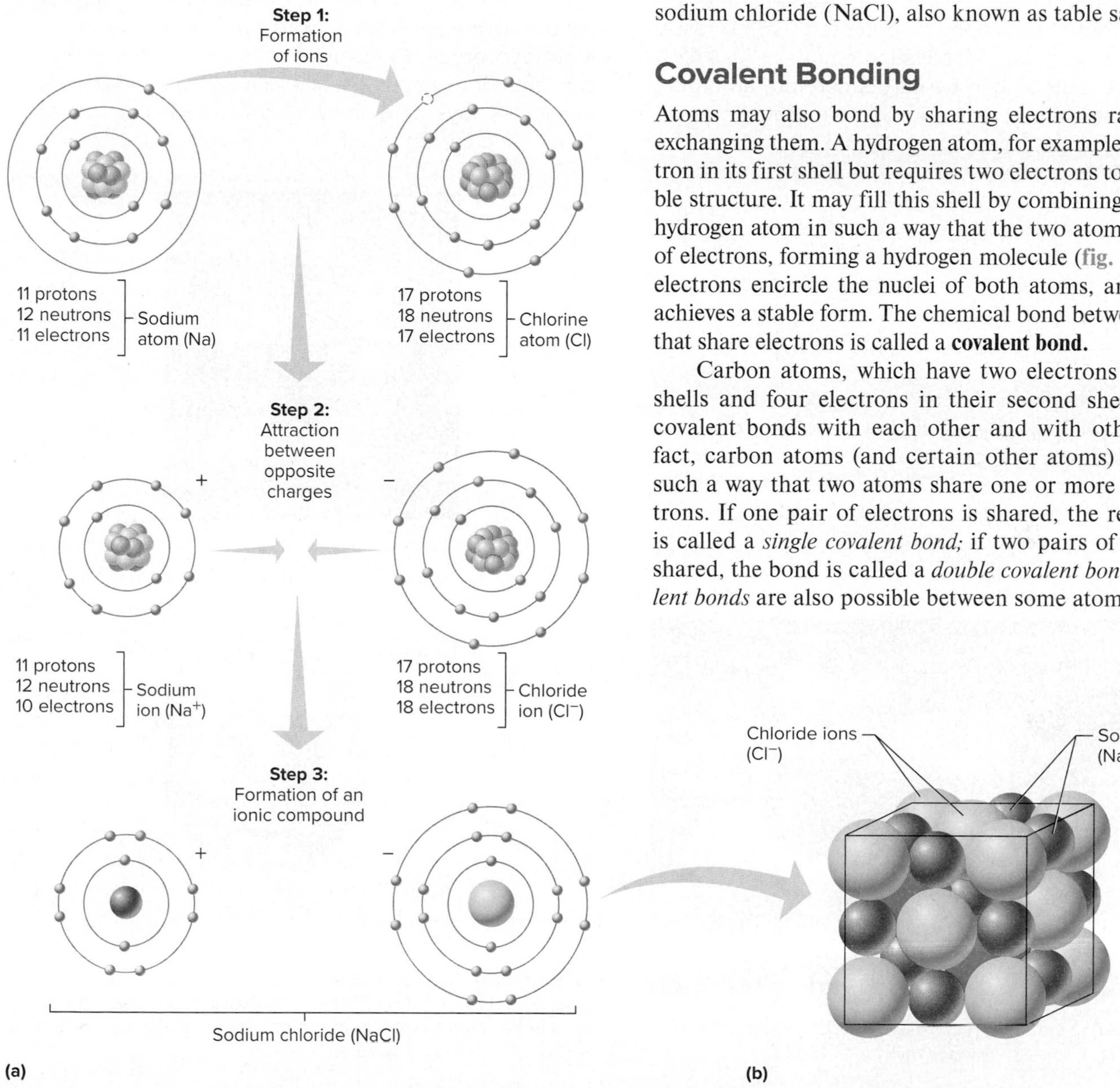

Figure 2.4 An ionic bond forms when one atom loses and another atom gains electrons **(a).** Ionically bonded atoms may form crystals **(b).**

Because oppositely charged ions attract, sodium and chloride ions react to form a type of chemical bond called an **ionic bond** (electrovalent bond) (fig. 2.4*a*). Ionically bound substances do not form discrete **molecules** (mol′ĕ-kūlz) (the term for when two or more atoms bond and form a new kind of particle)—instead, they form arrays, such as crystals of sodium chloride (NaCl), also known as table salt (fig. 2.4*b*).

Covalent Bonding

Atoms may also bond by sharing electrons rather than by exchanging them. A hydrogen atom, for example, has one electron in its first shell but requires two electrons to achieve a stable structure. It may fill this shell by combining with another hydrogen atom in such a way that the two atoms share a pair of electrons, forming a hydrogen molecule (fig. 2.5). The two electrons encircle the nuclei of both atoms, and each atom achieves a stable form. The chemical bond between the atoms that share electrons is called a **covalent bond.**

Carbon atoms, which have two electrons in their first shells and four electrons in their second shells, can form covalent bonds with each other and with other atoms. In fact, carbon atoms (and certain other atoms) may bond in such a way that two atoms share one or more pairs of electrons. If one pair of electrons is shared, the resulting bond is called a *single covalent bond;* if two pairs of electrons are shared, the bond is called a *double covalent bond. Triple covalent bonds* are also possible between some atoms.

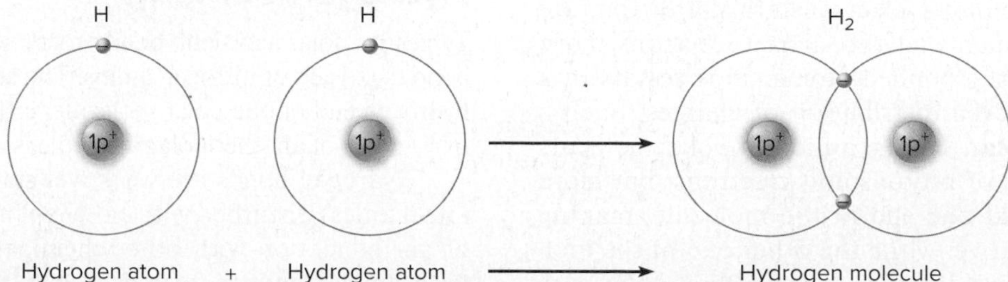

Figure 2.5 A hydrogen molecule forms when two hydrogen atoms share a pair of electrons. A covalent bond forms between the atoms.

CLINICAL APPLICATION 2.1

Radioactive Isotopes: Helpful and Harmful

Radioactive chemicals are useful in studying life processes and in diagnosing and treating some diseases. Atomic radiation is detected with special equipment, such as a scintillation counter (fig. 2A). A radioactive isotope can be introduced into an organism and then traced as it enters into metabolic activities. For example, the human thyroid gland is unique in using the element iodine in its metabolism. Therefore, radioactive iodine-131 is used to study thyroid functions and to evaluate thyroid disease (fig. 2B). Doctors use thallium-201, which has a half-life of 73.5 hours, to assess heart conditions, and gallium-67, with a half-life of 78 hours, to detect and monitor the progress of certain cancers and inflammatory diseases.

Atomic radiation also can change chemical structures and in this way alter vital cellular processes. For this reason, doctors sometimes use radioactive isotopes, such as cobalt-60, to treat cancers. The radiation from the cobalt preferentially kills the rapidly dividing cancer cells.

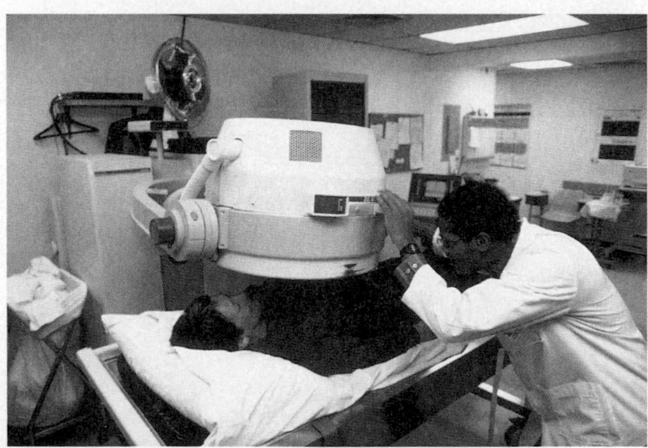

FIGURE 2A Scintillation counters detect radioactive isotopes. Mark Antman/The Image Works

Exposure to radiation can cause disease, such as certain cancers. The transfer of energy as radiation is emitted damages DNA in ways that kill cells or make them cancerous. Exposure to ultraviolet radiation in sunlight, for example, causes skin cancer, and excess medical X rays or gamma rays increase the risk of developing cancer in certain body parts.

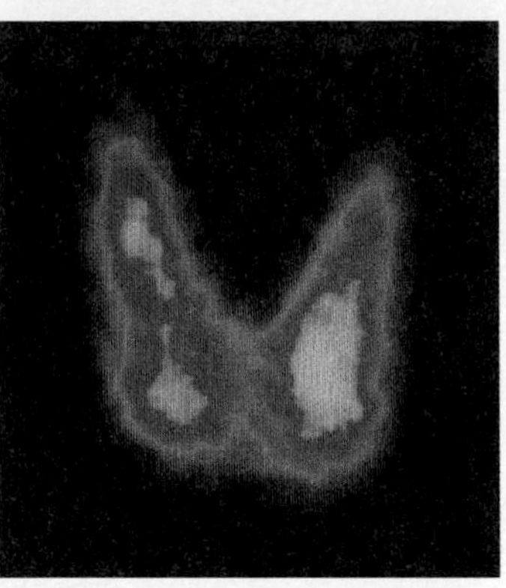

(a)

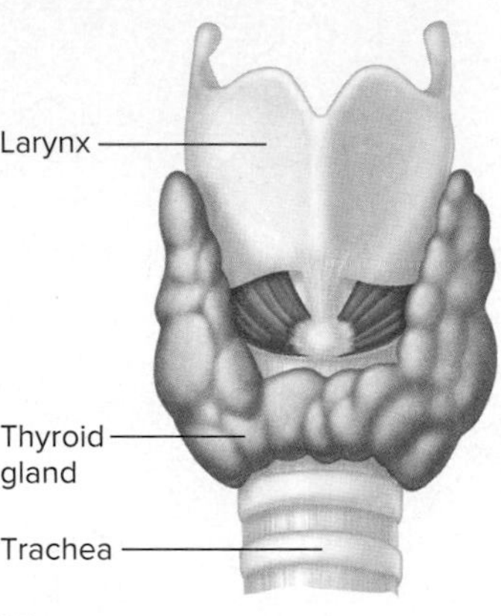

(b)

FIGURE 2B **(a)** Scan of the thyroid gland 24 hours after the patient received radioactive iodine. Note how closely the scan in **(a)** resembles the shape of the thyroid gland, shown in **(b)**. (a): Chris Priest/Science Source

In a *polar covalent bond,* electrons are shared, but are not shared equally, such that the shared electrons move more toward one of the bonded atoms. This results in a molecule with an uneven distribution of charges. Such a molecule is called **polar.** Unlike an ion, a polar molecule has an equal number of protons and electrons, but more of the electrons are at one end of the molecule, making that end slightly negative, while the other end of the molecule is slightly positive. Water is an important polar molecule (fig. 2.6*a*).

Hydrogen Bonding

Typically, polar covalent bonds form where hydrogen atoms bond to oxygen or nitrogen atoms. The attraction of the positive hydrogen end of one polar molecule to the negative nitrogen or oxygen end of another polar molecule is called a **hydrogen bond.**

Hydrogen bonds are weak, yet stable. That is, they stay intact unless disturbed with either an increase in temperature or an interaction with other chemicals. A water molecule, for example, tends to form hydrogen bonds with other water molecules. Above 0°C, however (body temperature is 37°C),

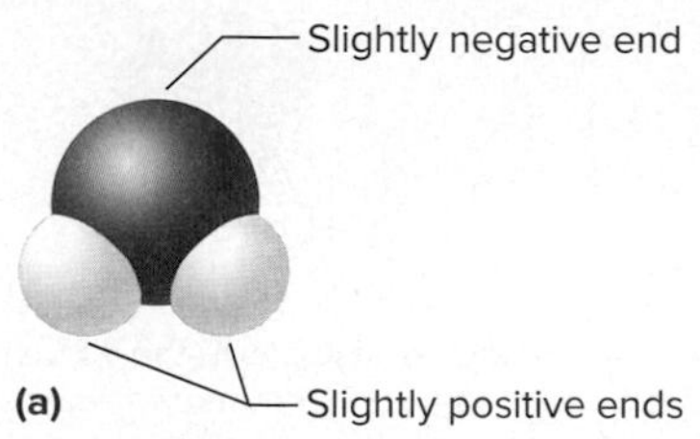

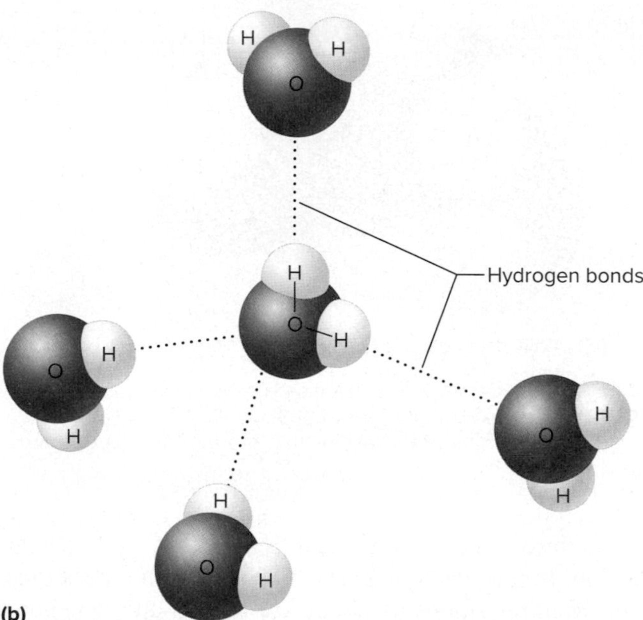

Figure 2.6 Water is a polar molecule. **(a)** Water molecules have equal numbers of electrons and protons but are polar because the electrons are shared unequally, creating slightly negative ends and slightly positive ends. **(b)** Hydrogen bonding connects water molecules. APR

increased molecular movement breaks the hydrogen bonds, and water is a liquid. In contrast, below 0°C the hydrogen bonds between the water molecules shown in figure 2.6*b* are strong enough to form ice.

In some cases, many hydrogen bonds form between polar regions of different parts of a single, very large molecule. Together, these individually weak bonds provide substantial strength. The contribution and importance of the stability of hydrogen bonds to protein and nucleic acid structure is described in section 2.6, Chemical Constituents of Cells (see figs. 2.18 and 2.21*b*).

PRACTICE 2.3

1. What is an ion?
2. Describe two ways that atoms bond with other atoms.
3. Distinguish between an ion and a polar molecule.

2.4 | Molecules, Compounds, and Chemical Reactions

LEARN

1. Explain how molecular and structural formulas symbolize the composition of compounds.
2. Describe three types of chemical reactions.

If atoms of the same element bond, they produce molecules of that element. Gases of hydrogen, oxygen, and nitrogen consist of such molecules (see fig. 2.5).

When atoms of different elements bond, they form molecules called **compounds.** Two atoms of hydrogen, for example, can bond with one atom of oxygen to produce a molecule of the compound water (H_2O) (fig. 2.7). Table sugar (*sucrose*), baking soda, natural gas, beverage alcohol, and most drugs are compounds.

A molecule of a compound consists of definite kinds and numbers of atoms. A molecule of water, for instance, always has two hydrogen atoms and one oxygen atom. If two hydrogen atoms bond with two oxygen atoms, the compound formed is not water, but hydrogen peroxide (H_2O_2).

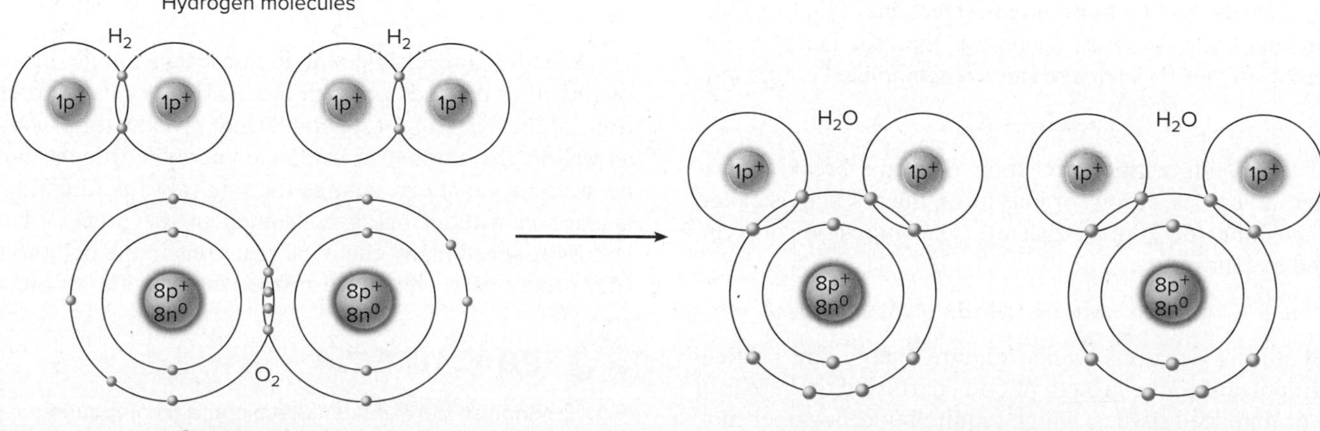

Figure 2.7 Hydrogen molecules can combine with oxygen molecules, forming water molecules. The overlapping shells represent the shared electrons of covalent bonds.

H—H	O=O	H\O/H	O=C=O
H_2	O_2	H_2O	CO_2

Figure 2.8 Structural and molecular formulas for molecules of hydrogen, oxygen, water, and carbon dioxide. Note the double covalent bonds. (Triple covalent bonds are also possible between some atoms.)

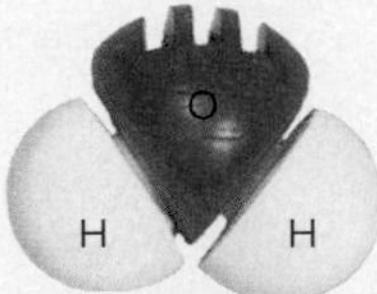

(a) A water molecule (H_2O), with the white parts depicting hydrogen atoms and the red part representing oxygen.

(b) A glucose molecule ($C_6H_{12}O_6$), in which the black parts represent carbon atoms.

Figure 2.9 Three-dimensional molecular models depict spatial relationships of the atoms in a molecule. (a, b): Courtesy John W. Hole, Jr.

Formulas

A **molecular formula** (mo-lek'ū-lar for'mū-lah) represents the numbers and types of atoms in a molecule. Such a formula displays the symbols for the elements in the molecule and the number of atoms of each element. For example, the molecular formula for water is H_2O, which means that each water molecule consists of two atoms of hydrogen and one atom of oxygen (fig. 2.8). The molecular formula for the sugar *glucose* is $C_6H_{12}O_6$, indicating that each glucose molecule consists of six atoms of carbon, twelve atoms of hydrogen, and six atoms of oxygen.

The atoms of each element in a molecule typically form a specific number of covalent bonds, represented by solid lines. Hydrogen atoms form single bonds, oxygen atoms form two bonds, nitrogen atoms form three bonds, and carbon atoms form four bonds. Symbols and lines depict bonds as follows:

— H —   — O —   —N— (three bonds)   —C— (four bonds)

These representations show how atoms are joined and arranged in molecules. Single lines represent single bonds, and double lines represent double bonds. Illustrations of this type are called **structural formulas** (struk'cher-al for'mū-lahz) (fig. 2.8). Three-dimensional models of molecules use different colors for the different kinds of atoms (fig. 2.9).

Chemical Reactions

Chemical reactions form or break bonds between atoms, ions, or molecules, generating new chemical combinations. For example, when two or more atoms (reactants) bond to form a more complex structure (product), the reaction is called **synthesis** (sin'thĕ-sis). Such a reaction is symbolized in this way:

$$A + B \rightarrow AB$$

If the bonds within a reactant molecule break so that simpler molecules, atoms, or ions form, the reaction is called **decomposition** (de"kom-po-zish'un). Decomposition is symbolized as follows:

$$AB \rightarrow A + B$$

Synthesis reactions, which require energy, are particularly important in the growth of body parts and the repair of worn or damaged tissues, which require building larger molecules from smaller ones. In contrast, decomposition reactions occur when food molecules are digested into smaller ones that can be absorbed.

A third type of chemical reaction is an **exchange reaction.** In this reaction, parts of two different types of molecules trade positions as bonds are broken and new bonds are formed. The reaction is symbolized as follows:

$$AB + CD \rightarrow AD + CB$$

An example of an exchange reaction is when an **acid** reacts with a **base,** producing water and a **salt.** Acids and bases are described in the next section.

Many chemical reactions are reversible. This means that the product (or products) of the reaction can change back to the reactant (or reactants) that originally underwent the reaction. A **reversible reaction** is symbolized with a double arrow:

$$A + B \leftrightharpoons AB$$

Whether a reversible reaction proceeds in one direction or the other depends on such factors as the relative proportions of the reactant (or reactants) and product (or products), as well as the amount of available energy. Particular atoms or molecules that can change the rate (not the direction) of a reaction without being consumed in the process, called **catalysts,** speed many chemical reactions in the body so that they proceed fast enough to sustain the activities of life.

PRACTICE 2.4

1. Distinguish between a molecule and a compound.
2. What is a molecular formula? A structural formula?
3. Describe three kinds of chemical reactions.

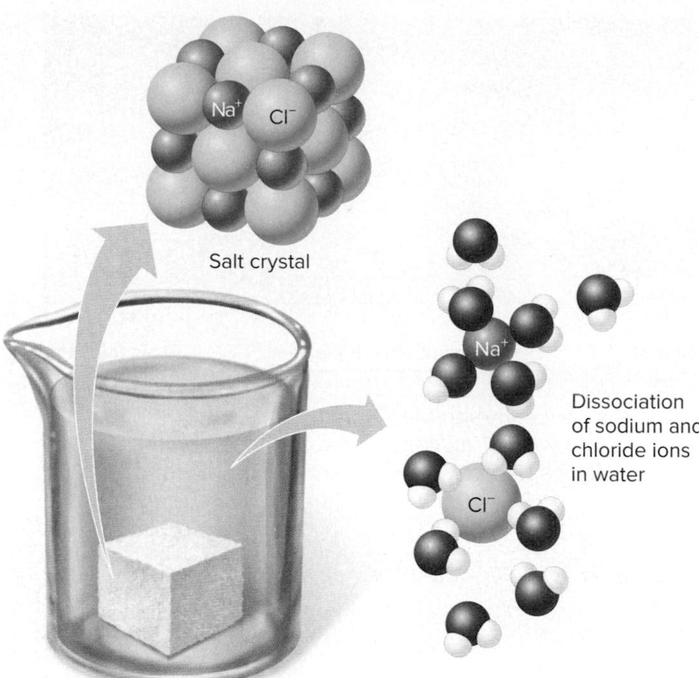

Figure 2.10 The polar nature of water molecules dissociates sodium chloride (NaCl), releasing sodium ions (Na^+) and chloride ions (Cl^-). APR

2.5 | Acids and Bases

LEARN

1. Define *acids, bases,* and *buffers.*
2. Define *pH* and be able to use the pH scale.

When ionically bound substances are placed in water, the slightly negative and positive ends of the water molecules cause the ions to leave each other (dissociate) and interact with the water molecules instead. For example, the salt sodium chloride (NaCl) releases sodium ions (Na^+) and chloride ions (Cl^-) when it is placed in water:

$$NaCl \rightarrow Na^+ + Cl^-$$

In this way, the polarity of water dissociates salts in any water solution, such as the internal environment (fig. 2.10). Because the resulting solution contains electrically charged particles (ions), it will conduct an electric current. Substances that release ions in water are therefore called **electrolytes** (e-lek′tro-lītz).

Acids (from *acere,* meaning sour) are electrolytes that release hydrogen ions (H^+) in water. For example, in water, the compound hydrochloric acid (HCl) releases hydrogen ions (H^+) and chloride ions (Cl^-):

$$HCl \rightarrow H^+ + Cl^-$$

Electrolytes that release ions that bond with hydrogen ions are called **bases.** For example, the compound sodium hydroxide (NaOH) releases hydroxide ions (OH^-) when placed in water. (Note: Some ions, such as OH^-, consist of two or more atoms.)

$$NaOH \rightarrow Na^+ + OH^-$$

The hydroxide ions, in turn, can bond with hydrogen ions to form water; thus, sodium hydroxide is a base. Many bases are present in the body fluids, but because of the way they react in water, the concentration of hydroxide ions is a good estimate of the total base concentration.

The concentrations of hydrogen ions (H^+) and hydroxide ions (OH^-) in body fluids greatly affect the chemical reactions that control certain physiological functions, such as blood pressure and breathing rate. Because their concentrations are inversely related (if one goes up, the other goes down), keeping track of one of them provides information on the other as well.

The pH Scale

A value called **pH** is a measure of hydrogen ion concentration. *pH* stands for **potential hydrogenation,** the potential for the system to accept H^+ ions. The pH scale ranges from 0 to 14. A solution with a pH of 7.0, the midpoint of the scale, contains equal numbers of hydrogen and hydroxide ions and is said to be **neutral.** A solution that contains more hydrogen ions than hydroxide ions has a pH less than 7.0 and is **acidic.** This is why the low end of the scale indicates a strong acid. It is already saturated with H^+ ions and has little potential to accept more (figure 2.11). A solution with fewer hydrogen ions than hydroxide ions has a pH greater than 7.0 and is **basic (alkaline).**

Figure 2.11 indicates the pH values of some common substances. Each whole number on the pH scale represents a tenfold difference in hydrogen ion concentration, and as the hydrogen ion concentration increases, the pH number decreases. Thus, a solution with a pH of 6 has ten times the hydrogen ion concentration of a solution with a pH of 7. This means that relatively small changes in pH can reflect large changes in hydrogen ion concentration.

Buffers are chemicals that resist pH change. They combine with hydrogen ions when these ions are in excess, or they donate hydrogen ions when these ions are depleted. Buffers and the regulation of the hydrogen ion concentration in body fluids are discussed further in section 18.5, Acid-Base Balance.

The pH of human blood is about 7.4, and ranges from 7.35 to 7.45 (see fig. 2.11). If the pH drops below 7.35, the person has *acidosis;* if it rises above 7.45, the condition is *alkalosis.* A substantial pH drop below 6.9 can induce a coma and a pH rise above 7.8 can have adverse effects on muscle contractions. Without medical intervention these extremes can cause death within a few hours. Homeostatic mechanisms such as those described in section 1.5, Maintenance of Life, regulate pH of the internal environment through the action of buffers.

PRACTICE 2.5

1. Compare the characteristics of an acid with those of a base.
2. What does pH measure?
3. What is a buffer?

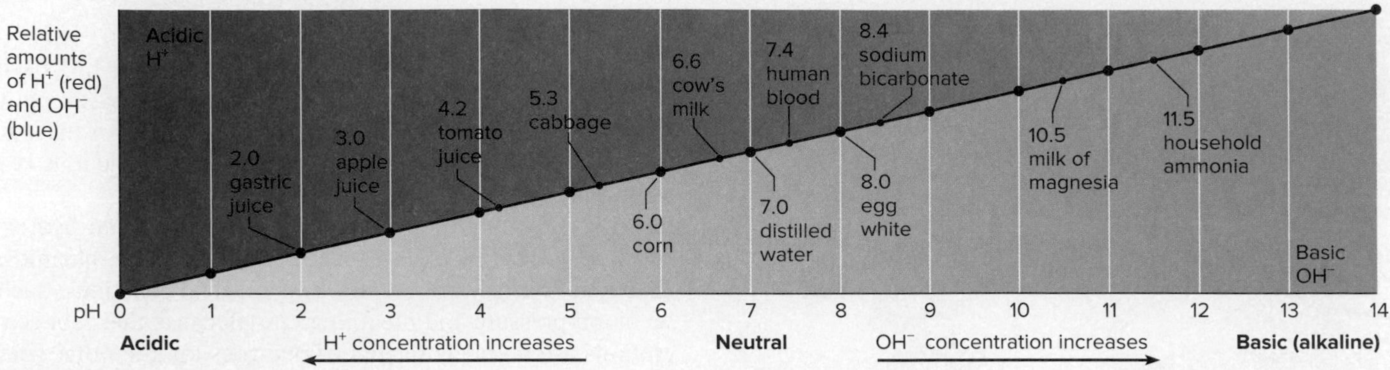

Figure 2.11 The pH scale measures hydrogen ion (H^+) concentration. As the concentration of H^+ increases, a solution becomes more acidic, and the pH value decreases. As the concentration of ions that bond with H^+ (such as hydroxide ions) increases, a solution becomes more basic (alkaline), and the pH value increases. The pH values of some common substances are shown.

PRACTICE FIGURE 2.11

How does the hydrogen ion concentration compare between a solution at pH 6.4 and a solution at pH 8.4?

Answer can be found in Appendix E.

2.6 | Chemical Constituents of Cells

LEARN

1. List the major inorganic chemicals in cells and identify the functions of each.
2. Describe the general functions of the four main groups of organic chemicals in cells.

Chemicals, including those that enter into metabolic reactions or are produced by them, can be divided into two large groups. Compounds that include carbon and hydrogen atoms are called **organic** (or-gan'ik). The rest are **inorganic** (in″or-gan'ik).

Many organic compounds dissolve in water, but most of these do not release ions. Substances that do not release ions are **nonelectrolytes.** Most inorganic substances dissolve in water and dissociate to release ions; those that do release ions are **electrolytes.**

Inorganic Substances

Among the inorganic substances common in cells are water, oxygen, carbon dioxide, and a group of compounds called salts.

Water

Water is the most abundant compound in living material and accounts for about two-thirds of the weight of an adult human. It is the major component of blood and other body fluids.

A substance in which other substances dissolve is a **solvent.** Water is often called the "universal" solvent because many substances readily dissolve in it. A substance dissolved in a liquid such as water is called a **solute.** When a solute dissolves, it is broken down into smaller and smaller pieces, eventually to molecular-sized particles, which may be ions. If two or more types of solutes are dissolved, they are much more likely to react with one another than were the original large pieces. Consequently, most metabolic reactions occur in water.

Moving chemicals in the body is another role for water. For example, blood, which is mostly water, carries many vital substances, such as oxygen, sugars, salts, and vitamins, from the organs of respiration and digestion to the body cells.

Water can absorb and carry heat. Blood, which is mostly water, carries heat released from muscle cells during exercise from deeper parts of the body to the surface, where it may be lost to the outside.

Oxygen

Molecules of oxygen (O_2) enter the body through the respiratory organs and are transported throughout the body by the blood. The red blood cells bind and carry most of the oxygen. Other cells then use the oxygen to release energy from the sugar glucose and other nutrients. The released energy drives the cell's metabolic activities.

Carbon Dioxide

Carbon dioxide (CO_2) is a simple, carbon-containing compound of the inorganic group. It is produced as a waste product when certain metabolic processes release energy, and it is then exhaled from the lungs.

Salts

A **salt** is a compound composed of oppositely charged ions, such as sodium (Na^+) and chloride (Cl^-), which form the familiar table salt NaCl. Salts are abundant in tissues and fluids. They dissociate to provide many necessary ions, including sodium (Na^+), chloride (Cl^-), potassium (K^+), calcium (Ca^{+2}), magnesium (Mg^{+2}), phosphate (PO_4^{-3}), carbonate (CO_3^{-2}), bicarbonate (HCO_3^-), and sulfate (SO_4^{-2}). These ions are important in metabolic processes, including the transport of substances into and out of cells, muscle contraction, and impulse conduction in nerve cells. Table 2.2 summarizes the functions of some of the inorganic substances in the body.

TABLE 2.2 Inorganic Substances Common in the Body

Substance	Symbol or Formula	Function(s)
I. Inorganic molecules		
Water	H_2O	Medium in which most biochemical reactions occur (section 1.5, Maintenance of Life); major component of body fluids (section 2.6, Chemical Constituents of Cells); helps regulate body temperature (section 6.4, Skin Functions); transports chemicals (section 12.1, Introduction)
Oxygen	O_2	Used in energy release from glucose molecules (section 4.4, Energy for Metabolic Reactions)
Carbon dioxide	CO_2	Waste product that results from metabolism (section 4.4, Energy for Metabolic Reactions); reacts with water to form carbonic acid (section 16.6, Gas Transport)
II. Inorganic ions		
Bicarbonate ions	HCO_3^-	Helps maintain acid-base balance (section 18.5, Acid-Base Balance)
Calcium ions	Ca^{+2}	Necessary for bone tissue (section 7.2, Bone Structure), muscle contraction (section 8.3, Skeletal Muscle Contraction), and blood clotting (blood coagulation) (section 12.4, Hemostasis)
Carbonate ions	CO_3^{-2}	Component of bone tissue (section 7.3, Bone Function)
Chloride ions	Cl^-	Major extracellular negatively charged ion (section 18.2, Distribution of Body Fluids)
Hydrogen ions	H^+	pH of the internal environment (section 18.5, Acid-Base Balance)
Magnesium ions	Mg^{+2}	Component of bone tissue (section 7.3, Bone Function); required for certain metabolic processes (section 15.11, Nutrition and Nutrients)
Phosphate ions	PO_4^{-3}	Required for synthesis of ATP, nucleic acids, and other vital substances (section 4.4, Energy for Metabolic Reactions; section 4.5, DNA (Deoxyribonucleic Acid)); component of bone tissue (section 7.3, Bone Function); helps maintain polarization of cell membranes (section 9.5, Charges Inside a Cell)
Potassium ions	K^+	Required for polarization of cell membranes (section 9.5, Charges Inside a Cell)
Sodium ions	Na^+	Required for polarization of cell membranes (section 9.5, Charges Inside a Cell); helps maintain water balance (section 11.7, Adrenal Glands)
Sulfate ions	SO_4^{-2}	Helps maintain polarization of cell membranes (section 9.5, Charges Inside a Cell)

PRACTICE 2.6

1. How do inorganic and organic molecules differ?
2. How do electrolytes and nonelectrolytes differ?
3. Name the inorganic substances common in body fluids.

Organic Substances

Important groups of organic chemicals in cells include carbohydrates, lipids, proteins, and nucleic acids.

Carbohydrates

Carbohydrates (kar″bō-hīdrātz) provide much of the energy that cells require. They supply materials to build certain cell structures and often are stored as reserve energy supplies.

Carbohydrate molecules consist of atoms of carbon, hydrogen, and oxygen. These molecules usually have twice as many hydrogen as oxygen atoms—the same ratio of hydrogen to oxygen as in water molecules (H_2O). This ratio is easy to see in the molecular formula of the carbohydrate glucose ($C_6H_{12}O_6$).

The number of carbon atoms in a carbohydrate molecule varies with the type of carbohydrate. Among the smallest carbohydrates are **sugars.**

Sugars with 5 carbon atoms (pentoses) or 6 carbon atoms (hexoses) are examples of **simple sugars,** or **monosaccharides** (mon″o-sak′ah-rīdz). The simple sugars include the 6-carbon sugars glucose, fructose, and galactose, as well as the 5-carbon sugars ribose and deoxyribose. Figure 2.12 illustrates the structural formulas of glucose.

(a) Some glucose molecules ($C_6H_{12}O_6$) have a straight chain of carbon atoms.

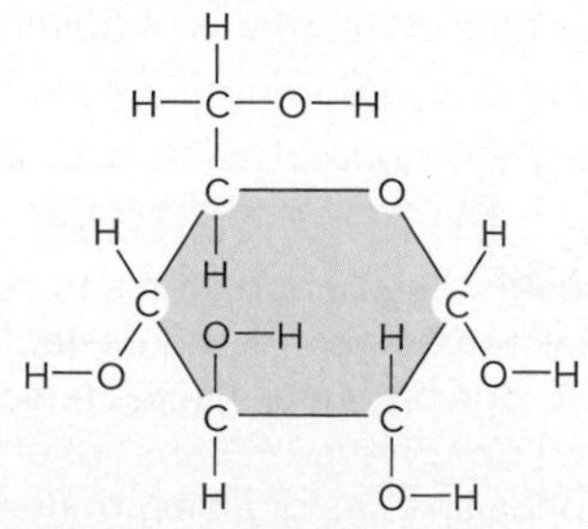

(b) More commonly, glucose molecules form a ring structure.

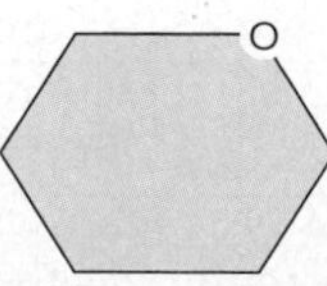

(c) This shape symbolizes the ring structure of a glucose molecule.

Figure 2.12 Structural formulas depict a molecule of glucose ($C_6H_{12}O_6$).

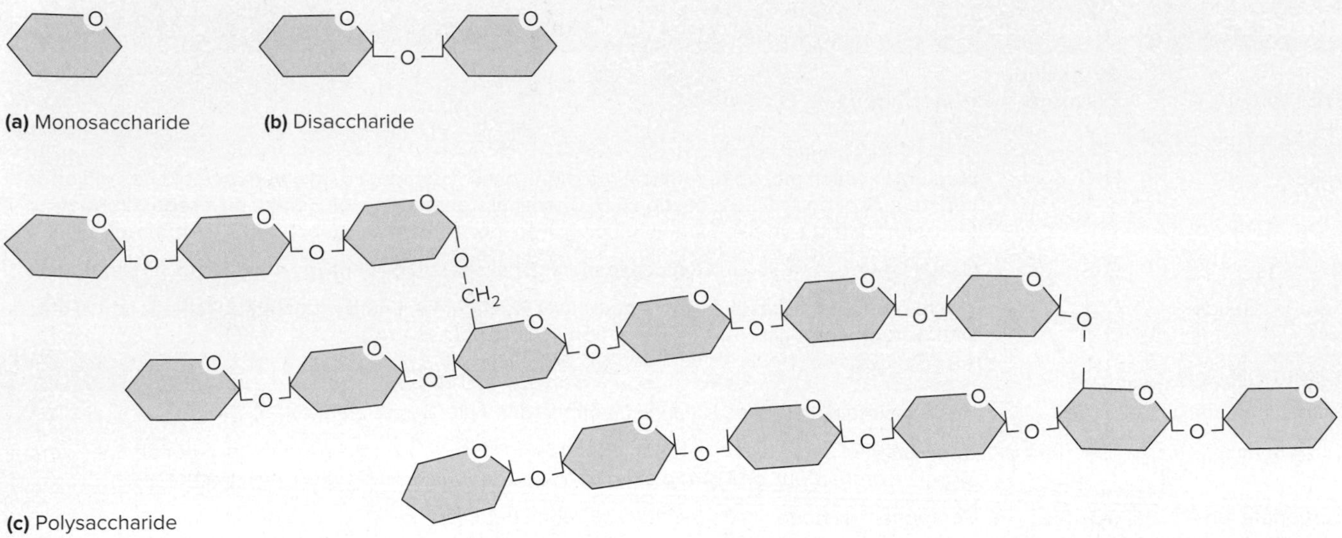

Figure 2.13 Carbohydrate molecules vary in size. **(a)** A monosaccharide molecule consists of one building block with 6 carbon atoms. **(b)** A disaccharide molecule consists of two of these building blocks. **(c)** A polysaccharide molecule consists of many such building blocks.

A number of simple sugar molecules may be linked to form molecules of different sizes (fig. 2.13). Some carbohydrates, such as sucrose (table sugar) and lactose (milk sugar), are *double sugars,* or **disaccharides** (dī-sak′ah-rīdz), whose molecules each consist of two simple sugar building blocks. Other carbohydrates are made up of many simple sugar units joined to form **polysaccharides** (pol″ē-sak′ah-rīdz), such as plant starch. Humans synthesize a polysaccharide similar to starch called *glycogen.*

Lipids

Lipids (lip′idz) are organic substances that are generally insoluble in water (hydrophobic) but soluble in certain organic solvents, such as ether and chloroform. Lipids include a variety of compounds—triglycerides, phospholipids, and steroids—that have vital functions in cells. The most abundant lipids are the triglycerides.

Triglycerides (fat) are used primarily to store energy for cellular activities. Triglyceride molecules can supply more energy, gram for gram, than carbohydrate molecules.

Like carbohydrates, triglyceride molecules are composed of carbon, hydrogen, and oxygen atoms. However, triglycerides have a much smaller proportion of oxygen atoms than do carbohydrates. The formula for the triglyceride tristearin, $C_{57}H_{110}O_6$, illustrates these characteristic proportions.

The building blocks of triglyceride molecules are **fatty acids** and **glycerol** (glis′er-ol). Each glycerol molecule bonds with three fatty acid molecules to produce a single triglyceride molecule (fig. 2.14).

The glycerol portions of all triglyceride molecules are identical, but triglycerides are diverse because there are many kinds of fatty acids. Fatty acid molecules differ in the lengths of their carbon atom chains, but most have an even number of carbon atoms. The chains also vary in the way the carbon atoms bond. In some molecules, the carbon atoms all join by single carbon-carbon bonds. This type of fatty acid is **saturated;** that is, each carbon atom is bound to as many hydrogen atoms as possible and is thus saturated with hydrogen atoms. Saturated triglycerides, or fats, are more abundant in fatty foods that are solids at room temperature, such as butter, lard, and most animal fats (figure 2.14*a*).

Other fatty acid chains are not bound to the maximum number of hydrogen atoms. These fatty acids, therefore, have one or more double bonds between carbon atoms, because a carbon atom must form four bonds to be stable. Fatty acid molecules with double bonds are **unsaturated,** and those with many double-bonded carbon atoms are polyunsaturated. Unsaturated fats are in fatty foods that are liquid at room temperature, such as soft margarine and seed oils (corn, grape, sesame, soybean, sunflower, and peanut). See figure 2.14*b*.

A **phospholipid** (fos″fō-lip′id) molecule is similar to a triglyceride molecule in that it consists of a glycerol portion and fatty acid chains (fig. 2.15*a,b*). A phospholipid, however, has only two fatty acid chains; in place of the third is a portion that includes a phosphate group. The phosphate portion is soluble in water (hydrophilic) and forms the "head" of the molecule, whereas the fatty acid portion is hydrophobic and forms a "tail" (fig. 2.15*c*). Phospholipids are important in cellular structures.

Steroid (ste′roid) molecules are complex structures that include four connected rings of carbon atoms (fig. 2.16). One of the more important steroids is cholesterol, which is in all body cells and is used to synthesize other steroids: hormones such as estrogen and progesterone from the ovaries, testosterone from the testes, and several hormones from the adrenal glands. Chapters 11 and 19 discuss these steroids. Table 2.3 lists the three important groups of lipids and their characteristics.

OF INTEREST Manufacturers add hydrogen atoms to certain vegetable oils to make them harder and easier to use. This process, called hydrogenation, produces fats that are partially unsaturated and also "trans." ("Trans" refers to atoms in a molecule on opposite sides of a backbone-like structure, like stores on opposite sides of a street. Atoms on the same side—like stores on the same side of a street—are called "cis.")

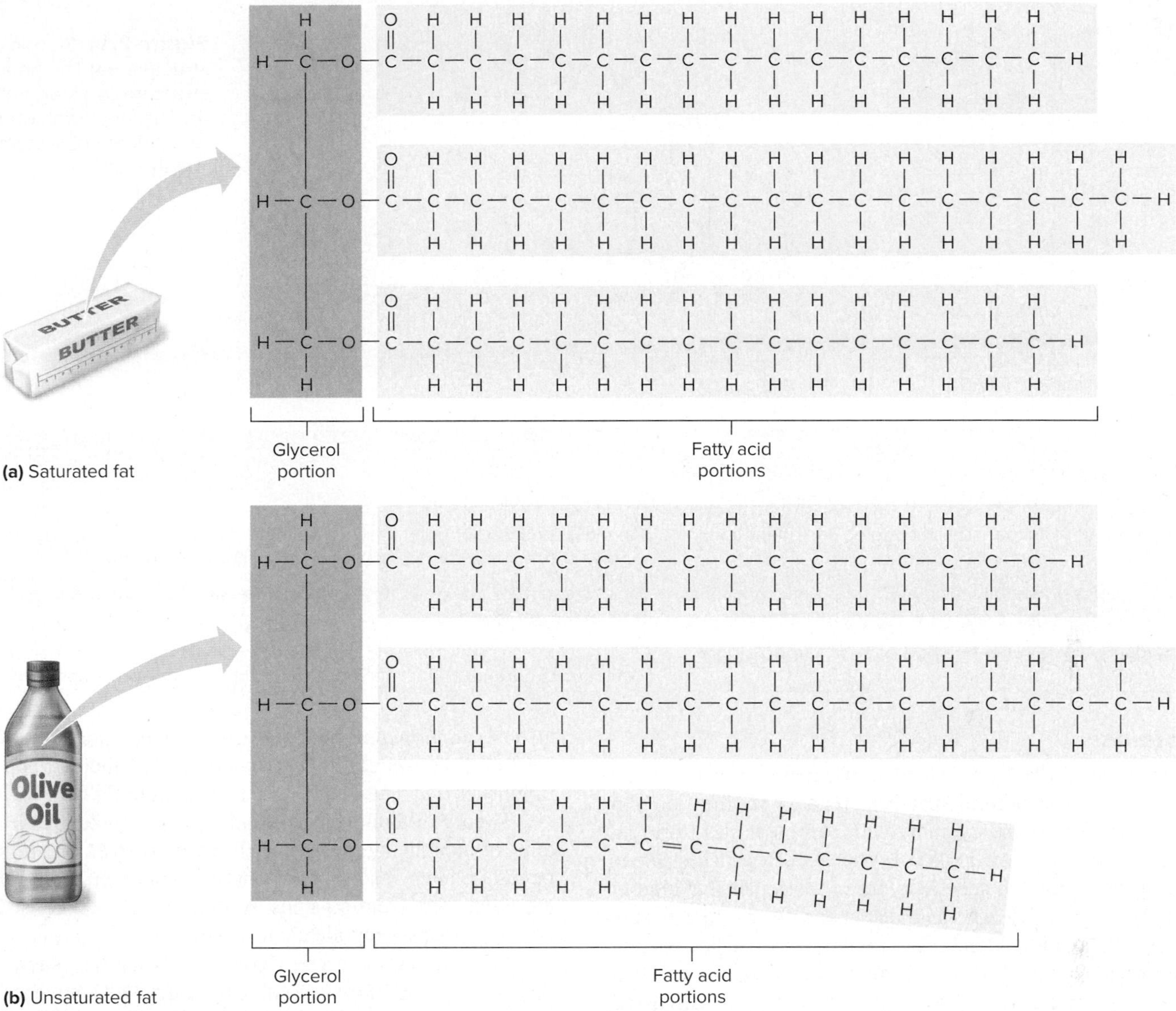

Glycerol portion

Fatty acid portions

(a) Saturated fat

Glycerol portion

Fatty acid portions

(b) Unsaturated fat

Figure 2.14 A triglyceride molecule (fat) consists of a glycerol portion and three fatty acid portions. A saturated fat contains only single bonds between all carbon atoms **(a)**. An unsaturated fat exhibits at least one double bond between two carbon atoms; here it is shown in red **(b)**.

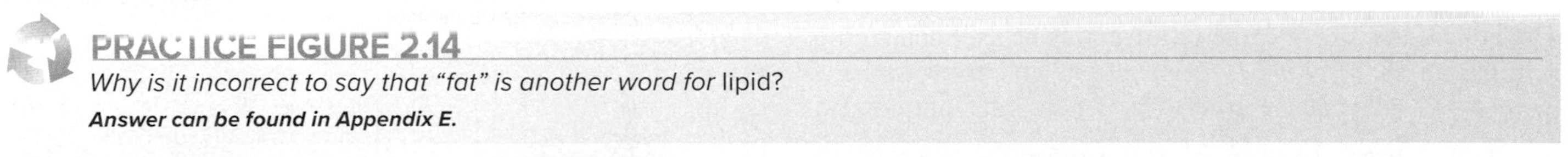

PRACTICE FIGURE 2.14

Why is it incorrect to say that "fat" is another word for lipid?

Answer can be found in Appendix E.

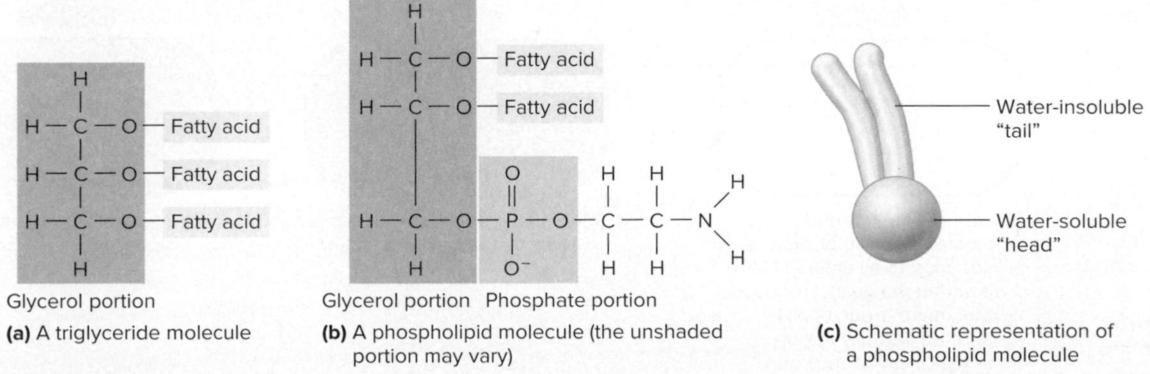

(a) A triglyceride molecule

(b) A phospholipid molecule (the unshaded portion may vary)

(c) Schematic representation of a phospholipid molecule

Figure 2.15 Fats and phospholipids. **(a)** A triglyceride molecule (fat) consists of a glycerol and three fatty acids. **(b)** In a phospholipid molecule, a phosphate-containing group replaces one fatty acid. The unshaded portion may vary. **(c)** Schematic representation of a phospholipid.

(a) General structure of a steroid

(b) Cholesterol

Figure 2.16 Steroid structure. **(a)** The general structure of a steroid. **(b)** The structural formula for cholesterol, a steroid widely distributed in the body.

TABLE 2.3 Important Groups of Lipids

Group	Basic Molecular Structure	Characteristics
Triglycerides	Three fatty acid molecules bound to a glycerol molecule	The most common lipids in the body; stored in fat tissue as an energy supply; fat tissue also provides thermal insulation beneath the skin
Phospholipids	Two fatty acid molecules and a phosphate group bound to a glycerol molecule	Used as structural components in cell membranes; abundant in liver and parts of the nervous system
Steroids	Four connected rings of carbon atoms	Widely distributed in the body and have a variety of functions; include cholesterol, hormones of adrenal cortex, and hormones from the ovaries and testes

Proteins

Proteins (prō′tēnz) have a variety of functions in the body. Many serve as structural materials, energy sources, or hormones. Some proteins combine with carbohydrates (forming glycoproteins) and function as *receptors* on cell surfaces, allowing cells to respond to specific types of molecules that bind to them. Proteins called *antibodies* detect and destroy foreign substances in the body. Metabolism could not occur fast enough to support life were it not for proteins called *enzymes,* which catalyze specific chemical reactions. (Enzymes are discussed in more detail in section 4.3, Control of Metabolic Reactions.)

Like carbohydrates and lipids, proteins are composed of atoms of carbon, hydrogen, and oxygen. In addition, all proteins contain nitrogen atoms, and some contain sulfur atoms. The building blocks of proteins are **amino acids.** Each amino acid has an *amino group* (—NH_2) at one end and a *carboxyl group* (—COOH) at the other (fig. 2.17*a*). Each amino acid also has a *side chain,* or *R group* ("R" may be thought of as the "rest of the molecule"). The composition of the R group distinguishes one type of amino acid from another (fig. 2.17*b,c*).

Twenty different types of amino acids make up the proteins of most living organisms. The amino acids join in polypeptide chains that vary in length from fewer than 100 to more than 5,000 amino acids. A protein molecule consists of one or more polypeptide chains.

Proteins have several levels of structure, shown in figure 2.18*a–c*. *Primary structure* is the amino acid sequence, the order in which particular amino acids occur in the polypeptide chain. *Secondary structure* results from hydrogen bonds between amino acids that are close together in the polypeptide chain.

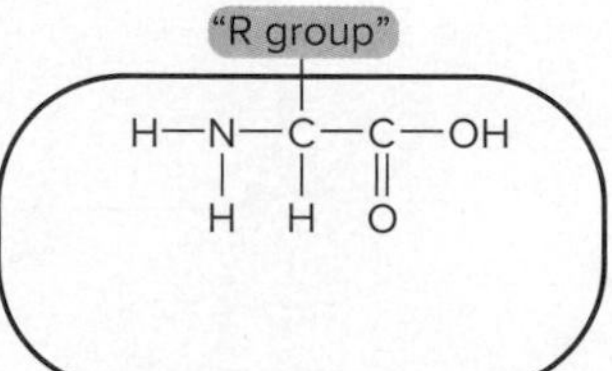

(a) General structure of an amino acid. The portion common to all amino acids is within the oval. It includes the amino group (—NH_2) and the carboxyl group (—COOH). The "R group," or the "rest of the molecule," varies and is what makes each type of amino acid unique.

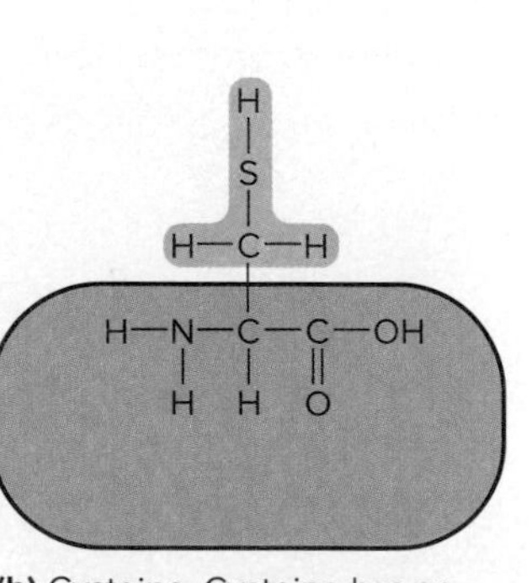

(b) Cysteine. Cysteine has an R group that contains sulfur.

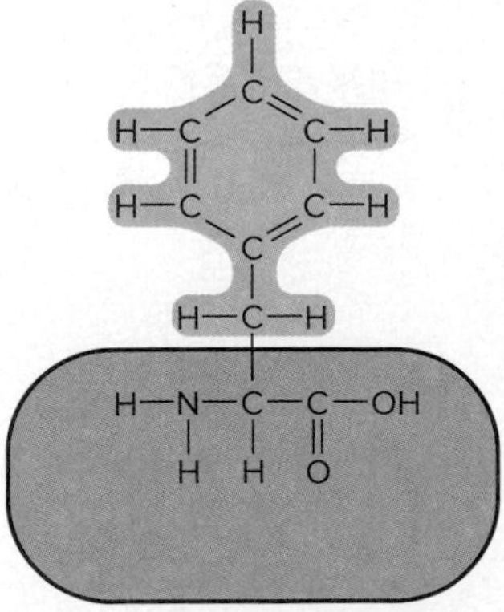

(c) Phenylalanine. Phenylalanine has a complex R group.

Figure 2.17 Amino acid structure. **(a)** An amino acid has an amino group, a carboxyl group, and a hydrogen atom that are common to all amino acid molecules, and a specific R group. **(b)** *and* **(c)** Two representative amino acids and their structural formulas. Each type of amino acid molecule has a particular shape due to its R group.

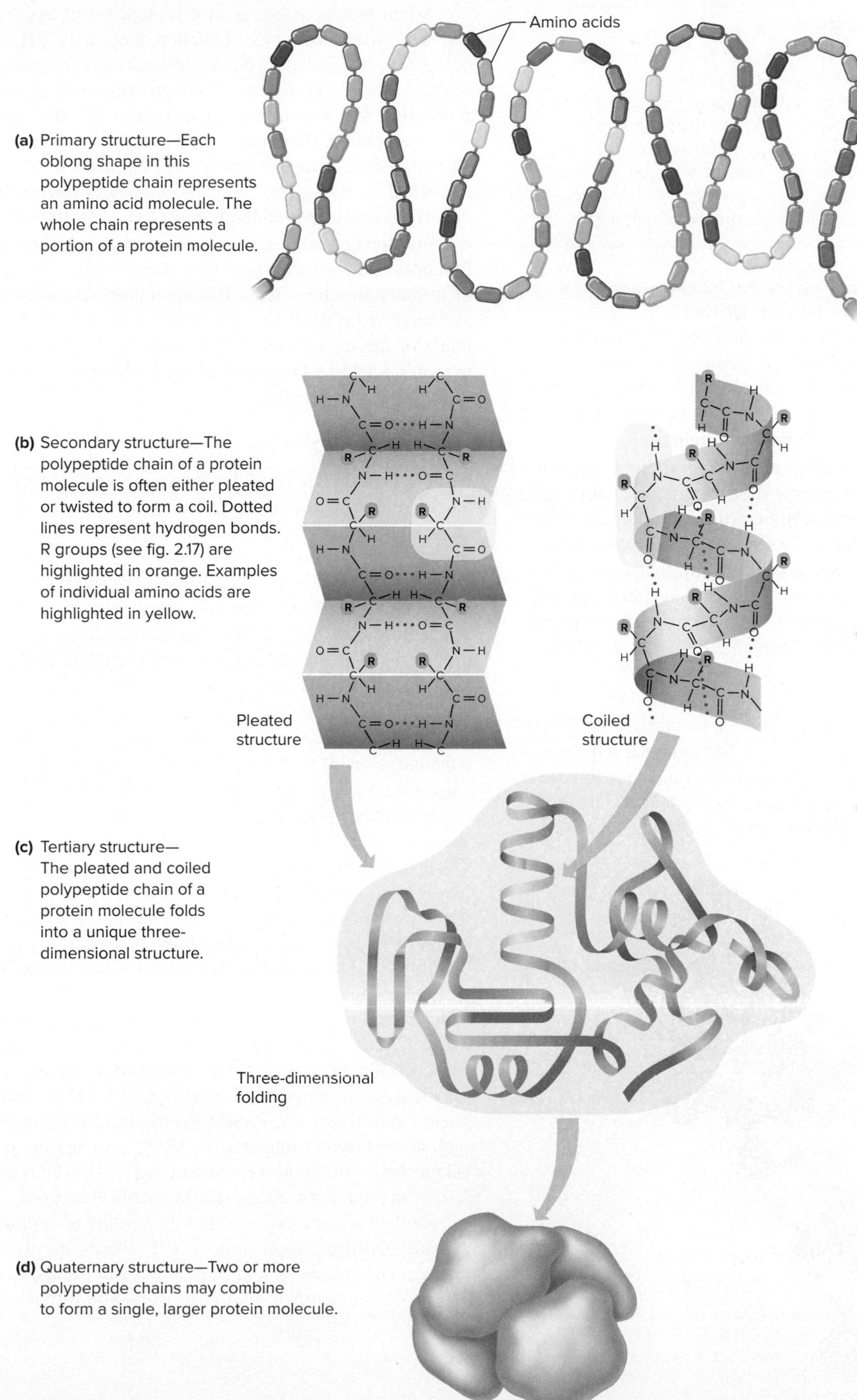

Amino acids
(a) Primary structure—Each oblong shape in this polypeptide chain represents an amino acid molecule. The whole chain represents a portion of a protein molecule.
(b) Secondary structure—The polypeptide chain of a protein molecule is often either pleated or twisted to form a coil. Dotted lines represent hydrogen bonds. R groups (see fig. 2.17) are highlighted in orange. Examples of individual amino acids are highlighted in yellow.
Pleated structure
Coiled structure
(c) Tertiary structure—The pleated and coiled polypeptide chain of a protein molecule folds into a unique three-dimensional structure.
Three-dimensional folding
(d) Quaternary structure—Two or more polypeptide chains may combine to form a single, larger protein molecule.

Figure 2.18 The levels of protein structure determine the overall, three-dimensional conformation, which is vital to the protein's function.

OF INTEREST A human body has more than 200,000 types of proteins, but only an estimated 20,000–25,000 genes, which are the instructions for production of particular polypeptides. The numbers are not the same because parts of some genes encode sequences of amino acids found in more than one type of protein. The relationship between protein number and gene number is a little like assembling a large and diverse wardrobe from a few basic pieces of clothing.

Tertiary structure introduces folds due to attractions involving amino acids far apart in the polypeptide chain.

Proteins with more than one polypeptide chain have a fourth level of conformation, the *quaternary structure.* The constituent polypeptides are connected, forming a very large protein (see fig. 2.18*d*). Hemoglobin is a quaternary protein made up of four separate polypeptide chains.

Hydrogen bonding and covalent bonding between atoms in different parts of the polypeptide chain (or chains) give the final protein a distinctive three-dimensional shape, called its **conformation** (fig. 2.19). The conformation of a protein determines its function. Some proteins are long and fibrous, such as the keratin proteins that form hair, or fibrin, the protein whose threads form a blood clot. Many proteins are globular and function as enzymes, ion channels, carrier proteins, or receptors. Myoglobin and hemoglobin, which transport oxygen in muscle and blood, respectively, are globular.

For some proteins, slight, reversible changes in conformation are part of their normal function. For example, some of the proteins involved in muscle contraction exert a pulling force as a result of such a shape change, leading to movement. The reversibility of these changes enables the protein to function repeatedly.

Figure 2.19 A model of a portion of the protein collagen. The complex shape of a protein is characteristic of that protein and determines its functional properties. Courtesy John W. Hole, Jr.

When hydrogen bonds in a protein break as a result of exposure to excessive heat, radiation, electricity, pH changes, or certain chemicals, a protein's unique shape may be changed dramatically, or *denatured.* Such proteins lose their special properties. For example, heat denatures the protein in egg white (albumin), changing it from a liquid to a solid. This is an irreversible change–a hard-boiled egg cannot return to its uncooked, runny state. Similarly, cellular proteins that are denatured may be permanently altered and lose their functions.

For most proteins, the conformation, which determines its function, is always the same for a given amino acid sequence or primary structure. Thus, it is the amino acid sequence that ultimately determines the role of a protein in the body. Genes, made of the nucleic acid DNA, contain the information for the amino acid sequences of all the body's proteins in a form that the cell can decode.

Nucleic Acids

Nucleic acids (nū-klā″ik as′idz) form genes and take part in protein synthesis. These molecules are generally very large. They include atoms of carbon, hydrogen, oxygen, nitrogen, and phosphorus, which form building blocks called **nucleotides.** Each nucleotide consists of a 5-carbon *sugar* (ribose or deoxyribose), a *phosphate group,* and one of several *nitrogenous* (nitrogen-containing) *bases* (fig. 2.20). A nucleic acid molecule consists of a chain of many nucleotides (polynucleotide chain). There are five different nitrogenous bases found in nucleic acids: adenine, thymine, guanine, cytosine and uracil.

Nucleic acids are of two types. One type–**RNA (ribonucleic acid)** (rī″bō-nū-klā′ik as′id)–is composed of molecules whose nucleotides have ribose. Most RNA molecules are single-stranded polynucleotide chains, but they can fold into shapes that enable them to interact with DNA (fig. 2.21*a*). The second type of nucleic acid–**DNA (deoxyribonucleic acid)** (dē-ok′sĭ-rībō-nū-klā″ik as′id)–has deoxyribose and forms a double polynucleotide chain. The two chains are held together by hydrogen bonds (fig. 2.21*b*). Adenine (A) always pairs with thymine (T). Cytosine (C) always pairs with guanine (G).

DNA molecules store information in a type of molecular code created by the sequences of the four types of nitrogenous bases. Cells use this information to synthesize protein molecules. RNA molecules carry out protein synthesis. Nucleic acids are discussed in more detail in sections 4.5, DNA (Deoxyribonucleic Acid), and 4.6, Protein Synthesis. Certain nucleotides, such as adenosine triphosphate (ATP), have another role, providing energy to chemical reactions (fig. 2.22). ATP is discussed further in section 4.4, Energy for Metabolic Reactions.

Table 2.4 summarizes the four groups of organic compounds. Clinical Application 2.2 discusses the use of biomarkers (both organic and inorganic compounds) in disease diagnosis, indicators of toxin exposure, and forensics.

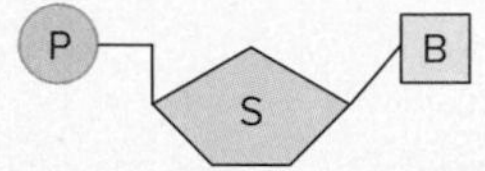

Figure 2.20 A nucleotide consists of a 5-carbon sugar (S = sugar), a phosphate group (P = phosphate), and a nitrogenous base (B = base).

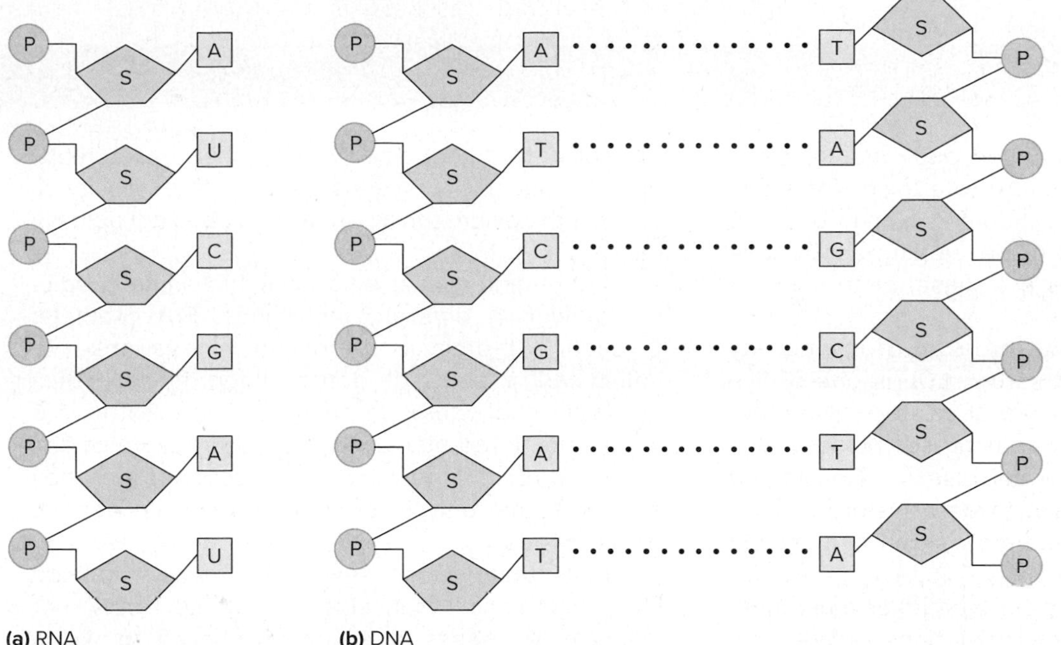

Figure 2.21 A schematic representation of nucleic acid structure. A nucleic acid molecule consists of **(a)** one (RNA) or **(b)** two (DNA) polynucleotide chains. DNA chains are held together by hydrogen bonds (dotted lines), and they twist, forming a double helix.

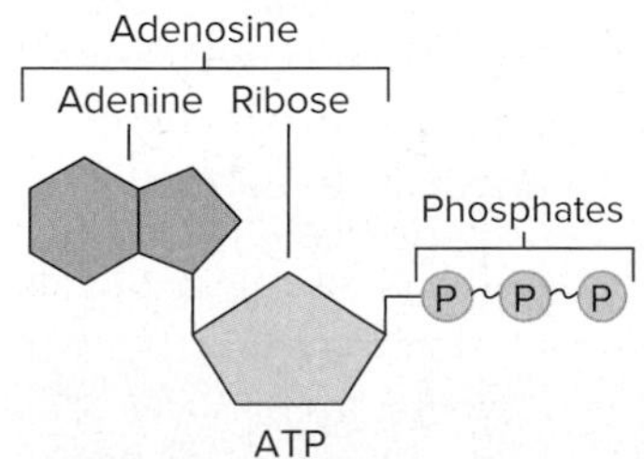

Figure 2.22 An ATP (adenosine triphosphate) molecule consists of an adenine, a ribose, and three phosphates. The wavy lines connecting the last two phosphates represent high-energy chemical bonds. When broken, these bonds release energy, which the cell uses for metabolic processes.

TABLE 2.4 Organic Compounds in Cells

Compound	Elements Present	General Form	Functions	Examples
Carbohydrates	C,H,O	Monosaccharide Disaccharide Polysaccharide	Provide energy, cell structure	Glucose Sucrose Glycogen
Lipids	C,H,O (often P)	Triglyceride Phospholipid Steroid	Provide energy, cell structure	Fats Cholesterol
Proteins	C,H,O,N (often S)	Polypeptide chain	Provide cell structure, enzymes, energy	Albumins, hemoglobin
Nucleic acids	C,H,O,N,P	Polynucleotide chain	Store information for protein synthesis; control cell activities	RNA, DNA

OF INTEREST Recall that water molecules are polar. Many larger molecules, including carbohydrates, proteins, and nucleic acids, have polar regions and dissolve easily in water as a result. Unlike electrolytes, however, these molecules do not dissociate when they dissolve in water—they remain intact. Such water-soluble molecules are said to be hydrophilic (they "like" water).

Molecules that lack polar regions, such as triglycerides and steroids, do not dissolve in water ("oil and water don't mix"). Such molecules do dissolve in lipid and are said to be lipophilic (they "like" lipid).

Water solubility and lipid solubility are important factors in drug delivery and in the movement of substances throughout the body. Much of this is discussed further in section 3.3, Movements Into and Out of the Cell.

PRACTICE 2.6

4. Compare the chemical composition of carbohydrates, lipids, proteins, and nucleic acids.
5. How does an enzyme affect a chemical reaction?
6. What is the chemical basis of the great diversity of proteins?
7. What are the functions of nucleic acids?

CLINICAL APPLICATION 2.2

Biomarkers

A biomarker is a chemical in the body that indicates a disease process or exposure to a toxin. Tests that measure levels of specific biomarkers may be used to indicate increased risk of developing a particular disease, to help diagnose a disease, or to select a treatment.

A familiar biomarker is the level of low-density lipoprotein (LDL) in blood serum. LDL is one of five sizes of lipoproteins that carry cholesterol and other lipids. Elevated LDL indicates increased risk of developing heart and blood vessel disease. Considered along with other factors such as family history, the biomarker test results may lead a physician to prescribe an LDL-lowering drug.

A clinical test based on a biomarker must meet certain criteria. Practical considerations include how easy the biomarker is to obtain and measure, and the cost of the biomarker test compared to the cost for treating disease detected at a more advanced stage. A blood or urine test, for example, is a simple way to sample a biomarker. Sensitivity and specificity are also important criteria for biomarkers. Sensitivity is the ability to detect disease only when it is actually present. Specificity is the biomarker test's ability to exclude the disease in a patient who does not actually have it.

A biomarker must also have the same predictive power in different individuals. For example, C-reactive protein is produced in response to inflammation, and high-density lipoproteins (HDLs) are large cholesterol-carrying particles. Both were once considered biomarkers for coronary heart disease, but recently researchers discovered that they are predictive only in people with certain genetic backgrounds.

Tests that include several biomarkers provide more information than single biomarker tests. For example, to assess exposure to tobacco smoke, a biomarker panel measures carbon monoxide and biochemicals that the body produces as it breaks down carcinogens in cigarette smoke. Biomarker panels are also valuable in selecting drugs most likely to work to treat a specific subtype of cancer in a particular individual.

ASSESS

CHAPTER ASSESSMENTS

2.1 Introduction

1. Define *chemistry*.

2.2 Fundamentals of Chemistry

2. Define *matter*.
3. Explain the relationship between elements and atoms.
4. List the four most abundant elements in the human body.
5. Describe the parts of an atom and where they are found within the atom.
6. Explain why a complete atom is electrically neutral.
7. Define *atomic number, atomic weight,* and *isotope*.
8. Explain how electrons are distributed within the electron shells of an atom.

2.3 Bonding of Atoms

9. An ionic bond forms when ________.
   a. atoms share electrons
   b. positively charged and negatively charged parts of polar covalent molecules attract
   c. ions with opposite electrical charges attract
   d. two atoms exchange protons

2.4 Molecules, Compounds, and Chemical Reactions

10. Explain the relationship between molecules and compounds.
11. Show the difference between a molecular formula and a structural formula.
12. The formula $C_6H_{12}O_6$ means ________.
13. Three major types of chemical reactions are ________, ________, and ________.
14. Explain what a reversible reaction is.
15. Define *catalyst*.

2.5 Acids and Bases

16. Define *acid* and *base*.
17. Explain what pH measures, and describe the pH scale.
18. Define *buffer*.

2.6 Chemical Constituents of Cells

19. Distinguish between inorganic and organic substances.
20. Distinguish between electrolytes and nonelectrolytes.
21. Describe the roles water and oxygen play in the human body.
22. List several ions in body fluids.
23. Describe the general characteristics of carbohydrates.
24. Distinguish between simple sugars and complex carbohydrates.
25. Describe the general characteristics of lipids, and list the three main kinds of lipids.

26. A triglyceride molecule consists of ________.
    a. 3 fatty acids and 1 glycerol
    b. 3 monosaccharides
    c. 3 amino acids
    d. 3 glycerols and 1 fatty acid
27. Explain the difference between saturated and unsaturated fats.
28. A hydrophilic molecule dissolves in ________.
    a. lipid but not water
    b. water but not lipid
    c. neither lipid nor water
    d. both lipid and water
29. List at least three functions of proteins.
30. Describe four levels of protein structure.
31. Explain how protein molecules may denature.
32. Describe the structure of nucleic acids.
33. Explain the major functions of nucleic acids.

ASSESS

INTEGRATIVE ASSESSMENTS/CRITICAL THINKING

Outcome 2.2

1. An advertisement for a cosmetic powder claims that the product is "chemical-free." Explain why this is impossible.

2. The thyroid gland metabolizes iodine, the most common form of which has a molecular weight of 127 (^{127}I). A physician wants to use a radioactive isotope of iodine (^{123}I) to test whether a patient's thyroid gland is metabolizing normally. Based on what you know about how atoms react, do you think this physician's plan makes sense or not?

Outcome 2.3

3. A student in your study group says "Water is considered a 'universal solvent,' which means that anything can dissolve in it." How would you correct this student's statement?

Outcomes 2.4, 2.5, 2.6

4. The bicarbonate buffer system, as shown in the following equation, is a chemical buffer found in extracellular and intracellular fluids. What is the type of chemical reaction shown? Carbonic anhydrase catalyzes this reaction. What specific chemical constituent is carbonic anhydrase? When a person has an asthma attack, where the bronchioles constrict and reduce airflow, what direction will the equation move (to the left *or* to the right) in response? Why?

$$\underset{\text{carbon dioxide}}{CO_2} + \underset{\text{water}}{H_2O} \leftrightarrow \underset{\text{carbonic acid}}{H_2CO_3} \leftrightarrow \underset{\text{hydrogen ion}}{H^+} + \underset{\text{bicarbonate ion}}{HCO_3^-}$$

Outcomes 2.5, 2.6

5. What acidic and basic substances do you encounter in your everyday activities? What acidic foods do you eat regularly? What basic foods do you eat?

Outcome 2.6

6. A topping for ice cream contains fructose, hydrogenated soybean oil, salt, and cellulose. What types of chemicals are in it?

7. At a restaurant, a server recommends a sparkling carbonated beverage, claiming that it contains no carbohydrates. The product label lists water and fructose as ingredients. Is the server correct?

8. How would you explain the dietary importance of amino acids and proteins to a person who is following a diet composed primarily of carbohydrates?

9. A friend, while frying some eggs, points to the change in the egg white (which contains a protein called albumin) and explains that if the conformation of a protein changes, it will no longer have the same properties and will lose its ability to function. Do you agree or disagree with this statement?

Chapter Summary

2.1 Introduction

Chemistry describes the composition of substances and how chemicals react with each other. The human body is composed of chemicals.

2.2 Fundamentals of Chemistry

1. Elements and atoms
   a. **Matter** is composed of elements.
   b. Some elements occur in pure form, but many are found combined with other elements.
   c. **Elements** are composed of atoms, which are the smallest complete units of elements.
   d. **Atoms** of different elements have characteristic sizes, weights, and ways of interacting.
2. Atomic structure
   a. An atom consists of one or more **electrons** surrounding a **nucleus,** which contains one or more **protons** and usually one or more **neutrons.**
   b. Electrons are negatively charged, protons are positively charged, and neutrons are uncharged.
   c. An atom is electrically neutral.
   d. If an atom gains or loses electrons, it becomes charged and is called an ion.
   e. An element's **atomic number** is equal to the number of protons in each atom. The **atomic weight** is equal to the number of protons plus the number of neutrons in each atom.
   f. **Isotopes** are atoms with the same atomic number but different atomic weights.
   g. Some isotopes are **radioactive.**

2.3 Bonding of Atoms

1. When atoms form **chemical bonds,** they gain, lose, or share electrons.
2. Electrons occupy shells around a nucleus.
3. Atoms with completely filled outer shells are **inert,** but atoms with incompletely filled outer shells tend to gain, lose, or share electrons and thus achieve stable structures.
4. Atoms that lose electrons become positively charged **ions.** Atoms that gain electrons become negatively charged ions.
5. Ions with opposite electrical charges attract and form **ionic bonds.** Atoms that share electrons form **covalent bonds.**
6. A **polar** covalently bonded molecule has an uneven distribution of charges.
7. The attraction of the positive hydrogen end of a polar molecule to the negative nitrogen or oxygen end of another polar molecule is called a **hydrogen bond.**

2.4 Molecules, Compounds, and Chemical Reactions

Two or more atoms of the same element may bond to form a molecule of that element. Atoms of different elements may bond to form a molecule of a compound. Molecules consist of definite kinds and numbers of atoms.

1. Formulas
   a. A **molecular formula** represents the numbers and types of atoms in a molecule.
   b. A **structural formula** depicts the arrangement of atoms within a molecule.
2. Chemical reactions
   a. A chemical reaction breaks or forms bonds between atoms, ions, or molecules.
   b. Three types of chemical reactions are: **synthesis,** in which larger molecules form from smaller particles; **decomposition,** in which larger molecules are broken down into smaller particles; and **exchange reactions,** in which the parts of two different molecules trade positions.
   c. Many reactions are **reversible.** The direction of a reaction depends on the proportions of reactants and end products and the energy available.

2.5 Acids and Bases

1. Compounds that release ions in water are **electrolytes.**
2. Electrolytes that release hydrogen ions are **acids,** and those that release hydroxide or other ions that react with hydrogen ions are **bases.**
3. A value called **pH** represents a solution's concentration of hydrogen ions (H^+) and hydroxide ions (OH^-).
4. A solution with equal numbers of H^+ and OH^- is **neutral** and has a pH of 7.0. A solution with more H^+ than OH^- is **acidic** and has a pH less than 7.0. A solution with fewer H^+ than OH^- is **basic (alkaline)** and has a pH greater than 7.0.
5. Each whole number on the pH scale represents a tenfold difference in the hydrogen ion concentration.
6. **Buffers** are chemicals that resist pH change.
7. The pH in the internal environment is regulated.

2.6 Chemical Constituents of Cells

Molecules that have carbon and hydrogen atoms are organic and are usually nonelectrolytes. Other molecules are inorganic and are usually electrolytes.

1. Inorganic substances
   a. Water is the most abundant compound in the body and is a **solvent** in which chemical reactions occur. Water transports chemicals and heat.
   b. Oxygen releases energy from glucose and other nutrients. This energy drives metabolism.
   c. Carbon dioxide is produced when metabolism releases energy.
   d. **Salts** provide a variety of ions that metabolic processes require.
2. Organic substances
   a. **Carbohydrates** provide much of the energy that cells require and also contribute to cell structure. Their basic building blocks are simple **sugar** molecules.
   b. **Lipids,** such as **triglycerides, phospholipids,** and **steroids,** supply energy and build cell parts. The basic building blocks of triglycerides are molecules of **glycerol** and **fatty acids.**
   c. **Proteins** serve as structural materials, energy sources, hormones, cell surface receptors, and enzymes.
      (1) The building blocks of proteins are **amino acids.**
      (2) Proteins vary in the numbers, types, and sequences of their amino acids.
      (3) Primary structure is the amino acid sequence. Secondary structure comes from attractions between amino acids that are close together in the primary structure. Tertiary structure reflects attractions of far-apart amino acids and folds the molecule.
      (4) The amino acid chain of a protein molecule folds into a complex shape **(conformation)** that is maintained largely by hydrogen bonds.
      (5) Excessive heat, radiation, electricity, altered pH, or various chemicals can denature proteins.
   d. **Nucleic acids** are the genetic material and control cellular activities.
      (1) Nucleic acid molecules are composed of **nucleotides.**
      (2) The two types of nucleic acids are **RNA** and **DNA.**
      (3) DNA molecules store information that cell parts use to construct specific protein molecules. RNA molecules play a role in the reactions of protein synthesis.

Cells 3

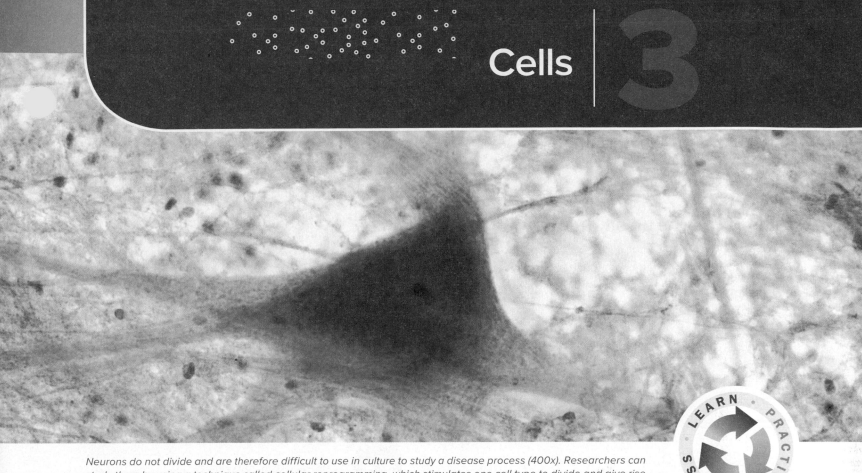

Neurons do not divide and are therefore difficult to use in culture to study a disease process (400x). Researchers can study them by using a technique called cellular reprogramming, which stimulates one cell type to divide and give rise to another cell type. Alvin Telser/McGraw-Hill Education

LEARN • PRACTICE • ASSESS

Reprogramming a cell. The first signs of amyotrophic lateral sclerosis (ALS), also known as Lou Gehrig's disease, are subtle—a foot drags, clothing feels heavy on the body, or an exercise usually done with ease becomes difficult. An actor was fired from a starring television role because of his slurred speech, and a teacher retired when he could no longer hold chalk or pens. Usually within five years of noticing these first signs, failure of the motor neurons that stimulate muscles becomes so widespread that breathing unaided becomes impossible.

ALS currently has no treatment. Part of the reason is that neurons do not divide, so it is impossible to maintain enough of a supply of them growing in a laboratory dish to test new drugs. A technology called cellular reprogramming, however, can take a specialized cell type back to a stage at which it can specialize in any of several ways. Then, by adding certain chemicals to the dish, researchers can guide the specialization toward the cell type, such as a neuron, that is affected in a certain disease. Cells can be reinvented in this way because they all contain a complete set of genes. Such a reprogrammed cell is like a stem cell, but it does not come from an embryo—it comes from a patient.

For ALS, cells taken from skin on the arms of two women in their eighties who have mild cases were reprogrammed to specialize as motor neurons. Researchers can now observe the very first inklings of the disease and use the findings to identify new drug targets and develop new drugs.

ALS was the first of hundreds of diseases now represented by reprogrammed cells, including inherited immune deficiencies, diabetes, blood disorders, and Parkinson disease. In the future, reprogrammed cells might be used therapeutically to replace abnormal cells. First, though, researchers must learn how to control the integration of reprogrammed cells into tissues and organs in the body.

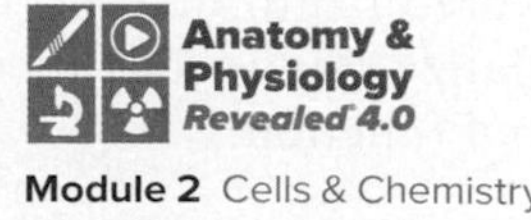

Module 2 Cells & Chemistry

LEARNING OUTLINE

After studying this chapter, you should be able to complete the "Learning Outcomes" that follow the major headings throughout the chapter.

AIDS TO UNDERSTANDING WORDS

cyt- [cell] *cyt*oplasm: fluid (cytosol) and organelles located between the cell membrane and the nuclear envelope.

endo- [within] *endo*plasmic reticulum: complex of membranous structures within the cytoplasm.

hyper- [above] *hyper*tonic: solution that has greater concentration of solutes (greater osmotic pressure) than the cytoplasm of a cell.

hypo- [below] *hypo*tonic: solution that has lower concentration of solutes (lower osmotic pressure) than the cytoplasm of a cell.

inter- [between] *inter*phase: stage between the end of one cell division and the beginning of the next.

iso- [equal] *iso*tonic: solution that has the same concentration of solutes (same osmotic pressure) as the cytoplasm of a cell.

mit- [thread] *mit*osis: process of cell division when threadlike chromosomes become visible within a cell.

phag- [to eat] *phag*ocytosis: process by which a cell takes in solid particles.

pino- [to drink] *pino*cytosis: process by which a cell takes in tiny droplets of liquid.

-som [body] ribo*som*e: tiny, spherical structure that consists of protein and RNA and functions in protein synthesis.

(Appendix A has a complete list of Aids to Understanding Words.)

3.1 | Introduction

LEARN

1. Explain how cells differ from one another.

Cells are the smallest, most basic units of life. That is, they are the smallest units that exhibit all of the characteristics of life that were discussed in chapter 1. The human body is a multicellular organism containing over 30 trillion cells. To fully appreciate this staggering number, a trillion is a thousand billion or, 1,000,000,000,000! Put another way, if you had a trillion pennies, you would be worth 10 billion dollars!

These trillions of cells all consist of the same basic structures, but they come in a large variety of sizes, shapes, and functions (fig. 3.1). For instance, nerve cells have long, threadlike extensions that conduct electrical impulses. Epithelial cells in the skin are small and flattened. They are tightly packed together to provide protection. Muscle cells are filled with proteins that are arranged so that they can generate forces.

A cell continually carries out activities essential for life, as well as more specialized functions, and adapts to changing conditions. Genes control a cell's actions and responses. Nearly all cells have a complete set of genetic instructions (the genome), yet they use only some of this information. Like a person accessing only a small part of the Internet, a cell accesses only some of the vast store of information in the genome to survive and specialize. This accounts for the great diversity in structure and function.

CAREER CORNER
Cytotechnologist

The woman has been tired for months, but blames her fatigue on frequent respiratory infections. When she notices several bruises, she visits her primary care physician, who sends a sample of her blood to a laboratory for analysis.

The cytotechnologist prepares a microscope slide from the blood sample and examines it. She finds too many white blood cells and too few red blood cells and platelets, suggesting that the patient has leukemia—a blood cancer.

A cytotechnologist recognizes signs of illness in cells. The job requires a bachelor's degree in biology, chemistry, or a related field that includes one to two years of an accredited program in cytotechnology. Further certification is available for Specialist in Cytotechnology and Technologist in Molecular Pathology, which may open up jobs in education and management. Cytotechnologists work independently but may consult with pathologists.

Cytotechnologists work in hospitals, doctors' offices, outpatient clinics, and diagnostic laboratories. Some tests performed on cells use automated equipment to identify cancerous or infected cells. Other tests may require the cytotechnologist's judgment, based on experience with and knowledge of cell biology.

A career in cytotechnology can be very rewarding, because detecting disease on the cellular level is critical to the accurate diagnosis of many conditions. Cytotechnologists also work in research settings or sell laboratory equipment.

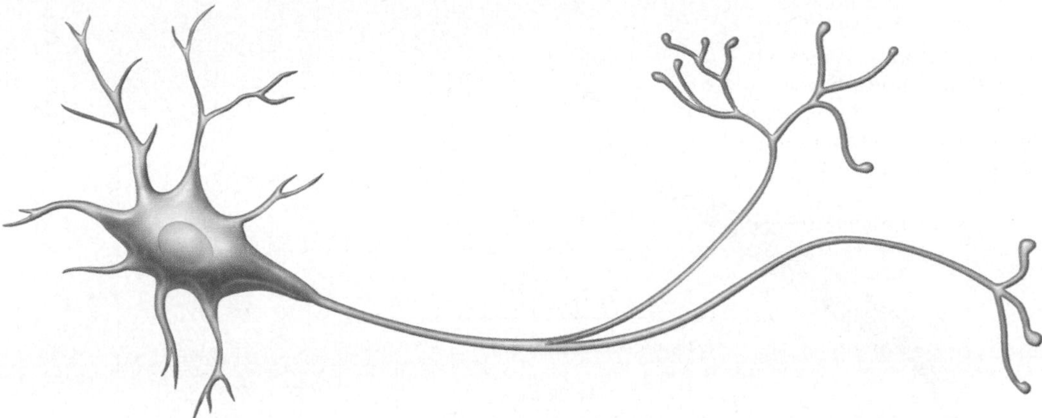

(a) A nerve cell's long extensions enable it to conduct electrical impulses from one body part to another.

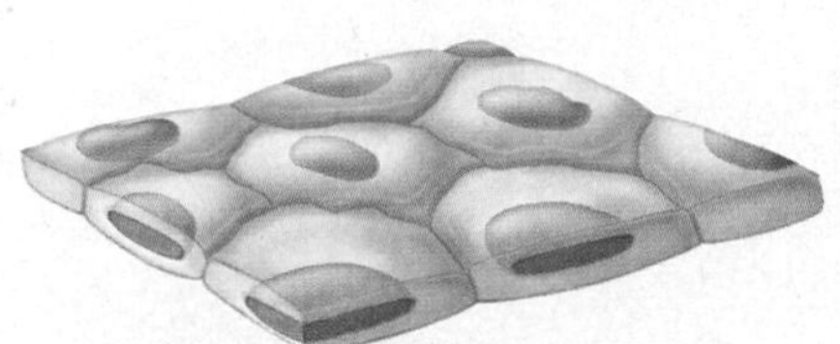

(b) The sheetlike organization of epithelial cells enables them to protect underlying cells.

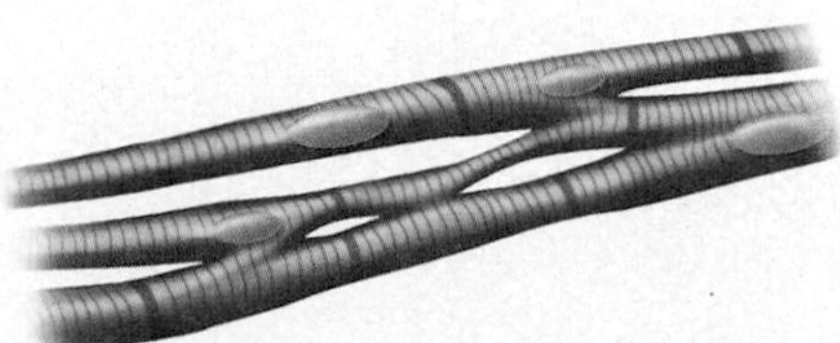

(c) The alignment of contractile proteins within muscle cells enables them to generate forces, pulling closer together the structures to which they attach.

Figure 3.1 Cells vary in size, shape, and function.

PRACTICE 3.1

Answers to the Practice questions can be found in the eBook.

1. What is a cell?
2. Give three examples of how a cell's shape makes possible the cell's function.

3.2 | Composite Cell

LEARN

1. Explain how the structure of a cell membrane makes possible its functions.
2. Describe each type of organelle, and explain its function.
3. Describe the structure and function of cilia and flagella.
4. Describe the cell nucleus and its parts.

Cells differ greatly in size, shape, content, and function. Therefore, describing a "typical" cell is challenging. The cell shown in figure 3.2 and described in this chapter is a composite cell that includes the "usual" cell structures. In a living organism, any given cell has most, but perhaps not all, of these structures, and cells have differing numbers of some of them.

Under the light microscope, a properly applied stain reveals three basic cell parts: the **cell membrane** (sel mem′-brān) that encloses the cell, the **nucleus** (nū′klē-us) that houses the genetic material (DNA) and controls cellular activities, and the **cytoplasm** (sī′tō-plazm) that fills out the cell.

Within the cytoplasm are specialized structures called **organelles** (or-gan-elz′), which can be seen clearly only under the higher magnification of an electron microscope. Organelles are suspended in a liquid called *cytosol.* They are not static and still, as figure 3.2 might suggest. Some organelles move within the cell. Even organelles that appear not to move are the sites of ongoing biochemical activities.

Organelles perform specific functions. They partition off biochemicals that might harm other cell parts, dismantle debris, process secretions, and extract energy from nutrients. Organelles interact, creating vast networks throughout the cell.

PRACTICE 3.2

1. Name the three major parts of a cell and their functions.
2. Define *organelles* and explain their general functions in a cell.

Cell Membrane

The cell membrane (also called the *plasma membrane*) is more than a simple boundary holding in the cellular contents. It is an actively functioning part of the living material. The cell membrane regulates movement of substances in and out of the cell and is the site of much biological activity. Many of a cell's actions that enable it to survive and to interact with other cells use a molecular communication process called *signal transduction.* A series of molecules that are part of the cell membrane form pathways that detect signals from outside the cell and transmit them inward, where yet other molecules orchestrate

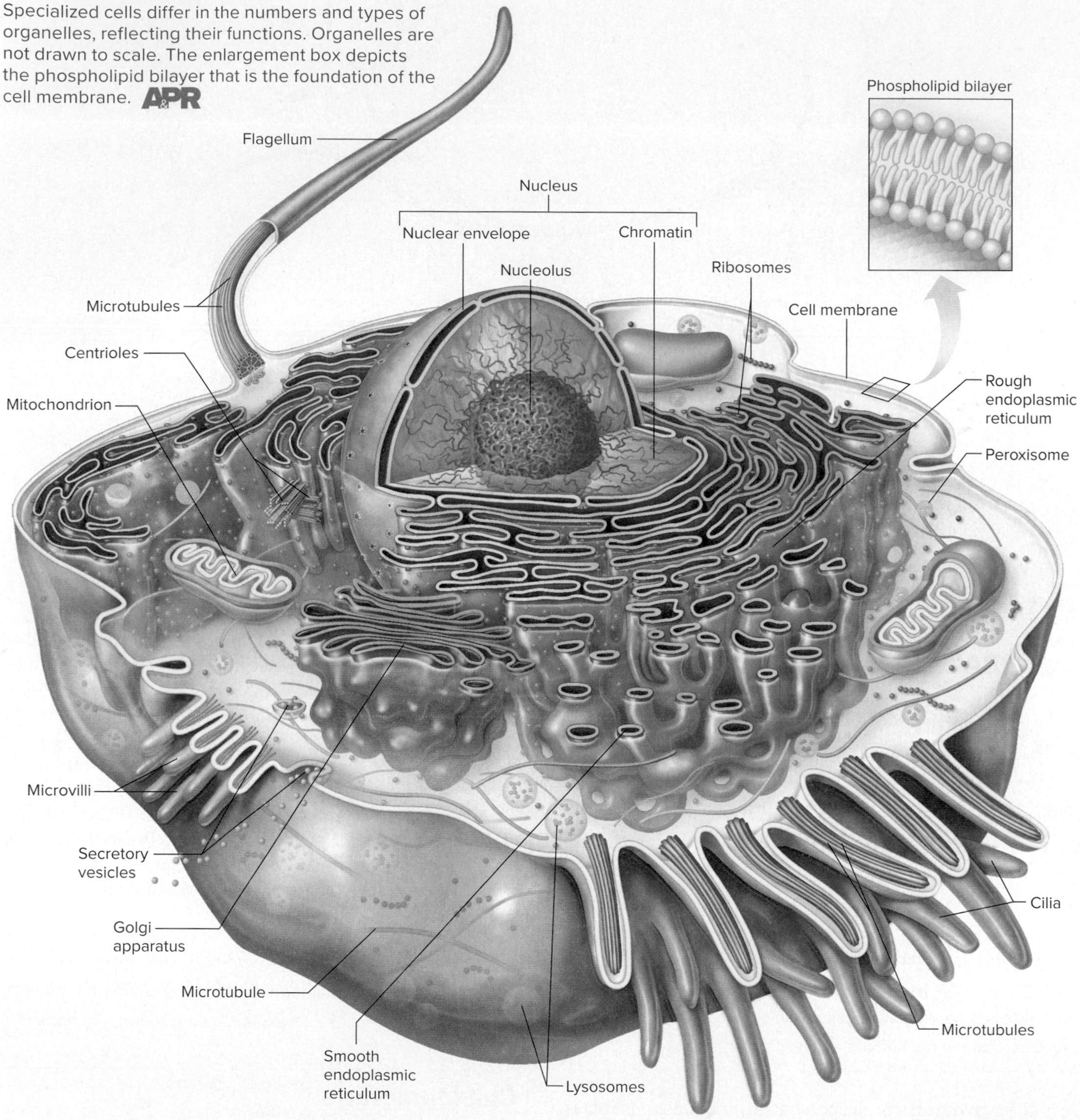

Figure 3.2 A composite cell illustrates the organelles and other structures found in cells. Specialized cells differ in the numbers and types of organelles, reflecting their functions. Organelles are not drawn to scale. The enlargement box depicts the phospholipid bilayer that is the foundation of the cell membrane. APR

the cell's response. The cell membrane also helps cells attach to certain other cells, which is important in forming tissues.

Cell Membrane Structure and Function

A cell membrane is thin and flexible and is composed mainly of lipids and proteins, with a few carbohydrates (fig. 3.3). The cell membrane maintains the structural integrity of the cell and is selectively permeable, or semipermeable, which means only certain substances can enter or leave a cell. Its basic framework is a double layer, or *bilayer,* of phospholipid molecules (see fig. 3.2). Each phospholipid molecule includes a phosphate group and two fatty acids bound to a glycerol molecule (see section 2.6, Chemical Constituents of Cells). The water-soluble phosphate "heads" form the intracellular and extracellular surfaces of the membrane, and the water-insoluble fatty acid "tails" make up the interior (the middle) of the membrane. The lipid molecules can move sideways within the plane of the membrane. The membrane layers form a soft and flexible, but stable, fluid film.

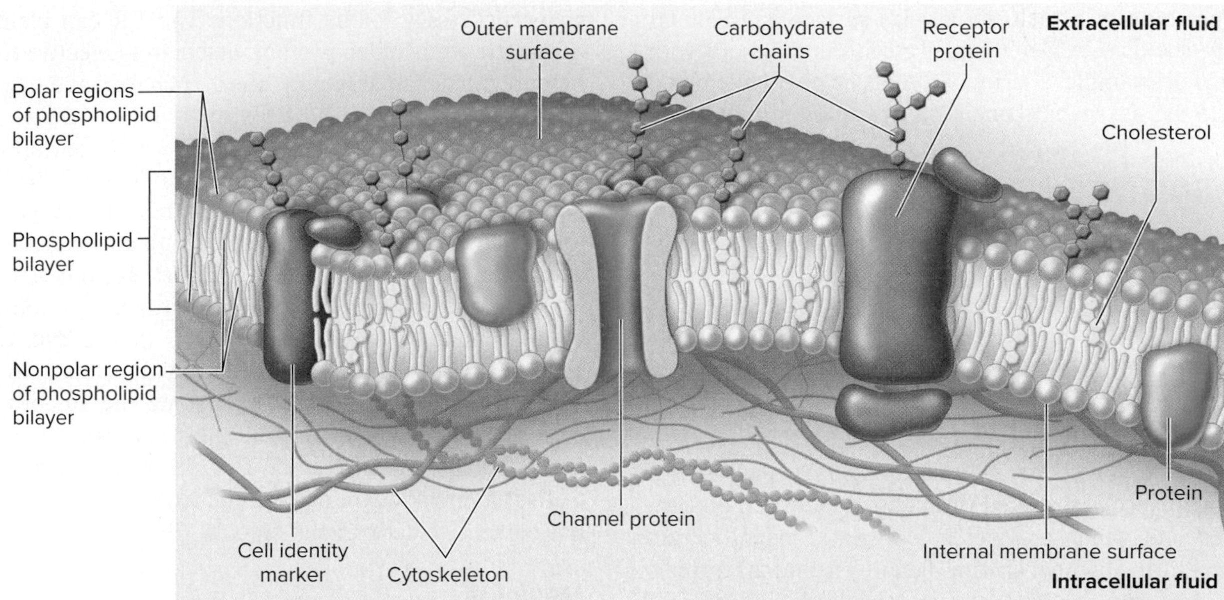

Figure 3.3 Structure of the cell membrane. The cell membrane is composed of two layers of phospholipid molecules with scattered embedded protein and cholesterol molecules. The hydrophilic heads of the phospholipids face the extracellular and intracellular fluids, and the hydrophobic tails form the middle of the membrane. APR

The cell membrane's middle portion is oily because it consists largely of the fatty acid tails of the phospholipid molecules. Molecules that are soluble in lipids, such as oxygen and carbon dioxide, can pass through the phospholipid bilayer easily. However, the bilayer is impermeable to water-soluble molecules, which include amino acids, sugars, proteins, nucleic acids, and certain ions. Cholesterol molecules embedded in the cell membrane's interior help make the membrane less permeable to water-soluble substances. The relatively rigid structure of the cholesterol molecules stabilizes the cell membrane.

A cell membrane includes a few types of lipid molecules, but many kinds of proteins. The proteins provide specialized functions of the membrane. Membrane proteins are classified according to their positions in relation to the phospholipid bilayer. Integral proteins extend through the lipid bilayer and may protrude from one or both faces (sides). Integral proteins that go through the lipid bilayer and extend through both faces of the membrane are called transmembrane proteins. In contrast, peripheral membrane proteins associate outside one side of the bilayer.

Membrane proteins also vary in shape—they may be globular, rodlike, or fibrous. The cell membrane is called a "fluid mosaic" because its proteins are embedded in an oily background and therefore can move, like ships on a sea.

Membrane proteins have a variety of functions. Some form receptors on the cell surface that bind incoming hormones or growth factors, starting signal transduction. Receptors are structures that have specific shapes that fit and hold certain molecules. Many receptors are partially embedded in the cell membrane. Other proteins transport ions or molecules across the cell membrane. Some membrane proteins form ion channels in the phospholipid bilayer that allow only particular ions to enter or leave the cell. Ion channels are specific for calcium (Ca^{+2}), sodium (Na^{+}), potassium (K^{+}), or chloride (Cl^{-}). A cell membrane may have a few thousand ion channels specific for each of these ions.

OF INTEREST Ten million or more ions can pass through an ion channel in one second!

Many ion channels open or close like a gate under specific conditions, such as a change in electrical forces across the membrane of a nerve cell, or receiving biochemical messages from inside or outside the cell.

Abnormal ion channels affect health. In cystic fibrosis, for example, abnormal chloride channels in cells lining the lung passageways and ducts of the pancreas cause the symptoms of impaired breathing and digestion. Sodium channels also malfunction. The overall result: Salt trapped inside cells draws moisture into the cells and thickens surrounding mucus. A drug for cystic fibrosis works by refolding a misfolded protein that forms chloride channels, allowing the channels to function properly.

Proteins with portions that extend inward from the inner face of the cell membrane anchor it to the protein rods and tubules that support the cell from within. Proteins that have portions that extend from the outer surface of the cell membrane mark the cell as part of a particular tissue or organ belonging to a particular person. This identification as "self" is important for the functioning of the immune system (see section 14.7, Immunity: Adaptive [Specific] Defenses). Many of these proteins that extend from the outsides of cells are attached to carbohydrates, forming glycoproteins.

Another type of protein on a cell's surface is a cellular adhesion molecule (CAM). It guides a cell's interactions with other cells. For example, a series of CAMs helps a white blood cell move to the site of an injury, such as a splinter in the skin.

PRACTICE 3.2

3. What is a *selectively permeable membrane?*
4. Describe the chemical structure of a cell membrane.

Cytoplasm

The cytoplasm is the gel-like material that includes the cellular organelles—it makes up most of a cell's volume. When viewed through a light microscope, cytoplasm usually appears as a clear jelly with specks scattered throughout. However, an electron microscope, which provides much greater magnification and the ability to distinguish fine detail (resolution), reveals that the cytoplasm contains vast and complex networks of membranes and organelles suspended in the more-liquid *cytosol*. Cytoplasm also includes abundant protein rods and tubules (*microfilaments* and *microtubules*) that form a framework, or **cytoskeleton** (sī″to-skel′e-ten), meaning "cell skeleton."

Organelles

Most cell activities occur in the cytoplasm, where nutrients are received, processed, and used. The following organelles have specific functions in carrying out these activities:

Ribosomes

Ribosomes (rī′bo-sōmz) are tiny, spherical structures composed of protein and RNA. They provide a structural support and enzymatic activity to link amino acids to synthesize proteins (see section 4.6, Protein Synthesis). Unlike many of the other organelles, ribosomes are not composed of or contained in membranes. They are scattered in the cytoplasm and also bound to the endoplasmic reticulum, another organelle (see fig. 3.2). Clusters of ribosomes in the cytoplasm, called *polysomes,* enable a cell to quickly manufacture proteins required in large amounts.

Endoplasmic Reticulum

Endoplasmic reticulum (en′do-plaz′mik rĕ-tik′u-lum) **(ER)** is a complex organelle composed of membrane-bound, flattened sacs, cylinders, and fluid-filled, bubblelike sacs called **vesicles.** These membranous parts are interconnected and communicate with the cell membrane, the nuclear envelope, and other organelles. The ER winds from the nucleus out toward the cell membrane. It provides a vast tubular network that transports molecules from one cell part to another.

The ER participates in the synthesis of protein and lipid molecules. These molecules may leave the cell as secretions or be used within the cell for such functions as producing new ER or cell membrane as the cell grows. The ER acts as a quality control center for the cell. Its chemical environment enables a forming protein to start to fold into the shape necessary for its function. The ER can identify and dismantle a misfolded protein, much as a defective toy might be pulled from an assembly line at a factory and discarded.

In many places, the ER's outer membrane is studded with ribosomes, which give the ER a textured appearance when viewed with an electron microscope (fig. 3.4*a*,*b*). These parts of the ER that have ribosomes are called *rough ER*. Proteins being synthesized move through ER tubules to another organelle, the Golgi apparatus, for further processing.

As the ER nears the cell membrane, it becomes more cylindrical and ribosomes become sparse and then are no longer associated with the ER. This section of the ER is called *smooth ER* (fig. 3.4*c*). Along the smooth ER are enzymes that are important in lipid synthesis, absorption of fats from the digestive tract, and the metabolism of drugs. Cells that break down drugs and alcohol, such as liver cells, have extensive networks of smooth ER.

Vesicles

Vesicles (ves′ĭ-klz) are membranous sacs that store or transport substances within a cell and between cells. Larger vesicles that contain mostly water form when part of the cell membrane folds inward and pinches off, bringing solid material from outside the cell into the cytoplasm. Smaller vesicles shuttle material from the rough ER to the Golgi apparatus as part of the process of secretion (see fig. 3.2).

Golgi Apparatus

A **Golgi apparatus** (gol′jē ap″ah-ra′tus) is a stack of five to eight flattened, membranous sacs that resemble pancakes (figs. 3.2 and 3.5). This organelle refines, packages, and transports proteins synthesized on ribosomes associated with the ER. A cell may have several Golgi apparatuses. Proteins arrive at the Golgi apparatus enclosed in vesicles composed of membrane from the ER. Sugar molecules were attached to some of the proteins in the ER, forming glycoproteins. These vesicles fuse with the membrane at the innermost end of the Golgi apparatus, which is specialized to receive glycoproteins. As the glycoproteins pass from layer to layer through the Golgi stacks, they are modified chemically. Some sugars may be added or removed, and proteins shortened. When the glycoproteins reach the outermost layer, they are packaged in bits of Golgi membrane, which bud off and form transport vesicles. A transport vesicle may then move to and fuse with the cell membrane, releasing its contents to the outside as a secretion (figs. 3.2 and 3.5). This

OF INTEREST Vesicles that deliver proteins and lipids to other cells are called *exosomes*. These vesicles remove debris, transport immune system molecules from cell to cell, and provide a vast communication network among cells. Because exosomes carry proteins, sugars, and nucleic acids from their cells of origin, and are found in all body fluids, researchers are investigating how to use them in diagnosing disease.

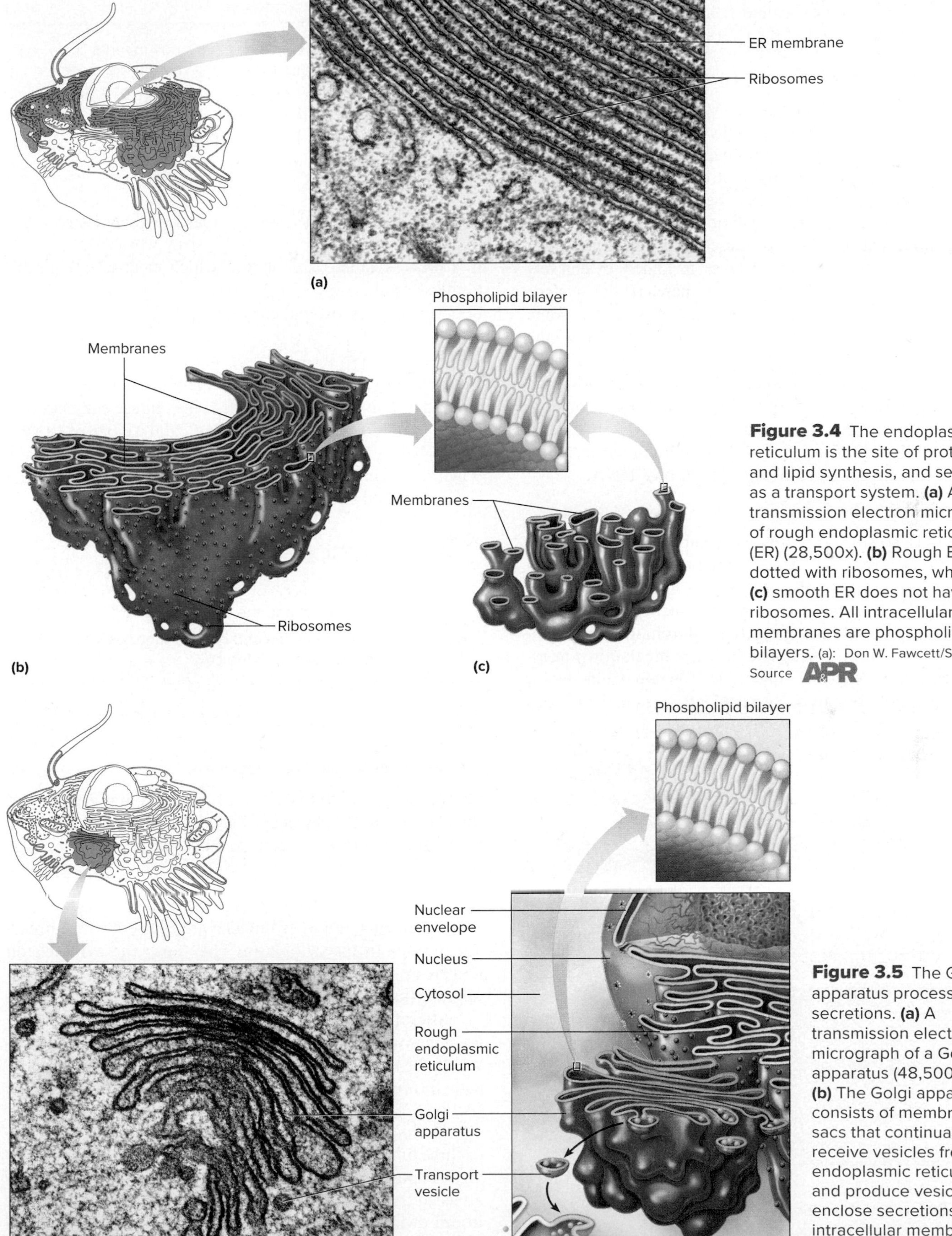

Figure 3.4 The endoplasmic reticulum is the site of protein and lipid synthesis, and serves as a transport system. **(a)** A transmission electron micrograph of rough endoplasmic reticulum (ER) (28,500x). **(b)** Rough ER is dotted with ribosomes, whereas **(c)** smooth ER does not have ribosomes. All intracellular membranes are phospholipid bilayers. (a): Don W. Fawcett/Science Source APR

Figure 3.5 The Golgi apparatus processes secretions. **(a)** A transmission electron micrograph of a Golgi apparatus (48,500x). **(b)** The Golgi apparatus consists of membranous sacs that continually receive vesicles from the endoplasmic reticulum and produce vesicles that enclose secretions. All intracellular membranes are phospholipid bilayers. (a): Biophoto Associates/Science Source APR

is an example of a process called *exocytosis* (see section 3.3, Movements Into and Out of the Cell).

Mitochondria

Mitochondria (mī″to-kon′drē-ah; *sing.* mī″to-kon′drē-on) are elongated, fluid-filled sacs that house most of the biochemical reactions that extract energy from the nutrients in digested food. These organelles are generally oblong, but vary somewhat in size and shape. They move slowly through the cytoplasm and reproduce by dividing.

A mitochondrion has an outer and an inner layer (figs. 3.2 and 3.6). The inner layer is folded extensively into partitions called *cristae,* which increase the surface area. Connected to the cristae are enzymes that control some of the chemical reactions that release energy from nutrients in a process called cellular respiration. Mitochondria store this energy in the chemical bonds of adenosine triphosphate (ATP) (see section 4.4, Energy for Metabolic Reactions). A cell can easily use energy stored as ATP. Very active cells, such as muscle cells, have thousands of mitochondria. Mitochondria contain a small amount of their own DNA.

Lysosomes

Lysosomes (lī′so-sōmz) are tiny membranous sacs that house enzymes that dismantle debris. They bud off of sections of Golgi membranes (see fig. 3.2). Lysosomes maintain the acidic pH that enables the enzymes to function; at the same time, lysosomes shield the rest of the cell from the acidic conditions. The powerful lysosomal enzymes break down nutrient molecules or ingested materials. Certain white blood cells, for example, engulf bacteria, which lysosomal enzymes

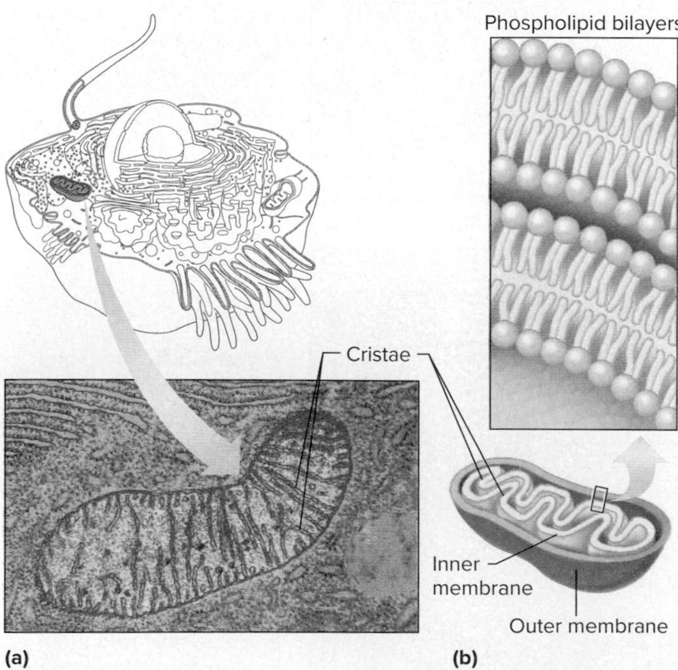

Figure 3.6 A mitochondrion is a major site of energy reactions. **(a)** A transmission electron micrograph of a mitochondrion (28,000x). **(b)** Cristae partition this saclike organelle. All intracellular membranes are phospholipid bilayers. (a): Bill Longcore/Science Source APR

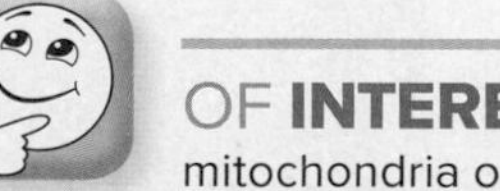

OF INTEREST A person inherits mitochondria only from the mother, because these organelles are excluded from the part of a sperm cell that enters an egg cell. Therefore, diseases that result from mutations in mitochondrial genes are passed only from the mother.

digest. In liver cells, lysosomes break down cholesterol, toxins, and drugs. Lysosomes also destroy worn cellular parts in a process called "autophagy," which means "eating self." Clinical Application 3.1 describes disorders that result from deficiencies of lysosomal enzymes. APR

Peroxisomes

Peroxisomes (pĕ-roks′ĭ-sōmz) are membranous sacs that are abundant in liver and kidney cells. They house enzymes (different from those in lysosomes) that catalyze (speed) a variety of biochemical reactions. Peroxisomal enzymes break down hydrogen peroxide (a by-product of metabolism) and fatty acids, and detoxify alcohol.

PRACTICE 3.2

5. What are the functions of ribosomes, the endoplasmic reticulum, vesicles, and the Golgi apparatus?
6. Explain how organelles and other structures interact to secrete substances from the cell.
7. What is the function of mitochondria?
8. What are the functions of lysosomes and peroxisomes?

Other Cellular Structures

Several other structures are not organelles, but are parts of the cytoskeleton. They occupy significant space in cells and are vitally important for cells to move and to divide.

Microfilaments and Microtubules

Microfilaments and microtubules are types of thin, threadlike strands in the cytoplasm. They form the cytoskeleton and are also part of certain structures (*centrosomes, cilia,* and *flagella*) that have specialized activities.

Microfilaments are tiny rods of the protein **actin.** They form meshworks or bundles, and provide cell motility (movement). In muscle cells, for example, microfilaments aggregate, forming *myofibrils,* which help these cells contract (see section 8.2, Structure of a Skeletal Muscle).

Microtubules are long, slender tubes with diameters two or three times those of microfilaments (fig. 3.7). Microtubules are composed of molecules of the globular protein *tubulin,* attached in a spiral to form a long tube. They are important in cell division. Intermediate filaments lie between microfilaments and microtubules in diameter, and are made of different proteins in different cell types. They are abundant in skin cells and neurons, but scarce in other cell types.

CLINICAL APPLICATION 3.1
Lysosomal Storage Diseases

The little boy was born in 1997. At first he cried frequently and had difficulty feeding, and his limbs were stiff. As time went on he became less alert, as his mental and motor skill development slowed and then stopped. At nine months old, he was diagnosed with Krabbe disease. He had inherited it from his parents, who are carriers. His lysosomes could not make an enzyme that is necessary to produce myelin, a mixture of lipids and proteins that insulates neurons. As a result, toxic biochemicals called galactolipids built up and damaged his brain.

By the time of diagnosis, damage to the boy's nervous system was already advanced. He ceased moving and responding, lost hearing and vision, and had to be tube fed. The boy lived for eight years. Had he been born today, he would have been tested for Krabbe disease along with dozens of other such "inborn errors of metabolism" with a few drops of blood taken from his heel shortly after birth. A stem cell transplant from a donor's umbilical cord blood may have prevented his symptoms.

Lysosomes house 43 types of enzymes, and if any one of them is abnormal, a "lysosomal storage disease" results. Each enzyme must be present within a certain concentration range in order for the cell to function properly. Although each of the 43 lysosomal storage diseases is rare, together they affect about 10,000 people worldwide.

Some lysosomal storage diseases can be treated by replacing the enzyme, using a drug to reduce the biochemical buildup, or using a drug that can unfold and correctly refold a misfolded enzyme, enabling it to function. For all lysosomal storage diseases, physical and occupational therapies and use of adaptive equipment can be helpful with activities of daily living.

Centrosome

A **centrosome** (sen'tro-sōm) is a structure near a Golgi apparatus and the nucleus. It is nonmembranous and consists of two hollow cylinders, called **centrioles,** that lie at right angles to each other. Each centriole is composed of nine groups of three microtubules (figs. 3.2 and 3.8). During mitosis, the centrioles distribute chromosomes to newly forming cells.

Cilia and Flagella

Cilia and flagella are motile structures that extend from the surfaces of certain cells. They are composed of microtubules in a "9 + 2" array, similar to centrioles but with two additional microtubules in the center. Cilia and flagella differ mainly in length and abundance.

Cilia fringe the free surfaces of some cells. A cilium is hairlike and anchored beneath the cell membrane (see fig. 3.2). Cilia form in precise patterns. They move in a coordinated

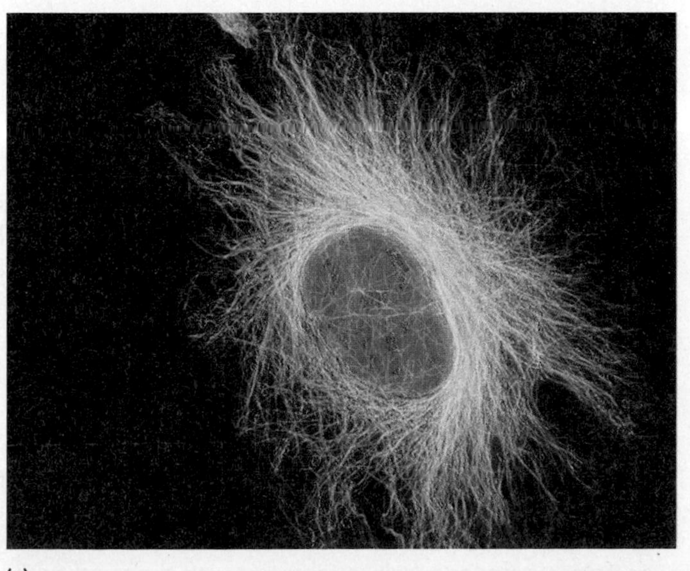

(a)

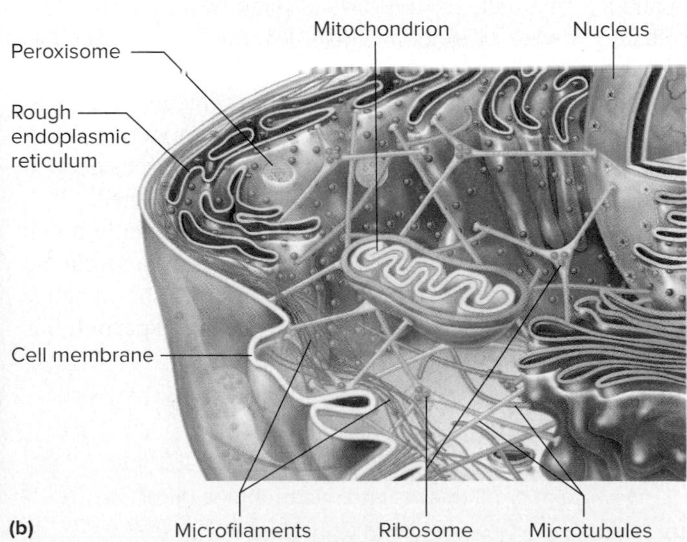

(b)

Figure 3.7 The cytoskeleton provides an inner framework for a cell. **(a)** In this falsely colored micrograph, the cytoskeleton is yellow and red (3,000x). The membrane is not visible. **(b)** Microtubules built of tubulin and microfilaments built of actin help maintain the shape of a cell by forming a scaffolding beneath the cell membrane and in the cytoplasm. (a): Dr. Gopal Murti/Science Source APR

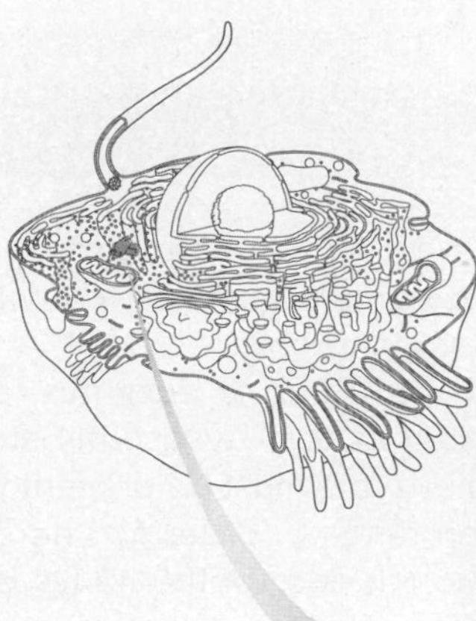

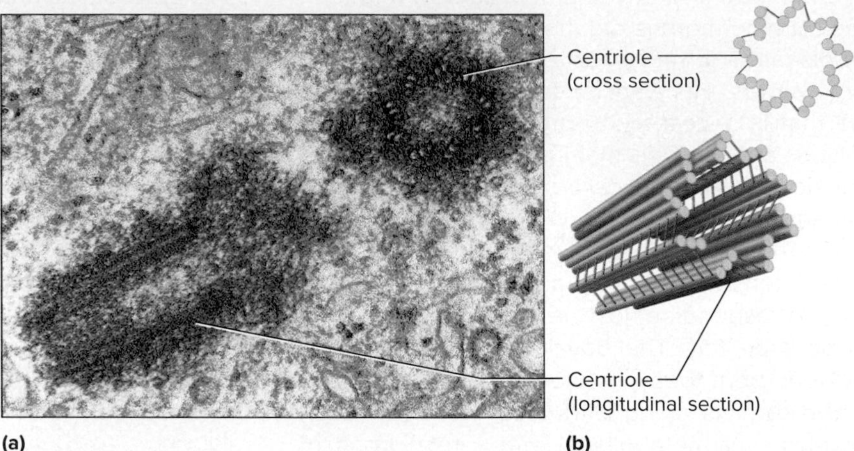

Figure 3.8 Centrioles are built of microtubules and form the structures (spindle fibers) that separate chromosome sets as a cell divides. **(a)** A transmission electron micrograph of the two centrioles in a centrosome (120,000x). **(b)** The centrioles lie at right angles to one another. (a): Don W. Fawcett/Science Source APR

OF **INTEREST** The cilia that wave secretions out of the respiratory system or move an egg toward the uterus are called motile cilia. Many motile cilia fringe certain cells. Another subtype of this organelle, which may be one per cell, called a primary or non-motile cilium, functions as a "cellular antenna," sensing signals and sending them into cells to control growth and maintain tissues. Evidence that primary cilia are important is that the cells of nearly all species have them, nearly all human cell types have them, and diseases in which they are abnormal typically affect several organ systems. Such conditions are called ciliopathies—"sick cilia."

"to-and-fro" manner, so that rows of them beat in succession, producing a wave of motion. This wave moves fluids, such as mucus, over the surface of certain tissues (fig. 3.9*a*). Early in development, beating cilia control the movements of cells as they join to form organs. Some cilia have receptors that detect molecules that signal sensations to cells. Cilia on cells deep in the nasal cavity, for example, assist in the sense of smell.

Flagella are much longer than cilia, and usually a cell has only one (see fig. 3.2). A flagellum moves in an undulating wave, which begins at its base. The tail of a sperm cell is a flagellum that enables the cell to "swim." The sperm tail is the only flagellum in humans (fig. 3.9*b*).

PRACTICE 3.2

9. How do microfilaments and microtubules differ?
10. What is a centrosome and what does it do?
11. Locate cilia and flagella and explain what they are composed of and what they do.

Cell Nucleus

The nucleus houses the genetic material (DNA), which directs all cell activities (figs. 3.2 and 3.10). It is a large, roughly spherical structure enclosed in a double-layered **nuclear envelope,** which consists of inner and outer lipid bilayer membranes. The nuclear envelope has protein-lined channels called *nuclear pores* that allow certain molecules to exit the nucleus. A nuclear pore is not just a hole, but a complex opening formed from 100 or so types of proteins. A nuclear pore is large enough to let out the RNA molecules that carry genes' messages, but not large enough to let out the DNA itself, which must remain in the nucleus to maintain the genetic information.

The nucleus contains a fluid, called *nucleoplasm,* in which the following structures are suspended:

Nucleolus

A **nucleolus** (nū-klē′o-lus) ("little nucleus") is a small, dense body composed largely of RNA and protein. It has no surrounding membrane and forms as specialized regions of certain chromosomes come together. Ribosomes form in the nucleolus and then migrate through nuclear pores to the cytoplasm.

Chromatin

Chromatin consists of loosely coiled fibers of DNA and protein. When the cell begins to divide, chromatin fibers coil tightly. They condense to form the individual **chromosomes** (krō′mo-sōmz) that become visible when stained and viewed under a light microscope. At times other than when the cell is dividing, chromatin unwinds locally to permit access to the information in certain genes (DNA sequences).

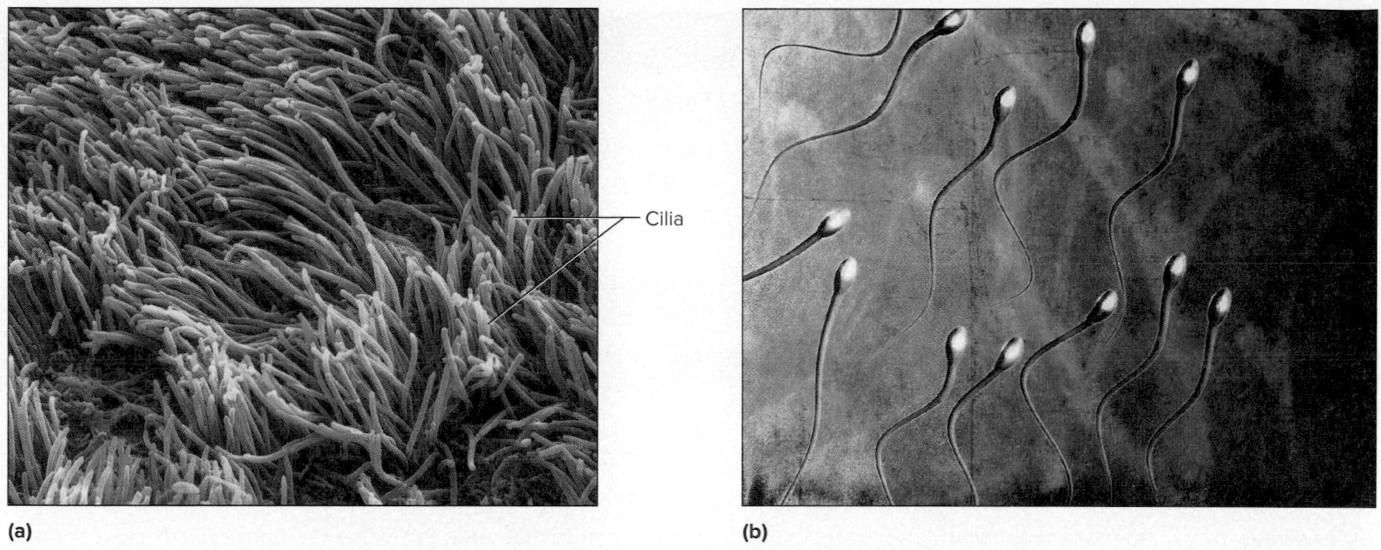

Figure 3.9 Cilia and flagella provide movement. **(a)** Cilia are motile, hairlike extensions that fringe the surfaces of certain cells, including those that form the inner lining of the respiratory tubes (5,800x). Cilia remove debris from the respiratory tract with their sweeping, to-and-fro movement. **(b)** Flagella form the tails of these human sperm cells, enabling them to "swim" (840x). (a): Oliver Meckes/Science Source; (b): Brand X Pictures/Getty Images

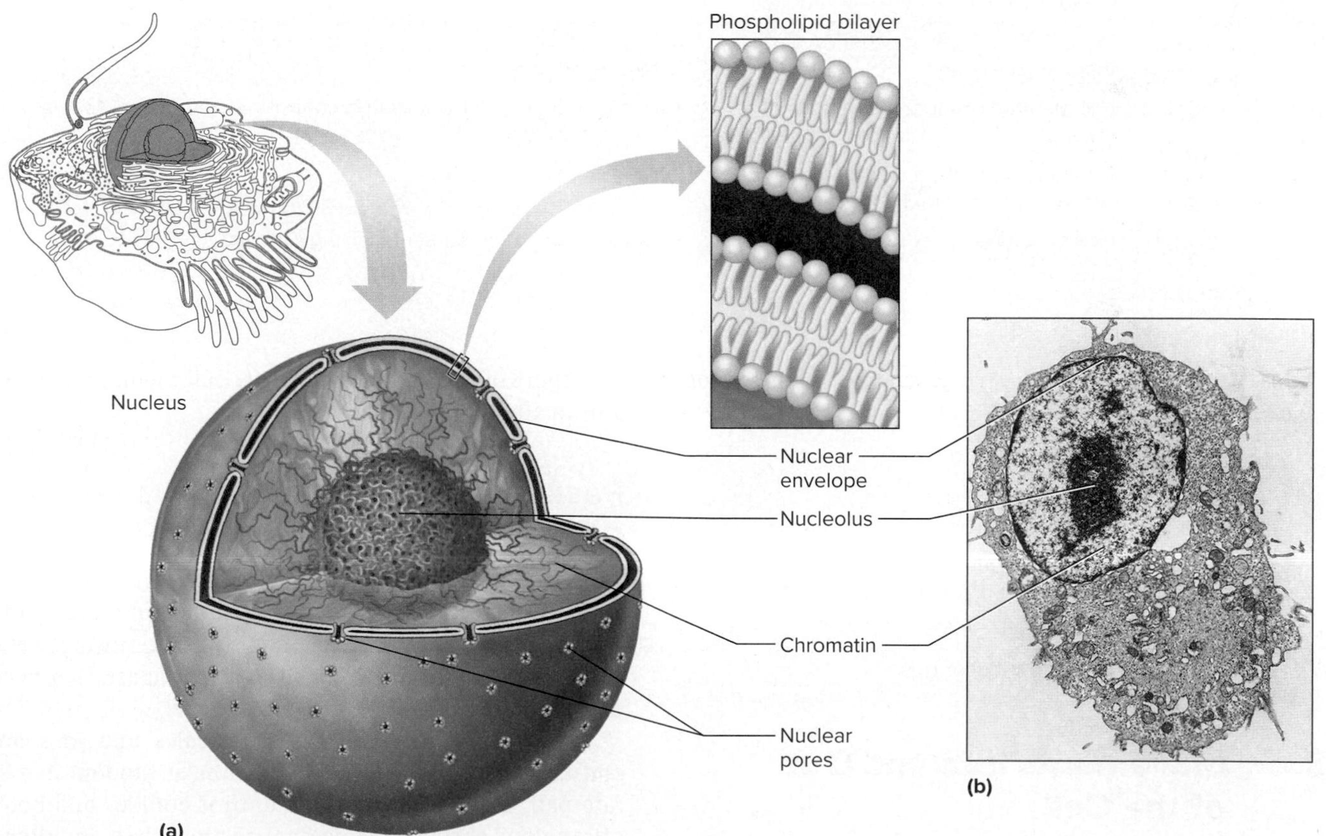

Figure 3.10 The nucleus. **(a)** The nuclear envelope is selectively permeable and allows certain substances to pass between the nucleus and the cytoplasm. Nuclear pores are more complex than depicted here. The nuclear envelope, like all intracellular membranes, consists of phospholipid bilayers. **(b)** A transmission electron micrograph of a cell nucleus (7,500x). It contains a nucleolus and masses of chromatin. (b): Dr. Gopal Murti/Science Source APR

PRACTICE FIGURE 3.10

What structures are inside the nucleus of a cell?

Answer can be found in Appendix E.

TABLE 3.1 Structures and Functions of Cell Parts

Cell Part(s)	Structure	Function(s)
Cell membrane	Membrane composed of protein and lipid molecules	Maintains integrity of cell and controls passage of materials into and out of cell
Ribosomes	Particles composed of protein and RNA molecules	Synthesize proteins
Endoplasmic reticulum	Complex of interconnected membrane-bounded sacs and canals	Transports materials within the cell, provides attachment for ribosomes, and synthesizes lipids
Vesicles	Membranous sacs	Contain and transport various substances
Golgi apparatus	Stack of flattened, membranous sacs	Packages protein molecules for transport and secretion
Mitochondria	Membranous sacs with inner partitions	Release energy from nutrient molecules and change energy into a usable form
Lysosomes	Membranous sacs	Digest worn cellular parts or substances that enter cells
Peroxisomes	Membranous sacs	House enzymes that catalyze diverse reactions, including breakdown of hydrogen peroxide and fatty acids, and alcohol detoxification
Microfilaments and microtubules	Thin rods and tubules	Support the cytoplasm and help move substances and organelles within the cytoplasm
Centrosome	Nonmembranous structure composed of two rodlike centrioles	Helps distribute chromosomes to new cells during cell division
Cilia and flagella	Motile projections attached beneath the cell membrane	Cilia propel fluid over cellular surfaces, and a flagellum enables a sperm cell to move
Nuclear envelope	Double membrane that separates the nuclear contents from the cytoplasm	Maintains integrity of nucleus and controls passage of materials between nucleus and cytoplasm
Nucleolus	Dense, nonmembranous body composed of protein and RNA	Site of ribosome synthesis
Chromatin	Fibers composed of protein and DNA	Contains information for synthesizing proteins

Table 3.1 summarizes the structures and functions of cell parts.

PRACTICE 3.2

12. Identify the structure that separates the nuclear contents from the cytoplasm.
13. What is produced in the nucleolus?
14. Describe chromatin and how it changes.

3.3 Movements Into and Out of the Cell

LEARN

1. Explain how substances move into and out of cells.

The cell membrane is a selective barrier that controls which substances enter and leave the cell. Movements of substances into and out of cells include passive mechanisms that do not require cellular energy (diffusion, facilitated diffusion, osmosis, and filtration) and active mechanisms that use cellular energy (active transport, endocytosis, and exocytosis). Transport of substances across the cell membrane is important in all body systems.

Passive Mechanisms

Diffusion

Diffusion (dĭ-fū′zhun) (also called *simple diffusion*) is the tendency of molecules or ions in a liquid solution or air to move from regions of higher concentration to regions of lower concentration. As the particles move farther apart, they become more evenly distributed, or more *diffuse.*

Diffusion occurs because molecules and ions are in constant motion. Each particle moves at random in a separate path along a straight line until it collides and bounces off another particle, changing direction, then colliding and changing direction once more. Molecules and ions exhibit a universal quality where they seek to reduce collisions and spread out as much as possible. You might say they want "personal space." This is true in living and nonliving systems.

Collisions are less likely if a solution has fewer particles, so there is a net movement of particles from a region of higher concentration to a region of lower concentration. The difference in concentration of particles from high to low establishes what is called a *concentration gradient.* Diffusion

Figure 3.11 A dissolving sugar cube illustrates diffusion. **(1–3)** A sugar cube placed in water slowly disappears as the sugar molecules dissolve and then diffuse from regions where they are more concentrated toward regions where they are less concentrated. **(4)** Eventually the sugar molecules are distributed evenly throughout the water.

occurs down a concentration gradient, which is beneficial for cells because it is "passive." That is, it requires no energy. With time, the concentration of a given substance becomes uniform throughout a solution. This state is called *diffusional equilibrium* (e″kwĭ-lib′re-um). Random movements continue, but there is no further net movement, and the concentration of a substance remains uniform throughout the solution.

A cube of sugar (a solute) put into a glass of water (a solvent) illustrates diffusion (fig. 3.11). At first, the sugar remains highly concentrated at the bottom of the glass. Diffusion moves the sugar molecules from the area of high concentration and disperses them into solution among the moving water molecules. Eventually, the sugar molecules become uniformly distributed in the water.

Diffusion of a substance across a cell membrane can happen only if (1) the cell membrane is permeable to that substance, (2) they are small and without an electrical charge, and (3) a concentration gradient exists such that the substance is at a higher concentration on one side of the cell membrane or the other (fig. 3.12). Consider oxygen and carbon dioxide, which are soluble in the lipid that forms much of the cell membrane. In the body, oxygen diffuses into cells and carbon dioxide diffuses out of cells, but equilibrium is never reached. Intracellular oxygen is always low because oxygen is constantly used up in metabolic reactions. Extracellular oxygen is maintained at a high level by homeostatic mechanisms in the respiratory and cardiovascular systems. Thus, a concentration gradient always allows oxygen to diffuse into cells.

The level of carbon dioxide is always higher inside cells because it is a waste product of metabolism. However, homeostasis maintains a lower extracellular carbon dioxide level. As a result, a concentration gradient is established that always favors carbon dioxide diffusing out of cells (fig. 3.13). Clinical Application 3.2 discusses the use of diffusion in hemodialysis.

Facilitated Diffusion

Substances that are too large such as a glucose molecule, or that have an electrical charge are not able to pass through the lipid bilayer of a cell membrane. To move down the concentration gradient they need the help of a membrane protein. This process is called **facilitated diffusion** (fah-sil″ĭ-tāt′ed dĭ-fu′zhun) (fig. 3.14). It is also referred to as facilitated transport and, as with simple diffusion, does not require an expenditure of energy. One form of facilitated diffusion uses the ion channels described in section 3.2, Composite Cell. Molecules such as

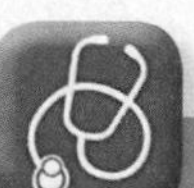

CLINICAL APPLICATION **3.2**

Diffusion in Hemodialysis

Dialysis is a technique that uses diffusion to separate small molecules from larger ones in a liquid. The artificial kidney uses a variant of this process—*hemodialysis*—to treat patients suffering from kidney damage or failure. An artificial kidney (dialyzer) passes blood from a patient through long, coiled tubing composed of porous cellophane. The size of the pores allows smaller molecules carried in the blood, such as the waste material urea, to exit through the tubing. Meanwhile, larger molecules, such as those of blood proteins, remain inside the tubing. The tubing is submerged in a tank of dialyzing fluid (wash solution), which contains varying concentrations of different chemicals. Altering the concentrations of molecules in the dialyzing fluid can control which molecules diffuse out of blood and which remain in it.

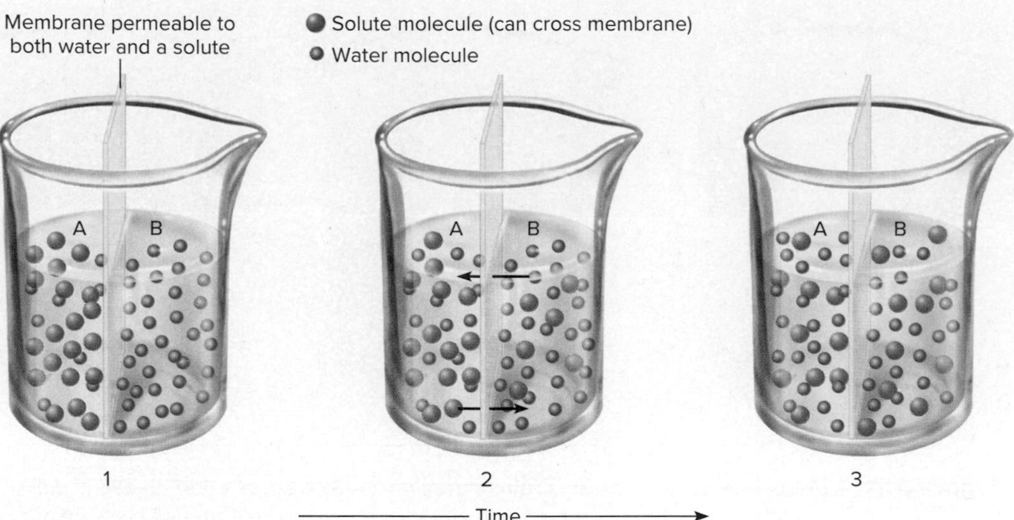

Figure 3.12 Diffusion is a passive movement of molecules. **(1)** A membrane permeable to water and solute molecules separates a container into two compartments. Compartment A contains both types of molecules, while compartment B contains only water molecules. **(2)** As a result of molecular motions, solute molecules tend to diffuse from compartment A into compartment B. Water molecules tend to diffuse from compartment B into compartment A. **(3)** Eventually, equilibrium is reached. APR

glucose and amino acids, that are not lipid-soluble and are too large to pass through ion channels, enter cells by another form of facilitated diffusion that uses a carrier molecule. For example, a glucose molecule outside a cell combines with a special protein carrier molecule at the surface of the cell membrane. The union of the glucose and the carrier molecule changes the shape of the carrier, enabling it to move glucose to the other side of the membrane. The carrier releases the glucose and then returns to its original shape and picks up another glucose molecule. The hormone *insulin,* discussed in section 11.8, Pancreas, promotes facilitated diffusion of glucose through the membranes of certain cells. The number of carrier molecules in the cell membrane limits the rate of facilitated diffusion. You can think of this as the same as having only so many doors to allow for entrance into a building. Facilitated diffusion is specific because a carrier molecule binds only a certain solute, such as glucose.

Osmosis

Osmosis (oz-mo′sis) is the movement of water across a selectively permeable membrane. It is often referred to as the "diffusion of water." This is true, but the whole truth is a bit more interesting and complex. Although water does move along a gradient from high to low concentration, it does so based upon the presence of a variety of dissolved solutes such as sodium and potassium ions or proteins. If there is a difference in solute concentration across a cell membrane, water will flow to the side with the higher solute concentration. This movement occurs because the higher the solute concentration, the less concentrated the water. Therefore, movement is from a solution with a higher concentration of water to a solution with a lower concentration of water (figure 3.15).

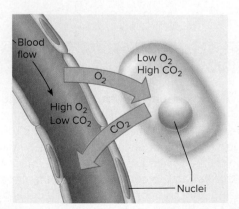

Figure 3.13 Diffusion enables oxygen to enter cells and carbon dioxide to leave.

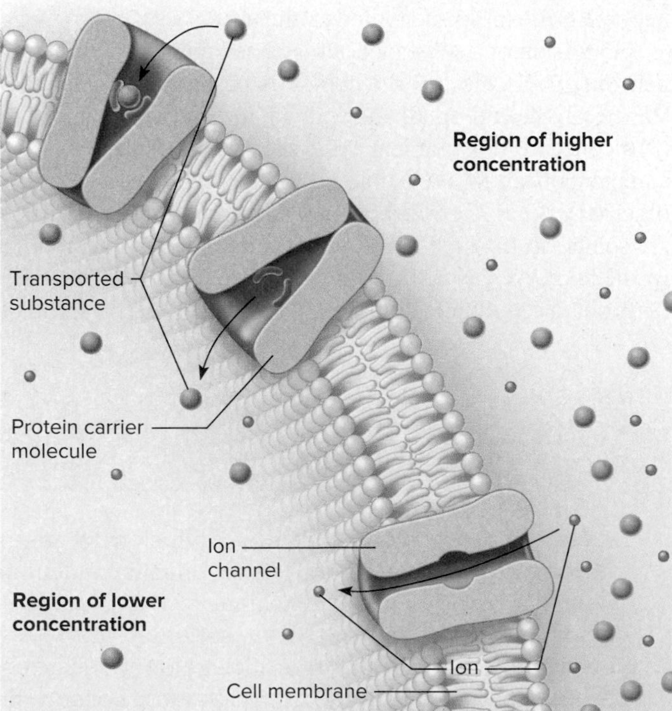

Figure 3.14 Facilitated diffusion uses carrier molecules to transport some substances into or out of cells, from a region of higher concentration to one of lower concentration. APR

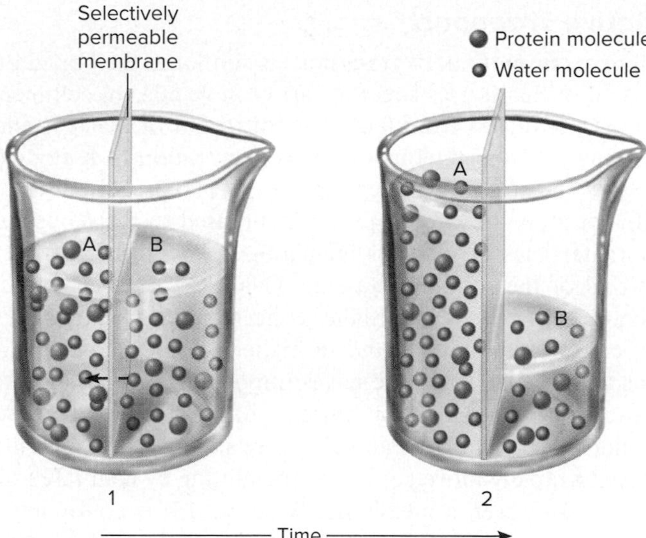

Figure 3.15 Osmosis. **(1)** A selectively permeable membrane separates the container into two compartments. At first, compartment A contains a higher concentration of protein than compartment B. Water moves by osmosis from compartment B into compartment A. **(2)** The membrane is impermeable to proteins, so equilibrium can be reached only by movement of water. As water accumulates in compartment A, the water level on that side of the membrane rises. APR

Osmotic Pressure

In figure 3.15, the selectively permeable membrane is permeable to water (the solvent) and impermeable to protein (the solute in this example). The protein concentration is greater in compartment A. Therefore, water moves from compartment B across the selectively permeable membrane and into compartment A by osmosis. Protein, on the other hand, cannot move out of compartment A, because the selectively permeable membrane is impermeable to it. Note in figure 3.15 that as osmosis occurs, the water level on side A rises. This ability of osmosis to generate enough pressure to lift a volume of water is called **osmotic pressure.**

The greater the concentration of impermeant solute particles in a solution, the greater the osmotic pressure of that solution. Water always tends to move toward solutions of greater osmotic pressure. That is, water moves by osmosis toward regions of trapped solute—whether in a laboratory exercise or in the body.

Tonicity

Tonicity refers to the composition of a solution outside of a cell. It can be potentially confusing because the terms refer to the dissolved solutes, but water is what is moving into and out of the cells. Cell membranes are generally permeable to water, so water equilibrates by osmosis throughout the body, and the concentration of water and solutes everywhere in the intracellular and extracellular fluids is essentially the same. Therefore, the osmotic pressure of the intracellular and extracellular fluids is the same. Any solution that has the same concentration of solutes (same osmotic pressure) as the cytoplasm is called **isotonic** (iso-, the same). See figure 3.16*b*.

Solutions that have a higher osmotic pressure than body fluids are called **hypertonic.** If cells are put into a hypertonic solution, water moves by osmosis out of the cells into the surrounding solution, and the cells shrink (fig. 3.16*c*). Conversely, cells put into a **hypotonic** solution, which has a lower osmotic pressure than body fluids, gain water by osmosis, and they swell (fig. 3.16*a*).

Filtration

Molecules move through membranes by diffusion because of random movements. In other instances, the process of **filtration** (fil-trā′shun) forces molecules through membranes by exerting pressure.

Filtration is commonly used outside the body to separate solids from water. One method is to pour a mixture of solids and water onto filter paper in a funnel. The

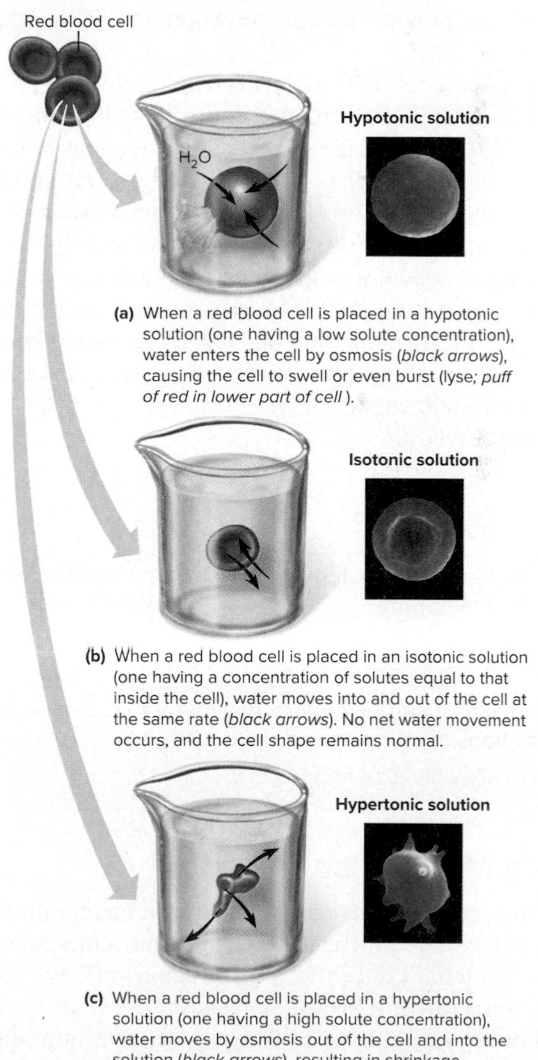

(a) When a red blood cell is placed in a hypotonic solution (one having a low solute concentration), water enters the cell by osmosis (*black arrows*), causing the cell to swell or even burst (lyse; *puff of red in lower part of cell*).

(b) When a red blood cell is placed in an isotonic solution (one having a concentration of solutes equal to that inside the cell), water moves into and out of the cell at the same rate (*black arrows*). No net water movement occurs, and the cell shape remains normal.

(c) When a red blood cell is placed in a hypertonic solution (one having a high solute concentration), water moves by osmosis out of the cell and into the solution (*black arrows*), resulting in shrinkage (crenation).

Figure 3.16 Effects of hypotonic, isotonic, and hypertonic solutions on red blood cells. The shape of a cell may change when it is placed in a new solution. (a, b, c): David M. Phillips/Science Source APR

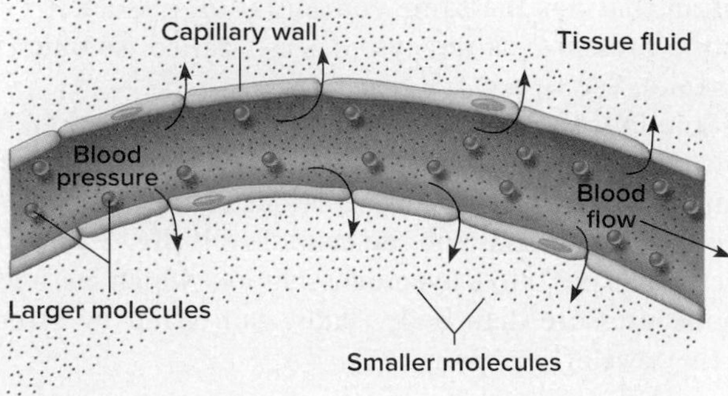

Figure 3.17 In filtration in the body, blood pressure forces smaller molecules through tiny openings in the capillary wall. The larger molecules remain inside.

paper is a porous membrane through which the small water molecules can pass, leaving the larger solid particles behind. *Hydrostatic pressure,* created by the weight of water due to gravity, forces the water molecules through to the other side.

In the body, tissue fluid forms when water and small dissolved substances are forced out through the thin, porous walls of blood capillaries (the smallest diameter blood vessels), but larger particles, such as blood protein molecules, are left inside (fig. 3.17). The force for this movement comes largely from blood pressure, generated mostly by heart action. Blood pressure is greater inside a blood vessel than outside it. However, the large proteins left inside capillaries oppose filtration by drawing water into blood vessels by osmosis. This action prevents the formation of excess tissue fluid, a condition called *edema.* Filtration also helps the kidneys cleanse blood.

PRACTICE 3.3

1. What types of substances diffuse most readily through a cell membrane?
2. Explain the differences among diffusion, facilitated diffusion, and osmosis.
3. Distinguish among hypertonic, hypotonic, and isotonic solutions.
4. How does filtration happen in the body?

Active Mechanisms

When molecules or ions pass through cell membranes by diffusion or facilitated diffusion, their net movements are from regions of higher concentration to regions of lower concentration. Sometimes, however, particles move from a region of lower concentration to one of higher concentration (against the concentration gradient). This movement requires energy. It comes from cellular metabolism and, specifically, from a molecule called **adenosine triphosphate (ATP).** Requiring energy to cross a cell membrane is a little like needing a push to enter a crowded room.

Active Transport

Active transport (ak′tiv trans′port) is similar to facilitated diffusion in that it uses specific carrier molecules in cell membranes. It differs from facilitated diffusion in that particles are moved from regions of low concentration to regions of high concentration, and energy from ATP is required. Up to 40% of a cell's energy supply may be used to actively transport particles through its membranes. The classic example is that of the sodium-potassium (Na^+/K^+) pump shown in figure 3.18. Na^+ is in a higher concentration on the outside of cells, while K^+ is found in higher concentration on the inside of cells. This gradient is important for physiological processes, especially for muscle contraction and nervous conduction (chapters 8 and 9), where sodium and potassium have to rapidly move across the membrane by facilitated diffusion. However, a metabolically active cell is continuously disrupting this gradient through its various uses of these ions. The sodium-potassium pump moves three Na^+ ions out of the cell for every two K^+ ions it brings in, thereby ensuring maintenance of the gradient.

The carrier molecules in active transport are proteins with binding sites that combine with the particles being transported.

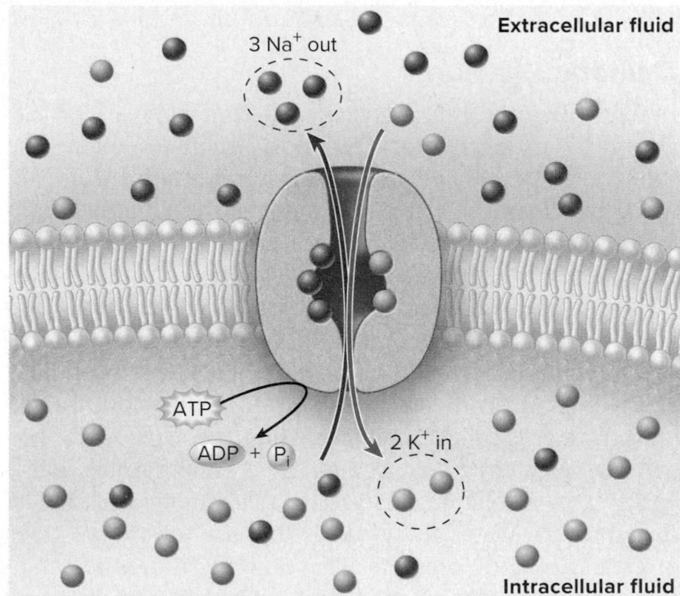

Figure 3.18 The sodium-potassium pump (Na^+–K^+ ATPase). In each cycle of action, this membrane carrier removes three sodium ions (Na^+) from the cell, brings two potassium ions (K^+) into the cell, and hydrolyzes one molecule of ATP. APR

PRACTICE FIGURE 3.18

What are two requirements for active transport to occur?

Why would the Na^+ –K^+ pump, but not osmosis, cease to function after a cell dies?

Answers can be found in Appendix E.

Such a union triggers the release of energy, and this alters the shape of the carrier protein. As a result, the "passenger" particles move through the membrane. Once on the other side, the transported particles are released, and the carriers can accept other passenger molecules at their binding sites.

Particles that are actively transported across cell membranes also include sugars and amino acids, as well as ions of calcium, and hydrogen. Some of these substances are actively transported into cells, and others are actively transported out.

Endocytosis and Exocytosis

Two processes use cellular energy to move substances into or out of a cell without actually crossing the cell membrane. In **endocytosis** (en″dō-sī-tō′sis), molecules or other particles that are too large to enter a cell by diffusion or active transport are conveyed in a vesicle that forms from a portion of the cell membrane. In **exocytosis** (ek″sō-sī-tō′sis), the reverse process secretes a substance stored in a vesicle from the cell. Nerve cells use exocytosis to release the neurotransmitter chemicals that signal other nerve cells, muscle cells, or glands.

Endocytosis happens in three ways: pinocytosis, phagocytosis, and receptor-mediated endocytosis. In **pinocytosis** (pi″no-si-to′sis), meaning "cell drinking," cells take in droplets of liquid from their surroundings (fig. 3.19). A small portion of the cell membrane indents, forming a tubelike area whose open end seals off and produces a small vesicle, which detaches from the membrane surface and moves into the cytoplasm. Eventually the vesicle's membrane breaks down and the liquid inside becomes part of the cytoplasm. In this way, a cell can take in water and the molecules dissolved in it, such as proteins, that otherwise might be too large to enter.

Phagocytosis (fag″o-sī-tō′sis), meaning "cell eating," is similar to pinocytosis, but the cell takes in solids rather than liquids. Certain types of white blood cells are called *phagocytes* because they can take in solid particles such as bacteria and cellular debris. When a particle outside the cell touches the cell membrane of a phagocyte, a portion of the membrane projects outward, surrounding the particle and slowly drawing it inside the cell. The part of the cell membrane surrounding the particle then detaches from the cell's surface, forming a vesicle containing the particle (fig. 3.19).

Once a particle has been phagocytized by a cell, a lysosome combines with the vesicle containing the particle (phagosome) to form a phagolysosome. Here, digestive enzymes decompose the contents. The products of this decomposition then leave the phagolysosome and enter the cytoplasm, where they may be used as raw materials in metabolic processes. Exocytosis may expel any remaining residue.

Pinocytosis and phagocytosis engulf any molecules in the vicinity of the cell membrane. In contrast, **receptor-mediated endocytosis** moves very specific kinds of particles into the cell. In this process, protein molecules extend through the cell membrane to the outer surface. Here, these proteins function as receptors to which only specific molecules from outside the cell can bind. Molecules that bind to specific receptors are called *ligands* (fig. 3.20). Cholesterol molecules enter cells by receptor-mediated endocytosis. Table 3.2 summarizes the types of movements into and out of cells.

OF INTEREST Endocytosis and exocytosis can act together, bringing a particle into a cell, and escorting it, in a vesicle, out of the cell at another place in the cell membrane. This process is called transcytosis. HIV can enter the body this way.

PRACTICE 3.3

5. Explain the mechanism that maintains unequal concentrations of ions on opposite sides of a cell membrane.
6. How are facilitated diffusion and active transport similar and different?
7. How do endocytosis and exocytosis differ?
8. Explain how receptor-mediated endocytosis is more specific than pinocytosis or phagocytosis.

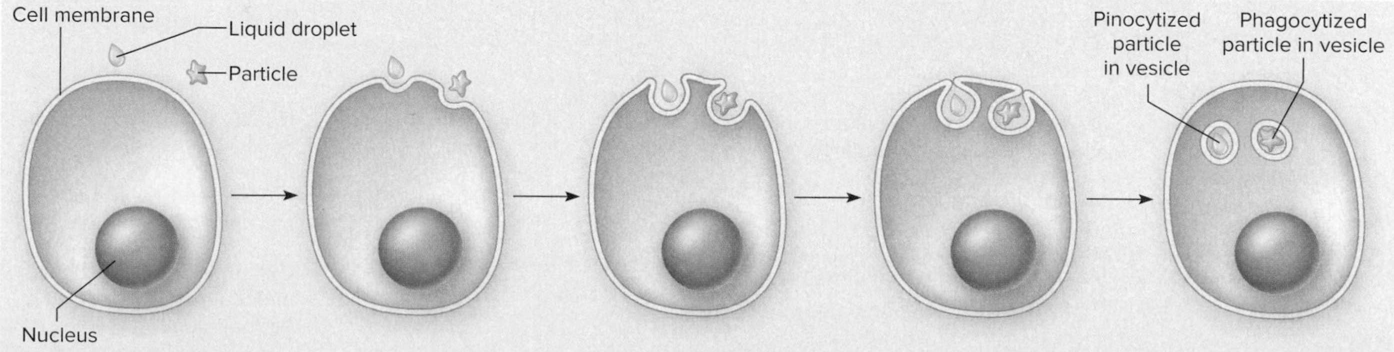

Figure 3.19 A cell may use pinocytosis to take in a fluid droplet and phagocytosis to take in a solid particle from its surroundings. APR

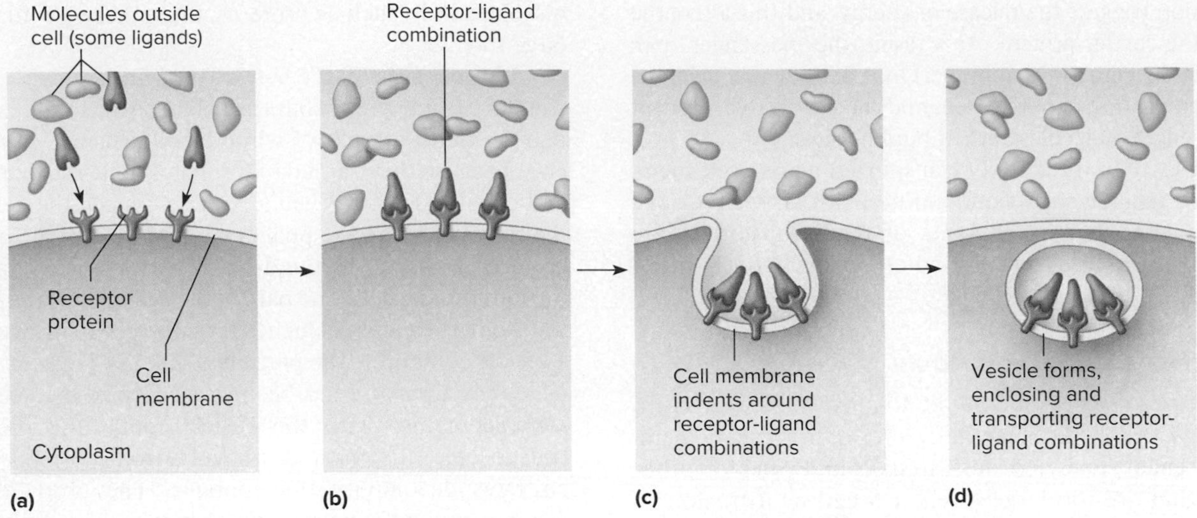

Figure 3.20 Receptor-mediated endocytosis brings specific molecules into a cell. **(a, b)** A specific molecule (ligand) binds to a receptor protein, forming a receptor-ligand combination. **(c)** The binding of the ligand to the receptor protein stimulates the cell membrane to indent. **(d)** Continued indentation forms a vesicle, which encloses and then transports the molecule into the cytoplasm.

TABLE 3.2 Movements Through Cell Membranes

Process	Characteristics	Source of Energy	Example
Passive mechanisms			
Diffusion	Molecules move through the phospholipid bilayer from regions of higher concentration to regions of lower concentration.	Molecular motion	Exchange of oxygen and carbon dioxide in the lungs
Facilitated diffusion	Ions move through channels, or molecules move by carrier proteins, across the membrane from a region of higher concentration toward one of lower concentration.	Molecular motion	Movement of glucose through a cell membrane
Osmosis	Water molecules move through a selectively permeable membrane toward the solution with more impermeant solute (greater osmotic pressure).	Molecular motion	Distilled water entering a cell
Filtration	Smaller molecules are forced through porous membranes from regions of higher pressure to regions of lower pressure.	Hydrostatic pressure	Molecules leaving blood capillaries
Active mechanisms			
Active transport	Carrier molecules transport molecules or ions through membranes from regions of lower concentration toward regions of higher concentration.	Cellular energy (ATP)	Movement of various ions, sugars, and amino acids through membranes
Endocytosis			
Pinocytosis	Membrane engulfs droplets containing dissolved molecules from surroundings.	Cellular energy	Uptake of water and solutes by all body cells
Phagocytosis	Membrane engulfs particles from surroundings.	Cellular energy	White blood cell engulfing bacterial cell
Receptor-mediated endocytosis	Membrane engulfs selected molecules combined with receptor proteins.	Cellular energy	Cell removing cholesterol molecules from its surroundings
Exocytosis	Vesicle fuses with membrane and releases contents outside of the cell.	Cellular energy	Neurotransmitter release

3.4 | The Cell Cycle

LEARN

1. Explain why regulation of the cell cycle is important to health.
2. Describe the cell cycle.
3. Explain how stem cells and progenitor cells make possible the growth and repair of tissues.
4. Explain how two differentiated cell types can have the same genetic information, but different appearances and functions.

The series of changes that a cell undergoes, from the time it forms until it divides, is called the **cell cycle** (fig. 3.21). This cycle may seem simple: A newly formed cell grows for a time and then divides to form two new cells, which in turn may grow and divide. Yet the phases and timing of the cycle are quite complex. The major phases are interphase, mitosis, and cytoplasmic division (cytokinesis). The resulting "daughter" cells may undergo further changes that make them specialize.

Groups of special proteins interact at certain times in the cell cycle, called *checkpoints,* in ways that control whether the cell cycle progresses. Of particular importance is the "restriction checkpoint" that determines a cell's fate. At the restriction checkpoint, the cell may continue in the cell cycle and divide, move into a nondividing stage as a specialized cell, or die. Checkpoints also ensure that cellular parts have been duplicated, and that chromosomes are distributed evenly into the forming daughter cells.

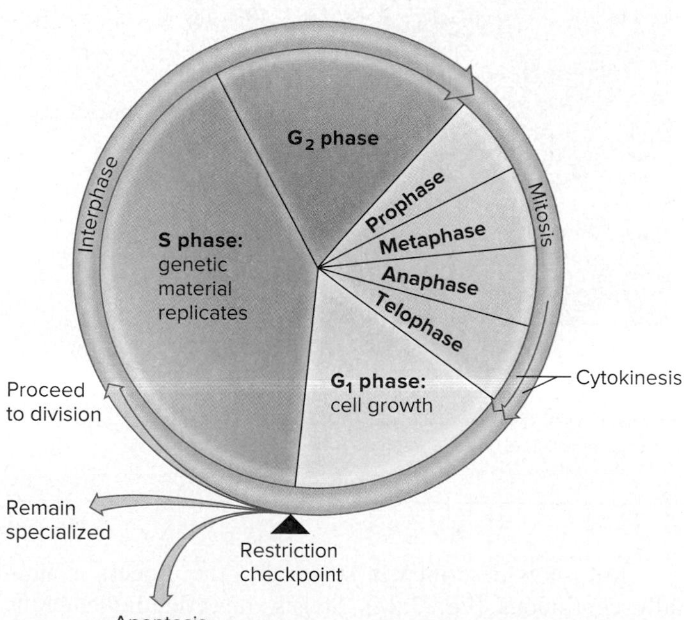

Figure 3.21 The cell cycle is divided into interphase, when cellular components duplicate, and cell division (mitosis and cytokinesis), when the cell splits in two, distributing its contents into two daughter cells. Interphase is divided into two gap phases (G_1 and G_2) when specific molecules and structures duplicate, and a synthesis phase (S), when the genetic material (DNA) replicates. Mitosis can be considered in stages—prophase, metaphase, anaphase, and telophase. The restriction checkpoint is when the cell "chooses" to die (apoptosis), remain specialized and exit the cell cycle, or continue in the cycle and divide again.

OF INTEREST Studies show that chronic psychological or emotional stress, clinical depression, obesity, and persistently elevated blood glucose levels can accelerate telomere shortening. This accelerates the aging process.

The cell cycle is very precisely regulated. Stimulation from a hormone or growth factor may trigger cell division. Such stimulation occurs, for example, when the breasts develop into milk-producing glands during pregnancy. Disruption of the cell cycle can affect health. If cell division is too infrequent, a wound cannot replace damaged cells and heal. If cell division is too frequent, an abnormal growth such as cancer forms. Clinical Application 3.3 discusses cancer.

Normally, most cells do not divide continually. If grown in a glass dish in a laboratory setting, most types of human cells divide only forty to sixty times. Presumably, such controls of cell division operate in the body too. Some cells, such as those that line the small intestine, may divide the maximum number of times. Others, such as nerve cells, normally do not divide.

A cell "knows" when to stop dividing because of a built-in "clock" in the form of the chromosome tips. These chromosome regions, called *telomeres,* shorten with each cell division. When the telomeres shorten to a certain length, the cell no longer divides. An enzyme called telomerase keeps telomeres long in cell types that must continually divide, such as certain cells in bone marrow.

Interphase

Before a cell actively divides, it must grow and duplicate much of its contents, so that two daughter cells can form from one. This period of preparedness is called **interphase.**

Once thought to be a time of rest, interphase is actually a time of great synthetic activity, when the cell obtains and utilizes nutrients to manufacture new living material and maintain its routine functions. The cell duplicates membranes, ribosomes, lysosomes, peroxisomes, and mitochondria. Perhaps most importantly, the cell in interphase takes on the tremendous task of replicating its genetic material. DNA replication is important so that each of the two new cells will have a complete set of genetic instructions (see section 4.5, DNA [Deoxyribonucleic Acid]).

Interphase is considered in phases based on the sequence of activities. DNA is replicated during the S (or synthesis) phase. Two gap (or growth) phases, called G_1 and G_2, bracket the S phase. Structures other than DNA are duplicated during the gap phases. Cellular growth occurs during the gap phases too.

Cell Division

There are two types of cell division. Meiosis is the formation of egg and sperm cells (chapter 19).

CLINICAL APPLICATION 3.3
Cancer

Cancer is a group of closely related diseases that can affect many different organs. The lifetime risk of developing cancer is one in two for males and one in three for females. These diseases result from changes in genes (mutations) that alter the cell cycle in somatic cells (cells other than sperm or eggs). Cancers share the following characteristics:

1. **Hyperplasia** is uncontrolled cell division. Normal cells divide a set number of times, signaled by the shortening of chromosome tips. Cancer cells make *telomerase,* which keeps chromosome tips long and silences signals that would stop division.
2. **Dedifferentiation** is loss of the specialized structures and functions of the normal type of cell from which the cancer cells descend (fig. 3A).
3. **Invasiveness** is the ability of cancer cells to break through boundaries, called *basement membranes,* that separate cell layers.
4. **Angiogenesis** is the ability of cancer cells to induce extension of nearby blood vessels. This blood supply nourishes the cells and removes wastes, enabling the cancer to grow.
5. **Metastasis** is the spread of cancer cells to other tissues, through the bloodstream or lymphatic system.

Mutations in certain genes cause cancer. Such a mutation may activate a cancer-causing oncogene (which normally controls mitotic rate) or inactivate a protective tumor-suppressor gene. A person may inherit one abnormal cancer-causing gene variant, present in all cells, that imparts susceptibility to developing cancer. The disease begins when a mutation disrupts the second copy of that gene in a cell of the affected organ. This second mutation may be a response to an environmental trigger. Most cancers do not result from such an inherited susceptibility, and instead occur when two mutations occur in both copies of a gene in the same somatic cell.

Cancers that affect different body parts may be caused by mutations in the same genes. A precision medicine approach to cancer treatment is to select drugs based on which genes have causative mutations, rather than by body part alone.

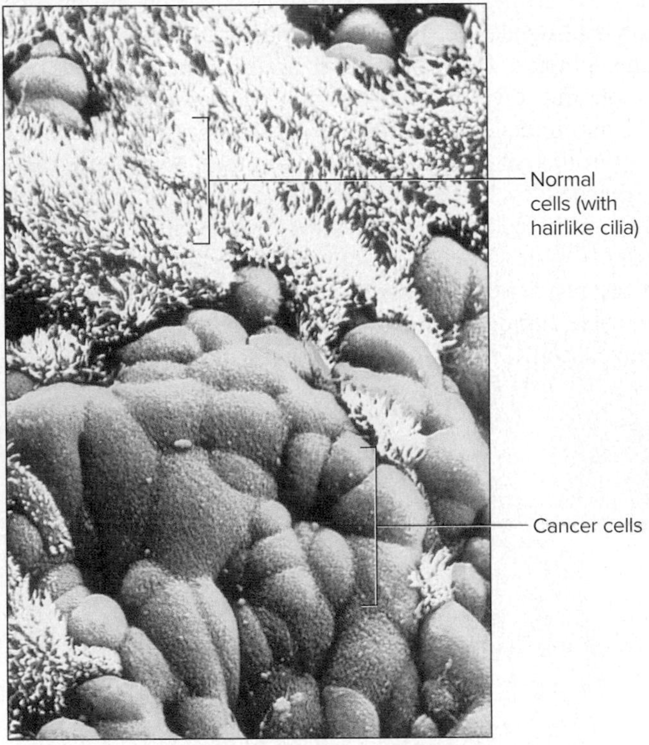

Figure 3A The absence of cilia on these cancer cells, compared to the nearby cilia-fringed cells from which they arose, is one sign of their dedifferentiation (2,250x).
Dr. Tony Brain/SPL/Science Source

The other, much more common form of cell division increases cell number, which is necessary for growth and development and for wound healing. It consists of two separate processes: (1) division of the nucleus, called **mitosis** (mī-tōsis), and (2) division of the cytoplasm, called **cytokinesis** (sī″tō-ki-nē′sis).

Division of the nucleus must be very precise, because it contains the DNA. Each new cell resulting from mitosis must have a complete and accurate copy of this information to survive. DNA replicates during interphase, but it is equally distributed into two cells during mitosis.

Mitosis is described in stages, but the process is actually continuous (fig. 3.22). Stages, however, indicate the sequence of major events. The stages are:

Prophase

One of the first signs that a cell is going to divide is that the chromosomes become visible in the nucleus when stained. The chromosomes become visible because the DNA coils very tightly, holding particles of stain. The cell has gone through S phase, so each prophase chromosome is composed of two identical structures (chromatids). They

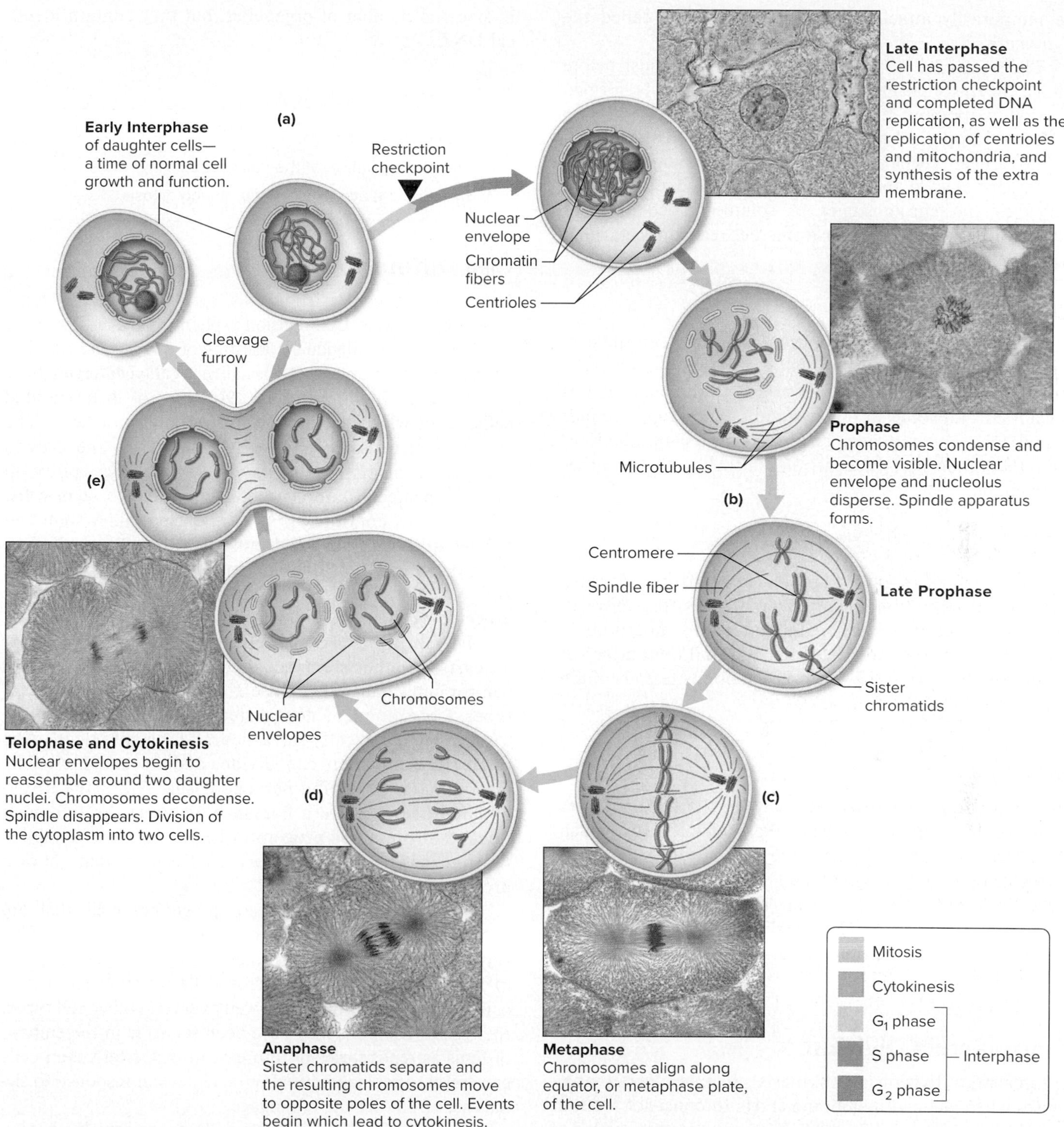

Late Interphase
Cell has passed the restriction checkpoint and completed DNA replication, as well as the replication of centrioles and mitochondria, and synthesis of the extra membrane.
Early Interphase
of daughter cells—
a time of normal cell growth and function.
(a)
Restriction checkpoint
Nuclear envelope
Chromatin fibers
Centrioles
Cleavage furrow
Prophase
Chromosomes condense and become visible. Nuclear envelope and nucleolus disperse. Spindle apparatus forms.
Microtubules
(b)
(e)
Centromere
Spindle fiber
Late Prophase
Sister chromatids
Chromosomes
Nuclear envelopes
Telophase and Cytokinesis
Nuclear envelopes begin to reassemble around two daughter nuclei. Chromosomes decondense. Spindle disappears. Division of the cytoplasm into two cells.
(d)
(c)
Anaphase
Sister chromatids separate and the resulting chromosomes move to opposite poles of the cell. Events begin which lead to cytokinesis.
Metaphase
Chromosomes align along equator, or metaphase plate, of the cell.
Mitosis
Cytokinesis
G1 phase
S phase
G2 phase
Interphase

Figure 3.22 Mitosis and cytokinesis produce two cells from one. **(a)** During interphase, before mitosis, chromosomes are visible only as chromatin fibers. A single pair of centrioles is present, but not visible at this magnification. **(b)** In prophase, as mitosis begins, chromosomes have condensed and are easily visible when stained. The centrioles have replicated, and each pair moves to an opposite end of the cell. The nuclear envelope and nucleolus disappear, and spindle fibers associate with the centrioles and the chromosomes. **(c)** In metaphase, the chromosomes line up midway between the centrioles. **(d)** In anaphase, the centromeres are pulled apart by the spindle fibers, and the chromatids, now individual chromosomes, move in opposite directions. **(e)** In telophase, chromosomes complete their migration and unwind to become chromatin fibers, the nuclear envelope re-forms, and microtubules disassemble. Cytokinesis, which actually began during anaphase, continues during telophase. Not all chromosomes are shown in these drawings. (Micrographs 360x) (a–e): Ed Reschke APR

are temporarily attached at a region on each called the *centromere.*

The centrioles of the centrosome replicate just before mitosis begins. During prophase, the two newly formed centriole pairs move to opposite ends of the cell. Soon the nuclear envelope and the nucleolus break up, disperse, and are no longer visible. Microtubules are assembled from tubulin proteins in the cytoplasm and associate with the centrioles and chromosomes. A spindle-shaped array of microtubules (spindle fibers) forms between the centrioles as they move apart.

Metaphase

The chromosomes line up about midway between the centrioles as a result of microtubule activity. The chromosomes align after spindle fibers attach to the centromeres of each chromosome so that a spindle fiber from one pair of centrioles contacts one centromere, and a spindle fiber from the other pair of centrioles attaches to the other centromere.

Anaphase

The centromeres are pulled apart. As the chromatids separate, they become individual chromosomes that move in opposite directions, once again guided by microtubule activity. The spindle fibers shorten and pull their attached chromosomes toward the centrioles at opposite ends of the cell.

Telophase

The final stage of mitosis begins when the chromosomes complete their migration toward the centrioles. It is much like the reverse of prophase. As the chromosomes approach the centrioles, they elongate and unwind from rods into threadlike chromatin. A nuclear envelope forms around each chromosome set, and nucleoli appear within the new nuclei. Finally, the microtubules disassemble into free tubulin molecules.

Cytoplasmic Division

Cytoplasmic division (cytokinesis) begins during anaphase, when the cell membrane starts to constrict around the middle of the cell (fig. 3.22). This constriction, called a *cleavage furrow,* continues to tighten throughout telophase. Contraction of a ring of microfilaments, which assemble in the cytoplasm and attach to the inner surface of the cell membrane, divides the cytoplasm. The contractile ring forms at right angles to the microtubules that pulled the chromosomes to opposite sides of the cell during mitosis. The ring pinches inward, separating the two newly formed nuclei and distributing about half of the organelles into each new cell. The newly formed cells may differ slightly in size and number of organelles, but they contain identical DNA.

PRACTICE 3.4

1. Outline the cell cycle.
2. Explain regulation of the cell cycle.
3. Describe the events that occur during mitosis.

Cell Differentiation

Cells come from preexisting cells, by the processes of mitosis and cytokinesis. Cell division explains how a fertilized egg develops into an individual consisting of trillions of cells. The process that enables cells to specialize is called **differentiation.**

An adult has more than 290 types of differentiated cells, 14 of which are unique to the embryo or fetus. The ability to generate new cells is essential to the growth and repair of tissues. Cells that retain the ability to divide repeatedly without specializing, called **stem cells,** allow for this continual growth and renewal. A stem cell divides mitotically to yield either two daughter cells like itself (stem cells), or one daughter cell that is a stem cell and one that becomes partially specialized, termed a **progenitor cell.**

The ability of a stem cell to divide and give rise to at least one other stem cell is called self-renewal. A progenitor cell's daughter cells can become any of a few cell types. For example, a neural stem cell divides to give rise to another stem cell and a neural progenitor cell. The progenitor cell then can divide, and its daughter cells differentiate, becoming nervous tissue. All of the differentiated cell types in a human body arise through such lineages of stem and progenitor cells. Figure 3.23 depicts the lineages of skin and nervous tissue, which share a progenitor cell.

Many organs have stem or progenitor cells that are stimulated to divide when injury or illness occurs. This action replaces cells, promoting healing. For example, one in 10,000 to 15,000 bone marrow cells is a stem cell, which can give rise to blood cells, as well as several other cell types. Stem cells in organs may have been set aside in the embryo or fetus as repositories of future healing. Certain stem cells can travel to replace injured or dead cells in response to signals sent from the site of damage.

Throughout development, cells progressively specialize by utilizing different parts of the complete genetic instructions, or genome, that are present in each cell. That is, some genes are "turned on" in certain cells, and other genes are turned on in other cell types. In this way, for example, an immature bone cell forms from a progenitor cell by synthesizing proteins that bind bone mineral and an enzyme required for bone formation. An immature muscle cell, in contrast, forms from a muscle progenitor cell by synthesizing contractile proteins. The bone

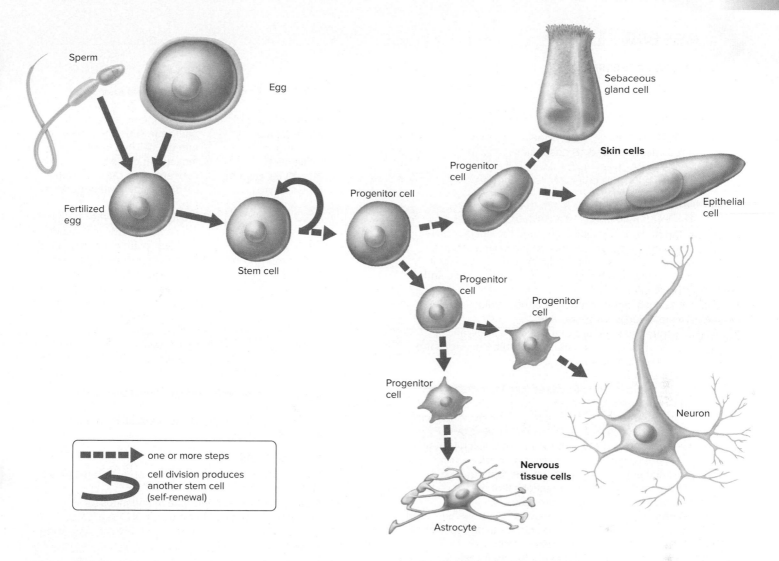

Figure 3.23 Cells specialize along cell lineage pathways. All cells in the human body ultimately descend from stem cells, through the processes of mitosis and differentiation. This simplified view depicts two pathways. A progenitor cell divides, yielding two progenitor daughter cells, one of which leads to skin cells and the other to nervous tissue cells. Imagine the complexity of the lineages of the more than 290 human cell types!

progenitor does not produce contractile proteins, nor does the muscle progenitor produce mineral-binding proteins. The final differentiated cell is like a database from which only some information is accessed.

Cell Death

A cell that does not divide or specialize has another option—it may die. **Apoptosis** (ap″o-tō′sis), programmed cell death, is a normal part of development, rather than the result of injury or disease. Apoptosis sculpts organs from naturally overgrown tissues. In the fetus, for example, apoptosis carves away webbing between developing fingers and toes, removes extra brain cells, and preserves only those immune system cells that recognize the body's cell surfaces. After birth, apoptosis occurs after a sunburn—it peels away damaged skin cells that might otherwise turn cancerous.

A cell in the throes of apoptosis goes through characteristic steps. It rounds up and bulges, the nuclear membrane breaks down, chromatin condenses, and enzymes cut the chromosomes into many equal-size pieces of DNA. Finally, the cell shatters into membrane-enclosed fragments, and a scavenger cell engulfs and destroys them. Apoptosis is a little like packing up the contents of a very messy room into plastic garbage bags, which are then removed.

PRACTICE 3.4

4. Why must cells divide and specialize?
5. Distinguish between a stem cell and a progenitor cell.
6. How are new cells generated and how do they specialize?
7. How is cell death a normal part of development?

ASSESS

CHAPTER ASSESSMENTS

3.1 Introduction

1. An adult human body consists of about __________ cells.
   a. 2 billion
   b. 50 billion
   c. 30 trillion
   d. 8 quadrillion
2. Define *cell.*
3. Discuss how cells differ from one another.

3.2 Composite Cell

4. The three major parts of a cell are __________.
   a. the nucleus, the nucleolus, and the nuclear envelope
   b. the nucleus, cytoplasm, and the cell membrane
   c. a nerve cell, an epithelial cell, and a muscle cell
   d. the endoplasmic reticulum, Golgi apparatus, and ribosomes
5. Explain the general function of organelles.
6. Define *selectively permeable.*
7. Describe the structure of a cell membrane and explain how this structural organization provides the membrane's function.
8. List three functions of membrane proteins.
9. Match the following structures with their definitions:
   (1) Golgi apparatus
   (2) mitochondria
   (3) peroxisomes
   (4) cilia
   (5) endoplasmic reticulum
   (6) cytoskeleton
   (7) vesicles
   (8) ribosomes
   A. sacs that contain enzymes that catalyze a variety of specific biochemical reactions
   B. structures on which protein synthesis occurs
   C. structures that house the reactions that release energy from nutrients
   D. a network of microfilaments and microtubules that supports and shapes a cell
   E. a structure that adds sugars to certain proteins and processes them for secretion
   F. membrane-bounded sacs
   G. a network of membranous channels and sacs where lipids and proteins are synthesized
   H. hairlike structures that extend from certain cell surfaces and wave about
10. List the parts of the nucleus and explain why each is important.

3.3 Movements Into and Out of the Cell

11. Distinguish between active and passive mechanisms of movement across cell membranes.
12. Match the transport mechanisms with their descriptions.
    (1) diffusion
    (2) facilitated diffusion
    (3) filtration
    (4) active transport
    (5) endocytosis
    (6) exocytosis
    A. the cell membrane engulfs a particle or substance, drawing it into the cell in a vesicle
    B. movement down the concentration gradient with a carrier protein, without energy input
    C. movement down the concentration gradient without a carrier protein or energy input
    D. a particle or substance leaves a cell in a vesicle that merges with the cell membrane
    E. movement against the concentration gradient with energy input
    F. hydrostatic pressure forces substances through membranes
13. Define *osmosis.*
14. Distinguish among hypertonic, hypotonic, and isotonic solutions.
15. Explain how phagocytosis differs from receptor-mediated endocytosis.

3.4 The Cell Cycle

16. Explain why it is important for the body to regulate the cell cycle.
17. Distinguish between interphase and mitosis.
18. The period of the cell cycle when DNA replicates is __________.
    a. G_1 phase
    b. G_2 phase
    c. S phase
    d. prophase
19. Explain how meiosis differs from mitosis.
20. __________ occur simultaneously.
    a. G_1 phase and G_2 phase
    b. Interphase and mitosis
    c. Cytokinesis and telophase
    d. Prophase and metaphase
21. Describe the events of mitosis in sequence.
22. Define *differentiation.*
23. A stem cell __________.
    a. forms from a progenitor cell
    b. self-renews
    c. is differentiated
    d. gives rise only to fully differentiated daughter cells
24. Describe the steps of apoptosis.

ASSESS

INTEGRATIVE ASSESSMENTS/CRITICAL THINKING

Outcomes 2.3, 2.6, 3.2

1. Given the structure of the cell membrane, which characteristics must a transmembrane protein have to extend through both faces of the membrane?
2. Why does a muscle cell contain many mitochondria, and why does a white blood cell contain many lysosomes?
3. Organelles compartmentalize a cell, much as a department store displays related items together. What advantage does such compartmentalization offer a large cell? Cite two examples of organelles and the activities they compartmentalize.
4. Exposure to tobacco smoke immobilizes cilia, and they eventually disappear. How might this effect explain why smokers have an increased incidence of coughing and respiratory infections?

Outcomes 3.2, 3.3

5. Which characteristic of cell membranes may explain why fat-soluble substances, such as chloroform and ether, rapidly affect cells?
6. Which process—diffusion, osmosis, or filtration—is utilized in the following situations?
   a. Injection of a drug that is hypertonic to the tissues stimulates pain.
   b. The urea concentration in the dialyzing fluid of an artificial kidney is decreased.

Outcome 3.4

7. New treatments for several conditions are being developed using stem cells in medical waste, such as biopsy material, teeth, menstrual blood, umbilical cords, and fatty tissue removed in liposuction. For example, fat samples from injured horses are used to grow stem cells to treat tendon injuries. Explain how the two defining characteristics of stem cells enable them to be used to replace damaged or diseased tissue, so that the new tissue functions as opposed to forming a scar.
8. Explain how the cell cycle of a cancer cell is abnormal.

Chapter Summary

3.1 Introduction

Cells vary considerably in size, shape, and function. The shapes of cells make possible their functions.

3.2 Composite Cell

A cell includes a ***cell membrane, cytoplasm,*** *and a* ***nucleus. Organelles*** *perform specific functions. The nucleus controls overall cell activities because it contains DNA, the genetic material.*

1. Cell membrane
   a. The cell membrane forms the outermost limit of the living material.
   b. It is a **selectively permeable** passageway that controls the entrance and exit of substances. Its molecules transmit signals.
   c. The cell membrane includes protein, lipid, and carbohydrate molecules.
   d. The cell membrane's framework is mainly a bilayer of phospholipid molecules.
   e. Molecules that are soluble in lipids pass through the cell membrane easily, but water-soluble molecules do not.
   f. Proteins function as receptors on membrane surfaces and form channels for the passage of ions and molecules.
   g. Patterns of surface carbohydrates associated with membrane proteins enable certain cells to recognize one another.
2. Cytoplasm
   a. Cytoplasm contains networks of membranes, organelles, and the rods and tubules of the **cytoskeleton,** suspended in cytosol.
   b. **Ribosomes** function in protein synthesis.
   c. The **endoplasmic reticulum** is a tubular communication system in the cytoplasm that transports lipids and proteins.
   d. **Vesicles** transport substances within and between cells.
   e. The **Golgi apparatus** adds sugars to certain proteins and processes them for secretion.
   f. **Mitochondria** contain enzymes that catalyze reactions that release energy from nutrient molecules.
   g. **Lysosomes** contain digestive enzymes that decompose substances.

h. **Peroxisomes** house enzymes that catalyze the breakdown of hydrogen peroxide and fatty acids, and detoxification of alcohol.
i. **Microfilaments** (actin) and **microtubules** (tubulin) aid cellular movements and support and stabilize the cytoplasm and organelles. Together they form the cytoskeleton. Microtubules also form centrioles, cilia, and flagella.
j. The centrosome contains **centrioles** that aid in distributing chromosomes during cell division.
k. **Cilia** and **flagella** are motile extensions from cell surfaces.

3. Cell nucleus
   a. The nucleus is enclosed in a double-layered **nuclear envelope.**
   b. It contains a **nucleolus,** where ribosomes are produced.
   c. It contains **chromatin,** which is composed of loosely coiled fibers of DNA and protein. As chromatin fibers condense during cell division, chromosomes hold stain and can be visualized using a microscope.

3.3 Movements Into and Out of the Cell

The cell membrane is a barrier through which substances enter and leave a cell.

1. Passive transport mechanisms do not require cellular energy.
   a. Diffusion
      (1) **Diffusion** is the movement of molecules or ions from regions of higher concentration to regions of lower concentration.
      (2) In the body, diffusion exchanges oxygen and carbon dioxide.
   b. Facilitated diffusion
      (1) In **facilitated diffusion,** ion channels and special carrier molecules move substances through the cell membrane.
      (2) This process moves substances only from regions of higher concentration to regions of lower concentration.
   c. Osmosis
      (1) **Osmosis** is the movement of water across a selectively permeable membrane into a compartment containing solute that cannot cross the same membrane.
      (2) Osmotic pressure increases as the number of impermeant particles dissolved in a solution increases.
      (3) A solution is isotonic to a cell when it has the same osmotic pressure as the cell.
      (4) Cells lose water when placed in hypertonic solutions and gain water when placed in hypotonic solutions.
   d. Filtration
      (1) **Filtration** is the movement of molecules through membranes from regions of higher hydrostatic pressure to regions of lower hydrostatic pressure.
      (2) Blood pressure causes filtration through porous capillary walls, forming tissue fluid.
2. Active mechanisms require cellular energy.
   a. Active transport
      (1) **Active transport** moves particles through membranes from a region of lower concentration to a region of higher concentration.
      (2) It requires cellular energy from adenosine triphosphate and carrier molecules in the cell membrane.
   b. Endocytosis and exocytosis
      (1) **Endocytosis** conveys large particles into a cell. **Exocytosis** is the reverse of endocytosis.
      (2) In **pinocytosis,** a cell membrane engulfs droplets of liquid.
      (3) In **phagocytosis,** a cell membrane engulfs solid particles.
      (4) **Receptor-mediated endocytosis** moves specific types of particles into cells.

3.4 The Cell Cycle

The cell cycle includes interphase, mitosis, and cytoplasmic division. It is highly regulated.

1. Interphase
   a. During **interphase,** a cell duplicates membranes, ribosomes, organelles, and DNA.
   b. Interphase terminates when mitosis begins.
2. Cell division
   a. Meiosis is a form of cell division that forms sex cells.
   b. **Mitosis** is the division and distribution of complete sets of genetic material to new cells, increasing cell number.
   c. The stages of mitosis are **prophase, metaphase, anaphase,** and **telophase.**
3. Cytoplasmic division
   a. Cytoplasmic division distributes cytoplasm into two portions.
   b. It begins during anaphase.
4. Differentiation
   a. **Differentiation** is cell specialization.
   b. **Stem cells** provide new cells for growth and tissue repair. A stem cell self-renews, yielding another stem cell and a **progenitor cell,** whose daughters follow a restricted number of fates.
5. Cell death
   a. A cell that does not divide or differentiate may undergo **apoptosis.**
   b. Apoptosis is a form of cell death that is a normal part of development.

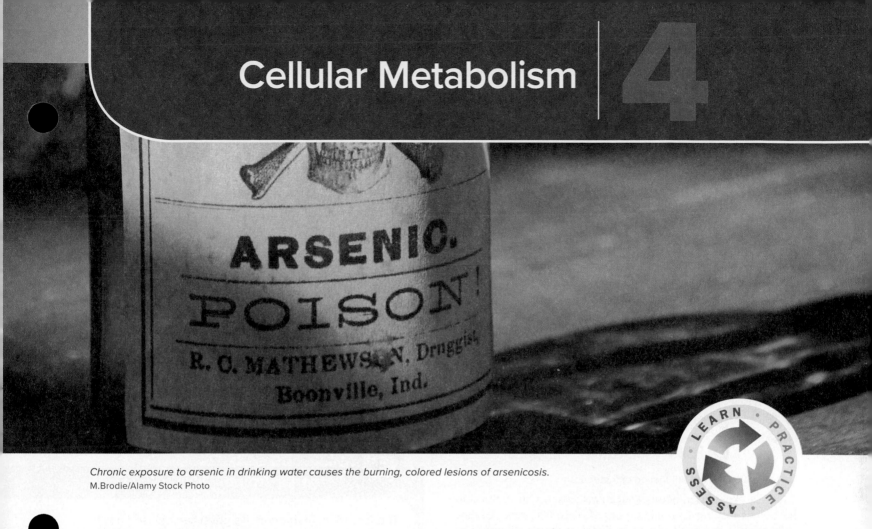

Cellular Metabolism 4

Chronic exposure to arsenic in drinking water causes the burning, colored lesions of arsenicosis.
M.Brodie/Alamy Stock Photo

Arsenic poisoning. Disrupting the body's ability to extract energy from nutrients can have serious effects on health. Arsenic is a chemical element that, if present in the body in excess, shuts down metabolism. It can do so suddenly or gradually.

Given in one large dose, arsenic causes chest pain, vomiting, diarrhea, shock, coma, and death. In contrast, a series of many small doses causes dark skin lesions that feel as if they are burning, numb hands and feet, and skin cancer. The condition may progress to paralysis and organ failure. Such gradual poisoning, called arsenicosis, may occur from contact with pesticides or environmental pollutants. The world's largest outbreak of arsenicosis, however, is due to a natural exposure.

When the World Bank and UNICEF began tapping into aquifers in India and Bangladesh in the late 1960s, they were trying to supply clean water to areas ravaged by sewage and industrial waste released from rivers during cycles of floods and droughts. Millions of people had already perished from diarrheal diseases due to the poor sanitation. But digging wells to provide clean water backfired when workers unwittingly penetrated a layer of sediment naturally rich in arsenic. The chemical has been leaching into the water in at least 2 million wells in Bangladesh alone ever since, reaching levels 30 times the safety limit set by the World Health Organization. Effects on health took several years to show up. When they did, the people thought arsenicosis was contagious. In addition to their physical pain, affected individuals bore the psychological pain of being shunned.

Arsenic damages the body by binding to bonds between sulfur atoms in proteins. The effects on metabolism largely stem from impairment of an enzyme that helps the breakdown products of glucose enter the mitochondria, where energy is extracted. Cells run out of energy.

Today, UNICEF is helping the people of India and Bangladesh avoid arsenic poisoning. People with arsenicosis are being diagnosed and treated, and given tanks to store rainwater. A vast education campaign has helped quell the stigma of arsenicosis. One program in West Bengal teaches people how to recognize arsenic-tainted sediments and gives them kits to test the water in new wells being drilled.

Module 2 Cells & Chemistry

LEARNING OUTLINE

After studying this chapter, you should be able to complete the "Learning Outcomes" that follow the major headings throughout the chapter.

AIDS TO UNDERSTANDING WORDS

an- [without] *an*aerobic respiration: respiratory process that does not require oxygen.

ana- [up] *ana*bolism: chemical reactions that require energy to build large molecules from smaller ones.

cata- [down] *cata*bolism: chemical reactions that break large molecules into smaller ones while releasing energy.

mut- [change] *mut*ation: change in genetic information.

-zym [causing to ferment] en*zym*e: protein that speeds a chemical reaction without itself being consumed.

(Appendix A has a complete list of Aids to Understanding Words.)

4.1 | Introduction

LEARN

1. Explain the overall function of metabolism.

A cell is like a little factory that never closes. Our multicellular bodies play host to many types of these factories. All day, they oversee chemical reactions. Some of the reactions are to maintain the cells' own life and integrity, and some of them will help the body as a whole survive. You can think of this as working for the good of the community. Like any productive factory, a cell needs a constant supply of energy. It also needs a "boss" to give orders. This boss is genetic material in the form of DNA, which carries the instructions for all of the processes that occur in the factory.

Metabolism is the set of all chemical reactions in a cell. You can think of a cell's energy supply as "currency," or money. Some reactions will make energy, whereas others will require energy to be spent. Special proteins called **enzymes** (en′zīmz) are vital to all of these activities because they allow chemical reactions in the body to proceed fast enough to sustain life. Enzymes control all of the interrelated reactions of cellular metabolism.

PRACTICE 4.1

Answers to the Practice questions can be found in the eBook.

1. What is cellular metabolism?
2. What are enzymes?

4.2 | Metabolic Reactions

LEARN

1. Compare and contrast anabolism and catabolism.

Metabolism is comprised of two classes of reactions. The reactions of **anabolism** (ah-nab′o-lizm) synthesize (build) larger

CAREER CORNER
Personal Trainer

The forty-five-year-old man's physician advised him to start an exercise program to lose weight, so the man joined a gym. But he hadn't been to one in years, and the rows of machines looked daunting. So he hired a personal trainer to develop an exercise routine that would be just what the doctor ordered.

A personal trainer assesses a client's fitness level, guides in the use of specific machines, and teaches how to do mat exercises. The trainer offers advice on which weight-lifting machines to use, how many repetitions to begin with, and how often to increase repetitions. The trainer might advise the client to lift weights only every other day, so that muscles will have time to recover from the microscopic tears caused by weight lifting. Using a mat, a trainer might lead a client through a series of exercises to strengthen the core abdominal muscles.

The personal trainer is part coach and part cheerleader, encouraging exercisers to push their limits. Trainers work in athletic clubs, at corporate fitness centers, in senior centers, and at other types of facilities that have exercise equipment. Some personal trainers work with clients at the clients' homes.

Minimal requirements to become a personal trainer include a high school diploma, cardiopulmonary resuscitation (CPR) training, and completion of a personal trainer course, which generally takes from a few months to a year. Passing a certification exam is required. Personal trainers tend to be outgoing, friendly people who enjoy helping others become physically fit.

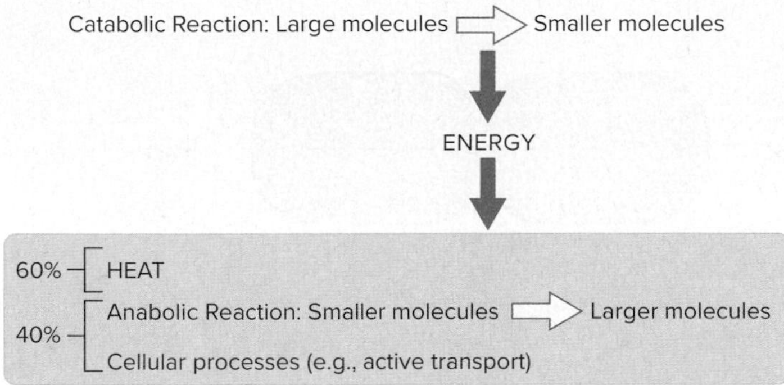

Figure 4.1. Catabolic reactions release energy as molecules are broken down. Some energy escapes as heat. The remainder will fuel anabolic reactions and cellular processes such as active transport.

molecules from smaller ones, requiring input of energy. The reactions of **catabolism** (kă-tab′o-lizm) decompose (break down) larger molecules into smaller ones, releasing energy. Reactions are often coupled together where the energy released from a catabolic reaction is used to fuel an anabolic reaction (fig. 4.1).

Anabolism

Anabolism provides the biochemicals required for cell growth and repair. A type of anabolic reaction called **dehydration synthesis** (dē″hī-drā′shun sin′the-sis), for example, joins many simple sugar molecules (monosaccharides) to form larger molecules of glycogen. When monosaccharides join, an –OH (hydroxyl group) from one monosaccharide molecule and an –H (hydrogen atom) from an –OH group of another monosaccharide molecule are removed. As the –H and –OH react to produce a water molecule (H_2O), the monosaccharides are joined by a shared oxygen atom (fig. 4.2). As this process repeats, the molecular chain extends.

Another example of dehydration synthesis is the linking of glycerol and fatty acid molecules in fat (adipose) cells to form triglyceride (fat) molecules. In this case, three hydrogen atoms are removed from a glycerol molecule and an –OH group is removed from each of three fatty acid molecules (fig. 4.3). The result is three water molecules and a single triglyceride molecule. Shared oxygen atoms bind the glycerol and fatty acid portions.

Yet another example of dehydration synthesis in cells joins amino acid molecules, forming protein molecules. When two amino acid molecules unite, an –OH from one and an –H from the –NH_2 group of another are removed. A water molecule forms, and the amino acid molecules are joined by a bond between a carbon atom and a nitrogen atom, called a *peptide bond* (fig. 4.4). Two bound amino acids form a *dipeptide,* and many linked in a chain form a *polypeptide.* Generally, a polypeptide that has a specific function and consists of 100 or more amino acids is considered a *protein.* Some protein molecules consist of more than one polypeptide.

Nucleic acids are also formed by dehydration synthesis joining nucleotides. This process is described later in the chapter.

Catabolism

Catabolic reactions break down larger molecules into smaller ones. An example of a catabolic reaction is **hydrolysis**

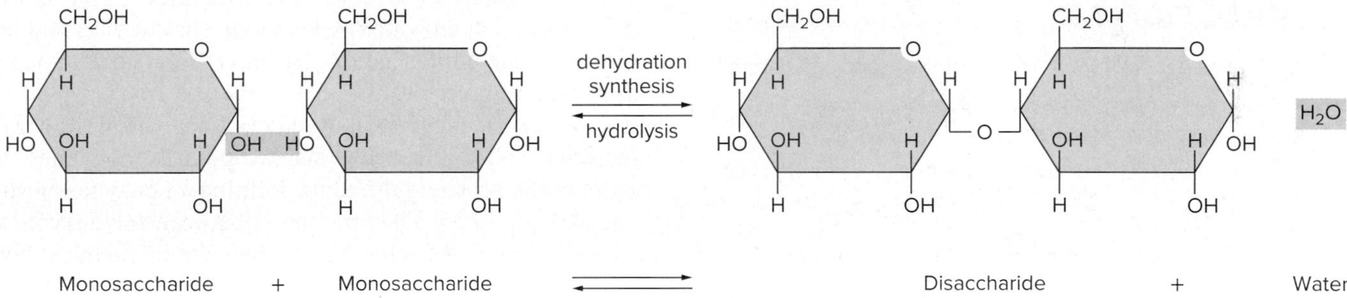

Figure 4.2 Building up and breaking down molecules. A disaccharide is formed from two monosaccharides in a dehydration synthesis reaction (arrows pointing to the right). In the reverse reaction, hydrolysis, a disaccharide is broken down into two monosaccharides (arrows pointing to the left).

Figure 4.3 Forming a triglyceride. A glycerol molecule and three fatty acid molecules participate in a dehydration synthesis reaction, yielding a triglyceride (fat) molecule (arrows pointing to the right). In the reverse reaction, hydrolysis, a triglyceride is broken down into three fatty acids and a glycerol (arrows pointing to the left).

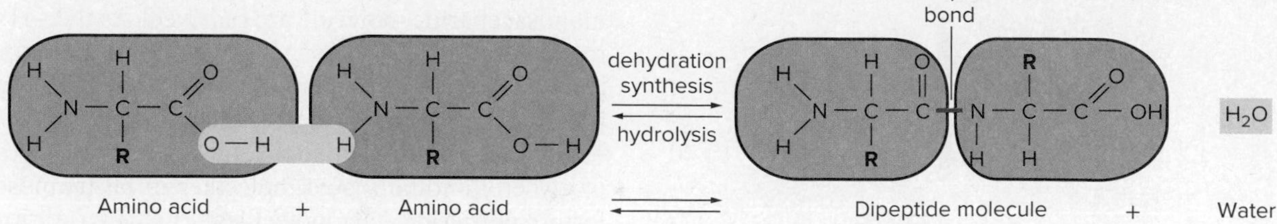

Figure 4.4 Peptide bonds link amino acids. When dehydration synthesis unites two amino acid molecules, a peptide bond forms between a carbon atom and a nitrogen atom, resulting in a dipeptide molecule (arrows pointing to the right). In the reverse reaction, hydrolysis, a dipeptide molecule is broken down into two amino acids (arrows pointing to the left).

(hi-drol'ĭ-sis), which breaks down carbohydrates, lipids, and proteins, and splits a water molecule in the process. For instance, hydrolysis of a disaccharide such as sucrose yields two monosaccharides (glucose and fructose) as a molecule of water splits:

$$\underset{\text{(Sucrose)}}{C_{12}H_{22}O_{11}} + \underset{\text{(Water)}}{H_2O} \rightarrow \underset{\text{(Glucose)}}{C_6H_{12}O_6} + \underset{\text{(Fructose)}}{C_6H_{12}O_6}$$

In this case, the bond between the simple sugars that form sucrose breaks. Then, the water supplies a hydrogen atom to one sugar and a hydroxide group to the other. Hydrolysis is the opposite of dehydration synthesis (see figs. 4.2, 4.3, and 4.4).

Hydrolysis is responsible for digestion. Specifically, it breaks down carbohydrates into monosaccharides, fats into glycerol and fatty acids, proteins into amino acids, and nucleic acids into nucleotides. (Chapter 15, sections 15.3 to 15.10, discuss digestion in detail.)

PRACTICE 4.2

1. What are the general functions of anabolism and catabolism?
2. What are the products of the anabolism of monosaccharides, glycerol and fatty acids, and amino acids?
3. Distinguish between dehydration synthesis and hydrolysis.

4.3 | Control of Metabolic Reactions

LEARN

1. Describe how enzymes control metabolic reactions.
2. Describe a metabolic pathway.

Specialized cells, such as nerve, muscle, or blood cells, carry out distinctive chemical reactions. However, all cells perform certain basic chemical reactions, such as the buildup and breakdown of carbohydrates, lipids, proteins, and nucleic acids. These reactions include hundreds of specific chemical changes that occur rapidly—yet in a coordinated fashion—thanks to enzymes.

Enzyme Action

Metabolic reactions require energy to proceed, as do all chemical reactions. The temperature conditions in cells, however, usually do not enable chemical reactions to proceed fast enough to support life. Enzymes make these reactions possible by lowering the amount of energy, called the *activation energy,* required to start these reactions. In this way, enzymes speed metabolic reactions. This acceleration is called *catalysis,* and an enzyme is a **catalyst.** Enzyme molecules are not consumed in the reactions they catalyze and can function repeatedly. Therefore, a few enzyme molecules can have a powerful effect.

Each type of enzyme is specific, acting only on a particular type of molecule, called its **substrate** (sub'strāt). Many enzymes are named after their substrates, with *-ase* as a suffix. A lipase, for example, catalyzes a reaction that breaks down a lipid. Another example of an enzyme is *catalase.* Its substrate is hydrogen peroxide, which is a toxic by-product of certain metabolic reactions. Catalase speeds breakdown of hydrogen peroxide into water and oxygen, preventing accumulation of hydrogen peroxide, which can damage cells.

Each enzyme must recognize its specific substrate. This ability of an enzyme to identify its substrate arises from the three-dimensional shape, or conformation, of the enzyme molecule. Each enzyme's polypeptide chain twists and coils into a unique conformation that fits the particular shape of its substrate molecule.

During an enzyme-catalyzed reaction, part of the enzyme molecule called the **active site** temporarily combines with parts of the substrate molecule, forming an enzyme-substrate complex (fig. 4.5). The interaction between the enzyme and the substrate at the active site strains certain chemical bonds in the substrate, altering its orientation in a way that lowers the amount of energy required to react. The reaction proceeds, the product forms, and the enzyme is released in its original conformation, able to bind another substrate molecule.

Many enzyme-catalyzed reactions are reversible, and in some cases the same enzyme catalyzes the reaction in both directions. An enzyme-catalyzed reaction can be summarized as follows:

$$\textbf{Substrate molecules} + \textbf{Enzyme molecule} \rightarrow \textbf{Enzyme-substrate complex} \rightarrow \textbf{Product (changed substrates)} + \textbf{Enzyme molecule}$$

The rate of an enzyme-catalyzed reaction depends partly on the number of enzyme and substrate molecules in the cell. A reaction is faster if the concentrations of the enzyme or the substrate increase. Enzyme efficiency varies greatly. Some enzymes can catalyze only a few reactions per second, whereas others can catalyze hundreds of thousands.

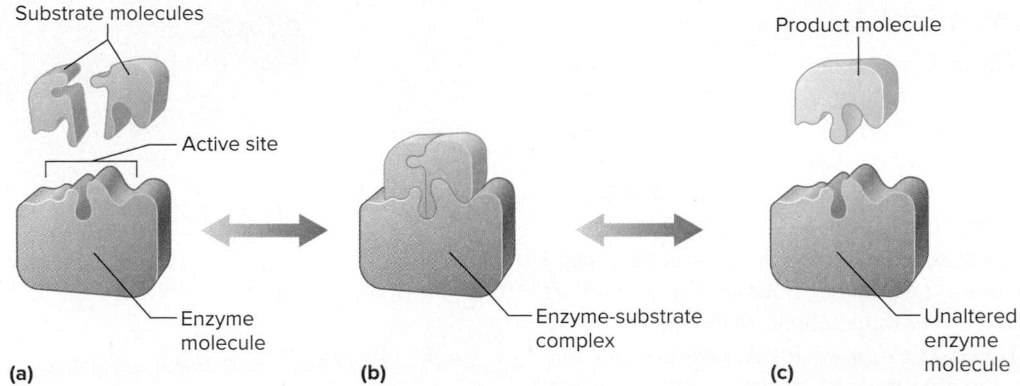

Figure 4.5 An enzyme-catalyzed reaction. As depicted here, many enzyme-catalyzed reactions are reversible. In the forward reaction (dark-shaded arrows), **(a)** the shapes of the substrate molecules fit the shape of the enzyme's active site. **(b)** When the substrate molecules temporarily combine with the enzyme, a chemical reaction proceeds. **(c)** The result is a product molecule and an unaltered enzyme. The active site changes shape as the substrate binds. Formation of the enzyme-substrate complex is more like a hand fitting into a glove, which has some flexibility, than a key fitting into a lock. APR

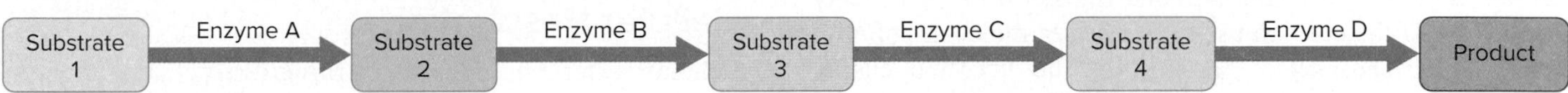

Figure 4.6 A metabolic pathway consists of a series of enzyme-controlled reactions leading to formation of a product.

PRACTICE 4.3

1. How do enzymes enable chemical reactions to proceed quickly?
2. How does an enzyme recognize its substrate?
3. List factors that affect the rate of an enzyme-controlled reaction.

Enzymes and Metabolic Pathways

Specific enzymes catalyze each of the hundreds of different chemical reactions that constitute cellular metabolism. Sequences of enzyme-controlled reactions, called **metabolic pathways,** lead to synthesis or breakdown of particular biochemicals (fig. 4.6). Every cell contains hundreds of different types of enzymes.

For some pathways, the enzymes are positioned in a cell in the exact same sequence as that of the reactions they control. The rate of a metabolic pathway is often determined by a *regulatory enzyme* that catalyzes one of its steps. The number of molecules of a regulatory molecule is limited. Consequently, the enzymes can become saturated if the substrate concentration exceeds a certain level. Once the enzyme is saturated, increasing the number of substrate molecules no longer affects the reaction rate.

A regulatory enzyme that controls an entire pathway is called a **rate-limiting enzyme,** and it is generally the first enzyme in a series. This position is important because if an enzyme at some other point in the sequence were rate-limiting, an intermediate chemical in the pathway might accumulate.

Factors That Alter Enzymes

Some enzymes become active only when they combine with a nonprotein component called a **cofactor.** A cofactor may be an ion of an element, such as copper, iron, or zinc, or a small organic molecule, called a **coenzyme** (kō-en'zīm). Many coenzymes are, or resemble, vitamin molecules. An example of a coenzyme is coenzyme A, which takes part in cellular respiration, discussed in section 4.4, Energy for Metabolic Reactions.

Almost all enzymes are proteins. Like other proteins, exposure to heat, radiation, electricity, certain chemicals, or fluids with extreme pH values can denature or otherwise inactivate enzymes. Radiation denatures some enzymes. Cyanide binds the active site of a key respiratory enzyme, preventing cells from releasing energy from nutrients, causing death.

PRACTICE 4.3

4. What is a metabolic pathway?
5. What is a rate-limiting enzyme?
6. What does a cofactor do?
7. List factors that can denature enzymes.

4.4 | Energy for Metabolic Reactions

LEARN

1. Identify the source of biological energy.
2. Describe how energy in the form of ATP becomes available for cellular activities.
3. Explain how cellular respiration releases chemical energy.

Energy is the capacity to change something; it is the ability to do work. We recognize energy by what it can do. Common forms of energy are heat, light, sound, electrical energy, mechanical energy, and chemical energy. Most metabolic reactions use chemical energy.

Release of Chemical Energy

Chemical energy is held in the bonds between the atoms of molecules and is released when these bonds break. Burning is an example of an intervention that can break the bonds of chemicals in the external environment (outside the body). Burning begins by applying heat. As the chemical burns, bonds break, and energy escapes as heat and light.

Cells "burn" glucose molecules in a process called **oxidation** (ok″sĭ-dā′shun). The energy released from breaking the bonds of glucose powers the anabolic reactions that build molecules in cells. However, oxidation inside cells and burning outside cells differ. Burning requires a large input of energy, most of which escapes as heat or light. In cells, enzymes reduce the activation energy required for the oxidation that occurs in the reactions of **cellular respiration.** These reactions release the energy in the bonds of nutrient molecules. Cells can capture about 40% of the energy released from breaking chemical bonds in cellular respiration and transfer it to high-energy electrons that the cell can use to synthesize molecules of **ATP (adenosine triphosphate).** The rest of the liberated energy escapes as heat, which helps maintain body temperature.

ATP Molecules

Each ATP molecule includes a chain of three linked chemical groups called phosphates (fig. 4.7). As energy is released during cellular respiration, some of it is captured in the bond that attaches the end (terminal) phosphate to the rest of the ATP molecule. This attachment is described as a "high-energy" bond. When energy is required for a metabolic reaction, the terminal phosphate bond breaks, releasing the stored energy. (The second phosphate bond is high-energy too.) The cell uses ATP for many functions, including active transport and synthesis of various compounds (anabolism).

An ATP molecule that has lost its terminal phosphate becomes an ADP (adenosine diphosphate) molecule. The ADP can be converted back into ATP by the addition of energy and a third phosphate. Thus, as figure 4.8*a* shows, ATP and ADP molecules shuttle back and forth between the energy-releasing reactions of cellular respiration and the energy-utilizing reactions of the cell. Figure 4.8*b* shows the coupling of an anabolic and catabolic reaction in cellular respiration.

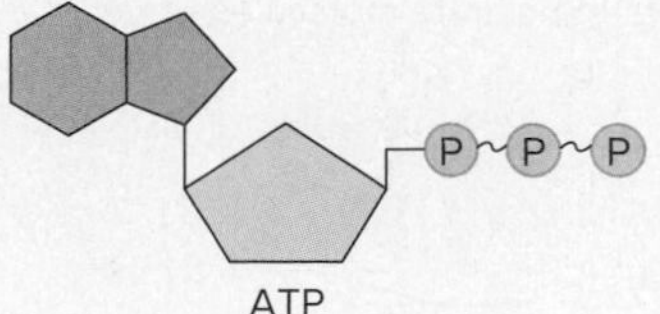

Figure 4.7 Phosphate bonds contain the energy stored in ATP.

PRACTICE 4.4

1. Define *energy.*
2. Explain how oxidation inside cells differs from burning in the outside environment.
3. What is the general function of ATP in metabolism?

Cellular Respiration

Cellular respiration consists of three distinct, yet interconnected, series of reactions: *glycolysis,* the *citric acid cycle,* and the *electron transport chain.* Glycolysis and the electron transport chain are pathways, similar to the general pathway shown in figure 4.6. Figure 4.9 depicts how a metabolic pathway can form a cycle, such as the citric acid cycle, if the product reacts to re-form the original substrate. Glucose and oxygen are required for cellular respiration. The following is the overall reaction of cellular respiration:

$$C_6H_{12}O_6 + O_2 \longrightarrow 6CO_2 + 6H_2O + \text{energy}$$

Glycolysis

The first series of reactions of cellular respiration is **glycolysis.** The word *glycolysis* means "the breaking of glucose." These reactions break a molecule of the 6-carbon sugar glucose into two molecules of 3-carbon **pyruvic acid.** Glycolysis takes place in the cytosol (the liquid portion of the cytoplasm). Because glycolysis does not directly require oxygen, it is also called the **anaerobic** (an″uh-rō′bik) phase of cellular respiration. Figure 4.10 depicts glycolysis and the reactions of cellular respiration that follow it if oxygen is present.

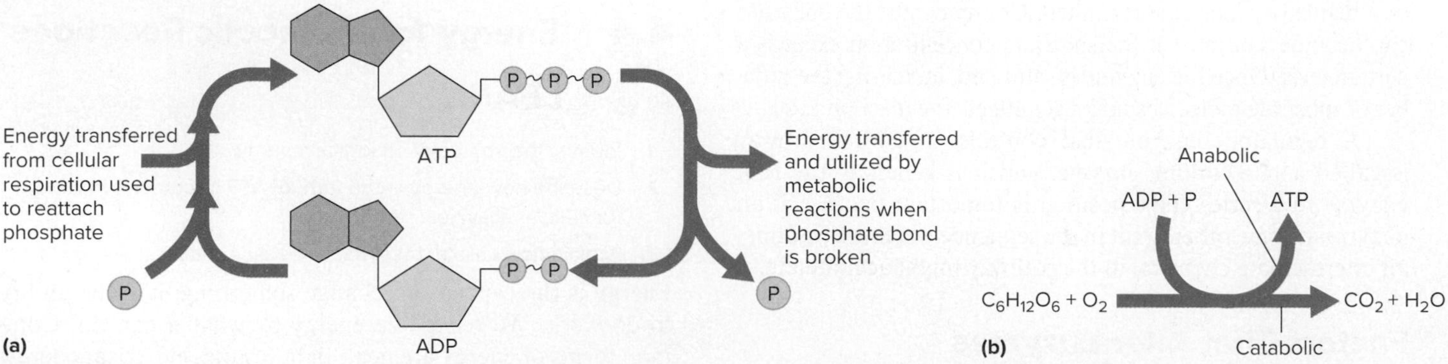

Figure 4.8 **(a)** ATP provides energy for metabolic reactions. **(b)** Cellular respiration couples the catabolic reaction of breaking down glucose in the presence of oxygen with the anabolic reaction of building ATP.

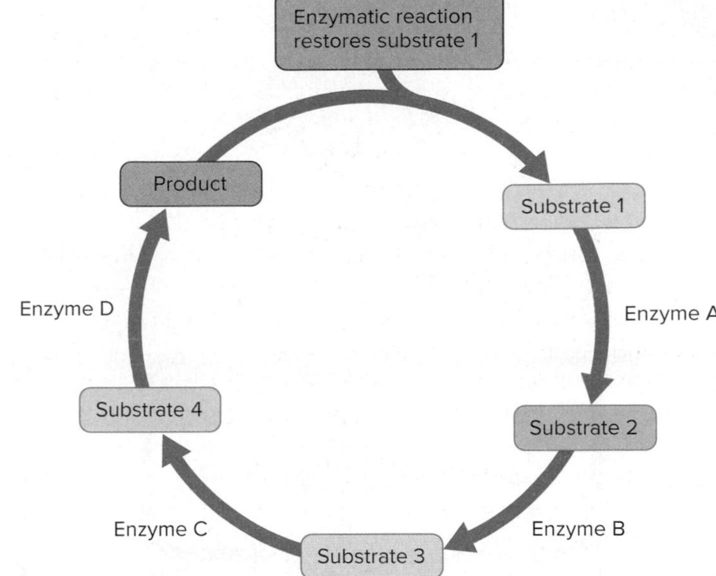

Figure 4.9 A metabolic cycle. A metabolic pathway (such as depicted in figure 4.6) can circularize, forming a metabolic cycle, when the product reacts and reforms the initial substrate.

Figure 4.10 Glycolysis takes place in the cytosol and does not require oxygen. Aerobic respiration takes place in the mitochondria and only in the presence of oxygen. The products of glycolysis and aerobic respiration include ATP, heat, carbon dioxide, and water. The relative size of the mitochondrion in comparison to the surrounding cytosol is greatly exaggerated to show the metabolic pathways. APR

PRACTICE FIGURE 4.10

Where in a cell does glycolysis occur?

Answer can be found in Appendix E.

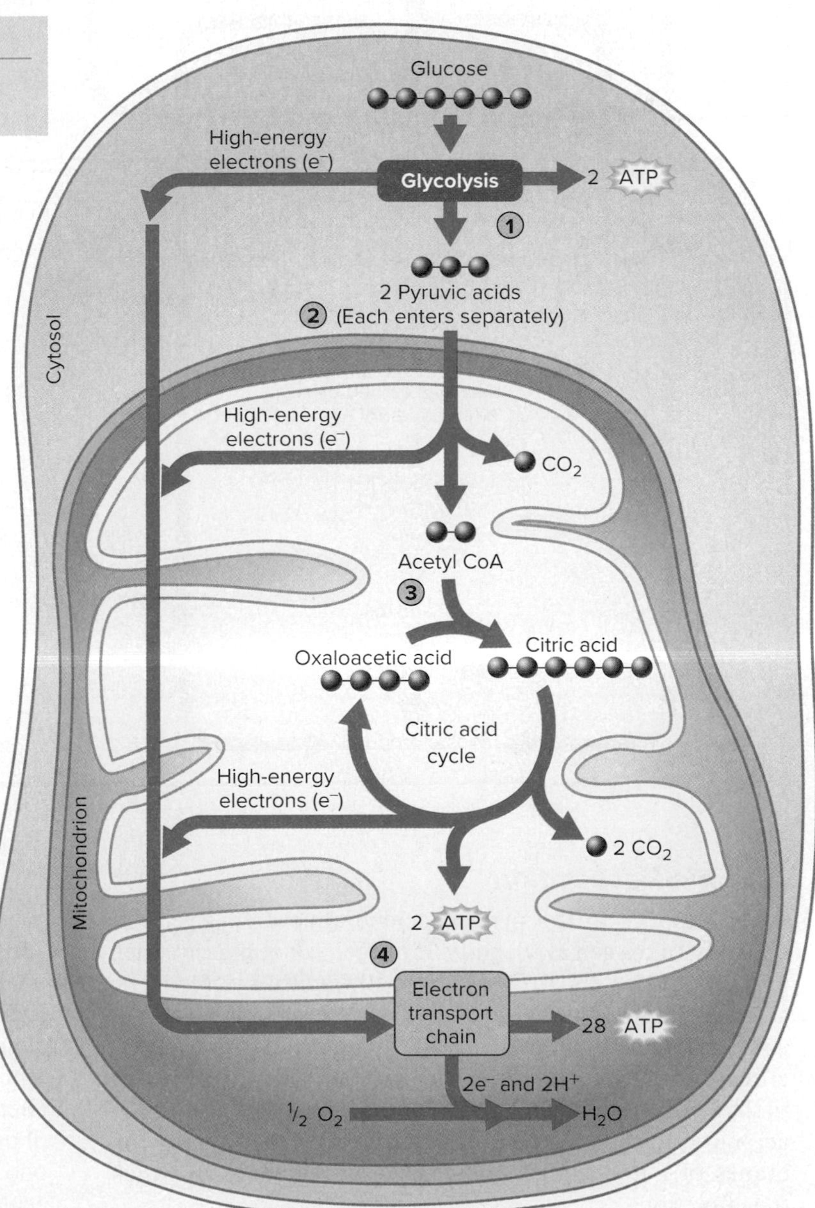

Glycolysis

① The 6-carbon sugar glucose is broken down in the cytosol into two 3-carbon pyruvic acid molecules with a net gain of 2 ATP and the release of high-energy electrons.

Citric Acid Cycle

② The 3-carbon pyruvic acids generated by glycolysis enter the mitochondria separately. Each loses a carbon (generating CO_2) and is combined with a coenzyme to form a 2-carbon acetyl coenzyme A (acetyl CoA). More high-energy electrons are released.

③ Each acetyl CoA combines with a 4-carbon oxaloacetic acid to form the 6-carbon citric acid, for which the cycle is named. For each citric acid, a series of reactions removes 2 carbons (generating 2 CO_2's), synthesizes 1 ATP, and releases more high-energy electrons. The figure shows 2 ATP resulting directly from 2 turns of the cycle per glucose molecule that enters glycolysis.

Electron Transport Chain

④ The high-energy electrons still contain most of the chemical energy of the original glucose molecule. Special carrier molecules bring the high-energy electrons to a series of enzymes that transfer much of the remaining energy to more ATP molecules. The electrons eventually combine with hydrogen ions and an oxygen atom to form water. The function of oxygen as the final electron acceptor in this last step is why the overall process is called aerobic respiration.

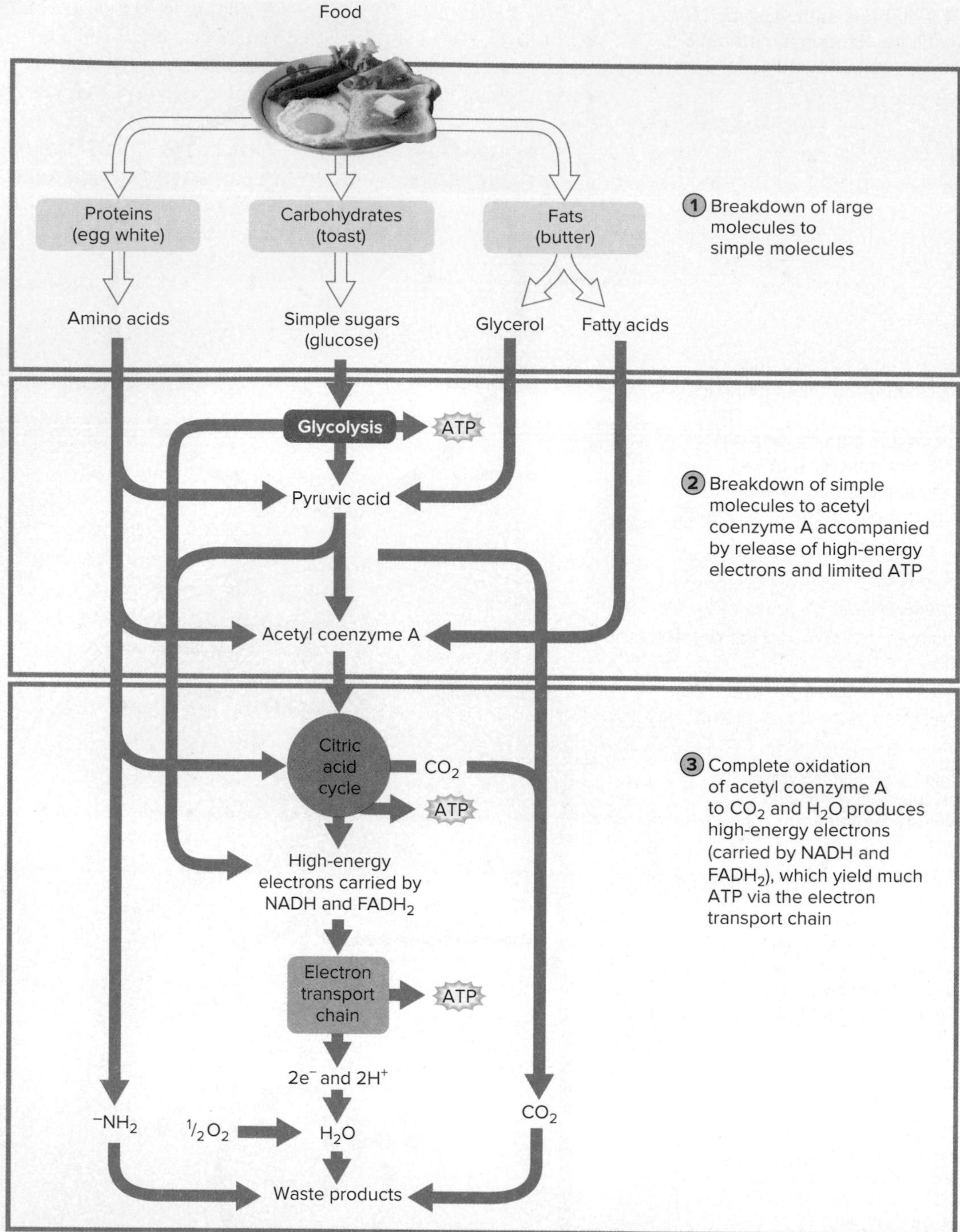

Figure 4.11 A summary of the breakdown (catabolism) of proteins, carbohydrates, and triglycerides (fats).
Red Tiger/Shutterstock

Aerobic Respiration

What happens to the pyruvic acid produced in glycolysis depends on oxygen availability. If oxygen is not present, then pyruvic acid enters an anaerobic pathway that yields lactic acid and limited energy (see fig. 8.10). If oxygen is present at a sufficient level, the pyruvic acid can enter the more energy-efficient pathways of **aerobic respiration** (a″er-o′bik res″pĭ-ra′shun) in the mitochondria. The enzymes responsible for aerobic respiration reside on the cristae of the inner membranes of mitochondria, aligned in the sequence in which they function.

In a mitochondrion, each molecule of pyruvic acid loses a carbon atom and binds a coenzyme to form a molecule of acetyl CoA, which can then combine with a 4-carbon compound to enter the **citric acid cycle** and then the **electron transport chain.** The final acceptor of the electrons passed along the electron transport chain is oxygen. The important role of oxygen is why the pathway is called "aerobic." (Because the reactions of the electron transport chain add phosphates to form ATP, they are also known as oxidative phosphorylation.) The complete oxidation of a glucose molecule yields an estimated theoretical maximum of 32 ATP molecules (fig. 4.10). Glycolysis generates 2 ATP, the citric acid cycle generates 2 ATP, and the electron transport chain generates 28 ATP molecules.

The theoretical maximum yield of 32 ATP molecules from cellular respiration of one glucose molecule is lower than has been presented in some textbooks in the past. New understanding of the complexities and the energy cost of these reactions is in large part the reason for the difference. Also, cells differ slightly in their exact mechanisms of cellular respiration. For example, the cells of skeletal muscle and the brain have a slightly lower theoretical ATP yield.

About 40 percent of the energy released in cellular respiration is captured to synthesize ATP. The rest is released as heat. Complete oxidation of glucose also produces carbon dioxide and water. The carbon dioxide is eventually exhaled, and the water becomes part of the internal environment.

In humans, metabolism does not generate enough water to meet daily needs, so we must drink water to survive. In contrast, a small desert rodent, the kangaroo rat, can survive almost entirely on the water produced by aerobic respiration.

This section has dealt with the metabolism of glucose, which is a carbohydrate. The other macronutrients, so named because they are required in large amounts, are fats and proteins. They are also broken down to release energy for ATP synthesis. For all three types of macronutrients, the final process is aerobic respiration. The most common entry point for aerobic respiration is into the citric acid cycle as acetyl coenzyme A (acetyl CoA) (fig. 4.11). Chapter 15, section 15.11, Nutrition and Nutrients, describes these metabolic pathways and their regulation further.

PRACTICE 4.4

4. Describe what happens during glycolysis.
5. What is the function of oxygen in cellular respiration?
6. What are the final products of cellular respiration?

4.5 | DNA (Deoxyribonucleic Acid)

LEARN

1. Describe how DNA molecules store genetic information.
2. Describe how DNA molecules are replicated.

Enzymes control essential metabolic reactions. Therefore, cells must have instructions for producing enzymes as well as other types of proteins. The sequences of building blocks of **DNA (deoxyribonucleic acid)** molecules hold the information to manufacture proteins in the form of a *genetic code.*

Genetic Information

DNA molecules pass from parents to offspring when a sperm fertilizes an egg. As an offspring grows and develops, mitosis passes the information in the DNA sequences of the chromosomes to new cells.

A complete set of genetic instructions for an individual constitutes the **genome** (jē′nōm). All cells except the sex cells contain two copies of the genome, one inherited from each parent. Segments of the genome that encode proteins are called **genes** (jēnz). Only about 1.5% of the human genome encodes protein and it is called the **exome** (ek-sōm). Much of the rest of the genome controls when and where genes become active to guide protein synthesis. Clinical Application 4.1 describes how exome sequencing aids disease diagnosis.

DNA Molecules

Recall from section 2.6, Chemical Constituents of Cells, that the building blocks of nucleic acids are nucleotides (see fig. 2.20). They are joined so that the sugars and phosphates alternate, forming a long "backbone" to the polynucleotide chain (see fig. 2.21*b*). DNA has two such polynucleotide chains.

In a DNA molecule, the nitrogenous bases project from the backbone and bind weakly to the bases of the other strand (fig. 4.12). The resulting structure is like a ladder, where the uprights represent the alternating sugar and phosphate backbones of the two strands, and the rungs represent the nitrogenous bases. In a nucleotide, the DNA base may be one of four types: *adenine* (A), *thymine* (T), *cytosine* (C), or *guanine* (G). A gene is a sequence of nucleotide bases along one DNA strand that specifies a particular protein's amino acid sequence.

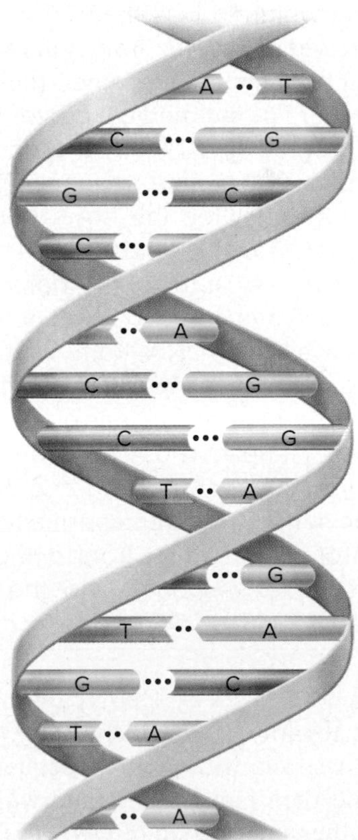

Figure 4.12 DNA structure. The molecular "ladder" of a double-stranded DNA molecule twists into a double helix. The ladder's "rungs" consist of complementary base pairs held together by hydrogen bonds—A with T (or T with A) and G with C (or C with G). APR

CLINICAL APPLICATION 4.1

Exome Sequencing

The exome is the part of the genome that encodes protein. It is only 1.5% of the genome—45 million of the 3 billion DNA bases—but includes 85% of the mutations known to affect health.

Consulting the DNA base sequence of the exome of a patient with an unusual set of symptoms can aid diagnosis by revealing what is abnormal at the molecular level. Thousands of patients, mostly children, have been properly diagnosed using exome sequencing technology. The approach can both identify rare disorders and reveal the molecular causes of more common conditions, such as autism. One of the first cases in which exome sequencing finally ended what patients call "the diagnostic odyssey" is described below.

A Boy with Dissolving Intestines

By the time he was four years old, the boy had already had 100 surgeries, including removal of his colon. His gastrointestinal (digestive) tract was riddled with holes, starting at age two with an abscess in his rectum. Feces leaked from his intestines. He was fed by tube, and weighed less than 20 pounds. Doctors thought he was suffering from inflammatory bowel disease, but his symptoms seemed much too severe.

Researchers at the medical center where the boy was being treated had received funding to perform one of the very first exome-sequencing experiments, planned for 2014. When the boy's doctors implored the researchers to help find out what was causing the boy's terrible symptoms, the date for the exome sequencing was moved up to 2009, to help him.

The analysis, on DNA from white blood cells, uncovered more than 16,000 variations in gene sequences. The researchers narrowed these down to 32 "candidate genes" whose functions could explain the boy's symptoms. A mutation in only one of those genes made sense. That gene normally prevents cells of the gastrointestinal tract from dying during infection. A disease corresponding to the mutation was already known, but it affects the immune system, not the digestive system. The boy's case was unusual, an "atypical presentation." Fortunately, a cord-blood stem cell transplant could treat the known condition, but until the exome sequencing, nobody knew that the boy had it. A stem cell transplant from an anonymous donor worked, and the boy rapidly gained weight. Today, he is much improved.

Speeding Diagnosis

Exome sequencing has been helpful in diagnosing common combinations of symptoms, such as intellectual disability, developmental delay, and birth defects. After tests for single gene disorders and chromosome abnormalities do not provide an answer, exome sequencing is increasingly the next step. It can be lifesaving. For example, exome sequencing revealed that a two-year-old boy who had severe feeding difficulties and poor weight gain had a mutation known to cause Marfan syndrome, although his symptoms were atypical. One symptom of Marfan syndrome is an enlarged aorta, the largest artery. An ultrasound revealed that the boy indeed had a bulge in the wall of the aorta near the heart, which could have caused sudden death. He was successfully treated.

Sequencing an exome to put a name to a patient's symptoms can also detect other, "unsolicited" (also called secondary or incidental) findings, such as an adult-onset cancer susceptibility, or a disease that has not yet affected health. The American College of Medical Genetics and Genomics recommends that health-care providers ask patients and parents before sequencing if they want to know about unsolicited findings. The organization maintains a list of several dozen conditions that it recommends reporting because they have treatments.

As the cost of exome sequencing decreases, it may become a standard part of diagnostic medicine. Studies are under way to sequence exomes to better understand the physiology behind many diseases.

The nitrogenous bases of DNA pair in specific ways: adenine only to thymine, and cytosine only to guanine. For example, a DNA strand with the base sequence A, C, G, C lies opposite and then binds to a strand with the sequence T, G, C, G (see the upper region of DNA in fig. 4.12). These combinations—A with T, and G with C—are called *complementary base pairs.*

The long DNA molecule forms a double helix. It folds very tightly to fit inside a cell's nucleus. The two strands of a DNA double helix run in opposite orientations ("head-to-tail").

A molecule of DNA is typically millions of base pairs long. The great length of DNA molecules may seem quite a challenge to copy (replicate) when a cell divides, but a set of special enzymes accurately and rapidly carries out this process.

DNA Replication

When a cell divides, each newly formed cell must have a copy of the original cell's genetic information (DNA) so it will be able to synthesize the proteins to build cellular parts

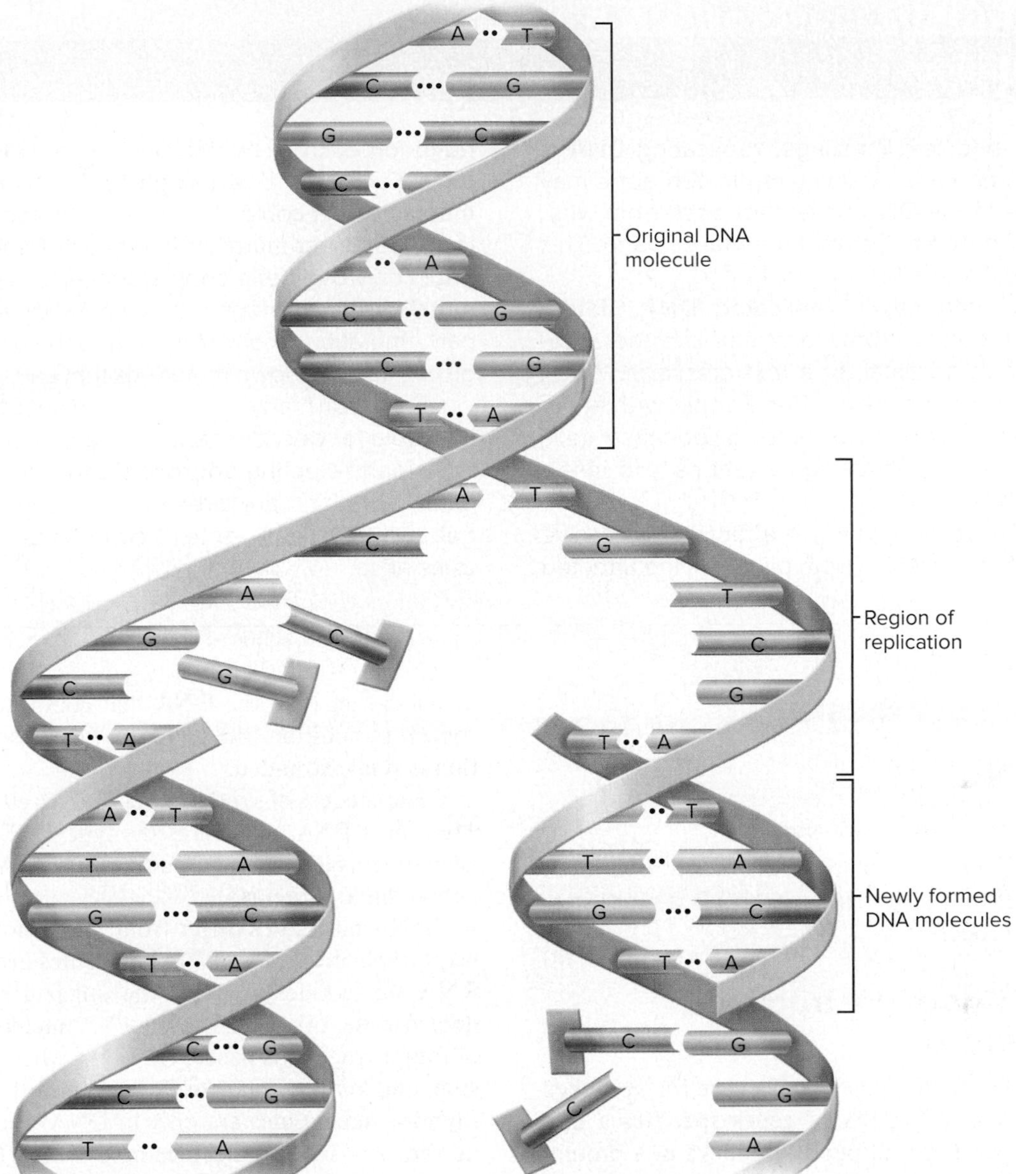

Figure 4.13 DNA replication forms two double helices from one. When a DNA molecule replicates, its original strands separate locally. A new strand of complementary nucleotides forms along each original strand, using free nucleotides in the cell. APR

and metabolize. This copying of DNA is called **replication** (rep″lĭ-kā′shun). It takes place during interphase of the cell cycle.

The double-stranded structure of DNA makes replication possible. As replication begins, hydrogen bonds between complementary base pairs in each DNA molecule break (fig. 4.13). The double helix unwinds and the two strands pull apart, exposing the nitrogenous bases. Then, an enzyme called DNA polymerase brings in new DNA nucleotides, which form complementary pairs with the exposed bases on the separated strand. Other enzymes knit together the sugar-phosphate backbone. In this way, a new strand of complementary nucleotides forms along each of the old strands.

DNA replication produces two complete DNA molecules, each with one old strand of the original molecule and one new strand. During mitosis, the two DNA molecules that form the two chromatids of each of the chromosomes separate so that one of these two DNA molecules passes to each of the new cells. Clinical Application 4.2 describes errors in DNA replication that change the sequence, generating a mutation.

PRACTICE 4.5

1. Distinguish between *genome* and *gene*.
2. Why must DNA molecules replicate?
3. List the steps of DNA replication.

CLINICAL APPLICATION 4.2
Mutations

Like typing a long text message, replicating DNA is an error-prone process. A newly replicated gene may have a change in the DNA base sequence from what is dictated in the original, complementary strand. This error is a type of mutation.

Cells can scan newly replicated DNA, detect mutations, and correct some of them, like the autocorrect feature in an email or a text message. When DNA repair fails, health may suffer. People with faulty DNA repair tend to develop cancers, because cells cannot remove cancer-causing mutations and insert the correct bases.

Mutations have two types of effects. A recessive mutation must be present in both copies of the affected gene (on each of two chromosomes) to cause symptoms. Each of us has a single copy of about 400 such mutations that could harm health if present in two copies. A dominant mutation exerts an effect even if present in only one gene copy. Databases organize human mutations into categories, such as by affected body part, mutations common in a particular population group, and all known mutations that cause a disease.

Tests for many single-gene diseases have been available for decades, and they are important in diagnosis and selecting appropriate treatments. Because most mutations are present from conception, some genetic tests may detect conditions before symptoms arise.

4.6 | Protein Synthesis

LEARN

1. Describe the steps of protein synthesis.

DNA provides the genetic instructions that a cell requires to synthesize proteins. Manufacturing proteins is a multi-step, enzyme-catalyzed process.

The Genetic Code—Instructions for Making Proteins

Cells can synthesize specific proteins because the sequence of nucleotide bases in the DNA of genes specifies a particular sequence of amino acid building blocks in a protein molecule. This correspondence of gene and protein building-block sequence is called the **genetic code.**

Recall from section 2.6, Chemical Constituents of Cells, that a biological protein may be built of twenty types of amino acids. Each amino acid is represented in a DNA molecule by a particular sequence of three nucleotides. The DNA sequence G, G, T represents one type of amino acid; G, C, A represents another; and T, T, A a different one. Other nucleotide sequences encode the instructions for beginning or ending the synthesis of a protein molecule. Thus, the sequence of nucleotides in a DNA molecule denotes the order of amino acids in a protein molecule, as well as where to start or stop synthesis of that protein.

Transcription

DNA molecules stay in the nucleus of a cell, maintaining the genetic information. However, protein synthesis occurs in the cytoplasm. The genetic information reaches the cytoplasm by being copied into molecules of RNA (ribonucleic acid). RNA can exit the nucleus because the molecules are much shorter than the DNA that composes entire chromosomes. In addition, the RNA that carries a gene's information is single-stranded.

The process of synthesizing RNA is called **transcription. Messenger RNA (mRNA)** is the type of RNA that carries a gene's message out of the nucleus. Other types of RNA also help to build proteins.

RNA molecules differ from DNA molecules in several ways (table 4.1). Most RNA molecules are single-stranded. RNA nucleotides include the sugar ribose rather than deoxyribose. Like DNA, each RNA nucleotide includes one of four nitrogenous bases. However, whereas adenine, cytosine, and guanine nucleotides are in both DNA and RNA, thymine nucleotides are only in DNA. In place of thymine nucleotides, RNA molecules have *uracil* (U) nucleotides.

The enzyme *RNA polymerase* synthesizes mRNA following the rules of complementary base pairing. For example, the DNA sequence A, T, G, C, G specifies the complementary mRNA bases U, A, C, G, C (table 4.2). Specific DNA sequences outside genes signal which of the two DNA strands contains the information to build a protein. RNA polymerase also recognizes sequences in the DNA that indicate where transcription of a gene begins, where it stops, and the correct direction to read the DNA, like a sentence. When the RNA polymerase reaches the end of the gene, it releases the newly formed mRNA. Transcription is complete.

Translation

Each amino acid in a protein is specified by three contiguous bases in the DNA sequence. Those bases, in the correct order, are represented by a three-base sequence, called a **codon** (kō′don), in mRNA (table 4.3). In addition to the mRNA codons that specify amino acids, the sequence AUG represents the "start" of a gene, whereas three other mRNA base sequences indicate "stop."

TABLE 4.1	A Comparison of DNA and RNA Molecules	
	DNA	**RNA**
Main location	Part of chromosomes, in nucleus	Cytoplasm
5-carbon sugar	Deoxyribose	Ribose
Basic molecular structure	Double-stranded	Single-stranded
Nitrogenous bases included	Adenine, thymine, cytosine, guanine	Adenine, uracil, cytosine, guanine
Major functions	Replicates prior to cell division; contains information for protein synthesis	mRNA carries transcribed DNA information to cytoplasm and acts as template for synthesis of protein molecules; tRNA carries amino acids to mRNA

TABLE 4.2 Complementary Base Pairs

Complementary strands of a DNA molecule		Transcribed strand of DNA	Complementary strand of mRNA
A	T	A	U
T	A	T	A
C	G	C	G
G	C	G	C

To guide protein synthesis, an mRNA molecule must leave the nucleus and associate with a ribosome in the cytoplasm. There the series of codons on mRNA is translated from the "language" of nucleic acids to the "language" of amino acids. This process of protein synthesis is appropriately called **translation.**

Building a protein molecule requires ample supplies of the correct amino acids in the cytoplasm and positioning them in the order specified along a strand of mRNA. A second kind of RNA molecule called **transfer RNA (tRNA)** correctly aligns amino acids, which are then linked by enzymatic action to form proteins (fig. 4.14). Like mRNA, tRNA is synthesized in the nucleus and sent into the cytoplasm, where it assists in constructing a protein molecule.

Because twenty different types of amino acids form biological proteins, at least twenty different types of tRNA molecules must be available, one for each type of amino acid. Each type of tRNA has a region at one end that consists of three nucleotides that form complementary base pairs with a specific mRNA codon. The three nucleotides in the tRNA are called an **anticodon** (an″tĭ-kō′don). In this way, tRNA carries its amino acid to a correct position on an mRNA strand. The temporary meeting of tRNA and rRNA takes place on a ribosome (fig. 4.14).

TABLE 4.3 Codons (mRNA Three-Base Sequences)

FIRST LETTER	SECOND LETTER U		SECOND LETTER C		SECOND LETTER A		SECOND LETTER G		THIRD LETTER
U	UUU	phenylalanine (phe)	UCU	serine (ser)	UAU	tyrosine (tyr)	UGU	cysteine (cys)	U
	UUC		UCC		UAC		UGC		C
	UUA	leucine (leu)	UCA		UAA	STOP	UGA	STOP	A
	UUG		UCG		UAG	STOP	UGG	tryptophan (trp)	G
C	CUU	leucine (leu)	CCU	proline (pro)	CAU	histidine (his)	CGU	arginine (arg)	U
	CUC		CCC		CAC		CGC		C
	CUA		CCA		CAA	glutamine (gln)	CGA		A
	CUG		CCG		CAG		CGG		G
A	AUU	isoleucine (ile)	ACU	threonine (thr)	AAU	asparagine (asn)	AGU	serine (ser)	U
	AUC		ACC		AAC		AGC		C
	AUA		ACA		AAA	lysine (lys)	AGA	arginine (arg)	A
	AUG	START methionine (met)	ACG		AAG		AGG		G
G	GUU	valine (val)	GCU	alanine (ala)	GAU	aspartic acid (asp)	GGU	glycine (gly)	U
	GUC		GCC		GAC		GGC		C
	GUA		GCA		GAA	glutamic acid (glu)	GGA		A
	GUG		GCG		GAG		GGG		G

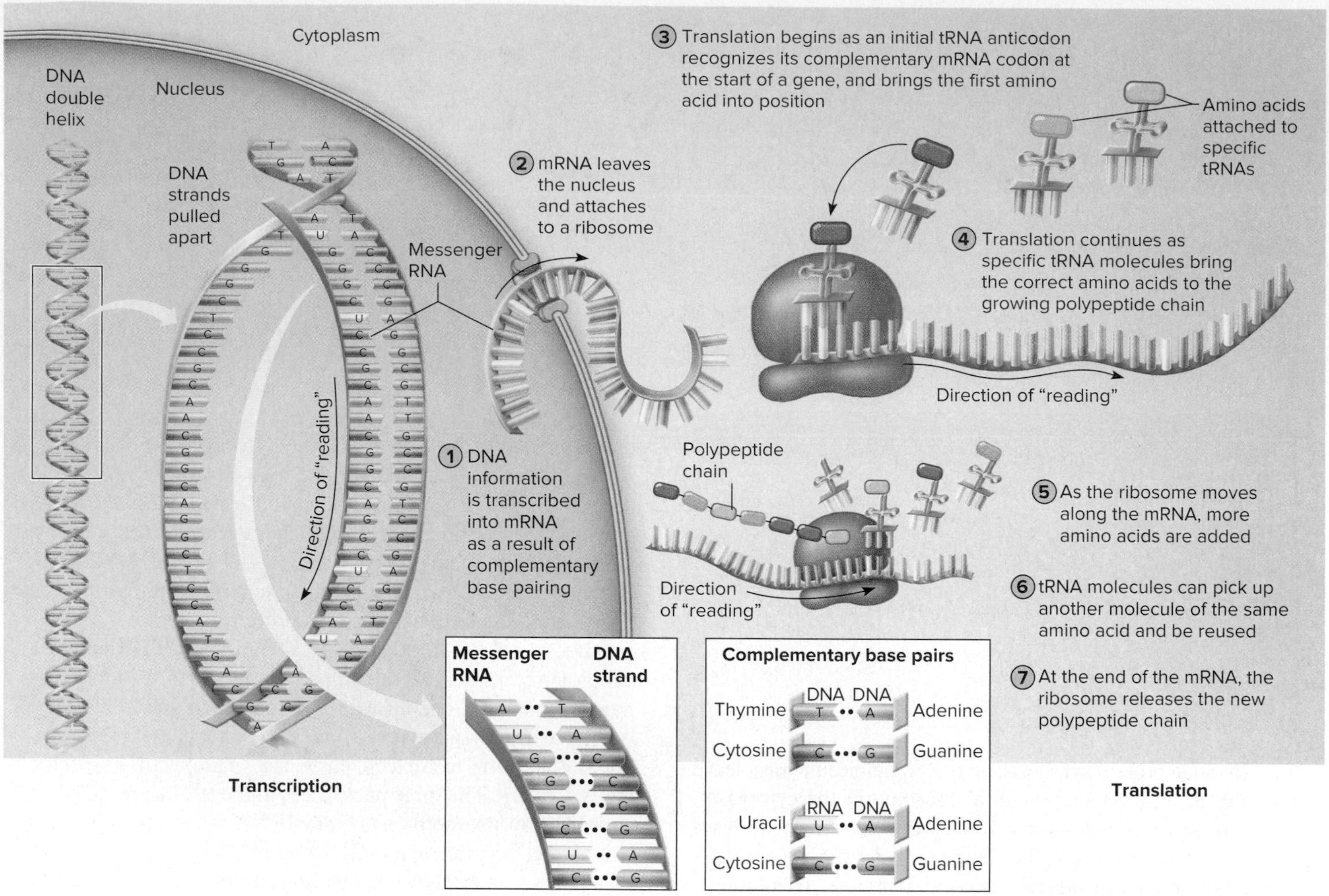

Figure 4.14 Protein synthesis. DNA information is transcribed into mRNA, which in turn is translated into a sequence of amino acids. A protein molecule consists of one or more polypeptides. The inset shows complementary base pairs in DNA and RNA. APR

PRACTICE FIGURE 4.14

What is the name of the molecule that carries DNA information so that it can be translated into protein?

Answer can be found in Appendix E.

There are 64 possible types of tRNA anticodons, because there are this many possible triplets. Some amino acids may bind to more than one type of anticodon.

All of the parts of the cell's protein synthetic machinery come together as translation of a particular mRNA into a polypeptide begins. First, a ribosome binds to an mRNA molecule at a specific start point. A tRNA molecule with the complementary anticodon holding its amino acid forms hydrogen bonds with the first mRNA codon. A second tRNA then binds the next codon, bringing its amino acid to an adjacent site on the ribosome. Then, a peptide bond forms between the two amino acids, beginning a chain. The first tRNA molecule is released from its amino acid and is recycled to the cytoplasm (fig. 4.15). This process of aligning and attaching amino acids repeats as the ribosome moves along the mRNA molecule. The amino acids delivered by the tRNA molecules are added one at a time to the elongating polypeptide chain.

As the protein molecule forms, the amino acid chain folds into its unique conformation and is then released, becoming a separate, functional molecule. Correct protein folding is essential to health. In cells, misfolded proteins are threaded through spool-shaped structures called *proteasomes.* Here, they are either refolded into the functional conformation, or destroyed if they are too abnormal.

A gene that is transcribed and translated into a protein is said to be *expressed.* The types and amounts of proteins in a cell, which can change with changing conditions, largely determine the function a cell performs in the body. Gene expression is the basis for cell differentiation, described in section 3.4, The Cell Cycle.

PRACTICE 4.6

1. Define *genetic code.*
2. What is the function of DNA?
3. How is genetic information carried from the nucleus to the cytoplasm?
4. List the steps of protein synthesis.

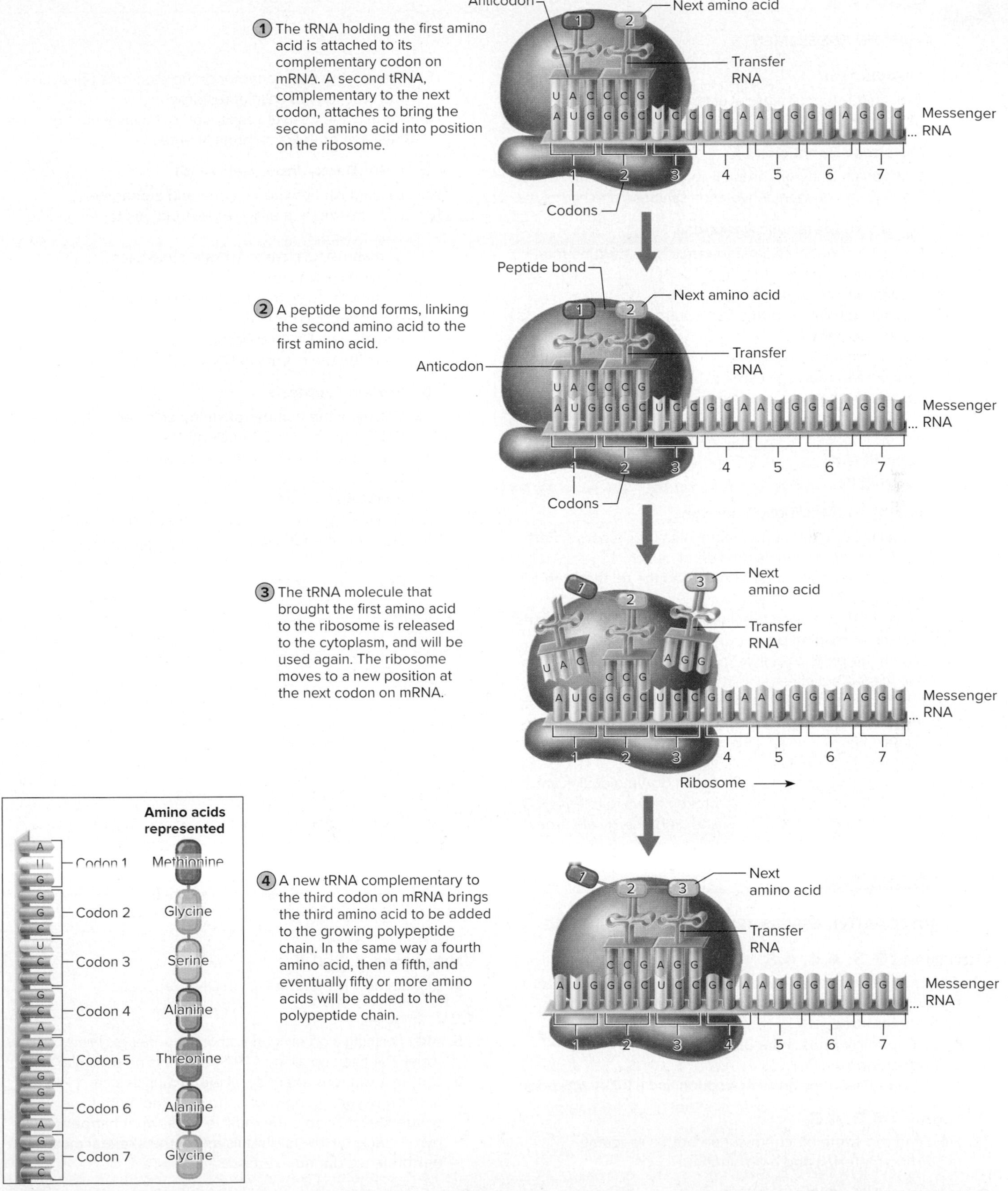

(1) The tRNA holding the first amino acid is attached to its complementary codon on mRNA. A second tRNA, complementary to the next codon, attaches to bring the second amino acid into position on the ribosome.
Anticodon
Next amino acid
Transfer RNA
Messenger RNA
Codons
1 2 3 4 5 6 7
U A C C C G
A U G G G C U C C G C A A C G G C A G G C
(2) A peptide bond forms, linking the second amino acid to the first amino acid.
Peptide bond
Next amino acid
Anticodon
Transfer RNA
Messenger RNA
Codons
(3) The tRNA molecule that brought the first amino acid to the ribosome is released to the cytoplasm, and will be used again. The ribosome moves to a new position at the next codon on mRNA.
Next amino acid
Transfer RNA
Messenger RNA
Ribosome
(4) A new tRNA complementary to the third codon on mRNA brings the third amino acid to be added to the growing polypeptide chain. In the same way a fourth amino acid, then a fifth, and eventually fifty or more amino acids will be added to the polypeptide chain.
Next amino acid
Transfer RNA
Messenger RNA
Amino acids represented
Codon 1 Methionine
Codon 2 Glycine
Codon 3 Serine
Codon 4 Alanine
Codon 5 Threonine
Codon 6 Alanine
Codon 7 Glycine

Figure 4.15 Protein synthesis occurs on ribosomes. Molecules of transfer RNA (tRNA) attach to and carry specific amino acids, aligning them in the sequence determined by the codons of mRNA. These amino acids, connected by peptide bonds, form a polypeptide chain of a protein molecule. The inset shows the correspondence between mRNA codons and the specific amino acids they encode.

ASSESS

CHAPTER ASSESSMENTS

4.1 Introduction

1. Explain the function of cellular metabolism.
2. Explain why enzymes are important in the body.

4.2 Metabolic Reactions

3. Distinguish between anabolism and catabolism.
4. Distinguish between dehydration synthesis and hydrolysis.

4.3 Control of Metabolic Reactions

5. Describe how an enzyme interacts with its substrate.
6. Define *active site.*
7. Define *metabolic pathway.*
8. Explain how one enzyme can control the rate of a metabolic pathway.
9. Define *cofactor.*
10. Exposure to radiation can __________ an enzyme and change its shape to the point where it becomes inactive.
    a. catalyze
    b. decompose
    c. synthesize
    d. denature

4.4 Energy for Metabolic Reactions

11. Explain how oxidation of molecules inside cells differs from burning materials outside of cells.
12. Explain the importance of ATP, and the relationship of ATP to ADP.
13. Distinguish between the anaerobic and aerobic phases of cellular respiration.
14. Match the parts of cellular respiration to their associated activities.
    (1) electron transport chain
    (2) glycolysis
    (3) citric acid cycle
    A. glucose molecules are broken down into pyruvic acid
    B. the final electron acceptor is oxygen
    C. acetyl CoA molecules are broken down to release CO_2 and high-energy electrons
15. Identify the final acceptor of the electrons released in the reactions of cellular respiration.
16. Identify the cellular respiration pathway where glucose, fats, and proteins commonly enter.

4.5 DNA (Deoxyribonucleic Acid)

17. Distinguish between a gene and a genome.
18. DNA information provides instructions for the cell to __________.
    a. manufacture carbohydrate molecules
    b. synthesize lipids
    c. manufacture RNA from amino acids
    d. synthesize protein molecules
19. Explain why DNA replication is essential.
20. Describe the events of DNA replication.

4.6 Protein Synthesis

21. Calculate the number of amino acids encoded by a DNA sequence of 27 nucleotides.
22. If a strand of DNA has the sequence A T G C G A T C C G C, then the sequence of an mRNA molecule transcribed from it is __________.
23. Distinguish between transcription and translation.
24. Describe the function of a ribosome in protein synthesis.
25. Define *gene expression.*

ASSESS

INTEGRATIVE ASSESSMENTS/CRITICAL THINKING

Outcomes 2.3, 4.4, 4.6

1. The chapter discusses several specific types of chemical bonds. Describe each of the following, and explain why each is important.
   a. High-energy phosphate bond
   b. Peptide bond
   c. The bond between an mRNA codon and a tRNA anticodon

Outcomes 4.2, 4.3

2. How can the same biochemical be both a reactant (a starting material) and a product?

Outcomes 4.3, 4.4

3. What effect might changes in the pH of body fluids or body temperature that accompany illness have on enzymes?

Outcomes 4.3, 4.5

4. Explain how proteins assist in DNA replication.

Outcome 4.4

5. After finishing a grueling marathon, a runner exclaims, "Whew, I think I've used up all my ATP!" Could this be possible?
6. During strenuous exercise, skeletal muscles can't keep up their oxygen supply with the demand. With lower oxygen levels within the muscle cells, what happens to pyruvic acid? How might this affect the skeletal muscle performance during exercise?

Outcome 4.6

7. Explain the advantage of polyribosomes in the translation process of protein synthesis.

Chapter Summary

4.1 Introduction

A cell continuously carries on thousands of metabolic reactions that maintain life and enable cell specialization. Cellular metabolism acquires, stores, and releases energy.

4.2 Metabolic Reactions

1. Anabolism
   a. **Anabolism** builds large molecules from smaller molecules.
   b. In **dehydration synthesis,** water forms as smaller molecules join by sharing atoms.
   c. Carbohydrates are synthesized from monosaccharides, triglycerides from glycerol and fatty acids, proteins from amino acids, and nucleic acids from nucleotides.
2. Catabolism
   a. **Catabolism** breaks down larger molecules into smaller ones.
   b. In **hydrolysis,** a water molecule is split as an enzyme breaks the bond between two parts of a molecule.
   c. Hydrolysis breaks down carbohydrates into monosaccharides, triglycerides into glycerol and fatty acids, proteins into amino acids, and nucleic acids into nucleotides.

4.3 Control of Metabolic Reactions

Enzymes control metabolic reactions, which include many specific chemical changes.

1. Enzyme action
   a. **Enzymes** lower the amount of energy required to start chemical reactions.
   b. Enzymes are molecules that promote metabolic reactions without being consumed (catalysis).
   c. An enzyme acts upon a specific **substrate** molecule.
   d. The shape of an enzyme molecule fits the shape of its substrate at the **active site.**
   e. When an enzyme combines with its substrate, the substrate changes, a product forms, and the enzyme is released in its original form.
   f. The rate of an enzyme-controlled reaction depends partly upon the numbers of enzyme and substrate molecules and the enzyme's efficiency.
2. Enzymes and metabolic pathways
   a. A sequence of enzyme-controlled reactions constitutes a **metabolic pathway.**
   b. Regulatory enzymes in limited numbers set rates of metabolic pathways.
   c. Regulatory enzymes become saturated when substrate concentrations exceed a certain level.
3. Factors that alter enzymes
   a. A **cofactor** is a nonprotein necessary for a specific enzyme to function. It may be a **coenzyme,** which is organic and can be a vitamin.
   b. Most enzymes are proteins. Harsh conditions undo, or denature, their specific shapes.
   c. Heat, radiation, electricity, certain chemicals, and extreme pH values denature enzymes.

4.4 Energy for Metabolic Reactions

***Energy** is the capacity to do work. Common forms of energy include heat, light, sound, electrical energy, mechanical energy, and chemical energy.*

1. Release of chemical energy
   a. Most metabolic processes use chemical energy released when chemical bonds break in **oxidation** reactions.
   b. The energy released from glucose breakdown during **cellular respiration** drives the reactions of cellular metabolism.
2. ATP molecules
   a. Energy is captured in the bond of the terminal phosphate of each molecule of **adenosine triphosphate (ATP).**
   b. When a cell requires energy, the terminal phosphate bond of an ATP molecule breaks, releasing stored energy.
   c. An ATP molecule that loses its terminal phosphate becomes ADP.
   d. An ADP molecule that captures energy and a phosphate becomes ATP.
3. Cellular respiration
   a. Glycolysis
      (1) **Glycolysis** is the series of reactions that break down a molecule of glucose into two molecules of **pyruvic acid.** This first phase of glucose decomposition does not directly require oxygen **(anaerobic).**
      (2) Some of the energy released is transferred to ATP.
   b. Aerobic respiration
      (1) The second phase of glucose decomposition requires oxygen and is thus called **aerobic respiration.**
      (2) Many more ATP molecules form during aerobic respiration than during the anaerobic phase.
      (3) The final products of glucose breakdown are carbon dioxide, water, and energy (ATP and heat).

4.5 DNA (Deoxyribonucleic Acid)

DNA molecules contain information that instructs a cell how to synthesize enzymes and other proteins.

1. Genetic information
   a. Inherited traits result from DNA information passed from parents to offspring.
   b. A complete set of genetic instructions is a **genome.** The part that encodes protein is the **exome.**
   c. A **gene** is a DNA sequence that contains the information for making a particular protein.
2. DNA molecules
   a. A DNA molecule consists of two strands of nucleotides wound into a double helix.
   b. The nucleotides of a DNA strand are linked in a specific sequence.
   c. The nucleotides of each strand pair with those of the other strand in a complementary fashion (A with T, and G with C).

3. DNA replication
   a. When a cell divides, each new cell requires a copy of the older cell's genetic information.
   b. DNA molecules **replicate** during interphase of the cell cycle.
   c. Each new DNA molecule has one old strand and one new strand.

4.6 Protein Synthesis

Genes provide instructions for making proteins, which take part in many aspects of cell function.

1. The **genetic code**—instructions for making proteins
   a. A sequence of DNA nucleotides encodes a sequence of amino acids.
   b. RNA molecules transfer genetic information from the nucleus to the cytoplasm.
   c. Transcription
      (1) **Transcription** is the copying of a DNA sequence into an RNA sequence. RNA has ribose instead of deoxyribose and uracil instead of thymine. Most RNA molecules are single-stranded.
      (2) **Messenger RNA (mRNA)** molecules consist of nucleotide sequences that are complementary to those of exposed strands of DNA. A **codon** is an mRNA triplet that specifies a particular amino acid.
      (3) Messenger RNA molecules associate with ribosomes and provide patterns for the synthesis of protein molecules.
   d. Translation
      (1) **Translation** is protein synthesis. It starts as a ribosome binds to an mRNA molecule.
      (2) Molecules of **transfer RNA (tRNA)** position amino acids along a strand of mRNA. A 3-base sequence in a tRNA called the **anticodon** complementary base pairs with the codon in the mRNA. The amino acids that the tRNAs bring to the mRNA form peptide bonds. A polypeptide chain grows and folds into a unique shape.

Tissues 5

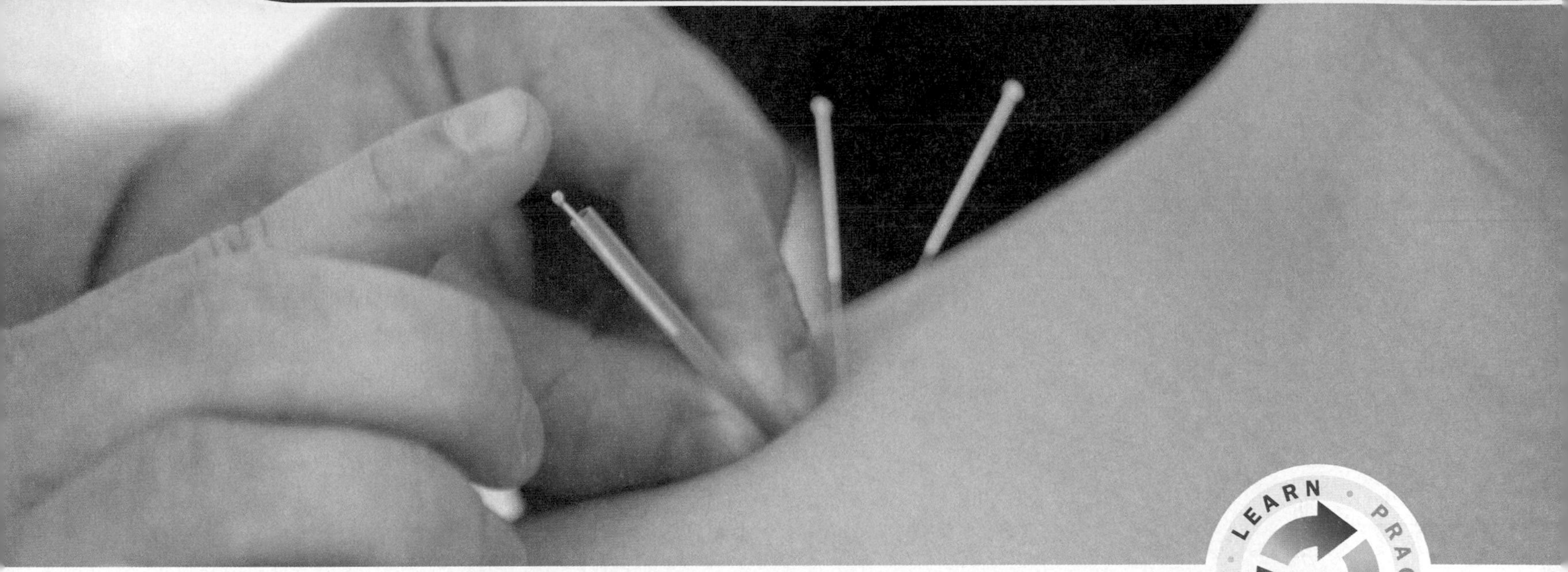

An acupuncturist inserting needles in a woman's shoulder, a therapy that complements conventional medicine. FogStock/Alamy

Just the thought of needles being poked into her skin made her feel woozy; however, following the suggestion of her physician, Jenny decided to try acupuncture to help relieve her musculoskeletal pain. A practice of traditional Chinese medicine that is several thousand years old, acupuncture has been integrated into Western medicine as a complementary therapy that enhances and supports conventional medicine.

Qi, in traditional Chinese medicine, is a "force" of energy that should be freely and robustly flowing throughout the body like a river. Sitting close to the body's surface are "meridians" that pass through and connect organs and body parts, acting as channels for qi to flow. It is believed that energy flow may be impeded when there is physical or emotional illness, such as stress. The goal of acupuncture is to clear these obstacles in the meridians to restore qi. Researchers have studied the role that connective tissues play in the body to help understand the mechanism of acupuncture. Connective tissues interact with and connect all body systems, including the immune, muscle, skeletal, and nervous systems. One theory of acupuncture is that it transmits signals through the body's fascia network. Fascia are layers of connective tissue surrounding muscles and certain other organs. These myofascial chains may explain why the placement of needles affects areas of the body that are far from where the needles are inserted. Helene Langevin, MD, led a study at the University of Vermont where researchers mapped acupuncture points on cadavers and examined the tissue beneath the points and found that about 80 percent of the points were located where connective tissue planes or networks converged, and she proposed that these planes were the channels described in traditional Chinese medicine.

Connective tissues contain an abundance of extracellular matrix, with few cells scattered among the matrix. Fibroblasts are cells that synthesize the extracellular matrix of connective tissues and control the amount of collagen and the composition of the connective tissue. Collagen fibers are strong and have the ability to stretch. Researchers have found that fibroblasts respond to the stretch by changing shape and expanding. The change in fibroblasts is mimicked by the stretch created with the up and down or twisted manipulation of acupuncture needles as they are inserted. Pulling on the tissue from the inside prompts a response by the fibroblasts. Fibroblasts release molecules like ATP, and it is thought that ATP may have an analgesic effect in relieving pain. Adhesion complexes on the surface of fibroblasts alter themselves in response to the stretching, resulting in relaxation and reduced muscle tension of the tissue. Because connective tissues surround nerves, blood vessels, lymphatic vessels, and muscle groups, less tension may affect how these structures function.

There are still many unknowns about the physiology of acupuncture, but it is obvious that the body's connective tissues play a key role.

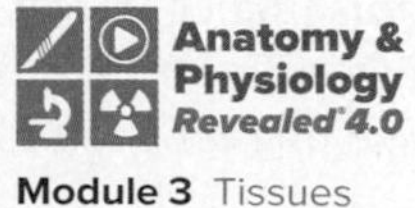

Module 3 Tissues

LEARNING OUTLINE

After studying this chapter, you should be able to complete the "Learning Outcomes" that follow the major headings throughout the chapter.

AIDS TO UNDERSTANDING WORDS

adip- [fat] *adip*ose tissue: tissue that stores fat.

chondr- [cartilage] *chondro*cyte: cartilage cell.

-cyt [cell] osteo*cyte*: bone cell.

epi- [upon] *epi*thelial tissue: tissue that covers all free body surfaces.

-glia [glue] neuro*glia:* cells that support neurons; part of nervous tissue.

inter- [between] *inter*calated disc: band between adjacent cardiac muscle cells.

macr- [large] *macro*phage: large phagocytic cell.

os- [bone] *os*seous tissue: bone tissue.

pseud- [false] *pseudo*stratified epithelium: tissue with cells that appear to be in layers, but are not.

squam- [scale] *squam*ous epithelium: tissue with flattened or scalelike cells.

strat- [layer] *strat*ified epithelium: tissue with cells in layers.

(Appendix A has a complete list of Aids to Understanding Words.)

5.1 | Introduction

LEARN

1. List the four major tissue types, and indicate a function of each type.

Cells, the basic units of structure and function in the human body, are organized into groups called **tissues** (tish'ūz). Each type of tissue is composed of similar cells with a common function.

The tissues of the human body are of four major types: *epithelial, connective, muscle,* and *nervous.* Epithelial tissues form protective coverings and function in secretion and absorption. Connective tissues support soft body parts and bind structures together. Muscle tissues produce body movements, and nervous tissues conduct impulses that help control and coordinate all body activities. Table 5.1 compares the four major tissue types.

Keep in mind that tissues are three-dimensional structures. Micrographs are photos of extremely thin slices (sections) of prepared tissue specimens, and may show more than one tissue type. The advantage of observing thin slices is that light passes through them more readily, making structures easier to see and photograph under magnification. However, the relationship of the structures in three dimensions may not be obvious. Figure 5.1 shows how a tubular structure would appear (a) in an oblique section and (b) in a longitudinal section (see also fig. 1.11).

Throughout this chapter, three-dimensional line drawings (for example, fig. 5.2*a*) are included with the micrograph (for example, fig. 5.2*b,c*) to emphasize the orientation and distinguishing characteristics of the specific tissues. Locator icons (as in fig. 5.2) indicate where in the body that particular tissue might be found.

PRACTICE 5.1

Answers to the Practice questions can be found in the eBook.

1. What is a tissue?
2. List the four major types of tissues.

CAREER CORNER
Medical Transcriptionist

The busy pediatrician reads her young patient's recent medical history on her handheld device just minutes before the appointment, and learns that after an anxious few moments when the fetal heart rate dipped, the birth had gone well. The clear writing of the medical transcriptionist informed the physician.

A medical transcriptionist combines excellent language and keyboarding skills with knowledge of the terminology of pathology, anatomy, and physiology, including many acronyms and abbreviations. In addition to excellence in spelling, grammar, and editing, a medical transcriptionist may also correct and edit documents generated from voice-recognition software.

Medical transcriptionists work in physicians' offices, in hospitals and clinics, for insurance companies, and for hospice and home health-care organizations. Some medical transcriptionists work from home. Medical transcriptionists create and maintain physical exam reports, autopsy reports, test result interpretations, referral letters, medical history notes, discharge summaries, and descriptions of surgical procedures.

Most medical transcriptionists have a two-year associate's degree or have completed a one-year certificate program. It is an excellent career choice for those who are interested in health care and have an ease with language.

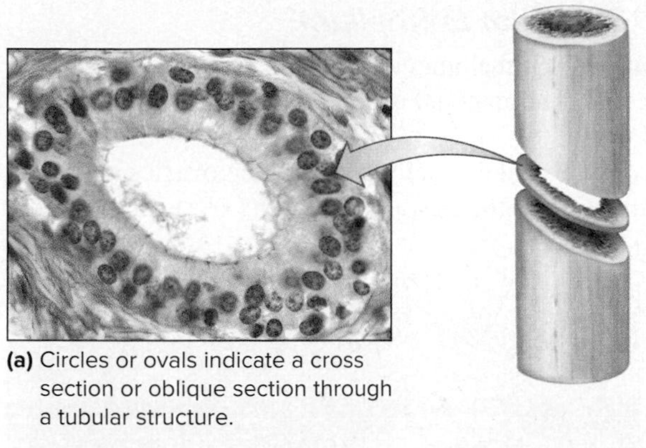

(a) Circles or ovals indicate a cross section or oblique section through a tubular structure.

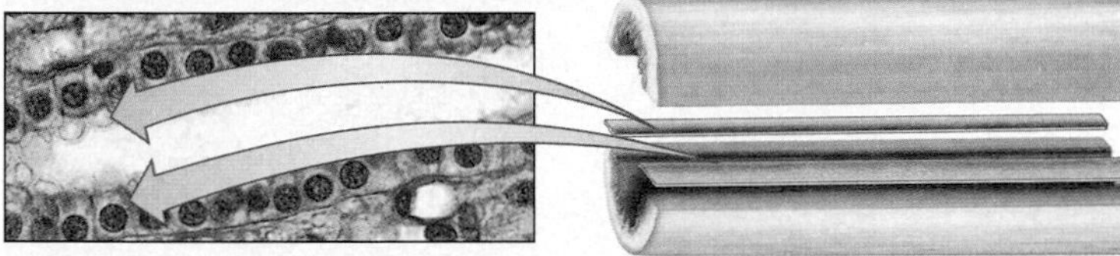

(b) Rows of cells with a space in between indicate a longitudinal section through a tube.

Figure 5.1 Tissue sections. **(a)** Oblique section (600x). **(b)** Longitudinal section (165x). (a): Al Telser/McGraw-Hill Education; (b): Victor P. Eroschenko

5.2 | Epithelial Tissues APR

LEARN

1. Describe the general characteristics and functions of epithelial tissue.
2. Name the types of epithelium and identify an organ in which each is found.
3. Explain how glands are classified.

General Characteristics and Categories

Epithelial (ep″ĭ-the′le-al) **tissues** are found throughout the body. Epithelium covers the body surface and organs, forms the inner lining of body cavities, lines hollow organs, and composes glands. It always has a *free (apical) surface* exposed to the outside or exposed internally to an open space. The underside of epithelial tissue is anchored to connective tissue by a thin, nonliving layer called the **basement membrane.**

As a rule, epithelial tissues lack blood vessels. However, nutrients diffuse to epithelium from underlying connective tissues, which have abundant blood vessels.

Epithelial cells readily divide. As a result, injuries heal rapidly as new cells replace lost or damaged ones. For example, skin cells and cells that line the stomach and intestines are continually damaged and replaced.

Epithelial cells are tightly packed. Consequently, these cells form protective barriers in such structures as the outer layer of the skin and the lining of the mouth. Other epithelial functions include secretion, absorption, and excretion.

Epithelial tissues are classified according to cell shape and the number of cell layers. Epithelial tissues composed of thin, flattened cells are *squamous epithelium;* those with cube-shaped cells are *cuboidal epithelium;* and those with elongated cells are *columnar epithelium.* Epithelium composed of a single layer of cells is called *simple,* whereas epithelium with two or more layers of cells is called *stratified.* In the following descriptions, note that the free surfaces of epithelial cells are modified in ways that reflect their specialized functions.

PRACTICE 5.2

1. List the general characteristics of epithelial tissue.
2. Describe the classification of epithelium in terms of cell shape and number of cell layers.

Simple Squamous Epithelium

Simple squamous (skwa′mus) **epithelium** consists of a single layer of thin, flattened cells. These cells fit tightly together, somewhat like floor tiles, and their nuclei are usually broad and thin (fig. 5.2).

TABLE 5.1 Tissues

Type	Function	Location	Distinguishing Characteristics
Epithelial	Protection, secretion, absorption, excretion	Cover body surface, cover and line internal organs, compose glands	Lack blood vessels, readily divide; cells are tightly packed
Connective	Bind, support, protect, fill spaces, store fat, produce blood cells	Widely distributed throughout body	Mostly have good blood supply, with at least two distinct cell types; cells are farther apart than epithelial cells with extracellular matrix in between
Muscle	Movement	Attached to bones, in the walls of hollow internal organs, heart	Able to generate forces in response to specific stimuli
Nervous	Conduct impulses for coordination, regulation, integration, and sensory reception	Brain, spinal cord, nerves	Cells communicate with each other and other body parts by receiving and sending stimuli

Substances pass rather easily through simple squamous epithelium, which is common at sites of diffusion and filtration. For instance, simple squamous epithelium lines the air sacs (alveoli) of the lungs where oxygen and carbon dioxide are exchanged. It also forms the walls of capillaries, lines the insides of blood and lymph vessels, and is part of the membranes that line body cavities and cover the viscera. However, because simple squamous epithelium is so thin and delicate, it is easily damaged.

Simple Cuboidal Epithelium

Simple cuboidal epithelium consists of a single layer of cube-shaped cells. These cells usually have centrally located, spherical nuclei (fig. 5.3).

Simple cuboidal epithelium covers the ovaries and lines most of the kidney tubules and the ducts of certain glands.

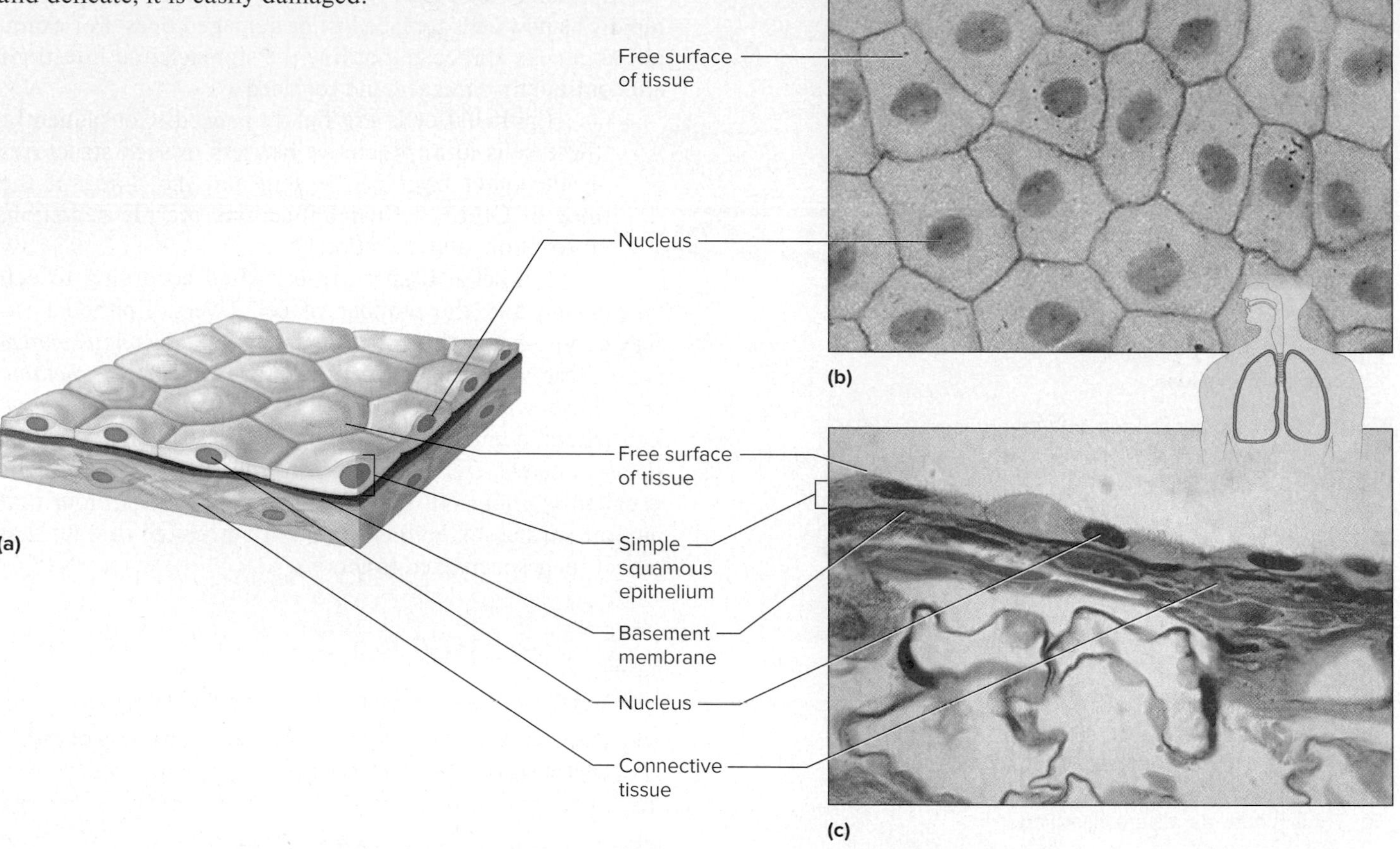

Figure 5.2 Simple squamous epithelium consists of a single layer of tightly packed, flattened cells. **(a)** Idealized representation of simple squamous epithelium. Micrographs of **(b)** a surface view (250x) and **(c)** a side view of a section through simple squamous epithelium (250x). (b): Al Telser/McGraw-Hill Education; (c): Ed Reschke/Photolibrary/Getty Images APR

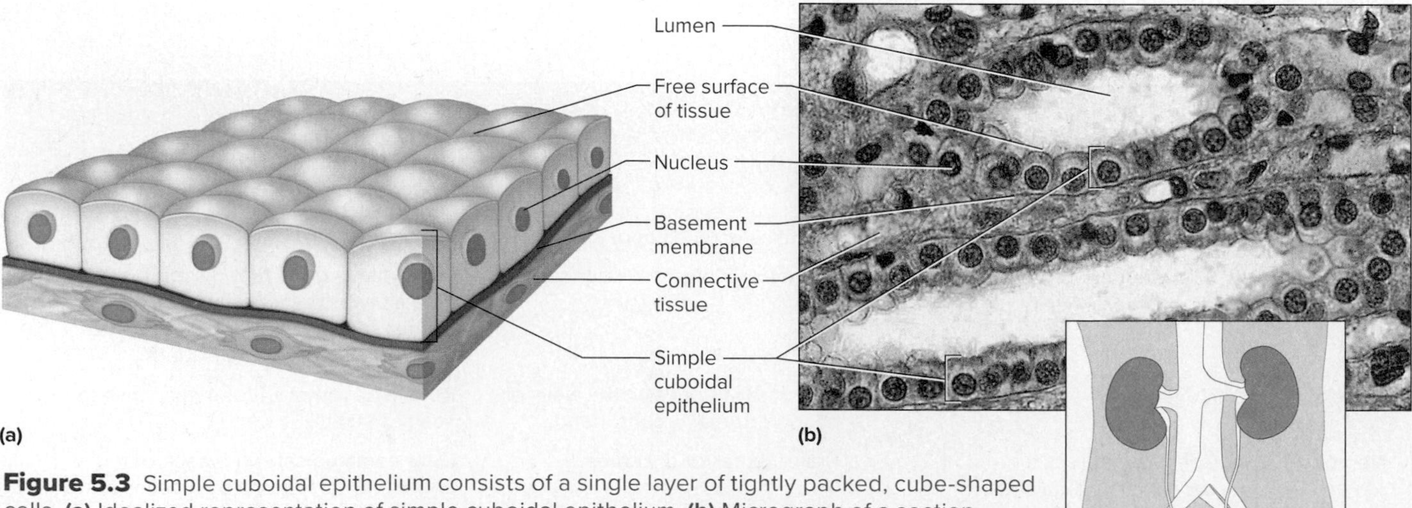

Figure 5.3 Simple cuboidal epithelium consists of a single layer of tightly packed, cube-shaped cells. **(a)** Idealized representation of simple cuboidal epithelium. **(b)** Micrograph of a section through simple cuboidal epithelium (165x). (b): Victor P. Eroschenko APR

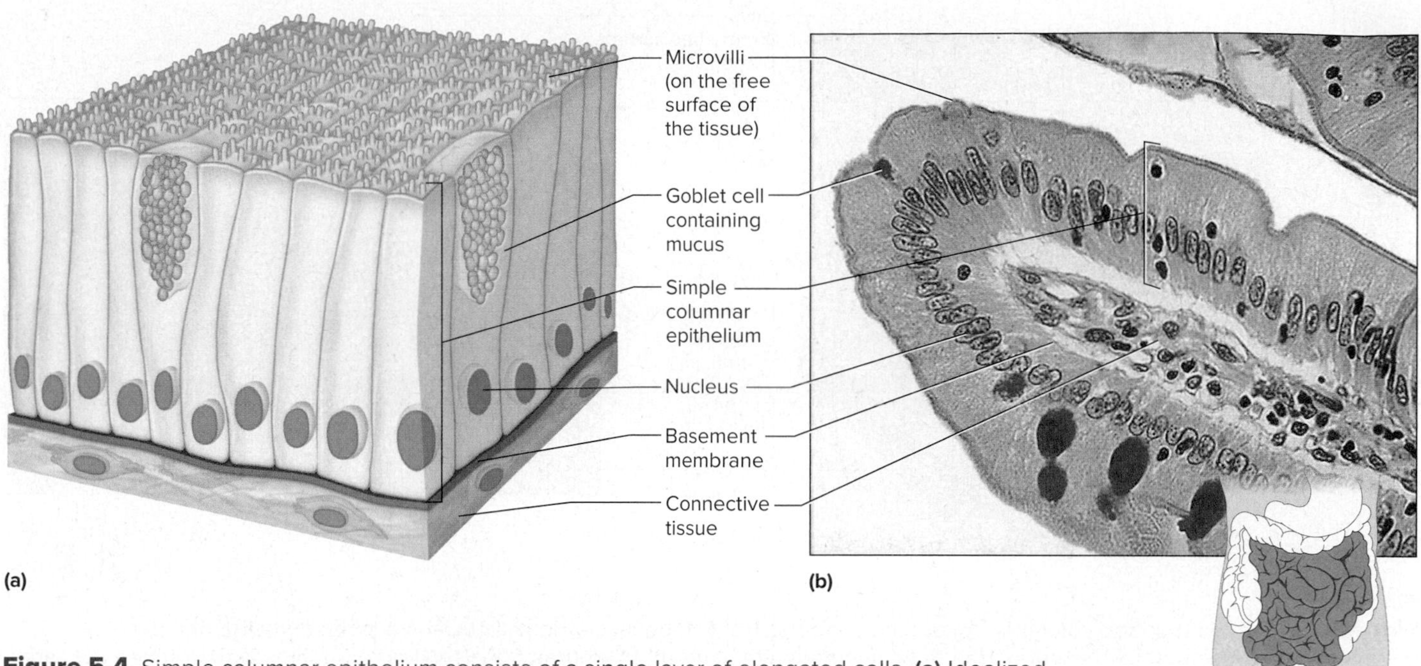

Figure 5.4 Simple columnar epithelium consists of a single layer of elongated cells. **(a)** Idealized representation of simple columnar epithelium. **(b)** Micrograph of a section through simple columnar epithelium (400x). (b): Victor P. Eroschenko APR

In these tubules and ducts, the free surface faces the hollow channel or **lumen.** In the kidneys, this tissue functions in the formation of urine; in glands, it secretes glandular products.

Simple Columnar Epithelium

The cells of **simple columnar epithelium** are distinguished by being taller than they are wide. This tissue is composed of a single layer of cells with elongated nuclei located at about the same level, near the basement membrane (fig. 5.4). The cells of this tissue can be ciliated or nonciliated. **Cilia** extend from the free surfaces of the cells and move constantly (see section 3.2, Composite Cell). In the female reproductive tract, cilia aid in moving the egg cell through the uterine tube to the uterus.

Nonciliated simple columnar epithelium lines the uterus and portions of the digestive tract, including the stomach and the small and large intestines. Because its cells are tall, this tissue is thick, which enables it to protect underlying tissues. Simple columnar epithelium also secretes digestive fluids and absorbs nutrients from digested food.

Simple columnar epithelial cells, specialized for absorption, often have many tiny, cylindrical processes, called **microvilli,** extending from their surfaces. These increase the surface area of the cell membrane where it is exposed to substances being absorbed (fig. 5.4).

Typically, specialized flask-shaped glandular cells are scattered among the columnar cells of simple columnar epithelium. These cells, called **goblet cells,** secrete a protective fluid, called *mucus,* onto the free surface of the tissue (fig. 5.4).

Pseudostratified Columnar Epithelium

The cells of **pseudostratified** (soo″dō-strat′ĭ-fīd) **columnar epithelium** appear to be stratified or layered, but they are not. A layered effect occurs because the nuclei lie at two or more levels in the row of aligned cells. However, the cells, which vary in shape, all reach the basement membrane, even though some of them may not contact the free surface.

Pseudostratified columnar epithelial cells commonly have cilia, which extend from their free surfaces. Goblet cells scattered throughout this tissue secrete mucus, which the cilia sweep away (fig. 5.5).

Pseudostratified columnar epithelium lines the passages of the respiratory system. Here, the mucus-covered linings are sticky and trap dust and microorganisms that enter with the air. The cilia move the mucus and its captured particles upward and out of the airways.

Stratified Squamous Epithelium

The many cell layers of **stratified squamous epithelium** make this tissue relatively thick. Cells divide in the deeper layers, and newer cells push older ones farther outward, where they flatten (fig. 5.6). Stratified epithelial tissues are named for the shape of their top layer of cells.

Stratified squamous epithelium forms the superficial layer of the skin (*epidermis*). As older skin cells are pushed outward, they accumulate proteins called *keratins,* and then harden and die. This "keratinization" produces a covering of dry, tough, protective material that prevents water and other substances from escaping underlying tissues and blocks chemicals and microorganisms from entering (see fig. 6.2).

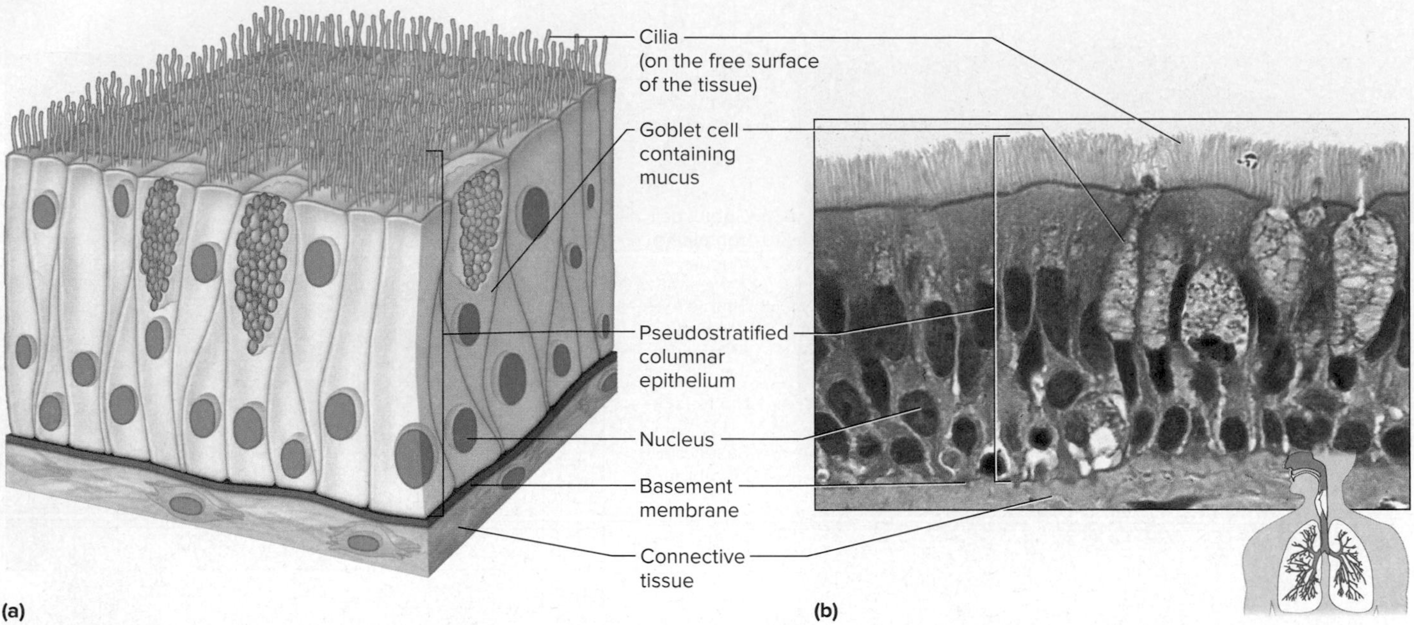

Figure 5.5 Pseudostratified columnar epithelium appears stratified because the cell nuclei are located at different levels. **(a)** Idealized representation of pseudostratified columnar epithelium. **(b)** Micrograph of a section through pseudostratified columnar epithelium (1,000x). (b): Dennis Strete/McGraw-Hill Education APR

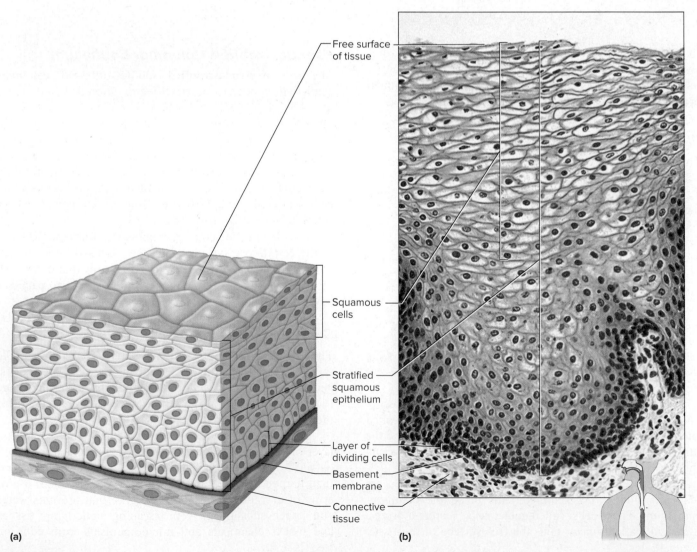

Figure 5.6 Stratified squamous epithelium consists of many layers of cells. **(a)** Idealized representation of stratified squamous epithelium. **(b)** Micrograph of a section through stratified squamous epithelium (65x). (b): Al Telser/McGraw-Hill Education APR

Stratified squamous epithelium also lines the oral cavity, esophagus, vagina, and anal canal. In these parts, the tissue is not keratinized; it stays soft and moist, and the cells on its free surfaces remain alive.

Stratified Cuboidal Epithelium

Stratified cuboidal epithelium consists of two or three layers of cuboidal cells that form the lining of a lumen (fig. 5.7). The layering of the cells provides more protection than the single layer affords.

Stratified cuboidal epithelium lines the ducts of the mammary glands, sweat glands, salivary glands, and pancreas. It also forms the lining of developing ovarian follicles and seminiferous tubules, which are parts of the female and male reproductive systems, respectively.

Stratified Columnar Epithelium

Stratified columnar epithelium consists of several layers of cells (fig. 5.8). The superficial cells are columnar, whereas the basal layers consist of cuboidal cells. Stratified columnar epithelium is found in part of the male urethra and lining the larger ducts of glands.

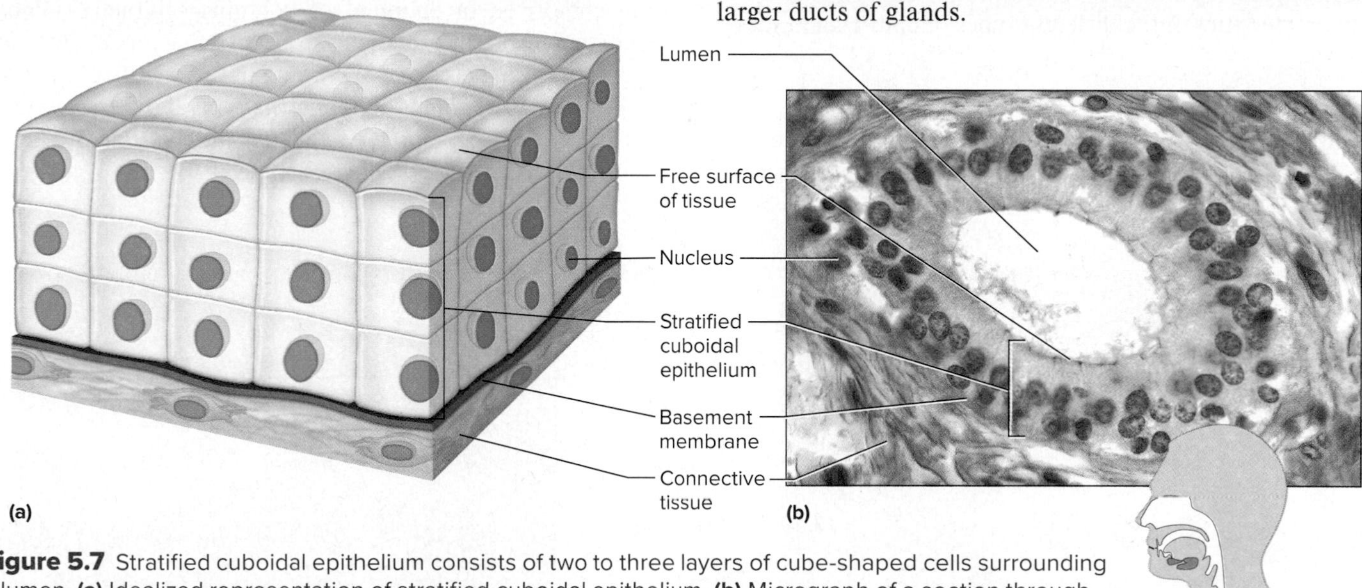

Figure 5.7 Stratified cuboidal epithelium consists of two to three layers of cube-shaped cells surrounding a lumen. **(a)** Idealized representation of stratified cuboidal epithelium. **(b)** Micrograph of a section through stratified cuboidal epithelium (600x). (b): Al Telser/McGraw-Hill Education APR

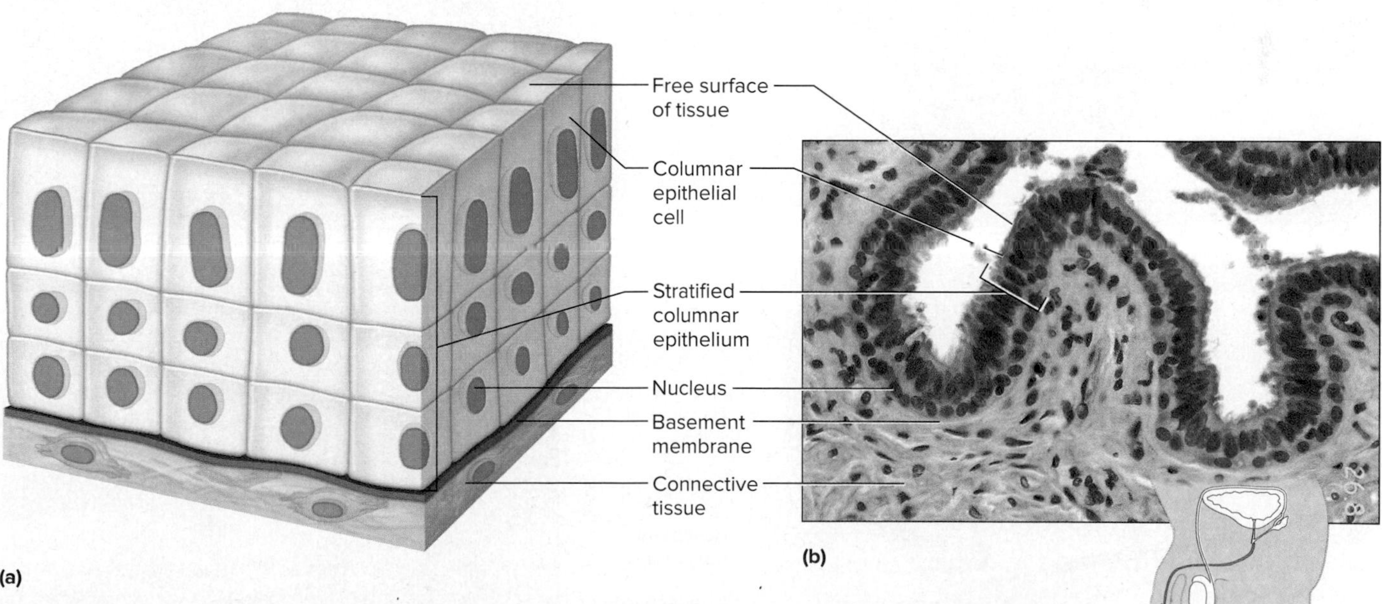

Figure 5.8 Stratified columnar epithelium consists of a superficial layer of columnar cells overlying several layers of cuboidal cells. **(a)** Idealized representation of stratified columnar epithelium. **(b)** Micrograph of a section through stratified columnar epithelium (230x). (b): McGraw-Hill Education/Al Telser, photographer APR

PRACTICE FIGURE 5.8

In the micrograph (b), is this section through the urethra a cross section or a longitudinal section?

Answer can be found in Appendix E.

Transitional Epithelium

Transitional epithelium (uroepithelium) is specialized to change in response to increased tension. It forms the inner lining of the urinary bladder and lines the ureters and the superior urethra. When the wall of one of these organs contracts, the tissue consists of several layers of irregular-shaped cells; when the organ is distended, however, the tissue stretches, and the cells elongate (fig. 5.9). In addition to providing an expandable lining, transitional epithelium forms a barrier that helps prevent the contents of the urinary tract from diffusing back into the internal environment. Table 5.2 summarizes the characteristics of the different types of epithelial tissues.

Glands

Glands are organs made of epithelial tissues. They produce and secrete their substances into ducts that lead to a specific target, or into the blood or other body tissues. Most glands are comprised of cuboidal or columnar epithelia. Glands that secrete their products into ducts that open onto surfaces, such as the skin or the lining of the digestive tract, are called **exocrine glands.** Glands that secrete their products into tissue fluid or blood are called **endocrine glands.** (Endocrine glands are discussed in chapter 11.)

> **OF INTEREST** Close to 90% of all human cancers are *carcinomas*, growths that originate in epithelium, due to the high rate of mitotic activity. Most carcinomas begin on surfaces that contact the external environment, such as skin, linings of the airways, or linings of the stomach or intestine. Carcinomas may also arise internally, as in a duct in a breast or in the prostate gland.

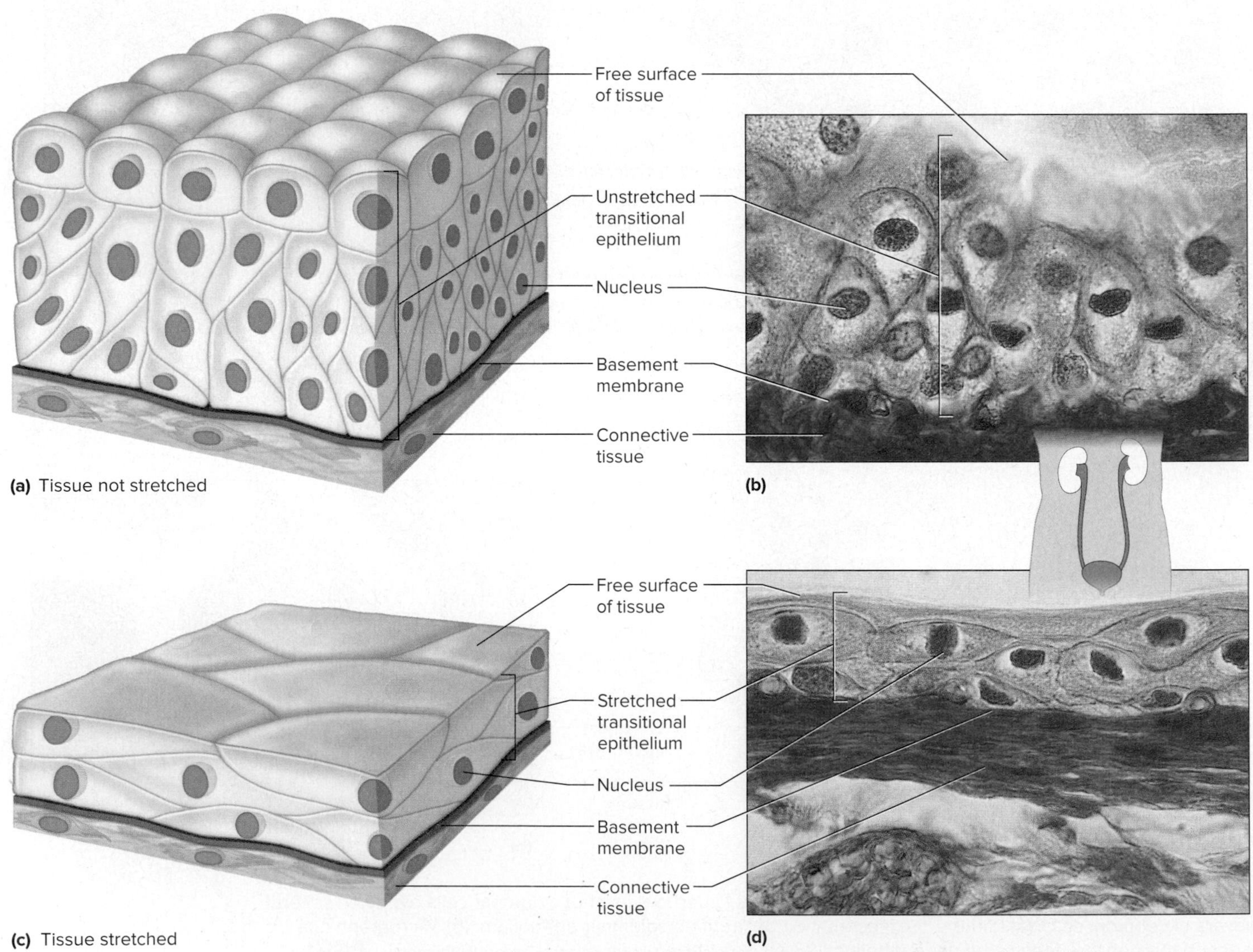

Figure 5.9 Transitional epithelium. **(a and c)** Idealized representation of transitional epithelium. **(b and d)** Micrographs of sections through transitional epithelium (675x). When the smooth muscle in the organ wall contracts, transitional epithelium is unstretched and consists of many layers. When the organ is distended, the tissue stretches and appears thinner. (b, d): Ed Reschke APR

TABLE 5.2 Epithelial Tissues

Type	Function	Location
Simple squamous epithelium	Filtration, diffusion, osmosis; covering of surfaces	Air sacs of the lungs, walls of capillaries, linings of blood and lymph vessels, part of the membranes lining body cavities and covering viscera
Simple cuboidal epithelium	Protection, secretion	Surface of ovaries, linings of kidney tubules, and linings of ducts of certain glands
Simple columnar epithelium	Protection, secretion, absorption	Linings of uterus, stomach, and intestines
Pseudostratified columnar epithelium	Protection, secretion, movement of mucus	Linings of respiratory passages
Stratified squamous epithelium	Protection	Superficial layer of skin, and linings of oral cavity, throat, vagina, and anal canal
Stratified cuboidal epithelium	Protection	Linings of ducts of mammary glands, sweat glands, salivary glands, and pancreas
Stratified columnar epithelium	Protection, secretion	Part of the male urethra and linings of larger ducts of excretory glands
Transitional epithelium	Stretchability, protection	Inner lining of urinary bladder and linings of ureters and part of urethra

Exocrine glands are classified according to the ways these glands secrete their products (fig. 5.10). Glands that release fluid by exocytosis are **merocrine** (mer′o-krin) **glands,** also called *eccrine glands.* Glands that lose small portions of their glandular cell bodies during secretion are called **apocrine** (ap′o-krin) **glands.** Glands that release entire cells that disintegrate to release cell secretions are called **holocrine** (ho′lo-krin) **glands.** Table 5.3 summarizes these glands and their secretions.

Most exocrine secretory cells are merocrine, and they can be further subclassified based on their secretion of serous fluid or mucus. *Serous fluid* typically is watery and slippery. **Serous cells** secreting this fluid, which lubricates, are commonly associated with the visceral and parietal membranes of the thoracic and abdominopelvic cavities. The thicker fluid, *mucus,* is rich in the glycoprotein *mucin.* Cells in the inner linings of the digestive, respiratory, and reproductive systems secrete abundant mucus, which protects underlying tissue. **Mucous cells** and goblet cells secrete mucus, but in different parts of the body.

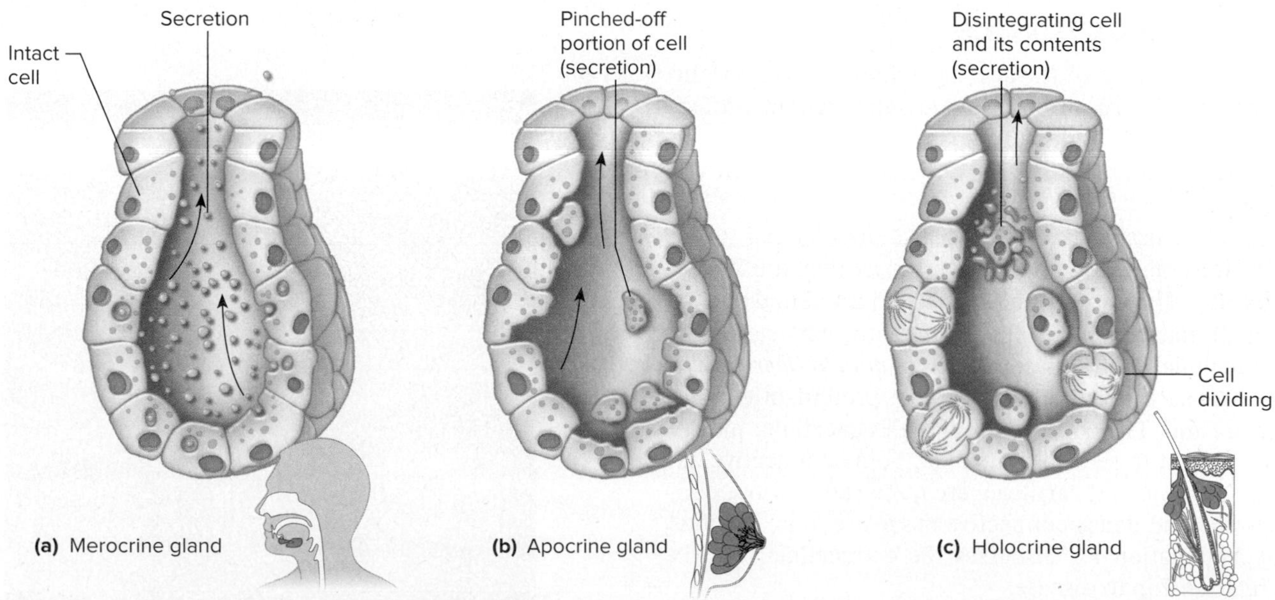

Figure 5.10 Types of exocrine glands. **(a)** APR Merocrine glands release secretions without losing cytoplasm. **(b)** APR Apocrine glands lose small portions of their cell bodies during secretion. **(c)** APR Holocrine glands release entire cells filled with secretory products.

TABLE 5.3 Exocrine Glandular Secretions

Type of Gland	Description of Secretion	Example
Merocrine glands	A fluid product released through the cell membrane by exocytosis	Salivary glands, pancreatic glands, sweat glands of the skin
Apocrine glands	Cellular product and portions of the free ends of glandular cells pinch off during secretion	Mammary glands, ceruminous glands lining the external ear canal
Holocrine glands	Disintegrated entire cells filled with secretory products	Sebaceous glands of the skin

PRACTICE 5.2

3. Describe the special functions of each type of epithelium.
4. Distinguish between exocrine glands and endocrine glands.
5. Explain how exocrine glands are classified.
6. Distinguish between serous fluid and mucus.

5.3 | Connective Tissues

LEARN

1. Compare and contrast the ground substance, cells, and fibers in different types of connective tissue.
2. Describe the major functions of each type of connective tissue.
3. Identify where each type of connective tissue is found.

General Characteristics

Connective (kŏ-nek′tiv) **tissues** have many roles in the body. They bind structures, provide support and protection, serve as frameworks, fill spaces, store fat, produce blood cells, protect against infections, and help repair tissue damage.

Connective tissues have at least two distinct cell types. The cells are not tightly bound together as they are in epithelial tissues. They are spread apart with an abundance of extracellular material, called the matrix, lying between them. This extracellular matrix is composed of *protein fibers,* and a *ground substance* consisting of nonfibrous protein, other molecules, and fluid. The consistency of the extracellular matrix varies from fluid to semisolid to solid. Some connective tissues, such as bone and cartilage, are quite rigid. Loose connective tissue and dense connective tissue are more flexible. Clinical Application 5.1 discusses the extracellular matrix and its relationship to disease.

Most connective tissue cells can divide. Connective tissues typically have good blood supplies and are well nourished, but the density of blood vessels (vascularity) varies with the type of connective tissue.

Major Cell Types

Connective tissues include a variety of cell types. Some cells are called *fixed cells* because they reside in the tissue for an extended period of time. These include fibroblasts and mast cells. Other cells, such as macrophages, are *wandering cells.* They move through and appear in tissues temporarily, usually in response to an injury or infection.

Fibroblasts (fi′bro-blasts) are the most common type of fixed cell in connective tissue. These large, star-shaped cells produce fibers by secreting proteins into the extracellular matrix (fig. 5.11).

Macrophages (mak′ro-fājez), or histiocytes, originate as white blood cells (see section 12.2, Formed Elements). They are almost as numerous as fibroblasts in some connective tissues. Macrophages are specialized to carry on phagocytosis (see section 3.3, Movements Into and Out of the Cell). They can move about and clear foreign particles from tissues (fig. 5.12).

Mast cells are large and widely distributed in connective tissues. They are usually near blood vessels (fig. 5.13). Mast

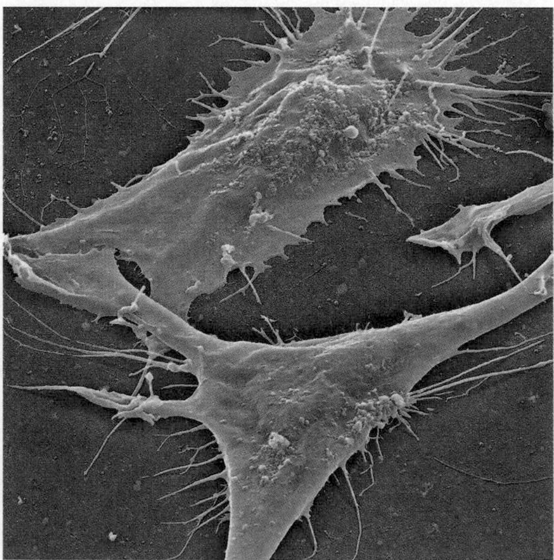

Figure 5.11 Scanning electron micrograph of fibroblasts, the most abundant cell type of connective tissue (1,500x).
Juergen Berger/Science Source

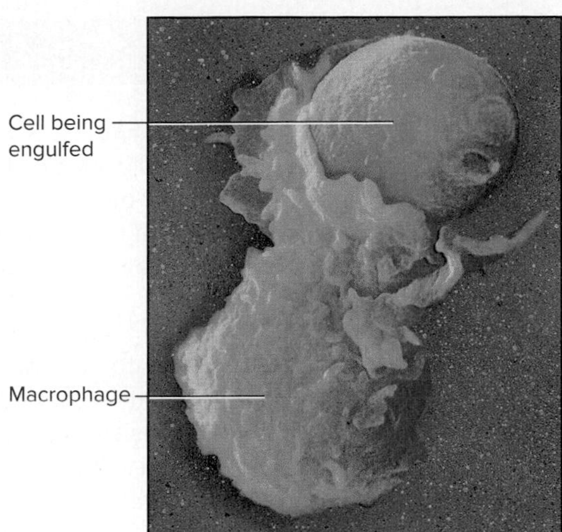

Figure 5.12 Macrophages are scavenger cells common in connective tissues. This scanning electron micrograph shows a macrophage engulfing a bacterium (7,500x). Biology Pics/Science Source

cells release *heparin,* a compound that prevents blood clotting, and *histamine,* a substance that promotes some of the reactions associated with inflammation and allergies (see section 14.7, Immunity: Adaptive (Specific) Defenses).

Connective Tissue Fibers

Fibroblasts produce three types of connective tissue fibers: collagen fibers, elastic fibers, and reticular fibers. Collagen and elastic fibers are the most abundant.

Collagen (kol'ah-jen) **fibers** are thick threads of the protein *collagen.* They are grouped in long, parallel bundles, and are flexible but only slightly elastic. More importantly, they have great tensile strength—that is, they resist considerable pulling force. Thus, collagen fibers are important components of body parts that hold structures together; they include **ligaments** (which connect bones to other bones) and **tendons** (which connect muscles to bones).

Tissue containing abundant collagen fibers is called *dense connective tissue.* It appears white, and for this reason collagen fibers are sometimes called *white fibers.* In contrast, *loose connective tissue* has fewer collagen fibers.

Elastic fibers are composed of a springlike protein called *elastin.* These thin fibers branch, forming complex networks. Elastic fibers are weaker than collagen fibers, but they are easily stretched or deformed and will resume their original lengths and shapes when the force acting on them is removed. Elastic fibers are abundant in body parts normally subjected to stretching, such as the vocal cords. They are also called *yellow fibers* because tissues well supplied with them appear yellowish.

Reticular fibers are thin collagen fibers. They are highly branched and form delicate supporting networks in a variety of tissues, including those of the spleen. Table 5.4 summarizes the components of connective tissue, and their functions.

When skin is exposed to prolonged and intense sunlight, connective tissue fibers lose elasticity, and the skin stiffens and becomes leathery. In time, the skin may sag and wrinkle. Collagen injections may temporarily smooth out wrinkles. However, collagen applied as a cream to the skin does not combat wrinkles because collagen molecules are far too large to penetrate the skin.

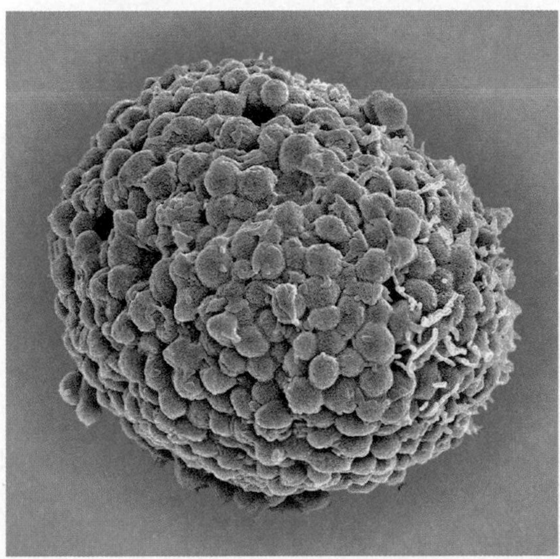

Figure 5.13 Scanning electron micrograph of a mast cell, which releases heparin and histamine (1,750x).
Steve Gschmeissner/Science Source

TABLE 5.4 General Components of Connective Tissue

Component	Function
Cellular	
Fibroblasts	Produce fibers
Macrophages	Carry on phagocytosis
Mast cells	Secrete heparin and histamine
Extracellular (matrix)	
Collagen fibers	Hold structures together with great tensile strength
Elastic fibers	Stretch easily
Reticular fibers	Lend delicate support
Ground substance	Fills in spaces around cells and fibers

CLINICAL APPLICATION 5.1

The Body's Glue: The Extracellular Matrix (ECM)

The extracellular matrix (ECM) is a complex and changing mix of molecules that modifies tissues to suit different organs and conditions. The ECM forms a scaffolding to organize cells into tissues. It also relays the biochemical signals that control cell division, cell differentiation, tissue repair, and cell migration.

The ECM has two basic components: the basement membrane that anchors epithelium to underlying connective tissue, and the rest of the material between cells, called the interstitial matrix. The basement membrane is mostly tightly packed collagen fibers from which large, cross-shaped glycoproteins called laminins extend. The laminins (and other glycoproteins such as fibronectin, the proteoglycans, and tenascin) traverse the interstitial matrix and contact receptors, called integrins, on other cells (fig. 5A). In this way, the ECM connects cells into tissues. At least twenty types of collagen and precursors of hormones, enzymes, growth factors, and immune system biochemicals (cytokines) compose versions of the ECM. The precursor molecules are activated under certain conditions.

The components of the ECM are always changing. As cells synthesize new proteins, others are degraded. The balance of components is important to maintaining and repairing organ structure. Disrupt the balance, and disease can result, as the following examples illustrate.

Cancer

Fibroblasts normally contract as they close a wound, and are then replaced with normal epithelium. A spreading cancer can control fibroblasts by emitting chemical signals that make the fibroblasts more contractile, and more like cancer cells. At the same time, alterations in laminins loosen the connections of the fibroblasts to surrounding cells. This abnormal flexibility enables the changed fibroblasts to migrate, helping the cancer spread.

Liver Fibrosis

In fibrosis, which is a part of all chronic liver diseases, collagen deposition increases so that the ECM exceeds its normal 3% of the organ. Healthy liver ECM sculpts a framework that supports the epithelial and vascular tissues of the organ. In response to a damaging agent such as a virus, alcohol, or a toxic drug, hepatic stellate cells secrete collagen fibers where the epithelium and blood vessels meet. Such limited fibrosis seals off the affected area, preventing spread of the damage. But if the process continues—if an infection is not treated or the noxious stimulus is not removed—the ECM grows and eventually blocks the interaction between liver cells and the bloodstream. Scarring of liver tissue occurs, with loss of normal function, resulting in a dangerous condition called *cirrhosis*.

Heart Failure and Atherosclerosis

The heart's ECM organizes cells into a three-dimensional network that coordinates their contractions into the rhythmic heartbeat necessary to pump blood. This ECM consists of collagen, fibronectin, laminin, and elastin surrounding cardiac muscle cells and myofibroblasts, and is also in the walls of arteries. Some forms of heart failure reflect imbalances of collagen production and degradation. In heart muscle, as in the liver, the natural response of ECM buildup is to wall off an area where circulation is blocked, but if it continues, the extra scaffolding stiffens the heart, which can lead to heart failure. During a myocardial infarction (heart attack), collagen synthesis and deposition increase in affected and nonaffected heart parts, which is why damage can continue even after pain stops. In atherosclerosis, excess ECM accumulates on the interior linings of arteries, blocking blood flow. Collagen accumulation is also associated with atherosclerosis.

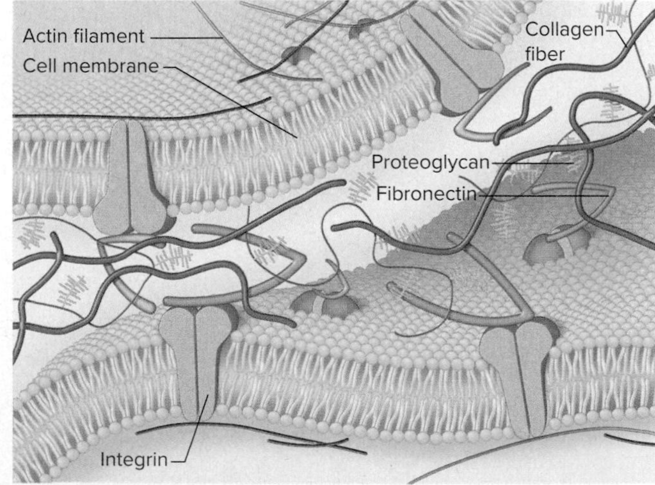

Figure 5A The extracellular matrix (ECM) is a complex and dynamic meshwork of various proteins and glycoproteins. Collagen is abundant. Other common components include integrins that anchor the ECM to cells, proteoglycans, and fibronectin. The ECM may also include precursors of growth factors, hormones, enzymes, and cytokines. It is vital to maintaining the specialized characteristics of tissues and organs.

Categories of Connective Tissue

Connective tissues are classified based upon the relative amounts of cells, protein fibers, and type of matrix, and are divided into two major categories. *Connective tissue proper* includes loose connective tissue and dense connective tissue. The *specialized connective tissues* include cartilage, bone, and blood. All of these tissues will have their own unique cells.

Loose Connective Tissue

Loose connective tissues, as the term implies, exhibit a loose arrangement of protein fibers that are not tightly bound together. These include areolar tissue, adipose tissue, and reticular connective tissue. **Areolar** (ah-rē′o-lar) **tissue** forms delicate, thin membranes throughout the body. The cells of this tissue, mainly fibroblasts, are located some distance apart and are separated by a gel-like ground substance that contains collagen and elastic fibers that fibroblasts secrete (fig. 5.14). Areolar tissue binds the skin to the underlying organs and fills spaces between muscles. It lies beneath most layers of epithelium, where its many blood vessels nourish nearby epithelial cells.

Adipose (ad′ĭ-pōs) **tissue,** or **fat,** develops when certain cells (adipocytes) store fat as droplets in their cytoplasm and enlarge (fig. 5.15). When such cells become so abundant that they crowd other cell types, they form adipose tissue. Adipose tissue lies beneath the skin, in spaces between muscles, around the kidneys, behind the eyeballs, in certain abdominal membranes, on the surface of the heart, and around certain joints. Adipose tissue cushions joints and some organs, such as the kidneys. It also insulates beneath the skin, and it stores energy in fat molecules.

Adipose tissue can be described as white adipose tissue (white fat) or brown adipose tissue (brown fat). White fat stores nutrients for nearby cells to use in the production of energy. Brown fat cells have many mitochondria that can break down nutrients to generate heat to warm the body. Infants have brown fat mostly on their backs to warm them. Adults have brown fat in the armpits, neck, and around the kidneys.

Reticular connective tissue is composed of thin, reticular fibers in a three-dimensional network. It helps provide the framework of certain internal organs, such as the liver and spleen.

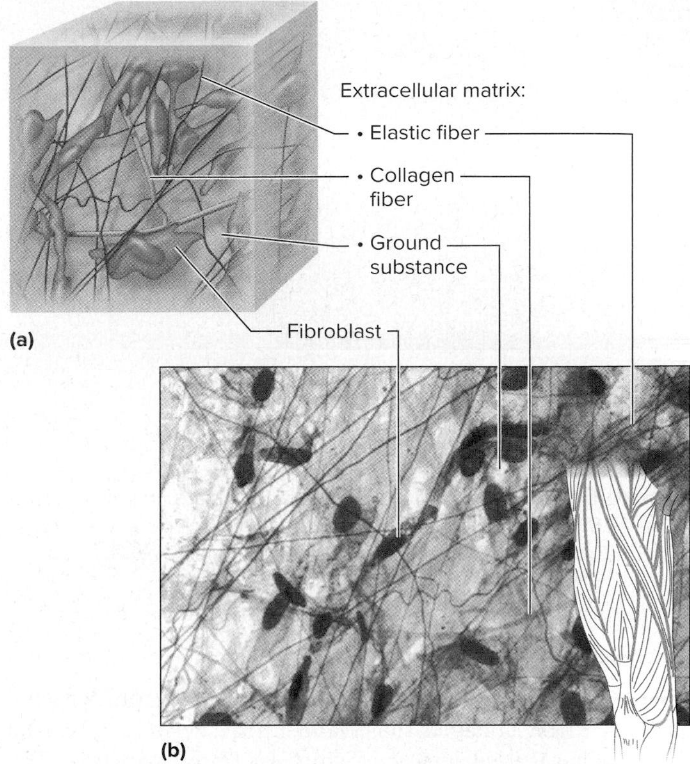

Figure 5.14 Areolar tissue contains numerous fibroblasts that produce collagen and elastic fibers (800x). (b): Dennis Strete/McGraw-Hill Education APR

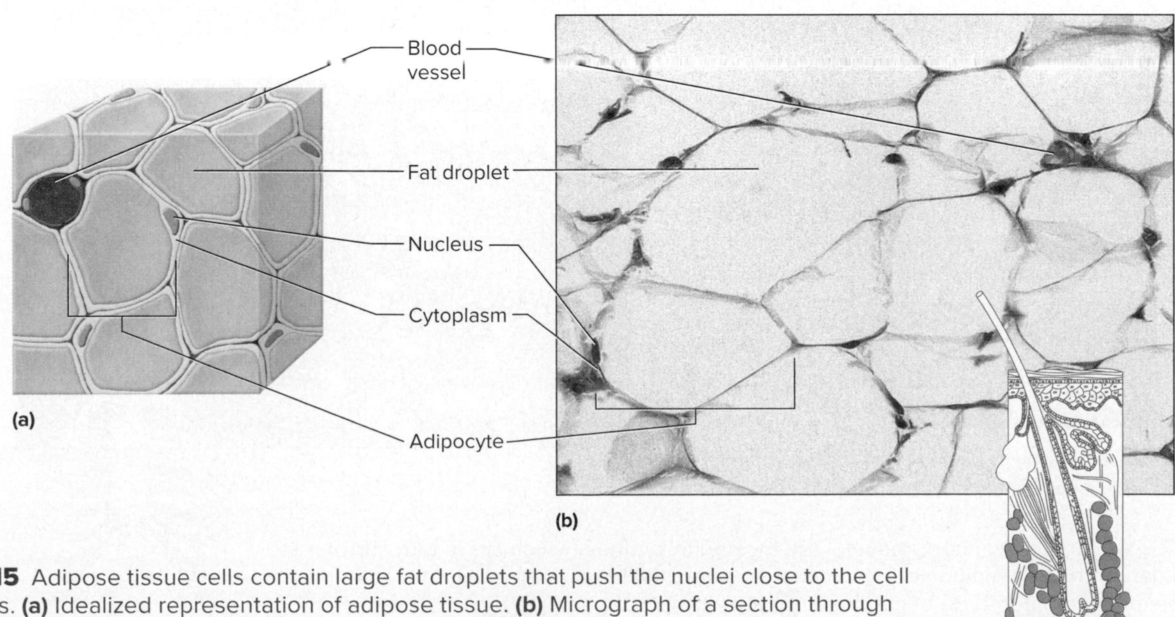

Figure 5.15 Adipose tissue cells contain large fat droplets that push the nuclei close to the cell membranes. **(a)** Idealized representation of adipose tissue. **(b)** Micrograph of a section through adipose tissue (400x). (b): Alvin Telser/McGraw-Hill Education APR

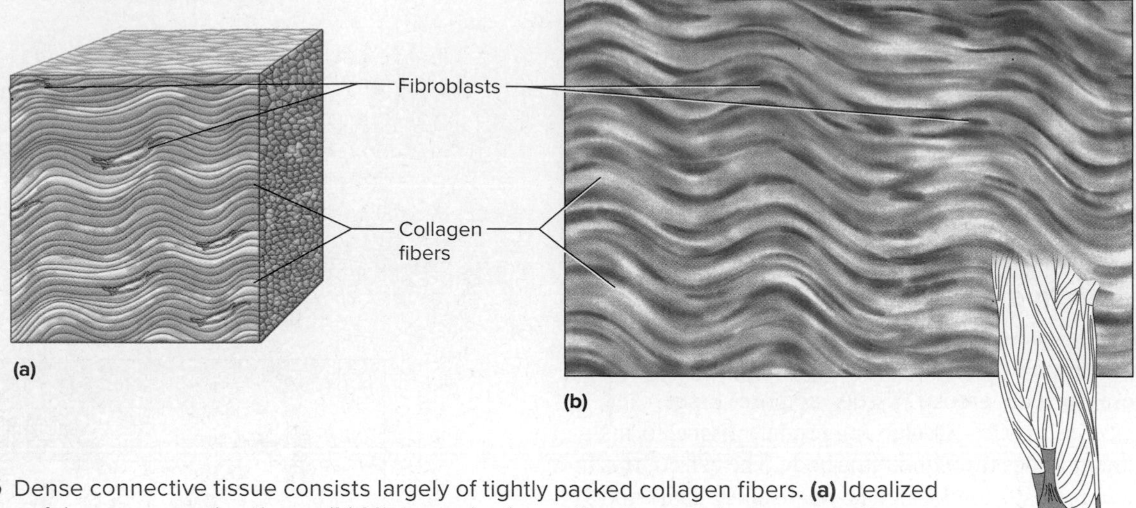

Figure 5.16 Dense connective tissue consists largely of tightly packed collagen fibers. **(a)** Idealized representation of dense connective tissue. **(b)** Micrograph of a section through dense connective tissue (500x). (b): Dennis Strete/McGraw-Hill Education APR

Dense Connective Tissue

Dense connective tissue consists of many closely, or "densely," packed, thick, collagen fibers and a fine network of elastic fibers. It has few cells, most of which are fibroblasts (fig. 5.16).

Collagen fibers of dense connective tissue are very strong, enabling the tissue to withstand pulling forces. It often binds body structures as part of tendons and ligaments. This type of tissue is also in the protective white layer of the eyeball and in the deep skin layer. The blood supply to dense connective tissue is poor, resulting in slower repair of damaged tissue.

PRACTICE 5.3

1. What are the general characteristics of connective tissues?
2. What are the characteristics of collagen and elastin?
3. What feature distinguishes adipose tissue from other connective tissues?
4. Explain the difference between loose connective tissue and dense connective tissue.

Cartilage

Cartilage (kar'ti-lij) is a rigid connective tissue. It provides support, frameworks, and attachments; protects underlying tissues; and forms structural models for many developing bones.

Cartilage extracellular matrix is abundant and is largely composed of collagen fibers embedded in a gel-like ground substance. Cartilage cells, called **chondrocytes** (kon'dro-sītz), occupy small chambers called *lacunae* and lie completely within the extracellular matrix (fig. 5.17).

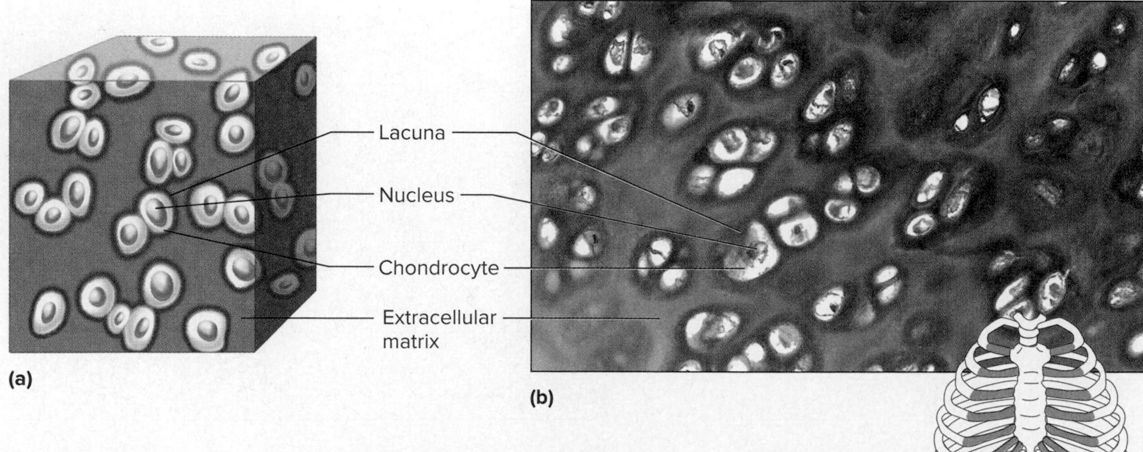

Figure 5.17 Cartilage cells (chondrocytes) are located in lacunae, which are in turn surrounded by extracellular matrix containing very fine collagen fibers. **(a)** Idealized representation of hyaline cartilage, the most common type of cartilage. **(b)** Micrograph of a section through hyaline cartilage (160x). (b): Al Telser/McGraw-Hill Education APR

A cartilaginous structure is enclosed in a covering of connective tissue called the **perichondrium.** Nutrients diffuse to cartilage cells from blood vessels in the perichondrium. Because this tissue does not have a direct blood supply, cartilage heals slowly and chondrocytes do not divide frequently.

Different types of extracellular matrix distinguish three types of cartilage. **Hyaline cartilage,** the most common type, has very fine collagen fibers in its extracellular matrix and looks somewhat like white glass (fig. 5.17). It is found on the ends of bones in many joints, in the soft part of the nose, and in the supporting rings of the respiratory passages. Hyaline cartilage is also important in the development and growth of most bones (see section 7.4, Bone Development, Growth, and Repair).

Elastic cartilage has a dense network of elastic fibers and thus is more flexible than hyaline cartilage (fig. 5.18). It provides the framework for the external ears and for parts of the larynx.

Fibrocartilage, a very tough tissue, has many collagen fibers (fig. 5.19). It is a shock absorber for structures that are subjected to pressure. For example, fibrocartilage forms pads (intervertebral discs) between the individual bones (vertebrae) of the spinal column. It also cushions bones in the knees and in the pelvic girdle.

OF INTEREST Between ages thirty and seventy, a person's nose may lengthen and widen by as much as half an inch, and the ears may lengthen by a quarter inch, because the cartilage in these areas continues to grow as we age.

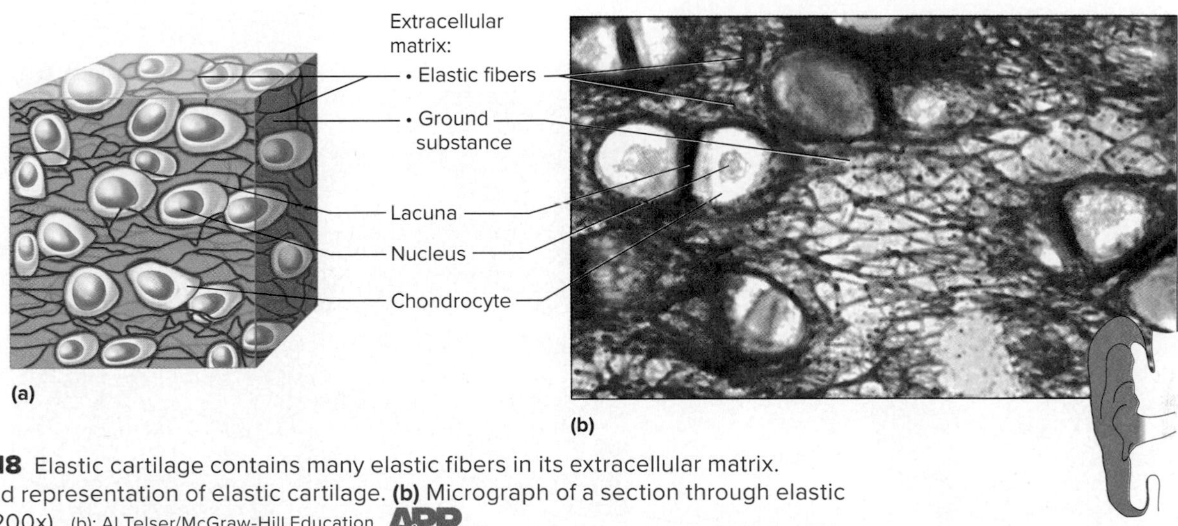

Figure 5.18 Elastic cartilage contains many elastic fibers in its extracellular matrix. **(a)** Idealized representation of elastic cartilage. **(b)** Micrograph of a section through elastic cartilage (200x). (b): Al Telser/McGraw-Hill Education APR

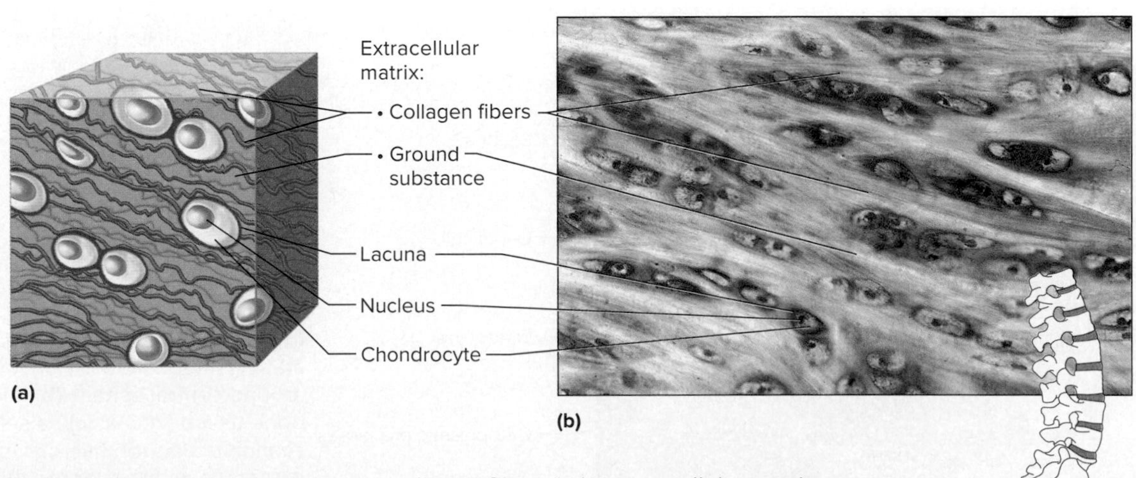

Figure 5.19 Fibrocartilage contains many large collagen fibers in its extracellular matrix. **(a)** Idealized representation of fibrocartilage. **(b)** Micrograph of a section through fibrocartilage (100x). (b): Victor P. Eroschenko APR

Bone

Bone (osseous tissue) is the most rigid connective tissue. Its hardness is largely due to mineral salts, such as calcium phosphate and calcium carbonate, between cells. This extracellular matrix also has many collagen fibers, which are flexible and reinforce the mineral components of bone.

Bone internally supports body structures. It protects vital parts in the cranial and thoracic cavities, and is an attachment for muscles. Bone also contains red marrow, which forms blood cells. It stores and releases inorganic chemicals such as calcium and phosphorus.

There are two types of bone tissue, compact and spongy bone, which will be discussed in further detail in section 7.2, Bone Structure. In compact bone, bony matrix is deposited in thin layers called *lamellae,* which form concentric patterns around tiny longitudinal tubes called *central canals,* or Haversian canals, which contain blood vessels (fig. 5.20). Bone cells, called **osteocytes** (os'tē-o-sītz), are located in lacunae, which are evenly spaced within the lamellae.

In compact bone, the osteocytes and layers of extracellular matrix, which are concentrically clustered around a central canal, form a cylinder-shaped unit called an **osteon** (os'te-on), or Haversian system. Many osteons cemented together make up the solid-appearing compact bone.

Central canals contain blood vessels and nerves. All bone cells have cytoplasmic processes that pass through very small tubes in the extracellular matrix called *canaliculi.* A network of canaliculi connects all bone cells. This allows for the rapid transport of nutrients and waste materials between blood vessels and bone cells. Thus, in spite of its hard, inert appearance, bone is a very metabolically active tissue that heals much more rapidly than does injured cartilage.

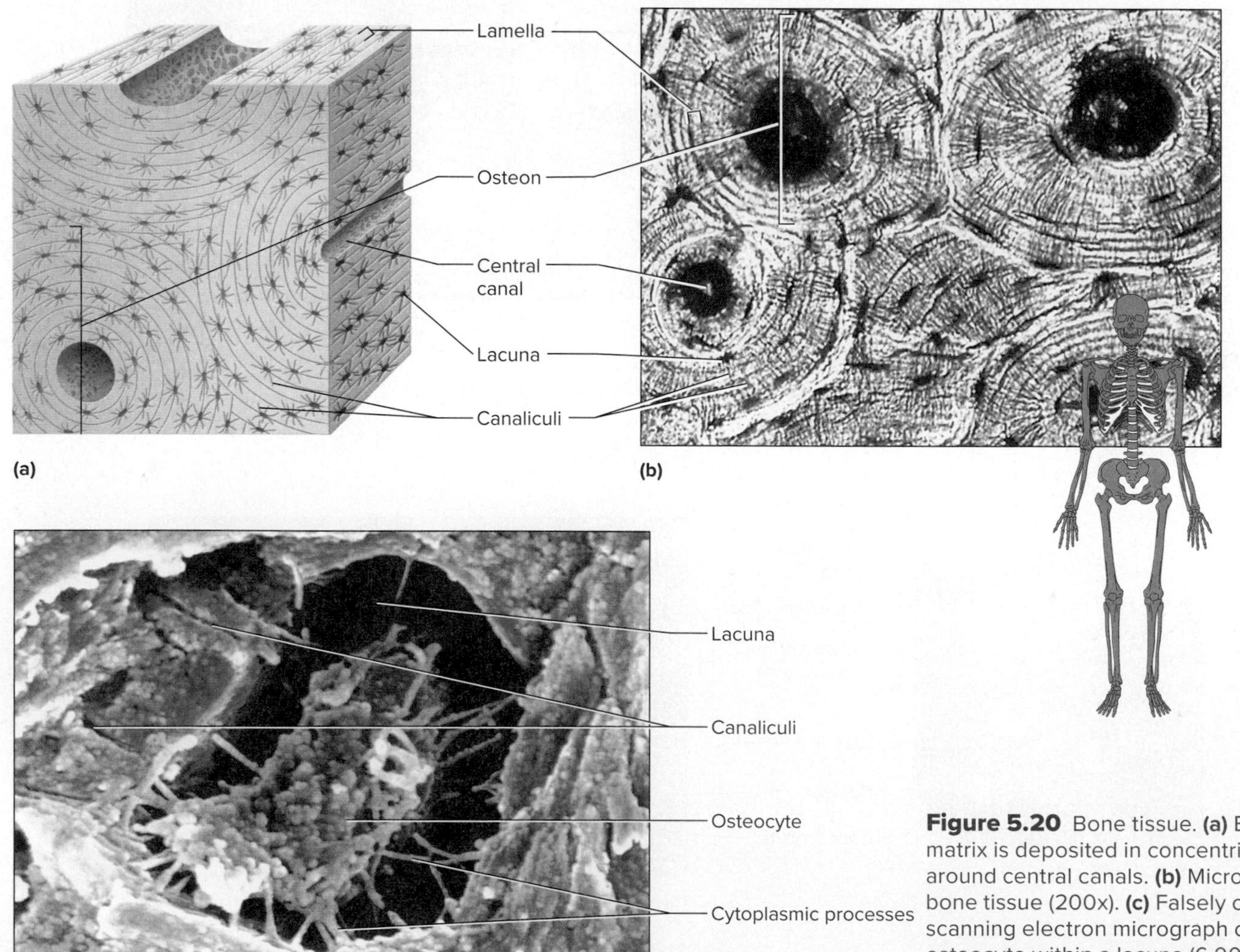

Figure 5.20 Bone tissue. **(a)** Bone matrix is deposited in concentric layers around central canals. **(b)** Micrograph of bone tissue (200x). **(c)** Falsely colored scanning electron micrograph of an osteocyte within a lacuna (6,000x).

(b): Dennis Strete/McGraw-Hill Education; (c): Prof. P. Motta/Univ. "La Sapienza"/Science Source APR

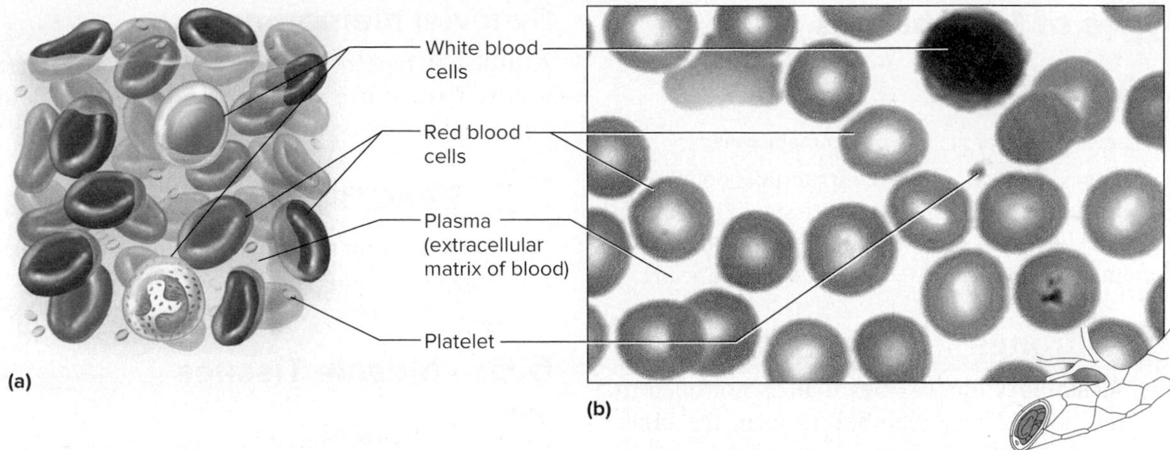

Figure 5.21 Blood tissue consists of red blood cells, white blood cells, and platelets suspended in plasma. **(a)** Idealized representation of a sample of blood. **(b)** Micrograph of a sample of blood (1,000x). (b): Al Telser/McGraw-Hill Education APR

PRACTICE FIGURE 5.21

What is the consistency of the extracellular matrix of blood?

Answer can be found in Appendix E.

Blood

Blood transports a variety of materials between interior body cells and those that exchange substances with the external environment. In this way, blood helps maintain stable internal environmental conditions.

Blood is composed of *formed elements* suspended in a fluid extracellular matrix called *blood plasma.* The formed elements include *red blood cells, white blood cells,* and cell fragments called *platelets* (fig. 5.21). Most blood cells form in red marrow within the hollow parts of certain long bones. Chapter 12 describes blood in detail. Table 5.5 lists the characteristics of the connective tissues.

PRACTICE 5.3

5. Describe the general characteristics of cartilage.
6. Explain why injured bone heals more rapidly than injured cartilage.
7. What are the major components of blood?

TABLE 5.5 Connective Tissues

Type	Function	Location
Loose connective tissue		
Areolar tissue	Binds organs	Beneath skin, between muscles, beneath epithelial tissues
Adipose tissue	Protects, insulates, stores fat	Beneath skin, around kidneys, behind eyeballs, on surface of heart
Reticular connective tissue	Supports	In walls of liver and spleen
Dense connective tissue	Binds body parts	Tendons, ligaments, deep layer of skin
Specialized connective tissue		
Hyaline cartilage	Supports, protects, provides framework	Ends of bones, nose, rings in the walls of respiratory passages
Elastic cartilage	Supports, protects, provides flexible framework	Framework of external ear and parts of larynx
Fibrocartilage	Supports, protects, absorbs shock	Between bony parts of spinal column, parts of pelvic girdle and knee
Bone	Supports, protects, provides framework	Bones of skeleton
Blood	Transports substances, helps maintain stable internal environment	Throughout body within a closed system of blood vessels and heart chambers

5.4 | Types of Membranes

LEARN

1. Distinguish among the four major types of membranes.

Epithelial membranes are thin, sheetlike structures composed of epithelium and underlying connective tissue. They cover body surfaces and line body cavities. The three major types of epithelial membranes are *serous, mucous,* and *cutaneous.*

Serous Membranes

Serous (se′rus) **membranes** line cavities that do not open to the outside of the body. These membranes form the inner linings of the thorax (parietal pleura) and abdomen (parietal peritoneum), and they cover the organs in these cavities (visceral pleura and visceral peritoneum, respectively, as shown in figs. 1.10 and 1.11). A serous membrane consists of a layer of simple squamous epithelium and a thin layer of areolar connective tissue. The cells of a serous membrane secrete watery *serous fluid,* which helps lubricate membrane surfaces.

Mucous Membranes

Mucous (mu′cus) **membranes** line cavities and tubes that open to the outside of the body, including the oral and nasal cavities and the tubes of the digestive, respiratory, urinary, and reproductive systems. A mucous membrane consists of epithelium overlying a layer of areolar connective tissue. Goblet cells within a mucous membrane secrete *mucus.*

The Cutaneous Membrane

Another epithelial membrane is the **cutaneous** (ku-ta′ne-us) **membrane,** more commonly called *skin.* It is described in detail in chapter 6.

Synovial Membrane

A different type of membrane, composed entirely of connective tissues, is a **synovial** (si-nō′vē-al) **membrane.** It lines joints and is discussed further in section 7.13, Joints.

PRACTICE 5.4

1. Name the four types of membranes, and explain how they differ.

5.5 | Muscle Tissues

LEARN

1. Distinguish among the three types of muscle tissues.

General Characteristics and Categories

Muscle (mus′el) **tissues** are able to generate forces. Their elongated cells, also called *muscle fibers,* can shorten (contract) and thicken. When stimulated by the nervous system, muscle fibers pull at their attached ends, which moves body parts.

Approximately 40% of the body, by weight, is skeletal muscle, and almost another 10% is smooth muscle and cardiac muscle combined. The three types of muscle tissue—skeletal, smooth, and cardiac—are introduced here and discussed in more detail in chapter 8.

Skeletal Muscle Tissue

Skeletal muscle tissue forms muscles that typically attach to bones and can be controlled by conscious effort. For this reason, it is often called *voluntary* muscle tissue. The long, threadlike cells of skeletal muscle have alternating light and dark cross-markings called *striations.* Each cell has many nuclei (fig. 5.22).

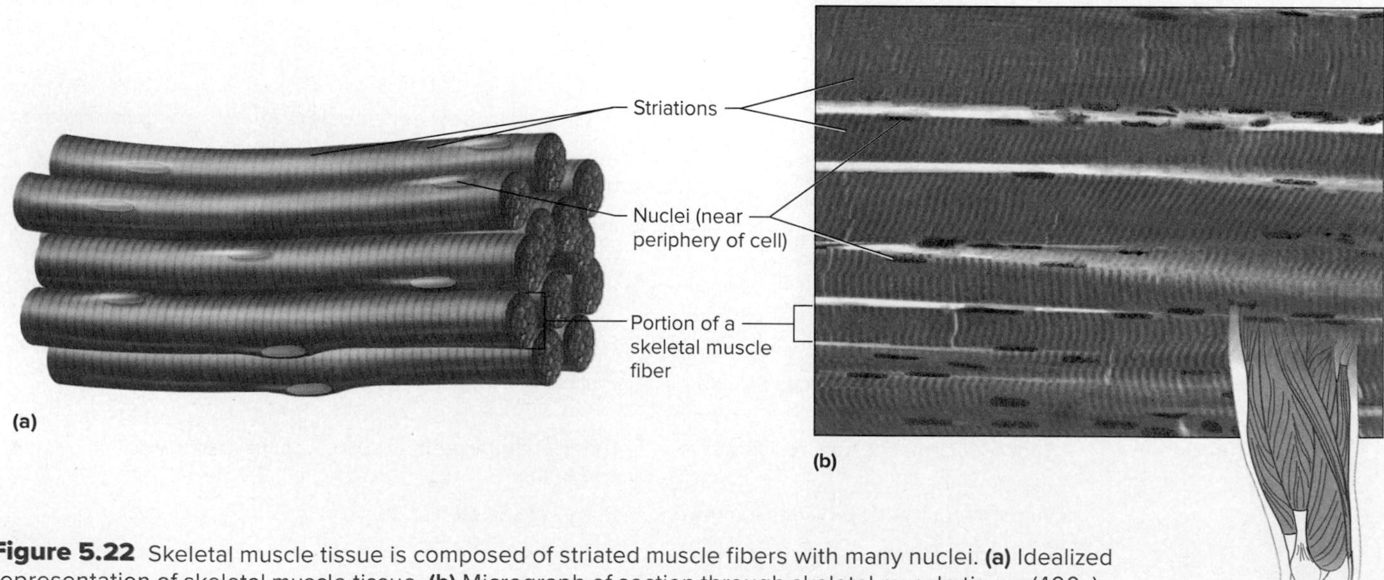

Figure 5.22 Skeletal muscle tissue is composed of striated muscle fibers with many nuclei. **(a)** Idealized representation of skeletal muscle tissue. **(b)** Micrograph of section through skeletal muscle tissue (400x).
(b): Al Telser/McGraw-Hill Education APR

The muscles built of skeletal muscle tissue move the head, trunk, and limbs. They enable us to make facial expressions, write, talk, sing, chew, swallow, and breathe.

Smooth Muscle Tissue

Smooth muscle tissue is so called because its cells do not have striations. Smooth muscle cells are shorter than skeletal muscle cells and are spindle-shaped, each with a single, centrally located nucleus (fig. 5.23). This tissue composes the walls of hollow internal organs, such as the stomach, intestines, urinary bladder, uterus, and blood vessels. Unlike skeletal muscle, smooth muscle cannot be stimulated to contract by conscious effort. Thus, its actions are *involuntary.* For example, smooth muscle tissue moves food through the digestive tract, constricts blood vessels, and empties the urinary bladder.

Cardiac Muscle Tissue

Cardiac muscle tissue is found only in the heart. Its cells, which are striated and branched, are joined end to end, forming complex networks. Each cardiac muscle cell has a single nucleus (fig. 5.24). Where one cell touches another cell is a specialized intercellular junction called an **intercalated disc,** discussed further in section 8.6, Cardiac Muscle.

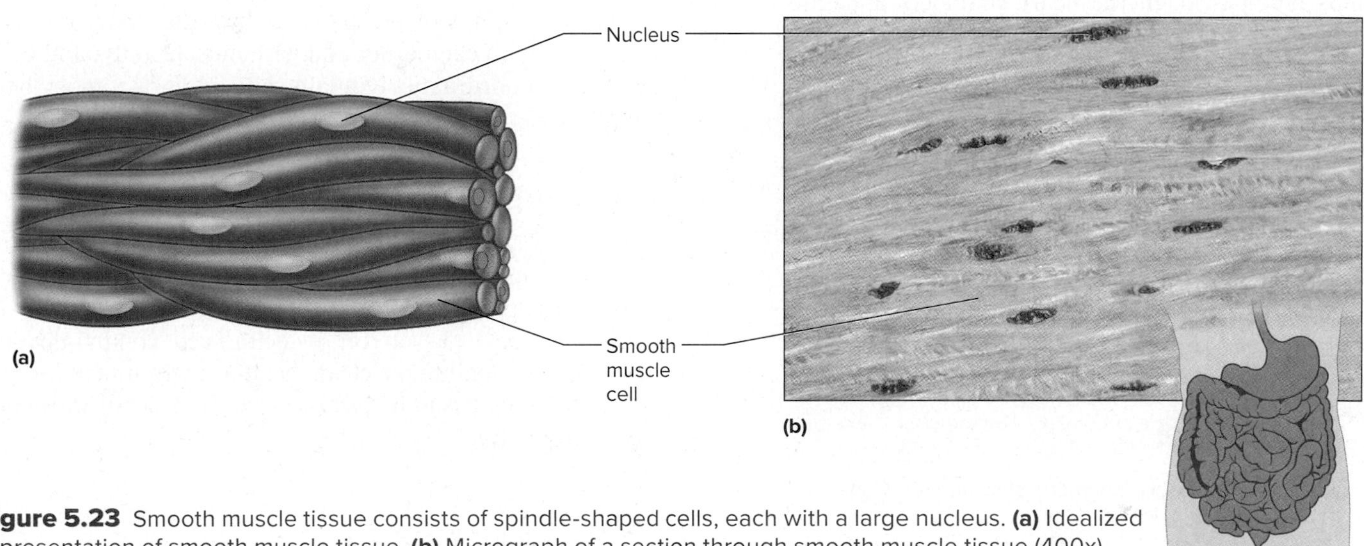

Figure 5.23 Smooth muscle tissue consists of spindle-shaped cells, each with a large nucleus. **(a)** Idealized representation of smooth muscle tissue. **(b)** Micrograph of a section through smooth muscle tissue (400x). Note that the plane of section does not always reveal the nucleus. (b): Al Telser/McGraw-Hill Education APR

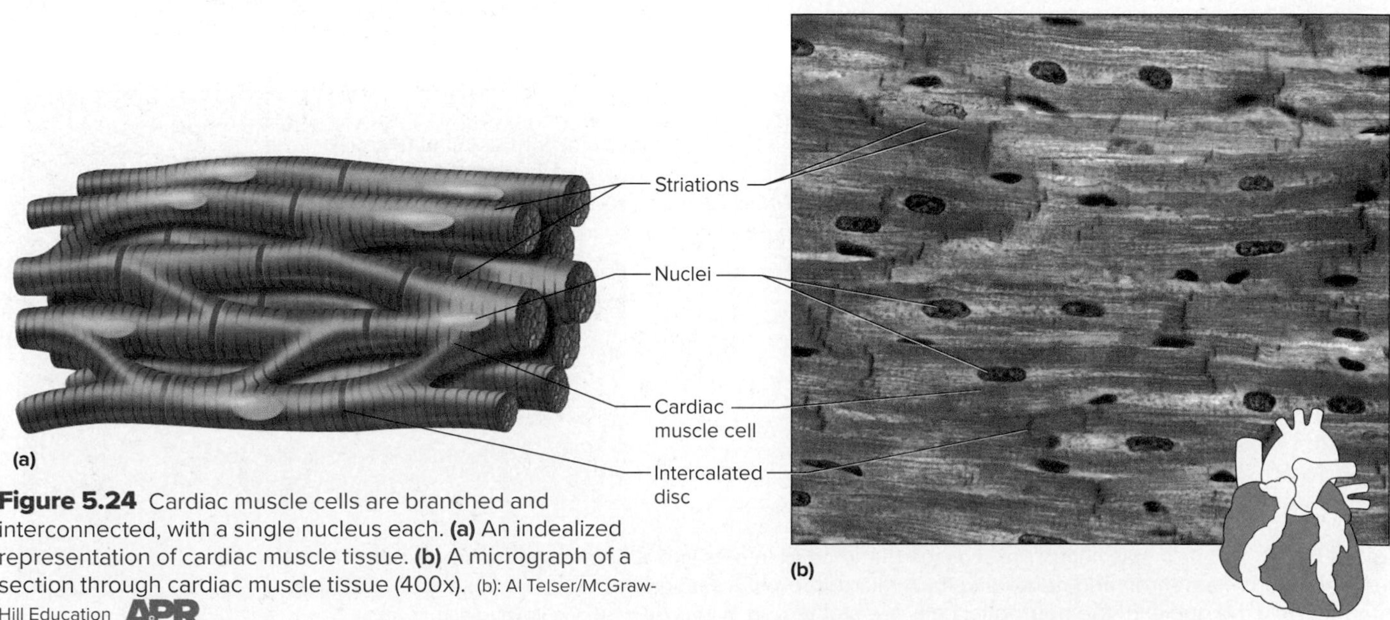

Figure 5.24 Cardiac muscle cells are branched and interconnected, with a single nucleus each. **(a)** An indealized representation of cardiac muscle tissue. **(b)** A micrograph of a section through cardiac muscle tissue (400x). (b): Al Telser/McGraw-Hill Education APR

TABLE 5.6 Muscle Tissues

Type	Function	Location
Skeletal muscle tissue (striated)	Voluntary movements of skeletal parts	Muscles usually attached to bones
Smooth muscle tissue (lacks striations)	Involuntary movements of internal organs	Walls of hollow internal organs
Cardiac muscle tissue (striated)	Contraction for heartbeats	Heart muscle

Cardiac muscle, like smooth muscle, is controlled involuntarily. This tissue makes up the bulk of the heart and pumps blood through the heart chambers and into blood vessels. Table 5.6 summarizes the general characteristics of muscle tissues.

PRACTICE 5.5

1. List the general characteristics of muscle tissues.
2. Distinguish among skeletal, smooth, and cardiac muscle tissues.

5.6 | Nervous Tissue

LEARN

1. Describe the general characteristics and functions of nervous tissue.

Nervous (ner′vus) **tissue** is found in the brain, spinal cord, and nerves. The cells responsible for the actions of the nervous system are called **neurons** (nu′ronz), or nerve cells (fig. 5.25). In addition to neurons, nervous tissue includes **neuroglia** (nu-ro′glē-ah), or "nerve glue," shown in figure 5.25. They lend support to, and are crucial to, the functioning of neurons.

Neurons

Neurons are different than all other cells because they can receive and send information, or stimuli. A neuron has a unique structure, with three parts, that allows for its specialized function (fig. 5.25). The cell body, or soma, looks like a generalized cell that we studied in chapter 3. However, it has two types of processes that look like wires hanging off of it. On the side of a neuron there are usually short but numerous processes called dendrites. These are the structures that will receive stimuli. The longer "wire" is called an axon and functions to send signals to other cells. As a result of the patterns by which neurons communicate with each other and with muscle and gland cells, they can coordinate, regulate, and integrate most body functions.

Neuroglial Cells

Neuroglia support and bind the components of nervous tissue, carry on phagocytosis, and help supply growth factors and nutrients to neurons by connecting them to blood vessels. They also play a role in cell-to-cell communication. Overall, they maintain a clean, healthy environment for neurons. Nervous tissue is discussed in more detail in section 9.1, Introduction.

PRACTICE 5.6

1. Describe the general characteristics of nervous tissues.
2. Distinguish between neurons and neuroglia.

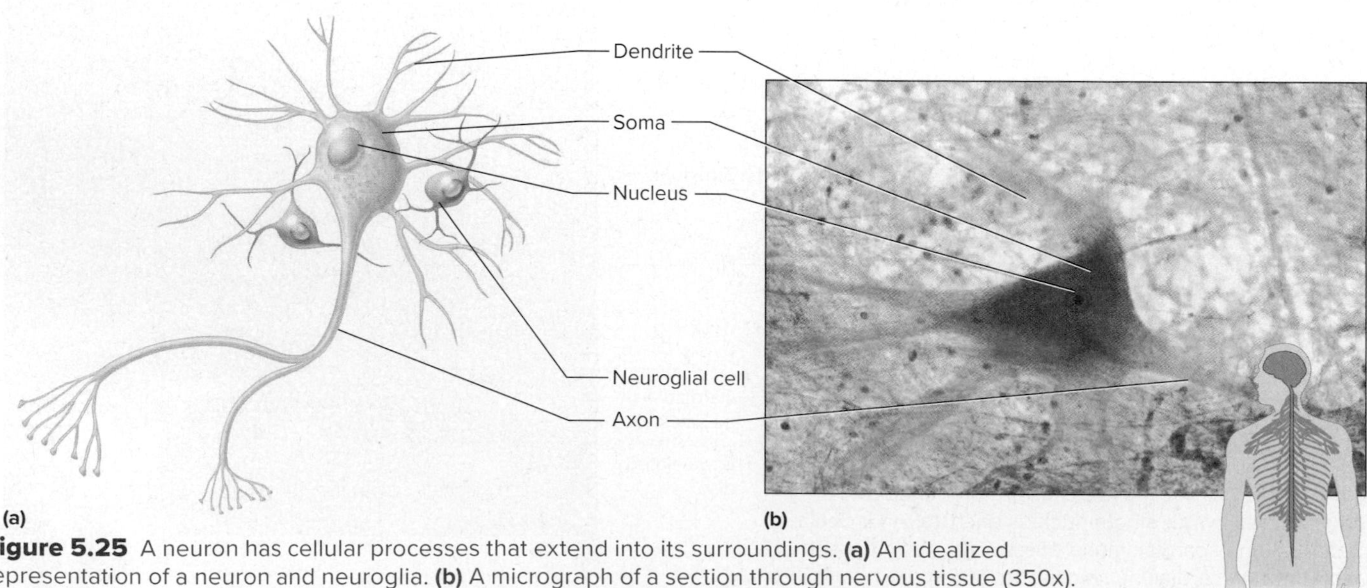

Figure 5.25 A neuron has cellular processes that extend into its surroundings. **(a)** An idealized representation of a neuron and neuroglia. **(b)** A micrograph of a section through nervous tissue (350x). Note that only the nuclei of the neuroglial cells are stained. (b): Alvin Telser/McGraw-Hill Education

ASSESS

CHAPTER ASSESSMENTS

5.1 Introduction

1. Which of the following is a major tissue type in the body?
   a. epithelial
   b. connective
   c. muscle
   d. all of the above
2. Which is the type of tissue that functions in secretion and absorption?
   a. epithelial
   b. connective
   c. muscle
   d. nervous

5.2 Epithelial Tissues

3. A general characteristic of epithelial tissues is that ________.
   a. numerous blood vessels are present
   b. cells are spaced apart
   c. cells readily divide
   d. there is much extracellular matrix between cells
4. Explain how the structure of epithelial tissues provides for the functions of epithelial tissues.
5. Match the epithelial tissue to an organ in which the tissue is found.

(1) simple squamous epithelium	A. lining of intestines
(2) simple cuboidal epithelium	B. lining of ducts of mammary glands
(3) simple columnar epithelium	C. lining of urinary bladder
(4) pseudostratified columnar epithelium	D. salivary glands
(5) stratified squamous epithelium	E. air sacs of lungs
(6) stratified cuboidal epithelium	F. respiratory passages
(7) stratified columnar epithelium	G. parts of male urethra
(8) transitional epithelium	H. lining of kidney tubules
(9) glandular epithelium	I. superficial layer of skin

6. Distinguish between exocrine glands and endocrine glands.
7. A gland that secretes substances out of cells by exocytosis is a(n) ________.
   a. merocrine gland
   b. apocrine gland
   c. holocrine gland
   d. mammary gland

5.3 Connective Tissues

8. Define *extracellular matrix.*
9. Describe three major types of connective tissue cells.
10. Distinguish between collagen and elastin.
11. Compare and contrast the different types of loose connective tissue.
12. Describe dense connective tissue.
13. Explain why injured dense connective tissue and cartilage are usually slow to heal.
14. Name the types of cartilages and describe their differences and similarities.
15. Describe how bone cells are organized in bone tissue.
16. The fluid extracellular matrix of blood is called ________.
    a. mucus
    b. serous fluid
    c. synovial fluid
    d. plasma

5.4 Types of Membranes

17. Identify the locations of four types of membranes in the body and indicate the types of tissues making up each membrane.

5.5 Muscle Tissues

18. Compare and contrast skeletal, smooth, and cardiac muscle tissues in terms of location, cell appearance, and control.

5.6 Nervous Tissue

19. Distinguish between the functions of neurons and neuroglia.

ASSESS

INTEGRATIVE ASSESSMENTS/CRITICAL THINKING

Outcomes 3.2, 3.4, 5.1, 5.2, 5.3, 5.5, 5.6

1. Tissue engineering combines living cells with synthetic materials to create functional substitutes for human tissues. What characteristics in the materials, structure, or actual components would you consider using to engineer replacement (a) skin, (b) bone, (c) muscle, and (d) blood?

2. Cancer-causing agents (carcinogens) usually act on dividing cells. Which of the four major types of tissues would carcinogens most influence, and why?

Outcomes 5.2, 5.4

3. Mucous cells may secrete excess mucus in response to irritants. What symptoms might this produce in the (a) digestive tract or (b) respiratory tract?

Outcomes 3.2, 3.3, 5.3

4. Collagen and elastin are added to many beauty products. What major type of tissues are they normally part of? Topically applied to the skin, are these proteins absorbed? Explain.

Outcome 5.4

5. Joints such as the elbow, shoulder, and knee contain considerable amounts of cartilage and dense connective tissue. Given the composition of these tissues, should a person who suffers a joint injury expect a quick recovery? Explain.

6. Disorders of collagen are characterized by deterioration of connective tissues. Why would such diseases produce widely varying symptoms?

Outcome 5.5

7. Identify the mistake made in each of these statements:

"I do crunches when I work out in order to strengthen my stomach."

"Because blood vessels are an extension of the heart, their walls contain cardiac muscle."

"When the doctor told me to push to deliver my son, I made my uterus contract."

Chapter Summary

5.1 Introduction

***Tissues** are groups of cells with specialized structural and functional roles. The four major types of human tissue are epithelial, connective, muscle, and nervous.*

5.2 Epithelial Tissues

1. General characteristics and categories
   a. **Epithelial tissue** covers organs, lines cavities and hollow organs, and is the major tissue of glands.
   b. Epithelium is anchored to connective tissue by a **basement membrane,** lacks blood vessels, consists of tightly packed cells, and is replaced continuously.
   c. Epithelial tissue protects, secretes, absorbs, and excretes.
   d. Epithelial tissues are classified according to cell shape and number of cell layers.
2. **Simple squamous epithelium**
   a. This tissue consists of a single layer of thin, flattened cells.
   b. It functions in gas exchange in the lungs, lines blood and lymph vessels, and is part of the membranes lining body cavities and covering viscera.
3. **Simple cuboidal epithelium**
   a. This tissue consists of a single layer of cube-shaped cells.
   b. It carries on secretion and absorption in the kidneys and various glands.
4. **Simple columnar epithelium**
   a. This tissue is composed of elongated cells whose nuclei are near the basement membrane.
   b. It lines the uterus and digestive tract.
   c. Many absorbing cells have microvilli.
   d. This tissue has goblet cells that secrete mucus.
5. **Pseudostratified columnar epithelium**
   a. Nuclei located at two or more levels give this tissue a stratified appearance.
   b. Cilia that are part of this tissue move mucus over the surface of the tissue.
   c. It lines respiratory passages.
6. **Stratified squamous epithelium**
   a. This tissue is composed of many layers of cells.
   b. It protects underlying cells.
   c. It forms the superficial layer of the skin and lines the oral cavity, esophagus, vagina, and anal canal.
7. **Stratified cuboidal epithelium**
   a. This tissue is composed of two or three layers of cube-shaped cells.
   b. It lines the ducts of the mammary glands, sweat glands, salivary glands, and pancreas.
   c. It protects.
8. **Stratified columnar epithelium**
   a. The top layer of cells in this tissue are column-shaped. Cube-shaped cells make up the bottom layers.
   b. It is in part of the male urethra and the lining of the larger ducts of exocrine glands.
   c. This tissue protects and secretes.
9. **Transitional epithelium**
   a. This tissue is specialized to stretch.
   b. It lines the urinary bladder, ureters, and superior urethra.
10. **Glands**
    a. Glands are composed of cells that are specialized to secrete substances.
    b. A gland consists of one or more cells.
       (1) **Exocrine glands** secrete into ducts.
       (2) **Endocrine glands** secrete into tissue fluid or blood.
    c. Exocrine glands are classified according to the composition of their secretions.
       (1) **Merocrine glands** secrete fluid without loss of cytoplasm.
          (a) Serous cells secrete a watery fluid.
          (b) Mucous cells secrete mucus.
       (2) **Apocrine glands** lose portions of their cells during secretion.
       (3) **Holocrine glands** release cells filled with secretory products.

5.3 Connective Tissues

1. General characteristics and categories
   a. **Connective tissue** connects, supports, protects, provides frameworks, fills spaces, stores fat, produces blood cells, protects against infection, and helps repair damaged tissues.
   b. Connective tissue cells usually have considerable **extracellular matrix** between them.
   c. This extracellular matrix consists of fibers and ground substance.
   d. Major cell types
      (1) **Fibroblasts** produce collagen and elastic fibers.
      (2) **Macrophages** are phagocytes.
      (3) **Mast cells** may release heparin and histamine, and usually are near blood vessels.
   e. Connective tissue fibers
      (1) **Collagen fibers** have great tensile strength.
      (2) **Elastic fibers** are composed of elastin and are very elastic.
      (3) **Reticular fibers** are thin collagen fibers.
2. Categories of connective tissue
   Connective tissue proper includes loose connective tissue and dense connective tissue. Specialized connective tissue includes cartilage, bone, and blood.
   a. **Loose connective tissue**
      (1) **Areolar tissue** forms thin membranes between organs and binds them. It is beneath the skin and between muscles.

(2) **Adipose tissue** stores fat, cushions, and insulates. It is found beneath the skin, in certain abdominal membranes, behind the eyeballs, and around the kidneys, heart, and various joints.

(3) **Reticular connective tissue** is composed of thin, collagen fibers. It helps provide the framework of the liver and spleen.

b. **Dense connective tissue**

(1) This tissue is largely composed of strong, collagen fibers.

(2) It is found in the tendons, ligaments, white portions of the eyes, and the deep skin layer.

c. **Cartilage**

(1) Cartilage provides a supportive framework for various structures.

(2) Its extracellular matrix is composed of fibers and a gel-like ground substance.

(3) Cartilaginous structures are enclosed in a perichondrium, which contains blood vessels.

(4) Cartilage lacks a direct blood supply and is slow to heal.

(5) Major types are **hyaline cartilage, elastic cartilage,** and **fibrocartilage.**

d. **Bone**

(1) The extracellular matrix of bone contains mineral salts and collagen.

(2) The cells of compact bone are usually organized in concentric circles around central canals. Canaliculi connect the cells.

(3) Bone is an active tissue that heals rapidly.

e. **Blood**

(1) Blood transports substances and helps maintain a stable internal environment.

(2) Blood is composed of red blood cells, white blood cells, and platelets suspended in plasma.

(3) Blood cells develop in red marrow in the hollow parts of long bones.

5.4 Types of Membranes

1. Epithelial membranes are composed of epithelium and underlying connective tissue. Serous, mucous, and cutaneous membranes are epithelial membranes.
2. **Serous membranes,** composed of epithelium and areolar connective tissue, are membranes that line body cavities that do not open to the outside and cover the organs in these cavities. The cells of a serous membrane secrete serous fluid to help lubricate membrane surfaces.
3. **Mucous membranes,** composed of epithelium and areolar connective tissue, are membranes that line body cavities that open to the outside. Goblet cells within these membranes secrete mucus.
4. Another epithelial membrane, the **cutaneous membrane,** is the external body covering commonly called skin.
5. **Synovial membranes,** composed entirely of connective tissues, line joints.

5.5 Muscle Tissues

1. General characteristics and categories
   a. Muscle cells are also called muscle fibers.
   b. **Muscle tissues** contract, moving structures that are attached to them.
   c. The three types are skeletal, smooth, and cardiac muscle tissues.
2. **Skeletal muscle tissue**
   a. Muscles containing this tissue usually are attached to bones and controlled by conscious effort.
   b. Cells are long and threadlike, with many nuclei and striations.
   c. Muscle cells contract when stimulated by nerve cells, and then relax when stimulation stops.
3. **Smooth muscle tissue**
   a. This tissue is in the walls of hollow internal organs.
   b. It is involuntarily controlled.
4. **Cardiac muscle tissue**
   a. This tissue is found only in the heart.
   b. Cells are joined by intercalated discs and form branched networks.

5.6 Nervous Tissue

1. **Nervous tissue** is in the brain, spinal cord, and peripheral nerves.
2. **Neurons** (nerve cells)
   a. Neurons sense changes and respond by conducting electrical impulses to other neurons or to muscles or glands.
   b. They coordinate, regulate, and integrate body activities.
3. **Neuroglia**
   a. Some of these cells bind and support nervous tissue.
   b. Others carry on phagocytosis.
   c. Still others connect neurons to blood vessels.
   d. Some are involved in cell-to-cell communication.

6 UNIT 2 SUPPORT AND MOVEMENT

Integumentary System

Administering a tattoo. Jeffrey Coolidge/Iconica/Getty Images

Tattoos. With their dramatic rise in popularity and being more socially acceptable than ever, you might even have one yourself. If not, you probably can't walk very far on campus without seeing tattoos on other students, and maybe even on some faculty. Often referred to as body art, the canvas for these illustrations is your skin.

The outer layer of the skin, the epidermis, is what we usually see. But the ink used for tattooing resides in the deeper layer called the dermis. Initially, the ink is injected into the region between the epidermis and dermis. The damaged epidermis is shed. Trapped by white blood cells as part of an immune response, the pigments remain in the upper part of the dermis. This becomes the long-term image, or art, that is visible. The fading of a tattoo over time is the result of the pigments migrating into deeper layers of the dermis.

Tools for tattooing recovered from some archeological sites are as old as 12,000 years. This long history of tattoos reveals ancient knowledge of anatomy and the healing arts. Otzi the iceman was found preserved in 1991 in the Italian Alps. He lived about 5,300 years ago and possesses the oldest surviving tattoos, 61 in all. It is thought that most of his tattoos had medical relevance, where either the tattoos themselves served as treatment, or they indicated areas of focus for medical intervention.

Contemporary tattoos are usually decorative, and most often are a form of self-expression. They exhibit a wide range of images in various categories. Animals such as tigers, lions, and family pets are popular standards. Hearts, roses, and skulls also seem to be quite numerous. Equally prevalent is fantasy art including dragons and the grim reaper, as well as Bible verses and important personal dates. Others are known as "functional tattoos." Alzheimer patients often get them to recall important information such as their name and home address.

Tattoos have also been used to mark criminals and prisoners of war. The most notorious was the tattooing of prisoners in Nazi concentration camps during the Second World War.

The American Academy of Dermatology warns against tattoos due to the possibility of infection, scarring, and the desire to have a tattoo removed later in one's life. To minimize potential danger, it recommends employing only certified and licensed tattoo artists.

Module 4 Integumentary System

LEARNING OUTLINE

After studying this chapter, you should be able to complete the "Learning Outcomes" that follow the major headings throughout the chapter.

AIDS TO UNDERSTANDING WORDS

cut- [skin] sub*cut*aneous: beneath the skin.

derm- [skin] *derm*is: inner layer of the skin.

epi- [upon] *epi*dermis: outer layer of the skin.

follic- [small bag] hair *follic*le: tubelike depression in which a hair develops.

kerat- [horn] *kerat*in: protein produced as epidermal cells die and harden.

melan- [black] *melan*in: dark pigment produced by certain cells.

seb- [grease] *seb*aceous gland: gland that secretes an oily substance.

sudor- [sweat] *sudor*iferous glands: exocrine glands that secrete sweat.

(Appendix A has a complete list of Aids to Understanding Words.)

6.1 | Introduction

LEARN

1. Describe what constitutes an organ, and name the large organ of the integumentary system.

The skin receives a great deal of care and medical attention. This is mostly because it is what the rest of the world sees. That is, it composes a great part of our outward appearance. You only need to look at the amount of time and money we spend caring for skin, hair, and nails to understand this. Dermatology is the medical specialty that investigates and treats issues with the skin. Changes in skin can also indicate problems with other body systems.

Two or more tissue types that are structurally connected and perform shared, specialized functions constitute an **organ** (see fig. 1.3). Likewise, two or more organs working toward the same goals are considered an organ system. The two main tissues of the skin—epithelial and connective—constitute the largest organ in the body by weight. The skin and its associated accessory structures and organs (hair, nails, muscles, sensory receptors, and glands) make up the **integumentary** (in-teg-u-men'tar-e) **system.** The integument forms a barrier between ourselves and the outside world. It is a strong, yet flexible covering for our bodies, with a variety of functions.

OF INTEREST The surface area of the skin of a 150-pound person is approximately 20 square feet.

PRACTICE 6.1

Answers to the Practice questions can be found in the eBook.

1. What constitutes an organ?

CAREER CORNER
Massage Therapist

The woman feels something give way in her left knee as she lands from a jump in her dance class. She limps away between her classmates, in great pain. At home, she uses "RICE"—rest, ice, compression, elevation—then has a friend take her to an urgent care clinic, where a physician diagnoses patellar tendinitis, or "jumper's knee." Frequent jumping followed by lateral movements caused the injury.

Three days later, at her weekly appointment with a massage therapist for stress relief, the woman mentions the injury. Over the next few weeks, the massage therapist applies light pressure to the injured area to stimulate circulation, and applies friction in a transverse pattern to break up scar tissue and relax the muscles. She also massages the muscles to improve flexibility.

A massage therapist manipulates soft tissues, using combinations of stroking, kneading, compressing, and vibrating, to relieve pain and reduce stress. Training includes 300 to 1,000 hours of class time, hands-on practice, and continuing education. Specialties include pediatrics, sports injuries, and even massage techniques for racehorses.

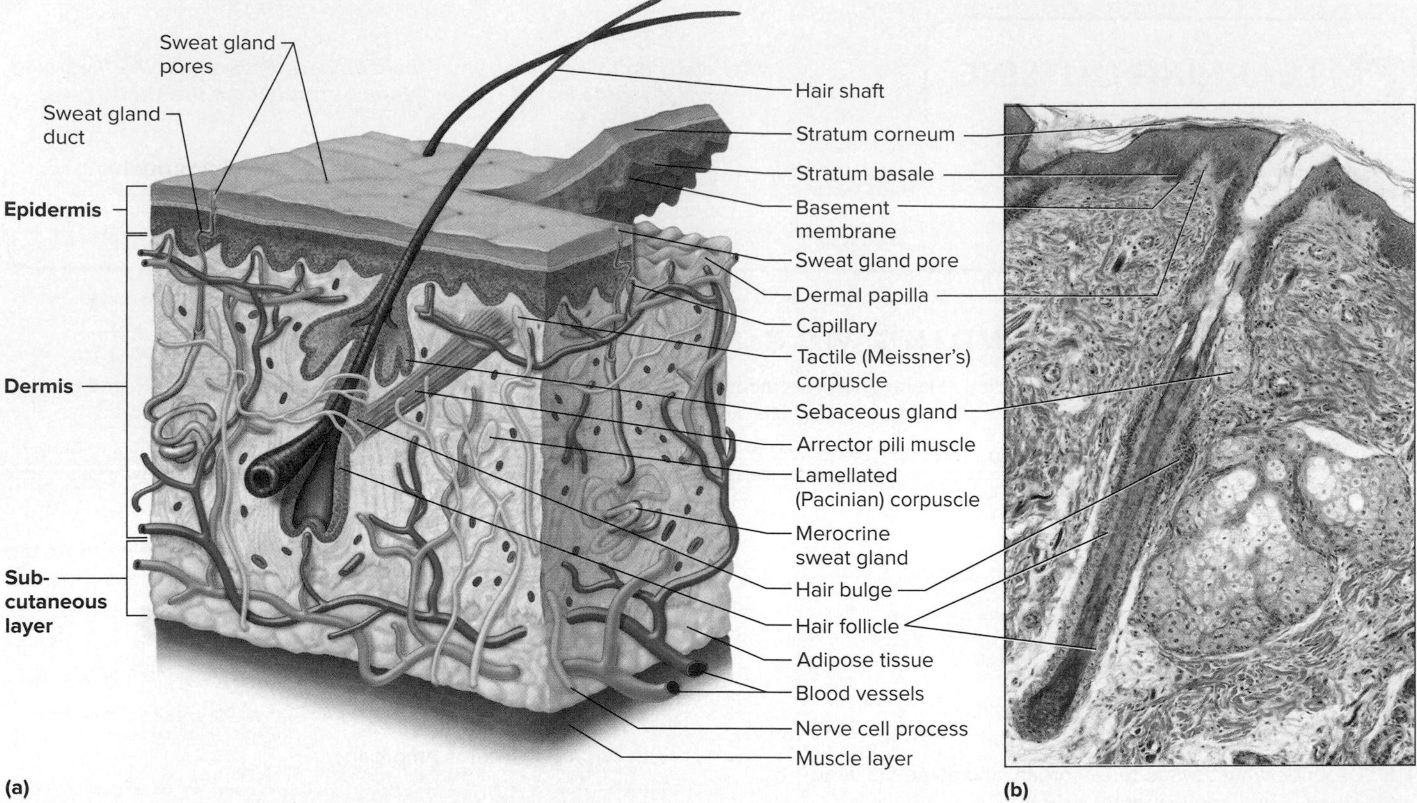

Figure 6.1 Skin. **(a)** The skin is an organ that includes two layers, the epidermis and dermis, that lie atop a subcutaneous ("beneath the skin") layer. **(b)** This light micrograph depicts the layered structure of the skin (75x). (b): Al Telser/McGraw-Hill Education APR

6.2 | Layers of the Skin

LEARN

1. Describe the structure of the layers of the skin.
2. Summarize the factors that determine skin color.

The skin includes two distinct layers (fig. 6.1). The outer layer, called the **epidermis** (ep″ĭ-der′mis), is composed of stratified squamous epithelium. The inner layer, or **dermis** (der′mis), is thicker than the epidermis. It is composed of connective tissue consisting of collagen and elastic fibers, along with smooth muscle tissue, nervous tissue, and blood. A *basement membrane* anchors the epidermis to the dermis and separates these two skin layers.

Beneath the dermis are masses of areolar tissue and adipose tissue that bind the skin to the underlying organs, forming the **subcutaneous** (sub″kū-tā′nē-us) **layer** (hypodermis). As its name indicates, this layer is beneath the skin and not a true layer of the skin. The collagen and elastic fibers of the subcutaneous layer are continuous with those of the dermis. Most of these fibers run parallel to the surface of the skin, extending in all directions. As a result, no sharp boundary separates the dermis and the subcutaneous layer. The adipose tissue of the subcutaneous layer insulates, helping to conserve body heat. The subcutaneous layer also contains the major blood vessels that supply the skin and underlying adipose tissue. Clinical Application 6.1 discusses the administration of treatments and drugs within the skin and subcutaneous layer.

PRACTICE 6.2

1. List the layers of the skin.
2. Name the tissues in the outer and inner layers of the skin.
3. Name the tissues in the subcutaneous layer beneath the skin.
4. What are the functions of the subcutaneous layer?

Epidermis

The epidermis lacks blood vessels because it is composed entirely of stratified squamous epithelium. However, the deepest layer of epidermal cells, called the **stratum basale** (stra′tum ba′sal), or stratum germinativum, is close to the dermis and is nourished by dermal blood vessels (fig. 6.1*a*). As the cells (basal cells) of this layer divide and grow, the older epidermal cells (keratinocytes) are pushed away from the dermis toward the skin surface. The farther the cells move away from the dermis toward the skin surface, the poorer their nutrient supply becomes, and in time they die.

The keratinocytes harden in a process called **keratinization** (ker″ah-tin″ĭ-zā′shun). The cytoplasm fills with strands of tough, fibrous, waterproof **keratin** proteins. As a result, many layers of tough, tightly packed dead cells accumulate in the outermost layer of the epidermis, called the **stratum corneum** (kor′nē-um). These dead cells are eventually shed.

The thickness of the epidermis varies among different body regions. In most areas, only four layers can be

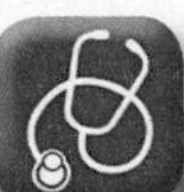

CLINICAL APPLICATION 6.1
Administering Treatments and Drugs

Various treatments temporarily smooth facial wrinkles. "Botox" is an injection of a very dilute solution of botulinum toxin. Produced by the bacterium *Clostridium botulinum,* the toxin causes food poisoning. It also blocks nerve activation of certain muscle cells, including the facial muscles that control smiling, frowning, and squinting. After three months, though, the facial nerves contact the muscles at different points, and the wrinkles return. Other anti-wrinkle treatments include chemical peels and dermabrasion to reveal new skin surface, as well as collagen injections, and transplants of subcutaneous fat from other body locations to the face.

Intradermal injections are administered into the skin. Subcutaneous injections are administered through a hollow needle into the subcutaneous layer beneath the skin. Subcutaneous injections and intramuscular injections, administered into muscles, are also called hypodermic injections.

Some drugs are introduced through the skin by means of an adhesive transdermal patch that includes a small reservoir containing the substance. The drug passes from the reservoir through a permeable membrane at a known rate. It then diffuses into the epidermis and enters the blood vessels of the dermis. Transdermal patches deliver drugs that protect against motion sickness, alleviate chest pain associated with heart disease, and lower blood pressure. A transdermal patch that delivers nicotine may be used to help people stop smoking.

distinguished (from deepest layer to superficial layer): the *stratum basale, stratum spinosum* (spi-no'sum), *stratum granulosum* (gran"u-lo'sum), and *stratum corneum.* An additional layer, the *stratum lucidum* (loo'sid-um), is in the thickened and hairless (glabrous) skin of the palms and soles (fig. 6.2).

Healthy skin does not completely wear away, because the production of epidermal cells in the stratum basale closely balances the loss of dead cells from the stratum corneum. However, the rate of cell division increases where the skin is rubbed or pressed regularly. This action causes growth of thickened areas called *calluses* on the palms and soles, and keratinized conical masses on the toes called *corns.*

The epidermis has important protective functions. It shields the moist underlying tissues against excess water loss, mechanical injury, and the effects of harmful chemicals. Intact epidermis also keeps out disease-causing microorganisms (pathogens).

Specialized cells in the epidermis called **melanocytes** produce **melanin** (mel'ah-nin), a pigment that provides skin

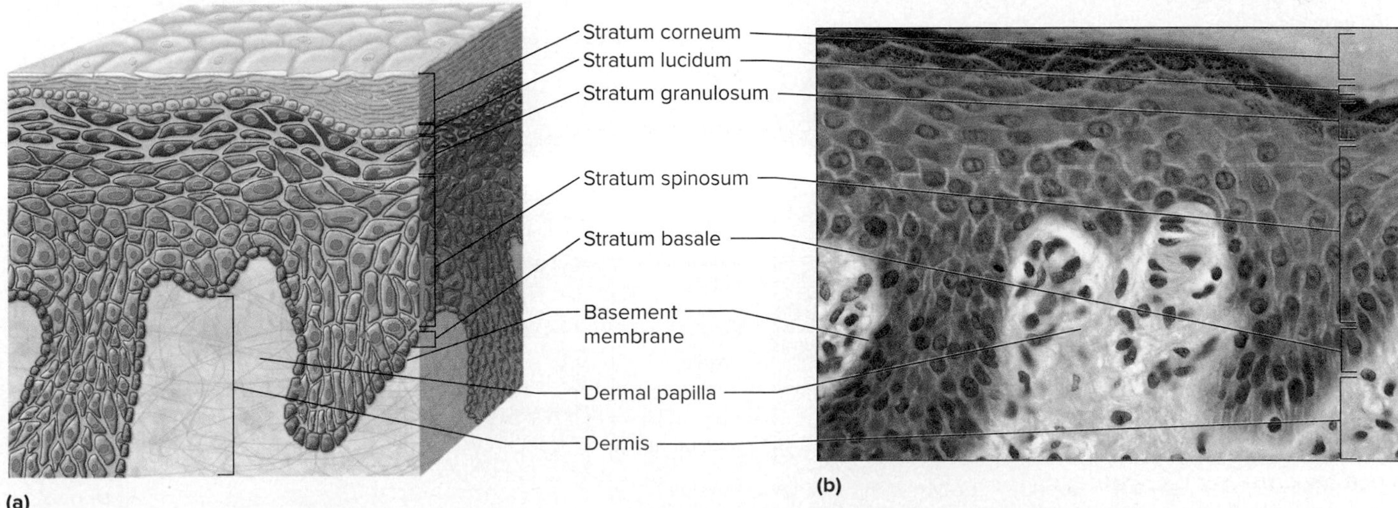

Figure 6.2 Epidermis of thick skin. **(a)** The layers of the epidermis are distinguished by changes in cells as they are pushed toward the surface of the skin. **(b)** Light micrograph of skin (500x). (b): Al Telser/McGraw-Hill Education APR

PRACTICE FIGURE 6.2

Where is thick skin found on the body?

Answer can be found in Appendix E.

color, in organelles called *melanosomes* (fig. 6.3*a*). The more melanin, the darker the skin. Melanin also absorbs ultraviolet radiation in sunlight. Without the protection of melanin, ultraviolet radiation could cause mutations in the DNA of skin cells and other damaging effects. Melanocytes lie in the deepest portion of the epidermis. They are the only cells that can produce melanin, but the pigment may also appear in nearby epidermal cells. This spreading of the pigment is possible because melanocytes have long, pigment-containing cellular extensions that pass upward between neighboring epidermal cells (fig. 6.3*b*). These extensions transfer melanin granules into neighboring cells by a process called **cytocrine secretion,** darkening the cells. Neighboring epidermal cells may contain more melanin than the melanocytes (fig. 6.3*b*). Clinical Application 6.2 discusses skin cancer arising from melanocytes and other epidermal cells.

Skin color is due largely to melanin, primarily the brownish-black **eumelanin** (ū-mel′ah-nin). The reddish-yellow **pheomelanin** (fē″ō-mel′ah-nin) is found in certain locations, such as the lips. All people have about the same number of melanocytes in their skin. Differences in skin color result from differences in the amount of melanin that melanocytes produce and in the distribution and size of the pigment granules. Skin color is mostly genetically determined. If genes instruct melanocytes to produce abundant melanin, the skin is dark.

OF **INTEREST** More than a hundred genes affect pigmentation of the skin, hair, and irises of the eyes.

Environmental and physiological factors also influence skin color, in addition to the effects of genes. Sunlight, ultraviolet light from sunlamps, and X rays darken existing melanin granules and stimulate production of more melanin. Blood in the dermal vessels may affect skin color as physiological changes occur. When blood is oxygen rich, the blood pigment hemoglobin is bright red, making the skin of light-complexioned people appear pinkish. When the blood oxygen concentration is low, hemoglobin is dark red, and the skin appears bluish—a condition called *cyanosis.*

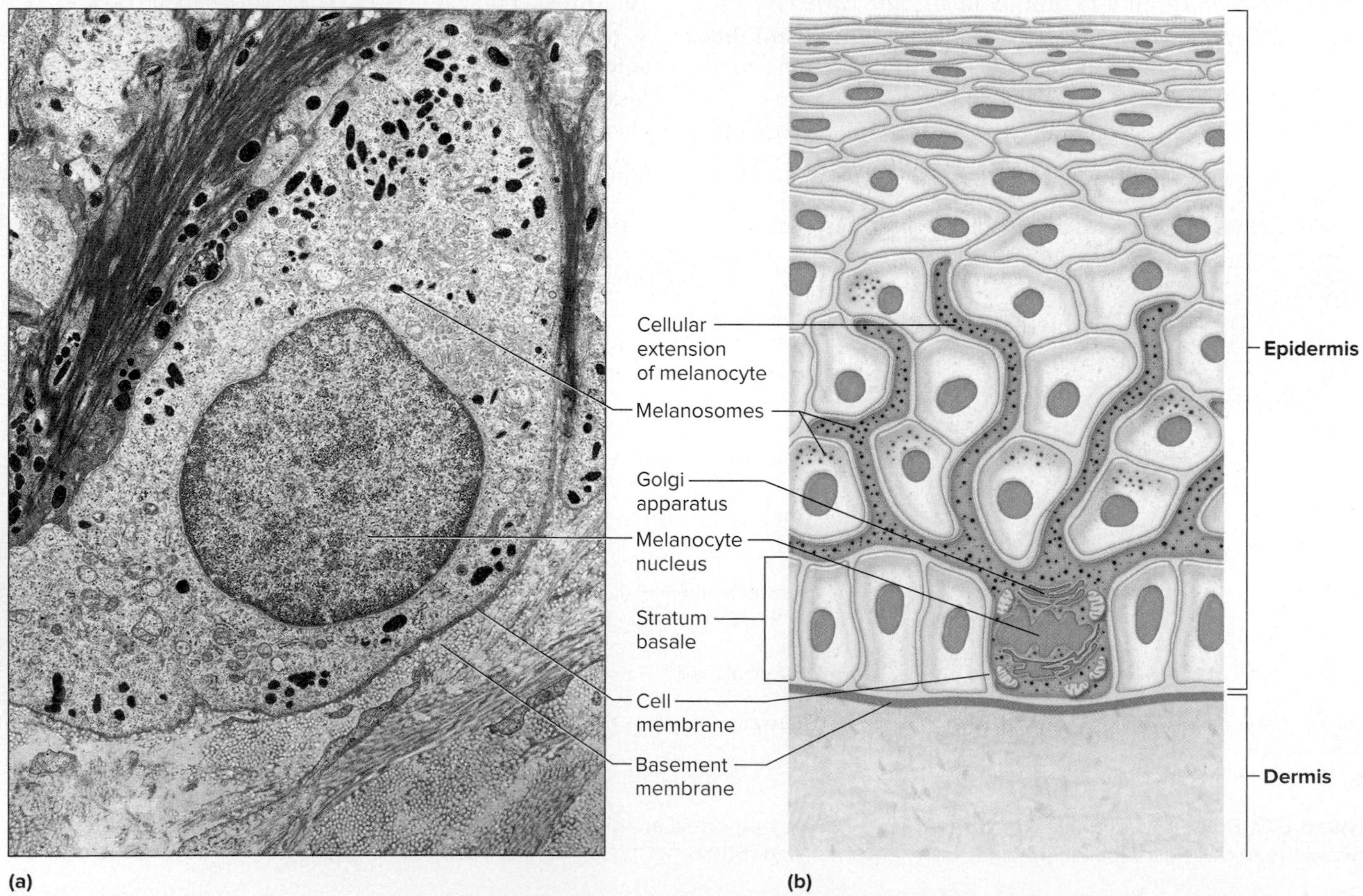

Figure 6.3 Melanocytes produce melanin. **(a)** Transmission electron micrograph of a melanocyte with pigment-containing melanosomes (4,500x). **(b)** A melanocyte may have pigment-containing extensions that pass between epidermal cells and transfer melanin into them. Note that much of the melanin is deposited above the nucleus, where the pigment can absorb UV radiation from outside before the DNA is damaged. (a): Don W. Fawcett/Science Source

CLINICAL APPLICATION 6.2
Skin Cancer

Skin cancer arises in nonpigmented epithelial cells in the deep layer of the epidermis, or from melanocytes. Skin cancers originating from epithelial cells are called *cutaneous carcinomas* (squamous cell carcinoma and basal cell carcinoma); those arising from melanocytes are *cutaneous melanomas* (melanocarcinomas or malignant melanomas) (fig. 6A). Statistically, one in five people in the United States will develop skin cancer at some point.

Cutaneous carcinomas are the most common type of skin cancer, affecting mostly light-skinned people over forty years of age regularly exposed to sunlight. Cutaneous carcinomas typically develop from hard, dry, scaly growths that have reddish bases. They may be flat or raised and usually firmly adhere to the skin. They are most common on the neck, face, or scalp. Cutaneous carcinomas grow slowly and are usually cured with surgical removal or radiation treatment.

Cutaneous melanomas are pigmented with melanin, often with a variety of colored areas, such as variegated brown, black, gray, or blue. Melanomas usually have irregular rather than smooth outlines, and may feel bumpy. The "ABCDE" rule provides a checklist for melanoma: A for asymmetry; B for border (irregular); C for color (more than one); D for diameter (more than 6 millimeters); and E for evolution or change.

People of any age may develop cutaneous melanomas. These cancers are caused by short, intermittent exposure to high-intensity sunlight, such as when a person who usually stays indoors occasionally sustains a blistering sunburn.

A cutaneous melanoma usually appears in the skin on the back or limbs, arising from normal-appearing skin or from a mole (nevus). The lesion spreads horizontally through the skin, but eventually may thicken and grow downward, invading deeper tissues. Surgical removal during the horizontal growth phase arrests the cancer in six of every seven cases. However, once the lesion thickens and deepens, it becomes difficult to treat, and the survival rate is low.

Both UVA and UVB types of ultraviolet radiation, which are different wavelengths of energy, can cause the mutations that trigger skin cancer. According to the American Cancer Society, the risk of developing skin cancer can be reduced by avoiding exposing the skin to high-intensity sunlight, using sunscreens and sunblocks that reflect or repel UV wavelengths, and examining the skin regularly and reporting any "ABCDE" lesions to a physician.

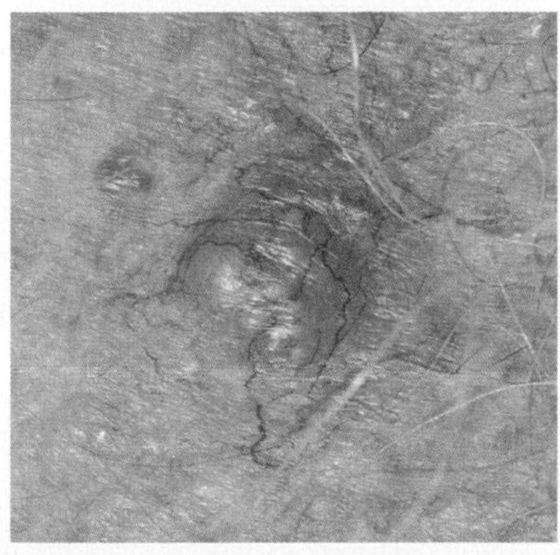

(a) Basal cell carcinoma

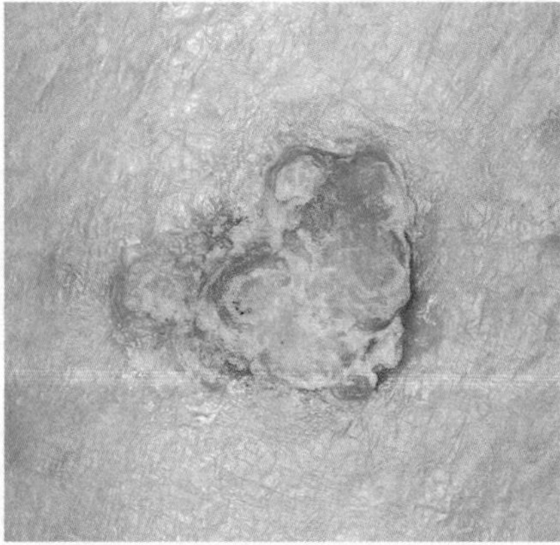

(b) Squamous cell carcinoma

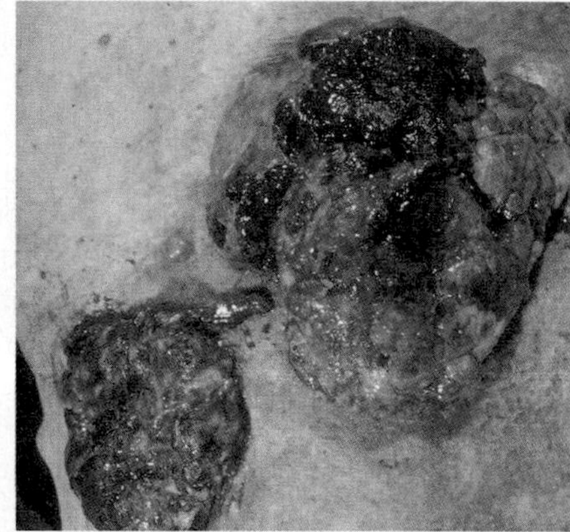

(c) Malignant melanoma

Figure 6A Skin cancer. (a): Dr. P. Marazzi/Science Photo Library/Getty Images; (b): Science Photo Library/Alamy Stock Photo; (c): McGraw-Hill Education

Diet and disease affect skin color. For example, a diet high in yellow vegetables may turn skin orange-yellow, because these foods are rich in a pigment called *carotene.* This pigment accumulates in the stratum corneum and adipose tissue of the dermis and subcutaneous layers. Skin color can also be affected by disease. A pathological cause of a yellowish skin tone is *jaundice,* which may indicate liver malfunction.

Epidermal cells can die if their blood supply from the dermis, which brings nutrients, is blocked. For example, when a person lies in one position for a prolonged period, the weight of the body pressing against the bed blocks the skin's blood supply. If cells die (necrosis), the tissues begin to break down, and a pressure ulcer (also called a decubitus ulcer or bedsore) may appear.

Pressure ulcers usually form in the skin overlying bony projections, such as on the hip, heel, elbow, or shoulder. Frequently changing body position or massaging the skin to stimulate blood flow in regions associated with bony prominences can prevent pressure ulcers.

PRACTICE 6.2

5. Explain how the epidermis is formed.
6. Distinguish between the stratum basale and the stratum corneum.
7. What is the function of melanin?
8. Which factors influence skin color?

Dermis

The boundary between the epidermis and dermis is uneven because epidermal ridges project inward and conical projections of dermis, called *dermal papillae,* extend into the spaces between the ridges (see fig. 6.1). Dermal papillae are found in the skin all over the body, but they are most abundant in the hands and feet. The ridges formed by the dermal papillae leave a patterned impression when a finger is pressed against a surface—this is a fingerprint.

Genes determine general fingerprint patterns. Skin ridges are altered slightly as a fetus presses them against the uterine wall. Because no two fetuses move exactly in the same ways, even the fingerprints of identical twins are not exactly alike.

The dermis binds the epidermis to underlying tissues (see fig. 6.1*a*). It is composed of areolar tissue along with dense connective tissue that includes tough collagen fibers and elastic fibers within a gel-like ground substance. Networks of these fibers give the skin toughness and elasticity.

Dermal blood vessels supply nutrients to all skin cells. These vessels also help regulate body temperature, as explained later in this chapter in section 6.4, Skin Functions.

Nerve cell processes are scattered throughout the dermis. Motor cell processes conduct impulses out from the brain or spinal cord to dermal muscles and glands. Sensory cell processes conduct impulses away from specialized sensory receptors, such as touch receptors in the dermis, and into the brain or spinal cord. Specialized sensory receptors are discussed in chapter 10 (see section 10.3, General Senses). The dermis also contains accessory structures, including hair follicles, sebaceous (oil-producing) glands, and sweat glands (see fig. 6.1).

PRACTICE 6.2

9. What types of tissues make up the dermis?
10. What are the functions of these tissues?

6.3 | Accessory Structures of the Skin: Epidermal Derivatives

LEARN

1. Describe the accessory structures associated with the skin.

Some of the accessory structures of the skin are actually made of the same epithelial tissues as the epidermis. Therefore, they are known as epidermal derivatives. They include nails, hair, and glands. If you follow the basement membrane at the bottom of the epidermis, you will see that it also surrounds glands and hair follicles (fig. 6.1).

Nails

Nails are protective coverings on the ends of the fingers and toes. Each nail consists of a *nail plate* that overlies a surface of skin called the *nail bed.* Specialized epithelial cells continuous with the epithelium of the skin produce the nail bed. The whitish, thickened, half-moon-shaped region (lunula) at the base of a nail plate is the most actively growing region. The epithelial cells here divide, and the newly formed cells become keratinized. This gives rise to tiny, keratinized scales that become part of the nail plate, pushing it forward over the nail bed. The keratin of nails is harder than that produced by the epidermal stratum corneum. In time, the nail plate extends beyond the end of the nail bed and with normal use gradually wears away (fig. 6.4).

Hair Follicles

Hair is present on all skin surfaces except the palms, soles, lips, nipples, and parts of the external reproductive organs. Each hair develops from a group of stem cells at the base of a tubelike depression called a **hair follicle** (hār fol′i-kl) (figs. 6.1 and 6.5). These stem cells originate from a region near the bottom of the hair follicle known as the *hair bulge,* and migrate downward. The follicle contains the *hair root,* which can extend from the surface through the dermis into the subcutaneous layer. The deepest portion of the hair root, located at the base of the hair follicle, is the *hair bulb.* It is composed of epithelial cells that are nourished from dermal blood vessels in a projection of connective tissue (hair papilla). As these epithelial cells divide and grow, they push older cells toward the surface. The cells that move upward and away from their nutrient supply become keratinized and die. Their remains constitute the structure of a developing *hair shaft* that extends outward, away from the skin surface (fig. 6.6). In other words, a hair is composed of dead epithelial cells.

Genes determine hair color by directing the type and amount of pigment that epidermal melanocytes produce. Dark hair has more of the brownish-black eumelanin, while blonde hair and red hair have more of the reddish-yellow pheomelanin. The white hair of a person with the

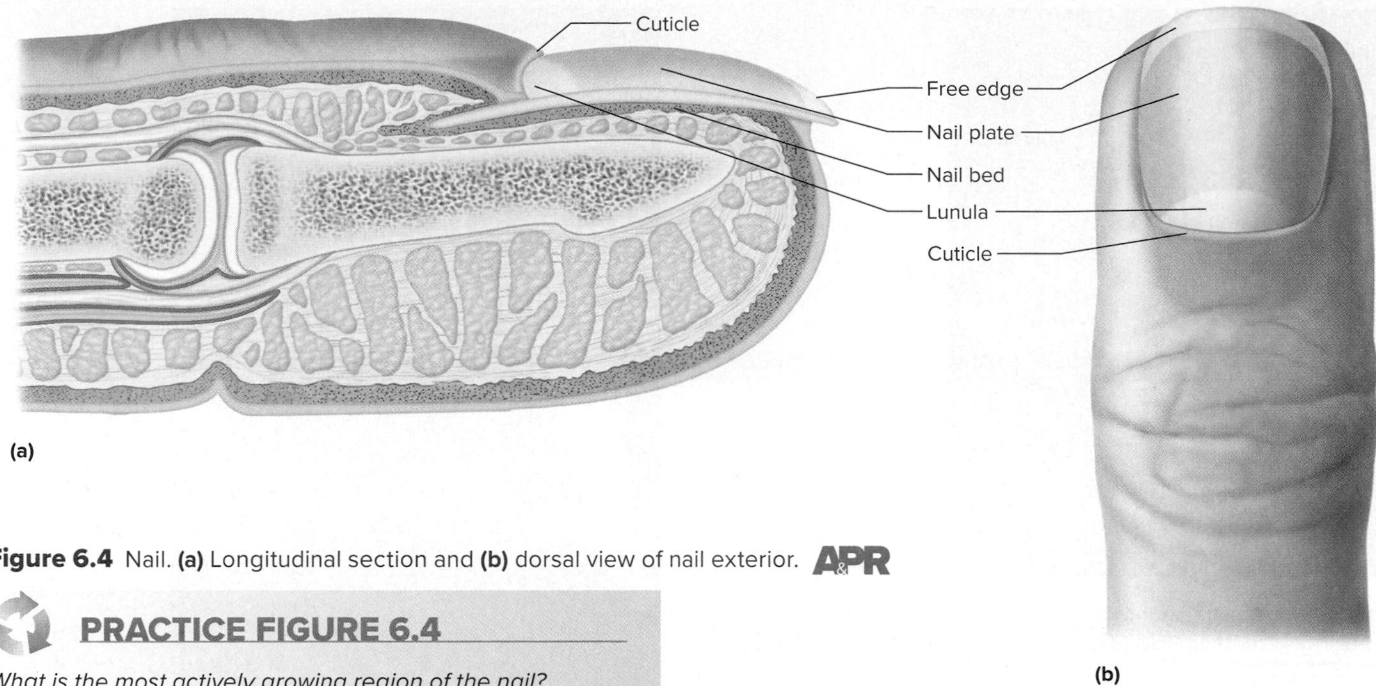

Figure 6.4 Nail. **(a)** Longitudinal section and **(b)** dorsal view of nail exterior. APR

PRACTICE FIGURE 6.4

What is the most actively growing region of the nail?

Answer can be found in Appendix E.

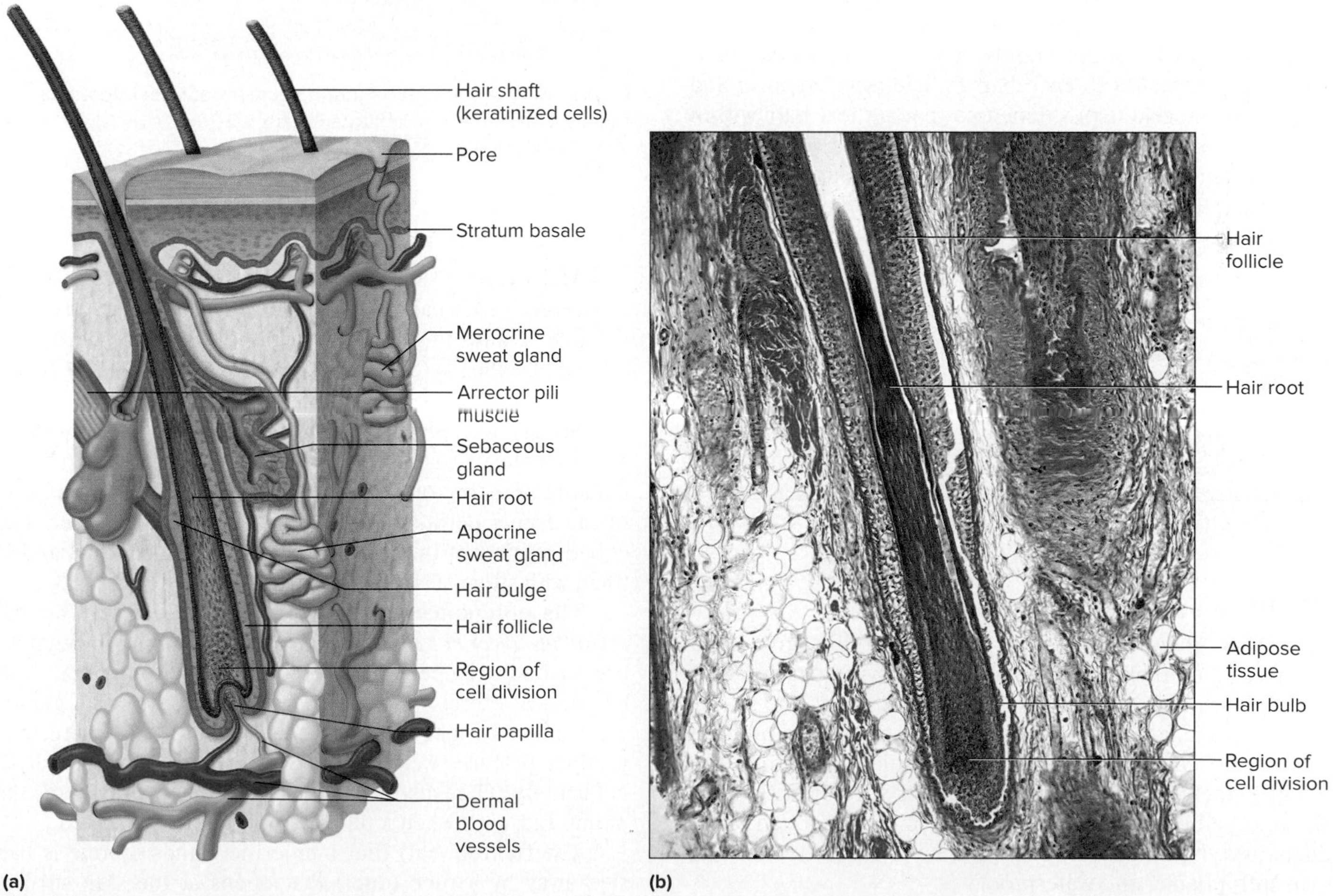

Figure 6.5 Hair follicle. **(a)** A hair grows from the base of a hair follicle when epithelial cells divide and older cells move outward and become keratinized. **(b)** Light micrograph of a hair follicle (175x). (b): Al Telser/McGraw-Hill Education

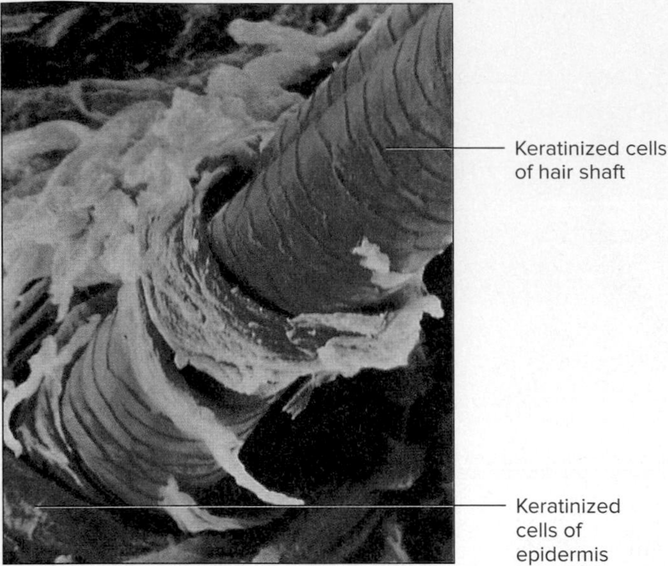

Figure 6.6 This scanning electron micrograph shows a hair emerging from its follicle (875x). CNRI/SPL/Science Source APR

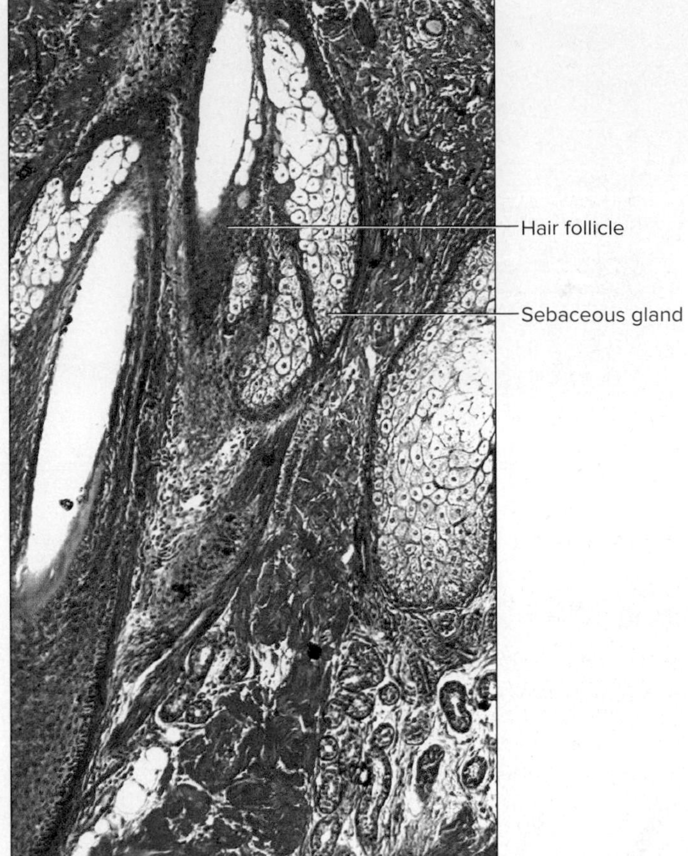

Figure 6.7 A sebaceous gland secretes sebum into a hair follicle, shown here in oblique section (200x). Alvin Telser/McGraw-Hill Education APR

inherited condition *albinism* lacks melanin altogether. A mixture of pigmented and unpigmented hairs appears gray.

A bundle of smooth muscle cells, forming the **arrector pili muscle,** attaches to each hair follicle (see figs. 6.1*a* and 6.5*a*). This muscle is positioned so that a short hair within the follicle stands on end when the muscle contracts. If a person is emotionally upset or very cold, nervous stimulation may cause the arrector pili muscles to contract, producing gooseflesh, or goose bumps.

OF **INTEREST** Stem cells from the hair bulge can give rise to epidermal cells of hair or skin. The first clue to the existence of these stem cells was the observation that new skin in burn patients arises from hair follicles. Manipulating these stem cells could someday treat baldness (alopecia) or extreme hairiness (hirsutism).

Glands

Sebaceous glands (se-bā′shus glandz) contain groups of specialized epithelial cells and are usually associated with hair follicles (figs. 6.5*a* and 6.7). They are holocrine glands (see section 5.2, Epithelial Tissues) and their cells produce globules of fatty material that accumulate, swelling and bursting the cells. The resulting oily mixture of fatty material and cellular debris called **sebum** moves through small ducts into the hair follicles. Sebum helps keep the hair and skin soft, pliable, and waterproof.

Many teens are all too familiar with a disorder of the sebaceous glands called *acne* (acne vulgaris). Overactive and inflamed glands in some body regions become plugged, producing blackheads (comedones), or are surrounded by small, red elevations producing pimples (pustules).

Sweat (swet) **glands,** also called sudoriferous glands, are exocrine glands that are widespread in the skin. Each gland consists of a tiny tube that originates as a ball-shaped coil in the deeper dermis or superficial subcutaneous layer. The coiled portion of the gland is closed at its deep end and is lined with sweat-secreting epithelial cells.

The most numerous and widespread sweat glands are the *merocrine (eccrine) sweat glands.* They respond throughout life to body temperature elevated by environmental heat or physical exercise (see fig. 6.5*a*). Merocrine sweat glands are abundant on the forehead, neck, and back, where they produce profuse sweat on hot days or during intense physical activity. They also release moisture that appears on the palms and soles when a person is emotionally stressed.

The fluid (sweat) that merocrine glands secrete is carried away by a tube (duct) that opens at the skin surface as a pore. Sweat is mostly water, but it also contains small amounts of salt and wastes, such as urea and uric acid. Thus, sweating is also an excretory function.

Other sweat glands, known as *apocrine sweat glands,* become active at puberty. Although they are called apocrine, these glands secrete by exocytosis, usually when a person is emotionally upset, frightened, in pain, or sexually aroused. The apocrine sweat glands are most numerous in the axillary regions and the groin. Ducts of these sweat glands open into hair follicles. The secretions of apocrine sweat glands include proteins and lipids that produce body odor when metabolized by skin bacteria.

Some sweat glands are structurally and functionally modified to secrete specific fluids. Such modified sweat glands include the *ceruminous glands* of the external ear canal that secrete earwax and the female *mammary glands* that secrete milk (see section 20.3, Pregnancy and the Prenatal Period).

OF INTEREST The average square inch (6.45 square centimeters) of skin holds 650 sweat glands, 20 blood vessels, 60,000 melanocytes, and more than 1,000 nerve endings.

PRACTICE 6.3

1. Describe the structure of the nail bed.
2. Explain how a hair forms.
3. What is the function of the sebaceous glands?
4. Distinguish between merocrine sweat glands and apocrine sweat glands.

6.4 | Skin Functions

LEARN

1. List various skin functions and explain how the skin helps regulate body temperature.
2. Describe wound healing.

The integumentary system has many functions. The two most obvious functions are protection and the sense of touch. However, it also has roles essential for maintaining homeostasis and tissue repair.

Protection and Touch

The epidermis serves as a waterproof protective barrier. It keeps out microorganisms such as bacteria and fungi. As mentioned in section 6.2, specialized cells called melanocytes produce melanin that protects us from ultraviolet radiation in sunlight.

Our sense of touch arises from the nerves in the skin, which contains sensory receptors that can detect pressure, changes in temperature, and pain. See chapter 10.

Vitamin D Production

The skin plays a role in the production of vitamin D, which is necessary for normal bone and tooth development. Some skin cells produce a vitamin D precursor (dehydrocholesterol), which when exposed to sunlight changes to an inactive form of vitamin D (cholecalciferol). In the liver and kidneys, the inactive form is modified and becomes active vitamin D (calcitriol).

Temperature Regulation

Regulation of body temperature is vitally important because even slight shifts can disrupt the rates of metabolic reactions. Normally the temperature of deeper body parts remains close to a set point of 37°C (98.6°F). Maintenance of a stable temperature requires that the amount of heat the body loses be balanced by the amount it produces. The skin plays a key role in the homeostatic mechanism that regulates body temperature.

Heat is a product of cellular metabolism; thus, the more active cells of the body are the major heat producers. Examples of cell types that release a great deal of heat are skeletal muscle cells, cardiac muscle cells, and cells of the liver.

When body temperature rises above the set point, the nervous system stimulates structures in the skin and other organs to release heat. For example, during physical exercise, active muscles release heat, which the blood carries away. The warmed blood reaches the part of the brain (the hypothalamus) that controls the body's temperature set point, which then signals smooth muscle in the walls of dermal blood vessels to relax. As these vessels dilate (vasodilation), more blood enters them, and some of the heat in the blood is released to lower body temperature.

At the same time as the skin loses heat, the nervous system stimulates the merocrine sweat glands to become active and to release sweat onto the skin surface. As this fluid evaporates (changes from a liquid to a gas), it carries heat away from the surface, cooling the skin.

When body temperature drops below the set point, as may occur in a very cold environment, the brain triggers different responses in the skin structures than those responses that occur with exposure to heat. Smooth muscle in the walls of dermal blood vessels is stimulated to contract; this decreases the flow of heat-carrying blood through the skin and helps reduce heat loss. Also, the merocrine sweat glands remain inactive, decreasing heat loss by evaporation. If body temperature continues to drop, the nervous system may stimulate muscle cells in the skeletal muscles throughout the body to contract slightly. This action requires an increase in the rate of cellular respiration, which releases heat as a by-product. If this response does not raise body temperature to normal, small groups of muscles may rhythmically contract with greater force, causing the person to shiver, generating more heat. Chapter 1 introduced this homeostatic mechanism (see fig. 1.6).

Deviation from the normal range for body temperature may have dangerous consequences. People with severe spinal cord injuries can no longer control body temperature, which fluctuates depending upon the environment. The very young have immature nervous systems, and thus have difficulty regulating their body temperature. The very old may have less adipose tissue in the subcutaneous layer beneath the skin (less insulation), and are less able to retain body heat.

In hypothermia, core body temperature falls below 95°F. Hypothermia initially produces shivering and a feeling of coldness. Symptoms of worsening hypothermia include a gradual loss of coordination, stiffening muscles, confusion, fatigue, and slow, shallow breathing. When core temperature falls to 87.8°F, the skin turns a bluish-gray, weakness intensifies, and consciousness fades.

In hyperthermia, core body temperature exceeds 101°F. The skin becomes hot, dry, and flushed, and the person becomes weak, dizzy, and nauseous, with headaches and a rapid, irregular pulse.

PRACTICE 6.4

1. List the functions of the skin.
2. How does the body lose excess heat?
3. Which actions help the body conserve heat?

Healing of Wounds

When a wound and the area surrounding it become red and painfully swollen, this is the result of the process of **inflammation,** which is a normal response to injury or stress. Blood vessels in affected tissues dilate and become more permeable, allowing fluids to leak into the damaged tissues. Inflamed skin may become reddened, warm, swollen, and painful to touch (table 6.1). However, the dilated blood vessels provide the tissues with more nutrients and oxygen, which aids healing.

The specific events in the healing process depend on the nature and extent of the injury. If a break in the skin is shallow, epithelial cells along its margin are stimulated to divide more rapidly than usual, and the newly formed cells fill the gap.

TABLE 6.1 Inflammation

Symptom	Cause
Redness	Vasodilation, more blood in area
Heat	Large amount of blood accumulating in area and as a by-product of increased metabolic activity in tissue
Swelling	Increased permeability of blood vessels, fluids leaving blood go into tissue spaces (edema)
Pain	Injury to neurons and increased pressure from edema

If the injury extends into the dermis or subcutaneous layer, blood vessels break, and the released blood forms a clot in the wound. The blood clot and the dried tissue fluids form a *scab* that covers and protects underlying tissues. Before long, fibroblasts migrate into the injured region and begin secreting collagen fibers that bind the edges of the wound. Suturing (stitching) or otherwise closing a large break in the skin speeds this process.

As healing continues, blood vessels extend into the area beneath the scab. Phagocytic cells remove dead cells and other debris. Eventually, the damaged tissues are replaced, and the scab sloughs off. Extensive production of collagen fibers may form an elevation above the normal epidermal surface, called a *scar.*

In large, open wounds, healing may be accompanied by formation of small, rounded masses called *granulations* that develop in the exposed tissues. A granulation consists of a new branch of a blood vessel and a cluster of collagen-secreting fibroblasts that the vessel nourishes. In time, some of the blood vessels are resorbed, and the fibroblasts move away, leaving a scar largely composed of collagen fibers. Figure 6.8 shows the stages in the healing of a wound. Clinical Application 6.3 describes healing of burned tissue.

PRACTICE 6.4

4. What is the tissue response to inflammation?
5. Distinguish between the activities necessary to heal a wound in the epidermis and those necessary to heal a wound in the dermis.
6. Explain the role of phagocytic cells in wound healing.
7. Define *granulation.*

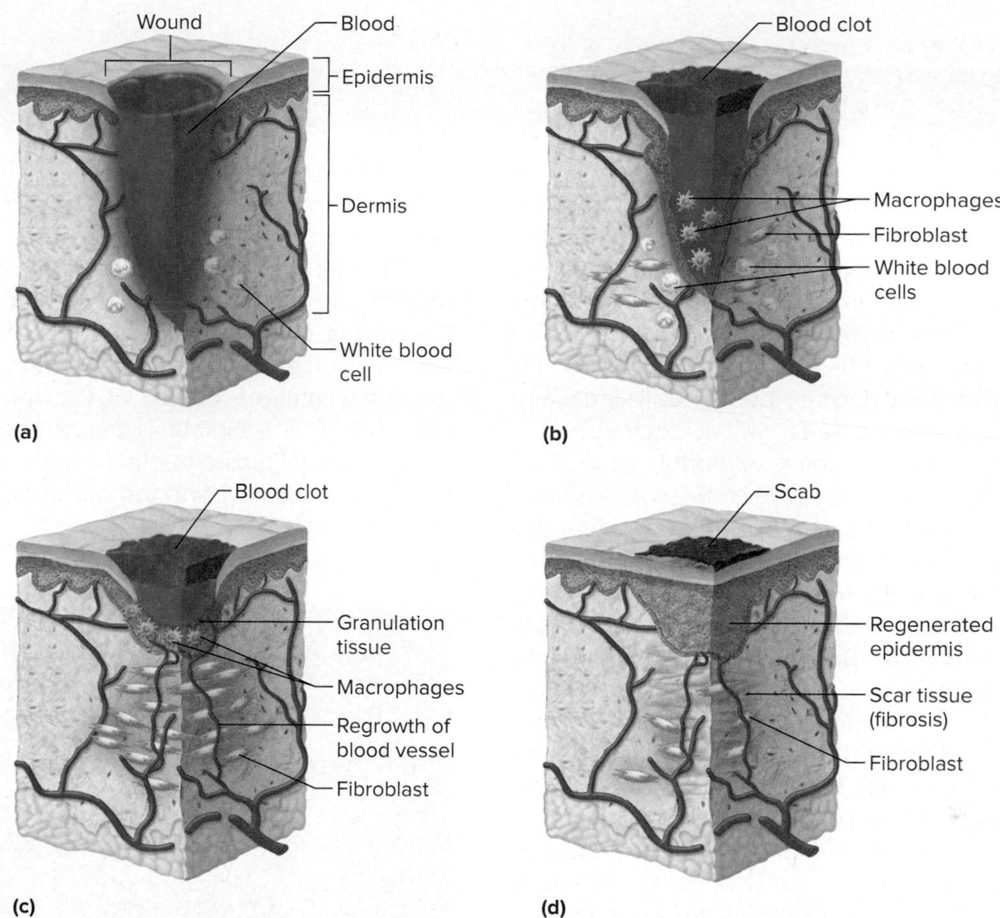

Figure 6.8 Healing of a wound. **(a)** When skin is injured, blood escapes from dermal blood vessels, and **(b)** a blood clot soon forms. **(c)** Blood vessels send out branches, and fibroblasts migrate into the area. The fibroblasts produce collagen fibers. **(d)** The scab formed by the blood clot and dried tissue fluid protects the damaged region until the skin is mostly repaired. Then, the scab sloughs off. Scar tissue continues to form, elevating the epidermal surface. (Note: The cells are not drawn to scale.)

COMMON SKIN DISORDERS

acne (ak'nē) Disease of the sebaceous glands that produces blackheads and pimples.

alopecia (al″o-pē'shē-ah) Hair loss, usually sudden.

athlete's foot (ath'-lētz foot) Fungus (*Tinea pedis*) infection usually in the skin of the toes and soles.

birthmark (berth' mark) Congenital blemish or spot on the skin, visible at birth or soon after.

boil (boil) Bacterial infection (furuncle) of the skin, produced when bacteria enter a hair follicle.

carbuncle (kar'bung-kl) Bacterial infection, similar to a boil, that spreads into the subcutaneous tissues.

cyst (sist) Liquid-filled sac or capsule.

dermatitis (der″mah-tī'tis) Inflammation of the skin.

eczema (ek'zĕ-mah) Noncontagious skin rash that produces itching, blistering, and scaling.

erythema (er″ĭ-thē'mah) Reddening of the skin due to dilation of dermal blood vessels in response to injury or inflammation.

herpes (her'pēz) Infectious disease of the skin, caused by the herpes simplex virus and characterized by recurring formations of small clusters of vesicles.

impetigo (im″pĕ-ti'go) Contagious disease of bacterial origin, characterized by pustules that rupture and become covered with loosely held crusts.

keloid (kē'loid) Elevated, enlarging fibrous scar usually initiated by an injury.

mole (mōl) Benign skin tumor (nevus) that is usually pigmented; colors range from brown to black.

pediculosis (pĕ-dik″ū-lō'sis) Disease produced by an infestation of lice.

pruritus (proo-rī'tus) Itching of the skin.

psoriasis (so-rī'ah-sis) Chronic skin disease characterized by red patches covered with silvery scales.

pustule (pus'tūl) Elevated, pus-filled area on the skin.

scabies (skā'bēz) Disease resulting from an infestation of mites.

seborrhea (seb″o-rē'ah) Hyperactivity of the sebaceous glands, causing greasy skin and dandruff.

ulcer (ul'ser) Open sore.

urticaria (ur″tĭ-ka'rē-ah) Allergic reaction of the skin that produces reddish, elevated patches (hives).

vitiligo (vit'ĭ-lī'go) Loss of melanocytes in parts of the epidermis, producing whitened areas of skin.

wart (wort) Flesh-colored, raised area caused by a viral infection.

CLINICAL APPLICATION 6.3

Burns

A few hours outside on a sunny summer day, without use of sunscreen, may result in a minor sunburn. The slightly burned skin becomes inflamed, warming and reddening (erythema) as dermal blood vessels dilate. Mild edema may swell the exposed tender skin, and a few days later the surface layer of skin may peel. Any burn injuring only the epidermis is a *superficial partial-thickness* (first degree) *burn*. Healing usually takes a few days to two weeks, with no scarring.

A burn that destroys some epidermis as well as some underlying dermis is a *deep partial-thickness* (second degree) *burn*. Fluid escapes from damaged dermal capillaries, accumulating beneath the outer layer of epidermal cells, forming blisters. The injured region becomes moist and firm and may vary from dark red to waxy white. Such a burn most commonly occurs as a result of exposure to hot objects, hot liquids, flames, or burning clothing.

The extent of healing of a deep partial-thickness burn depends upon stem cells that are associated with accessory structures of the skin. These structures include hair follicles, sweat glands, and sebaceous glands. They survive the injury because although they are derived from the epidermis, they extend into the dermis. During healing, the stem cells divide, and their daughter cells grow out onto the surface of the exposed dermis, spread over it, and differentiate as new epidermis.

A burn that destroys the epidermis, the dermis, and the accessory structures of the skin is a *full-thickness* (third degree) *burn*. The injured skin becomes dry and leathery, and it may vary in color from red to black to white. A full-thickness burn usually results from immersion in hot liquids or prolonged exposure to hot objects, flames, or corrosive chemicals. Because most of the epithelial cells in the affected region are destroyed, healing occurs as epithelial cells from the margin of the burn grow inward. If the injured area is extensive, a transplant may be necessary. A skin transplant transfers a thin layer of skin from an unburned region of the body (an autograft) or from a cadaver (a homograft) to the damaged area. Artificial skin consisting of a collagen framework seeded with a patient's own cells provides a skin substitute for transplant (**fig. 6B**).

The treatment of a burn patient requires estimating the extent of the body's affected surface. Physicians use the "rule of nines," subdividing the skin's surface into regions, each accounting for 9% (or some multiple of 9%) of the total surface area (**fig. 6C**). This estimate is important in planning to replace body fluids and electrolytes lost from injured tissues and for covering the burned area with skin or skin substitutes.

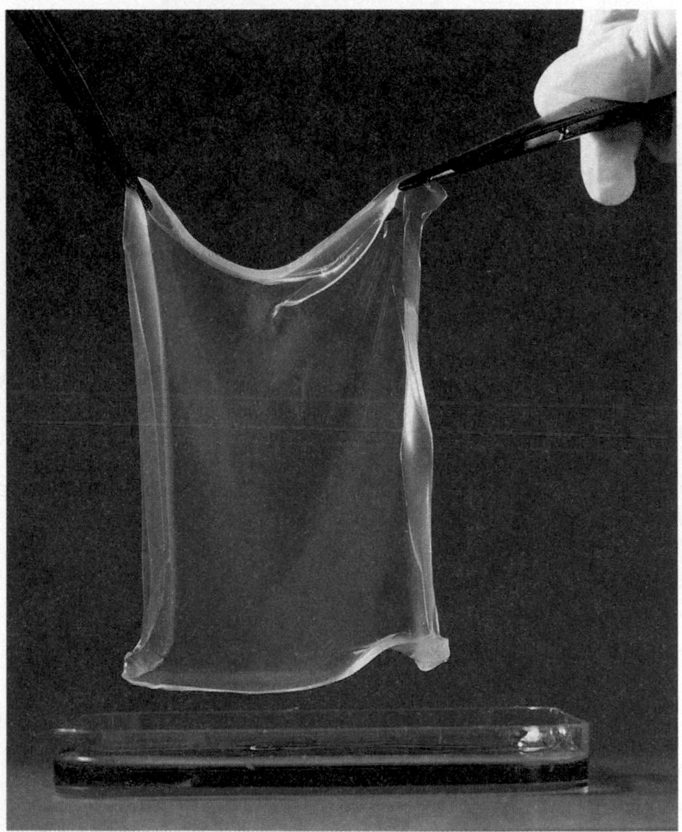

Figure 6B Tissue-engineered (artificial) skin. Mauro Fermariello/Science Source

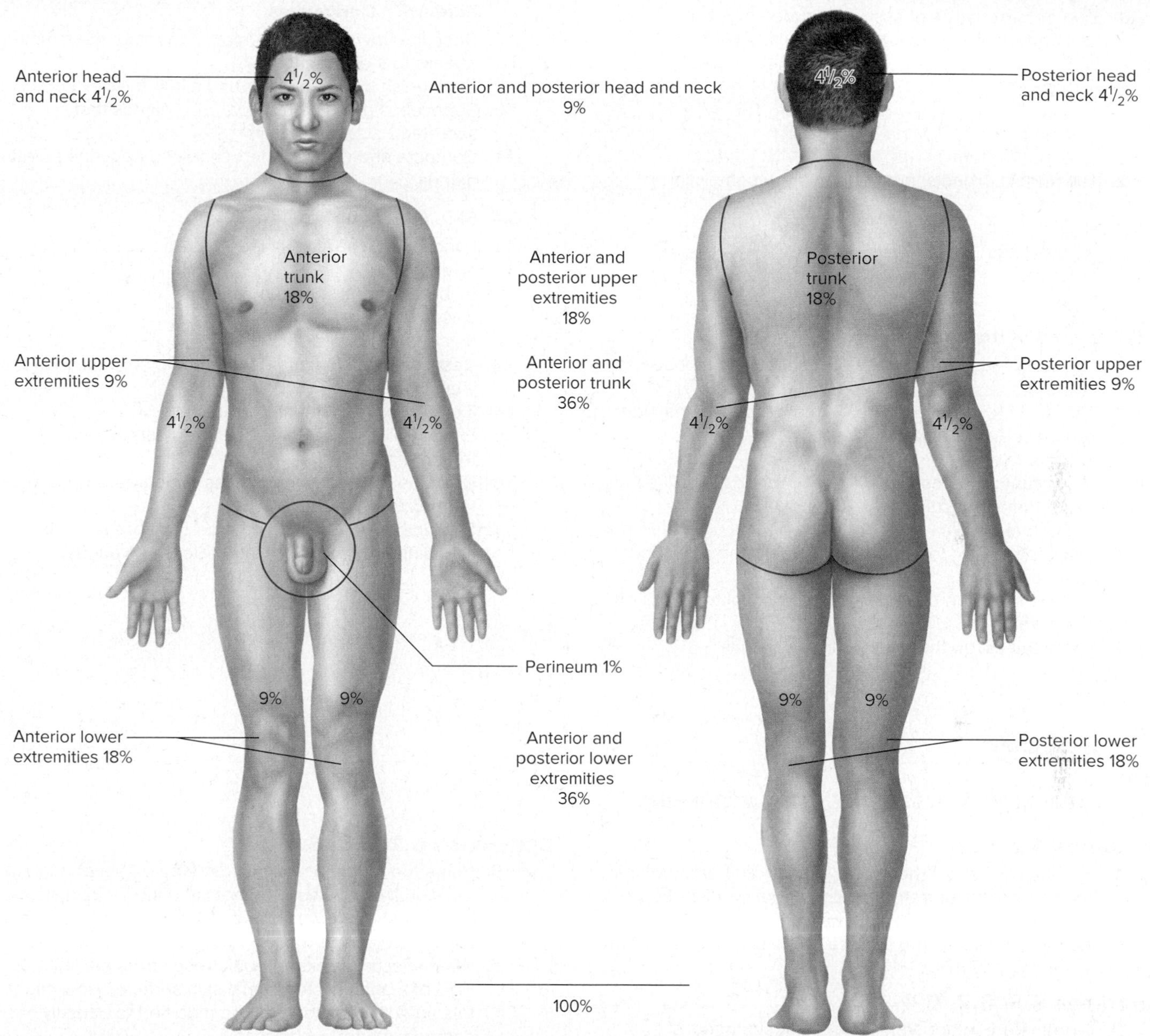

Figure 6C As an aid for estimating the extent of damage burns cause, the body is subdivided into regions, each representing 9% (or some multiple of 9%) of the total skin surface area.

ASSESS

CHAPTER ASSESSMENTS

6.1 Introduction

1. Two or more types of tissues grouped together and performing specialized functions defines a(n) ________.
   a. organelle
   b. cell
   c. organ
   d. organ system
2. The largest organ(s) in the body, by weight, is (are) the ________.
   a. liver
   b. intestines
   c. lungs
   d. skin

6.2 Layers of the Skin

3. The epidermis on the soles of the feet is composed of layers of ________ tissue.
4. The ________ layer of epidermal cells contains older keratinized cells and dead cells.
   a. stratum corneum
   b. stratum granulosum
   c. stratum spinosum
   d. stratum basale
5. Discuss the function of melanin, other than providing color to the skin.
6. List and describe the influence of each factor affecting skin color.
7. Name the tissue that makes up the dermis.

6.3 Accessory Structures of the Skin: Epidermal Derivatives

8. Describe how nails are formed, and relate the structure of nails to their function.
9. Distinguish between a hair and a hair follicle.
10. Sebaceous glands are ________ glands that secrete ________.
11. Compare and contrast merocrine and apocrine sweat glands.

6.4 Skin Functions

12. Functions of the skin include ________.
    a. excretion
    b. body temperature regulation
    c. sensory reception
    d. all of the above
13. Describe how skin plays a role in the production of vitamin D.
14. Explain how body heat is produced.
15. Explain how sweat glands help regulate body temperature.
16. Describe the body's responses to decreasing body temperature.
17. Explain how the healing of superficial breaks in the skin differs from the healing of deeper wounds.

ASSESS

INTEGRATIVE ASSESSMENTS/CRITICAL THINKING

Outcomes 5.2, 6.2

1. A child who fell off the jungle gym asked her mom why she was bleeding from only one of her wounds but not the other. Her mom told her that the non-bleeding wound is closer to the surface of the skin. Explain what she means by this.

Outcomes 5.3, 6.2, 6.4

2. A premature infant typically lacks subcutaneous adipose tissue. Also, the surface area of an infant's body is relatively large compared to its volume. How do these factors affect the ability of a premature infant to regulate its body temperature?

Outcome 6.2

3. Which of the following would result in the more rapid absorption of a drug: a subcutaneous injection or an intradermal injection? Why?

4. Everyone's skin contains about the same number of melanocytes, even though people have many different skin colors. How is this possible?

Outcomes 6.2, 6.3, 6.4

5. Using the rule of nines, estimate the extent of damage for an individual whose body, clad only in a long-sleeve, short nightgown, was burned when the nightgown caught fire as she was escaping from a burning room. What special problems would result from the loss of 50% of a person's functional skin surface? How might this person's environment be modified to partially compensate for such a loss?

Outcomes 6.2, 6.4

6. How is it protective for skin to peel after a severe sunburn?

7. As a rule, a superficial partial-thickness burn is more painful than one involving deeper tissues. How would you explain this observation?

Outcomes 6.3, 6.4

8. As a person ages, sweat glands become less active. What considerations should an elderly man make as he prepares to mow his lawn on a hot afternoon?

Integumentary System

SKELETAL SYSTEM

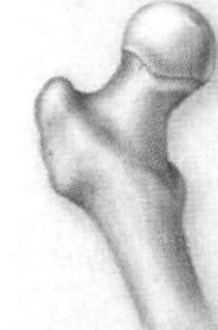

Vitamin D, production of which begins in the skin, helps provide calcium needed for bone matrix.

MUSCULAR SYSTEM

Rhythmic muscle contractions (shivering) work with the skin to control body temperature. Muscles act on facial skin to create expressions.

NERVOUS SYSTEM

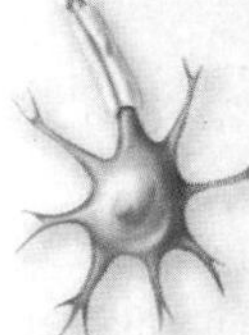

Sensory receptors provide information about the outside world to the nervous system. Nerves control the activity of sweat glands.

ENDOCRINE SYSTEM

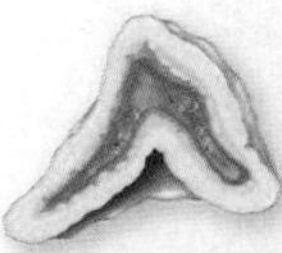

Hormones help to increase skin blood flow during exercise. Other hormones stimulate either the synthesis or the decomposition of subcutaneous fat.

CARDIOVASCULAR SYSTEM

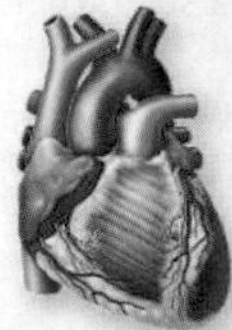

Skin blood vessels play a role in regulating body temperature.

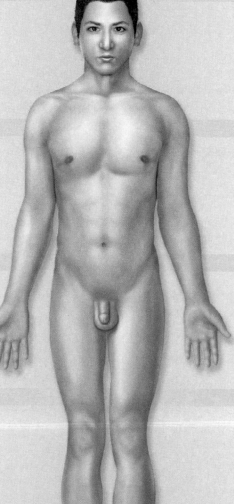

LYMPHATIC SYSTEM

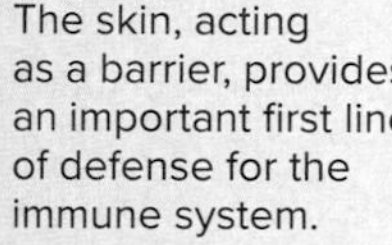

The skin, acting as a barrier, provides an important first line of defense for the immune system.

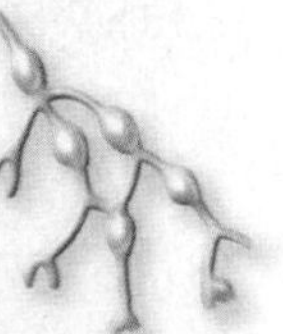

DIGESTIVE SYSTEM

Excess calories may be stored as subcutaneous fat. Vitamin D, production of which begins in the skin, stimulates dietary calcium absorption.

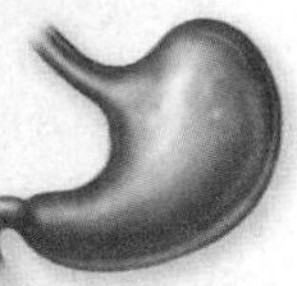

RESPIRATORY SYSTEM

Stimulation of skin receptors may alter respiratory rate.

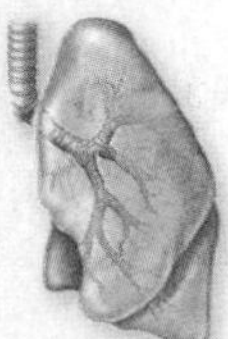

URINARY SYSTEM

The kidneys help compensate for water and electrolytes lost in sweat.

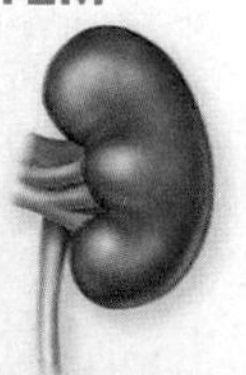

REPRODUCTIVE SYSTEM

Sensory receptors play an important role in sexual activity and in the suckling reflex.

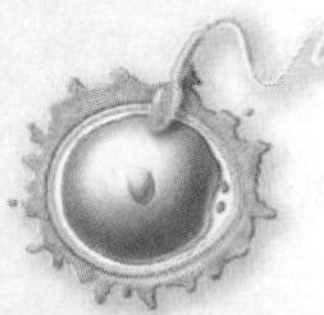

The skin provides protection, contains sensory receptors, and helps control body temperature.

Chapter Summary

6.1 Introduction

*An **organ** is formed by two or more tissue types grouped together and performing specialized functions. The skin, the largest organ in the body by weight, is part of the **integumentary system.***

6.2 Layers of the Skin

*Skin is composed of an **epidermis** and a **dermis** separated by a basement membrane. Beneath the skin is the **subcutaneous layer** that binds the skin to underlying organs, stores fat, and contains blood vessels that supply the skin.*

1. Epidermis
   a. The epidermis is stratified squamous epithelium that lacks blood vessels.
   b. The deepest layer of the epidermis, called the stratum basale, contains cells that divide.
   c. Epidermal cells undergo **keratinization** as they mature and are pushed toward the surface.
   d. The outermost layer, called the stratum corneum, is composed of dead epidermal cells.
   e. The epidermis protects underlying tissues against water loss, mechanical injury, and the effects of harmful chemicals.
   f. **Melanin,** a pigment that provides skin color, protects underlying cells from the effects of ultraviolet light.
   g. Melanocytes transfer melanin to nearby epidermal cells.
   h. Skin color is largely determined by the amount of melanin in the epidermis.
      (1) All people have about the same number of melanocytes.
      (2) Skin color is due largely to the amount of melanin and the distribution and size of pigment granules in the epidermis.
      (3) Environmental and physiological factors, as well as genes, influence skin color.
2. Dermis
   a. The dermis binds the epidermis to underlying tissues.
   b. Dermal blood vessels supply nutrients to all skin cells and help regulate body temperature.
   c. Nerve cell processes are scattered throughout the dermis.
      (1) Some dermal nerve cell processes conduct impulses to muscles and glands of the skin.
      (2) Other dermal nerve cell processes are associated with sensory receptors in the skin, and conduct impulses to the brain and spinal cord.
   d. The dermis also has hair follicles, sebaceous glands, and sweat glands.

6.3 Accessory Structures of the Skin: Epidermal Derivatives

1. Nails
   a. **Nails** are protective covers on the ends of fingers and toes.
   b. Specialized epidermal cells that are keratinized make up nails.
   c. The keratin of nails is harder than that produced by the skin's epidermal cells.
2. Hair follicles
   a. Each hair develops from epidermal cells at the base of a tubelike **hair follicle.**
   b. As newly formed cells develop and grow, older cells are pushed toward the surface and undergo keratinization.
   c. Hair color is determined by genes that direct the amount of eumelanin or pheomelanin produced by melanocytes associated with hair follicles.
   d. A bundle of smooth muscle cells is attached to each hair follicle.
3. Skin glands
   a. **Sebaceous glands** are usually associated with hair follicles.
   b. Sebaceous glands secrete sebum, which helps keep the skin and hair soft and waterproof.
   c. Each **sweat gland** is a coiled tube.
   d. Merocrine sweat glands, whose ducts open at the skin surface, respond to elevated body temperature, and produce sweat that is primarily water but also contains salts and wastes.
   e. Apocrine sweat glands, whose ducts open into hair follicles, respond to emotional upset and produce sweat containing proteins and lipids.

6.4 Skin Functions

The skin is vital in maintaining homeostasis.

1. The skin is a protective covering, slows water loss, and houses sensory receptors.
2. The skin produces vitamin D precursor.
3. The skin helps regulate body temperature.
   a. When body temperature rises above the normal set point, dermal blood vessels dilate and sweat glands secrete sweat.
   b. When body temperature drops below the normal set point, dermal blood vessels constrict and sweat glands become inactive.
   c. If body temperature continues to drop, skeletal muscles rhythmically contract.

Skin injuries trigger inflammation. The affected area becomes red, warm, swollen, and tender.

4. Dividing epithelial cells fill in shallow cuts in the epidermis.
5. Clots close deeper cuts, sometimes leaving a scar where connective tissue replaces skin.
6. Granulations form in large, open wounds as part of the healing process.

Skeletal System

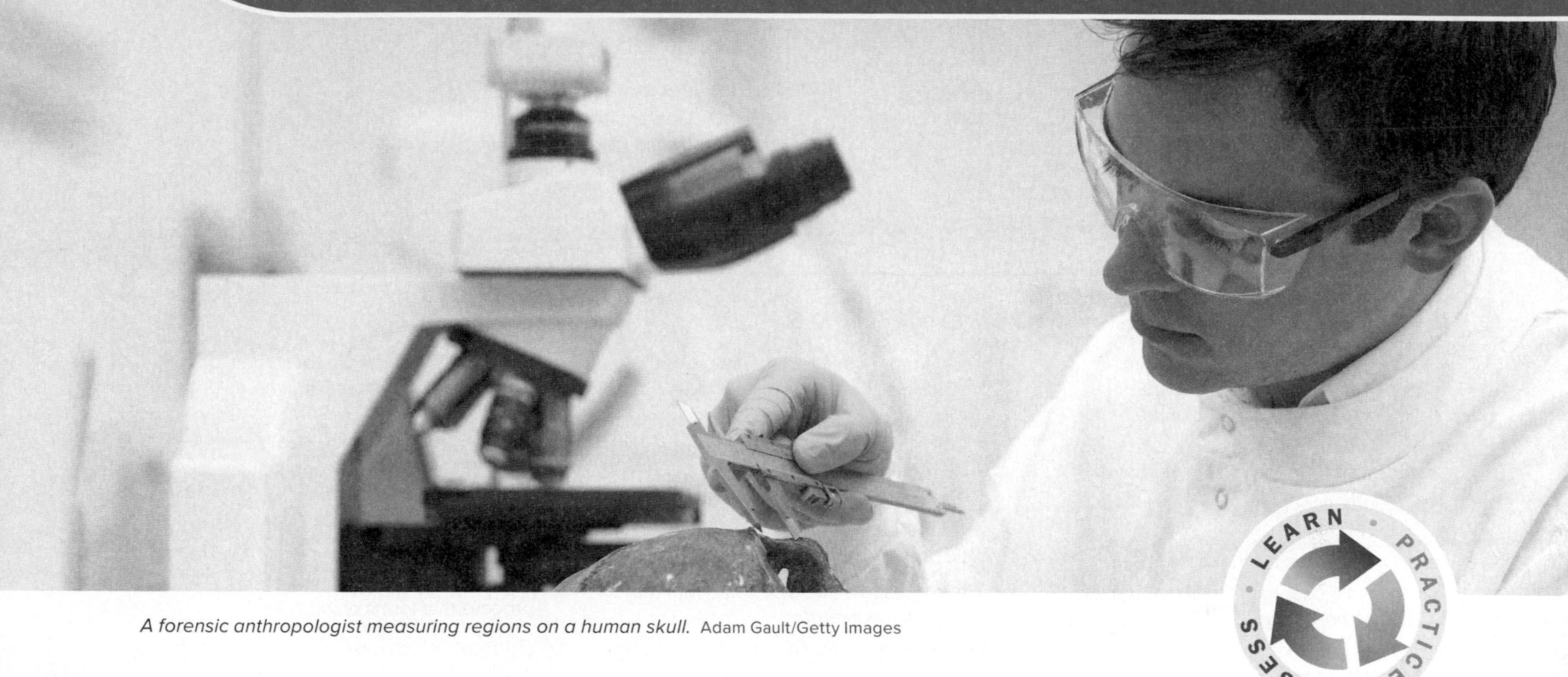

A forensic anthropologist measuring regions on a human skull. Adam Gault/Getty Images

When human skeletal remains are found with no positive identification, a forensic anthropologist might be consulted to determine the gender, age at the time of the death, stature, and possibly the cause of death.

As a skeleton ages, there are numerous prominent changes. The hip bone begins as three separate bones: the ilium, ischium, and pubis. They fuse together in the mid to late teens, but do not completely ossify until the early twenties. The ends of the long bones also fuse to the shaft at various ages. For the femur, this is between fourteen and twenty-one years, and for the clavicle it occurs between twenty and thirty years of age. The sutures in the skull close and often disappear. This begins in the late twenties and might continue into the fifties. The vertebrae in the lower back will show signs of wear and osteoarthritis. This usually begins in the sixties but can begin in the thirties. The teeth can also provide clues about age. If the secondary molars are not yet erupted, it indicates a pre-adolescent. Severe wear on the teeth is often associated with a middle-aged or elderly individual.

The region of the hip bones is the most reliable to determine gender. In general, they are tall and narrow in males and short and broad in females. In addition, there are a few differences in some of the landmarks on the hips. The skull also holds some clues. A female chin is often oval in shape, whereas males tend to have a more squared off chin and jaw. The area on the occipital bone for attachment of neck muscles is more pronounced in males, and the parietal bones in females exhibit a certain swelling.

If bones were broken earlier in life, signs of healing will be evident. If a broken bone is part of the trauma that caused death, signs of healing will be absent. Rheumatoid arthritis and gout will cause wear and erosion of joint surfaces.

Stature and size can also be estimated. There are calculations used based on the length of bones such as the femur and humerus to estimate height. Markings and ridges on bones can indicate the size of the large muscles of the arms and legs.

Analysis of the facial bones can often yield an approximation of what the person "looked like." This includes facial features such as the length and width of the nose and the distance between the eyes.

Determination of ancestry and geographic origin is possible but somewhat controversial and less reliable than aging and sexing a skeleton.

After a full examination, with all factors considered, a fairly accurate assessment of "who" the skeleton was in life is possible.

Module 5 Skeletal System

LEARNING OUTLINE

After studying this chapter, you should be able to complete the "Learning Outcomes" that follow the major headings throughout the chapter.

AIDS TO UNDERSTANDING WORDS

acetabul- [vinegar cup] *acetabul*um: depression of the hip bone that articulates with the head of the femur.

ax- [axis] *ax*ial skeleton: bones of the skull, vertebral column, ribs, and sternum.

-blast [bud] osteo*blast:* cell that will form bone tissue.

carp- [wrist] *carp*als: wrist bones.

-clast [break] osteo*clast:* cell that breaks down bone tissue.

condyl- [knob] *condyl*e: rounded, bony process.

corac- [a crow's beak] *corac*oid process: beaklike process of the scapula.

cribr- [sieve] *cribr*iform plate: portion of the ethmoid bone with many small openings.

crist- [crest] *crist*a galli: bony ridge that projects upward into the cranial cavity.

fov- [pit] *fov*ea capitis: pit in the head of a femur.

glen- [joint socket] *glen*oid cavity: depression in the scapula that articulates with the head of a humerus.

inter- [among, between] *inter*vertebral disc: structure between vertebrae.

intra- [inside] *intra*membranous bone: bone that forms within sheetlike masses of connective tissue.

meat- [passage] external acoustic *meat*us: canal of the temporal bone that leads inward to parts of the ear.

odont- [tooth] *odont*oid process: toothlike process of the second cervical vertebra.

poie- [make, produce] hemato*poie*sis: process that forms blood cells.

(Appendix A has a complete list of Aids to Understanding Words.)

7.1 | Introduction

LEARN

1. List the active tissues found in a bone.

The term *skeleton* has evolved over time, but combining elements from the Greek and Latin meanings we get "dried framework." However, the dry specimens you study in the lab can easily give you the wrong impression of a bone's true nature. A living bone is wrapped in connective tissues and has a blood and nerve supply just as all organs do. Bones have many functions, including support, movement, protection, blood cell formation, and the storage of minerals.

The human skeletal "system" is much like those of our pet dogs and cats, where it is strong enough to bear weight yet light enough to allow movement. But the human skeleton is uniquely adapted for walking on two legs and for holding things in the hands, or "grasping." Made of not only bones, the skeletal system also includes tendons, ligaments, and cartilage that often work together as joints. Two major divisions are recognized in the skeletal system: the axial skeleton and the appendicular skeleton.

PRACTICE 7.1

Answers to the Practice questions can be found in the eBook.

1. Name the living tissues in bone.

7.2 | Bone Structure

LEARN

1. Describe the macroscopic and microscopic structure of a long bone, and list the functions of these parts.

The adult human skeleton has 206 bones that vary greatly in size and shape (reference table 7.1 later in section 7.5). However, they are similar in structure, development, and function.

Bone Classification

Bones may be classified according to their shapes—long, short, flat, or irregular.

- **Long bones** have long longitudinal axes and expanded ends. Examples of long bones are the forearm and thigh bones.
- **Short bones** have roughly equal lengths and widths. The bones of the wrists and ankles are this type. A special type of short bone is a **sesamoid bone** or **round bone.** This type of bone is typically small and nodular and develops within a tendon or adjacent to a joint. The kneecap (patella) is a sesamoid bone.
- **Flat bones** are platelike structures with broad surfaces, such as the ribs, the scapulae, and some bones of the skull.

- **Irregular bones** have a variety of shapes, and most are connected to several other bones. Irregular bones include the vertebrae that compose the backbone and many facial bones.

Structure of a Long Bone

The femur, the bone in the thigh, illustrates the structure of a long bone (fig. 7.1). At each end of a long bone is an expanded portion called an **epiphysis** (e-pif′ĭ-sis) (plural, *epiphyses*), which articulates (forms a joint) with another bone. The epiphysis that is nearest the attachment to the trunk of the body is called the proximal epiphysis. The epiphysis that is farthest from the trunk of the body is called the distal epiphysis. The outer surface of the articulating portion of the epiphysis is coated with a layer of hyaline cartilage called **articular cartilage** (ar-tik′u-lar kar′tĭ-lij). The shaft of the bone, between the epiphyses, is called the **diaphysis** (dī-af′ĭ-sis). The epiphyseal line is the region where the epiphysis and diaphysis fuse together during growth. While the bone is still growing, it is called the epiphyseal plate, or growth plate.

A tough covering of dense connective tissue called the **periosteum** (per″ē-os′tē-um) completely encloses the bone, except for the articular cartilage on the bone's ends. The periosteum is firmly attached to the bone, and periosteal fibers are continuous with the connecting ligaments and tendons. The periosteum also helps form and repair bone tissue.

A bone's shape makes possible the bone's functions. For example, bony projections called *processes* provide sites where ligaments and tendons attach; grooves and openings form passageways for blood vessels and nerves; and a depression of one bone may articulate with a process of another.

The wall of the diaphysis is mainly composed of tightly packed tissue called **compact bone** (kom′pakt bōn), also called cortical bone. Compact bone has a continuous extracellular matrix with no gaps. The epiphyses, in contrast, are composed largely of **spongy bone** (spun′jē bōn), also called cancellous bone, with thin layers of compact bone on their surfaces. Spongy bone consists of numerous branching bony plates called **trabeculae** (trah-bek′u-lē). Irregular connecting spaces between these plates help reduce the bone's weight (fig. 7.2). The bony plates are most highly developed in the regions of the epiphyses that are subjected to compressive forces. Both compact bone and spongy bone are strong and resist bending.

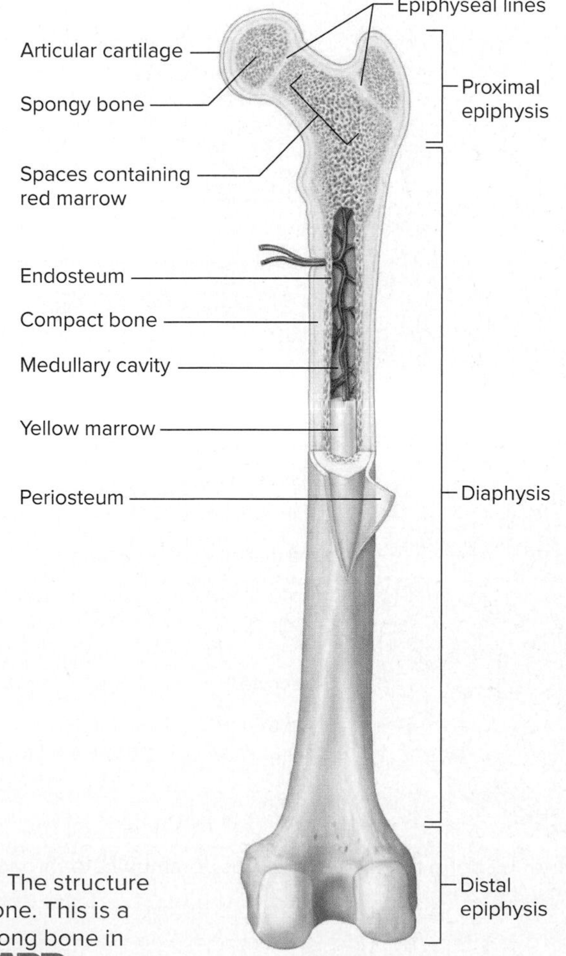

Figure 7.1 The structure of a long bone. This is a femur, the long bone in the thigh. APR

Radiology Technologist

At age fifty-two the woman is younger than most of the others having their bone mineral density measured. She had been advised by her gynecologist to have a baseline test to assess the health of her skeleton because her parents had osteoporosis.

A radiology technologist conducts the test. She explains the test to the patient, then positions her on her back on a padded table, fully clothed. The scanner passes painlessly over the patient's hip and lower spine, emitting low-dose X rays that form images of the bones. Spaces on the scan indicate osteopenia, the low bone mineral density that may be a prelude to osteoporosis.

Radiology technologists administer medical imaging tests, such as ultrasound and magnetic resonance imaging (MRI), as well as mammography and the X-ray cross sections of computerized tomography (CT). They protect patients from radiation with drapes. By positioning the patients and operating scanning devices, they produce images from which a radiologist can diagnose an illness or injury.

A registered radiology technologist completes two years of training at a hospital or a two- or four-year program at a college or university, and must pass a national certification exam.

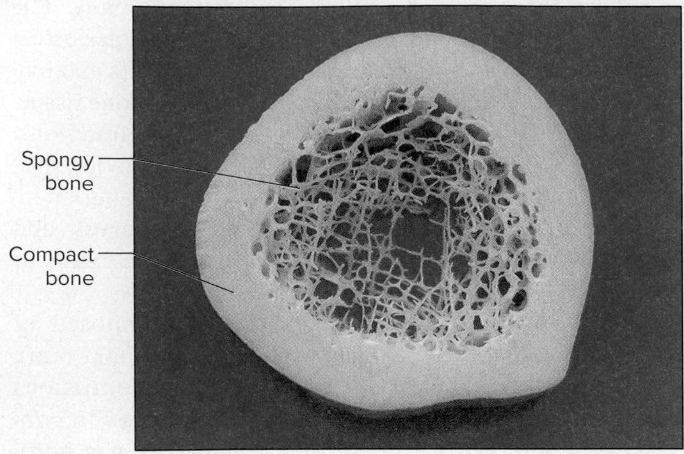

Figure 7.2 This cross section of a long bone reveals a layer of spongy bone beneath a layer of compact bone. Ed Reschke APR

Compact bone in the diaphysis of a long bone forms a tube with a hollow chamber called the **medullary cavity** (med′u-lār″e kav′ĭ-tē) that is continuous with the spaces of the spongy bone. A thin layer of cells called the **endosteum** (en-dos′te-um) lines the medullary cavity, as well as the spaces within spongy bone, and a specialized type of soft connective tissue called **marrow** (mar′ō) fills them.

Microscopic Structure

Recall from chapter 5 (see section 5.3, Connective Tissues) that bone cells called *osteocytes* occupy very small, bony chambers called **lacunae.** The lacunae are within the bony matrix of the **lamellae,** which form concentric circles around *central canals* (Haversian canals). Osteocytes exchange substances with nearby cells by means of cellular processes passing through **canaliculi** (fig. 7.3; see fig. 5.20). The extracellular

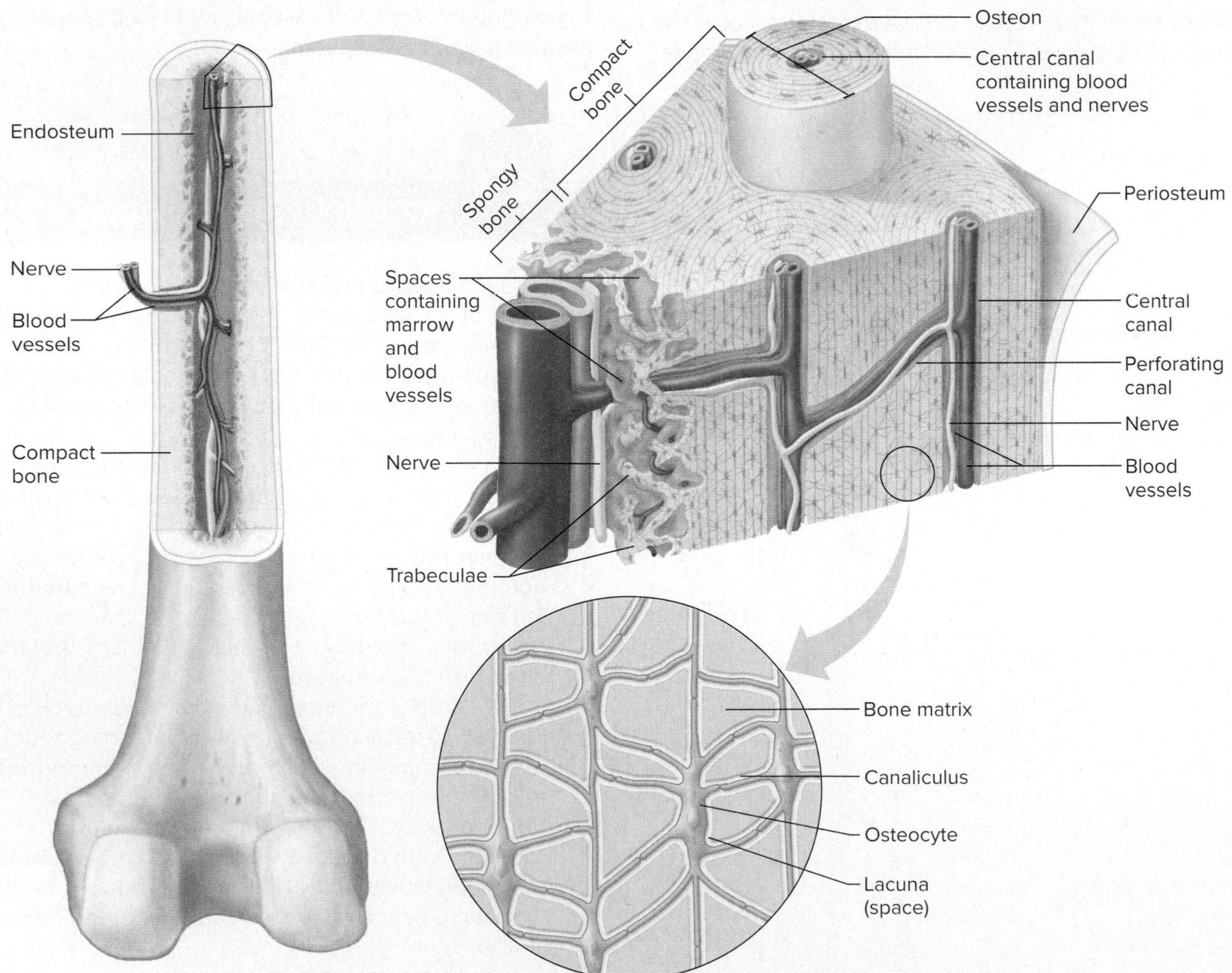

Figure 7.3 Compact bone is composed of osteons cemented together by bone matrix. Osteocytes communicate through tunnel-like extensions called canaliculi. (Note: Drawings are not to scale.) APR

matrix of bone tissue is largely collagen and inorganic salts (calcium phosphate). Collagen gives bone its strength and resilience, and inorganic salts make it hard and resistant to crushing.

In compact bone, the osteocytes and layers of extracellular matrix concentrically clustered around a central canal form a cylinder-shaped unit called an *osteon* (Haversian system). Many osteons together form the substance of compact bone.

Each central canal contains blood vessels and nerve fibers surrounded by loose connective tissue. Blood in these vessels nourishes bone cells associated with the central canal.

Central canals extend longitudinally through bone tissue, and transverse *perforating canals* (Volkmann's canals) connect them. Perforating canals contain larger blood vessels and nerves by which the smaller blood vessels and nerves in central canals communicate with the surface of the bone and the medullary cavity (fig. 7.3).

Spongy bone is also composed of osteocytes and extracellular matrix, but the bone cells do not aggregate around central canals. Instead, the cells lie in the *trabeculae* and get nutrients from substances diffusing into canaliculi that lead to the surface of these thin, bony plates.

PRACTICE 7.2

1. Explain how bones are classified by shape.
2. List five major parts of a long bone.
3. How do compact and spongy bone differ in structure?
4. Describe the microscopic structure of compact bone.

7.3 | Bone Function

LEARN

1. Discuss the major functions of bones.

Bones shape, support, and protect body structures. They also aid body movements, house tissue that produces blood cells, and store inorganic salts.

Support, Protection, and Movement

Bones give shape to structures such as the head, face, thorax, and limbs. They are also strong enough to lend support and protection. For example, the bones of the lower limbs, pelvis, and backbone support the body's weight. The bones of the skull protect the eyes, ears, and brain. Bones of the rib cage and shoulder girdle protect the heart and lungs, whereas the bones of the pelvic girdle protect the lower abdominal and internal reproductive organs. Bones are not only strong, they are light enough to work with muscles to bring about movement of the limbs and other body parts.

Blood Cell Formation

The process of blood cell formation, called **hematopoiesis** (he″mă-to-poi-ē′sis), begins in the *yolk sac,* which lies outside the human embryo (see section 20.3, Pregnancy and the Prenatal Period). Later in development, blood cells are manufactured in the liver and spleen, and still later they form in bone marrow.

Marrow is a soft, netlike mass of connective tissue within the medullary cavities of long bones, in the irregular spaces of spongy bone, and in the larger central canals of compact bone tissue. It is of two kinds: red and yellow. **Red marrow** functions in the formation of red blood cells (erythrocytes), white blood cells (leukocytes), and blood platelets. The color comes from the oxygen-carrying pigment **hemoglobin** in the red blood cells.

In an infant, red marrow occupies the cavities of most bones. As a person ages, **yellow marrow,** which stores fat, replaces much of the red marrow. Yellow marrow is not active in blood cell production. In an adult, red marrow is primarily found in the spongy bone of the skull, ribs, breastbone (sternum), collarbones (clavicles), backbones (vertebrae), and hip bones. If the supply of blood cells is deficient, some yellow marrow may become red marrow, which then reverts to yellow marrow when the deficiency is corrected. Chapter 12 (see section 12.2, Formed Elements) describes blood cell formation in more detail.

Storage of Inorganic Salts

Vital metabolic processes require calcium. The extracellular matrix of bone tissue is rich in calcium salts, mostly in the form of calcium phosphate. When the blood is low in calcium, *parathyroid hormone* stimulates osteoclasts to break down bone tissue, releasing calcium salts from the extracellular matrix into the blood. A high blood calcium level inhibits osteoclast activity, and *calcitonin,* a hormone from the thyroid gland, stimulates osteoblasts to form bone tissue, storing excess calcium in the extracellular matrix (fig. 7.4). Chapter 11 (see sections 11.5, Thyroid Gland, and 11.6, Parathyroid Glands) describes the details of this homeostatic mechanism. Maintaining sufficient blood calcium levels is important in muscle contraction, nerve cell function, blood clotting, and other physiological processes.

In addition to storing calcium and phosphorus, bone tissue contains smaller amounts of magnesium, sodium, potassium, and carbonate ions. Bones also accumulate certain harmful metallic elements such as lead, radium, or strontium. These elements are not normally present in the body but are sometimes accidentally ingested.

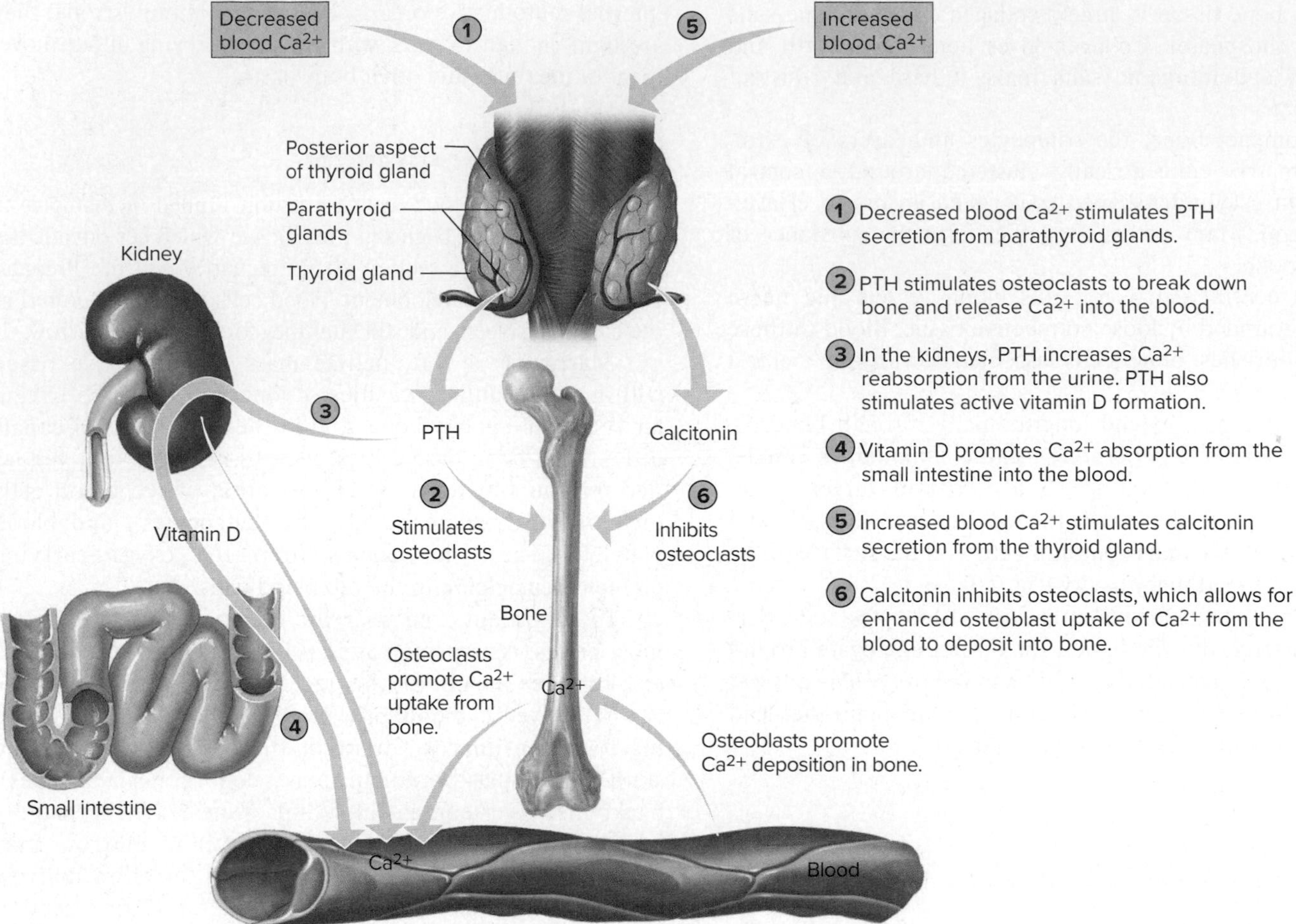

Decreased blood Ca^{2+}
Increased blood Ca^{2+}
Posterior aspect of thyroid gland
Parathyroid glands
Thyroid gland
Kidney
PTH
Calcitonin
Stimulates osteoclasts
Inhibits osteoclasts
Vitamin D
Bone
Osteoclasts promote Ca^{2+} uptake from bone.
Ca^{2+}
Osteoblasts promote Ca^{2+} deposition in bone.
Small intestine
Ca^{2+}
Blood
1 Decreased blood Ca^{2+} stimulates PTH secretion from parathyroid glands.
2 PTH stimulates osteoclasts to break down bone and release Ca^{2+} into the blood.
3 In the kidneys, PTH increases Ca^{2+} reabsorption from the urine. PTH also stimulates active vitamin D formation.
4 Vitamin D promotes Ca^{2+} absorption from the small intestine into the blood.
5 Increased blood Ca^{2+} stimulates calcitonin secretion from the thyroid gland.
6 Calcitonin inhibits osteoclasts, which allows for enhanced osteoblast uptake of Ca^{2+} from the blood to deposit into bone.

Figure 7.4 Hormonal regulation of bone calcium deposition and resorption.

PRACTICE FIGURE 7.4

What three components of a homeostatic mechanism (see fig. 1.5) are shown in this figure?

Answer can be found in Appendix E.

PRACTICE 7.3

1. Name the major functions of bones.
2. Distinguish between the functions of red marrow and yellow marrow.
3. List the substances normally stored in bone tissue.

7.4 Bone Development, Growth, and Repair

LEARN

1. Distinguish between intramembranous and endochondral bones, and explain how such bones develop and grow.

Parts of the skeletal system begin to form during the first few weeks of prenatal development, and bony structures continue to develop and grow into adulthood. Bones form by replacing existing connective tissues in either of two ways: (1) intramembranous bones originate between sheetlike layers of connective tissues; and (2) endochondral bones begin as masses of hyaline cartilage that are later replaced by bone tissue (fig. 7.5). The formation of bone is called **ossification** (os″ĭ-fi-kā′shun).

Intramembranous Bones

The flat bones of the skull are **intramembranous bones** (in″trah-mem′brah-nus bōnz). They develop in the fetus from membranelike layers of unspecialized, or relatively undifferentiated, connective tissues at the sites of the future bones. Dense networks of blood vessels are contained within these connective tissues. Some of the partially differentiated progenitor cells enlarge and further differentiate into bone-forming cells called **osteoblasts** (os′tē-o-blasts). The osteoblasts deposit bony matrix around themselves, forming spongy bone tissue in all directions within the layers of connective tissues. When an extracellular matrix completely surrounds osteoblasts, they are called **osteocytes.** Eventually, cells of

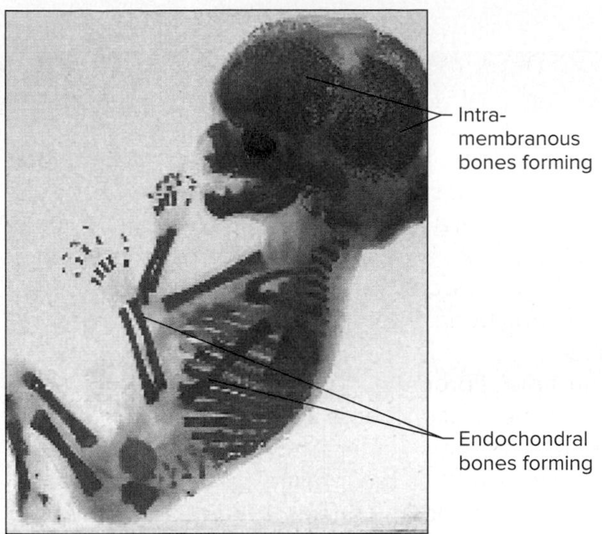

Figure 7.5 Intramembranous bones in the fetus form by replacing unspecialized connective tissue. Endochondral bones form from hyaline cartilage "models" that are gradually replaced with the harder tissue of bone. Note the stained, developing bones of this fourteen-week fetus. Biophoto Associates/Science Source

the membranous tissues that persist outside the developing bone give rise to the periosteum. Osteoblasts on the inside of the periosteum form a layer of compact bone over the surface of the newly formed spongy bone.

Endochondral Bones

Most of the bones of the skeleton are **endochondral bones** (en″do-kon′dral bōnz). They develop in the fetus from masses of hyaline cartilage shaped like future bony structures. These cartilaginous models grow rapidly for a time and then begin to change extensively.

In a long bone, changes begin in the center of the diaphysis, where the cartilage slowly breaks down and disappears (fig. 7.6*a*,*b*). At about the same time, a periosteum forms from connective tissue that encircles the developing diaphysis. Blood vessels and osteoblasts from the periosteum invade the disintegrating cartilage, and spongy bone forms in its place. This region of bone formation is called the *primary ossification center,* and bone tissue develops from it toward the ends of the cartilaginous structure (fig. 7.6*c*). Meanwhile, osteoblasts from the periosteum deposit a thin layer of compact bone around the primary ossification center.

The epiphyses of the developing bone remain cartilaginous and continue to grow. Later, *secondary ossification centers* appear in the epiphyses, and spongy bone forms in all directions from them (fig. 7.6*d*). As spongy bone is deposited in the diaphysis and in the epiphysis, a band of cartilage called the **epiphyseal plate** (ep″ĭ-fiz′ē-al plāt) remains between these two ossification centers (fig. 7.6*e*).

The cartilaginous tissue of the epiphyseal plate includes layers of young cells that are undergoing mitosis and producing new cells. As these cells enlarge and an extracellular matrix forms around them, the cartilaginous plate thickens, lengthening the bone. At the same time, calcium salts accumulate in the extracellular matrix adjacent to the oldest cartilaginous cells. As the extracellular matrix calcifies, the cartilage cells begin to die.

In time, bone-resorbing cells called **osteoclasts** (os′tē-o-klasts) break down the calcified extracellular matrix. These large, multinucleated cells originate in bone marrow when certain single-nucleated white blood cells (monocytes) fuse. Osteoclasts secrete an acid that dissolves the inorganic component of the calcified matrix, and their lysosomal enzymes digest the organic components. After osteoclasts remove the extracellular matrix, bone-building osteoblasts invade the region and deposit new bone tissue in place of the calcified cartilage.

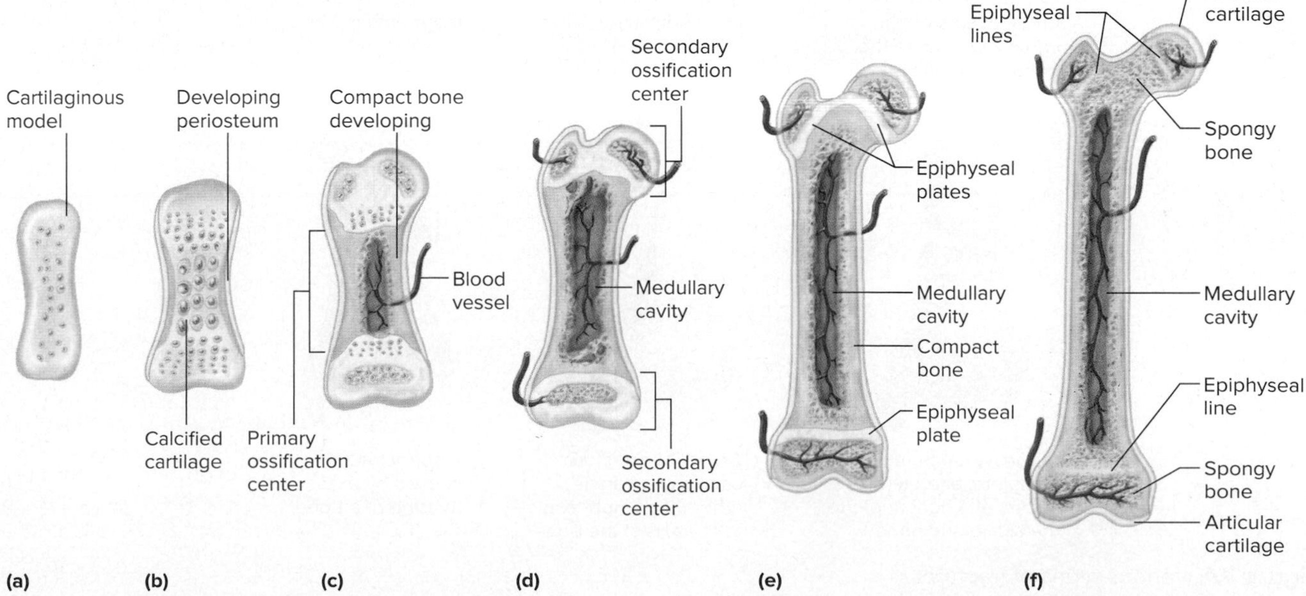

Figure 7.6 Major stages **(a–d** fetal, **e** child**)** in the development of an endochondral bone. In an **(f)** adult, when bone growth ceases, an epiphyseal line is what remains of the epiphyseal plate. (Note: Relative bone sizes are not to scale.)

CLINICAL APPLICATION 7.1

Bone Fractures

A *fracture* is a break in a bone. A fracture is classified by its cause as a traumatic, spontaneous, or pathologic fracture, and by the nature of the break as a greenstick, fissured, comminuted, transverse, oblique, or spiral fracture (**fig. 7A**). A broken bone exposed to the outside by an opening in the skin is termed a compound (open) fracture.

When a bone breaks, blood vessels in it rupture, and the periosteum is likely to tear. Blood from the broken vessels spreads through the damaged area and soon forms a blood clot, or *hematoma.* Vessels in surrounding tissues dilate, swelling and inflaming the tissues.

Within days or weeks, developing blood vessels and large numbers of osteoblasts originating in the periosteum invade the hematoma. The osteoblasts rapidly divide in the regions close to the new blood vessels, building spongy bone nearby. Granulation tissue develops, and in regions farther from a blood supply, fibroblasts produce masses of fibrocartilage. Meanwhile, phagocytic cells begin to remove the blood clot, as well as any dead or damaged cells in the affected area. Osteoclasts also appear and resorb bone fragments, aiding in "cleaning up" debris.

In time, fibrocartilage fills the gap between the ends of the broken bone. This mass, a *cartilaginous soft callus,* is later replaced by bone tissue in much the same way that the hyaline cartilage of a developing endochondral bone is replaced. That is, the cartilaginous callus breaks down, blood vessels and osteoblasts invade the area, and a *hard bony callus* fills the space.

Typically, more bone is produced at the site of a healing fracture than is necessary to replace the damaged tissues. Osteoclasts remove the excess,

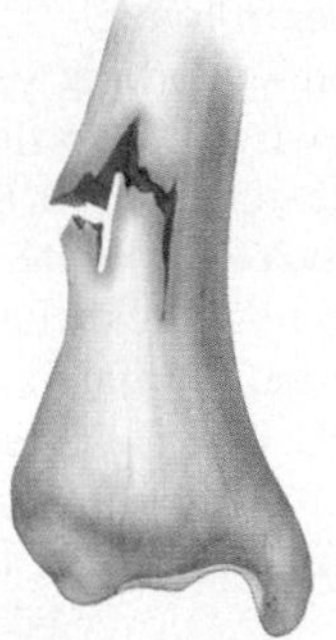

A *greenstick* fracture is incomplete, and the break occurs on the convex surface of the bend in the bone.

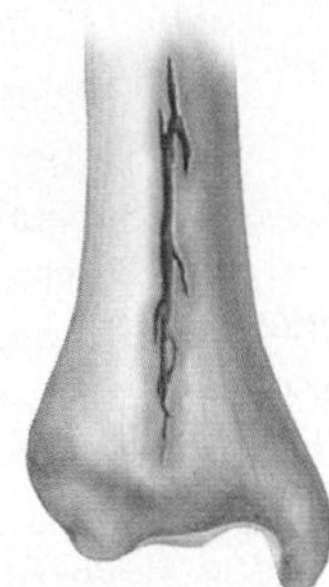

A *fissured* fracture is an incomplete longitudinal break.

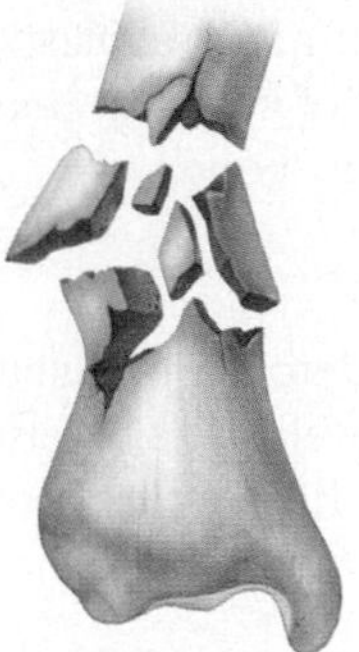

A *comminuted* fracture is complete and fragments the bone.

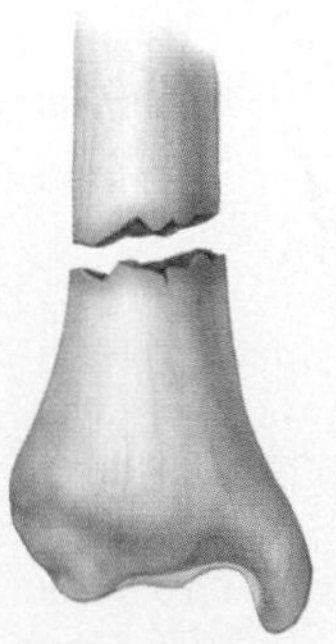

A *transverse* fracture is complete, and the break occurs at a right angle to the axis of the bone.

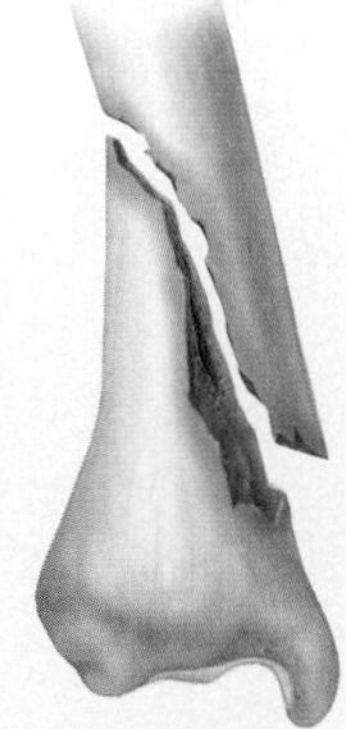

An *oblique* fracture occurs at an angle other than a right angle to the axis of the bone.

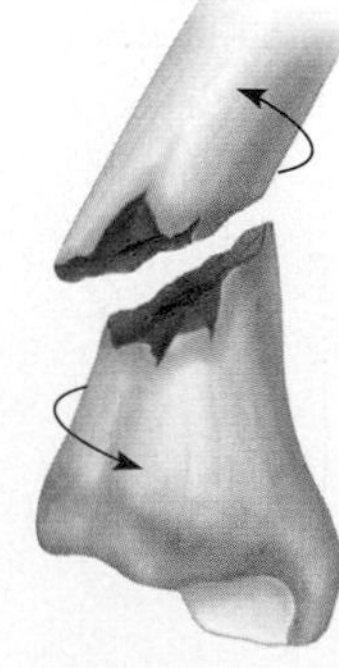

A *spiral* fracture is caused by excessive twisting of a bone.

Figure 7A Various types of fractures.

and the result is a bone shaped much like the original (fig. 7B).

Several techniques are used to help the bone-healing process. The first casts to immobilize fractured bones were introduced in 1876, and soon after, doctors began using screws and plates internally to align healing bone parts. Today, orthopedic surgeons also use rods, wires, and nails. These devices have become lighter and smaller; many are built of titanium. A device called a hybrid fixator treats a broken leg using metal pins internally to align bone pieces. The pins are anchored to a metal ring device worn outside the leg. Experimental approaches to helping bones heal include cartilage grafts and infusions of stem cells taken from a patient's own bone marrow.

If an epiphyseal plate is damaged as a result of a fracture before it ossifies, elongation of the long bone may prematurely cease, or if growth continues, it may be uneven. For this reason, injuries to the epiphyses of a young person's bones are of special concern. Surgery may be used on an epiphysis to equalize the growth of bones developing at very different rates.

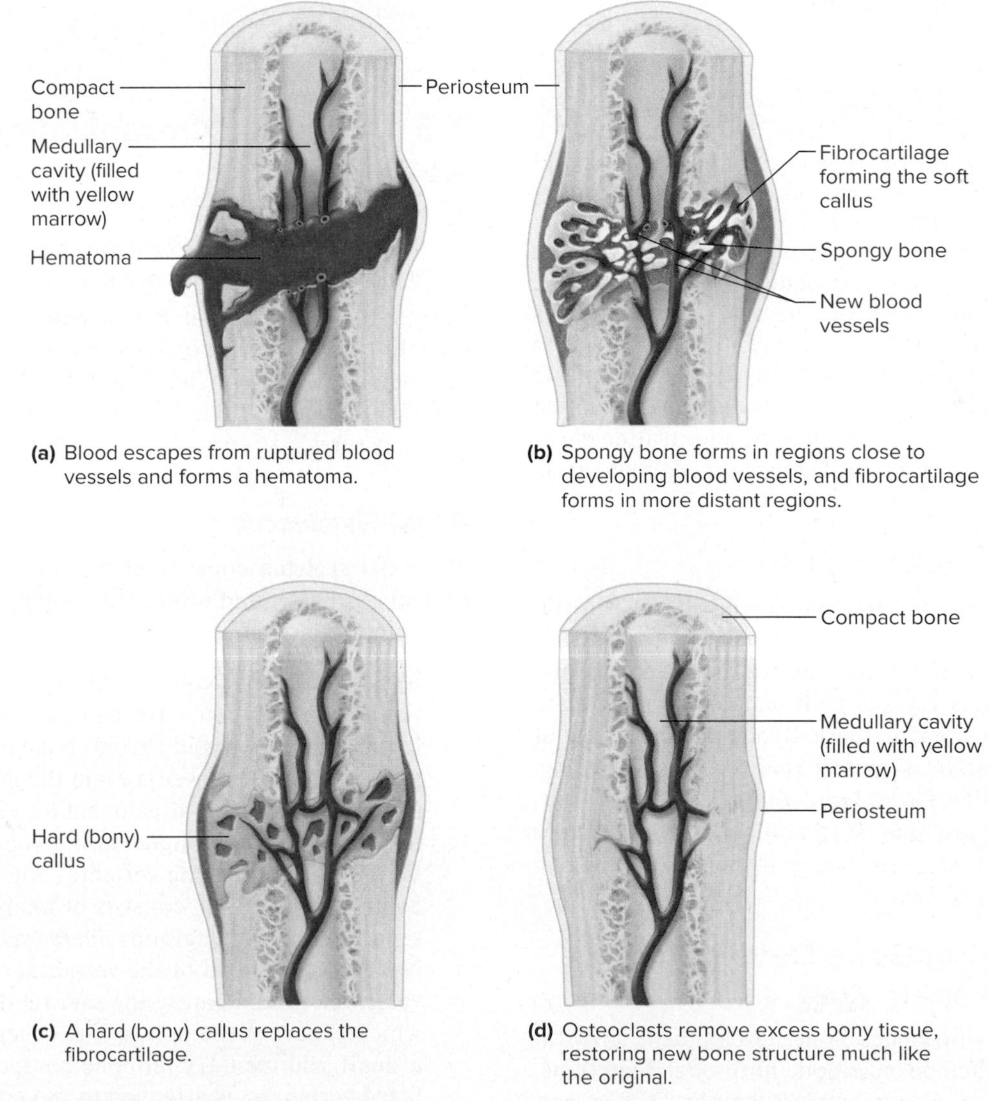

(a) Blood escapes from ruptured blood vessels and forms a hematoma.

(b) Spongy bone forms in regions close to developing blood vessels, and fibrocartilage forms in more distant regions.

(c) A hard (bony) callus replaces the fibrocartilage.

(d) Osteoclasts remove excess bony tissue, restoring new bone structure much like the original.

Figure 7B Major steps **(a–d)** in repair of a fracture.

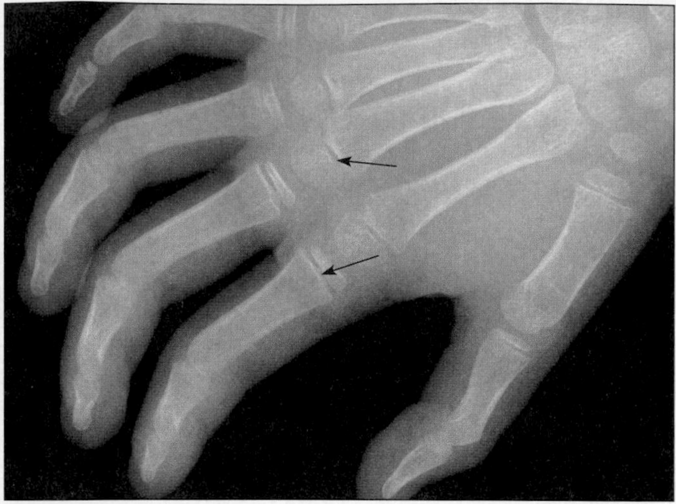

Figure 7.7 Radiograph of the right hand. Epiphyseal plates (arrows) in a child's bones indicate that the bones are still lengthening. Ed Reschke/Getty Images APR

A long bone continues to lengthen while the cartilaginous cells of the epiphyseal plates are active (fig. 7.7). However, once the ossification centers of the diaphysis and epiphyses meet and the epiphyseal plates ossify, lengthening is no longer possible in that end of the bone (see fig. 7.6*f*).

A developing long bone thickens as compact bone is deposited on the outside, just beneath the periosteum. As this compact bone forms on the surface, osteoclasts erode other bone tissue on the inside. The resulting space becomes the medullary cavity of the diaphysis, which later fills with marrow. The bone in the central regions of the epiphyses and diaphysis remains spongy, and hyaline cartilage on the ends of the epiphyses persists throughout life as articular cartilage.

Homeostasis of Bone Tissue

After the intramembranous and endochondral bones form, the actions of osteoclasts and osteoblasts continually remodel them. Throughout life, osteoclasts resorb bone matrix and osteoblasts replace it. Hormones that regulate blood calcium help control these opposing processes of *resorption* and *deposition* of matrix (see sections 11.5, Thyroid Gland, and 11.6, Parathyroid Glands). As a result, the total mass of bone tissue of an adult skeleton normally remains nearly constant, even though 3% to 5% of bone calcium is exchanged each year.

Factors Affecting Bone Development, Growth, and Repair APR

A number of factors influence bone development, growth, and repair. These include nutrition, hormonal secretions, and physical exercise. For example, vitamin D is necessary for proper absorption of calcium in the small intestine. In the absence of this vitamin, dietary calcium is poorly absorbed, and the inorganic salt portion of bone matrix will lack calcium, softening and thereby deforming bones. Growth hormone secreted by the pituitary gland stimulates division of the cartilage cells in the epiphyseal plates. Sex hormones stimulate ossification of the epiphyseal plates. Physical exercise pulling on muscular attachments to bones stresses the bones, stimulating the bone tissue to thicken and strengthen. Clinical Application 7.1 describes repair of a fractured bone.

PRACTICE 7.4

1. Describe the development of an intramembranous bone.
2. Explain how an endochondral bone develops.
3. Explain how osteoclasts and osteoblasts remodel bone.
4. Explain how nutritional factors, hormones, and physical exercise affect bone development and growth.

7.5 | Skeletal Organization

LEARN

1. Distinguish between the axial and appendicular skeletons, and name the major parts of each.

For purposes of study, it is convenient to divide the skeleton into two major portions—an axial skeleton and an appendicular skeleton (fig. 7.8). Table 7.1 lists the bones of the adult skeleton. Table 7.2 lists terms that describe skeletal structures.

Axial Skeleton

The **axial skeleton** consists of the bony and cartilaginous parts that support and protect the organs of the head, neck, and trunk. These parts include:

- **Skull.** The skull is composed of the **cranium** (krā′nē-um), or brain case, and the *facial bones.*
- **Hyoid bone.** The hyoid (hī′oid) bone is located in the neck between the lower jaw and the larynx. It supports the tongue and is an attachment for certain muscles that help move the tongue during swallowing.
- **Vertebral column.** The vertebral column, or spinal column (backbone), consists of many vertebrae separated by cartilaginous *intervertebral discs.* Near the distal end of the vertebral column, five vertebrae fuse, forming the **sacrum** (sa′krum), which is part of the pelvis. The **coccyx** (kok′siks), a small, rudimentary tailbone composed of four fused vertebrae, is attached to the end of the sacrum.
- **Thoracic cage.** The thoracic cage protects the organs of the thoracic cavity and the upper abdominal cavity.

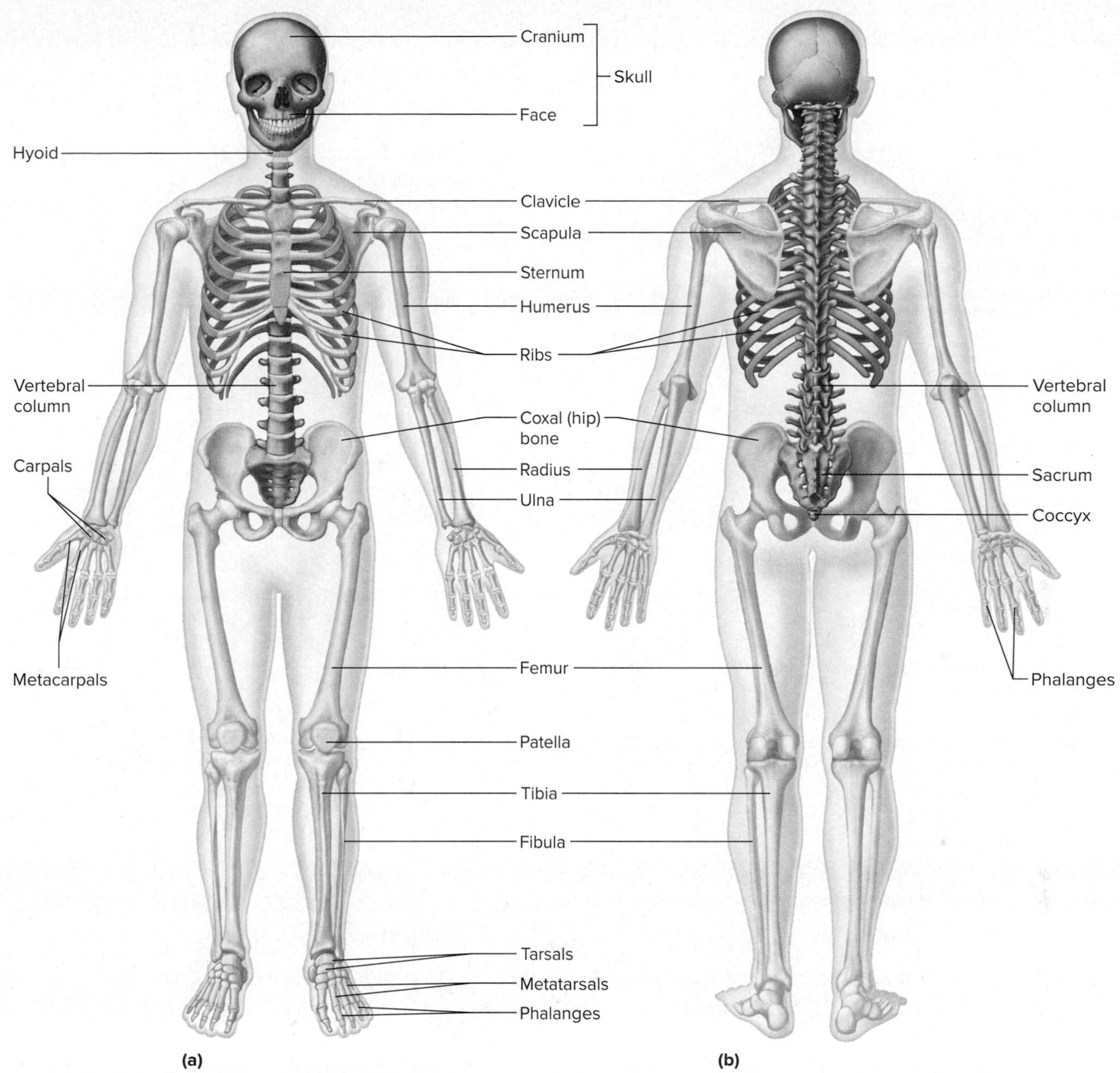

Figure 7.8 Major bones of the skeleton. **(a)** Anterior view. **(b)** Posterior view. The axial portion is shown in orange, and the appendicular portions are shown in yellow. APR

It is composed of twelve pairs of **ribs,** which articulate posteriorly with thoracic vertebrae. The thoracic cage also includes the **sternum** (ster′num), or breastbone, to which most of the ribs attach anteriorly.

Appendicular Skeleton

The **appendicular skeleton** consists of the bones of the upper and lower limbs and the bones that anchor the limbs to the axial skeleton. It includes:

- **Pectoral girdle**. The pectoral (pek′to-ral) girdle is formed by a **scapula** (scap′u-lah), or shoulder blade, and a **clavicle** (klav′ĭ-k′l), or collarbone, on both sides of the body. The pectoral girdle connects the bones of the upper limbs to the axial skeleton and aids in upper limb movements.
- **Upper limbs.** Each upper limb consists of a **humerus** (hū′mer-us), or arm bone, two forearm bones—a **radius** (ra′de-us) and an **ulna** (ul′nah)—and a hand. The humerus, radius, and ulna articulate with each other at the elbow joint. At the distal end of the radius and ulna is the hand. There are eight **carpals** (kar′pals), or wrist bones. The five bones of the palm are called **metacarpals** (met″ah-kar′pals), and the fourteen bones of the fingers are called **phalanges** (fah-lan′jēz; singular, *phalanx,* fa′lanks).

TABLE 7.1 Bones of the Adult Skeleton

1. Axial Skeleton
   a. Skull
      8 cranial bones
      frontal 1, temporal 2
      parietal 2, sphenoid 1
      occipital 1, ethmoid 1
      14 facial bones
      maxilla 2, lacrimal 2
      zygomatic 2, nasal 2
      palatine 2, vomer 1
      inferior nasal concha 2
      mandible 1 — 22 bones
   b. Middle ear bones
      malleus 2
      incus 2
      stapes 2 — 6 bones
   c. Hyoid
      hyoid bone 1 — 1 bone
   d. Vertebral column
      cervical vertebra 7
      thoracic vertebra 12
      lumbar vertebra 5
      sacrum 1
      coccyx 1 — 26 bones
   e. Thoracic cage
      rib 24
      sternum 1 — 25 bones
2. Appendicular Skeleton
   a. Pectoral girdle
      scapula 2
      clavicle 2 — 4 bones
   b. Upper limbs
      humerus 2
      radius 2
      ulna 2
      carpal 16
      metacarpal 10
      phalanx 28 — 60 bones
   c. Pelvic girdle
      coxal (hip) bone 2 — 2 bones
   d. Lower limbs
      femur 2
      tibia 2
      fibula 2
      patella 2
      tarsal 14
      metatarsal 10
      phalanx 28 — 60 bones

Total 206 bones

TABLE 7.2 Terms Used to Describe Skeletal Structures

Term	Definition	Examples
Condyle (kon′dīl)	Rounded process that usually articulates with another bone	Occipital condyle of the occipital bone (fig. 7.12)
Crest (krest)	Narrow, ridgelike projection	Iliac crest of the ilium (fig. 7.27)
Epicondyle (ep″ĭ-kon′dīl)	Projection situated above a condyle	Medial epicondyle of the humerus (fig. 7.23)
Facet (fas′et)	Small, nearly flat surface	Costal facet of the thoracic vertebra (fig. 7.16)
Fontanel (fon″tah-nel′)	Soft spot in the skull where membranes cover the space between bones	Anterior fontanel between the frontal and parietal bones (fig. 7.15)
Foramen (fo-ra′men)	Opening through a bone that usually is a passageway for blood vessels, nerves, or ligaments	Foramen magnum of the occipital bone (fig. 7.12)
Fossa (fos′ah)	Relatively deep pit or depression	Olecranon fossa of the humerus (fig. 7.23)
Fovea (fo′ve-ah)	Tiny pit or depression	Fovea capitis of the femur (fig. 7.29)
Head (hed)	Enlargement on the end of a bone	Head of the humerus (fig. 7.23)
Meatus (me-a′tus)	Tubelike passageway within a bone	External acoustic meatus of the temporal bone (fig. 7.11)
Process (pros′es)	Prominent projection on a bone	Mastoid process of the temporal bone (fig. 7.11)
Sinus (si′nus)	Cavity within a bone	Frontal sinus of the frontal bone (fig. 7.14)

TABLE 7.2 (continued)

Term	Definition	Examples
Spine (spīn)	Thornlike projection	Spine of the scapula (fig. 7.22)
Sulcus (sul'kus)	Furrow or groove	Intertubercular sulcus of the humerus (fig. 7.23)
Suture (soo'cher)	Interlocking line of union between bones	Lambdoid suture between the occipital and parietal bones (fig. 7.11)
Trochanter (tro-kan'ter)	Relatively large process	Greater trochanter of the femur (fig. 7.29)
Tubercle (tu'ber-kl)	Small, knoblike process	Greater tubercle of the humerus (fig. 7.23)
Tuberosity (tu"bĕ-ros'ĭ-te)	Knoblike process usually larger than a tubercle	Radial tuberosity of the radius (fig. 7.24)

- **Pelvic girdle.** The pelvic girdle is formed by two hip bones attached to each other anteriorly and to the sacrum posteriorly. The hip bones, the sacrum, and the coccyx form the **pelvis.**
- **Lower limbs.** Each lower limb consists of a **femur** (fē'mur), or thigh bone, two leg bones—a large **tibia** (tib'ē-ah) and a slender **fibula** (fib'ū-lah)—and a foot. The femur and tibia articulate with each other at the knee joint, where the **patella** (pah-tel'ah), also called the kneecap, covers the anterior surface. At the distal ends of the tibia and fibula is the foot. There are seven **tarsals** (tahr'sals), or ankle bones. The five bones of the instep are called **metatarsals** (met"ah-tahr'sals), and the fourteen bones of the toes (like the fingers) are called **phalanges.**

OF INTEREST The skeleton of an average 160-pound body weighs about 29 pounds.

PRACTICE 7.5

1. Distinguish between the axial and appendicular skeletons.
2. List the bones of the axial skeleton and of the appendicular skeleton.

7.6 | Skull APR

LEARN

1. Locate and identify the bones and the major features of the bones that compose the skull.

A human skull typically consists of twenty-two bones. Except for the lower jaw, the skull bones are firmly interlocked along immovable joints called *sutures* (soo'cherz) (fig. 7.9). Eight of these interlocked bones make up the cranium, and fourteen form the facial skeleton. (Three other bones in each middle ear are discussed in chapter 10, section 10.7, Sense of Hearing.) Reference plates 8, 9, 10, and 11 show the human skull and its parts.

Cranium

The **cranium** encloses and protects the brain, and its surface provides attachments for muscles that make chewing and head movements possible. Some of the cranial bones contain air-filled cavities called *paranasal sinuses* that are lined with mucous membranes and are connected by passageways to the nasal cavity (fig. 7.10). Sinuses reduce the skull's weight and increase the intensity of the voice by serving as resonant sound chambers.

The eight bones of the cranium, shown in figures 7.9 and 7.11, are:

- **Frontal bone.** The frontal (frun'tal) bone forms the anterior portion of the skull above the eyes. On the upper margin of each orbit (the bony socket of the eye), the frontal bone is marked by a *supraorbital foramen* (or *supraorbital notch* in some skulls), through which blood vessels and nerves pass to the tissues of the forehead. Within the frontal bone are two *frontal sinuses,* one above each eye near the midline (see fig. 7.10).
- **Parietal bones.** One parietal (pah-rī'ĕ-tal) bone is located on each side of the skull just behind the frontal bone (fig. 7.11). Together, the parietal bones form the bulging sides and roof of the cranium. They are joined at the midline along the *sagittal suture,* and they meet the frontal bone along the *coronal suture.*
- **Occipital bone.** The occipital (ok-sip'ĭ-tal) bone joins the parietal bones along the *lambdoid* (lam'doid) *suture* (figs. 7.11 and 7.12). It forms the back of the skull and the base of the cranium. A large opening on its lower surface is the **foramen magnum,** where nerve fibers from the brain enter the vertebral canal to become part of the spinal cord. Rounded processes called *occipital condyles,* located on each side of the foramen magnum, articulate with the first vertebra (atlas) of the vertebral column.
- **Temporal bones.** A temporal (tem'po-ral) bone on each side of the skull joins the parietal bone along a *squamous suture* (see figs. 7.9 and 7.11). The temporal bones form parts of the sides and the base of the cranium. Located near the inferior margin is an opening, the *external acoustic meatus,* which leads

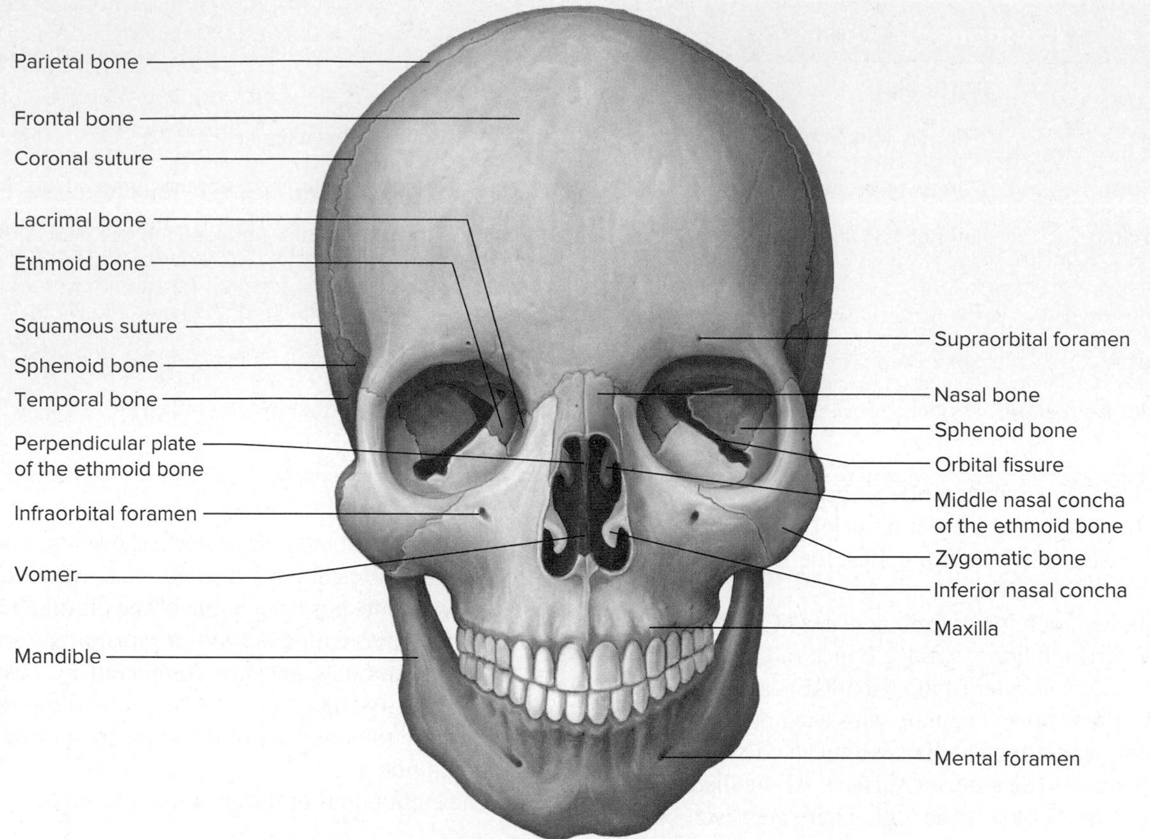

Figure 7.9 Anterior view of the skull. APR

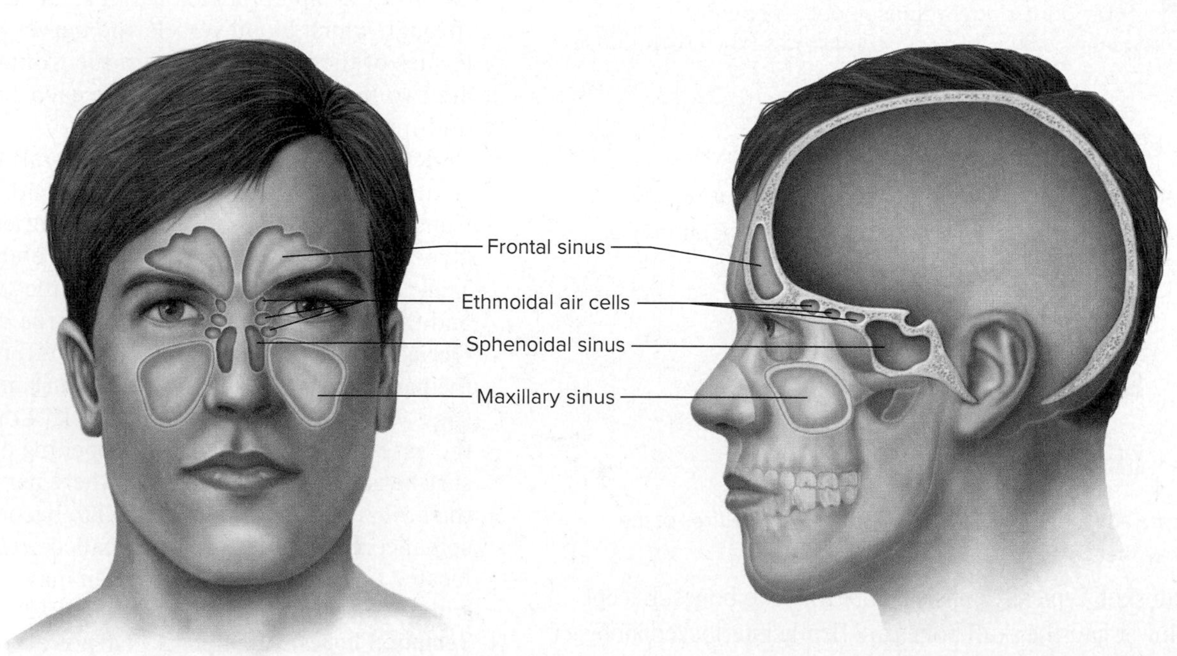

Figure 7.10 Locations of the paranasal sinuses.

inward to parts of the ear. The temporal bones have depressions called the *mandibular fossae* that articulate with condyles of the mandible.

Below each external acoustic meatus are two projections—a rounded *mastoid process* and a long, pointed *styloid process.* The mastoid process provides an attachment for certain muscles of the neck, whereas the styloid process anchors muscles associated with the tongue and pharynx. A *zygomatic process* projects anteriorly from the temporal bone, joins the *zygomatic bone,* and helps form the prominence of the cheek.

- **Sphenoid bone.** The sphenoid (sfe'noid) bone is wedged between several other bones in the anterior portion of the cranium (figs. 7.11 and 7.12). This bone helps form the base of the cranium, the sides of the skull, and the floors and sides of the orbits. Along the midline within the cranial cavity, a portion of the sphenoid bone indents, forming the saddle-shaped *sella turcica* (sel'ah tur'si-ka). The pituitary gland occupies this depression. The sphenoid bone also contains two *sphenoidal sinuses* (see fig. 7.10).
- **Ethmoid bone.** The ethmoid (eth'moid) bone is located in front of the sphenoid bone (figs. 7.11 and 7.13). It consists of two masses, one on each side of the nasal cavity, which are joined horizontally by thin *cribriform* (krib'rĭ-form) *plates.* These plates form part of the roof of the nasal cavity (fig. 7.13).

Projecting upward into the cranial cavity between the cribriform plates is a triangular process of the ethmoid bone called the *crista galli* (kris'tă gal'lī) (crest of the rooster). Membranes that enclose the brain attach to this process (figs. 7.13 and 7.14). Portions of the ethmoid bone also form sections of the cranial floor, the orbital walls, and the nasal cavity walls. A *perpendicular plate* extends inferiorly between the cribriform plates and forms most of the nasal septum (fig. 7.14).

Delicate scroll-shaped plates called the *superior nasal conchae* (kong'ke) and the *middle nasal conchae* project inward from the lateral portions of the ethmoid bone toward the perpendicular plate (see fig. 7.9). The lateral portions of the ethmoid bone contain many small spaces, the *ethmoidal air cells,* that together form the ethmoidal sinus (see fig. 7.10).

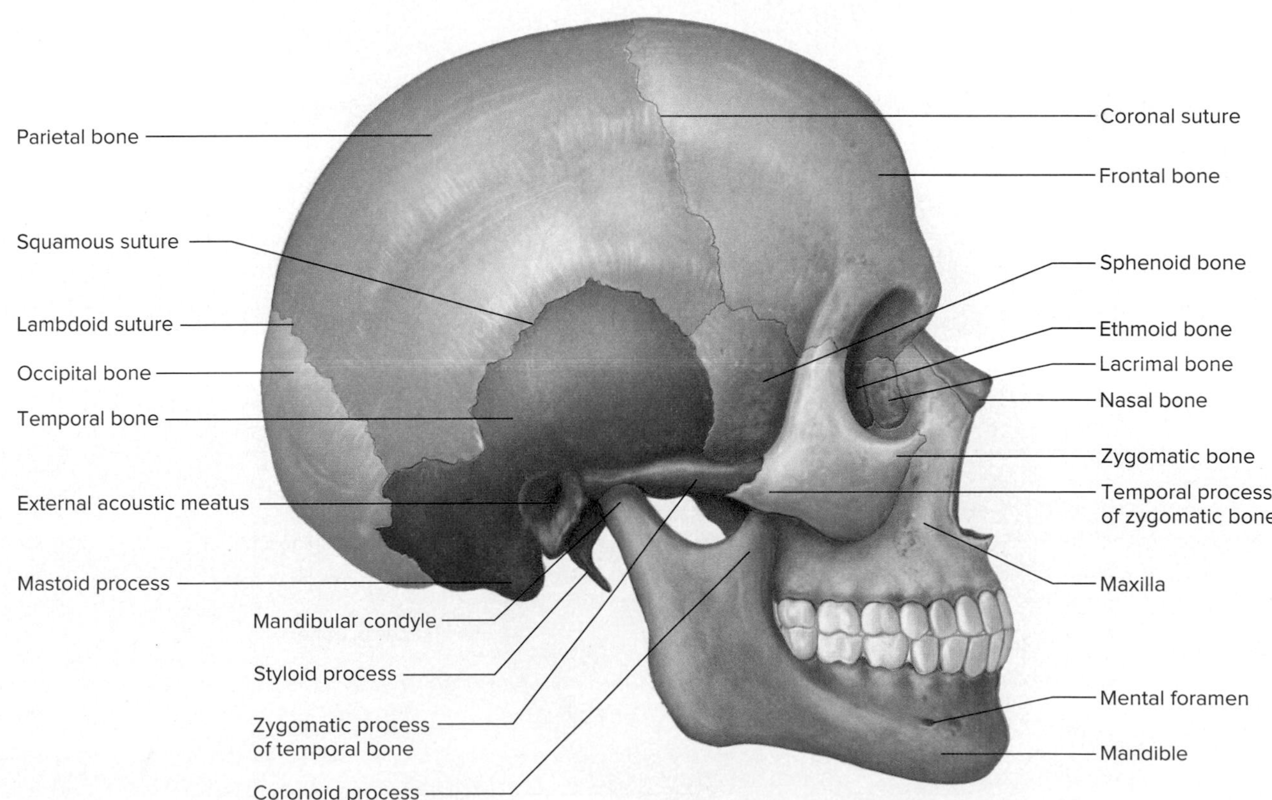

Figure 7.11 Right lateral view of the skull. APR

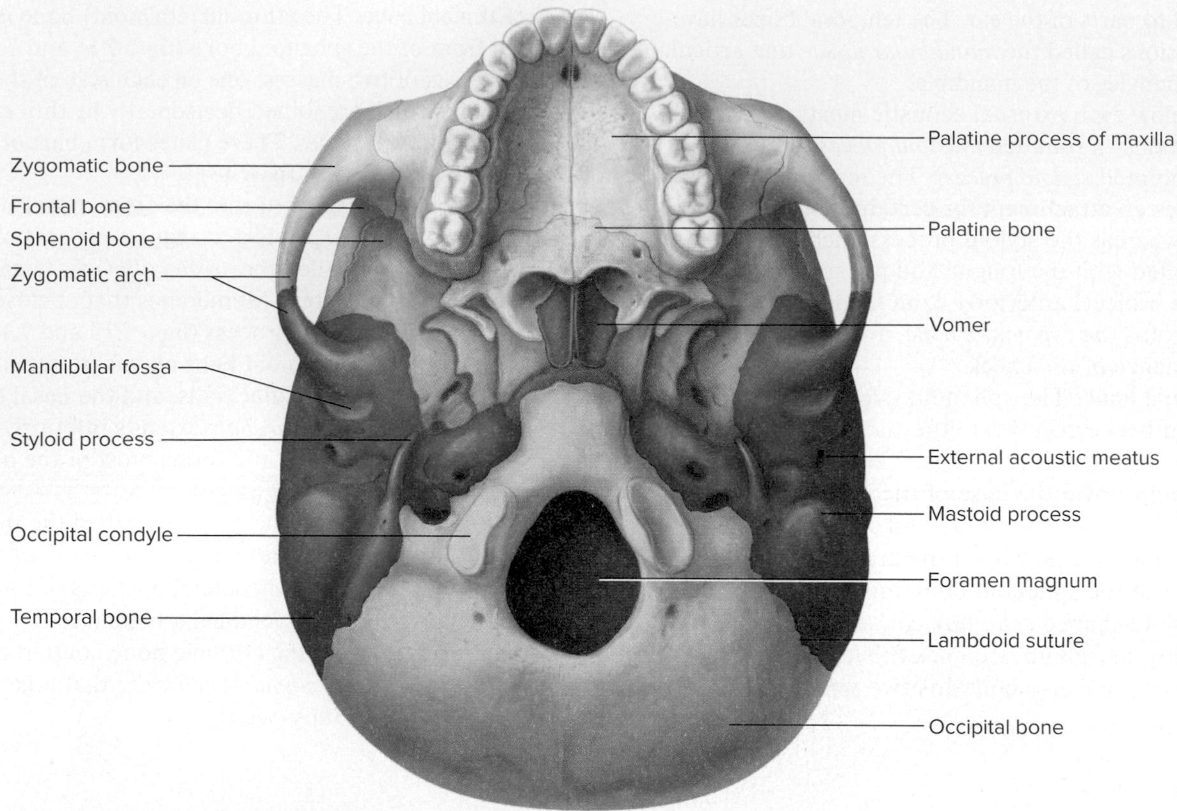

Figure 7.12 Inferior view of the skull. APR

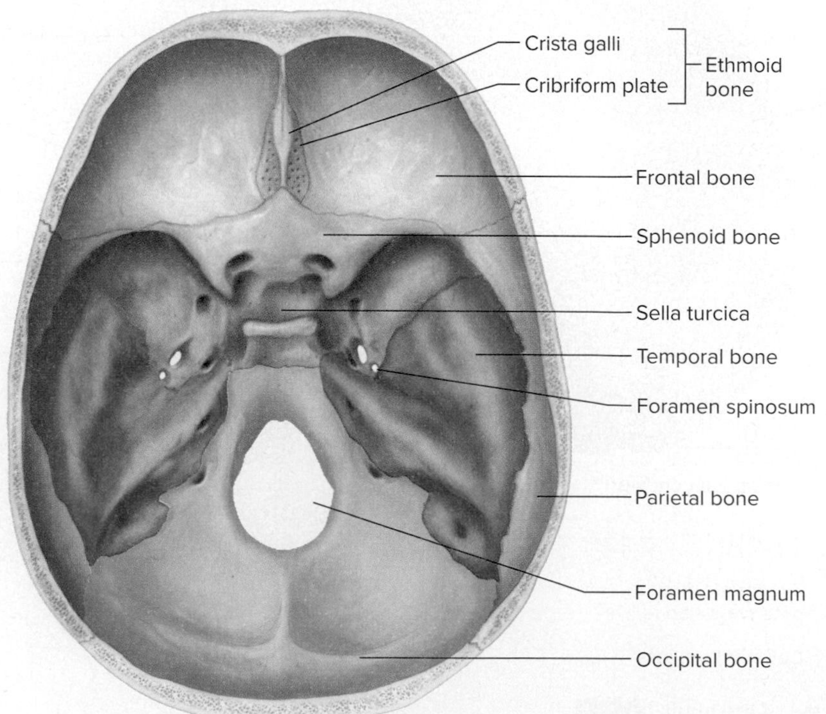

Figure 7.13 Floor of the cranial cavity, viewed from above. APR

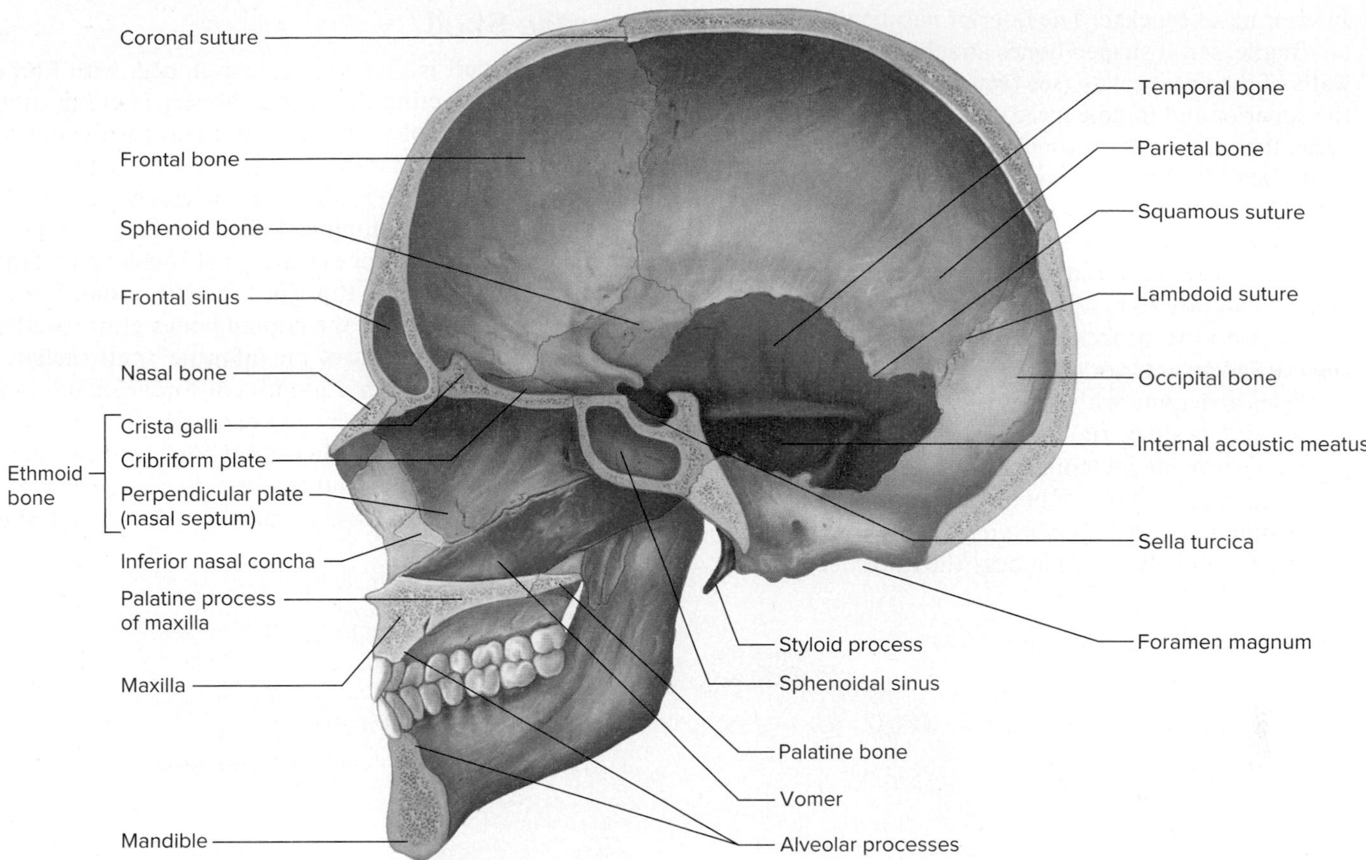

Figure 7.14 Sagittal section of the skull. APR

Facial Skeleton

The **facial skeleton** consists of thirteen immovable bones and a movable lower jawbone. These bones form the basic shape of the face and provide attachments for muscles that move the jaw and control facial expressions.

The bones of the facial skeleton are:

- **Maxillae.** The maxillae (mak-sil′e; singular, *maxilla,* mak-sil′ah) form the upper jaw (see figs. 7.11 and 7.12). Portions of these bones compose the anterior roof of the mouth (*hard palate*), the floors of the orbits, and the sides and floor of the nasal cavity. They also contain the sockets of the upper teeth. Inside the maxillae, lateral to the nasal cavity, are *maxillary sinuses,* the largest of the sinuses (see fig. 7.10).

  During development, portions of the maxillae called *palatine processes* grow together and fuse along the midline forming the anterior section of the hard palate. The inferior border of each maxillary bone projects downward, forming an *alveolar* (al-vē′o-lar) *process* (fig. 7.14). Together, these processes form a horseshoe-shaped *alveolar arch* (dental arch). Teeth occupy cavities (dental alveoli) in this arch. Dense connective tissue binds teeth to the bony sockets.
- **Palatine bones.** The L-shaped palatine (pal′ah-tīn) bones are located behind the maxillae (see figs. 7.12 and 7.14). The horizontal portions form the posterior section of the hard palate and the floor of the nasal cavity. The perpendicular portions help form parts of the lateral walls of the nasal cavity.
- **Zygomatic bones.** The zygomatic (zi″go-mat′ik) bones form the prominences of the cheeks below and to the sides of the eyes (see figs. 7.11 and 7.12). These bones also help form the lateral walls and the floors of the orbits. Each bone has a *temporal process,* which extends posteriorly, joining the zygomatic process of a temporal bone. Together, a temporal process and a zygomatic process form a *zygomatic arch.*
- **Lacrimal bones.** A lacrimal (lak′rĭ-mal) bone is a thin, scalelike structure located in the medial wall of each orbit between the ethmoid bone and the maxilla (see figs. 7.9 and 7.11).
- **Nasal bones.** The nasal (nā′zal) bones are long, thin, and nearly rectangular (see figs. 7.9 and 7.11). They lie side by side and are fused at the midline, where they form the bridge of the nose.
- **Vomer.** The thin, flat vomer (vō′mer) is located along the midline within the nasal cavity (see figs. 7.9 and 7.14). Posteriorly, it joins the perpendicular plate of the ethmoid bone, and together they form the nasal septum.

- **Inferior nasal conchae.** The inferior nasal conchae are fragile, scroll-shaped bones attached to the lateral walls of the nasal cavity (see figs. 7.9 and 7.14). Like the superior and middle nasal conchae of the ethmoid bone, the inferior nasal conchae support mucous membranes in the nasal cavity.
- **Mandible.** The mandible (man′dĭ-b′l) is also called the lower jawbone. It has a horizontal, horseshoe-shaped body with a vertical, flat portion projecting upward at each end (see figs. 7.9 and 7.11). This projection is divided into two processes—a posterior *mandibular condyle* and an anterior *coronoid process.* The mandibular condyles articulate with the mandibular fossae of the temporal bones (see fig. 7.12), whereas the coronoid processes provide attachments for muscles used in chewing. A curved bar of bone on the superior border of the mandible, the *alveolar arch,* contains the hollow sockets (dental alveoli) that bear the lower teeth.

Infantile Skull

At birth the skull is incompletely developed, with fibrous membranes connecting the cranial bones. These membranous areas of incomplete intramembranous ossification are called **fontanels** (fon″tah-nelz′) or, more commonly, soft spots (fig. 7.15). They permit some movement between the bones, so that the developing skull is partially compressible and can slightly change shape. This enables an infant's skull to more easily pass through the birth canal. Eventually the fontanels close as the cranial bones grow together.

Other characteristics of an infantile skull include a relatively small face with a prominent forehead and large orbits. The jaw and nasal cavity are small, the sinuses are incompletely formed, and the frontal bone is in two parts. The skull bones are thin, but they are also somewhat flexible and thus are less easily fractured than adult skull bones.

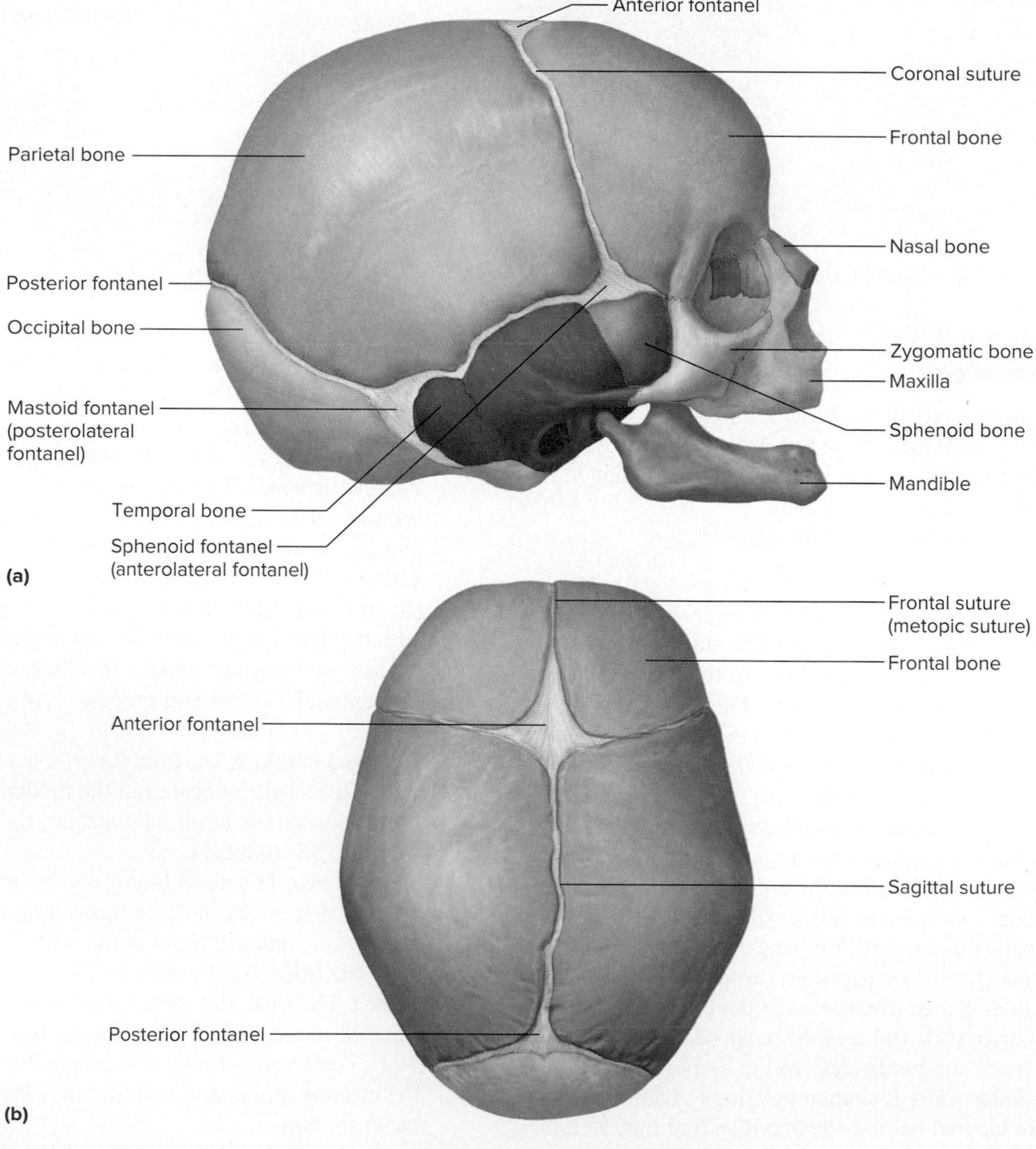

Figure 7.15 Fontanels. **(a)** Right lateral view and **(b)** superior view of the infantile skull.

PRACTICE 7.6

1. Locate and name each of the bones of the cranium.
2. Locate and name each of the facial bones.
3. Explain how an adult skull differs from that of an infant.

7.7 | Vertebral Column

LEARN

1. Locate and identify the bones and the major features of the bones that compose the vertebral column.

The **vertebral column** extends from the skull to the pelvis and forms the vertical axis of the skeleton. It is composed of many bony parts, called **vertebrae** (ver'tĕ-bre), that are separated by pads of fibrocartilage called **intervertebral discs** and are connected to one another by ligaments (fig. 7.16). The vertebral column supports the head and trunk of the body, and protects the spinal cord.

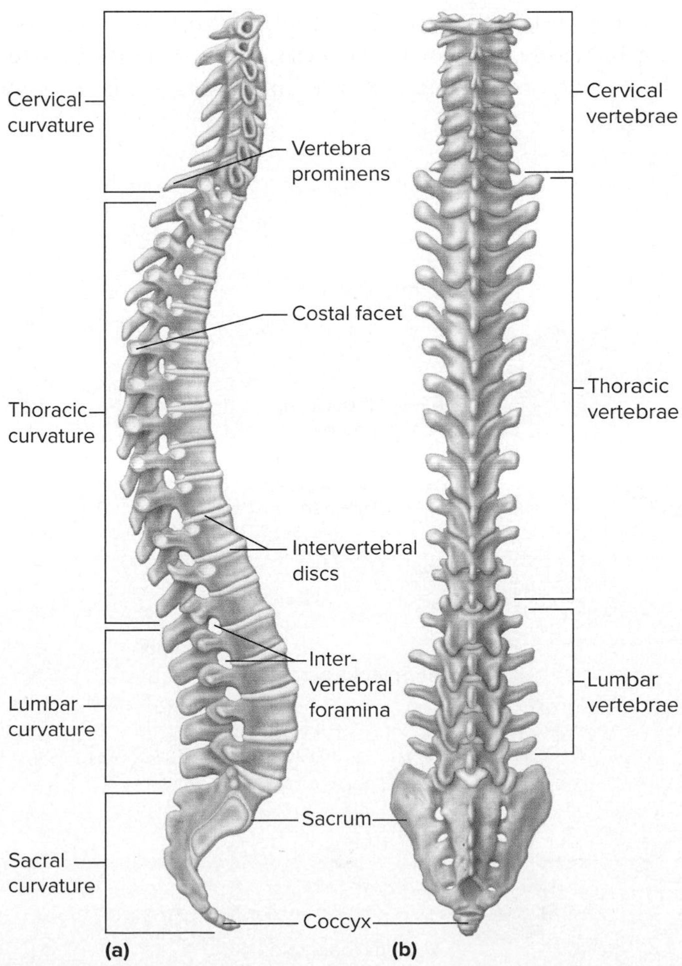

Figure 7.16 The curved vertebral column consists of many vertebrae separated by intervertebral discs. **(a)** Right lateral view. **(b)** Posterior view. APR

A Typical Vertebra

The vertebrae in different regions of the vertebral column have special characteristics, but they also have features in common. A typical vertebra has a drum-shaped *body,* which forms the thick, anterior portion of the bone (fig. 7.17). A longitudinal row of these vertebral bodies supports the weight of the head and trunk. The intervertebral discs, which separate adjacent vertebral bodies, cushion and soften the forces generated by such movements as walking and jumping.

Projecting posteriorly from each vertebral body are two short stalks called *pedicles* (ped'ĭ-k'lz). Two plates called *laminae* (lam'i-ne) arise from the pedicles and fuse in the back to become a *spinous process.* The pedicles, laminae, and spinous process together complete a bony *vertebral arch* around the *vertebral foramen.* The vertebral foramina within the bones of the vertebral column form a *vertebral canal,* through which the spinal cord passes.

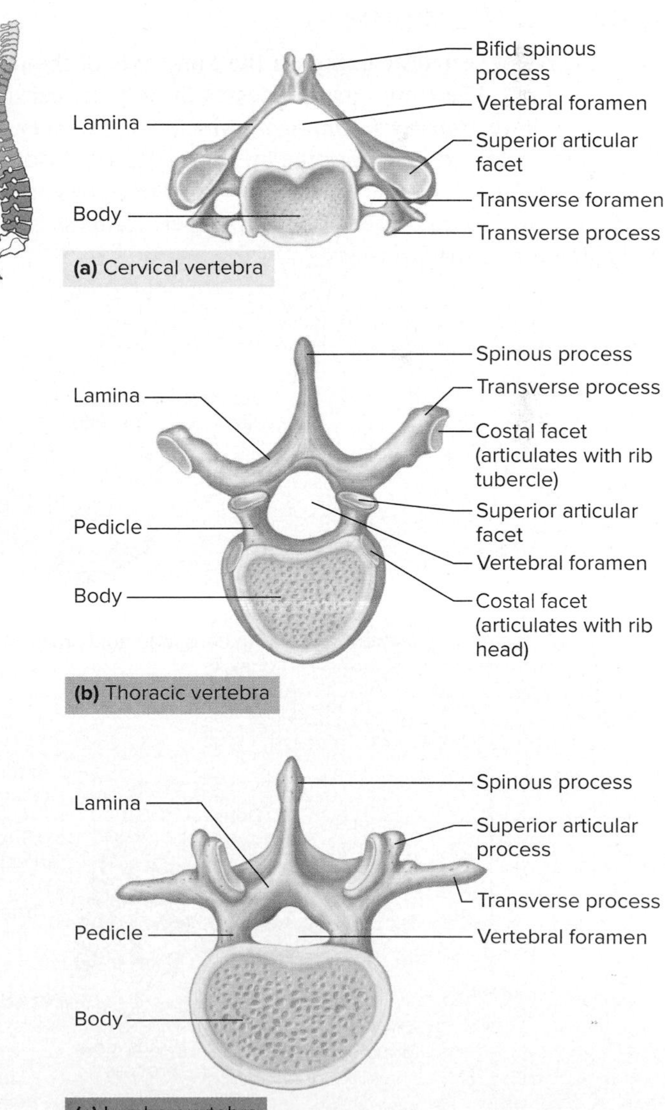

Figure 7.17 Superior view of **(a)** a cervical vertebra, **(b)** a thoracic vertebra, and **(c)** a lumbar vertebra. APR

If the laminae of the vertebrae do not unite during development, the vertebral arch remains incomplete, causing a condition called *spina bifida.* The contents of the vertebral canal protrude outward. This condition occurs most frequently in the lumbosacral region. Spina bifida is associated with folic acid deficiency in certain genetically susceptible individuals.

Between the pedicles and laminae of a typical vertebra are the *transverse processes,* which project laterally and posteriorly. Ligaments and muscles are attached to the dorsal spinous process and the transverse processes. Projecting upward and downward from each vertebral arch are *superior* and *inferior articular processes.* These processes bear cartilage-covered facets by which each vertebra is joined to the one above and the one below it.

On the lower surfaces of the vertebral pedicles are notches that align with adjacent vertebrae to help form openings called *intervertebral foramina* (in″ter-ver′tĕ-bral fo-ram′ĭ-nah) (see fig. 7.16). These openings provide passageways for spinal nerves.

Cervical Vertebrae

Seven **cervical vertebrae** compose the bony axis of the neck (see fig. 7.16). The transverse processes of only the cervical vertebrae have *transverse foramina,* which are passageways for blood vessels to and from the brain (fig. 7.17*a*). Also, the spinous processes of the second through the sixth cervical vertebrae are uniquely forked (bifid). These processes provide attachments for muscles.

The spinous process of the seventh cervical vertebra is called the vertebra prominens. It gets this name because it is more "prominent" than those of the first six cervical vertebrae. It is easily found at the base of the neck and serves as a landmark for locating other vertebral parts.

Two of the cervical vertebrae are of special interest: the atlas and the axis (fig. 7.18). The first vertebra, or **atlas** (at′las), supports the head. It has practically no body or spinous process and appears as a bony ring with two transverse processes. On its superior surface are two kidney-shaped *facets* that articulate with the occipital condyles.

The second cervical vertebra, or **axis** (ak′sis), bears a toothlike *dens* (odontoid process) on its body. This process projects upward and lies in the ring of the atlas. As the head is turned from side to side, the atlas pivots around the dens.

Thoracic Vertebrae

The twelve **thoracic vertebrae** are larger than the cervical vertebrae (see fig. 7.16). Each thoracic vertebra has a long, pointed spinous process which slopes downward. Thoracic vertebrae except for T10–T12 have two costal facets that articulate with ribs (see fig. 7.17*b*).

Beginning with the third thoracic vertebra and moving inferiorly, the bodies of these bones increase in size. Thus, they are adapted to bear increasing loads of body weight.

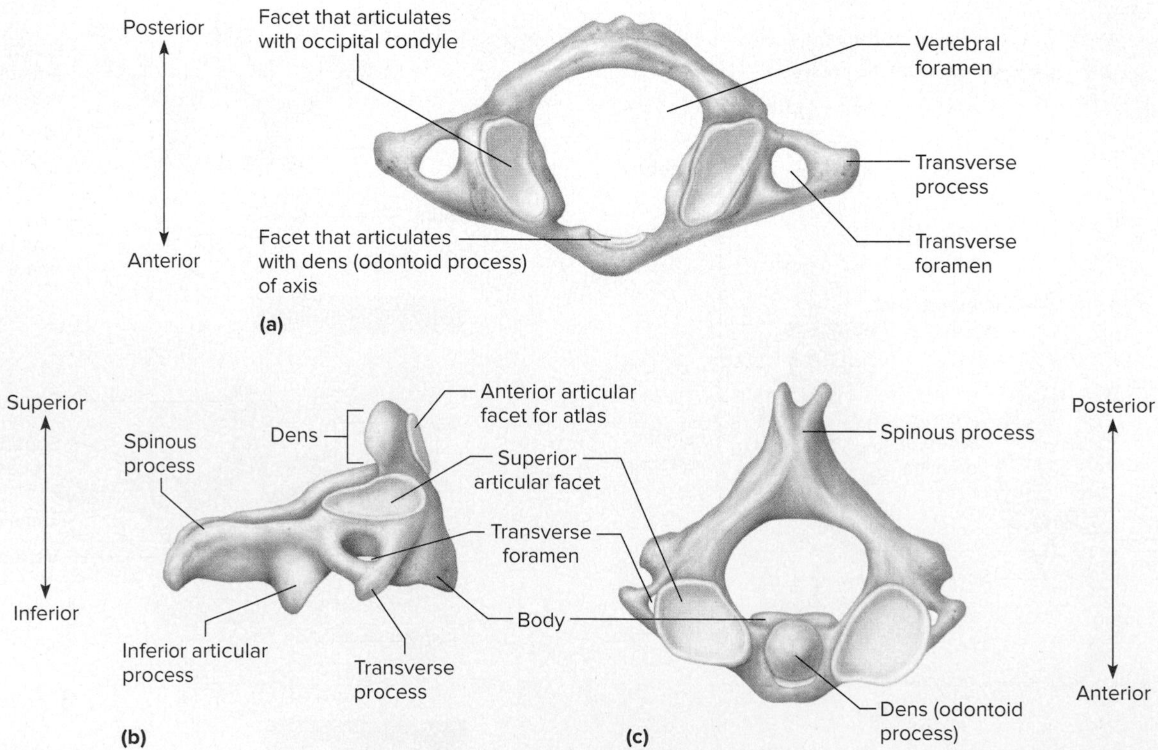

Figure 7.18 Atlas and axis. **(a)** Superior view of the atlas. **(b)** Right lateral view and **(c)** superior view of the axis. 

CLINICAL APPLICATION 7.2

Disorders of the Vertebral Column

Changes in the intervertebral discs may cause painful problems. Each disc is composed of a tough, outer layer of fibrocartilage (annulus fibrosus) and an elastic central mass (nucleus pulposus). With age, these discs degenerate. The central masses lose firmness, and the outer layers thin, weaken, and crack. Taking a fall or lifting a heavy object can exert enough pressure to break the outer layers of the discs and squeeze out the central masses. Such a rupture may press on the spinal cord or on spinal nerves that branch from it. This condition, called a *ruptured* or *herniated* (protruding) *disc,* may cause inflammation with back pain and numbness, or loss of muscular function in the parts innervated by the affected spinal nerves.

The pain of a herniated disc in the lower back decreases in about one month for half of affected individuals and for most others within six months. Pain medications and physical therapy may be very helpful to strengthen muscles surrounding the injured area.

For about 10% of people with herniated discs, a surgical procedure called a laminectomy may relieve the pain by removing a portion of the posterior arch of a vertebra, which reduces pressure on nerve tissues. Then a surgeon may perform a microdiscectomy to remove the herniated disc material.

The painful condition spondylolisthesis results from a vertebra that slips out of place over the vertebra beneath it. Most commonly the fifth lumbar vertebra slides forward over the body of the sacrum. Spondylolisthesis may be present at birth, be a consequence of small stress fractures of the vertebra between the superior and inferior articular processes (spondylolysis), or, most commonly, be associated with the loss of moisture in intervertebral discs that accompanies aging. Gymnasts, football players, and other athletes and dancers who flex, extend, or rotate their vertebral columns excessively and forcefully are at elevated risk of developing spondylolisthesis.

Poor posture, injury, or disease can affect the curvatures of the vertebral column. An exaggerated thoracic curvature causes rounded shoulders and a hunchback. This condition, called *kyphosis,* is seen in adolescents who undertake strenuous athletic activities. Unless corrected before bone growth completes, the condition can permanently deform the vertebral column.

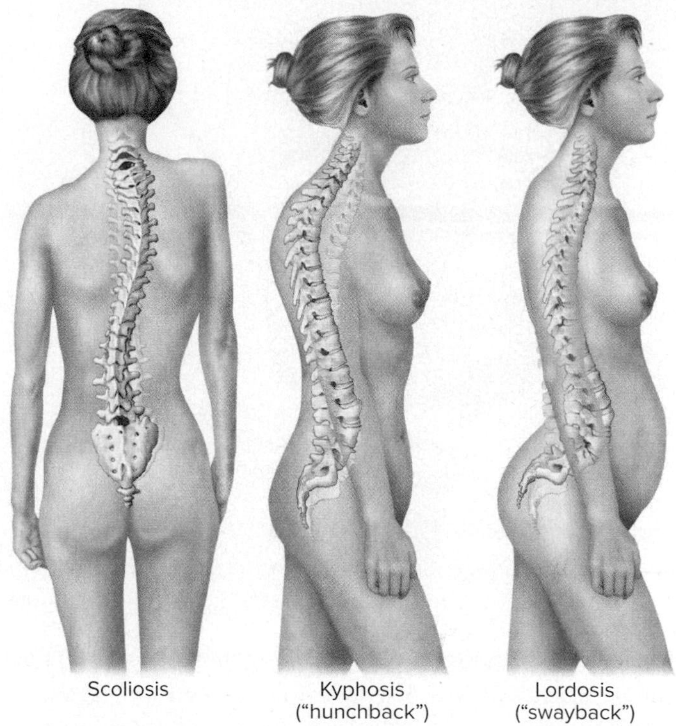

The vertebral column may develop an abnormal lateral curvature, placing one hip or shoulder lower than the other, which may displace or compress the thoracic and abdominal organs. This condition, called *scoliosis,* is most common in adolescent females. It may accompany poliomyelitis, rickets, or tuberculosis, or have an unknown cause. An accentuated lumbar curvature is called *lordosis,* or swayback.

As a person ages, the intervertebral discs shrink and become more rigid, and compression is more likely to fracture the vertebral bodies. Consequently, height may decrease, and the thoracic curvature of the vertebral column may become accentuated (kyphosis), bowing the back.

Lumbar Vertebrae

Five **lumbar vertebrae** are in the small of the back (loin) (see fig. 7.16). These vertebrae are adapted with larger and stronger bodies to support more weight than the vertebrae above them (see fig. 7.17*c*).

Sacrum

The **sacrum** (sa'krum) is a triangular structure, composed of five fused vertebrae, that forms the base of the vertebral column (fig. 7.19). The spinous processes of these fused bones form a ridge of *tubercles.* To the sides of the tubercles

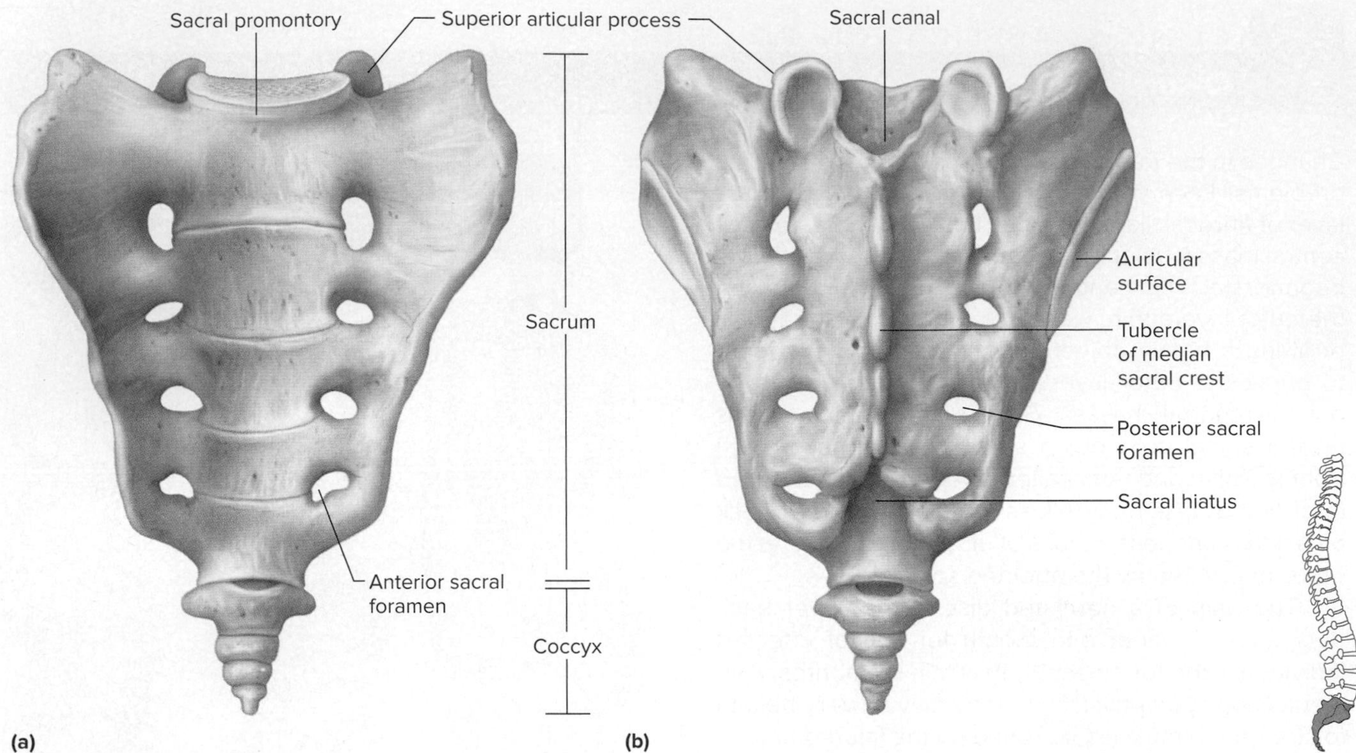

Figure 7.19 Sacrum and coccyx. **(a)** Anterior view and **(b)** posterior view. APR

are rows of openings, the *posterior sacral foramina,* through which nerves and blood vessels pass.

The vertebral foramina of the sacral vertebrae form the *sacral canal,* which continues through the sacrum to an opening of variable size at the tip, called the *sacral hiatus* (sa′kral hī-a′tus). On the ventral surface of the sacrum, four pairs of *anterior sacral foramina* provide passageways for nerves and blood vessels.

Coccyx

The **coccyx** (kok′siks), or tailbone, is the lowest part of the vertebral column and is typically composed of four fused vertebrae (fig. 7.19). Ligaments attach it to the margins of the sacral hiatus. Clinical Application 7.2 details some common vertebral column disorders.

PRACTICE 7.7

1. Describe the structure of the vertebral column.
2. Describe a typical vertebra.
3. Explain how the structures of cervical, thoracic, and lumbar vertebrae differ.

7.8 | Thoracic Cage

LEARN

1. Locate and identify the bones and the major features of the bones that compose the thoracic cage.

The **thoracic cage** includes the ribs, the thoracic vertebrae, the sternum, and the costal cartilages that attach the ribs to the sternum (fig. 7.20). These bones support the pectoral girdle and upper limbs, protect the viscera in the thoracic and upper abdominal cavities, and play a role in breathing.

Ribs

The usual number of **ribs** is twenty-four—one pair attached to each of the twelve thoracic vertebrae. The first seven rib pairs, *true ribs* (vertebrosternal ribs), join the sternum directly by their costal cartilages. The remaining five pairs are called *false ribs* because their cartilages do not reach the sternum directly. Instead, the cartilages of the upper three false ribs (vertebrochondral ribs) join the cartilages of the seventh rib. The last two (or sometimes three) rib pairs are called *floating ribs* (vertebral ribs) because they have no cartilaginous attachments to the sternum.

A typical rib has a long, slender shaft, which curves around the chest and slopes downward. On the posterior end is an enlarged *head* by which the rib articulates with a *facet* on the body of its own vertebra and with the body of the next higher vertebra. A *tubercle,* close to the head of the rib, articulates with the transverse process of the vertebra.

Sternum

The **sternum,** or breastbone, is located along the midline in the anterior portion of the thoracic cage (fig. 7.20). This flat, elongated bone develops in three parts—an upper *manubrium* (mah-nu′bre-um), a middle *body,* and a lower *xiphoid* (zīf′oid) *process* that projects downward. The manubrium articulates with the clavicles by facets on its superior border.

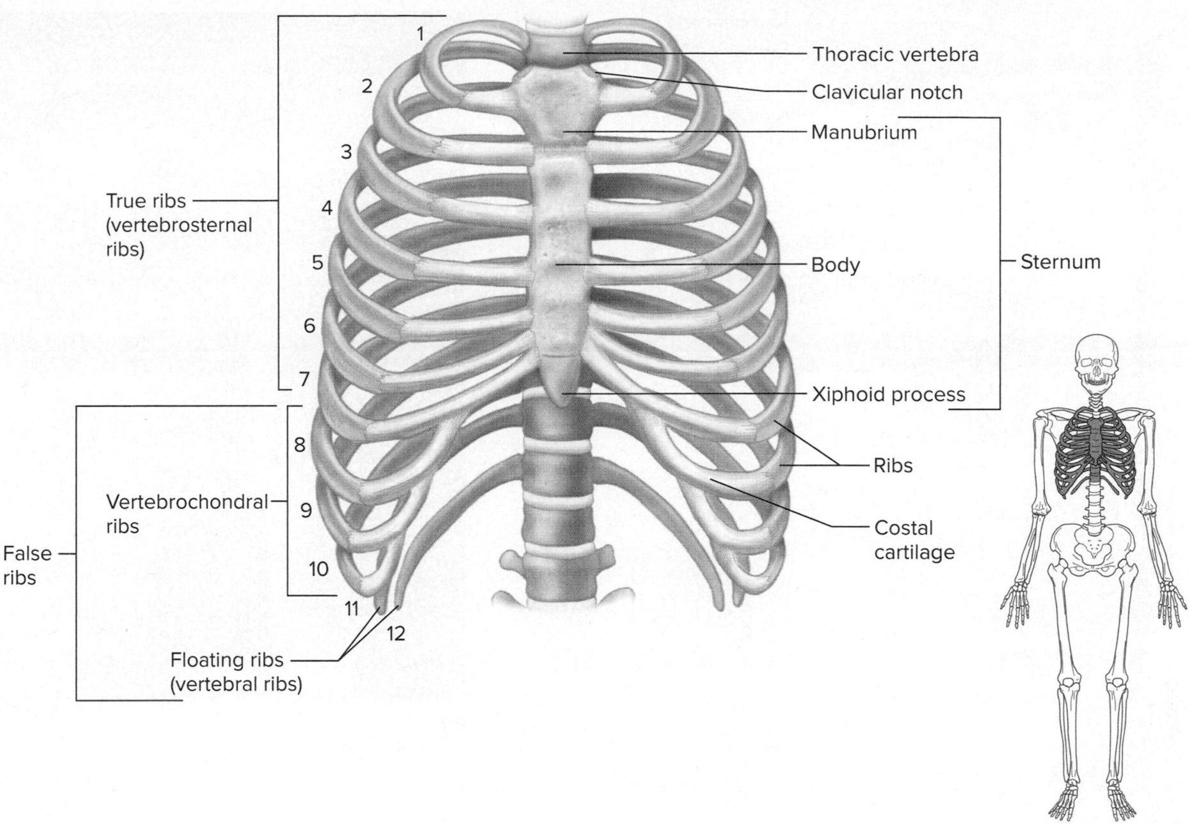

Figure 7.20 The thoracic cage includes the ribs, the thoracic vertebrae, the sternum, and the costal cartilages that attach the ribs to the sternum. APR

PRACTICE 7.8

1. Which bones compose the thoracic cage?
2. What are the differences among true, false, and floating ribs?
3. Name the three parts of the sternum.

7.9 | Pectoral Girdle

LEARN

1. Locate and identify the bones and the major features of the bones that compose the pectoral girdle.

Each **pectoral girdle**, or shoulder girdle, is composed of two parts—a clavicle and a scapula (fig. 7.21). Although the word *girdle* suggests a ring-shaped structure, the pectoral girdle is an incomplete ring. It is open in the back between the scapulae, and the sternum separates its bones in front. The pectoral girdle supports the upper limbs and is an attachment for several muscles that move them.

Clavicles

The **clavicles,** or collarbones, are slender, rodlike bones with elongated S shapes (fig. 7.21). Located at the base of the neck, they run horizontally between the manubrium and the scapulae.

The clavicles brace the freely movable scapulae, helping to hold the shoulders in place. They also provide attachments for muscles of the upper limbs, chest, and back.

Scapulae

The **scapulae** (skap'u-le), or shoulder blades, are broad, somewhat triangular bones located on either side of the upper back (figs. 7.21 and 7.22). A *spine* divides the posterior surface of each scapula into unequal portions. This spine leads to an *acromion* (ah-kro'me-on) *process* that forms the tip of the shoulder. The acromion process articulates with the clavicle and provides attachments for muscles of the upper limb and chest. A *coracoid* (kor'ah-koid) *process* curves anteriorly and inferiorly to the clavicle. The coracoid process also provides attachments for upper limb and chest muscles. On the lateral surface of the scapula and between these processes is a depression called the *glenoid cavity* (glenoid fossa of the scapula) that articulates with the head of the humerus.

PRACTICE 7.9

1. Which bones form the pectoral girdle?
2. What is the function of the pectoral girdle?

Acromion process
Acromial end
Sternal end
Clavicle
Head of humerus
Coracoid process
Scapula
Rib
Costal cartilage
Humerus
Ulna
Radius
(a)

Acromion process
Head of humerus
Coracoid process
Sternum
Humerus
Rib
(b)

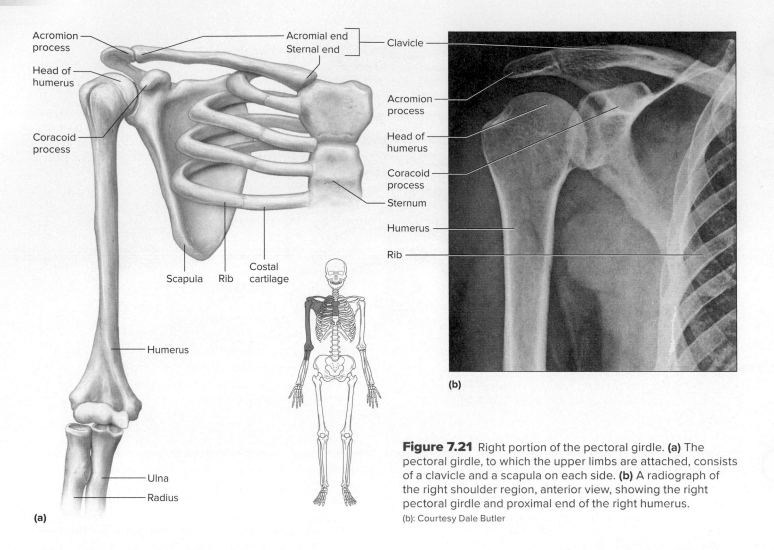

Figure 7.21 Right portion of the pectoral girdle. **(a)** The pectoral girdle, to which the upper limbs are attached, consists of a clavicle and a scapula on each side. **(b)** A radiograph of the right shoulder region, anterior view, showing the right pectoral girdle and proximal end of the right humerus.
(b): Courtesy Dale Butler

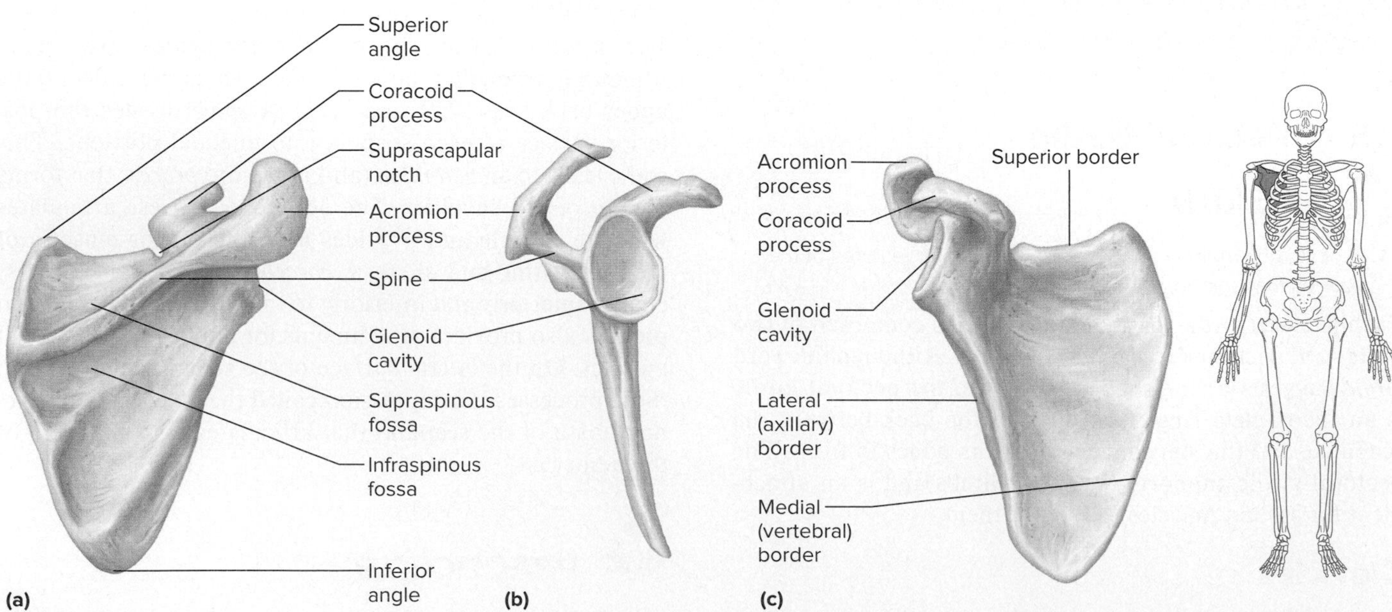

Figure 7.22 Right scapula. **(a)** Posterior surface. **(b)** Lateral view showing the glenoid cavity, which articulates with the head of the humerus. **(c)** Anterior surface. APR

7.10 | Upper Limb

LEARN

1. Locate and identify the bones and the major features of the bones that compose the upper limb.

The bones of the upper limb form the framework of the arm, forearm, and hand. They also provide attachments for muscles and interact with muscles to move limb parts. These bones include a humerus, a radius, an ulna, carpals, metacarpals, and phalanges (see fig. 7.8).

Humerus

The **humerus** is a long bone that extends from the scapula to the elbow (fig. 7.23). At its upper end is a smooth, rounded *head* that fits into the glenoid cavity of the scapula. Just below the head are two processes—a *greater tubercle* on the lateral side and a *lesser tubercle* on the anterior side. These tubercles provide attachments for muscles that move the upper limb at the shoulder. Between them is a narrow furrow, the *intertubercular sulcus* (intertubercular groove).

The narrow depression along the lower margin of the humerus head that separates it from the tubercles is called the *anatomical neck.* Just below the head and the tubercles is a tapering region called the *surgical neck,* so named because fractures commonly occur there. Near the middle of the bony shaft on the lateral side is a rough, V-shaped area called the *deltoid tuberosity.* It provides an attachment for the muscle (deltoid) that raises the upper limb horizontally to the side.

At the lower end of the humerus are two smooth *condyles,* a lateral *capitulum* that articulates with the radius and a medial *trochlea* that articulates with the ulna. Above the condyles on either side are *epicondyles,* which provide attachments for muscles and ligaments of the elbow. Between the epicondyles anteriorly is a depression, the *coronoid* (kor′o-noid) *fossa,* that receives a process of the ulna (coronoid process) when the elbow bends. Another depression on the posterior surface, the *olecranon* (ō″lek′ra-non) *fossa,* receives a different ulnar process, the olecranon process, when the elbow straightens.

Radius

The **radius,** located on the thumb side of the forearm, extends from the elbow to the wrist. It crosses over the ulna when the hand is turned so that the palm faces backward (fig. 7.24). A thick, disclike *head* at the upper end of the radius articulates with the capitulum of the humerus and a notch of the ulna (radial notch). This arrangement allows the radius to rotate.

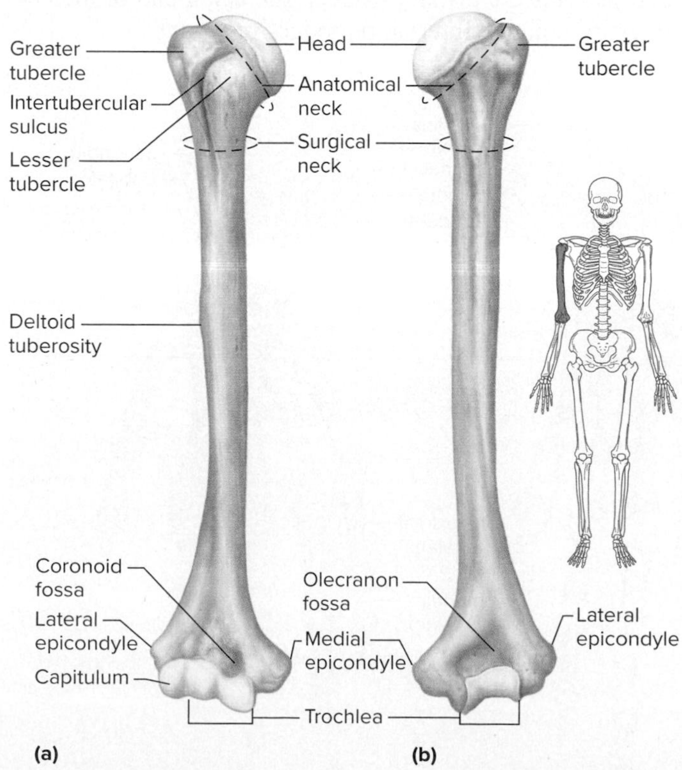

Figure 7.23 Right humerus. **(a)** Anterior surface. **(b)** Posterior surface. APR

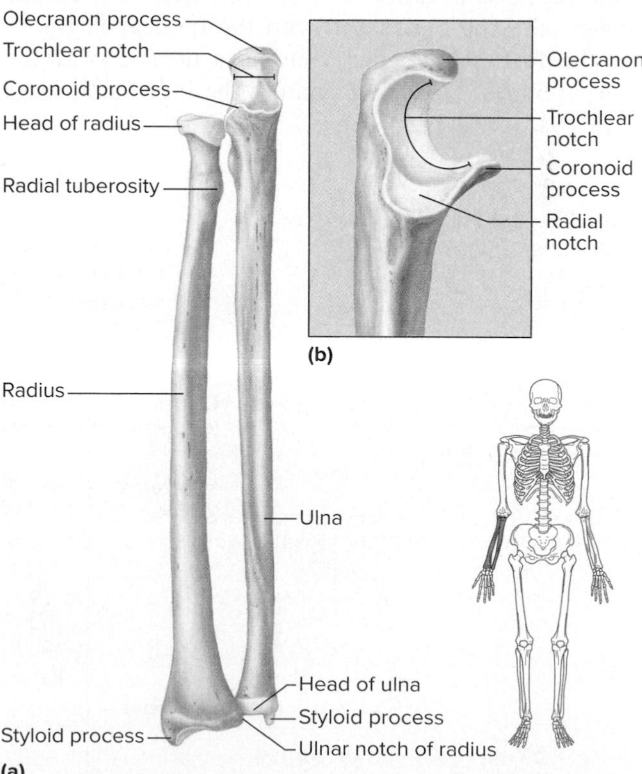

Figure 7.24 Right radius and ulna. **(a)** The head of the radius articulates with the radial notch of the ulna, and the head of the ulna articulates with the ulnar notch of the radius. **(b)** Lateral view of the proximal end of the ulna. APR

On the radial shaft just below the head is a process called the *radial tuberosity.* It is an attachment for a muscle (biceps brachii) that bends the upper limb at the elbow. At the distal end of the radius, a lateral *styloid* (stī'loid) *process* provides attachments for ligaments of the wrist.

Ulna

The **ulna** is medial to the radius. It is longer than the radius and overlaps the end of the humerus posteriorly (fig. 7.24). At its proximal end, the ulna has a wrenchlike opening, the *trochlear* (trok'lē-ar) *notch,* that articulates with the trochlea of the humerus. Two processes above and below this notch, the *olecranon process* and the *coronoid process,* provide attachments for muscles.

At the distal end of the ulna, its knoblike *head* articulates laterally with a notch of the radius (ulnar notch) and with a disc of fibrocartilage inferiorly. This disc, in turn, joins one of the wrist bones (triquetrum). A medial *styloid process* at the distal end of the ulna provides attachments for wrist ligaments.

Hand

The hand is made up of the wrist, palm, and fingers. The skeleton of the wrist consists of eight small **carpal bones** firmly bound in two rows of four bones each. The resulting compact mass is called a *carpus* (kar'pus). The carpus articulates with the radius and with the fibrocartilaginous disc on the ulnar side. Its distal surface articulates with the metacarpal bones. Figure 7.25 names the individual bones of the carpus.

Five **metacarpal bones,** one in line with each finger, form the framework of the palm or *metacarpus* (met"ah-kar'pus) of the hand. These bones are cylindrical, with rounded distal ends that form the knuckles of a clenched fist. They are numbered 1–5, beginning with the metacarpal of the thumb (fig. 7.25). The metacarpals articulate proximally with the carpals and distally with phalanges.

The **phalanges** are the finger bones. Each finger has three phalanges—a proximal, a middle, and a distal phalanx—except the thumb, which has two (it does not have a middle phalanx).

PRACTICE 7.10

1. Locate and name each of the bones of the upper limb.
2. Explain how the bones of the upper limb articulate.

7.11 | Pelvic Girdle

LEARN

1. Locate and identify the bones and the major features of the bones that compose the pelvic girdle.

The **pelvic girdle** consists of two coxal bones (hip bones, pelvic bones, or innominate bones) that articulate with each other anteriorly and with the sacrum posteriorly. The sacrum, coccyx, and pelvic girdle together form the bowl-shaped pelvis, or pelvic region (fig. 7.26). The pelvic girdle supports the trunk, provides attachments for the lower limbs, and protects the urinary bladder, the distal end of the large intestine, and the internal reproductive organs.

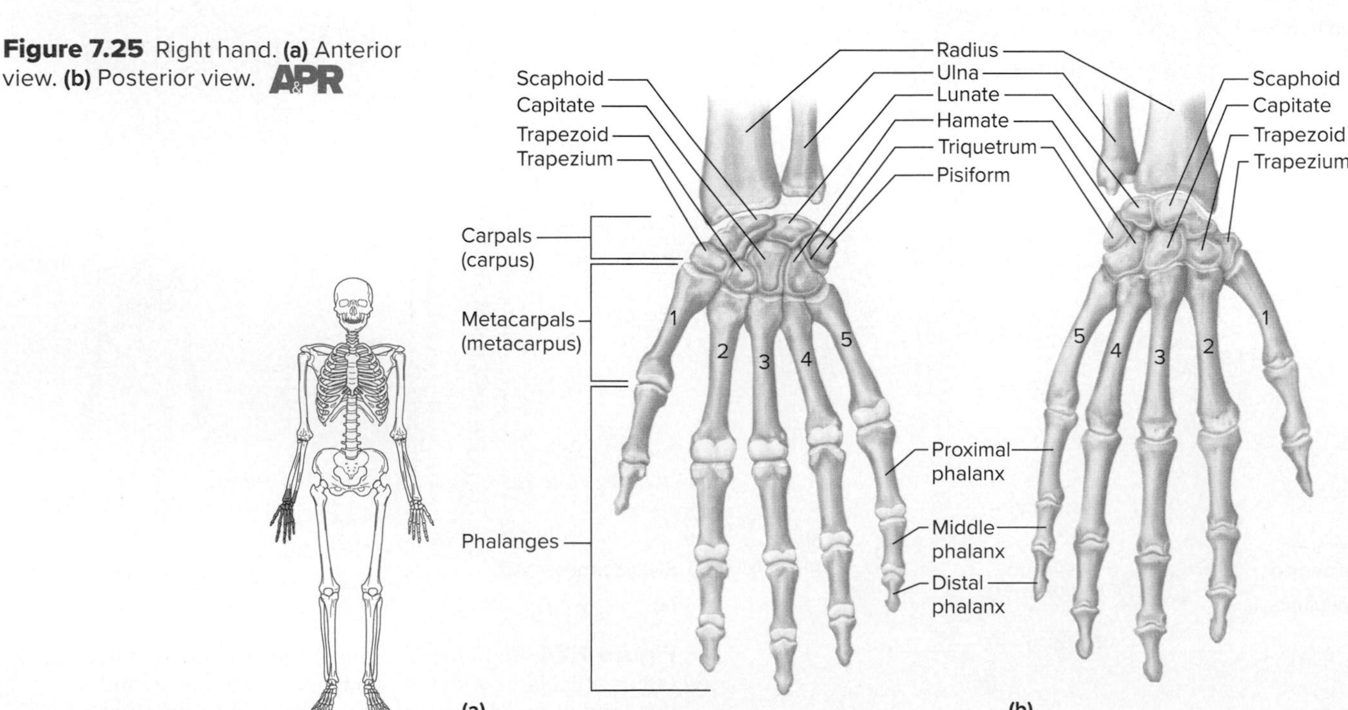

Figure 7.25 Right hand. **(a)** Anterior view. **(b)** Posterior view. APR

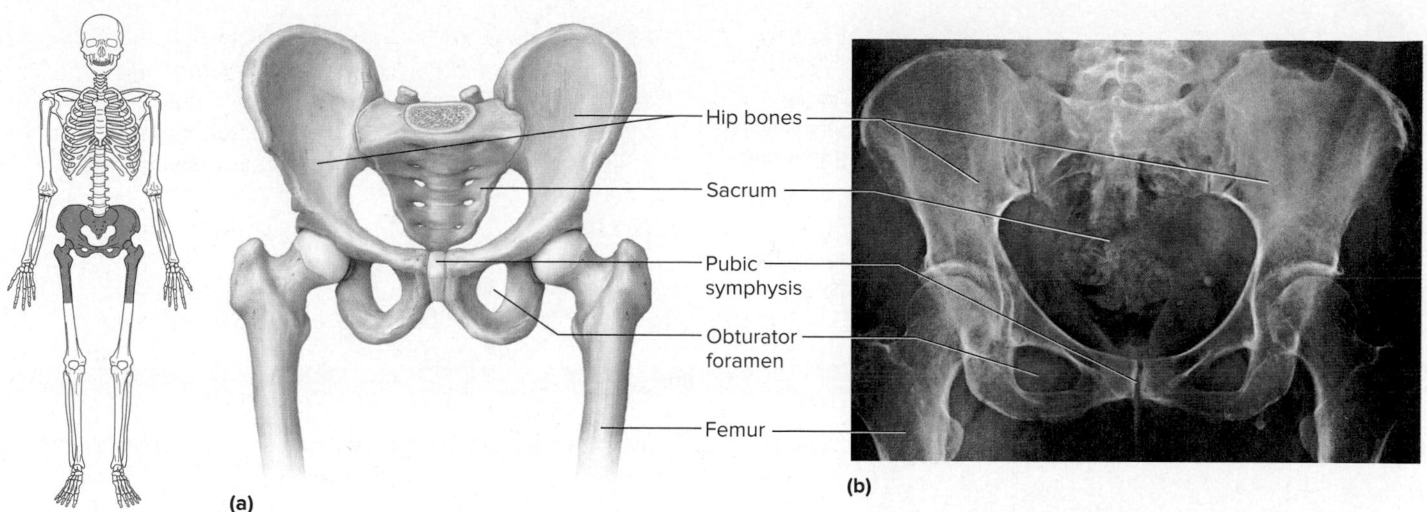

Figure 7.26 Pelvic girdle, anterior view. **(a)** The pelvic girdle is formed by two hip bones. The pelvis includes the pelvic girdle, as well as the sacrum and the coccyx. **(b)** A radiograph of the pelvic girdle showing the sacrum, coccyx, and proximal ends of the femurs. (b): Courtesy Dale Butler APR

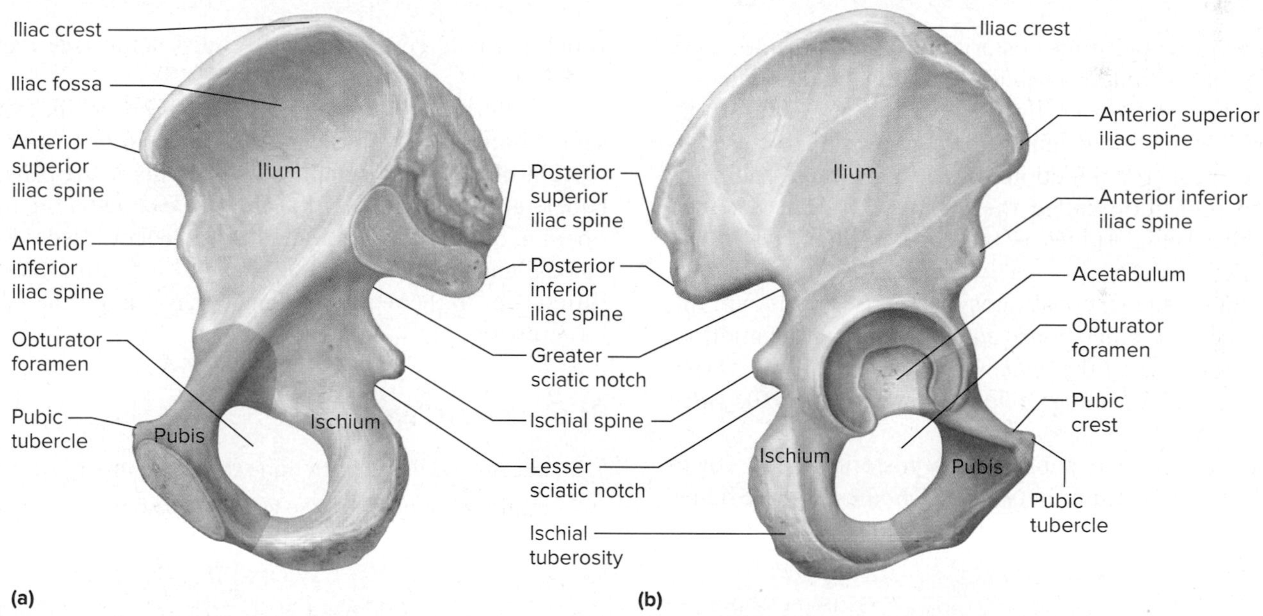

Figure 7.27 Right coxal (hip) bone. **(a)** Medial surface. **(b)** Lateral view. APR

Each coxal bone develops from three parts—an ilium, an ischium, and a pubis (fig. 7.27). These parts don't fuse until a person's late teens and don't fully ossify until their early twenties. The fusion occurs in the region of a cup-shaped cavity called the *acetabulum* (as″ĕ-tab′u-lum). This depression, on the lateral surface of the coxal bone, receives the rounded head of the femur (thigh bone).

The **ilium** (il′ē-um), which is the largest and uppermost portion of the hip bone, flares outward, forming the prominence of the hip. The margin of this prominence is called the *iliac crest.*

Posteriorly, the ilium joins the sacrum at the *sacroiliac* (sa″kro-il′e-ak) *joint.* Anteriorly, a projection of the ilium, the *anterior superior iliac spine,* can be felt lateral to the groin and provides attachments for ligaments and muscles.

The **ischium** (is′kē-um), which forms the lowest portion of the hip bone, is L-shaped, with its angle, the

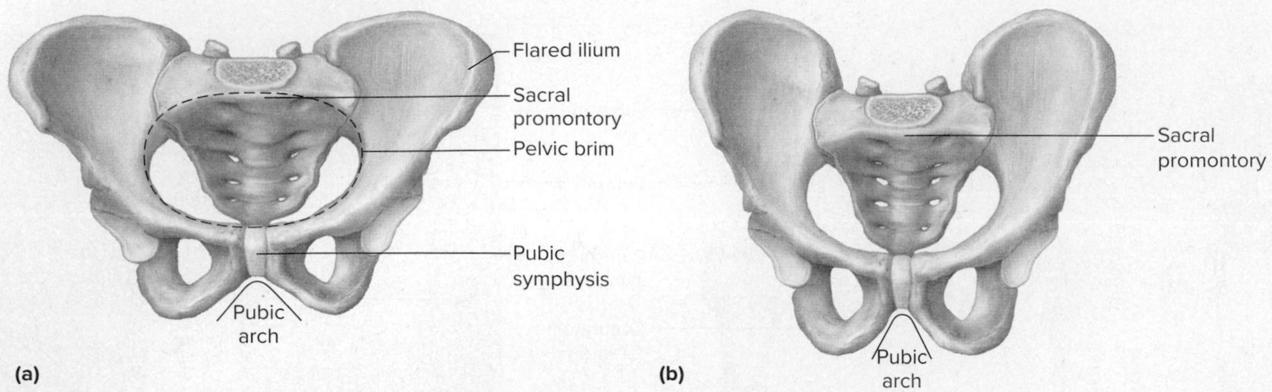

Figure 7.28 The female pelvis is usually wider in all diameters and roomier than that of the male. **(a)** Female pelvis. **(b)** Male pelvis.

PRACTICE FIGURE 7.28

What are some of the specific differences between the female and male pelves?

Answer can be found in Appendix E.

ischial tuberosity, pointing posteriorly and downward. This tuberosity has a rough surface that provides attachments for ligaments and lower limb muscles. It also supports the weight of the body during sitting. Above the ischial tuberosity, near the junction of the ilium and ischium, is a sharp projection called the *ischial spine.* The distance between the ischial spines is the shortest diameter of the pelvic outlet.

The **pubis** (pū'bis) constitutes the anterior portion of the hip bone. The two pubic bones come together at the midline, forming a joint called the *pubic symphysis* (pu'bik sim'fi-sis). The angle these bones form below the symphysis is the *pubic arch* (**fig. 7.28**).

A portion of each pubis passes posteriorly and downward, joining an ischium. Between the bodies of these bones on either side is a large opening, the *obturator foramen,* which is the largest foramen in the skeleton (see figs. 7.26 and 7.27).

A line drawn along each side of the pelvis from the sacral promontory downward and anteriorly to the upper margin of the pubic symphysis would mark the *pelvic brim* (linea terminalis) (fig. 7.28). The *pelvic outlet* is the inferior opening of the pelvis, bounded by the coccyx, ischial tuberosities, and pubic symphysis. **Table 7.3** summarizes some differences in the female and male pelves and other skeletal structures.

PRACTICE 7.11

1. Locate and name each bone that forms the pelvis.
2. Name the bones that fuse to form a hip bone.

TABLE 7.3 Differences Between the Female and Male Pelvic Regions

Part	Differences
Pelvic girdle	Female hip bones are lighter, thinner, and have less evidence of muscular attachments. The female obturator foramina are triangular, whereas the male's are oval. The female acetabula are smaller and the pubic arch is wider than corresponding structures of a male.
Pelvic cavity	Female pelvic cavity is wider in all diameters and is shorter, roomier, and less funnel-shaped. The distances between the female ischial spines and ischial tuberosities are greater than in a male.
Sacrum	Female sacrum is wider, and the sacral curvature is bent more sharply posteriorly than in a male.
Coccyx	Female coccyx is more movable than that of a male.

7.12 | Lower Limb

LEARN

1. Locate and identify the bones and the major features of the bones that compose the lower limb.

Bones of the lower limb form the framework of each thigh, leg, and foot. They include a femur, a tibia, a fibula, tarsals, metatarsals, and phalanges (see fig. 7.8).

Femur

The **femur,** or thigh bone, is the longest bone in the body and extends from the hip to the knee (fig. 7.29). A large, rounded *head* at its proximal end projects medially into the acetabulum of the hip bone. On the head, a pit called the *fovea capitis* marks the attachment of a ligament (ligamentum capitis). Just below the head are a constriction, or *neck,* and two large processes—a superior, lateral *greater trochanter* and an inferior, medial *lesser trochanter.* These processes provide attachments for muscles of the lower limbs and buttocks.

OF INTEREST The strongest bone in the body, the femur, has greater pressure tolerance and bearing strength than a rod of equivalent weight made of cast steel.

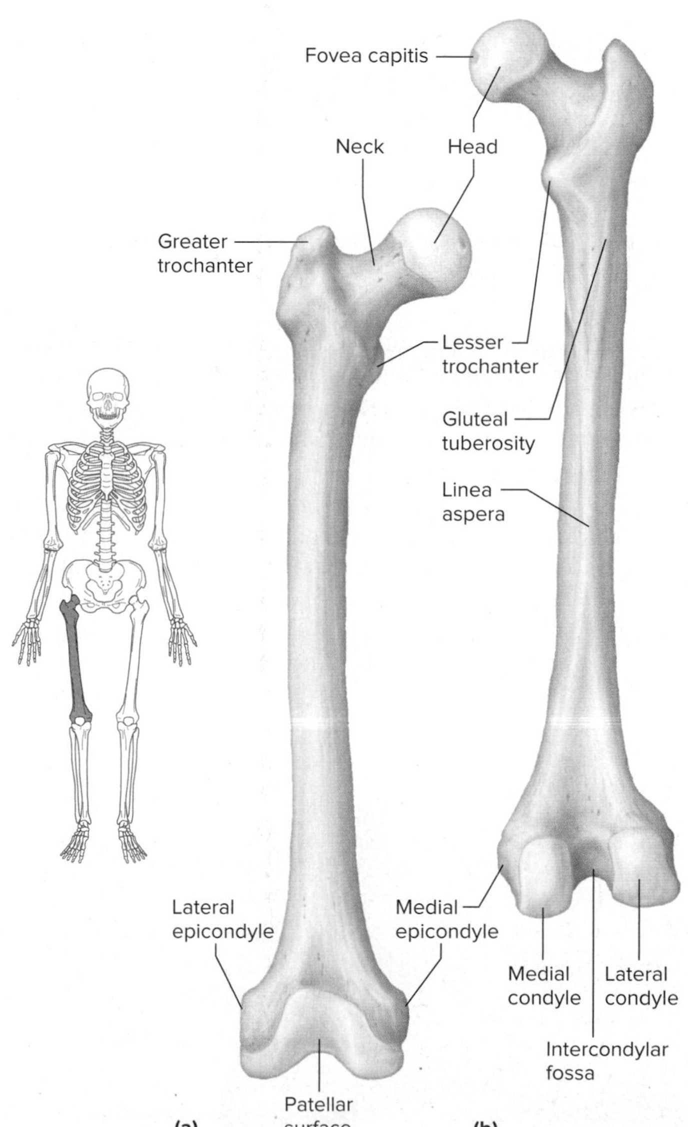

Figure 7.29 Right femur. **(a)** Anterior surface. **(b)** Posterior surface. APR

At the distal end of the femur, two rounded processes, the *lateral* and *medial condyles,* articulate with the tibia of the leg. A **patella,** or kneecap, also articulates with the femur on its distal anterior surface (see fig. 7.8). The patella is located in a tendon that passes anteriorly over the knee.

Tibia

The **tibia,** or shin bone, is the larger of the two leg bones and is located on the medial side (fig. 7.30). Its proximal end is expanded into *medial* and *lateral condyles,* which have concave surfaces and articulate with the condyles of the femur. Below the condyles, on the anterior surface, is a process called the *tibial tuberosity.* It provides an attachment for the *patellar ligament,* which is a continuation of the patellar tendon.

At its distal end, the tibia expands, forming a prominence on the inner ankle called the *medial malleolus* (mah-le'o-lus), which is an attachment for ligaments. On its lateral side is a depression that articulates with the fibula. The inferior surface of the tibia's distal end articulates with a large bone (the talus) in the ankle.

Fibula

The **fibula** is a long, slender bone located on the lateral side of the tibia (fig. 7.30). Its ends are slightly enlarged into a proximal *head* and a distal *lateral malleolus.* The head articulates with the tibia just below the lateral condyle; however, it does not enter into the knee joint and does not bear any body weight. The lateral malleolus articulates with the ankle and protrudes on the lateral side.

Foot

The foot is made up of the ankle, the instep, and the toes. The ankle, or *tarsus* (tahr'sus), is composed of seven **tarsal bones** (figs. 7.31 and 7.32). One of these bones, the **talus** (ta'lus), can move freely where it joins the tibia and fibula. Figure 7.32 names the individual bones of the tarsus.

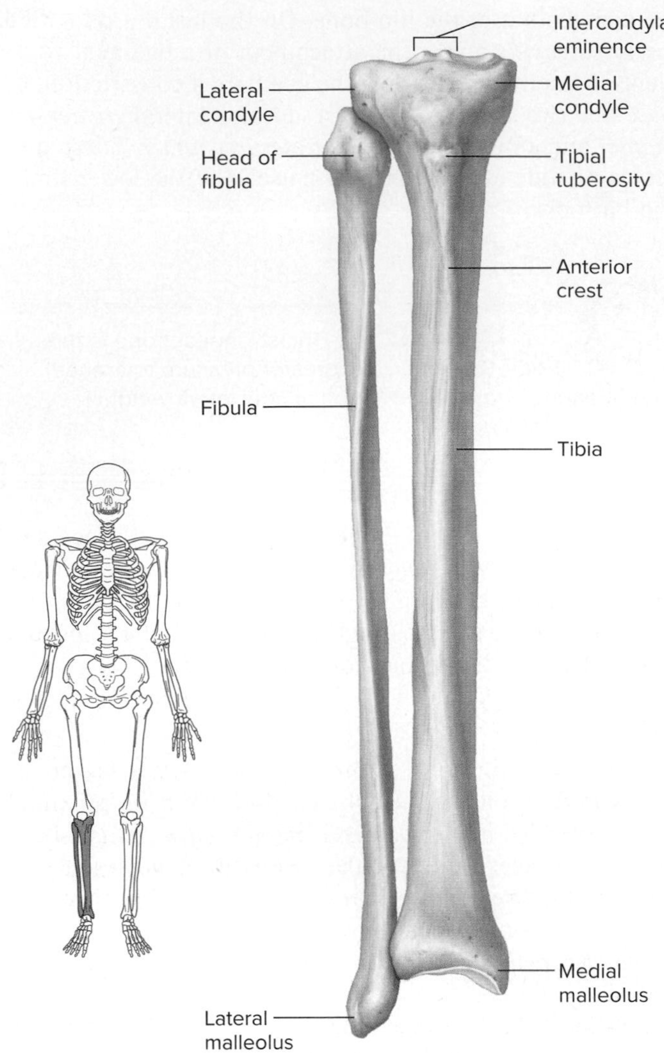

Figure 7.30 Right tibia and fibula, anterior view. APR

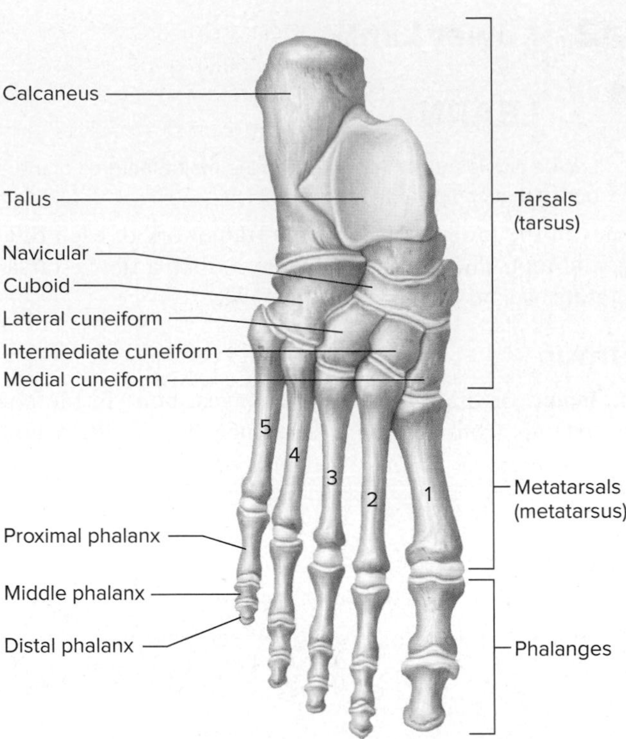

Figure 7.32 Right foot, viewed superiorly. APR

The largest of the tarsals, the **calcaneus** (kal-kā′nē-us), or heel bone, is below the talus, where it projects backward, forming the base of the heel. The calcaneus helps support body weight and provides an attachment for the muscles that move the foot (see fig. 7.31).

The instep, or *metatarsus* (met″ah-tar′sus), consists of five elongated **metatarsal bones** that articulate with the tarsus. They are numbered 1–5, beginning on the medial side

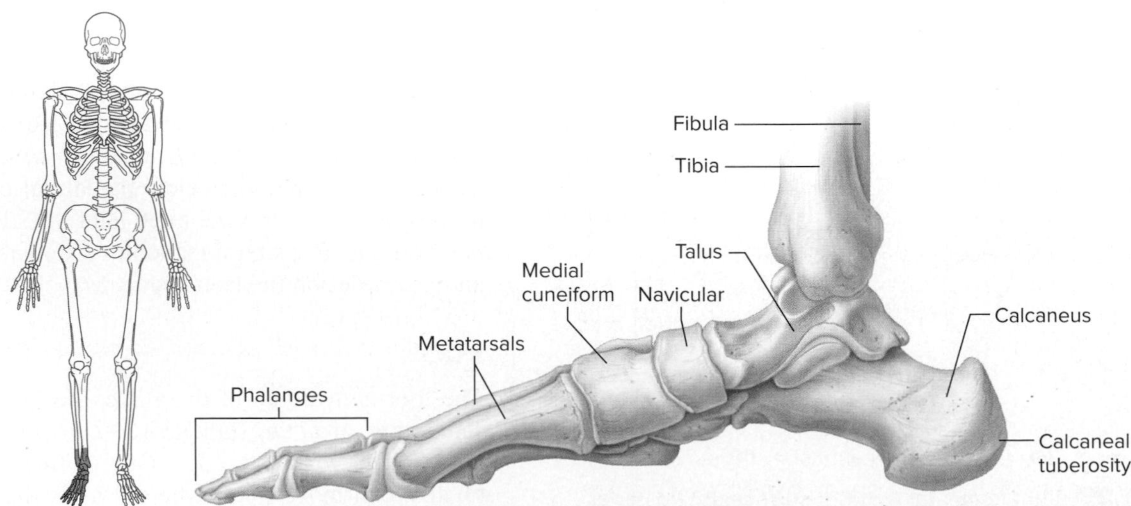

Figure 7.31 Right foot, medial view. The talus moves freely where it articulates with the tibia and fibula.

(fig. 7.32). The heads at the distal ends of these bones form the ball of the foot. The tarsals and metatarsals are bound by ligaments in a way that forms the arches of the foot. A longitudinal arch extends from the heel to the toe, and a transverse arch stretches across the foot. These arches provide a stable, springy base for the body. Sometimes, however, the tissues that bind the metatarsals weaken, producing fallen arches, or flat feet.

The proximal **phalanges** of the toes, which are similar to those of the fingers, align and articulate with the metatarsals. Each toe has three phalanges—a proximal, a middle, and a distal phalanx—except the great toe, which has only a proximal and a distal phalanx.

PRACTICE 7.12

1. Locate and name each of the bones of the lower limb.
2. Explain how the bones of the lower limb articulate.
3. Describe how the foot is adapted to support the body.

7.13 | Joints

LEARN

1. Classify joints according to the type of tissue binding the bones, describe the different joint characteristics, and name an example of each joint type.
2. List six types of synovial joints, and describe the actions of each.
3. Explain how skeletal muscles produce movements at joints, and identify several types of joint movements.

Joints (articulations) are functional junctions between bones. They bind parts of the skeletal system, make possible bone growth, permit parts of the skeleton to change shape during childbirth, and enable the body to move in response to skeletal muscle contractions.

The number of joints changes in the human body from birth until old age. However, typically there are about 230 joints in the average adult. Joints may be classified according to the degree of movement they make possible. Joints can be immovable (synarthrotic), slightly movable (amphiarthrotic), or freely movable (diarthrotic). They vary considerably in function and structure. Joints also can be grouped according to the type of tissue (fibrous, cartilaginous, or synovial) that binds the bones at each junction. Currently, this structural classification by tissue type is most commonly used.

Fibrous Joints

Fibrous (fī′brus) **joints** lie between bones that closely contact one another and are held together by a thin layer of dense connective tissue. An example of such a joint is a *suture* between a pair of flat bones of the skull (fig. 7.33).

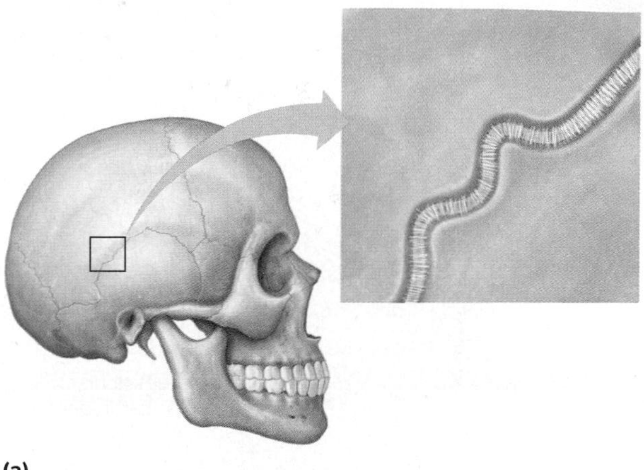

(a)

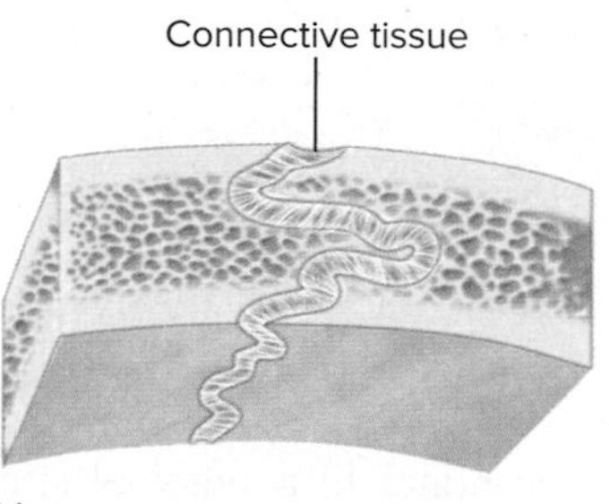

(b)

Figure 7.33 Fibrous joints. **(a)** The fibrous joints between the bones of the skull are immovable and are called sutures. **(b)** A thin layer of connective tissue connects the bones at the suture.

Generally, no appreciable movement (synarthrotic) takes place at a fibrous joint. Some fibrous joints, such as the joint in the leg between the distal ends of the tibia and fibula, have limited movement (amphiarthrotic).

Cartilaginous Joints

Hyaline cartilage, or fibrocartilage, connects the bones of **cartilaginous** (kar″tĭ-lah′jin-us) **joints.** For example, joints of this type separate the vertebrae of the vertebral column. Each intervertebral disc is composed of a band of fibrocartilage (annulus fibrosus) surrounding a pulpy or gelatinous core (nucleus pulposus). The disc absorbs shocks and helps equalize pressure between vertebrae when the body moves (see fig. 7.16).

Due to the slight flexibility of the discs, cartilaginous joints allow limited movement (amphiarthrotic), as when the back is bent forward or to the side or is twisted. Other examples of cartilaginous joints include the pubic symphysis and the first rib with the sternum.

Synovial Joints APR

Most joints in the skeletal system are **synovial** (sĭ-nō′vē-al) **joints,** which allow free movement (diarthrotic). They are more complex structurally than fibrous or cartilaginous joints.

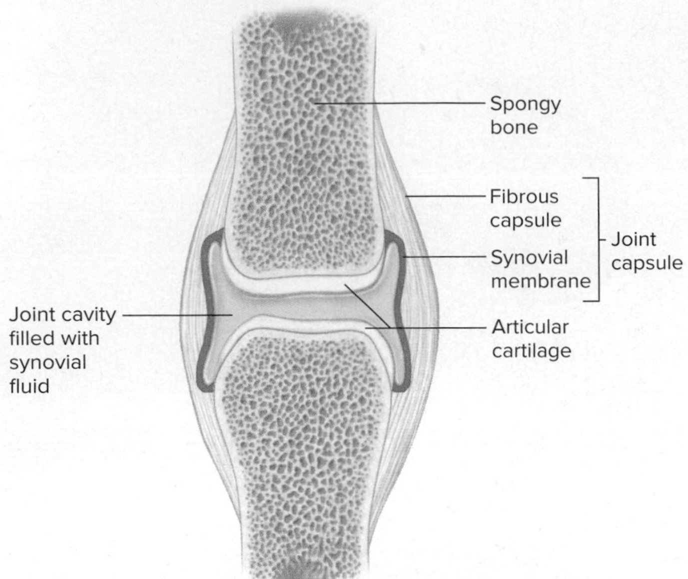

Figure 7.34 The generalized structure of a synovial joint.

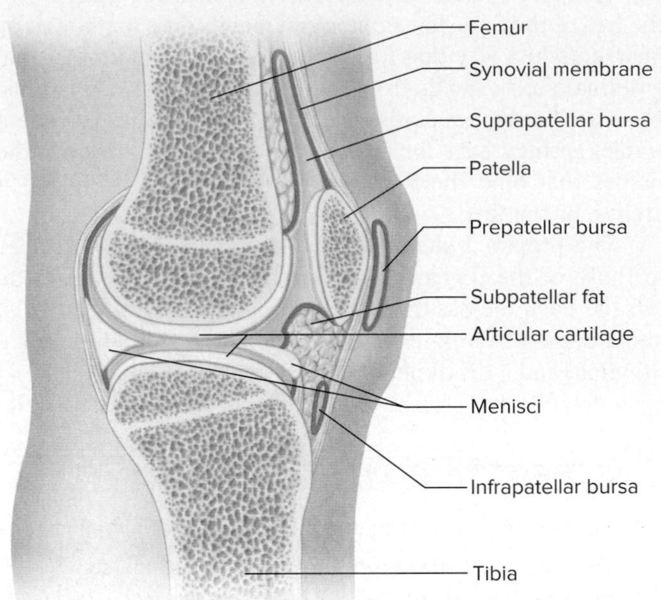

Figure 7.35 Menisci separate the articulating surfaces of the femur and tibia. Several bursae are associated with the knee joint. Synovial spaces in this and other line drawings of joints in this chapter are exaggerated.

The articular ends of the bones in a synovial joint are covered with a thin layer of hyaline cartilage (articular cartilage). A tubular *joint capsule* holds the bones of a synovial joint together. Each joint capsule is composed of an outer fibrous layer of dense connective tissue and an inner lining of *synovial membrane,* which secretes *synovial fluid* (fig. 7.34). Ligaments in the fibrous layer prevent the bones from being pulled apart. Synovial fluid has a consistency similar to uncooked egg white, and it lubricates joints.

Some synovial joints have flattened, shock-absorbing pads of fibrocartilage called **menisci** (mĕ-nis′kī) (singular, *meniscus*) between the articulating surfaces of the bones (fig. 7.35). Such joints may also have fluid-filled sacs called **bursae** (ber′se) associated with them. Each bursa is lined with synovial membrane, which may be continuous with the synovial membrane of a nearby joint cavity. Bursae are commonly located between tendons and underlying bony prominences, as in the patella of the knee or the olecranon process of the elbow. They aid the movement of tendons that glide over these bony parts or over other tendons. Figure 7.35 shows and names some of the bursae associated with the knee.

Based upon the shapes of their parts and the movements they permit, synovial joints are classified as follows:

- A **ball-and-socket joint,** or **spheroidal joint,** consists of a bone with a globular or slightly egg-shaped head that articulates with the cup-shaped cavity of another bone. Such a joint allows the widest range of motion, permitting movements in all planes (*multiaxial movement*), including rotational movement around a central axis. The shoulder and hip have joints of this type (fig. 7.36*a*).
- In a **condylar joint,** or **ellipsoidal joint,** the ovoid condyle of one bone fits into the elliptical cavity of another bone, as in the joints between the metacarpals and phalanges (fig. 7.36*b*). This type of joint permits back and forth and side to side movement in two planes (*biaxial movement*), but not rotation.
- The articulating surfaces of **plane joints,** or **gliding joints,** are nearly flat or slightly curved. Most of the joints in the wrist and ankle, as well as those between the articular processes of adjacent vertebrae, belong to this group (fig. 7.36*c*). They allow sliding and twisting movements (*nonaxial movement*). The sacroiliac joints and the joints formed by ribs 2–7 connecting with the sternum are also plane joints.
- In a **hinge joint,** the convex surface of one bone fits into the concave surface of another, as in the elbow and the joints between the phalanges (fig. 7.36*d*). Such a joint resembles the hinge of a door in that it permits movement in one plane only (*uniaxial movement*).
- In a **pivot joint,** or **trochoid joint,** the cylindrical surface of one bone rotates within a ring formed of bone and ligament. Movement is limited to the rotation around a central axis (*uniaxial movement*). The joint between the atlas and the dens of the axis is of this type (fig. 7.36*e*).

- A **saddle joint,** or **sellar joint,** forms between bones whose articulating surfaces have both concave and convex regions. The surface of one bone fits the complementary surface of the other. This physical relationship permits a variety of movements, mainly in two planes (*biaxial movement*), as in the joint between the carpal (trapezium) and metacarpal of the thumb (fig. 7.36*f*).

Table 7.4 summarizes the types of joints. Clinical Application 7.3 discusses injuries and conditions that affect the joints.

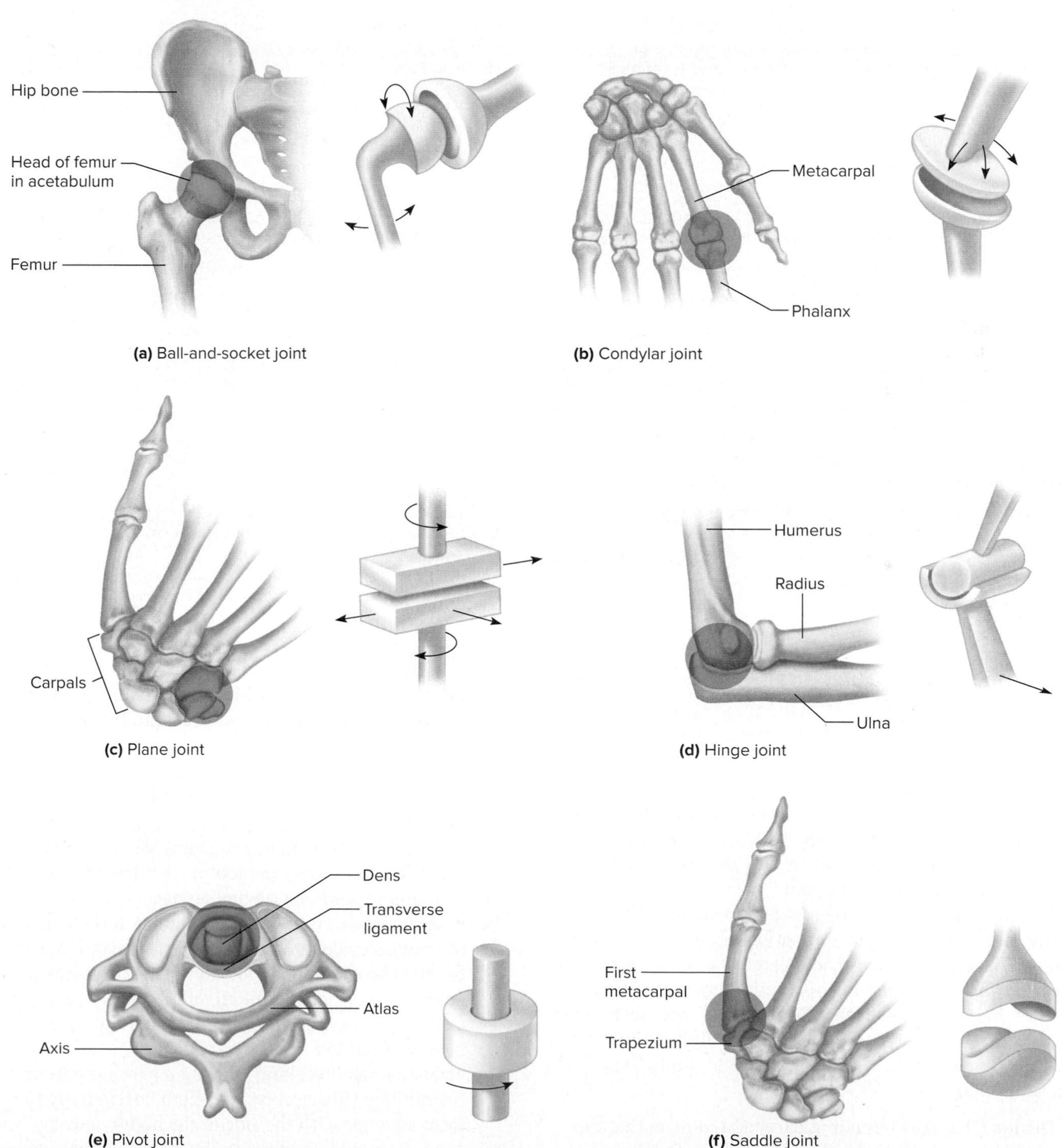

Figure 7.36 Types and examples of synovial (freely movable) joints **(a–f)**. APR

TABLE 7.4 Types of Joints

Type of Joint	Description	Possible Movements	Examples
Fibrous	Articulating bones are fastened together by a thin layer of dense connective tissue.	None or slight twisting	Suture between bones of skull, joint between the distal ends of tibia and fibula
Cartilaginous	Articulating bones are connected by hyaline cartilage or fibrocartilage.	Limited movement, as when the back is bent or twisted	Joints between the bodies of vertebrae, pubic symphysis
Synovial	Articulating ends of bones are surrounded by a joint capsule of ligaments and synovial membranes; ends of articulating bones are covered by hyaline cartilage and separated by synovial fluid.	Allow free movement (see the following list, "Types of Joint Movements")	
Ball-and-socket	Ball-shaped head of one bone articulates with cup-shaped cavity of another.	Movements in all planes, including rotation	Shoulder, hip
Condylar	Oval-shaped condyle of one bone articulates with elliptical cavity of another.	Variety of movements in two planes, but no rotation	Joints between the metacarpals and phalanges
Plane	Articulating surfaces are nearly flat or slightly curved.	Sliding or twisting	Joints between various bones of wrist and ankle, sacroiliac joints, joints between ribs 2–7 and sternum
Hinge	Convex surface of one bone articulates with concave surface of another.	Flexion and extension	Elbow, joints of phalanges
Pivot	Cylindrical surface of one bone articulates with ring of bone and ligament.	Rotation around a central axis	Joint between the atlas and dens of the axis
Saddle	Articulating surfaces have both concave and convex regions; the surface of one bone fits the complementary surface of another.	Variety of movements, mainly in two planes	Joint between the carpal and metacarpal of thumb

Types of Joint Movements

Skeletal muscle action produces movements at synovial joints. Typically, one end of a muscle is attached to a less movable or relatively fixed part on one side of a joint, and the other end of the muscle is fastened to a more movable part on the other side. When the muscle contracts, its fibers pull its movable end (*insertion*) toward its fixed end (*origin*) and a movement occurs at the joint.

Note that in the anatomical position some actions have already occurred, such as supination of the forearm and hand, extension of the elbow, and extension of the knee. The muscle actions described in this section consider the entire range of movement at each joint, and do not necessarily presume that the starting point is the anatomical position. The following terms describe movements at joints (figs. 7.37, 7.38, and 7.39).

- **flexion** (flek′shun) Bending parts at a joint so that the angle between them decreases and the parts come closer together (bending the knee).
- **extension** (ek-sten′shun) Moving parts at a joint so that the angle between them increases and the parts move farther apart (straightening the knee).
- **dorsiflexion** (dor″sĭ-flek′shun) Movement at the ankle that brings the foot closer to the shin (rocking back on one's heels).
- **plantar flexion** (plan′tar flek′shun) Movement at the ankle that moves the foot farther from the shin (walking or standing on one's toes).
- **hyperextension** (hī″per-ek-sten′shun) A term sometimes used to describe the extension of the parts at a joint beyond the anatomical position (bending the head back beyond the upright position); often used to describe an abnormal extension beyond the normal range of motion, resulting in injury.
- **abduction** (ab-duk′shun) Moving a part away from the midline (lifting the upper limb horizontally to form an angle with the side of the body) or away from the axial line of the limb (spreading the fingers or toes). Abduction of the head and neck and

bending of the trunk to the side may be termed *lateral flexion.*

- **adduction** (ah-duk′shun) Moving a part toward the midline (returning the upper limb from the horizontal position to the side of the body) or toward the axial line of the limb (moving the fingers and toes closer together).
- **rotation** (rō-tā′shun) Moving a part around an axis (twisting the head from side to side). Medial (internal) rotation is the turning of a limb on its longitudinal axis so its anterior surface moves toward the midline, whereas lateral (external) rotation is the turning of a limb on its longitudinal axis in the opposite direction.
- **circumduction** (ser″kum-duk′shun) Moving a part so that its end follows a circular path (moving the finger in a circular motion without moving the hand).
- **pronation** (prō-nā′shun) Rotation of the forearm so the palm is downward or facing posteriorly (in anatomical position). Prone refers to the body lying face down.
- **supination** (soo″pĭ-nā′shun) Rotation of the forearm so the palm is upward or facing anteriorly (in anatomical position). Supine refers to the body lying face up.
- **eversion** (e-ver′zhun) Turning the foot so the plantar surface faces laterally.
- **inversion** (in-ver′zhun) Turning the foot so the plantar surface faces medially.

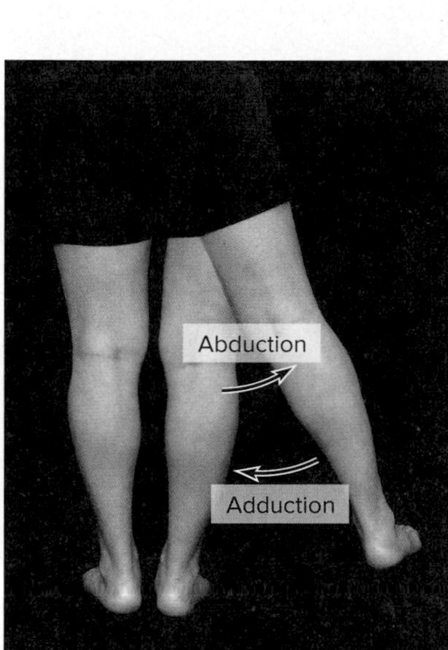

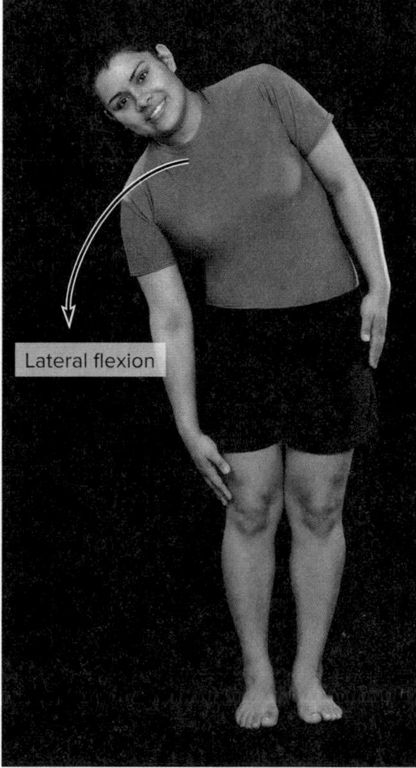

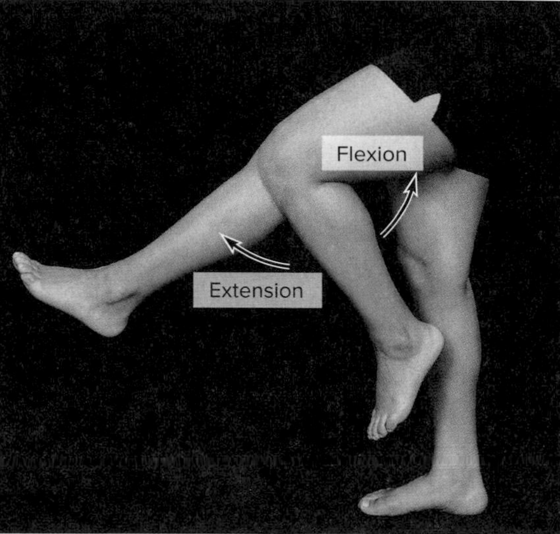

Figure 7.37 Joint movements illustrating abduction, adduction, lateral flexion, extension, and flexion. J and J Photography APR

CLINICAL APPLICATION 7.3

Joint Disorders and Injuries

Joints must support weight, must provide a variety of body movements, and are used frequently. Trauma, overuse, infection, a misdirected immune system attack, or degeneration can injure joints. Consider some common joint problems.

Sprains

Sprains result from overstretching or tearing the connective tissues, including cartilage, ligaments, and tendons, associated with a joint. However, sprains do not dislocate the articular bones. Usually a forceful wrenching or twisting sprains a wrist or ankle. For example, inverting an ankle too far can sprain it by stretching the ligaments on its lateral surface. Severe injuries may pull these structures loose from their attachments.

A sprained joint is painful and swollen, restricting movement. Immediate treatment of a sprain is rest; more serious cases require medical attention. However, immobilization of a joint, even for a brief period, causes bone resorption and weakens ligaments. Consequently, exercise may help strengthen the joint.

Bursitis

Overuse of a joint or stress on a bursa may cause *bursitis,* an inflammation of a bursa. The bursa between the calcaneus (heel bone) and the Achilles tendon may become inflamed as a result of a sudden increase in physical activity using the feet. Bursitis is treated with rest. Medical attention may be necessary.

Arthritis

Arthritis causes inflamed, swollen, and painful joints. More than a hundred different types of arthritis affect millions of people worldwide. The most common types of arthritis are rheumatoid arthritis (RA), osteoarthritis (OA), and Lyme arthritis.

Rheumatoid Arthritis (RA)

Rheumatoid arthritis, an autoimmune disorder (a condition in which the immune system attacks the body's healthy tissues), is painful and debilitating. The synovial membrane of a joint becomes inflamed and thickened. Then the articular cartilage is damaged, and fibrous tissue infiltrates, interfering with joint movements. Over time, the joints may ossify, fusing the articulating bones. RA may affect many joints or only a few. It is often accompanied by muscular atrophy, fatigue, and other symptoms.

Osteoarthritis (OA)

Osteoarthritis, a degenerative disorder, may result from aging or a poorly healed injury, or it may be inherited. Articular cartilage softens and disintegrates gradually, roughening the articular surfaces. Joints become painful, with restricted movement. OA usually affects the most active joints, such as those of the fingers, hips, knees, and the lower vertebral column.

If a person with osteoarthritis is overweight or obese, the first treatment is usually an exercise and dietary program to lose weight. Nonsteroidal anti-inflammatory drugs (NSAIDs) such as aspirin and ibuprofen have been used for many years to control osteoarthritis symptoms. NSAIDs called COX-2 inhibitors relieve inflammation without the gastrointestinal side effects of earlier drugs, but the FDA (Food and Drug Administration) advises that they be prescribed only to people who do not have risk factors for cardiovascular disease, to which some of these drugs are linked.

Lyme Arthritis

Lyme disease, a bacterial infection passed in a tick bite, causes intermittent arthritis of several joints, usually weeks after the initial symptoms of rash, fatigue, and flu-like aches and pains. *Lyme arthritis* was first observed in Lyme, Connecticut, where an astute woman alerted a prominent rheumatologist to the fact that many of her young neighbors had what appeared to be the very rare juvenile form of rheumatoid arthritis. Researchers then traced the illness to a tick-borne bacterial infection. Antibiotic treatment that begins as soon as the early symptoms are recognized may prevent Lyme arthritis. Other types of bacteria can cause arthritis, too.

Injuries to the elbow, shoulder, and knee are commonly diagnosed and treated using a procedure called *arthroscopy.* Arthroscopy enables a surgeon to visualize the interior of a joint and perform procedures, guided by the image on a video screen. An arthroscope is a thin, tubular instrument about 25 centimeters long containing optical fibers that transmit an image. The surgeon inserts the device through a small incision in the joint capsule. Arthroscopy is much less invasive than conventional surgery. Some runners have undergone uncomplicated arthroscopy and raced several weeks later.

- **retraction** (re-trak'shun) Moving a part backward (pulling the head backward).
- **protraction** (prō-trak'shun) Moving a part forward (thrusting the head forward).
- **elevation** (el"ĕ-vā'shun) Raising a part (shrugging the shoulders).
- **depression** (dē-presh'un) Lowering a part (drooping the shoulders).

PRACTICE 7.13

1. Describe the characteristics of the three major types of joints.
2. Name six types of synovial joints.
3. What terms describe movements possible at synovial joints?

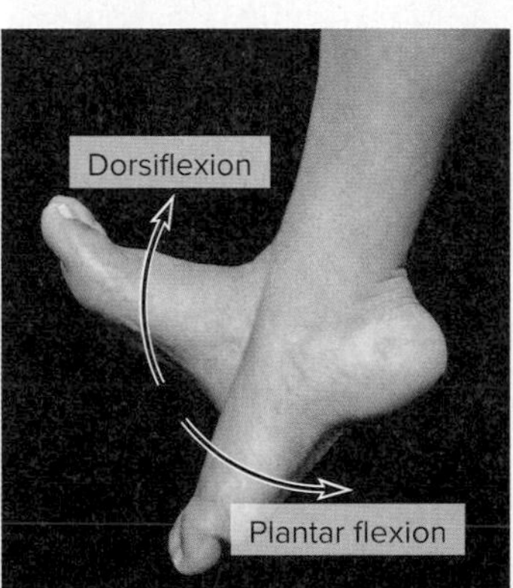

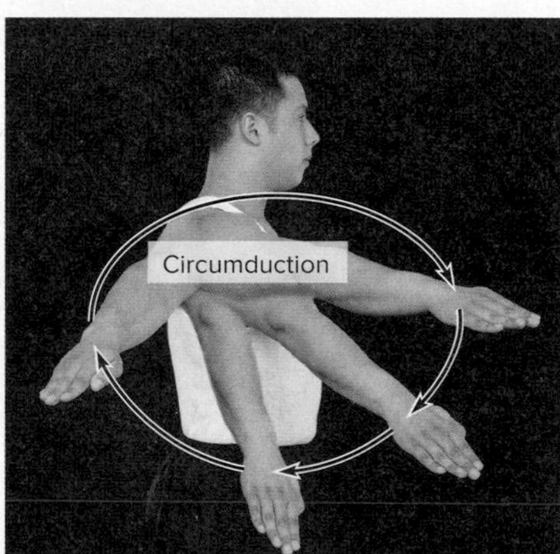

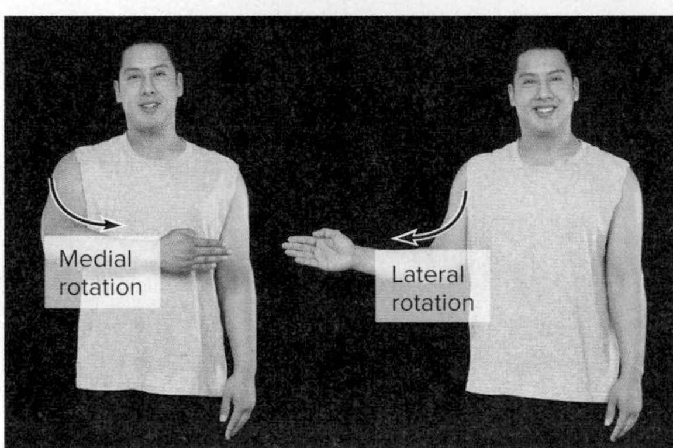

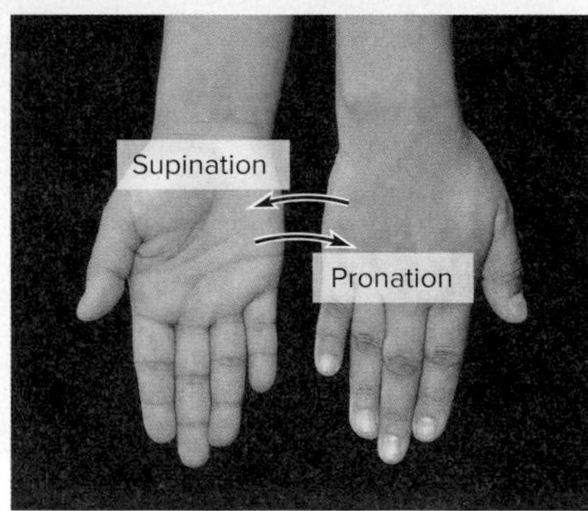

Figure 7.38 Joint movements illustrating dorsiflexion, plantar flexion, circumduction, rotation, supination, and pronation.
J and J Photography APR

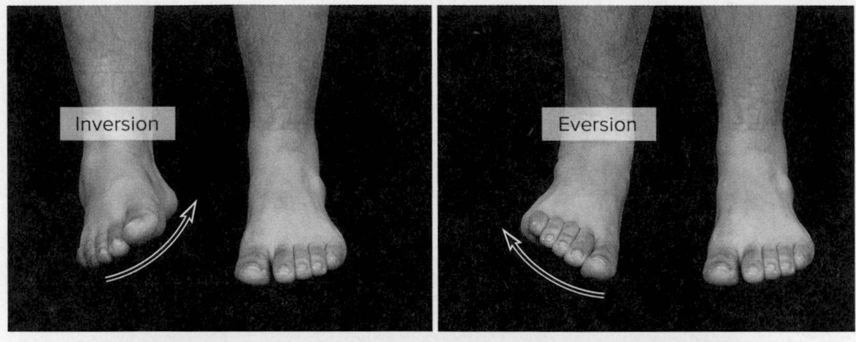

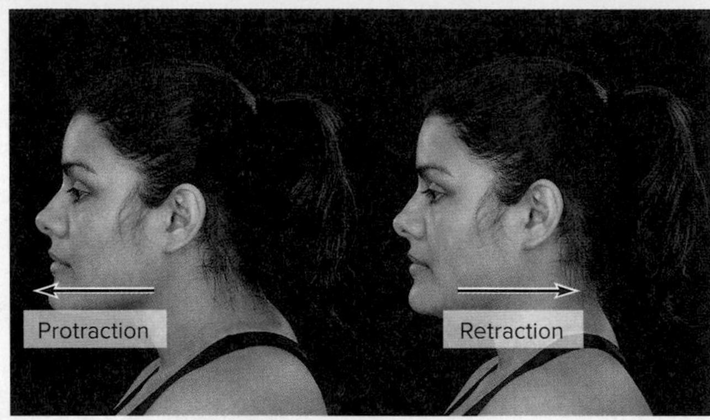

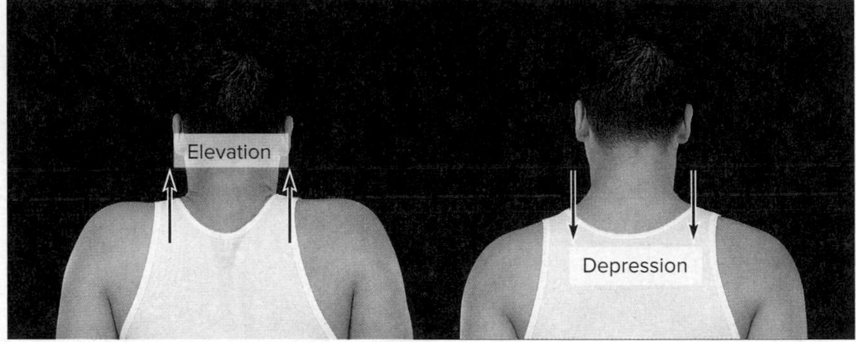

Figure 7.39 Joint movements illustrating inversion, eversion, protraction, retraction, elevation, and depression.
J and J Photography APR

Skeletal System

INTEGUMENTARY SYSTEM

Vitamin D, production of which begins in the skin, plays a role in calcium absorption and availability for bone matrix.

LYMPHATIC SYSTEM

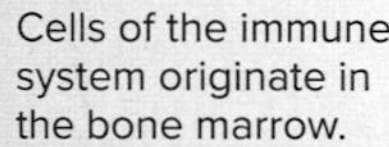

Cells of the immune system originate in the bone marrow.

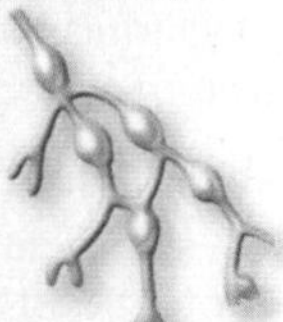

MUSCULAR SYSTEM

Muscles pull on bones to cause movement.

DIGESTIVE SYSTEM

Absorption of dietary calcium provides material for bone matrix.

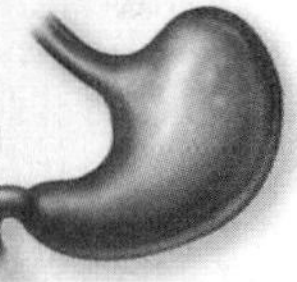

NERVOUS SYSTEM

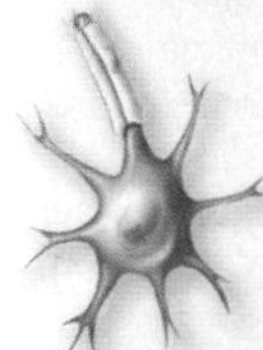

Proprioceptors sense the position of body parts. Pain receptors warn of trauma to bone. Bones protect the brain and spinal cord.

RESPIRATORY SYSTEM

Ribs and muscles work together in breathing.

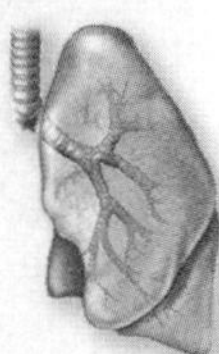

ENDOCRINE SYSTEM

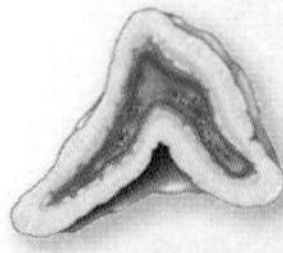

Some hormones act on bone to help regulate blood calcium levels.

URINARY SYSTEM

The kidneys and bones work together to help regulate blood calcium levels.

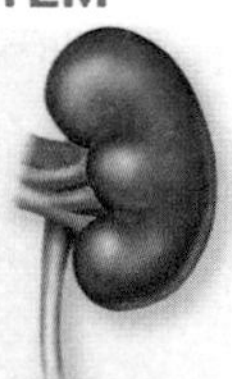

CARDIOVASCULAR SYSTEM

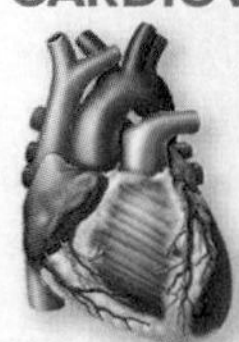

Blood transports nutrients to bone cells. Bone helps regulate plasma calcium levels, important to heart function.

REPRODUCTIVE SYSTEM

The pelvis helps support the uterus during pregnancy. Bones provide a source of calcium during lactation.

Bones provide support, protection, and movement and also play a role in calcium balance.

ASSESS

CHAPTER ASSESSMENTS

7.1 Introduction

1. Active, living tissues found in bone include ______.
   a. blood
   b. nervous tissue
   c. bone tissue
   d. all of the above

7.2 Bone Structure

2. Sketch a typical long bone, and label its epiphyses, diaphysis, medullary cavity, periosteum, and articular cartilages. On the sketch, designate the locations of compact and spongy bone.
3. Discuss the functions of the parts labeled in the sketch you made for question 2.
4. Distinguish between the microscopic structure of compact bone and spongy bone.

7.3 Bone Function

5. Give several examples of how bones support and protect body parts.
6. List and describe other functions of bones.

7.4 Bone Development, Growth, and Repair

7. Explain how the development of intramembranous bone differs from that of endochondral bone.
8. ______ are mature bone cells, whereas ______ are bone-forming cells and ______ are bone-resorbing cells.
9. Explain the function of an epiphyseal plate.
10. Physical exercise pulling on muscular attachments to bones stimulates ______.

7.5 Skeletal Organization

11. Bones of the head, neck, and trunk compose the ______ skeleton; bones of the limbs and their attachments compose the ______ skeleton.

7.6–7.12 (Skull–Lower Limb)

12. Name the bones of the cranium and the facial skeleton.
13. Describe a typical vertebra, and distinguish among the cervical, thoracic, and lumbar vertebrae.
14. Name the bones that compose the thoracic cage.
15. The clavicle and scapula form the ______ girdle, whereas the hip bones form the ______ girdle.
16. Name the bones of the upper and lower limbs.
17. Match the parts listed on the left with the bones listed on the right.

(1) foramen magnum	A. maxilla
(2) mastoid process	B. occipital bone
(3) palatine process	C. temporal bone
(4) sella turcica	D. femur
(5) deltoid tuberosity	E. fibula
(6) greater trochanter	F. humerus
(7) lateral malleolus	G. radius
(8) medial malleolus	H. sternum
(9) radial tuberosity	I. tibia
(10) xiphoid process	J. sphenoid bone

7.13 Joints

18. Describe and give an example of a fibrous joint, a cartilaginous joint, and a synovial joint.
19. Name an example of each type of synovial joint, and describe the parts of the joint as they relate to the movement(s) allowed by that particular joint.
20. Joint movements occur when a muscle contracts and the muscle fibers pull the muscle's movable end of attachment to the bone, the ______, toward its fixed end, the ______.
21. Match the movement on the left with the appropriate description on the right.

(1) rotation	A. turning the palm upward
(2) supination	B. decreasing the angle between parts
(3) extension	C. moving a part forward
(4) eversion	D. moving a part around an axis
(5) protraction	E. moving a part toward midline
(6) flexion	F. turning the foot so plantar surface faces laterally
(7) pronation	G. increasing the angle between parts
(8) abduction	H. lowering a part
(9) depression	I. turning the palm downward
(10) adduction	J. moving a part away from midline

ASSESS

INTEGRATIVE ASSESSMENTS/CRITICAL THINKING

Outcomes 5.3, 7.2, 7.6

1. How does the structure of a bone make it strong yet lightweight?

Outcomes 5.3, 7.13

2. How would you explain to an athlete why damaged joint ligaments and cartilages are so slow to heal following an injury?

Outcome 7.3

3. Vitamin D is converted to the hormone calcitriol by three body organs: the skin, liver, and kidneys. Calcitriol enhances the absorption of calcium in the small intestine and acts to prevent the elimination of calcium in urine. Which hormone does calcitriol work with to raise blood calcium? Explain.

Outcomes 7.3, 7.4, 7.10, 7.12

4. When a child's bone is fractured, growth may be stimulated at the epiphyseal plate of that bone. What problems might this extra growth cause in an upper or lower limb before the growth of the other limb compensates for the difference in length?

Outcomes 7.4, 7.11

5. Suppose archaeologists discover human skeletal remains in Ethiopia. Examination of the bones suggests that the remains represent four individuals. Two of the skeletons have a broad pubic bone and a pelvis with a wide pubic arch. Within the two groups defined by pelvic differences, smaller skeletons have bones with evidence of epiphyseal plates, but larger bones have only a thin line where the epiphyseal plates should be. Give the age group and sex of the four individuals in the find.

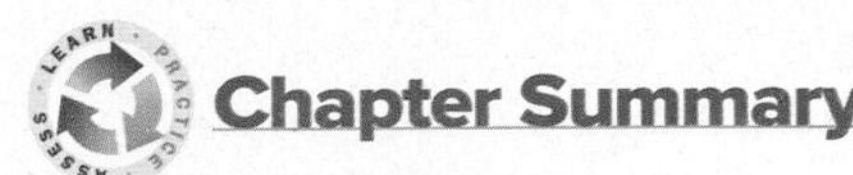

Chapter Summary

7.1 Introduction

Individual bones are the organs of the ***skeletal system.*** *A bone contains active tissues.*

7.2 Bone Structure

Bone structure reflects its function.

1. Bones are classified according to their shapes—long, short, flat, or irregular.
2. Parts of a **long bone**
   a. **Epiphyses** at each end are covered with **articular cartilage** and articulate with other bones.
   b. The shaft of a bone is called the **diaphysis.**
   c. A bone is covered by a **periosteum,** except for the articular cartilage.
   d. **Compact bone** has a continuous extracellular matrix with no gaps.
   e. **Spongy bone** has irregular interconnecting spaces between bony plates, **trabeculae,** that reduce the weight of bone.
   f. Both compact and spongy bone are strong and resist bending.
   g. The diaphysis contains a **medullary cavity** filled with **marrow.**
3. Microscopic structure
   a. Compact bone contains osteons cemented together.
   b. Central canals contain blood vessels that nourish the cells of osteons.
   c. Diffusion from the surface of the thin, bony plates nourishes the cells of spongy bone.

7.3 Bone Function

1. Support and protection
   a. Bones shape and form body structures.
   b. Bones support and protect softer underlying tissues.
2. Body movement
   Bones and muscles interact to move body parts.
3. Blood cell formation
   a. At different ages, hematopoiesis occurs in the yolk sac, liver and spleen, and red bone marrow.
   b. Red marrow houses developing red blood cells, white blood cells, and blood platelets. Yellow marrow stores fat.
4. Storage of inorganic salts
   a. Bones store calcium in the extracellular matrix of bone tissue, which contains large quantities of calcium phosphate.
   b. When blood calcium is low, osteoclasts break down bone, releasing calcium salts. When blood calcium is high, osteoblasts form bone tissue and store calcium salts.
   c. Bones store small amounts of magnesium, sodium, potassium, and carbonate ions.

7.4 Bone Development, Growth, and Repair

1. Intramembranous bones
   a. **Intramembranous bones** develop from sheetlike layers of unspecialized connective tissues.
   b. **Osteoblasts** within the membranous layers form bone tissue.
   c. Mature bone cells are called **osteocytes.**
2. Endochondral bones
   a. **Endochondral bones** develop as hyaline cartilage that is later replaced by bone tissue.
   b. The primary ossification center appears in the diaphysis, whereas secondary ossification centers appear in the epiphyses.
   c. An **epiphyseal plate** remains between the primary and secondary ossification centers.
   d. The epiphyseal plates are responsible for lengthening.
   e. Long bones continue to lengthen until the epiphyseal plates are ossified.
   f. Growth in thickness is due to ossification beneath the periosteum.
3. Homeostasis of bone tissue
   a. **Osteoclasts** break down bone matrix, and osteoblasts deposit bone matrix to continually remodel bone.
   b. The total mass of bone remains nearly constant.
4. Factors affecting bone development, growth, and repair include nutrition, hormonal secretions, and physical exercise.

7.5 Skeletal Organization

1. The skeleton can be divided into axial and appendicular portions.
2. The **axial skeleton** consists of the **skull, hyoid bone, vertebral column,** and **thoracic cage.**
3. The **appendicular skeleton** consists of the **pectoral girdle, upper limbs, pelvic girdle,** and **lower limbs.**

7.6 Skull

The ***skull*** *consists of twenty-two bones: eight cranial bones and fourteen facial bones.*

1. Cranium
   a. The **cranium** encloses and protects the brain.
   b. Some cranial bones contain air-filled paranasal sinuses.
   c. Cranial bones include the **frontal bone, parietal bones, occipital bone, temporal bones, sphenoid bone,** and **ethmoid bone.**
2. Facial skeleton
   a. The **facial skeleton** forms the basic shape of the face and provides attachments for muscles.
   b. Facial bones include the **maxillae, palatine bones, zygomatic bones, lacrimal bones, nasal bones, vomer bone, inferior nasal conchae,** and **mandible.**
3. Infantile skull
   a. **Fontanels** connect incompletely developed bones.
   b. The proportions of the infantile skull are different from those of an adult skull.

7.7 Vertebral Column

The ***vertebral column*** *extends from the skull to the pelvis and protects the spinal cord. It is composed of* ***vertebrae*** *separated by intervertebral discs.*

1. A typical vertebra
   a. A typical vertebra consists of a body and a bony vertebral arch, which surrounds the spinal cord.
   b. Notches on the upper and lower surfaces provide intervertebral foramina through which spinal nerves pass.
2. Cervical vertebrae
   a. Transverse processes of **cervical vertebrae** bear transverse foramina.
   b. The **atlas** (first vertebra) supports and balances the head.
   c. The dens of the **axis** (second vertebra) provides a pivot for the atlas when the head is turned from side to side.
3. Thoracic vertebrae
   a. **Thoracic vertebrae** are larger than cervical vertebrae.
   b. Facets on the sides articulate with the ribs.
4. Lumbar vertebrae
   a. The vertebral bodies of **lumbar vertebrae** are large and strong.
   b. They support more body weight than other vertebrae.
5. Sacrum
   a. The **sacrum** is a triangular structure formed of five fused vertebrae.
   b. Vertebral foramina form the sacral canal.
6. Coccyx
   a. The **coccyx,** composed of four fused vertebrae, forms the lowest part of the vertebral column.
   b. It acts as a shock absorber when a person sits.

7.8 Thoracic Cage

*The **thoracic cage** includes the **ribs**, thoracic vertebrae, **sternum**, and costal cartilages. It supports the pectoral girdle and upper limbs, protects viscera, and functions in breathing.*

1. Ribs
   a. Twelve pairs of ribs attach to the twelve thoracic vertebrae.
   b. Costal cartilages of the true ribs join the sternum directly. Those of the false ribs join it indirectly or not at all.
   c. A typical rib has a shaft, a head, and tubercles that articulate with the vertebrae.
2. Sternum
   a. The sternum consists of a manubrium, body, and xiphoid process.
   b. It articulates with the clavicles.

7.9 Pectoral Girdle

*The **pectoral girdle** is composed of two **clavicles** and two **scapulae.** It forms an incomplete ring that supports the upper limbs and provides attachments for muscles that move the upper limbs.*

1. Clavicles
   a. Clavicles are rodlike bones located between the sternum and the scapulae.
   b. They hold the shoulders in place and provide attachments for muscles.
2. Scapulae
   a. The scapulae are broad, triangular bones.
   b. They articulate with the humerus of each upper limb and provide attachments for muscles of the upper limbs and chest.

7.10 Upper Limb

Bones of the upper limb form the framework, provide the attachments for muscles, and interact with muscles to move the limb and its parts.

1. Humerus
   a. The **humerus** extends from the scapula to the elbow.
   b. It articulates with the radius and ulna at the elbow.
2. Radius
   a. The **radius** is located on the thumb side of the forearm between the elbow and the wrist.
   b. It articulates with the humerus, ulna, and wrist.
3. Ulna
   a. The **ulna** is longer than the radius and overlaps the humerus posteriorly.
   b. A styloid process at the distal end of the ulna provides attachments for wrist ligaments.
4. Hand
   a. The wrist is composed of eight **carpal bones** that form a carpus.
   b. The palm or metacarpus includes five **metacarpal bones** and fourteen **phalanges** compose the fingers.

7.11 Pelvic Girdle

*The **pelvic girdle** consists of two coxal (hip) bones that articulate with each other anteriorly and with the sacrum posteriorly.*

1. The sacrum, coccyx, and pelvic girdle form the bowl-shaped **pelvis.**
2. Each coxal bone consists of an **ilium, ischium,** and **pubis,** which are fused in the region of the acetabulum.
   a. The ilium
      (1) The ilium is the largest portion of the hip bone.
      (2) It joins the sacrum at the sacroiliac joint.
   b. The ischium
      (1) The ischium is the lowest portion of the hip bone.
      (2) It supports the body weight when sitting.
   c. The pubis
      (1) The pubis is the anterior portion of the hip bone.
      (2) The pubic bones are joined anteriorly at the pubic symphysis.

7.12 Lower Limb

Bones of the lower limb provide a framework for the lower limb and provide attachments for muscles that move the lower limb.

1. Femur
   a. The **femur** extends from the hip to the knee.
   b. The patella articulates with the femur's anterior surface.
2. Tibia
   a. The **tibia** is located on the medial side of the leg.
   b. It articulates proximally with the femur and distally with the talus of the ankle.
3. Fibula
   a. The **fibula** is located on the lateral side of the tibia.
   b. It articulates proximally with the tibia and distally with the ankle. It does not bear body weight.
4. Foot
   a. The ankle or tarsus consists of the talus and six other **tarsal bones.**
   b. The instep or metatarsus includes five **metatarsal bones,** and fourteen **phalanges** compose the toes.

7.13 Joints

***Joints** can be classified according to degree of movement as well as according to the type of tissue that binds the bones.*

1. Fibrous joints
   a. Bones at **fibrous joints** are tightly joined by a layer of dense connective tissue.
   b. Limited movement (amphiarthrotic) or no movement (synarthrotic) occurs at a fibrous joint.
2. Cartilaginous joints
   a. A layer of cartilage joins the bones of **cartilaginous joints.**
   b. Such joints allow limited movement (amphiarthrotic).
3. Synovial joints
   a. The bones of a **synovial joint** are covered with hyaline cartilage and held together by a fibrous joint capsule.
   b. The joint capsule consists of an outer layer of ligaments and an inner lining of synovial membrane.
   c. Pads of fibrocartilage, **menisci,** act as shock absorbers in some synovial joints.
   d. **Bursae** are located between tendons and underlying bony prominences.
   e. Synovial joints that allow free movement (diarthrotic) include **ball-and-socket, condylar, plane, hinge, pivot,** and **saddle joints.**
4. Types of joint movements
   a. Muscles fastened on either side of a joint produce the movements of synovial joints.
   b. Joint movements include **flexion, extension, dorsiflexion, plantar flexion, hyperextension, abduction, adduction, rotation, circumduction, pronation, supination, eversion, inversion, retraction, protraction, elevation,** and **depression.**

REFERENCE PLATES

HUMAN SKULL

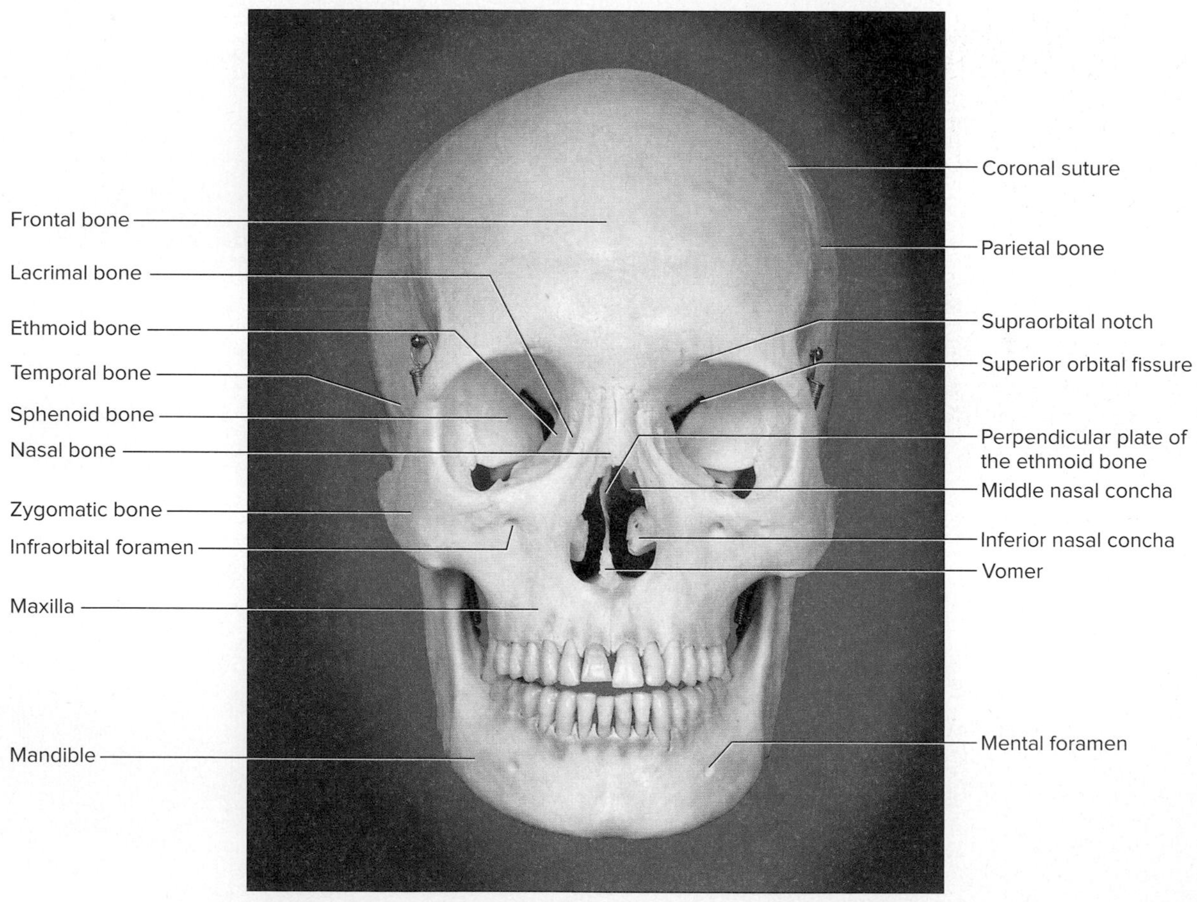

PLATE EIGHT

The skull, anterior view. J and J Photography

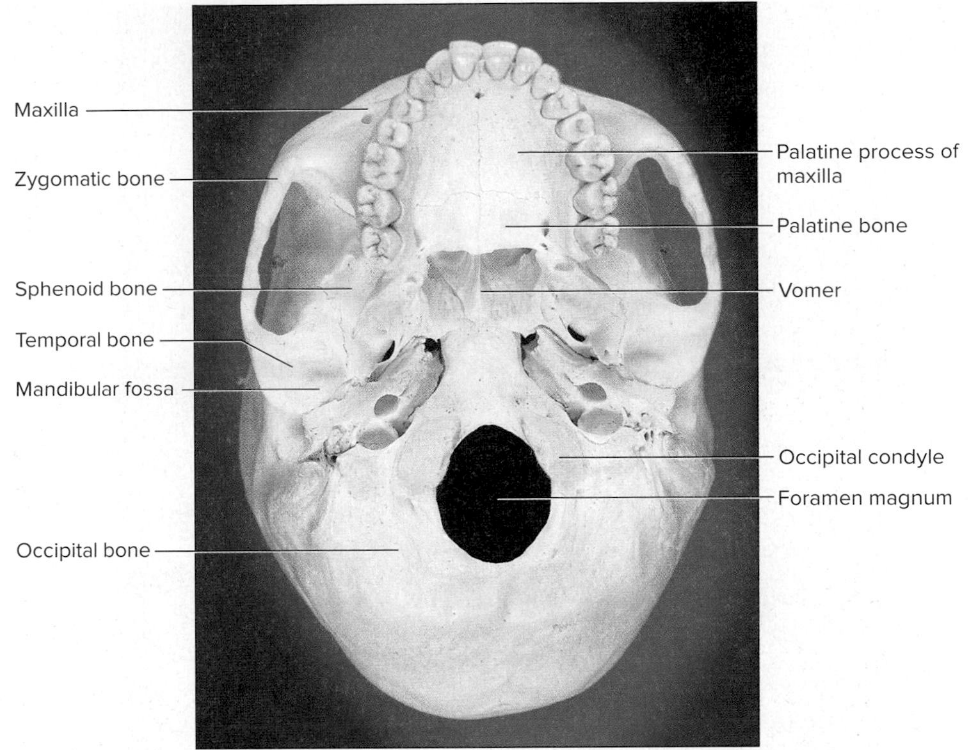

PLATE NINE

The skull, inferior view. J and J Photography

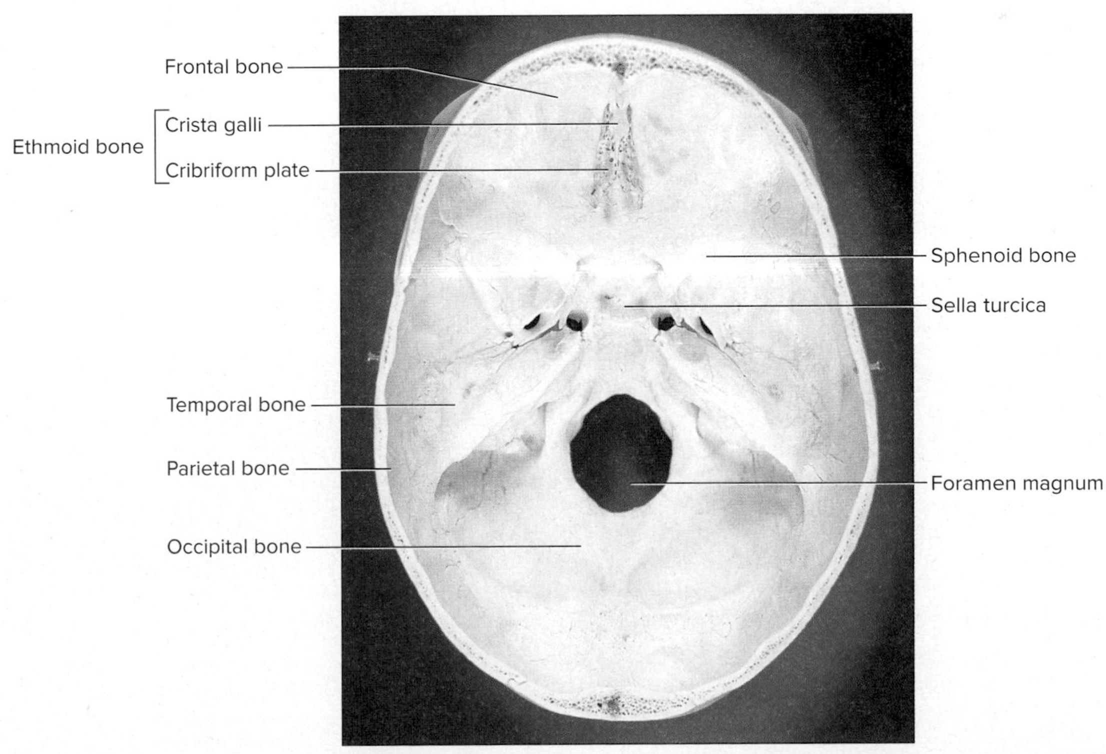

PLATE TEN

The skull, floor of the cranial cavity. J and J Photography

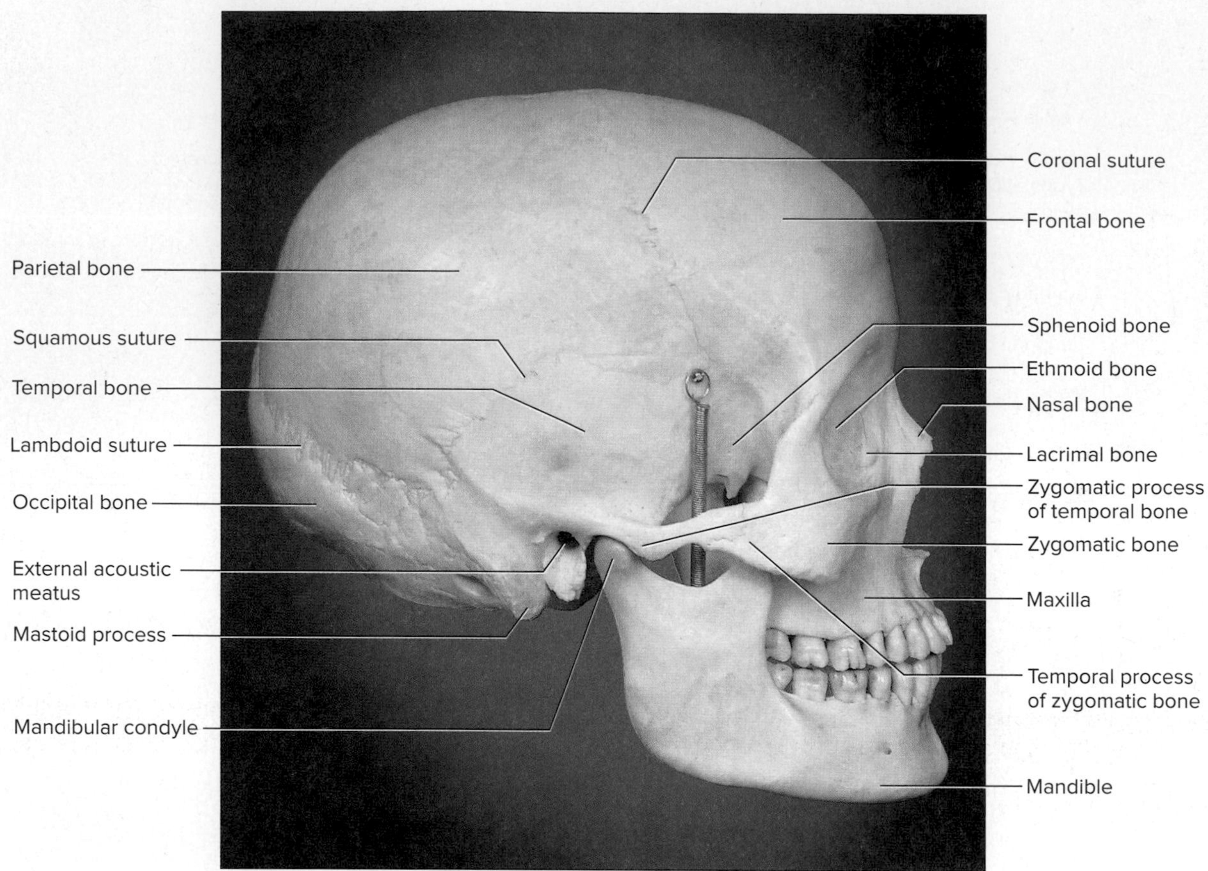

PLATE ELEVEN

The skull, right lateral view. J and J Photography

Muscular System

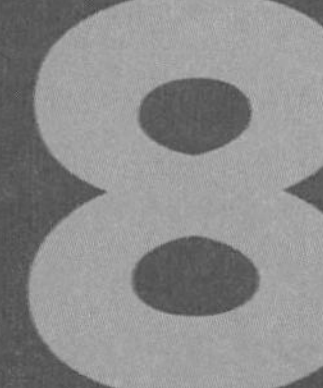

Regular resistance training (weight training) can strengthen muscles. pixs4u/Shutterstock

LEARN • PRACTICE • ASSESS •

Double the muscle. The newborn had an astonishing appearance—his prominent arm and thigh muscles looked as if he'd been weightlifting in the womb. When the child reached five years of age, his muscles were twice normal size, and he could lift weights heavier than many adults could lift. He also had half the normal amount of body fat.

The boy's muscle cells cannot produce a protein called myostatin, which normally stops stem cells from developing into muscle cells. In this boy, a mutation turned off this genetic brake, and as a result his muscles bulged, their cells both larger and more numerous than those in the muscles of an unaffected child. The boy is healthy so far, but because myostatin is also normally made in cardiac muscle, he may develop heart problems.

Other species with myostatin mutations are well known. Naturally "double-muscled" cattle and sheep are valued for their high weights early in life. Chicken breeders lower myostatin production to yield meatier birds, and "mighty mice" with silenced myostatin genes are used in basic research to study muscle overgrowth. In clinical applications, researchers are investigating ways to block myostatin activity to stimulate muscle growth to reverse muscle-wasting from AIDS, cancer, and muscular dystrophy. Myostatin could also be abused to enhance athletic performance.

Apart from double-muscle mutations, resistance (weight) training can increase the ratio of muscle to fat in our bodies, which offers several benefits. Because muscle cells burn calories at three times the rate of fat cells, a lean body is more energetically efficient. Weight training increases muscle strength and bone density; lowers blood pressure; decreases the risks of developing arthritis, osteoporosis, and diabetes mellitus; and is even associated with improved self-esteem and fewer sick days.

Anatomy & Physiology *Revealed* 4.0

Module 6 Muscular System

LEARNING OUTLINE

After studying this chapter, you should be able to complete the "Learning Outcomes" that follow the major headings throughout the chapter.

AIDS TO UNDERSTANDING WORDS

calat- [something inserted] inter*calat*ed disc: dense band that connects cardiac muscle cells.

erg- [work] syn*erg*ist: muscle that works with an agonist to produce a movement.

hyper- [over, more] muscular *hyper*trophy: enlargement of muscle fibers.

inter- [between] *inter*calated disc: dense band that connects cardiac muscle cells.

laten- [hidden] *laten*t period: time between application of a stimulus and the beginning of a muscle contraction.

myo- [muscle] *myo*fibril: contractile structure within a muscle cell.

sarco- [flesh] *sarco*plasm: material (cytoplasm) within a muscle fiber.

syn- [together] *syn*ergist: muscle that works with an agonist to produce a movement.

tetan- [stiff] *tetan*ic contraction: sustained muscular contraction.

-troph [well fed] muscular hyper*trophy*: enlargement of muscle fibers.

(Appendix A has a complete list of Aids to Understanding Words.)

8.1 | Introduction

LEARN

1. List various outcomes of muscle actions.

Muscles are organs composed of specialized cells that generate forces, allowing all types of movement. These actions include walking, speaking, breathing, pumping blood, and moving food through the digestive tract.

Muscle is of three types—skeletal muscle, smooth muscle, and cardiac muscle, as described in section 5.5, Muscle Tissues. This chapter focuses mostly on skeletal muscle, which attaches to bones and is mostly under conscious or voluntary control. It also functions to maintain posture and balance, as well as generate heat through "shivering." Smooth muscle and cardiac muscle are discussed briefly.

8.2 | Structure of a Skeletal Muscle

LEARN

1. Identify the structures that make up a skeletal muscle.
2. Identify the major parts of a skeletal muscle fiber, and the function of each.
3. Discuss nervous stimulation of a skeletal muscle.

The human body has more than 600 distinct skeletal muscles. The face alone includes 60 muscles, more than 40 of which are used to frown, and 20 to smile. Thinner than a thread and barely visible, the stapedius in the middle ear is the body's smallest muscle. In contrast is the gluteus maximus, the largest muscle, located in the buttock. Averaging

CAREER CORNER
Physical Therapy Assistant

The forty-eight-year-old man has joined an over-thirty basketball league, and he cannot keep up. He lies on the gym floor, in pain, after an overambitious jump shot. He had felt a sudden twinge in his knee, and now the area is red and swelling.

In the emergency department of the nearest hospital, an orthopedist sends the man for an MRI scan, which reveals a small tear in the anterior cruciate ligament (ACL). The patient is willing to give up basketball and doesn't want surgery, so he sees a physical therapist twice a week for a month. A physical therapy assistant works with the man, leading him through a series of exercises that may restore full mobility.

The therapy, under the supervision of the physical therapist, begins with stepping, squats, and using a single-leg bicycle that isolates and builds up the muscles of the injured limb. Therapy progresses to deep knee bends and lifting weights to further build up muscles around the injured joint. The PT assistant gives her patient exercises to do daily at home. The therapy builds the muscles around the knee to compensate for the hurt ACL, restoring full range of motion.

A physical therapy assistant must complete a two-year college program and pass a certification exam. PT assistants work in hospitals, skilled nursing facilities, private homes, schools, fitness centers, and workplaces.

about 18 inches in length, the sartorius, found in the thigh, is the longest muscle in the body. However, at the microscopic level, all skeletal muscles are built from the same tissues.

Connective Tissue Coverings

Layers of connective tissue enclose and separate all parts of a skeletal muscle. Dense connective tissue called **fascia** (fash′e-ah) separates an individual skeletal muscle from adjacent muscles and holds it in position (fig. 8.1). Fascia blends with the **epimysium** (ep″i-mis′ē-um), a layer of connective tissue that closely surrounds each skeletal muscle (fig. 8.1). Other layers of connective tissue, called **perimysium** (per″i-mis′ē-um), extend inward from the epimysium and separate the muscle tissue into small sections called **fascicles** (fas′ĭ-k′lz). Fascicles are bundles of skeletal muscle fibers. Each muscle fiber within a fascicle lies within a layer of connective tissue in the form of a thin covering called **endomysium** (en″do-mis′ē-um). This organization allows the parts to move somewhat independently. Many blood vessels and nerves pass through these layers.

The connective tissue layers may project beyond the muscle's end to form a cordlike tendon. Fibers in a tendon may intertwine with those in a bone's periosteum, attaching the muscle to the bone. In other cases, the connective tissue forms broad fibrous sheets called **aponeuroses** (ap″o-nū-rōsēz), which may attach to bone, skin, or to the connective tissue of adjacent muscles.

In *tendinitis,* a tendon (the attachment of a muscle to a bone) becomes painfully inflamed and swollen following injury or the repeated stress of athletic activity. If rest, physical therapy, and anti-inflammatory drugs do not alleviate tendinitis, then ultrasound can be applied to break up scar tissue. In *tenosynovitis,* the connective tissue sheath of the tendon (the tenosynovium) is inflamed. The tendons most commonly affected are those associated with the joint capsules of the shoulder, elbow, and hip, and those that move the hand, thigh, and foot.

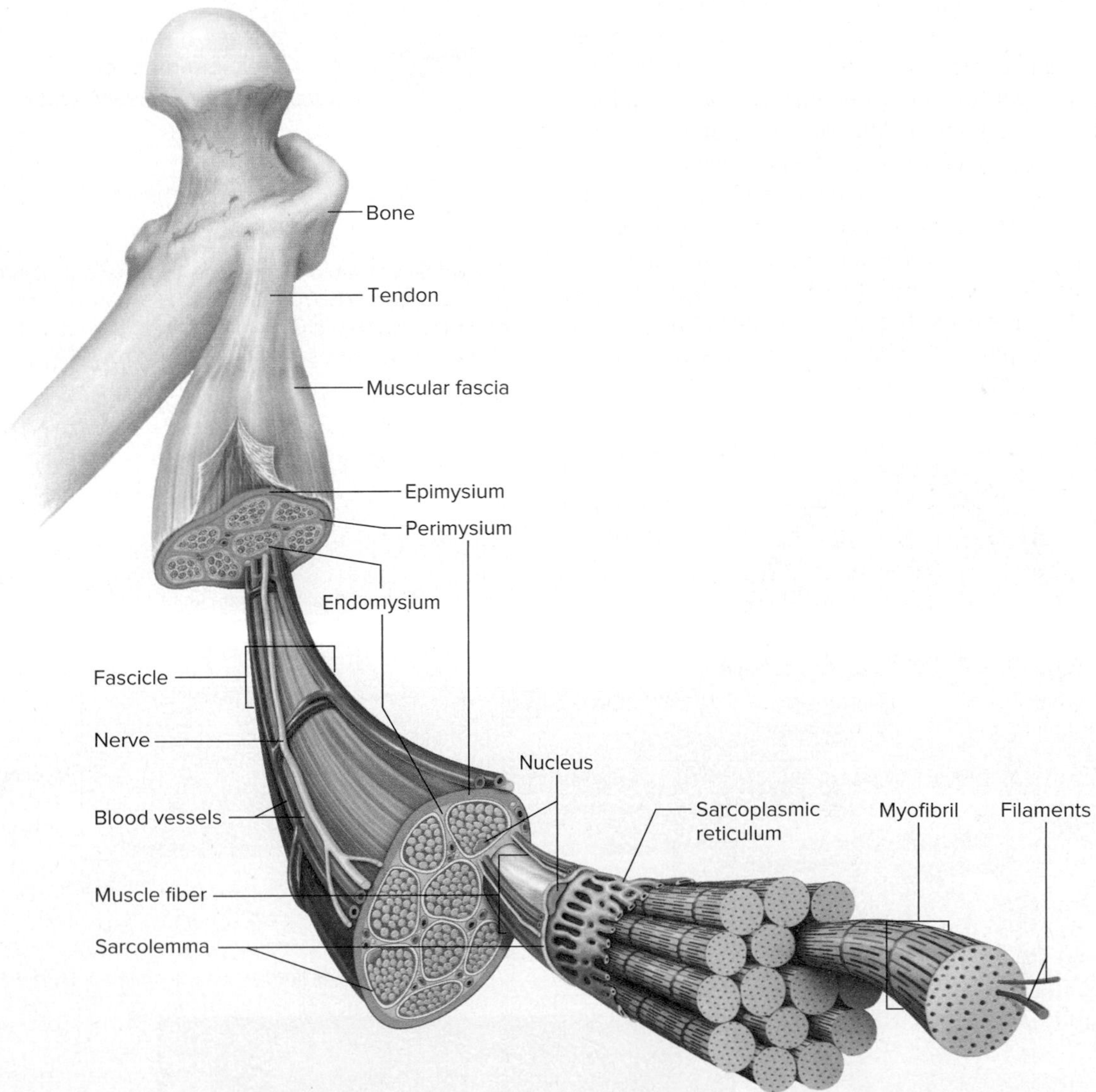

Figure 8.1 A skeletal muscle is composed of a variety of tissues, including layers of connective tissue. Fascia covers the surface of the muscle, epimysium lies beneath the fascia, and perimysium extends into the structure of the muscle where it groups muscle cells into fascicles. Endomysium separates individual muscle fibers within fascicles. APR

Skeletal Muscle Fibers

A skeletal muscle fiber is a single cell that contracts (exerts a pulling force) in response to stimulation and then relaxes when the stimulation ends. Each skeletal muscle fiber is a long, thin, cylinder with rounded ends. It may extend the full length of the muscle. Just beneath its cell membrane (or *sarcolemma*), the cytoplasm (or *sarcoplasm*) of the fiber has many small, oval nuclei and mitochondria (fig. 8.1). The sarcoplasm also contains many threadlike **myofibrils** (mī″o-fī′brilz) that lie parallel to one another.

Myofibrils play a fundamental role in muscle contraction. They consist of two kinds of protein filaments (myofilaments)—thick filaments composed of the protein **myosin** (mī′o-sin) and thin filaments composed mainly of the protein **actin** (ak′tin) (fig. 8.2). (Two other thin filament proteins, troponin and tropomyosin, are discussed later in section 8.3, Skeletal Muscle Contraction.) The organization of these filaments produces the characteristic alternating light and dark *striations,* or bands, of a skeletal muscle fiber.

The striations of skeletal muscle result from a repeating pattern of units called **sarcomeres** (sar′kō-mĕrz) within each muscle fiber. The myofibrils are essentially sarcomeres joined end-to-end (figs. 8.2 and 8.3). Muscle fibers, and in a way muscles themselves, may be thought of as a collection of sarcomeres. Sarcomeres are discussed later as the functional units of muscle contraction (section 8.3, Skeletal Muscle Contraction).

The striation pattern of skeletal muscle fibers has two main parts. The first, the *I bands* (the light bands), are composed of thin filaments directly attached to structures called *Z lines.* The second part of the striation pattern consists of

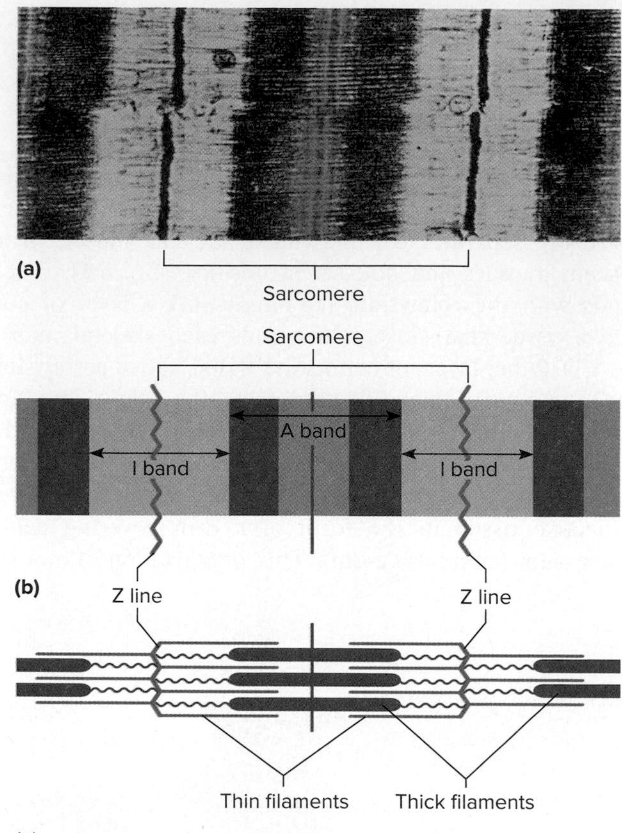

Figure 8.3 A sarcomere is the functional unit of muscle contraction. **(a)** Micrograph (16,000x). **(b, c)** The spatial relationship of thin and thick filaments in a sarcomere is the basis for the repeating pattern of striations in skeletal muscle. APR (a): ©H.E. Huxley

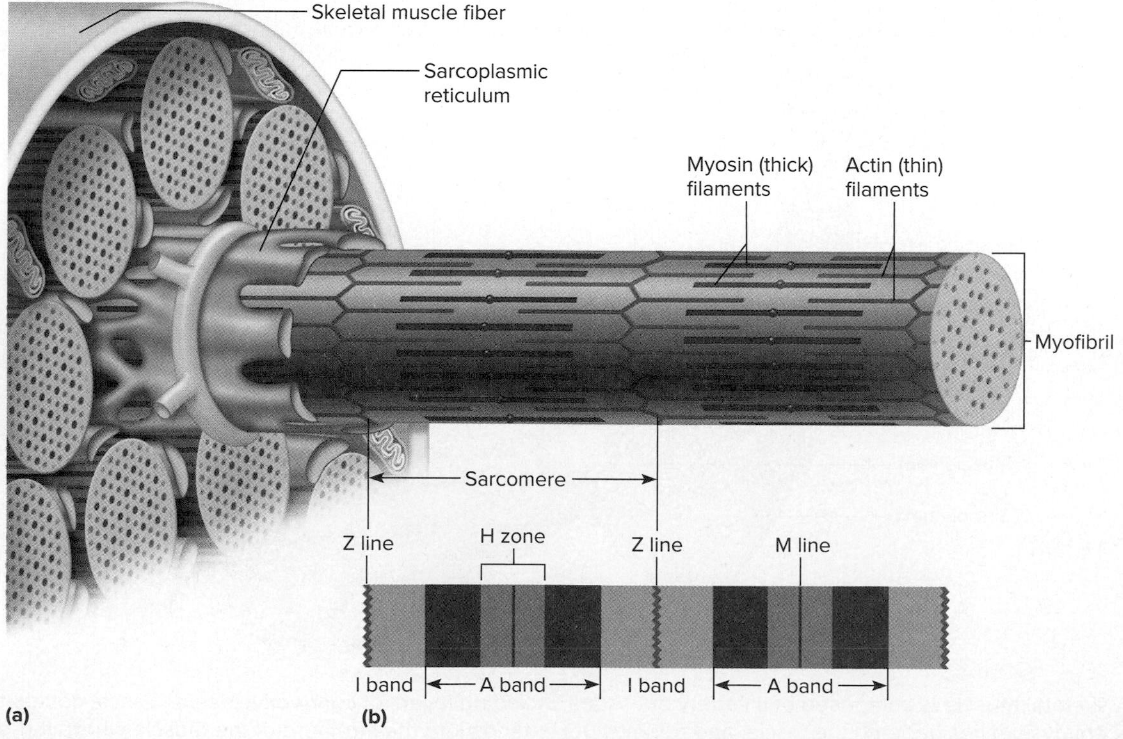

Figure 8.2 Skeletal muscle fiber. **(a)** A skeletal muscle fiber contains many myofibrils, each consisting of **(b)** repeating units called sarcomeres. The characteristic striations of a sarcomere reflect the organization of actin (thin) and myosin (thick) filaments. APR

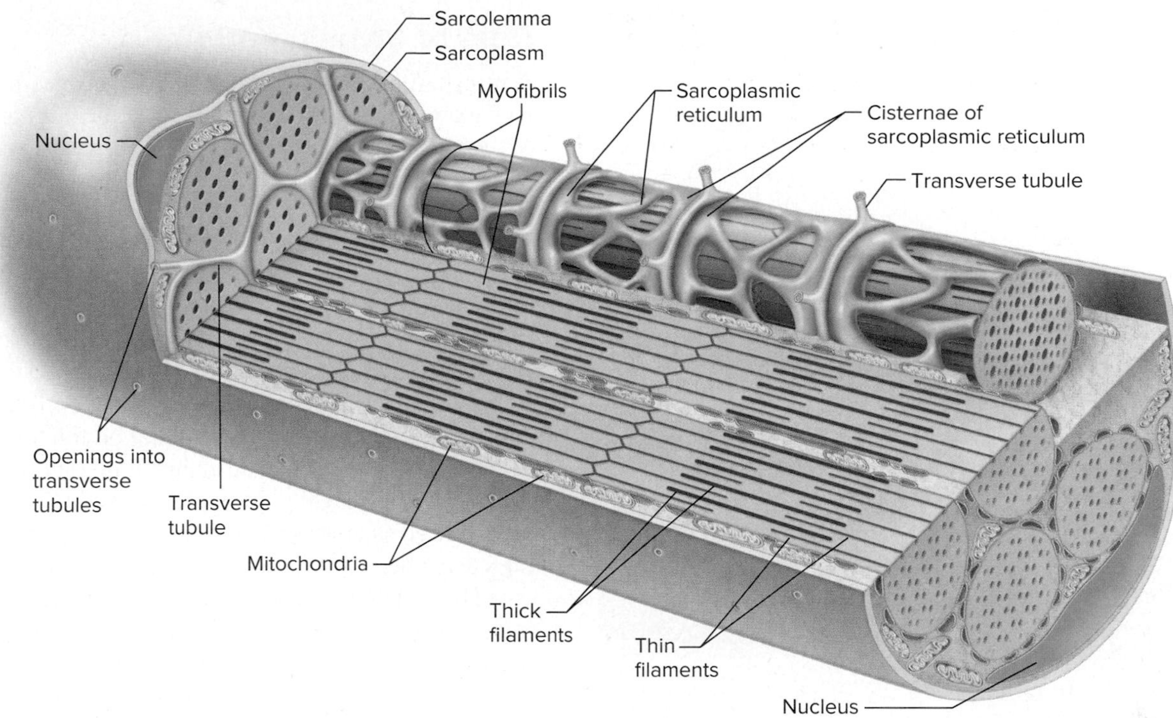

Figure 8.4 The sarcoplasm of a skeletal muscle fiber contains a network of sarcoplasmic reticulum and a system of transverse tubules.

the *A bands* (the dark bands), which extend the length of the thick filaments. The A bands have a central region (*H zone*), consisting only of thick filaments, located between two regions where the thick and thin filaments overlap. A thickening known as the *M line* is located down the center of the A band (see fig. 8.2). The M line consists of proteins that help hold the thick filaments in place. A sarcomere extends from one Z line to the next (see figs. 8.2 and 8.3).

Within the sarcoplasm of a muscle fiber is a network of membranous channels that surrounds each myofibril and runs parallel to it (fig. 8.4). These membranes form the **sarcoplasmic reticulum,** which corresponds to the endoplasmic reticulum of other types of cells. Another set of membranous channels, called **transverse tubules** (T tubules), extends inward as invaginations from the fiber's membrane and passes all the way through the fiber. Thus, each transverse tubule opens to the outside of the muscle fiber and contains extracellular fluid. Furthermore, each transverse tubule lies between two enlarged portions of the sarcoplasmic reticulum called **cisternae,** near the region where the thick and thin filaments overlap. The sarcoplasmic reticulum and transverse tubules play important roles in activating the contraction mechanism when the muscle fiber is stimulated.

PRACTICE 8.2

Answers to the Practice questions can be found in the eBook.

1. Describe how connective tissue is part of a skeletal muscle.
2. Describe the general structure of a skeletal muscle fiber.
3. Explain why skeletal muscle fibers appear striated.
4. Explain the relationship between the sarcoplasmic reticulum and the transverse tubules.

Neuromuscular Junction

Recall from section 5.6, Nervous Tissue, that neurons (nerve cells) play a role in communication within the body by conducting electrical impulses. Neurons that control effectors (such as muscles) are called **motor neurons.** Normally, a skeletal muscle fiber contracts only when stimulated by a motor neuron. The opening vignette to chapter 3 describes amyotrophic lateral sclerosis (ALS, or Lou Gehrig's disease), which impairs the motor neurons that control skeletal muscles.

Each skeletal muscle fiber is functionally (but not physically) connected to the axon of a motor neuron that passes outward from the brain or the spinal cord. This is much like the functional connection whereby you can talk into a cell phone although your mouth is not in direct physical contact with it. The functional connection between a neuron and another cell is called a **synapse** (sin'aps). Neurons communicate with the cells that they control by releasing chemicals, called **neurotransmitters** (nu″ro-trans'mit-erz), at synapses.

The synapse between a motor neuron and the muscle fiber that it controls is called a **neuromuscular junction.** Here, the muscle fiber membrane is specialized to form a **motor end plate.** At the motor end plate, nuclei and mitochondria are abundant. The sarcolemma has indentations and is extensively folded. The end of the motor neuron extends fine projections into the indentations of the muscle fiber membrane (fig. 8.5).

A small gap called the **synaptic cleft** separates the membrane of the neuron and the membrane of the muscle fiber. The cytoplasm at the distal ends of these motor neuron axons is rich in mitochondria and contains many tiny vesicles (synaptic vesicles) that store neurotransmitter molecules (fig. 8.5).

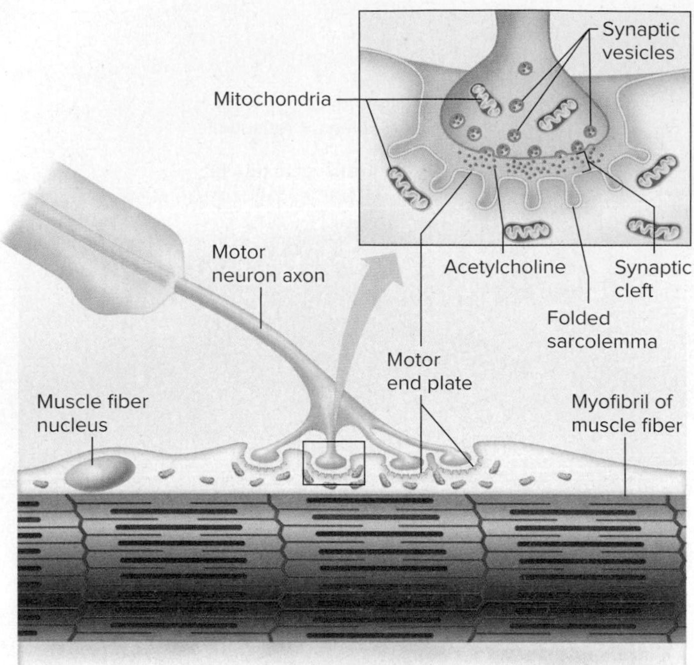

Figure 8.5 A neuromuscular junction includes the end of a motor neuron and the motor end plate of a muscle fiber. APR

PRACTICE FIGURE 8.5

How does acetylcholine released into the synaptic cleft reach the muscle fiber membrane?

Answer can be found in Appendix E.

PRACTICE 8.2

5. Which two structures approach each other at a neuromuscular junction?
6. Describe a motor end plate.
7. What is the function of a neurotransmitter?

8.3 | Skeletal Muscle Contraction

LEARN

1. Identify the major events of skeletal muscle fiber contraction.
2. List the energy sources for muscle fiber contraction.
3. Describe how oxygen debt develops.
4. Describe how a muscle may become fatigued.
5. Distinguish between muscle fiber types.
6. Describe effects of skeletal muscle use and disuse.

A muscle fiber contraction involves an interaction of organelles and molecules in which myosin binds to actin and exerts a pulling force. The result is a movement within the myofibrils in which the filaments of actin and myosin slide past one another, increasing the area of overlap. This action shortens (contracts) the muscle fiber, which then pulls on the body part that it moves.

Role of Myosin and Actin

A myosin molecule is composed of two twisted protein strands with globular parts called heads projecting from one end. Many of these molecules together compose a thick filament. An actin molecule is a globular structure with a binding site to which the myosin heads can attach. Many actin molecules twist into a double strand (helix), forming a thin filament. The proteins **troponin** and **tropomyosin** are also part of the thin filament (fig. 8.6).

The sarcomere is considered the functional unit of skeletal muscles because the contraction of an entire skeletal muscle can be described in terms of the shortening of the sarcomeres within its muscle fibers. The force that shortens the sarcomeres comes from the myosin heads pulling on the thin filaments. A myosin head can attach to an actin binding site, forming a *cross-bridge,* and bend slightly, pulling on the actin filament. Then the myosin head can release, straighten, combine with another binding site further down the actin filament, and pull again (fig. 8.7).

The **sliding filament model** of muscle contraction includes all of these actin-myosin interactions and is named for how the sarcomeres shorten. Thick and thin filaments do not change length. Rather, they slide past one another, with the thin filaments moving toward the center of the sarcomere from both ends (fig. 8.8).

The myosin heads contain an enzyme, **ATPase,** which catalyzes the breakdown of ATP to ADP and phosphate (see section 4.4, Energy for Metabolic Reactions). This reaction provides energy that straightens the myosin head into a "cocked" position, much like pulling back on the moving part of a spring-operated toy. The cocked myosin head stays in this position until it binds to actin, forming a cross-bridge. When this happens, the "spring" is released and the cross-bridge pulls on the thin filament. Another ATP binding to the myosin breaks the cross-bridge, releasing the myosin head from the actin, but not breaking down the ATP. The ATPase then catalyzes the breakdown of ATP to ADP and phosphate, putting the myosin head in a "cocked" position again. This cycle repeats as long as ATP is available as an energy source and as long as the muscle fiber is stimulated to contract.

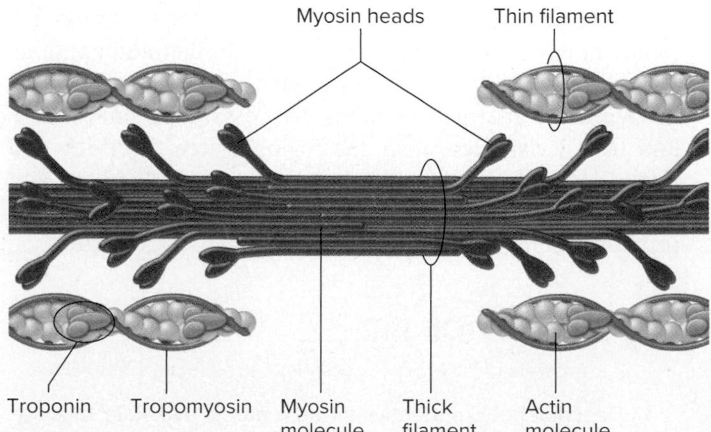

Figure 8.6 Thick filaments are composed of the protein myosin, and thin filaments are composed primarily of the protein actin. Myosin molecules have globular heads that extend toward nearby thin filaments.

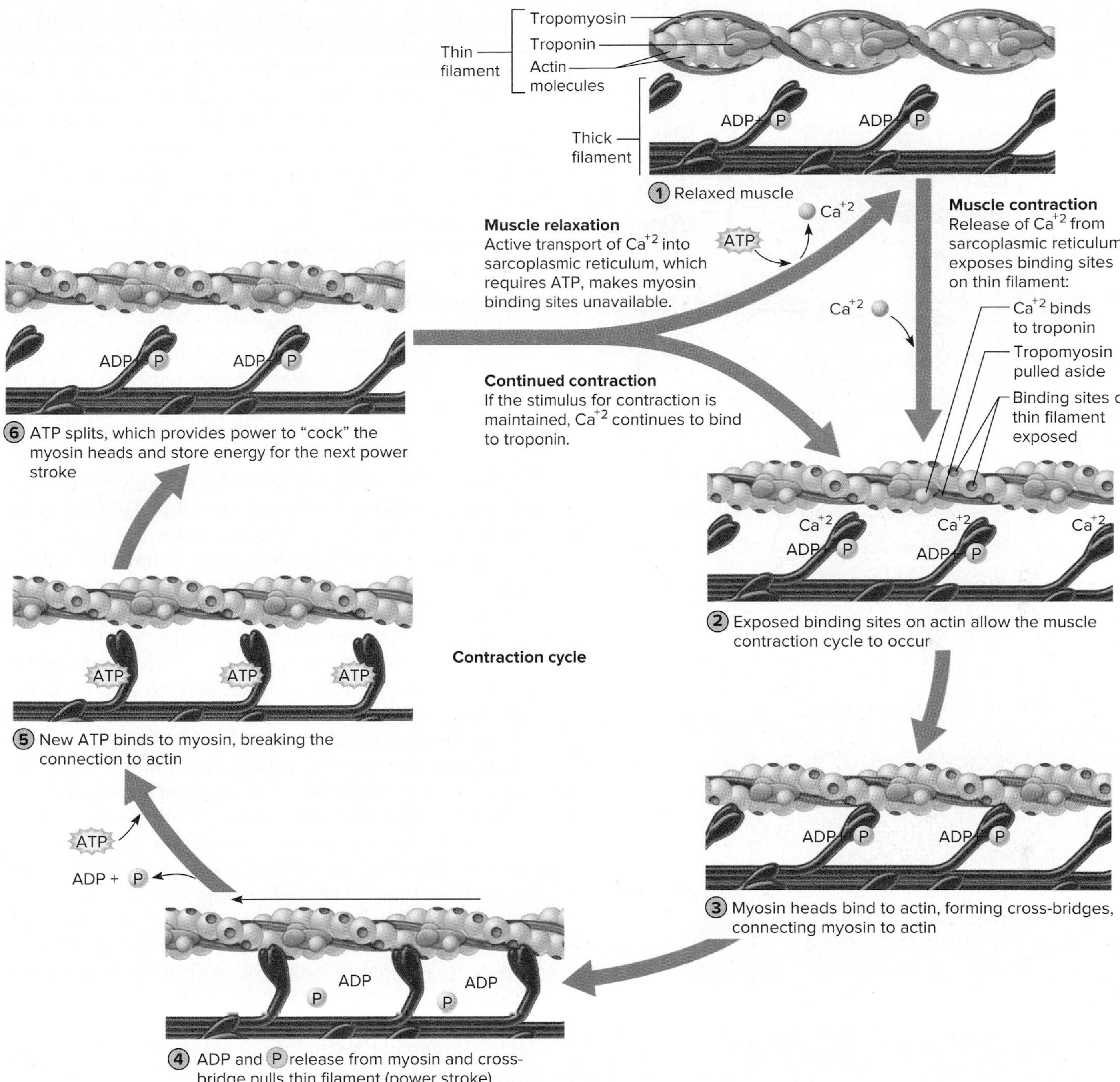

Figure 8.7 The sliding filament model. **(1)** Relaxed muscle. **(2)** When the calcium ion concentration in the cytosol rises, binding sites on thin filaments become exposed, and **(3)** myosin heads bind to the actin, forming cross-bridges. **(4)** Upon binding to actin, myosin heads spring from the cocked position and the cross-bridges pull on thin filaments. **(5)** ATP binds (but is not yet broken down), causing the myosin heads to release from the thin filament, thus breaking the cross-bridges. **(6)** ATP breakdown provides energy to again "cock" the unattached myosin heads. As long as ATP and calcium ions are present, the cycle continues. When the calcium ion concentration in the cytosol is low, the muscle is relaxed. APR

Several hours after death, skeletal muscles partially contract and become rigid, fixing the joints in place. This condition, *rigor mortis,* may continue for 72 hours or more. It results from an increase in membrane permeability to calcium ions, which allows cross-bridges to form, and a decrease in ATP in muscle fibers, which prevents relaxation. The actin and myosin filaments of the muscle fibers remain linked until the proteins begin to decompose.

Stimulus for Contraction APR

A skeletal muscle fiber normally does not contract until the neurotransmitter **acetylcholine** (as″ĕ-til-kō′lēn) stimulates it. This neurotransmitter is synthesized in the cytoplasm of the motor neuron and stored in vesicles at the distal end of the motor neuron axons. When an impulse (see section 9.5, Charges Inside a Cell, and section 9.6, Impulse Conduction) reaches the end of a

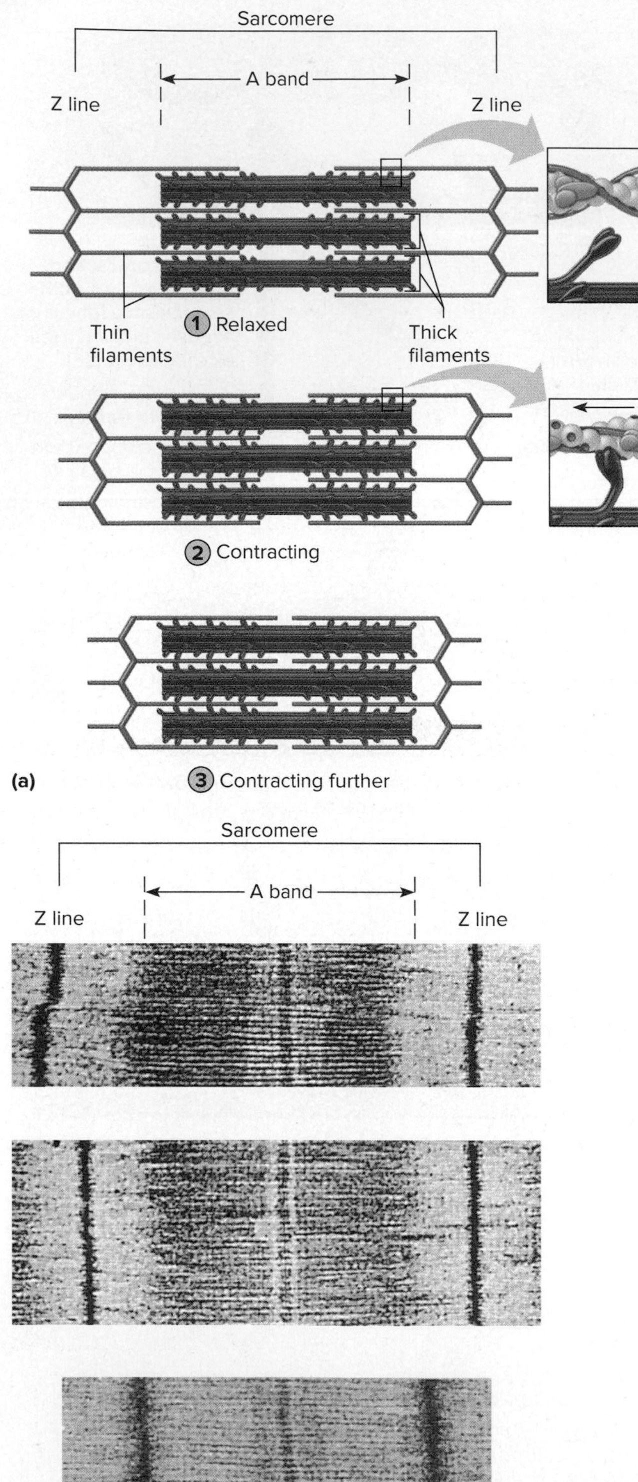

Figure 8.8 When a skeletal muscle contracts, **(a)** individual sarcomeres shorten as thin filaments slide past thick filaments toward the center of the sarcomere. **(b)** This transmission electron micrograph shows a sarcomere shortening during muscle contraction (23,000x). (b): ©H. E. Huxley APR

PRACTICE FIGURE 8.8

What happens to the length of the thick and thin filaments during contraction?

Answer can be found in Appendix E.

motor neuron axon, some of the vesicles release their acetylcholine into the synaptic cleft, the space between the motor neuron axon and the motor end plate (see fig. 8.5).

Acetylcholine diffuses rapidly across the synaptic cleft and binds to specific protein molecules (acetylcholine receptors) in the muscle fiber membrane at the motor end plate, increasing membrane permeability to sodium ions. Entry of these charged particles into the muscle cell stimulates an electrical impulse much like the impulse on the motor neuron. The impulse passes in all directions over the surface of the muscle fiber membrane and travels through the transverse tubules, deep into the fiber, until it reaches the sarcoplasmic reticulum (see fig. 8.4).

The sarcoplasmic reticulum contains a high concentration of calcium ions. In response to a muscle impulse, the membranes of the sarcoplasmic reticulum become more permeable to these ions, and the calcium ions diffuse into the cytosol of the muscle fiber.

When a high concentration of calcium ions is in the cytosol, troponin and tropomyosin interact in a way that exposes binding sites on actin where myosin heads can attach. As a result, cross-bridge linkages form between the thick and thin filaments, and the muscle fiber contracts (see figs. 8.7 and 8.8). The contraction, which requires ATP, continues as long as the motor neuron releases acetylcholine.

When nervous stimulation ceases, three events lead to muscle relaxation. First, the acetylcholine that stimulated the muscle fiber is rapidly decomposed by the enzyme **acetylcholinesterase** (as′ee-til-kō″lin-es′ter-ās). This enzyme is present at the neuromuscular junction on the membrane of the motor end plate. Without acetylcholine binding to its receptors, the impulse on the muscle fiber ceases.

The second event in muscle relaxation takes place once acetylcholine is broken down and the stimulus to the muscle fiber ceases. Using ATP as an energy source, calcium ions are actively transported back into the sarcoplasmic reticulum, which decreases the calcium ion concentration of the cytosol. The third event involves ATP binding to myosin, breaking the cross-bridge linkages between thin and thick filaments, and consequently relaxing the muscle fiber. Table 8.1 summarizes the major events leading to muscle contraction and relaxation.

PRACTICE 8.3

1. Explain how an impulse on a motor neuron can trigger a muscle contraction.
2. Explain how the filaments of a myofibril interact during muscle contraction.

Energy Sources for Contraction

ATP molecules supply the energy for muscle fiber contraction. However, when a contraction starts, a muscle fiber has only enough ATP to enable it to contract for a very short time. Therefore, when a fiber is active, ATP must be regenerated from ADP and phosphate. The molecule that initially makes this possible is **creatine phosphate** (krē′ah-tin fos′fāt).

TABLE 8.1 Major Events of Muscle Contraction and Relaxation

Muscle Fiber Contraction
1. An impulse travels down a motor neuron axon.
2. The motor neuron releases the neurotransmitter acetylcholine (ACh).
3. ACh binds to ACh receptors in the muscle fiber membrane.
4. The sarcolemma is stimulated. An impulse travels over the surface of the muscle fiber and deep into the fiber through the transverse tubules.
5. The impulse reaches the sarcoplasmic reticulum, and calcium channels open.
6. Calcium ions diffuse from the sarcoplasmic reticulum into the cytosol and bind to troponin molecules.
7. Tropomyosin molecules move and expose specific sites on actin where myosin heads can bind.
8. Cross-bridges form, linking thin and thick filaments.
9. Thin filaments are pulled toward the center of the sarcomere by pulling of the cross-bridges.
10. The muscle fiber exerts a pulling force on its attachments as a contraction occurs.
Muscle Fiber Relaxation
1. Acetylcholinesterase decomposes acetylcholine, and the muscle fiber membrane is no longer stimulated.
2. Calcium ions are actively transported into the sarcoplasmic reticulum.
3. ATP breaks cross-bridge linkages between actin and myosin filaments without breakdown of the ATP itself.
4. Breakdown of ATP "cocks" the myosin heads.
5. Troponin and tropomyosin molecules block the interaction between myosin and actin filaments.
6. The muscle fiber remains relaxed, yet ready, until stimulated again.

Like ATP, creatine phosphate contains high-energy phosphate bonds. When ATP supply is sufficient, an enzyme in the mitochondria (creatine phosphokinase) catalyzes the synthesis of creatine phosphate, which stores excess energy in its phosphate bonds (fig. 8.9).

Creatine phosphate is four to six times more abundant in muscle fibers than ATP, but it cannot directly supply energy to a cell's energy-utilizing reactions. Instead, as ATP decomposes, the phosphate from creatine phosphate can be transferred to ADP molecules, converting them back into ATP. Active muscle fibers, however, rapidly exhaust the supply of creatine phosphate. When this happens, the muscle fibers use cellular respiration of glucose as an energy source for synthesizing ATP.

Oxygen Supply and Cellular Respiration

Glycolysis can take place in the cytosol in the absence of oxygen (anaerobic), as discussed in section 4.4, Energy for Metabolic Reactions. The more complete breakdown of glucose occurs in the mitochondria and requires oxygen. The blood carries the oxygen from the lungs to body cells to support this aerobic respiration. Red blood cells carry the oxygen, loosely bound to molecules of **hemoglobin,** the protein responsible for the red color of blood.

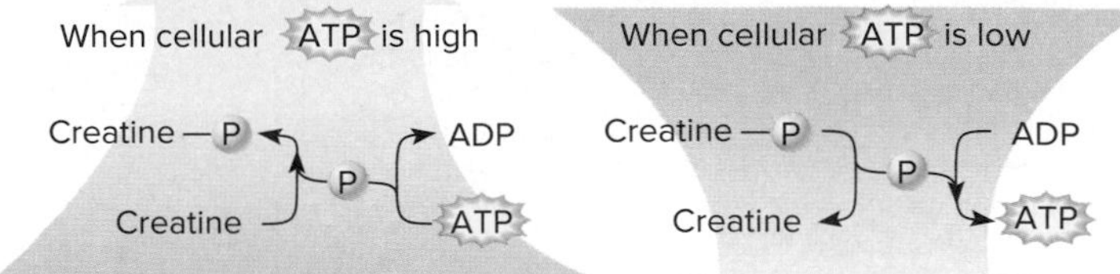

Figure 8.9 Creatine phosphate is synthesized when ATP levels in a muscle cell are high. Creatine phosphate may be used to replenish ATP when ATP levels in a muscle cell are low.

Another protein, **myoglobin,** is synthesized in muscle cells and imparts the reddish-brown color of skeletal muscle tissue. Like hemoglobin, myoglobin can combine loosely with oxygen. Myoglobin's ability to temporarily store oxygen increases the amount of oxygen available in the muscle cells to support aerobic respiration (fig. 8.10).

Oxygen Debt

When a person is resting or is moderately active, the respiratory and cardiovascular systems can usually supply sufficient oxygen to skeletal muscles to support aerobic respiration. This may not be the case when skeletal muscles are used strenuously for even a minute or two. In this situation, muscle fibers increasingly rely on anaerobic respiration to obtain energy.

In anaerobic respiration, glucose molecules are broken down by glycolysis, yielding *pyruvic acid,* which would enter the citric acid cycle under aerobic conditions (see section 4.4, Energy for Metabolic Reactions). Because the oxygen supply is low, however, the pyruvic acid reacts to produce **lactic acid** (fig. 8.10). Lactic acid dissociates rapidly to form lactate ion (lactate) and hydrogen ion. Lactate leaves muscle cells by facilitated diffusion, enters the bloodstream, and eventually reaches the liver. In liver cells, reactions requiring ATP synthesize glucose from lactate.

During strenuous exercise, available oxygen is used primarily to synthesize the ATP the muscle fiber requires to contract, rather than to make ATP for synthesizing glucose from lactate. Consequently, as lactate accumulates, a person develops an **oxygen debt** that must be repaid. Oxygen debt (also called *excess post-exercise oxygen consumption,* or EPOC) equals the amount of oxygen that liver cells require to convert the accumulated lactate into glucose, plus the amount muscle cells require to restore ATP and creatine phosphate to their original concentrations and to return blood and tissue oxygen levels to normal. The conversion of lactate back into glucose is slow. Repaying an oxygen debt following vigorous exercise may take several hours.

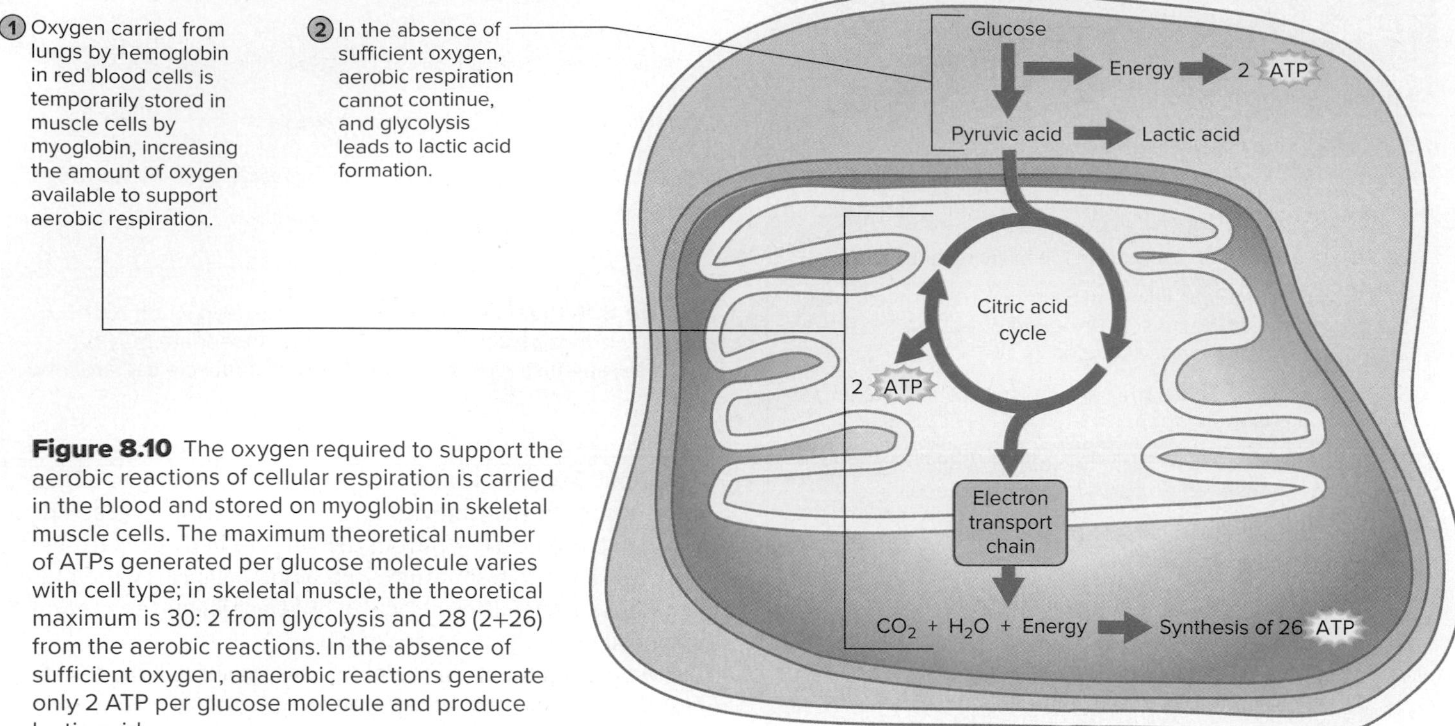

① Oxygen carried from lungs by hemoglobin in red blood cells is temporarily stored in muscle cells by myoglobin, increasing the amount of oxygen available to support aerobic respiration.
② In the absence of sufficient oxygen, aerobic respiration cannot continue, and glycolysis leads to lactic acid formation.
Glucose
Energy
2 ATP
Pyruvic acid
Lactic acid
Citric acid cycle
2 ATP
Electron transport chain
CO_2 + H_2O + Energy
Synthesis of 26 ATP

Figure 8.10 The oxygen required to support the aerobic reactions of cellular respiration is carried in the blood and stored on myoglobin in skeletal muscle cells. The maximum theoretical number of ATPs generated per glucose molecule varies with cell type; in skeletal muscle, the theoretical maximum is 30: 2 from glycolysis and 28 (2+26) from the aerobic reactions. In the absence of sufficient oxygen, anaerobic reactions generate only 2 ATP per glucose molecule and produce lactic acid.

The metabolic capacity of a muscle may change with physical training. With high-intensity exercise, which depends more on glycolysis for ATP, a muscle synthesizes more glycolytic enzymes, and its capacity for glycolysis increases. With aerobic exercise, more capillaries and mitochondria form, and the muscle's capacity for aerobic respiration increases. Table 8.2 summarizes muscle metabolism, and Clinical Application 8.1 discusses abuse of steroid drugs to enhance muscle performance.

Heat Production

Less than half of the energy released in cellular respiration is transferred to ATP, and the rest becomes heat. Although all active cells generate heat, muscle tissue is a major heat source because muscle is such a large proportion of the total body mass. Blood transports heat generated in muscle to other tissues, which helps maintain body temperature. In cold temperatures, skeletal muscles can also generate body heat through involuntary contractions commonly called shivering.

Muscle Fatigue

A muscle exercised strenuously for a prolonged period may have a decreased ability to contract, a condition called *fatigue*. The obvious cause of muscle fatigue is the depletion of the energy source, ATP. Although not completely understood, fatigue more than likely involves a complex combination of other factors, including a drop in pH due to a buildup of lactic acid. Fluctuations in pH can have adverse effects on the physiology of a muscle fiber. Electrolyte imbalance and central nervous system exhaustion have also been proposed as potential factors.

Occasionally, a muscle becomes fatigued and cramps at the same time. A cramp is a painful condition in which a muscle undergoes a sustained involuntary contraction. Cramps are thought to occur when changes in the extracellular fluid surrounding the muscle fibers and their motor neurons somehow trigger uncontrolled stimulation of the muscle.

TABLE 8.2 Muscle Metabolism

Type of Exercise	Pathway Used	ATP Production	Waste Product
Low to moderate intensity: Blood flow provides sufficient oxygen for cellular requirements	Glycolysis, leading to pyruvic acid formation and aerobic respiration	30 ATP per glucose for skeletal muscle	Carbon dioxide is exhaled
High intensity: Oxygen supply is not sufficient for cellular requirements	Glycolysis, leading to lactic acid formation	2 ATP per glucose	Lactic acid accumulates

CLINICAL APPLICATION 8.1

Steroids and Athletes—An Unhealthy Combination

It seems that not a year goes by without a few famous athletes confessing to, or being caught using, steroid hormones to bulk up their muscles to improve performance. High school and college athletes have abused steroids too. Athletes who abuse steroids seek the hormone's ability to increase muscular strength. They are caught when the steroids or their breakdown products are detected in urine or when natural testosterone levels plummet in a negative feedback response to the outside hormone supply (fig. 8A).

Improved performance today due to steroid use may have consequences tomorrow. Steroids hasten adulthood, stunting height and causing early hair loss. In males, excess steroid hormones lead to breast development, and in females to a deepened voice, hairiness, and a male physique. The drugs may damage the kidneys, liver, and heart. Atherosclerosis may develop because steroids raise LDL cholesterol. In males, the body mistakes the synthetic steroids for the natural hormone and lowers its own production of testosterone. Infertility may result. Steroids can also cause psychiatric symptoms, including delusions, depression, and violence.

Anabolic steroids have been used for medical purposes since the 1930s, to treat underdevelopment of the testes and the resulting testosterone deficiency, anemia, and muscle-wasting disorders. Today, they are used to treat wasting associated with AIDS.

Figure 8A Sprinter Ben Johnson ran away with the gold medal in the 100-meter race at the 1988 Summer Olympics—but then had to return the award when traces of a steroid drug showed up in his urine. Drug abuse continues to be a problem among amateur as well as professional athletes. Mike Powell/Getty Images

Types of Muscle Fibers and Muscle Use

Not all muscle fibers are the same. *Fast fibers* are built for rapid movements, whereas *slow fibers* allow for endurance activities. All whole muscles are comprised of a unique combination of these two fiber types.

Fast Fibers

Fast fibers make up the majority of muscle fibers in the body. They are called fast fibers because they are able to generate their maximum force rapidly. Large in diameter, they also provide fairly powerful contractions. They contain a large amount of glycogen and relatively few mitochondria, making them mostly dependent on anaerobic energy production (glycolysis). Therefore, they are not ideal for prolonged activities. Whole skeletal muscles constructed primarily from fast fibers are known as white muscles. Examples include most of the muscles in the hand, the biceps brachii in the arm, and those that move the eyeball. Chicken breast is called "white meat" because it is mostly made of fast fibers for quick, short bursts of flight.

Slow Fibers

Slow fibers are small in diameter. Taking longer to reach peak tension, slow fibers have the ability to provide contraction for prolonged periods. Densely packed with mitochondria, capillaries, and myosin, they are custom built for aerobic energy production. Therefore, they are much more resistant to fatigue than fast fibers. Skeletal muscles with a high percentage of slow fibers are often called red muscles because of the high concentration of myoglobin. Examples include the muscles of the back and vertebral column that maintain posture and many of the muscles in the legs for standing and walking. "Dark meat" in chicken is found in the legs and thighs where the muscles contain mostly slow, red fibers.

Exercise and Muscle Use

Skeletal muscles are very responsive to an increase or decrease in activity. Forcefully exercised muscles enlarge, which is called *muscular hypertrophy*. Conversely, an unused muscle undergoes *atrophy*, decreasing in size and strength.

The way a muscle responds to use also depends on the type of exercise. A muscle contracting with lower intensity, during swimming or running, activates slow fibers. With use, these specialized muscle fibers develop more mitochondria, and more extensive capillary networks envelop them. Such changes increase the slow fibers' ability to resist fatigue during prolonged exercise, although their sizes and strengths may remain unchanged.

Forceful exercise, such as weightlifting, in which a muscle exerts more than 75% of its maximum tension, utilizes fast fibers. In response to strenuous exercise, these fibers produce new filaments of actin and myosin, the diameters of the muscle fibers increase, and the entire muscle enlarges. That is, there is no increase in the number of muscle fibers; the existing fibers become larger.

The strength of a muscular contraction is directly proportional to the diameter of the activated muscle fibers. Consequently, an enlarged muscle can produce stronger contractions than before. Such a change, however, does not increase the muscle's ability to resist fatigue during activities like swimming or running.

If regular exercise stops, the capillary networks shrink, and the number of mitochondria within the muscle fibers drops. The number of actin and myosin filaments decreases, and the entire muscle atrophies. Such atrophy commonly occurs when accidents or diseases block motor impulses from reaching muscle fibers. An unused muscle may shrink to less than half its usual size within a few weeks.

The fibers of muscles whose motor neurons are severed not only shrink, but also may fragment and, in time, be replaced by fat or fibrous connective tissue. However, reinnervation within the first few months following an injury may restore muscle function.

Astronauts experience muscle atrophy and impaired performance with long-term exposure to the microgravity environment of space. Customized workouts using special resistance equipment can minimize the changes in muscle structure and function.

PRACTICE 8.3

3. Which chemicals provide the energy to regenerate ATP?
4. What are the sources of oxygen for aerobic respiration?
5. How are lactic acid and oxygen debt related?
6. What is the relationship between cellular respiration and heat production?
7. What are the causes of skeletal muscle fatigue?
8. How do fast fibers and slow fibers differ?
9. How does skeletal muscle respond to different types of exercise and to no exercise?

8.4 | Muscular Responses

LEARN

1. Distinguish among a twitch, recruitment, and a sustained contraction.
2. Explain how muscular contractions move body parts and help maintain posture.
3. Distinguish among the types of contractions.

One way to observe muscle contraction is to remove a single muscle fiber from a skeletal muscle and connect it to a device that records changes in the fiber's length. Such experiments usually require an electrical device that can produce stimuli of varying strengths and frequencies.

Threshold Stimulus

When an isolated muscle fiber in the laboratory is exposed to a series of stimuli of increasing strength, the fiber remains unresponsive until a certain strength of stimulation called the *threshold stimulus* is applied. Once threshold is reached, an electrical impulse is generated that spreads throughout the muscle fiber, releasing enough calcium ions from the sarcoplasmic reticulum to activate cross-bridge binding and contract that fiber. In the body, a single impulse in a motor neuron normally releases enough ACh at the neuromuscular junction to bring a muscle fiber to threshold.

Recording of a Muscle Contraction

The contractile response of a single muscle fiber to a single impulse is called a **twitch.** A twitch consists of a period of contraction, during which pulling force increases, followed by a period of relaxation, during which the pulling force declines. These events can be recorded in a pattern called a *myogram* (fig. 8.11). Note that a twitch has a brief delay between the time of stimulation and the beginning of contraction. This is the **latent period,** which in human muscle is approximately 2 milliseconds. During this time, calcium ions, actin, and myosin are preparing to interact as described in section 8.3, Skeletal Muscle Contraction.

A muscle fiber brought to threshold under a given set of conditions tends to contract completely, such that each twitch generates the maximum force of a single muscle fiber. This phenomenon has been termed an *all-or-none* response: either the muscle fiber contracts or it does not. There is no partial contraction of a single muscle fiber. The myogram of twitch contractions allows us to understand and visualize the various phases of muscle contraction. However, a twitch contraction of a single fiber is of no use in the overall action of a whole muscle.

Contractions of whole muscles enable us to perform everyday activities, but the force generated by those contractions

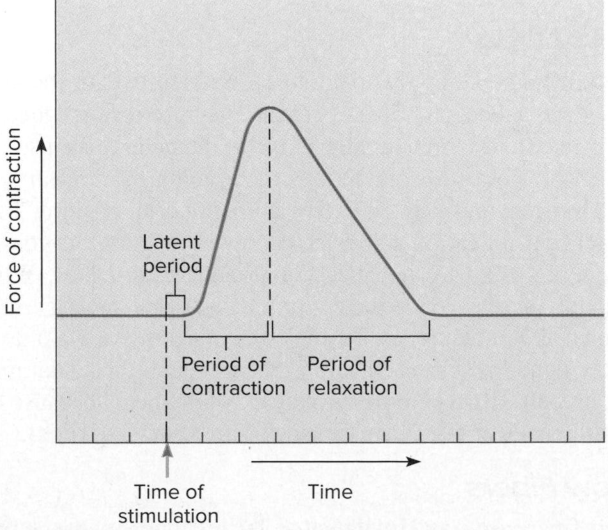

Figure 8.11 A myogram of a single muscle twitch.

must be controlled. For example, holding a paper coffee cup firmly enough that it does not slip through our fingers, but not so forcefully as to crush it, requires precise control of contractile force. In the whole muscle, the degree of tension developed reflects (1) the frequency at which individual muscle fibers are stimulated and (2) how many fibers take part in the overall contraction of the muscle.

Summation

A muscle fiber exposed to a series of stimuli of increasing frequency reaches a point when it is unable to completely relax before the next stimulus in the series arrives. When this happens, the force of individual twitches combines by the process of **summation.**

At higher frequencies of stimulation, the time spent in relaxation becomes very brief. A condition called partial tetany results. If the frequency of contraction is so rapid that the fiber doesn't relax at all, it is called a **complete tetanic** (tĕ-tan′ik) **contraction,** or tetanus (fig. 8.12). Partial tetanic contractions occur frequently in skeletal muscles during everyday activities. Complete tetany does not occur in the body, but can be demonstrated in the laboratory.

Recruitment of Motor Units

Summation increases the force of contraction of a single muscle fiber, but a whole muscle can generate more force if more muscle fibers participate in the contraction. A muscle fiber typically has a single motor end plate. The axons of motor neurons, however, are densely branched, which enables one such axon to control many muscle fibers. A motor neuron and the muscle fibers that it controls constitute a **motor unit** (mō′tor ū′nit) (fig. 8.13). Each motor unit is a functional unit because an impulse in its motor neuron will cause all of the muscle fibers in that motor unit to contract at the same time.

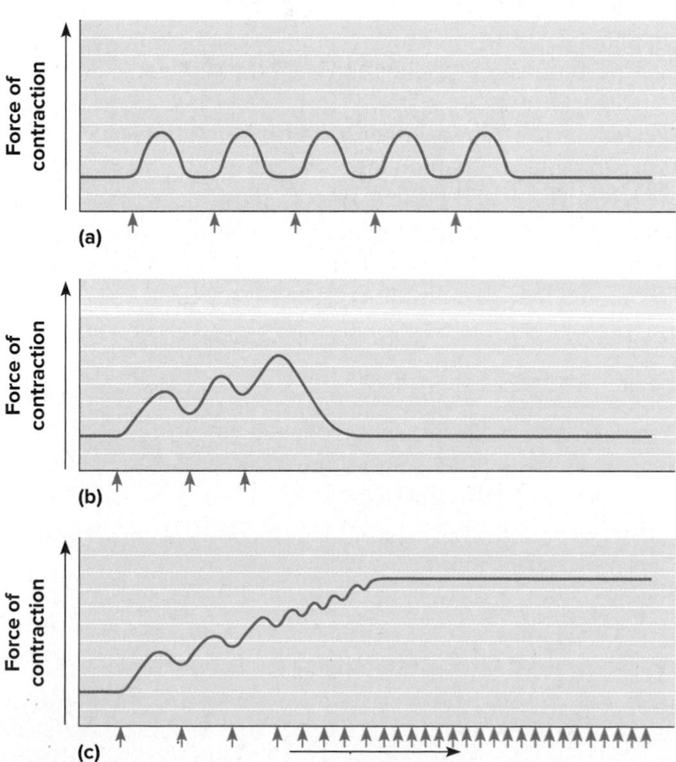

Figure 8.12 Myograms of **(a)** a series of twitches, **(b)** summation, and **(c)** a tetanic contraction. Note that stimulation frequency increases from one myogram to the next.

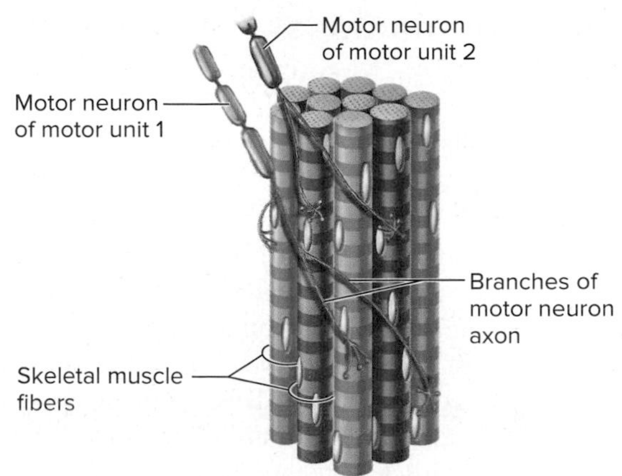

Figure 8.13 Portions of two motor units. Muscle fibers within a motor unit are innervated by a single neuron and may be distributed throughout the muscle.

A whole muscle is composed of many motor units controlled by different motor neurons. Like muscle fibers, motor neurons must be brought to threshold before an impulse is generated. It turns out that some motor neurons are more easily brought to threshold than others. If only the more sensitive motor neurons reach threshold, few motor units contract. At higher intensities of stimulation, other motor neurons are brought to threshold, and more motor units are activated. An increase in the number of motor units activated during a contraction is called **recruitment.** As the intensity of stimulation increases, recruitment of motor units continues until, finally, all motor units in that muscle are activated and the muscle contracts with maximal tension.

Sustained Contractions and Muscle Tone

Summation and recruitment together can produce a *sustained contraction* of increasing strength. Sustained contractions of whole muscles enable us to perform everyday activities. Such contractions are responses to a rapid series of impulses transmitted from the brain and spinal cord on motor neuron axons.

Even when a muscle appears to be at rest, its fibers undergo some sustained contraction. This is called **muscle tone.** Muscle tone is a response to nervous stimulation that originates repeatedly from the spinal cord and stimulates only a few muscle fibers at a time. Muscle tone is particularly important in maintaining posture. If muscle tone is suddenly lost, as happens when a person loses consciousness, the body collapses.

Types of Contractions

The term *contraction* can be misleading. Muscles do not have to "shorten" to generate a force. The type of contraction associated with movement that actually involves the shortening of the muscle is called an **isotonic contraction** (*iso* means the "same," and *tonic* in this context refers to "tension"). As the muscle shortens and generates a force, the tension throughout the movement remains the same. An example would be going to the gym and using a ten-pound weight for the "biceps curl." You start with the weight at your side and then lift it up to your shoulder. You can see and feel your biceps become short and squat. In an **isometric contraction** (*metric* refers to a measurement), the muscle generates a force without shortening. This occurs when holding a weight at your side. The muscles have to generate enough force to resist overstretching and oppose the gravitational pull of the weight. Although the muscle does not shorten, the tension increases as time passes and the weight appears to get heavier. Most movements employ various combinations of isotonic and isometric contractions.

PRACTICE 8.4

1. Define *threshold stimulus.*
2. What is a motor unit?
3. Distinguish between a twitch and a sustained contraction.
4. What is recruitment?
5. How is muscle tone maintained?
6. How do isotonic and isometric contractions differ?

8.5 | Smooth Muscle

LEARN

1. Distinguish between the structures and functions of multiunit smooth muscle and visceral smooth muscle.
2. Compare the contraction mechanisms of skeletal and smooth muscle.

The contractile mechanism of smooth muscle is essentially the same as for skeletal muscle. The cells of smooth muscle, however, have some important structural and functional differences from the other types of muscle.

Smooth Muscle Cells

Recall from section 5.5, Muscle Tissues, that smooth muscle cells are elongated, with tapering ends. Smooth muscle cells contain thick and thin filaments, but these filaments are organized differently and more randomly than those in skeletal muscle. Therefore, smooth muscle cells lack striations (and appear "smooth" under the microscope). The sarcoplasmic reticulum in these cells is not well developed.

The two major types of smooth muscle are multiunit and visceral. In **multiunit smooth muscle,** the muscle cells are separate rather than organized into sheets. Smooth muscle of this type is found in the irises of the eyes and in the walls of blood vessels. Typically, multiunit smooth muscle tissue contracts only in response to stimulation by neurons or certain hormones.

Visceral smooth muscle is composed of sheets of spindle-shaped cells in close contact with one another (see fig. 5.23). This more common type of smooth muscle is found in the walls of hollow organs, such as the stomach, intestines, urinary bladder, and uterus.

Visceral smooth muscle displays *rhythmicity,* a pattern of repeated contractions. Rhythmicity is due to self-exciting cells that deliver spontaneous impulses periodically into surrounding muscle tissue. When one cell is stimulated, the impulse may excite adjacent cells, which in turn stimulate still others. These two features—rhythmicity and transmission of impulses from cell to cell—are largely responsible for the wavelike motion, called **peristalsis** (per″ĭ-stal′sis), that helps force the contents of certain tubular organs along their lengths. Peristalsis occurs, for example, in the intestines.

Smooth Muscle Contraction

Smooth muscle contraction resembles skeletal muscle contraction in a number of ways. Both mechanisms include reactions of actin and myosin, both are triggered by membrane impulses and an increase in intracellular calcium ions, and both use energy from ATP. However, these two types of muscle tissue also have significant differences.

Recall that acetylcholine is the neurotransmitter in skeletal muscle. Two neurotransmitters commonly affect smooth muscle—acetylcholine and norepinephrine. Each of these neurotransmitters stimulates contractions in some smooth muscle and inhibits contractions in other smooth muscle (see section 9.16, Autonomic Nervous System). Also, a number of hormones affect smooth muscle, stimulating contractions in some cases and altering the degree of response to neurotransmitters in others.

Smooth muscle is slower to contract and to relax than skeletal muscle. On the other hand, smooth muscle can maintain a forceful contraction longer with a given amount of ATP. Also, unlike skeletal muscle, smooth muscle cells can change length without changing tautness. As a result, smooth muscle in the stomach and intestinal walls can stretch as these organs fill, yet maintain a constant pressure inside these organs.

PRACTICE 8.5

1. Describe two major types of smooth muscle.
2. What special characteristics of visceral smooth muscle make peristalsis possible?
3. How does smooth muscle contraction differ from skeletal muscle contraction?

8.6 | Cardiac Muscle

LEARN

1. Compare the contraction mechanisms of cardiac and skeletal muscle.

Cardiac muscle is found only in the heart. Its mechanism of contraction is essentially the same as that of skeletal and smooth muscle, but with some important differences.

Cardiac Muscle Cells and Contraction

Cardiac muscle is composed of branching, striated cells interconnected in three-dimensional networks (see fig. 5.24). Each cell has many filaments of actin and myosin, organized similarly to those in skeletal muscle. A cardiac muscle cell also has a sarcoplasmic reticulum, many mitochondria, and a system of transverse tubules. However, the sarcoplasmic reticulum of cardiac muscle cells is less well developed and stores less calcium than that of skeletal muscle, and the transverse tubules of cardiac muscle are larger. Many calcium ions released into the cytosol in response to muscle impulses come from the extracellular fluid through these large transverse tubules. This mechanism causes cardiac muscle twitches to last longer than skeletal muscle twitches.

The opposing ends of cardiac muscle cells are connected by structures called **intercalated discs.** These are elaborate junctions between cardiac muscle cell membranes. Intercalated discs allow impulses to pass freely so that they travel rapidly from cell to cell, triggering contraction. The discs help to join cells and to transmit the force of contraction from cell to cell. Thus, when one portion of the cardiac muscle network is stimulated, the resulting impulse passes to the other parts of the network, and the whole structure contracts as a functional unit.

Cardiac muscle is also self-exciting and rhythmic. Consequently, a pattern of contraction and relaxation repeats, causing the rhythmic contractions of the heart.

Table 8.3 summarizes the characteristics of the three types of muscle tissue. Clinical Application 8.2 considers several inherited diseases that affect the muscular system.

PRACTICE 8.6

1. How is cardiac muscle similar to smooth muscle?
2. How is cardiac muscle similar to skeletal muscle?
3. What is the function of intercalated discs?
4. What characteristic of cardiac muscle contracts the heart as a unit?

8.7 | Skeletal Muscle Actions

LEARN

1. Explain how the attachments, locations, and interactions of skeletal muscles make different movements possible.

Skeletal muscles provide a variety of body movements, as described in section 7.13, Joints. Each muscle's movement depends largely on the kind of joint it is associated with and the way the muscle attaches on either side of that joint.

Origin and Insertion

One end of a skeletal muscle usually attaches to a relatively immovable or fixed part on one side of a movable joint, and the other end attaches to a movable part on the other side of that joint, such that the muscle crosses the joint. The less movable end of the muscle is called its **origin** (or′ĭ-jin), and the more movable end is its **insertion** (in-ser′shun). When a muscle contracts, its insertion is pulled toward its origin.

Some muscles have more than one origin or insertion. *Biceps brachii* in the arm, for example, has two origins. This is reflected in the name *biceps,* which means "two heads." (Note: The head of a muscle is the part nearest its origin.) One head of biceps brachii attaches to the coracoid process of the scapula, and the other head arises from a tubercle above the glenoid cavity of the scapula. The muscle runs along the anterior surface of the humerus and is inserted by means of a tendon on the radial tuberosity of the radius. When biceps brachii contracts, its insertion is pulled toward its origin, and the forearm flexes at the elbow (fig. 8.14).

Muscle Movements

Whenever limbs or other body parts move, bones and muscles interact as simple mechanical devices called **levers** (lev′erz). A lever has four basic components: (1) a rigid bar

TABLE 8.3 Types of Muscle Tissue

	Skeletal	Smooth	Cardiac
Major Location	Skeletal muscles	Walls of hollow viscera, blood vessels	Wall of the heart
Major Function	Movement of bones at joints, maintenance of posture	Movement of viscera, peristalsis, vasoconstriction	Pumping action of the heart
Cellular Characteristics			
Striations	Present	Absent	Present
Nucleus	Many nuclei	Single nucleus	Single nucleus
Special features	Well-developed transverse tubule system	Lacks transverse tubules	Well-developed transverse tubule system; intercalated discs separating adjacent cells
Mode of Control	Voluntary	Involuntary	Involuntary
Contraction Characteristics	Contracts and relaxes rapidly when stimulated by a motor neuron	Contracts and relaxes slowly; single unit type is self-exciting; rhythmic	Network of cells contracts as a unit; self-exciting; rhythmic

or rod, (2) a fulcrum or pivot on which the bar turns, (3) an object moved against resistance, and (4) a force that supplies energy for the movement of the bar. The actions of bending and straightening the upper limb at the elbow illustrate bones and muscles functioning as levers.

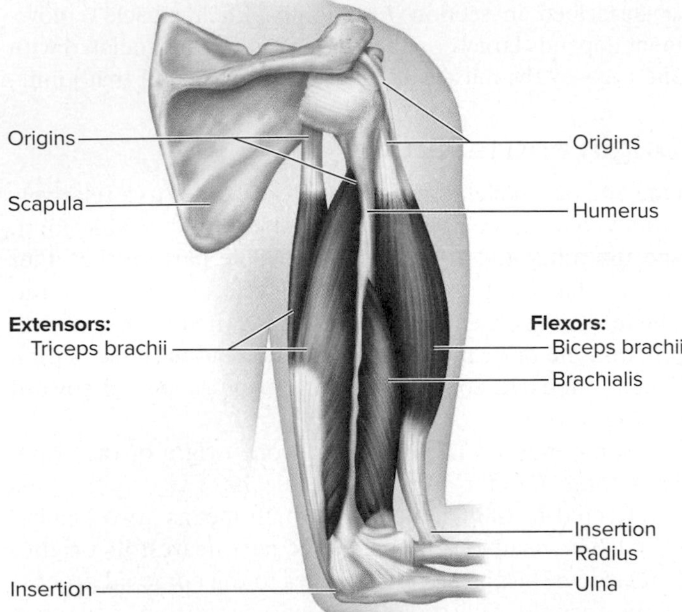

Figure 8.14 The biceps brachii has two heads that originate on the scapula. A tendon inserts this muscle on the radius.

When the upper limb bends, the forearm bones represent the rigid bar, the elbow joint is the fulcrum, the hand is moved against the resistance provided by the weight, and the force is supplied by muscles on the anterior side of the arm (fig. 8.15*a*). One of these muscles, the biceps brachii, is attached by a tendon to a projection on a bone (radius) in the forearm, a short distance distal to the elbow.

When the upper limb straightens at the elbow, the forearm bones again serve as the rigid bar, the elbow joint serves as the fulcrum, and the hand moves against the resistance by pulling on the rope to raise the weight (fig. 8.15*b*). However, in this case the triceps brachii, a muscle located on the posterior side of the arm, supplies the force. A tendon of this muscle attaches to a projection on a forearm bone (ulna) at the point of the elbow.

Sometimes the less movable and more movable bones are reversed based upon the action of the muscle in question. Therefore, an alternate way to describe muscle attachments is to use the directional terms *proximal* and *distal* for the attachments of appendicular muscles and the terms *superior* and *inferior* for axial muscles. Thus, the proximal attachments of biceps brachii are on the coracoid process and the tubercle above the glenoid cavity of the scapula. The distal attachment is on the radial tuberosity (see table 8.9). Similarly, the superior attachment of rectus abdominis is on the costal cartilage and xiphoid process of the sternum, and the inferior attachment is on the pubic crest and pubic symphysis (see table 8.11). The tables throughout this chapter that describe muscle actions

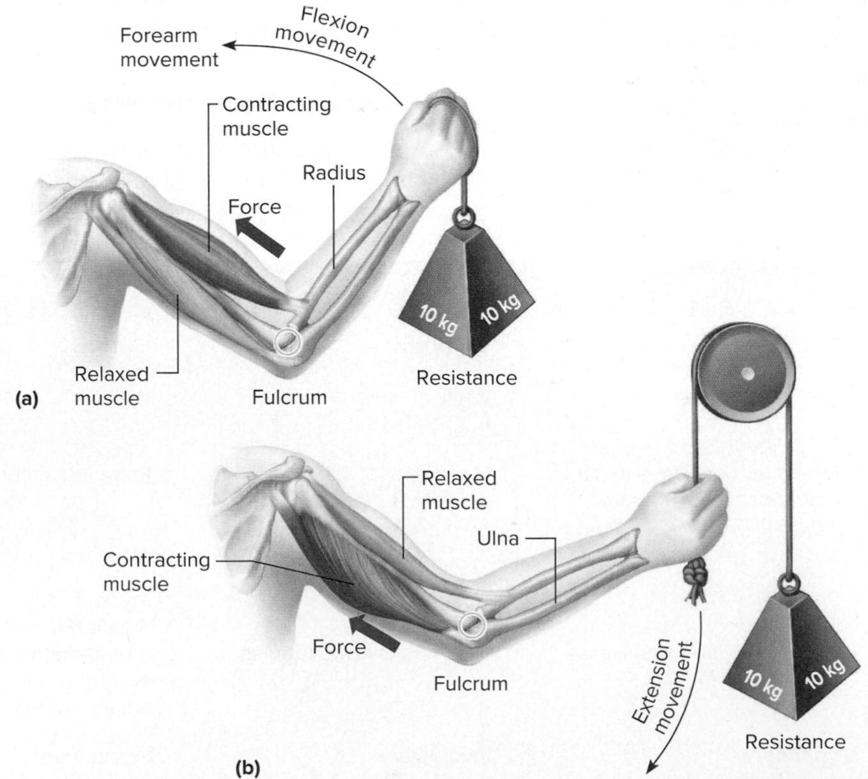

Figure 8.15 Levers and movement. **(a)** When the upper limb bends at the elbow or **(b)** when the upper limb straightens at the elbow, the bones and muscles function as levers.

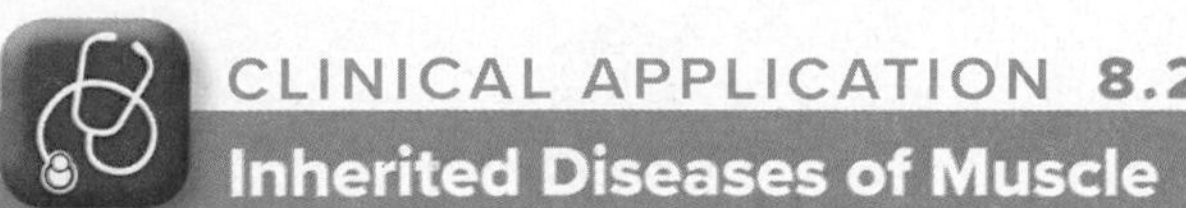

Several inherited conditions affect muscle tissue. These disorders differ in the nature of the genetic defect, the type of protein that is abnormal in form or function, and the muscles that are impaired.

The Muscular Dystrophies—Missing Proteins

A muscle cell is packed with filaments of actin and myosin. Much less abundant, but no less important, is a protein called *dystrophin.* It holds skeletal muscle cells together by linking actin in the cell to glycoproteins in the cell membrane, which helps attach the cell to the extracellular matrix. Missing or abnormal dystrophin or the glycoproteins cause muscular dystrophies. These illnesses vary in severity and age of onset, but in all cases, muscles weaken and degenerate. Eventually fat and connective tissue replace muscle.

Figure 8B These twins are both affected with Duchenne muscular dystrophy because it is an inherited genetic disease.

Duchenne muscular dystrophy (DMD) is the most severe type of the illness (fig. 8B). Symptoms begin by age five and affect only boys. By age thirteen, the person cannot walk, and by early adulthood he usually dies from failure of the respiratory muscles. In DMD, dystrophin is absent or shortened. In Becker muscular dystrophy, symptoms begin in early adulthood, are less severe, and result from underproduction of dystrophin. An experimental genetic therapy produces nearly full-length dystrophin by skipping over the part of the gene that includes the mutation.

Charcot-Marie-Tooth Disease—A Duplicate Gene

Charcot-Marie-Tooth disease causes a slowly progressing weakness in the muscles of the hands and feet and a decrease in tendon reflexes in these parts. In this illness, an extra gene impairs the insulating sheath around affected nerve cells, so that nerve cells cannot adequately stimulate muscles. Physicians perform two tests—electromyography and nerve conduction velocity—to diagnose Charcot-Marie-Tooth disease. It is also possible to test for gene mutations to confirm a diagnosis based on symptoms.

Hereditary Idiopathic Dilated Cardiomyopathy—A Tiny Glitch

This very rare inherited form of heart failure usually begins in a person's forties and is lethal in 50% of cases within five years of diagnosis, unless a heart transplant can be performed. The condition is caused by a mutation in a gene that encodes a form of actin found only in cardiac muscle. The mutation disturbs actin's ability to anchor to the Z lines in heart muscle cells, preventing actin from effectively transferring the force of contraction. As a result, the heart chambers enlarge and eventually fail.

identify the traditional origin and insertions, but the alternative terms described here can be easily determined from that information.

Muscle Relationships

The terms **flexion** and **extension** describe opposing movements and changes in the angle between bones that meet at a joint. For example, flexion of the elbow refers to a movement of the forearm that bends the elbow, or decreases the angle. In general, flexion refers to bringing the bones closer together (fig 8.15*a*). Extension of the elbow widens the angle, increasing the distance between the bones (fig. 8.15*b*).

Skeletal muscles almost always function in groups. A number of terms describe the roles of muscles in performing particular actions. The **prime mover,** or **agonist,** generates the majority of the force during a desired action (fig. 8.14). A **synergist** aids the prime mover in the desired action or by inhibiting the opposing action. In the example of elbow flexion, the biceps brachii is the prime mover for flexion, and the synergist is the brachialis. The triceps brachii is the **antagonist** to the biceps brachii and brachialis because it brings about the opposite action, extension. Note that the role of a muscle is dependent on the movement; in the example of elbow extension, triceps brachii is the prime mover and biceps brachii and brachialis are the antagonists.

Relationships between muscles depend on the action in question and can be complex. For example, pectoralis major, a chest muscle, and latissimus dorsi, a back muscle, are synergistic for medial rotation of the arm. However, they are antagonistic to each other for flexion and extension of the shoulder. Similarly, two muscles on the lateral forearm, flexor carpi radialis and extensor carpi radialis longus, are synergistic for abduction of the hand, yet they are antagonistic for flexion and extension of the wrist. Thus, any role of a muscle must be learned in the context of a particular movement.

Because students (and patients) often find it helpful to think of movements in terms of the specific actions of the muscles involved, we may also describe flexion and extension in these terms. Thus, the action of biceps brachii may be described as "flexion of the forearm at the elbow," and the action of the quadriceps group as "extension of the leg at the knee." We believe this occasional departure from strict anatomical terminology may facilitate learning.

PRACTICE 8.7

1. Distinguish between the origin and the insertion of a muscle.
2. Define *agonist.*
3. What is the function of a synergist? An antagonist?

8.8 | Major Skeletal Muscles

LEARN

1. Identify and locate the major skeletal muscles of each body region.
2. Identify the actions of the major skeletal muscles of each body region.

This section discusses the locations, actions, and attachments of some of the major skeletal muscles. (Figs. 8.16 and 8.17 and reference plates 1 and 2 show the locations of the superficial skeletal muscles–those near the surface.)

Naming Muscles

The names of these muscles often describe them. A name may indicate a muscle's relative size, shape, location, action, or number of attachments, or the direction of its fibers, as in the following examples:

pectoralis major Large (major) and located in the pectoral region (chest).
deltoid Shaped like a delta or triangle.
extensor digitorum Extends the digits (fingers or toes).

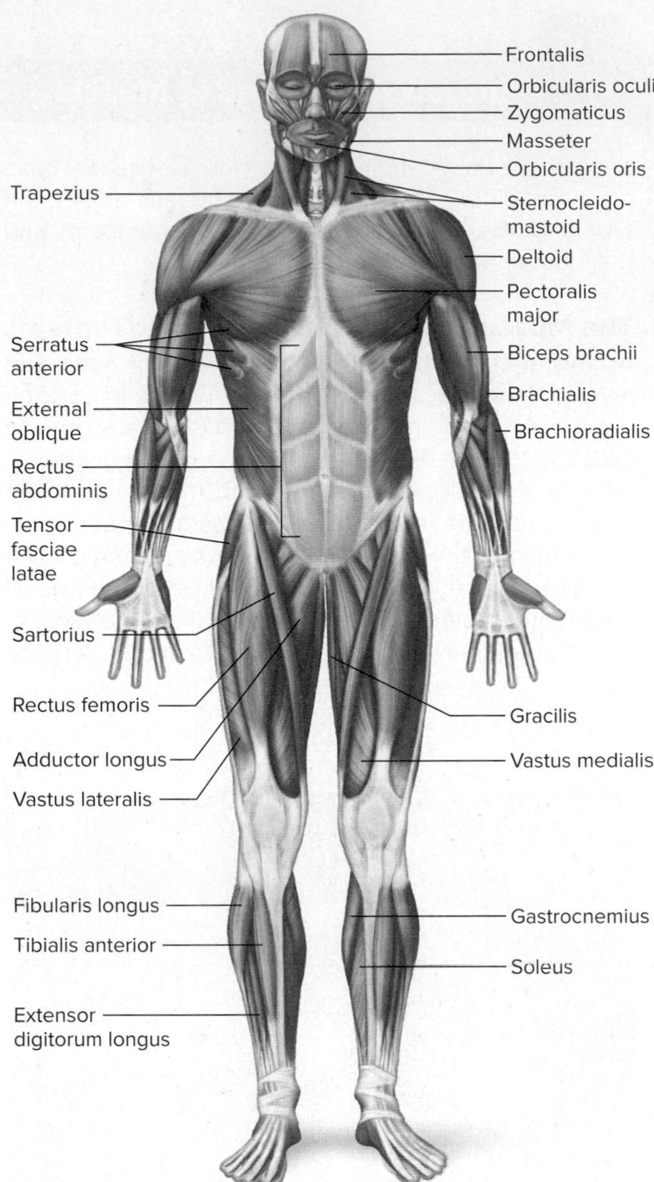

Figure 8.16 Anterior view of superficial skeletal muscles.

biceps brachii Having two heads (biceps) or points of origin and located in the brachium (arm).
sternocleidomastoid Attached to the sternum, clavicle, and mastoid process.
external oblique Located near the outside, with fibers that run obliquely (in a slanting direction).

Note that in the anatomical position some actions have already occurred, such as supination of the forearm and hand, extension of the elbow, and extension of the knee. The muscle actions described in the following section consider the entire range of movement at each joint, and do not presume that the starting point is the anatomical position.

Some muscles have more than one origin or more than one insertion. The wide range of attachments of some of the larger muscles has the effect of giving those muscles

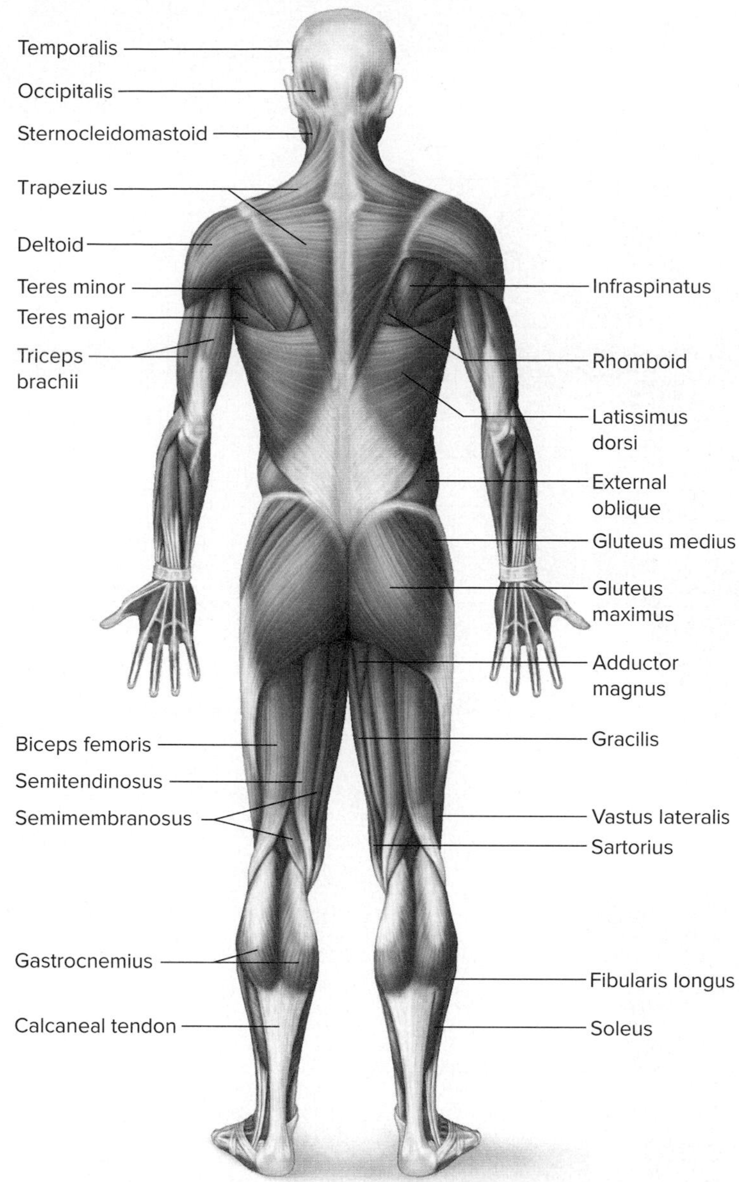

Figure 8.17 Posterior view of superficial skeletal muscles.

different, sometimes opposing actions, depending on which portion of the muscle is active. Many of these cases are identified in the appropriate sections and tables.

You may have noticed some discrepancies between anatomical terminology and the terms the general public uses when referring to body parts. For example, someone with a bruised thigh may complain of a sore leg, instead of correctly referring to the thigh. In this book, we have used accepted anatomical terminology when referring to body parts. On the other hand, you will likely be dealing not only with colleagues (who rely on the precision of correct terminology when communicating) but also with patients (who simply, sometimes desperately, want to communicate). You must become a master of both ways of communicating. The bottom line is being able to communicate accurately with colleagues and effectively with patients.

Muscles of Facial Expression

A number of small muscles that lie beneath the skin of the face and scalp enable us to communicate feelings through facial expression (fig. 8.18*a*). Many of these muscles, located around the eyes and mouth, are responsible for such expressions as surprise, sadness, anger, fear, disgust, and pain.

Some of the muscles of facial expression join the bones of the skull to connective tissue in various regions of the overlying skin. They include:

epicranius (ep″ĭ-krā′nē-us) Composed of two parts, *frontalis* (frun-ta′lis) and *occipitalis* (ok-sip″ĭ-ta′lis)
orbicularis oculi (or-bik′u-la-rus ok′u-lī)
orbicularis oris (or-bik′u-la-rus o′ris)
buccinator (buk′sĭ-nā″tor)
zygomaticus (zī″gō-mat′ik-us)
platysma (plah-tiz′mah)

Table 8.4 lists the origins, insertions, and actions of the muscles of facial expression. Section 10.9, Sense of Sight, describes the muscles that move the eyes.

TABLE 8.4 Muscles of Facial Expression APR

Muscle	Origin	Insertion	Action
Epicranius	Occipital bone	Skin around eye	Elevates eyebrow
Orbicularis oculi	Maxilla and frontal bone	Skin around eye	Closes eye
Orbicularis oris	Muscles near the mouth	Skin of lips	Closes and protrudes lips
Buccinator	Alveolar processes of maxilla and mandible	Orbicularis oris	Compresses cheeks
Zygomaticus	Zygomatic bone	Skin and muscle at corner of mouth	Elevates corner of mouth
Platysma	Fascia in upper chest	Skin and muscles below mouth	Depresses lower lip and angle of mouth

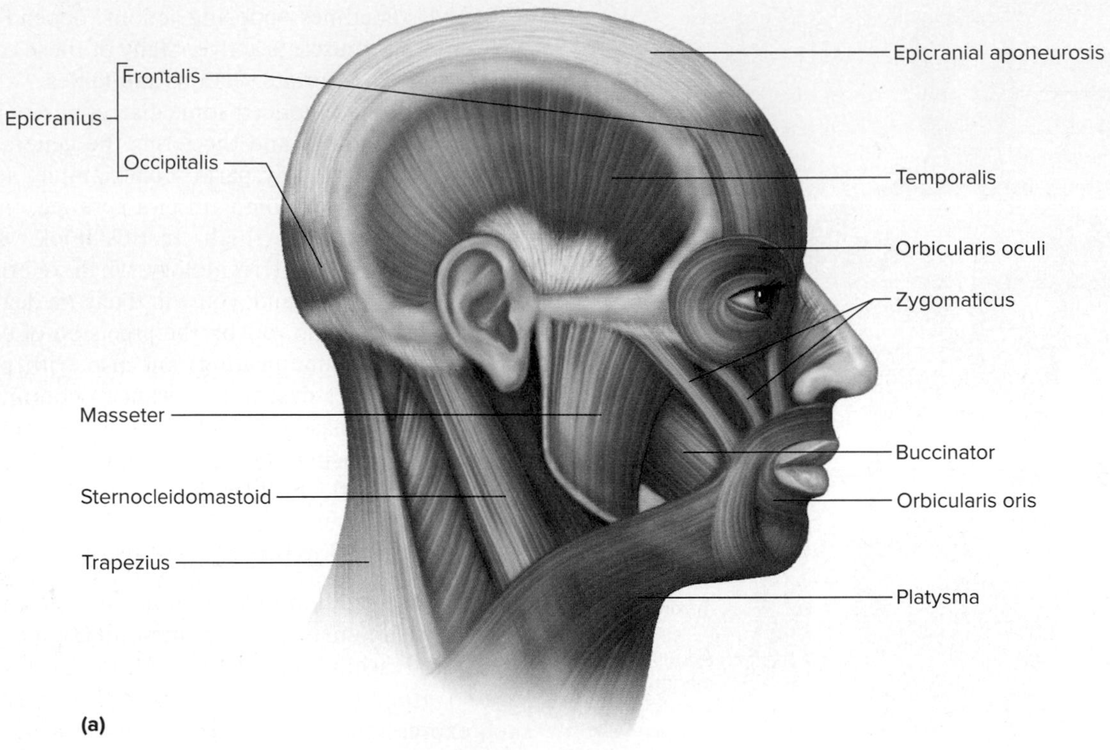

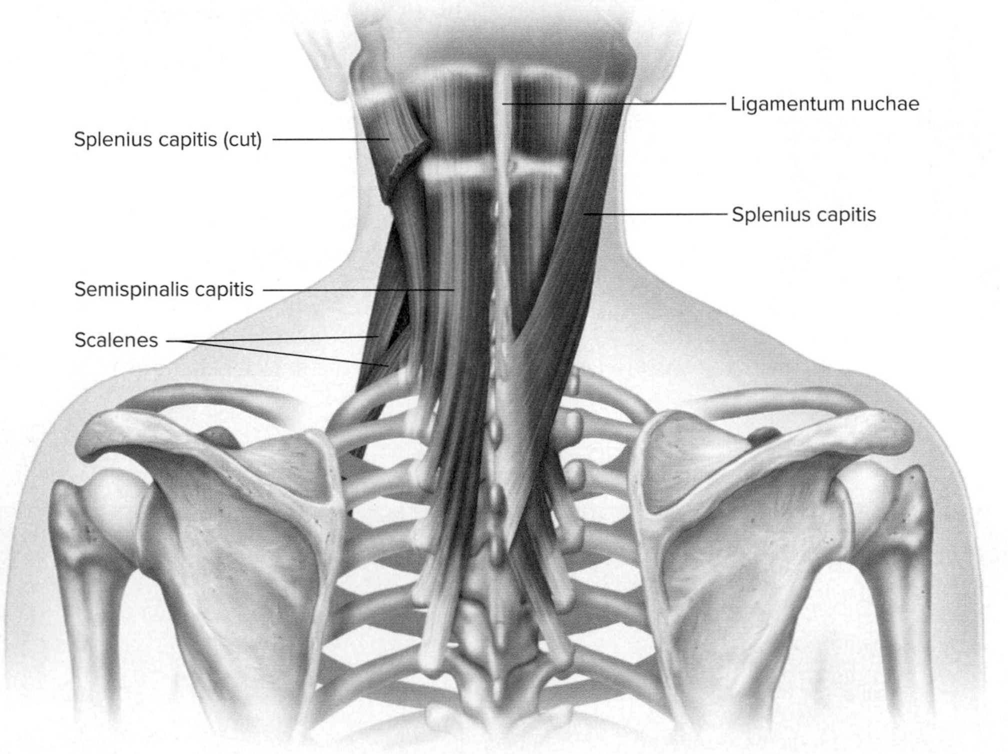

Figure 8.18 Muscles of the face and neck. **(a)** Lateral view including muscles of facial expression and mastication. **(b)** Posterior view of muscles that move the head. APR

Muscles of Mastication

Muscles attached to the mandible produce chewing movements. Two pairs of these muscles elevate the mandible, a motion used in biting. These muscles are the *masseter* (mas-se'ter) and the *temporalis* (tem-po-ra'lis) (fig. 8.18*a*). Table 8.5 lists the origins, insertions, and actions of the muscles of mastication.

Grinding the teeth, a common response to stress, may strain the temporomandibular joint through the excessive force generated by the masseter and temporalis muscles. This condition, called temporomandibular joint syndrome (TMJ syndrome), may produce headache, earache, and pain in the jaw, neck, or shoulder.

Muscles that Move the Head

Head movements result from the actions of paired muscles in the neck and upper back. These muscles flex, extend, and rotate the head. They are aided by an elastic ligament (*ligamentum nuchae*), which limits flexion of the neck and helps to hold the head upright. They include (see figs. 8.18*b*, 8.19, and 8.20):

sternocleidomastoid (ster″nō-klī″do-mas'toid)
splenius capitis (sple'nē-us kap'ĭ-tis)
semispinalis capitis (sem″ē-spi-na'lis kap'ĭ-tis)
scalenes (skā'lēnz)

Table 8.6 lists the origins, insertions, and actions of muscles that move the head.

Muscles that Move the Pectoral Girdle

The muscles that move the pectoral girdle are closely associated with those that move the arm. A number of these chest and shoulder muscles attach from the scapula to nearby bones and move the scapula in various directions. They include (figs. 8.19 and 8.20):

trapezius (trah-pē'zē-us)
rhomboid (rom-boid') *major*
levator scapulae (le-va'tor scap'u-lē)
serratus anterior (ser-ra'tus an-te'rē-or)
pectoralis (pek″to-ra'lis) *minor*

Table 8.7 lists the origins, insertions, and actions of the muscles that move the pectoral girdle.

TABLE 8.5 Muscles of Mastication APR

Muscle	Origin	Insertion	Action
Masseter	Zygomatic arch	Posterior lateral surface of mandible	Elevates and protracts mandible
Temporalis	Temporal bone	Coronoid process of mandible	Elevates and retracts mandible

TABLE 8.6 Muscles that Move the Head APR

Muscle	Origin	Insertion	Action
Sternocleidomastoid	Manubrium of sternum and medial clavicle	Mastoid process of temporal bone	Individually: laterally flexes head and neck to the same side, rotates head to the opposite side Together: pull the head forward and down; also aid in forceful inhalation by elevating sternum and first ribs
Splenius capitis	Ligamentum nuchae; spinous processes of 7th cervical and upper thoracic vertebrae	Occipital bone and mastoid process of temporal bone	Individually: rotates head to the same side Together: bring head into an upright position
Semispinalis capitis	Below the articular facets of lower cervical vertebrae; transverse processes of upper thoracic vertebrae	Occipital bone	Individually: rotates head to the opposite side Together: extend head and neck
Scalenes	Transverse processes of cervical vertebrae	Superior and lateral surfaces of first two ribs	Individually: laterally flexes head and neck to the same side Together: elevate first two ribs during forceful inhalation

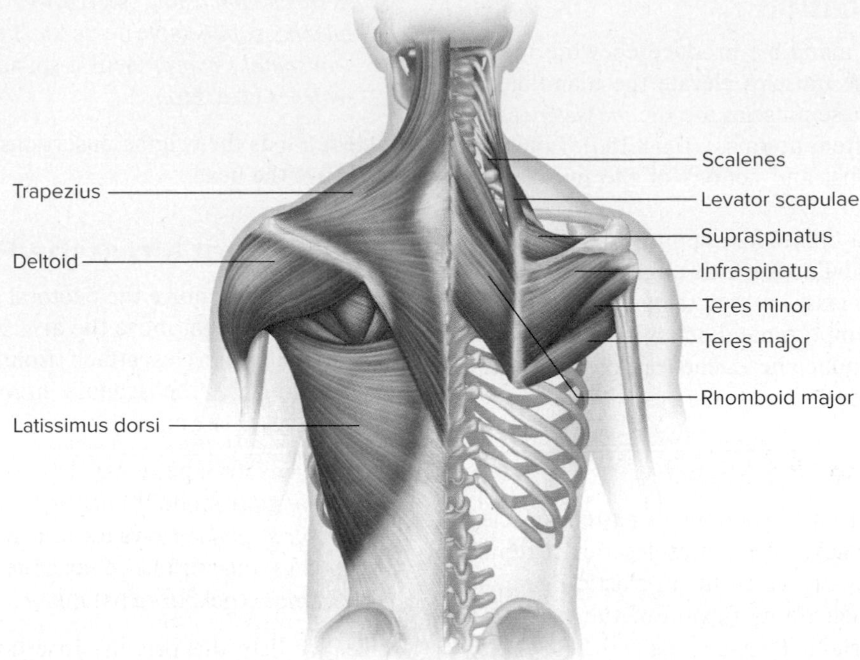

Figure 8.19 Muscles of the posterior shoulder. The right trapezius is removed to show underlying muscles. APR

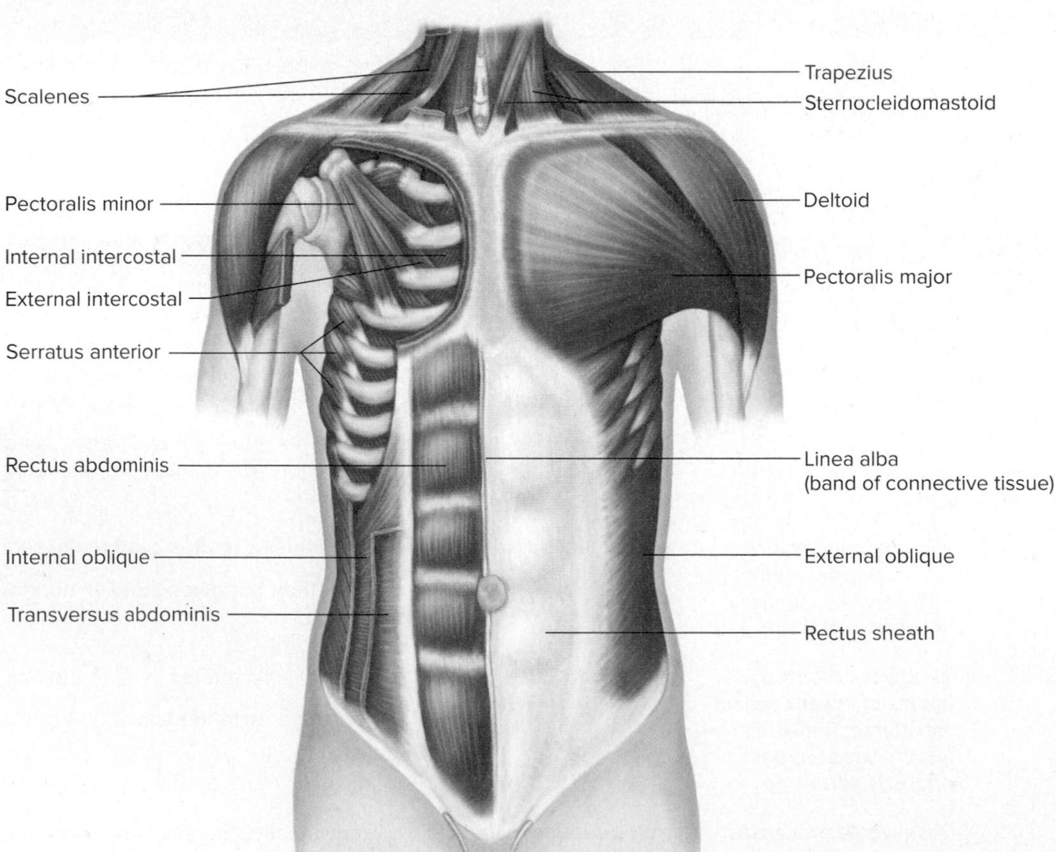

Figure 8.20 Muscles of the anterior chest and abdominal wall. The right pectoralis major is removed to show the pectoralis minor. APR

TABLE 8.7 Muscles that Move the Pectoral Girdle APR

Muscle	Origin	Insertion	Action
Trapezius	Occipital bone, ligamentum nuchae, and spinous processes of 7th cervical and all thoracic vertebrae	Clavicle; spine and acromion process of scapula	Rotates and retracts scapula Superior portion elevates scapula Inferior portion depresses scapula
Rhomboid major	Spinous processes of upper thoracic vertebrae	Medial border of scapula	Elevates and retracts scapula
Levator scapulae	Transverse processes of cervical vertebrae	Superior angle and medial border of scapula	Elevates scapula
Serratus anterior	Anterior surfaces of ribs 1–10	Medial border of scapula	Protracts and rotates scapula
Pectoralis minor	Anterior surfaces of ribs 3–5	Coracoid process of scapula	Depresses and protracts scapula, elevates ribs during forceful inhalation

Muscles that Move the Arm

The arm is one of the more freely movable parts of the body. Muscles that attach from the humerus to various regions of the pectoral girdle, ribs, and vertebral column make these movements possible (figs. 8.19, 8.20, 8.21, and 8.22). These muscles can be grouped according to their primary actions—flexion, extension, abduction, and rotation—as follows:

Flexors
coracobrachialis (kor″ah-kō-brā′kē-al-is)
pectoralis (pek″to-ra′lis) *major*

Extensors
teres (te′rēz) *major*
latissimus dorsi (lah-tis′ĭ-mus dor′sī)

Abductors
supraspinatus (sū″prah-spī′nā-tus)
deltoid (del′toid)

Rotators
subscapularis (sub-scap′u-lar-is)
infraspinatus (in″frah-spi′na-tus)
teres (te′rēz) *minor*

The movements of flexion and extension of the shoulder may be less obvious than at other joints. Movements of the arm forward and upward flex the shoulder, and the opposite movements extend it. Table 8.8 lists the origins, insertions, and actions of muscles that move the arm.

TABLE 8.8 Muscles that Move the Arm APR

Muscle	Origin	Insertion	Action
Coracobrachialis	Coracoid process of scapula	Medial midshaft of humerus	Flexes arm at shoulder, adducts arm
Pectoralis major	Clavicle, sternum, and costal cartilages of upper ribs	Intertubercular sulcus of humerus	Flexes arm at shoulder, adducts and medially rotates arm
Teres major	Lateral border of scapula	Intertubercular sulcus of humerus	Extends arm at shoulder, adducts and medially rotates arm
Latissimus dorsi	Spinous processes of lower thoracic and lumbar vertebrae, iliac crest, and lower ribs	Intertubercular sulcus of humerus	Extends arm at shoulder, adducts and medially rotates arm
Supraspinatus	Supraspinous fossa of scapula	Greater tubercle of humerus	Abducts arm
Deltoid	Acromion process, spine of scapula, and clavicle	Deltoid tuberosity of humerus	Lateral portion abducts arm Anterior portion flexes arm at shoulder Posterior portion extends arm at shoulder
Subscapularis	Anterior surface of scapula	Lesser tubercle of humerus	Medially rotates arm
Infraspinatus	Infraspinous fossa of scapula	Greater tubercle of humerus	Laterally rotates arm
Teres minor	Lateral border of scapula	Greater tubercle of humerus	Laterally rotates arm

Muscles that Move the Forearm

Muscles that attach from the radius or ulna to the humerus or pectoral girdle produce most of the forearm movements. A group of muscles located along the anterior surface of the humerus flexes the elbow, and a single posterior muscle extends this joint. Other muscles move the radioulnar joint and rotate the forearm.

Muscles that move the forearm include (figs. 8.21, 8.22, and 8.23):

Flexors
biceps brachii (bī'seps brā'kē-ī)
brachialis (brā'kē-al-is)
brachioradialis (brā''kē-o-rā''dē-a'lis)

Extensor
triceps brachii (tri'seps brā'kē-i)

Rotators
supinator (su'pĭ-nā-tor) (Note: This deep muscle is not shown in these figures, but can be found in APR.)
pronator teres (pro-nā'tor te'rēz)
pronator quadratus (pro-nā'tor kwod-ra'tus)

Table 8.9 lists the origins, insertions, and actions of muscles that move the forearm.

Muscles that Move the Hand

Many muscles move the hand. They originate from the distal end of the humerus and from the radius and ulna. The two major groups of these muscles are flexors on the anterior side of the forearm and extensors on the posterior side. These muscles include (figs. 8.23 and 8.24):

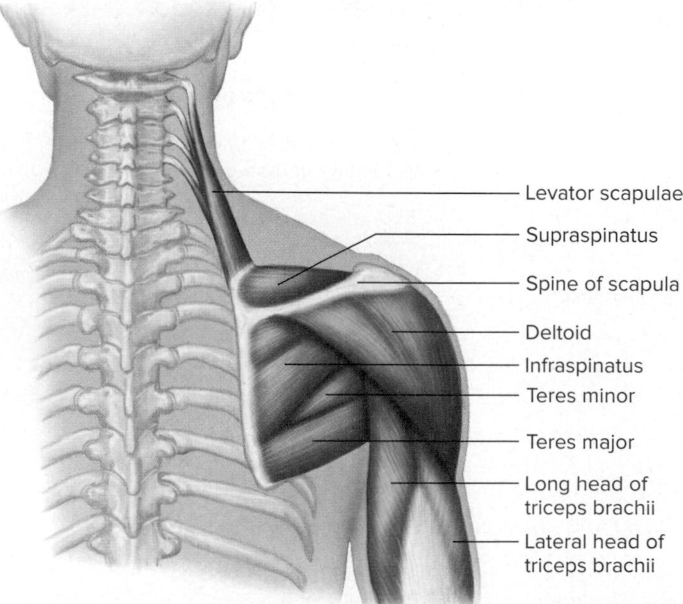

Figure 8.21 Muscles of the posterior surface of the scapula and arm. (Note: The medial head of triceps brachii is not visible.) APR

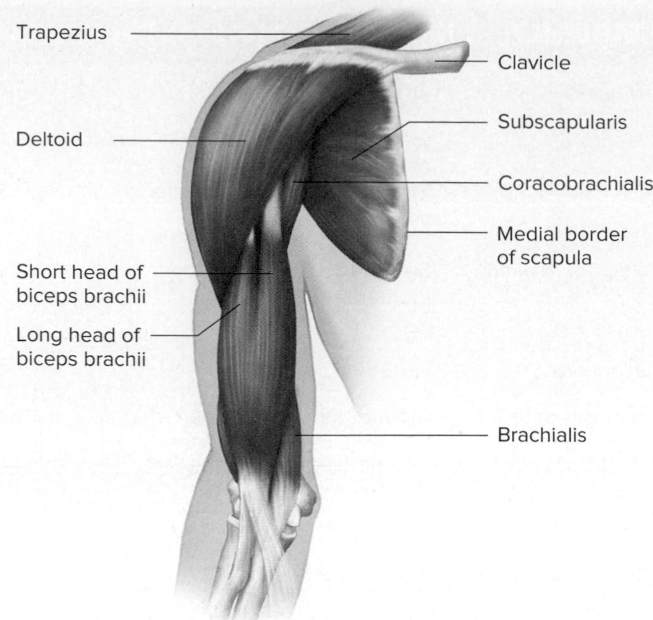

Figure 8.22 Muscles of the anterior shoulder and arm. 

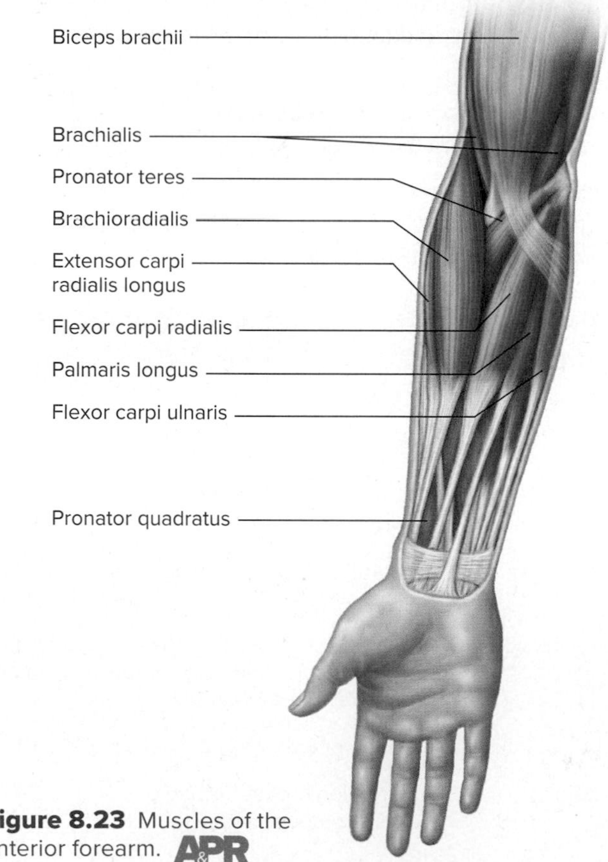

Figure 8.23 Muscles of the anterior forearm. APR

Flexors
flexor carpi radialis (flex'sor kar-pī' rā''dē-a'lis)
flexor carpi ulnaris (flex'sor kar-pī' ul-na'ris)
palmaris longus (pal-ma'ris long'gus)
flexor digitorum profundus (flex'sor dij''ĭ-to'rum pro-fun'dus)

TABLE 8.9 Muscles that Move the Forearm APR

Muscle	Origin	Insertion	Action
Biceps brachii	Coracoid process (short head); tubercle above glenoid cavity of scapula (long head)	Radial tuberosity	Flexes forearm at elbow, supinates forearm and hand
Brachialis	Anterior surface of humerus	Coronoid process of ulna	Flexes forearm at elbow
Brachioradialis	Distal lateral end of humerus	Lateral surface of radius above styloid process	Flexes forearm at elbow
Triceps brachii	Tubercle below glenoid cavity of scapula (long head); lateral surface of humerus (lateral head); posterior surface of humerus (lateral and medial heads)	Olecranon process of ulna	Extends forearm at elbow
Supinator	Lateral epicondyle of humerus and proximal ulna	Anterior and lateral surface of radius	Supinates forearm and hand
Pronator teres	Medial epicondyle of humerus and coronoid process of ulna	Lateral surface of radius	Pronates forearm and hand
Pronator quadratus	Anterior distal end of ulna	Anterior distal end of radius	Pronates forearm and hand

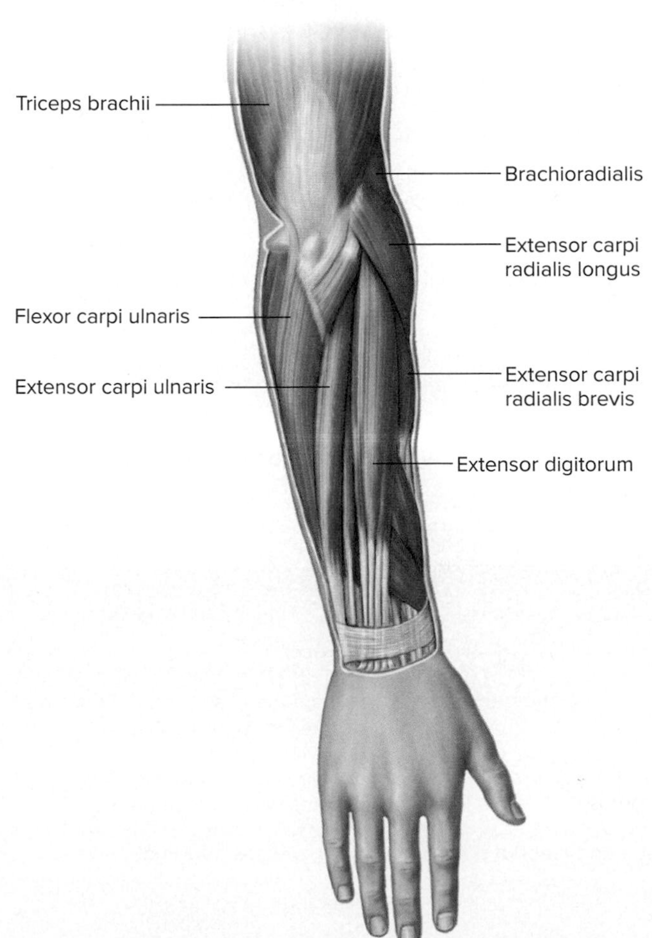

Figure 8.24 Muscles of the posterior forearm. 

Extensors

extensor carpi radialis longus (eks-ten'sor kar-pī' rā''dē-a'lis long'gus)
extensor carpi radialis brevis (eks-ten'sor kar-pī' rā''dē-a'lis brev'ĭs)
extensor carpi ulnaris (eks-ten'sor kar-pī' ul-na'ris)
extensor digitorum (eks-ten'sor dij''ĭ-to'rum)

Table 8.10 lists the origins, insertions, and actions of muscles that move the hand.

Muscles of the Abdominal Wall

Bone supports the walls of the chest and pelvic regions, but not those of the abdomen. Instead, the anterior and lateral walls of the abdomen are composed of layers of broad, flattened muscles. These muscles connect the rib cage and vertebral column to the pelvic girdle. A band of tough connective tissue called the **linea alba** extends from the xiphoid process of the sternum to the pubic symphysis (see fig. 8.20). It is an attachment for some of the abdominal wall muscles. Another important attachment is the inguinal ligament, which extends from the anterior superior iliac spine to the pubis near the pubic symphysis.

Contraction of these muscles decreases the size of the abdominal cavity and increases the pressure inside. These actions help press air out of the lungs during forceful exhalation and aid in the movements of defecation, urination, vomiting, and childbirth.

The abdominal wall muscles include (see fig. 8.20):

external oblique (eks-ter'nal o-blēk')
internal oblique (in-ter'nal o-blēk')
transversus abdominis (trans-ver'sus ab-dom'ĭ-nis)
rectus abdominis (rek'tus ab-dom'ĭ-nis)

Table 8.11 lists the origins, insertions, and actions of muscles of the abdominal wall.

Muscles of the Pelvic Floor

The inferior outlet of the pelvis is closed off by two muscular sheets—a deeper **pelvic diaphragm** and a more superficial **urogenital diaphragm.** Together they form the floor of the pelvis. The pelvic diaphragm spans the outlet of the pelvic cavity, and the urogenital diaphragm fills the space within the pubic arch (see fig. 7.28). Just anterior to the anal canal, a deep central tendon serves as an attachment for a number of these muscles. The muscles of the male and female pelvic floors include (fig. 8.25):

Pelvic diaphragm
levator ani (le-va'tor ah-ni')
coccygeus (kok-sij'e-us)

Urogenital diaphragm
superficial transversus perinei (su"per-fish'al trans-ver'sus per"ĭ-ne'i)
bulbospongiosus (bul"bo-spon"je-o'sus)
ischiocavernosus (is"ke-o-kav"er-no'sus)

Table 8.12 lists the origins, insertions, and actions of muscles of the pelvic floor.

Muscles that Move the Thigh

Muscles that move the thigh are attached to the femur and to some part of the pelvic girdle. These muscles are in anterior, medial, and posterior groups. Muscles of the anterior group primarily flex the hip; those of the medial group adduct the thigh; those of the posterior group extend the hip, abduct the thigh, or rotate the thigh. The muscles in these groups include (figs. 8.26, 8.27, and 8.28):

Anterior group
psoas (so'as) *major*
iliacus (il'ĭ-ak-us)

TABLE 8.10 Muscles that Move the Hand

Muscle	Origin	Insertion	Action
Flexor carpi radialis	Medial epicondyle of humerus	Base of second and third metacarpals	Flexes wrist, abducts hand
Flexor carpi ulnaris	Medial epicondyle of humerus and olecranon process of ulna	Carpal bones and fifth metacarpal bone	Flexes wrist, adducts hand
Palmaris longus	Medial epicondyle of humerus	Fascia of palm	Flexes wrist
Flexor digitorum profundus	Anterior and medial surface of ulna	Distal phalanges of fingers 2–5	Flexes wrist and joints of fingers
Extensor carpi radialis longus	Lateral distal end of humerus	Base of second metacarpal	Extends wrist, abducts hand
Extensor carpi radialis brevis	Lateral epicondyle of humerus	Base of third metacarpal	Extends wrist, abducts hand
Extensor carpi ulnaris	Lateral epicondyle of humerus and proximal, posterior ulna	Base of fifth metacarpal	Extends wrist, adducts hand
Extensor digitorum	Lateral epicondyle of humerus	Posterior surface of phalanges in fingers 2–5	Extends wrist and joints of fingers

TABLE 8.11 Muscles of the Abdominal Wall

Muscle	Origin	Insertion	Action
External oblique	Outer surfaces of lower 8 ribs	Outer lip of iliac crest and linea alba	Compresses abdomen, flexes and rotates vertebral column
Internal oblique	Iliac crest and inguinal ligament	Lower 3–4 ribs, linea alba, and crest of pubis	Compresses abdomen, flexes and rotates vertebral column
Transversus abdominis	Costal cartilages of lower 6 ribs, processes of lumbar vertebrae, lip of iliac crest, and inguinal ligament	Linea alba and crest of pubis	Compresses abdomen
Rectus abdominis	Crest of pubis and pubic symphysis	Xiphoid process of sternum and costal cartilages of ribs 5–7	Compresses abdomen, flexes vertebral column

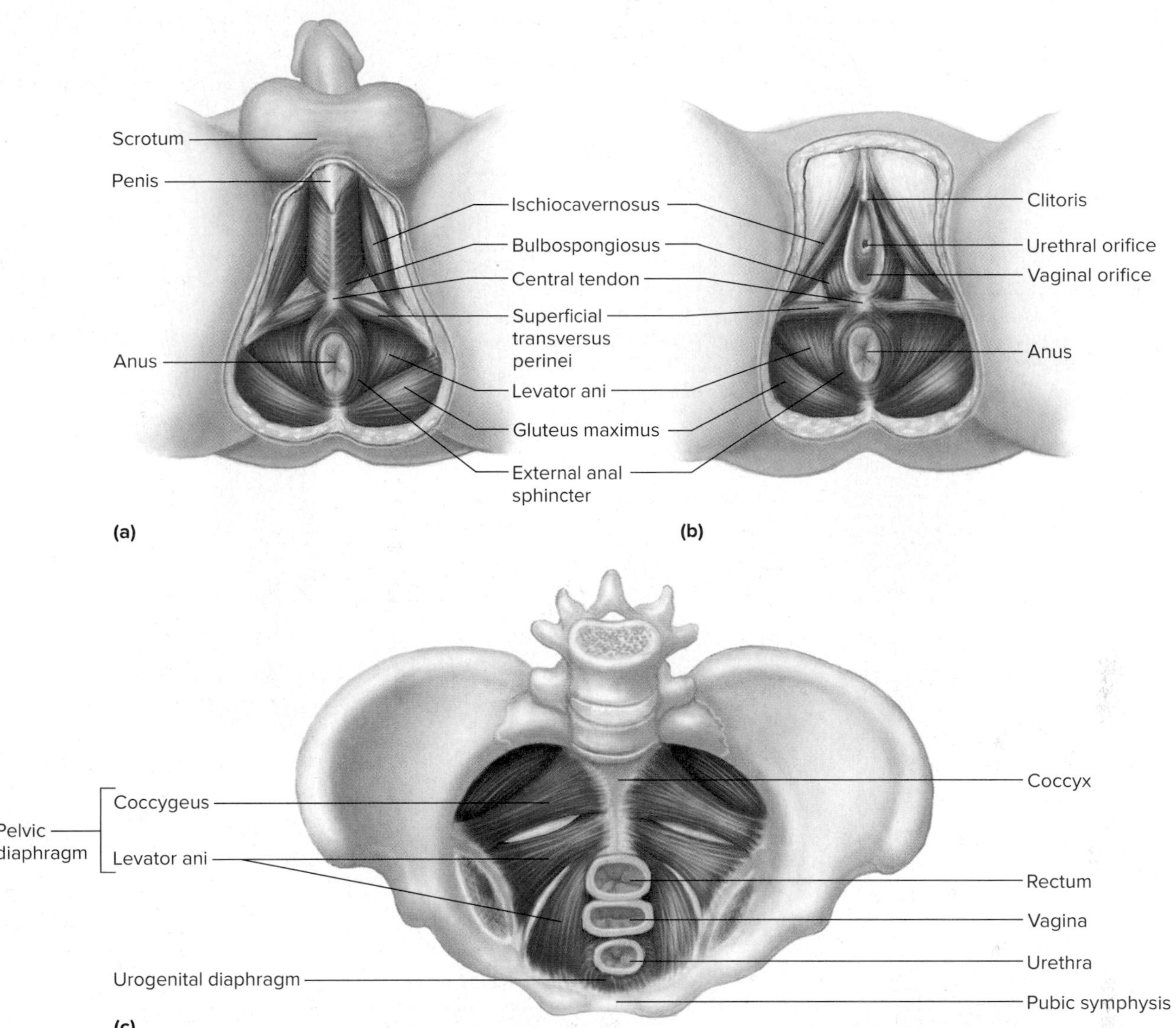

Figure 8.25 External view of **(a)** the male pelvic floor and **(b)** the female pelvic floor. **(c)** Internal view of the female pelvic and urogenital diaphragms. APR

TABLE 8.12 Muscles of the Pelvic Floor

Muscle	Origin	Insertion	Action
Levator ani	Pubis and ischial spine	Coccyx	Supports pelvic viscera, compresses anal canal
Coccygeus	Ischial spine	Sacrum and coccyx	Supports pelvic viscera, compresses anal canal
Superficial transversus perinei	Ischial tuberosity	Central tendon	Supports pelvic viscera
Bulbospongiosus	Central tendon	Males: Corpus cavernosa of penis Females: Corpus cavernosa of clitoris	Males: Assists emptying of urethra, assists erection of penis Females: Constricts vagina, assists erection of clitoris
Ischiocavernosus	Ischial tuberosity	Males: Corpus cavernosa of penis Females: Corpus cavernosa of clitoris	Males: Contributes to erection of the penis Females: Contributes to erection of the clitoris

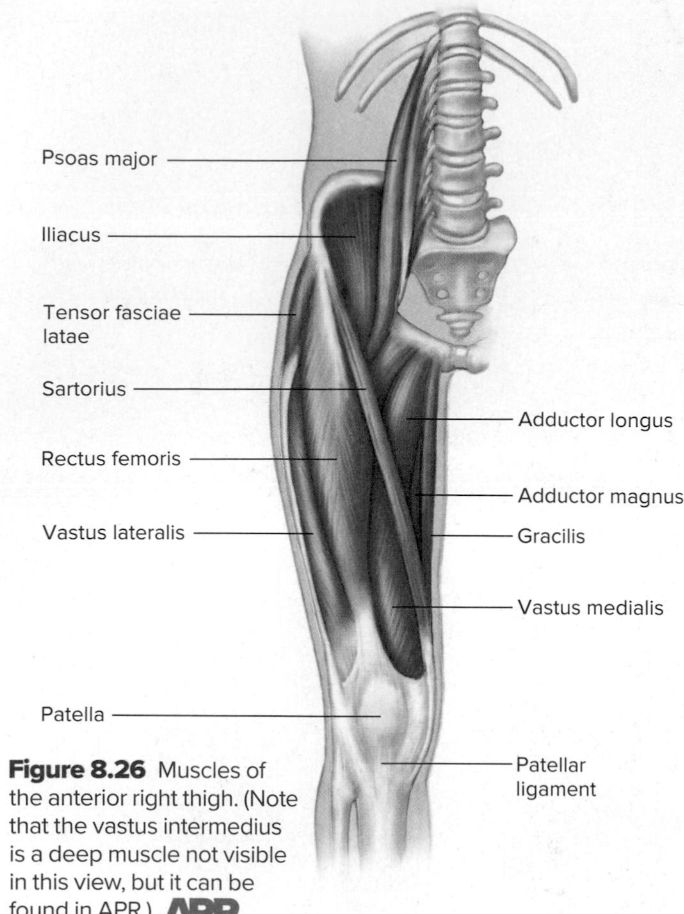

Figure 8.26 Muscles of the anterior right thigh. (Note that the vastus intermedius is a deep muscle not visible in this view, but it can be found in APR.) APR

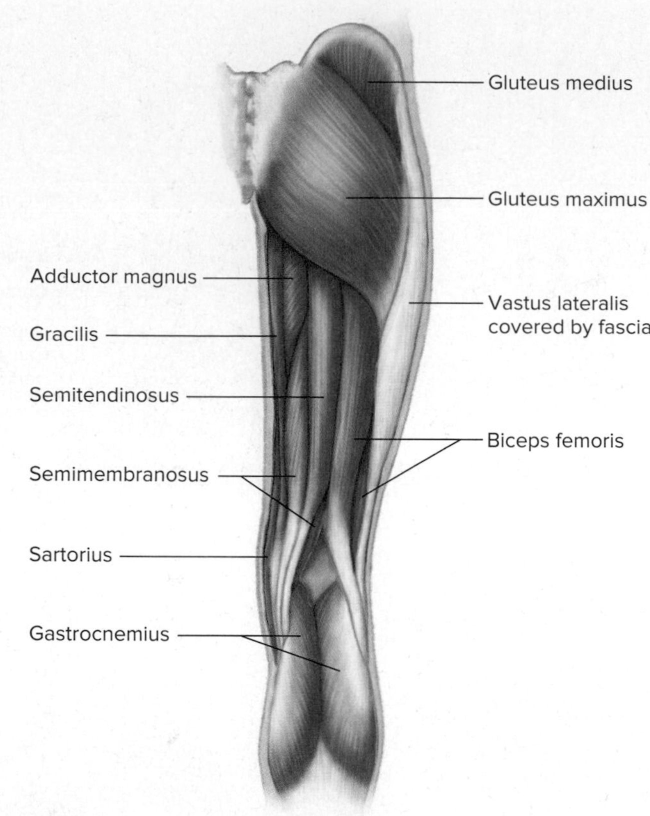

Figure 8.28 Muscles of the posterior right thigh. APR

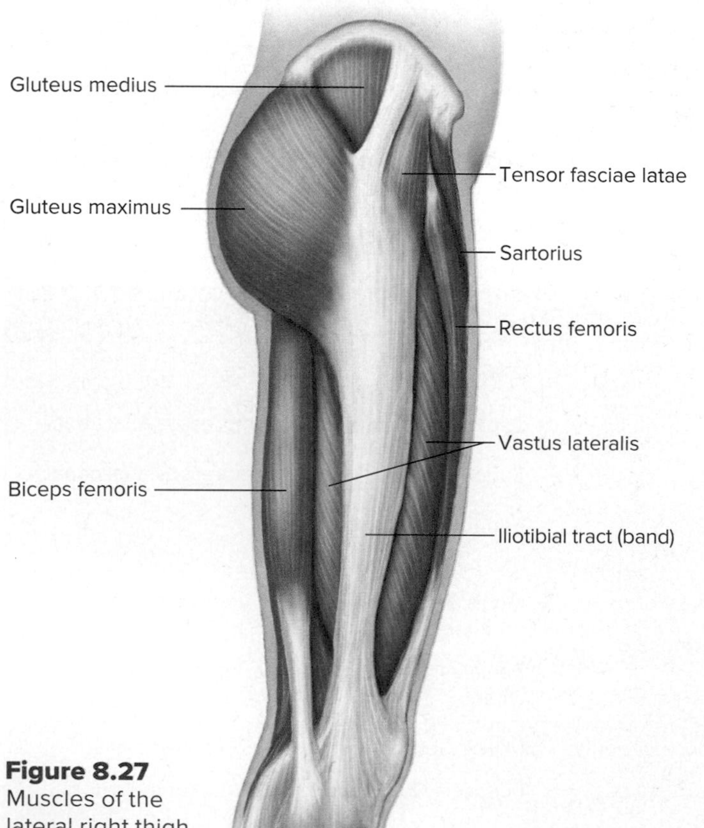

Figure 8.27 Muscles of the lateral right thigh. APR

Posterior group
gluteus maximus (gloo′tē-us mak′si-mus)
gluteus medius (gloo′tē-us me′de-us)
gluteus minimus (gloo′tē-us min′ĭ-mus)
tensor fasciae latae (ten′sor fash′ē-e lah-tē)

Medial Group
adductor longus (ah-duk′tor long′gus)
adductor magnus (ah-duk′tor mag′nus)
gracilis (gras′il-is)

Table 8.13 lists the origins, insertions, and actions of muscles that move the thigh.

Muscles that Move the Leg

Muscles that move the leg attach from the tibia or fibula to the femur or to the pelvic girdle. They can be separated into two major groups—those that flex the knee and those that extend the knee. Muscles that move the leg include the hamstring group and the quadriceps femoris group (see figs. 8.26, 8.27, and 8.28):

Flexors
hamstring group
biceps femoris (bī′seps fem′or-is)
semitendinosus (sem″ē-ten′dĭ-nō-sus)
semimembranosus (sem″ē-mem′brah-nō-sus)
sartorius (sar-to′rē-us)

TABLE 8.13 Muscles that Move the Thigh

Muscle	Origin	Insertion	Action
Psoas major	Bodies and transverse processes of lumbar vertebrae	Lesser trochanter of femur	Flexes thigh at hip
Iliacus	Iliac fossa of ilium	Lesser trochanter of femur	Flexes thigh at hip
Gluteus maximus	Sacrum, coccyx, and posterior surface of ilium	Posterior surface of femur and fascia of thigh	Extends thigh at hip, laterally rotates thigh
Gluteus medius	Lateral surface of ilium	Greater trochanter of femur	Abducts thigh, medially rotates thigh
Gluteus minimus	Lateral surface of ilium	Greater trochanter of femur	Abducts thigh, medially rotates thigh
Tensor fasciae latae	Anterior iliac crest	Fascia of thigh	Abducts thigh, medially rotates thigh
Adductor longus	Pubic bone near pubic symphysis	Posterior surface of femur	Adducts thigh, flexes thigh at hip
Adductor magnus	Pubis and ischial tuberosity	Posterior surface of femur	Adducts thigh, extends thigh at hip
Gracilis	Lower edge of pubis	Proximal medial surface of tibia	Adducts thigh, flexes thigh at hip, medially rotates thigh and leg

TABLE 8.14 Muscles that Move the Leg

Muscle	Origin	Insertion	Action
Sartorius	Anterior superior iliac spine	Proximal medial surface of tibia	Flexes leg at knee, flexes thigh at hip, abducts thigh, laterally rotates thigh, medially rotates leg
Hamstring group			
Biceps femoris	Ischial tuberosity and posterior surface of femur	Head of fibula	Flexes leg at knee, extends thigh at hip
Semitendinosus	Ischial tuberosity	Proximal medial surface of tibia	Flexes leg at knee, extends thigh at hip
Semimembranosus	Ischial tuberosity	Medial condyle of tibia	Flexes leg at knee, extends thigh at hip
Quadriceps femoris group			
Rectus femoris	Anterior inferior iliac spine and margin of acetabulum	Patella, by the tendon which continues as patellar ligament to tibial tuberosity	Extends leg at knee, flexes thigh at hip
Vastus lateralis	Greater trochanter and posterior surface of femur	Patella, by the tendon which continues as patellar ligament to tibial tuberosity	Extends leg at knee
Vastus medialis	Medial surface of femur	Patella, by the tendon which continues as patellar ligament to tibial tuberosity	Extends leg at knee
Vastus intermedius	Anterior and lateral surfaces of femur	Patella, by the tendon which continues as patellar ligament to tibial tuberosity	Extends leg at knee

Extensors

quadriceps femoris group (kwod′rĭ-seps fem′or-is)
- *rectus femoris* (rek′tus fem′or-is)
- *vastus lateralis* (vas″tus lat″er-a′lis)
- *vastus medialis* (vas″tus mē″de-a′lis)
- *vastus intermedius* (vas″tus in″ter-mē′dē-us)

Table 8.14 lists the origins, insertions, and actions of muscles that move the leg.

Muscles that Move the Foot

A number of muscles that move the foot are in the leg. They attach from the femur, tibia, and fibula to bones of the foot, move the foot upward (dorsiflexion) or downward (plantar flexion), and turn the sole of the foot medial (inversion) or lateral (eversion). These muscles include (figs. 8.29, 8.30, and 8.31):

Dorsiflexors

tibialis anterior (tib″ē-a′lis an-te′rē-or)
fibularis (peroneus) tertius (fib″ū-la′ris ter′shus)
extensor digitorum longus (eks-ten′sor dij″ĭ-tō′rum long′gus)

Plantar flexors

gastrocnemius (gas″trok-nē′mē-us)
soleus (sō′lē-us)
flexor digitorum longus (flek′sor dij″ĭ-tō′rum long′gus)

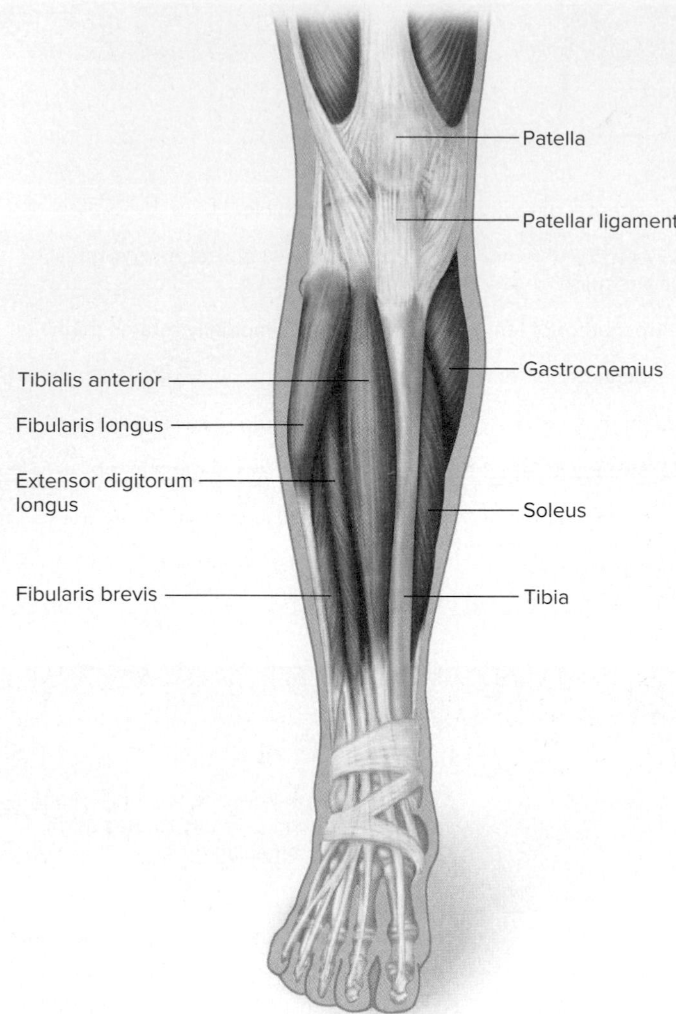

Figure 8.29 Muscles of the anterior right leg. 

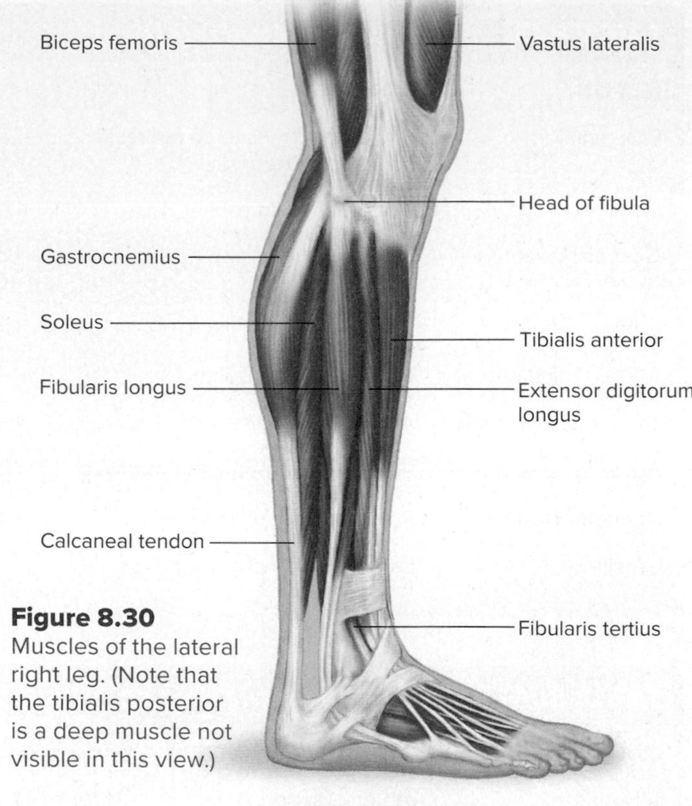

Figure 8.30 Muscles of the lateral right leg. (Note that the tibialis posterior is a deep muscle not visible in this view.)

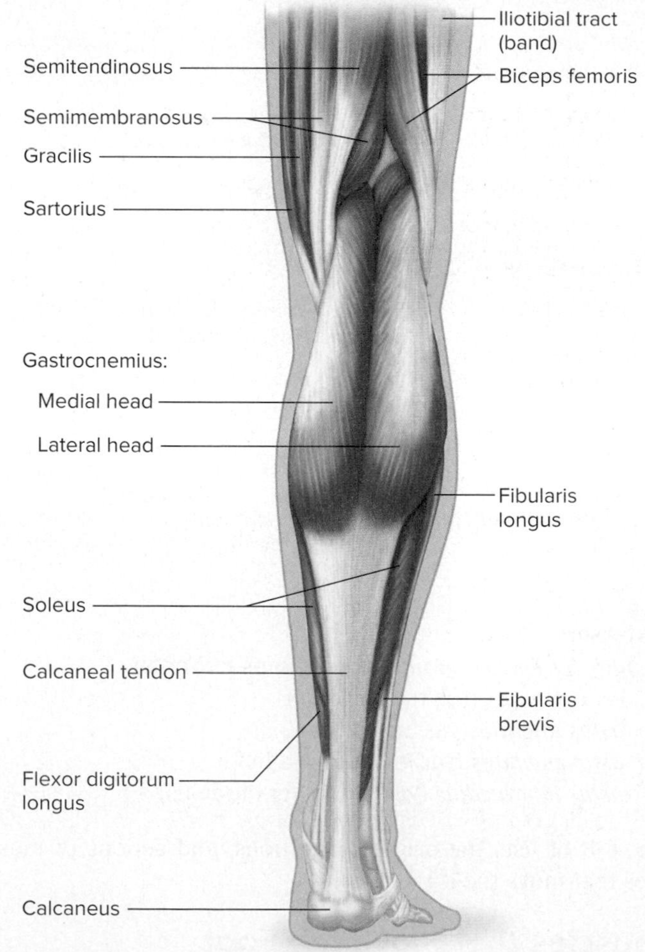

Figure 8.31 Muscles of the posterior right leg. 

Invertor
tibialis posterior (tib″ē-a′lis pos-tēr′ē-or)

Evertor
fibularis (peroneus) longus (fib″ū-la′ris long′gus)
fibularis (peroneus) brevis (fib″ū-la′ris bre′vis)

Table 8.15 lists the origins, insertions, and actions of muscles that move the foot.

PRACTICE 8.8

1. What information is imparted in a muscle's name?
2. Which muscles provide facial expressions? The ability to chew? Head movements?
3. Which muscles move the pectoral girdle? Abdominal wall? Pelvic outlet? The arm, forearm, and hand? The thigh, leg, and foot?

TABLE 8.15 Muscles that Move the Foot APR

Muscle	Origin	Insertion	Action
Tibialis anterior	Lateral condyle and lateral surface of tibia	Tarsal bone (medial cuneiform) and first metatarsal	Dorsiflexion and inversion of foot
Fibularis tertius	Anterior surface of fibula	Dorsal surface of fifth metatarsal	Dorsiflexion and eversion of foot
Extensor digitorum longus	Lateral condyle of tibia and anterior surface of fibula	Dorsal surfaces of middle and distal phalanges of the four lateral toes	Dorsiflexion of foot, extension of four lateral toes
Gastrocnemius	Lateral and medial condyles of femur	Posterior surface of calcaneus	Plantar flexion of foot, flexion of leg at knee
Soleus	Head and shaft of fibula and posterior surface of tibia	Posterior surface of calcaneus	Plantar flexion of foot
Flexor digitorum longus	Posterior surface of tibia	Distal phalanges of the four lateral toes	Flexion of the four lateral toes
Tibialis posterior	Lateral condyle and posterior surface of tibia, and posterior surface of fibula	Tarsal and metatarsal bones	Inversion and plantar flexion of foot
Fibularis longus	Lateral condyle of tibia and head and shaft of fibula	Tarsal bone (medial cuneiform) and first metatarsal	Eversion and plantar flexion of foot; also supports arch
Fibularis brevis	Lower lateral surface of fibula	Base of fifth metatarsal	Eversion and plantar flexion of foot

ASSESS

CHAPTER ASSESSMENTS

8.1 Introduction

1. The three types of muscle tissue are ______, ______, and ______.

8.2 Structure of a Skeletal Muscle

2. Describe the difference between a tendon and an aponeurosis.
3. Describe how connective tissue associates with skeletal muscle.
4. List the major parts of a skeletal muscle fiber, and describe the function of each part.
5. Describe a neuromuscular junction.
6. A neurotransmitter ______.
   a. binds actin filaments, causing them to slide
   b. diffuses across a synapse from a neuron to a muscle cell
   c. carries ATP across a synapse
   d. travels across a synapse from a muscle cell to a neuron

8.3 Skeletal Muscle Contraction

7. List the major events of muscle fiber contraction and relaxation.
8. Describe how ATP and creatine phosphate interact.
9. Describe how muscles obtain oxygen.
10. Describe how an oxygen debt may develop.
11. Explain how muscles may become fatigued.
12. Explain how skeletal muscle function affects the maintenance of body temperature.
13. Describe the characteristics of a muscle primarily composed of fast fibers to a muscle primarily composed of slow fibers.
14. Explain the causes of skeletal muscle hypertrophy and atrophy.

8.4 Muscular Responses

15. Define *threshold stimulus.*
16. Sketch a myogram of a single muscular twitch, and identify the latent period, period of contraction, and period of relaxation.
17. Define *motor unit.*
18. Which of the following describes the addition of muscle fibers to take part in a contraction?
    a. summation
    b. recruitment
    c. tetany
    d. twitch
19. Explain how skeletal muscle stimulation produces a sustained contraction.
20. Distinguish between tetanic contraction and muscle tone.
21. Compare isotonic and isometric contractions.

8.5 Smooth Muscle

22. Distinguish between multiunit and visceral smooth muscle cells.
23. Compare smooth and skeletal muscle contractions.

8.6 Cardiac Muscle

24. Make a table comparing contraction mechanisms of cardiac and skeletal muscle.

8.7 Skeletal Muscle Actions

25. Distinguish between a muscle's origin and its insertion.
26. Define *agonist, antagonist,* and *synergist.*

8.8 Major Skeletal Muscles

27. Match the muscles to their descriptions and functions.

(1) buccinator	A. inserted on coronoid process of mandible
(2) epicranius	B. elevates corner of mouth
(3) orbicularis oris	C. elevates scapula
(4) platysma	D. brings head into an upright position
(5) rhomboid major	E. elevates eyebrow
(6) splenius capitis	F. compresses cheeks
(7) temporalis	G. fascia in upper chest is origin
(8) zygomaticus	H. closes lips
(9) biceps brachii	I. extends forearm at elbow
(10) brachialis	J. extends arm at shoulder
(11) deltoid	K. abducts arm
(12) latissimus dorsi	L. inserted on radial tuberosity
(13) pectoralis major	M. flexes arm at shoulder
(14) pronator teres	N. pronates forearm
(15) teres minor	O. inserted on coronoid process of ulna
(16) triceps brachii	P. rotates arm laterally
(17) biceps femoris	Q. inverts foot
(18) external oblique	R. member of quadriceps femoris group
(19) gastrocnemius	S. plantar flexor of foot
(20) gluteus maximus	T. compresses contents of abdominal cavity
(21) gluteus medius	U. extends thigh at hip
(22) gracilis	V. hamstring muscle
(23) rectus femoris	W. adducts thigh
(24) tibialis anterior	X. abducts thigh

28. Which muscles can you identify in the body of this model?

Corbis

ASSESS

INTEGRATIVE ASSESSMENTS/CRITICAL THINKING

Outcomes 4.4, 8.3

1. As lactate and other substances accumulate in an active muscle, they stimulate pain receptors and the muscle may feel sore. How might the application of heat or substances that dilate blood vessels relieve such soreness?

Outcomes 5.3, 8.2

2. Discuss how connective tissue is part of the muscular system.

Outcomes 5.5, 8.2

3. What purpose is served by skeletal muscle cells being multinucleated?

Outcomes 8.2, 8.8

4. Organophosphates are widely used pesticides in agriculture and gardens at home. Ingestion or inhalation may result in poisoning the body by inhibiting the enzyme acetylcholinesterase. Explain how this would affect the neuromuscular junction and overall function of skeletal muscles, particularly those muscles that are necessary to sustain life.

Outcomes 8.3, 8.4

5. A woman takes her daughter to a sports medicine specialist and asks the specialist to determine the percentage of fast- and slow-twitch fibers in the girl's leg muscles. The parent wants to know if the healthy girl should try out for soccer or cross-country running. Do you think this is a valid reason to test muscle tissue? Why or why not?

6. Following an injury to a nerve, the muscle it innervates may become paralyzed. How would you explain to a patient the importance of moving the disabled muscles passively or contracting them using electrical stimulation?

Outcomes 8.3, 8.6

7. Make an argument as to why cardiac muscle is suitable for the wall of the heart, while skeletal muscle is not.

Outcomes 8.4, 8.8

8. What steps might be taken to minimize atrophy of the skeletal muscles in patients confined to bed for prolonged periods of time?

Chapter Summary

8.1 Introduction

The three types of muscle tissue are skeletal, smooth, and cardiac.

8.2 Structure of a Skeletal Muscle

Individual muscles are the organs of the muscular system. They include skeletal muscle tissue, nervous tissue, blood, and connective tissues.

Muscular System

INTEGUMENTARY SYSTEM

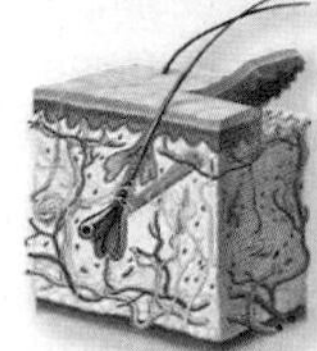

The skin increases heat loss during skeletal muscle activity.

SKELETAL SYSTEM

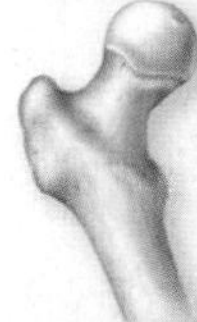

Bones provide attachments that allow skeletal muscles to cause movement.

NERVOUS SYSTEM

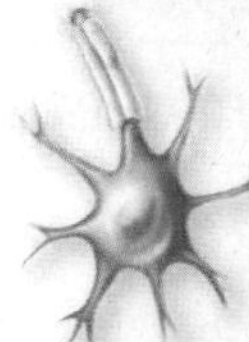

Neurons control muscle contractions.

ENDOCRINE SYSTEM

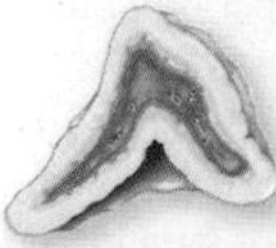

Hormones help increase blood flow to exercising skeletal muscles.

CARDIOVASCULAR SYSTEM

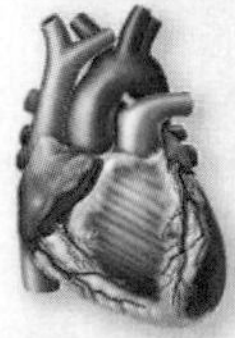

The heart pumps as a result of cardiac muscle contraction. Blood flow delivers oxygen and nutrients and removes wastes.

LYMPHATIC SYSTEM

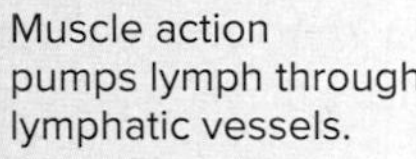

Muscle action pumps lymph through lymphatic vessels.

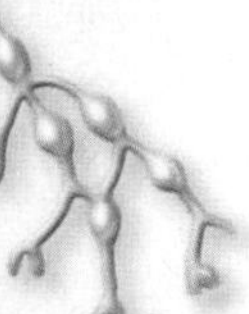

DIGESTIVE SYSTEM

Skeletal muscles are important in swallowing. The digestive system absorbs nutrients needed for muscle contraction.

RESPIRATORY SYSTEM

Breathing depends on skeletal muscles. The lungs provide oxygen for body cells and excrete carbon dioxide.

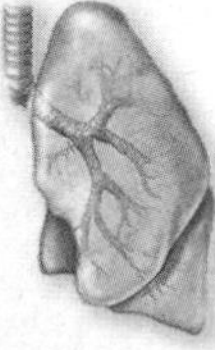

URINARY SYSTEM

Skeletal muscles help control expulsion of urine from the urinary bladder.

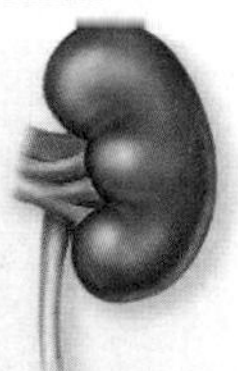

REPRODUCTIVE SYSTEM

Skeletal muscles are important in sexual activity.

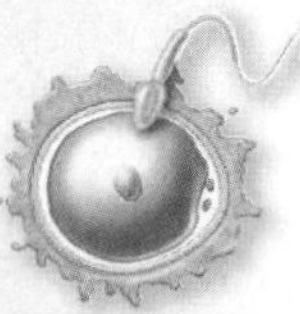

Muscles provide the force for moving body parts.

1. Connective tissue coverings
   a. **Fascia** covers skeletal muscles.
   b. Other connective tissues attach muscles to bones or to other muscles.
   c. A network of connective tissue extends throughout the muscular system.
2. Skeletal muscle fibers
   a. Each skeletal muscle fiber is a single muscle cell.
   b. The cytoplasm contains mitochondria, sarcoplasmic reticulum, and **myofibrils** of actin and myosin.
   c. The organization of **actin** and **myosin** filaments produces striations.
   d. **Transverse tubules** extend inward from the cell membrane and associate with the **sarcoplasmic reticulum.**
3. **Neuromuscular junction**
   a. **Motor neurons** stimulate muscle fibers to contract.
   b. In response to an impulse, the end of a motor neuron axon secretes a **neurotransmitter,** which stimulates the muscle fiber to contract.

8.3 Skeletal Muscle Contraction

Muscle fiber contraction results from a sliding movement of actin and myosin filaments.

1. Role of myosin and actin
   a. Heads of myosin filaments form cross-bridge linkages with actin filaments.
   b. The reaction between actin and myosin filaments generates the force of contraction.
2. Stimulus for contraction
   a. **Acetylcholine** released from the distal end of a motor neuron axon stimulates a skeletal muscle fiber.
   b. Acetylcholine causes the muscle fiber to conduct an impulse over the surface of the fiber that reaches deep within the fiber through the transverse tubules.
   c. The impulse signals the sarcoplasmic reticulum to release calcium ions.
   d. Cross-bridge linkages form between actin and myosin, and the cross-bridges pull on actin filaments, shortening the fiber.
   e. The muscle fiber relaxes when myosin heads release from actin, breaking the cross-bridges (ATP is needed, but is not broken down) and when calcium ions are actively transported (requiring ATP breakdown) back into the sarcoplasmic reticulum.
   f. **Acetylcholinesterase** breaks down acetylcholine.
3. Energy sources for contraction
   a. ATP supplies the energy for muscle fiber contraction.
   b. **Creatine phosphate** stores energy that can be used to synthesize ATP.
   c. ATP is needed for muscle relaxation.
4. Oxygen supply and cellular respiration
   a. Aerobic respiration requires oxygen.
   b. Red blood cells carry oxygen to body cells.
   c. **Myoglobin** in muscle cells helps maintain oxygen availability.
5. **Oxygen debt**
   a. During rest or moderate exercise, muscles receive enough oxygen to respire aerobically.
   b. During strenuous exercise, oxygen deficiency may cause lactic acid to be produced. Lactic acid dissociates to form lactate.
   c. Oxygen debt is the amount of oxygen required to convert lactate to glucose and to restore supplies of ATP and creatine phosphate.
6. Heat production
   a. More than half of the energy released in cellular respiration is lost as heat.
   b. Muscle action is an important source of body heat.
7. Muscle fatigue
   a. A fatigued muscle loses its ability to contract.
   b. Muscle fatigue may be due in part to increased production of lactic acid.
8. Types of muscle fibers and muscle use
   a. The **fast fibers** are large in diameter, contain abundant glycogen and few mitochondria, produce most ATP by glycolysis, and produce more forceful contractions.
   b. The **slow fibers** are small in diameter, contain abundant mitochondria and blood capillaries, are fatigue-resistant, and use aerobic respiration to produce ATP.
   c. New myofilaments are produced during high intensity exercise, resulting in muscle hypertrophy.
   d. Stopping regular exercise results in muscle atrophy.

8.4 Muscular Responses

1. Threshold stimulus is the minimal stimulus required to elicit a muscular contraction.
2. Recording a muscle contraction
   a. A **twitch** is a single contraction reflecting stimulation of a muscle fiber.
   b. A myogram is a recording of an electrically stimulated isolated muscle.
   c. The **latent period,** the time between stimulus and responding muscle contraction, is followed by a period of contraction and a period of relaxation.
3. **Summation**
   a. A rapid series of stimuli may produce summation of twitches.
   b. Very rapid stimulation can lead to partial or **complete tetanic contraction.**
4. **Recruitment** of motor units
   a. One motor neuron and the muscle fibers associated with it constitute a **motor unit.**
   b. All the muscle fibers of a motor unit contract together.
   c. Recruitment increases the number of motor units being activated in a whole muscle.
   d. The many motor units in a whole muscle are controlled by different motor neurons which respond to different thresholds of stimulation.
   e. At a low intensity of stimulation, small numbers of motor units contract.
   f. At increasing intensities of stimulation, other motor units are recruited until the muscle contracts with maximal force.
5. Sustained contractions and muscle tone
   a. Summation and recruitment together can produce a sustained contraction of increasing strength.
   b. Even when a muscle is at rest, its fibers usually remain partially contracted. This is called **muscle tone.**
6. Types of contractions
   a. Shortening of a muscle while force generated remains the same throughout movement is **isotonic contraction.**
   b. In **isometric contraction,** there is no shortening, although muscle generates a force against gravity.

8.5 Smooth Muscle

The contractile mechanism of smooth muscle is similar to that of skeletal muscle.

1. Smooth muscle cells
   a. Smooth muscle cells contain filaments of actin and myosin, which are less organized than those in skeletal muscle.
   b. Types include **multiunit smooth muscle** and **visceral smooth muscle.**
   c. Visceral smooth muscle displays rhythmicity and is self-exciting.
2. Smooth muscle contraction
   a. Two neurotransmitters—acetylcholine and norepinephrine—and hormones affect smooth muscle function.
   b. Smooth muscle can maintain a contraction longer with a given amount of energy than can skeletal muscle.
   c. Smooth muscle can change length without changing tension.

8.6 Cardiac Muscle

1. Like skeletal muscle cells, cardiac muscle cells have actin and myosin filaments that are well-organized and striated.
2. Cardiac muscle twitches last longer than skeletal muscle twitches.
3. Intercalated discs connect cardiac muscle cells.
4. A network of cells contracts as a unit.
5. Cardiac muscle is self-exciting and rhythmic.

8.7 Skeletal Muscle Actions

The type of movement a skeletal muscle produces depends on the way the muscle attaches on either side of a joint.

1. **Origin** and **insertion**
   a. The relatively immovable end of a skeletal muscle is its origin, and the relatively movable end is its insertion.
   b. Some muscles have more than one origin.
2. Interaction of skeletal muscles
   a. Skeletal muscles function in groups.
   b. An **agonist** causes a movement.
   c. **Antagonists** are muscles that oppose a movement.
   d. Muscles that work together to assist a movement are **synergists.**
   e. An agonist doing most of the work to cause a movement is a **prime mover.**
   f. Smooth movements result from agonists and antagonists working together.

8.8 Major Skeletal Muscles

1. Muscles of facial expression
   a. These muscles lie beneath the skin of the face and scalp and are used to communicate feelings through facial expression.
   b. They include the epicranius, orbicularis oculi, orbicularis oris, buccinator, zygomaticus, and platysma.
2. Muscles of mastication
   a. These muscles attach to the mandible and are used in chewing.
   b. They include the masseter and temporalis.
3. Muscles that move the head
   a. Muscles in the neck and upper back move the head.
   b. They include the sternocleidomastoid, splenius capitis, semispinalis capitis, and the scalenes.
4. Muscles that move the pectoral girdle
   a. Most of these muscles connect the scapula to nearby bones and closely associate with muscles that move the arm.
   b. They include the trapezius, rhomboid major, levator scapulae, serratus anterior, and pectoralis minor.
5. Muscles that move the arm
   a. These muscles connect the humerus to various regions of the pectoral girdle, ribs, and vertebral column.
   b. They include the coracobrachialis, pectoralis major, teres major, latissimus dorsi, supraspinatus, deltoid, subscapularis, infraspinatus, and teres minor.
6. Muscles that move the forearm
   a. These muscles connect the radius and ulna to the humerus or pectoral girdle.
   b. They include the biceps brachii, brachialis, brachioradialis, triceps brachii, supinator, pronator teres, and pronator quadratus.
7. Muscles that move the hand
   a. These muscles arise from the distal end of the humerus and from the radius and ulna.
   b. They include the flexor carpi radialis, flexor carpi ulnaris, palmaris longus, flexor digitorum profundus, extensor carpi radialis longus, extensor carpi radialis brevis, extensor carpi ulnaris, and extensor digitorum.
8. Muscles of the abdominal wall
   a. These muscles connect the rib cage and vertebral column to the pelvic girdle.
   b. They include the external oblique, internal oblique, transversus abdominis, and rectus abdominis.
9. Muscles of the pelvic floor
   a. These muscles form the floor of the pelvic cavity and fill the space within the pubic arch.
   b. They include the levator ani, coccygeus, superficial transversus perinei, bulbospongiosus, and ischiocavernosus.
10. Muscles that move the thigh
    a. These muscles attach to the femur and to some part of the pelvic girdle.
    b. They include the psoas major, iliacus, gluteus maximus, gluteus medius, gluteus minimus, tensor fasciae latae, adductor longus, adductor magnus, and gracilis.
11. Muscles that move the leg
    a. These muscles connect the tibia or fibula to the femur or pelvic girdle.
    b. They include the hamstring group (biceps femoris, semitendinosus, semimembranosus), sartorius, and the quadriceps femoris group (rectus femoris, vastus lateralis, vastus medialis, vastus intermedius).
12. Muscles that move the foot
    a. These muscles attach the femur, tibia, and fibula to bones of the foot.
    b. They include the tibialis anterior, fibularis tertius, extensor digitorum longus, gastrocnemius, soleus, flexor digitorum longus, tibialis posterior, fibularis longus, and fibularis brevis.

9

UNIT 3 INTEGRATION AND COORDINATION

Nervous System

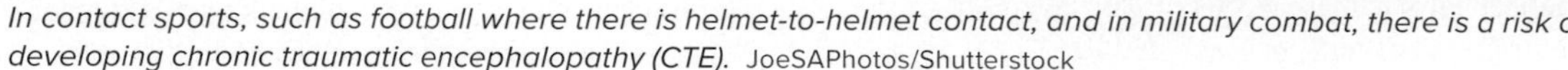

In contact sports, such as football where there is helmet-to-helmet contact, and in military combat, there is a risk of developing chronic traumatic encephalopathy (CTE). JoeSAPhotos/Shutterstock

Jason had played both sides of the line, offensive and defensive, since he was in third grade, and most recently, he had played high school varsity football as defensive tackle until he received a blow to the side of his head that was diagnosed as a concussion. However, the months of rehabilitation he spent to improve his memory, cognition, vision, and balance was likely not the result of a single blow to his head. Given Jason's history of playing football, he may be at risk of developing chronic traumatic encephalopathy (CTE), a brain condition associated with repeated blows or impacts to the head.

An associate professor at Boston University's School of Medicine and Engineering, Dr. Lee Goldstein, MD, PhD, clarified the terms *concussion, traumatic brain injury (TBI),* and *CTE.* A concussion is a syndrome that occurs when a person is hit in the head once every couple of years based on the neurological signs and symptoms. A TBI is an injury, not a syndrome, that involves damage to the brain tissue. A person can have a TBI without a concussion, and vice versa. CTE is a neurodegenerative disease that progresses independently of future hits. This means that focusing only on concussions does not prevent the development of CTE, and CTE can occur even without signs of concussion. Evidence suggests that head impact, not concussion, causes CTE. Whether it be by headers in soccer, tackles in football, or blasts in military combat, smaller hits or impacts have a cumulative effect, labeled as "the bobblehead effect." The majority of hits are subconcussive, below concussion threshold. It is particularly damaging when cumulative subconcussive hits occur when the brain is not fully healed, especially in young people whose peak brain maturation occurs between ages 9 and 12. According to findings shared by Dr. Goldstein, about 20% of athletes with CTE never suffer a diagnosed concussion.

A key biomarker of CTE in the brain is a tau protein. Head trauma can boost normal tau protein levels in the brain, resulting in an abnormal buildup that causes brain cell death, and CTE spreads in the brain as other neurodegenerative diseases do. Researchers have hypothesized that early CTE may result from damaged blood vessels that become leaky, so that blood proteins escape into the brain tissue and provoke inflammation. The brains of mice that were subjected to head impact had a brain scan called dynamic contrast-enhanced magnetic resonance imaging (DCE-MRI) that detected leaky blood vessels. It was also found that head impact caused persistent changes in electrical functions in the brain, which may be a contributor to cognitive impairment that affects some people after head impact injury.

Research is being done to find how and why CTE manifests itself, who is at risk, and why. Studies have been done on former NFL players and other athletes, as well as on military veterans. CTE can be confirmed with an autopsy at death; however, researchers are working toward identifying a way to diagnose CTE in the living. For example, scientists at Texas Christian University have studied a possible biomarker, neurofilament light (NF-L), to help identify and prevent CTE through blood tests. The goal of Dr. Lee Goldstein's and his colleagues' work is to help people affected by head injuries or who are at risk for CTE by facilitating the development of new diagnostics, therapeutics, protective equipment, and preventive measures.

Module 7 Nervous System

LEARNING OUTLINE

After studying this chapter, you should be able to complete the "Learning Outcomes" that follow the major headings throughout the chapter.

AIDS TO UNDERSTANDING WORDS

ax- [axis] *ax*on: cylindrical process that carries impulses away from a neuron cell body.

dendr- [tree] *dendr*ite: branched process that serves as a receptor surface of a neuron.

funi- [small cord or fiber] *funi*culus: major neural tract or bundle of myelinated nerve cell axons in the spinal cord.

gangli- [a swelling] *gangli*on: mass of neuron cell bodies.

-lemm [rind or peel] neuri*lemm*a: sheath that surrounds the myelin of a nerve cell axon.

mening- [membrane] *mening*es: membranous coverings of the brain and spinal cord.

moto- [moving] *moto*r neuron: neuron that stimulates a muscle to contract or a gland to secrete.

peri- [around] *peri*pheral nervous system: portion of the nervous system that consists of neural structures outside the brain and spinal cord.

plex- [interweaving] choroid *plex*us: mass of specialized capillaries and neuroglia associated with spaces in the brain.

sens- [feeling] *sens*ory neuron: neuron that conducts impulses into the brain or spinal cord.

syn- [together] *syn*apse: junction between two neurons.

ventr- [belly or stomach] *ventr*icle: fluid-filled space in the brain.

(Appendix A has a complete list of Aids to Understanding Words.)

9.1 | Introduction

LEARN

1. Distinguish between the two types of cells that compose nervous tissue.
2. Name the two major groups of nervous system organs.

Driving into campus, you see a traffic light turn from green to yellow. You make a "decision" to either step on the brake or keep your foot on the accelerator. Although this is a somewhat simplified example, it contains the three major aspects of the nervous system: ***sensory input, integration and processing*** (decision making), and a ***motor output*** (response). See figure 9.1. Employing this flow of information, the nervous system is the ultimate command center for all body functions. This includes thinking, movement, and all internal activities collectively called physiology.

Recall from chapter 5 that nervous tissue consists of bundles of **neurons,** also called nerve cells. They conduct signals, or **impulses.** Information is then passed to another neuron or other tissues. Biological messenger molecules called **neurotransmitters** (nur″ō-trans′-miterz) are the sources of this information. The nervous system also includes cells called **neuroglia** that provide protection, support, insulation, and nutrients for neurons.

The **endocrine system** is a secondary regulatory system. Under control of the nervous system, it helps maintain homeostasis by using other biological messengers called **hormones** (chapter 11).

PRACTICE 9.1

Answers to the Practice questions can be found in the eBook.

1. What are the three major aspects of the nervous system?
2. What are the two major types of cells that form nervous tissue?

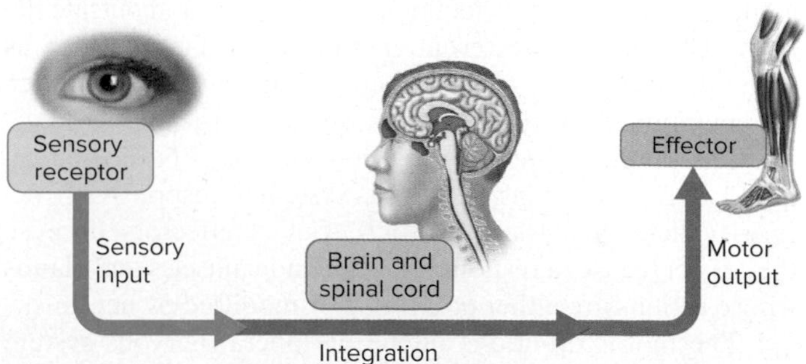

Figure 9.1 The flow of information in the nervous system.

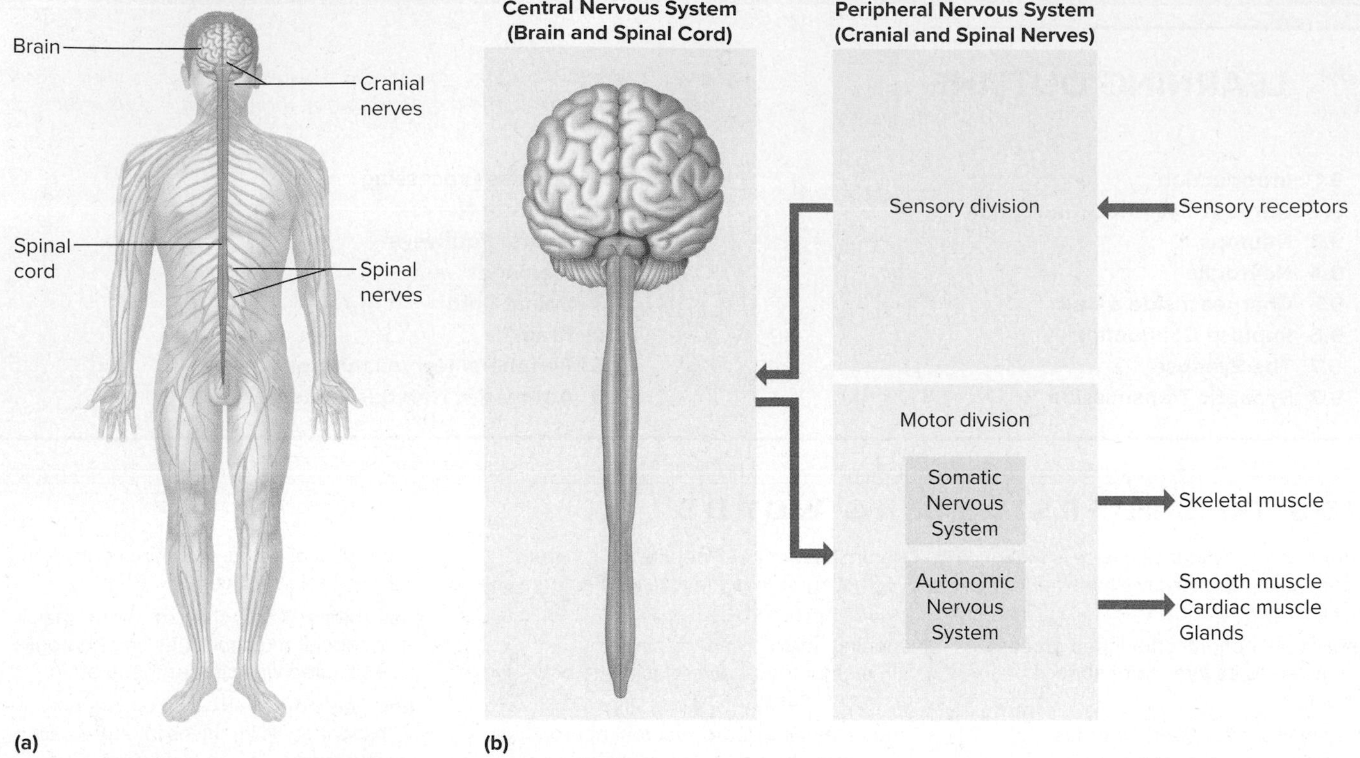

Figure 9.2 Nervous system. **(a)** The nervous system includes the central nervous system (brain and spinal cord) and the peripheral nervous system (cranial nerves and spinal nerves). **(b)** The nervous system receives information from sensory receptors and initiates responses through effector organs (muscles and glands).

9.2 | Nervous System Organization

LEARN

1. Explain the general functions of the nervous system.

The organs of the nervous system can be divided into two groups. The **central nervous system (CNS)** consists of the brain and spinal cord, while the **peripheral nervous system (PNS)** includes the cranial and spinal nerves that connect the central nervous system to other body parts (fig. 9.2*a*). In the CNS, the signals are integrated; that is, they are brought together for processing. As a result of this *integrative function,* we make conscious or subconscious decisions, and then we use *motor functions* to act on them.

The PNS is separated into the sensory and motor divisions (fig. 9.2*b*). The sensory, or ***afferent*** (to go into), division brings information into the central nervous system. Sensory receptors gather information by detecting changes inside and outside the body. They monitor external environmental factors, such as light and sound intensities, and conditions of the body's internal environment, such as temperature and oxygen level.

The motor (***efferent***) division of the PNS conducts impulses from the central nervous system to responsive structures called **effectors** (e-fek′torz). Called effectors because they effect (cause) a response, they include muscles and glands whose actions are either controlled or modified by neurons.

The motor functions of the peripheral nervous system fall into two categories: *voluntary* and *involuntary.* Those that are under voluntary (conscious) control involve the **somatic nervous system,** which controls skeletal muscle. In contrast, the **autonomic nervous system** controls effectors that are involuntary, such as cardiac muscle, smooth muscle, and various glands (fig. 9.2*b*).

PRACTICE 9.2

1. What are the two major subdivisions of the nervous system?
2. How do sensory receptors collect information?
3. How does the central nervous system integrate incoming information?
4. What are the two types of motor functions of the nervous system?

9.3 | Neurons

LEARN

1. Describe the general structure of a neuron.
2. Explain how differences in structure and function are used to classify neurons.

Neuron Structure

Neurons, or nerve cells, are the functional units of the nervous system. Responsible for generating and conducting impulses, they are the true "communicators" in the CNS and PNS. Neurons vary considerably in size and shape, but they all share common features. They all have three distinct regions: a **cell body** (soma), **dendrites,** and an **axon** (fig. 9.3*a*). To picture the relative sizes of a typical neuron's parts,

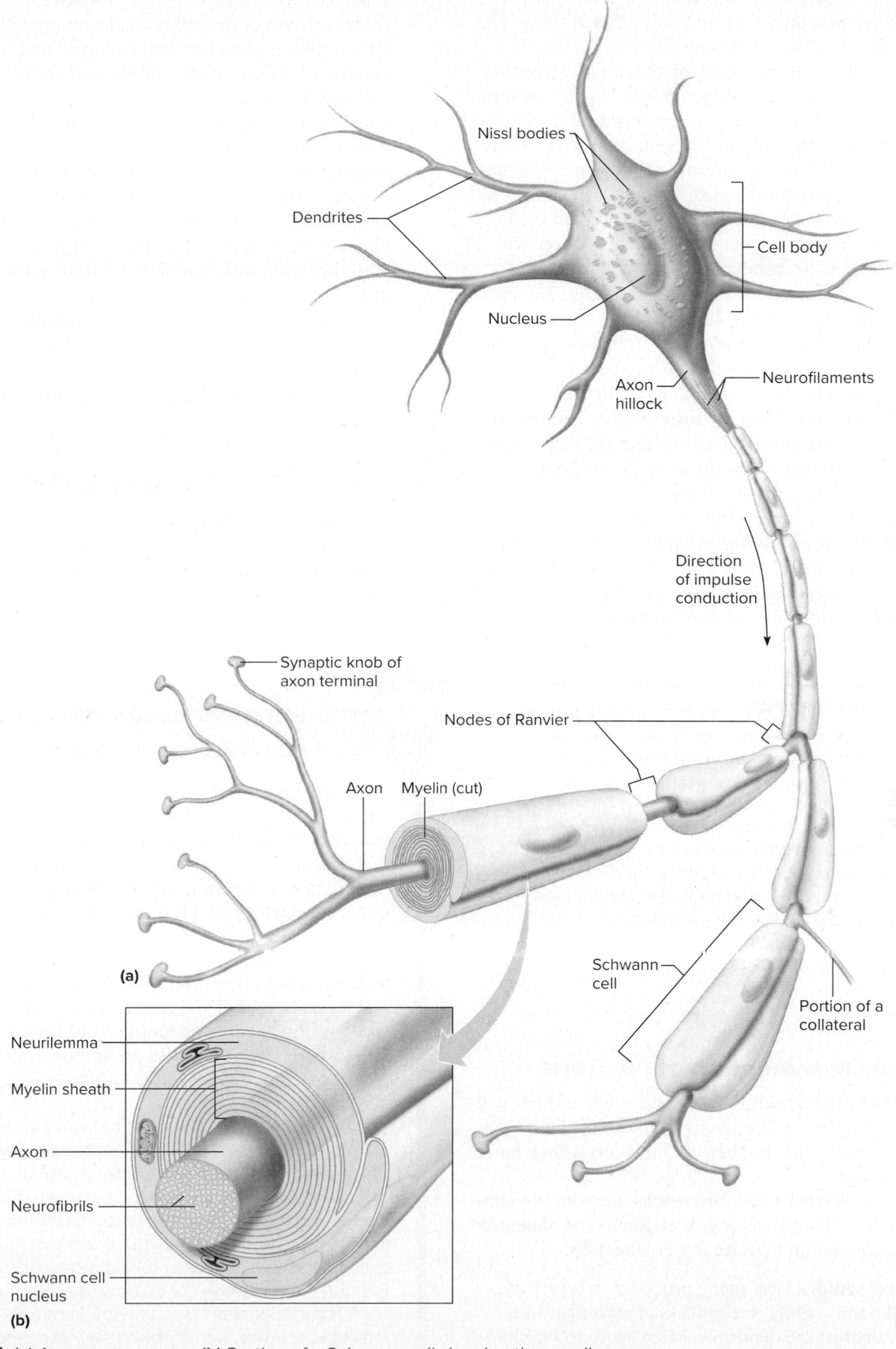

Figure 9.3 **(a)** A common neuron. **(b)** Portion of a Schwann cell showing the neurilemma.

imagine that the cell body is the size of a tennis ball. The axon would then be a mile long and half an inch thick. The dendrites would fill a large bedroom.

The cell body contains most of the usual organelles, such as the golgi apparatus, mitochondria, and lysosomes. The rough endoplasmic reticulum, together with free ribosomes, form **Nissl bodies,** which function in protein synthesis and give the cytoplasm a "grainy" appearance. The cell body also contains protein fibers called neurofilaments that extend into the axon. They function to maintain the shape of the cell and to ensure the proper diameter of the axon, which is important for impulse conduction.

Dendrites (dendro: Greek for *tree*) are highly branched processes arising from the cell body. They provide tremendous surface area to receive information from other neurons. Axons are the long processes that arise from the side of the cell body opposite the dendrites. They function to send impulses away from the cell body toward other neurons and cells. In most neurons, the axon arises from the cell body as a cone-shaped thickening called the *axon hillock*. Some axons also have side branches called collaterals. The terminal end of an axon can split into many fine extensions that contact the dendrites of other neurons at the synapse, a tiny gap between neurons (see section 9.7). Larger axons of peripheral neurons are enclosed in *myelin sheaths.* Narrow gaps in the myelin are called **nodes of Ranvier** (nōdz uv ron′vee-ay) (fig. 9.3).

OF **INTEREST** Myelin begins to form on axons during the fourteenth week of prenatal development. Yet many of the axons in newborns are not completely myelinated. As a result, an infant's nervous system cannot function as effectively as that of an older child or adult. Infants' responses to stimuli are coarse and undifferentiated, and may involve the whole body. Normally, all myelinated axons begin to develop sheaths by the time a child starts to walk, and myelination continues into adolescence. Deficiencies of essential nutrients during the developmental years may limit myelin formation, which may impair nervous system function later in life.

Structural Classification of Neurons

Neurons differ in the structure, size, and shape of their cell bodies. They also vary in the length and size of their axons and dendrites and in the number of connections they make with other neurons.

On the basis of structural differences, neurons are classified into three major groups (fig. 9.4). Each type of neuron is specialized to send an impulse in one direction:

- **Multipolar neurons** have many processes arising from their cell bodies. Only one process of each neuron is an axon; the rest are dendrites. Most neurons whose cell bodies lie within the brain or spinal cord are multipolar.
- **Bipolar neurons** have only two processes, one arising from each end of the cell body. These processes are structurally similar, but one is an axon and the other a dendrite. Neurons in specialized parts of the eyes, nose, and ears are bipolar.
- **Unipolar neurons** (also called pseudounipolar neurons because of the way they develop) have a single process extending from the cell body. A short distance from the cell body, this process divides into two branches, which really function as a single axon. One branch (the peripheral process) is associated with dendrites near a peripheral body part. The other branch (the central process) enters the brain or spinal cord. The cell bodies of unipolar neurons are found in some of the specialized masses of nervous tissue called **ganglia** (gang′gle-ah) (singular, *ganglion*), which are located outside the brain and spinal cord.

Functional Classification of Neurons

Neurons also vary in function. Different neurons may conduct impulses into the brain or spinal cord, conduct impulses from one area of the brain or spinal cord to another, or

CAREER CORNER

Occupational Therapist

The man with amyotrophic lateral sclerosis (ALS, or Lou Gehrig's disease) had been growing frustrated with his increasing inability to carry out the activities of daily living. He couldn't use his hands, and his wrists were growing weaker. A visit from an occupational therapist greatly improved both his independence and his spirit.

The occupational therapist showed the man how to continue to use a bathroom sink by supporting his weight on his arms, and how to use mirrors to see, when his neck could no longer turn. The therapist was comforting and practical as he demonstrated how to repurpose metal salad tongs to hold toilet paper to care for bathroom needs.

An occupational therapist helps a person maintain normal activities while struggling with a disease, injury, disability, or other limitation. The therapist evaluates the patient's situation and how it is likely to change, sets goals, researches and presents interventions and adaptive equipment that may help, and assesses results. The therapist may also instruct family members and caregivers on how to assist the patient.

Occupational therapists work in health-care facilities, schools, home health services, and nursing homes. They must have a master's degree in occupational therapy, and state licensure.

conduct impulses out of the brain or spinal cord. On the basis of functional differences, neurons are grouped as follows (fig. 9.5):

- **Sensory neurons** (afferent neurons) conduct impulses from peripheral body parts into the brain or spinal cord. Sensory neurons either have specialized *receptor ends* at the tips of their dendrites, or they have dendrites that are closely associated with *receptor cells* in the skin or in sensory organs.

  Changes that occur inside or outside the body stimulate receptor ends or receptor cells, triggering sensory impulses. The impulses travel along the sensory neuron axons, which lead to the brain or spinal cord, where other neurons can process the impulses. Most sensory neurons are unipolar; some are bipolar.
- **Interneurons** (also called *association* or *internuncial neurons*) lie entirely within the brain or spinal cord. They are multipolar and link other neurons. Interneurons conduct impulses from one part of the brain or spinal cord to another. That is, they may direct incoming sensory impulses to appropriate parts of the CNS for processing and interpreting. Other impulses are transferred to motor neurons. The cell bodies of some interneurons aggregate in specialized masses of nervous tissue called **nuclei** (singular, *nucleus*). Nuclei are similar to ganglia, but are within the CNS.
- **Motor neurons** (efferent neurons) are multipolar and conduct impulses out of the brain or spinal cord to effectors. Motor impulses control muscle contraction and the secretions of glands.

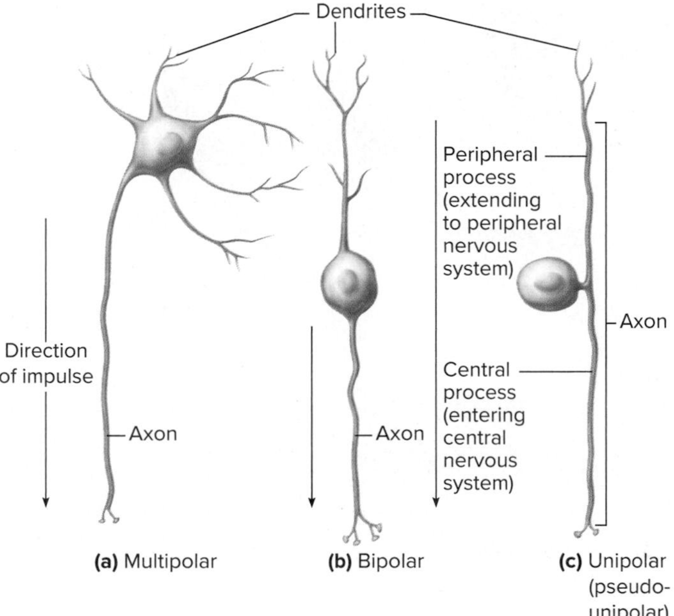

Figure 9.4 Structural types of neurons include **(a)** multipolar neurons, **(b)** bipolar neurons, and **(c)** unipolar neurons.

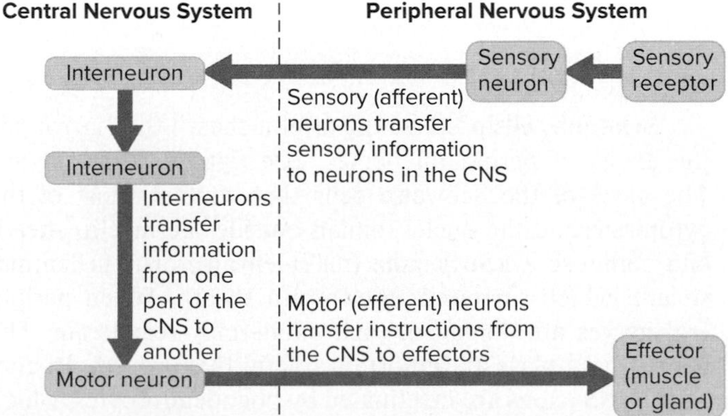

Figure 9.5 Neurons are classified by function as well as structure. Sensory (afferent) neurons carry information into the central nervous system (CNS), interneurons are completely within the CNS, and motor (efferent) neurons carry instructions to effectors.

PRACTICE 9.3

1. Distinguish between a dendrite and an axon.
2. Describe the components of a neuron.
3. Name three groups of neurons based on structure and three groups based on function.

9.4 | Neuroglia

LEARN

1. State the functions of neuroglia in the central nervous system.
2. Distinguish among the types of neuroglia in the central nervous system.
3. Describe the Schwann cells of the peripheral nervous system.

Neuroglia ("nerve glue") is a collective term for six types of cells: four in the CNS and two in the PNS. They are also called glial cells and have quite diverse functions. These include protection, insulation, and general support.

Glial cells far outnumber the neurons in the CNS and are of the following types (fig. 9.6): **Microglia** are small "spider-shaped" cells scattered throughout the central nervous system. They develop from white blood cells called monocytes. Acting as phagocytes, they remove bacterial cells and cellular debris. They also help to form scars in areas of damage.

Oligodendrocytes are aligned along axons of neurons. They produce insulating layers of myelin, called a myelin sheath. This sheath increases the speed of conduction in neurons in the brain and spinal cord.

Astrocytes (star-shaped cells) are the most numerous of all glial cells and can make up to 90% of all the cells in certain regions of the brain. Preventing capillary leakage, they play a major role in the maintenance of the *blood-brain barrier,* including the regulation of ions and nutrients in neurons.

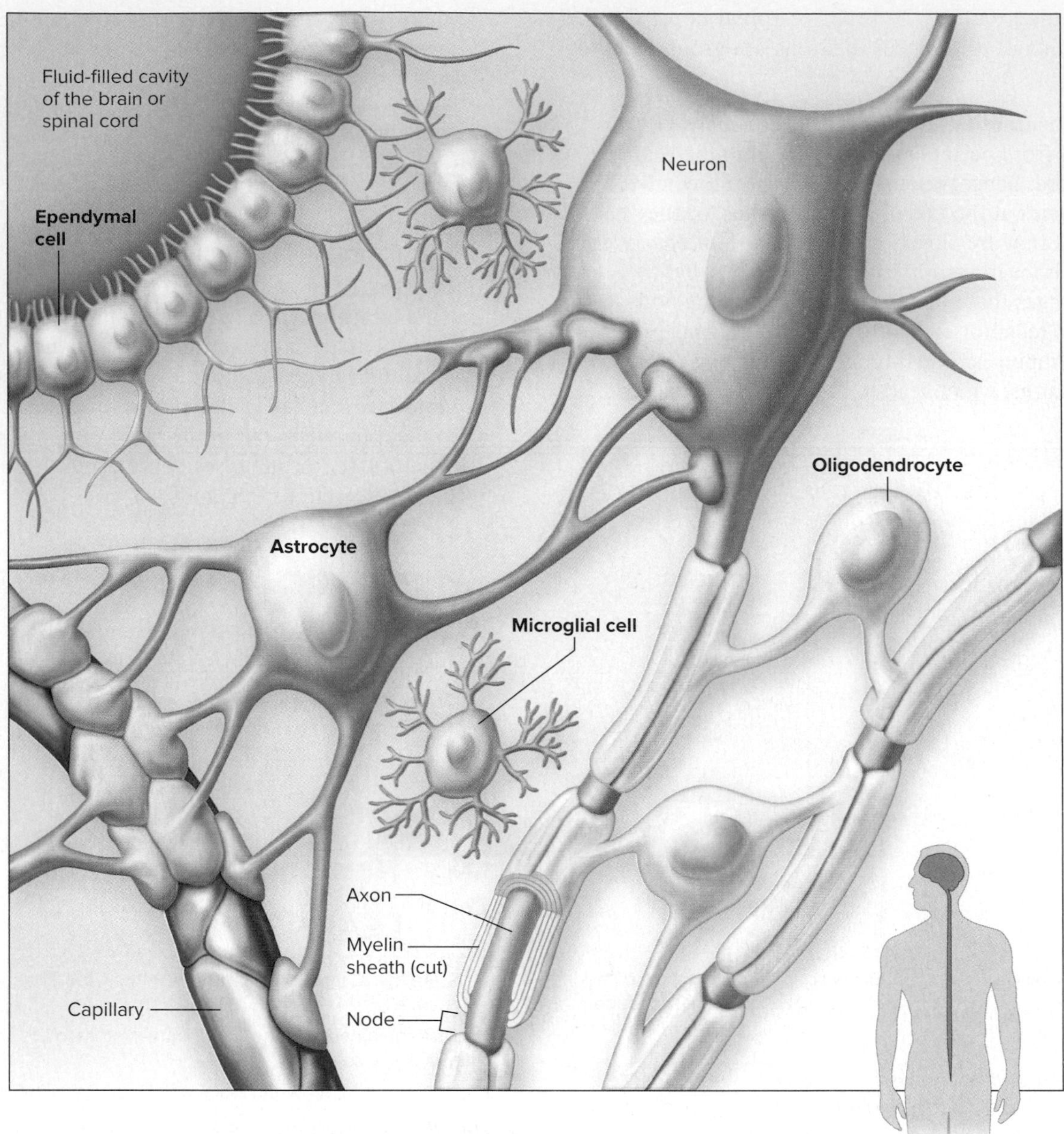

Figure 9.6 Types of neuroglia in the central nervous system include the microglial cell, the oligodendrocyte, the astrocyte, and the ependymal cell.

Knowledge of the blood-brain barrier is of particular importance for the development of drugs that can enter the CNS.

Ependymal cells form an epithelial-like membrane that lines the cavities in the brain called ventricles and the central canal in the spinal cord. They produce the cerebrospinal fluid (CSF) (sections 9.12, 9.13, and 9.14) that fills these ventricles and the central canal.

Glial cells resemble neurons but do not generate or conduct impulses. Most glial cells can still divide, making more cells. Most mature neurons lack centrioles that are needed for cell division, and therefore lose this ability. Because the creation of tumors is a result of overactive cell division, brain tumors are usually gliomas.

Schwann cells produce the myelin sheath that surrounds the axons of peripheral nerves. See figures 9.3b and 9.7. The parts of the Schwann cells that contain most of the cytoplasm and the nuclei remain outside the myelin sheath and compose a **neurilemma** (nū″rĭ-lem′ah), or neurilemmal sheath, which surrounds the myelin sheath. When peripheral nerves are damaged, their axons may regenerate. The neurilemma plays an important role in this process. In contrast, CNS axons are myelinated by oligodendrocytes, which do not provide a neurilemma. Consequently, damaged CNS axons usually do not regenerate.

Satellite cells provide a protective coat around the cell bodies of peripheral neurons (fig. 9.7).

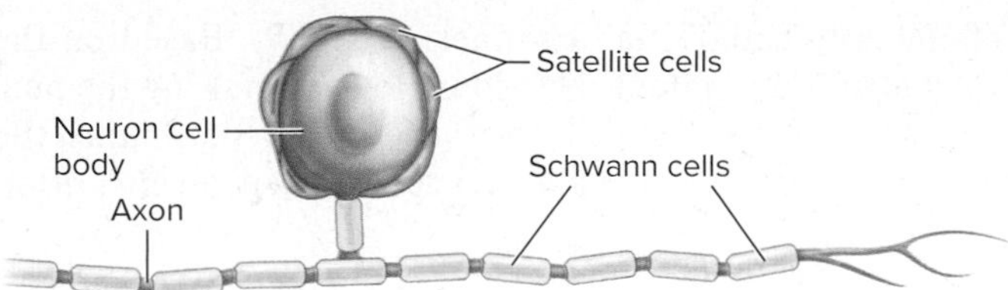

Figure 9.7 Satellite cells and Schwann cells in the PNS.

PRACTICE 9.4

1. Distinguish among the types of neuroglia in the central nervous system.
2. Provide the functions of the glial cells that support neurons in the central nervous system.
3. What is the function of Schwann cells and satellite cells in the peripheral nervous system?
4. Explain why axons of peripheral nerves can regenerate, but axons of central nervous system nerves cannot.

9.5 | Charges Inside a Cell

LEARN

1. Explain how a membrane becomes polarized.
2. Describe the events that lead to the generation of an action potential.

Most cells are negatively charged on the inside, whereas the outside (the extracellular space) is positively charged. Exhibiting a difference in charges in a given area is known as **polarity.** An example of this is the polarity in a water molecule, where there are negatively and positively charged regions (see chapter 2). Neurons and muscle cells are known as "excitable" because they can respond to stimuli. That is, unlike most other cells, they have the ability to rapidly and dramatically change the internal charge, sending it well into the positive range. This sets off a cascade of events, allowing neurons to communicate with each other.

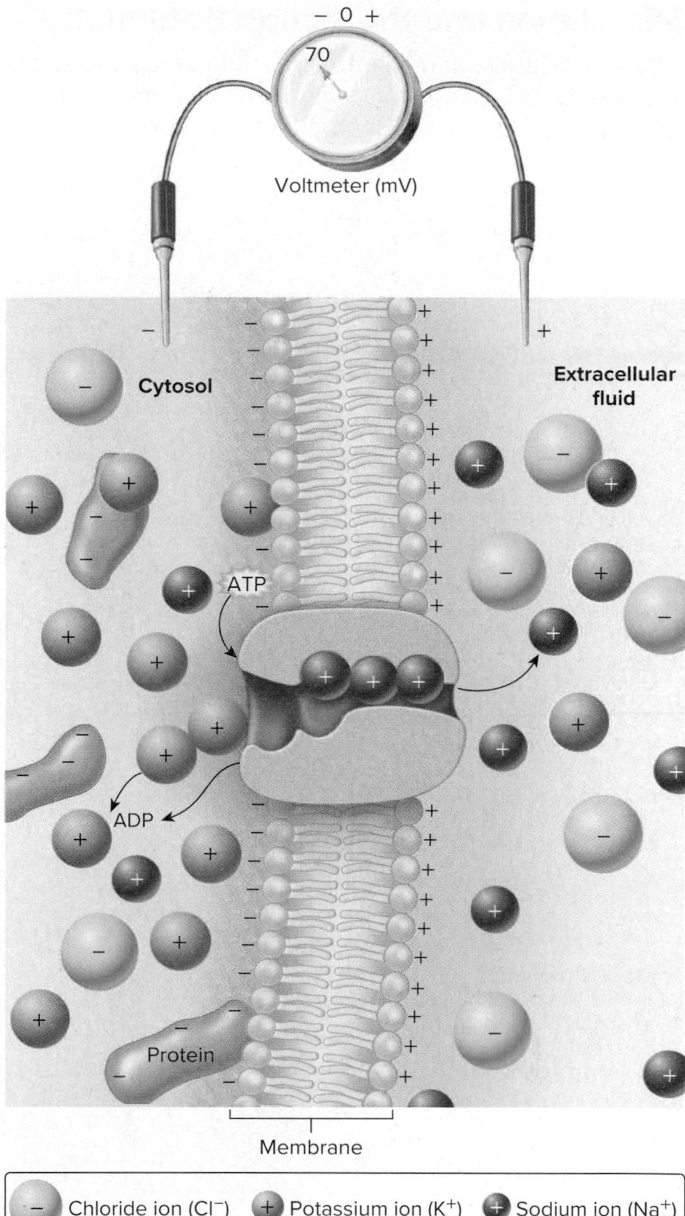

Figure 9.8 Resting membrane potential.

Membrane Potential and Distribution of Ions

A charge inside of a cell is known as a **membrane** potential (fig. 9.8). The term indicates that there is the potential for charges to move across the cell membrane. When charges flow from one area to another, they create an electrical current. The **resting membrane potential** in a neuron is about –70 mV (millivolts). This is when the neuron is at rest or inactive. Compared to the outlets in your house that carry a potential of 110 volts, this is very small, but it is quite a big deal to a neuron.

Charges in cells arise mostly from ions (fig. 9.8). Potassium (K^+) ions are found in higher concentration on the inside of a neuron. However, the overall charge on the inside is due to the many negatively charged proteins and large phosphate and sulfate anions (negatively charged ions). These are known as "fixed" anions because they cannot cross the cell membrane. Sodium (Na^+) ions are in higher concentration on the outside of a neuron. The concentration gradients of K^+ and Na^+ ions set the stage for a neuron to be excited and rapidly get back to rest.

Stimulation and the Action Potential

A neuron will stay at rest until it is stimulated. A stimulus is anything that can change the resting potential of –70 mV in either direction. An excitatory stimulus will open "chemically gated" Na^+ channels (fig. 9.9). Based on the concentration gradient, Na^+ ions flow in, making the neuron less negative. A special type of excitatory stimulus, the **threshold stimulus,** usually a specific neurotransmitter,

1 **Resting membrane potential.** Na^+ channels (*purple*) and most, but not all, K^+ channels (*pink*) are closed. The outside of the cell membrane is positively charged compared to the inside.

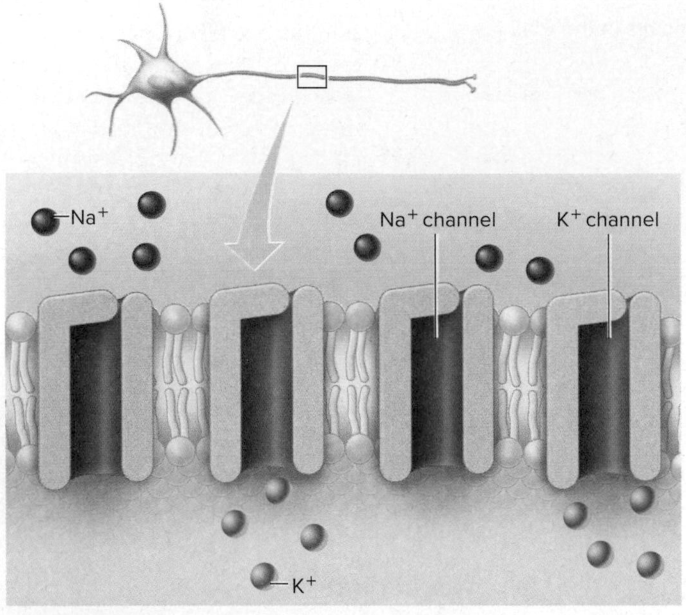

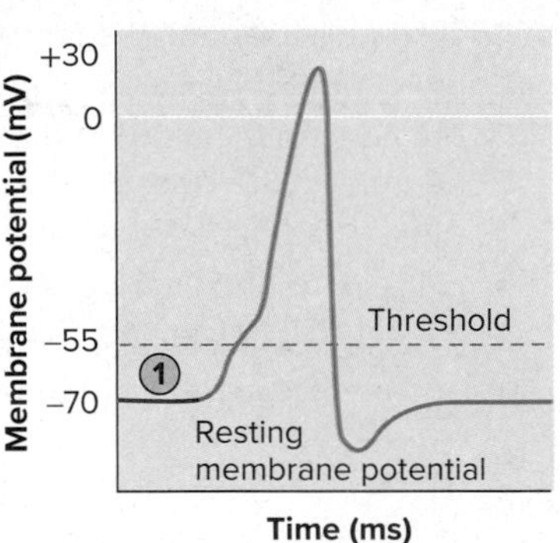

2 **Depolarization.** Na^+ channels open. K^+ channels begin to open. Depolarization results because the inward movement of Na^+ makes the inside of the membrane positive.

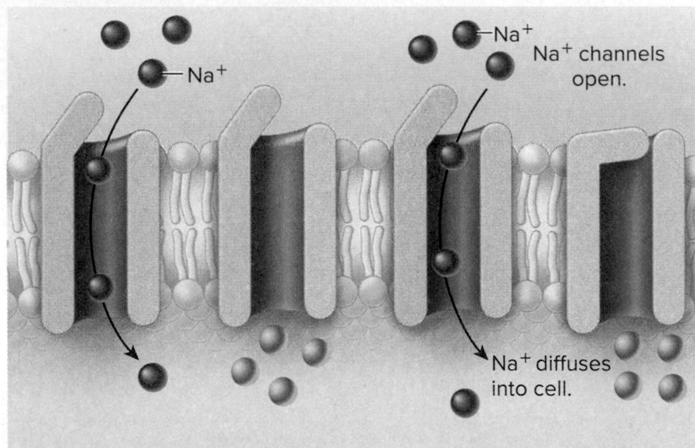

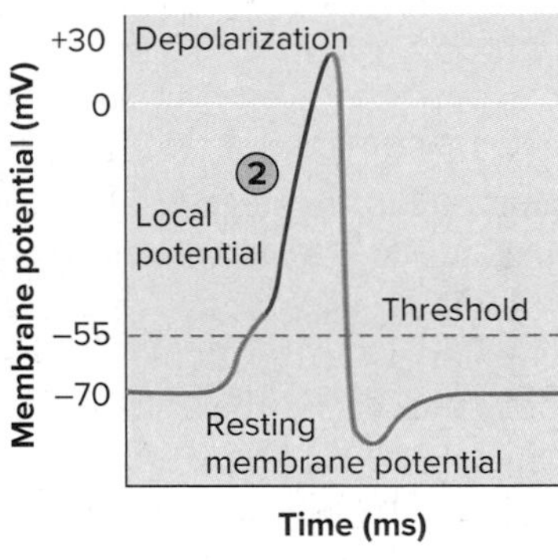

3 **Repolarization.** Na^+ channels close and additional K^+ channels open. Na^+ movement into the cell stops, and K^+ movement out of the cell increases, causing repolarization.

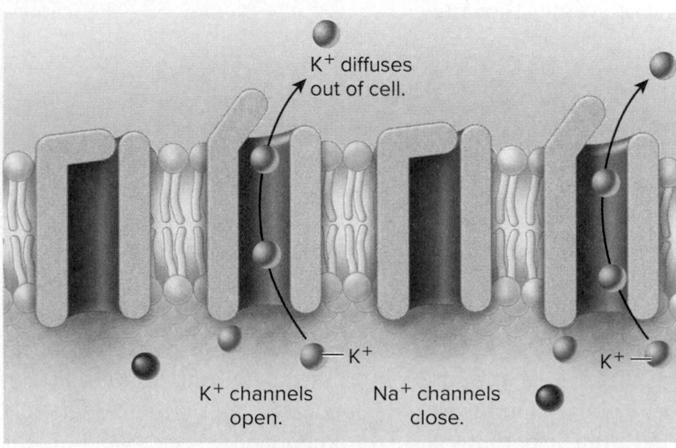

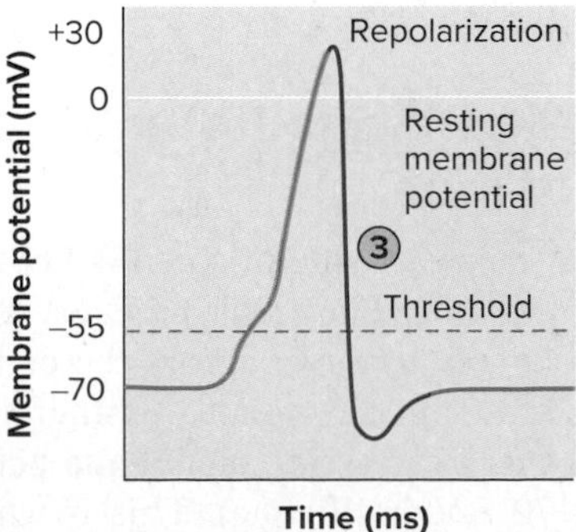

Figure 9.9 Ion channels and the action potential. 

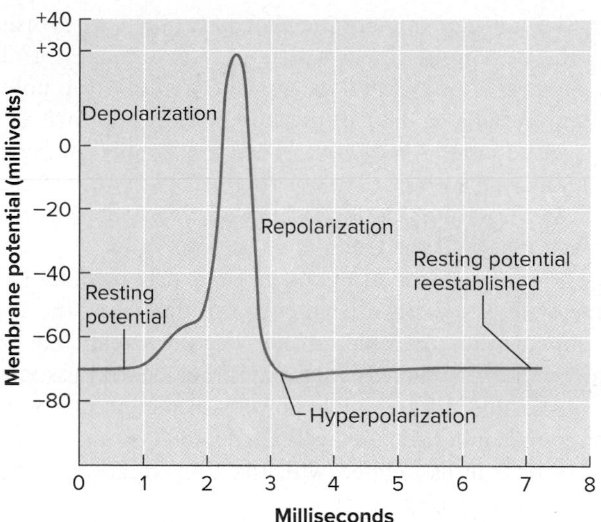

Figure 9.10 A recording of an action potential.

will cause Na^+ to flow in long enough to change the potential to −55 mV. This is called the **threshold potential.** At this point, "voltage-gated" sodium channels open and the charge rapidly rises to about +30 mV, known as the **action potential.** This change from a negative to a positive charge is known as **depolarization,** because the inside and outside of the neuron now each possess a positive charge. Reaching the action potential is often referred to as an **"all-or-none"** response, because either it is achieved or it is not. That is, if the neuron does not make it to −55 mV, nothing much happens, but once it does, there is no turning back. This is when the neuron takes action, or "fires," beginning the process of sending the signal along the axon.

After the action potential is reached, there is a quick return to the resting membrane potential. This is accomplished through the opening of channels that allow potassium to rush out. This is known as **repolarization,** because the polarity between the inside and the outside of the neuron is reestablished. At the end of this process, there is a bit of an "overshoot" called **hyperpolarization,** where the potential temporarily dips a bit below −70 mV (fig. 9.10). The Na^+/K^+ pumps then clean up the "mess" by moving Na^+ ions back out and K^+ ions back in.

PRACTICE 9.5

1. Describe the ion distribution that creates the resting membrane potential in an inactive neuron.
2. List the major events of an action potential.
3. Explain the all-or-none response.

9.6 | Impulse Conduction

LEARN

1. Explain how an impulse is conducted in unmyelinated neurons; in myelinated neurons.

An action potential (an impulse) at the trigger zone of an axon causes an electric current to flow to the adjacent region of the axon membrane. This *local current* stimulates the adjacent axon membrane to its threshold level and triggers another action potential. The new action potential, in turn, stimulates the next adjacent region of the axon. As this pattern repeats, a series of action potentials (impulses) occurs along the axon to the axon terminal (fig. 9.11). This is called *impulse conduction.* Table 9.1 summarizes impulse conduction.

Action potential movement along an unmyelinated axon is called continuous conduction, because it flows uninterrupted along its entire length. A myelinated axon functions differently, because myelin insulates and prevents

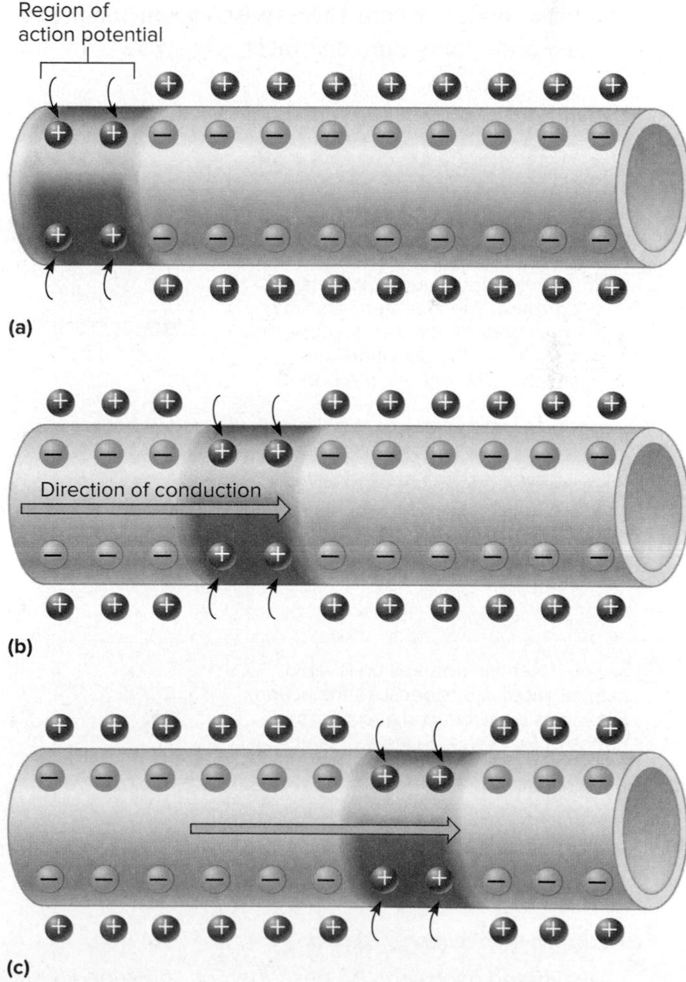

Figure 9.11 Impulse conduction. **(a)** An action potential in one region stimulates the adjacent region. **(b)** and **(c)** A series of action potentials occurs along the axon. APR

TABLE 9.1 Impulse Conduction
1. Neuron membrane maintains resting potential.
2. Threshold stimulus is received.
3. Sodium channels in the trigger zone of the axon open.
4. Sodium ions diffuse inward, depolarizing the axon membrane.
5. Potassium channels in the axon membrane open.
6. Potassium ions diffuse outward, repolarizing the axon membrane.
7. The resulting action potential causes a local electric current that stimulates the adjacent portions of the axon membrane.
8. A series of action potentials occurs along the axon.

almost all ion movement through the axon membrane it encloses. The myelin sheath would prevent impulse conduction altogether if the sheath were continuous. However, nodes of Ranvier interrupt the sheath. Action potentials occur at these nodes, where the exposed axon membrane has sodium and potassium channels (fig. 9.12). In this case, the adjacent membrane that is brought to threshold is at the next node down the axon. An impulse traveling along a myelinated axon thus appears to jump from node to node, eventually to the axon terminal. This is known as saltatory conduction (*saltation* is Latin for "jump" or "leap"). It is many times faster than conduction on an unmyelinated axon. Saltatory conduction occurs on myelinated axons in both the PNS and the CNS.

The speed of impulse conduction is proportional to the diameter of the axon—the greater the diameter, the faster the impulse. For example, an impulse on a relatively thick myelinated axon, such as that of a motor neuron associated with a skeletal muscle, might travel 120 meters per second. An impulse on a thin, unmyelinated axon, such as that of a sensory neuron associated with the skin, might move only 0.5 meter per second.

Following an action potential, a threshold stimulus will not trigger another action potential on that portion of the axon. This brief period, called the *refractory period,* limits the frequency of action potentials and also ensures that the impulses progress in only one direction—down the axon. This is because the area upstream from where the action potential has just occurred is still in the refractory period from the previous action potential.

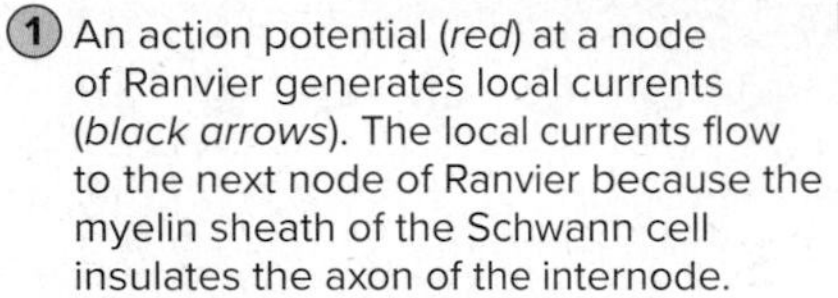

(1) An action potential (*red*) at a node of Ranvier generates local currents (*black arrows*). The local currents flow to the next node of Ranvier because the myelin sheath of the Schwann cell insulates the axon of the internode.

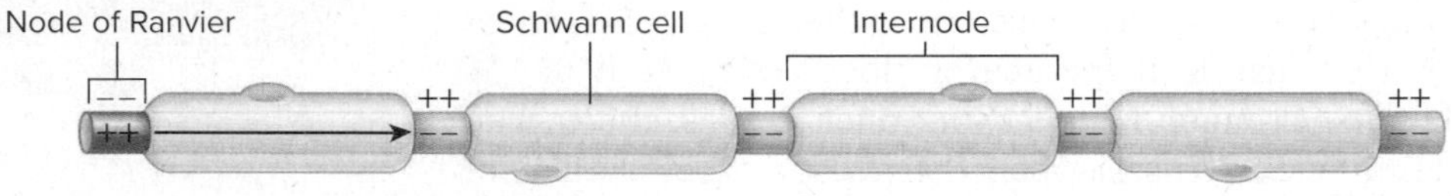

(2) When the depolarization caused by the local currents reaches threshold at the next node of Ranvier, a new action potential is produced (*red*).

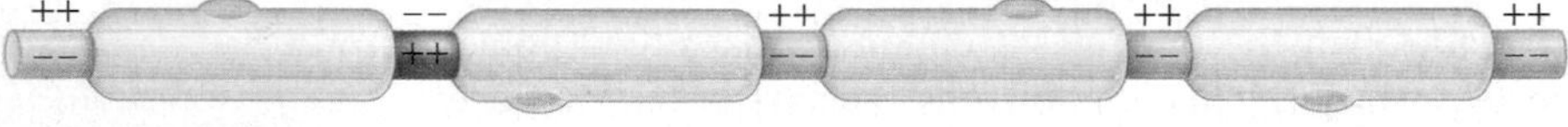

(3) Action potential propagation is rapid in myelinated axons because the action potentials are produced at successive nodes of Ranvier (*1–5*) instead of at every part of the membrane along the axon.

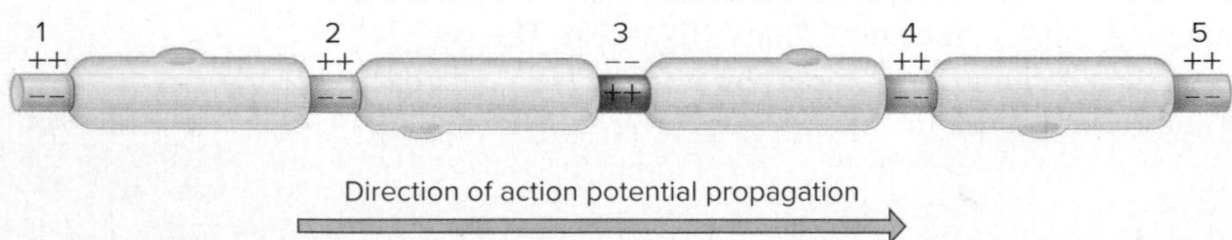

Figure 9.12 Saltatory conduction: Action potential conduction in a myelinated axon. Saltatory conduction speeds up the action potential propagation along the axon.

PRACTICE FIGURE 9.12

Myelinated axons can be described as "functionally shorter" axons compared to unmyelinated axons. Explain what is meant by "functionally shorter."

Answer can be found in Appendix E.

PRACTICE 9.6

1. What is the relationship between action potentials and impulses?
2. Explain how impulse conduction differs in myelinated and unmyelinated nerve fibers.
3. What is the refractory period, and what purpose does it serve?

9.7 | The Synapse

LEARN

1. Explain how information passes from one neuron to another.

As in the case of a motor neuron and a skeletal muscle fiber, the functional connection between two neurons is called a **synapse.** The neurons at a synapse are not in direct physical contact, but are separated by a gap called a **synaptic cleft.** Communication along a neural pathway must cross these gaps (fig. 9.13).

When you get a text message, the person texting is the sender and you are the receiver. Similarly, the neuron conducting the impulse to the synapse is the sender, or *presynaptic neuron.* The neuron that receives input at the synapse is the receiver, or *postsynaptic neuron.* The mechanism whereby this message crosses the synaptic cleft is called *synaptic transmission.* It is a one-way process, from presynaptic neuron to postsynaptic neuron (or to another postsynaptic cell, such as a skeletal muscle fiber). Clinical Application 9.1 discusses some factors that affect synaptic transmission.

Chemicals called **neurotransmitters** carry out synaptic transmission. The distal ends of axons have one or more extensions called **synaptic knobs,** which contain many membranous sacs called *synaptic vesicles* (dendrites do not have synaptic knobs). When an impulse reaches the synaptic knob of a presynaptic neuron, some of the synaptic vesicles release neurotransmitter molecules by exocytosis (figs. 9.14 and 9.15). The neurotransmitter molecules diffuse across the synaptic cleft and react with specific receptors on the membrane of the postsynaptic cell.

Once the neurotransmitter molecules bind to receptors on a postsynaptic cell, the effect is either excitatory

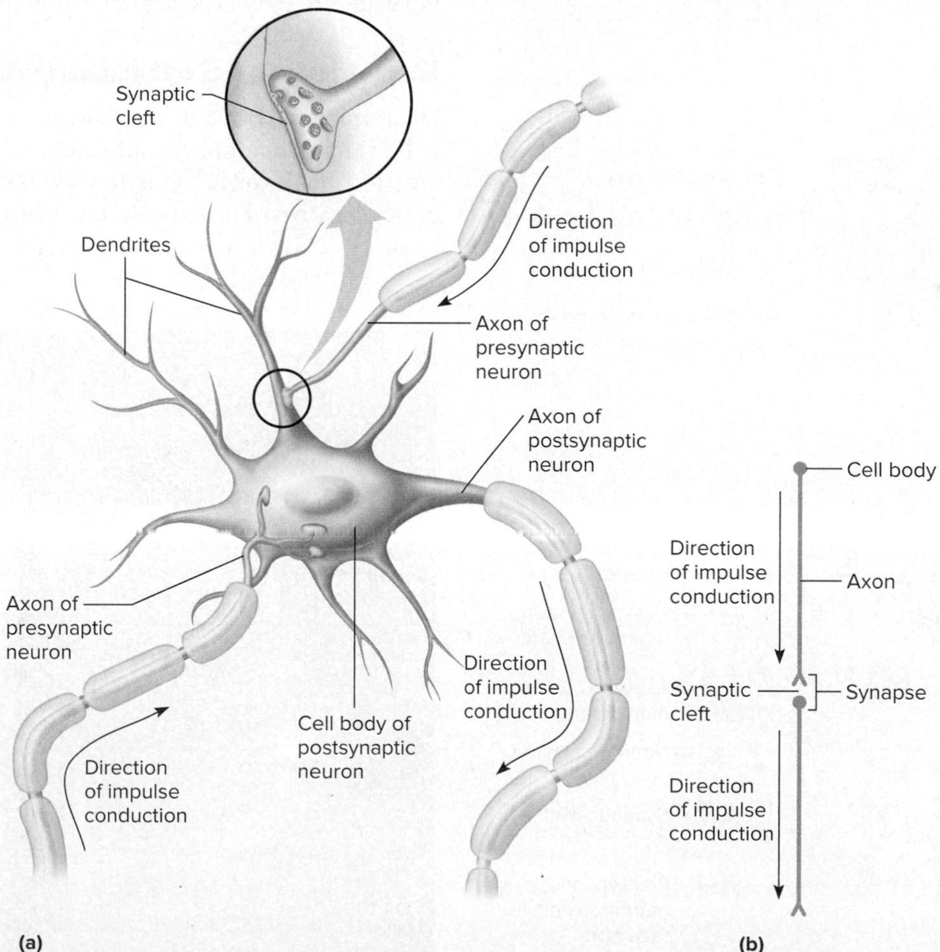

Figure 9.13 Synapses separate neurons. **(a)** For an impulse to continue on a postsynaptic neuron, the neurotransmitter must cross the synapse and stimulate the postsynaptic neuron. Most synapses are between an axon and a dendrite or between an axon and a cell body. **(b)** A schematic representation of presynaptic and postsynaptic neurons. APR

CLINICAL APPLICATION 9.1

Factors Affecting Synaptic Transmission

Many chemicals affect synaptic transmission. A drug called Dilantin (diphenylhydantoin) treats seizure disorders by blocking gated sodium channels, thereby limiting the frequency of action potentials reaching the axon terminal. Caffeine in coffee, tea, and cola drinks stimulates nervous system activity by lowering the thresholds at synapses. As a result, postsynaptic neurons are more easily excited. Antidepressants called "selective serotonin reuptake inhibitors" keep the neurotransmitter serotonin in synapses longer, compensating for a still little-understood decreased serotonin release that presumably causes depression.

(stimulating an impulse) or inhibitory (preventing an impulse). The net effect on the postsynaptic cell depends on the combined effect of the excitatory and inhibitory inputs from as few as 1 to as many as 10,000 presynaptic neurons.

PRACTICE 9.7

1. Describe the events that occur at a synapse.

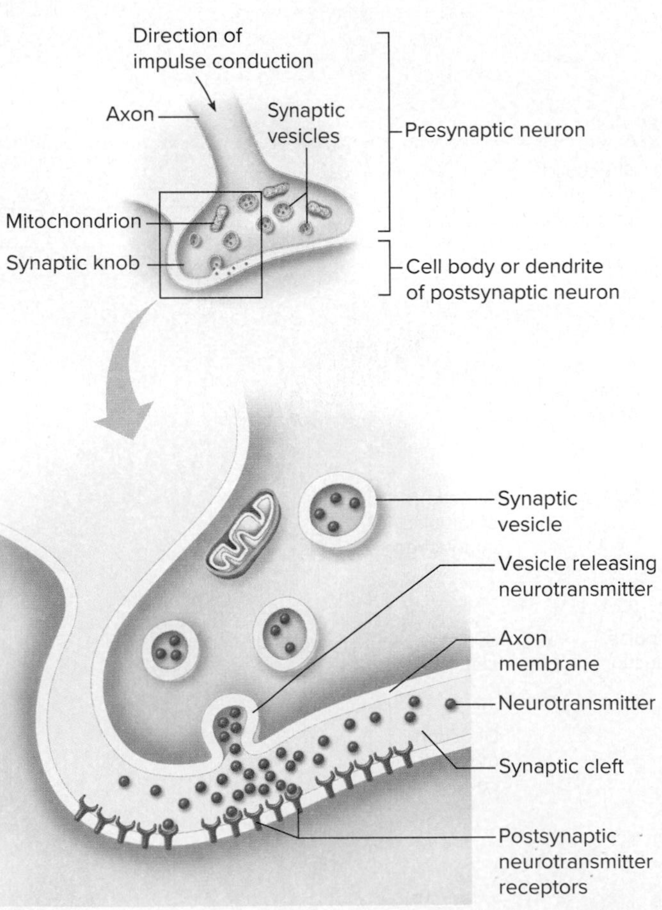

Figure 9.14 Synaptic transmission. When an impulse reaches the synaptic knob at the end of an axon, synaptic vesicles release neurotransmitter molecules that diffuse across the synaptic cleft and bind to specific receptors on the membrane of the postsynaptic cell. APR

9.8 | Synaptic Transmission

LEARN

1. Identify the changes in membrane potential associated with excitatory and inhibitory neurotransmitters.

Neurotransmitters have various effects when they diffuse across the synaptic cleft and react with specific receptor molecules in the postsynaptic neuron membrane.

Excitatory and Inhibitory Actions

Neurotransmitters are not always excitatory; often they will inhibit a neuron. An inhibitory stimulus will cause Cl^- ions to flow in or K^+ ions to flow out, making the charge more negative. This makes the neuron less likely to get

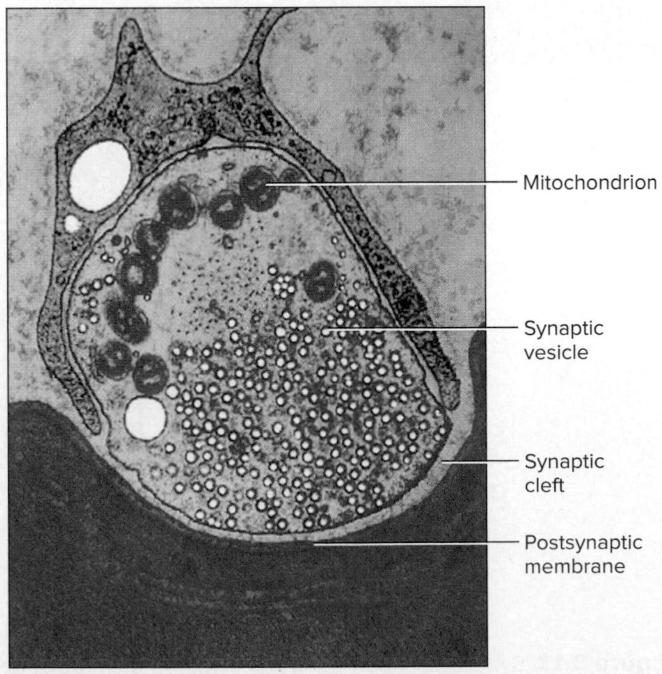

Figure 9.15 This transmission electron micrograph of a synaptic knob shows abundant synaptic vesicles, which are filled with neurotransmitter molecules (37,500x). Don W. Fawcett/ Science Source

excited (fig. 9.16). A complex combination of ***excitation*** and ***inhibition*** throughout the nervous system is crucial for proper functioning. You can think of this like the electrical circuits in your home. Most often, there are some turned off, while others are on.

The synaptic knobs of a thousand or more neurons may communicate with the dendrites and cell body of a single postsynaptic neuron. Neurotransmitters released by some of these presynaptic neurons have an excitatory action, while those from others have an inhibitory action. The overall effect on the postsynaptic neuron depends on which presynaptic neurons are releasing neurotransmitters from moment to moment. If more excitatory than inhibitory neurotransmitters are released, the postsynaptic neuron's threshold may be reached, and an action potential triggered. Conversely, if most of the neurotransmitters released are inhibitory, threshold may not be reached.

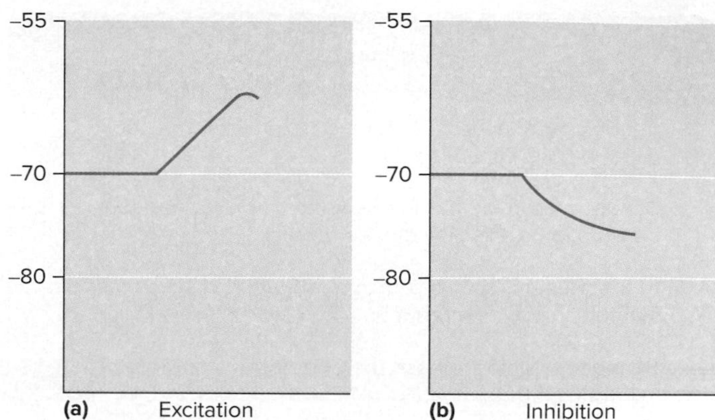

Figure 9.16 Excitatory **(a)** stimulus versus an inhibitory **(b)** stimulus.

Neurotransmitters

Most neurotransmitter molecules are synthesized in the cytoplasm of the synaptic knobs and stored in the synaptic vesicles. More than 100 different types of neurotransmitters have been identified in the nervous system. The different neurotransmitters include *acetylcholine,* which stimulates skeletal muscle contractions (see section 8.3, Skeletal Muscle Contraction); a group of compounds called *monoamines* (such as epinephrine, norepinephrine, dopamine, and serotonin), which form from modified amino acids; several *amino acids* (such as glycine, glutamic acid, aspartic acid, and gamma-aminobutyric acid—GABA); and more than 50 *neuropeptides,* which are short chains of amino acids. The action of a neurotransmitter depends on the receptors at a particular synapse. Some neurons release only one type of neurotransmitter, whereas others release two or three types. Table 9.2 lists some neurotransmitters and their actions.

TABLE 9.2 Some Neurotransmitters and Representative Actions

Neurotransmitter	Location	Major Actions
Acetylcholine	CNS	Controls skeletal muscle actions
	PNS	Stimulates skeletal muscle contraction at neuromuscular junctions; may excite or inhibit autonomic nervous system actions, depending on receptors
Monoamines		
Norepinephrine	CNS	Creates a sense of feeling good; low levels may lead to depression
	PNS	May excite or inhibit autonomic nervous system actions, depending on receptors
Dopamine	CNS	Creates a sense of feeling good; deficiency in some brain areas is associated with Parkinson disease
	PNS	Limited actions in autonomic nervous system; may excite or inhibit, depending on receptors
Serotonin	CNS	Primarily inhibitory; leads to sleepiness; action is blocked by LSD, enhanced by selective serotonin reuptake inhibitor drugs (SSRIs)
Histamine	CNS	Release in hypothalamus promotes alertness
Amino acids		
GABA	CNS	Generally inhibitory
Glutamic acid	CNS	Generally excitatory
Neuropeptides		
Substance P	PNS	Excitatory; pain perception
Endorphins, enkephalins	CNS	Generally inhibitory; reduce pain by inhibiting substance P release
Gases		
Nitric oxide	PNS	Vasodilation
	CNS	May play a role in memory

TABLE 9.3	Events Leading to the Release of a Neurotransmitter
1. Action potential passes along an axon and over the surface of its synaptic knob.	
2. Synaptic knob membrane becomes more permeable to calcium ions, and they diffuse inward.	
3. In the presence of calcium ions, synaptic vesicles fuse to synaptic knob membrane.	
4. Synaptic vesicles release their neurotransmitter into synaptic cleft.	

When an action potential reaches the membrane of a synaptic knob, it increases the membrane's permeability to calcium ions by opening calcium ion channels in the membrane. Calcium ions diffuse inward, and in response some synaptic vesicles fuse with the membrane and release their contents, neurotransmitter molecules, into the synaptic cleft. The neurotransmitter molecules then diffuse across the synaptic cleft, and may bind specific receptors on the postsynaptic cell. Table 9.3 summarizes the events leading to the release of a neurotransmitter.

A released neurotransmitter is either decomposed or otherwise removed from the synaptic cleft. This prevents a released neurotransmitter from acting on postsynaptic neurons continuously. Some neurotransmitters are decomposed by enzymes. For example, the enzyme *acetylcholinesterase* breaks down acetylcholine and is present in the synapse and on the postsynaptic membrane of neuromuscular junctions, which control skeletal muscle contraction. Other neurotransmitters are transported back into the synaptic knob that released them (a process called reuptake) or into nearby neurons or neuroglia.

PRACTICE 9.8

1. Distinguish between the actions of excitatory and inhibitory neurotransmitters.
2. What types of chemicals function as neurotransmitters?
3. What are possible fates of neurotransmitters?

9.9 | Impulse Processing

LEARN

1. Describe the general ways in which the nervous system processes information.

The way the nervous system processes and responds to impulses reflects, in part, the organization of neurons and their axons in the brain and spinal cord.

Neuronal Pools

Neurons in the CNS are organized into **neuronal pools.** These are groups of neurons that make hundreds of synaptic connections with each other and perform a common function. Each pool receives input from neurons, which may be part of other pools. Each pool generates output. Neuronal pools may have excitatory or inhibitory effects on other pools or on peripheral effectors.

A neuron in a neuronal pool may receive excitatory and inhibitory input as a result of incoming impulses and neurotransmitter release. If the net effect of the input is excitatory, threshold may be reached, and an outgoing impulse triggered. If the net effect is excitatory but subthreshold, an impulse is not triggered.

Facilitation

Repeated impulses on an excitatory presynaptic neuron may cause that neuron to release more neurotransmitter in response to a single impulse, making it more likely to bring the postsynaptic cell to threshold. This phenomenon is called **facilitation** (fah-sil″ĭ-ta′shun).

Convergence

Any single neuron in a neuronal pool may receive input from two or more incoming axons. Axons originating from different parts of the nervous system and leading to the same neuron exhibit **convergence** (kon-ver′jens) (fig. 9.17*a*).

Convergence makes it possible for impulses arriving from different sources to have an additive effect on a neuron. For example, if a neuron receives subthreshold stimulation from one input neuron, it may reach threshold if it receives additional stimulation from a second input neuron at the same time. As a result, an impulse may occur in the postsynaptic cell, travel to a particular effector, and cause a response.

Incoming impulses often bring information from several sensory receptors that detect changes. Convergence allows

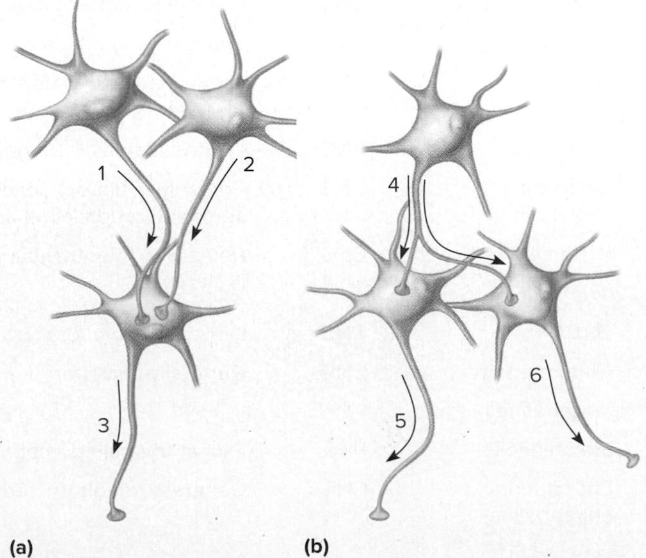

Figure 9.17 Impulse processing in neuronal pools. **(a)** Axons of neurons 1 and 2 converge to the cell body of neuron 3. **(b)** The axon of neuron 4 diverges to the cell bodies of neurons 5 and 6.

the nervous system to collect a variety of kinds of information, process it, and respond to it in a specific way.

Divergence

A neuron of a neuronal pool may exhibit **divergence** (di-ver'jens) by synapsing with several other neurons (see fig. 9.17*b*). For example, an impulse from one neuron may stimulate two others; each of these, in turn, may stimulate several others, and so forth. Divergence can amplify an impulse—that is, spread it to more neurons in the pool. As a result of divergence, the effect of a single neuron in the CNS may be amplified so that impulses reach enough motor units within a skeletal muscle to cause forceful contraction (see section 8.4, Muscular Responses). Similarly, an impulse originating from a sensory receptor may diverge and reach several different regions of the CNS, where the resulting impulses are processed and acted upon.

PRACTICE 9.9

1. Define *neuronal pool.*
2. Distinguish between convergence and divergence.

9.10 | Types of Nerves

LEARN

1. Describe how nerves are classified.

Nerves are bundles of axons located in the PNS. Nerves that conduct impulses to the brain or spinal cord are called **sensory nerves,** and those that conduct impulses to muscles or glands are termed **motor nerves.** Most nerves include axons of both sensory and motor neurons and are called **mixed nerves.**

An axon is often referred to as a *nerve fiber.* The axons that bring sensory information into the CNS may be called **sensory fibers,** or **afferent fibers.** In contrast, **motor fibers** or **efferent fibers** conduct impulses from the CNS to effectors (muscles or glands). Thus, sensory nerves contain only sensory fibers, motor nerves contain only motor fibers, and mixed nerves contain both sensory and motor fibers. The nerve fibers that compose a nerve are bundled within layers of connective tissue (fig. 9.18).

Recall from figure 8.1 that epimysium, perimysium, and endomysium connective tissue separates muscle tissue into compartments. Similarly, a nerve is defined by an outer *epineurium,* with *perineurium* surrounding a nerve fascicle within the nerve, and *endoneurium* surrounding an individual nerve fiber.

PRACTICE 9.10

1. What is a nerve?
2. How does a mixed nerve differ from a sensory nerve? From a motor nerve?

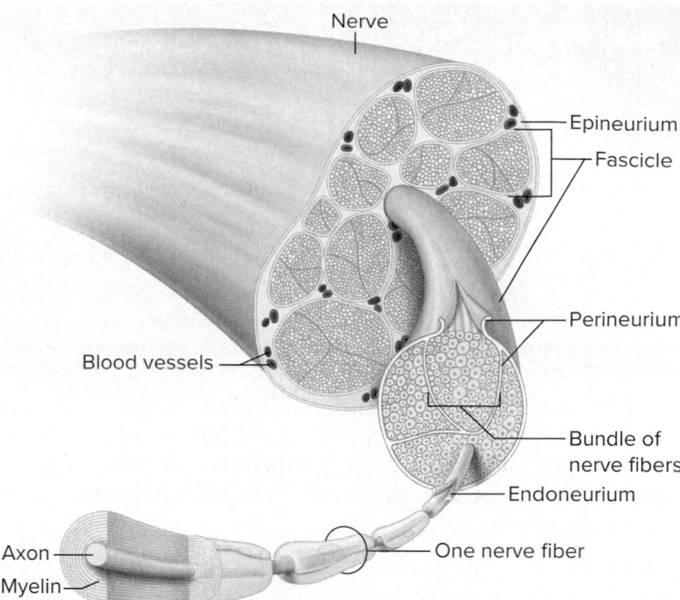

Figure 9.18 Connective tissue binds a bundle of nerve fibers, forming a fascicle. Many fascicles are bundled together to form a nerve.

9.11 | Neural Pathways

LEARN

1. Describe the function of each part of a reflex arc, and name two reflex examples.

The routes impulses follow as they travel through the nervous system are called *neural pathways.* The simplest of these pathways includes only a few neurons and is called a **reflex** (rē'fleks) **arc.** It constitutes the structural and functional basis for involuntary actions called **reflexes.**

Reflex Arcs

A reflex arc begins with a receptor at the end of a sensory (or afferent) neuron. This neuron usually leads to several interneurons in the CNS, which serve as a processing center, or *reflex center.* These interneurons communicate with motor (or efferent) neurons, whose axons pass outward from the CNS to effectors, such as muscles or glands (fig. 9.19). The

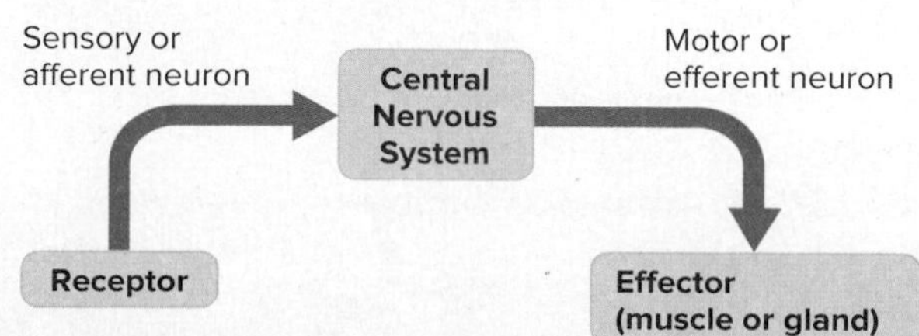

Figure 9.19 A reflex arc is the simplest neural pathway. It involves a sensory neuron that sends a message to the CNS, interneurons within the CNS, and a motor neuron that sends the message from the CNS to a muscle or gland. APR

TABLE 9.4 Parts of a Reflex Arc

Part	Description	Function
Receptor	Receptor end of a dendrite or a specialized receptor cell in a sensory organ	Senses specific type of internal or external change
Sensory neuron	Dendrite, cell body, and axon of a sensory neuron	Carries information from receptor into brain or spinal cord
Interneuron	Dendrite, cell body, and axon of a neuron within the brain or spinal cord	Carries information from sensory neuron to motor neuron
Motor neuron	Dendrite, cell body, and axon of a motor neuron	Carries instructions from brain or spinal cord out to effector
Effector	Muscle or gland	Responds to stimulation (or inhibition) by motor neuron and produces reflex or behavioral action

interneurons can also connect with interneurons in other parts of the nervous system. Table 9.4 summarizes the parts of a reflex arc.

Reflex Behavior

Reflexes are automatic responses to changes (stimuli) within or outside the body. They help maintain homeostasis by controlling many involuntary processes, such as heart rate, breathing rate, blood pressure, and digestion. Reflexes also carry out the automatic actions of swallowing, sneezing, coughing, and vomiting.

The *patellar reflex* (knee-jerk reflex) is an example of a simple reflex involving a pathway of only two neurons—a sensory neuron communicating directly with a motor neuron. Striking the patellar ligament just below the patella initiates this reflex. The quadriceps femoris muscle group, which is attached to the patella by a tendon, is pulled slightly, stimulating stretch receptors in these muscles. The receptors, in turn, trigger impulses that pass along the axon of a sensory neuron into the spinal cord. Within the spinal cord, the sensory axon synapses with a motor neuron. An impulse is then triggered along the axon of the motor neuron and travels back to the quadriceps femoris group. The muscle group contracts in response, and the reflex is completed as the knee extends (fig. 9.20).

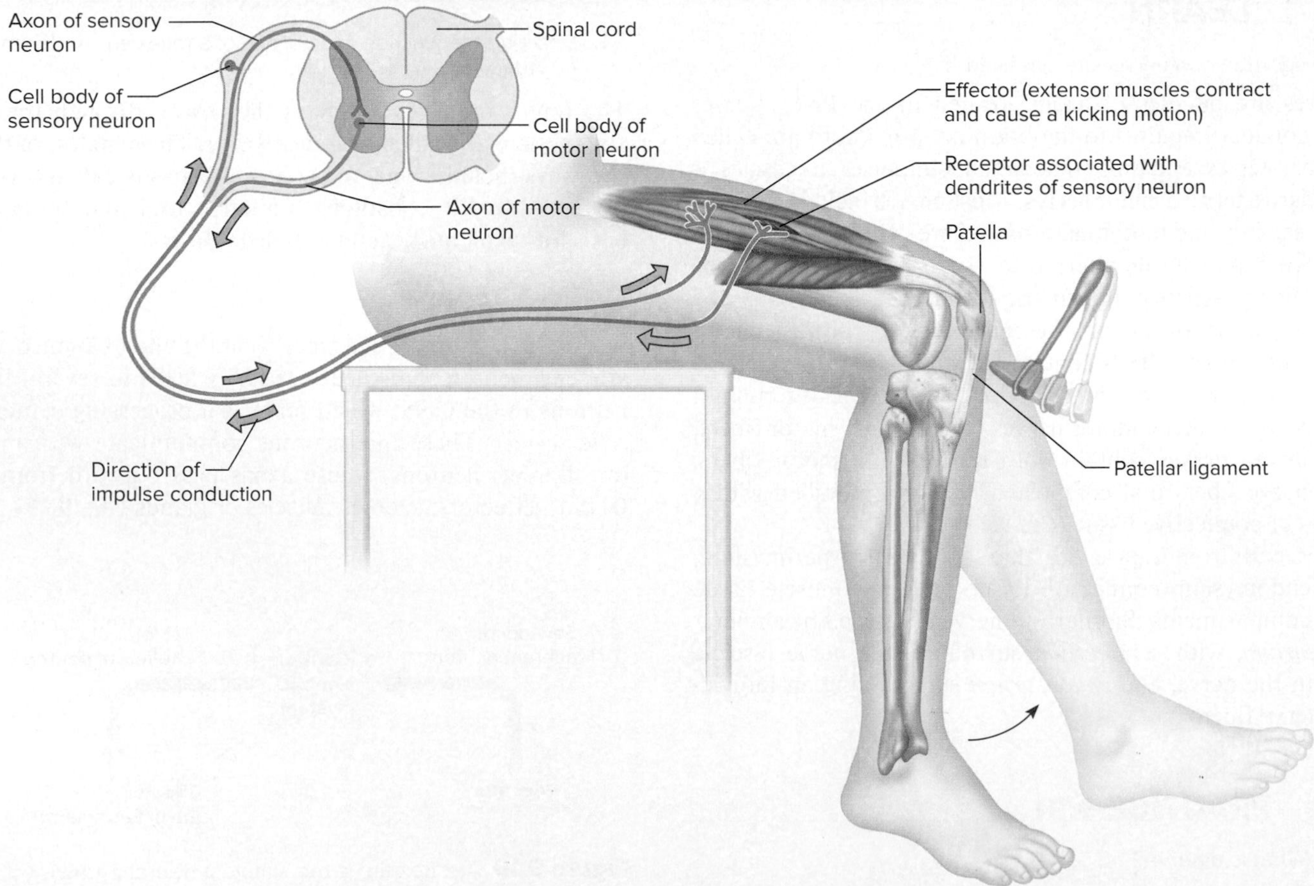

Figure 9.20 The patellar reflex involves a sensory neuron and a motor neuron. Note the single synapse within the spinal cord.

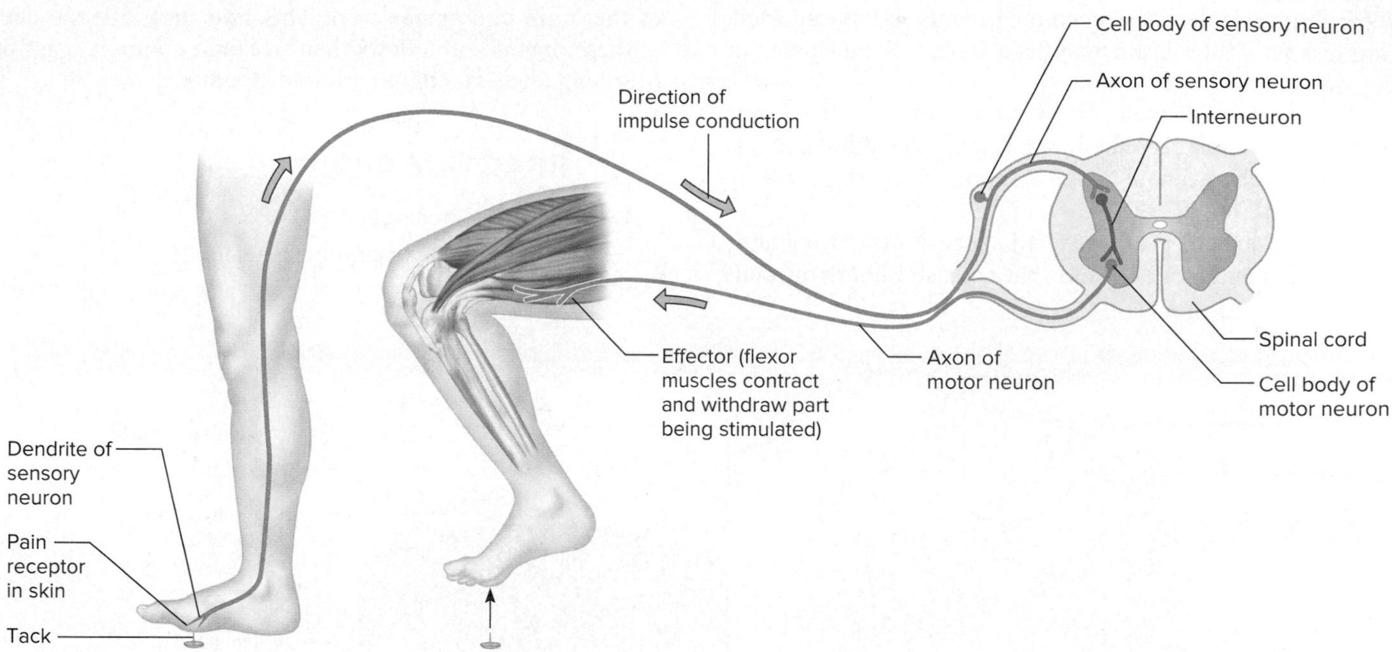

Figure 9.21 A withdrawal reflex involves a sensory neuron, an interneuron, and a motor neuron.

The patellar reflex helps maintain upright posture. If the knee begins to bend from the force of gravity when a person is standing still, the quadriceps femoris group is stretched, the reflex is triggered, and the leg straightens again.

Another type of reflex, called a *withdrawal reflex,* occurs when a person unexpectedly touches a body part to something painful, such as stepping on a tack. This activates skin receptors and sends sensory impulses to the spinal cord. There, the impulses pass to the interneurons of a reflex center and are directed to motor neurons. The motor neurons activate fibers in the flexor muscles of the leg and thigh, which contract in response, pulling the foot away from the painful stimulus. At the same time, the antagonistic extensor muscles are inhibited. This inhibition of antagonists allows the flexor muscles to effectively withdraw the affected part. Concurrent with the withdrawal reflex, other interneurons carry sensory impulses to the brain and the person becomes aware of the experience and may feel pain (fig. 9.21). A withdrawal reflex is protective because it may limit tissue damage caused by touching something harmful.

An anesthesiologist may try to initiate a reflex in a patient being anesthetized to determine how well the anesthetic drug is affecting nerve functions. A neurologist may test reflexes to determine the location and extent of damage from a nervous system injury.

PRACTICE 9.11

1. What is a neural pathway?
2. List the parts of a reflex arc.
3. Define *reflex.*
4. List the actions that occur during a withdrawal reflex.

9.12 | Meninges

LEARN

1. Describe the coverings of the brain and spinal cord.

Bones, membranes, and fluid surround the organs of the CNS. The brain lies in the cranial cavity of the skull, and the spinal cord occupies the vertebral canal in the vertebral column. Layered membranes called **meninges** (mĕ-nin′jēz) (singular, *meninx*) lie between these bony coverings and the soft tissues of the CNS, protecting the brain and spinal cord (fig. 9.22*a*).

The meninges have three layers—dura mater, arachnoid mater, and pia mater (fig. 9.22*b*). The **dura mater** (du′rah ma′ter) is the outermost layer. It is composed primarily of tough, white, fibrous connective tissue and contains many blood vessels and nerves. The dura mater attaches to the inside of the cranial cavity and forms the internal periosteum of the surrounding skull bones. In some regions, the dura mater extends inward between lobes of the brain and forms partitions that support and protect these parts.

The dura mater continues into the vertebral canal as a strong, tubular sheath that surrounds the spinal cord. It ends as a closed sac below the tip of the cord. The membrane around the spinal cord is not attached directly to the vertebrae but is separated by an **epidural space,** which lies between the dural sheath and the bony walls (fig. 9.23). This space contains loose connective and adipose tissues, which pad the spinal cord.

A blow to the head may break some blood vessels associated with the brain, and escaping blood may collect beneath the dura mater. Such a *subdural hematoma* increases pressure between the rigid bones of the skull and the soft tissues of the brain. Unless the accumulating blood is removed, compression of the brain may lead to functional losses or even death.

The **arachnoid mater** is a thin, weblike membrane without blood vessels that lies between the dura and pia maters. A **subarachnoid space** contains the clear, watery **cerebrospinal fluid (CSF).**

The **pia mater** (pī′ah mā′ter) is very thin and contains many nerves and blood vessels that nourish underlying cells of the brain and spinal cord. This layer hugs the surfaces of these organs and follows their irregular contours, passing over high areas and dipping into depressions.

PRACTICE 9.12

1. Describe the meninges.
2. State the location of cerebrospinal fluid.

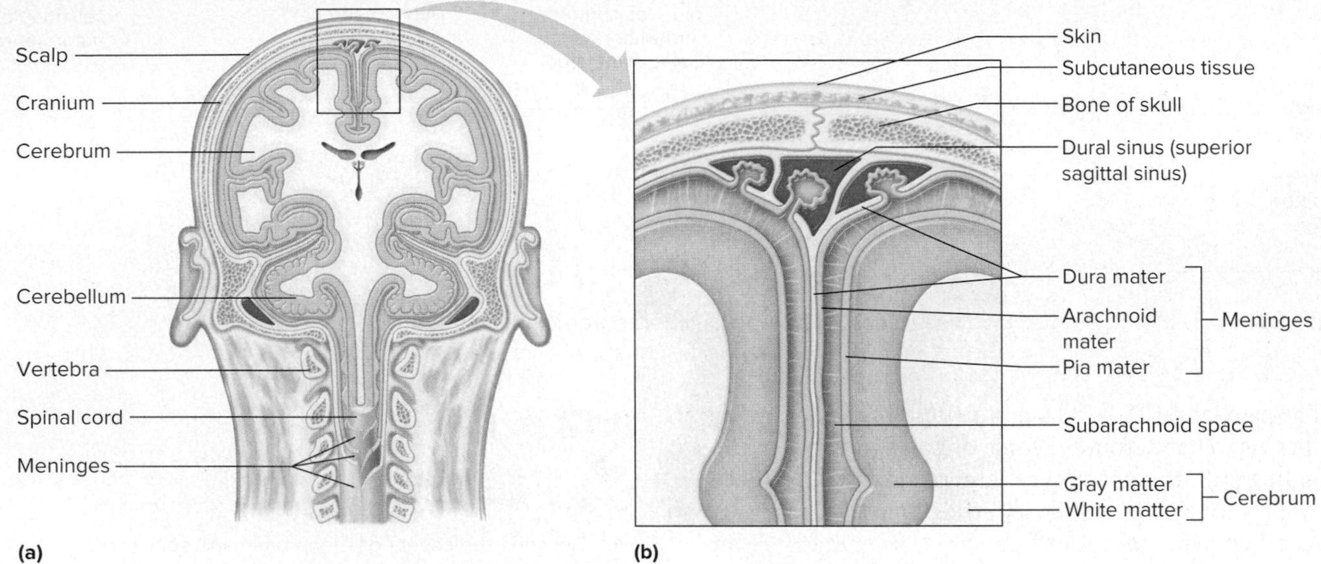

Figure 9.22 Meninges. **(a)** Membranes called meninges enclose the brain and spinal cord. **(b)** The meninges include three layers: dura mater, arachnoid mater, and pia mater.

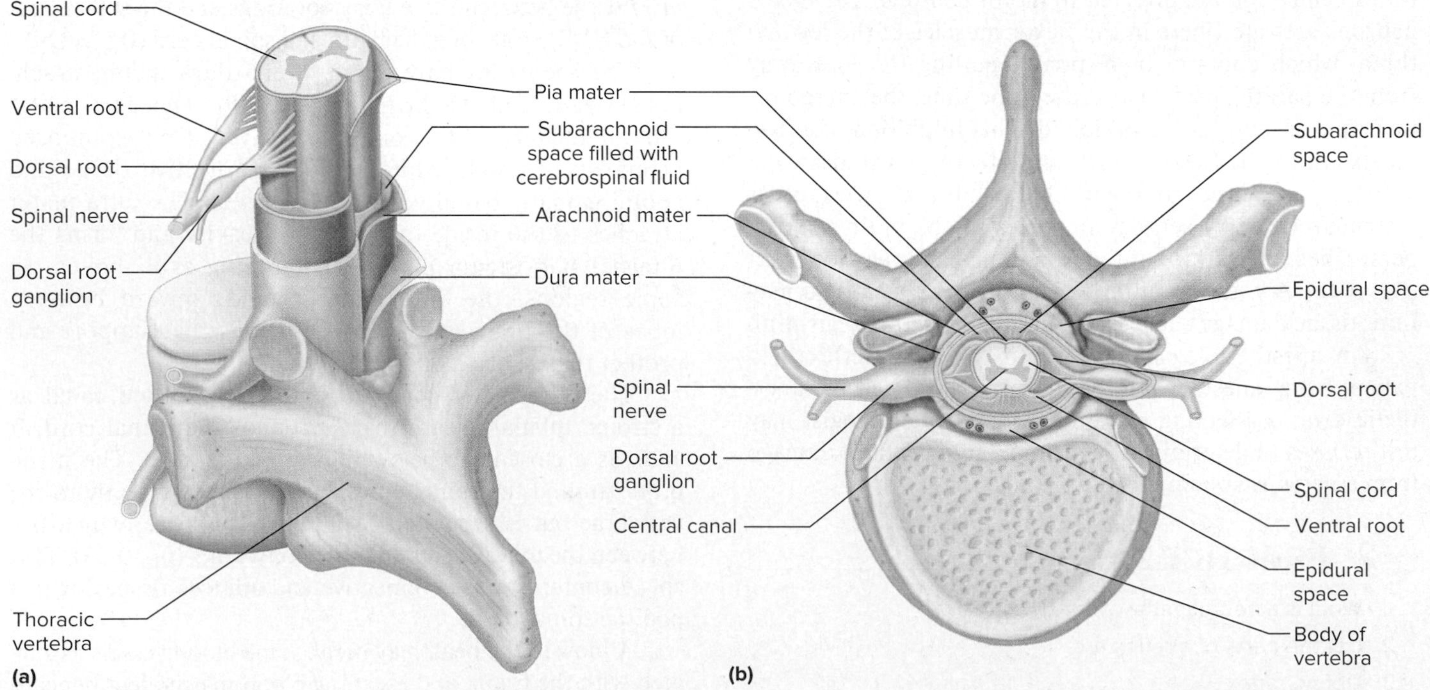

Figure 9.23 Meninges of the spinal cord. **(a)** The dura mater encloses the spinal cord. **(b)** Tissues forming a protective pad around the cord fill the epidural space between the dural sheath and the bone of the vertebra. APR

9.13 | Spinal Cord

LEARN

1. Describe the structure of the spinal cord and its major functions.

The **spinal cord** is a slender column of nervous tissue that passes downward from the brain into the vertebral canal. Although continuous with the brain, the spinal cord begins where nervous tissue leaves the cranial cavity at the level of the foramen magnum.

In the neck region, a thickening in the spinal cord, called the *cervical enlargement,* gives rise to nerves to the upper limbs. A similar thickening in the lower back, the *lumbar enlargement,* gives rise to nerves to the lower limbs. The spinal cord tapers to a point and terminates near the intervertebral disc that separates the first and second lumbar vertebrae. From this point, nervous tissue, including axons of both motor and sensory neurons, extends downward to become spinal nerves at the remaining lumbar and sacral levels forming a structure called the *cauda equina* (horse's tail) (fig. 9.24).

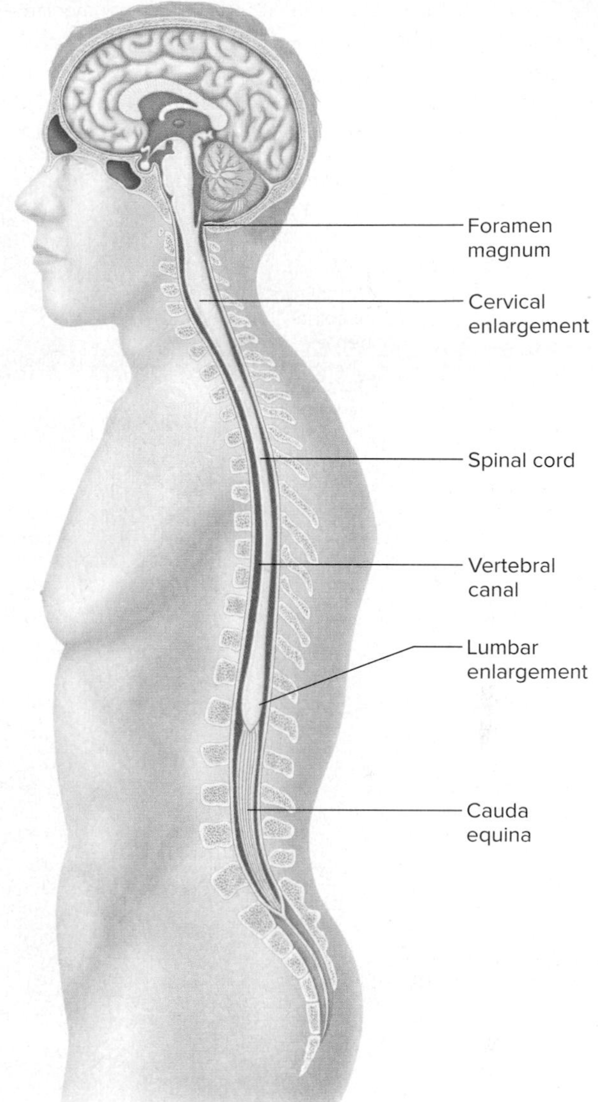

Figure 9.24 The spinal cord begins at the level of the foramen magnum and ends near the intervertebral disc between the first and second lumbar vertebrae. APR

Structure of the Spinal Cord

The spinal cord consists of thirty-one segments, each of which gives rise to a pair of **spinal nerves.** These nerves (part of the peripheral nervous system) branch to various body parts and connect them with the CNS (see fig. 9.36).

Two grooves, a deep *anterior median fissure* and a shallow *posterior median sulcus,* extend the length of the spinal cord, dividing it into right and left halves (fig. 9.25). A cross section of the cord reveals a core of gray matter (mostly cell bodies and dendrites) within white matter (axons). The pattern of gray matter roughly resembles a butterfly with its wings spread. The posterior and anterior wings of gray matter are called the *posterior horns* and *anterior horns,* respectively. Between them on either side in the thoracic and upper lumbar segments is a protrusion of gray matter called the *lateral horn.*

Neurons with large cell bodies located in the anterior horns give rise to motor fibers that pass out through spinal nerves to skeletal muscles. However, the majority of neurons in the gray matter of the spinal cord are interneurons.

The sensory neurons that are associated with the spinal cord have cell bodies that are not in the spinal cord at all. Rather, the cell bodies of these unipolar neurons are clustered in the *dorsal root ganglia* (singular, *ganglion*) associated with the spinal nerves at each segment of the spinal cord (fig. 9.25).

Gray matter divides the white matter of the spinal cord into three regions on each side—the *anterior, lateral,* and *posterior funiculi* (fig. 9.25). Each funiculus consists of longitudinal bundles of myelinated axons that comprise major neural pathways. In the central nervous system, such bundles of axons are called **tracts.**

A horizontal bar of gray matter in the middle of the spinal cord, the *gray commissure,* connects the wings of the gray matter on the right and left sides. This bar surrounds the **central canal,** which contains cerebrospinal fluid.

Functions of the Spinal Cord

The spinal cord has two major functions—conducting impulses to and from the brain, and serving as a center for spinal reflexes. The tracts of the spinal cord consist of axons that provide a two-way communication system between the brain and the body parts outside the nervous system. The tracts that carry sensory information to the brain are called **ascending tracts** (fig. 9.26); those that carry motor instructions from the brain to muscles and glands are called **descending tracts** (fig. 9.27).

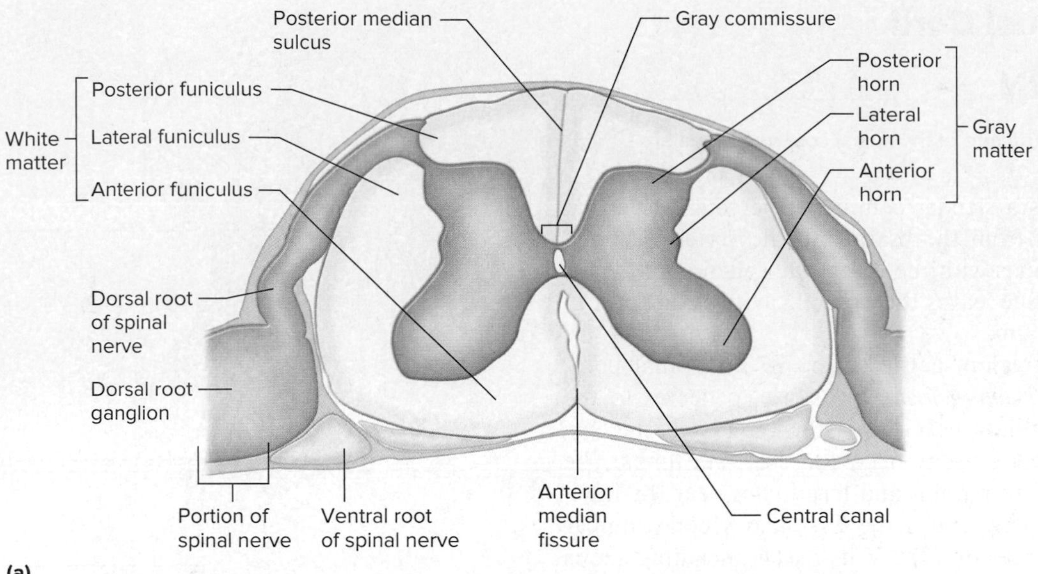

(a)

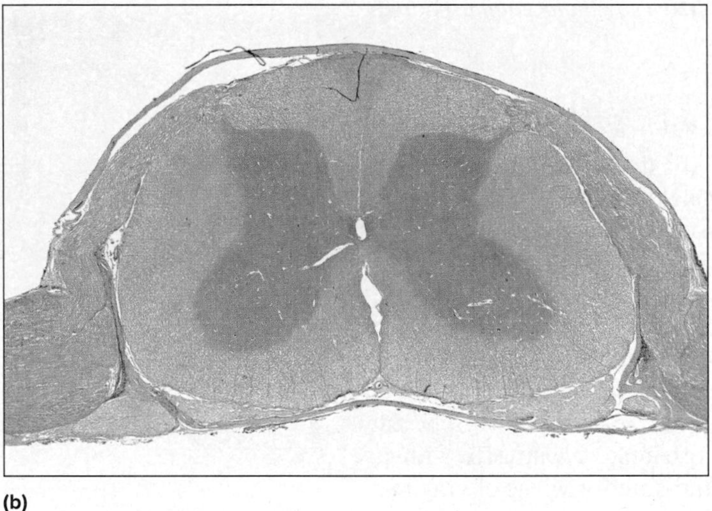

(b)

Figure 9.25 The spinal cord. **(a)** A cross section of the spinal cord. **(b)** A micrograph of a cross section of the spinal cord (10x). APR (b): Carolina Biological Supply Company/Phototake

All the axons in a given tract typically originate from neuron cell bodies in the same part of the nervous system and terminate together in another part. The names that identify tracts often reflect these common origins and terminations. For example, a *spinothalamic tract* begins at the various levels of the spinal cord and conducts sensory impulses associated with the sensations of pain, touch, and temperature to the thalamus of the brain. A *corticospinal tract* originates in the cortex of the brain and conducts motor impulses downward to spinal nerves at various levels of the spinal cord. These impulses control skeletal muscle movements.

Corticospinal tracts are also called *pyramidal tracts* after the pyramid-shaped areas in the medulla oblongata of the brain through which they pass. Other descending tracts, called *extrapyramidal tracts,* control motor activities associated with maintaining balance and posture.

Some axons extend to the base of the spinal cord from the toes. If you stub your toe, a sensory message reaches the spinal cord in under a hundredth of a second.

In addition to providing a pathway for tracts, the spinal cord functions in many reflexes, including the patellar and withdrawal reflexes described previously. These are called **spinal reflexes** because their reflex arcs pass through the spinal cord.

PRACTICE 9.13

1. Describe the structure of the spinal cord.
2. Describe the general functions of the spinal cord.
3. Distinguish between an ascending and a descending tract.

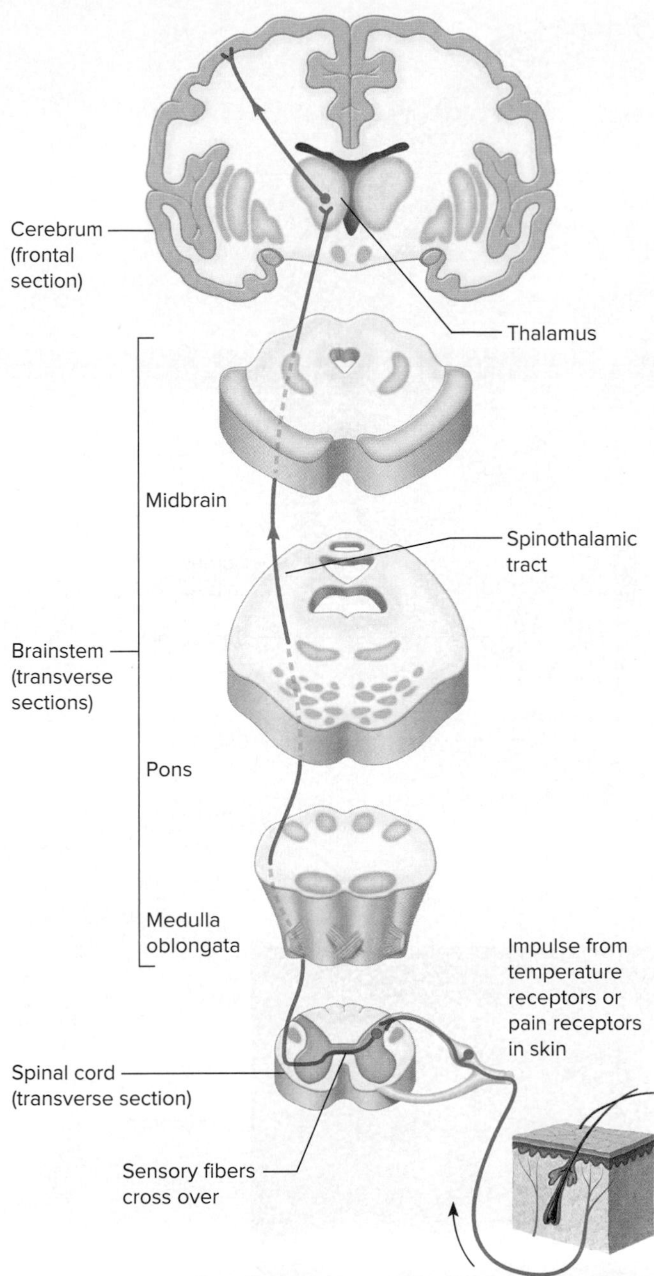

Figure 9.26 Ascending tracts. Some sensory tracts bringing information from skin receptors cross over in the spinal cord and ascend to the brain. Other sensory tracts cross over in the medulla oblongata (not shown).

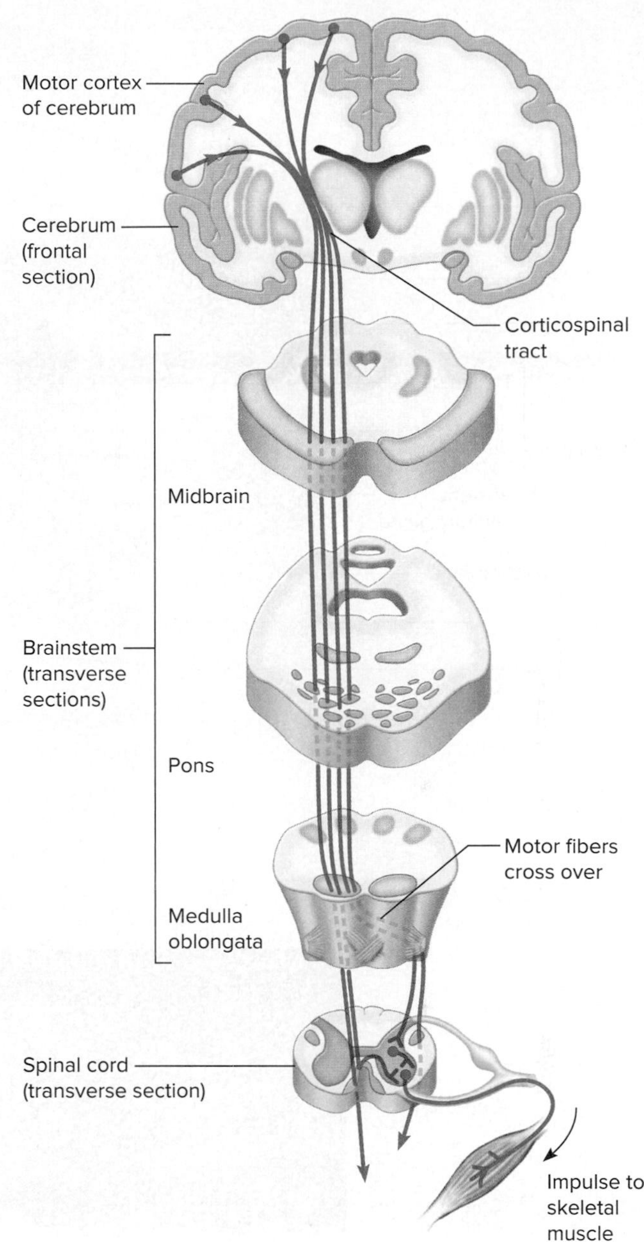

Figure 9.27 Descending tracts. Most of the axons of the corticospinal tract originate in the cerebral cortex, cross over in the medulla oblongata, and descend in the spinal cord. There, they synapse with motor neurons whose axons lead to the spinal nerves that supply skeletal muscles. Some axons in this tract cross over in the spinal cord.

9.14 | Brain

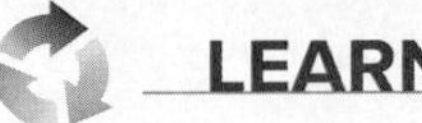

LEARN

1. Name the major parts of the brain and their functions.
2. Distinguish among sensory, association, and motor areas of the cerebral cortex.
3. Describe the location, formation, and function of cerebrospinal fluid.

The **brain** is structured in reverse fashion to the spinal cord. The gray matter is on the outside, surrounding the deeper white matter. It is composed of about 100 billion (10^{11}) multipolar neurons, which communicate with one another and with neurons in other parts of the nervous system. The brain also includes neuroglia, which outnumber the neurons. As figure 9.28 shows, the brain can be divided into four major portions—the cerebrum, the diencephalon, the brainstem, and the cerebellum. The *cerebrum,* the largest part, includes centers associated with sensory and motor functions and provides higher mental functions, including memory and reasoning. The *diencephalon* also processes sensory information. Neural pathways in the *brainstem* connect

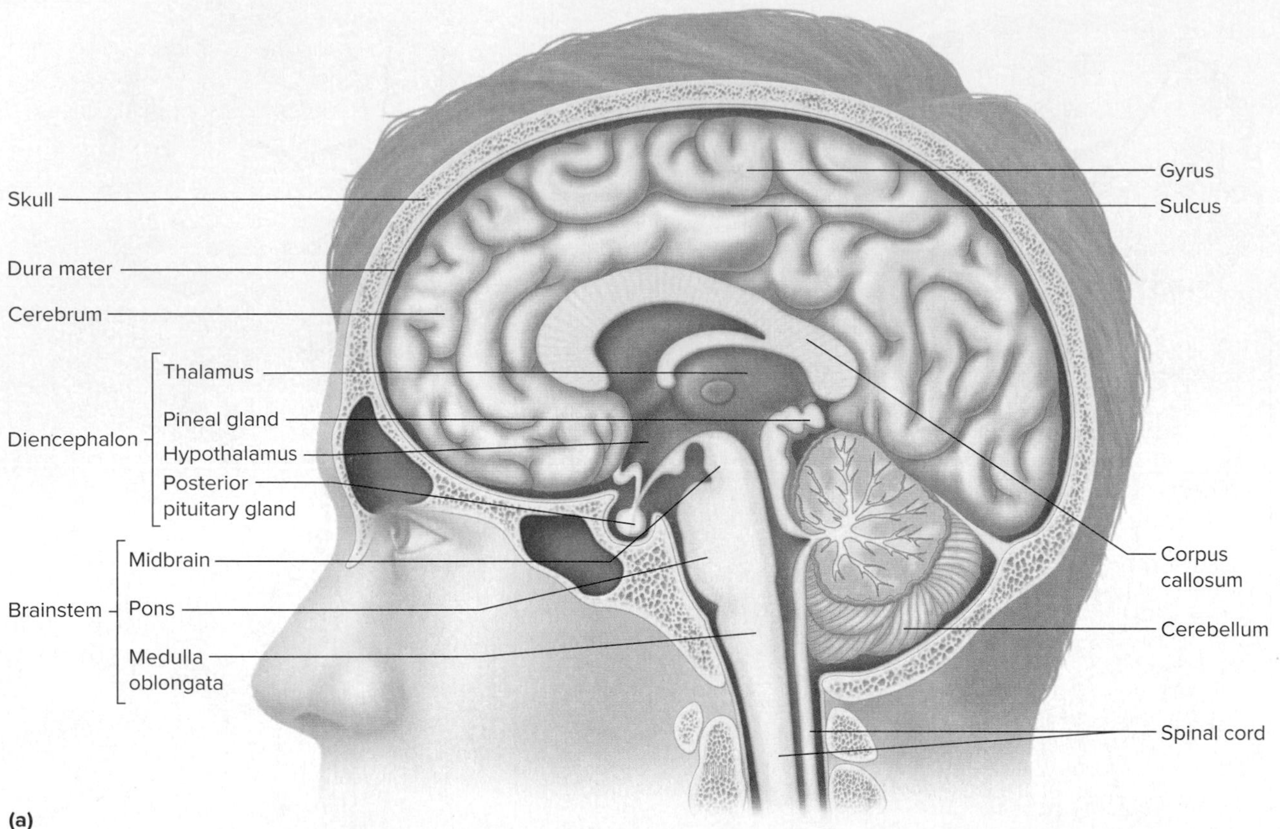

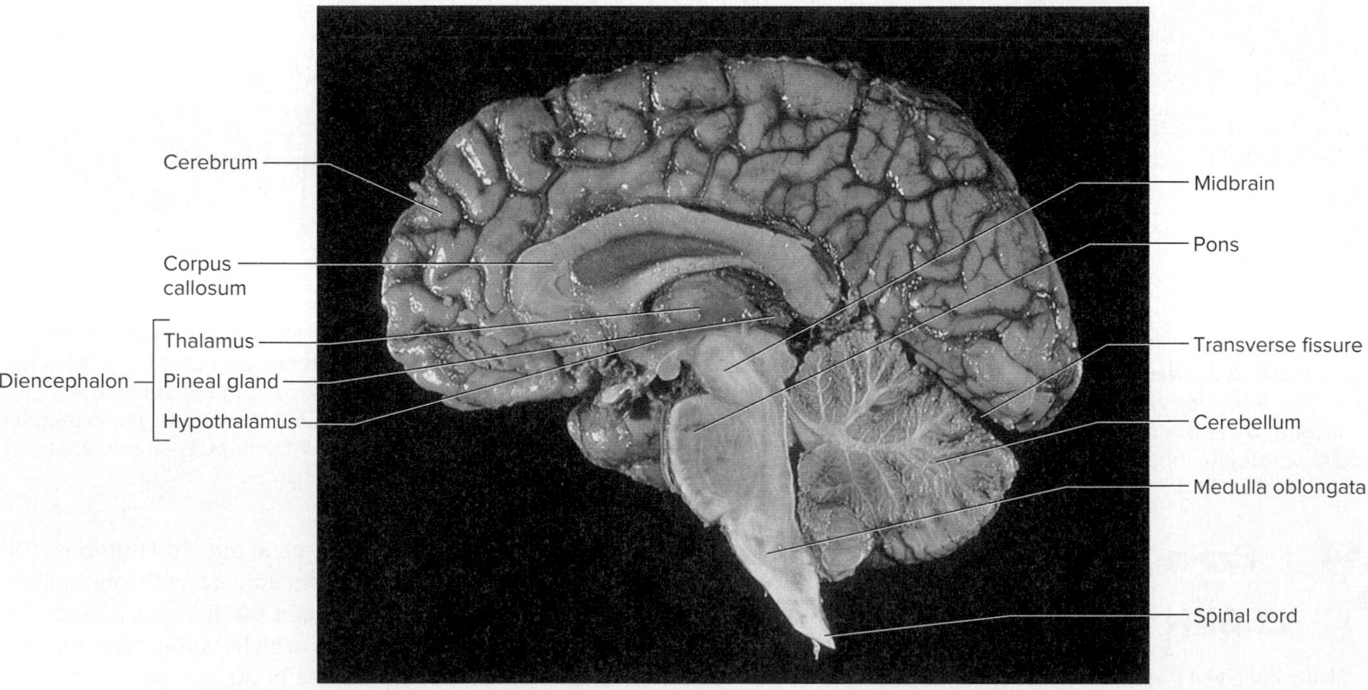

Figure 9.28 Midsagittal section of the brain and spinal cord, including the uncut medial surface of the right cerebral hemisphere. **(a)** The major portions of the brain are the cerebrum, the diencephalon, the brainstem, and the cerebellum. **(b)** Photo of a midsagittal section through a human brain. APR (b): Martin Rotker/Science Source

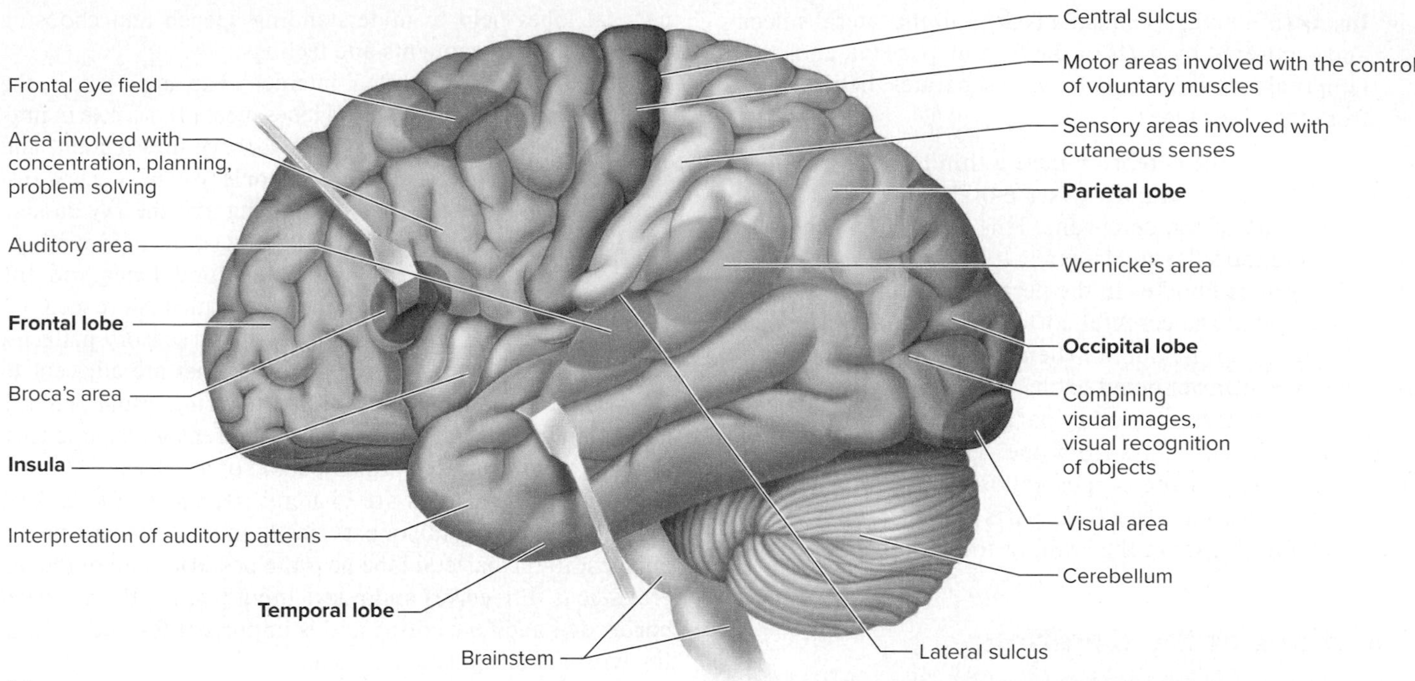

parts of the nervous system and regulate certain visceral activities. The *cerebellum* includes centers of gray matter that coordinate voluntary muscular movements.

Structure of the Cerebrum

The **cerebrum** (ser′ĕ-brum) consists of two large masses called the left and right **cerebral hemispheres** (ser″ĕ-bral hem′ĭ-sfērz), which are essentially mirror images of each other. A broad, flat bundle of axons called the **corpus callosum** (kor′pus kah-lo′sum) connects the cerebral hemispheres. A layer of dura mater (falx cerebri) separates them.

The surface of the cerebrum has many ridges (convolutions) or **gyri** (jahy′-rahy) (singular, *gyrus*), separated by grooves. A shallow groove is called a **sulcus** (sul′kus), and a deep groove is called a **fissure.** The structural organization of these elevations and depressions is complex, but they form similar patterns in all normal brains. For example, a *longitudinal fissure* separates the right and left cerebral hemispheres, a *transverse fissure* separates the cerebrum from the cerebellum, and several sulci divide each hemisphere into lobes.

In a condition called lissencephaly, which means "smooth brain," sulci and gyri are absent. Lissencephaly is associated with intellectual disability, developmental delay, and seizures.

The lobes of the cerebral hemispheres are named after the skull bones they underlie (fig. 9.29). They include:

- **Frontal lobe** The frontal lobe forms the anterior part of each cerebral hemisphere. It is bordered posteriorly by a *central sulcus,* which extends from the longitudinal fissure at a right angle, and inferiorly by a *lateral sulcus,* which extends from the undersurface of the brain along its sides.
- **Parietal lobe** The parietal lobe is posterior to the frontal lobe and separated from it by the central sulcus.

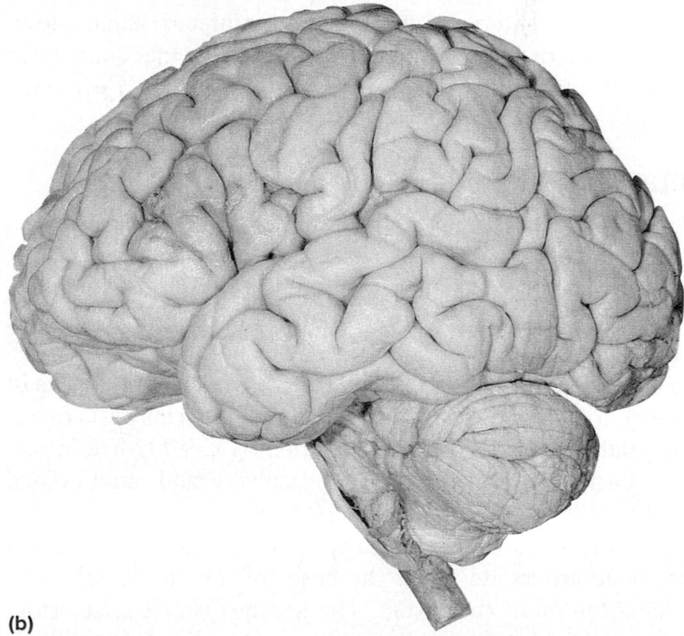
(b)

Figure 9.29 Lateral view of the brain. **(a)** Some sensory, association, and motor areas of the left cerebral cortex. **(b)** Photo of a human brain and part of the spinal cord. APR
(b): Rebecca Gray/McGraw-Hill Education

- **Temporal lobe** The temporal lobe lies below the frontal and parietal lobes and is separated from them by the lateral sulcus.
- **Occipital lobe** The occipital lobe forms the posterior part of each cerebral hemisphere and is separated from the cerebellum by a shelflike extension of dura mater (tentorium cerebelli). The boundary between the occipital lobe and the parietal and temporal lobes is not distinct.

- **Insula** (in'su-lah) The insula is deep in the lateral sulcus and is covered by parts of the frontal, parietal, and temporal lobes. A *circular sulcus* separates the insula from the other lobes.

All lobes of the cerebrum have a thin layer of gray matter called the **cerebral cortex** (ser"ĕ-bral kor'teks). It is the outermost part of the cerebrum. This layer covers the gyri and dips into the sulci and fissures. It contains nearly 75% of all the neuron cell bodies in the nervous system.

Just beneath the cerebral cortex is a mass of white matter that makes up the bulk of the cerebrum. This mass contains bundles of myelinated axons that connect neuron cell bodies of the cortex with other parts of the nervous system. Some of these fibers pass from one cerebral hemisphere to the other by way of the corpus callosum, and others carry sensory or motor impulses from parts of the cortex to areas of gray matter deeper in the brain or to the spinal cord.

Functions of the Cerebrum

The cerebrum provides higher brain functions. It has centers for interpreting sensory impulses arriving from sense organs and centers for initiating voluntary muscular movements. The cerebrum stores the information that constitutes memory and utilizes it to reason. Intelligence and personality also stem from cerebral activity.

Functional Areas of the Cerebral Cortex

Specific regions of the cerebral cortex perform specific functions. Although functions overlap among regions, the cortex can be divided into sensory, association, and motor areas.

Sensory areas in several lobes of the cerebrum interpret impulses that arrive from sensory receptors, producing feelings or sensations. For example, sensations from all parts of the skin (cutaneous senses) arise in the anterior parts of the parietal lobes along the central sulcus (fig. 9.29). The posterior parts of the occipital lobes receive visual input (visual area), and the superior posterior temporal lobes contain the centers for hearing (auditory area). The sensory areas for taste are located near the bases of the lateral sulci and include parts of the insula. The sense of smell arises from centers deep in the temporal lobes.

Sensory fibers from the peripheral nervous system cross over either in the spinal cord (see fig. 9.26) or in the brainstem. Thus, the centers in the right cerebral hemisphere interpret impulses originating from the left side of the body, and vice versa.

Association areas are neither primarily sensory nor primarily motor. They connect with one another and with other brain structures. Association areas analyze and interpret sensory experiences and oversee memory, reasoning, verbalizing, judgment, and emotion. Association areas occupy the anterior portions of the frontal lobes and are widespread in the lateral parts of the parietal, temporal, and occipital lobes (fig. 9.29).

The association areas of the frontal lobes control a number of higher intellectual processes. These include concentrating, planning, complex problem solving, and judging the possible consequences of behavior. Association areas of the parietal lobes help in understanding speech and choosing words to express thoughts and feelings.

Association areas often interact. The area where the occipital, parietal, and temporal lobes meet plays a role in integrating visual, auditory, and other sensory information, and then interpreting a situation. For example, you hear leaves rustling, look up and see branches swaying and the sky darkening, feel the temperature drop, and realize a storm is coming.

The association areas of the temporal lobes and the regions of the posterior ends of the lateral sulcus store memory of visual scenes, music, and other complex sensory patterns. Association areas of the occipital lobes that are adjacent to the visual centers are important in analyzing visual patterns and combining visual images with other sensory experiences, as when you recognize another person or an object.

Not all brain areas are bilateral. *Wernicke's* (ver'nĭ-kēz) *area* is in the temporal lobe, typically in the left hemisphere, adjacent to the parietal lobe near the posterior end of the lateral sulcus. It receives and relays input from both the visual cortex and auditory cortex and is important for understanding written and spoken language.

The primary **motor areas** of the cerebral cortex lie in the frontal lobes, just in front of the central sulcus (fig. 9.29). The nervous tissue in these regions contains many large *pyramidal cells,* named for their pyramid-shaped cell bodies. These cells are also termed *upper motor neurons,* because of their location.

Impulses from the pyramidal cells travel downward through the brainstem and into the spinal cord on the corticospinal tracts (see fig. 9.27). Here they form synapses with *lower motor neurons* whose axons leave the spinal cord and reach skeletal muscle fibers. Most of the axons in these tracts cross over from one side of the brain to the other within the brainstem. As a result, the motor area of the right cerebral hemisphere generally controls skeletal muscles on the left side of the body, and vice versa.

In addition to the primary motor areas, certain other regions of the frontal lobe affect motor functions. For example, a region called the **motor speech area,** or *Broca's* (bro'kahz) *area,* is in the frontal lobe, typically in the left hemisphere, just anterior to the primary motor cortex and superior to the lateral sulcus. This area provides motor instructions to muscles necessary for speech (fig. 9.29).

In the superior part of the frontal lobe is a region called the *frontal eye field.* The motor cortex in this area controls voluntary movements of the eyes and eyelids. Another region just anterior to the primary motor area controls learned movement patterns that make skills such as writing possible (fig. 9.29).

The functions of the insula are not as well known as those of the other lobes, because its location deep within the cerebrum makes it impossible to study with surface electrodes. However, studies that use functional MRI scanning suggest that the insula serves as a crossroads for translating sensory information into appropriate emotional responses, such as feeling disgust at the sight of something unpleasant, or a feeling of joy when hearing a symphony or when biting into a slice of pizza. Some researchers hypothesize that in some complex way the insula is responsible for some of the qualities that make us

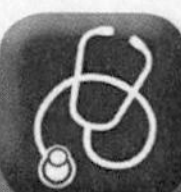

CLINICAL APPLICATION 9.2
Cerebral Cortex Injuries

The effects of injuries to the cerebral cortex depend on the location and extent of the damage. For example, injury to the motor areas of one frontal lobe causes partial or complete paralysis on the opposite side of the body. Damage to the association areas of the frontal lobe may impair concentration on complex mental tasks, making a person appear disorganized and easily distracted. Damage to association areas of the temporal lobes may impair recognition of printed words or the ability to arrange words into meaningful thoughts.

human. Clinical Application 9.2 discusses some injuries to the association areas of the cerebral cortex.

PRACTICE 9.14

1. List the major divisions of the brain.
2. Describe the cerebral cortex.
3. Describe the major functions of the cerebrum.
4. Locate the major functional areas of the cerebral cortex.

Hemisphere Dominance

Both cerebral hemispheres participate in basic functions, such as receiving and analyzing sensory impulses, controlling skeletal muscles, and storing memory. However, in most individuals, one side of the cerebrum is the **dominant hemisphere,** controlling the ability to use and understand language.

In most people, the left hemisphere is dominant for the language-related activities of speech, writing, and reading, and for complex intellectual functions requiring verbal, analytical, and computational skills. In others, the right hemisphere is dominant for language-related abilities, or the hemispheres are equally dominant. Broca's area in the dominant hemisphere controls the muscles that function in speaking.

In addition to carrying on basic functions, the nondominant hemisphere specializes in nonverbal functions, such as motor tasks that require orientation of the body in space, understanding and interpreting musical patterns, and nonverbal visual experiences. The nondominant hemisphere also controls emotional and intuitive thinking.

Nerve fibers of the corpus callosum, which connect the cerebral hemispheres, allow the dominant hemisphere to control the motor cortex of the nondominant hemisphere (see fig. 9.28). These fibers also transfer sensory information reaching the nondominant hemisphere to the dominant one, where the information can be used in decision making.

Basal Nuclei

Deep within each cerebral hemisphere are several *nuclei* (regions of gray matter) called **basal nuclei** (or *basal ganglia*) (fig. 9.30). Technically, however, a ganglion is a cluster of neuron cell bodies in the peripheral nervous system. They

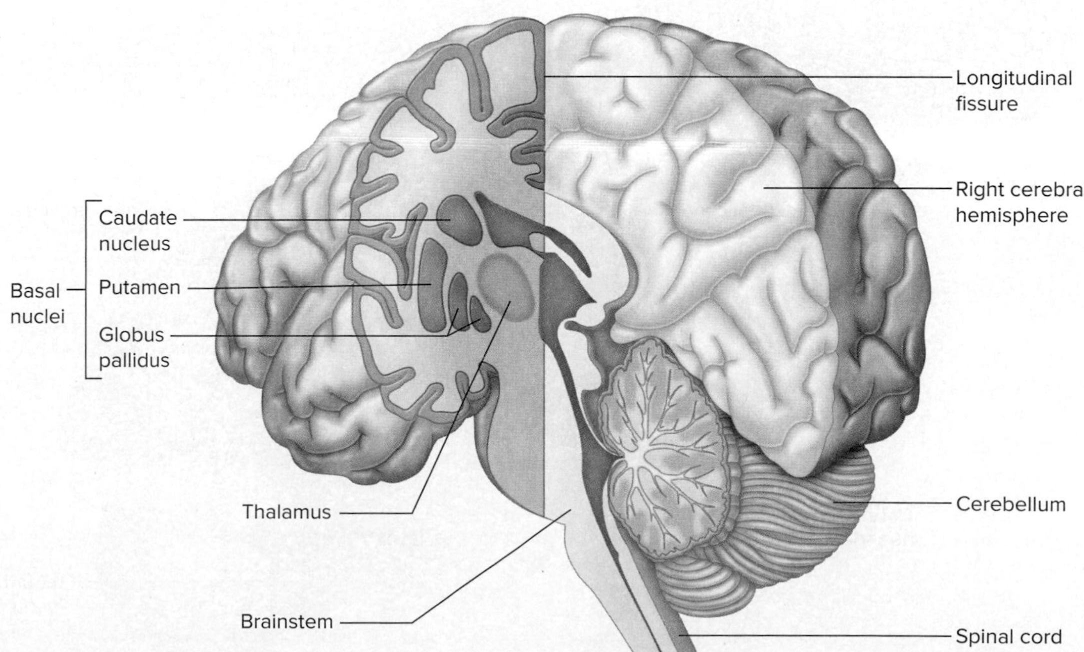

Figure 9.30 A frontal (coronal) section of the left cerebral hemisphere (posterior view) reveals some of the basal nuclei. (Note that the hypothalamus is more anterior and is not visible in this section.) APR

include the *caudate nucleus,* the *putamen,* and the *globus pallidus.* The basal nuclei produce the inhibitory neurotransmitter *dopamine.* The neurons of the basal nuclei interact with other brain areas, including the motor cortex, thalamus, and cerebellum. These interactions, through a combination of stimulation and inhibition, facilitate voluntary movement.

The signs of Parkinson disease and Huntington disease result from altered activity of basal nuclei neurons. In Parkinson disease, nearby neurons release less dopamine, and the basal nuclei become overactive, inhibiting movement. In Huntington disease, basal nuclei neurons gradually deteriorate, resulting in unrestrained movement.

Ventricles and Cerebrospinal Fluid

Interconnected cavities called **ventricles** lie within the cerebral hemispheres and brainstem (fig. 9.31). These spaces are continuous with the central canal of the spinal cord, and similar to it, they contain cerebrospinal fluid (CSF).

The largest ventricles are the *lateral ventricles* (first and second ventricles), which extend into the cerebral hemispheres and occupy parts of the frontal, temporal, and occipital lobes. A narrow space that constitutes the *third ventricle* is in the midline of the brain, beneath the corpus callosum. This ventricle communicates with the lateral ventricles through openings (interventricular foramina) in its anterior end. The *fourth ventricle* is in the brainstem just anterior to the cerebellum. A narrow canal, the *cerebral aqueduct,* connects it to the third ventricle and passes lengthwise through the brainstem. The fourth ventricle is continuous with the central canal of the spinal cord and has openings in its wall that lead into the subarachnoid space of the meninges.

Tiny, reddish, cauliflower-like masses containing specialized capillaries from the pia mater, called **choroid plexuses** (plek'sus-ez), secrete cerebrospinal fluid (fig. 9.32). These structures project into all four ventricles, but the lateral ventricles produce most of the cerebrospinal fluid. From there, it circulates slowly into the third and fourth ventricles. Small amounts enter the central canal of the spinal cord, but most of the cerebrospinal fluid circulates through the subarachnoid space of both the brain and the spinal cord by passing through the openings in the wall of the fourth ventricle near the cerebellum. It completes its circuit by being reabsorbed into the blood.

Cerebrospinal fluid completely surrounds the brain and spinal cord because it occupies the subarachnoid space of the meninges. In effect, these organs float in the fluid, which supports and protects them by absorbing forces that might

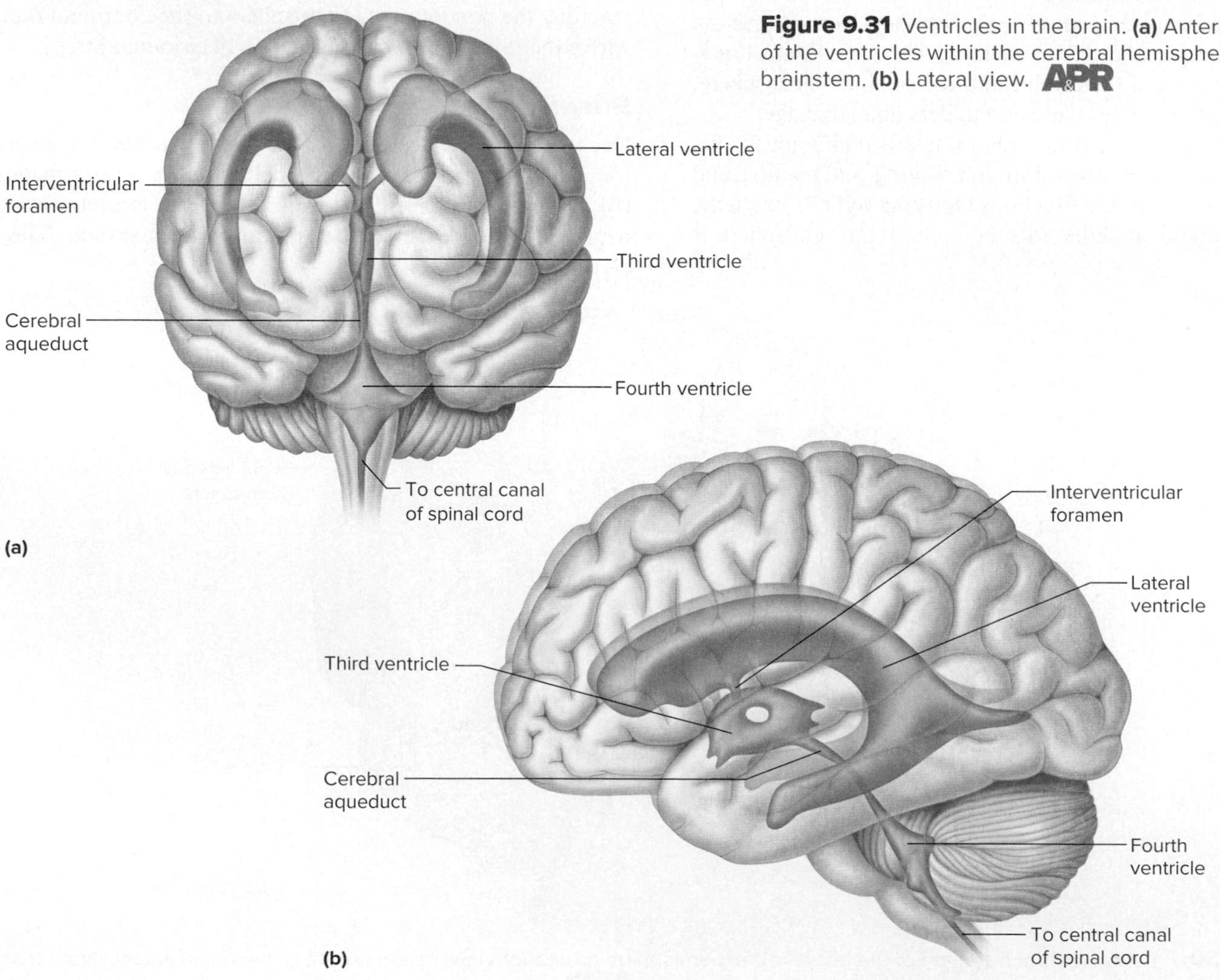

Figure 9.31 Ventricles in the brain. **(a)** Anterior view of the ventricles within the cerebral hemispheres and brainstem. **(b)** Lateral view. APR

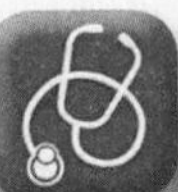

CLINICAL APPLICATION 9.3

Impaired Cerebrospinal Fluid Circulation

The fluid pressure in the ventricles normally remains relatively constant, because cerebrospinal fluid is secreted and reabsorbed continuously and at equal rates. An infection, a tumor, or a blood clot can interfere with fluid circulation, increasing pressure in the ventricles and thus in the cranial cavity (intracranial pressure). This buildup of pressure can injure the brain by forcing it against the rigid skull or partially through the foramen magnum at the base of the skull, causing a *herniation*.

A *lumbar puncture* (spinal tap) measures the pressure of cerebrospinal fluid. A very thin hollow needle is inserted into the subarachnoid space between the third and fourth or between the fourth and fifth lumbar vertebrae and an instrument called a *manometer* measures the pressure.

otherwise jar and damage them. Cerebrospinal fluid also maintains a stable ionic concentration in the CNS and provides a pathway to the blood for wastes. Clinical Application 9.3 discusses impaired cerebrospinal fluid circulation.

PRACTICE 9.14

5. What is hemisphere dominance?
6. What are the major functions of the dominant hemisphere? The nondominant one?
7. Where are the ventricles of the brain?
8. Describe the circulation of cerebrospinal fluid.

Diencephalon

The **diencephalon** (di″en-sef′ah-lon) is located between the cerebral hemispheres and above the midbrain. It surrounds the third ventricle and is composed largely of gray matter. Within the diencephalon, a dense mass called the **thalamus** bulges into the third ventricle from each side (see fig. 9.30). Another region of the diencephalon that includes many nuclei (masses of gray matter) is the **hypothalamus.** It lies below the thalamus and forms the lower walls and floor of the third ventricle.

The thalamus is a selective gateway for sensory impulses ascending from other parts of the nervous system

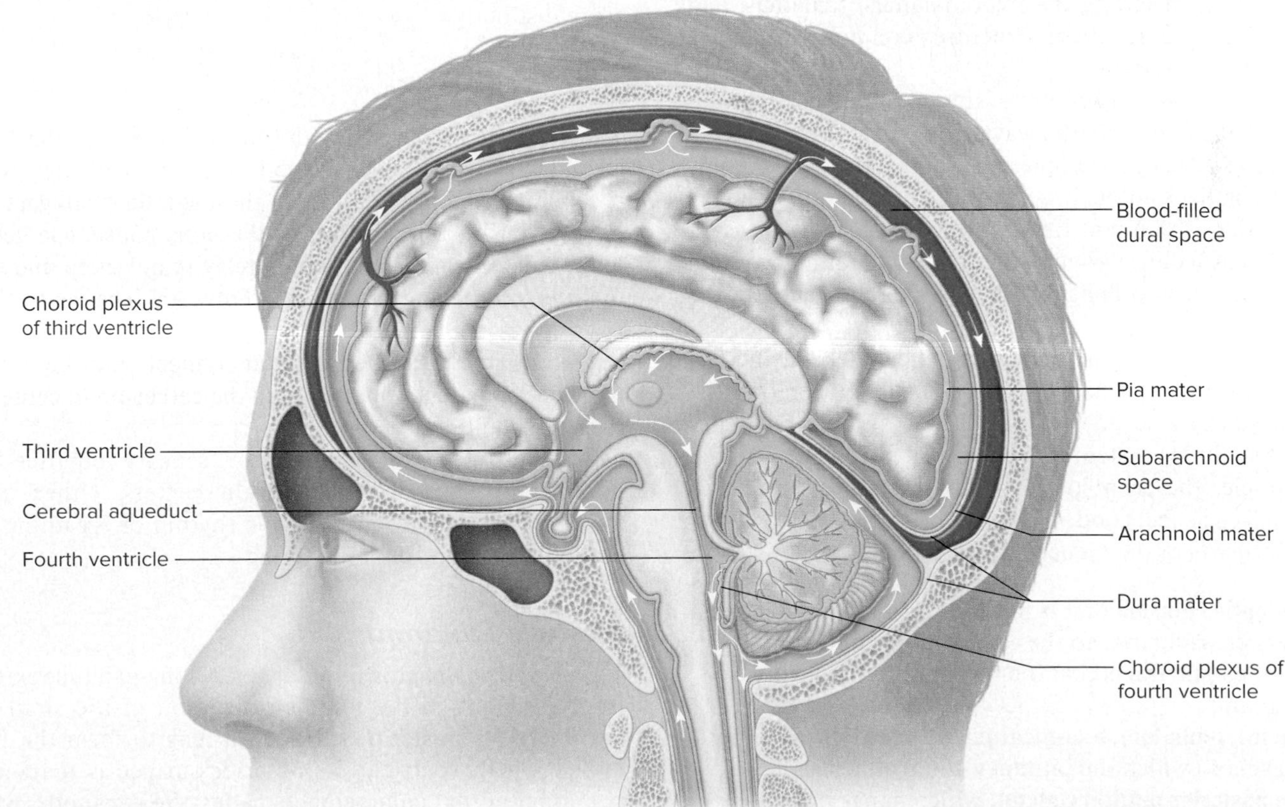

Figure 9.32 The choroid plexuses in the walls of the ventricles secrete cerebrospinal fluid. The fluid circulates through the ventricles and central canal, enters the subarachnoid space, and is reabsorbed into the blood. APR

to the cerebral cortex. It receives all sensory impulses (except those associated with the sense of smell) and channels them to the appropriate regions of the cortex for interpretation. The thalamus produces a general awareness of certain sensations, such as pain, touch, and temperature, and the cerebral cortex pinpoints the origin of the sensory stimulation.

Various pathways connect the hypothalamus to the cerebral cortex, thalamus, and other parts of the brainstem so that it can receive impulses from them and send impulses to them. The hypothalamus helps maintain homeostasis by regulating a variety of visceral activities and by linking the nervous and endocrine systems.

The hypothalamus regulates:

- Heart rate and arterial blood pressure
- Body temperature
- Water and electrolyte balance
- Control of hunger and body weight
- Control of movements and glandular secretions of the stomach and intestines
- Production of hormones that stimulate the pituitary gland to secrete pituitary hormones
- Sleep and wakefulness

Structures in the general region of the diencephalon also control emotional responses. For example, regions of the cerebral cortex in the medial parts of the frontal and temporal lobes interconnect with a number of deep masses of gray matter, including the hypothalamus, thalamus, and basal nuclei. Together these structures compose a complex called the **limbic system.**

The limbic system controls emotional experience and expression. It can modify the way a person acts by producing such feelings as fear, anger, pleasure, and sorrow. The limbic system recognizes upsets in a person's physical or psychological condition that might threaten life. By causing pleasant or unpleasant feelings about experiences, the limbic system guides a person into behavior that is likely to increase the chance of survival.

A whiff of a certain scent may elicit vivid memories, because sensory information from olfactory receptors (the sense of smell) also goes to the limbic system. Olfactory input to the limbic system is also why odors can alter mood. For example, the scent of just-mowed grass or an ocean breeze makes us feel good.

Other parts of the diencephalon include:

- The **optic chiasma** that is formed by some optic nerve fibers crossing over to the opposite side of the brain and the **optic tracts** that then lead to the visual areas of the brain;
- The **infundibulum,** a conical process behind the optic chiasma to which the pituitary gland attaches;
- The **posterior pituitary gland,** which hangs from the floor of the hypothalamus;
- The **mammillary bodies,** which appear as two rounded structures behind the infundibulum; and
- The **pineal gland** (pin'e-al gland), a cone-shaped structure attached to the upper part of the diencephalon (see section 11.9, Pineal, Thymus, and Other Glands).

Brainstem

The **brainstem** is a bundle of nervous tissue that connects the cerebrum, diencephalon, and cerebellum to the spinal cord. It consists of many tracts and several nuclei. The parts of the brainstem include the midbrain, pons, and medulla oblongata (see figs. 9.28 and 9.33).

Midbrain

The **midbrain** is a short section of the brainstem between the diencephalon and the pons (see fig. 9.28). It contains bundles of myelinated axons that join lower parts of the brainstem and spinal cord with higher parts of the brain. Two prominent bundles of axons in the anterior of the midbrain are the corticospinal tracts. They are the main motor pathways between the cerebrum and lower parts of the nervous system.

The midbrain includes several masses of gray matter that serve as reflex centers. For example, the midbrain contains the centers for certain visual reflexes, such as those responsible for allowing the eyes to look at a stationary object as the head turns. It also contains the auditory reflex centers that enable a person to move the head to hear sounds more distinctly.

Pons

The **pons** (ponz) occupies the full thickness of the brainstem, but is most visible anteriorly as a rounded bulge, where it separates the midbrain from the medulla oblongata (see fig. 9.28). The anterior part of the pons consists largely of longitudinal nerve fibers, which relay impulses to and from the medulla oblongata and the cerebrum. The ventral part of the pons also has large, transverse bundles of nerve fibers that wrap around to the back and connect with the cerebellum. They conduct impulses from the cerebrum to centers in the cerebellum.

Several nuclei of the pons relay sensory impulses from peripheral nerves to higher brain centers. Other nuclei may contribute to controlling the rhythm of breathing (see section 16.4, Control of Breathing).

Medulla Oblongata

The **medulla oblongata** (mĕ-dul'ah ob"long-gah'tah) extends from the pons to the foramen magnum of the skull (see fig. 9.28). Its posterior surface flattens to form the floor of the fourth ventricle. Its anterior surface is marked by two longitudinal enlargements called the pyramids, which contain the corticospinal tracts. Most of the fibers of the corticospinal tracts cross over at this level (see figs. 9.27 and 9.33).

(a) Ventral view

(b) Dorsal view

Regions of the brainstem

Diencephalon | Midbrain | Pons | Medulla oblongata

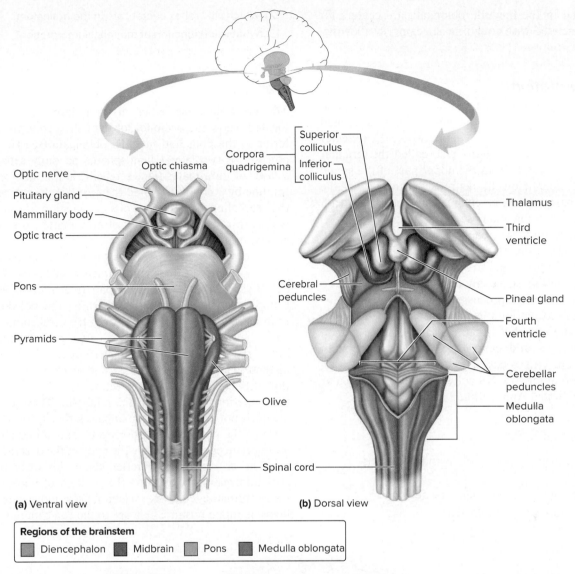

Figure 9.33 The brainstem. **(a)** Ventral view of the brainstem. **(b)** Dorsal view of the brainstem with the cerebellum removed, exposing the fourth ventricle. APR

PRACTICE FIGURE 9.33

What is the relative position of the fourth ventricle to the third ventricle?

Answer can be found in Appendix E.

All of the ascending and descending nerve fibers connecting the brain and spinal cord must pass through the medulla oblongata because of its location. In the spinal cord, the white matter surrounds a central mass of gray matter. Here in the medulla oblongata, however, nerve fibers separate the gray matter into nuclei, some of which relay ascending impulses to the other side of the brainstem and then on to higher brain centers. Other nuclei in the medulla oblongata control vital visceral activities. These centers include:

- The **cardiac center.** Impulses originating in the cardiac center are conducted to the heart on peripheral nerves, altering heart rate.
- The **vasomotor center.** Certain neurons of the vasomotor center initiate impulses that travel to smooth muscle in the walls of certain blood vessels and stimulate the smooth muscle to contract. This constricts the blood vessels (vasoconstriction), raising blood pressure. Other neurons of the vasomotor center produce the opposite effect—dilating blood vessels (vasodilation) and consequently dropping blood pressure.
- The **respiratory center.** Groups of neurons in the respiratory center maintain breathing rhythm and adjust the rate and depth of breathing.

Still other nuclei in the medulla oblongata are centers for the reflexes associated with coughing, sneezing, swallowing, and vomiting.

Reticular Formation

Scattered throughout the medulla oblongata, pons, and midbrain is a complex network of nerve fibers associated with tiny islands of gray matter. This network, the **reticular formation** (rĕ-tik′u-lar for-ma′shun), also called the reticular activating system, extends from the upper part of the spinal cord into the diencephalon. Its neurons join centers of the hypothalamus, basal nuclei, cerebellum, and cerebrum with all of the major ascending and descending tracts.

When sensory impulses reach the reticular formation, it responds by activating the cerebral cortex into a state of wakefulness. Without this arousal, the cortex remains unaware of stimulation and cannot interpret sensory information or carry on thought processes. Thus, decreased activity in the reticular formation results in sleep. If the reticular formation is injured so that it cannot function, the person remains unconscious and cannot be aroused, even with strong stimulation. This is called a comatose state. Barbiturate drugs, which dampen CNS activity, affect the reticular formation (see Clinical Application 9.4).

PRACTICE 9.14

9. What are the major functions of the thalamus? The hypothalamus?
10. How may the limbic system influence behavior?
11. List the structures of the brainstem.
12. Which vital reflex centers are in the brainstem?
13. What is the function of the reticular formation?

Cerebellum

The **cerebellum** (ser″ĕ-bel′um) is a large mass of tissue located below the occipital lobes of the cerebrum and posterior to the pons and medulla oblongata (see fig. 9.28). It consists of two lateral hemispheres partially separated by a layer of dura mater (falx cerebelli) and connected in the midline by a structure called the *vermis.* Like the cerebrum, the cerebellum is composed of a deep region of white matter with a thin, convoluted layer of gray matter, the **cerebellar cortex,** on its surface.

The cerebellum communicates with other parts of the CNS by means of three pairs of tracts called *cerebellar peduncles* (figs. 9.33 and 9.34). One pair (the inferior peduncles) brings sensory information concerning the position of the limbs, joints, and other body parts to the cerebellum. Another pair (the middle peduncles) conducts signals from the cerebral cortex to the cerebellum concerning the desired positions of these parts. After integrating and analyzing this information, the cerebellum sends correcting impulses via a third pair (the superior peduncles) to the midbrain. These corrections are incorporated into motor impulses that travel downward through the pons, medulla oblongata, and spinal cord in the appropriate patterns to move the body in the desired way.

The cerebellum is a reflex center for integrating sensory information concerning the position of body parts and for coordinating complex skeletal muscle movements. It also helps maintain posture. Damage to the cerebellum is likely to

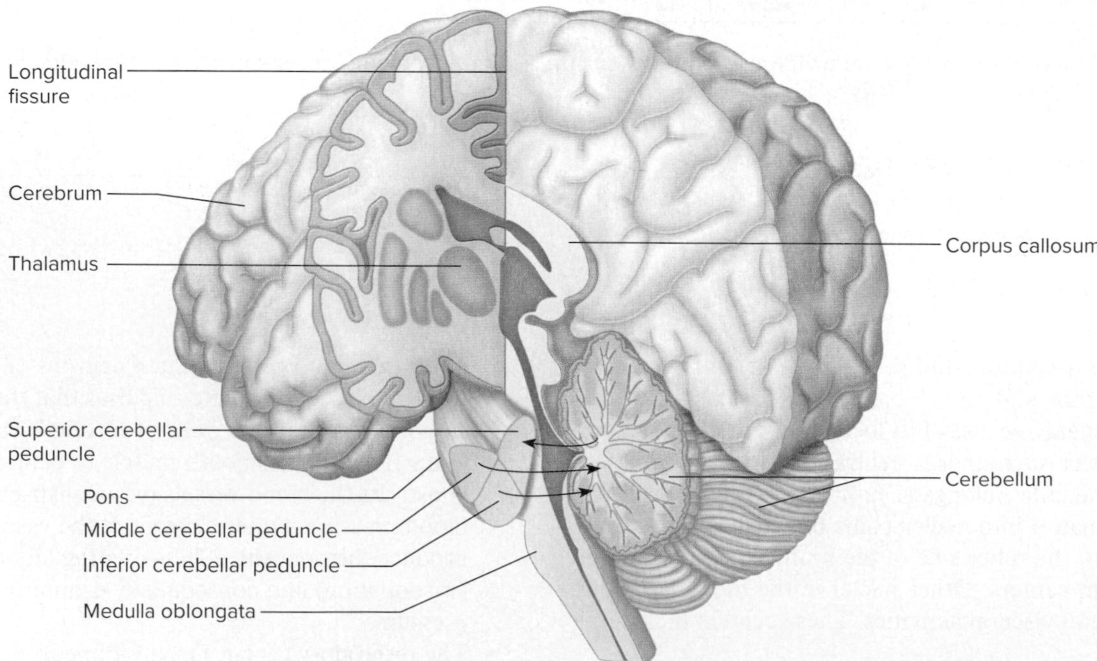

Figure 9.34 A midsagittal section through the cerebellum, which is located below the occipital lobes of the cerebrum. The cerebellum communicates with other parts of the nervous system by means of the cerebellar peduncles. APR

result in tremors, inaccurate movements of voluntary muscles, loss of muscle tone, a staggering walk, and loss of balance.

PRACTICE 9.14

14. Where is the cerebellum located?
15. What are the major functions of the cerebellum?

9.15 | Peripheral Nervous System

LEARN

1. List the major parts of the peripheral nervous system.
2. Name the cranial nerves, and list their major functions.
3. Describe the structure of a spinal nerve.

The peripheral nervous system (PNS) consists of **nerves** that branch from the CNS and connect it to other body parts. The PNS includes the cranial nerves, which arise from the brain, and the spinal nerves, which arise from the spinal cord. The PNS includes both sensory and motor divisions (see fig. 9.2).

The motor portion of the PNS can be subdivided into the somatic and autonomic nervous systems. Generally, the **somatic** (sō-mat′ik) **nervous system** consists of the cranial and spinal nerve fibers that connect the CNS to the skin and skeletal muscles; it plays a role in conscious activities. The **autonomic** (aw″to-nom′ik) **nervous system** includes cranial and spinal nerve fibers that connect the CNS to viscera, such as the heart, stomach, intestines, and glands; it controls subconscious activities. Table 9.5 outlines the subdivisions of the nervous system.

TABLE 9.5 Subdivisions of the Nervous System

1. Central nervous system (CNS)
   a. Brain
   b. Spinal cord
2. Peripheral nervous system (PNS)
   a. Cranial nerves arising from the brain and brainstem
      (1) Sensory fibers connecting peripheral sensory receptors to the CNS
      (2) Somatic fibers connecting to skin and skeletal muscles
      (3) Autonomic fibers connecting to viscera
   b. Spinal nerves arising from the spinal cord
      (1) Sensory fibers connecting peripheral sensory receptors to the CNS
      (2) Somatic fibers connecting to skin and skeletal muscles
      (3) Autonomic fibers connecting to viscera

Cranial Nerves

Twelve pairs of **cranial nerves** are located on the underside of the brain (fig. 9.35). The first pair is associated with the cerebrum and the second pair with the thalamus. All of the other cranial nerves are associated with the brainstem (cranial nerve XI is an exception, described later in this section). They pass from their sites of origin through foramina of the skull and lead to parts of the head, neck, and trunk.

Most of the cranial nerves are mixed nerves containing both sensory and motor nerve fibers. However, some cranial nerves associated with special senses, such as smell and vision, contain only sensory fibers. Other cranial nerves that affect muscles and glands are composed primarily of motor fibers. Table 9.6 identifies which cranial nerves are sensory, motor, or mixed.

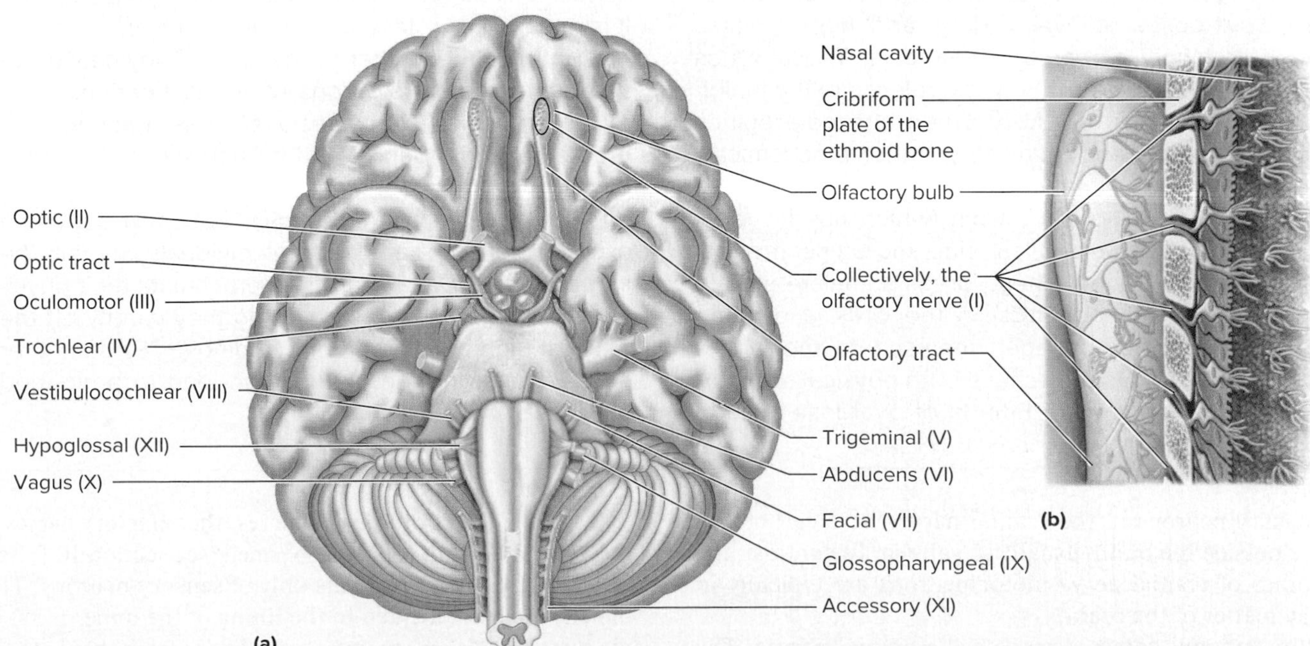

Figure 9.35 The cranial nerves, except for the first two pairs, arise from the brainstem (cranial nerve XI is an exception, described in the text). **(a)** They are identified by numbers indicating their order, by their function, or by the general distribution of their fibers. **(b)** The collection of axons passing through the cribriform plate comprise the olfactory nerve (I). APR

CLINICAL APPLICATION 9.4
Drug Abuse

Drug addiction is defined by the National Institute on Drug Abuse as "a chronic, relapsing brain disease that is characterized by compulsive drug seeking and use, despite harmful consequences." Stopping drug use while addicted causes intense, unpleasant withdrawal symptoms. Prolonged and repeated abuse of a drug may also result in *drug tolerance,* in which the physiological response to a particular dose of the drug becomes less intense over time. Drug tolerance results as the drug increases synthesis of certain liver enzymes, which metabolize the drug more rapidly, so that the addicted individual needs the next dose sooner. Drug tolerance also arises from physiological changes that lessen the drug's effect on its target cells. The most commonly abused drugs are CNS depressants ("downers"), CNS stimulants ("uppers"), hallucinogens, and anabolic steroids (see Clinical Application 8.1).

CNS depressants include barbiturates, benzodiazepines, opiates, and cannabinoids. *Barbiturates* act uniformly throughout the brain, but the reticular formation is particularly sensitive to their effects. CNS depression occurs due to inhibited secretion of certain excitatory and inhibitory neurotransmitters. Effects range from mild calming of the nervous system (sedation) to sleep, loss of sensory sensations (anesthesia), respiratory distress, cardiovascular collapse, and death.

The *benzodiazepines,* such as diazepam, depress activity in the limbic system and the reticular formation. Low doses relieve anxiety, and higher doses cause sedation, sleep, or anesthesia. These drugs increase either the activity or the release of the inhibitory neurotransmitter GABA. When benzodiazepines are metabolized, they may form other biochemicals that have depressing effects.

The *opiates* include heroin (which has no legal use in the United States), codeine, morphine, meperidine, and methadone. These drugs stimulate certain receptors (opioid receptors) in the CNS, and when taken in prescribed dosages, they sedate and relieve pain (analgesia). Opiates cause both physical and psychological dependence. Effects of overdose include a feeling of well-being (euphoria), respiratory distress, convulsions, coma, and possible death. On the other hand, these drugs are very important in treating chronic, severe pain. For example, cancer patients find pain relief with oxycodone, which is taken twice daily in a timed-release pill.

The *cannabinoids* include marijuana and hashish, both derived from the hemp plant. Hashish is several times more potent than marijuana. These drugs depress higher brain centers and release lower brain centers from the normal inhibitory influence of the higher centers. This induces an anxiety-free state, characterized by euphoria and a distorted perception of time and space. *Hallucinations* (sensory perceptions that have no external stimuli), respiratory distress, and vasomotor depression may occur with higher doses.

CNS stimulants include cocaine (one form is "crack") and amphetamines. These drugs have great abuse potential and may quickly produce psychological dependence. Cocaine, especially when smoked or inhaled, produces euphoria but may also change personality, cause seizures, and constrict certain blood vessels, leading to sudden death from stroke or cardiac arrhythmia. Cocaine's very rapid effect, and perhaps its addictiveness, reflect its rapid entry and metabolism in the brain. Cocaine arrives at the basal nuclei in four to six minutes and is mostly cleared within thirty minutes. The drug inhibits transporter molecules that remove dopamine from synapses after it is released. "Ecstasy" is an amphetamine.

Hallucinogens alter perceptions. They cause *illusions,* which are distortions of vision, hearing, taste, touch, and smell; *synesthesia,* such as "hearing" colors or "feeling" sounds; and hallucinations. The most commonly abused and most potent hallucinogen is lysergic acid diethylamide (LSD). LSD may act as an excitatory neurotransmitter. Individuals under the influence of LSD may greatly overestimate their physical capabilities, such as believing they can fly off the top of a high building. Phencyclidine (PCP) is another abused hallucinogen. Its use can lead to prolonged psychosis.

Sensory neurons in the cranial nerves have cell bodies that are outside the brain, usually in ganglia. In contrast, the cell bodies of cranial nerve motor neurons are typically in the gray matter of the brain.

Numbers and names designate the cranial nerves. The numbers indicate the order, superior to inferior, in which the nerves arise from the brain, and the names describe their primary functions or the general distribution of their fibers (fig. 9.35).

The first pair of cranial nerves, the **olfactory nerves (I),** are associated with the sense of smell (see section 10.5, Sense of Smell) and contain axons only of sensory neurons. These bipolar neurons, located in the lining of the upper nasal cavity, serve as *olfactory receptor cells.* Axons from these receptor cells pass upward through the cribriform plates of the ethmoid bone. They conduct impulses to the olfactory neurons in the *olfactory bulbs,* which are extensions of the cerebral cortex just beneath the frontal lobes (see fig. 10.4). Sensory

TABLE 9.6 Functions of the Cranial Nerves APR

Nerve	Type	Function
I Olfactory	Sensory	Sensory fibers conduct impulses associated with the sense of smell.
II Optic	Sensory	Sensory fibers conduct impulses associated with the sense of vision.
III Oculomotor	Primarily motor	Motor fibers conduct impulses to muscles that raise eyelids, move eyes, adjust the amount of light entering the eyes, and focus lenses. Some sensory fibers conduct impulses associated with the condition of muscles.
IV Trochlear	Primarily motor	Motor fibers conduct impulses to muscles that move the eyes. Some sensory fibers conduct impulses associated with the condition of muscles.
V Trigeminal	Mixed	
Ophthalmic division		Sensory fibers conduct impulses from the surface of the eyes, tear glands, scalp, forehead, and upper eyelids.
Maxillary division		Sensory fibers conduct impulses from the upper teeth, upper gum, upper lip, lining of the palate, and skin of the face.
Mandibular division		Sensory fibers conduct impulses from the skin of the jaw, lower teeth, lower gum, and lower lip. Motor fibers conduct impulses to muscles of mastication and to muscles in the floor of the mouth.
VI Abducens	Primarily motor	Motor fibers conduct impulses to muscles that move the eyes. Some sensory fibers conduct impulses associated with the condition of muscles.
VII Facial	Mixed	Sensory fibers conduct impulses associated with taste receptors of the anterior tongue. Motor fibers conduct impulses to muscles of facial expression, tear glands, and salivary glands.
VIII Vestibulocochlear	Sensory	
Vestibular branch		Sensory fibers conduct impulses associated with the sense of equilibrium.
Cochlear branch		Sensory fibers conduct impulses associated with the sense of hearing.
IX Glossopharyngeal	Mixed	Sensory fibers conduct impulses from the pharynx, tonsils, posterior tongue, and carotid arteries. Motor fibers conduct impulses to muscles of the pharynx used in swallowing and to salivary glands.
X Vagus	Mixed	Somatic motor fibers conduct impulses to muscles associated with speech and swallowing; autonomic motor fibers conduct impulses to the heart, smooth muscle, and glands in the thorax and abdomen. Sensory fibers conduct impulses from the pharynx, larynx, esophagus, and viscera of the thorax and abdomen.
XI Accessory	Primarily motor	
Cranial branch		Motor fibers conduct impulses to muscles of the soft palate, pharynx, and larynx.
Spinal branch		Motor fibers conduct impulses to muscles of the neck and back.
XII Hypoglossal	Primarily motor	Motor fibers conduct impulses to muscles that move the tongue.

Note: The cranial nerves described as *primarily motor* do have some sensory fibers associated with specialized receptors (proprioceptors) that give information about length and force of contraction of skeletal muscles. Because this information is part of motor control, these nerves are still considered motor nerves.

impulses are conducted from the olfactory bulbs along *olfactory tracts* to cerebral centers, where they are interpreted.

The second pair of cranial nerves, the **optic nerves (II),** lead from the eyes to the brain and are associated with vision. The neurons associated with this nerve are called ganglion cells, and their cell bodies are in *ganglion cell layers* in the retinas of the eyes. The axons of these neurons pass in the optic nerve through the *optic foramina* of the orbits and continue into the visual neural pathways of the brain (see section 10.9, Sense of Sight). Sensory impulses conducted on the optic nerves are interpreted in the visual cortices of the occipital lobes.

The third pair of cranial nerves, the **oculomotor nerves (III),** arise from the midbrain and pass into the orbits of the eyes. One component of each nerve connects to the voluntary muscles that raise the eyelid and to four of the six muscles that move the eye. A second component of each oculomotor nerve is part of the autonomic nervous system and innervates involuntary muscles in the eye. These muscles adjust the amount of light entering the eye and focus the lens (see section 10.9, Sense of Sight).

The fourth pair of cranial nerves, the **trochlear nerves (IV),** arise from the midbrain and are the smallest cranial nerves. Each nerve conducts motor impulses to a fifth voluntary muscle, not innervated by the oculomotor nerve, that moves the eye (see section 10.9, Sense of Sight).

The fifth pair of cranial nerves, the **trigeminal nerves (V),** are the largest cranial nerves and arise from the pons. They are mixed nerves, with the sensory parts more extensive than the motor parts. Each sensory component includes three large branches, called the ophthalmic, maxillary, and mandibular divisions.

The *ophthalmic division* of the trigeminal nerves consists of sensory fibers that conduct impulses to the brain from the surface of the eyes, the tear glands, and the skin of the anterior scalp, forehead, and upper eyelids. The fibers of the *maxillary division* conduct sensory impulses from the upper teeth, upper gum, and upper lip, as well as from the mucous lining of the palate and the skin of the face. The *mandibular division* includes both motor and sensory fibers. The sensory branches conduct impulses from the scalp behind the ears, the skin of the jaw, the lower teeth, the lower gum, and the lower lip. The motor branches innervate the muscles of mastication and certain muscles in the floor of the mouth.

The sixth pair of cranial nerves, the **abducens nerves (VI),** are quite small and originate from the pons near the medulla oblongata. Each nerve enters the orbit of the skull and innervates the remaining muscle that moves the eye (see section 10.9, Sense of Sight).

The seventh pair of cranial nerves, the **facial nerves (VII),** arise from the lower part of the pons and emerge on the sides of the face. Their sensory branches are associated with taste receptors on the anterior two-thirds of the tongue, and some of their motor fibers conduct impulses to the muscles of facial expression. Still other motor fibers of these nerves function in the autonomic nervous system and stimulate secretions from tear glands and salivary glands.

The eighth pair of cranial nerves, the **vestibulocochlear nerves (VIII),** are sensory nerves that arise from the medulla oblongata. Each of these nerves has two distinct parts—a vestibular branch and a cochlear branch.

The neuron cell bodies associated with the *vestibular branch* fibers are located in ganglia in parts of the inner ears. These parts contain the receptors involved with reflexes that help maintain equilibrium (see section 10.8, Sense of Equilibrium). The neuron cell bodies of the *cochlear branch* fibers are located in another ganglion in the part of the inner ear that houses the hearing receptors. Information from these branches reaches the medulla oblongata and midbrain on its way to the temporal lobes, where it is interpreted (see section 10.7, Sense of Hearing).

The ninth pair of cranial nerves, the **glossopharyngeal nerves (IX),** are associated with the tongue and pharynx (see sections 15.3, Mouth, and 15.5, Pharynx and Esophagus). These mixed nerves arise from the medulla oblongata, with predominantly sensory fibers. Their sensory fibers conduct impulses from the linings of the pharynx, tonsils, and posterior third of the tongue to the brain. Fibers in the motor component innervate muscles of the pharynx that function in swallowing.

The tenth pair of cranial nerves, the **vagus nerves (X),** originate in the medulla oblongata and extend downward through the neck into the chest and abdomen. These nerves are mixed. They also contain both somatic and autonomic branches, with autonomic fibers predominant. Certain somatic motor fibers conduct impulses to muscles of the larynx that are associated with speech and swallowing. Other motor fibers of the vagus nerves are autonomic. They innervate the heart, smooth muscle, and glands in the thorax and abdomen.

The eleventh pair of cranial nerves, the **accessory nerves (XI),** originate in the spinal cord, but have both cranial and spinal branches. Because of the cranial branches, the accessory nerves are included as cranial nerves. Each *cranial branch* passes upward into the cranial cavity and joins a vagus nerve. This branch conducts impulses to muscles of the soft palate, pharynx, and larynx. The *spinal branch* descends into the neck and innervates motor fibers to the trapezius and sternocleidomastoid muscles.

The twelfth pair of cranial nerves, the **hypoglossal nerves (XII),** arise from the medulla oblongata and pass into the tongue. They include motor fibers that conduct impulses to muscles that move the tongue in speaking, chewing, and swallowing. Table 9.6 summarizes the functions of the cranial nerves.

The consequences of a cranial nerve injury depend on the injury's location and its extent. Damage to one member of a nerve pair will not cause a total loss of function, but injury to both nerves may. If a nerve is severed completely, functional loss of that nerve is total; if the cut is incomplete, loss may be partial.

PRACTICE 9.15

1. Define *peripheral nervous system.*
2. Distinguish between somatic and autonomic nerve fibers.
3. Name the cranial nerves, and list the major functions of each.

Spinal Nerves

Thirty-one pairs of **spinal nerves** originate from the spinal cord (fig. 9.36). All but the first pair are mixed nerves that provide two-way communication between the spinal cord and parts of the upper and lower limbs, neck, and trunk.

Spinal nerves are named according to the level from which they arise. Each pair of nerves is numbered in sequence. On each vertebra, the vertebral notches (the major parts of the intervertebral foramina) are associated with the inferior part of their respective vertebrae. For this reason, each spinal nerve, as it passes through the intervertebral foramen, is associated with the vertebra above it. The cervical spinal nerves are an exception, because spinal nerve C1 passes superior to the vertebra C1. Thus, although there are seven cervical vertebrae, there are eight pairs of *cervical nerves* (numbered C1 to C8). There are twelve pairs of *thoracic nerves* (numbered T1 to T12), five pairs of *lumbar nerves* (numbered L1 to L5), five pairs of *sacral nerves* (numbered S1 to S5), and one pair of *coccygeal nerves* (Co).

The adult spinal cord ends at the level between the first and second lumbar vertebrae. Nervous tissue that becomes

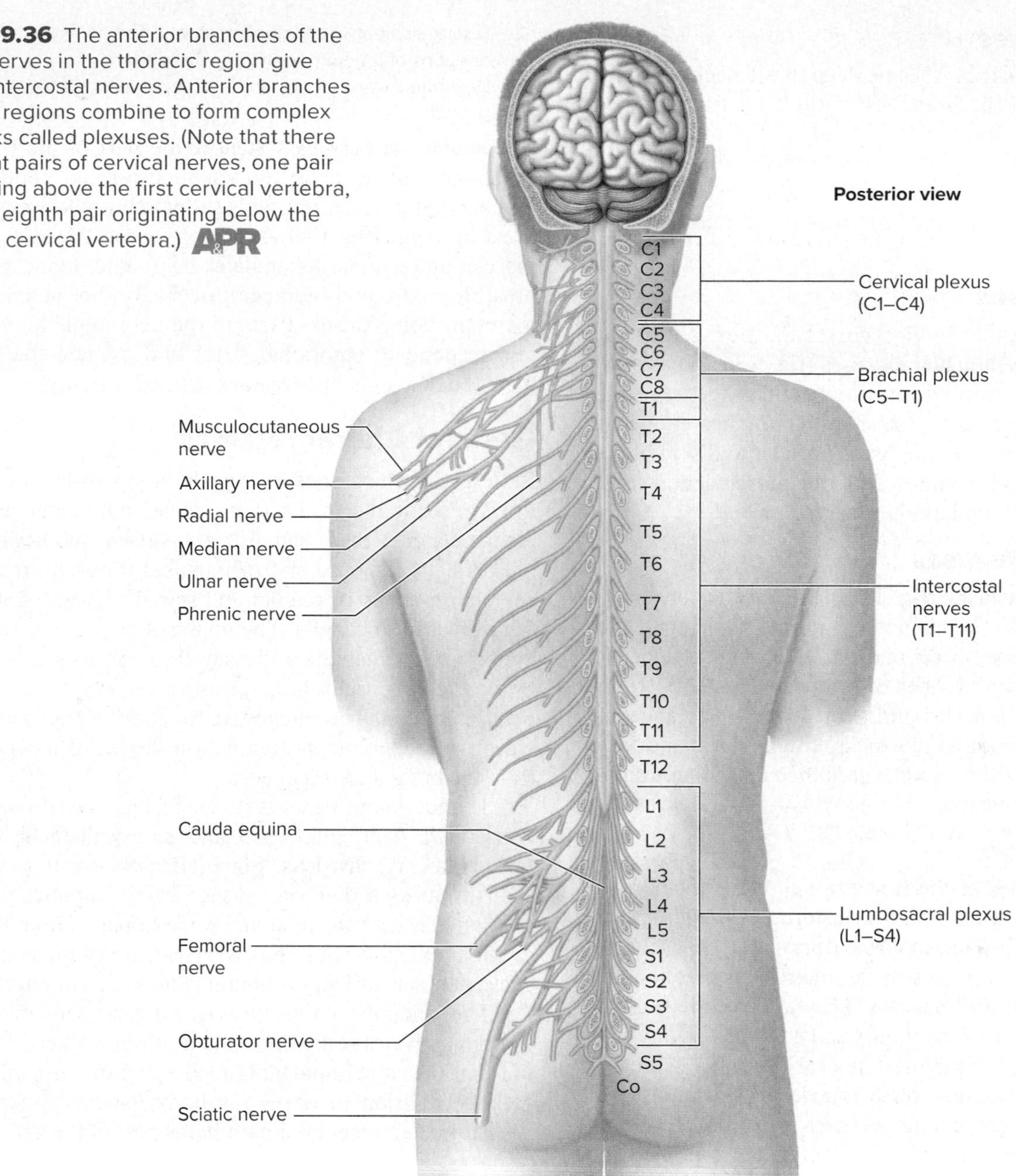

Figure 9.36 The anterior branches of the spinal nerves in the thoracic region give rise to intercostal nerves. Anterior branches in other regions combine to form complex networks called plexuses. (Note that there are eight pairs of cervical nerves, one pair originating above the first cervical vertebra, and the eighth pair originating below the seventh cervical vertebra.) APR

the lumbar, sacral, and coccygeal nerves descends beyond the end of the cord, forming a structure called the *cauda equina* (horse's tail) (see fig. 9.24).

Each spinal nerve emerges from the cord by two short branches, or *roots,* which lie within the vertebral canal. The **dorsal root** (posterior or sensory root) can be identified by an enlargement called the *dorsal root ganglion* (see fig. 9.23*a*). This ganglion contains the cell bodies of the unipolar sensory neurons whose axons have two functional parts. The *peripheral processes* of these axons conduct impulses inward from the peripheral body parts. The *central processes* of these axons extend through the dorsal root and into the spinal cord, where they form synapses with dendrites of other neurons (see fig. 9.4). The **ventral root** (anterior or motor root) of each spinal nerve consists of axons from the motor neurons whose cell bodies are within the gray matter of the cord.

A ventral root and a dorsal root unite to form a spinal nerve, which extends outward from the vertebral canal through an *intervertebral foramen* (see fig. 7.16). Just beyond its foramen, each spinal nerve divides into several parts.

The main parts of several spinal nerves combine to form complex networks called **plexuses** (fig. 9.36). In a plexus, instead of continuing directly to peripheral body parts, spinal nerve axons are sorted and recombined. As a result, axons that originate from different spinal nerves can reach a peripheral body part in the same peripheral nerve.

Cervical Plexuses

The **cervical plexuses** (fig. 9.36) lie deep in the neck on either side and form from the branches of the first four cervical nerves. Axons from these plexuses supply the muscles and skin of the neck. In addition, axons from the third, fourth, and fifth cervical nerves pass into the right and left **phrenic nerves,** which conduct motor impulses to the muscle fibers of the diaphragm.

Brachial Plexuses

Branches of the lower four cervical nerves and the first thoracic nerve give rise to the **brachial plexuses** (fig. 9.36). These networks of axons are deep within the shoulders between the neck and axillae (armpits). The major branches emerging from the brachial plexuses supply the muscles and skin of the arm, forearm, and hand, and include the **musculocutaneous, ulnar, median, radial,** and **axillary nerves.**

Lumbosacral Plexuses

The **lumbosacral plexuses** (fig. 9.36) are formed on either side by the five lumbar spinal nerves and the first four sacral spinal nerves. These networks of axons extend into the pelvic cavity, giving rise to a number of motor and sensory axons associated with the muscles and skin of the lower abdominal wall, external genitalia, buttocks, thighs, legs, and feet. The major branches of these plexuses include the **obturator, femoral,** and **sciatic nerves.**

Intercostal Nerves

The anterior branches of the first through eleventh thoracic spinal nerves do not enter a plexus. Instead, they enter spaces between the ribs and become **intercostal nerves** (fig. 9.36). These nerves supply motor impulses to the intercostal muscles and the upper abdominal wall muscles. They also receive sensory impulses from the skin of the thorax and abdomen.

Spinal nerves may be injured in a variety of ways, including stabs, gunshot wounds, birth injuries, dislocations and fractures of the vertebrae, and pressure from tumors in surrounding tissues. For example, a sudden extension followed by flexion of the neck, called *whiplash,* can occur during rear-end automobile collisions and may stretch the superficial nerves of the cervical plexuses. Whiplash may cause continuing headaches and pain in the neck and skin, which the cervical nerves supply.

PRACTICE 9.15

4. How are spinal nerves grouped?
5. Describe how a spinal nerve joins the spinal cord.
6. Name and locate the major nerve plexuses.

9.16 | Autonomic Nervous System

LEARN

1. Describe the functions of the autonomic nervous system.
2. Distinguish between the sympathetic and parasympathetic divisions of the autonomic nervous system.
3. Describe a sympathetic and a parasympathetic nerve pathway.

The **autonomic nervous system** is the part of the PNS that functions independently (autonomously) and continuously without conscious effort. This system controls visceral functions by regulating the actions of smooth muscle, cardiac muscle, and glands. It regulates heart rate, blood pressure, breathing rate, body temperature, and other activities that maintain homeostasis. Parts of the autonomic nervous system respond to emotional stress and prepare the body to meet the demands of strenuous physical activity.

General Characteristics

Reflexes in which sensory signals originate from receptors in the viscera and in the skin regulate autonomic activities. Axons conduct these signals to centers in the brain or spinal cord. In response, motor impulses travel out from these centers on axons in cranial and spinal nerves. These axons typically lead to ganglia. The impulses they conduct are integrated in these ganglia and relayed to effectors that respond by contracting (muscles), releasing secretions (glands), or being inhibited. The integrative function of the ganglia provides the autonomic system with a degree of independence from the brain and spinal cord.

The autonomic nervous system includes two divisions—the **sympathetic** (sim″pah-thet′ik) and **parasympathetic** (par″ah-sim″pah-thet′ik) **divisions.** Some effectors are innervated by axons from each division. In such cases, impulses from one division may activate an organ, while impulses from the other division inhibit it. Thus, the divisions may act antagonistically, alternately activating or inhibiting the actions of effectors.

The functions of the autonomic divisions are mixed; that is, each activates some organs and inhibits others. However, the divisions have important functional differences. The sympathetic division prepares the body for energy-expending, stressful, or emergency situations, as part of the *fight-or-flight* response. Conversely, the parasympathetic division is most active under ordinary, restful conditions, such as after a meal. It is often described as *rest and digest.* It also counterbalances the effects of the sympathetic division and restores the body to a resting state following a stressful experience. For example, during an emergency the sympathetic division increases heart rate; following the emergency, the parasympathetic division decreases heart rate.

Autonomic Neurons

The neurons of the autonomic nervous system are motor neurons. However, unlike the motor pathways of the somatic nervous system, which usually include a single neuron between the brain or spinal cord and a skeletal muscle, those of the autonomic system include two neurons (fig. 9.37). The cell body of the first neuron, the preganglionic neuron, is located in the brain or spinal cord. Its axon, the **preganglionic fiber** (prē″gang-glē-on′ik fī′ber), leaves the CNS and synapses with one or more neurons whose cell bodies are located in the PNS

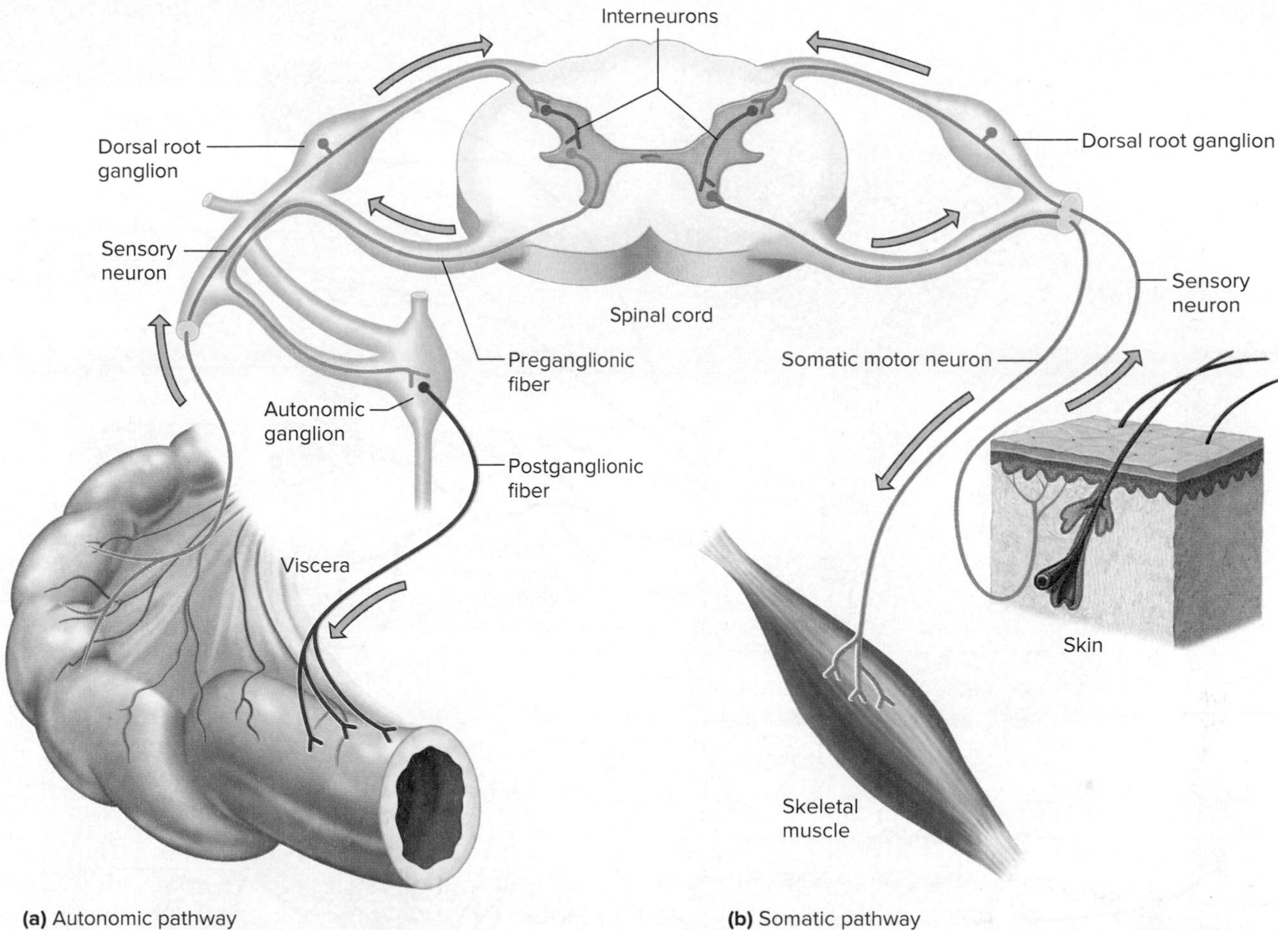

Figure 9.37 Motor pathways. **(a)** Autonomic pathways include two neurons between the CNS and an effector. **(b)** Somatic pathways usually have a single neuron between the CNS and an effector. In both cases, the motor fibers pass through the ventral root of the spinal cord. (Both autonomic and somatic pathways are bilateral, but each is shown on only one side for simplicity.)

within an autonomic ganglion. The axon of such a second neuron, or postganglionic neuron, is called a **postganglionic fiber** (pōst″gang-glē-on′ik fī′ber), and it extends to a visceral effector.

Sympathetic Division

In the sympathetic division, the preganglionic fibers originate from neurons in the gray matter of the spinal cord (fig. 9.38). Their axons leave the cord through the ventral roots of spinal nerves in the first thoracic through the second lumbar segments. These fibers extend a short distance, then leave the spinal nerves, and each enters a member of a chain of sympathetic ganglia (*paravertebral ganglia*). There are two of these sympathetic chains, one extending longitudinally along each side of the vertebral column (fig. 9.38).

In paravertebral ganglia, preganglionic fibers form synapses with second neurons. The axons of these neurons, the postganglionic fibers, typically return to spinal nerves and extend to visceral effectors. In some cases, the preganglionic fibers pass through the paravertebral ganglia and form synapses within *collateral ganglia,* which are found partway between the sympathetic chain ganglia and the target organs. One exception to this pattern involves certain hormone-secreting cells of the adrenal gland, which are innervated directly by preganglionic neurons (fig. 9.38).

Parasympathetic Division

The preganglionic fibers of the parasympathetic division arise from the brainstem and sacral region of the spinal cord (fig. 9.39). From there, they lead outward in cranial or sacral nerves to *terminal ganglia* located near or in various viscera. The relatively short postganglionic fibers continue from the ganglia to specific muscles or glands in these viscera.

PRACTICE 9.16

1. Describe the parts of the autonomic nervous system.
2. Distinguish between the divisions of the autonomic nervous system.
3. Describe a sympathetic nerve pathway and a parasympathetic nerve pathway.

Autonomic Neurotransmitters

The preganglionic fibers of the sympathetic and parasympathetic divisions all secrete **acetylcholine** and are therefore called **cholinergic fibers** (kō″lin-er′jik fī′berz). The parasympathetic postganglionic fibers are also cholinergic. (One exception is parasympathetic neurons that secrete nitric

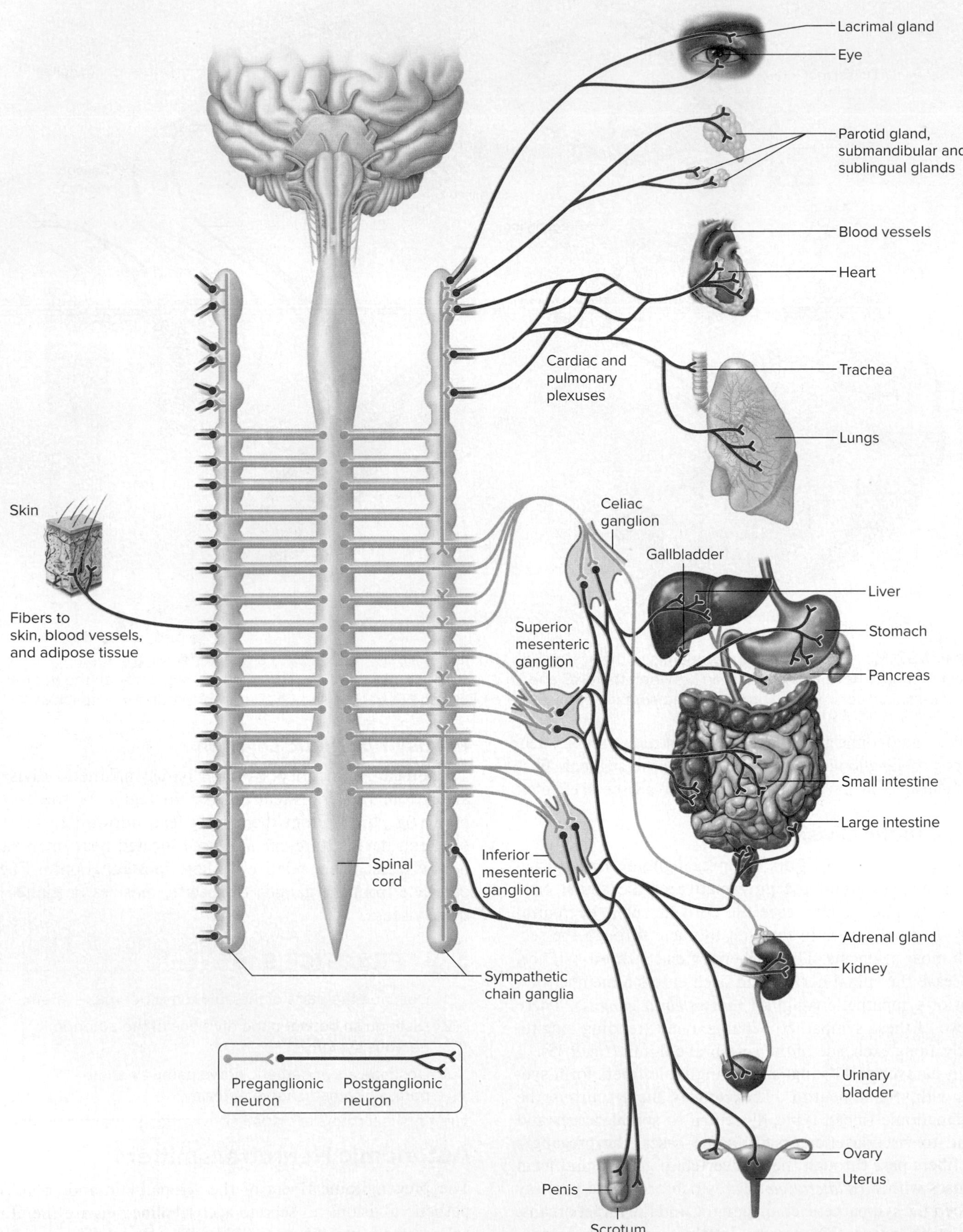

Figure 9.38 The preganglionic fibers of the sympathetic division of the autonomic nervous system arise from the thoracic and lumbar regions of the spinal cord (T1–L2). Note that the adrenal medulla is innervated directly by a preganglionic fiber. APR

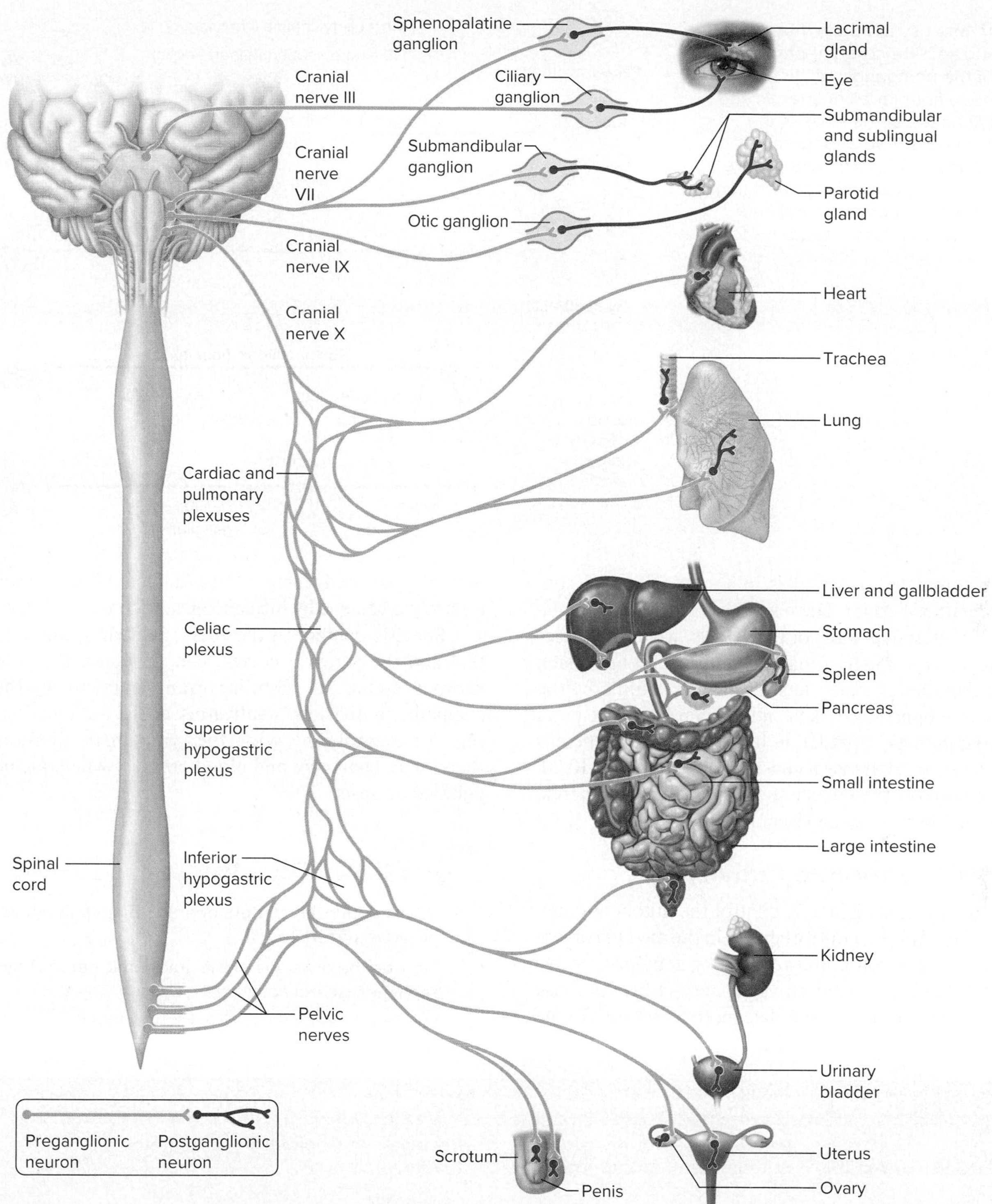

Figure 9.39 The preganglionic fibers of the parasympathetic division of the autonomic nervous system arise from the brainstem and sacral region of the spinal cord. APR

oxide; these are described in section 19.2, Organs of the Male Reproductive System.) Most sympathetic postganglionic neurons secrete **norepinephrine** (noradrenalin) and are called **adrenergic fibers** (ad″ren-ur′jik fī′berz) (fig. 9.40). The different postganglionic neurotransmitters cause the different effects that the sympathetic and parasympathetic divisions have on their effector organs.

Most organs receive innervation from both sympathetic and parasympathetic divisions, usually with opposing actions. For example, parasympathetic activity increases activity of the digestive system, whereas sympathetic activity decreases it. Similarly, sympathetic stimulation increases heart rate, but parasympathetic action slows heart rate.

Some viscera are controlled primarily by one division or the other. That is, the divisions are not always actively antagonistic. For example, the sympathetic division regulates the diameter of most blood vessels, which lack parasympathetic innervation. Smooth muscle in the walls of these vessels is

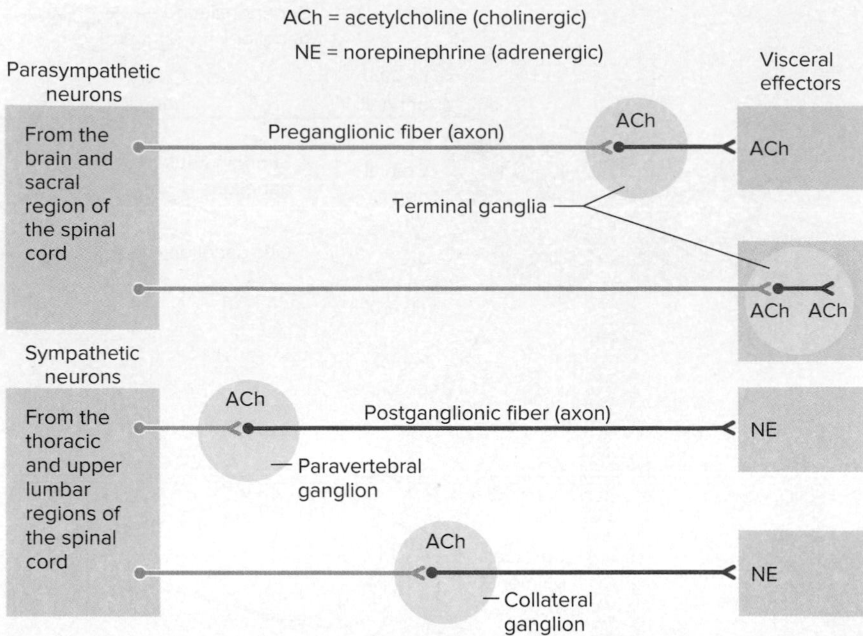

Figure 9.40 Most sympathetic fibers are adrenergic and secrete norepinephrine at the ends of the postganglionic fiber; parasympathetic fibers are cholinergic and secrete acetylcholine at the ends of the postganglionic fibers. Two arrangements of parasympathetic postganglionic fibers are found in both the cranial and sacral portions. Similarly, sympathetic paravertebral and collateral ganglia are found in both the thoracic and lumbar portions.

continuously stimulated and thus is in a state of partial contraction (sympathetic tone). Decreasing sympathetic stimulation relaxes the muscular walls of the vessels, which increases the diameter of the vessels (dilates). Conversely, increasing sympathetic stimulation constricts the vessels. Similarly, the two systems have opposing effects on the diameter of the pupil of the eye. Sympathetic stimulation dilates the pupil, whereas parasympathetic signals cause a constriction. See figure 10.20. Table 9.7 summarizes the effects of stimulation by adrenergic and cholinergic fibers on some visceral effectors.

Control of Autonomic Activity

The brain and spinal cord largely control the autonomic nervous system. For example, control centers in the medulla oblongata for cardiac, vasomotor, and respiratory activities receive sensory impulses from viscera on vagus nerve fibers and use autonomic nerve pathways to stimulate motor responses in the heart, blood vessels, and lungs. Similarly, the hypothalamus helps regulate body temperature, hunger, thirst, and water and electrolyte balance by influencing autonomic pathways.

Specific centers in the brain, including the limbic system and the cerebral cortex, can influence the autonomic nervous system. A familiar example involving the parasympathetic division is salivating at the anticipation of eating. An example involving the sympathetic division is the increase in heart rate and blood pressure when one becomes agitated or upset.

PRACTICE 9.16

4. Which neurotransmitters operate in the autonomic nervous system?
5. How do the divisions of the autonomic nervous system regulate visceral activities?
6. How are autonomic activities controlled?

TABLE 9.7 Effects of Neurotransmitter Substances on Visceral Effectors or Actions

Visceral Effector or Action	Response to Adrenergic Stimulation (Sympathetic)	Response to Cholinergic Stimulation (Parasympathetic)
Pupil of the eye	Dilation	Constriction
Heart rate	Increases	Decreases
Bronchioles of lungs	Dilation	Constriction
Muscle of intestinal wall	Slows peristaltic action	Speeds peristaltic action
Intestinal glands	Secretion decreases	Secretion increases
Blood distribution	More blood to skeletal muscles; less blood to digestive organs	More blood to digestive organs; less blood to skeletal muscles
Blood glucose concentration	Increases	Decreases
Salivary glands	Secretion decreases	Secretion increases
Tear glands	No action	Secretion
Muscle of gallbladder wall	Relaxation	Contraction
Muscle of urinary bladder wall	Relaxation	Contraction

Nervous System

INTEGUMENTARY SYSTEM

Sensory receptors provide the nervous system with information about the outside world.

SKELETAL SYSTEM

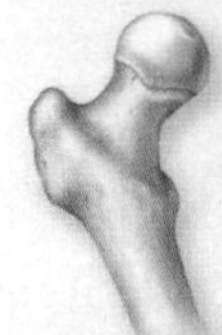

Bones protect the brain and spinal cord and help maintain plasma calcium, which is important to neuron function.

MUSCULAR SYSTEM

The nervous system controls movement and processes information about the position of body parts.

ENDOCRINE SYSTEM

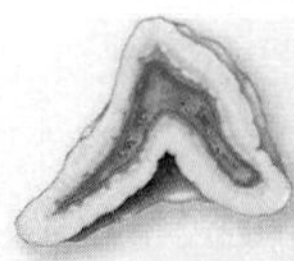

The hypothalamus controls secretion of many hormones.

CARDIOVASCULAR SYSTEM

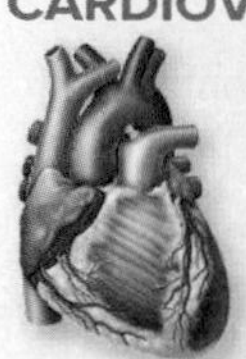

The nervous system helps control blood flow and blood pressure.

LYMPHATIC SYSTEM

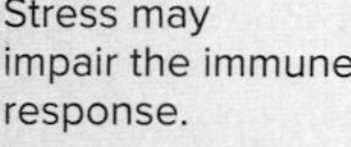

Stress may impair the immune response.

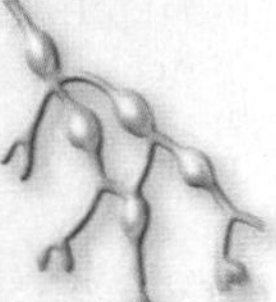

DIGESTIVE SYSTEM

The nervous system can influence digestive function.

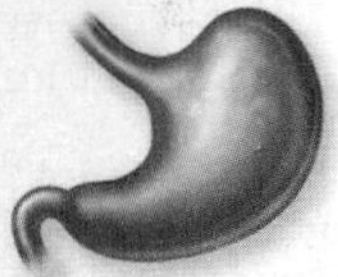

RESPIRATORY SYSTEM

The nervous system alters respiratory activity to control oxygen levels and blood pH.

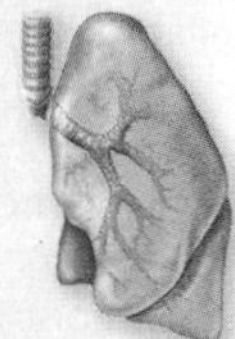

URINARY SYSTEM

The nervous system plays a role in urine production and elimination.

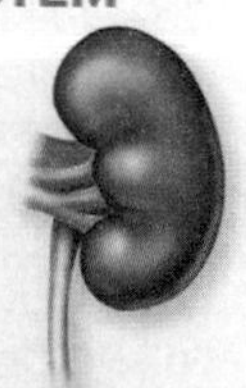

REPRODUCTIVE SYSTEM

The nervous system plays a role in egg and sperm formation, sexual pleasure, childbirth, and nursing.

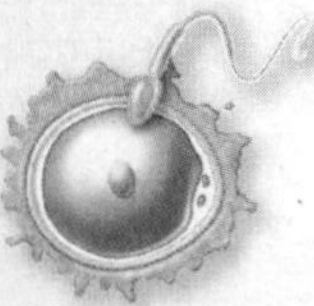

Neurons conduct impulses that allow body systems to communicate.

ASSESS

CHAPTER ASSESSMENTS

9.1 Introduction

1. The general function of neurons is to ________, whereas the general functions of neuroglia are to ________.
2. List the general functions of the nervous system.

9.2 Nervous System Organization

3. Explain the relationship between the CNS and the PNS.

9.3 Neurons

4. Match the neuron part on the left to its description on the right.

(1) dendrite	A. a cell process that sends information
(2) axon	B. one of usually several cell processes that receive information
(3) cell body	C. the rounded part of a neuron

5. Describe three structures found in neurons that are also in other cell types, and describe two structures that are unique to neurons.
6. Distinguish among multipolar, bipolar, and unipolar neurons.
7. Distinguish among sensory neurons, interneurons, and motor neurons.

9.4 Neuroglia

8. Match the types of neuroglia to their functions.

(1) ependymal cells	A. form a myelin sheath around peripheral nerves
(2) oligodendrocytes	B. phagocytize cellular debris and bacteria
(3) astrocytes	C. line inner parts of ventricles and spinal cord
(4) Schwann cells	D. form scar tissue and regulate ion and nutrient concentrations in the CNS
(5) microglial cells	E. form a myelin sheath around neurons in the CNS

9. The part of a Schwann cell that contributes to the myelin sheath is the ________, and the part that contributes to the neurilemma is the ________.

9.5 Charges Inside a Cell

10. Explain how a membrane becomes polarized.
11. Describe how ions associated with nerve cell membranes are distributed.
12. Define *resting potential.*
13. Explain the relationship between threshold potential and an action potential.
14. List the events that occur during an action potential.
15. "All-or-none" response in impulse conduction means that ________.

9.6 Impulse Conduction

16. Choose the correct sequence of events along an axon:
    a. Resting potentials are propagated along a stimulated axon, causing a very small action potential.
    b. A threshold stimulus opens K^+ channels and the ions diffuse in, depolarizing the cell membrane. Then Na^+ channels open, Na^+ exits, and the cell membrane repolarizes, generating an action potential that stimulates adjacent cell membrane, forming the impulse.
    c. A threshold stimulus opens Na^+ channels and the ions diffuse in, depolarizing the cell membrane. Then K^+ channels open, K^+ exits, and the cell membrane repolarizes, generating an action potential that stimulates adjacent cell membrane, forming the impulse.
    d. A threshold stimulus opens Na^+ channels and the ions diffuse in, depolarizing the cell membrane. Then K^+ channels open, K^+ exits, and the cell membrane repolarizes, generating an action potential that inhibits adjacent cell membrane, forming the impulse.
17. Distinguish between myelinated and unmyelinated axons.
18. Explain why a myelin sheath covering an entire axon (with no nodes of Ranvier) would inhibit conduction of an impulse.
19. The ________ period ensures that action potentials progress in one direction toward the axon terminal.

9.7 The Synapse

20. Define *synapse.*
21. Explain how information passes from one neuron to another.

9.8 Synaptic Transmission

22. Distinguish between excitatory and inhibitory actions of neurotransmitters.
23. Neurotransmitters are synthesized in ________ and are stored in ________.
24. Match the neurotransmitter to its description on the right.

(1) biogenic amine	A. short chains of amino acids
(2) acetylcholine	B. a modified amino acid
(3) neuropeptide	C. an amino acid
(4) GABA	D. stimulates skeletal muscle contraction

25. Explain what happens to neurotransmitters after they are released.

9.9 Impulse Processing

26. Describe the components of a neuronal pool.
27. "Facilitation in a neuronal pool" refers to ________.
28. Distinguish between convergence and divergence in a neuronal pool.

9.10 Types of Nerves

29. Describe how sensory, motor, and mixed nerves differ.

9.11 Neural Pathways

30. Distinguish between a reflex arc and a reflex.
31. Describe the components of a reflex arc and their functions.
32. List three body functions that reflexes control.

9.12 Meninges

33. Match each layer of the meninges to its description.

(1) dura mater	A. the thin, innermost layer, containing blood vessels and nerves
(2) arachnoid mater	B. the tough, outermost layer, consisting mostly of connective tissue
(3) pia mater	C. the lacy membrane, lacking blood vessels, sandwiched between the other two layers

9.13 Spinal Cord

34. Describe the structure of the spinal cord.
35. Distinguish between the ascending and descending tracts of the spinal cord.

9.14 Brain

36. Name the four major parts of the brain and describe their general functions.
37. The area of the brain that contains centers controlling visceral activities is the ________.
    a. cerebrum
    b. cerebellum
    c. brainstem
    d. diencephalon
38. The structure that connects the cerebral hemispheres is the ________.
39. Distinguish between a sulcus and a fissure.
40. Relate the lobes of the cerebral hemispheres to the skull bones.
41. Locate the sensory, association, and motor areas of the cerebral cortex, and describe the general functions of each.
42. Define *hemisphere dominance.*
43. Distinguish between ganglia and nuclei.
44. The function of the basal nuclei is to ________.
45. Locate the ventricles in the brain.
46. Explain how cerebrospinal fluid is produced and how it functions.
47. The part of the diencephalon that regulates hunger, weight, water and electrolyte balance, sleep and wakefulness, temperature, arterial blood pressure, heart rate, production of substances that stimulate the pituitary gland, and movement and secretion in areas of the digestive tract is the ________.
    a. thalamus
    b. pineal gland
    c. infundibulum
    d. hypothalamus
48. Define *limbic system,* and explain its functions.
49. The parts of the brainstem are the ________, ________, and ________.
50. List the functions of the three parts of the brainstem.
51. Vomiting is controlled by ________.
    a. the reticular formation
    b. the medulla oblongata
    c. the midbrain
    d. the pons
52. Describe what happens to the body when the reticular formation receives sensory impulses, and what happens when it does not receive stimulation.
53. Describe the functions of the cerebellum.

9.15 Peripheral Nervous System

54. Distinguish between cranial nerves and spinal nerves.
55. Distinguish between the somatic nervous system and the autonomic nervous system.
56. Match the cranial nerves to the body parts or functions that they affect. More than one nerve pair may correspond to the same structure or function.

(1) olfactory nerves (I)
(2) optic nerves (II)
(3) oculomotor nerves (III)
(4) trochlear nerves (IV)
(5) trigeminal nerves (V)
(6) abducens nerves (VI)
(7) facial nerves (VII)
(8) vestibulocochlear nerves (VIII)
(9) glossopharyngeal nerves (IX)
(10) vagus nerves (X)
(11) accessory nerves (XI)
(12) hypoglossal nerves (XII)

A. vision
B. hearing and equilibrium
C. muscles of the larynx, pharynx, soft palate, sternocleidomastoid and trapezius muscles
D. heart, various smooth muscles and glands in the thorax and abdomen
E. taste, facial expressions, secretion of tears and saliva
F. sense of smell
G. tongue movements and swallowing
H. face and scalp
I. eye movements

57. Explain how the spinal nerves are classified and numbered.
58. Describe the structure of a spinal nerve.
59. Define *plexus,* and locate the major plexuses of the spinal nerves.

9.16 Autonomic Nervous System

60. Describe the general functions of the autonomic nervous system.
61. Distinguish between the sympathetic and parasympathetic divisions of the autonomic nervous system.
62. Distinguish between preganglionic and postganglionic neurons.
63. The effects of the sympathetic and parasympathetic autonomic divisions differ because ________.
64. List two ways in which the CNS controls autonomic activities.

ASSESS

INTEGRATIVE ASSESSMENTS/CRITICAL THINKING

Outcomes 3.4, 9.3, 9.4

1. State two reasons why rapidly growing brain cancers are composed of neuroglia rather than neurons.

Outcomes 8.2, 8.5, 8.6, 9.2, 9.16

2. Skeletal muscle, visceral smooth muscle, and cardiac muscle are all innervated by nerves. However, only one of these muscle types depends solely on nerve stimulation for contraction to occur. What is this muscle type? Explain what might stimulate the other two muscle types to contract.

Outcomes 9.2, 9.11

3. While driving home at night, Chris suddenly sees a deer run out in front of his car, so he reacts and abruptly turns the steering wheel to just barely miss hitting the animal. Explain the role of the central nervous system and peripheral nervous system in this incident.

Outcomes 9.3, 9.4, 9.6, 9.13, 9.14

4. In multiple sclerosis, nerve fibers in the CNS lose their myelin. Explain why this loss affects skeletal muscle function.

Outcomes 9.3, 9.11, 9.13, 9.14

5. List four skills encountered in everyday life that depend on nervous system function, and list the part of the nervous system responsible for each.

Outcomes 9.5, 9.8, 9.15

6. The vagus nerve of the parasympathetic nervous system secretes the neurotransmitter acetylcholine (ACh) into the synapse that is received by cholinergic receptors on sinoatrial node cells (pacemaker) of the heart, resulting in potassium (K^+) gated channels to open. Would this result in an excitation or an inhibition? Would this speed up or slow down the heart rate? Explain.

Outcome 9.6

7. When you lie on your arm and it "goes to sleep," what causes this and why does it cause an unpleasant feeling of prickly pins?

Outcomes 9.11, 9.13

8. The biceps-jerk reflex is carried out by motor neurons that exit the spinal cord in the fifth spinal nerve (C5). The triceps-jerk reflex uses motor neurons in the seventh spinal nerve (C7). Describe how these reflexes might be tested to help pinpoint damage in a patient with a neck injury.

Outcomes 9.11, 9.14

9. Describe the roles of the cerebrum and cerebellum in athletics.

Outcomes 9.13, 9.14

10. Describe expected functional losses in a patient who has suffered injury to the right occipital lobe of the cerebral cortex compared to injury in the right temporal lobe of the cerebral cortex.

Chapter Summary

9.1 Introduction

1. Three major aspects of the nervous system are **sensory input, integration and processing,** and **motor output.**
2. Nervous tissue includes **neurons,** which are the structural and functional units of the nervous system, and **neuroglia.**

9.2 Nervous System Organization

1. Organs of the nervous system are divided into the **central nervous system** and **peripheral nervous system.**
2. Sensory functions involve **sensory receptors** that detect internal and external changes.
3. Integrative functions collect sensory information and make decisions that motor functions carry out.
4. Motor functions stimulate **effectors** to respond.

9.3 Neurons

1. A neuron includes a cell body, dendrites, and an axon.
2. Dendrites and the cell body provide receptive surfaces.
3. A single axon arises from the cell body and may be enclosed in a myelin sheath with gaps called **nodes of Ranvier.**
4. Classification of neurons
   a. Neurons are classified structurally as **multipolar, bipolar,** or **unipolar.**
   b. Neurons are classified functionally as **sensory neurons, interneurons,** or **motor neurons.**

9.4 Neuroglia

1. Neuroglia in the central nervous system include **microglial cells, oligodendrocytes, astrocytes,** and **ependymal cells.**
2. In the peripheral nervous system, **Schwann cells** form **myelin** sheaths, and **satellite cells** protect cell bodies.

9.5 Charges Inside a Cell

A cell membrane is usually polarized as a result of unequal ion distribution.

1. Distribution of ions
   a. Channels in cell membranes that allow passage of some ions but not others set up differences in the concentrations of specific ions inside and outside a neuron.
   b. Potassium ions pass more easily through cell membranes than do sodium ions.
2. **Resting potential**
   a. A high concentration of sodium ions is outside a cell membrane, and a high concentration of potassium ions is inside.
   b. Many negatively charged ions are inside a cell.
   c. In a resting cell, more positive ions leave than enter, so the outside of the cell membrane develops a positive charge, while the inside develops a negative charge.
3. Potential changes
   a. Stimulation of a cell membrane affects the membrane's resting potential.
   b. When its resting potential becomes less negative, a membrane becomes depolarized.
   c. Potential changes are graded.
   d. Achieving **threshold potential** triggers an **action potential.**
4. Action potential
   a. At threshold, sodium channels open, and sodium ions diffuse inward, depolarizing the membrane.
   b. At almost the same time, potassium channels open, and potassium ions diffuse outward, repolarizing the membrane.
   c. This rapid sequence of depolarization and repolarization is an action potential.
   d. Many action potentials can occur in a neuron without disrupting the ion concentrations. Active transport contributes to maintaining these concentrations.
5. All-or-none response
   a. An action potential occurs in an all-or-none manner whenever a stimulus of threshold intensity is applied to an axon.

b. All of the action potentials triggered on an axon are of the same strength.

9.6 Impulse Conduction

1. Impulse conduction (action potential progression down the axon)
   a. Unmyelinated axons conduct action potentials uninterrupted along their entire lengths.
   b. Myelinated axons conduct impulses more rapidly than unmyelinated axons.
   c. Axons with larger diameters conduct impulses faster than those with smaller diameters.
   d. The refractory period limits impulse frequency and ensures that impulses move in one direction.

9.7 The Synapse

A synapse is a junction between two neurons.

1. A presynaptic neuron conducts an impulse into a synapse; a postsynaptic neuron or other cell responds.
2. Axons have synaptic knobs at their distal ends, which secrete **neurotransmitters.**
3. A neurotransmitter is released when an impulse reaches the synaptic knob at the end of an axon.
4. A neurotransmitter reaching the postsynaptic neuron membrane is either excitatory or inhibitory.

9.8 Synaptic Transmission

1. Excitatory and inhibitory actions
   a. Neurotransmitters that may trigger impulses are **excitatory.** Those that inhibit impulses are **inhibitory.**
   b. The net effect of synaptic knobs communicating with a neuron depends on which knobs are activated from moment to moment.
2. Neurotransmitters
   a. The nervous system produces many different neurotransmitters.
   b. Neurotransmitters include acetylcholine, biogenic amines, amino acids, and peptides.
   c. A synaptic knob releases neurotransmitters when an action potential increases membrane permeability to calcium ions.
   d. After being released, neurotransmitters are decomposed or removed from synaptic clefts.

9.9 Impulse Processing

How the nervous system processes and responds to information reflects the organization of neurons in the brain and spinal cord.

1. **Neuronal pools**
   a. Neurons form pools in the central nervous system.
   b. Each pool receives impulses, processes them, and conducts impulses away.
2. **Facilitation**
   a. Each neuron in a pool may receive excitatory and inhibitory stimuli.
   b. A neuron is facilitated when repeated impulse conduction increases its release of neurotransmitter in response to a single impulse.
3. **Convergence**
   a. Impulses from two or more incoming axons may converge on a single neuron.
   b. Convergence enables impulses from different sources to have an additive effect on a neuron.
4. **Divergence**
   a. Impulses occurring on a neuron of a neuronal pool often synapse with several other neurons.
   b. As a result of divergence, the action of a single neuron may be amplified to affect more postsynaptic cells.

9.10 Types of Nerves

1. Nerves are cordlike bundles (fascicles) of nerve fibers (axons).
2. Nerves are **sensory, motor,** or **mixed,** depending on which type of fibers they contain.

9.11 Neural Pathways

*A **neural pathway** is the route that information follows through the nervous system.*

1. A **reflex arc** usually includes a sensory neuron, a reflex center composed of interneurons, and a motor neuron.
2. Reflex behavior
   a. **Reflexes** are automatic, subconscious responses to changes.
   b. Reflexes help maintain homeostasis.
   c. Two neurons carry out the patellar reflex. It is therefore monosynaptic.
   d. Withdrawal reflexes are protective.

9.12 Meninges

1. Bone and **meninges** surround the brain and spinal cord.
2. The meninges are the **dura mater, arachnoid mater,** and **pia mater.**
3. **Cerebrospinal fluid** fills the space between the arachnoid and pia maters.

9.13 Spinal Cord

*The **spinal cord** is a column of nervous tissue that extends from the brain into the vertebral canal.*

1. Structure of the spinal cord
   a. Each of the spinal cord's thirty-one segments gives rise to a pair of **spinal nerves** (two pairs are associated with C1).
   b. The spinal cord has a cervical enlargement and a lumbar enlargement.
   c. A central core of gray matter lies within white matter.
   d. White matter consists of bundles of myelinated axons called **tracts.**
2. Functions of the spinal cord
   a. The spinal cord provides a two-way communication system between the brain and other body parts and serves as a center for **spinal reflexes.**
   b. **Ascending tracts** conduct sensory impulses from sensory receptors to the brain. **Descending tracts** conduct motor impulses from the brain to muscles and glands.

9.14 Brain

*The **brain** is subdivided into the cerebrum, diencephalon, brainstem, and cerebellum.*

1. Structure of the **cerebrum**
   a. The cerebrum consists of two **cerebral hemispheres** connected by the **corpus callosum.**

   b. The cerebral cortex is a thin layer of gray matter near the surface.
   c. White matter consists of myelinated axons that connect neurons in the nervous system and communicate with other body parts.
2. Functions of the cerebrum
   a. The cerebrum provides higher brain functions.
   b. The cerebral cortex consists of **sensory, association,** and **motor** areas.
   c. One cerebral hemisphere is usually dominant for certain intellectual functions.
3. Ventricles and cerebrospinal fluid
   a. **Ventricles** are interconnected cavities within the cerebral hemispheres and brainstem.
   b. Cerebrospinal fluid fills the ventricles.
   c. The **choroid plexuses** in the walls of the ventricles secrete cerebrospinal fluid.
4. Diencephalon
   a. The **diencephalon** contains the **thalamus,** which is a central relay station for incoming sensory impulses, and the **hypothalamus,** which maintains homeostasis.
   b. The **limbic system** produces emotions and modifies behavior.
5. Brainstem
   a. The **brainstem** consists of the **midbrain, pons,** and **medulla oblongata.**
   b. The midbrain contains reflex centers associated with eye and head movements.
   c. The pons relays impulses between the cerebrum and other parts of the nervous system and contains centers that may help regulate breathing.
   d. The medulla oblongata relays all ascending and descending impulses and contains neural centers that control heart rate, blood pressure, and respiratory function.
   e. The **reticular formation** filters incoming sensory impulses, arousing the cerebral cortex into wakefulness when significant input arrives.
6. Cerebellum
   a. The **cerebellum** consists of two hemispheres.
   b. It functions primarily as a reflex center for integrating sensory information required in the coordination of skeletal muscle movements and the maintenance of equilibrium.

9.15 Peripheral Nervous System

The peripheral nervous system consists of cranial and spinal nerves that branch from the brain and spinal cord to all body parts. It is also subdivided into the ***somatic*** *and* ***autonomic nervous systems.***

1. Cranial nerves
   a. Twelve pairs of **cranial nerves** connect the brain to parts in the head, neck, and trunk.
   b. Most cranial nerves are mixed, but some are purely sensory, and others are primarily motor.
   c. The names of the cranial nerves indicate their primary functions or the general distributions of their fibers.
2. Spinal nerves
   a. Thirty-one pairs of **spinal nerves** originate from the spinal cord.
   b. All but the first pair are mixed nerves that provide a two-way communication system between the spinal cord and parts of the upper and lower limbs, neck, and trunk.
   c. Spinal nerves are grouped according to the levels from which they arise, and they are numbered in sequence.
   d. Each spinal nerve emerges by a **dorsal root** and a **ventral root.**
   e. Each spinal nerve divides into several branches just beyond its foramen.
   f. Most spinal nerves combine to form **plexuses** in which nerve fibers are sorted and recombined so that those fibers associated with a peripheral body part reach it together.

9.16 Autonomic Nervous System

The ***autonomic nervous system*** *functions without conscious effort. It regulates the visceral activities that maintain homeostasis.*

1. General characteristics
   a. Autonomic functions are reflexes controlled from centers in the brain and spinal cord.
   b. The autonomic nervous system consists of two divisions—the **sympathetic** and the **parasympathetic.**
   c. The sympathetic division responds to stressful and emergency conditions.
   d. The parasympathetic division is most active under ordinary conditions.
2. Autonomic nerve fibers
   a. Autonomic nerve fibers are motor fibers.
   b. Sympathetic fibers leave the spinal cord and synapse in paravertebral ganglia.
   c. Parasympathetic fibers begin in the brainstem and sacral region of the spinal cord and synapse in ganglia near viscera.
3. Autonomic neurotransmitters
   a. Sympathetic and parasympathetic **preganglionic fibers** secrete acetylcholine.
   b. Parasympathetic **postganglionic fibers** secrete acetylcholine. Sympathetic postganglionic fibers secrete norepinephrine.
   c. The different effects of the autonomic divisions are due to the different neurotransmitters the postganglionic fibers release.
   d. The two divisions usually have opposite actions.
4. Control of autonomic activity
   a. The autonomic nervous system is controlled by the brain and spinal cord, but is somewhat independent.
   b. Control centers in the medulla oblongata and hypothalamus utilize autonomic nerve pathways.
   c. The limbic system and cerebral cortex affect the autonomic system's actions.

The Senses 10

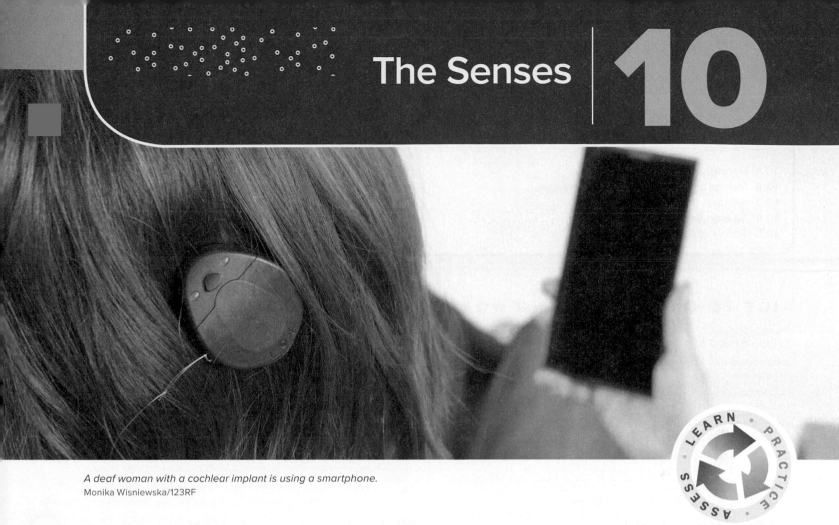

A deaf woman with a cochlear implant is using a smartphone.
Monika Wisniewska/123RF

Imagine not being able to hear your favorite song, your beloved movie or TV show, the voice on the other end of a phone call, the alarm that awakens you in the morning, or conversation among your family and friends. Even the most powerful hearing aids that focus on making sounds louder may not help you understand what is being sung or said. It's like listening to an out-of-tune radio station. Damage to parts of the inner ear cochlea may result in sensorineural hearing loss, or nerve deafness, where there is a problem with the conduction of sound signals to the brain. An electronic medical device called a cochlear implant may be surgically placed to bypass the damaged inner ear and directly stimulate the auditory nerve, so that the person can hear sounds and understand speech again.

The external part of a cochlear implant contains a *microphone* that picks up environmental sounds, as well as a *speech processor* that picks up the sounds from the microphone, analyzes and digitizes the signal, and then sends it to the *transmitter.* The transmitter sends coded signals to the implant's internal part, the *receiver,* which is implanted under the skin. The receiver converts the signals to electrical impulses and transmits them through a *stimulator* to electrodes within the cochlea. The auditory nerve is stimulated by the electrodes, and the brain interprets these signals as sound.

Hearing loss affects more than just a person's ability to hear or process sounds. It has been linked to dementia, accidents, and depression. New technologies have been developed that connect the sound processors to smartphones and other audio devices, which may help reduce risks and improve quality of life. In June 2017, the Federal Drug Administration (FDA) approved a device called the Nucleus 7 sound processor that links cochlear implants with phones or tablets, a collaboration between Apple and Cochlear. This technology allows a person to get phone calls that are directly routed to the implant, and to also stream music, TV shows, podcasts, and audio books. Because streaming high-quality audio and voice quickly drains batteries, a Bluetooth wireless technology called Bluetooth Low-Energy Audio (BLEA) was engineered by Apple to extend battery life. The iPhone or iPod Touch is paired with the cochlear implant upon activation. This system supports hearing from an implant for the ear on one side, and hearing from a conventional hearing aid on the ear's other side, called a "bimodal" setup. A person can then drop the volume on the side where there is more background noise to focus on conversation with someone on the other side. Wireless technology accessories allow hands-free use of the iPhone, while also improving the signal-to-noise ratio because the signal bypasses the microphone and enters the sound processor directly.

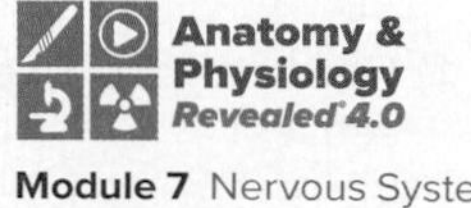

Module 7 Nervous System

LEARNING OUTLINE

After studying this chapter, you should be able to complete the "Learning Outcomes" that follow the major headings throughout the chapter.

AIDS TO UNDERSTANDING WORDS

choroid [skinlike] *choroid* coat: middle, vascular layer of the eye.

cochlea [snail] *cochlea:* coiled tube in the inner ear.

iris [rainbow] *iris:* colored, muscular part of the eye.

labyrinth [maze] *labyrinth:* complex system of connecting chambers and tubes of the inner ear.

lacri- [tears] *lacri*mal gland: tear gland.

macula [spot] *macula* lutea: yellowish spot on the retina.

olfact- [to smell] *olfact*ory: pertaining to the sense of smell.

scler- [hard] *scler*a: tough, outer protective layer of the eye.

tympan- [drum] *tympan*ic membrane: eardrum.

vitre- [glass] *vitre*ous humor: clear, jellylike substance within the eye.

(Appendix A has a complete list of Aids to Understanding Words.)

10.1 | Introduction

LEARN

1. Distinguish between general senses and special senses.

You are sitting in your favorite restaurant looking at its appetizing menu, with the smell of food all around. Your favorite band is playing through the sound system. After your meal, you exit the restaurant with a certain discomfort associated with a very full stomach. All of these experiences, pleasurable and otherwise, are orchestrated by our senses. These senses are necessary for us not only to enjoy life, but to indicate potential problems regarding our survival. They derive from structures called sensory receptors that detect environmental and internal changes. These trigger impulses that travel on sensory pathways into our central nervous system for processing and a possible response.

Senses fall into two major categories. *General senses* include touch, temperature, pressure, and pain. The *special senses* are taste, smell, vision, and hearing and balance. Receptors for the special senses are parts of complex, specialized organs found in the head.

PRACTICE 10.1

Answers to the Practice questions can be found in the eBook.

1. How are special senses different from general senses?

10.2 | Receptors, Sensations, and Perception

LEARN

1. Name five kinds of receptors, and explain their functions.
2. Explain how a sensation arises.

Our awareness of different sensory events arises from receptors that respond to specific stimuli and the ability of different parts of the brain to interpret the resulting impulses.

Types of Receptors

Sensory receptors are diverse but share certain features. Each type of receptor is particularly sensitive to a distinct kind of environmental change and is much less sensitive to other forms of stimulation. Sensory receptors are categorized into five types according to their sensitivities: **Chemoreceptors** (kē″mō-re-sep′torz) are stimulated by changes in the concentration of certain chemicals; **pain receptors** (pān re-sep′torz) by tissue damage; **thermoreceptors** (ther′mō-re-sep′torz) by changes in temperature; **mechanoreceptors** (mek″ah-no-re-sep′torz) by changes in pressure or movement; and **photoreceptors** (fō″tō-re-sep′torz) by light.

Sensation and Perception

A **sensation** occurs when sensory receptors are stimulated and send the information to the brain. A **perception** is the conscious awareness of stimuli. Thus, pain is a sensation, but realizing that you have just stepped on a tack is a perception.

When impulses travel away from sensory receptors into the central nervous system, the resulting sensation depends on which region of the brain receives the impulses. For example, when you rub your eye, you might "see" flashes of light. This is known as the *labeled line principle* because any kind of stimulus to the optic nerve is carried to the visual cortex in the brain. That is, it is labeled for vision.

In a process called **projection,** the brain projects the sensation back to its apparent source. Projection allows a person to perceive the region of stimulation. For example, a specific region of the body that might be experiencing pain.

Sensory Adaptation

The brain must prioritize the sensory input it receives, or incoming unimportant information would be overwhelming. For example, until this sentence prompts you to think about it, you are probably unaware of the pressure of your clothing against your skin, or the background noise in the room. This ability of the nervous system to become less responsive to a maintained stimulus is called **sensory adaptation** (ad″ap-tā′shun). It may result from receptors becoming unresponsive or inhibition along the central nervous system pathways leading to the respective sensory regions of the cerebral cortex.

PRACTICE 10.2

1. List five general types of sensory receptors.
2. Explain how a perception is different from a sensation.
3. What is sensory adaptation?

10.3 | General Senses

LEARN

1. Describe the receptors associated with the senses of touch, pressure, temperature, body position, movement, stretch, and pain.
2. Describe how the sense of pain is produced.

General senses are widespread, and are associated with receptors in the skin, muscles, joints, and viscera. They include the senses of touch and pressure, temperature, body position, movement, stretch, and pain.

Touch and Pressure Senses

The senses of touch and pressure derive from three kinds of receptors (fig. 10.1). These receptors respond to mechanical forces that deform or displace tissues. Touch and pressure receptors include:

- **Free nerve endings** These receptors are common in epithelial tissues, where the ends of dendrites branch and extend between epithelial cells. They are responsible for the sensation of itching, as well as other sensations described later.
- **Tactile (Meissner's) corpuscles** These are small, oval masses of flattened connective tissue cells in connective tissue sheaths. Two or more sensory nerve fibers branch into each corpuscle and end in it as tiny knobs. Tactile corpuscles are abundant in hairless areas of skin, such as the lips, fingertips, palms, soles, nipples, and external genital organs. They respond to the motion of objects that barely contact the skin. The brain interprets impulses from them as the sensation of light touch.
- **Lamellated (Pacinian) corpuscles** These sensory bodies are relatively large structures composed of connective tissue fibers and cells, with a single sensory nerve fiber branch extending into each. They are common in the deeper dermal and subcutaneous tissues and in the connective tissue capsule of synovial joints. Lamellated corpuscles respond to heavy pressure and are associated with the sensation of deep pressure.

Temperature Senses

Temperature sensation depends on two types of free nerve endings in the skin. Those that respond to warmer temperatures are called *warm receptors,* and those that respond to colder temperatures are called *cold receptors.*

Warm receptors are most sensitive to temperatures above 25°C (77°F) and become unresponsive at

CAREER CORNER
Optometrist

The woman squints at the medicine bottle, just barely making out the letters of the label. Since turning fifty, she's noticed difficulty reading small print, and she finally acknowledges that she may need glasses. She visits an optometrist, who uses a series of tests to detect the woman's farsightedness. The optometrist assesses visual acuity (reading an eye chart), peripheral vision, color and depth perception, focusing ability, and coordination of the eyes. He also checks for signs of disease, including macular degeneration, glaucoma, cataracts, conjunctivitis, and diabetic retinopathy. The optometrist assures the patient that her eyes are healthy, but eyeglasses or contact lenses will enable her to easily read the fine print.

An optometrist completes four years in optometry school postcollege. Optometrists with advanced training can provide vision therapy (exercises to better coordinate the eyes) and help patients with low vision use devices to supplement lenses and surgery. An optometrist can be the first health-care professional to report a rare syndrome that has ocular symptoms, such as thin retinal layers or abnormal blood vessels.

An optometrist may work with an ophthalmologist to care for a patient before and after a procedure, such as cataract surgery. In some states, optometrists can prescribe certain medications related to vision and eye care. Optometrists work in private practice and in vision care centers.

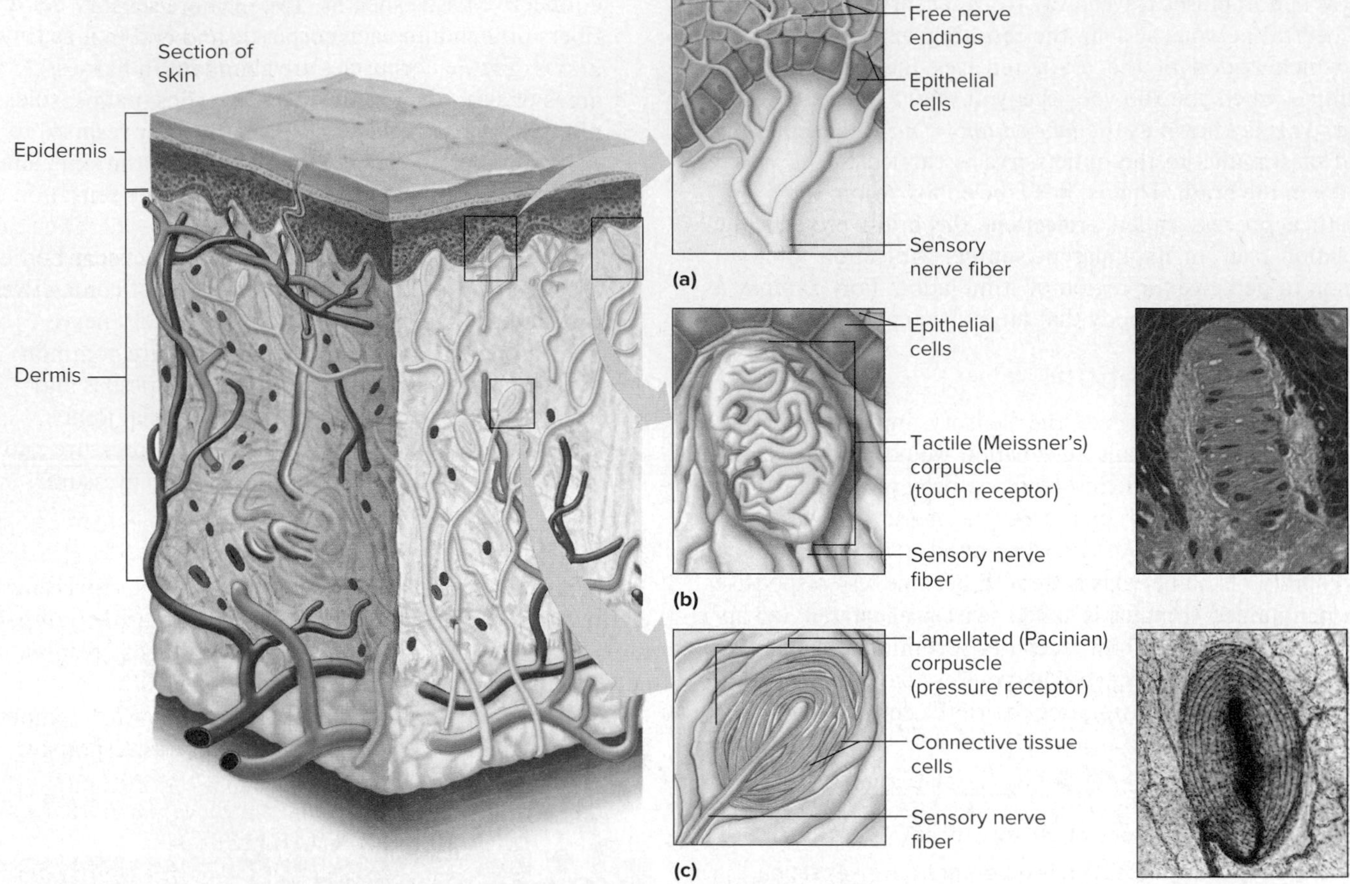

Figure 10.1 Touch and pressure receptors include **(a)** free ends of sensory nerve fibers, **(b)** tactile corpuscles (with 225x micrograph), and **(c)** lamellated corpuscles (with 50x micrograph). APR photos: (b): Ed Reschke; (c): Ed Reschke/Getty Images

temperatures above 45°C (113°F). Temperatures near and above 45°C also stimulate pain receptors, producing a burning sensation.

Cold receptors are most sensitive to temperatures between 10°C (50°F) and 20°C (68°F). Temperatures below 10°C also stimulate pain receptors, producing a freezing sensation.

Both warm and cold receptors adapt rapidly. Within about a minute of continuous stimulation, the sensation of warmth or cold begins to fade.

Body Position, Movement, and Stretch Receptors

We need a continual "sense" of where we are in space and time. This is known as **proprioception.** Proprioceptors are associated with skeletal muscle and provide information about body position. **Muscle spindles** are specialized fibers buried in skeletal muscles that monitor the state of contraction. This is achieved through the action of sensory neurons wrapped around the spindle (fig. 10.2). When a muscle

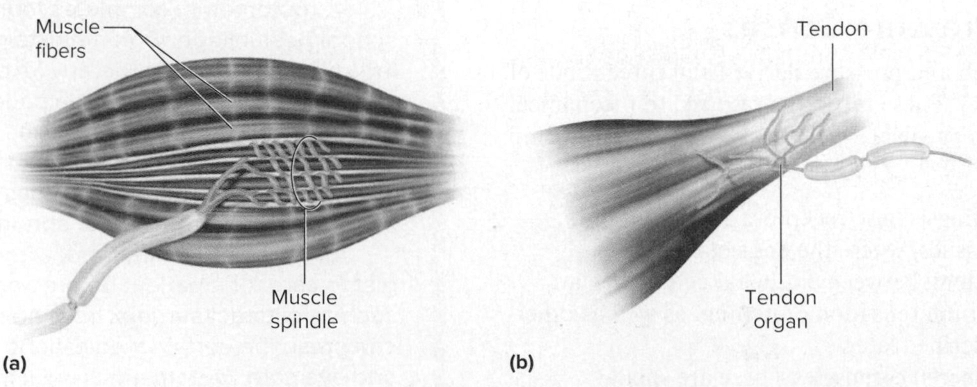

Figure 10.2 Proprioceptors. **(a)** Muscle spindle. **(b)** Tendon organ.

contracts, so do the spindle fibers, allowing the sensory neurons to continuously feed information to the brain and spinal cord. The knee-jerk reflex discussed in chapter 9 is the result of stretching the spindle when the patellar tendon is struck. The counteracting response is a contraction of the extensor muscles, resulting in the leg "jerking" or "kicking" out.

Golgi tendon organs detect how much a tendon is stretched during muscle contraction. When a tendon reaches the maximum amount of stretch it can endure without potential injury, the Golgi organ sends signals that inhibit any further stimulation of the associated muscles.

Even a seemingly simple movement, such as walking, requires a highly integrated and coordinated effort: Some muscles will relax, while others have to generate the appropriate amount of force. Therefore, spindles and Golgi organs must work with the brain and spinal cord to orchestrate all movement.

A gene called *PIEZO2* has been identified as instrumental in proprioception and the function of some touch-related mechanoreceptors. A mutation of this gene was first discovered in mice, and then in human subjects. These subjects suffered a delay in the development of motor ability associated with crawling and walking, and have never been able to run or jump because of the precise input needed from proprioceptors.

The blood vessels close to the heart contain stretch receptors called **baroreceptors** that monitor blood pressure (see chapter 13).

Sense of Pain

Receptors that transmit impulses perceived as pain are called nociceptors (although a rough and nuanced translation, the term comes from the Latin *nocere,* which means "to hurt" or "to injure"). Some of these receptors are free nerve endings found in the skin and other tissues and are stimulated by actual or potential tissue damage. Overstimulation of other receptors, such as the cold and warm receptors discussed earlier, can send signals perceived as pain.

Nociceptors communicate with other pain pathway sensory neurons using two main neurotransmitters: substance P and glutamate. These carry the messages through the spinal cord and to the brain, respectively. When tissue is damaged, prostaglandins are released, increasing the sensitivity of nociceptors and subsequently the intensity of pain. The analgesic effects of ibuprofen and aspirin are a function of their ability to inhibit the production of prostaglandins.

We produce natural "painkillers" called endorphins and enkephalins. These neurotransmitters inhibit the release of substance P by binding to opiate receptors in the spinal cord. This same mechanism is responsible for the powerful analgesic effect of narcotics such as morphine and heroin. Pharmacological manipulation of glutamate has proven more challenging as a potential pain inhibitor because of its role in learning, memory, and anxiety disorders.

Pain is actually a good thing because it indicates danger. There is a medical condition known as congenital insensitivity to pain (CIP), in which affected people do not respond to tissue damage, as they cannot feel pain. In most cases it is caused by malfunctioning Na^+ channels. But some cases also involve an overproduction of endorphins. More common is *peripheral neuropathy,* in which the hands and/or feet become numb due to too few tactile corpuscles. In one study, people with normal pain sensation had an average of 12 corpuscles per square millimeter of skin, whereas people with peripheral neuropathy had fewer than 3. The most common causes of peripheral neuropathy are diabetes mellitus, cancer treatments, vitamin deficiency, and HIV infection.

Visceral Pain

As a rule, pain receptors are the only receptors in viscera whose stimulation produces sensations. Pain receptors in these organs respond differently to stimulation than those associated with surface tissues. For example, localized damage to intestinal tissue during surgical procedures may not elicit pain sensations, even in a conscious person. However, when visceral tissues are subjected to more widespread stimulation, such as when intestinal tissues are stretched or smooth muscle in intestinal walls undergoes a spasm, a strong pain sensation may follow. Once again, the resulting pain seems to stem from stimulation of mechanoreceptors and from chemoreceptors. The chemoreceptors respond to decreased blood flow accompanied by lower tissue oxygen concentration and accumulation of pain-stimulating chemicals.

Visceral pain may feel as if it is coming from a part of the body other than the part being stimulated, a phenomenon called **referred pain.** For example, pain originating in the heart may be referred to the left shoulder or left upper limb (fig. 10.3*a*). Referred pain may arise from common nerve pathways that conduct sensory impulses from skin areas as well as from viscera. Pain impulses from the heart travel over the same nerve pathways as those from the skin of the left shoulder and left upper limb (fig. 10.3*b*). Consequently, during a heart attack the cerebral cortex may incorrectly interpret the source of the pain impulses as the left shoulder or left upper limb, rather than the heart.

PRACTICE 10.3

1. Describe the three types of touch and pressure receptors.
2. Describe the receptors that sense temperature.
3. Explain the function of muscle spindles and Golgi tendon organs.
4. What types of stimuli excite pain receptors?
5. Describe how pain may be pharmacologically controlled.
6. What is referred pain?

Pain Nerve Fibers

The fibers (axons) that conduct impulses away from pain receptors are of two main types: fast pain fibers and slow pain fibers. *Fast pain fibers* are myelinated. They conduct impulses rapidly and are associated with the immediate

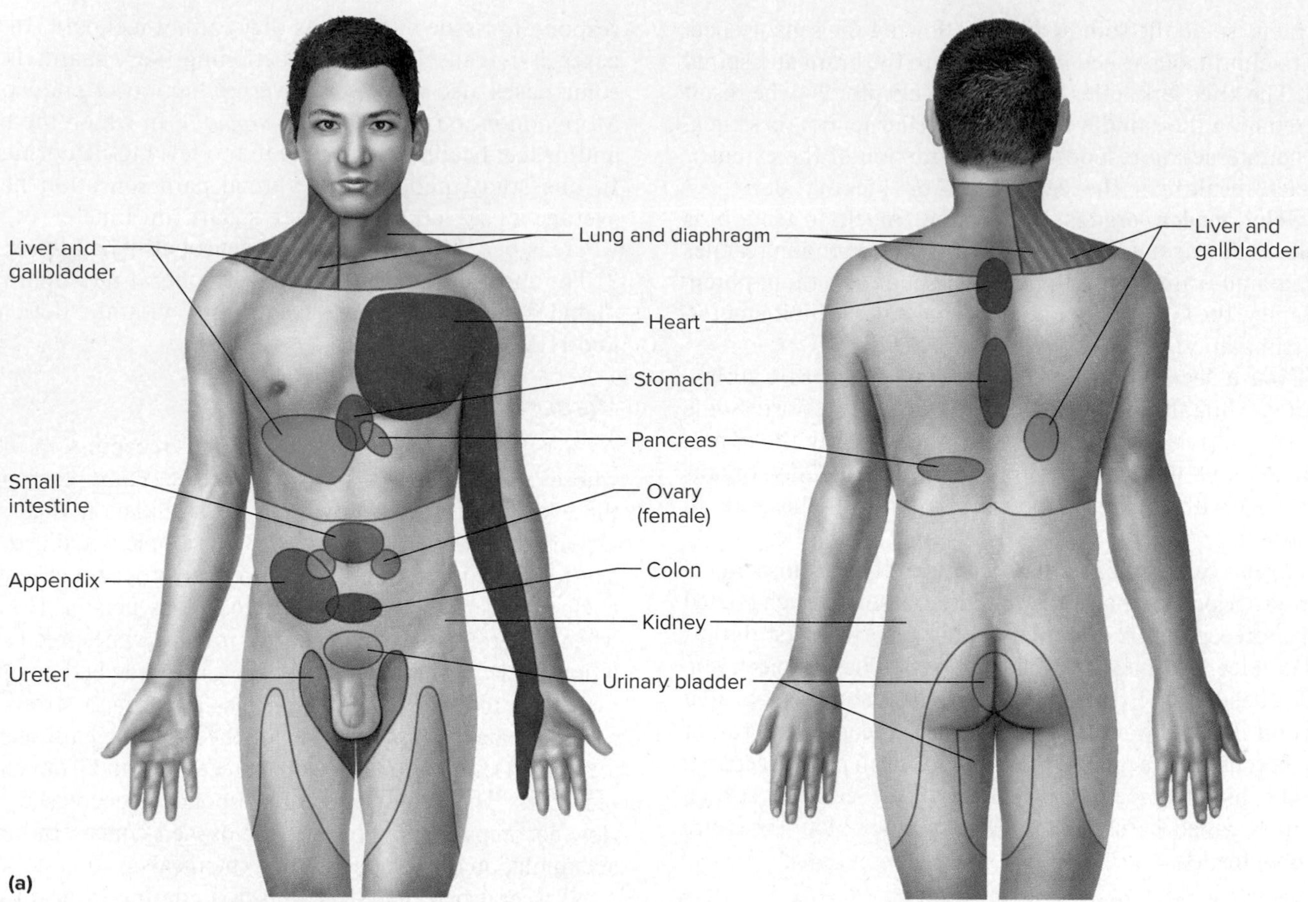

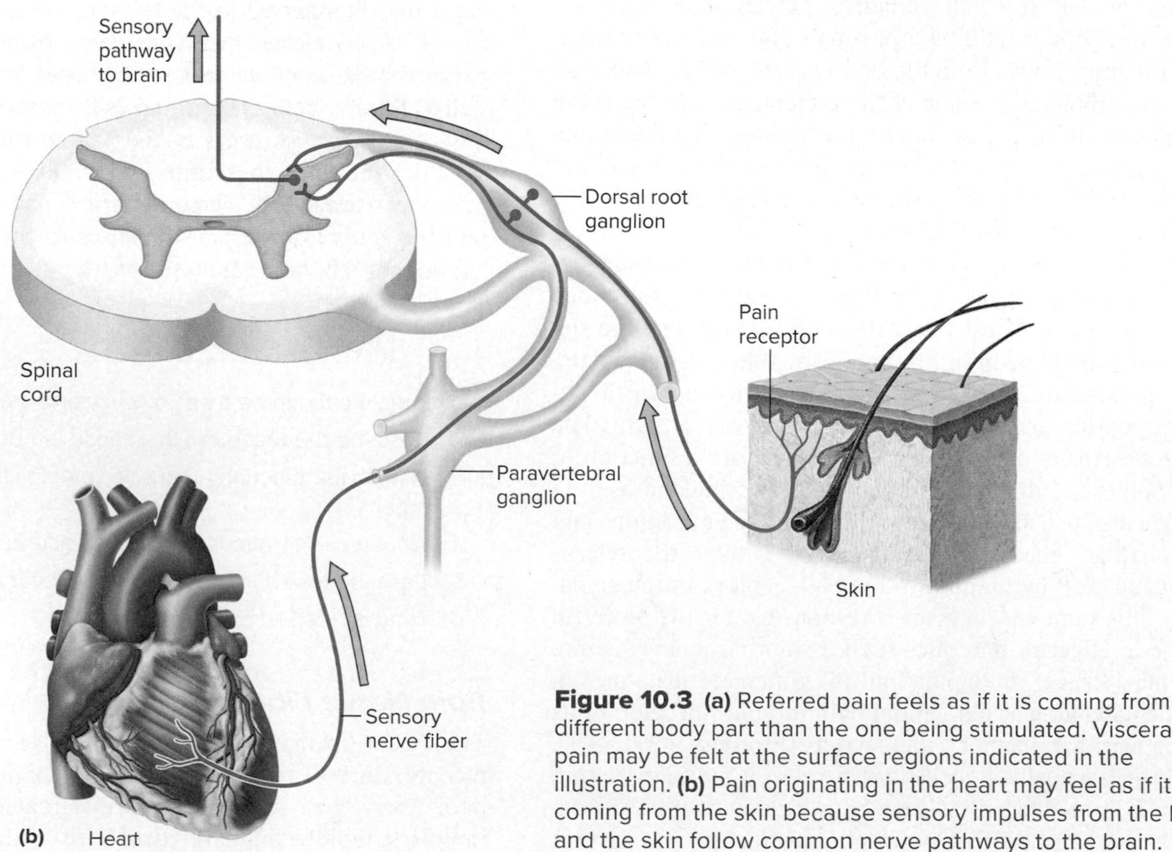

Figure 10.3 **(a)** Referred pain feels as if it is coming from a different body part than the one being stimulated. Visceral pain may be felt at the surface regions indicated in the illustration. **(b)** Pain originating in the heart may feel as if it is coming from the skin because sensory impulses from the heart and the skin follow common nerve pathways to the brain.

sensation of sharp pain, which typically originates from a restricted area of the skin and seldom continues after the pain-producing stimulus stops. *Slow pain fibers* are unmyelinated. They conduct impulses more slowly and produce a delayed, dull, aching sensation that may be diffuse and difficult to pinpoint. Such pain may continue for some time after the original stimulus ceases. Immediate pain is usually sensed as coming only from the skin; delayed pain is felt in deeper tissues as well.

An event that stimulates pain receptors usually triggers impulses on both fast and slow pain fibers. The result is a dual sensation—a sharp, pricking pain, followed shortly by a dull, aching pain that is usually more intense and may worsen with time. Chronic pain can cause prolonged suffering.

Pain Pathways

Pain impulses that originate from the head reach the brain on sensory fibers of cranial nerves. All other pain impulses travel on the sensory fibers of spinal nerves. They pass into the spinal cord by way of the dorsal roots of the spinal nerves. Within the spinal cord, neurons process pain impulses in the gray matter of the posterior horns. Impulses are then conducted to the brain.

In the brain, most pain fibers terminate in the thalamus, the reticular formation, or the limbic system (see section 9.14, Brain). From there, other neurons conduct impulses to the hypothalamus and cerebral cortex. The cerebral cortex determines pain intensity, locates the pain source, and carries out motor responses to the pain. The emotional response to pain involves the limbic system.

PRACTICE 10.3

7. Describe two types of pain fibers.
8. How do acute pain and chronic pain differ?
9. What parts of the brain interpret pain impulses?

10.4 | Special Senses

LEARN

1. Identify the locations of the receptors associated with the special senses.

Special senses are those whose sensory receptors are in large, complex sensory organs in the head. These senses and their respective organs include the following:

Clinical Application 10.1 describes a condition in which some of these assignments of senses to sense organs are altered.

PRACTICE 10.4

1. Where in the body are the sensory organs of the special senses located?

10.5 | Sense of Smell

LEARN

1. Compare the receptor cells involved with the senses of smell and taste.
2. Explain the mechanism for smell.

The sense of smell is associated with complex sensory structures in the upper region of the nasal cavity.

Olfactory Receptors

Olfactory (smell) receptors and taste receptors are chemoreceptors, which means that chemicals dissolved in liquids stimulate them. Smell and taste function closely together and aid in food selection because we usually smell food at the same time as we taste it.

Olfactory Organs

The **olfactory organs** are yellowish-brown masses of epithelium about the size of postage stamps that cover the upper parts of the nasal cavity, the superior nasal conchae, and a part of the nasal septum. The olfactory organs contain **olfactory receptor cells,** which are bipolar neurons surrounded by columnar epithelial cells (fig. 10.4). Hairlike cilia cover tiny knobs at the distal ends of these neurons' dendrites. In any particular such neuron, the cilia harbor many copies of one type of olfactory receptor membrane protein.

Chemicals that are inhaled, called odorant molecules, stimulate various sets of olfactory receptor proteins, and therefore various sets of olfactory receptor cells, to send a signal of a detected odor to the brain. Odorant molecules enter the nasal cavity but must dissolve at least partially in the watery fluids that surround the cilia before receptors can detect them. Foods, flowers, and many other things release odorant molecules.

Olfactory Pathways

Stimulated olfactory receptor cells send impulses along their axons (which together form the first pair of cranial nerves). These axons synapse with neurons located in enlargements called **olfactory bulbs.** The olfactory bulbs lie within the cranial cavity on either side of the crista galli of the ethmoid bone (see fig. 7.13). In the olfactory bulbs, the impulses are processed, and as a result additional impulses are conducted along the **olfactory tracts.** The major interpreting areas for these impulses are deep within the temporal lobes and at the bases of the frontal lobes, anterior to the hypothalamus. Impulses conducted on the olfactory tracts also reach the limbic system (see section 9.14, Brain), which is why one can have an emotional response to an odor.

CLINICAL APPLICATION 10.1

Synesthesia: Connected Senses

"The song was full of glittering orange diamonds."
"The paint smelled blue."
"The sunset was salty."
"The pickle tasted like a rectangle."
About 1 in 2,000 people have a condition called synesthesia ("joined sensation"), in which sensation and perception mix. The brain perceives a stimulus to one sense as coming from another, or links a sense to something that isn't a sense. Most commonly, letters, numbers, or periods of time evoke specific colors. These associations are involuntary, are very specific, and persist over a lifetime. For example, a person might report that 3 is always mustard yellow or Thursday is a very dark, shiny brown.

Synesthesia runs in families, and geneticists have associated the condition with inheriting variants in any of four different genes. Female "synesthetes" outnumber males six to one. Creative individuals are overrepresented among those with the condition. They include Franz Liszt, architect Frank Lloyd Wright, and physicist Richard Feynman, who used to include the hues with which he visualized chemical equations on the chalkboard, to the amusement of his students. One of the co-authors of this book has it—to her, days are colors. The earliest recorded mention of synesthesia is an essay from John Locke in 1690. People with synesthesia are recognizing that their peculiar talent has a name, thanks to Internet groups devoted to the condition.

Researchers hypothesize that mixed senses are present in all babies, but synesthesia develops in individuals who do not lose as many synapses as others as they age. (A loss of 20 billion synapses a day is normal for adults.) Imaging studies and animal experiments have localized the neurons that convey synesthetic connections to the general area where the temporal, parietal, and occipital lobes meet. Synesthesia is increasingly viewed as an enhancement to learning—and a fuller way of enjoying our sensual worlds.

OF INTEREST Canines' excellent sense of smell is the basis of using service dogs to detect impending health problems in their owners. The dogs sense subtle odors that people emit when becoming ill in certain ways. Service dogs are used to sense imminent seizures, drops in blood glucose and heart rate, and lung, breast, and thyroid cancers. Experiments confirm that dogs are especially sensitive to odorant molecules on the skin or in the sweat of sick people.

Olfactory Stimulation

When odorant molecules bind to olfactory receptor proteins in olfactory receptor cell membranes, a chemical pathway is activated that culminates in a depolarizing influx of sodium ions. This may trigger an action potential if the depolarization reaches threshold. The action potentials from this and other olfactory receptor cells travel to the olfactory bulbs in the brain, where the sensation of smell arises.

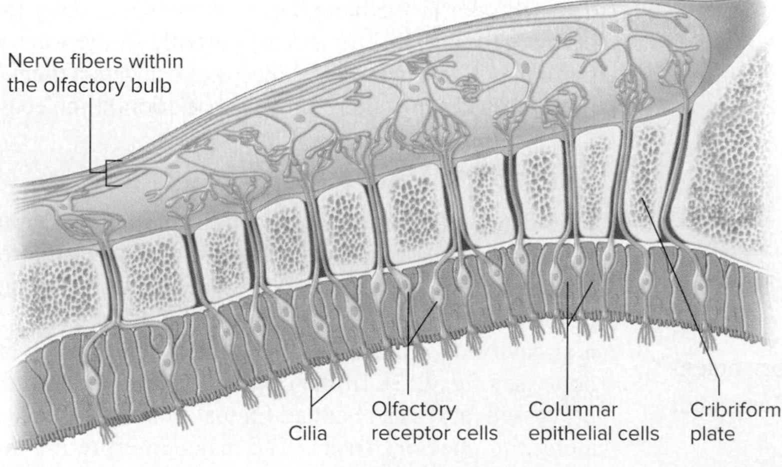

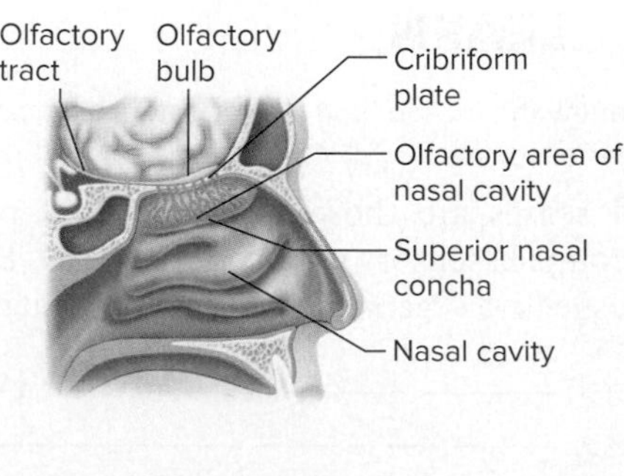

Figure 10.4 Olfactory receptors convey the sense of smell. **(a)** Columnar epithelial cells support olfactory receptor cells, which have cilia at their distal ends. The actual olfactory receptors, which are membrane proteins, are on the cilia. Binding of odorants to these receptors in distinctive patterns conveys the information that the brain interprets as an odor. **(b)** The olfactory area is associated with the superior nasal concha. APR

The several hundred types of olfactory receptor cells can code for many thousands of odors when they signal the brain in groups. That is, an odorant molecule stimulates a distinct set of receptor types. An olfactory receptor cell has only one type of olfactory receptor, but that receptor can bind several types of odorant molecules. In addition, any one odorant molecule can bind several types of receptors. The brain interprets this binding information as a combinatorial olfactory code. In a simplified example, if there are ten odor receptors, parsley might stimulate receptors 3, 4, and 8, while chocolate might stimulate receptors 1, 5, and 10.

A person may have to sniff and force air up to the receptor areas to smell a faint odor, because the olfactory organs are high in the nasal cavity above the usual pathway of inhaled air. Astronauts on the first space flights reportedly could not smell their food because they had to squeeze the food from tubes directly into their mouths—odorant molecules were not inhaled as they usually are.

The sense of smell adapts rapidly, and the smell quickly fades. However, adaptation to one scent will not diminish sensitivity to new odors. For example, a person visiting a fish market might at first be acutely aware of the fishy smell, but then that odor fades. If a second person enters the fish market wearing a strong perfume, the person already there, who has become accustomed to the stinky fish, will nonetheless detect the flowery scent of the perfume.

Partial or complete loss of smell is called *anosmia.* It may result from inflammation of the nasal cavity lining due to a respiratory infection, tobacco smoking, or use of certain drugs, such as cocaine.

PRACTICE 10.5

1. Where are olfactory receptors located?
2. Trace the pathway of an olfactory impulse from a receptor to the cerebrum.

10.6 | Sense of Taste

LEARN

1. Explain the mechanism for taste.

Taste buds are the special organs of taste (fig. 10.5). The 10,000 or so taste buds are located mostly on the surface of the tongue and are associated with tiny elevations called *papillae.* About 1,000 taste buds are scattered in the roof of the mouth and walls of the throat.

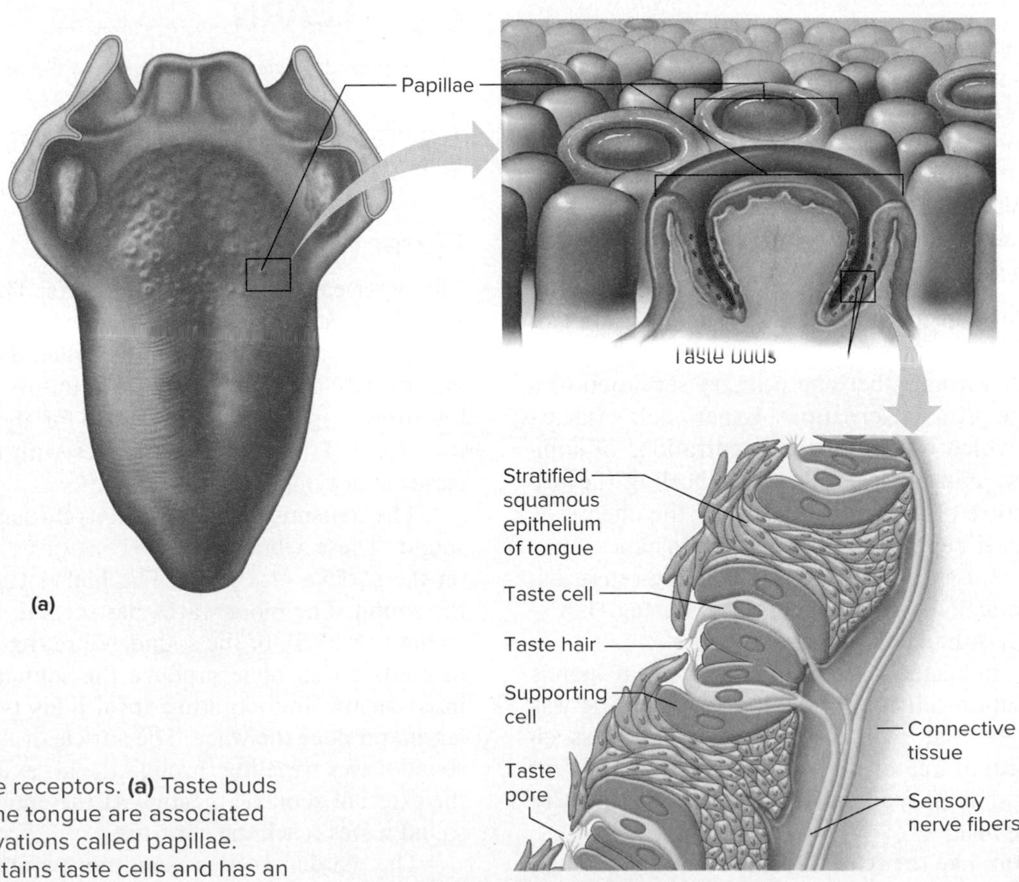

Figure 10.5 Taste receptors. **(a)** Taste buds on the surface of the tongue are associated with nipplelike elevations called papillae. **(b)** A taste bud contains taste cells and has an opening, the taste pore, at its free surface.

Taste Receptors

Each taste bud includes 50 to 150 modified epithelial cells, the **taste cells** (gustatory cells), which function as sensory receptors. Each taste cell is replaced every ten days, on average. The taste bud also includes epithelial supporting cells. The entire structure is spherical, with an opening, the **taste pore,** on its free surface. Tiny projections called **taste hairs** protrude from the outer ends of the taste cells and extend from the taste pore. These taste hairs are the sensitive parts of the receptor cells.

A particular chemical must dissolve in the watery fluid surrounding the taste buds in order for it to be tasted. The salivary glands provide this fluid. Food molecules bind to specific receptor proteins embedded in taste hairs on the taste cells. A particular taste sensation results from the pattern of specific taste receptor cells that bind food molecules.

Interwoven among the taste cells and wrapped around them is a network of sensory fibers. Stimulation of a taste receptor cell triggers an impulse on a nearby fiber, and the impulse is then conducted to the brain for interpretation of the sensation.

The taste cells in all taste buds look alike microscopically, but are of at least five types. Each type is most sensitive to a particular kind of chemical stimulus, producing at least five primary taste (gustatory) sensations.

Taste Sensations

The five primary taste sensations are:

1. *Sweet,* such as table sugar
2. *Sour,* such as a lemon
3. *Salty,* such as table salt
4. *Bitter,* such as caffeine or quinine
5. Umami (savory) (a Japanese term meaning "delicious"), a response to certain amino acids associated with meats and their chemical relatives, such as monosodium glutamate (MSG)

Some investigators recognize other taste sensations—*alkaline* and *metallic.*

A flavor results from either one primary sensation or a combination of the primary sensations. Experiencing flavors involves tasting, which reflects the concentrations of stimulating chemicals, as well as smelling and feeling the texture and temperature of foods. Furthermore, the chemicals in some foods—chili peppers and ginger, for instance—may stimulate pain receptors, which cause a burning sensation. In fact, the chemical in chili peppers that tastes "hot"—capsaicin—actually stimulates warm receptors.

Experiments indicate that each taste cell responds to one taste sensation only, with distinct receptors. It was historically thought that each taste sensation was associated with a particular area of the tongue. However, current research suggests that all types are scattered over the entire surface of the tongue.

Taste sensation, like the sense of smell, undergoes adaptation rapidly. Moving bits of food over the surface of the tongue to stimulate different receptors at different moments keeps us from losing taste due to sensory adaptation.

Our varied taste sensations may help us stay healthy. Sweet tastes direct us to energy-providing carbohydrates, sour tastes compel us to avoid foods containing dangerous acids, salty foods entice us to eat sodium, bitter tastes help us avoid poisons, and savory umami taste sensations increase our protein intake.

Taste Pathways

Sensory impulses from taste receptor cells in the tongue travel on fibers of the facial, glossopharyngeal, and vagus nerves into the medulla oblongata. From there, the impulses ascend to the thalamus and are directed to the gustatory cortex, in the parietal lobe of the cerebrum, along a deep part of the lateral sulcus (see fig. 9.29).

PRACTICE 10.6

1. Why is saliva necessary for the sense of taste?
2. Name the five primary taste sensations.
3. Trace a sensory impulse from a taste receptor to the cerebral cortex.

10.7 | Sense of Hearing

LEARN

1. Explain the function of each part of the ear.

The organ of hearing, the ear, has outer, middle, and inner parts. The ear also functions in the sense of equilibrium.

Outer (External) Ear

The outer ear consists of three parts. The first is an outer, funnel-like structure called the **auricle** (aw′ri-kl) or pinna. The second is an S-shaped tube called the **external acoustic meatus** (mē-ā′tus), or external auditory canal, that leads inward through the temporal bone for about 2.5 centimeters (fig. 10.6). The meatus terminates with the third part, the **eardrum** or tympanic membrane.

The transmission of vibrations through matter produces sound. These vibrations travel in waves, much like ripples on the surface of a pond. The higher the wave, the louder the sound. The more waves per second, the higher the frequency, or pitch, of the sound. Vibrating strings on a guitar or reeds on an oboe produce the sounds of these musical instruments, and vibrating vocal folds (vocal cords) in the larynx produce the voice. The auricle of the ear helps collect sound waves traveling through the air and directs them into the external acoustic meatus. At the end of the meatus, the sound waves reach the eardrum.

The eardrum is a semitransparent membrane covered by a thin layer of skin on its outer surface and by a

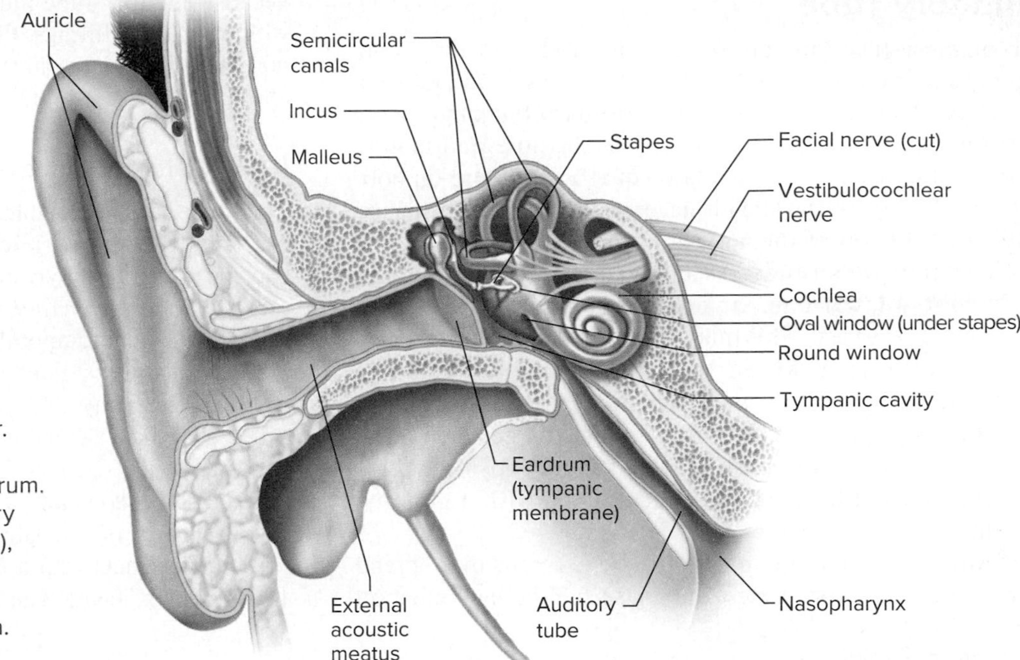

Figure 10.6 Major parts of the ear. The outer ear includes the auricle, external acoustic meatus, and eardrum. The middle ear includes the auditory ossicles (malleus, incus, and stapes), the oval window, and the round window. The inner ear includes the semicircular canals and the cochlea.
APR

PRACTICE FIGURE 10.6

Which cranial bone is shown in this figure?

Answer can be found in Appendix E.

mucous membrane on the inside. It has an oval margin and is cone-shaped, with the apex of the cone directed inward. The attachment of one of the auditory ossicles (the malleus) maintains the eardrum's cone shape. Sound waves that enter the external acoustic meatus change the pressure on the eardrum, which vibrates back and forth in response and thus reproduces the vibrations of the sound-wave source.

Middle Ear

The middle ear, or *tympanic cavity,* is an air-filled space in the temporal bone. It contains three small bones called **auditory ossicles** (aw'di-to"rē os'i-klz): the *malleus,* the *incus,* and the *stapes* (fig. 10.7). Tiny ligaments attach them to the wall of the tympanic cavity, and they are covered by a mucous membrane. These bones bridge the eardrum and the inner ear, transferring vibrations between these parts. Specifically, the malleus attaches to the eardrum, and when the eardrum vibrates, the malleus vibrates in unison. The malleus causes the incus to vibrate, and the incus passes the movement on to the stapes. An oval ligament holds the stapes to an opening in the wall of the tympanic cavity called the **oval window,** which leads into the inner ear. Vibration of the stapes at the oval window moves a fluid in the inner ear, which stimulates the hearing receptors.

The auditory ossicles help increase (amplify) the force of vibrations as they pass from the eardrum to the oval window, in addition to transferring vibrations. The vibrational force concentrates as it moves from the outer to the inner ear because the ossicles transmit vibrations from the relatively large surface of the eardrum to a much smaller area at the oval window. As a result, the pressure (per square millimeter) that the stapes applies on the oval window is many times greater than the pressure that sound waves exert on the eardrum.

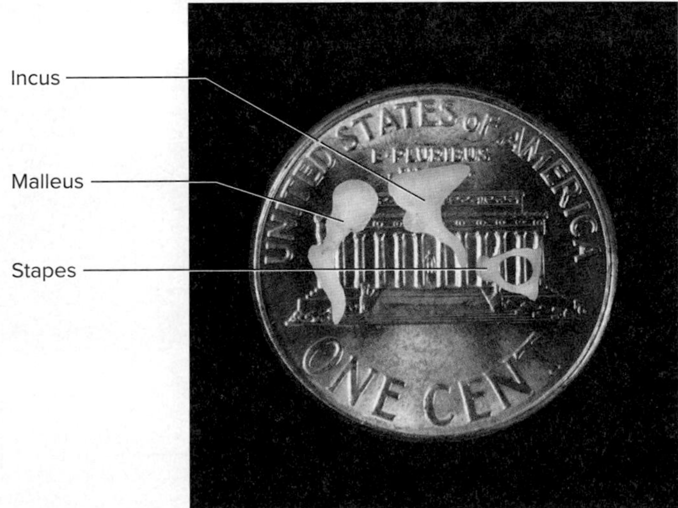

Figure 10.7 The auditory ossicles—the malleus, incus, and stapes—are bones that bridge the eardrum and the inner ear (2.5x). Comparison to a penny emphasizes their tiny size.
J and J Photography

Auditory Tube

An **auditory tube** (aw′di-to″rē tūb), or eustachian tube, connects each middle ear to the back of the nasal cavity (nasopharynx). This tube conducts air between the tympanic cavity and the outside of the body by way of the nose and mouth. The auditory tube helps maintain equal air pressure on both sides of the eardrum, which is necessary for normal hearing.

The function of the auditory tube is noticeable during rapid changes in altitude. As a person moves from a higher altitude to a lower one, air pressure on the outside of the eardrum increases. This may push the eardrum inward, impairing hearing. When the air pressure difference is great enough, air movement through the auditory tube equalizes the pressure on both sides of the eardrum, and the membrane moves back into its regular position. This restores normal hearing and is associated with a popping sound.

Mucous membrane infections of the throat may spread through the auditory tubes and cause middle ear infection because auditory tube mucous membranes connect directly with middle ear linings. Pinching a nostril when blowing the nose may force material from the throat up the auditory tube and into the middle ear.

Inner (Internal) Ear

The inner ear is a complex system of communicating chambers and tubes called a **labyrinth** (lab′ĭ-rinth). Each ear has two parts to the labyrinth—the *bony (osseus) labyrinth* and the *membranous labyrinth* (fig. 10.8). The bony labyrinth is a cavity within the temporal bone. The membranous labyrinth is a tube of similar shape that lies within the bony labyrinth. Between the bony and membranous labyrinths is a fluid called **perilymph,** which is secreted by cells in the wall of the bony labyrinth. The membranous labyrinth contains another fluid, called **endolymph.**

The parts of the labyrinths include three membranous semicircular ducts within three bony **semicircular canals,** and a **cochlea** (kok′lē-ah). The semicircular canals and associated

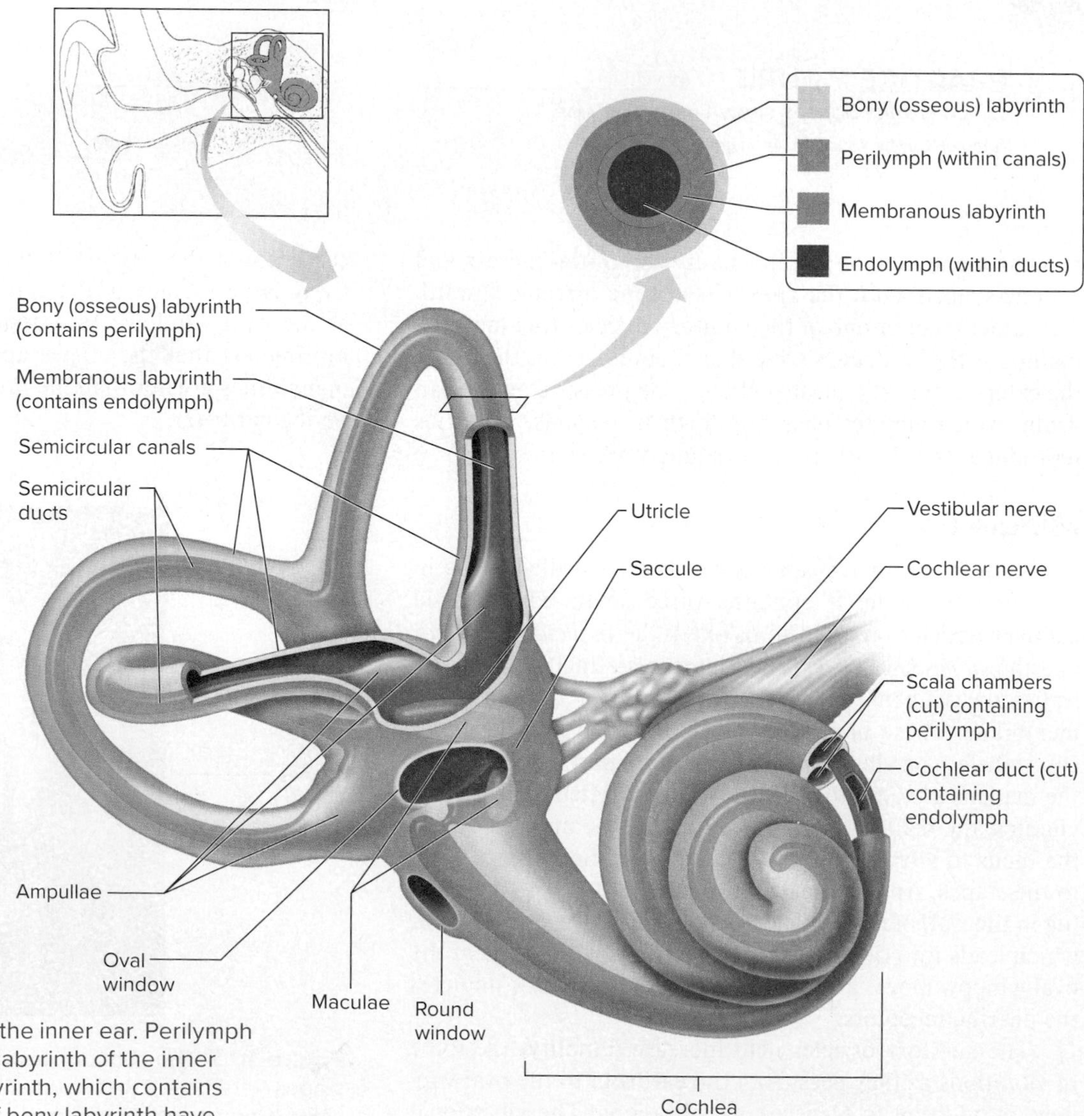

Figure 10.8 A closer look at the inner ear. Perilymph separates the bony (osseous) labyrinth of the inner ear from the membranous labyrinth, which contains endolymph. Note that areas of bony labyrinth have been removed to reveal underlying structures.

structures provide a sense of equilibrium (discussed in section 10.8, Sense of Equilibrium). The cochlea functions in hearing.

The cochlea has a bony core and a thin, bony shelf that extends out from the core and coils around it. The shelf divides the bony labyrinth of the cochlea into upper and lower compartments. The upper compartment, called the *scala vestibuli,* leads from the oval window to the tip of the cochlea. The lower compartment, the *scala tympani,* extends from the tip of the cochlea to a membrane-covered opening in the wall of the middle ear called the **round window** (see figs. 10.6 and 10.8).

The part of the membranous labyrinth within the cochlea is called the *cochlear duct.* It lies between the two bony compartments and ends as a closed sac near the tip of the cochlea. The cochlear duct is separated from the scala vestibuli by a *vestibular membrane* (Reissner's membrane) and from the scala tympani by a *basilar membrane* (**fig. 10.9**).

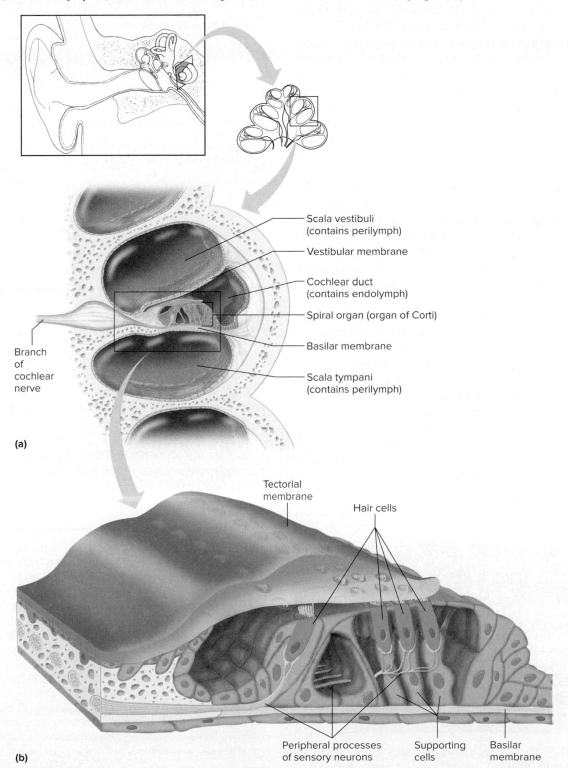

Figure 10.9 The cochlea. **(a)** Cross section of the cochlea. **(b)** The spiral organ and the tectorial membrane. 

The basilar membrane has many thousands of elastic fibers, allowing it to move in response to sound vibrations. Sound vibrations entering the perilymph at the oval window travel along the scala vestibuli and pass through the vestibular membrane and into the endolymph of the cochlear duct, where they move the basilar membrane.

After passing through the basilar membrane, the vibrations enter the perilymph of the scala tympani. Their forces are dissipated into the air in the tympanic cavity by movement of the membrane covering the round window.

The **spiral organ** (organ of Corti) contains the hearing receptors. It is located on the upper surface of the basilar membrane and stretches from the apex to the base of the cochlea (fig. 10.9). The receptor cells, called *hair cells,* are organized in rows and have many hairlike processes that project into the endolymph of the cochlear duct. Above these hair cells is a *tectorial membrane* attached to the bony shelf of the cochlea, passing over the receptor cells and contacting the tips of their hairs.

As sound vibrations move the basilar membrane, the hairs move back and forth against the tectorial membrane, and the resulting mechanical deformation of the hairs stimulates the hair cells (figs. 10.9 and 10.10). Hair cells at different locations along the length of the cochlear duct respond to different frequencies (pitch) of sound vibrations. This enables us to hear sounds of different pitch simultaneously.

Hair cells are epithelial but function somewhat like neurons. For example, when a hair cell is at rest, its membrane is polarized. When it is stimulated, selective ion channels open, depolarizing the membrane and making it more permeable to calcium ions. The hair cell has no axon or dendrites, but it has neurotransmitter-containing vesicles near its base. As calcium ions diffuse into the cell, some of these vesicles fuse with the cell membrane and release a neurotransmitter by exocytosis. The neurotransmitter stimulates the dendrites of nearby sensory neurons. In response, these neurons send action potentials along the cochlear branch of the vestibulocochlear nerve to the auditory cortex of the temporal lobe of the brain.

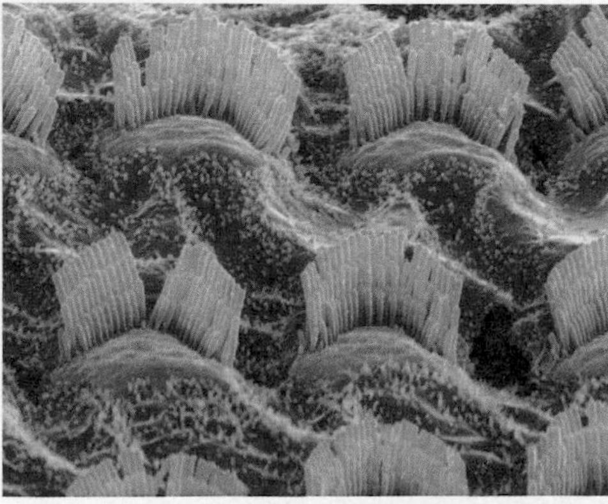

Figure 10.10 Scanning electron micrograph of hair cells in the spiral organ (5,000x). Clouds Hill Imaging Ltd./Science Source

TABLE 10.1	Steps in the Generation of Sensory Impulses from the Ear
1.	Sound waves enter external acoustic meatus.
2.	Sound waves cause eardrum to reproduce vibrations coming from sound source.
3.	Auditory ossicles amplify and transfer vibrations to end of stapes.
4.	Movement of stapes at oval window transfers vibrations to perilymph in scala vestibuli.
5.	Vibrations pass through vestibular membrane and enter endolymph of cochlear duct, where they move the basilar membrane.
6.	Different frequencies of vibration of basilar membrane stimulate different sets of receptor cells.
7.	As a receptor cell depolarizes, its membrane becomes more permeable to calcium ions.
8.	Inward diffusion of calcium ions causes vesicles at the base of the receptor cell to release neurotransmitter.
9.	Neurotransmitter stimulates dendrites of nearby sensory neurons.
10.	Sensory impulses are triggered on fibers of the cochlear branch of vestibulocochlear nerve.
11.	Auditory cortices of temporal lobes interpret sensory impulses.

The ear of a young person with normal hearing can detect sound waves with frequencies ranging from 20 to more than 20,000 vibrations per second. The range of greatest sensitivity is 2,000 to 3,000 vibrations per second. Table 10.1 summarizes the steps of hearing.

Recall from section 9.6, Impulse Conduction, that action potentials are all-or-none. More-intense stimulation of the hair cells causes more action potentials per second to reach the auditory cortex, and we sense a louder sound.

OF **INTEREST** Units called *decibels* (dB) measure sound intensity on a logarithmic scale. The decibel scale begins at 0 dB, which is the intensity of the sound that is least perceptible by a normal human ear. A sound of 10 dB is 10 times as intense as the least perceptible sound; a sound of 20 dB is 100 times as intense; and a sound of 30 dB is 1,000 times as intense. A whisper has an intensity of about 40 dB, normal conversation measures 60 to 70 dB, and heavy traffic produces about 80 dB. A sound of 120 dB, common at a rock concert, produces discomfort, and a sound of 140 dB, such as that emitted by a jet plane at takeoff, causes pain. Frequent or prolonged exposure to sounds with intensities above 85 dB can damage hearing receptors and cause permanent hearing loss.

Auditory Pathways

The nerve fibers associated with hearing enter the auditory pathways, which pass into the auditory cortices of the temporal lobes of the cerebrum. Here they are interpreted. On the way, some of these fibers cross over, so that impulses arising from each ear are interpreted on both sides of the brain. Consequently, damage to a temporal lobe on one side of the brain does not necessarily cause complete hearing loss in the ear on that side.

Several factors cause partial or complete hearing loss. Interference with the transmission of vibrations to the inner ear is called *conductive hearing loss.* Conductive hearing loss may be due to plugging of the external acoustic meatus or to changes in the eardrum or auditory ossicles. For example, the eardrum may harden as a result of disease and become less responsive to sound waves, or disease or injury may tear or perforate the eardrum.

Damage to the cochlea, auditory nerve, or auditory pathways can cause *sensorineural hearing loss.* Loud sounds, tumors in the central nervous system, brain damage from vascular accidents, or use of certain drugs can also cause sensorineural hearing loss.

PRACTICE 10.7

1. How are sound waves transmitted through the outer, middle, and inner ears?
2. Distinguish between the osseous and membranous labyrinths.
3. Describe the spiral organ.
4. Distinguish between conductive and sensorineural hearing loss.

10.8 | Sense of Equilibrium

LEARN

1. Distinguish between static and dynamic equilibrium.

The sense of equilibrium (balance) is really two senses—static equilibrium and dynamic equilibrium—that come from different sensory organs. The organs of **static equilibrium** (stat′ik e″kwĭ-lib′re-um) sense the position of the head, maintaining balance, stability, and posture when the head and body are still. When the head and body suddenly move or rotate, the organs of **dynamic equilibrium** (di-nam′ik e″kwĭ-lib′re-um) detect such motion and aid in maintaining balance.

Static Equilibrium

The organs of static equilibrium are in the **vestibule,** a bony chamber between the semicircular canals and the cochlea. The membranous labyrinth inside the vestibule consists of two expanded chambers—a **utricle** (yoo′trĭ-kl) and a **saccule** (sak′ūl) (see fig. 10.8).

The saccule and utricle each have a tiny structure called a **macula** (mak′u-lah). Maculae have many hair cells, which serve as sensory receptors. The hairs of the hair cells project into a mass of gelatinous material, which has grains of calcium carbonate (otoliths) embedded in it. These particles add weight to the gelatinous structure.

Bending the head forward, backward, or to either side tilts the gelatinous masses of the maculae, and as they sag in response to gravity, the hairs projecting into them bend. This action causes the hair cells to signal the sensory neurons associated with them in a manner similar to that of hair cells associated with hearing. The resulting action potentials are conducted into the central nervous system on the vestibular branch of the vestibulocochlear nerve, informing the brain of the head's new position. The brain responds by adjusting the pattern of motor impulses to skeletal muscles, which contract or relax to maintain balance (fig. 10.11).

Dynamic Equilibrium

The organs of dynamic equilibrium are the three semicircular canals in the labyrinth. They detect motion of the head and aid in balancing the head and body during sudden movement. These canals lie at right angles to each other (see fig. 10.8).

Suspended in the perilymph of the bony portion of each semicircular canal is a membranous semicircular duct that ends in a swelling called an **ampulla** (am-pul′ah), which houses the sensory organs of the semicircular canals. Each of these sensory organs, called a **crista ampullaris** (kris′tah am-pul′ar-is), contains a number of sensory hair cells and supporting cells. Like the hairs of the maculae, the hair cells of the crista ampullaris extend upward into a dome-shaped, gelatinous mass called the *cupula* (fig. 10.12).

Rapid movement of the head or body stimulates the hair cells of the crista ampullaris (fig. 10.13). At such times, the semicircular canals move with the head or body, but the fluid inside the membranous ducts remains stationary. (Imagine turning rapidly while holding a full glass of water.) This action bends the cupula in one or more of the canals in a direction opposite that of the head or body movement, and the hairs embedded in it also bend. The stimulated hair cells signal their associated neurons, which conduct impulses to the brain. The brain interprets these impulses as a movement in a particular direction.

Parts of the cerebellum are particularly important in interpreting impulses from the semicircular canals. Analysis of such information allows the brain to predict the consequences of rapid body movements. By modifying signals to appropriate skeletal muscles, the cerebellum can maintain balance.

Other sensory structures aid in maintaining equilibrium. For example, certain mechanoreceptors (proprioceptors), particularly those associated with the joints of the

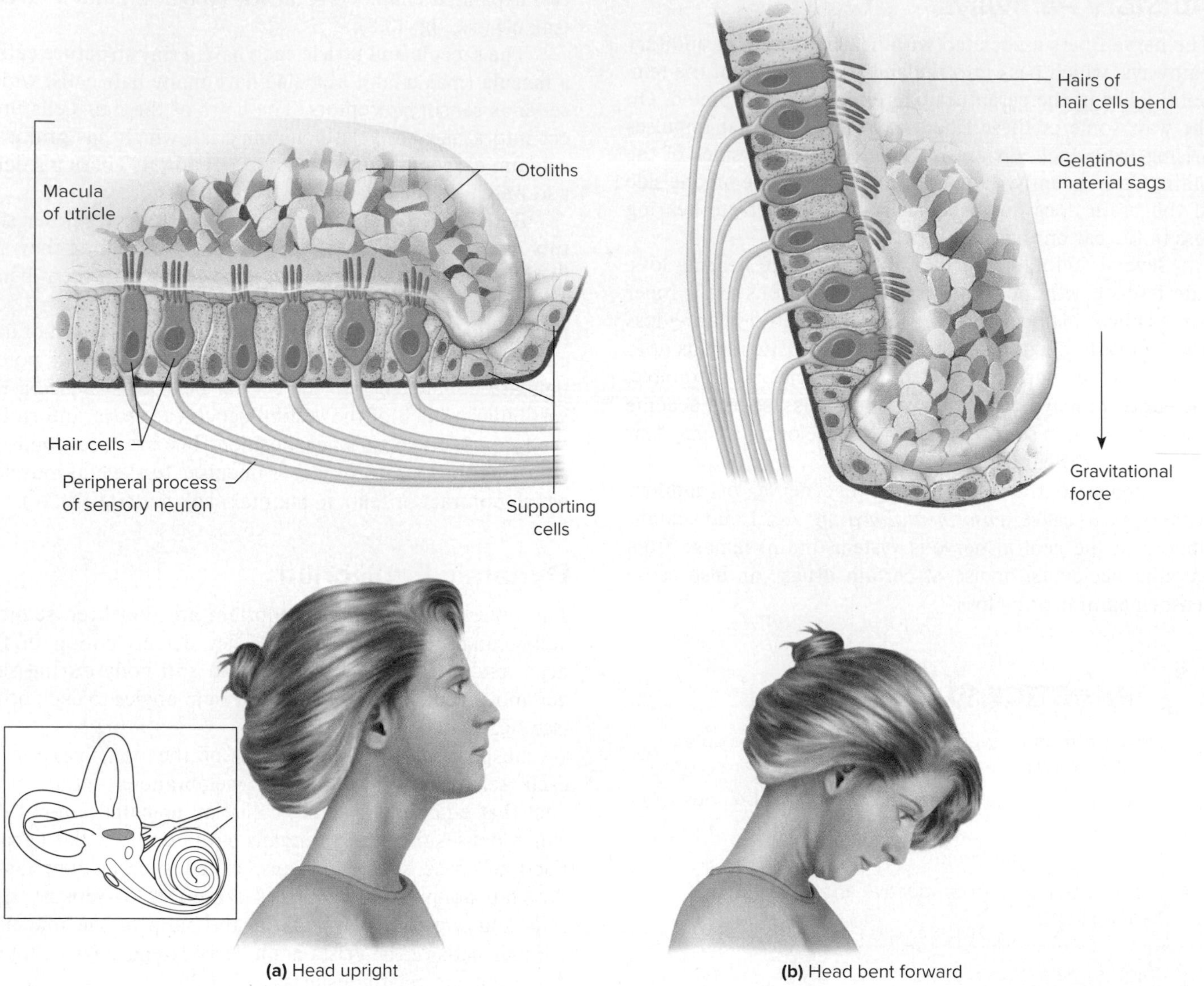

(a) Head upright

(b) Head bent forward

Figure 10.11 The maculae respond to changes in head position. **(a)** Macula of the utricle with the head in an upright position. **(b)** Macula of the utricle with the head bent forward.

neck, inform the brain about the position of body parts. In addition, the eyes detect changes in position that result from body movements. Such visual information is so important that even if the organs of equilibrium are damaged, a person may be able to maintain normal balance by keeping the eyes open and moving slowly.

The nausea, vomiting, dizziness, and headache of *motion sickness* arise from sensations that don't make sense. The eyes of a person reading in a moving car, for example, signal the brain that the person is stationary, because the print doesn't move. However, receptors in the skin detect bouncing, swaying, starting, and stopping as the inner ear detects movement. The contradiction triggers the symptoms. Similarly, in a passenger of an airplane flying through heavy turbulence, receptors in the skin and inner ear register the chaos outside, but the eyes focus on the immobile seats and surroundings.

To prevent or lessen the misery of motion sickness, focus on the horizon or an object in the distance ahead. Medications are available by pill (diphenhydramine and dimenhydrinate) and, for longer excursions, in a skin patch (scopolamine).

PRACTICE 10.8

1. Distinguish between static and dynamic equilibrium.
2. Which structures provide the sense of static equilibrium? Of dynamic equilibrium?
3. How does sensory information from other receptors help maintain equilibrium?

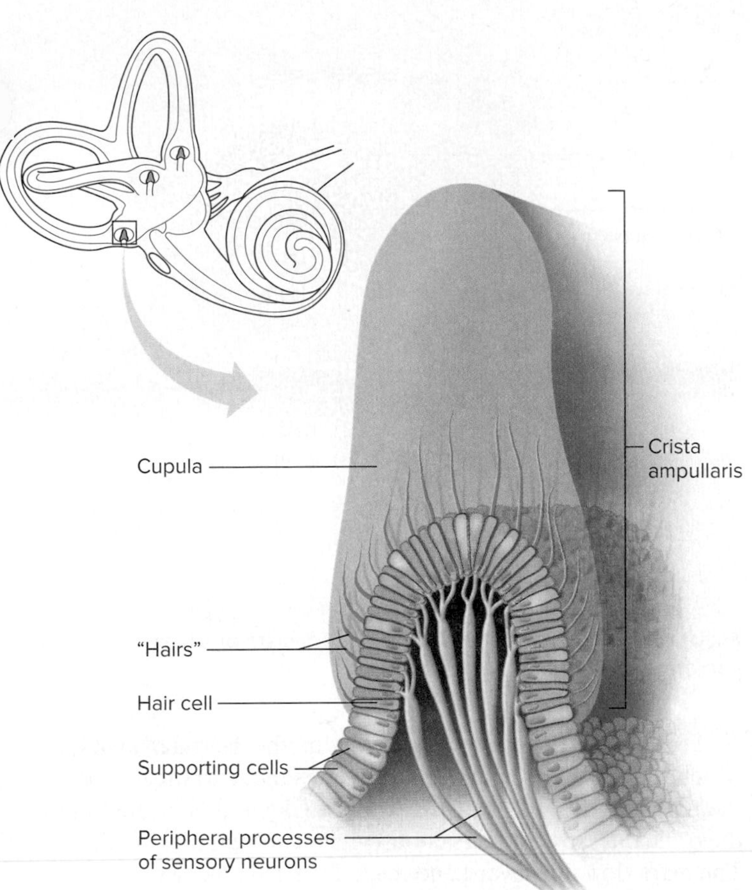

Figure 10.12 A crista ampullaris is located within the ampulla of each semicircular duct.

(a) Head in still position

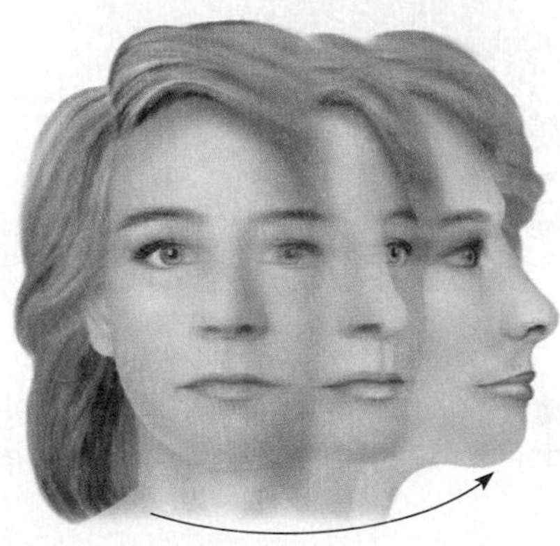

(b) Head rotating

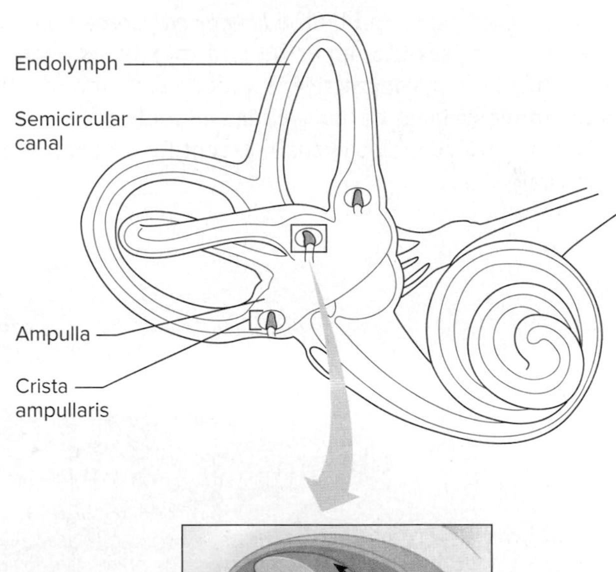

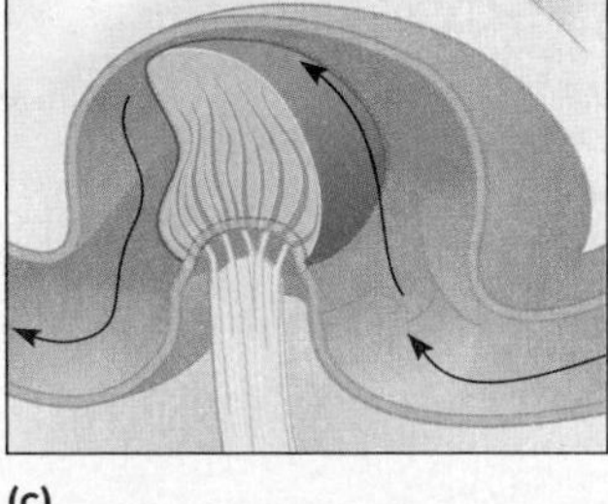

(c)

Figure 10.13 Equilibrium. **(a)** When the head is stationary, the cupula of the crista ampullaris remains upright. **(b)** and **(c)** When the head is moving rapidly, the cupula bends opposite the motion of the head, stimulating sensory receptors.

10.9 | Sense of Sight

LEARN

1. Explain the function of each part of the eye.
2. Explain how the eye refracts light.
3. Describe the visual nerve pathway.

The eye, the organ containing visual receptors, provides vision, with the assistance of *accessory organs.* These accessory organs include the eyelids and lacrimal apparatus, which protect the eye, and a set of extrinsic muscles, which move the eye.

Visual Accessory Organs

The eye, lacrimal gland, and associated extrinsic muscles are housed in the orbital cavity, or orbit, of the skull. Each orbit is lined with the periosteum of various bones, and also contains fat, blood vessels, nerves, and connective tissues.

Each **eyelid,** both upper and lower, has four layers—skin, muscle, connective tissue, and conjunctiva. The skin of the eyelid, which is the thinnest skin of the body, covers the lid's outer surface and fuses with its inner lining near the margin

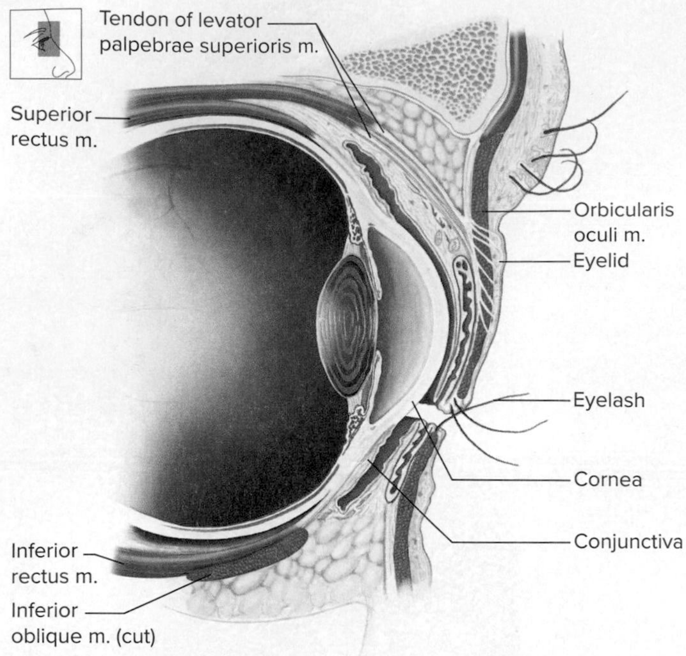

Figure 10.14 Sagittal section of the closed eyelids and anterior portion of the eye. (**m.** stands for **muscle.**)

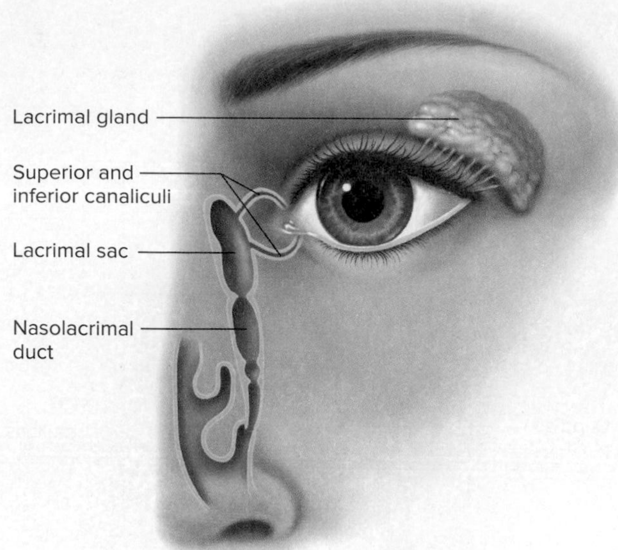

Figure 10.15 The lacrimal apparatus consists of a tear-secreting gland and a series of ducts.

of the lid. The eyelids are moved by the *orbicularis oculi* muscle (see fig. 8.18*a*), which acts as a sphincter and closes the lids when it contracts, and by the *levator palpebrae superioris* muscle, which raises the upper lid and thus helps open the eye (fig. 10.14). The **conjunctiva** is a mucous membrane that lines the inner surfaces of the eyelids and folds back to cover the anterior surface of the eyeball, except for its central portion (cornea).

The *lacrimal apparatus* consists of the **lacrimal gland,** which secretes tears, and a series of ducts that carry tears into the nasal cavity (fig. 10.15). The lacrimal gland is located in the orbit and secretes tears continuously through tiny tubules. The tears flow downward and medially across the eye.

Two small ducts (the superior and inferior canaliculi) collect tears, which then flow into the *lacrimal sac,* located in a deep groove of the lacrimal bone. From there, the tears flow into the *nasolacrimal duct,* which empties into the nasal cavity. Secretion by the lacrimal gland moistens and lubricates the surface of the eye and the lining of the lids. Tears

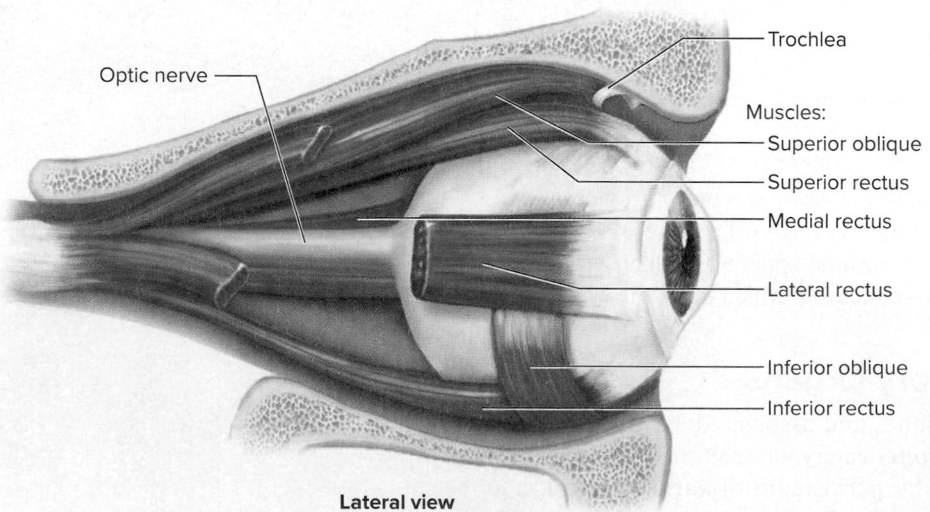

Figure 10.16 Extrinsic muscles of the right eye (lateral view). APR

PRACTICE FIGURE 10.16

Are the extrinsic muscles of the eye under voluntary control?

Answer can be found in Appendix E.

TABLE 10.2 Muscles Associated with the Eyelids and Eyes

Name	Innervation	Function
Muscles of the Eyelids		
Orbicularis oculi	Facial nerve (VII)	Closes eye
Levator palpebrae superioris	Oculomotor nerve (III)	Opens eye
Extrinsic Muscles of the Eyes		
Superior rectus	Oculomotor nerve (III)	Rotates eye upward and toward midline
Inferior rectus	Oculomotor nerve (III)	Rotates eye downward and toward midline
Medial rectus	Oculomotor nerve (III)	Rotates eye toward midline
Lateral rectus	Abducens nerve (VI)	Rotates eye away from midline
Superior oblique	Trochlear nerve (IV)	Rotates eye downward and away from midline
Inferior oblique	Oculomotor nerve (III)	Rotates eye upward and away from midline

also have an enzyme (*lysozyme*) that kills bacteria, reducing the risk of eye infections.

The **extrinsic muscles** of the eye arise from the bones of the orbit and insert by broad tendons on the eye's tough outer surface. Six extrinsic muscles move the eye in different directions. Any given eye movement may utilize more than one extrinsic muscle, but each muscle is associated with a primary action. Figure 10.16 illustrates the locations of these extrinsic muscles, and table 10.2 lists their functions, as well as the functions of the eyelid muscles.

PRACTICE 10.9

1. Explain how the eyelid moves.
2. Describe the conjunctiva.
3. What is the function of the lacrimal apparatus?

Structure of the Eye

The eye is a hollow, spherical structure about 2.5 centimeters in diameter. Its wall has three distinct layers—an outer (fibrous) layer, a middle (vascular) layer, and an inner (nervous) layer. The spaces within the eye are filled with fluids that help maintain its shape. Figure 10.17 shows the major parts of the eye.

Outer Layer

Light is often described as moving in straight lines called *rays* of light. The anterior sixth of the outer layer bulges forward as the transparent **cornea** (kor′nē-ah), the window of the eye. The cornea helps focus entering light rays. The cornea is composed largely of connective tissue with a thin surface layer of epithelium. It is transparent because it contains few cells and no blood vessels, and the cells and collagenous fibers form

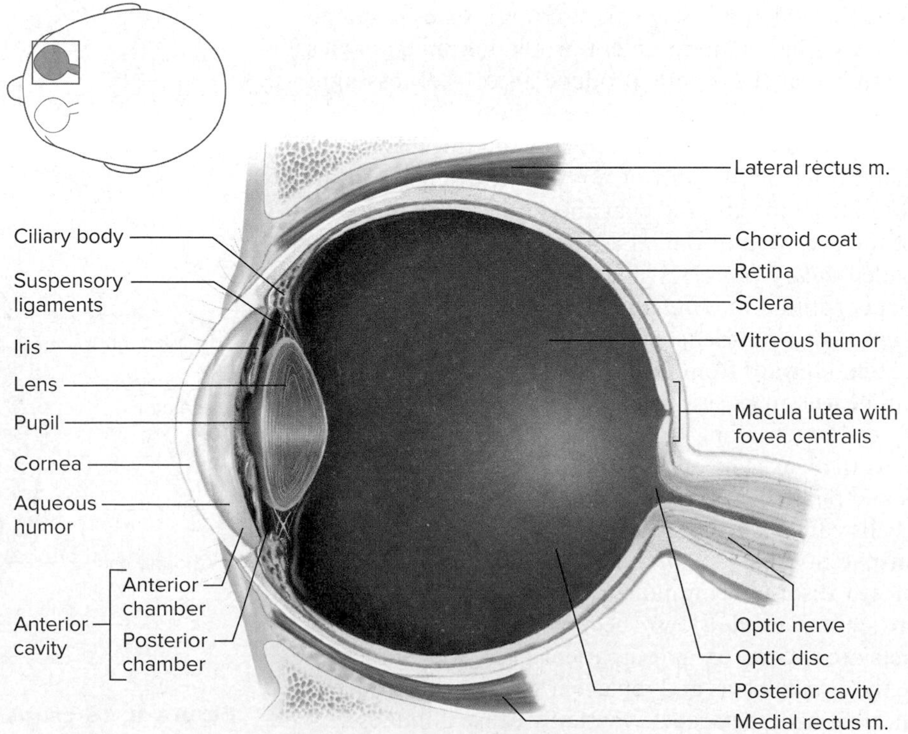

Figure 10.17 Transverse section of the right eye (superior view). APR

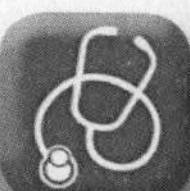

CLINICAL APPLICATION 10.2
Corneal Transplants

Worldwide, the most common cause of blindness is loss of transparency of the cornea. Each year, 40,000 corneal transplants are performed in the United States. Corneas are "immune privileged," not evoking an immune response, so anyone can donate to anyone. However, corneal transplants are effective only if the transplanted tissue includes stem cells normally found in a layer of cells, called the limbus, that separates the cornea from the conjunctiva. (The cornea itself does not contain stem cells.) Researchers performed limbal cell transplants on patients with corneal damage in one eye. They took the stem cells from the patients' healthy eyes, culturing the limbal cells into a bluish, translucent gel, which they then applied to the affected eyes. Vision returned. Limbal stem cell transplants may prove to be more effective than corneal transplants, which have been done since 1905.

unusually regular patterns. The use of stem cells in corneal transplants is discussed in Clinical Application 10.2.

The cornea is continuous with the **sclera** (skle'rah), the white portion of the eye. The sclera makes up the posterior five-sixths of the outer layer of the wall of the eye. It is opaque due to many large, disorganized, collagenous and elastic fibers. The sclera protects the eye and serves as an attachment for the extrinsic muscles. In the back of the eye, the **optic nerve** and certain blood vessels pierce the sclera. Clinical Application 10.3 describes headaches that include visual symptoms.

Middle Layer

The middle layer of the wall of the eye includes the choroid coat, ciliary body, and iris (fig. 10.17). The **choroid coat** (ko'roid kōt), in the posterior five-sixths of the globe of the eye, is loosely joined to the sclera and is honeycombed with blood vessels, which nourish surrounding tissues. The choroid coat also has many pigment-producing melanocytes. The melanin that these cells produce absorbs excess light, which helps keep the inside of the eye dark.

The **ciliary body** (sil'ē-er"ē bod'ē) is the thickest part of the middle layer of the wall of the eye. It extends forward and inward from the choroid coat and forms a ring inside the front of the eye. Within the ciliary body are many radiating folds called *ciliary processes* and groups of smooth muscle cells that constitute the *ciliary muscle.*

Many strong but delicate fibers, called *suspensory ligaments,* extend inward from the ciliary processes and hold the transparent **lens** in position (fig. 10.18). The distal ends of these fibers attach along the margin of a thin capsule that surrounds the lens. The body of the lens lies directly behind the iris and pupil and is composed of highly specialized epithelial cells called *lens fibers.* The cytoplasm of these cells is the transparent substance of the lens.

An eye disorder common in older people is *cataract.* The lens or its capsule slowly becomes cloudy and opaque. Cataracts are treated on an outpatient basis with surgery to remove the clouded lens and replace it with an artificial lens. Without treatment, cataracts eventually cause blindness.

The ciliary muscle and suspensory ligaments, along with the structure of the lens itself, enable the lens to adjust shape to facilitate focusing, a phenomenon called **accommodation** (ah-kom"o-dā'shun). The lens is enclosed by a clear capsule composed largely of elastic fibers. This elastic nature keeps the lens under constant tension, and enables it to assume a globular shape. The suspensory ligaments attached to the margin of the capsule are also under tension. When they pull outward, flattening the capsule and the lens inside, the lens focuses on distant objects (fig. 10.19*b*). However, if the tension on the suspensory ligaments relaxes, the elastic lens capsule rebounds, and the lens surface becomes more convex—focused for viewing closer objects (fig. 10.19*a*).

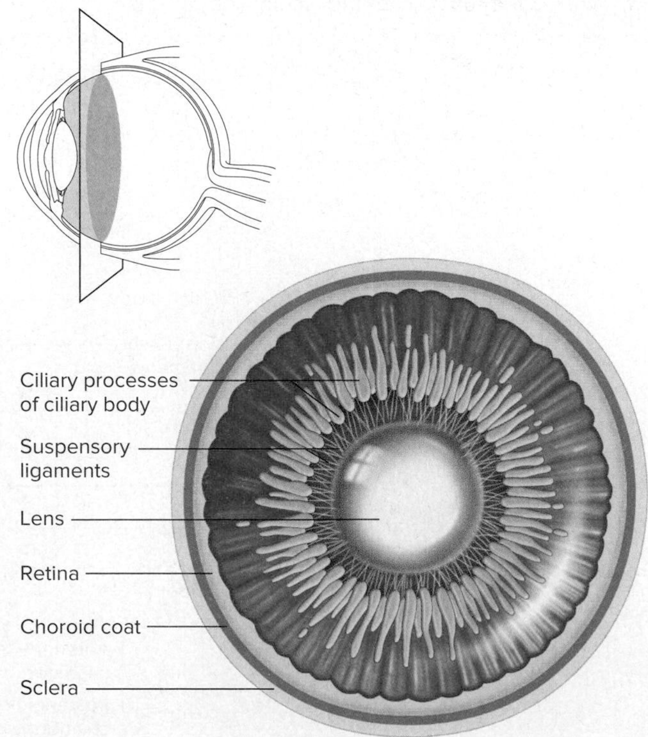

Figure 10.18 Lens and ciliary body viewed from behind.

CLINICAL APPLICATION 10.3

Headache

Headaches are common. The cells of the nervous tissue in the brain lack pain receptors, but nearly all the other tissues of the head, including the meninges and blood vessels, are richly innervated and can be the source of headache pain.

Many a *migraine* sufferer knows that an attack is imminent early in the morning, when an ominous dull throbbing begins, often on one side of the head. Migraine may be more than head pain—the person feels unwell in a general sense, and an attack can be disabling, lasting from a few hours to several days. In a migraine, certain cranial blood vessels constrict, producing a localized cerebral blood deficiency. When vasodilation quickly follows, a severe headache results. Several types of drugs, such as the triptans, effectively relieve migraines, but they are best taken at the first sign of illness. Several drugs may need to be tried to find an effective one. For some people, keeping a diary of events before an attack can reveal a trigger, such as eating chocolate or exposure to low-pressure weather systems.

In some individuals, migraine begins with an "aura" of shimmery bright lights in the peripheral vision. Often accompanying the aura is "photophobia," which is head pain when exposed to light. Photophobia arises from a group of brain neurons closely associated with the optic nerve (see fig. 10.26). Completely blind migraine sufferers, whose optic nerves do not function, do not experience photophobia. But migraine sufferers who are "legally blind," with partially degenerated retinas but intact and functional optic nerves, do experience pain from light.

Other headaches are associated with stressful life situations that cause fatigue, emotional tension, anxiety, or frustration. These conditions can trigger various physiological changes, such as prolonged contraction of the skeletal muscles in the forehead, sides of the head, or back of the neck, which stimulate pain receptors and produce a *tension headache*. More severe *vascular headaches* accompany constriction or dilation of the cranial blood vessels. For example, the throbbing headache of a "hangover" from drinking too much alcohol may be due to blood pulsating through dilated cranial vessels. Yet other causes of headaches include sensitivity to food additives, high blood pressure, increased intracranial pressure due to a tumor or to blood escaping from a ruptured vessel, decreased cerebrospinal fluid pressure following a lumbar puncture, and sensitivity to or withdrawal from certain drugs.

The ciliary muscle controls the actions of the suspensory ligaments in accommodation. One set of ciliary muscle cells extends back from fixed points in the sclera to the choroid coat. When the muscle contracts, the choroid coat is pulled forward and the ciliary body shortens. This action relaxes the suspensory ligaments, and the lens thickens in response (see fig. 10.19*a*). When the ciliary muscle relaxes, tension on the suspensory ligaments increases, and the lens becomes thinner and less convex again (see fig. 10.19*b*).

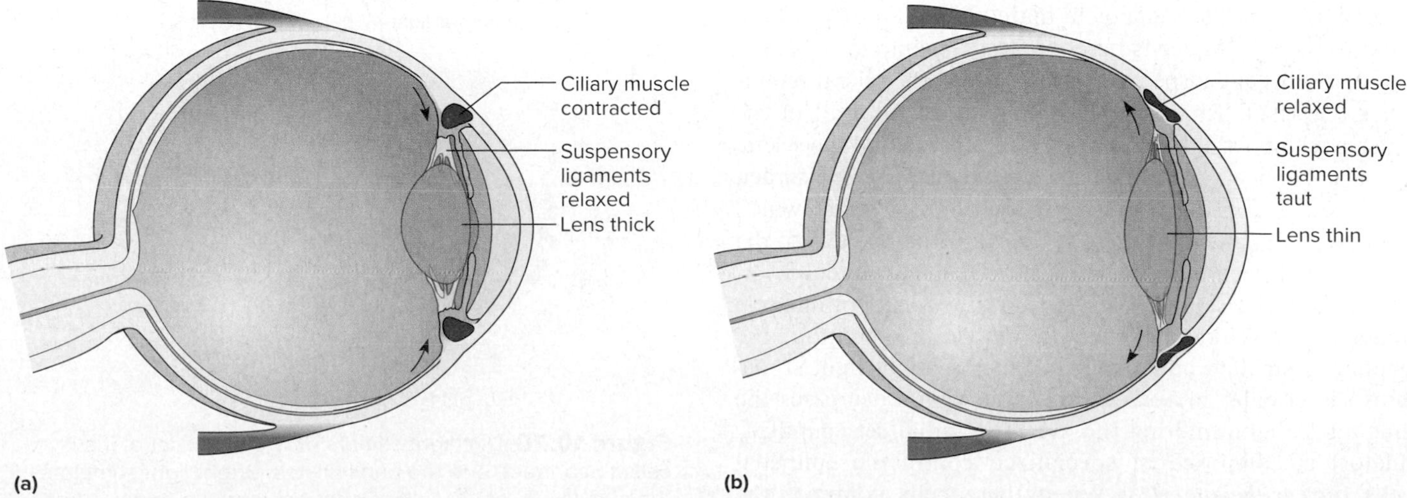

Figure 10.19 Accommodation. **(a)** The lens thickens as the ciliary muscle contracts. **(b)** The lens thins as the ciliary muscle relaxes.

CLINICAL APPLICATION 10.4
Glaucoma

An eye disorder called *glaucoma* develops when aqueous humor forms faster than it is removed. As fluid accumulates in the anterior chamber of the eye, fluid pressure rises and is transmitted to all parts of the eye. In time, the building pressure squeezes shut blood vessels that supply the receptor cells of the retina. Cells that are robbed of nutrients and oxygen in this way may die, and permanent blindness can result.

When diagnosed early, glaucoma can usually be treated successfully with drugs, laser surgery, or traditional surgery. All of these treatments promote the outflow of aqueous humor. In its early stages, glaucoma typically produces no symptoms, so discovery of the condition usually depends on measuring intraocular pressure (pressure inside the eye), using an instrument called a *tonometer.*

PRACTICE 10.9

4. Describe the outer and middle layers of the eye.
5. What factors contribute to the transparency of the cornea?
6. How does the shape of the lens change during accommodation?
7. Why would reading for a long time cause eye fatigue, while looking at a distant scene is restful?

The **iris** (i′ris) is a thin diaphragm composed mostly of connective tissue and smooth muscle fibers. From the outside, the iris is the colored part of the eye. The iris extends forward from the periphery of the ciliary body and lies between the cornea and lens (see fig. 10.17). The iris divides the space (anterior cavity) separating these parts into an *anterior chamber* (between the cornea and the iris) and a *posterior chamber* containing the lens (between the iris and the vitreous body).

The epithelium on the inner surface of the ciliary body secretes a watery fluid called **aqueous humor** (ā′kwē-us hu′mor) into the posterior chamber. The fluid circulates from this chamber through the **pupil** (pū′pil), a circular opening in the center of the iris, and into the anterior chamber. Aqueous humor fills the space between the cornea and lens, helps nourish these parts, and aids in maintaining the shape of the front of the eye. Aqueous humor leaves the anterior chamber through veins and a special drainage canal, the scleral venous sinus (canal of Schlemm). This sinus is in the wall of the anterior chamber at the junction of the cornea and the sclera. Clinical Application 10.4 discusses glaucoma, an eye disorder.

The iris controls the size of the pupil, through which light passes as it enters the eye. The contractile cells of the iris are organized into two groups: a *circular set* and a *radial set.* The circular set (pupillary constrictor) is smooth muscle and acts as a sphincter. When the muscle cells contract, the pupil gets smaller, and less light enters. Bright light stimulates the circular muscles to contract, which decreases the amount of light entering the eye. The radial set (pupillary dilator) is composed of specialized contractile epithelial cells (*myoepithelial cells*). When these cells contract, the pupil's diameter increases and more light enters (fig. 10.20).

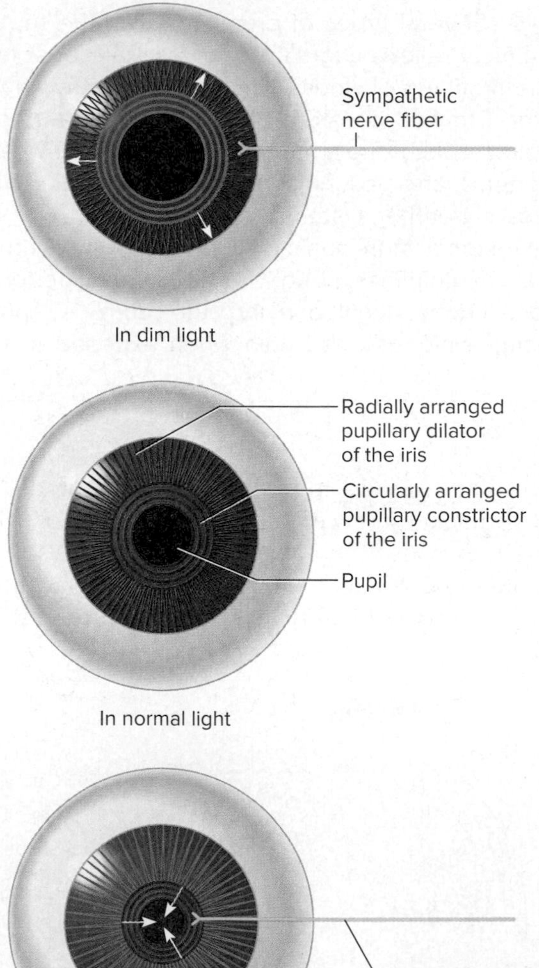

Figure 10.20 Dim light stimulates the radial set of iris cells to contract, and the pupil dilates. Bright light stimulates the circular set of iris cells to contract, and the pupil constricts.

Dim light stimulates the radial cells to contract, which increases the amount of light entering the eye.

Inner Layer

The inner layer of the wall of the eye consists of the **retina** (ret'ĭ-nah), which contains the visual receptor cells (photoreceptors). The retina is a nearly transparent sheet of tissue continuous with the optic nerve in the back of the eye and extending forward as the inner lining of the eyeball. The retina ends just behind the margin of the ciliary body.

The retina is thin and delicate, but its structure is quite complex. It has a number of distinct layers, as figures 10.21 and 10.22 illustrate.

In the central region of the retina is a yellowish spot called the **macula lutea.** A depression in its center, called the **fovea centralis** (fō'vē-ah sen-tral'is), is in the region of the retina that produces the sharpest vision (see figs. 10.17 and 10.23).

Just medial to the fovea centralis is an area called the **optic disc** (op'tik disk) (fig. 10.23). Here, nerve fibers from the retina leave the eye and form the optic nerve. Because the optic disc region does not have photoreceptors, it is commonly known as the *blind spot* of the eye. A central artery and vein also pass through the optic disc. These vessels are continuous with the capillary networks of the retina. Along with vessels in the underlying choroid coat, they supply blood to the cells of the inner layer.

The space bounded by the lens, ciliary body, and retina is the largest compartment of the eye and is called the *posterior cavity* (see fig. 10.17). It is filled with a transparent, jellylike fluid called **vitreous humor** (vit'rē-us hū'mor), which along with collagen fibers forms the *vitreous body.* The vitreous body supports the internal parts of the eye and helps maintain the eye's shape.

As a person ages, tiny, dense clumps of gel or deposits of crystal-like substances form in the vitreous humor. When these clumps cast shadows on the retina, the person sees small, moving specks in the field of vision, called *floaters.*

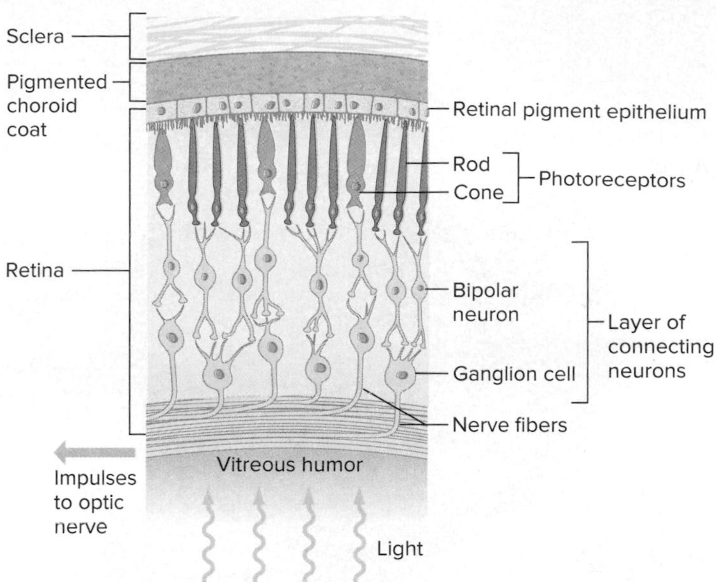

Figure 10.21 The retina consists of several cell layers. Light penetrates a layer of connecting neurons to reach the rods and cones, which are the photoreceptors. The retinal pigment epithelium absorbs stray light.

PRACTICE 10.9

8. Explain the source of aqueous humor, and trace its path through the eye.
9. How does the pupil respond to changes in light intensity?
10. Describe the structure of the retina.

Light Refraction and Corrective Lenses

When a person sees an object, either the object is giving off light or it is reflecting light from another source. Rays of light enter the eye, and an image of the object is focused

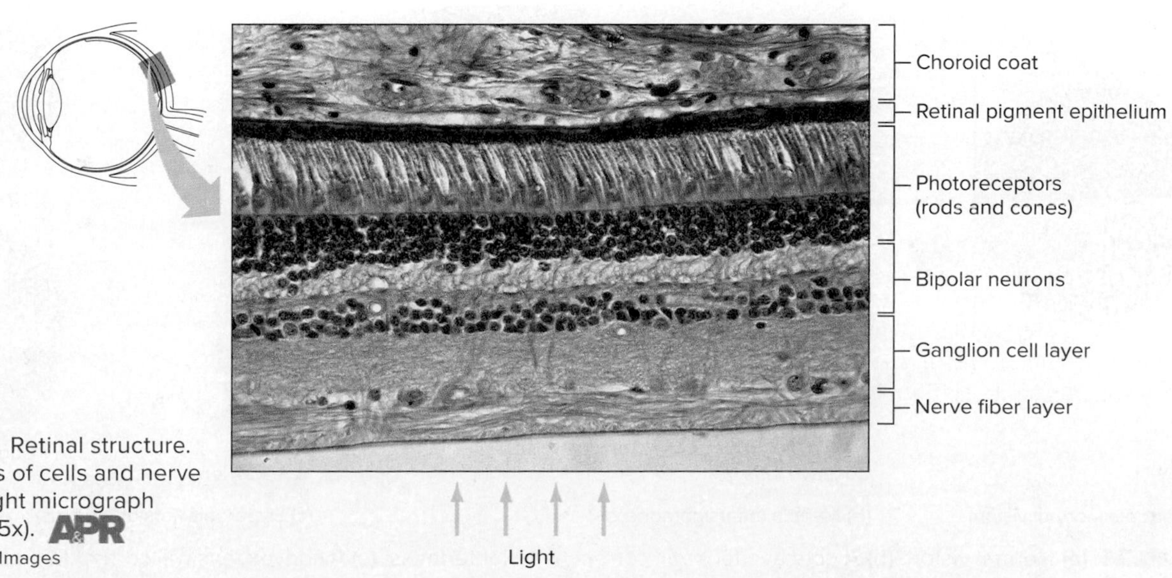

Figure 10.22 Retinal structure. Note the layers of cells and nerve fibers in this light micrograph of the retina (75x). APR
Ed Reschke/Getty Images

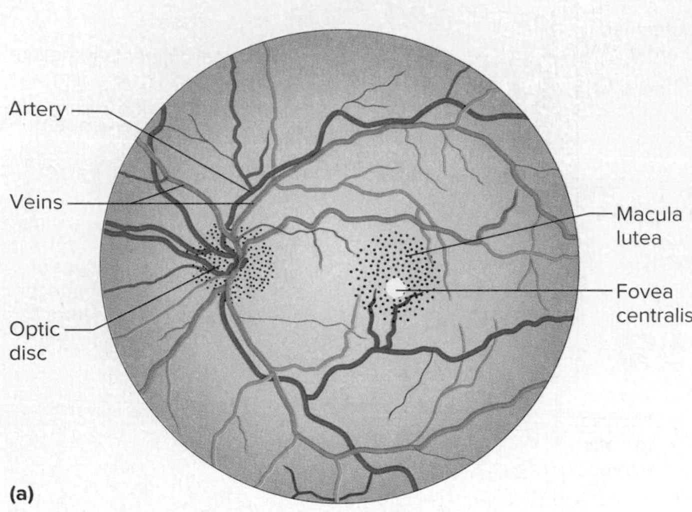

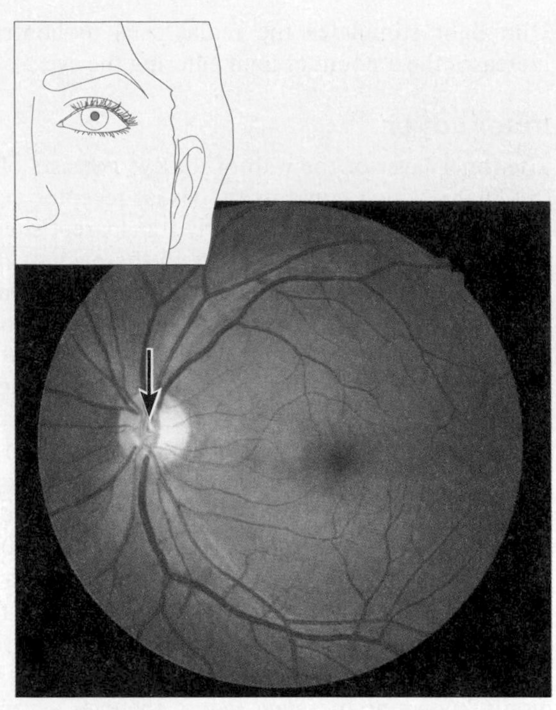

Figure 10.23 The retina. **(a)** Major features of the retina. **(b)** Nerve fibers leave the retina of the eye in the area of the optic disc (arrow) to form the optic nerve in this magnified view of the retina (53x). (b): Thalerngsak Mongkolsin/Shutterstock

on the retina. Focusing bends the light rays, a phenomenon called **refraction** (re-frak′shun), so the image falls on the fovea centralis of the retina.

In normal vision (emmetropia), light rays focus sharply on the retina, much as a motion picture image is focused on a screen for viewing. If the lens cannot properly focus the light rays on the retina, two common conditions are possible: **nearsightedness** (myopia) or **farsightedness** (hyperopia) (fig. 10.24). In nearsightedness, the light rays focus in front of the retina. They then scatter, resulting in a blurred image. The correction for this is a concave lens that causes the light rays to travel a bit further to focus on the retina. The corrective measure for farsightedness is the opposite. A convex lens will focus the light rays sooner.

PRACTICE 10.9

11. What is refraction?
12. What parts of the eye provide refracting surfaces?

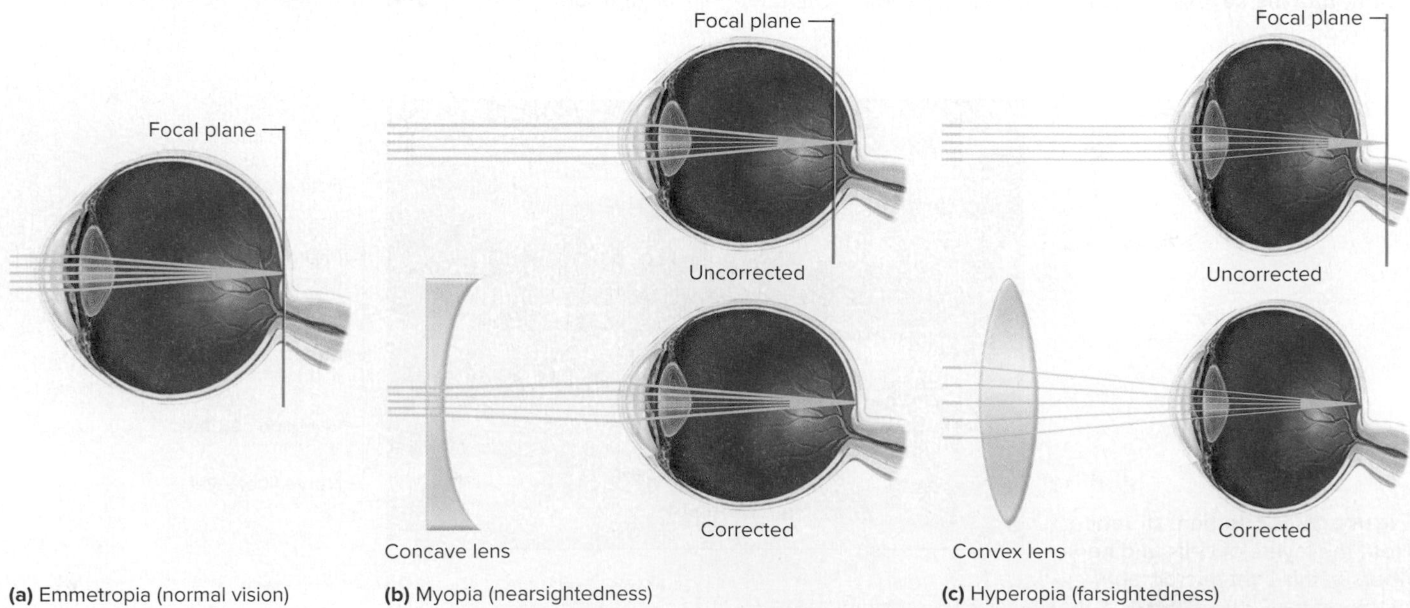

Figure 10.24 **(a)** Normal vision. **(b)** A concave lens will correct nearsightedness. **(c)** A convex lens will correct farsightedness.

Photoreceptors

Photoreceptors are modified neurons of two distinct kinds, as figure 10.21 illustrates. One group of photoreceptors, called **rods** (rodz), have long, thin projections at their ends, and provide vision without color, in shades of gray. The other group of photoreceptors, called **cones** (kōnz), have short, blunt projections, and provide color vision. There is only one type of rod, but there are three types of cones.

Rods and cones are in a deep part of the retina. They are closely associated with an adjacent layer of retinal pigment epithelium (RPE) that absorbs light the photoreceptors do not absorb. With the pigment of the choroid coat, the RPE keeps light from reflecting off surfaces inside the eye (see fig. 10.22).

Photoreceptors are stimulated only when light reaches them. A light image focused on an area of the retina stimulates some photoreceptors, which results in impulses traveling to the brain. However, the impulses leaving in response to each activated photoreceptor deliver only a fragment of the information required for the brain to interpret a complete scene.

Rods and cones contribute to different aspects of vision. Rods are hundreds of times more sensitive to light than cones and therefore can provide vision in dim light, but without color. Cones detect color but are less sensitive and do not respond in dim light.

OF INTEREST A human eye has 125 million rods and 7 million cones. A cat has three types of cone cells, but sees mostly pastels. A dog has two types of cone cells, and its visual world is much like that of a person with colorblindness. Researchers corrected colorblindness in monkeys by introducing the genes for human cone pigments into their eyes.

Rods and cones also differ in the sharpness of the perceived images, or visual acuity. Cones provide sharp images, and rods provide more general outlines of objects. Rods give less precise images because axon branches from rods undergo convergence (see section 9.9, Impulse Processing). Because of this, impulses from a group of rods are conducted to the brain on a single nerve fiber (fig. 10.25*a*). Thus, if a point of light stimulates a rod, the brain cannot tell which one of many receptors has been stimulated. Convergence of impulses is less common among cones. When a cone is stimulated, the brain can pinpoint the stimulation more accurately (fig. 10.25*b*).

The fovea centralis, the area of sharpest vision, does not have rods but contains densely packed cones with few or no converging fibers. Also in the fovea centralis, the overlying layers of the retina and the retinal blood vessels are displaced to the sides, more fully exposing photoreceptors to incoming

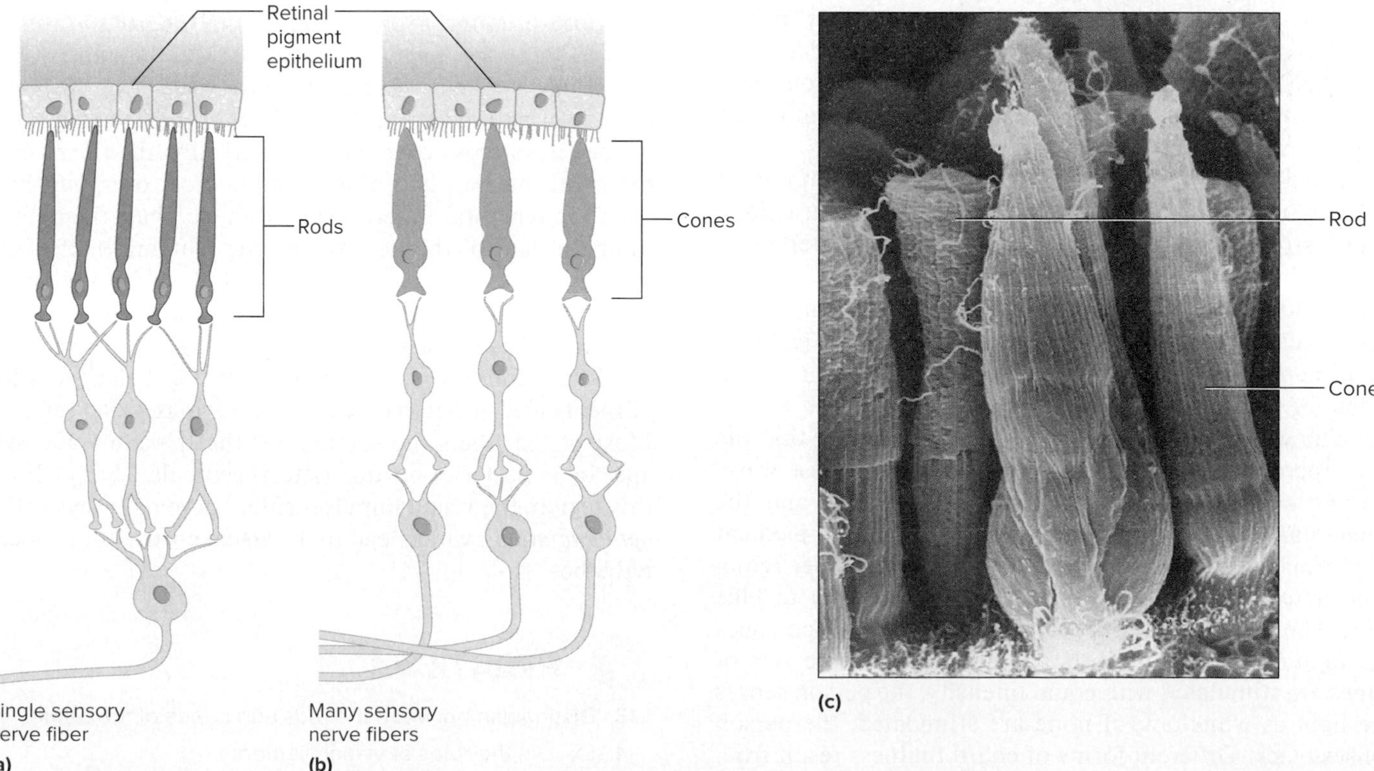

Figure 10.25 Rods and cones are photoreceptors. **(a)** A single sensory nerve fiber conducts impulses from several rods to the brain. **(b)** Separate sensory nerve fibers conduct impulses from cones to the brain. **(c)** Scanning electron micrograph of rods and cones (1,350x). (c): Frank S. Werblin, Ph.D.

light (see fig. 10.23). Consequently, to view something in detail, a person moves the eyes so that the important part of an image falls on the fovea centralis.

Photopigments

Both rods and cones contain light-sensitive pigments that decompose when they absorb light energy. The light-sensitive biochemical in rods is called **rhodopsin** (ro-dop'sin), or *visual purple.* In the presence of light, rhodopsin molecules are broken down into a colorless protein called *opsin* and a yellowish substance called *retinal* (retinene) that is synthesized from vitamin A.

Decomposition of rhodopsin molecules activates an enzyme that initiates a series of reactions altering the permeability of the rod cell membrane. As a result, a complex pattern of nervous stimulation originates in the retina. The impulses are conducted away from the retina along the optic nerve into the brain, where they are interpreted as vision.

In bright light, nearly all of the rhodopsin in the rods of the retina decomposes, greatly reducing rod sensitivity. In dim light, however, regeneration of rhodopsin from opsin and retinal is faster than rhodopsin breakdown. ATP provides the energy required for this regeneration (see section 4.4, Energy for Metabolic Reactions).

The light-sensitive pigments in cones, as in the rods, are composed of retinal and protein. In cones, however, three types of opsin proteins, different from the protein portion of rhodopsin found in rods, combine with retinal to form the three cone pigments. The three types of cones each contain one of these three photopigments.

Poor vision in dim light, called *nightblindness,* results from vitamin A deficiency. Lack of the vitamin reduces the supply of retinal, rhodopsin production falls, and rod sensitivity is low. Nightblindness is treated by supplementing the diet with vitamin A.

Daylight, as well as so-called white light from indoor lighting, is actually a mixture of light of different colors. An illustration of this is light passing through water vapor in the air after a storm and separating into its component colors to form a rainbow. These different colors can be described either by a physical property called a *wavelength,* or simply by the name of a color, such as red, green, or blue.

The wavelength of light determines the color that the brain perceives from it. For example, the shortest wavelengths of visible light are perceived as violet, and the longest are perceived as red. One type of cone pigment (erythrolabe) is most sensitive to red light, another (chlorolabe) to green light, and the third (cyanolabe) to blue light. The color a person perceives depends on which cones the light in a given image stimulates. If all three sets of cones are stimulated with equal intensity, the person senses the light as white, and if none are stimulated, the person senses black. Different forms of colorblindness result from the body's inability to produce different types of cone pigments.

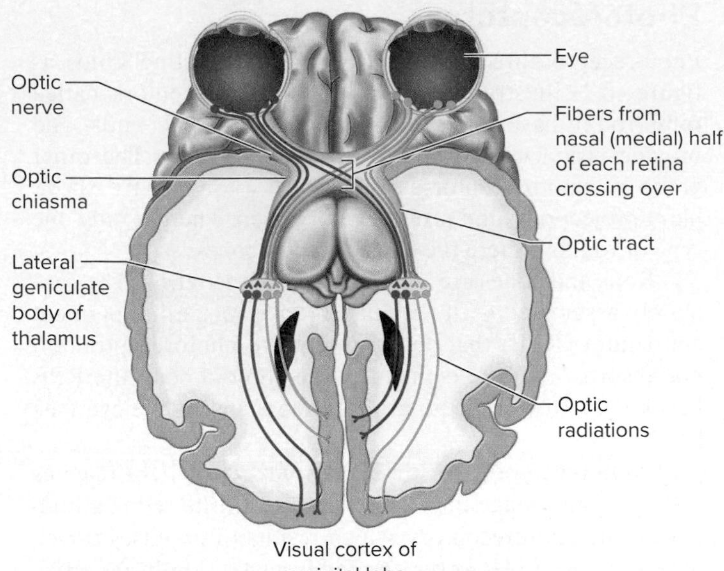

Figure 10.26 A visual pathway includes the optic nerve, optic chiasma, optic tract, and optic radiations. Blue lines indicate information originating in the right retina, and purple lines indicate information originating in the left retina (inferior view of the brain, transverse section). APR

Visual Pathways

Visual pathways conduct impulses from the retina to the visual cortex, where they are perceived as vision. The pathways begin as the axons of the retinal neurons leave the eyes to form the *optic nerves* (fig. 10.26). Just anterior to the pituitary gland, these nerves give rise to the X-shaped *optic chiasma* (op'tik kī-az'mah). Within the chiasma, some of the fibers cross over. More specifically, the fibers from the nasal (medial) half of each retina cross over, but those from the temporal (lateral) sides do not. Thus, fibers from the nasal half of the left eye and the temporal half of the right eye form the *right optic tract,* and fibers from the nasal half of the right eye and the temporal half of the left eye form the *left optic tract.*

Just before the nerve fibers reach the thalamus, a few of them enter nuclei that function in various visual reflexes. Most of the fibers, however, enter the thalamus and synapse in its posterior portion (lateral geniculate body). From this region, the visual impulses enter nerve pathways called *optic radiations,* which lead to the visual cortex of the occipital lobes.

PRACTICE 10.9

13. Distinguish between the rods and cones of the retina.
14. Explain the roles of visual pigments.
15. Trace an impulse from the retina to the visual cortex.

ASSESS

CHAPTER ASSESSMENTS

10.1 Introduction

1. Distinguish between general senses and special senses.

10.2 Receptors, Sensations, and Perception

2. Match each sensory receptor to the type of stimulus to which it is likely to respond.

(1) chemoreceptor	A. approaching headlights
(2) pain receptor	B. a change in blood pressure
(3) thermoreceptor	C. the smell of roses
(4) mechanoreceptor	D. an infected tooth
(5) photoreceptor	E. a cool breeze

3. Explain the difference between a sensation and a perception.
4. Explain the projection of a sensation.
5. You fill up the tub to take a hot bath, but the water is too hot to the touch. You try a second and third time, and within a few seconds it feels fine. Which of the following is the most likely explanation?
   a. The water has cooled down unusually quickly.
   b. Your ability to sense heat has adapted.
   c. Your nervous system is suddenly not functioning properly.
   d. Your ability to sense cold has adapted.

10.3 General Senses

6. Describe the functions of free nerve endings, tactile corpuscles, and lamellated corpuscles.
7. Identify the proprioceptor that is involved in causing the contracting muscle to relax after being stretched, preventing potential injury.
8. Explain why pain may be referred, and provide an example.
9. How does tissue damage increase nociceptor sensitivity?

10.4 Special Senses

10. Identify the location of the receptors for smell, taste, hearing, equilibrium, and sight.

10.5 Sense of Smell

11. Which two of the following are part of the olfactory organs?
    a. olfactory receptor cells
    b. columnar epithelial cells in the nasal mucosa
    c. the brain
    d. the eyes
12. Trace an impulse from the olfactory receptor cells to the interpreting center of the cerebrum.

10.6 Sense of Taste

13. Salivary glands are important in taste because ________.
    a. they provide the fluid in which food molecules dissolve
    b. the taste receptors are located in salivary glands
    c. salivary glands are part of the brain
    d. they produce enzymes to break down the food
14. Name the five primary taste sensations.
15. Trace the pathway of an impulse from a taste receptor to the interpreting center of the cerebrum.

10.7 Sense of Hearing

16. Match the ear area with the associated structure.

(1) outer ear	A. cochlea
(2) middle ear	B. eardrum
(3) inner ear	C. auditory ossicles

17. Trace the path of sound waves from the external acoustic meatus to the hearing receptors.
18. Describe the functions of the auditory ossicles.
19. The function of the auditory tube is to ____________.
    a. equalize air pressure on both sides of the eardrum
    b. transmit sound vibrations to the eardrum
    c. contain the hearing receptors
    d. connect the ears
20. Distinguish between the osseous and membranous labyrinths.
21. Describe the cochlea and its function.
22. Trace an impulse from the spiral organ to the interpreting centers of the cerebrum.
23. Which of the following best describes hearing receptor "hair cells"?
    a. They are neurons.
    b. They lack ion channels.
    c. They are epithelial, but function like neurons.
    d. They are built of the protein keratin.
24. Explain how a hearing receptor stimulates a sensory neuron.

10.8 Sense of Equilibrium

25. Contrast static equilibrium and dynamic equilibrium.
26. Describe the organs of static and dynamic equilibrium and their functions.

10.9 Sense of Sight

27. Match the visual accessory organ with its function.

(1) eyelid	A. moves the eye
(2) conjunctiva	B. covers the eye
(3) lacrimal gland	C. lines the eyelids
(4) extrinsic muscle	D. produces tears

28. Name the three layers of the eye wall and describe the functions of each layer.
29. Explain why looking at a close object causes fatigue, in terms of how accommodation is accomplished.
30. Explain the mechanisms of pupil constriction and pupil dilation.
31. All of the following are compartments within the eye. In which one is vitreous humor found?
    a. anterior chamber
    b. posterior chamber
    c. anterior cavity
    d. posterior cavity
32. Distinguish between the fovea centralis and the optic disc.
33. Explain how light is focused on the retina.
34. Distinguish between rods and cones.
35. Explain why cone vision is generally more acute than rod vision.
36. Describe the function of rhodopsin.
37. Explain why rod vision may be more important under dim light conditions.
38. Describe the relationship between light wavelength and color vision.
39. Trace an impulse from the retina to the visual cortex.

ASSESS

INTEGRATIVE ASSESSMENTS/CRITICAL THINKING

Outcomes 6.2, 10.2, 10.9

1. PET (positron emission tomography) scans of the brains of people who have been blind since birth reveal high neural activity in the visual centers of the cerebral cortex when these people read Braille. However, when sighted individuals run their fingers over the raised letters of Braille, the visual centers do not show increased activity. Explain these experimental results.

Outcomes 7.13, 8.8, 10.3

2. Jerry was feeling pain in his right shoulder around his acromioclavicular joint and mid-scapula. The pain was worse when lying down at night, but better during the day. He also complained of shortness of breath and abdominal bloating. An X ray showed no sign of shoulder problems other than mild osteoarthritis, but it did show his diaphragm was elevated on the right side. What do you think Jerry's diagnosis might be, and why might he have these symptoms?

Outcomes 8.2, 8.3, 10.9

3. Strabismus, a condition commonly called "cross-eyed," is an uncommon congenital problem that occurs when the extrinsic eye muscles pull unequally, resulting in the inability to fixate on an object with both eyes. Why might putting an eye patch on the stronger eye help correct this problem? Why might botulinum toxin (Botox), which blocks acetylcholine (ACh) release in the neuromuscular junction, be used in eye surgery to correct strabismus? If strabismus is not corrected, what may occur if it is allowed to continue?

Outcomes 9.14, 10.2, 10.5

4. Loss of the sense of smell often precedes the major symptoms of Alzheimer disease and Parkinson disease. What additional information is needed to use this association to prevent or treat these diseases?

Outcomes 10.2, 10.7, 10.8

5. People who are deaf due to cochlear damage may still suffer from motion sickness. Why?

Outcomes 10.2, 10.7, 10.8

6. Labyrinthitis is an inflammation of the inner ear. What symptoms would you expect in a patient with this disorder?

Outcomes 10.5, 10.6

7. Describe how the taste of a medicine might be modified from sour to sweet, so that children would be more willing to take it.

Outcome 10.9

8. Cheryl is forty-five years old. While at a restaurant with her husband, she realizes she has to move the menu further and further from her eyes in order to focus and see the words. What is Cheryl's condition, and why is it occurring at this point in her life?

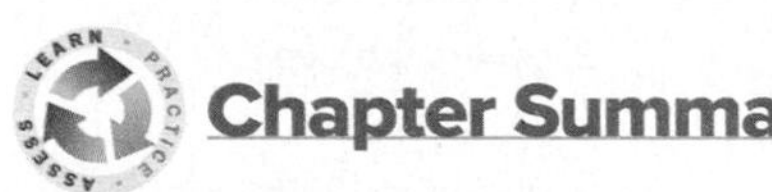

Chapter Summary

10.1 Introduction

Sensory receptors sense changes in their surroundings.

10.2 Receptors, Sensations, and Perception

1. Types of receptors
   a. Each type of receptor is most sensitive to a distinct type of stimulus.
   b. The major types of receptors are **chemoreceptors, pain receptors, thermoreceptors, mechanoreceptors,** and **photoreceptors.**
2. Sensations
   a. A **sensation** is the awareness of sensory stimulation.
   b. A particular part of the cerebral cortex interprets every impulse reaching it in a specific way.
   c. The cerebral cortex projects a sensation back to the region of stimulation.
3. **Sensory adaptation** may involve receptors becoming unresponsive or inhibition along the CNS pathways leading to the sensory regions of the cerebral cortex.

10.3 General Senses

General senses are associated with receptors in the skin, muscles, joints, and viscera.

1. Touch and pressure senses
   a. Free ends of sensory nerve fibers are receptors for the sensation of itching.
   b. **Tactile corpuscles** are receptors for the sensation of light touch.
   c. **Lamellated corpuscles** are receptors for the sensation of heavy pressure.
2. Temperature senses
   Temperature receptors include two sets of free nerve endings that are warm and cold receptors.
3. Body position, movement, and stretch receptors
4. Sense of pain
   a. Pain receptors are free nerve endings that tissue damage stimulates.
   b. Visceral pain
      (1) Pain receptors are the only receptors in viscera that provide sensations.

(2) Pain sensations produced from visceral receptors may feel as if they are coming from some other body part, called **referred pain.**
(3) Visceral pain may be referred because sensory impulses from the skin and viscera travel on common nerve pathways.

c. Pain nerve fibers
(1) The two main types of pain fibers are fast pain fibers and slow pain fibers.
(2) Fast pain fibers conduct impulses that reach the brain quickly. Impulses on slow pain fibers reach the brain with a slight delay.
(3) Pain impulses are processed in the gray matter of the spinal cord and ascend to the brain.
(4) Within the brain, most pain impulses pass through the thalamus, reticular formation, or limbic system before being conducted to the cerebral cortex.

10.4 Special Senses

Special senses have receptors within large, complex sensory organs of the head.

10.5 Sense of Smell

1. Olfactory receptors
   a. Olfactory receptors are chemoreceptors that are stimulated by chemicals dissolved in liquid.
   b. Olfactory receptors function with taste receptors and aid in food selection.
2. Olfactory organs
   a. **Olfactory organs** consist of receptors and supporting cells in the nasal cavity.
   b. **Olfactory receptor cells** are bipolar neurons with cilia.
3. Olfactory pathways
   Impulses travel from the olfactory receptor cells through the olfactory nerves, **olfactory bulbs,** and **olfactory tracts** to interpreting centers in the temporal and frontal lobes of the cerebrum.
4. Olfactory stimulation
   a. Olfactory impulses may result when odorant molecules bind cell surface olfactory receptors on cilia of receptor cells. The binding pattern encodes a specific odor, which is interpreted in the brain.
   b. The sense of smell adapts rapidly.

10.6 Sense of Taste

1. Taste receptors
   a. **Taste buds** consist of taste (receptor) cells and supporting cells.
   b. **Taste cells** have **taste hairs.**
   c. Taste hair surfaces have receptors to which chemicals bind, stimulating impulses.
2. Taste sensations
   a. The five primary taste sensations are sweet, sour, salty, bitter, and umami.
   b. Various taste sensations result from the stimulation of one or more types of taste receptors.
   c. A single taste receptor cell detects only one of the five tastes, but receptors corresponding to different tastes are scattered on the tongue.
3. Taste pathways
   a. Sensory impulses from taste receptors travel on fibers of the facial, glossopharyngeal, and vagus nerves.
   b. These impulses are conducted to the medulla oblongata and then ascend to the thalamus, from which they are conducted to the gustatory cortex in the parietal lobes of the cerebrum.

10.7 Sense of Hearing

1. Outer ear
   The outer ear collects sound waves of vibrating objects.
2. Middle ear
   Auditory ossicles of the middle ear conduct sound waves from the eardrum to the oval window of the inner ear.
3. Auditory tube
   The **auditory tube** connects the middle ear to the nasopharynx and helps maintain equal air pressure on both sides of the eardrum.
4. Inner ear
   a. The inner ear is a complex system of connected tubes and chambers—the bony and membranous **labyrinths.**
   b. The **spiral organ** contains hearing receptors that are stimulated by vibrations in the fluids of the inner ear.
   c. Different frequencies of vibrations stimulate different sets of receptor cells.
5. Auditory pathways
   a. Auditory nerve fibers conduct impulses to the auditory cortices of the temporal lobes.
   b. Some auditory nerve fibers cross over, so that impulses arising from each ear are interpreted on both sides of the brain.

10.8 Sense of Equilibrium

1. Static equilibrium
   Static equilibrium maintains the stability of the head and body when they are motionless.
2. Dynamic equilibrium
   a. **Dynamic equilibrium** balances the head and body when they are rotated or otherwise moved suddenly.
   b. Other structures that help maintain equilibrium include the eyes and mechanoreceptors associated with certain joints.

10.9 Sense of Sight

1. Visual accessory organs
   Visual accessory organs include the **eyelids,** lacrimal apparatus, and **extrinsic muscles** of the eyes.
2. Structure of the eye
   a. The wall of the eye has an outer (fibrous), a middle (vascular), and an inner (nervous) layer.
   (1) The outer layer is protective, and its transparent anterior portion **(cornea)** refracts light entering the eye.
   (2) The middle layer is vascular and contains melanin that keeps the inside of the eye dark.
   (3) The inner layer contains the photoreceptors.
   b. The lens is a transparent, elastic structure. The ciliary muscle controls its shape.
   c. The **lens** must thicken to focus on close objects.

   d. The **iris** is a muscular diaphragm that controls the amount of light entering the eye.
   e. Spaces within the eye are filled with fluids that help maintain its shape.
3. **Light refraction**
   The cornea and lens refract light rays to focus an image on the **fovea centralis** of the **retina.**
4. Photoreceptors
   a. Photoreceptors are rods and cones.
   b. Rods are responsible for colorless vision in dim light, and cones provide color vision.
5. Photopigments
   a. A light-sensitive pigment in rods decomposes in the presence of light and triggers a complex series of reactions that initiate impulses.
   b. Color vision comes from three sets of cones containing different light-sensitive pigments.
6. Visual pathways
   a. Nerve fibers from the retina form the optic nerves.
   b. Some fibers cross over in the optic chiasma.
   c. Most of the fibers enter the thalamus and synapse with others that continue to the visual cortex in the occipital lobes.

Endocrine System

The endocrine system produces hormones, which act within an individual. Humans may also produce pheromones, which affect other individuals and may play a role in mate selection, as they do in rodents and insects. rawpixel/123RF

Smelly T-shirts. The endocrine system produces hormones, which are biochemicals that send messages within an individual. Less well understood are pheromones, which are chemical signals sent between individuals of a species. In insects and rodents, pheromones stimulate mating behavior. Experiments suggest that this may be the case with humans, too.

Mice and rats choose mates that are dissimilar to themselves with respect to a group of genes that provide immunity. Their sense of smell helps them discern appropriate mates. Biologists hypothesize that choosing mates based on scent may protect offspring in two ways—it prevents close relatives from mating, and it may team immune systems with different strengths.

Researchers have traced mouse social and mating behavior to receptors in the olfactory epithelium, in the nasal cavity. Molecules in mouse urine that influence social behavior bind to these receptors, called trace-amine-associated receptors. The genes that encode the receptors are also in the human genome.

To test whether heterosexual humans use the sense of smell to respond to pheromones in mate selection as rodents do, researchers in Switzerland recruited forty-nine young women and forty-four young men. Each participant donated DNA, which was typed for human versions of genes that affect mating in rodents. The women used nasal spray for two weeks to clear their nasal passages. The men wore the same T-shirt on two consecutive days, using no deodorant or soap and avoiding contact with anything smelly that could linger. Each woman was then given three T-shirts from men genetically similar to her and three T-shirts from men genetically dissimilar to her, not knowing which shirts came from which men.

The women rated the shirts on intensity, pleasantness, and sexiness. Like the mice and rats, women preferred the sweaty T-shirts from the men least like them genetically.

Another experiment supported these findings. Women were given vials of fluid to sniff that either contained or did not contain a component of male sweat called androstadienone. Although they didn't know which samples they were sniffing, the women consistently reported mood elevation and sexual arousal when they smelled the androstadienone. In addition, their saliva had increased amounts of cortisol, a hormone that raises the blood sugar level, when they smelled the chemical, suggesting that it might be a human pheromone. Despite the mounting scientific evidence for human pheromones, a definitive human pheromone has not yet been described.

Anatomy & Physiology Revealed 4.0
Module 8 Cells & Chemistry

LEARNING OUTLINE

After studying this chapter, you should be able to complete the "Learning Outcomes" that follow the major headings throughout the chapter.

AIDS TO UNDERSTANDING WORDS

-crin [to secrete] endo*crine*: pertaining to internal secretions.

diure- [to pass urine] *diure*tic: substance that promotes urine production.

endo- [within] *endo*crine gland: gland that releases its secretion internally into a body fluid.

exo- [outside] *exo*crine gland: gland that releases its secretion to the outside through a duct.

hyper- [above] *hyper*thyroidism: condition resulting from an above-normal secretion of thyroid hormone.

hypo- [below] *hypo*thyroidism: condition resulting from a below-normal secretion of thyroid hormone.

para- [beside] *para*thyroid glands: set of glands on the posterior surface of the thyroid gland.

toc- [birth] oxy*toc*in: hormone that stimulates the uterine muscles to contract during childbirth.

-tropic [influencing] adrenocortico*tropic* hormone: hormone that influences secretions from the adrenal cortex.

(Appendix A has a complete list of Aids to Understanding Words.)

11.1 | Introduction

LEARN

1. Describe the secretions of the endocrine system.
2. Compare the nervous system and endocrine system in maintaining homeostasis.
3. Distinguish between endocrine, autocrine, and paracrine secretions.

Regulating the functions of the human body to maintain homeostasis is an enormous job. This is achieved by the partnering of the nervous system and the **endocrine system** to coordinate and execute communication throughout the entire body.

The endocrine system is unique in that the organs are not anatomically adjacent to each other. Scattered from the head to the genital region, functionally they are glands. The major endocrine glands are the pituitary gland, thyroid gland, parathyroid glands, adrenal glands, pancreas, pineal gland, thymus, and reproductive glands (testes and ovaries) (fig. 11.1).

Specialized cells in the liver, heart, kidneys, and gastrointestinal tract also produce hormones, although these organs are primarily parts of other systems.

Unlike exocrine glands (see chapter 5), endocrine glands secrete chemical messengers called **hormones** (hor'mōnz) into the extracellular fluid that then diffuse into the bloodstream. The term *hormone* has Greek origins and roughly means to "set in motion." Hormones flow through the blood and take action on tissues in other regions of the body, regulating metabolic activities, including water and electrolyte balance, growth, and much of the physiology in all body systems. This is possible because target cells will have specific receptors for the hormone.

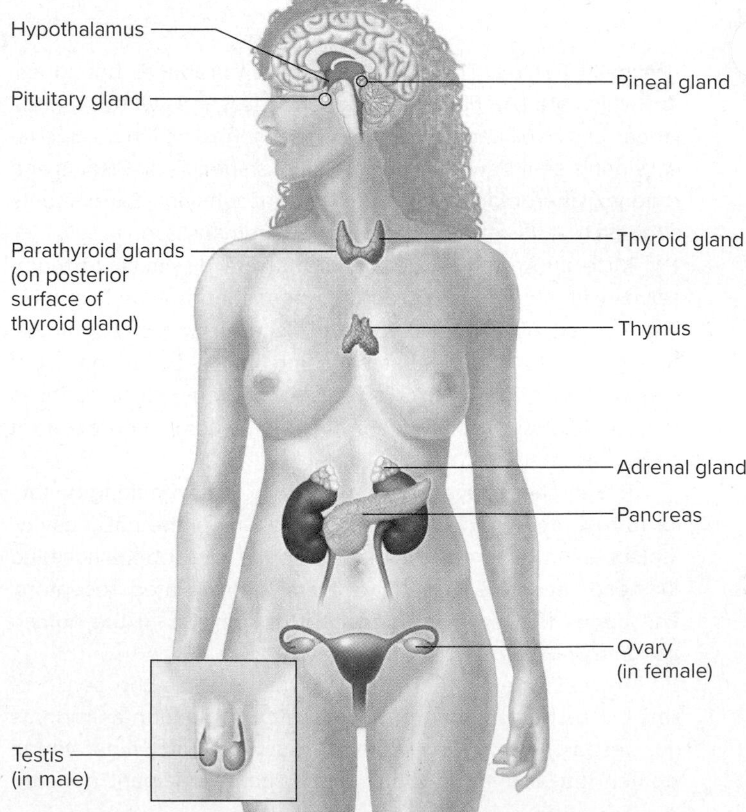

Figure 11.1 Locations of the major endocrine glands. The pituitary, thyroid, parathyroid, adrenal glands, and pancreas are the main topics of this chapter. The functions of the other glands are described in more detail in subsequent chapters.

TABLE 11.1 A Comparison Between the Nervous System and the Endocrine System

	Nervous System	Endocrine System
Cells	Neurons	Epithelial and others
Chemical signal	Neurotransmitter	Hormone
Specificity of response	Receptors on postsynaptic cell	Receptors on target cell
Speed of onset	Seconds	Seconds to hours
Duration of action	Very brief unless neuronal activity continues	May be brief or may last for days even if secretion ceases

The nervous system also oversees communication using chemical signals that bind to receptor molecules (fig. 11.2). In general, the nervous system employs fast-acting neurotransmitters released directly to the target, whereas hormones have longer-lasting effects. Table 11.1 summarizes some similarities and differences between the nervous and endocrine systems.

Some glands secrete substances into the interstitial fluid, but because these secretions are rapidly broken down, they do not reach the bloodstream and are not hormones according to the traditional definition. However, they do function similarly as messenger molecules and are sometimes referred to as "local hormones." **Paracrine** secretions affect only neighboring cells. An example of this is the release of histamine from white blood cells that dilate blood vessels. **Autocrine** secretions affect only the cell secreting the substance and includes liver cells stimulating themselves to release stored iron.

PRACTICE 11.1

Answers to the Practice questions can be found in the eBook.

1. What are the components of the endocrine system?
2. State some general functions of hormones.
3. What determines whether a cell is a target cell for a particular hormone?
4. Explain how the nervous and endocrine systems are alike and how they differ.
5. How do paracrine and autocrine secretions function differently than traditionally defined hormones?

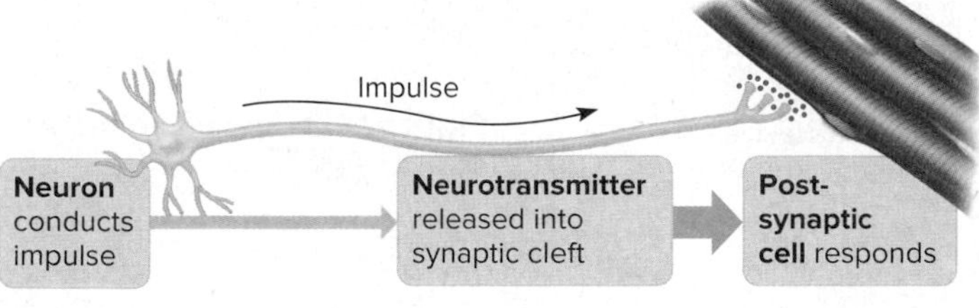

(a)

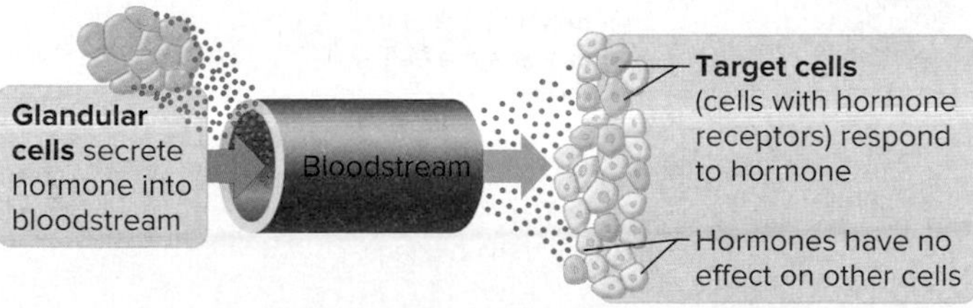

(b)

Figure 11.2 Chemical communication takes place in the nervous system and the endocrine system. In both cases, cells respond to chemicals released from other cells. **(a)** Neurons release neurotransmitters into a synapse, affecting postsynaptic cells. **(b)** Glands release hormones into the bloodstream. Blood carries hormone molecules throughout the body, but only target cells respond.

PRACTICE FIGURE 11.2

What do postsynaptic cells and target cells have in common that allow them to respond to secreted chemicals?

Answer can be found in Appendix E.

CAREER CORNER

Licensed Practical Nurse

The man has been suffering from cancer for many months and now is bedridden. He receives hospice care in his home. Three times a week, a licensed practical nurse (LPN) arrives to take the man's vital signs, check that he is receiving the proper doses of medications to stay comfortable, and help the family by answering their questions about caring for their loved one. The LPN might suggest or demonstrate ways to bathe or feed the patient, and change the bedding. Should the patient exhibit signs of active dying or require further medication, or if a change in status has occurred, the LPN reports the information to the registered nurse or physician assigned to the case.

An LPN is an entry-level nursing position. In the United States, becoming an LPN requires completing a post–high school program that typically takes one year. In some states, an LPN is called a licensed vocational nurse, or LVN. An LPN must pass a national licensure exam. LPNs work in hospitals, nursing homes, assisted living facilities, physicians' offices, and with home health-care agencies.

11.2 | Hormone Action

LEARN

1. Explain how steroid and nonsteroid hormones affect target cells.

Most hormones are of two general types. They are either steroids (or steroidlike substances) or nonsteroids. Steroids are synthesized from **cholesterol.** Nonsteroids are amines, peptides, polypeptides, proteins, or glycoproteins, all synthesized from amino acids (table 11.2).

TABLE 11.2 Types of Hormones

Type of Compound	Formed From	Examples
Steroids	Cholesterol	Estrogen, testosterone, aldosterone, cortisol
Amines	Amino acids	Norepinephrine, epinephrine, thyroid hormones
Peptides	Amino acids	Antidiuretic hormone, oxytocin, thyrotropin-releasing hormone
Polypeptides and proteins	Amino acids	Parathyroid hormone, growth hormone, prolactin
Glycoproteins	Protein and carbohydrate	Follicle-stimulating hormone, luteinizing hormone, thyroid-stimulating hormone

Steroid Hormones

Steroid molecules consist of complex rings of carbon and hydrogen atoms, and some oxygen atoms (see fig. 2.16). Steroids differ according to the types and numbers of atoms attached to these rings and the ways the atoms are joined.

Steroid hormones are poorly soluble in water. They are carried in the bloodstream weakly bound to plasma proteins in such a way that they are released from the proteins in sufficient quantity to act on their target cells.

Steroid hormones can diffuse into cells relatively easily, because lipids make up the bulk of cell membranes and steroid molecules are lipid-soluble. Steroid hormones may enter any cell in the body, but only target cells will respond. When a steroid hormone molecule enters a target cell, the following events occur (fig. 11.3):

1. The lipid-soluble steroid hormone diffuses through the cell membrane.

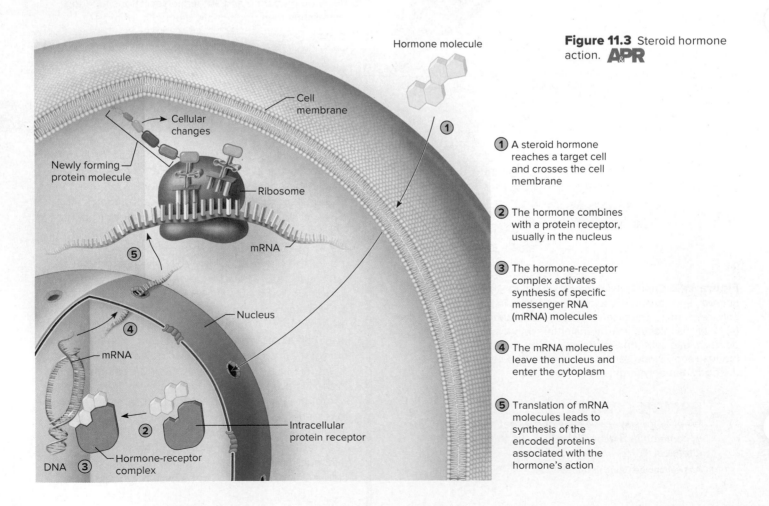

② The hormone combines with a protein receptor, usually in the nucleus

Figure 11.3 Steroid hormone action. APR

2. The steroid hormone binds a specific protein molecule—the receptor for that hormone.
3. The resulting hormone-receptor complex binds in the nucleus to specific sequences of the target cell's DNA, activating transcription of specific genes into messenger RNA (mRNA) molecules.
4. The mRNA molecules leave the nucleus and enter the cytoplasm.
5. Translation of mRNA molecules leads to the synthesis of specific proteins.

The newly synthesized proteins, which may be enzymes, transport proteins, or even hormone receptors, carry out the specific effects associated with the particular steroid hormone.

Nonsteroid Hormones

Nonsteroid hormones, such as amines, peptides, and proteins, usually bind receptors in target cell membranes. Each of these receptor molecules is a protein with a *binding site* and an *activity site.* A hormone molecule delivers its message to its target cell by uniting with the binding site of its receptor. This combination stimulates the receptor's activity site to interact with other membrane proteins. The hormone that triggers this first step, in what becomes a cascade of chemical activity, is called a *first messenger.* The chemicals in the cell that induce changes in response to the hormone's binding are called *second messengers.* The entire process of chemical communication, from outside cells to inside, is called **signal transduction.**

The second messenger associated with one group of hormones is *cyclic adenosine monophosphate*, also called **cyclic AMP (cAMP)** (sĭ'klik ay em pee). This mechanism works as follows (fig. 11.4):

1. A hormone binds to its receptor.
2. The resulting hormone-receptor complex activates a membrane protein called a *G protein.*
3. The G protein activates an enzyme called *adenylate cyclase,* which is a membrane protein.
4. In the cytoplasm, activated adenylate cyclase catalyzes the formation of cAMP from ATP.
5. cAMP activates another set of enzymes, called protein kinases, which transfer phosphate groups from ATP to their substrate molecules, which are specific proteins in the cell. This action, called phosphorylation, alters the shapes of these substrate molecules, thereby activating them.

The activated proteins then alter various cellular processes, bringing about the characteristic effect of the hormone.

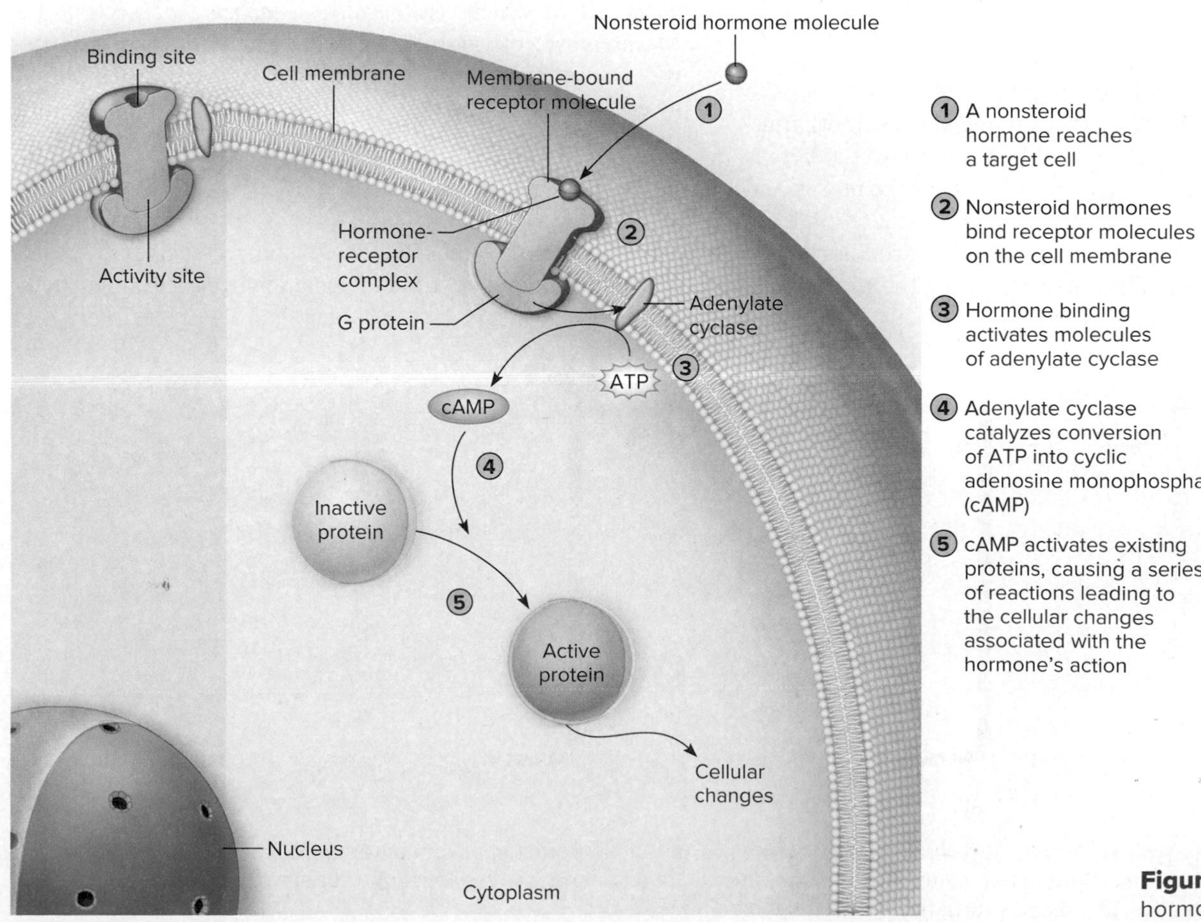
Nonsteroid hormone molecule
Binding site
Cell membrane
Membrane-bound receptor molecule
Activity site
Hormone-receptor complex
G protein
Adenylate cyclase
ATP
cAMP
Inactive protein
Active protein
Cellular changes
Nucleus
Cytoplasm
1
2
3
4
5
1 A nonsteroid hormone reaches a target cell
2 Nonsteroid hormones bind receptor molecules on the cell membrane
3 Hormone binding activates molecules of adenylate cyclase
4 Adenylate cyclase catalyzes conversion of ATP into cyclic adenosine monophosphate (cAMP)
5 cAMP activates existing proteins, causing a series of reactions leading to the cellular changes associated with the hormone's action

Figure 11.4 Nonsteroid hormone action. APR

The type of membrane receptors present and the specific protein substrate molecules in a cell determine the cell's response to a hormone. Such responses to second messenger activation include altering membrane permeabilities, activating enzymes, promoting synthesis of certain proteins, stimulating or inhibiting specific metabolic pathways, moving the cell, and initiating secretion of hormones or other substances.

Another enzyme (phosphodiesterase) quickly inactivates cAMP, so that its effect is short-lived. For this reason, a continuing response of a target cell requires a continuing signal from hormone molecules binding the target cell's membrane receptors.

A number of other second messengers work in much the same way. These include diacylglycerol (DAG) and inositol triphosphate (IP_3).

Abnormal or missing G proteins cause a variety of disorders, including colorblindness, precocious puberty, retinitis pigmentosa, and several thyroid problems.

Prostaglandins

Chemicals called **prostaglandins** (pros″tah-glan′dinz) also regulate cells. Prostaglandins are lipids synthesized from a fatty acid (arachidonic acid) in cell membranes. A great variety of cells produce prostaglandins, including those of the liver, kidneys, heart, lungs, thymus, pancreas, brain, and reproductive organs. Prostaglandins usually act more locally than hormones, often affecting only the organ where they are produced.

Prostaglandins are present in very small amounts but are potent. They are not stored in cells; instead they are synthesized just before release. They are rapidly inactivated.

Prostaglandins produce diverse and even opposite effects. Some prostaglandins, for example, relax smooth muscle in the airways of the lungs and in blood vessels, whereas others contract smooth muscle in the walls of the uterus and intestines. Prostaglandins stimulate hormone secretion from the adrenal cortex and inhibit secretion of hydrochloric acid from the stomach wall. They also influence the movements of sodium ions and water molecules in the kidneys, help regulate blood pressure, and have powerful effects on male and female reproductive physiology.

PRACTICE 11.2

1. How does a steroid hormone promote cellular changes? How does a nonsteroid hormone do the same?
2. What is a second messenger?
3. What are prostaglandins?
4. What are the effects of prostaglandins?

11.3 | Control of Hormonal Secretions

LEARN

1. Discuss how negative feedback mechanisms regulate hormonal secretions.
2. Explain how the nervous system controls secretion of hormones.

Hormones are continually excreted in the urine and broken down by various enzymes, primarily in the liver. Therefore, maintaining constant hormone levels in the blood requires ongoing hormone secretion. To increase or decrease the blood levels of a hormone, the body increases or decreases secretion. Not surprisingly, hormone secretion is precisely regulated.

Generally, hormone secretion is controlled in three ways, all of which use *negative feedback* (see section 1.5, Maintenance of Life). In each case, an endocrine gland or the system controlling it detects the concentration (blood level) of the hormone the gland secretes, a process the hormone controls, or an action the hormone has on the internal environment (fig. 11.5). The three mechanisms of hormone control are:

1. The hypothalamus, which constantly receives information about the internal environment, regulates

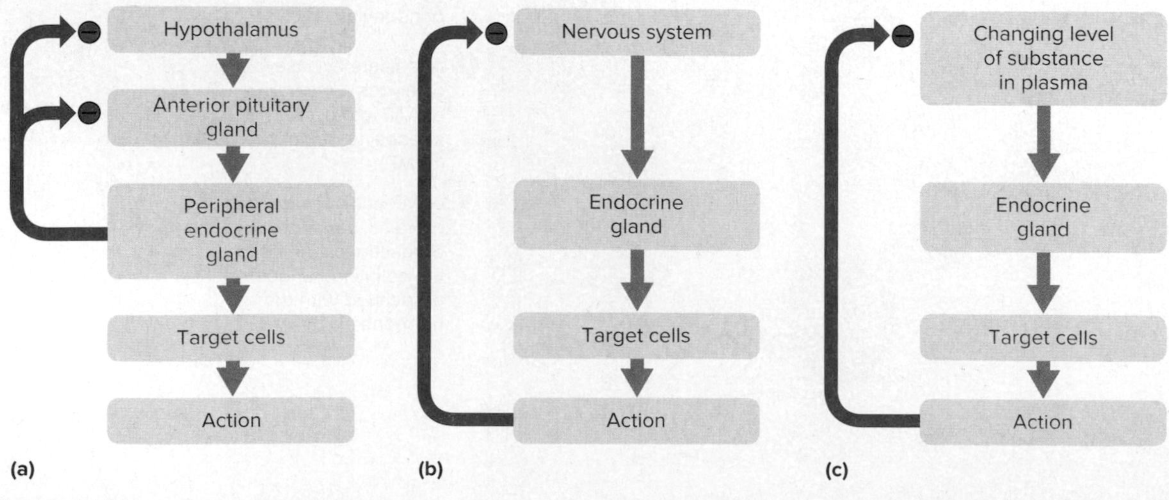

Figure 11.5 Control of the endocrine system occurs in three ways: **(a)** The hypothalamus and anterior pituitary stimulate other endocrine glands; **(b)** the nervous system stimulates a gland directly; or **(c)** changes in the level of a substance in the blood stimulates a gland directly. (⊖ indicates negative feedback inhibition.)

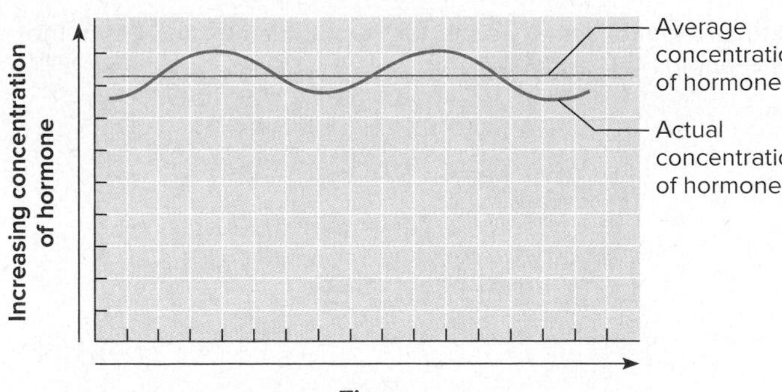

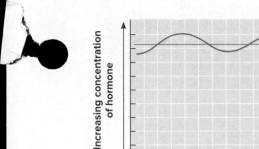

Figure 11.6 As a result of negative feedback, hormone concentrations remain relatively stable, although they may fluctuate slightly above and below average concentrations.

the anterior pituitary gland's release of hormones. Many anterior pituitary hormones affect the activity of other endocrine glands (fig. 11.5*a*).

2. The nervous system stimulates some glands directly. The adrenal medulla, for example, secretes its hormones in response to impulses from the sympathetic nervous system (fig. 11.5*b*).
3. Another group of glands responds directly to changes in the composition of the internal environment. For example, when the blood glucose level rises, the pancreas secretes insulin, and when the blood glucose level falls, it secretes glucagon (fig. 11.5*c*).

In each of these cases, as hormone levels rise in the blood and the hormone exerts its effects, negative feedback inhibits the system, and hormone secretion decreases. Then, as hormone levels in the blood decrease and the hormone's effects are no longer taking place, inhibition of the system is lifted, and secretion of that hormone increases again. As a result of negative feedback, hormone levels in the bloodstream remain relatively stable, tending to fluctuate slightly above and below an average value (fig. 11.6).

PRACTICE 11.3

1. Explain three examples of control of hormonal secretion.
2. Describe a negative feedback system that controls hormone secretion.

11.4 | Pituitary Gland

LEARN

1. Describe the locations of the anterior and posterior lobes of the pituitary gland, and list the hormones they secrete.
2. Describe the functions of the hormones that the pituitary gland secretes.
3. Explain how the secretion of each pituitary hormone is regulated.

The **pituitary gland** (hypophysis) is located at the base of the brain, where a pituitary stalk (infundibulum) attaches it to the hypothalamus. The gland is about 1 centimeter in diameter and consists of an **anterior pituitary** (pĭ-tū′ĭ-tar″ē), or anterior lobe, and a **posterior pituitary,** or posterior lobe (fig. 11.7).

OF INTEREST In the fetus, a narrow region develops between the anterior and posterior lobes of the pituitary gland. Called the *intermediate* lobe (pars intermedia), it produces melanocyte-stimulating hormone (MSH), which regulates the synthesis of melanin—the pigment in skin and in parts of the eyes and brain. In most adults, this intermediate lobe is no longer a distinct structure, but its secretory cells persist in the two remaining lobes.

The brain controls most of the pituitary gland's activity. In fact, the posterior pituitary is actually part of the nervous system. Certain neurons whose cell bodies are in the hypothalamus have axons that extend down into the posterior pituitary gland. Impulses on their axons trigger the secretion of chemicals from their axon terminals, which then enter the bloodstream as posterior pituitary hormones (fig. 11.8). Note that although these chemicals are released by neurons, they are not considered neurotransmitters. Because they enter the bloodstream, they are considered hormones.

The anterior pituitary consists of glandular cells rather than neurons, but it is still controlled by the brain. Axon terminals of another population of hypothalamic neurons secrete hormones called **releasing hormones** (and some release-inhibiting hormones) into a capillary network associated with the hypothalamus. The capillaries merge to form the *hypophyseal portal veins*, which pass downward along the pituitary stalk and give rise to a capillary network in the anterior pituitary (fig. 11.8). Thus, the hypothalamus secretes hormones that the blood carries directly to target cells in the anterior pituitary.

Upon reaching the anterior pituitary, each of the hypothalamic hormones acts on a specific population of target cells. Some of the resulting actions are inhibitory, but most stimulate the anterior pituitary to release hormones that stimulate secretions from peripheral endocrine glands. In many of these cases, negative feedback regulates hormone levels in the bloodstream.

PRACTICE 11.4

1. Where is the pituitary gland located?
2. Explain how the hypothalamus controls the secretory activity of the posterior and anterior lobes of the pituitary gland.

Anterior Pituitary Hormones

The anterior pituitary is enclosed in a capsule of dense connective tissue. It consists largely of epithelial tissue

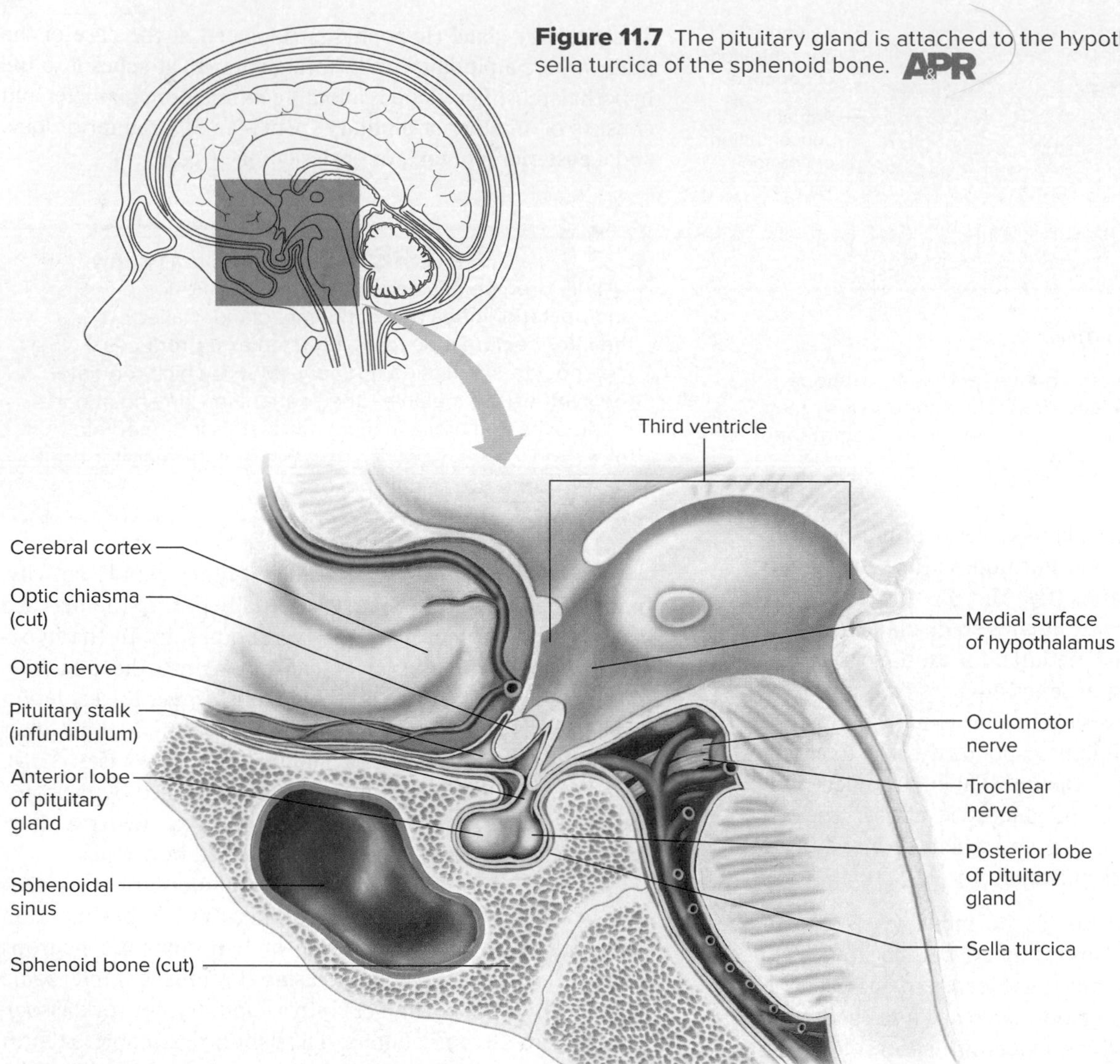

Figure 11.7 The pituitary gland is attached to the hypothalamus and lies in the sella turcica of the sphenoid bone. APR

organized in blocks around many thin-walled blood vessels. So far, researchers have identified five types of secretory cells in this epithelium. Four of these cell types each secrete a different hormone–growth hormone (GH), prolactin (PRL), thyroid-stimulating hormone (TSH), and adrenocorticotropic hormone (ACTH). The fifth type of cell secretes both follicle-stimulating hormone (FSH) and luteinizing hormone (LH). (In males, luteinizing hormone is also referred to as interstitial cell stimulating hormone, or ICSH.)

Growth hormone (GH) stimulates cells to enlarge and divide more frequently. It also enhances the movement of amino acids across cell membranes and speeds the rate at which cells utilize carbohydrates and fats. The hormone's effect on amino acids is important in stimulating growth.

Two hormones from the hypothalamus control GH secretion: *GH-releasing hormone (GHRH)* stimulates growth hormone secretion, and *GH-inhibiting hormone (GHIH,* also called *somatostatin)* inhibits growth hormone secretion.

Nutritional state also influences control of GH. For example, more GH is released in response to an abnormally low blood glucose concentration. Conversely, when the blood glucose concentration increases, GH secretion decreases. Increased levels of some amino acids stimulate growth hormone secretion. Clinical Application 11.1 discusses the insufficient secretion and oversecretion of growth hormone.

Prolactin (prō-lak′tin) **(PRL)** stimulates and sustains a woman's milk production following the birth of an infant (see section 20.3, Pregnancy and the Prenatal Period). The control of prolactin secretion involves a combination of stimulating and inhibiting hormones, similar to the control of growth hormone. No normal physiological role for PRL in human males has been firmly established. Abnormally elevated levels of PRL can disrupt sexual function in both sexes.

Thyroid-stimulating hormone (TSH) (also called *thyrotropin*) regulates thyroid hormone production and secretion by the thyroid gland (see section 11.5, Thyroid Gland). The hypothalamus stimulates TSH secretion by secreting

CLINICAL APPLICATION 11.1

Hyposecretion and Hypersecretion of Growth Hormone

Insufficient secretion of growth hormone (GH) during childhood limits growth, causing pituitary dwarfism. Body parts are normally proportioned, and mental development is normal—the individual is just very small. Typically, hormone therapy can stimulate some growth.

Oversecretion of GH during childhood causes gigantism, in which height may exceed 8 feet. This rare condition is usually a result of a pituitary gland tumor, which may also cause oversecretion of other pituitary hormones. As a result, a person with gigantism often has several metabolic disturbances.

Acromegaly is the overproduction of growth hormone in adulthood. The many symptoms attesting to the wide effects of this hormone include enlargement of the heart, the bones, the thyroid gland, facial features, the hands, the feet, and the head. Early symptoms include headache, joint pain, fatigue, and depression.

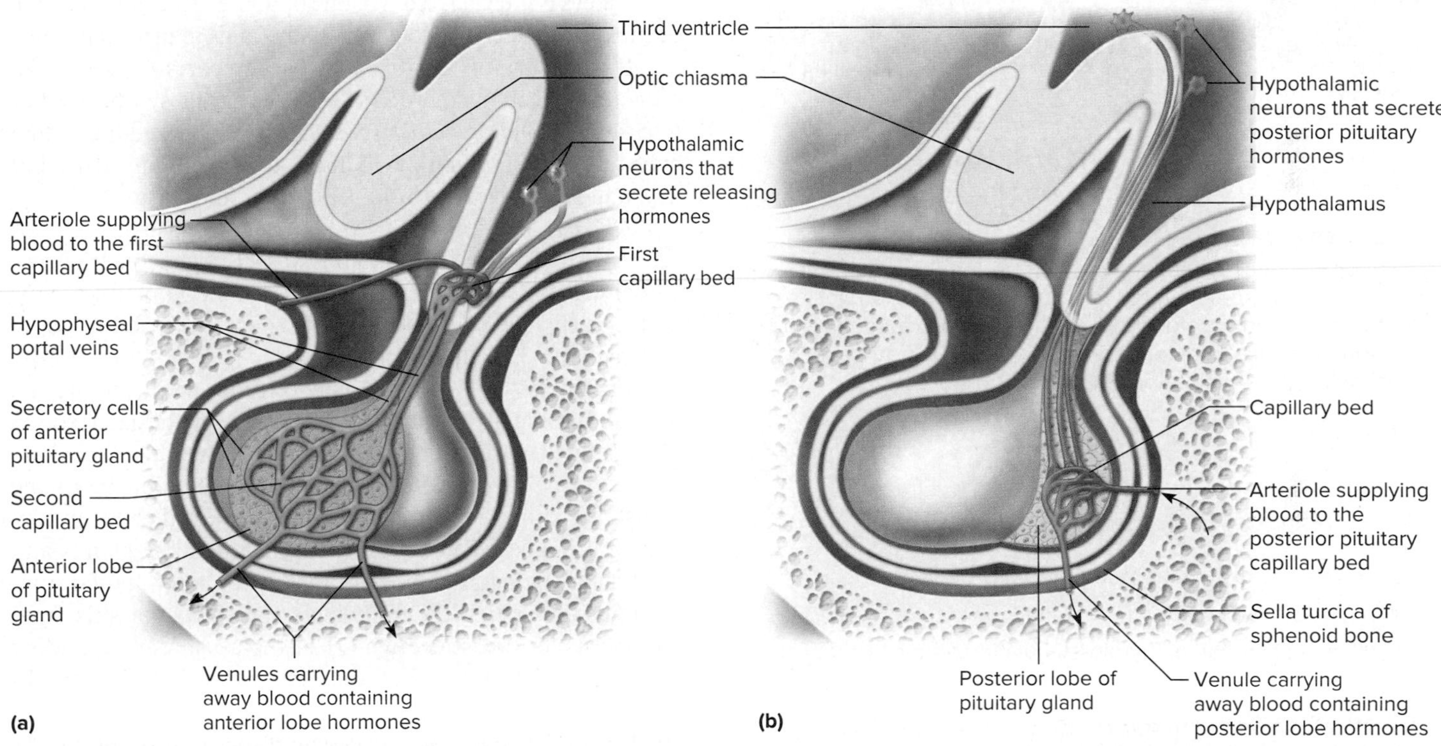

Figure 11.8 Secretion of pituitary hormones. **(a)** Releasing hormones from neurons in the hypothalamus stimulate secretory cells of the anterior lobe of the pituitary gland to secrete anterior pituitary hormones. **(b)** Other neurons in the hypothalamus release their hormones directly into capillaries of the posterior lobe of the pituitary gland as posterior pituitary hormones. APR

thyrotropin-releasing hormone (TRH) (fig. 11.9). Circulating thyroid hormones inhibit release of TRH and TSH. As the blood concentration of thyroid hormones increases, secretion of TRH and TSH decreases.

Adrenocorticotropic hormone (a-drē″nō-kor″te-kō-trō-pik′ hor′mōn) **(ACTH)** controls the manufacture and secretion of certain hormones from the outer layer, or *cortex*, of the adrenal gland. (These hormones are discussed in section 11.7, Adrenal Glands.) ACTH secretion is stimulated by *corticotropin-releasing hormone (CRH),* which the hypothalamus releases in response to decreased concentrations of adrenal cortical hormones. Stress may also stimulate the release of CRH, which results in increased ACTH secretion.

Follicle-stimulating hormone (fol′ĭ-kl stim′u-lā″ting hor′mōn) **(FSH)** and **luteinizing hormone** (lū′tē-in-īz″ing hor′mōn) **(LH)** are *gonadotropins,* which means they exert their actions on the gonads, or reproductive organs. Gonads are the **testes** (tes′tēz) in the male and the **ovaries** (ō′vah-rēz) in the female. (Section 19.4, Hormonal Control of Male Reproductive Functions, and section 19.6, Oogenesis and the Ovarian Cycle, discuss the functions of these gonadotropins and the ways they interact.)

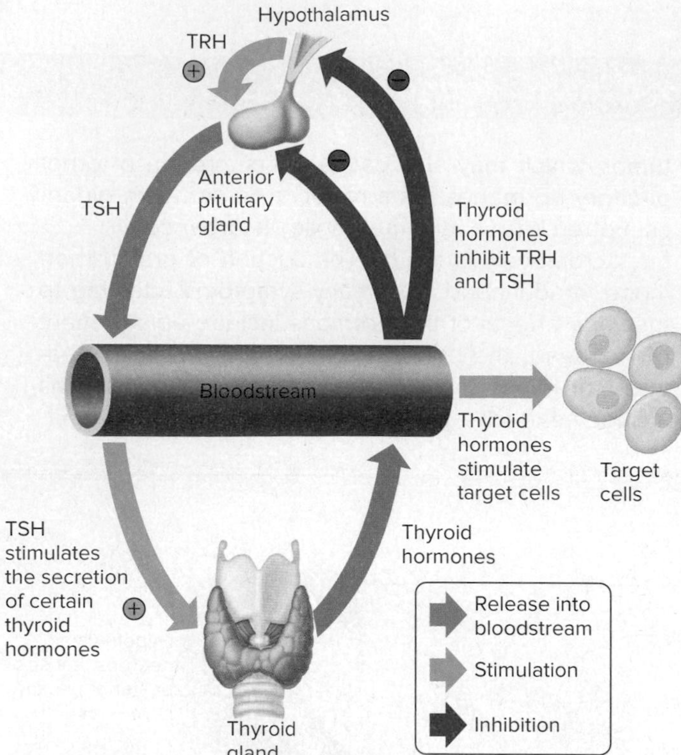

Figure 11.9 Thyrotropin-releasing hormone (TRH) from the hypothalamus stimulates the anterior pituitary gland to release thyroid-stimulating hormone (TSH), which stimulates the thyroid gland to release thyroid hormones. These thyroid hormones reduce the secretion of TSH and TRH by negative feedback. (⊕ = stimulation; ⊖ = inhibition)

PRACTICE 11.4

3. How does growth hormone affect protein synthesis?
4. What is the function of prolactin?
5. How is secretion of thyroid-stimulating hormone regulated?
6. What is the function of adrenocorticotropic hormone?
7. What is a gonadotropin?

Posterior Pituitary Hormones

The posterior pituitary consists mostly of axons and neuroglia, unlike the anterior pituitary, which is composed primarily of glandular epithelial cells. Neuroglia support the axons, which originate from neurons in the hypothalamus. The secretions of these neurons function not as neurotransmitters, but as hormones.

The hormones associated with the posterior pituitary are **antidiuretic hormone** (an″tī-dī″u-ret′ik hor′mōn) **(ADH)** and **oxytocin** (ok″sĭ-tō′sin) **(OT)**. These hormones are transported down axons through the pituitary stalk to the posterior lobe, and are stored in vesicles (secretory granules) near the ends of the axons. Impulses from the hypothalamus release the hormones into the blood. Thus, though synthesized in the hypothalamus, ADH and OT are considered posterior pituitary hormones because they enter the bloodstream from the posterior pituitary gland (fig. 11.8).

A *diuretic* is a chemical that increases urine production, whereas an *antidiuretic* decreases urine formation. ADH produces an antidiuretic effect by reducing the volume of water the kidneys excrete. In this way, ADH helps regulate the water concentration of body fluids.

The hypothalamus controls ADH secretion. Certain neurons in this part of the brain, called *osmoreceptors,* sense changes in the osmotic pressure of body fluids. Dehydration due to loss of water increasingly concentrates blood solutes. Osmoreceptors, sensing the resulting increase in osmotic pressure, signal the posterior pituitary to release ADH, which acts on target cells in the kidneys. As a result, the kidneys produce less urine, conserving water. On the other hand, drinking excess water dilutes body fluids, inhibiting ADH release. In response, the kidneys excrete a larger volume of dilute urine until the concentration of water and solutes in body fluids returns to normal.

If an injury or tumor damages any parts of the ADH-regulating mechanism, too little ADH may be synthesized or released, producing *diabetes insipidus.* An affected individual may produce as much as 15 liters of very dilute urine per day, and solute concentrations in body fluids rise.

ADH is also known as *vasopressin,* because at high levels it causes constriction of blood vessels (vasoconstriction). Under certain conditions, such as when blood pressure falls dangerously low, ADH may help to maintain blood pressure (see section 13.5, Blood Pressure).

In females, OT contracts smooth muscle in the uterine wall and stimulates uterine contractions in the later stages of childbirth. Stretching of uterine and vaginal tissues late in pregnancy triggers OT release during childbirth. In the breast, OT stimulates contraction of specialized cells (myoepithelial cells) associated with the milk-producing glands and their ducts. In lactating breasts, this action forces liquid from the milk glands into the milk ducts and ejects the milk from the breasts for breastfeeding. OT also plays a role in bonding between mother and infant and between sexual partners. In males, OT may play a role in the sexual response, including erection of the penis and movement of sperm. In addition, OT is an antidiuretic, but it is much weaker than ADH. Table 11.3 reviews the hormones of the pituitary gland.

If a pregnant woman near or at her "due date" has certain signs of the approaching birth, but is not yet experiencing uterine contractions (labor pains), she may be given a form of oxytocin to stimulate contractions. Oxytocin may also be administered to the mother following childbirth to contract the uterus sufficiently to squeeze broken blood vessels closed, lessening the risk of hemorrhage.

PRACTICE 11.4

8. What is the function of antidiuretic hormone?
9. How is secretion of antidiuretic hormone controlled?
10. What effects does oxytocin produce in females?

TABLE 11.3 Hormones of the Pituitary Gland

Hormone	Action	Source of Control
Anterior Lobe		
Growth hormone (GH)	Stimulates an increase in the size and division rate of body cells; enhances movement of amino acids across membranes	Secretion stimulated by growth hormone-releasing hormone from the hypothalamus. Secretion inhibited by growth hormone inhibiting hormone from hypothalamus
Prolactin (PRL)	Sustains milk production after birth	Secretion inhibited by prolactin inhibiting hormone from the hypothalamus. Secretion stimulated by prolactin-releasing factor from hypothalamus
Thyroid-stimulating hormone (TSH)	Controls secretion of hormones from thyroid gland	Thyrotropin-releasing hormone (TRH) from hypothalamus
Adrenocorticotropic hormone (ACTH)	Controls secretion of certain hormones from adrenal cortex	Corticotropin-releasing hormone (CRH) from hypothalamus
Follicle-stimulating hormone (FSH)	In females, responsible for the development of egg-containing follicles in ovaries and stimulates follicular cells to secrete estrogen; in males, stimulates production of sperm cells	Gonadotropin-releasing hormone from hypothalamus
Luteinizing hormone (LH)	Promotes secretion of sex hormones; plays a role in releasing an egg cell in females	Gonadotropin-releasing hormone from hypothalamus
*Posterior Lobe**		
Antidiuretic hormone (ADH)	Causes kidneys to conserve water; in high concentration constricts blood vessels	Hypothalamus in response to changes in water concentration in body fluids
Oxytocin (OT)	Contracts smooth muscle in the uterine wall; contracts myoepithelial cells associated with milk-secreting glands	Hypothalamus in response to stretching of uterine and vaginal walls and stimulation of breasts

*These hormones are synthesized in the hypothalamus, as explained in the text.

11.5 | Thyroid Gland

LEARN

1. Describe the location of the thyroid gland, and list the hormones it secretes.
2. Describe the functions of the hormones that the thyroid gland secretes.
3. Explain how the secretion of each thyroid hormone is regulated.

The **thyroid gland** (thī′roid gland) is a very vascular structure that consists of two large lobes connected by a broad *isthmus* (is′mus). The lobes are just inferior to the larynx and anterior and lateral to the trachea (fig. 11.10 and reference plate 4). The larynx is easily identified by the large thyroid cartilage or "Adam's apple" in the front of the neck. Of the three hormones from the thyroid gland, two help control how many calories the body consumes, and one plays a role in bone growth and maintenance of blood calcium levels.

Structure of the Gland

A capsule of connective tissue covers the thyroid gland. The gland is made up of many secretory parts called *follicles.* The follicles have cavities filled with a clear, viscous substance called *colloid* and are lined with a single layer of cuboidal epithelial cells. These **follicular cells** produce and secrete hormones that are then either stored in the colloid or released into the blood in nearby capillaries.

Hormones of the Thyroid Gland

The follicular cells of the thyroid gland synthesize two hormones—**thyroxine** (thī-rok′sin) and **triiodothyronine** (trī″ī-ō″dō-thī′ro-nēn)—collectively referred to as **thyroid hormone.** Thyroxine is also known as tetraiodothyronine, or T_4, because it contains four atoms of iodine. Triiodothyronine, or T_3, includes three atoms of iodine. Thyroxine and triiodothyronine have similar actions, although triiodothyronine is five times more potent. Thyroid hormone helps regulate the metabolism of carbohydrates, lipids, and proteins. It increases the rate at which cells release energy from carbohydrates, increases the rate of protein synthesis, and stimulates breakdown and mobilization of lipids. Thyroid hormone is the major factor determining how many calories the body must consume at rest in order to maintain life, which is known as the *basal metabolic rate (BMR).* Thyroid hormone is required for normal growth and development, and is essential to nervous system maturation.

Follicular cells require iodine salts (iodides) to produce thyroxine and triiodothyronine. Foods normally provide iodides. After the iodides have been absorbed from the intestine, blood transports them to the thyroid gland. An efficient active transport mechanism moves the iodides into

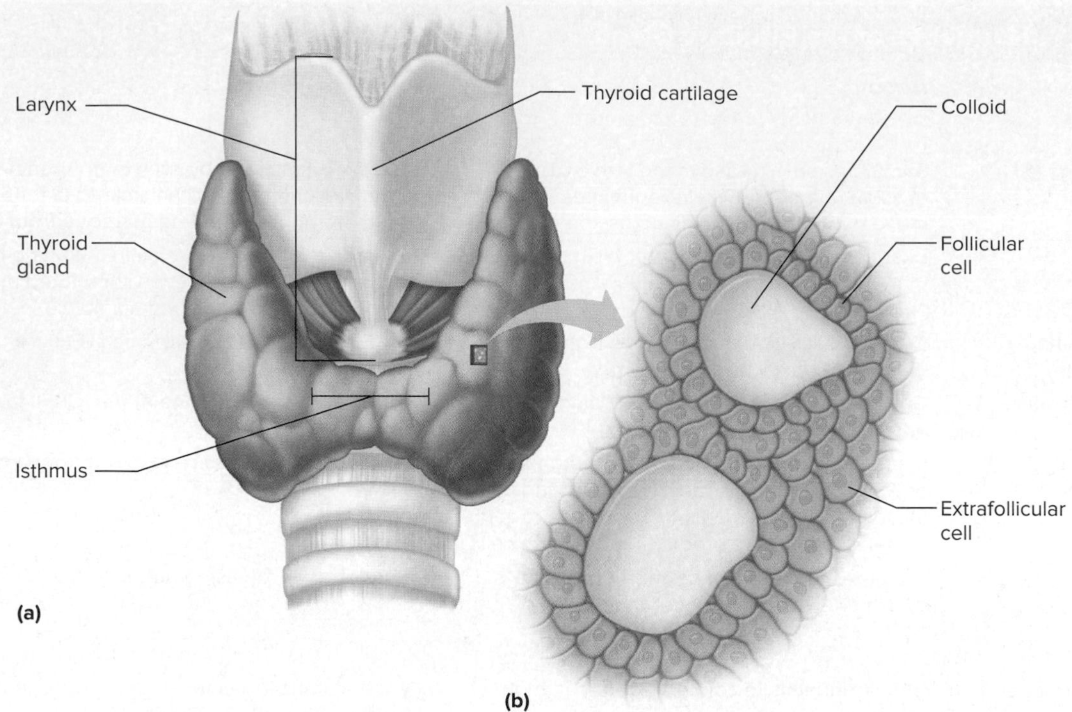

Figure 11.10 Thyroid gland. **(a)** The thyroid gland consists of two lobes connected anteriorly by an isthmus. **(b)** Follicular cells secrete the thyroid hormones: thyroxine and triiodothyronine. Extrafollicular cells secrete calcitonin. APR

the follicular cells, where they are used to synthesize the hormones. The hypothalamus and anterior pituitary gland control the synthesis and release of T_4 and T_3. Once in the blood, thyroxine and triiodothyronine combine with proteins (plasma proteins) and are transported to body cells.

A third hormone of the thyroid gland, **calcitonin** (kal′sĭ-tō′nin), is not referred to as "thyroid hormone" because it is produced by the thyroid's extrafollicular cells, separate from the gland's follicles. Along with parathyroid hormone (PTH) from the parathyroid glands, calcitonin regulates the concentrations of blood calcium and phosphate ions (see fig. 7.4).

The blood concentration of calcium ions controls calcitonin release. As the blood calcium concentration increases, so does calcitonin secretion. Calcitonin stimulates the bone-forming activity of osteoblasts, inhibits the bone-resorbing activity of osteoclasts (see section 7.4, Bone Development, Growth, and Repair), and increases the kidneys' excretion of calcium and phosphate ions—actions that lower the blood calcium and phosphate ion concentrations. Table 11.4 reviews the actions and controls of hormones of the thyroid gland.

Thyroid disorders may produce underactivity (*hypothyroidism*) or overactivity (*hyperthyroidism*) of the glandular cells. One form of hypothyroidism, called *cretinism,* affects newborns as a result of insufficient thyroid hormone during the pregnancy. If untreated shortly after birth, symptoms include stunted growth, abnormal bone formation, slowed mental development, low body temperature, and sluggishness. Hypothyroidism is also common among older adults, producing fatigue and weight gain.

Hyperthyroidism produces an elevated metabolic rate, restlessness, and overeating. The eyes may protrude (exophthalmia) because of swelling in the tissues behind them.

TABLE 11.4 Hormones of the Thyroid Gland APR

Hormone	Action	Source of Control
Thyroxine (T_4)	Increases rate of energy release from carbohydrates; increases rate of protein synthesis; accelerates growth; necessary for normal nervous system maturation	Thyroid-stimulating hormone from the anterior pituitary gland
Triiodothyronine (T_3)	Same as above, but five times more potent than thyroxine	Thyroid-stimulating hormone from the anterior pituitary gland
Calcitonin	Lowers blood calcium and phosphate ion concentrations by stimulating deposition of calcium and phosphate ions into bones, inhibiting release of these ions from bones, and by increasing excretion of these ions by kidneys	Blood calcium concentration

Depending on the specific disease, both hypothyroidism and hyperthyroidism may result in an enlarged thyroid gland. This typically is visible as a bulge in the neck called a *goiter.* The presence of a goiter does not by itself indicate the nature of the thyroid disorder.

PRACTICE 11.5

1. Where is the thyroid gland located?
2. Which hormones of the thyroid gland affect carbohydrate metabolism and protein synthesis?
3. How does the thyroid gland influence the concentrations of blood calcium and phosphate ions?

11.6 | Parathyroid Glands

LEARN

1. Describe the locations of the parathyroid glands, and identify the hormone they secrete.
2. Describe the functions of the hormone that the parathyroid glands secrete.
3. Explain how the secretion of parathyroid hormone is regulated.

The **parathyroid glands** (par″ah-thī′roid glandz) are on the posterior surface of the thyroid gland, as figure 11.11 shows. Most individuals have four parathyroid glands—a superior and an inferior gland associated with each of the thyroid's bilateral lobes. The hormone they secrete helps control blood calcium levels.

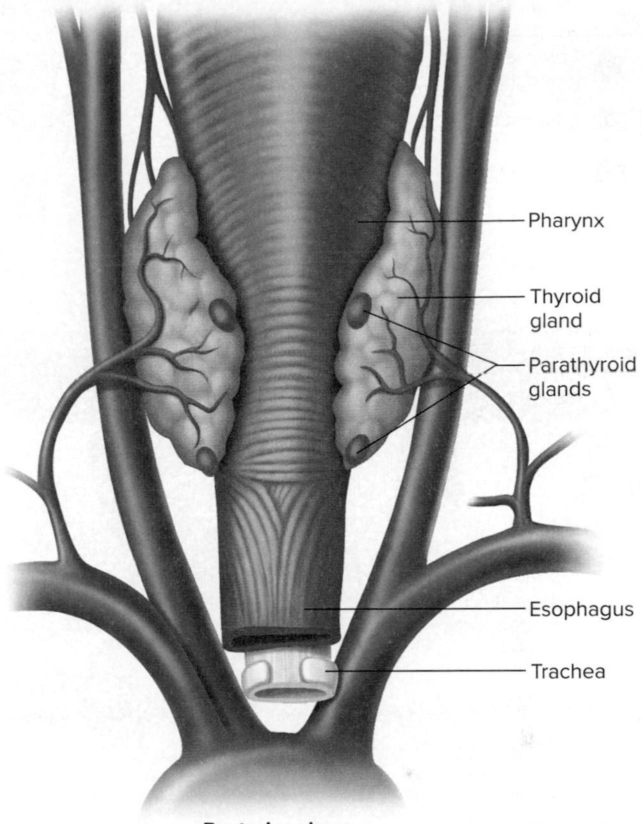

Figure 11.11 The parathyroid glands are embedded in the posterior surface of the thyroid gland.

Structure of the Glands

A thin capsule of connective tissue covers each small, yellowish-brown parathyroid gland. The body of the gland consists of many tightly packed secretory cells closely associated with capillary networks.

Parathyroid Hormone

The parathyroid glands secrete **parathyroid hormone (PTH),** which increases the blood calcium ion concentration and decreases the blood phosphate ion concentration. PTH affects the bones, kidneys, and intestine.

The extracellular matrix of bone tissue is rich in mineral salts, including calcium phosphate (see section 7.3, Bone Function). PTH inhibits the activity of osteoblasts and increases the activity of osteoclasts to resorb bone and release calcium and phosphate ions into the blood. At the same time, PTH causes the kidneys to conserve blood calcium and to excrete more phosphate ions in the urine. It also causes the kidneys to activate vitamin D, which stimulates calcium absorption from food in the intestine, further increasing the blood calcium concentration.

Negative feedback between the parathyroid glands and the blood calcium concentration regulates PTH secretion. As the blood calcium concentration drops, more PTH is secreted; as the blood calcium concentration rises, less PTH is released (fig. 11.12).

To summarize, calcitonin and PTH activities maintain a stable blood calcium concentration. Calcitonin decreases an above-normal blood calcium concentration, while PTH increases a below-normal blood calcium concentration without changing blood phosphate (see fig. 7.7).

Injury to the parathyroids or their surgical removal can cause *hypoparathyroidism,* in which decreased PTH secretion reduces osteoclast activity. Although the bones remain strong, the blood calcium concentration decreases. The nervous system may become abnormally excitable, triggering spontaneous impulses. As a result, muscles may undergo tetanic contractions, possibly leading to respiratory failure and death.

A tumor in a parathyroid gland may cause *hyperparathyroidism,* which increases PTH secretion. This stimulates osteoclast activity, and as bone tissue is resorbed, the bones soften, deform, and more easily fracture spontaneously. In addition, excess calcium and phosphate released into body fluids may be deposited in various places, causing new problems, such as kidney stones.

PRACTICE 11.6

1. Where are the parathyroid glands located?
2. How does parathyroid hormone help regulate concentrations of blood calcium and phosphate ions?

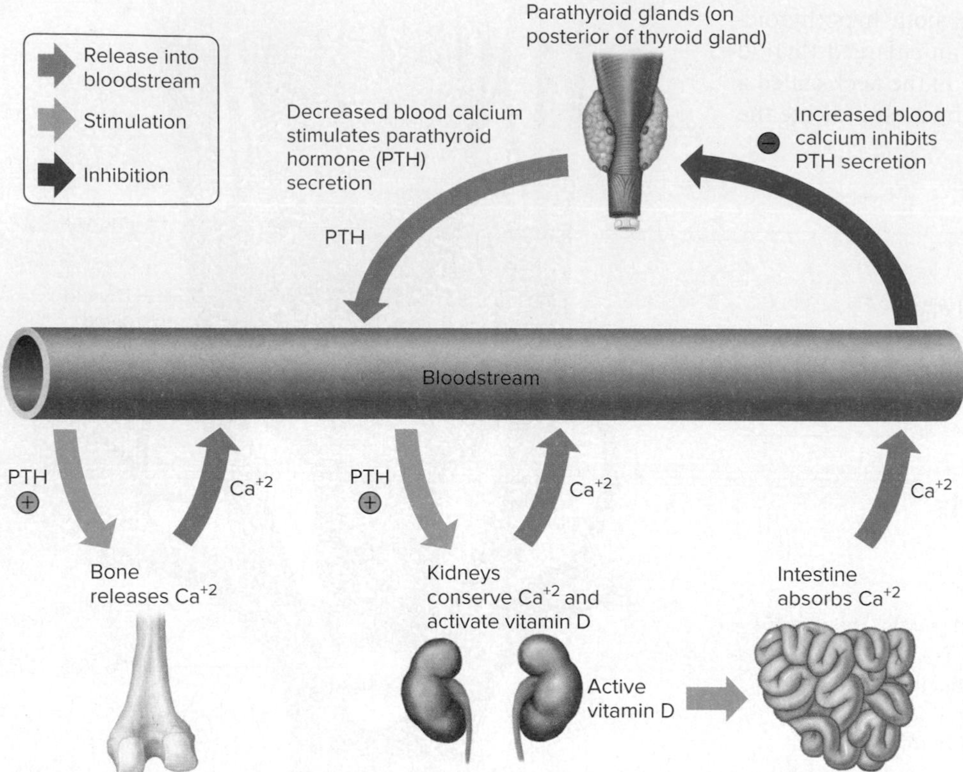

Figure 11.12 Parathyroid hormone (PTH) stimulates bone to release calcium (Ca^{+2}) and the kidneys to conserve calcium. It indirectly stimulates the intestine to absorb calcium. The resulting increase in blood calcium concentration inhibits secretion of PTH by negative feedback. (⊕ = stimulation; ⊖ = inhibition) APR

11.7 | Adrenal Glands

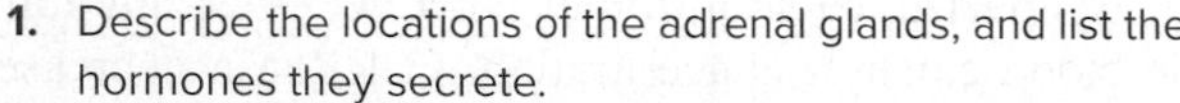

LEARN

1. Describe the locations of the adrenal glands, and list the hormones they secrete.
2. Describe the functions of the hormones that the adrenal glands secrete.
3. Explain how the secretion of each adrenal hormone is regulated.

The **adrenal glands** (ah-drē'nal glandz) are closely associated with the kidneys. A gland sits atop each kidney like a cap and is embedded in the mass of adipose tissue that encloses the kidney (fig. 11.13 and reference plate 6). Adrenal hormones play roles in maintaining blood sodium levels and responding to stress. They also include certain sex hormones.

Structure of the Glands

Each adrenal gland is very vascular and consists of two parts: The central portion is the **adrenal medulla** (ah-drē'nal me-dul'ah), and the outer part is the **adrenal cortex** (ah-drē'nal kor'teks). These regions are not sharply divided, but they are functionally distinct structures that secrete different hormones.

The adrenal medulla consists of irregularly shaped cells organized in groups around blood vessels. These cells are intimately connected with the sympathetic division of the autonomic nervous system. Adrenal medullary cells are actually modified postganglionic neurons. Preganglionic autonomic nerve fibers control their secretions (see section 9.16, Autonomic Nervous System).

The adrenal cortex, which makes up the bulk of the adrenal gland, is composed of closely packed masses of epithelial cells, organized in layers. These layers form the outer (glomerulosa), middle (fasciculata), and inner (reticularis) zones of the cortex (fig. 11.13*b*). As in the adrenal medulla, the cells of the adrenal cortex are well supplied with blood vessels.

PRACTICE 11.7

1. Where are the adrenal glands located?
2. Describe the two portions of an adrenal gland.

Hormones of the Adrenal Medulla

The cells of the adrenal medulla secrete two closely related hormones—**epinephrine** (ep″ĭ-nef'rin) (adrenaline) and **norepinephrine** (nor″ep-ĭ-nef'rin) (noradrenaline). These hormones have similar molecular structures and physiological functions. In fact, epinephrine, which makes up 80 percent of the adrenal medullary secretion, is synthesized from norepinephrine.

The effects of the adrenal medullary hormones resemble those of sympathetic neurons stimulating their effectors. The hormonal effects, however, last up to ten times longer than nervous stimulation because hormones are broken down more slowly than are neurotransmitters. Epinephrine and norepinephrine increase heart rate, the force of cardiac muscle contraction, and blood glucose level. They also dilate airways, which makes breathing easier, elevate blood pressure, and decrease digestive activity.

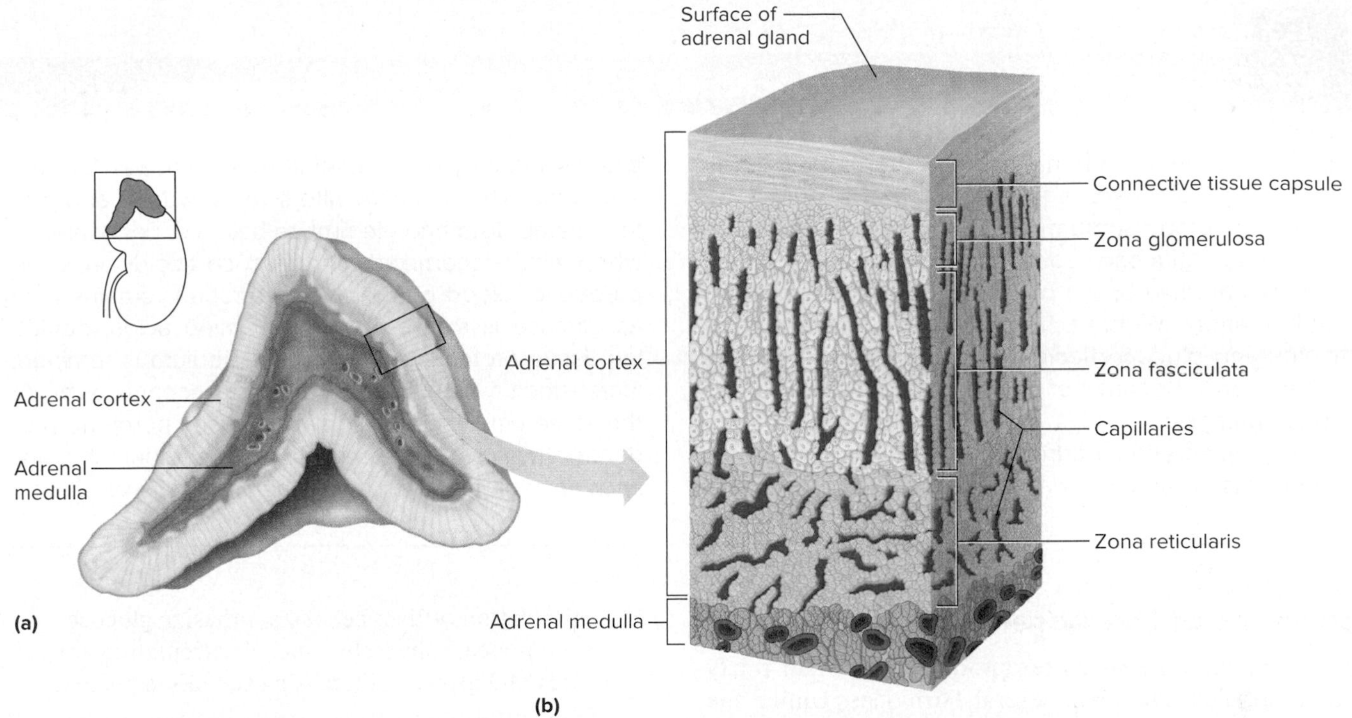

Figure 11.13 Adrenal glands. **(a)** An adrenal gland consists of an outer cortex and an inner medulla. **(b)** The cortex consists of three layers, or zones, of cells. 

Impulses arriving on sympathetic preganglionic nerve fibers stimulate the adrenal medulla to release its hormones at the same time that sympathetic impulses are stimulating other effectors. These sympathetic impulses originate in the hypothalamus in response to stress. In this way, adrenal medullary secretions function with the sympathetic division of the autonomic nervous system in preparing the body for energy-expending action, also called "fight-or-flight responses." Table 11.5 compares some of the effects of the adrenal medullary hormones.

Tumors in the adrenal medulla can increase hormonal secretion. Release of norepinephrine usually predominates, prolonging sympathetic responses—high blood pressure, increased heart rate, elevated blood sugar, and so forth. Surgical removal of the tumor corrects the condition.

PRACTICE 11.7

3. Name the hormones the adrenal medulla secretes.
4. What effects do hormones from the adrenal medulla produce?
5. What stimulates release of hormones from the adrenal medulla?

TABLE 11.5 Comparative Effects of Epinephrine and Norepinephrine

Structure or Function Affected	Epinephrine	Norepinephrine
Heart	Increases rate	Increases rate
	Increases force of contraction	Increases force of contraction
Blood vessels	Vasodilation, especially important in skeletal muscle at onset of fight-or-flight response	Vasoconstriction in skin and viscera shifts blood flow to other areas, such as exercising skeletal muscle
Systemic blood pressure	Some increase due to increased cardiac output	Some increase due to increased cardiac output and vasoconstriction (offset in some areas, such as exercising skeletal muscle, by local vasodilation due to other factors)
Airways	Dilation	Some dilation
Reticular formation of brainstem	Activated	Little effect
Liver	Promotes breakdown of glycogen to glucose, increasing blood sugar concentration	Little effect on blood glucose level
Metabolic rate	Increases	Increases

CLINICAL APPLICATION 11.2

Hyposecretion and Hypersecretion of Adrenal Cortical Hormones

Hyposecretion of adrenal cortical hormones leads to *Addison disease,* a condition characterized by decreased blood sodium, increased blood potassium, low blood glucose concentration (hypoglycemia), dehydration, low blood pressure, and increased skin pigmentation. Without treatment with mineralocorticoids and glucocorticoids, Addison disease can be lethal in days because of severe disturbances in electrolyte balance.

Hypersecretion of adrenal cortical hormones, which may be associated with an adrenal tumor or with the anterior pituitary oversecreting ACTH, causes *Cushing syndrome*. This condition alters carbohydrate and protein metabolism and electrolyte balance. For example, when mineralocorticoids and glucocorticoids are overproduced, blood glucose concentration remains high as glucose is synthesized from amino acids, depleting tissue protein. Also, too much sodium is retained, increasing tissue fluids, and the skin becomes puffy. At the same time, increases in adrenal sex hormone production may cause masculinizing effects in a female, such as beard growth and deepening of the voice.

Hormones of the Adrenal Cortex

The cells of the adrenal cortex produce more than thirty different steroids, including several hormones. Unlike the adrenal medullary hormones, without which a person can still survive, some adrenal cortical hormones are vital. Without extensive electrolyte therapy, a person who lacks these hormones will likely die within a week. The most important adrenal cortical hormones are aldosterone, cortisol, and certain sex hormones.

Aldosterone

Cells in the outer zone of the adrenal cortex synthesize **aldosterone** (al-dos′ter-ōn″). This hormone is called a **mineralocorticoid** (min″er-al-ō-kor′tĭ-koid) because it helps regulate the concentration of mineral electrolytes. More specifically, aldosterone causes the kidney to conserve sodium ions and excrete potassium ions. By conserving sodium ions, aldosterone stimulates water retention indirectly by osmosis, helping to maintain blood volume and blood pressure.

An increase in the blood concentration of potassium ions stimulates the cells that secrete aldosterone. The kidneys indirectly stimulate aldosterone secretion if blood pressure falls or the blood sodium ion concentration decreases.

Cortisol

Cortisol (kor′tĭ-sol) (hydrocortisone) is a **glucocorticoid** (gloo″kō-kor′tĭ-koid) hormone, which means it affects glucose metabolism. It is produced in the middle zone of the adrenal cortex and, like aldosterone, is a steroid. Cortisol also influences protein and fat metabolism.

The more important actions of cortisol include:

1. Inhibition of protein synthesis in tissues, increasing the blood concentration of amino acids.
2. Promotion of fatty acid release from adipose tissue, increasing the utilization of fatty acids and decreasing the use of glucose as energy sources.
3. Stimulation of liver cells to synthesize glucose from noncarbohydrates, such as circulating amino acids and glycerol, increasing the blood glucose concentration.

These actions of cortisol help keep blood glucose concentration within the normal range between meals. This control is important, because a few hours without food can exhaust the supply of liver glycogen, a major source of glucose.

Negative feedback controls cortisol release. This is much like control of thyroid hormones, involving the hypothalamus, anterior pituitary gland, and adrenal cortex. The hypothalamus secretes corticotropin-releasing hormone (CRH) into the hypophyseal portal veins, which carry CRH to the anterior pituitary, stimulating it to secrete ACTH. In turn, ACTH stimulates the adrenal cortex to release cortisol. Cortisol inhibits the release of CRH and ACTH, and as concentrations of these fall, cortisol production drops (fig. 11.14).

The set point of the feedback mechanism controlling cortisol secretion may change to meet the demands of changing conditions. For example, under stress—as from injury, disease, or emotional upset—information concerning the stressful condition reaches the brain. In response, brain centers signal the hypothalamus to release more CRH, elevating the blood cortisol concentration until the stress subsides (fig. 11.14).

Adrenal Sex Hormones

Cells in the inner zone of the adrenal cortex produce sex hormones. These hormones are male types (adrenal androgens), but some are converted to female hormones (estrogens) in the skin, liver, and adipose tissue. The amounts of adrenal sex hormones are very small compared to the supply of sex hormones from the gonads, but they may contribute to early development of reproductive organs. Table 11.6 summarizes the characteristics of the adrenal cortical hormones. Clinical Application 11.2 discusses the insufficient secretion and oversecretion of hormones of the adrenal cortex.

TABLE 11.6 Hormones of the Adrenal Cortex APR

Hormone	Action	Factor Regulating Secretion
Aldosterone	Helps regulate concentration of extracellular electrolytes by conserving sodium ions and excreting potassium ions	Blood sodium and potassium concentrations
Cortisol	Decreases protein synthesis, increases fatty acid release, and stimulates glucose synthesis from noncarbohydrates	Corticotropin-releasing hormone from the hypothalamus and adrenocorticotropic hormone (ACTH) from the anterior pituitary
Adrenal androgens	Supplement sex hormones from the gonads; may be converted to estrogens in females	Adrenocorticotropic hormone (ACTH) from the anterior pituitary plus unknown factors

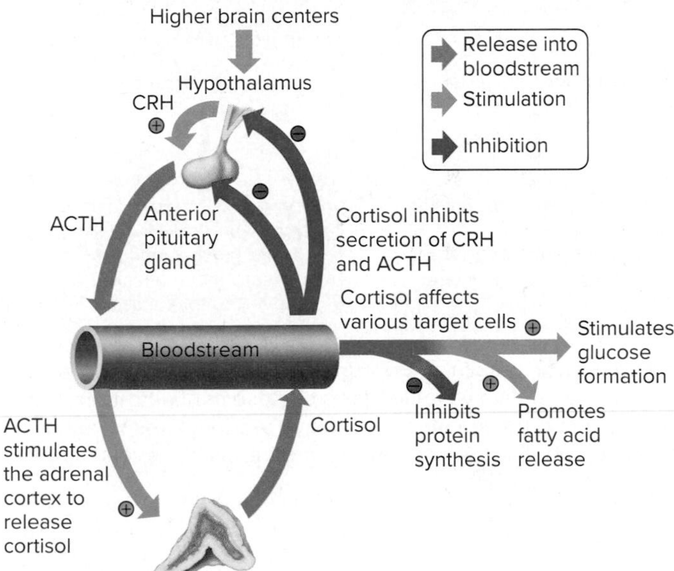

Figure 11.14 Negative feedback regulates cortisol secretion, similar to the regulation of thyroid hormone secretion. (⊕ = stimulation; ⊖ = inhibition)

PRACTICE 11.7

6. Name the most important hormones of the adrenal cortex.
7. What is the function of aldosterone?
8. What actions does cortisol produce?
9. How are the blood concentrations of aldosterone and cortisol regulated?

11.8 | Pancreas

LEARN

1. Describe the location of the pancreas, and list the hormones it secretes.
2. Describe the functions of the hormones that the pancreas secretes.
3. Explain how the secretion of each pancreatic hormone is regulated.

The **pancreas** (pan'krē-as) consists of two major types of secretory tissues. This organization reflects the pancreas's dual function as an exocrine gland that secretes digestive juice and an endocrine gland that releases hormones that control the blood glucose level (fig. 11.15 and reference plate 6). The dual nature of the pancreas begins in the embryo. First, ducts form whose walls harbor progenitor cells (see fig. 3.23). Some of the progenitor cells divide to yield daughter cells

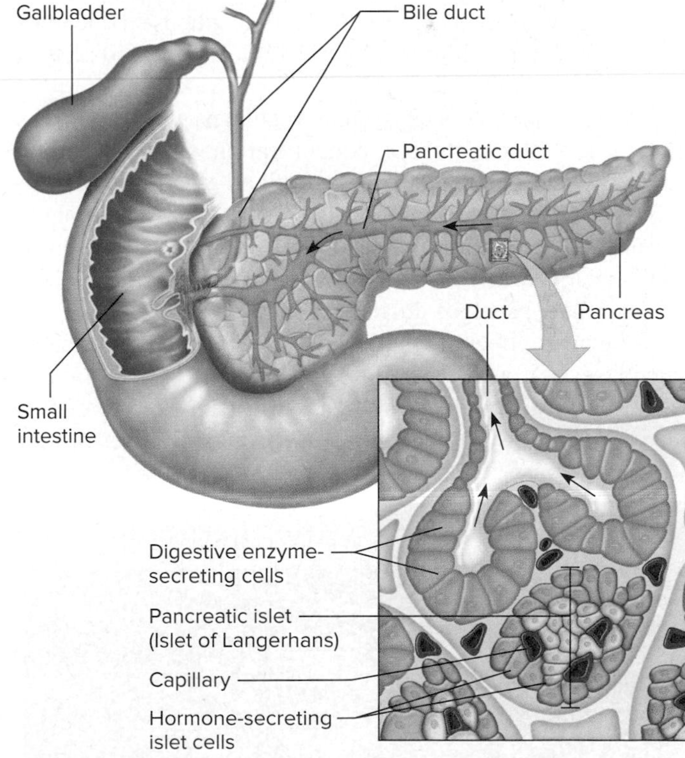

Figure 11.15 The hormone-secreting cells of the pancreas are grouped in clusters, or islets, that are near blood vessels. Other pancreatic cells secrete digestive enzymes into ducts leading to the small intestine. APR

PRACTICE FIGURE 11.15

What term describes the secretions that enter the pancreatic ducts?

Answer can be found in Appendix E.

that specialize as exocrine cells, and others divide to yield cells that differentiate into endocrine cells. The two functions are elaborated as the gland develops further.

Structure of the Gland

The pancreas is an elongated, somewhat flattened organ posterior to the stomach and partly posterior to the parietal peritoneum. A duct joins the pancreas to the duodenum (the first section of the small intestine). Digestive juice, the exocrine secretion of the pancreas, flows through this duct to the intestine. Section 15.7, Pancreas, discusses the digestive functions of the pancreas.

The endocrine part of the pancreas consists of groups of cells that are closely associated with blood vessels. These groups form "islands" of cells called *pancreatic islets* (islets of Langerhans) (figs. 11.15 and 11.16). The pancreatic islets include two distinct types of cells—alpha cells, which secrete the hormone glucagon, and beta cells, which secrete the hormone insulin.

Hormones of the Pancreatic Islets

Glucagon (gloo'kah-gon) raises the blood sugar concentration by stimulating the liver to break down glycogen and convert certain noncarbohydrates, such as amino acids, into glucose. These actions raise the blood glucose concentration. Glucagon much more effectively elevates blood glucose than does epinephrine.

A negative feedback system regulates glucagon secretion. A low blood glucose concentration stimulates alpha cells to release glucagon. When the blood glucose concentration rises, glucagon secretion falls (fig. 11.17). This control prevents hypoglycemia when the blood glucose concentration is relatively low, such as between meals, or when glucose is used rapidly, such as during exercise.

The main effect of **insulin** (in'su-lin) is to lower the blood glucose level, exactly opposite that of glucagon. Insulin does this in part by promoting facilitated diffusion of glucose into cells that have insulin receptors, for use in cellular respiration

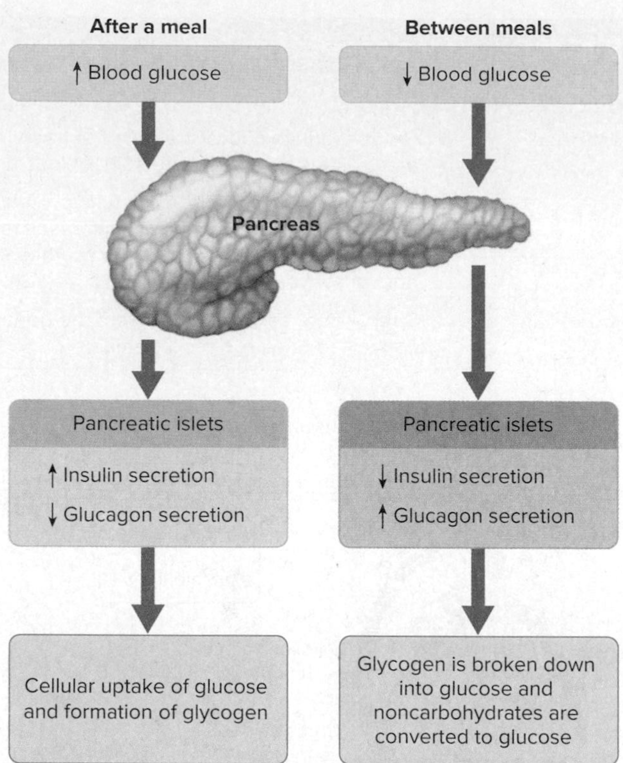

Figure 11.17 Insulin and glucagon function together to help maintain a relatively stable blood glucose concentration after a meal and between meals. Negative feedback responding to the blood glucose concentration controls the levels of both hormones. APR

(see section 3.3, Movements Into and Out of the Cell). Such cells include those of adipose tissue, liver, and skeletal muscle. (Glucose uptake by active skeletal muscle does not require insulin.) Insulin also stimulates the liver to form glycogen from glucose and inhibits conversion of noncarbohydrates into glucose. In addition, insulin promotes transport of amino acids into cells, increases the rate of protein synthesis, and stimulates adipose cells to synthesize and store fat.

A negative feedback system sensitive to the blood glucose concentration regulates insulin secretion. When the blood glucose concentration is high, such as after a meal, beta cells release insulin. Insulin helps prevent too high a blood glucose concentration by promoting glycogen formation in the liver and entrance of glucose into adipose and muscle cells. When glucose concentration falls, such as between meals or during the night, insulin secretion decreases (fig. 11.17).

As insulin secretion decreases, less glucose enters adipose and resting muscle cells. Cells that lack insulin receptors and are therefore not dependent on insulin, such as nerve cells, can still take up glucose from the blood. At the same time that insulin is decreasing, glucagon secretion is increasing. Insulin and glucagon are coordinated to maintain a relatively stable blood glucose concentration, despite great variation in the amount of carbohydrates a person eats (fig. 11.17). Clinical Application 11.3 discusses diabetes

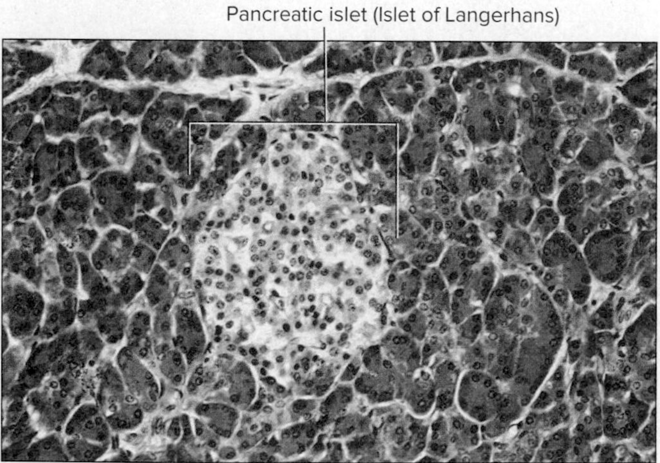

Figure 11.16 Light micrograph of a pancreatic islet (250x). APR Victor P. Eroschenko

CLINICAL APPLICATION 11.3
Diabetes Mellitus

Diabetes mellitus is a metabolic disease that arises from a deficiency of insulin or an inability of cells to recognize insulin. Persistent elevated blood glucose affects the eyes, heart, kidneys, and peripheral nerves. According to the Centers for Disease Control and Prevention, 29.1 million people in the United States have diabetes—that is 9.3% of the population.

Insulin deficiency, or an impaired insulin response, disturbs metabolism of carbohydrates, proteins, and fats. Because insulin helps glucose cross some cell membranes, diabetes impairs movement of glucose into adipose and resting skeletal muscle cells. At the same time, formation of glycogen, which is a long chain of glucose molecules, declines. As a result, the blood sugar concentration rises (hyperglycemia). When it reaches a certain level, the kidneys begin to excrete the excess. Glucose in the urine (glycosuria) raises the urine's osmotic pressure, and too much water is excreted. Excess water output causes dehydration and extreme thirst (polydipsia).

Diabetes mellitus also hampers synthesis of proteins and fats. Glucose-starved cells increasingly use proteins for energy. As a result, tissues waste away as weight drops, hunger increases, exhaustion becomes overwhelming, and wounds do not heal; children with this disorder stop growing. Changes in fat metabolism cause fatty acids and ketone bodies to accumulate in the blood, which lowers pH (acidosis). Dehydration and acidosis may harm brain cells, causing disorientation, coma, and eventually death.

The two most common forms of diabetes mellitus are type 1 (insulin-dependent or juvenile diabetes) and type 2 (non-insulin-dependent or maturity-onset diabetes).

Type 1 Diabetes Mellitus

Type 1 diabetes mellitus usually appears before age twenty. It is an autoimmune disease: the immune system destroys the beta cells of the pancreas (see section 14.7, Immunity: Adaptive [Specific] Defenses).

People with type 1 diabetes must carefully monitor their blood glucose levels. They do this in two ways. Every three months, a laboratory test checks the levels of hemoglobin molecules in the blood that have glucose bound to them. This measurement is called "A1C" and should be between 6 and 7. A1C represents the blood glucose level over the preceding three months, the life span of red blood cells, which transport hemoglobin. The second type of test is self-monitoring of blood glucose. The person uses a test kit to draw a drop of blood, applies it to a test strip, then uses a meter to read the concentration of glucose in the blood (in milligrams per deciliter). Normal plasma levels of glucose should range from 90 to 130 mg/dL before meals and less than 180 mg/dL one to two hours after meals. Most people with type 1 diabetes check their glucose this way two to four times a day. Smartphone apps record blood glucose levels with foods eaten and exercise done during the testing period.

Delivery of insulin to treat type 1 diabetes mimics normal pancreatic function (fig. 11A). People with type 1 diabetes inject insulin once or several times a day, or receive the hormone from an implanted insulin pump. A newer method is an inhaled form of insulin. For some, a wearable pump delivers insulin through a catheter placed under the skin. Another approach is pancreatic islet transplantation, which thousands of people have had.

Type 2 Diabetes Mellitus

About 85% to 90% of people with diabetes mellitus have type 2, in which the beta cells produce insulin but body cells lose the ability to recognize it. Most affected individuals are overweight when symptoms begin. Treatment includes a low-carbohydrate, high-protein diet; aerobic and weight-bearing exercise; and maintaining a desirable body weight. Several oral drugs can help control glucose levels, which can delay the onset of diabetes-related complications. Weight loss surgery, including procedures that remove access to part of the digestive tract, can stabilize blood glucose control for years.

People with any type of diabetes must monitor and regulate their blood glucose level to forestall complications, which include coronary artery disease, peripheral nerve damage, and retinal damage. Evidence suggests that complications may begin even before blood glucose level indicates disease. The American Diabetes Association recognizes "pre-diabetes" as blood glucose levels above the normal range but not yet indicative of type 2 diabetes. About 85 million people in the United States fall into this category.

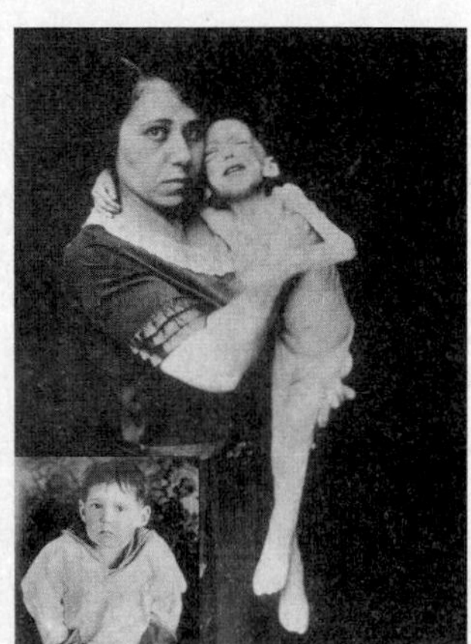

Figure 11A Before and after insulin treatment: The boy in his mother's arms is three years old but weighs only 15 pounds because of type 1 diabetes mellitus. The inset shows the same child after just two months of receiving insulin—his weight had doubled. (Both): Source Eli Lilly & Company Archives

mellitus, a disorder that affects the beta cells' ability to produce insulin or the body's ability to respond to it.

Nerve cells, including those of the brain, obtain glucose by a facilitated diffusion mechanism that does not require insulin, but rather depends only on the blood glucose concentration. For this reason, nerve cells are particularly sensitive to changes in blood glucose concentration. Conditions that cause such changes–for example, oversecretion of insulin leading to decreased blood glucose–are likely to affect brain functions.

PRACTICE 11.8

1. What is the endocrine portion of the pancreas called?
2. What is the function of glucagon?
3. What is the function of insulin?
4. How are glucagon and insulin secretion controlled?
5. Why are nerve cells particularly sensitive to changes in blood glucose concentration?

11.9 | Pineal, Thymus, and Other Glands

LEARN

1. Describe the locations of the pineal, thymus, and other endocrine glands covered in this section, and list the hormones they secrete.
2. Describe the functions of the hormones that the glands covered in this section secrete.
3. Explain how the secretion of each of the hormones described in this section is regulated.

The pineal and thymus glands are formally part of the endocrine system because their primary role is to produce and secrete hormones. The ovaries and testes are often considered part of the endocrine system proper, but they do have other functions, such as producing egg and sperm cells. Although their primary functions lie with other systems, the heart, kidneys, and digestive tract do produce hormones.

Pineal Gland

The **pineal gland** (pin'ē-al gland) is a small structure located deep between the cerebral hemispheres, where it is attached to the upper part of the thalamus near the roof of the third ventricle (see fig. 11.1). The pineal gland secretes the hormone **melatonin** (mel"ah-tō'nin) in response to changing light conditions outside the body. Melatonin helps to regulate **circadian rhythms** (ser"kad-ē'an rithmz), or daily sleep-wake cycles. Impulses originating in the retinas of the eyes are conducted along a complex pathway that eventually reaches the pineal gland. Melatonin secretion increases as levels of light drop. This usually makes you feel "sleepy." The fact that melatonin secretion responds to day length may explain why traveling across several time zones produces the temporary insomnia of jet lag. Clinical Application 11.4 discusses biological rhythms.

Thymus Gland

The **thymus** (thī'mus) lies in the mediastinum, posterior to the sternum and between the lungs. It is relatively large in young children but shrinks with age (see fig. 11.1). The thymus secretes a group of hormones called **thymosins** (thī'mo-sinz) that affect the production and differentiation of certain white blood cells (lymphocytes). In this way, the thymus plays an important role in immunity, discussed in section 14.4, Lymphatic Tissues and Organs.

Other Glands

The reproductive organs that secrete important hormones include the testes, which produce testosterone; the ovaries, which produce estrogens and progesterone; and the **placenta** (plah-sen'tah), which produces estrogens, progesterone, and gonadotropin. These glands and their secretions are discussed in sections 19.2, Organs of the Male Reproductive System, 19.5, Organs of the Female Reproductive System, and 20.3, Pregnancy and the Prenatal Period.

The digestive glands that secrete hormones are associated with the linings of the stomach and small intestine. Section 15.6, Stomach, and section 15.7, Pancreas, describe these structures and their secretions.

Other organs outside of the endocrine system produce hormones. The heart, for example, secretes *atrial natriuretic peptide,* a hormone that stimulates urinary sodium excretion (see section 17.3, Urine Formation). The kidneys and the liver secrete a red blood cell growth hormone called *erythropoietin* (see section 12.2, Formed Elements).

PRACTICE 11.9

1. Where is the pineal gland located?
2. What is the function of the pineal gland?
3. Where is the thymus located?
4. Which reproductive organs secrete hormones?
5. Which other organs secrete hormones?

11.10 | Stress and Health

LEARN

1. Describe how the body responds to stress.

Survival depends on the maintenance of homeostasis. Therefore, factors that change the body's internal environment can threaten life. When sensory receptors detect such changes, impulses to the hypothalamus trigger physiological responses that preserve homeostasis. These responses include increased activity in the sympathetic division of

CLINICAL APPLICATION 11.4
Biological Rhythms

Biological rhythms are changes that systematically recur in organisms. The period of any rhythm is the duration of one complete cycle. The frequency of a rhythm is the number of cycles per time unit. The study of biological rhythms is called *chronobiology.*

Three common types of rhythms in humans are ultradian, infradian, and circadian rhythms. *Ultradian rhythms* have periods shorter than 24 hours and include the cardiac cycle and the breathing cycle. Periods of *infradian rhythms,* such as the female reproductive cycle, are longer than 24 hours. Periods of *circadian rhythms,* such as the sleep-wake cycle, variation in body temperature, and changes in hormone secretion, are approximately 24 hours.

Both external (exogenous) and internal (endogenous) factors regulate human biological rhythms. Exogenous factors are environmental components, such as daily temperature changes and the light-dark cycle. Endogenous factors include "clock" genes. Many members of an extended family in Utah, for example, have "advanced sleep phase syndrome" due to a mutation in a gene called "period." The effect is striking—they promptly fall asleep at 7:30 each night and awaken suddenly at 4:30 a.m.

The sleep-wake cycle is the most obvious circadian rhythm in humans. It is largely controlled by the pattern of daylight and night, but under laboratory conditions of constant light or dark, the human body eventually follows an approximately 25-hour cycle. Using a backlit electronic device, such as a smartphone or tablet, can delay falling asleep long after the device is shut off. Experiments show that such light exposure decreases melatonin production by about 22 percent.

Body temperature is mostly endogenously regulated, but light exposure and physical activity help keep this rhythm on a 24- rather than 25-hour cycle. Body temperature is usually lowest between 4 and 6 a.m., and then increases and peaks between 5 and 11 p.m. It drops during the late evening hours and into the night.

Platelet cohesion, blood pressure, and pulse rate are typically highest 2 hours after awakening. This may explain why heart attacks and strokes are more likely to occur between 6 a.m. and noon than at other times.

Plasma cortisol surges and peaks at about 6 a.m., and then gradually declines to its minimum level in late evening before increasing again in the early morning. Growth hormone secretion peaks during the night. Antidiuretic hormone secretion is greater at night, when it decreases urine formation.

the autonomic nervous system, including increased secretion of adrenal hormones. A factor that can stimulate such a response is called a **stressor,** and the condition it produces in the body is called **stress.**

Types of Stress

Stressors include physical factors, such as exposure to extreme heat or cold, decreased oxygen concentration, infections, injuries, prolonged heavy exercise, and loud sounds. Stressors also include psychological factors, such as thoughts about real or imagined dangers, personal losses, and unpleasant social interactions. Feelings of anger, fear, grief, anxiety, depression, and guilt can also produce psychological stress. Sometimes, even pleasant stimuli, such as friendly social contact, feelings of joy and happiness, or sexual arousal, may be stressful.

Responses to Stress

Physiological responses to stress consist of a series of reactions called the *stress response* or *general adaptation syndrome,* which is under hypothalamic control. These reactions proceed through two stages: the immediate "alarm" stage and the slower to respond, but longer lasting, "resistance" stage.

In the immediate stage, the hypothalamus activates mechanisms that prepare the body for "fight or flight." These mechanisms include raising blood concentrations of glucose, glycerol, and fatty acids; increasing heart rate and blood pressure; dilating air passages; shunting blood from the skin and digestive organs to the skeletal muscles; and increasing epinephrine secretion from the adrenal medulla (fig. 11.18).

Other hormones whose secretions increase with acute stress include glucagon, growth hormone (GH), and antidiuretic hormone (ADH). Glucagon and GH mobilize energy sources, such as glucose, glycerol, fatty acids, and amino acids. ADH stimulates the kidneys to retain water, which helps to maintain blood volume—particularly important if a person is sweating heavily or bleeding.

In the resistance stage of the response, the hypothalamus releases CRH. This stimulates the anterior pituitary to secrete ACTH, which increases cortisol secretion. Cortisol increases blood amino acid concentration, fatty acid release, and glucose formation from noncarbohydrates. Thus, while the alarm stage responses prepare the body for physical action to alleviate stress, cortisol supplies cells with chemicals required during stress (fig. 11.18).

The stress response is beneficial in the face of a life-threatening event, but long-term stress can be harmful. Persistently increased cortisol secretion may make an individual more susceptible to infectious diseases and some cancers by decreasing the number of certain white blood cells (lymphocytes). Also, excess cortisol production may raise the risk of developing high blood pressure, atherosclerosis, and gastrointestinal ulcers.

PRACTICE 11.10

1. What is stress?
2. Distinguish between physical stress and psychological stress.
3. Describe the stress response.

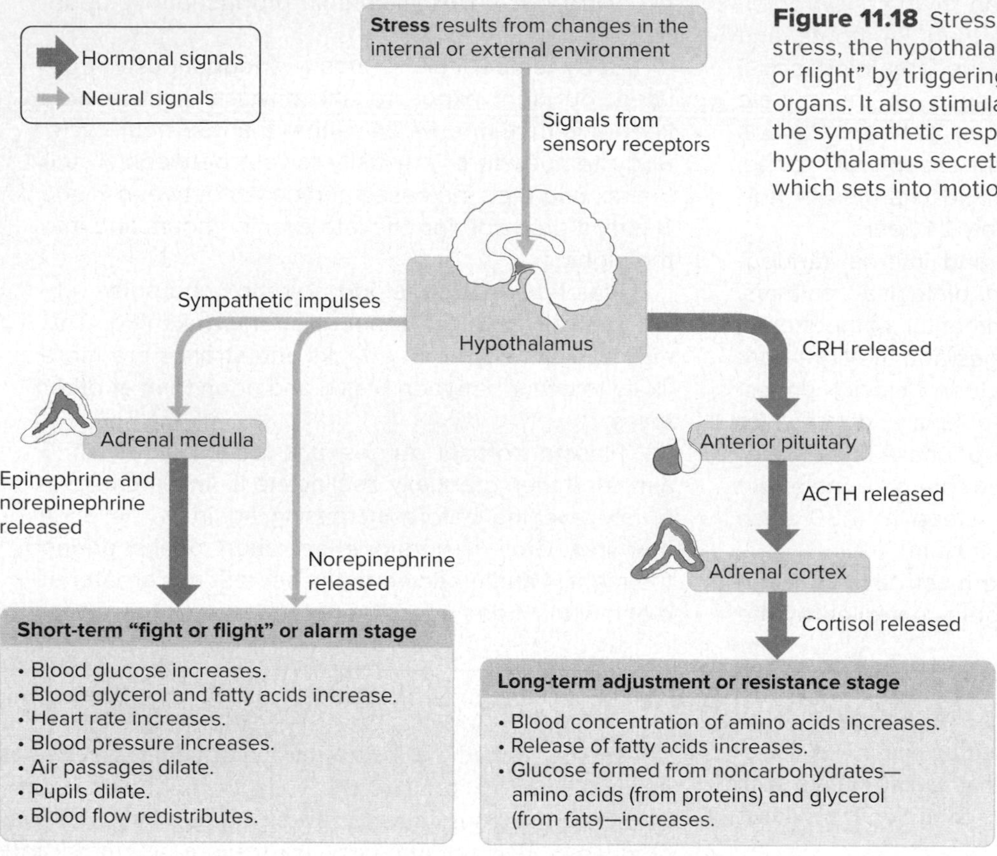

Figure 11.18 Stress response. During the alarm stage of stress, the hypothalamus helps prepare the body for "fight or flight" by triggering sympathetic impulses to various organs. It also stimulates epinephrine release, intensifying the sympathetic responses. In the resistance stage, the hypothalamus secretes corticotropin-releasing hormone, which sets into motion more lasting responses to stress.

Endocrine System

INTEGUMENTARY SYSTEM

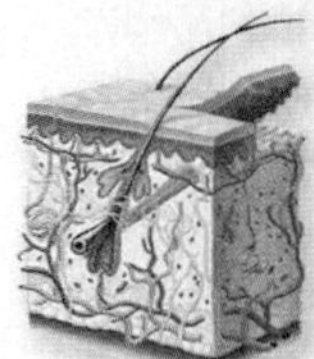

Melanocytes produce skin pigment in response to hormonal stimulation.

SKELETAL SYSTEM

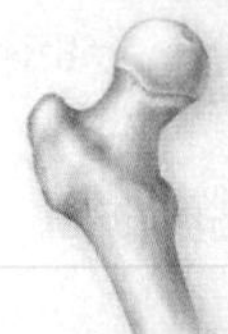

Hormones act on bones to control calcium balance.

MUSCULAR SYSTEM

Hormones help increase blood flow to exercising muscles.

NERVOUS SYSTEM

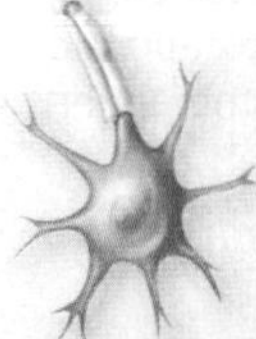

Neurons control the secretions of the anterior and posterior pituitary glands and the adrenal medulla.

CARDIOVASCULAR SYSTEM

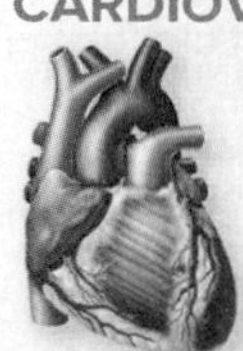

Hormones are carried in the bloodstream; some have direct actions on the heart and blood vessels.

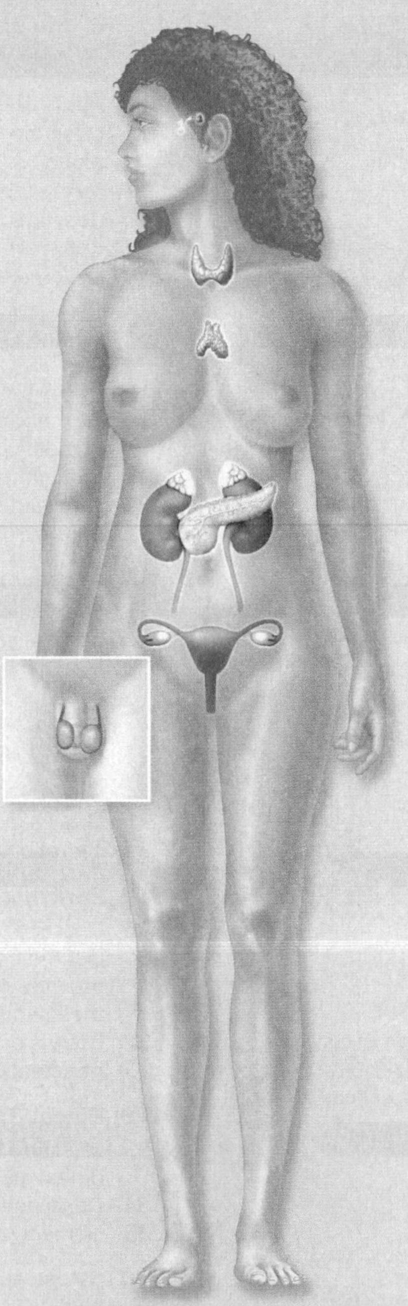

Glands secrete hormones that have a variety of effects on cells, tissues, organs, and organ systems.

LYMPHATIC SYSTEM

Hormones stimulate lymphocyte production.

DIGESTIVE SYSTEM

Hormones help control digestive system activity.

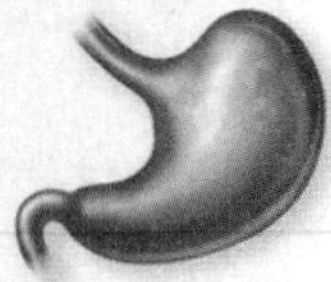

RESPIRATORY SYSTEM

Decreased oxygen causes hormonal stimulation of red blood cell production; red blood cells transport oxygen and carbon dioxide.

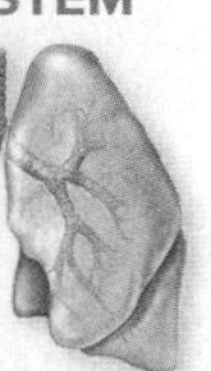

URINARY SYSTEM

Hormones act on the kidneys to help control water and electrolyte balance.

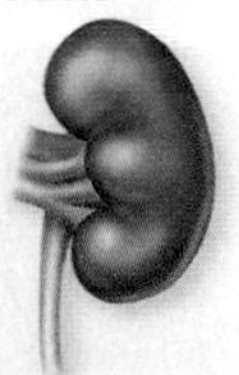

REPRODUCTIVE SYSTEM

Sex hormones play a major role in development of secondary sex characteristics, egg, and sperm.

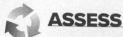

ASSESS

CHAPTER ASSESSMENTS

11.1 Introduction

1. Define *hormone* and *target cell*.
2. Contrast endocrine glands and exocrine glands.
3. Compare and contrast chemical communication in the nervous and endocrine systems.
4. Explain the specificity of a hormone for its target cell.
5. Functions of hormones include which of the following?
   a. control rates of certain chemical reactions
   b. transport substances across cell membranes
   c. help regulate water and electrolyte balances
   d. all of the above

11.2 Hormone Action

6. List the steps of steroid hormone action.
7. List the steps in the action of most nonsteroid hormones.
8. Explain how prostaglandins are similar to hormones and how they are different.

11.3 Control of Hormonal Secretions

9. Draw diagrams of the three mechanisms by which hormone secretion is controlled, including negative feedback.

11.4 Pituitary Gland

10. Describe the location and structure of the pituitary gland.
11. Explain the two ways in which the brain controls pituitary gland activity.
12. Releasing hormones come from which one of the following?
    a. thyroid gland
    b. anterior pituitary gland
    c. posterior pituitary gland
    d. hypothalamus
13. List the hormones secreted by the anterior pituitary.
14. Match the following hormones with their actions. More than one hormone can correspond to the same function.

(1) growth hormone	A. milk production
(2) thyroid-stimulating hormone	B. cell division
(3) prolactin	C. metabolic rate
(4) adrenocorticotropic hormone	D. exerts action on gonads
(5) follicle-stimulating hormone	E. controls secretion of adrenal cortex hormones
(6) luteinizing hormone	

15. Describe the control of growth hormone secretion.
16. Prolactin does which of the following?
    a. stimulates breast milk secretion
    b. stimulates breast milk production
    c. inhibits breast milk secretion
    d. inhibits breast milk production
17. Diagram the control of thyroid hormone secretion.
18. Describe the anatomical differences between the anterior and posterior lobes of the pituitary gland.
19. Describe the functions of the posterior pituitary hormones.
20. Under which of the following conditions would you expect an increase in antidiuretic hormone secretion?
    a. An individual ingests excess water.
    b. The posterior pituitary is removed from an individual because of a tumor.
    c. An individual is rescued after three days in the desert without food or water.
    d. An individual with normal kidneys is producing a large volume of urine.

11.5 Thyroid Gland

21. Describe the location and structure of the thyroid gland.
22. Match the hormones from the thyroid gland with their descriptions.

(1) thyroxine	A. most potent at controlling metabolism
(2) triiodothyronine	B. regulates blood calcium
(3) calcitonin	C. has four iodine atoms

23. List the source of control for each thyroid hormone.

11.6 Parathyroid Glands

24. Describe the location and structure of the parathyroid glands.
25. Explain the general function of parathyroid hormone.
26. Draw a diagram that shows how the secretion of parathyroid hormone is regulated.

11.7 Adrenal Glands

27. Distinguish between the adrenal medulla and the adrenal cortex.
28. Match the adrenal hormones with their source and actions.

(1) cortisol	A. cortex; sodium retention
(2) aldosterone	B. cortex; fatty acid release
(3) epinephrine	C. medulla; fight-or-flight response

29. Draw a diagram illustrating the regulation of cortisol secretion.

11.8 Pancreas

30. Describe the location and structure of the pancreas.
31. List the hormones secreted by the pancreatic islets, the type of cell that secretes each, and the actions of these hormones.
32. Draw a diagram that shows how the secretion of pancreatic hormones is regulated.

11.9 Pineal, Thymus, and Other Glands

33. Describe the location and general function of the pineal gland.
34. Describe the location and general function of the thymus.
35. Name five additional hormone-secreting organs.

11.10 Stress and Health

36. Define *stress*.
37. List the similarities and differences between the short-term alarm stage of stress and the long-term resistance stage.

ASSESS

INTEGRATIVE ASSESSMENTS/CRITICAL THINKING

Outcomes 1.5, 11.5

1. Amy noticed a slight enlargement in the anterior portion of her neck. Her doctor suspected a goiter, so she ordered some tests to try to determine the cause. The results showed that Amy's TRH levels were low, TSH levels were high, and T_3 and T_4 levels were high. Given these results, which endocrine organ is the problem? Is this condition an example of *hypersecretion* or *hyposecretion*? Why was Amy's neck enlarged?

Outcomes 2.2, 11.2, 11.3, 11.4, 11.5

2. When reactor 4 at the Chernobyl Nuclear Power Station in Ukraine exploded at 1:23 p.m. on April 26, 1986, a great plume of radioactive isotopes erupted into the air and spread for thousands of miles. Most of the isotopes emitted immediately following the blast were of the element iodine. Which of the glands of the endocrine system would be most seriously—and immediately—affected by the blast, and how do you think this would become evident in the nearby population?

Outcomes 4.4, 8.3, 11.4, 11.7, 11.8

3. Epinephrine, glucagon, and growth hormone are all hormones that have a glucose-sparing effect. What does this effect mean and what is its purpose?

Outcomes 7.3, 11.6

4. The parathyroid hormone (PTH) stimulates osteoclasts to release calcium phosphate salts from bone. Why is it important that PTH then stimulate the kidneys to eliminate phosphate?

Outcomes 7.4, 11.3, 11.4

5. Growth hormone is administered to people who have pituitary dwarfism. Parents wanting their normal children to be taller have requested the treatment for them. Do you think this is a wise request? Why or why not?

Outcomes 11.3, 11.4

6. What hormone supplements would an adult whose anterior pituitary has been removed require?

Outcomes 11.3, 11.4, 11.5, 11.10

7. How might a patient with hyperthyroidism modify their lifestyle to minimize the drain on body energy resources?

Outcomes 11.3, 11.7

8. The adrenal cortex of a patient who has lost a large volume of blood will increase secretion of aldosterone. What effect will this increased secretion have on the patient's blood concentrations of sodium and potassium ions?

Outcomes 11.3, 11.8

9. Why might oversecretion of insulin actually reduce glucose uptake by nerve cells?

Chapter Summary

11.1 Introduction

The endocrine and nervous systems maintain homeostasis. Like the nervous system, the endocrine system exerts precise effects in helping regulate metabolic processes.

1. The endocrine system is a network of glands that secrete **hormones,** which travel in the bloodstream and affect the functioning of **target cells.**
2. Both the endocrine system and the nervous system use chemical signals that bind to specific receptor molecules.
3. The endocrine system has a slower onset, but has longer lasting actions than the nervous system.
4. **Paracrine** secretions act locally, and **autocrine** secretions act on the cells that produce them.

11.2 Hormone Action

Endocrine glands secrete hormones that affect target cells with specific receptors. Hormones are very potent.

1. Chemically, hormones are steroids, amines, peptides, proteins, or glycoproteins.
2. Steroid hormones
   a. Steroid hormones enter a target cell and bind receptors, forming complexes in the nucleus.
   b. These complexes activate specific genes, so that specific proteins are synthesized.
3. Nonsteroid hormones
   a. Nonsteroid hormones bind receptors in the target cell membrane.

b. The hormone-receptor complex signals a G protein to stimulate a membrane protein, such as adenylate cyclase, to induce formation of second messenger molecules.
c. A second messenger, such as cyclic adenosine monophosphate **(cAMP),** diacylglycerol (DAG), or inositol triphosphate (IP_3), activates protein kinases.
d. Protein kinases activate protein substrate molecules, which in turn change a cellular process.

4. Prostaglandins
   a. **Prostaglandins** act on the cells of the organs that produce them.
   b. Prostaglandins are present in small amounts and have powerful hormonelike effects.

11.3 Control of Hormonal Secretions

The concentration of each hormone in body fluids is regulated.

1. Some endocrine glands secrete hormones in response to releasing hormones that the hypothalamus and anterior pituitary secrete.
2. Other glands secrete their hormones in response to impulses from neurons.
3. Some glands respond to levels of a substance in the bloodstream.
4. Negative feedback guides these control mechanisms.
   a. In a negative feedback system, a gland senses the concentration of a substance it regulates.
   b. When the concentration of the regulated substance reaches a certain point, it inhibits the gland.
   c. As the gland secretes less hormone, the amount of the regulated substance also decreases.
   d. Negative feedback systems maintain relatively stable hormone concentrations.

11.4 Pituitary Gland

*The **pituitary gland** has an anterior lobe and a posterior lobe. The hypothalamus controls most pituitary secretions.*

1. Anterior pituitary hormones
   a. The **anterior pituitary** secretes **growth hormone (GH), prolactin (PRL), thyroid-stimulating hormone (TSH), adrenocorticotropic hormone (ACTH), follicle-stimulating hormone (FSH),** and **luteinizing hormone (LH).**
   b. Growth hormone
      (1) GH stimulates cells to enlarge and divide more frequently.
      (2) GH-releasing hormone and GH release-inhibiting hormone from the hypothalamus control GH secretion.
   c. PRL stimulates and sustains milk production.
   d. Thyroid-stimulating hormone
      (1) TSH controls thyroid hormone secretion from the thyroid gland.
      (2) The hypothalamus secretes thyrotropin-releasing hormone (TRH), which regulates TSH secretion.
   e. Adrenocorticotropic hormone
      (1) ACTH controls secretion of hormones from the adrenal cortex.
      (2) The hypothalamus secretes corticotropin-releasing hormone (CRH), which regulates ACTH secretion.
   f. FSH and LH are gonadotropins, hormones which regulate the functions of the gonads.
2. Posterior pituitary hormones
   a. The **posterior pituitary** gland consists largely of neuroglia and nerve fibers.
   b. Neurons whose cell bodies are in the hypothalamus produce the hormones secreted by the posterior pituitary.
   c. **Antidiuretic hormone (ADH)**
      (1) ADH reduces the volume of water the kidneys excrete.
      (2) The hypothalamus regulates ADH secretion.
   d. **Oxytocin (OT)**
      (1) OT contracts smooth muscle in the uterine wall.
      (2) OT also contracts myoepithelial cells that secrete and eject milk.

11.5 Thyroid Gland

*The **thyroid gland** in the neck consists of two lobes.*

1. Structure of the gland
   a. The thyroid gland consists of many follicles.
   b. The follicles are fluid-filled and store hormones.
2. Hormones of the thyroid gland
   a. **Thyroxine** and **triiodothyronine** (collectively called **thyroid hormone**) increase the metabolic rate of cells, enhance protein synthesis, and stimulate lipid utilization.
   b. **Calcitonin** decreases blood calcium level and increases blood phosphate ion concentration.

11.6 Parathyroid Glands

*The **parathyroid glands** are on the posterior surface of the thyroid gland.*

1. Each parathyroid gland consists of secretory cells that are well supplied with capillaries.
2. **Parathyroid hormone (PTH)**
   a. PTH increases blood calcium level and decreases blood phosphate ion concentration.
   b. A negative feedback mechanism operates between the parathyroid glands and the blood.

11.7 Adrenal Glands

*The **adrenal glands** are located atop the kidneys.*

1. Structure of the glands
   a. Each gland consists of an **adrenal medulla** and an **adrenal cortex.**
   b. These parts are functionally distinct, and secrete different hormones.
2. Hormones of the adrenal medulla
   a. The adrenal medulla secretes **epinephrine** and **norepinephrine,** which have similar effects.
   b. Sympathetic impulses stimulate secretion of these hormones.
3. Hormones of the adrenal cortex
   a. The adrenal cortex produces several steroid hormones.
   b. **Aldosterone** is a mineralocorticoid that causes the kidneys to conserve sodium ions and water and to excrete potassium ions.

c. **Cortisol** is a glucocorticoid that affects carbohydrate, protein, and fat metabolism.
d. Adrenal sex hormones
    1. These hormones are of the male type but may be converted to female hormones.
    2. They may supplement the sex hormones the gonads produce.

11.8 Pancreas

*The **pancreas** secretes digestive juices as well as hormones.*

1. Structure of the gland
    a. The pancreas is attached to the small intestine.
    b. The pancreatic islets secrete glucagon and insulin.
2. Hormones of the pancreatic islets
    a. **Glucagon** stimulates the liver to produce glucose from glycogen and noncarbohydrates.
    b. **Insulin** moves glucose across some cell membranes, stimulates glucose and fat storage, and promotes protein synthesis.
    c. Nerve cells do not require insulin to take up glucose.

11.9 Pineal, Thymus, and Other Glands

1. **Pineal gland**
    a. The pineal gland is attached to the thalamus.
    b. It secretes **melatonin** in response to varying light conditions.
2. **Thymus**
    a. The thymus lies behind the sternum and between the lungs.
    b. It secretes **thymosins,** which affect the production of certain lymphocytes that function in immunity.
3. Reproductive organs
    a. The testes secrete testosterone.
    b. The ovaries secrete estrogens and progesterone.
    c. The **placenta** secretes estrogens, progesterone, and gonadotropin.
4. Digestive glands
    Certain glands of the stomach and small intestine secrete hormones.
5. Other hormone-producing organs
    Other organs, such as the heart and the kidneys, also produce hormones.

11.10 Stress and Health

***Stress** occurs when the body responds to stressors that threaten the maintenance of homeostasis. Stress responses include increased activity of the sympathetic nervous system and increased secretion of adrenal hormones.*

1. Types of stress
    a. Physical stress results from environmental factors that are harmful or potentially harmful to tissues.
    b. Psychological stress results from thoughts about real or imagined dangers.
2. Responses to stress
    a. Responses to stress maintain homeostasis.
    b. The hypothalamus controls the stress response.

12 UNIT 4 TRANSPORT

Blood

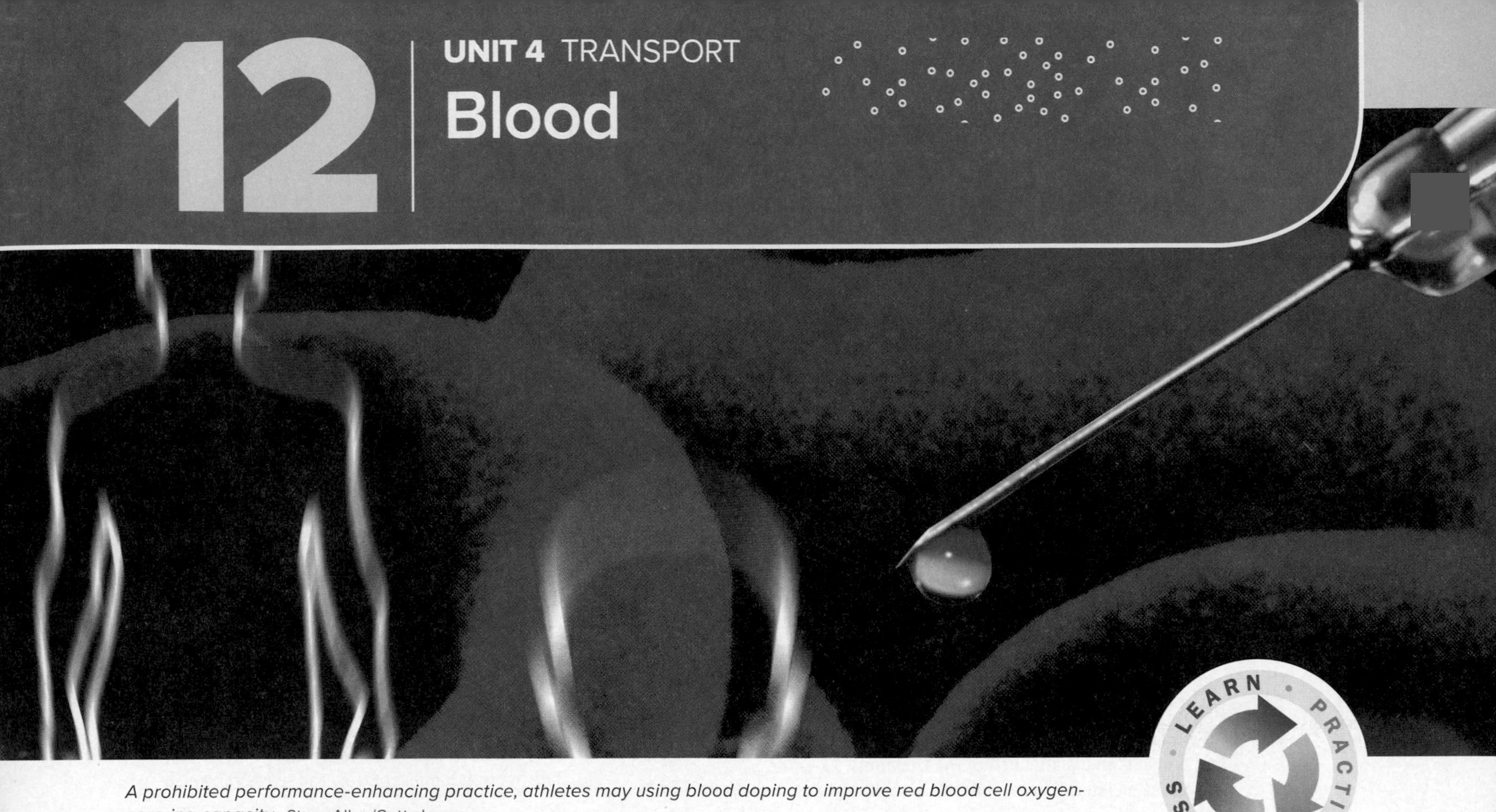

A prohibited performance-enhancing practice, athletes may using blood doping to improve red blood cell oxygen-carrying capacity. Steve Allen/Getty Images

In their pursuit to be the best, as well as to obtain financial and social rewards, elite athletes may have the mind-set to win at all costs, which includes cheating. These athletes recognize that the advantages of cheating, such as a valuable contract and a low risk of getting caught, greatly outweigh the penalties, which might be a one-year ban from competition. They live in a world where they are told that a performance-enhancing drug is dangerous, but they disregard that risk even if it means that their life may be shortened. Performance-enhancing drugs, like erythropoietin and anabolic steroids, mimic natural chemicals in the body, which makes detection difficult.

Hemoglobin is the oxygen-carrying molecule found in erythrocytes (red blood cells). The hormone erythropoietin (EPO) stimulates red blood cell production in red bone marrow. Enhancing red blood cell production, and therefore, hemoglobin concentration, increases oxygen-carrying capacity, with the intent to improve stamina and performance. Blood doping is defined by the World Anti-Doping Agency as the misuse of techniques or substances to increase red blood cell count, which is prohibited. One form of blood doping involves taking blood from the athlete and separating red blood cells from the blood and freezing them. Consequently, the kidneys and liver detect lower oxygen levels (hypoxemia) and act by secreting EPO, ultimately raising the red blood cell count back to normal. Then, just prior to competition, the red blood cells that were previously removed are thawed and injected back into the bloodstream. The effect is similar to an EPO injection with the same risks. An increase in the hematocrit value (the percentage of red blood cells) increases the viscosity of the blood, which puts an increased stress on the heart because it is having to pump against a greater resistance.

Athletes have come up with creative ways to avoid detection, such as using quantities just small enough to avoid detection, but large enough to have an effect (called microdosing). External injections of EPO can be detected through blood and urine tests. EPO being produced by the athlete's own cells was undetectable as a banned substance until recently when a Duke University research team discovered that during storage, large RNAs in red blood cells were cut by enzymes, creating a by-product called microRNA-70 (miR-70) that can be used as a biomarker for detecting stored red blood cells. To get around the illegal use of cortisol injections to calm inflammation or repair damaged tissue, a legal procedure used by athletes involves providing a sample of their blood, centrifuging it to aggregate the platelets, and then injecting the platelets back into the athlete at the site of inflammation or injury. Growth factors secreted by platelets speed up the repair process and reduce inflammation.

Other performance-enhancing practices may be used in the future, such as "poop doping" that involves fecal transplants of intestinal microorganisms, thought to enhance muscle recovery, that are not normally found in gut microbiomes except in elite athletes. New technologies such as "gene-editing" or "gene doping" may be used to modify cells or the body's DNA to improve gene expression through the use of vectors or CRISPR-Cas9. The World Anti-Doping Agency has already banned these technologies.

Anatomy & Physiology *Revealed 4.0*

Module 9 Cardiovascular System

LEARNING OUTLINE

After studying this chapter, you should be able to complete the "Learning Outcomes" that follow the major headings throughout the chapter.

AIDS TO UNDERSTANDING WORDS

agglutin- [to glue together] *agglutin*ation: clumping of red blood cells.

bil- [bile] *bil*irubin: pigment excreted in the bile.

embol- [stopper] *embol*ism: a mass lodging in and obstructing a blood vessel.

erythr- [red] *erythr*ocyte: red blood cell.

hema- [blood] *hema*tocrit: percentage of red blood cells in a given volume of blood.

hemo- [blood] *hemo*globin: red pigment responsible for the color of blood.

leuko- [white] *leuko*cyte: white blood cell.

-osis [abnormal condition] leukocyt*osis:* condition in which white blood cells are overproduced.

-poie [make, produce] erythro*poie*tin: hormone that stimulates the production of red blood cells.

-stasis [halt] hemo*stasis:* process that stops bleeding from damaged blood vessels.

thromb- [clot] *thromb*ocyte: blood platelet involved in the formation of a blood clot.

(Appendix A has a complete list of Aids to Understanding Words.)

12.1 | Introduction

LEARN

1. Describe the general characteristics of blood, and discuss its major functions.
2. Distinguish between the formed elements and the liquid portion of blood.

Essential to life, blood carries out many vital functions and contributes to homeostasis as it connects the internal and external environments of the body. The blood, heart, and blood vessels form the circulatory system. Blood is pumped by the heart and flows within tubes, called blood vessels. As blood circulates, it transports dissolved nutrients and oxygen to cells from the digestive system and lungs, respectively, and carbon dioxide and other waste products from cells to the lungs and kidneys to be removed from the body. Clotting proteins, hormones, ions, heat, and cells are also carried within the blood.

Blood is the only connective tissue with a fluid extracellular matrix, called **plasma.** Plasma is a mixture of substances dissolved or suspended in water. The **formed elements** are produced in red bone marrow and include red blood cells, white blood cells, and cell fragments called platelets (fig. 12.1). Red blood cells transport gases, white blood cells fight disease, and platelets help control blood loss.

Blood is three to four times more viscous (thick) than water, as well as slightly heavier, because of its many constituents. If blood is too viscous, blood flow becomes sluggish, and if viscosity is too low, blood flows too quickly. In either case, the heart has to work harder. The normal pH of blood is between 7.35 and 7.45, which is slightly alkaline. Blood volume varies with gender, body size, amount of adipose tissue, and changes in fluid and electrolyte concentrations. An average-sized adult female has 4–5 liters of blood, whereas an average-sized male has 5–6 liters.

In a laboratory, a blood draw may be done to determine possible abnormalities of the blood. As part of a complete blood count, a sample of blood is taken from the patient, and then the test tube is centrifuged to separate the whole blood into its liquid portion (plasma) and solid portion (formed elements). Red blood cells are dense, so they settle to the bottom of the test tube. The percentage of red blood cells in whole blood is called the **hematocrit (HCT),** which is normally about 45% of blood volume. The white blood cells and platelets account for less than 1% of blood volume and are seen in the "buffy coat." The remaining blood sample is the clear, straw-colored plasma that makes up about 55% of the blood volume (fig. 12.2). Among other causes, a low hematocrit value may indicate anemia or nutritional deficiencies, whereas a high hematocrit value may indicate excessive, abnormal production of blood cells or dehydration.

PRACTICE 12.1

Answers to the Practice questions can be found in the eBook.

1. What are the major components of blood?
2. What factors affect blood volume?
3. What might cause abnormal hematocrit values?

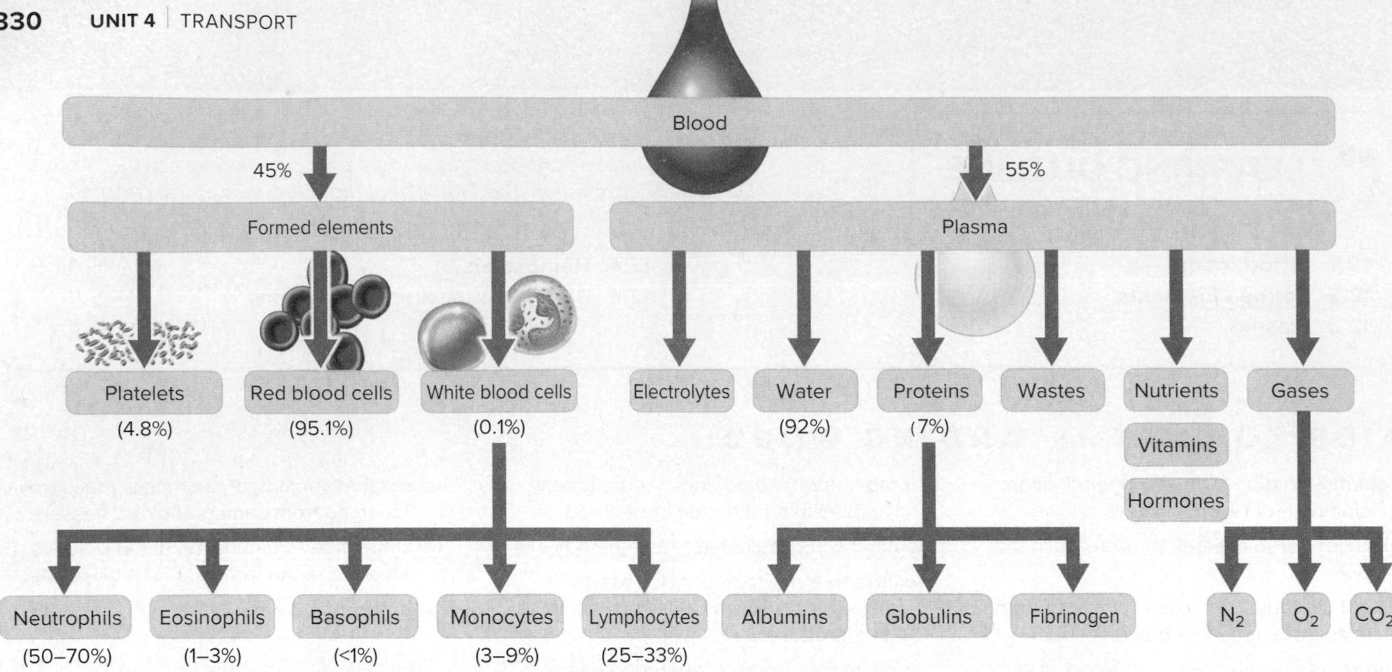

Figure 12.1 Blood composition. Blood is a complex mixture of formed elements in a liquid extracellular matrix, called plasma. Note that water and proteins account for 99% of the plasma.

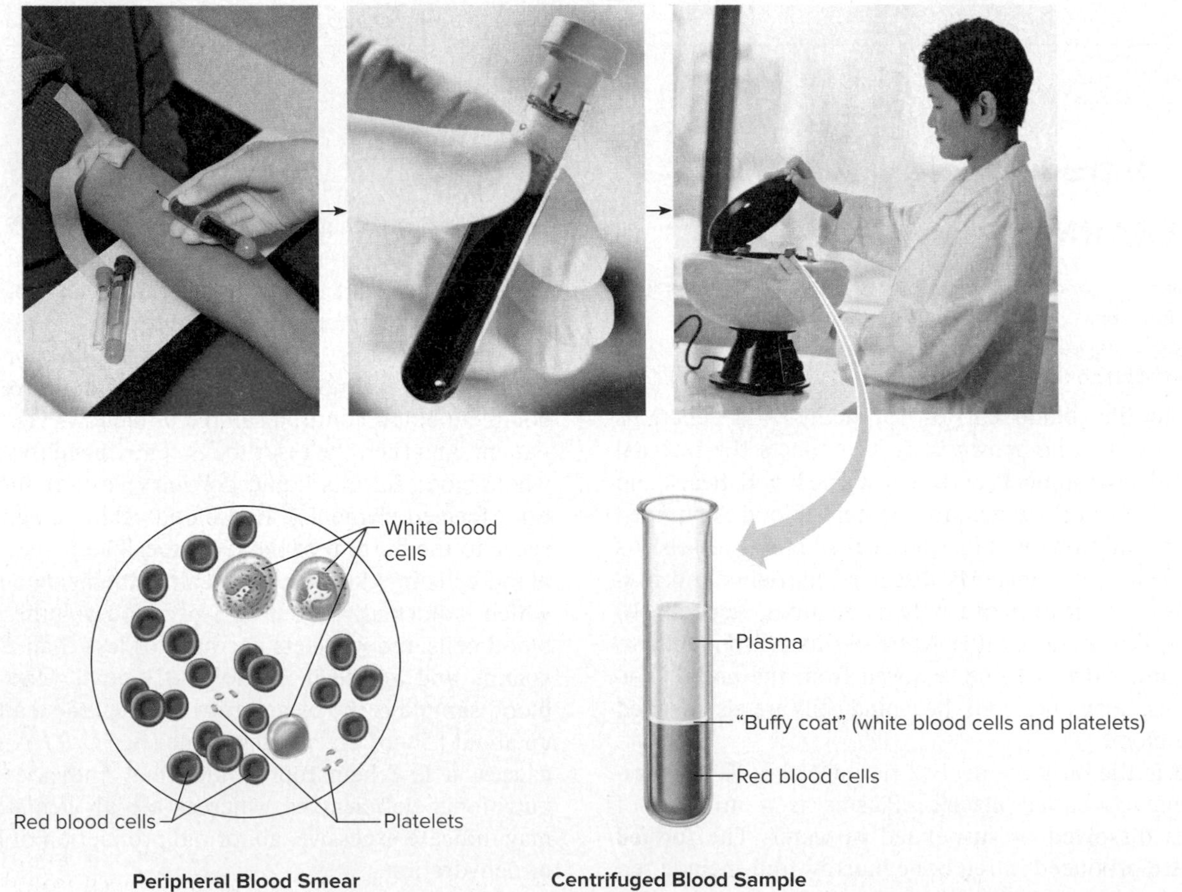

Figure 12.2 Blood consists of a liquid portion called plasma and a solid portion (the formed elements) that includes red blood cells, white blood cells, and platelets. Blood cells and platelets can be seen under a light microscope when a blood sample is smeared onto a glass slide. APR (photos: left and right): Keith Brofsky/Getty Images; (photo: center): Comstock/PunchStock

PRACTICE FIGURE 12.2

How would the percentage of red blood cells (hematocrit) change in a person who is dehydrated? Would the actual number of red blood cells be affected?

Answers can be found in Appendix E.

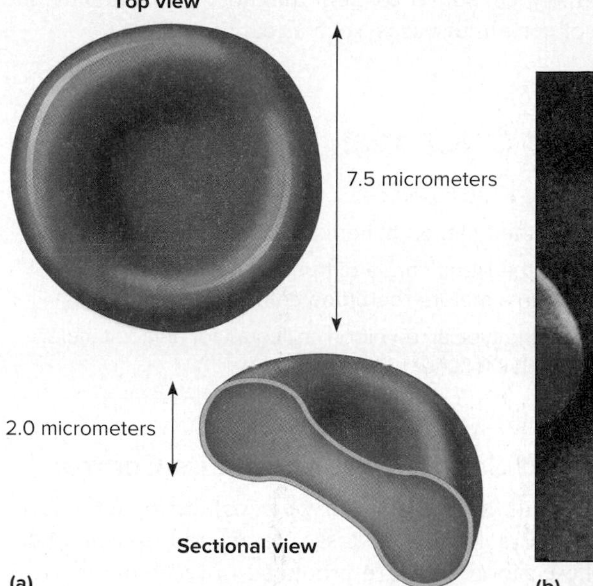

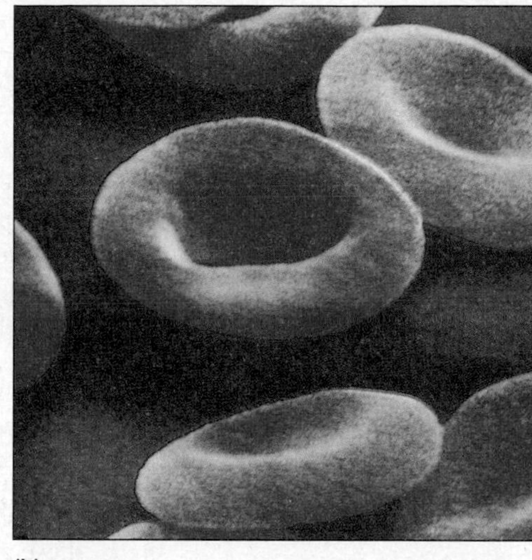

Figure 12.3 Red blood cells. **(a)** The biconcave shape of a red blood cell makes possible its function of transporting gases. **(b)** A falsely colored scanning electron micrograph of human red blood cells (5,000x). APR

(b): Bill Longcore/Science Source

12.2 | Formed Elements

LEARN

1. Explain the significance of red blood cell counts.
2. Summarize the control of red blood cell production.
3. Distinguish among the five types of white blood cells, and give the function(s) of each type.
4. Explain the significance of platelet counts.

Red Blood Cells

Red blood cells, also called **erythrocytes** (ĕ-ri-thro′-sītz), are shaped like biconcave discs as a result of losing most of their organelles, including the nucleus, during development. Some scientists suggest that the flattened shape with a depressed center increases the surface area relative to its volume, making it ideal for gas exchange (fig. 12.3). Occupying about one-third of a red blood cell's volume, **hemoglobin** (hē″mo-glō′bin), a globular protein that carries oxygen and some carbon dioxide, is oriented close to the plasma membrane so that the distance for diffusion is reduced. A normal, healthy concentration of hemoglobin in the blood is 13–18 g/100 mL in adult males and 12–16 g/100 mL in adult females.

The structure of the hemoglobin molecule includes four polypeptide chains (the globin portion), each with a heme pigment, and in the center of each heme is an iron that binds to oxygen (fig. 12.4). When oxygen is bound to hemoglobin, resulting in **oxyhemoglobin,** the color of blood is a bright red, and when the oxygen is released, the resulting deoxyhemoglobin makes blood a darker color.

Prolonged oxygen deficiency (hypoxia) causes the skin and mucous membranes to appear a bluish color due to an abnormally high blood concentration of **deoxyhemoglobin** in the superficial blood vessels, a condition called cyanosis. (The blood is not blue, however, but appears blue because of how the tissue of the skin absorbs and reflects light.) Exposure to low temperature may also result in cyanosis, which slows the blood flow by constricting superficial blood vessels. This gives additional time for more oxygen to diffuse out of the blood, and consequently increases the concentration of deoxyhemoglobin.

CAREER CORNER
Phlebotomy Technician

The phlebotomy technician is a regular visitor to hospital inpatient rooms. Morning and night, a technician arrives with a rack of tubes for blood collection and materials to puncture veins and collect blood. The phlebotomy technician can use several blood-drawing techniques and handle other body fluids.

The phlebotomy technician explains the aseptic blood-drawing procedure, palpates arm veins to locate the best to puncture, calms common fears of needles, and prepares samples for analysis. It takes practice to remove blood as painlessly as possible!

Phlebotomy technicians work not only in hospitals but also in stand-alone blood-drawing facilities and blood donation centers, and they may travel to nursing homes and assisted living facilities. Training programs require a high school diploma. Most training programs take two to three months, and include classroom instruction (which may be done online in some programs) and practical training. All candidates for phlebotomy technician jobs must pass a national certification exam.

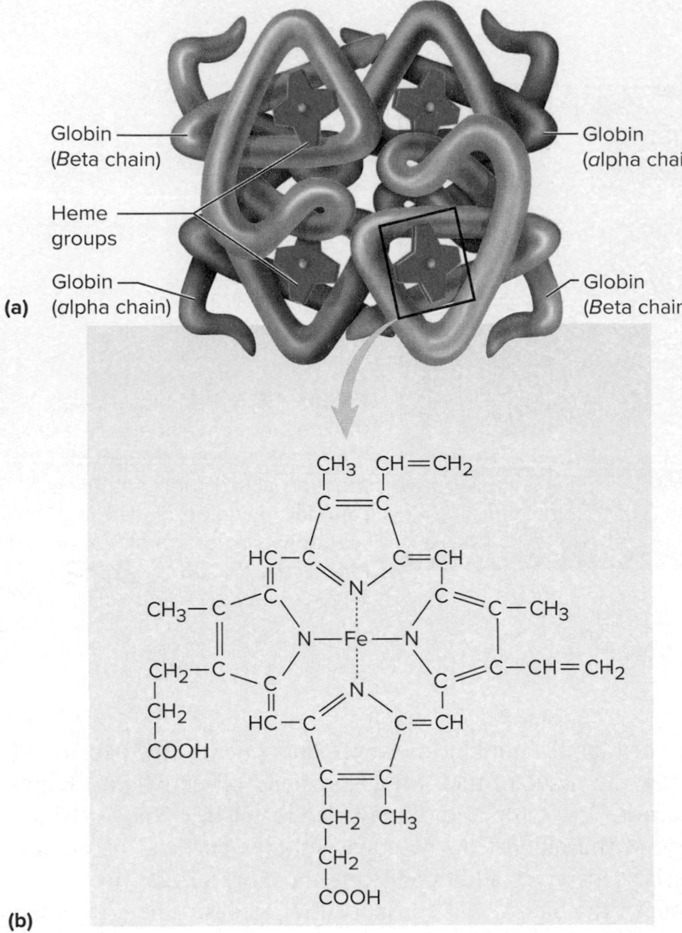

Figure 12.4 **(a)** Structure of hemoglobin molecule; **(b)** chemical structure of the heme portion that includes iron.

Because red blood cells do not have mitochondria, they use none of the oxygen they carry. Consequently, they can only produce ATP through glycolysis (see section 4.4, Energy for Metabolic Reactions). Without a nucleus, red blood cells cannot divide. Red blood cells, like all blood cells, are produced from stem cells in red bone marrow.

Red Blood Cell Counts

The number of red blood cells in a microliter (µL or mcL or 1 mm^3) of blood is called the *red blood cell count* (*RBCC* or *RCC*). This number varies from time to time even in healthy individuals. However, the typical range for adult males is 4,700,000 to 6,100,000 cells per microliter, and that for adult females is 4,200,000 to 5,400,000 cells per microliter.

An increase in the number of circulating red blood cells increases the blood's *oxygen-carrying capacity,* much as a decrease in the number of circulating red blood cells decreases the blood's oxygen-carrying capacity. Changes in this number may affect health. For this reason, red blood cell counts, as well as hematocrit and hemoglobin concentration, are routinely consulted to help diagnose and evaluate the courses of certain diseases, such as cancer.

PRACTICE 12.2

1. Describe a red blood cell.
2. What is the function of hemoglobin?
3. What is the significance of the organelles that are lacking in a mature red blood cell?
4. What is the typical red blood cell count for an adult male? For an adult female?

Red Blood Cell Production and Its Control

Red blood cell formation (*erythropoiesis*) during fetal development occurs in the yolk sac, liver, and spleen. After birth, these blood cells are produced in red bone marrow, a process called **hemopoiesis** (see section 7.3, Bone Function). Figure 12.5 illustrates the development of red blood cells from **hematopoietic** (he″mat-ō-poi-et′ik) **stem cells** (blood-forming cells), which are also called *hemocytoblasts.* As the red blood cells develop, they synthesize hemoglobin. When there is enough hemoglobin, most of the organelles are ejected, forming a *reticulocyte* that still contains some rough endoplasmic reticulum (RER). This young red blood cell enters circulation, ejects its remaining RER, and becomes a mature erythrocyte. It takes three to five days for development from a hemocytoblast to a mature red blood cell.

The average life span of a red blood cell is 120 days. Many of these cells are removed from the circulation each day, and yet the number of cells in the circulating blood remains relatively stable. This observation suggests a homeostatic control of the rate of red blood cell production.

The hormone **erythropoietin** (ĕ-rith″rō-poi′ĕ-tin) controls the rate of red blood cell formation through *negative feedback.* The kidneys, and to a lesser extent the liver, release erythropoietin in response to prolonged oxygen deficiency (fig. 12.6). At high altitudes, for example, where the amount of oxygen in the air is reduced, the blood oxygen level initially decreases. This drop in the blood oxygen level triggers the release of erythropoietin, which travels via the blood to the red bone marrow and stimulates red blood cell production.

After a few days of exposure to high altitudes, many newly formed red blood cells appear in the circulating blood. The increased rate of production continues until the number of erythrocytes in the circulation is sufficient to supply tissues with oxygen. When the availability of oxygen returns to normal, erythropoietin release decreases, and the rate of red blood cell production returns to normal as well. An excessive increase in red blood cells is called **polycythemia.** This condition increases blood viscosity, slowing blood flow and impairing circulation.

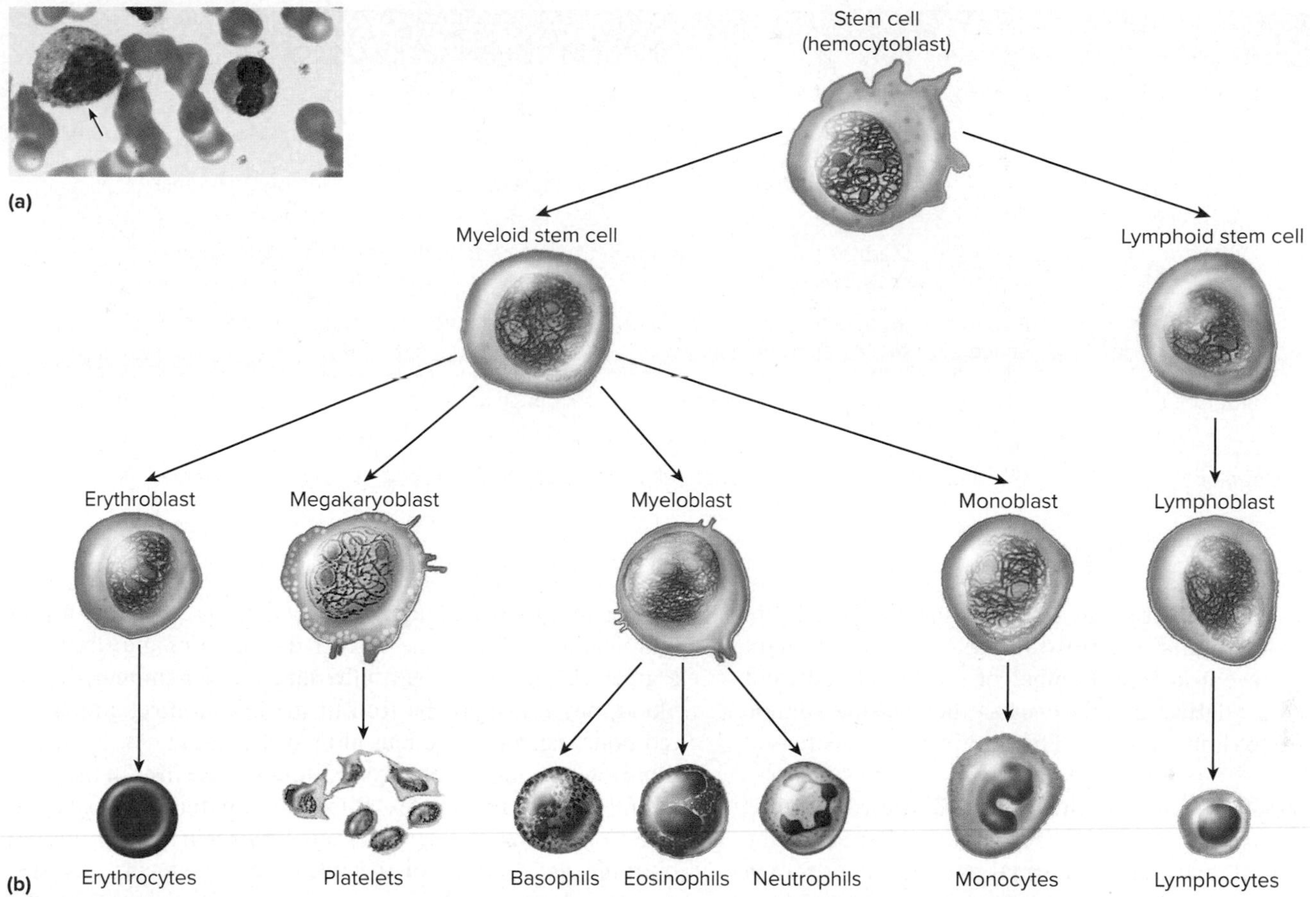

Figure 12.5 Blood cell formation (hemopoeisis). **(a)** A light micrograph of a hematopoietic stem cell (arrow) in red bone marrow (500x). **(b)** Development of red blood cells, platelets, and white blood cells from hematopoietic stem cells in bone marrow. APR
(a): Alvin Telser/McGraw-Hill Education

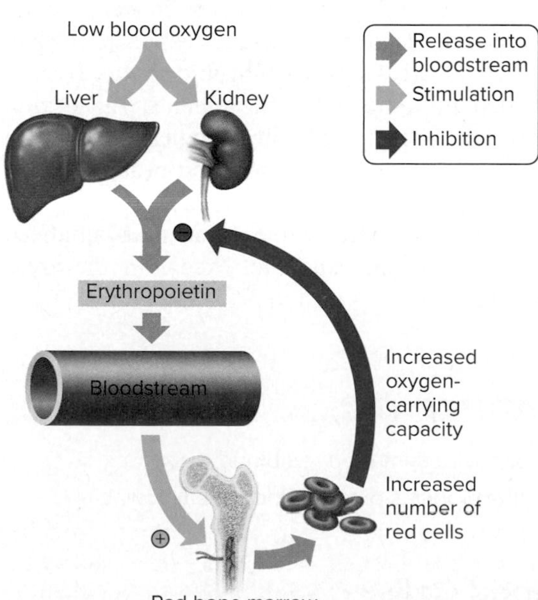

Figure 12.6 Low blood oxygen causes the kidneys and, to a lesser degree, the liver to release erythropoietin. Erythropoietin stimulates target cells in the red bone marrow to increase the production of red blood cells, which carry oxygen to tissues.

Dietary Factors Affecting Red Blood Cell Production

The availability of B-complex vitamins—*vitamin B_{12}* and *folic acid*—significantly influences red blood cell production. Because these vitamins are required for DNA synthesis, they are necessary for the growth and division of cells. Cell division is frequent in blood-forming (hematopoietic) tissue, so this tissue is especially vulnerable to a deficiency of either of these vitamins. Hemoglobin synthesis and normal red blood cell production also require iron. The small intestine absorbs iron slowly from food. The body reuses much of the iron released by the decomposition of hemoglobin from damaged red blood cells. Nonetheless, insufficient dietary iron can reduce hemoglobin synthesis.

A deficiency of red blood cells or a reduction in the amount of hemoglobin they contain results in a condition called **anemia.** This reduces the oxygen-carrying capacity of the blood, and the affected person may appear pale and

PRACTICE FIGURE 12.6

How might kidney disease affect red blood cell production and oxygen-carrying capacity?

Answer can be found in Appendix E.

TABLE 12.1 Types of Anemia

Primary Cause	Due to	Results in
Decreased RBC number	Hemorrhage	Hemorrhagic anemia
	Bacterial infections or blood transfusion incompatibilities destroy RBCs	Hemolytic anemia
	Deficiency of intrinsic factor from stomach causes inadequate vitamin B_{12} absorption	Pernicious anemia
	Destruction of bone marrow by radiation, certain medications, cancer, viruses, and certain poisons	Aplastic anemia
Decreased hemoglobin concentration	Dietary malnourishment, heavy menstruation, persistent bleeding ulcer	Iron-deficiency anemia
Abnormal hemoglobin	Genetic defect resulting in abnormal hemoglobin structure	Sickle cell anemia

lack energy. There are several types of anemia, which are briefly categorized and described in table 12.1. A pregnant woman may have a normal number of red blood cells, but she develops a relative anemia because her plasma volume increases due to fluid retention. This shows up as a decreased hematocrit.

In contrast to anemia, the inherited disorder called *hemochromatosis* results in the absorption of iron in the small intestine at ten times the normal rate. Iron builds up in organs, to toxic levels. Treatment is periodic blood removal, as often as every week. Clinical Application 12.1 discusses the use of bone marrow and stem cell transplants in sickle cell disease and certain cancers.

PRACTICE 12.2

5. Where are red blood cells produced?
6. How does a red blood cell change as it matures?
7. How is red blood cell production controlled?
8. Which vitamins are necessary for red blood cell production?
9. Why is iron required for the formation of red blood cells?

Death and Destruction of Red Blood Cells

Red blood cells are elastic and flexible, and they readily bend as they pass through small blood vessels. As the cells near the end of their three-month life span, however, they become more fragile. The cells may sustain damage simply passing through capillaries, particularly those in active muscles that must withstand strong forces. **Macrophages** phagocytize and destroy damaged red blood cells, primarily in the liver and spleen. Recall from section 5.3, Connective Tissues, that macrophages are large, phagocytic, wandering cells.

Hemoglobin molecules liberated from red blood cells break down, separating the heme groups from the four polypeptide "globin" chains. These chains are broken down further into individual amino acids that may be used in protein synthesis. The heme further decomposes into iron and a greenish pigment called **biliverdin.** Carrier proteins in the blood may transport the iron to the hematopoietic tissue in red bone marrow to be reused in synthesizing new hemoglobin. However, most of the iron is stored in the liver in the form of an iron-protein complex. Biliverdin eventually is converted to an orange-yellow pigment called **bilirubin**. Biliverdin and bilirubin are secreted in the bile as bile pigments (see section 15.8, Liver and Gallbladder). Figure 12.7 summarizes the life, death, and destruction of a red blood cell.

In jaundice (icterus), accumulation of bilirubin turns the skin and eyes yellowish. Newborns can develop *physiologic jaundice* a few days after birth. This condition may be the result of immature liver cells that ineffectively secrete bilirubin into the bile. Treatment includes exposure to fluorescent light, which breaks down bilirubin in the tissues, and feedings that promote bowel movements. In hospital nurseries, babies being treated for physiological jaundice lie under "bili lights," clad only in diapers and protective goggles. The healing effect of fluorescent light was discovered in the 1950s, when an astute nurse noted that jaundiced babies improved after sun exposure, except in the areas their diapers covered.

PRACTICE 12.2

10. What happens to damaged red blood cells?
11. What are the products of hemoglobin breakdown?

White Blood Cells

White blood cells, also called **leukocytes** (lū'kō-sītz), protect against disease. Blood transports leukocytes to sites of infection. Leukocytes can squeeze between the endothelial cells that form the walls of the smallest blood vessels. This movement, called *diapedesis* (di″ah-pĕ-de'sis), allows the white

CLINICAL APPLICATION 12.1

Bone Marrow and Stem Cell Transplants

In sickle cell disease, a single DNA base mutation changes one amino acid in the protein part of hemoglobin, causing hemoglobin to crystallize in a low-oxygen environment. The crystal formation bends the red blood cells containing the abnormal hemoglobin into a sickle shape. The sickled cells tend to get stuck and block the flow in small blood vessels, causing excruciating joint pain and damaging organs. Hospitalization for blood transfusions may be necessary to overcome painful sickling "crises" of blocked circulation.

A drug, hydroxyurea, is used to activate the body's production of a form of hemoglobin normally made only in the fetus. The fetal hemoglobin slows sickling, which enables the red blood cells to reach the lungs—where fresh oxygen restores the cells' normal shapes. A bone marrow transplant or an umbilical cord stem cell transplant from a donor may completely cure sickle cell disease. These procedures have a low risk of fatality.

Bone marrow transplants also enable people with certain cancers to tolerate high levels of chemotherapy drugs. In a bone marrow transplant, a hollow needle and syringe remove normal red bone marrow from the spongy bone of the donor, or stem cells are separated out from the donor's bloodstream. Stem cells from the umbilical cord of a newborn can be used and are less likely to stimulate an immune response in the recipient. The cells are injected into the bloodstream of the recipient. About 5% of recipients get what's called graft-versus-host disease, where their immune system rejects the transplant, resulting in death by infection. Autologous ("self") stem cell transplants may be done, which is a safer alternative.

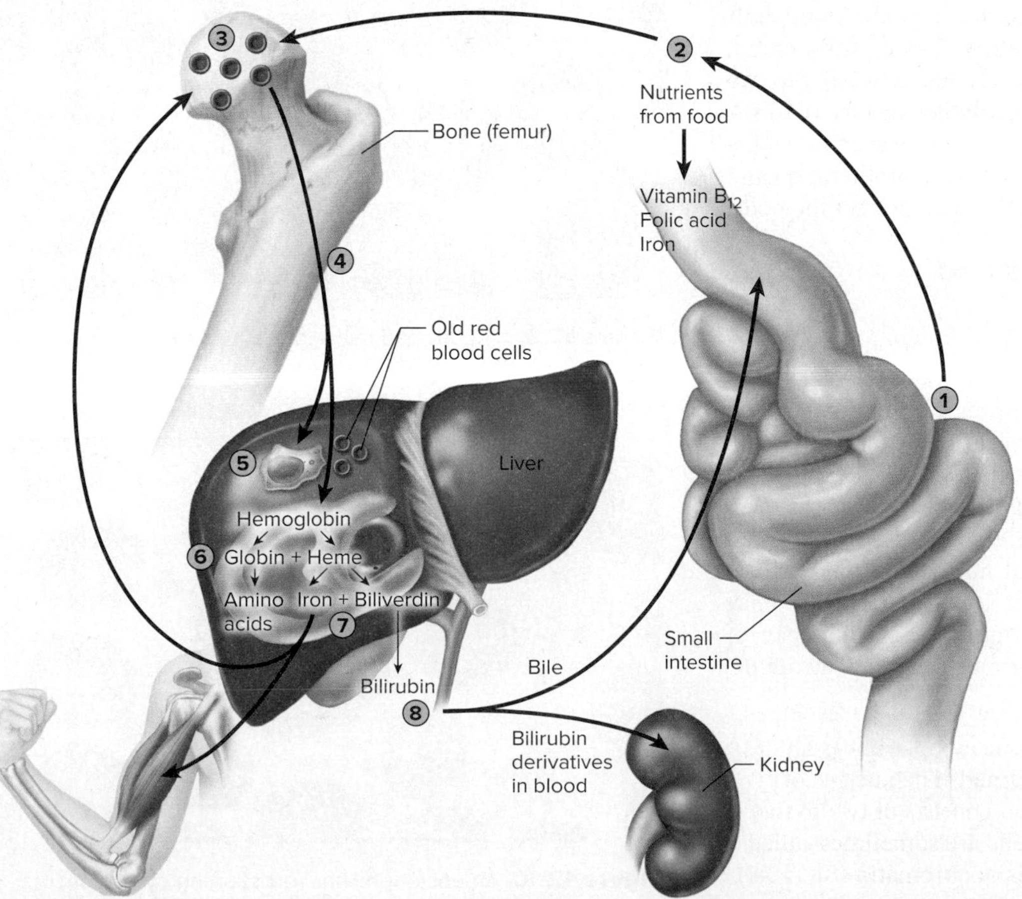

1. Nutrients are absorbed by the small intestine.
2. Blood transports absorbed nutrients.
3. Nutrients (and erythropoietin) are used by red bone marrow to produce red blood cells.
4. Red blood cells circulate in the bloodstream for about 120 days.
5. Macrophages in liver (and spleen) phagocytize and break down old red cells.
6. Hemoglobin is broken down into globin and heme. Globin is further degraded into amino acids that may be reused.
7. Heme is further broken down into iron, which is either stored, or recycled to produce hemoglobin and myoglobin. The pigment is converted to biliverdin.
8. Most biliverdin is converted into bilirubin. These pigments become part of bile or bilirubin derivatives are carried by the blood to the kidneys to be eliminated in urine.

Figure 12.7 The life, death, and destruction of a red blood cell. APR

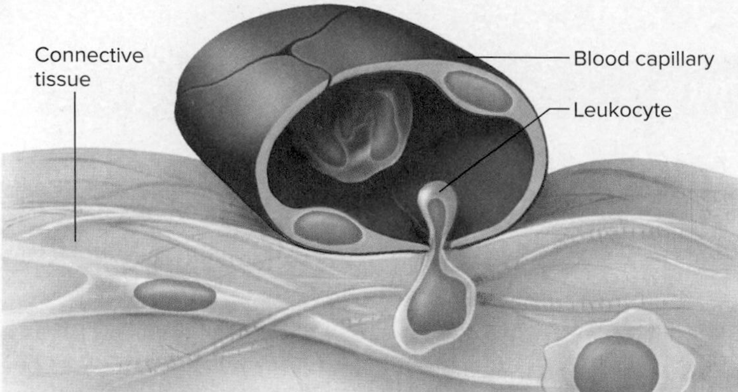

Figure 12.8 Leukocytes squeeze between the endothelial cells of a capillary wall and enter the tissue space outside the blood vessel, using a type of movement called diapedesis.

blood cells to leave circulation (fig. 12.8). Once outside the blood, they move through interstitial spaces using a form of self-propulsion called *amoeboid motion,* as they follow a chemical "trail" released by damaged tissues—an ability termed *positive chemotaxis.*

Like red blood cells, leukocytes have limited life spans, so they must constantly be replaced. Leukocytes develop from hematopoietic stem cells in the red bone marrow (see fig. 12.5) in response to hormones, much as red blood cells form from precursors upon stimulation from erythropoietin. The hormones that affect leukocyte development fall into two groups—**interleukins** and **colony-stimulating factors (CSFs)**—that stimulate red bone marrow to synthesize specific leukocytes. These are also involved in stimulating proliferation and activation of mature leukocytes, so they can protect the body.

Types of White Blood Cells and Their Functions

Normally, five types of white blood cells are in circulating blood. They differ in size, the nature of their cytoplasm, the shape of the nucleus, and their staining characteristics, and they are named for these distinctions. The two major groups are the **granulocytes** and **agranulocytes** (see fig. 12.5).

Leukocytes with specific granular cytoplasm, shown more distinctly when stained with Wright's stain, are called *granulocytes.* Consisting of a lobed nucleus, a typical granulocyte is about twice the size of a red blood cell. They develop in red bone marrow and have a short life span of about 12 hours. Members of this group include neutrophils, eosinophils, and basophils.

- **Neutrophils** (nū′tro-filz) have fine cytoplasmic granules that appear light pink when stained. The nucleus of an older neutrophil is lobed and consists of two to five sections (segments, so these cells are sometimes called *segs*) connected by thin strands of chromatin (fig. 12.9). Younger neutrophils are also called *bands* because their nuclei are C-shaped. Neutrophils are the most numerous, accounting for 50% to 70% of the leukocytes in a typical blood sample from an adult. Neutrophils are mobile and active phagocytes. They aggressively kill bacteria by creating a *respiratory burst,* a toxic chemical cloud of oxidizing agents such as hydrogen peroxide and bleach, around the bacteria. As the bacteria die, they are phagocytized by the neutrophils. Similar to monocytes, neutrophils contain many lysosomes that are filled with digestive enzymes that break down captured bacteria, nutrients, and worn-out organelles.
- **Eosinophils** (ē″o-sin′o-filz) contain coarse, uniformly sized cytoplasmic granules that appear deep red in acid stain (fig. 12.10). The nucleus usually has only two lobes (termed bilobed). Eosinophils make up 1% to 3% of the total number of circulating leukocytes. Eosinophils are only weakly phagocytic. They are attracted to and kill certain parasites, such as parasitic worms (tapeworms and roundworms), by producing toxic chemicals. Eosinophils also help control inflammation and allergic reactions by removing biochemicals associated with these reactions.

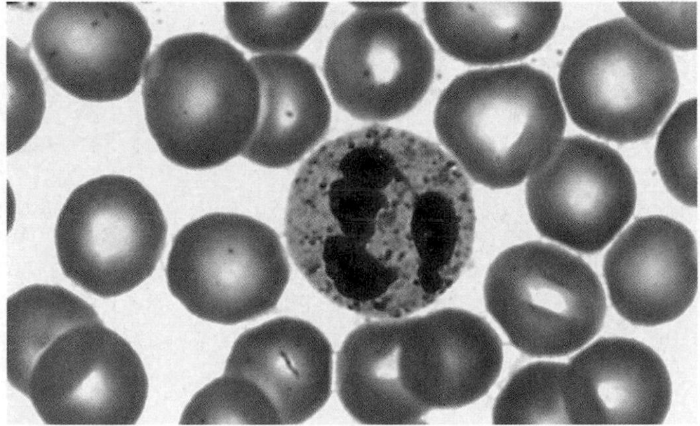

Figure 12.9 A neutrophil has a lobed nucleus with two to five segments (2,000x). APR Ed Reschke

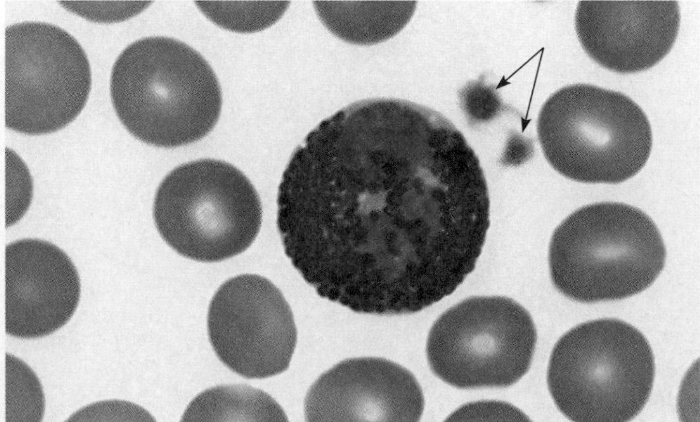

Figure 12.10 An eosinophil has red-staining cytoplasmic granules (2,000x), and a bilobed nucleus. Note the platelets indicated by the arrows. APR Ed Reschke

- **Basophils** (bā'so-filz) are similar to eosinophils in size and in the shape of their nuclei, but they have fewer, more irregularly shaped cytoplasmic granules that appear deep blue in basic stain (fig. 12.11). The rarest of the leukocytes, basophils, usually account for less than 1% of the circulating leukocytes. Basophils migrate to damaged tissues, where they release *heparin,* which inhibits blood clotting, and *histamine,* which promotes inflammation by widening the blood vessel diameter (vasodilation) to speed blood flow to the area. Basophils also play major roles in certain allergic reactions.

Lacking specific granules that stain conspicuously and distinctly, the leukocytes of the *agranulocyte* group include monocytes and lymphocytes. Monocytes and lymphocytes are produced in red bone marrow, as other formed elements are. However, some types of lymphocytes migrate to the thymus to mature.

- **Monocytes** (mon'o-sītz), the largest blood cells, are two to three times greater in diameter than red blood cells (fig. 12.12). Their nuclei are round, kidney-shaped, oval, or lobed. They usually make up 3% to 9% of the leukocytes in a blood sample and live for several weeks or even months. Monocytes migrate to certain tissues and transform into various types of macrophages, which are highly phagocytic cells that engulf bacteria, dead cells, and other debris in the tissues. Their number increases with viral infections and inflammation.
- **Lymphocytes** (lim'fo-sītz) are only slightly larger than red blood cells. Although there are different sizes, a typical lymphocyte has a large, round nucleus surrounded by a thin rim of cytoplasm (fig. 12.13). These cells are the second most abundant leukocyte, accounting for 25% to 33% of circulating leukocytes. Mature lymphocytes colonize in lymphatic organs, such as the lymph nodes and spleen (see section 14.4, Lymphatic Tissues and Organs) and may live for years. There are different categories of lymphocytes, and they are all important in *immunity.* Lymphocytes called cytotoxic T cells, for example, directly attack and destroy foreign cells. Other categories of lymphocytes are B cells and NK cells (see section 14.7, Immunity: Adaptive [Specific] Defenses).

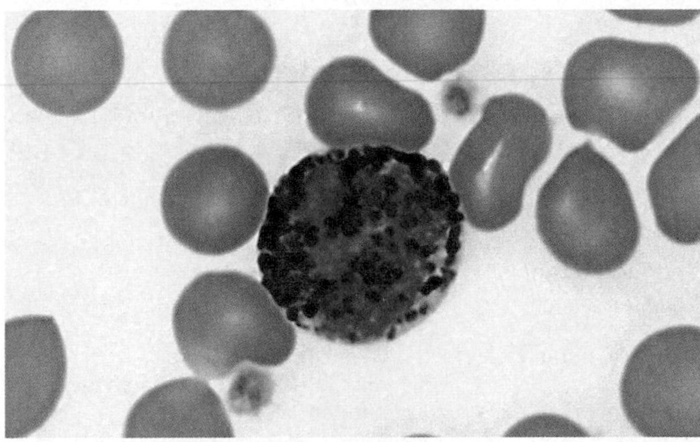

Figure 12.11 A basophil has a violet-stained nucleus and cytoplasmic granules that stain deep blue (2,000x). APR
Ed Reschke

PRACTICE 12.2

12. How do white blood cells reach microorganisms that are outside blood vessels?
13. Which hormones are necessary for differentiation of white blood cells from hematopoietic stem cells in red bone marrow?
14. Distinguish between granulocytes and agranulocytes.
15. List the five types of white blood cells, and explain how they differ from one another.
16. How do white blood cells fight infection?
17. Which white blood cells are the most active phagocytes?

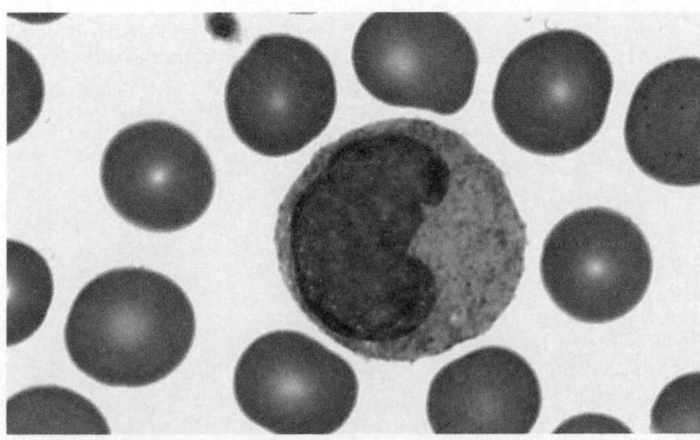

Figure 12.12 A monocyte is the largest of the blood cells (2,000x). APR Ed Reschke

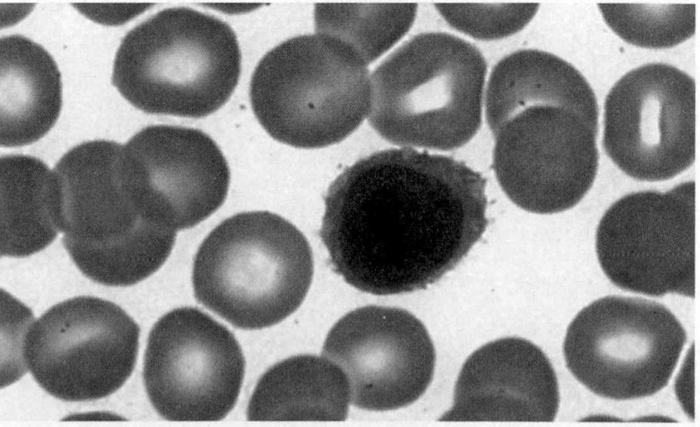

Figure 12.13 A lymphocyte, the smallest of the white blood cells, has a large, round nucleus that occupies much of the cytoplasm (2,000x). APR Ed Reschke

White Blood Cell Counts

The number of leukocytes in a microliter of human blood, called the *white blood cell count* (*WBCC* or *WCC*), normally is 3,500 to 10,500 cells. White blood cell counts are of clinical interest because their number may change in response to abnormal conditions. A total number of white blood cells exceeding 10,500 per microliter of blood constitutes **leukocytosis,** indicating acute infection, such as appendicitis. The white blood cell count is greatly elevated in leukemia, as Clinical Application 12.2 describes.

A total white blood cell count below 3,500 per microliter of blood is called **leukopenia.** Such a deficiency may accompany typhoid fever, influenza, measles, mumps, chickenpox, AIDS, or poliomyelitis.

A *differential white blood cell count* (*DIFF*) lists percentages of the various types of leukocytes in a blood sample. This test is useful because the relative proportions of white blood cells may change in particular diseases. The number of neutrophils, for instance, usually increases during bacterial infections. Eosinophils may become more abundant during certain parasitic infections and allergic reactions. In AIDS, the number of T lymphocytes (T cells) drops sharply.

PRACTICE 12.2

18. What is the normal human white blood cell count?
19. Distinguish between leukocytosis and leukopenia.
20. What is a differential white blood cell count?

Blood Platelets APR

Platelets (plāt′letz), or **thrombocytes** (throm′bo-sītz), are not complete cells (see fig. 12.10). They arise from very large cells in red bone marrow, called **megakaryocytes** (meg″ah-kar′ē-o-sītz), that fragment, releasing small pieces–platelets–into the circulation. The megakaryocytes develop long cellular extensions that break off in small sections to form platelets in the bone marrow. Megakaryocytes, and therefore platelets, develop from hematopoietic stem cells (see fig. 12.5) in response to the hormone thrombopoietin (throm″bō-poi′ĕ-tin).

Each platelet has a membrane but lacks a nucleus and is less than half the size of a red blood cell. It is capable of amoeboid movement and may live for about ten days. Platelets help close breaks in damaged blood vessels, as explained in section 12.4, Hemostasis. Table 12.2 summarizes the general characteristics of the formed elements.

TABLE 12.2 Cellular Components of Blood

Component	Description	Number Present	Function
Red blood cell (erythrocyte)	Biconcave disc without a nucleus; about one-third hemoglobin	4,700,000–6,100,000 per microliter, male; 4,200,000–5,400,000 per microliter, female	Transports oxygen and carbon dioxide
White blood cell (leukocyte)		3,500–10,500 per microliter	Destroys pathogenic microorganisms and parasites and removes worn cells
Granulocytes	About twice the size of red blood cells; cytoplasmic granules are present		
1. Neutrophil	Nucleus with two to five lobes; cytoplasmic granules stain light purple in neutral stain	50% to 70% of white blood cells present	Phagocytizes small particles
2. Eosinophil	Nucleus bilobed, cytoplasmic granules stain red in acid stain	1% to 3% of white blood cells present	Kills parasites and moderates allergic reactions
3. Basophil	Nucleus bilobed, cytoplasmic granules stain blue in basic stain	Less than 1% of white blood cells present	Releases heparin and histamine
Agranulocytes	Cytoplasmic granules are absent		
1. Monocyte	Two to three times larger than a red blood cell; nucleus shape varies from spherical to lobed	3% to 9% of white blood cells present	Phagocytizes large particles
2. Lymphocyte	Only slightly larger than a red blood cell; its nucleus nearly fills cell	25% to 33% of white blood cells present	Provides immunity
Platelet (thrombocyte)	Cellular fragment	150,000–350,000 per microliter	Helps control blood loss from broken vessels

CLINICAL APPLICATION 12.2

Leukemia

When the twenty-three-year-old had a routine physical examination, she expected reassurance that her healthy lifestyle had indeed been keeping her healthy. After all, she felt great. What she got, a few days later, was a shock. Instead of having 3,500 to 10,500 white blood cells per microliter of blood, she had more than ten times that number—and many of the cells were cancerous. She had chronic myeloid leukemia (CML). Her red bone marrow was flooding her circulation with too many granulocytes, most of them poorly differentiated (fig. 12A).

Another type of leukemia is lymphoid, in which the cancer cells are lymphocytes. Both myeloid and lymphoid leukemia can cause fatigue, headaches, nosebleeds, and other bleeding, frequent respiratory infections, fever, bone pain, bruising, and other signs of slow blood clotting.

The symptoms arise from the disrupted proportions of the blood's formed elements and their malfunction. Immature white blood cells increase the risk of infection. Leukemic cells crowd out red blood cells and their precursors in the red marrow, causing anemia and the resulting fatigue. Platelet deficiency (thrombocytopenia) increases clotting time, causing bruises and bleeding.

Leukemia is also classified as acute or chronic. An acute condition appears suddenly, symptoms progress rapidly, and without treatment, death occurs in a few months. Chronic forms begin more slowly and may remain undetected for months or even years or, in rare cases, decades. Without treatment, life expectancy after symptoms develop is about three years.

Traditional cancer treatments destroy any cell that is actively dividing. More specifically, the drug imatinib (Gleevec) specifically targets only the cancer cells by nestling into ATP-binding sites on a version of an enzyme called a tyrosine kinase that is found only in cancer cells. Inhibiting this enzyme prevents cancer cells from dividing. If cancer cells become resistant to Gleevec, even newer drugs are available that target the cancer cells in different ways. People with leukemia have other options. Bone marrow and stem cell transplants may cure the condition.

Another improvement in leukemia treatment is refinement of diagnosis, based on identifying the proteins that leukemia cells produce. This information is used to predict which drugs are most likely to be effective, and which will cause intolerable side effects, in particular individuals. For example, some people with acute lymphoblastic leukemia (ALL), diagnosed on the basis of the appearance of the cancer cells in a blood smear, do not respond to standard chemotherapy. However, DNA microarray (also called DNA chip) technology revealed that the cells of most patients who do not improve produce different proteins than the cancer cells of patients who do respond to the drugs used to treat ALL. The nonresponders actually have a different form of leukemia, called mixed-lineage leukemia. These patients respond to different drugs.

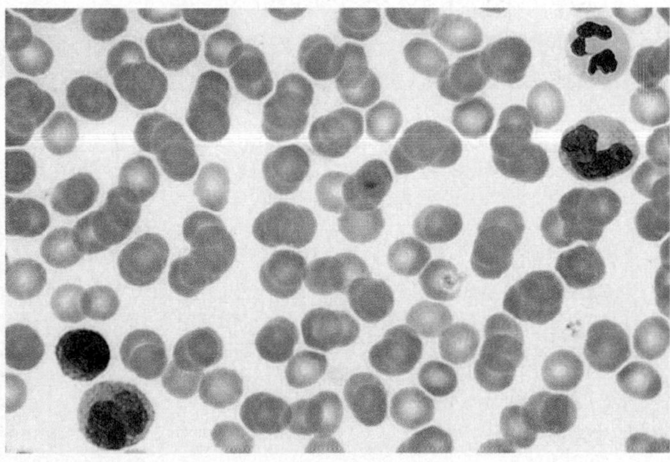

(a)

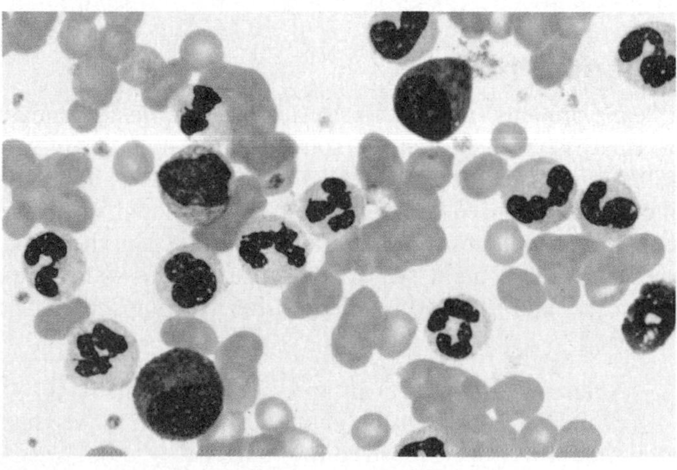

(b)

Figure 12A Leukemia and blood cells. **(a)** Normal blood cells (700×). **(b)** Blood cells from a person with granulocytic leukemia, a type of myeloid leukemia (700×). Note the increased number of leukocytes. (a): Al Telser/McGraw-Hill Education; (b): Joaquin Carrillo-Farga/Science Source

Platelet Counts

Under normal conditions, the platelet count varies from 150,000 to 400,000 per microliter. **Thrombocytosis** (also called thrombocythemia) is a high platelet count in the blood, which may be caused by a genetic defect or an infection. There is a high risk of blood clots and bleeding with this condition. When the platelet count falls below normal, it is called **thrombocytopenia.** An excessively low platelet count puts a person at risk for internal bleeding because there are not enough platelets to stop the bleeding. A decrease in the number of platelets may be due to a genetic condition, leukemia, radiation, certain medicines, exposure to toxic chemicals, heavy alcohol drinkers, or certain viruses or autoimmune diseases.

PRACTICE 12.2

21. What is the normal blood platelet count?
22. What is the function of blood platelets?
23. What is the risk of high platelet count? Of low platelet count?

12.3 | Plasma

LEARN

1. Describe the functions of each of the major components of plasma.

Plasma is the clear, straw-colored, liquid portion of the blood in which the cells and platelets are suspended. It is approximately 92% water and contains a complex mixture of organic and inorganic biochemicals. Functions of plasma include transporting gases, vitamins, and other nutrients; helping to regulate fluid and electrolyte balance; and maintaining a favorable pH.

Plasma Proteins

Plasma proteins (plaz'mah pro'tēnz) are the most abundant of the dissolved substances (solutes) in plasma. These proteins remain in the blood and interstitial fluids, and ordinarily are not used as energy sources. Although there are enzymes, hormones, and certain other proteins in the blood, there are three main types of plasma proteins–albumins, globulins, and fibrinogen–which differ in composition and function.

- **Albumins** (al-bu'minz) are the smallest plasma proteins, yet account for about 60% of them by weight. Albumins are synthesized in the liver.

  Recall from section 3.3, Movements Into and Out of the Cell, that an impermeant solute on one side of a selectively permeable membrane creates an osmotic pressure, and that water always moves toward a greater osmotic pressure. Plasma proteins are too large to pass through the capillary walls, so they are impermeant. They create an osmotic pressure that holds water in the capillaries, despite blood pressure forcing water out of capillaries by filtration. The term *colloid osmotic pressure* is used to describe this osmotic effect due to the plasma proteins. Because albumins are so plentiful, they are an important determinant of the colloid osmotic pressure of the plasma.

  By maintaining the colloid osmotic pressure of plasma, albumins and other plasma proteins help regulate water movement between the blood and the tissues. In doing so, the plasma proteins help control blood volume, which, in turn, directly affects blood pressure (see section 13.5, Blood Pressure).

  If the concentration of plasma proteins falls, tissues swell. This condition is called *edema.* As the concentration of plasma proteins drops, so does the colloid osmotic pressure. Water leaves the blood vessels and accumulates in the interstitial spaces, causing swelling. A low plasma protein concentration may result from starvation, a protein-deficient diet, or an impaired liver that cannot synthesize plasma proteins.
- **Globulins** (glob'ū-linz) make up about 36% of the plasma proteins. They can be further subdivided into *alpha, beta,* and *gamma globulins.* The liver synthesizes alpha and beta globulins, which have a variety of functions that include transporting lipids and fat-soluble vitamins. Within lymphatic tissues, B lymphocytes (B cells) differentiate into *plasma cells* that produce the gamma globulins, which are antibodies (see section 14.7, Immunity: Adaptive [Specific] Defenses).
- **Fibrinogen** (fī-brin'o-jen) constitutes about 4% of the plasma proteins, and functions in blood coagulation (clotting). It is discussed in section 12.4, Hemostasis. Synthesized in the liver, fibrinogen is the largest of the plasma proteins. Table 12.3 summarizes the characteristics of the plasma proteins.

TABLE 12.3 Plasma Proteins

Protein	Percentage of Total	Origin	Function
Albumins	60%	Liver	Help maintain colloid osmotic pressure
Globulins	36%		
Alpha globulins		Liver	Transport lipids and fat-soluble vitamins
Beta globulins		Liver	Transport lipids and fat-soluble vitamins
Gamma globulins		Lymphatic tissues	Constitute the antibodies of immunity
Fibrinogen	4%	Liver	Plays a key role in blood coagulation

PRACTICE 12.3

1. List three main types of plasma proteins.
2. How do albumins help maintain water balance between the blood and the tissues?
3. What are the functions of the globulins?
4. What is the role of fibrinogen?

Gases and Nutrients

The most important *blood gases* are oxygen and carbon dioxide. Plasma also contains a considerable amount of dissolved nitrogen, which ordinarily has no physiological function. Section 16.6, Gas Transport, discusses blood gases and their transport.

The *plasma nutrients* include amino acids, simple sugars, nucleotides, vitamins, minerals, and lipids, all absorbed from the digestive tract. For example, plasma transports glucose from the small intestine to the liver, where it may be stored as glycogen or converted to fat. If blood glucose concentration drops below the normal range, glycogen may be broken down into glucose, as described in section 11.8, Pancreas. Plasma also carries recently absorbed amino acids to the liver, where they may be used to manufacture proteins, or deaminated and used as an energy source (see section 15.11, Nutrition and Nutrients).

Plasma lipids include fats (triglycerides), phospholipids, and cholesterol. Because lipids are not water-soluble and plasma is almost 92% water, these lipids are carried in the plasma attached to proteins.

Nonprotein Nitrogenous Substances

Molecules that contain nitrogen atoms but are not proteins comprise a group called **nonprotein nitrogenous substances.** In plasma, this group includes amino acids, urea, uric acid, creatine (kre'ah-tin), and creatinine (kre-at'i-nin). Amino acids come from protein digestion and amino acid absorption. Urea and uric acid are products of protein and nucleic acid catabolism, respectively. Creatinine results from the metabolism of creatine. As discussed in section 8.3, Skeletal Muscle Contraction, creatine is part of *creatine phosphate* in muscle tissue, where it stores energy in phosphate bonds.

Plasma Electrolytes

Blood plasma contains *electrolytes* that are absorbed from the intestine or released as by-products of cellular metabolism. They include sodium, potassium, calcium, magnesium, chloride, bicarbonate, phosphate, and sulfate ions. Sodium and chloride ions are the most abundant. Bicarbonate ions are important in maintaining the pH of plasma. Like other plasma constituents, bicarbonate ions are regulated so that their blood concentrations remain relatively stable. Chapter 18 (see section 18.4, Electrolyte Balance) discusses these electrolytes in connection with water and electrolyte balance.

PRACTICE 12.3

5. Which gases are in plasma?
6. Which nutrients are in plasma?
7. What is a nonprotein nitrogenous substance?
8. What are the sources of plasma electrolytes?

12.4 | Hemostasis

LEARN

1. Define *hemostasis,* and explain the mechanisms that help achieve it.
2. Review the major steps in blood coagulation.

Hemostasis (he″mo-sta'sis) refers to the process that stops bleeding, which is vitally important when blood vessels are damaged. Following an injury to the blood vessels, several actions may help to limit or prevent blood loss. These overlapping events include vascular spasm, platelet plug formation, and blood coagulation.

Vascular Spasm

Cutting or breaking a small blood vessel stimulates **vascular spasm,** the contraction of smooth muscle in the wall of the blood vessel. Blood loss lessens almost immediately, and the ends of the severed vessel may close completely. This effect results from direct stimulation of the vessel wall, from reflexes triggered by pain receptors in the injured tissues, and from **serotonin** released from platelets that stimulates smooth muscle contraction of the blood vessel walls (called vasoconstriction) to reduce blood loss.

The reflex response of vascular spasm may last only a few minutes, but the effect of the direct stimulation usually continues for about 30 minutes. By then, a blockage called a *platelet plug* has formed, and blood is coagulating.

Platelet Plug Formation

Platelets are normally repelled by a smooth, intact endothelium lining the blood vessels; however, when a blood vessel breaks, "sticky" platelets adhere to the exposed collagen from connective tissues underlying the endothelium. Platelets also adhere to each other, forming a platelet plug in the vascular break. A plug may control blood loss from a small break, but a larger break may require a blood clot to halt bleeding. Figure 12.14 shows the steps in platelet plug formation.

PRACTICE 12.4

1. What is hemostasis?
2. How does a vascular spasm help control bleeding?
3. Describe the formation of a platelet plug.

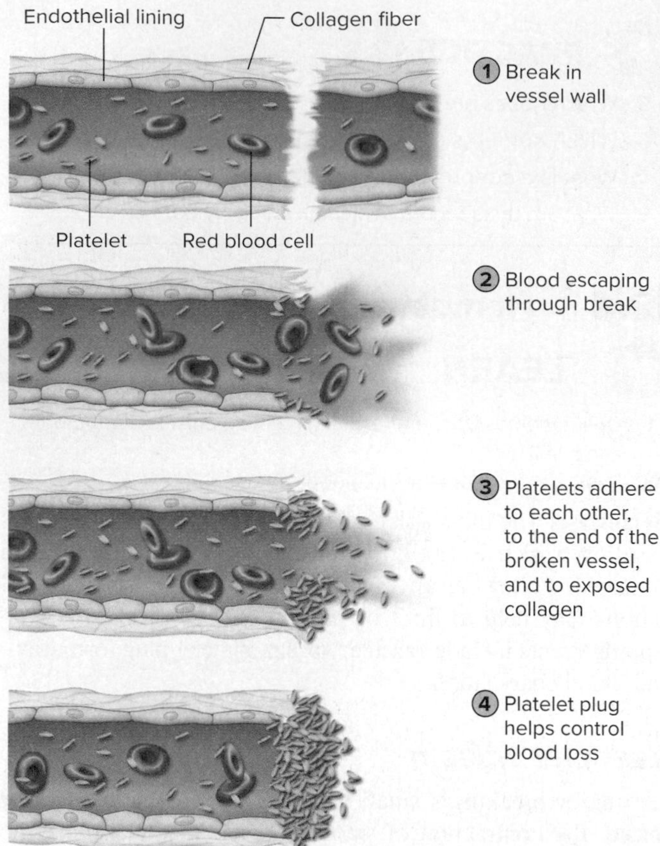

Figure 12.14 Steps in platelet plug formation.

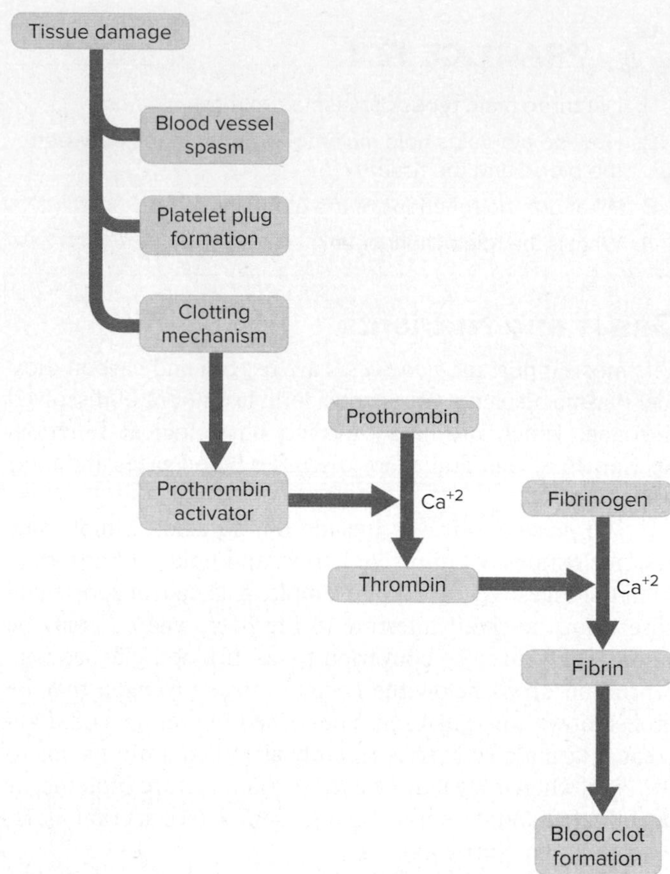

Figure 12.15 The three mechanisms of hemostasis: vascular spasm, platelet plug formation, and the steps involved in blood coagulation.

PRACTICE FIGURE 12.15

What is the major event in blood clot formation? How might the blood clotting mechanism be affected if a person has hypocalcemia (low blood Ca^{2+})?

Answers can be found in Appendix E.

Blood Coagulation

Coagulation (kō-ag″ū-lā′shun), the most effective hemostatic mechanism, forms a *blood clot* in a series of reactions, each one activating the next. Blood coagulation is complex and utilizes many biochemicals called *clotting factors.* Some of these factors promote coagulation, and others inhibit it. Whether or not blood coagulates depends on the balance between these two groups of factors. Normally, anticoagulants prevail, and the blood does not clot. However, as a result of injury (trauma), biochemicals that favor coagulation may increase in concentration, and the blood may coagulate.

The major event in blood clot formation is the conversion of the soluble plasma protein *fibrinogen* into insoluble threads of the protein **fibrin.** Formation of fibrin takes several steps, summarized as the clotting mechanism in figure 12.15. First, damaged tissues release *tissue thromboplastin,* initiating a series of reactions that results in the production of an enzyme called *prothrombin activator.* This series of changes requires calcium ions, as well as certain proteins and phospholipids. As its name suggests, prothrombin activator acts on prothrombin.

Prothrombin is an alpha globulin that the liver continually produces and is thus a normal constituent of plasma. Prothrombin activator converts prothrombin into **thrombin**, an enzyme which in turn catalyzes a reaction that joins fragments of fibrinogen into long threads of fibrin.

Once fibrin threads form, they stick to the exposed surfaces of damaged blood vessels, creating a meshwork that entraps blood cells and platelets (fig. 12.16). The resulting mass is a blood clot, which may block a vascular break and prevent further blood loss. *Fibroblasts* (see section 5.3, Connective Tissues) and the proteins they secrete form fibrous connective tissue in the walls of the ruptured vessels, which helps strengthen and seal the vascular breaks. The clear, yellow liquid that remains after the clot forms is called **serum.** Serum is plasma minus the clotting factors.

The amount of prothrombin activator in the blood is directly proportional to the degree of tissue damage. Once a blood clot begins to form, additional clotting occurs because thrombin also acts directly on blood clotting factors other than fibrinogen, causing prothrombin to form more thrombin. Blood clot formation is an example of a *positive feedback system,* in which the original action stimulates more of the same type of action. Such a positive feedback mechanism

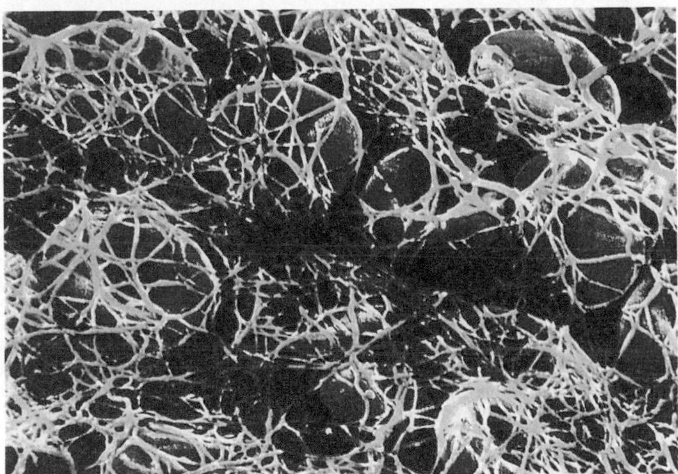

Figure 12.16 A scanning electron micrograph of red blood cells caught in a meshwork of fibrin threads (2,800x). SPL/Science Source

produces unstable conditions and can operate for only a short time without disrupting the stable internal environment (see section 1.5, Maintenance of Life). Normal clotting is a temporary response to injury that helps preserve the internal environment by minimizing blood loss.

Clot Dissolution

Many clots, including those that form in tissues as a result of blood leakage (hematomas), disappear in time. This clot dissolution, called *fibrinolysis,* requires conversion of a plasma protein, *plasminogen,* to *plasmin,* a protein-splitting enzyme that can digest fibrin threads and other proteins associated with blood clots. Plasmin formation may dissolve a whole clot; however, clots that fill large blood vessels are removed by drug action or surgically.

Abnormal Clot Formation

A blood clot abnormally forming in a vessel is a **thrombus** (throm'bus). A clot that dislodges, or a fragment of a clot that breaks loose and is carried away by the blood flow, is an **embolus** (em'bo-lus). Generally, emboli continue to move until they reach narrow places in vessels, where they may lodge and block blood flow (an embolism).

A blood clot forming in a vessel that supplies a vital organ, such as the heart (coronary thrombosis) or the brain (cerebral thrombosis), blocks blood flow and kills tissues the vessel serves (*infarction*) and may be fatal. A blood clot that travels and then blocks a vessel that supplies a vital organ, such as the lungs (pulmonary embolism), affects the portion of the organ the blocked blood vessel supplies. An embolism can be fatal as well. Clinical Application 12.3 discusses deep vein thrombosis.

Abnormal clot formations are often associated with conditions that change the endothelial linings of vessels. For example, in *atherosclerosis* (ath″er-ō″skle-rō'sis), accumulations of fatty deposits change arterial linings, sometimes initiating inappropriate clotting (fig. 12.17).

Normally, blood flow throughout the body prevents formation of a massive clot by rapidly carrying excess thrombin away; however, slow moving blood, or pooling of the blood, may result in undesirable coagulation. When clotting factors, such as thrombin, are not diluted and washed away, this provides more time for a clot to form.

Drugs based on "clot-busting" biochemicals can be lifesavers. Delivered by catheter to the clot location, *tissue plasminogen activator* (tPA) may quickly restore blocked cerebral circulation if given within no more than 3 to 4½ hours after a stroke, for example. Derived from bacteria, *streptokinase* is a drug that can be given intravenously to dissolve a clot. *Urokinase*, an enzyme produced by certain kidney cells, is another drug that is a plasminogen activator. Drugs that are

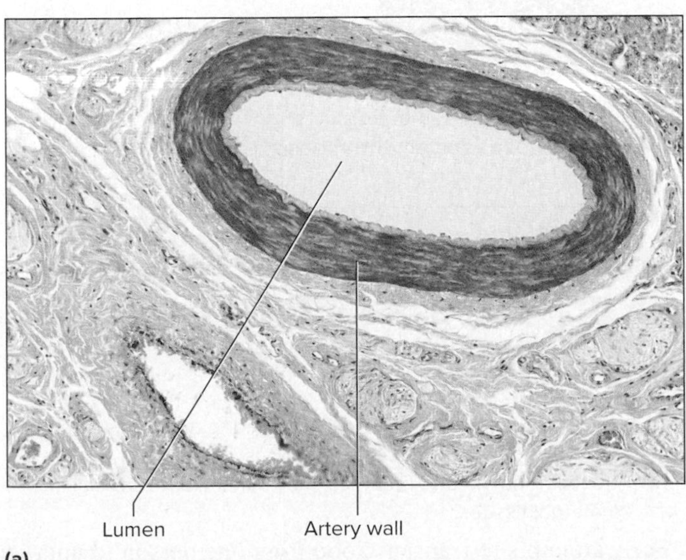

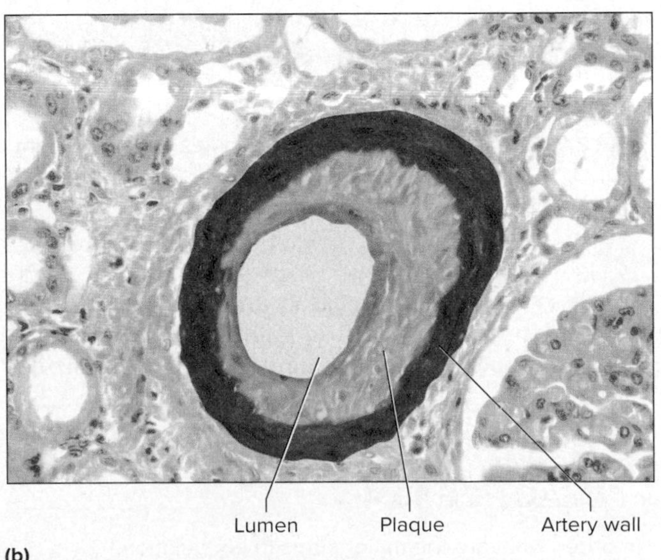

Figure 12.17 Artery cross sections, falsely colored light micrographs. **(a)** Normal artery (50x), and **(b)** the inner wall of an artery changed as a result of atherosclerosis (100x). Not only is blood flow impeded, but the uneven inner surface can snag platelets, triggering coagulation. (a, b): Al Telser/McGraw-Hill Education

CLINICAL APPLICATION 12.3
Deep Vein Thrombosis

After a transcontinental flight, a man complained of cramps behind his knees. He mentioned the unfamiliar sensations to his traveling companion, who urged him to call his doctor. The nurse who took the call told the man to seek immediate medical attention, but because he was already starting to feel better, he didn't take her advice. Instead he continued his trip, reaching his destination after a short bus ride. Three days later, the man suddenly collapsed and died from a pulmonary embolism, a complication of the condition called deep vein thrombosis (DVT).

The man had two key risk factors for DVT: sitting for long periods, and an inherited clotting disorder that he had not known about called factor V Leiden. Other risk factors for DVT are prolonged immobility due to surgery; oral contraceptive use; hormone replacement therapy (estrogen); pregnancy; recent surgery (of the abdomen, pelvis, or limbs); and cancer (including lymphoma and cancers of the ovaries, pancreas, colon, liver, and stomach).

In DVT, stagnant blood pools, leading to clot formation, typically in the femoral or popliteal veins or in the deep veins of the pelvis. Symptoms occur in half of all affected individuals. These include deep muscle pain, redness, swelling, and possibly discoloration and dilation of surface veins (phlebitis). Part of the clot may break off hours or days after it forms and follow the path of circulation, lodging in the pulmonary arteries. This is a pulmonary embolism, and it is life-threatening. Symptoms include chest pain, anxiety, racing pulse, sweating, cough with bloody sputum, and loss of consciousness. In the United States, approximately 2 million people a year develop DVT, and 200,000 die from a pulmonary embolism.

Guidelines from the American College of Chest Physicians recommend preventive measures against DVT for patients at higher risk. These actions include taking anticoagulants if immobilization is expected and wearing compression stockings that help keep blood flowing in the legs. Doing exercises while immobilized during travel is a good idea for everyone. Some airlines advise passengers on how to exercise on cramped flights, such as by curling and uncurling the toes and/or moving the feet up and down (fig. 12B).

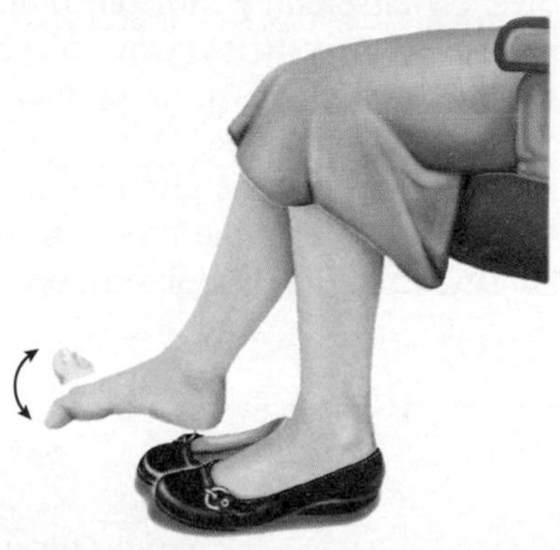

Figure 12B Exercising the toes and ankles on a long flight can lower the risk of deep vein thrombosis.

not clot dissolving include heparin, an anticoagulant that interferes with prothrombin activator formation, and coumadin, a drug that acts to prevent vitamin K from synthesizing clotting factors.

PRACTICE 12.4

4. In order, what are the major steps in coagulation?
5. What prevents the formation of massive clots throughout the cardiovascular system?
6. Distinguish between a thrombus and an embolus.

12.5 | Blood Groups and Transfusions

LEARN

1. Explain blood typing and how it is used to avoid adverse reactions following blood transfusions.
2. Describe how blood reactions may occur between fetal and maternal tissues.

Early attempts to transfer blood from one person to another produced varied results. Sometimes, the recipient improved. Other times, the recipient suffered a blood transfusion reaction in which the red blood cells clumped, obstructing vessels and producing great pain and organ damage.

Eventually, scientists determined that blood is of differing types and that only certain combinations of blood types are compatible. These discoveries led to the development of procedures for typing blood. Today, safe transfusions of whole blood depend on two blood tests—"type and cross match." First the recipient's ABO blood type and Rh status (discussion to follow) are determined. Following this, a "cross match" is done of the recipient's serum with a small amount of the donor's red blood cells that have the same ABO type and Rh status as the recipient. Compatibility is determined by examining the mixture under a microscope for **agglutination,** the clumping of red blood cells.

Antigens and Antibodies

An **antigen** (an′tĭ-jen) is any molecule that triggers an immune response, the body's reaction to invasion by a foreign substance or organism. When the immune system encounters an antigen not found on the body's own cells, it will respond by producing **antibodies** (an′tĭ-bod″ēz). In a transfusion reaction, antigens (*agglutinogens*) on the surface of the donated red blood cells react with antibodies (*agglutinins*) in the plasma of the recipient, resulting in the agglutination of the donated red blood cells.

A mismatched blood transfusion quickly produces telltale signs of agglutination—anxiety, breathing difficulty, facial flushing, headache, and severe pain in the neck, chest, and lumbar area. Red blood cells burst, releasing free hemoglobin. Liver cells and macrophages phagocytize the hemoglobin, converting it to bilirubin, which may accumulate to cause yellowing of the skin (jaundice). Free hemoglobin reaching the kidneys may ultimately cause them to fail.

Only a few of the 32 known antigens on red blood cell membranes can produce serious transfusion reactions. These include the antigens of the ABO group and those of the Rh group. Avoiding the mixture of certain kinds of antigens and antibodies prevents adverse transfusion reactions.

ABO Blood Group

The *ABO blood group* is based on the presence (or absence) of two major antigens on red blood cell membranes—antigen A and antigen B. A and B antigens are carbohydrates attached to glycolipids projecting from the red blood cell surface. A person's erythrocytes have on their surfaces one of four antigen combinations: only A, only B, both A and B, or neither A nor B.

A person with only antigen A has *type A blood.* A person with only antigen B has *type B blood.* An individual with both antigens A and B has *type AB blood.* A person with neither antigen A nor B has *type O blood.* Thus, all people have one of four possible ABO blood types—A, B, AB, or O. The resulting ABO blood type is inherited. It is the consequence of DNA encoding enzymes to synthesize A or B antigens on erythrocytes in one of these four combinations.

Antibodies that affect the ABO blood group antigens are present in the plasma about two to eight months following birth. Individuals with type A blood have anti-B antibody in their plasma; those with type B blood have anti-A antibody; those with type AB blood have neither antibody; and those with type O blood have both anti-A and anti-B antibodies (fig. 12.18 and table 12.4). The antibodies anti-A and anti-B are large and do not cross the placenta. Thus, a pregnant

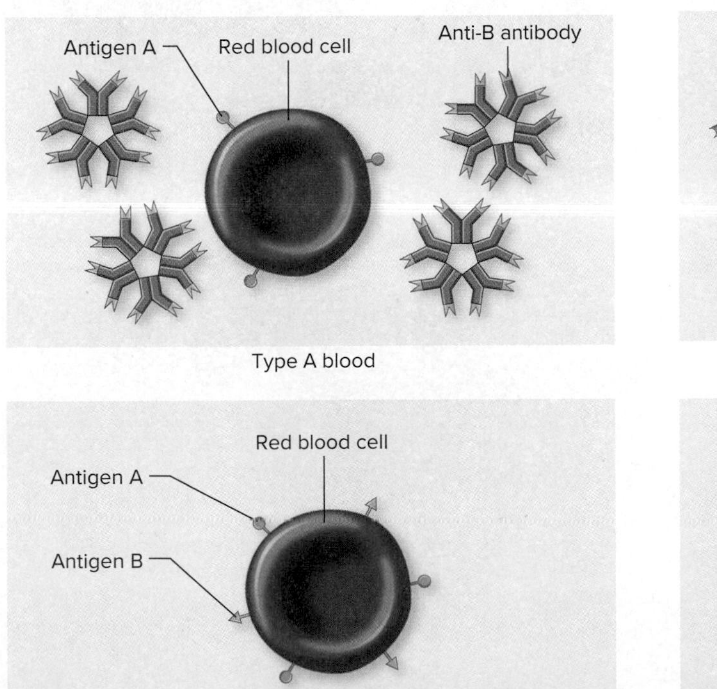

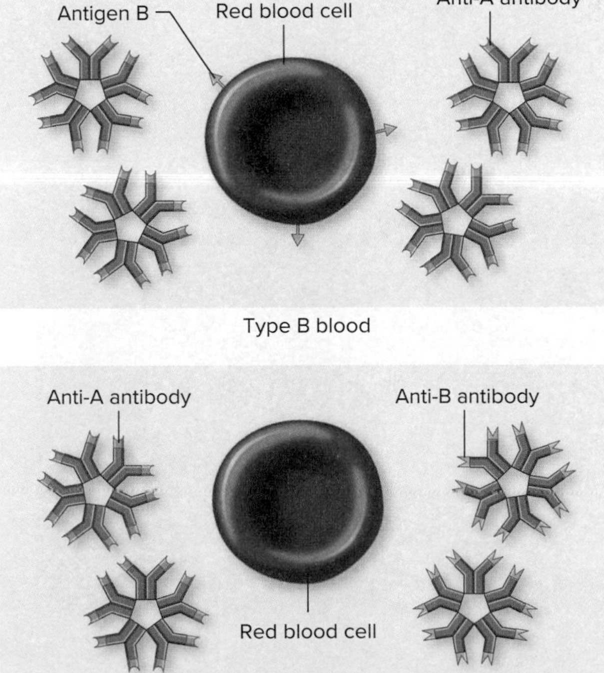

Figure 12.18 Different combinations of antigens and antibodies distinguish blood types. (Cells and antibodies not drawn to scale.)

TABLE 12.4 Antigens and Antibodies of the ABO Blood Group

Blood Type	Antigen	Antibody
A	A	Anti-B
B	B	Anti-A
AB	A and B	Neither anti-A nor anti-B
O	Neither A nor B	Both anti-A and anti-B

woman and her fetus may have different ABO blood types, but agglutination in the fetus will not occur.

An antibody of one type will react with an antigen of the same type and clump red blood cells (fig. 12.19); therefore, such combinations must be avoided. A major concern in blood transfusion procedures is that donated blood cells not clump to antibodies in the recipient's plasma. For this reason, a person with type A (anti-B) blood must not receive blood of type B or AB, either of which would clump in the presence of anti-B in the recipient's type A blood. Likewise, a person with type B (anti-A) blood must not receive type A or AB blood, and a person with type O (anti-A and anti-B) blood must not receive type A, B, or AB blood.

Type AB blood does not have anti-A or anti-B antibodies, so an AB person can receive a transfusion of blood of any other type. For this reason, individuals with type AB blood are called *universal recipients.* Type O blood has neither antigen A nor antigen B. Theoretically, this blood type could be transfused into persons with blood of any type. Individuals with type O blood are called *universal donors.*

Using donor blood of the same type as the recipient is best for all blood types (table 12.5). When donating a permissible blood type to the recipient, precautionary measures should be taken to reduce the chance of an adverse reaction, such as minimizing the plasma, and transfusing slowly so that the recipient's larger blood volume will dilute the donor's antibodies.

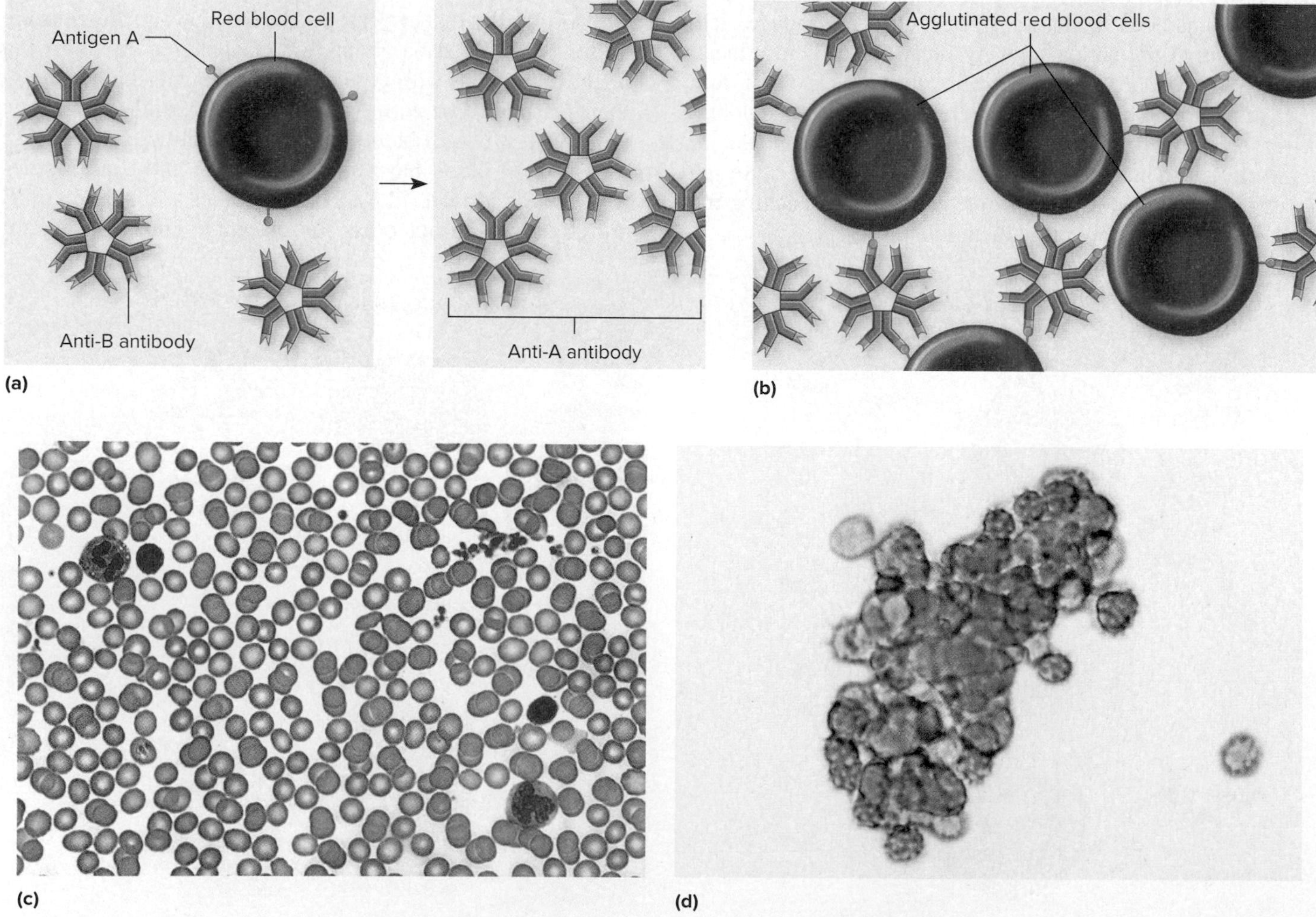

Figure 12.19 Agglutination. **(a)** If red blood cells with antigen A are added to blood containing anti-A antibody, **(b)** the antibodies react with the antigens, causing clumping (agglutination). (Cells and antibodies in [a] and [b] not drawn to scale.) Micrographs of **(c)** nonagglutinated blood (750x) and **(d)** agglutinated blood (260x). (c): Biophoto Associates/Science Source; (d): Ed Reschke/Getty Images

TABLE 12.5 Preferred and Permissible Blood Types for Transfusions

Blood Type of Recipient	Preferred Blood Type of Donor	Permissible Blood Types of Donor
A–	A–	O–
A+	A+	A–, O–, O+
B–	B–	O–
B+	B+	B–, O–, O+
AB–	AB–	A–, B–, O–
AB+	AB+	AB–, A–, A+ , B–, B+ , O–, O+
O–	O–	None
O+	O+	O–

In the United States, the most common ABO blood types are O (47%) and A (41%). Rarer are type B (9%) and type AB (3%). These percentages vary in subpopulations and over time, reflecting changes in the genetic structure of populations.

OF INTEREST Blood in the umbilical cord at birth is rich in stem cells that can be used to treat a variety of disorders, including leukemias, sickle cell disease and other hemoglobin abnormalities, and certain inborn errors of metabolism. For many illnesses, it is more effective to receive donor stem cells, because receiving one's own could reintroduce the disease.

PRACTICE 12.5

1. Distinguish between antigens and antibodies.
2. What is the main concern when blood is transfused from one individual to another?
3. Why is an individual with type AB blood called a universal recipient?
4. Why is an individual with type O blood called a universal donor?

Rh Blood Group

The *Rh blood group* was named after the rhesus monkey in which it was first studied. In humans, this group includes several Rh antigens (factors). The most prevalent of these is *antigen D.*

If the Rh antigens are present on the red blood cell membranes, the blood is said to be *Rh-positive.* Conversely, if the red blood cells do not have Rh antigens, the blood is called *Rh-negative.* The presence (or absence) of Rh antigens is an inherited trait, and only 15% of the U.S. population is Rh-negative. Anti-Rh antibodies (anti-Rh) form only in Rh-negative individuals in response to the presence of red blood cells with Rh antigens. This happens, for example, if an individual with Rh-negative blood receives a transfusion of Rh-positive blood. The Rh antigens stimulate the recipient to begin producing anti-Rh antibodies. Generally, this initial transfusion has no serious consequences, but if an individual with Rh-negative blood—who is now sensitized to Rh-positive blood—receives another transfusion of Rh-positive blood some months later, the donated red cells are likely to agglutinate.

A similar situation of Rh incompatibility arises when an Rh-negative woman is pregnant with an Rh-positive fetus. Her first pregnancy with an Rh-positive fetus would probably be uneventful. However, if at the time of the infant's birth (or if a miscarriage occurs) the placental membranes that separated the maternal blood from the fetal blood during the pregnancy tear, some of the infant's Rh-positive blood cells may enter the maternal circulation. These Rh-positive cells may then stimulate the maternal tissues to produce anti-Rh antibodies (fig. 12.20). If a woman who has already developed anti-Rh antibodies becomes pregnant with a second Rh-positive fetus, these antibodies, called hemolysins, cross the placental membrane and destroy the fetal red blood cells (fig. 12.20). The fetus then develops a condition called *erythroblastosis fetalis,* or hemolytic disease of the fetus and newborn.

Erythroblastosis fetalis is extremely rare today because obstetricians carefully track Rh status. An Rh-negative woman who might carry an Rh-positive fetus is given an injection of a drug called RhoGAM at week 28 of her pregnancy and after delivery of an Rh-positive baby. RhoGAM is a preparation of anti-Rh antibodies, which bind to and shield any Rh-positive fetal cells that might contact the woman's cells and sensitize her immune system. RhoGAM must be given within 72 hours of possible contact with Rh-positive cells—including giving birth, terminating a pregnancy, miscarrying, or undergoing amniocentesis (a prenatal test in which a needle is inserted into the uterus).

PRACTICE 12.5

5. What is the Rh blood group?
6. What are two ways that Rh incompatibility can arise?

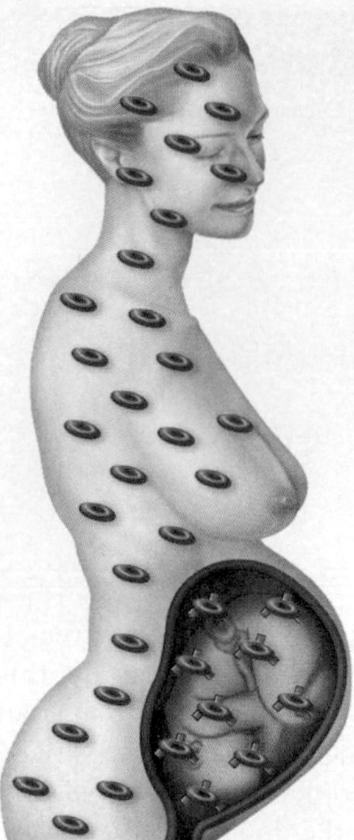

Rh-negative female with Rh-positive fetus

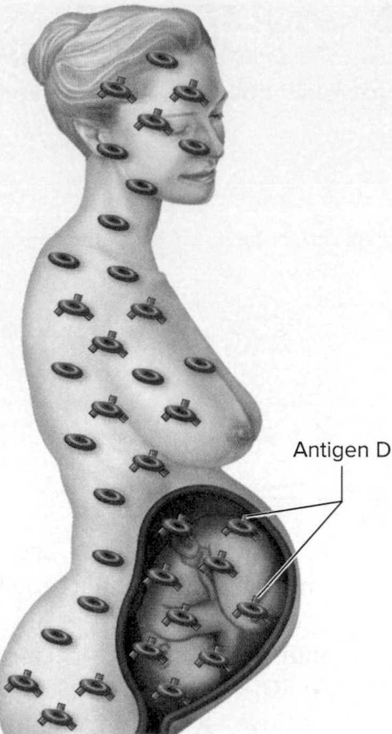

During childbirth, cells from the Rh-positive fetus may enter the female's bloodstream

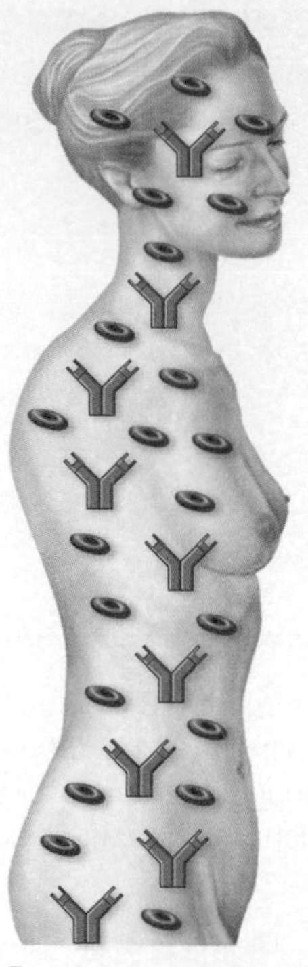

Female becomes sensitized—antibodies (Y) form against Rh-positive blood cells between pregnancies

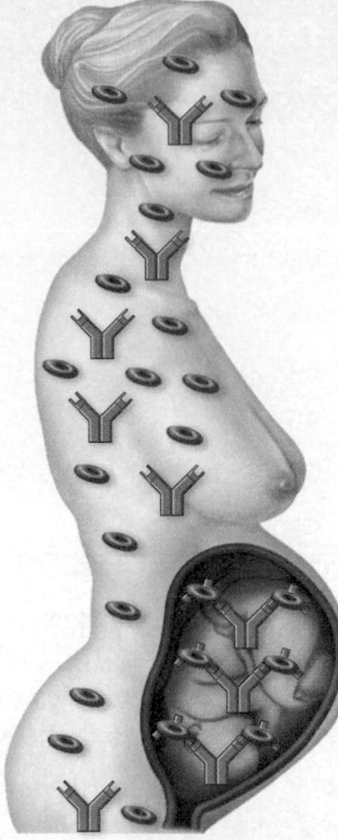

In the next Rh-positive pregnancy, maternal antibodies attack fetal red blood cells

Figure 12.20 Rh incompatibility. If a man who is Rh-positive and a woman who is Rh-negative conceive a child who is Rh-positive, the woman's body may manufacture antibodies that attack future Rh-positive offspring.

ASSESS

CHAPTER ASSESSMENTS

12.1 Introduction

1. Major functions of blood include ________.
   a. nutrient, hormone, and oxygen transport
   b. helping maintain the stability of interstitial fluid
   c. heat distribution
   d. all of the above
2. Formed elements in blood are ________, ________, and ________.
3. The liquid portion of blood is called ________.

12.2 Formed Elements

4. Describe a red blood cell.
5. Contrast oxyhemoglobin and deoxyhemoglobin.
6. Connect the significance of red blood cell counts with the function of red blood cells.
7. Describe the life cycle of a red blood cell, beginning with its production and ending with its death and destruction.
8. List dietary factors affecting red blood cell production.
9. Name five types of leukocytes, identifying which are granulocytes and which are agranulocytes, and list the major function(s) of each type.
10. ________ are fragments of megakaryocytes that function in ________.

12.3 Plasma

11. The most abundant component of plasma is ________.
    a. waste
    b. oxygen
    c. proteins
    d. water
12. Name three types of plasma proteins, and indicate the major function(s) of each type.
13. Name the gases and nutrients found in plasma.
14. Define *nonprotein nitrogenous substances,* and name those commonly present in plasma.
15. The most abundant plasma electrolytes are ________ and ________.

12.4 Hemostasis

16. ________ is the term for stoppage of bleeding.
17. Explain how vascular spasm is stimulated following an injury.
18. Platelets adhering to form a plug may control blood loss from a ________ break, but a larger break may require a ________ to halt bleeding.
19. Describe the major steps leading to the formation of a blood clot.

20. Distinguish between a thrombus and an embolus.
21. Describe ways to prevent abnormal blood clotting.

12.5 Blood Groups and Transfusions

22. An individual with B antigens and anti-A antibodies is ABO blood type ________.
23. Explain why the individual described in question 22 should not receive a transfusion with type AB blood.
24. Distinguish between Rh-positive and Rh-negative blood.
25. Describe *erythroblastosis fetalis,* and explain how this condition may develop.

ASSESS

INTEGRATIVE ASSESSMENTS/CRITICAL THINKING

Outcomes 3.3, 12.1, 12.3, 12.4

1. How might a low-protein diet affect blood volume and the ability of the blood to clot?

Outcomes 3.4, 12.2

2. If a patient with cancer is treated using a chemotherapy drug that reduces the rate of cell division, how might the patient's white blood cell count change? How might the patient's environment be modified to compensate for the effects of these changes?

Outcomes 8.3, 12.2

3. A week or so prior to participating in a competitive 13.1-mile run on the Oregon coast, Frank, who lives in Portland, Oregon, decided to fly to Denver, Colorado, to train for a few days. What advantages would Frank obtain by training in Denver?

Outcome 12.2

4. How would you explain to a patient with leukemia, who has a greatly elevated white blood cell count, the importance of avoiding bacterial infections?

Outcomes 12.2, 12.5

5. Why can an individual receive platelets donated by anyone, but must receive only a particular type of whole blood?

Outcomes 12.3, 12.4

6. Why do patients with liver diseases commonly develop blood clotting disorders?

Outcomes 12.4, 12.5

7. A student, Joanne, explains to another student that when a recipient receives an incompatible blood type from a donor, it causes coagulation. Explain the error in her explanation.

Outcome 12.5

8. Commercially available antiserum samples containing antibodies for antigens A, B, and D are used to determine the type of a particular patient's blood. The antiserum is mixed with the sample of blood, and if agglutination (clumping) occurs, that means the antigen with the same name as the antiserum is present on those RBCs. Indicate the blood type (both ABO and Rh) of this individual. What blood type(s) could safely receive blood from this individual? What blood type(s) could this individual safely receive? (Note: The control indicates the absence of agglutination, or clumping.)

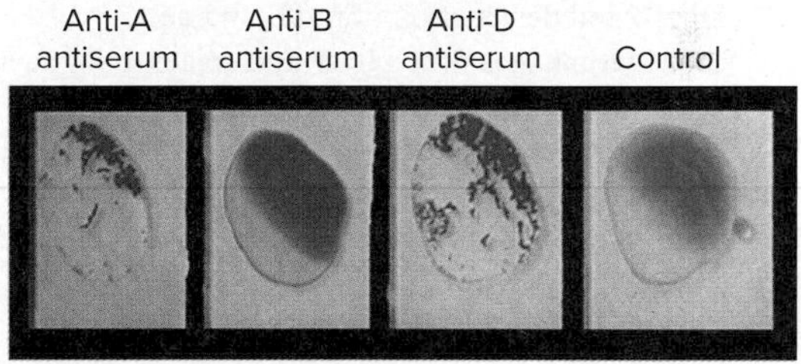

Image Source/Getty Images

9. Explain why Tammy, who has blood type B–, can receive blood from a donor who has blood type B– but not B+. Are there other blood types that would be an acceptable donor for Tammy?

Chapter Summary

12.1 Introduction

Blood is a type of connective tissue in which cells are suspended in a liquid extracellular matrix. It transports substances between body cells and the external environment, and helps maintain a stable internal environment.

1. Blood can be separated into formed elements and liquid portions.
   a. The formed elements portion consists of red blood cells, white blood cells, and platelets.
   b. The liquid **plasma** includes water, gases, nutrients, hormones, electrolytes, and cellular wastes.
2. Blood is more viscous than water and has a slightly alkaline pH.
3. Blood volume varies with body size, percent adipose tissue, and fluid and electrolyte balance. A complete blood count includes the percentage of red blood cells (the hematocrit value).

12.2 Formed Elements

1. Red blood cells
   a. **Red blood cells (erythrocytes)** are biconcave discs with shapes that increase surface area.

b. Red blood cells contain **hemoglobin,** which carries oxygen, and some carbon dioxide.
c. Prolonged oxygen deficiency may result in cyanosis.
d. Red blood cells depend on anaerobic mechanisms to synthesize ATP.

2. Red blood cell counts
   a. The red blood cell count equals the number of cells per microliter of blood.
   b. The average count ranges from approximately 4 to 6 million cells per microliter of blood.
   c. Red blood cell count and hemoglobin concentration determines the oxygen-carrying capacity of the blood. It is used to diagnose and evaluate the courses of certain diseases.
3. Red blood cell production and its control
   a. Red bone marrow produces red blood cells.
   b. In health, the number of red blood cells remains relatively stable.
   c. **Erythropoietin** controls the rate of red blood cell formation by negative feedback.
4. Dietary factors affecting red blood cell production
   a. Availability of vitamin B_{12} and folic acid influences red blood cell production.
   b. Hemoglobin synthesis requires iron.
5. Death and destruction of red blood cells
   a. **Macrophages** in the liver and spleen phagocytize damaged red blood cells.
   b. Hemoglobin molecules decompose, and nearly all of the iron they contain is recycled.
   c. **Biliverdin** and **bilirubin** are pigments, released from the heme (iron) portion, excreted in bile.
6. White blood cells
   a. White blood cells are able to leave circulation and move toward the injured area.
   b. **White blood cells (leukocytes)** develop from hematopoietic stem cells in red bone marrow, in response to **interleukins** and **colony-stimulating factors.**
7. Types of white blood cells and their functions
   a. **Granulocytes** include **neutrophils, eosinophils,** and **basophils.**
   b. **Neutrophils** are most numerous and have a three- to five-lobed nucleus. They are phagocytes that kill by creating a respiratory burst.
   c. **Eosinophils** have a bilobed nucleus, and granules stain them an orange-red color. They kill parasites and help control inflammation and allergic reactions.
   d. **Basophils** are the least numerous and have violet-stained granules. They release heparin, an anticoagulant, and histamine, a vasodilator, which promotes inflammation to increase blood flow to injured tissues.
   e. **Agranulocytes** include **monocytes** and **lymphocytes.**
   f. **Monocytes** are largest in size, with a U-shaped nucleus. They differentiate into macrophages that are highly phagocytic.
   g. **Lymphocytes** are the second-most numerous with a nucleus that takes up most of the cell. They are important in the immune response.
8. White blood cell counts
   a. Normal total white blood cell counts vary from 3,500 to 10,500 cells per microliter of blood.
   b. The number of white blood cells may change in response to abnormal conditions, such as infections, emotional disturbances, or excessive loss of body fluids.
   c. A differential white blood cell count indicates the percentages of various types of leukocytes.
9. Blood platelets
   a. Blood **platelets,** which develop in the red bone marrow in response to **thrombopoietin,** are fragments of giant cells.
   b. Platelets help close breaks in blood vessels.
10. Platelet count
    a. The normal platelet count varies from 150,000 to 350,000 platelets per microliter of blood.
    b. The number of platelets may change in response to abnormal conditions, such as infection, genetic mutations, toxins, or certain viruses or autoimmune diseases.

12.3 Plasma

Plasma transports gases and nutrients, helps regulate fluid and electrolyte balance, and helps maintain stable pH.

1. Plasma proteins
   a. **Plasma proteins** remain in blood and interstitial fluids, and are not normally used as energy sources.
   b. Three major types exist.
      (1) **Albumins** help maintain the colloid osmotic pressure.
      (2) Alpha and beta **globulins** transport lipids and fat-soluble vitamins, and gamma globulins are the antibodies that provide immunity.
      (3) **Fibrinogen** functions in blood clotting.
2. Gases and nutrients
   a. Gases in plasma include oxygen, carbon dioxide, and nitrogen.
   b. Plasma nutrients include simple sugars, amino acids, and lipids.
      (1) The liver stores glucose as glycogen and releases glucose whenever blood glucose concentration falls.
      (2) Amino acids are used to synthesize proteins and are deaminated for use as energy sources.
      (3) Lipoproteins function in the transport of lipids.
3. Nonprotein nitrogenous substances
   a. **Nonprotein nitrogenous substances** are composed of molecules that contain nitrogen atoms but are not proteins.
   b. They include amino acids, urea, uric acid, creatine, and creatinine.
4. Plasma electrolytes
   a. Plasma electrolytes include ions of sodium, potassium, calcium, magnesium, chloride, bicarbonate, phosphate, and sulfate.
   b. Bicarbonate ions are important in helping maintain the pH of plasma.

12.4 Hemostasis

Hemostasis is the stoppage of bleeding.

1. **Vascular spasm**
   a. Smooth muscle in blood vessel walls reflexively contract following injury.

b. Platelets release **serotonin,** which stimulates vasoconstriction and helps maintain vascular spasm.
2. Platelet plug formation
a. Platelets adhere to rough surfaces and exposed collagen.
b. Platelets adhere to each other at injury sites and form platelet plugs in broken vessels.
3. Blood coagulation
a. Blood clotting is the most effective means of hemostasis.
b. Clot formation depends on the balance between factors that promote clotting and those that inhibit clotting.
c. The basic event of **coagulation** is the conversion of soluble fibrinogen into insoluble **fibrin.**
d. Biochemicals that promote clotting include prothrombin activator, **prothrombin, thrombin,** and calcium ions.
4. Clot dissolution
a. **Plasmin** is an enzyme that dissolves clots.
5. Abnormal clot formation
a. A **thrombus** is an abnormal blood clot in a vessel. An **embolus** is a clot or fragment of a clot that moves in a vessel.
b. Abnormal clots are associated with atherosclerosis and immobilization that slows blood flow.
c. Drugs that dissolve clots or that prevent coagulation lower the risk of abnormal clot formation.

12.5 Blood Groups and Transfusions

Blood can be typed on the basis of cell surface antigens.

1. Antigens and antibodies
a. An **antigen** is any molecule that triggers an immune response.
b. **Antibodies** are protein molecules produced in an immune response to an encounter with an antigen not found on the body's own cells.
2. ABO blood group
a. Blood is grouped according to the presence or absence of antigens A and B attached to the surface of red blood cells.
b. Whenever antigen A is absent, anti-A antibody is present; whenever antigen B is absent, anti-B antibody is present.
c. Mixing red blood cells that contain an antigen with plasma that contains the corresponding antibody results in an **agglutination** reaction or adverse transfusion reaction in a patient.
3. Rh blood group
a. Rh antigens are present on the red blood cell membranes of Rh-positive blood. Rh antigens are absent in Rh-negative blood.
b. An Rh-negative person exposed to Rh-positive blood produces anti-Rh antibodies in response to the presence of Rh antigens.
c. Mixing Rh-positive red blood cells with plasma that contains anti-Rh antibodies agglutinates the positive cells.
d. Anti-Rh antibodies in maternal blood may cross the placental tissues and react with the red blood cells of an Rh-positive fetus.

13 Cardiovascular System

An implantable cardioverter defibrillator (ICD) delivers a shock to a heart whose ventricles are contracting wildly, restoring a normal heartbeat. Jochen Tack/Alamy Stock Photo

Cardiovascular defibrillators. A man rushing to catch a flight at a busy airport stops suddenly, looks about in confusion, and collapses. As people gather around him, a woman checks for a heartbeat and pulse and then tells another onlooker to bring her a device mounted on a nearby wall. It is an automated external defibrillator (AED), and looks like a laptop computer. The woman learned how to use it in a cardiopulmonary resuscitation class. She brings it over to the man, opens it, and places electrode pads over the man's chest, as indicated in a drawing on the inner cover of the defibrillator. Then the device speaks: "Analyzing heart rhythm," it declares, as a computer assesses the heart rhythm. After a short pause, the device says, "Charging, stand clear," and then "Push button." The woman does so, and the device delivers a shock to the man's chest. It assesses the heart rhythm again, and instructs the woman to deliver a second shock. Soon the man regains consciousness, just as emergency medical technicians (EMTs) arrive.

An AED in a public place can save the life of a person suffering *sudden cardiac arrest,* failure of the heart to circulate blood. One study conducted at airports found that they saved 64% of the people on whom they were used. Without defibrillation, only 5% to 7% of people survive sudden cardiac arrest. Each minute, the odds of survival drop by 10%, and after six minutes, brain damage is irreversible.

Sudden cardiac arrest can result from an abnormally accelerated heartbeat (tachycardia) or a chaotic and irregular contraction of the heart muscle (ventricular fibrillation). The bioelectrical malfunction that usually causes these conditions may result from an artery blocked with plaque or from buildup of scar tissue from a previous myocardial infarction (heart attack).

For people who know they have an inherited disorder that causes sudden cardiac arrest (by having suffered an event and having had genetic tests), a device called an implantable cardioverter defibrillator (ICD) can be placed under the skin of the chest in a one-hour procedure. Like the AED, the ICD monitors heart rhythm. When the telltale deviations of ventricular tachycardia or ventricular fibrillation begin, it delivers a shock, preventing cardiac arrest. ICDs have been so successful in preventing cardiac arrests that they may be recommended to people at high risk for this condition.

Anatomy & Physiology Revealed® 4.0

Module 9 Cardiovascular System

LEARNING OUTLINE

After studying this chapter, you should be able to complete the "Learning Outcomes" that follow the major headings throughout the chapter.

AIDS TO UNDERSTANDING WORDS

brady- [slow] *brady*cardia: abnormally slow heartbeat.

diastol- [dilation] *diastol*ic pressure: blood pressure when the ventricle of the heart is relaxed.

-gram [something written] electrocardio*gram:* recording of the electrical changes in the myocardium during a cardiac cycle.

papill- [nipple] *papill*ary muscle: small mound of muscle projecting into a ventricle of the heart.

syn- [together] *syn*cytium: mass of merging cells that act together.

systol- [contraction] *systol*ic pressure: blood pressure resulting from a single ventricular contraction.

tachy- [rapid] *tachy*cardia: abnormally fast heartbeat.

(Appendix A has a complete list of Aids to Understanding Words.)

13.1 | Introduction

LEARN

1. Discuss the functions of the organs of the cardiovascular system.

The heart is a muscular pump that generates the force required to move blood through the vessels. The heart and blood vessels together constitute the **cardiovascular system** (fig. 13.1). During normal daily activities, the heart pumps the entire volume of blood approximately every minute. On an expanded scale, it pumps about 7,000 liters of blood through the body each day, contracting about 2.5 billion times in an average lifetime.

Vessels carrying blood toward the heart are veins, whereas those carrying blood away from the heart are the arteries. Gas and nutrient exchange occur in the microscopic vessels that lie between arteries and veins, the **capillaries.**

The cardiovascular system has two closed pathways, or circuits, of blood flow. The **pulmonary** (pul′mo-ner″ē) **circuit** sends oxygen-poor blood to the lungs to pick up oxygen and unload carbon dioxide. The **systemic** (sis-tem′ik) **circuit** sends oxygen-rich blood and nutrients to all body cells and removes wastes.

PRACTICE 13.1

Answers to the Practice questions can be found in the eBook.

1. Name the parts of the cardiovascular system.
2. Distinguish between the pulmonary and systemic circuits of the cardiovascular system.

CAREER CORNER
Surgical Technician

A surgical technician enters the room of a woman scheduled to undergo heart surgery. The technician gently washes the patient. Meanwhile, another surgical technician checks that the operating room is clean, and that the surgical instruments are sterilized. She sets out the instruments on a tray in the order in which the surgeon will require them.

Surgical technicians assist doctors and nurses in gowning, gloving, and putting on masks after scrubbing. During the operation, surgical technicians hand instruments to the surgeon, and may prepare tissue samples, medications, and sterile dressings. An important task is to count the instruments and dressing materials before and after the procedure, to be certain none are left inside the patient. The technician may prepare the operating room for the next procedure, or accompany the patient to the recovery room. The job requires long periods of standing.

A surgical technician can specialize, such as in open heart or brain surgery. Jobs are in hospitals, dental surgical practices, and outpatient surgical facilities. Becoming a surgical technician requires an associate's (two-year) degree. Most states require certification and continuing education.

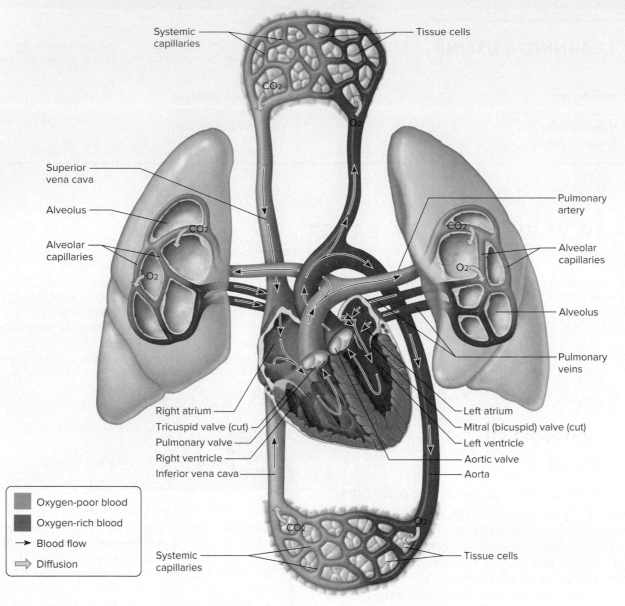

Figure 13.1 The cardiovascular system transports blood between the body cells and organs such as the lungs, intestines, and kidneys that communicate with the external environment. Vessels in the pulmonary circuit transport blood from the heart to the lungs and back to the heart, replenishing oxygen (O_2) and releasing the metabolic waste carbon dioxide (CO_2). Vessels of the systemic circuit supply all of the other cells. (Structures are not drawn to scale.) APR

13.2 | Structure of the Heart

LEARN

1. Distinguish between the coverings of the heart and the layers that compose the wall of the heart.
2. Identify and locate the major parts of the heart, and discuss the functions of each part.
3. Trace the pathway of blood through the heart and the vessels of coronary circulation.

The heart is strategically located close to the lungs. The outward appearance is often referred to as cone-shaped. Inside, the majority of the heart is actually hollow with a double pump, divided into left and right halves.

Size and Location of the Heart

The heart is typically about the size of a fist, obviously varying with body size. However, an average adult's heart is about 14 centimeters long and 9 centimeters wide.

In the mediastinum just superior to the diaphragm, the heart is bordered laterally by the lungs. The "cone" is actually upside down with the apex (peak) on the bottom, left of the sternum at about the level of the fifth rib (fig. 13.2). This is where a stethoscope is placed to listen to a heartbeat. The broad end of the heart, the base, is attached to large vessels, and is on the top beneath the second rib. The aorta lies behind the upper part of the sternum.

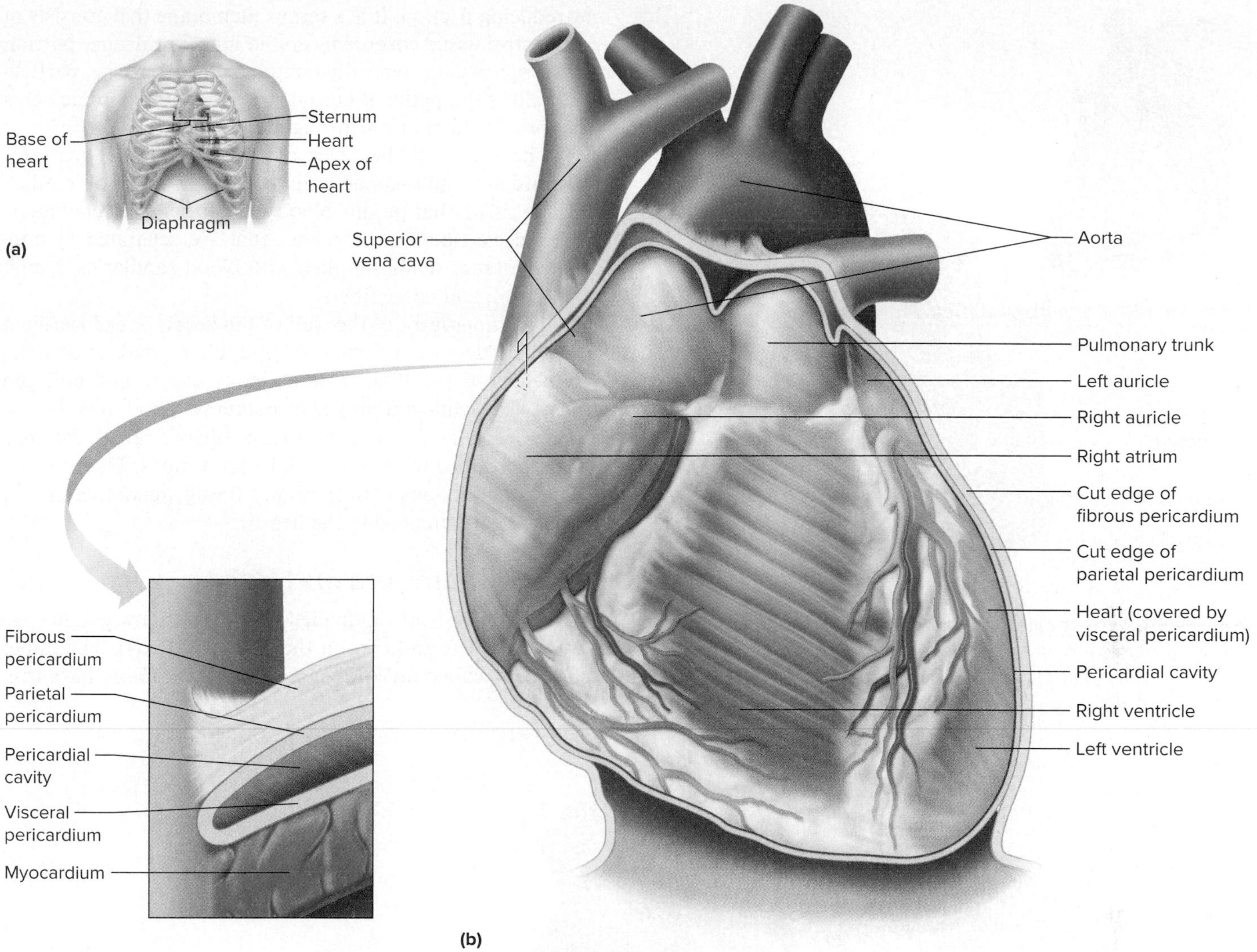

Figure 13.2 The heart is within the mediastinum of the thoracic cavity and is enclosed by a layered pericardium. APR

Coverings of the Heart

The **pericardium** (per″ĭ-kar′dē-um) is a membranous sac that encloses the heart and the proximal ends of the large blood vessels to which it attaches. The outer layer consists of a fibrous bag, the *fibrous pericardium,* composed of dense connective tissue. The fibrous pericardium is attached to the central part of the diaphragm, the posterior of the sternum, the vertebral column, and the large blood vessels associated with the heart.

The fibrous pericardium surrounds more delicate serous layers. The innermost layer, the **visceral pericardium** (epicardium), covers the heart. At the base of the heart, the visceral pericardium turns back upon itself to become the **parietal pericardium,** which covers the inner surface of the fibrous pericardium (see fig. 13.3 and reference plate 3). Between the parietal and visceral serous layers of the pericardium is a space, the **pericardial cavity,** that contains a small volume of serous fluid (fig. 13.3). This fluid reduces friction between the pericardial membranes as the heart moves within them.

In *pericarditis,* inflammation of the pericardium is often due to viral or bacterial infection. This painful condition interferes with heart movements.

PRACTICE 13.2

1. Where is the heart located?
2. Distinguish between the visceral pericardium and the parietal pericardium.

Wall of the Heart

The wall of the heart is composed of three distinct layers—an outer epicardium, a middle myocardium, and an inner endocardium (fig. 13.3). The **epicardium** (ep″ĭ-kar′dē-um), which corresponds to the visceral pericardium, protects the heart

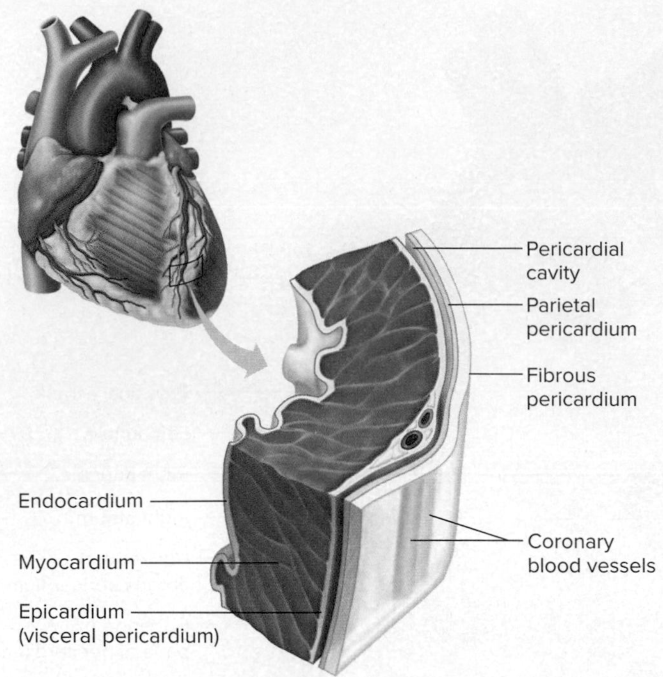

Figure 13.3 The heart wall has three layers: an endocardium, a myocardium, and an epicardium. APR

by reducing friction. It is a serous membrane that consists of connective tissue covered by epithelium. The deeper portion of the epicardium typically contains adipose tissue, particularly along the paths of coronary arteries and cardiac veins that provide blood flow through the myocardium.

The thick middle layer of the wall of the heart, or **myocardium** (mī″o-kar′dē-um), consists largely of cardiac muscle tissue that pumps blood out of the heart chambers. The muscle fibers lie in planes that are separated by connective tissues richly supplied with blood capillaries, lymph capillaries, and nerve fibers.

The inner layer of the wall of the heart, or **endocardium** (en″dō-kar′dē-um), consists of epithelium and underlying connective tissue that contains many elastic and collagen fibers. The endocardium also contains blood vessels and some specialized cardiac muscle fibers, called *Purkinje fibers,* described in section 13.3, Heart Actions. The endocardium is continuous with the inner linings (endothelium) of blood vessels attached to the heart.

Heart Chambers and Valves

Internally, the heart is divided into four hollow chambers—two on the left and two on the right (fig. 13.4). The upper chambers, called **atria** (ā′trē-ah; singular, *atrium*), have thin

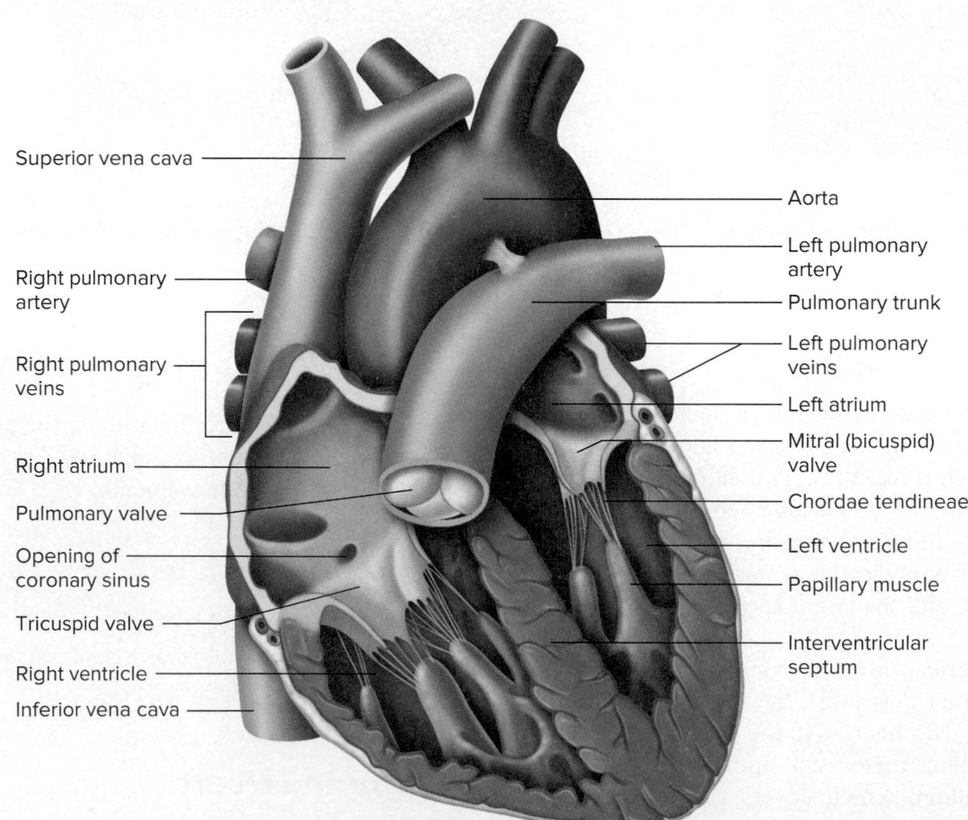

Figure 13.4 Frontal section of the heart showing the connection between the right ventricle and the pulmonary trunk, as well as the four hollow chambers. The plane of section does not show the aortic valve, and shows only parts of the tricuspid and mitral valves. APR

walls and receive blood returning to the heart. Small, earlike projections called *auricles* extend anteriorly from the atria. They serve to increase the blood volume capacity of the atria. The lower chambers, the **ventricles** (ven′trĭ-klz), receive blood from the atria and contract to force blood out of the heart into arteries.

A solid, wall-like **septum** separates the atrium and ventricle on the right side from their counterparts on the left. As a result, blood from one side of the heart never mixes with blood from the other side (except in the fetus, see section 20.3, Pregnancy and the Prenatal Period). An *atrioventricular valve* (AV valve) on the right and on the left ensures one-way blood flow between the atrium and the ventricle on each side.

The right atrium receives blood from the **vena cavae,** two large veins—the *superior vena cava* and the *inferior vena cava.* A smaller vein, the *coronary sinus,* also drains venous blood into the right atrium from the myocardium.

The large **tricuspid valve**, which has three tapered projections called *cusps* as its name implies, lies between the right atrium and the right ventricle (fig. 13.4). The valve permits blood to move from the right atrium into the right ventricle and prevents backflow.

Strong, fibrous strings called **chordae tendineae** (kor′de ten′dĭ-nē-a) attach to the cusps of the tricuspid valve on the ventricular side. These strings originate from small mounds of cardiac muscle tissue, the **papillary muscles,** that project inward from the walls of the ventricle. The papillary muscles contract when the ventricle contracts. As the tricuspid valve closes, these muscles pull on the chordae tendineae and prevent the cusps from swinging back (everting) into the atrium.

The right ventricle has a thinner muscular wall than the left ventricle (fig. 13.4). This right chamber pumps blood a fairly short distance to the lungs against a relatively low resistance to blood flow. The left ventricle, on the other hand, must force blood to all the other parts of the body against a much greater resistance to flow.

When the muscular wall of the right ventricle contracts, the blood inside its chamber is put under increasing pressure, and the tricuspid valve is pushed closed by the increased pressure. As a result, the only exit for the blood is through the *pulmonary trunk,* which divides to form the left and right *pulmonary arteries* that lead to the lungs. At the base of this trunk is a **pulmonary valve** which consists of three cusps. This valve allows blood to leave the right ventricle and prevents backflow into the ventricular chamber (fig. 13.5).

The left atrium receives blood from the lungs through four *pulmonary veins*—two from the right lung and two from the left lung. Blood passes from the left atrium into the left ventricle through the **mitral valve** (shaped like a mitre, a type of headpiece), also called the bicuspid valve, which prevents blood from flowing back into the left atrium from

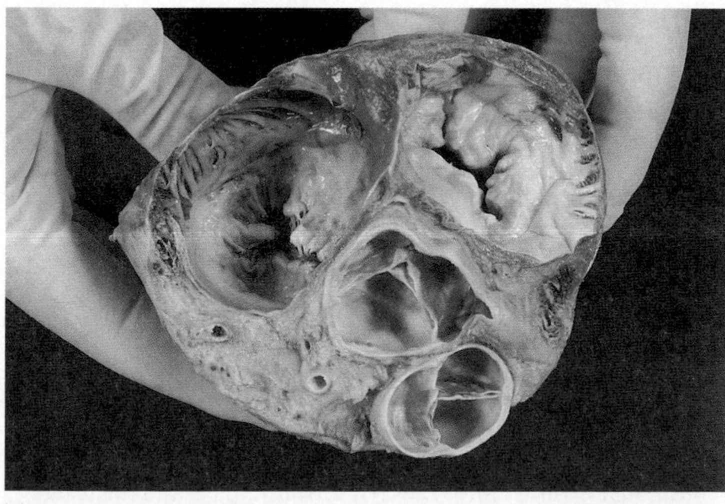

(a)

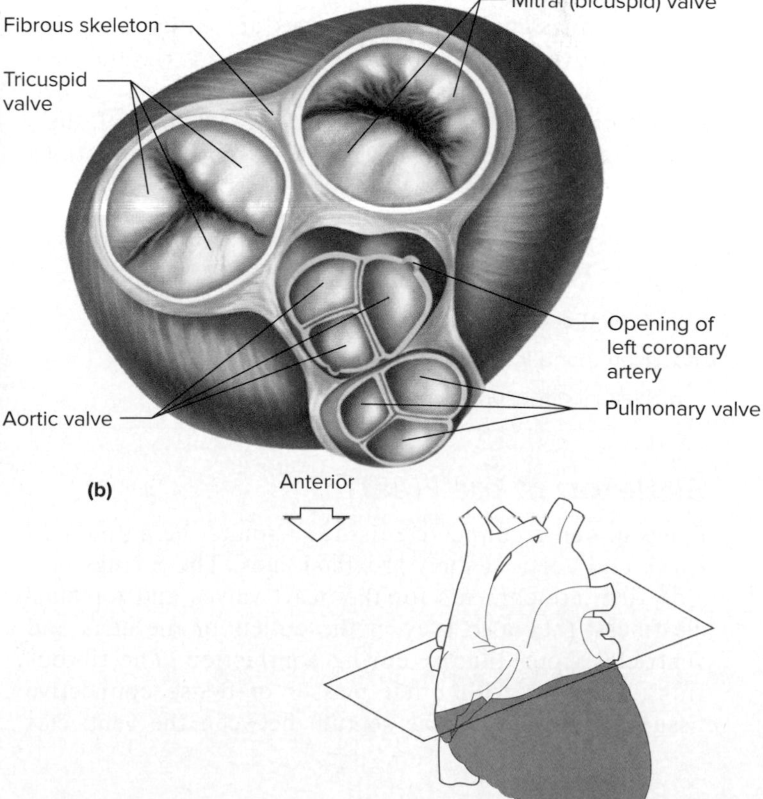

(b)

Figure 13.5 Heart valves. **(a)** Photograph of a transverse section through the heart, showing the four valves (superior view). **(b)** The skeleton of the heart consists of fibrous rings to which the heart valves are attached (superior view).

(a): Karl Rubin/McGraw-Hill Education

TABLE 13.1 Heart Valves

Valve	Location	Function
Tricuspid valve	Opening between right atrium and right ventricle	Prevents blood from moving from the right ventricle into the right atrium during ventricular contraction
Pulmonary valve	Entrance to pulmonary trunk	Prevents blood from moving from the pulmonary trunk into the right ventricle during ventricular relaxation
Mitral (bicuspid) valve	Opening between left atrium and left ventricle	Prevents blood from moving from the left ventricle into the left atrium during ventricular contraction
Aortic valve	Entrance to aorta	Prevents blood from moving from the aorta into the left ventricle during ventricular relaxation

the left ventricle (see fig. 13.4). As with the tricuspid valve, the papillary muscles and the chordae tendineae prevent the cusps of the mitral valve from swinging back into the left atrium.

When the left ventricle contracts, the mitral valve closes passively, and the only exit is through a large artery, the **aorta** (ā-or′tah). At the base of the aorta is the **aortic valve** (ā-or′tik valv), which consists of three cusps (fig. 13.5). The aortic valve opens and allows blood to leave the left ventricle as it contracts. When the ventricular muscles relax, this valve closes and prevents blood from backing up into the left ventricle.

The mitral and tricuspid valves are called atrioventricular valves because they are between atria and ventricles. The pulmonary and aortic valves are called "semilunar" because of the half-moon shapes of their cusps. Table 13.1 summarizes the locations and functions of the heart valves.

PRACTICE 13.2

3. Describe the layers of the heart wall.
4. Name and locate the four chambers of the heart.
5. Describe the function of each heart valve.

Skeleton of the Heart

Rings of dense connective tissue surround the pulmonary trunk and aorta at their proximal ends. These rings provide firm attachments for the heart valves and for muscle fibers; they also prevent the outlets of the atria and ventricles from dilating during contraction. The fibrous rings, together with other masses of dense connective tissue in the part of the septum between the ventricles (interventricular septum), constitute the *skeleton of the heart* (fig. 13.5*b*).

Blood Flow Through the Heart, Lungs, and Tissues

Blood flows in a continuous one-way circle. With regard to gas exchange, we recognize two subdivisions, or circuits (fig. 13.6). The exchange of blood between the heart and lungs is called the pulmonary circuit. The systemic circuit is the flow of blood between the heart and all body tissues. Because there is no beginning or ending in a circle, we could start anywhere and follow a drop of blood through the entire cardiovascular system until it returned to the same spot. But the right atrium provides a logical place to start. This chamber receives blood low in oxygen from the superior and inferior vena cavae. Pushing through the tricuspid (right AV) valve, blood enters the right ventricle. From here, blood passes through the pulmonary semilunar valve and into a large vessel called the pulmonary trunk. The pulmonary trunk branches into the left and right pulmonary arteries. They take blood to the left and right lungs, respectively. It is in the lungs that gas exchange occurs. The carbon dioxide brought back as metabolic waste from all cells leaves the blood and is exhaled. The newly inhaled oxygen is picked up by the blood. From the lungs, blood flows through the pulmonary veins into the left atrium, past the bicuspid (mitral) valve, and into the left ventricle. It then pushes through the aortic semilunar valve into the aorta. The aorta branches, carrying oxygen-rich blood to all tissue cells. The cells unload carbon dioxide into the blood. The circle is completed when the oxygen-poor blood is returned to the right atrium through the superior and inferior vena cavae. Clinical Application 13.1 discusses some complications that may arise in the cardiovascular system.

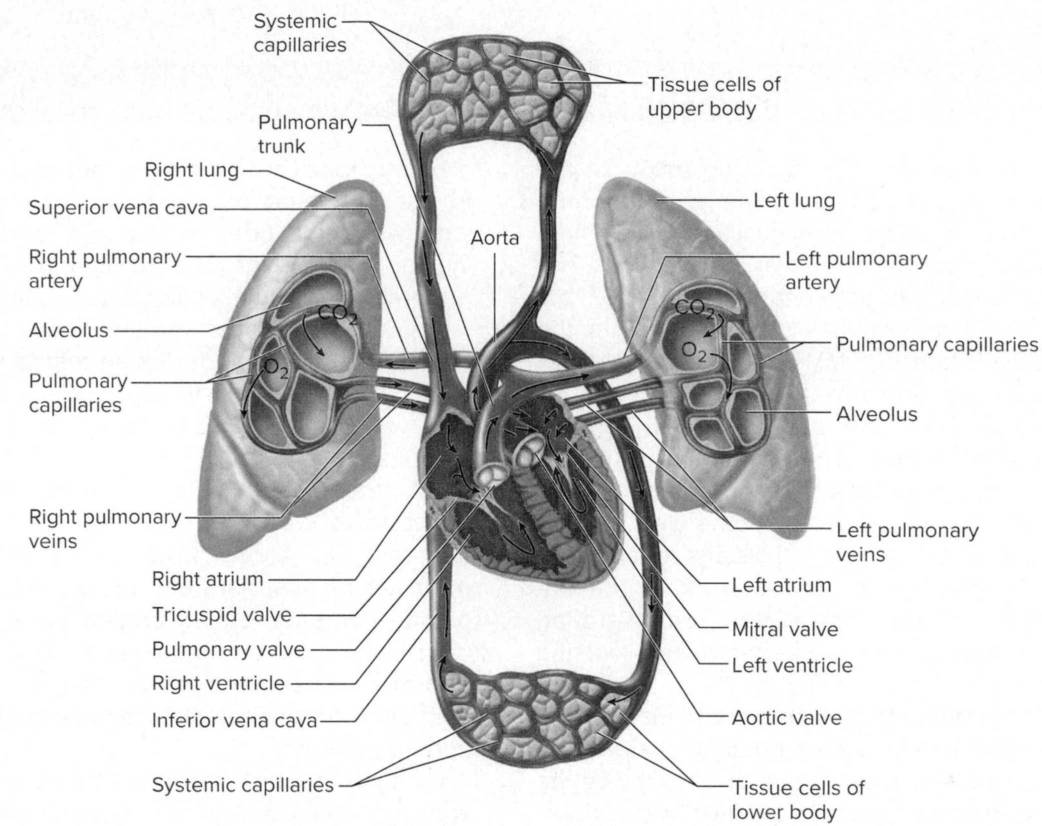

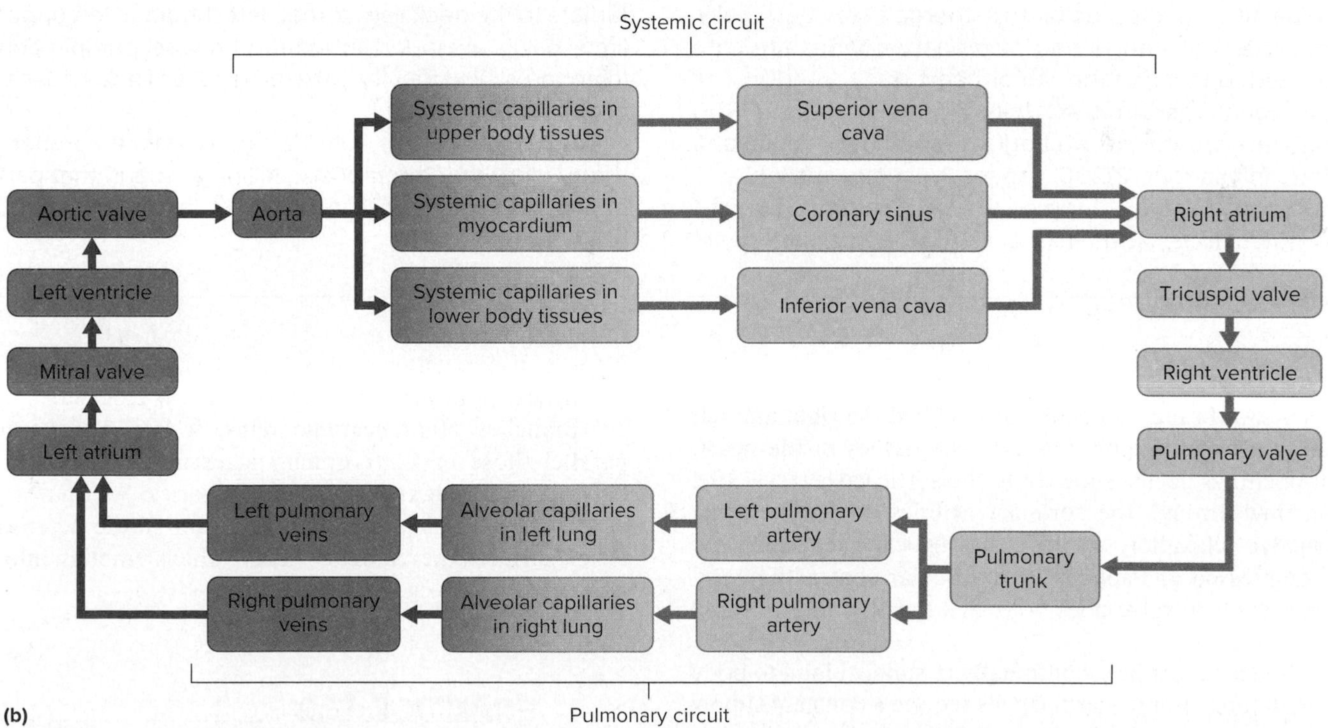

Figure 13.6 Pulmonary and systemic circuits. **(a)** The right ventricle forces blood to the lungs as part of the pulmonary circuit, whereas the left ventricle forces blood to all other body parts as part of the systemic circuit. (Structures are not drawn to scale.) **(b)** A flow chart showing the structures involved and direction of blood flow through the pulmonary and systemic circuits. APR

CLINICAL APPLICATION 13.1

Valvular Disease, Heart Transplant, Angina Pectoris, Myocardial Infarction

Mitral valve prolapse (MVP) affects up to 6% of the U.S. population. In this condition, one (or both) of the cusps of the mitral valve stretches and bulges into the left atrium during ventricular contraction. The valve usually continues to function adequately, but sometimes blood regurgitates (flows back) into the left atrium. Symptoms of MVP include chest pain, palpitations, fatigue, and anxiety.

People with MVP are particularly susceptible to infective endocarditis. This inflammation of the endocardium due to an infection appears as a plant-like growth on the mitral valve. Some people with MVP may be advised by their physicians to take antibiotics before undergoing dental work to prevent *Streptococcus* bacteria in the mouth from migrating through the bloodstream to the heart and causing this infection.

In heart transplantation, the recipient's failing heart is removed, except for the posterior walls of the right and left atria and their connections to the venae cavae and pulmonary veins. The donor heart is prepared similarly and is attached to the atrial structures remaining in the recipient's thoracic cavity. Finally, the recipient's aorta and pulmonary arteries are connected to those of the donor heart.

Each year in the United States, about 4,000 patients are on the waiting list for a heart transplant, but only about 2,200 organs become available. A solution for some people in need of a heart is a left ventricular assist device, or LVAD. The patient wears a battery pack with controls that attach by a cable to the implanted device. More than 6,000 people worldwide have been helped with LVADs while waiting for a transplant. The devices, now in their third generation, are smaller, lighter, and have fewer moving parts than the first models used in the 1980s. A fourth-generation model being tested replaces both ventricles and the aortic and mitral valves. Patients with LVADs report more energy, better sleep, and improved mood.

A thrombus or embolus that partially blocks or narrows a coronary artery branch causes a decrease in blood flow called ischemia. This deprives myocardial cells of oxygen, producing a painful condition called angina pectoris. The pain usually happens during physical activity, when oxygen demand exceeds oxygen supply. Pain lessens with rest. Emotional disturbance may also trigger angina pectoris.

Angina pectoris feels like heavy pressure, tightening, or squeezing in the chest, usually behind the sternum or in the anterior upper thorax. The pain may radiate to the neck, jaw, throat, left shoulder, left upper limb, back, or upper abdomen. Profuse perspiration (diaphoresis), difficulty breathing (dyspnea), nausea, or vomiting may occur.

A blood clot may completely obstruct a coronary artery (coronary thrombosis), killing tissue in that part of the heart. This is a myocardial infarction (MI) or heart attack.

Blood Supply to the Heart

The first two branches of the aorta, called the **right** and **left coronary arteries,** supply blood to the tissues of the heart. Their openings lie just superior to the aortic valve (fig. 13.7). Blood flow through the coronary arteries increases during ventricular relaxation because the myocardial vessels are not compressed as happens in ventricular contraction, and the closed aortic valve does not block the openings to these vessels.

The heart must beat continually to supply blood to body tissues. To do this, myocardial cells require a constant supply of oxygen-rich blood. Branches of the coronary arteries feed the many capillaries of the myocardium (fig. 13.8). The smaller branches of these arteries typically have connections (anastomoses) between vessels that provide alternate pathways for blood, called collateral circulation. These detours in circulation may supply oxygen and nutrients to the myocardium when a coronary artery is blocked.

Branches of the **cardiac veins,** whose paths roughly parallel those of the coronary arteries, drain blood that has passed through myocardial capillaries. As figure 13.8*b* shows, these veins join an enlarged vein on the heart's posterior surface—the **coronary sinus**—which empties into the right atrium (see fig. 13.4).

PRACTICE 13.2

6. Trace the path of blood through the heart.
7. Which vessels supply blood to the myocardium?
8. How does blood return from the cardiac tissues to the right atrium?

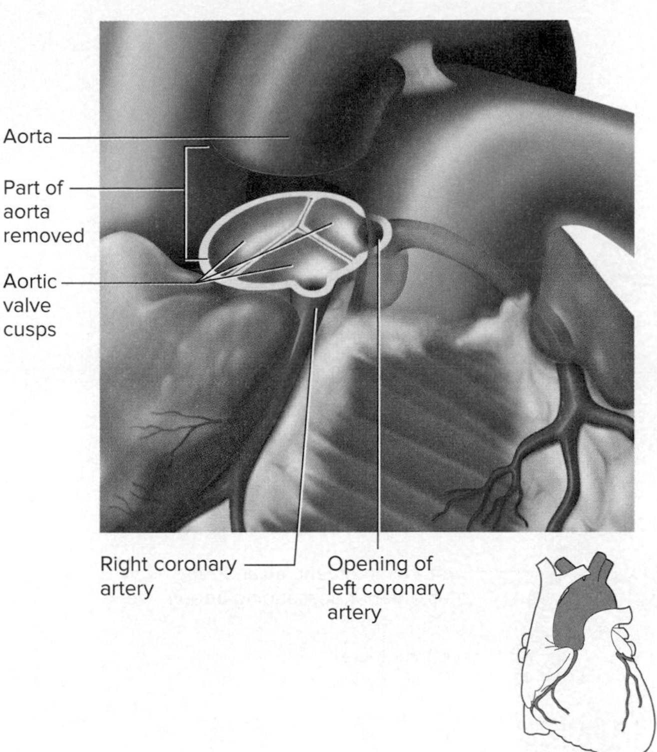

Figure 13.7 The openings of the coronary arteries lie just superior to the aortic valve.

13.3 | Heart Actions

LEARN

1. Describe the cardiac conduction system and the cardiac cycle.
2. Identify the parts of a normal ECG pattern, and discuss the significance of this pattern.
3. Explain control of the cardiac cycle.

The heart chambers function in a coordinated fashion. The structure of the cardiac muscle allows for regulations of their actions so that atria contract together, while ventricles relax. The ventricles then contract, while the atria relax. The state of contraction is known as **systole,** whereas **diastole** is the state of relaxation. Then, the atria and ventricles both relax for a brief interval. This series of events constitutes a complete heartbeat, or **cardiac cycle:** the contraction and subsequent relaxation of all four heart chambers. These actions can be measured and recorded in an ECG (electrocardiogram). There are also two sounds associated with a heartbeat.

Cardiac Muscle Cells

Cardiac muscle cells function like those of skeletal muscles, but the cells connect in branching networks. The intercalated discs join cardiac muscle cells, allowing action potentials to spread throughout a network of cells. As a result, cardiac muscle cells may contract as a unit.

A mass of cells that act as a unit is called a **functional syncytium** (sin-sish′e-um). Two such structures are in the heart—in the atrial walls and in the ventricular walls. Portions of the heart's fibrous skeleton separate these masses of cardiac muscle, except for a small area in the right atrial floor. In this region, specialized conduction fibers connect the *atrial syncytium* and the *ventricular syncytium.*

Cardiac Conduction System

Throughout the heart are clumps and strands of specialized cardiac muscle tissue whose cells contain only a few myofibrils. Instead of contracting, the cells in these areas initiate and distribute impulses throughout the myocardium. They form the **cardiac conduction system** (kar′dē-ak kon-duk′shun sis′tem), which coordinates the events of the cardiac cycle (fig. 13.9).

A key portion of the cardiac conduction system is the **SA node,** or **sinoatrial node,** which is a small elongated mass of specialized cardiac muscle tissue located just beneath the epicardium in the right atrium near the opening of the superior vena cava. Its cells are continuous with those of the atrial syncytium.

The cells of the SA node can reach threshold on their own, and their cell membranes contact one another. Without stimulation from nerve fibers or any other outside agents, the nodal cells initiate cardiac impulses that spread into the surrounding myocardium and stimulate cardiac muscle cells to contract.

SA node activity is rhythmic. Because it generates the heart's rhythmic contractions, the SA node is also called the **pacemaker.**

Figure 13.10 traces the path of a cardiac impulse. As a cardiac impulse from the SA node reaches the atrial syncytium, the right and left atria begin to contract almost simultaneously. The cardiac impulse is not conducted directly into the ventricular syncytium, which is separated from the atrial syncytium by the fibrous skeleton of the heart. Instead, the cardiac impulse passes along specialized cardiac muscle cells, called junctional fibers, of the cardiac conduction system. The junctional fibers lead to a mass of specialized cardiac muscle tissue called the AV node, or **atrioventricular node.** This node is in the inferior part of the septum that separates the atria (interatrial septum) and lies just beneath the endocardium. It provides the only normal conduction pathway between the atrial and ventricular syncytia.

The junctional fibers that conduct the cardiac impulse into the AV node have small diameters, and because small fibers conduct impulses slowly, they delay conduction of the impulse. The impulse is delayed further as it moves through the AV node. This allows time for the atria to contract

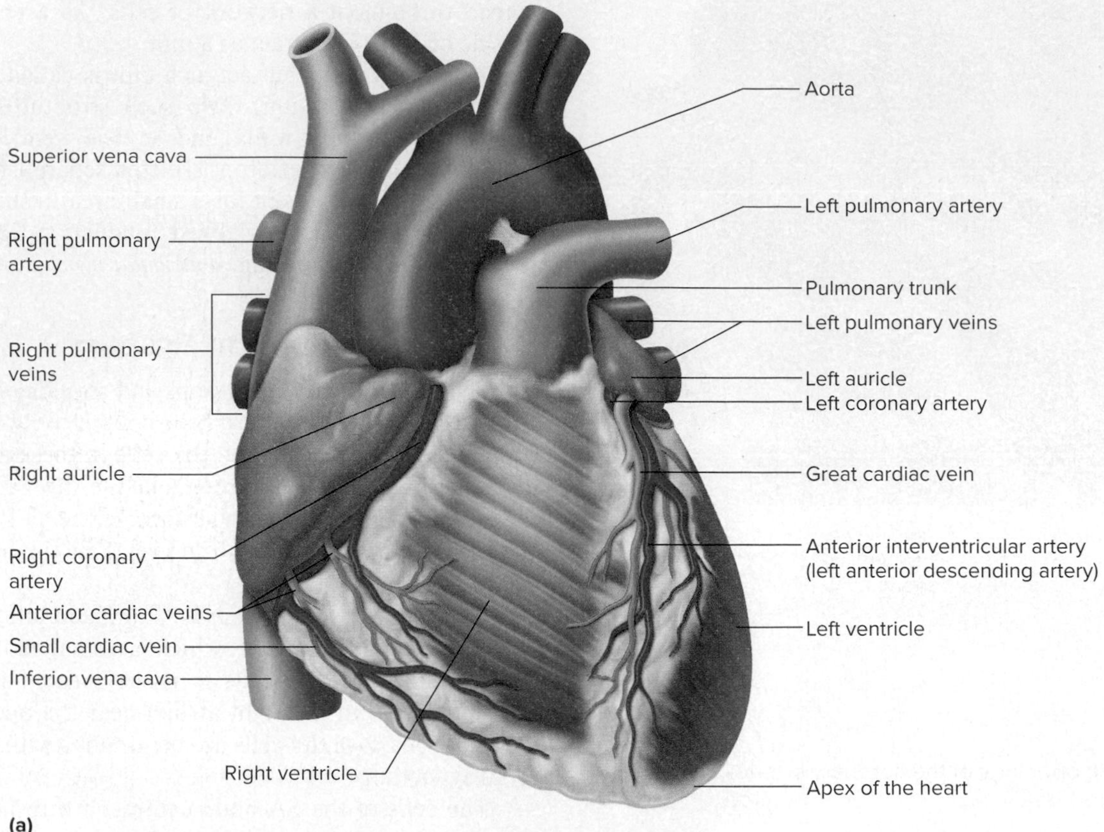

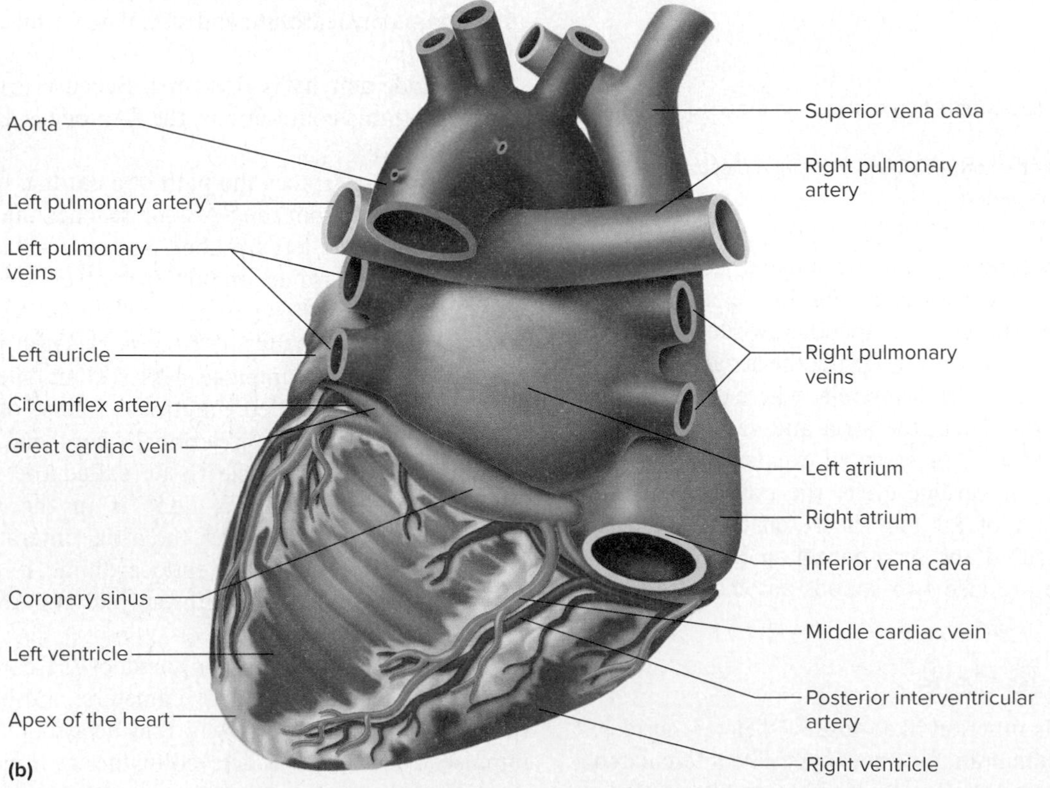

Figure 13.8 Blood vessels associated with the surface of the heart. **(a)** Anterior view. **(b)** Posterior view. APR

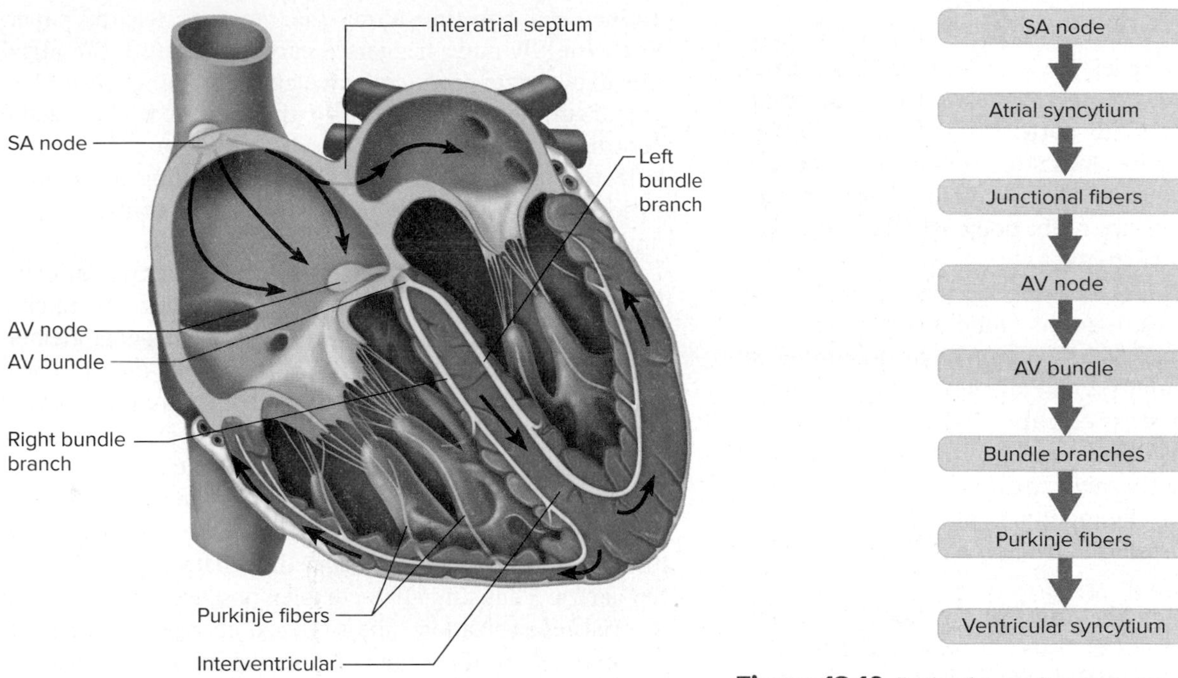

Figure 13.9 Components of the cardiac conduction system. APR

Figure 13.10 Path of a cardiac impulse.

completely so they empty most of their blood into the ventricles before the ventricles contract.

Once the cardiac impulse reaches the distal side of the AV node, it passes into a group of large conduction fibers that make up the **AV bundle** (bundle of His). The AV bundle enters the upper part of the interventricular septum and divides into right and left **bundle branches** that lie just beneath the endocardium. About halfway down the septum, the branches give rise to enlarged **Purkinje fibers** (pur-kin'jē fī'berz). The AV bundle and Purkinje fibers conduct the impulse rapidly.

Purkinje fibers spread from the interventricular septum into the papillary muscles, which project inward from ventricular walls. Then, the Purkinje fibers extend downward to the apex of the heart. There they curve around the tips of the ventricles and pass upward over the lateral walls of these chambers. Along the way, they branch from the interventricular septum and the ventricular walls into the papillary muscles, which project inward from the ventricular walls. The Purkinje fibers also give off many small branches which become continuous with cardiac muscle fibers.

The muscle fibers in ventricular walls form irregular whorls (spiral patterns). When impulses on the Purkinje fibers stimulate these muscle fibers, the ventricular walls contract with a twisting motion (fig. 13.11). This action squeezes blood out of the ventricular chambers and forces it into the aorta and pulmonary trunk.

PRACTICE 13.3

1. What series of events constitutes a single complete heartbeat?
2. What is a functional syncytium?
3. What types of tissues make up the cardiac conduction system?
4. How is a cardiac impulse initiated?
5. How is a cardiac impulse conducted from the right atrium to the other heart chambers?

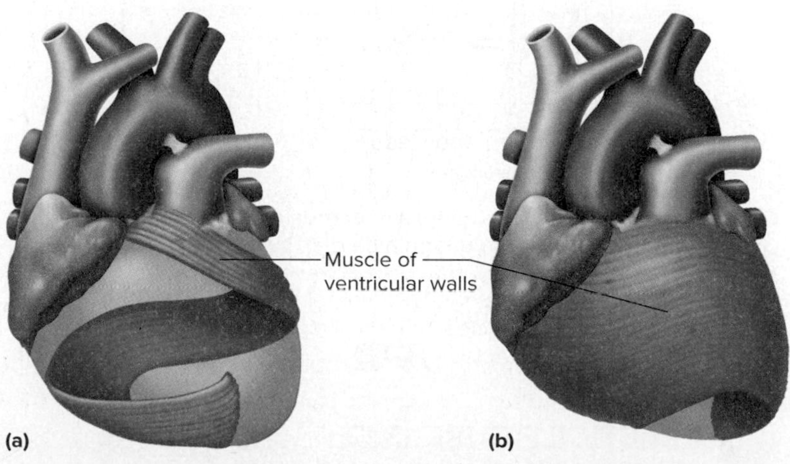

Figure 13.11 The muscle cells within the ventricular walls interconnect to form whorled networks. The networks of groups **(a)** and **(b)** surround both ventricles in these anterior views of the heart. APR

Electrocardiogram

An **electrocardiogram** (e-lek″trō-kar′dē-o-gram″), or **ECG,** is a recording of the electrical changes in the myocardium during a cardiac cycle. (This pattern is generated as action potentials stimulate cardiac muscle cells to contract, but it is not the same as individual action potentials.) These changes are detectable on the surface of the body because body fluids can conduct electrical currents.

To record an ECG, electrodes are placed on the skin and connected by wires to an instrument that responds to small electrical changes. These changes are recorded on an electronic device and may be displayed on a screen or printed on a moving strip of paper. Up-and-down movements, or deflections from the baseline, correspond to electrical changes in the myocardium. The paper moves at a known rate, so the distance between deflections indicates the time elapsing between phases of the cardiac cycle.

As figure 13.12*a* illustrates, a normal ECG pattern includes several deflections, or *waves,* during each cardiac cycle. Between cycles, the cardiac muscle cells remain polarized, with no detectable electrical changes, producing a baseline along the moving strip of paper. When the SA node triggers a cardiac impulse, the atrial cells depolarize, producing an electrical change. A deflection occurs, and at the end of the electrical change the recording returns to the baseline position. This first deflection produces a *P wave,* corresponding to depolarization of the atrial cells that will lead to contraction of the atria (fig. 13.12*b*).

When the cardiac impulse reaches the ventricular cells, they rapidly depolarize. The ventricular walls are thicker than those of the atria, so the electrical change is greater, and a greater deflection is recorded. When the electrical change ends, the recording returns to the baseline, leaving a pattern called the *QRS complex* (fig. 13.12*b*). The complex corresponds to depolarization of the ventricles just prior to the contraction of the ventricular walls. The record of atrial repolarization seems to be missing from the pattern because atrial fibers repolarize at the same time that ventricular fibers depolarize. Thus, the QRS complex obscures the recording of atrial repolarization. The electrical changes occurring as the ventricular muscle fibers repolarize produce a *T wave* as a deflection occurs again. This ends the ECG pattern for a given cardiac cycle (fig. 13.12*b*).

Physicians use ECG patterns to assess the heart's ability to conduct impulses. For example, the period between the beginning of a P wave and the beginning of a QRS complex, called the *PQ interval* (or if a Q wave is not visible, the PR interval), indicates the time for the cardiac impulse to travel from the SA node through the AV node. Ischemia or other problems affecting the fibers of the AV conduction pathways can prolong this PQ interval. Similarly, injury to the AV bundle can extend the QRS complex, because it may take longer for an impulse to spread throughout the ventricular walls (fig. 13.13).

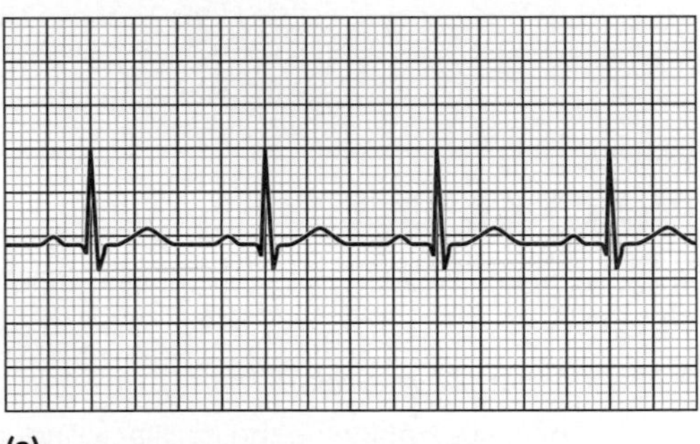

(a)

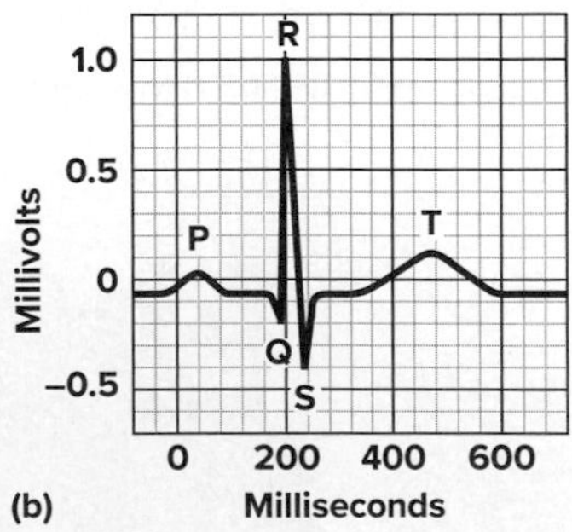

(b)

Figure 13.12 An electrocardiogram records electrical changes in the myocardium during a cardiac cycle. **(a)** A normal ECG. **(b)** In an ECG pattern, the P wave results from a depolarization of the atria, the QRS complex results from a depolarization of the ventricles, and the T wave results from a repolarization of the ventricles. APR

PRACTICE FIGURE 13.12

Which two electrical events occur during the QRS complex?

Answer can be found in Appendix E.

PRACTICE 13.3

6. What is an electrocardiogram?
7. Which cardiac events do the P wave, QRS complex, and T wave represent?

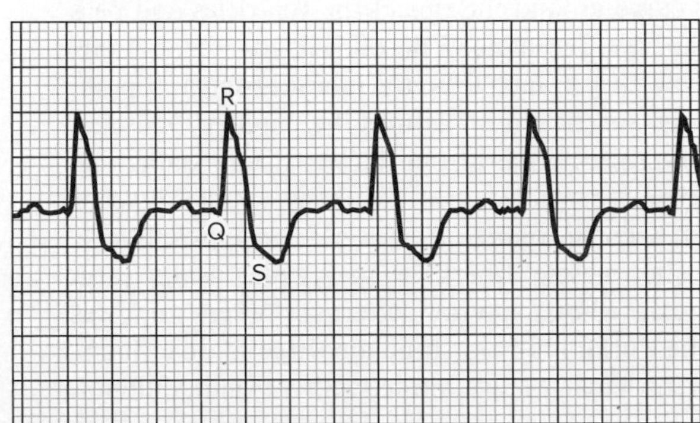

Figure 13.13 A prolonged QRS complex may result from damage to the AV bundle fibers.

Heart Sounds

A heartbeat heard through a stethoscope sounds like *lubb-dupp.* These sounds are due to vibrations in the heart tissues associated with the valves closing.

The first part of a heart sound (*lubb*) originates during ventricular systole, when the AV valves close. The second part (*dupp*) occurs during ventricular diastole, when the pulmonary and aortic valves close.

Heart sounds provide information about the condition of the heart valves. For example, inflammation of the endocardium (endocarditis) may erode the edges of the valvular cusps. As a result, the cusps may not close completely and some blood may leak back through the valve, producing an abnormal sound called a *murmur.* The seriousness of a murmur depends on the degree of valvular damage. Many heart murmurs are harmless. Open heart surgery may be needed to repair or replace severely damaged valves.

Pressure and Volume Changes During a Cardiac Cycle

During a cardiac cycle, the pressure in the heart chambers rises and falls (see fig. 13.14). These pressure changes open and close the valves, much like doors being blown open or closed by the wind. Early in ventricular diastole, the ventricular pressure is lower than the atrial pressure and the AV valves open. The ventricles fill. About 70% of the returning blood enters the ventricles prior to atrial contraction, and ventricular pressure gradually increases. Note that this occurs during the T wave. During atrial systole, the remaining 30% of returning blood is pushed into the ventricles, and ventricular pressure increases (fig. 13.15*a*). Then, as the ventricles contract, ventricular pressure rises sharply. As soon as the ventricular pressure exceeds the atrial pressure, the AV valves close. At the same time, the papillary muscles contract. By pulling on the chordae tendineae, they prevent the cusps of the AV valves from bulging too far into the atria.

During ventricular systole, the AV valves remain closed. The atria are now relaxed, and pressure in the atria is low, even lower than venous pressure. As a result, blood flows into the atria from the large, attached veins. That is, as the ventricles are contracting, the atria are filling, already preparing for the next cardiac cycle (fig. 13.15*b*).

As ventricular systole progresses, ventricular pressure continues to increase until it exceeds the pressure in the pulmonary trunk (right side) and aorta (left side). At this point, the pressure differences across the semilunar valves open the pulmonary and aortic valves, and blood is ejected from each valve's respective ventricle into these arteries. This also occurs during the T wave.

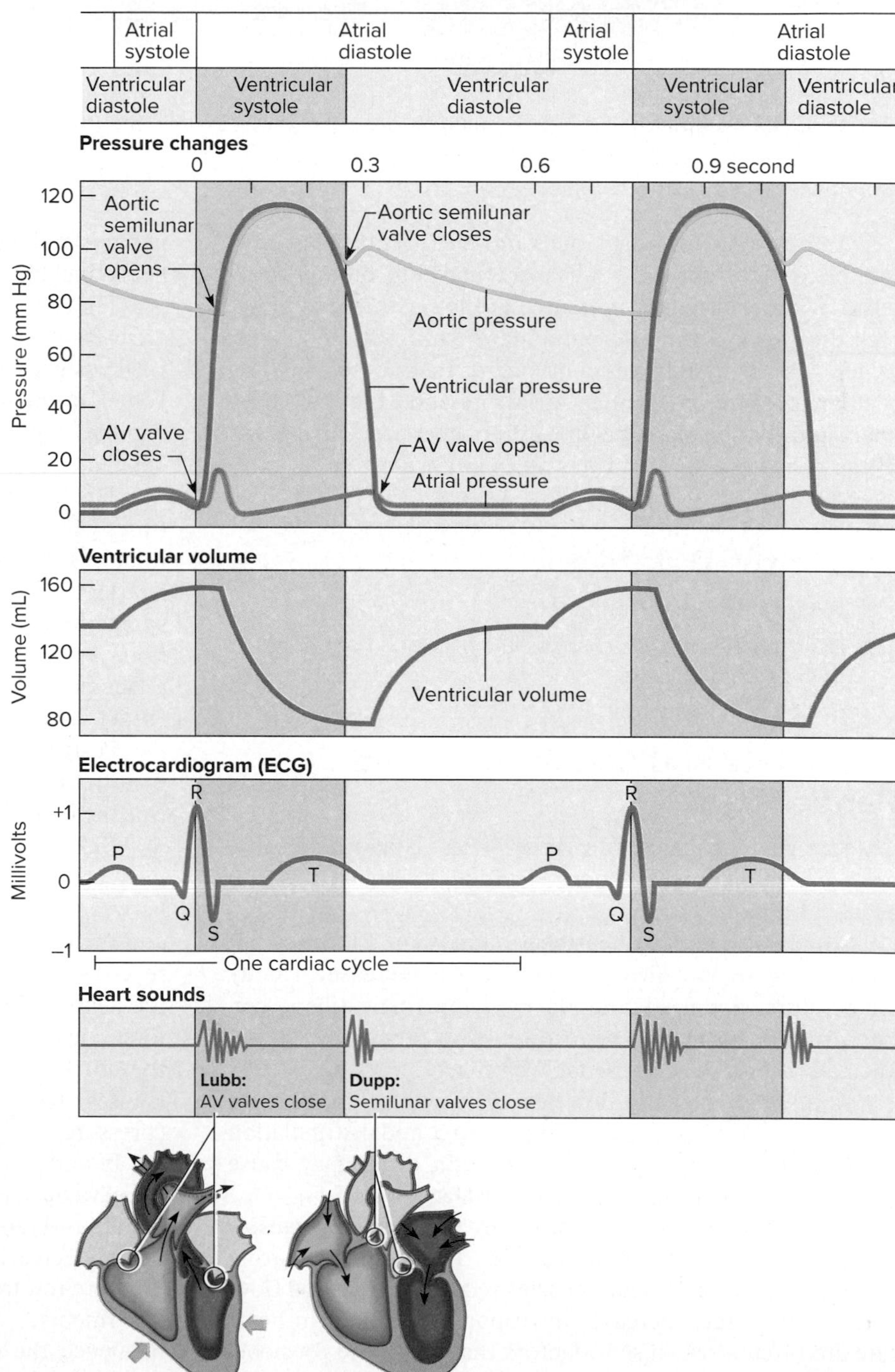

Figure 13.14 The graph depicts changes that occur during the cardiac cycle, with parts of the ECG pattern and corresponding heart sounds.

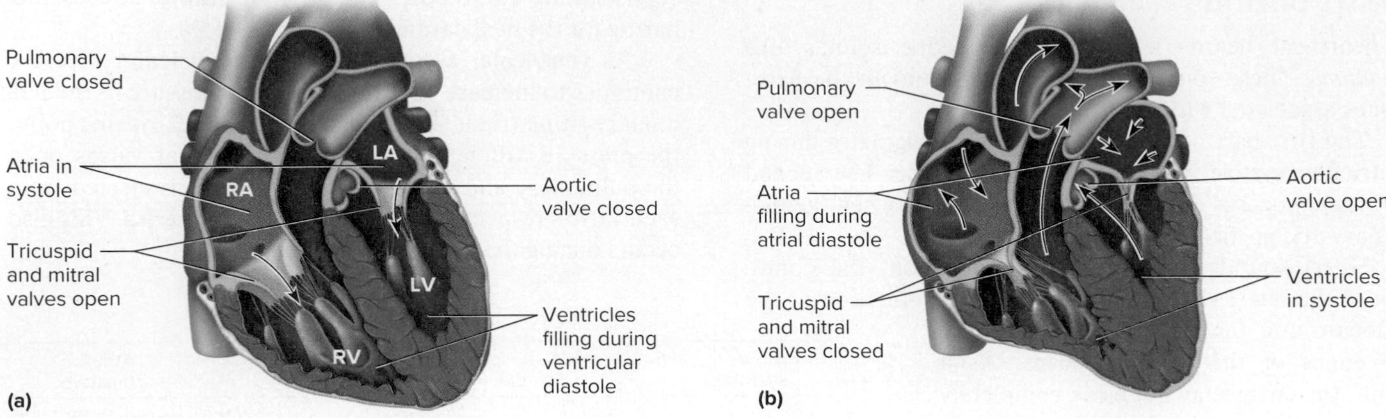

Figure 13.15 A cardiac cycle. The atria **(a)** empty during atrial systole and **(b)** fill with blood during atrial diastole. APR

As blood flows out of the ventricles, ventricular pressure begins to drop, and it falls even further as the ventricles relax. When ventricular pressures are lower than the blood pressure in the aorta and pulmonary trunk, the semilunar valves close. The ventricles continue to relax. As soon as ventricular pressure is less than atrial pressure, the AV valves open, and the ventricles begin to fill once more. During this filling phase, the atria and ventricles are in diastole.

PRACTICE 13.3

8. What causes heart sounds?
9. Describe the pressure changes in the atria and ventricles during a cardiac cycle.

Heart Rate and Regulation of the Cardiac Cycle

An average adult heart rate while at rest is about 70 to 75 beats per minute, with a normal range of about 60 to 100 beats per minute.

An abnormally fast heartbeat, more than 100 beats per minute at rest, is called *tachycardia.* An increase in body temperature, nodal stimulation by sympathetic fibers, certain drugs or hormones, heart disease, excitement, exercise, anemia, or shock can cause tachycardia.

Bradycardia means a slow heart rate, fewer than 60 beats per minute. Decreased body temperature, nodal stimulation by parasympathetic impulses, or certain drugs may cause bradycardia. It may also occur during sleep.

The volume of blood pumped by the heart changes to accommodate cellular requirements. For example, during strenuous exercise, skeletal muscles require more blood flow, and the heart rate increases in response. Changes in heart rate are often a response to factors that affect the SA node, such as the motor impulses conducted on the parasympathetic and sympathetic nerve fibers (see section 9.16, Autonomic Nervous System). Both divisions of the autonomic nervous system innervate various structures throughout the heart, including the SA node, AV node, and myocardium.

The parasympathetic fibers that innervate the heart arise from neurons in the medulla oblongata and synapse with postganglionic fibers in the wall of the heart (fig. 13.16). When stimulated, the postganglionic fibers release acetylcholine, which decreases SA nodal activity. As a result, the heart rate decreases.

The SA node on its own would send impulses about 100 times per minute. However, parasympathetic fibers conduct impulses continually to the SA node, "braking" to a resting heart rate of 60 to 80 beats per minute in an adult. Parasympathetic activity can change heart rate in either direction. An increase in the impulses slows the heart rate, and a decrease in the impulses releases the parasympathetic "brake" and increases the heart rate.

Postganglionic sympathetic neurons respond to stimulation by secreting the neurotransmitter norepinephrine. Norepinephrine increases the rate and force of myocardial contractions.

Reflexes called *baroreceptor reflexes* involving the *cardiac center* of the medulla oblongata maintain balance between the inhibitory effects of parasympathetic fibers and the excitatory effects of sympathetic fibers. This center receives sensory information and relays motor impulses to the heart and blood vessels in response. For example, receptors sensitive to stretch are located in certain regions of the aorta (aortic arch) and in the carotid arteries (carotid sinuses) (fig. 13.16). These receptors, called *baroreceptors* (pressoreceptors), can detect changes in blood pressure. Rising blood pressure stretches the receptors, and they signal the cardiac center in the medulla oblongata. In response, the medulla oblongata sends parasympathetic impulses to the heart, decreasing heart rate. This action helps lower blood pressure toward normal.

Impulses from the cerebrum or hypothalamus also influence the cardiac control center. These impulses may decrease heart rate, as occurs when a person faints following an emotional upset, or they may increase heart rate during a period of anxiety.

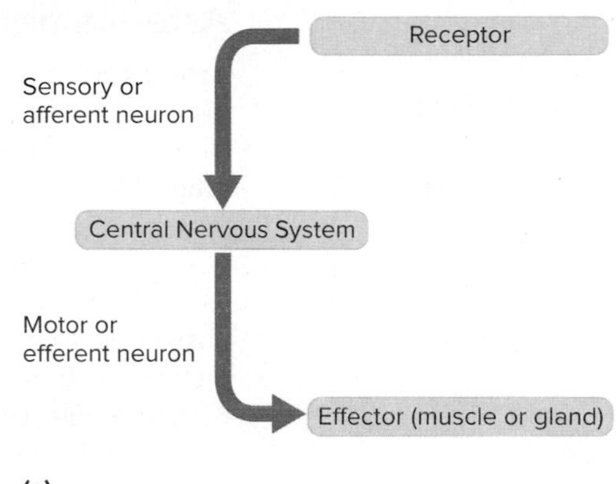

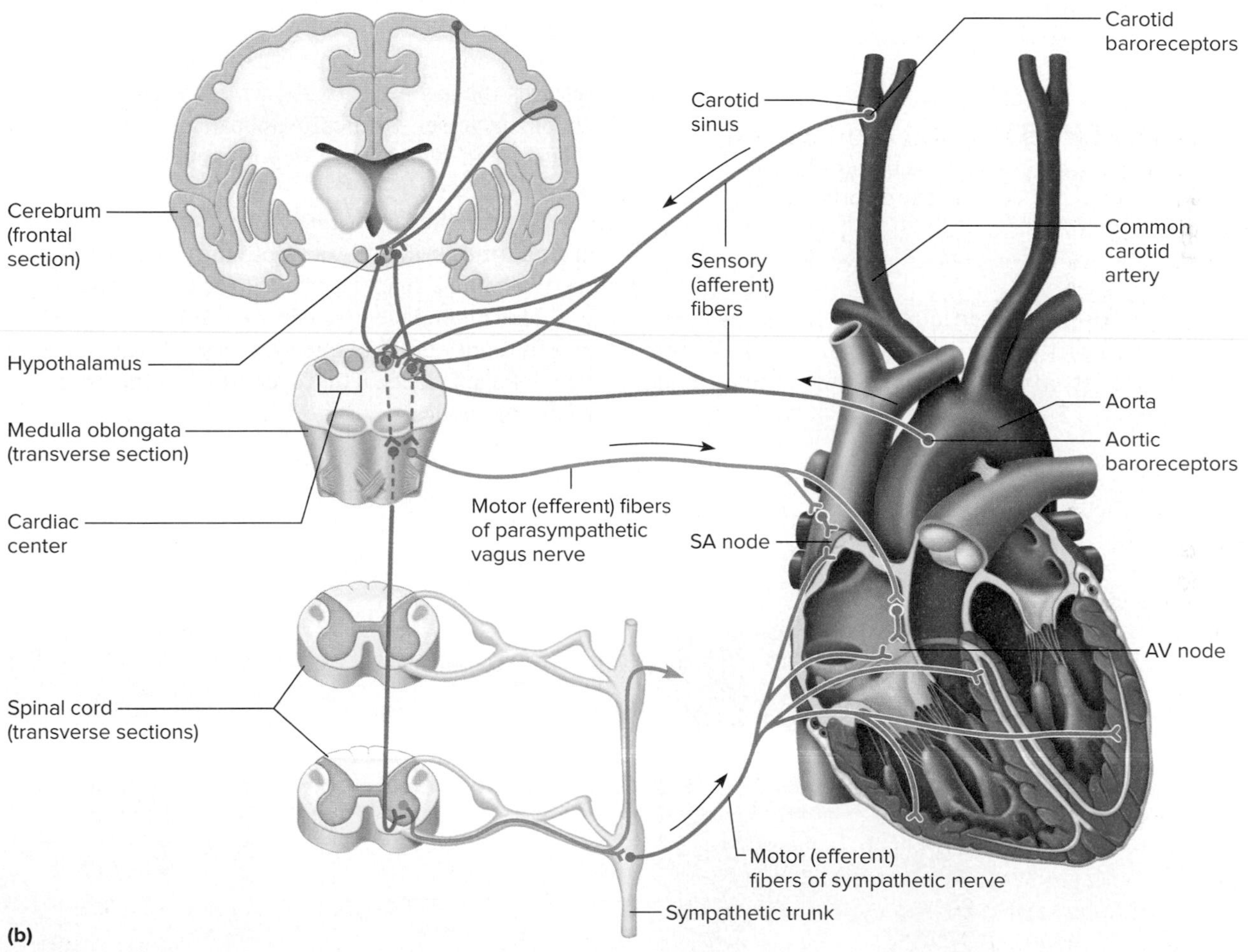

Figure 13.16 Baroreceptor reflex. **(a)** Schematic of a general reflex arc. Note the similarity to figure 1.5. **(b)** Autonomic impulses alter the activities of the SA and AV nodes. APR

Two other factors that influence heart rate are temperature change and certain ions. Rising body temperature increases heart action, which is why heart rate usually increases during fever. Abnormally low body temperature decreases heart action.

The most important ions that influence heart action are potassium (K^+) and calcium (Ca^{+2}). An excess of potassium ions in the blood (*hyperkalemia*) decreases the rate and force of myocardial contractions. If the potassium ion concentration in the blood drops below normal (*hypokalemia*), the heart may develop a potentially life-threatening abnormal rhythm (arrhythmia).

An excess of calcium ions in the blood (*hypercalcemia*) increases heart action, which can result in dangerously extended heart contractions. Conversely, a low blood calcium concentration (*hypocalcemia*) depresses heart action.

PRACTICE 13.3

10. How do parasympathetic and sympathetic impulses help control heart rate?
11. How do changes in body temperature affect heart rate?
12. Describe the effects on the heart of abnormal concentrations of potassium and calcium ions.

13.4 | Blood Vessels

LEARN

1. Compare the structures and functions of the major types of blood vessels.
2. Describe how substances are exchanged between blood in the capillaries and the tissue fluid surrounding body cells.

OF INTEREST The 62,000 or so miles of blood vessels in an average human body would wrap two and a half times around Earth if extended end-to-end.

The blood vessels form a closed circuit of tubes that carries blood from the heart to the body cells and back. These vessels include arteries, arterioles, capillaries, venules, and veins.

Arteries and Arterioles

Arteries are strong, elastic vessels adapted for transporting blood away from the heart under relatively high pressure. These vessels subdivide into progressively thinner tubes and eventually give rise to finer, branched **arterioles** (ar-te′rē-ōlz).

The wall of an artery consists of three distinct layers (fig. 13.17). The innermost layer (*tunica interna*) is composed of a layer of simple squamous epithelium, called **endothelium,** that rests on a connective tissue membrane that is rich in elastic and collagen fibers (see fig. 13.17*a*). Endothelium helps prevent blood clotting by providing a smooth surface that allows blood cells and platelets to flow through the vessel without being damaged and by secreting biochemicals that inhibit platelet aggregation. Endothelium also may help regulate local blood flow by secreting substances that dilate or constrict blood vessels. For example, endothelium releases the gas nitric oxide, which relaxes the smooth muscle of the vessel. Clinical Application 13.2 describes atherosclerosis, in which fatty deposits accumulate on the inner walls of arteries.

The middle layer (*tunica media*) makes up the bulk of the arterial wall. It includes smooth muscle cells, which encircle the tube, and a thick layer of elastic connective tissue.

The outer layer (*tunica externa*) is relatively thin and chiefly consists of connective tissue with irregular elastic and collagen fibers. This layer attaches the artery to the surrounding tissues.

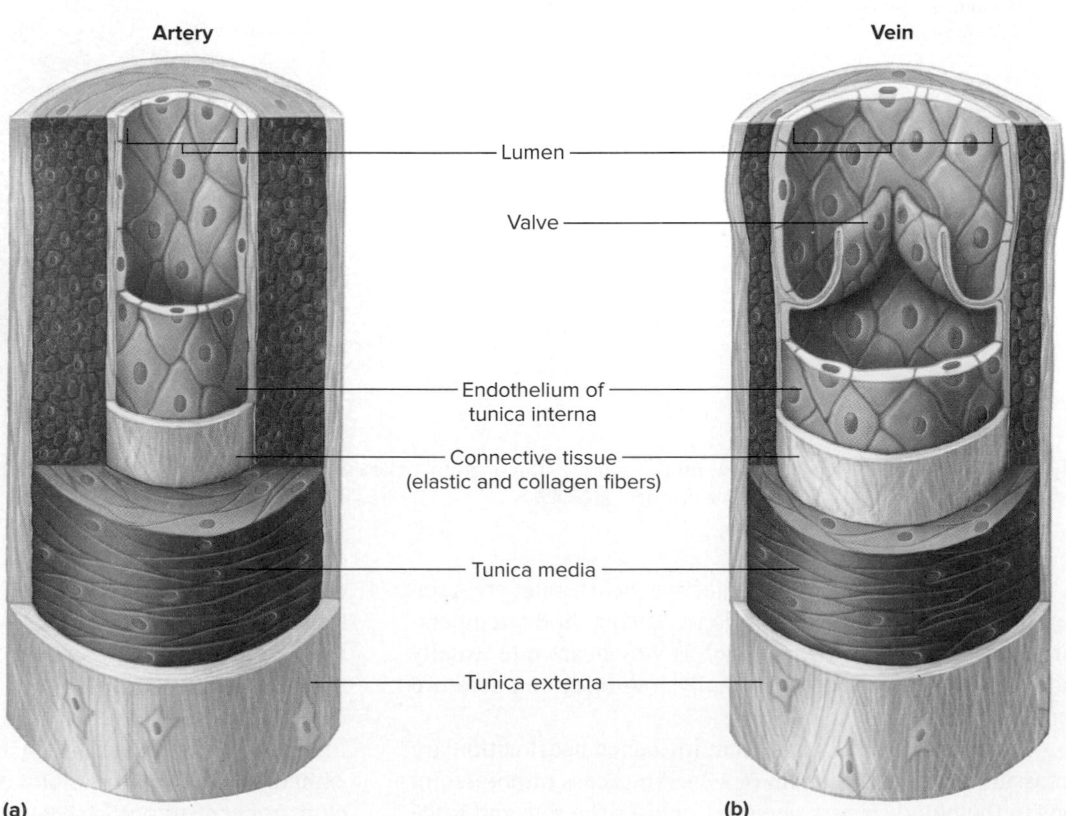

Figure 13.17 Blood vessels. **(a)** The wall of an artery. **(b)** The wall of a vein. APR

The sympathetic branches of the autonomic nervous system innervate smooth muscle in artery and arteriole walls. *Vasomotor fibers* stimulate the smooth muscle to contract, reducing the diameter of the vessel, lessening blood flow. This action is called **vasoconstriction** (vas″ō-kon-strik′shun). If vasomotor impulses are inhibited, the smooth muscle cells relax, and the diameter of the vessel increases, allowing greater blood flow. This response is called **vasodilation** (vās″ō-dī-lā′shun). Changes in the diameters of arteries and arterioles greatly influence blood flow and blood pressure (as described in section 13.5, Blood Pressure).

The walls of the larger arterioles have three layers, similar to those of arteries. These walls thin as the arterioles approach capillaries. The wall of a very small arteriole consists only of an endothelial lining and some smooth muscle cells, surrounded by a small amount of connective tissue (fig. 13.18).

PRACTICE 13.4

1. Describe the wall of an artery.
2. What is the function of smooth muscle in the arterial wall?
3. How is the structure of an arteriole different from that of an artery?

Capillaries

Capillaries (kap′ĭ-lar″ēz), the smallest-diameter blood vessels, connect the smallest arterioles and the smallest venules. Capillaries are extensions of the inner linings of arterioles in that their walls are endothelium (fig. 13.18). The thin walls of capillaries form the semipermeable layer through which substances in the blood are exchanged for substances in the tissue fluid surrounding body cells.

The openings in capillary walls are thin slits between the endothelial cells (fig. 13.19). The sizes of these openings and, consequently, the permeability of the capillary wall vary from tissue to tissue. For example, the openings are relatively smaller in capillaries of smooth, skeletal, and cardiac muscle than they are in capillaries associated with endocrine glands, the kidneys, and the lining of the small intestine.

Capillary density reflects tissues' rates of metabolism. Muscle and nervous tissues, which use abundant oxygen and nutrients, are richly supplied with capillaries. Tissues with slower metabolic rates have fewer capillaries, as in some cartilage, or lack them entirely, as in the cornea.

The spatial patterns of capillaries also differ in various body parts. For example, some capillaries pass directly from arterioles to venules, but others lead to highly branched networks (fig. 13.20).

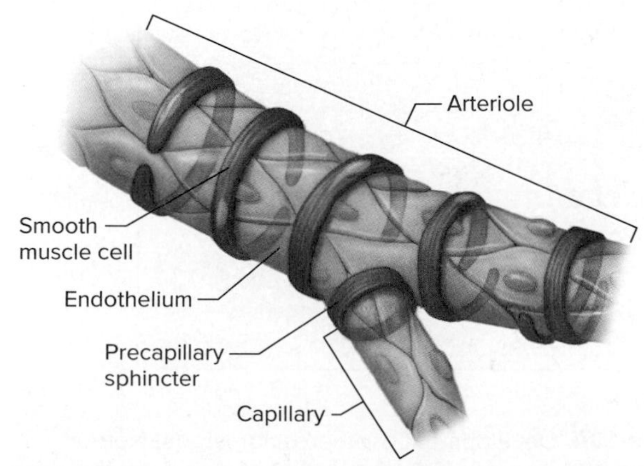

Figure 13.18 The smallest arterioles have only a few smooth muscle cells in their walls. Capillaries lack these smooth muscle cells.

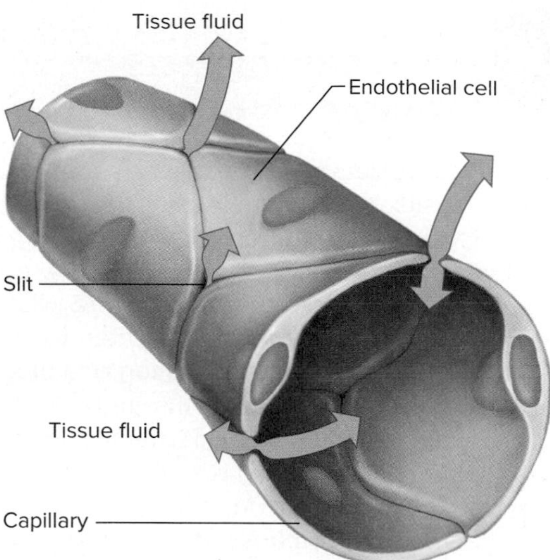

Figure 13.19 In capillaries, substances are exchanged by diffusion between the blood and tissue fluid through openings (slits) separating endothelial cells.

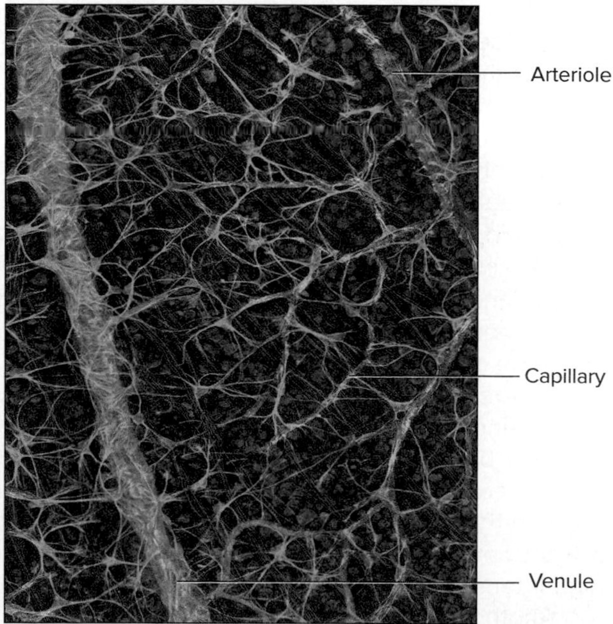

Figure 13.20 Falsely colored light micrograph of a capillary network. Thomas Deerinck, NCMIR/Science Source

CLINICAL APPLICATION 13.2

Atherosclerosis

In the arterial disease *atherosclerosis* (ath″er-ō-sklĕ-rō′sis), deposits of fatty materials, particularly cholesterol, form within and on the inner lining of the arterial walls. Such deposits, called *plaque,* protrude into the lumens of vessels and interfere with blood flow (fig. 13A). Furthermore, plaque often forms a surface texture that can initiate formation of a blood clot, increasing the risk of developing thrombi or emboli that result in inadequate blood flow (ischemia) downstream from the obstruction causing tissue death (necrosis).

Plaque accumulation in coronary arteries may cause a *coronary thrombosis* or *coronary embolism.* Similar changes in brain arteries increase the risk of a *cerebral vascular accident (CVA),* also called a *stroke,* due to cerebral thrombosis, embolism, or hemorrhage. In addition, the walls of affected arteries may degenerate, losing their elasticity and becoming hardened, or *sclerotic.* In this stage of the disease, a sclerotic vessel may rupture under the force of blood pressure.

Risk factors for developing atherosclerosis include a fatty diet, elevated blood pressure, tobacco smoking, obesity, and lack of physical exercise. Genetic factors may also increase the risk of developing atherosclerosis.

For millions of people who cannot control cholesterol with diet and exercise, drugs called statins can inhibit a liver enzyme that the body uses to synthesize cholesterol. Statins dramatically reduce LDL cholesterol and regulate triglyceride levels, lowering the risk of atherosclerosis.

Several invasive treatments attempt to clear clogged arteries. In *percutaneous transluminal angioplasty,* a thin, plastic catheter with a tiny deflated balloon at its tip is passed through an incision in the skin into a large artery in the arm or thigh, guided through other arteries and into the lumen of the affected blood vessel. Once in position at the blockage, the balloon is inflated for several minutes. The inflated balloon compresses the plaque against the arterial wall, widening the arterial lumen and restoring blood flow. In some cases, the catheter also introduces a stent, which is a coiled steel tube that widens as the balloon inflates, helping to keep the lumen open. However, blockage can recur if the underlying cause of cholesterol buildup is not addressed.

Lasers are also used to destroy atherosclerotic plaque and to channel through arterial obstructions to increase blood flow. In *laser angioplasty,* the laser is introduced through a catheter inserted into an incision made in the skin. The catheter is then advanced until it is in the lumen of an obstructed artery, where it delivers high-intensity light energy through optical fibers.

Another invasive procedure for treating arterial obstruction is *bypass graft surgery.* A surgeon uses a vein or an artery from another part of the patient's body to detour around the obstruction. The graft bypasses the narrowed region of the affected artery, supplying blood to the tissues downstream. If a vein is used, the vein is connected backward so that its valves do not impede blood flow. Another bypass graft technique connects the distal end of the left internal mammary artery (LIMA) to the coronary artery beyond the obstruction.

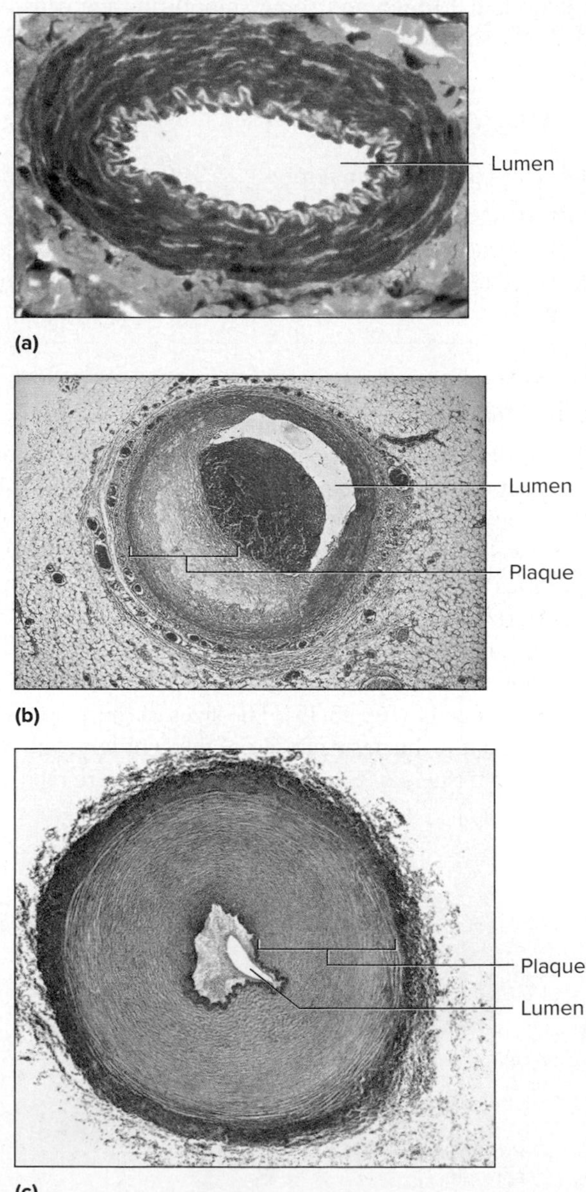

Figure 13A Development of atherosclerosis. **(a)** Normal arteriole (100x). **(b** and **c)** Accumulation of plaque on the inner wall of an arteriole (100x). (a): Biophoto Associates/Science Source; (b): Image Source/Getty Images; (c): Alfred Pasieka/Getty Images

Smooth muscle that encircles the capillary where it branches off from an arteriole regulates blood distribution in capillary pathways. These *precapillary sphincters* may close a capillary by contracting or open it by relaxing (see fig. 13.18). Both precapillary sphincters and arterioles respond to the demands of the cells that they supply. When these cells have low concentrations of oxygen and nutrients, the sphincter and arteriole relax and flow increases; when cellular requirements have been met, the sphincter and arteriole may contract again, decreasing flow to previous levels. In this way, blood flow can follow different pathways through a tissue to meet the changing cellular requirements.

Routing of blood flow to different parts of the body is due to vasoconstriction and vasodilation of arterioles and precapillary sphincters. During exercise, for example, blood enters the capillary networks of the skeletal muscles, where the cells have increased oxygen and nutrient requirements. At the same time, blood can bypass some of the capillary networks in the digestive tract tissues, where demand for blood is less immediate.

PRACTICE 13.4

4. Describe a capillary wall.
5. What is the function of a capillary?
6. What controls blood flow into capillaries?

Exchanges in Capillaries

Gases, nutrients, and metabolic by-products are exchanged between the blood in capillaries and the tissue fluid surrounding body cells. The substances exchanged move through capillary walls by diffusion, filtration, and osmosis (see section 3.3, Movements Into and Out of the Cell).

Because blood entering systemic capillaries carries high concentrations of oxygen and nutrients, these substances diffuse through the capillary walls and enter the tissue fluid. Conversely, the concentrations of carbon dioxide and other metabolic by-products are generally greater in the tissues, and such wastes diffuse into the capillary blood.

Plasma proteins generally remain in the blood because they are too large to diffuse through the membrane channels or the slitlike openings between the endothelial cells of most capillaries. Also, these proteins are not soluble in the lipid parts of capillary cell membranes.

Diffusion depends on concentration gradients, but filtration occurs as hydrostatic pressure pushes water and other small molecules through a membrane. In capillaries, the blood pressure generated when ventricle walls contract provides the force for filtration.

Blood pressure also moves blood through the lumen of the arteries and arterioles. This pressure decreases as the distance from the heart increases, because of friction (peripheral resistance) between the blood and the vessel walls. For this reason, blood pressure is greater in the arteries than in the arterioles, and greater in the arterioles than in the capillaries. Blood pressure is similarly greater at the arteriolar end of a capillary than at the venular end. Therefore, the filtration effect occurs primarily at the arteriolar ends of capillaries, whereas diffusion takes place along their entire lengths.

The presence of an impermeant solute on one side of a cell membrane creates an osmotic pressure. Plasma proteins trapped in the capillaries create an osmotic pressure that draws water into the capillaries. The term *colloid osmotic pressure* describes this osmotic effect due solely to the plasma proteins.

The effect of capillary blood pressure, which favors filtration, opposes the plasma colloid osmotic pressure, which favors reabsorption. At the arteriolar end of capillaries, the blood pressure is higher than the colloid osmotic pressure, so filtration predominates here. At the venular end, the colloid osmotic pressure is essentially unchanged, but the blood pressure has decreased due to peripheral resistance through the capillary, so reabsorption predominates (fig. 13.21).

Normally, more fluid leaves the capillaries than returns to them. Closed-ended vessels called lymphatic capillaries collect the excess tissue fluid and return it through lymphatic vessels to the venous circulation. Section 14.2, Lymphatic Pathways, discusses this mechanism.

Unusual events may increase blood flow to capillaries, sending excess fluid into the spaces between tissue cells. This may occur in response to certain chemicals, such as *histamine,* that vasodilate the arterioles near capillaries and increase capillary permeability. Enough fluid may leak out of the capillaries to overwhelm lymphatic drainage. Affected tissues become swollen (edematous) and painful.

PRACTICE 13.4

7. Which forces affect the exchange of substances between blood and tissue fluid?
8. Why is the fluid movement out of a capillary greater at its arteriolar end than at its venular end?

Venules and Veins

Venules (ven'ūlz) are the microscopic vessels that continue from the capillaries and merge to form **veins** (vānz). The veins, which transport blood back to the atria, follow pathways that roughly parallel those of the arteries.

The walls of veins are similar to those of arteries in that they are composed of three distinct layers (see fig. 13.17*b*). However, the middle layer of the venous wall is much thinner than that of the arterial wall. Consequently, veins have thinner walls that have less smooth muscle and less elastic connective tissue than those of comparable arteries. The lumens of veins have a greater diameter (fig. 13.22).

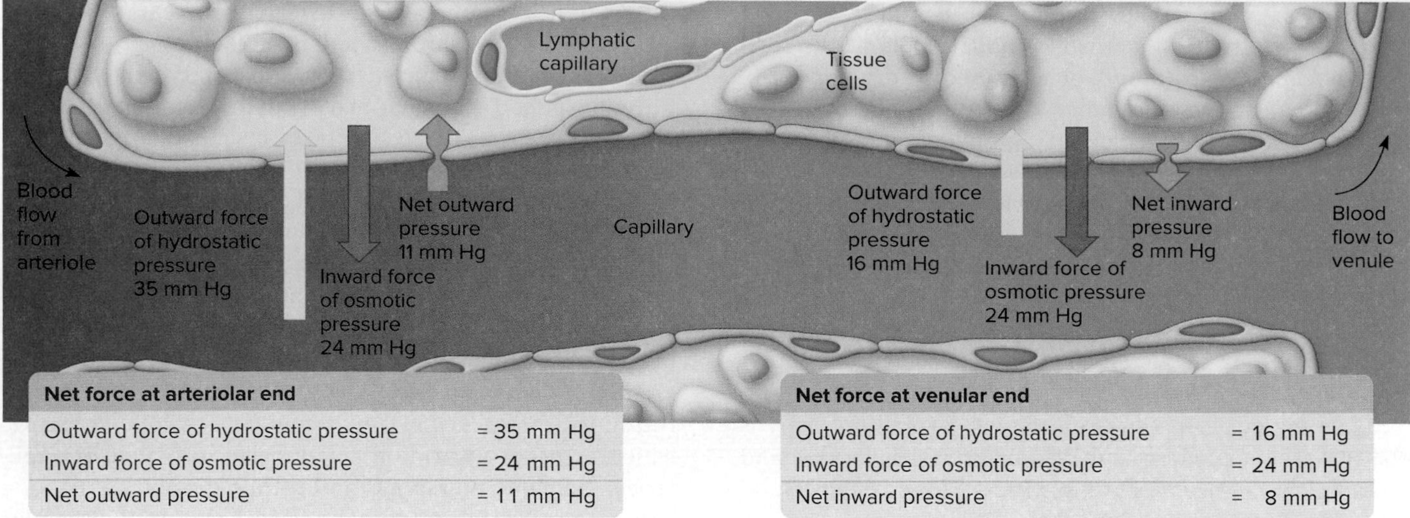

Net force at arteriolar end	
Outward force of hydrostatic pressure	= 35 mm Hg
Inward force of osmotic pressure	= 24 mm Hg
Net outward pressure	= 11 mm Hg

Net force at venular end	
Outward force of hydrostatic pressure	= 16 mm Hg
Inward force of osmotic pressure	= 24 mm Hg
Net inward pressure	= 8 mm Hg

Figure 13.21 Water and other substances leave capillaries because of a net outward pressure at the capillaries' arteriolar ends. Water enters at the capillaries' venular ends because of a net inward pressure. Substances move in and out along the length of the capillaries according to their respective concentration gradients. APR

PRACTICE FIGURE 13.21

Which substances do not leave the blood at the arteriolar end of the capillary and draw water by osmosis back into the capillary at the venular end of the capillary?

Answer can be found in Appendix E.

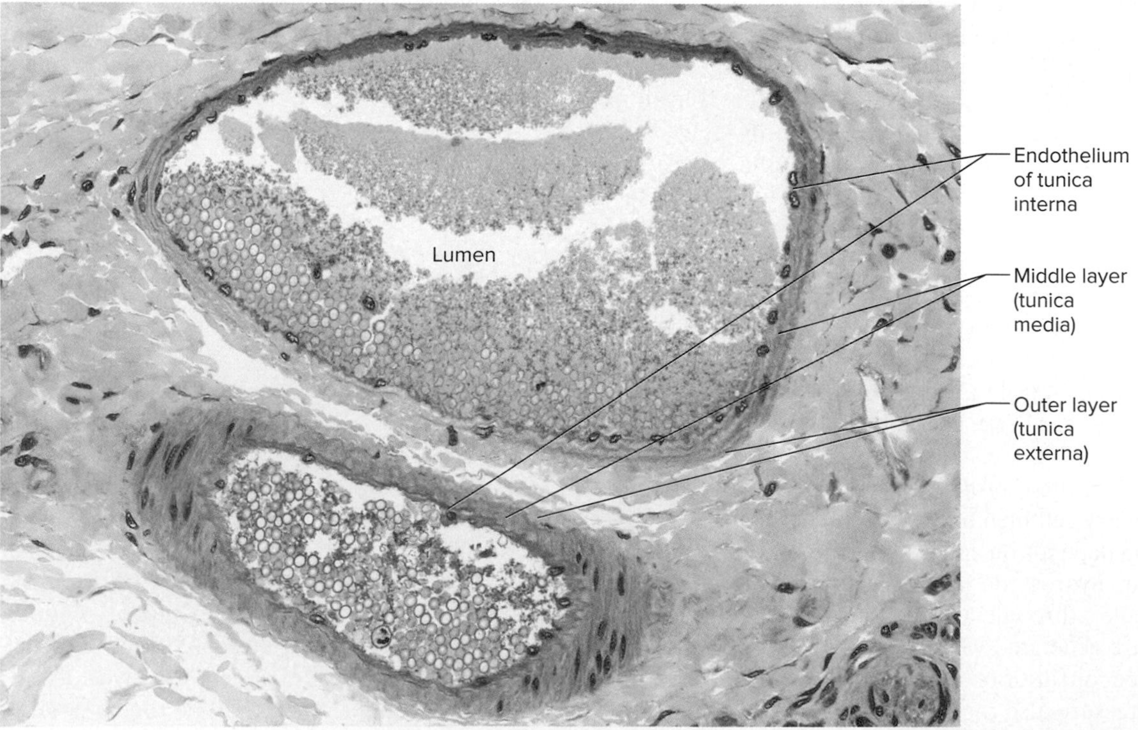

Figure 13.22 Note the structural differences in these cross sections of a venule (top) and an arteriole (bottom) (200x). 
Al Telser/McGraw-Hill Education

Many veins, particularly those in the upper and lower limbs, are buried in skeletal muscles and have *valves,* which project inward from their linings. Most valves are composed of two leaflets that close to prevent backflow of blood in a vein (fig. 13.23). The contraction of the muscles squeezes the vessels, pushing blood toward the heart. The valves also aid in returning blood to the heart because they open as long as the blood flow is toward the heart, but prevent flow in the opposite direction.

Veins also function as blood reservoirs. For example, in hemorrhage accompanied by a drop in arterial blood pressure, sympathetic impulses reflexively stimulate the muscular walls of the veins. The resulting constriction of the veins helps maintain blood pressure by returning more blood to be pumped by the heart. This mechanism helps ensure a nearly normal blood pressure even when as much as 25% of blood volume is lost. Table 13.2 summarizes the characteristics of blood vessels.

PRACTICE 13.4

9. How does the structure of a vein differ from that of an artery?
10. How does venous circulation help maintain blood pressure when hemorrhaging causes blood loss?

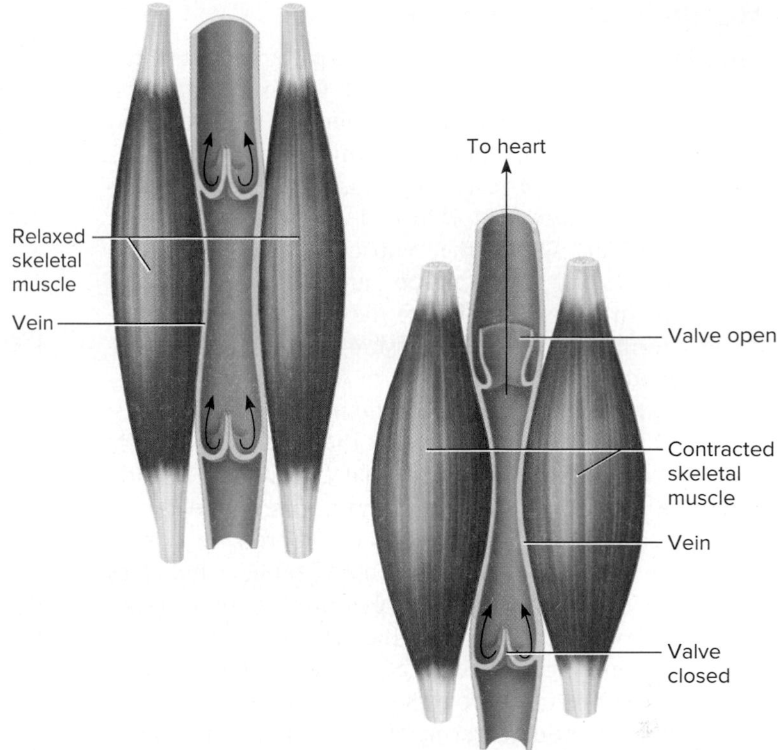

Figure 13.23 Muscular contractions and valves assist veins in returning blood back to the heart.

13.5 | Blood Pressure

LEARN

1. Explain how blood pressure is produced and controlled.
2. Describe the mechanisms that aid in returning venous blood to the heart.

Blood pressure is the force that blood exerts against the inner walls of the blood vessels. Although this force is present throughout the vascular system, the term *blood pressure* most commonly refers to pressure in the arteries supplied by branches of the aorta (systemic arteries). This is known as clinical blood pressure and is what is measured for medical applications.

OF **INTEREST** Contraction of the human heart can create enough pressure to squirt blood almost 10 feet into the air.

TABLE 13.2 Characteristics of Blood Vessels

Vessel	Type of Wall	Function
Artery	Thick, strong wall with three layers—an endothelial lining, a middle layer of smooth muscle and elastic connective tissue, and an outer layer of connective tissue	Transports blood under relatively high pressure from heart to arterioles
Arteriole	Thinner wall than an artery but with three layers; smaller arterioles have an endothelial lining, some smooth muscle tissue, and a small amount of connective tissue	Connects an artery to a capillary; helps control blood flow into a capillary by vasoconstricting or vasodilating
Capillary	Single layer of squamous epithelium	Allows nutrients, gases, and wastes to be exchanged between the blood and tissue fluid; connects an arteriole to a venule
Venule	Thinner wall than in an arteriole, less smooth muscle and elastic connective tissue	Connects a capillary to a vein
Vein	Thinner wall than an artery but with similar layers; the vein middle layer is much thinner; some veins have flaplike valves	Transports blood under relatively low pressure from a venule to the heart; valves prevent backflow of blood; serves as a blood reservoir

Arterial Blood Pressure

Arterial blood pressure rises and falls in a pattern corresponding to the phases of the cardiac cycle. That is, contraction of the ventricles (ventricular systole) squeezes blood out and into the pulmonary trunk and aorta, which sharply increases the pressures in these arteries. The maximum pressure during ventricular contraction is called the **systolic pressure** (sis-tol′ik presh′ur). When the ventricles relax (ventricular diastole), the arterial pressure drops, and the lowest pressure that remains in the arteries before the next ventricular contraction is termed the **diastolic pressure** (dī-a-stol′ik presh′ur). Clinical blood pressure is typically measured using the auscultatory method with a sphygmomanometer (blood pressure cuff) (fig. 13.24*a*). Wrapped around the arm, it is actually measuring the pressure in the brachial artery. This is fairly accurate because it is close to the aorta. It is recorded as a fraction, such as 127/74 (with normal blood pressure, at rest, being no greater than 120/80). The upper number indicates the arterial systolic pressure (SP) in mm of mercury (Hg), while the lower number indicates the arterial diastolic pressure (DP) in mm Hg. Figure 13.24*b* shows how these pressures decrease as distance from the left ventricle increases.

The surge of blood entering the arterial system during a ventricular contraction distends the elastic arterial walls, but the pressure begins to drop almost immediately as the contraction ends, and the arterial walls recoil. This alternate expanding and recoiling of the arterial wall can be felt as a **pulse** in an artery that runs close to the body surface. The radial artery is commonly used to take a person's pulse. Other sites where an arterial pulse is easily detected include the carotid, brachial, femoral, and dorsalis pedis arteries (see section 13.6, Arterial System).

PRACTICE 13.5

1. What is blood pressure?
2. Distinguish between systolic and diastolic blood pressure.
3. What causes a pulse in an artery?

Factors that Influence Arterial Blood Pressure

Arterial blood pressure depends on a variety of factors. These include cardiac output, blood volume, peripheral resistance, and blood viscosity (fig. 13.25).

Cardiac Output

In addition to producing blood pressure by forcing blood into the arteries, heart action determines how much blood enters the arterial system with each ventricular contraction. The volume of blood discharged from the ventricle with each contraction is called the **stroke volume** and equals about 70 milliliters in an average-weight male at rest. The volume discharged from the ventricle per minute is called the **cardiac output.** It is calculated by multiplying the stroke volume by the heart rate in beats per minute (cardiac output = stroke volume × heart rate). For example, if the stroke volume is 70 milliliters per beat and the heart rate is 72 beats per minute, the cardiac output is 5,040 milliliters per minute.

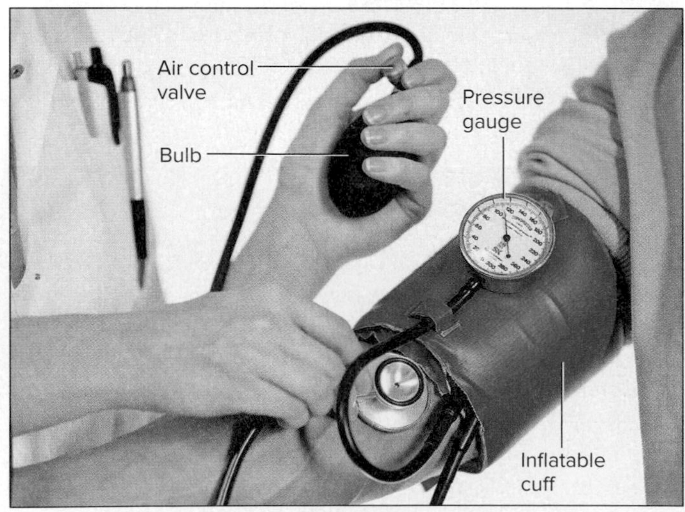

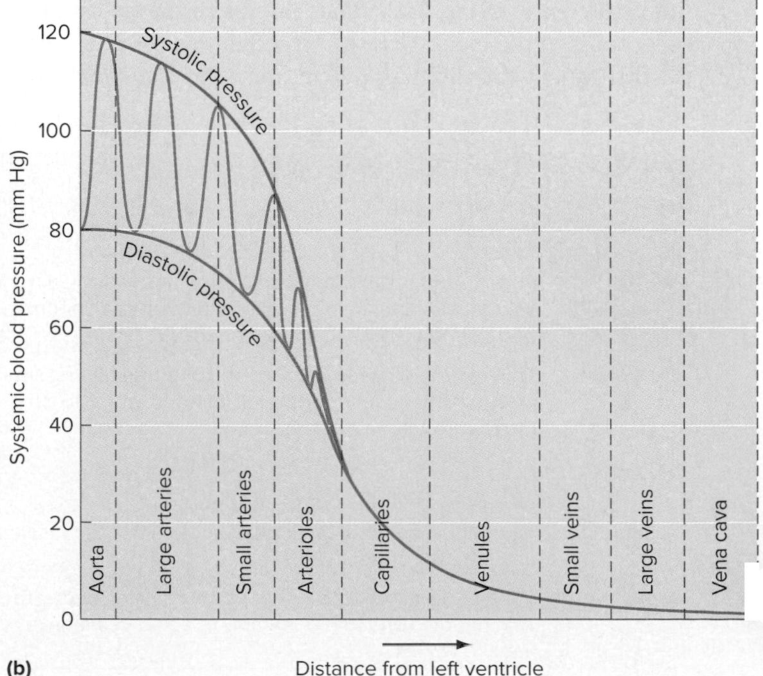

Figure 13.24 Measuring blood pressure. **(a)** A sphygmomanometer is used to measure blood pressure. **(b)** Blood pressure decreases as the distance from the left ventricle increases. Systolic pressure occurs during maximal ventricular contraction. Diastolic pressure occurs when the ventricles relax. (a): Johner/SuperStock

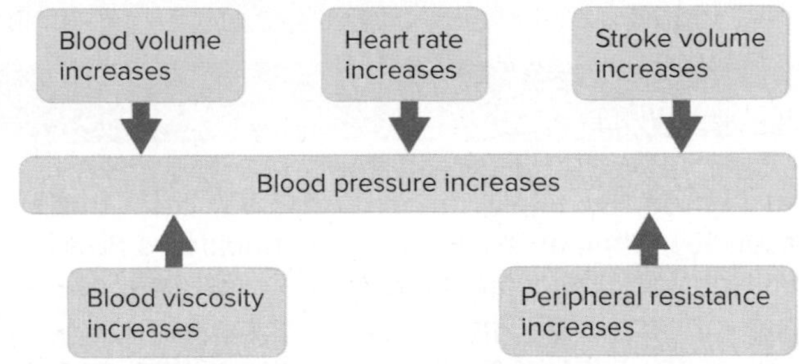

Figure 13.25 Some of the factors that influence arterial blood pressure.

Blood pressure varies with the cardiac output. If either the stroke volume or the heart rate increases, so does the cardiac output, and blood pressure increases. Conversely, if the stroke volume or the heart rate decreases, the cardiac output decreases, and blood pressure decreases.

Blood Volume

Blood volume equals the sum of the formed elements and plasma volumes in the vascular system. Although the blood volume varies somewhat with age, body size, and sex, it is usually about 5 liters for adults, or 8% of body weight in kilograms.

Normally, blood pressure is directly proportional to blood volume in the cardiovascular system. Thus, any changes in blood volume can initially alter the blood pressure. For example, if a hemorrhage reduces blood volume, blood pressure initially drops. If a transfusion restores normal blood volume, normal blood pressure may be reestablished. Blood volume can also fall if the fluid balance is upset, as happens in dehydration. Fluid replacement can reestablish normal blood volume and pressure.

Peripheral Resistance

Friction between the blood and the walls of blood vessels produces a force called **peripheral resistance** (pĕ-rif′er-al re-zis′tans), which hinders blood flow. Blood pressure must overcome peripheral resistance if the blood is to continue flowing. Factors that alter the peripheral resistance change blood pressure. For example, contracting smooth muscle in arteriolar walls increases the peripheral resistance by constricting these vessels. Blood backs up into the arteries supplying the arterioles, and the arterial pressure rises. Dilation of arterioles has the opposite effect—peripheral resistance decreases, and arterial blood pressure drops in response.

Blood Viscosity

Viscosity (vis-kos′ĭ-tē) is the difficulty with which the molecules of a fluid flow past one another. The greater the viscosity, the greater the resistance to flow.

Blood cells and plasma proteins increase blood viscosity. The greater the blood's resistance to flowing, the greater the force needed to move it through the vascular system. Thus, it is not surprising that blood pressure rises as blood viscosity increases and drops as viscosity decreases.

Clinical Application 13.3 discusses the effect of blood pressure in an aneurysm, varicose veins, and hypertension.

PRACTICE 13.5

4. How are cardiac output and blood pressure related?
5. How does blood volume affect blood pressure?
6. What is the relationship between peripheral resistance and blood pressure? Between blood viscosity and blood pressure?

Control of Blood Pressure

Blood pressure (BP) is determined by cardiac output (CO) and peripheral resistance (PR) according to this relationship:

$$BP = CO \times PR$$

Maintenance of normal arterial pressure therefore requires regulation of these two factors.

The volume of blood returning to the heart and entering the ventricles affects cardiac output. Entering blood mechanically stretches myocardial cells in the ventricular walls. Within limits, the greater the length of these cells prior to contraction, the greater the force with which they contract. The relationship between cell length (due to stretching of the cardiac muscle cell just before contraction) and force of contraction is called the *Frank-Starling law of the heart.* This relationship becomes important, for example, during exercise, when venous blood returns more rapidly to the heart. The more blood that enters the heart from the veins, the greater the ventricular distension, the stronger the contraction, the greater the amount of blood pumped in a single beat (stroke volume), and the greater the cardiac output. Conversely, the less blood that returns from the veins, the less the ventricle distends, the weaker the ventricular contraction, and the lesser the stroke volume and cardiac output. This mechanism ensures that the volume of blood discharged from the heart is equal to the volume entering its chambers.

Cardiac output and peripheral resistance are controlled in part by baroreceptor reflexes. **Baroreceptors** are sensory receptors in the aortic arch and carotid arteries that sense changes in blood pressure. If arterial pressure increases, impulses travel from the baroreceptors to the cardiac center of the medulla oblongata. This center relays parasympathetic impulses to the SA node in the heart, and the heart rate decreases in response. As a result of this *cardioinhibitor reflex,* cardiac output falls, and blood pressure decreases toward the normal level (fig. 13.26). Conversely, decreasing arterial blood pressure initiates the *cardioaccelerator reflex,* which sends sympathetic impulses to the SA node. As a result, the heart beats faster, increasing cardiac output and arterial pressure. Other factors that increase heart rate and blood pressure include exercise, a rise in body temperature, and emotional responses, such as fear and anger. Clinical Application 13.4 discusses the effects of exercise on cardiovascular functioning.

Peripheral resistance also controls blood pressure. Changes in arteriole diameters regulate peripheral resistance.

CLINICAL APPLICATION 13.3

Aneurysm, Varicose Veins, and Hypertension

If blood pressure dilates a weakened area of an artery wall, a bulge called an aneurysm may form and enlarge. If the lining of the vessel tears and allows blood to enter the middle layer of the arterial wall, it is called a dissecting aneurysm. An aneurysm may cause symptoms by pressing on nearby organs, or it may rupture and cause great blood loss, which is life-threatening.

Aneurysms may also result from trauma, high blood pressure, certain infections, inherited disorders such as Marfan syndrome, or congenital defects in blood vessels. Common sites of aneurysms include the thoracic and abdominal aorta and an arterial circle at the base of the brain (cerebral arterial circle). Some aneurysms may be treated with a stent (a small mesh tube that holds a vessel open), or by replacing the affected part of the vessel with a synthetic graft. A recent technique grafts veins from the patient's own thighs, which are less likely to cause infection or be rejected by the immune system.

Varicose veins are abnormal and irregular dilations in superficial veins, particularly in the legs. This condition is usually associated with prolonged, increased pressure in the affected vessels due to gravity, as when a person stands. Crossing the legs or sitting in a chair so that its edge presses against the area behind the knee can obstruct venous blood flow and aggravate varicose veins. Increased venous pressure stretches and widens the veins. The valves in these vessels lose their ability to block the backward flow of blood, and blood accumulates in the veins. The resulting increased venous pressure is accompanied by rising pressure in the venules and capillaries that supply the veins. Consequently, tissues in affected regions typically become edematous and painful.

Hypertension, or high blood pressure, is persistently elevated systemic arterial pressure. It is one of the more common diseases of the cardiovascular system.

High blood pressure with unknown cause is called *essential* (also primary or idiopathic) *hypertension.* Elevated blood pressure that is a consequence of another problem, such as kidney disease, is called *secondary hypertension.*

The consequences of prolonged, uncontrolled hypertension can be very serious. As the left ventricle works harder to pump blood at a higher pressure, the myocardium thickens, enlarging the heart. If coronary blood flow cannot support this growth, parts of the heart muscle die and fibrous tissue replaces them. Eventually, the enlarged and weakened heart fails to maintain adequate output for survival.

Exercising regularly, maintaining a healthy body weight, reducing stress, and limiting dietary sodium may control blood pressure. If necessary, medications may include diuretics and/or inhibitors of sympathetic nerve action.

Because blood vessels with smaller diameters offer a greater resistance to blood flow, factors that cause arteriole vasoconstriction increase peripheral resistance, which raises blood pressure, and factors causing vasodilation decrease peripheral resistance, lowering blood pressure.

The *vasomotor center* of the medulla oblongata continually sends sympathetic impulses to smooth muscle in the arteriole walls, keeping them in a state of tonic contraction. This action helps maintain the peripheral resistance associated with normal blood pressure. Because the vasomotor center responds to changes in blood pressure, it can increase peripheral resistance by increasing its outflow of sympathetic impulses, or it can decrease such resistance by decreasing its sympathetic outflow. In the latter case, the vessels vasodilate as sympathetic stimulation decreases.

Whenever arterial blood pressure suddenly increases, baroreceptors in the aorta and carotid arteries signal the vasomotor center, and the sympathetic outflow to the arterioles falls. The resulting vasodilation decreases peripheral resistance, and blood pressure lowers toward the normal level.

Certain chemicals, including carbon dioxide, oxygen, and hydrogen ions, also influence peripheral resistance by affecting precapillary sphincters and smooth muscle in arteriole walls. For example, increasing blood carbon dioxide, decreasing blood oxygen, and lowering blood pH relaxes smooth muscle in the systemic circulation. This increases local blood flow to tissues with high metabolic rates, such as exercising skeletal muscles. In addition, epinephrine and norepinephrine vasoconstrict many systemic vessels, increasing peripheral resistance.

PRACTICE 13.5

7. Describe how the volume of blood returning to the heart and entering the ventricles affects cardiac output.
8. What is the function of baroreceptors in the aortic arch and carotid arteries?
9. How does the vasomotor center control peripheral resistance?

CLINICAL APPLICATION 13.4

Exercise and the Cardiovascular System

The cardiovascular system adapts to aerobic exercise. The aerobically conditioned athlete or person who exercises regularly experiences increases in heart pumping efficiency, blood volume, blood hemoglobin concentration, and the number of mitochondria in muscle fibers. These adaptations improve oxygen delivery to, and utilization by, muscle tissue.

An athlete's heart typically changes in response to increased demands, and may enlarge 40% or more. Myocardial mass increases, the ventricular cavities expand, and the ventricle walls thicken. At rest, stroke volume increases, and heart rate and blood pressure decrease. To a physician unfamiliar with a conditioned cardiovascular system, a trained athlete may appear abnormal. For example, a prolonged QT interval on an electrocardiogram for a longtime distance runner is not necessarily an indication of "long QT syndrome," as it might be in a less physically fit individual.

The cardiovascular system responds beautifully to a slow, steady buildup in exercise frequency and intensity. It may not react well to sudden demands, such as when a person who never exercises suddenly shovels snow.

For exercise to benefit the cardiovascular system, the heart rate must be elevated to 70% to 85% of its "theoretical maximum" for 30 to 60 minutes, at least three to four times a week, according to the American Heart Association. To calculate your theoretical maximum, subtract your age from 220. If you are eighteen years old, your theoretical maximum is 202 beats per minute. Then, 70% to 85% of this value is 141 to 172 beats per minute. Some good activities for raising the heart rate are tennis, skating, skiing, handball, vigorous dancing, hockey, basketball, biking, and fast walking.

It is wise to consult a physician before starting an exercise program. People over age thirty are advised to have a stress test, which is an electrocardiogram taken while exercising. (The standard electrocardiogram is taken at rest.) An arrhythmia that appears only during exercise may indicate heart disease that has not yet produced symptoms. The American Heart Association suggests that after a physical exam, a sedentary person wishing to start an exercise program begin with 30 minutes of activity (perhaps broken into two 15-minute sessions at first) at least five times per week.

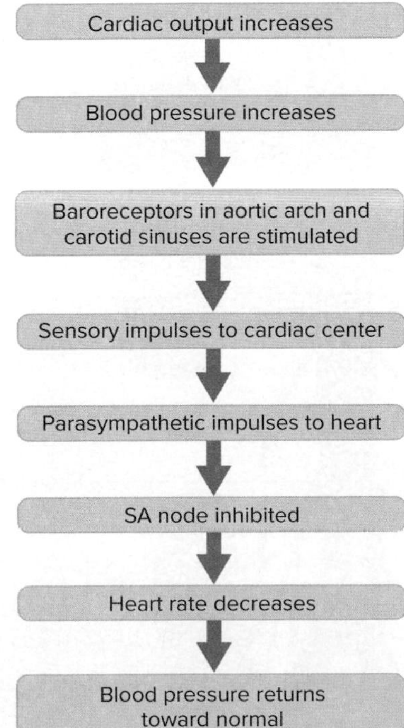

Figure 13.26 If blood pressure rises, baroreceptors initiate the cardioinhibitor reflex, which lowers the blood pressure.

Venous Blood Flow

Blood pressure decreases as blood moves through the arterial system and into the capillary networks, so that little pressure remains at the venular ends of capillaries (see fig. 13.24*b*). Instead, blood flow through the venous system is only partly the direct result of heart action and depends on other factors, such as skeletal muscle contraction, breathing movements, and vasoconstriction of veins (*venoconstriction*).

Contracting skeletal muscles press on veins moving blood from one valve section to another (see fig. 13.23). This massaging action of contracting skeletal muscles helps push blood through the venous system toward the heart.

Respiratory movements also move venous blood. During inspiration, the pressure within the thoracic cavity is reduced as the diaphragm contracts and the rib cage moves upward and outward. At the same time, the pressure within the abdominal cavity is increased as the diaphragm presses downward on the abdominal viscera. Consequently, blood is squeezed out of the abdominal veins and forced into thoracic veins. During exercise, these respiratory movements, along with skeletal muscle contractions, increase the return of venous blood to the heart.

Venoconstriction also returns venous blood to the heart. When venous pressure is low, sympathetic reflexes stimulate smooth muscle in the walls of veins to contract. The veins provide a blood reservoir that can adapt its capacity to changes in blood volume. If some blood is lost and blood pressure falls, venoconstriction can force blood out of this reservoir. By maintaining venous return to the heart, venoconstriction helps to maintain blood pressure.

PRACTICE 13.5

10. What is the function of venous valves?
11. How do skeletal muscles and respiratory movements affect venous blood flow?
12. What factors stimulate venoconstriction?

13.6 | Arterial System

LEARN

1. Identify and locate the major arteries.

The **aorta** is the largest-diameter artery in the body. It extends upward from the left ventricle, arches over the heart to the left, and descends just anterior and to the left of the vertebral column. Figure 13.27 shows the aorta and its main branches.

Principal Branches of the Aorta

The first part of the aorta is called the *ascending aorta.* At the root are the three cusps of the aortic valve, and opposite each cusp is a swelling in the aortic wall called an **aortic sinus.** The right and left coronary arteries arise from two of these sinuses (see fig. 13.7).

Three major arteries originate from the *aortic arch* (arch of the aorta): the **brachiocephalic** (brāk″ē-ō-sĕ-fal′ik) **trunk,** the left **common carotid** (kah-rot′id) **artery,** and the left **subclavian** (sub-klā′vē-an) **artery.**

The upper part of the *descending aorta* is left of the midline. It gradually extends medially and lies directly anterior to the vertebral column at the level of the twelfth thoracic vertebra. The part of the descending aorta above the diaphragm is the **thoracic aorta.** It gives off many small branches to the thoracic wall and the thoracic viscera.

Below the diaphragm, the descending aorta becomes the **abdominal aorta.** It gives off branches to the abdominal wall and abdominal organs. Branches to abdominal organs include: the **celiac** (sē′lē-ak) **artery,** which gives rise to the *gastric, splenic,* and *hepatic arteries;* the **superior** (supplies small intestine and superior part of large intestine) and **inferior** (supplies inferior part of large intestine) **mesenteric** (mes″en-ter′ik) **arteries;** and the **suprarenal** (soo″prah-rē′nal) **arteries, renal** (rē′nal) **arteries,** and **gonadal** (go′nad-al) **arteries,** which supply blood to the adrenal glands, kidneys, and ovaries or testes, respectively. The abdominal aorta ends near the brim of the pelvis, where it divides into right and left **common iliac** (il′ē-ak) **arteries.** These vessels supply blood to lower regions of the abdominal wall, the pelvic organs, and the lower extremities. Table 13.3 summarizes the major branches of the aorta.

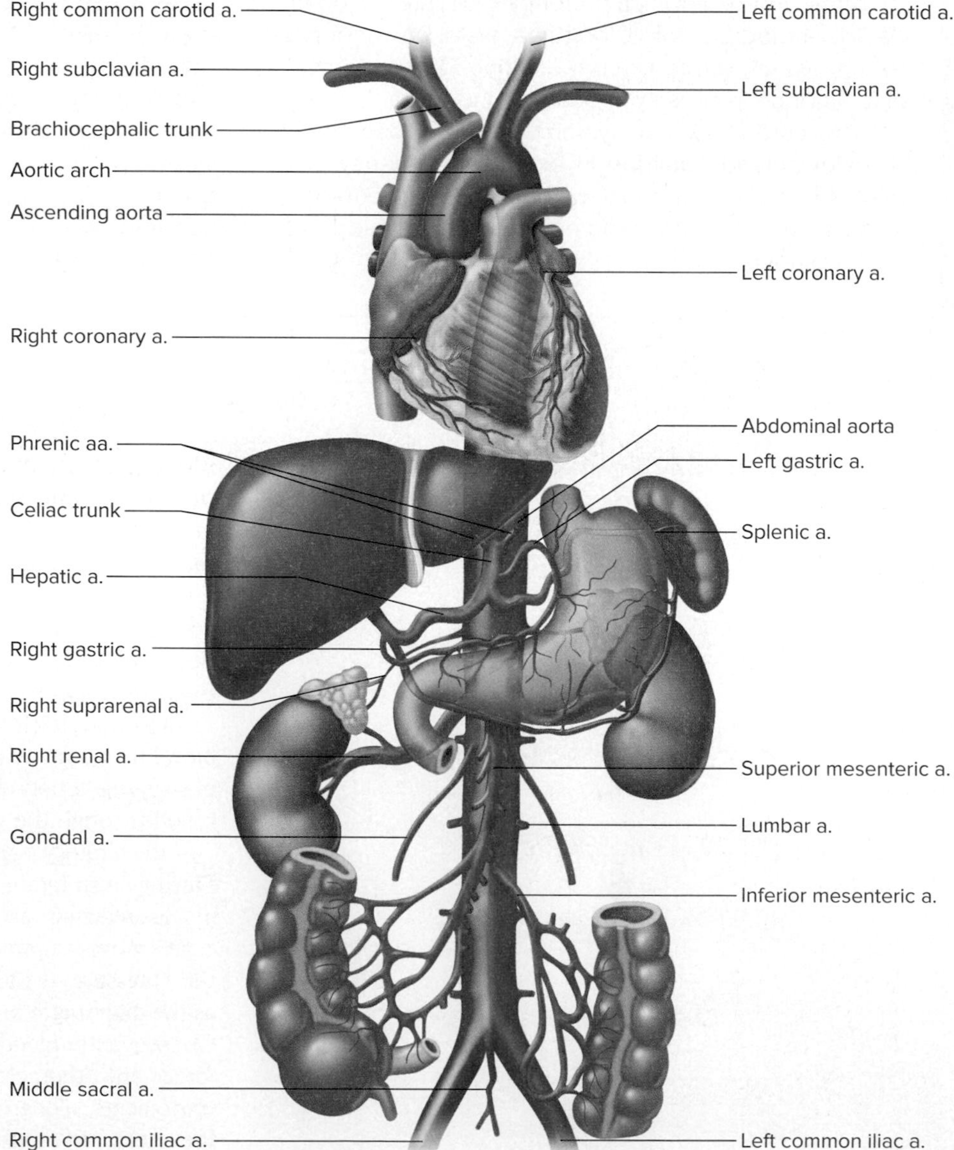

Figure 13.27 Major branches of the aorta. (**a.** stands for **artery, aa.** stands for **arteries.**) APR

TABLE 13.3 Major Branches of the Aorta

Portion of Aorta	Branch	General Regions or Organs Supplied
Ascending aorta	Right and left coronary arteries	Heart
Arch of the aorta	Brachiocephalic trunk Left common carotid artery Left subclavian artery	Right upper limb, right side of head Left side of head Left upper limb
Descending aorta		
Thoracic aorta	Bronchial artery Pericardial artery Esophageal artery Mediastinal artery Posterior intercostal artery	Bronchi Pericardium Esophagus Mediastinum Thoracic wall
Abdominal aorta	Celiac artery Phrenic artery Superior mesenteric artery Suprarenal artery Renal artery Gonadal artery Inferior mesenteric artery Lumbar artery Middle sacral artery Common iliac artery	Stomach, spleen, liver Diaphragm Portions of small and large intestines Adrenal gland Kidney Ovary or testis Lower portions of large intestine Posterior abdominal wall Sacrum and coccyx Lower abdominal wall, pelvic organs, and lower limb

Arteries to the Neck, Head, and Brain

Branches of the subclavian and common carotid arteries supply blood to structures in the neck, head, and brain (fig. 13.28). The main divisions of the subclavian artery to these regions include the vertebral and thyrocervical arteries. The common carotid artery communicates with these regions by means of the internal and external carotid arteries.

The **vertebral arteries** pass upward through the foramina of the transverse processes of the cervical vertebrae and enter the skull through the foramen magnum. These vessels supply blood to the brainstem, spinal cord, and to the vertebrae and their associated ligaments and muscles.

In the cranial cavity, the vertebral arteries unite to form a single *basilar artery.* This vessel passes along the ventral brainstem and gives rise to branches leading to the pons, midbrain, and cerebellum. The basilar artery ends by dividing into two *posterior cerebral arteries* that supply parts of the occipital and temporal lobes of the cerebrum. The posterior cerebral arteries also help form the **cerebral arterial circle** (circle of Willis) at the base of the brain, which connects the vertebral artery and internal carotid artery systems (fig. 13.29). The union of these systems provides alternate pathways for blood to circumvent blockages and reach brain tissues. It also equalizes blood pressure in the brain's blood supply.

The **thyrocervical** (thī″rō-ser′vĭ-kal) **arteries** are short vessels. These vessels branch to the thyroid gland, parathyroid glands, larynx, trachea, esophagus, and pharynx, as well as to muscles in the neck, shoulder, and back.

The left and right *common carotid arteries* diverge into the internal and external carotid arteries. The **external carotid artery** courses upward on the side of the head, giving off branches to structures in the neck, face, jaw, scalp, and base of the skull. The **internal carotid artery** begins lateral to the external carotid artery, then extends medially to follow a deep course upward along the pharynx to the base of the skull. Entering the cranial cavity, it provides the major blood supply to the brain. Near the base of each internal carotid artery is an enlargement called a **carotid sinus.** Like the aortic arch, these structures contain baroreceptors that control blood pressure. Table 13.4 summarizes the major branches of the external and internal carotid arteries.

Arteries to the Shoulder and Upper Limb

The subclavian artery, after branching toward the neck, continues into the arm (fig. 13.30). It passes between the clavicle and the first rib, and becomes the axillary artery. The **axillary artery** supplies branches to structures in the axilla and chest wall. It becomes the **brachial artery,** which courses along the humerus to the elbow. It gives rise to a *deep brachial artery* that curves posteriorly around the humerus and supplies the triceps brachii muscle. In the elbow, the brachial artery divides into an ulnar artery and a radial artery.

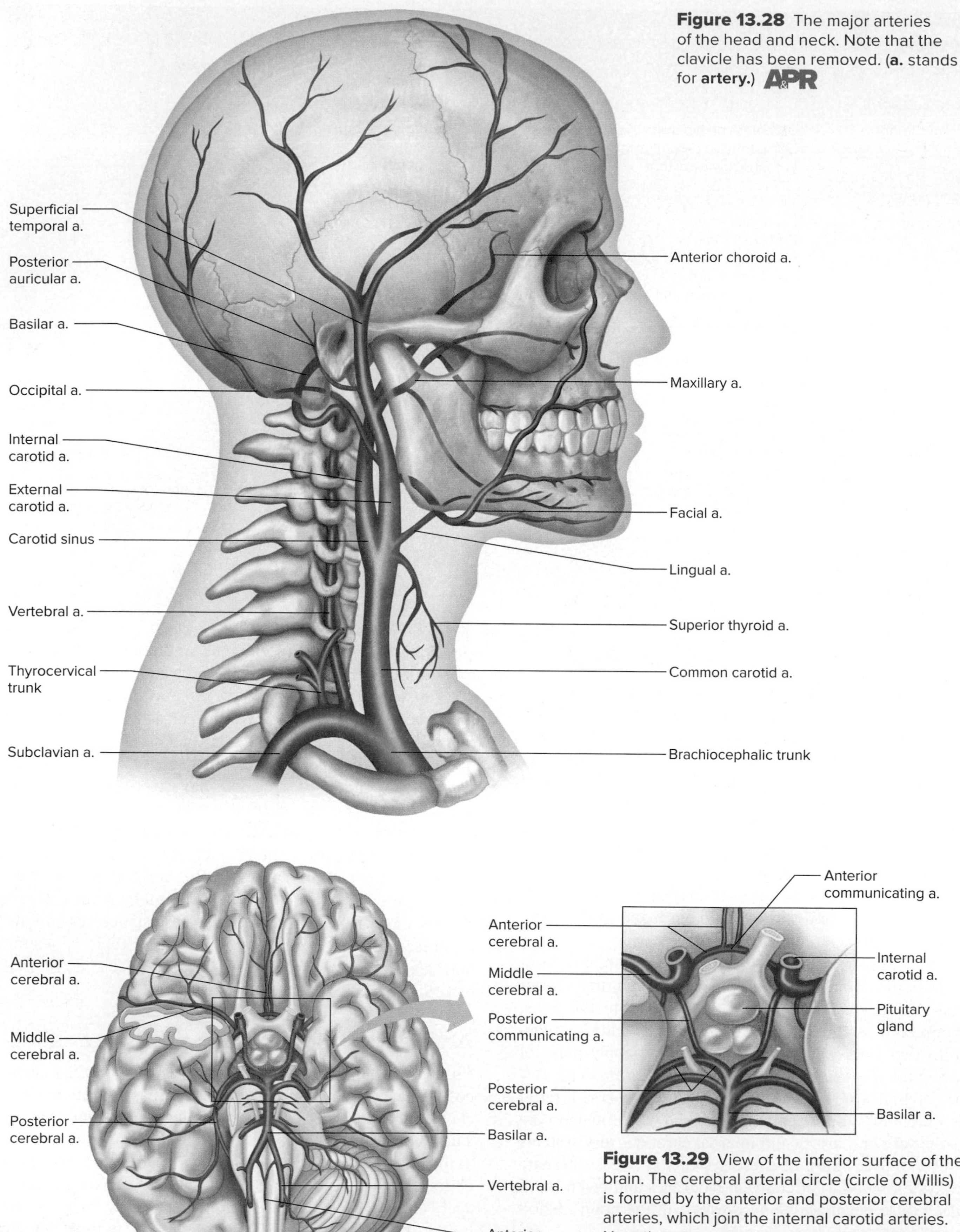

Figure 13.28 The major arteries of the head and neck. Note that the clavicle has been removed. (**a.** stands for **artery.**) APR

Figure 13.29 View of the inferior surface of the brain. The cerebral arterial circle (circle of Willis) is formed by the anterior and posterior cerebral arteries, which join the internal carotid arteries. Note that some structures have been removed from the right side of the figure to show underlying blood vessels. (**a.** stands for **artery.**) APR

TABLE 13.4 Major Branches of the External and Internal Carotid Arteries

Artery	Branch	General Region or Organs Supplied
External carotid artery	Superior thyroid artery	Larynx and thyroid gland
	Lingual artery	Tongue and salivary glands
	Facial artery	Pharynx, palate, chin, lips, and nose
	Occipital artery	Posterior scalp, meninges, and neck muscles
	Posterior auricular artery	Ear and lateral scalp
	Maxillary artery	Teeth, jaw, cheek, and eyelids
	Superficial temporal artery	Parotid salivary gland and surface of the face and scalp
Internal carotid artery	Ophthalmic artery	Eye and eye muscles
	Anterior choroid artery	Choroid plexus and brain
	Anterior cerebral artery	Frontal lobes of the brain
	Middle cerebral artery	Parietal lobes of the brain

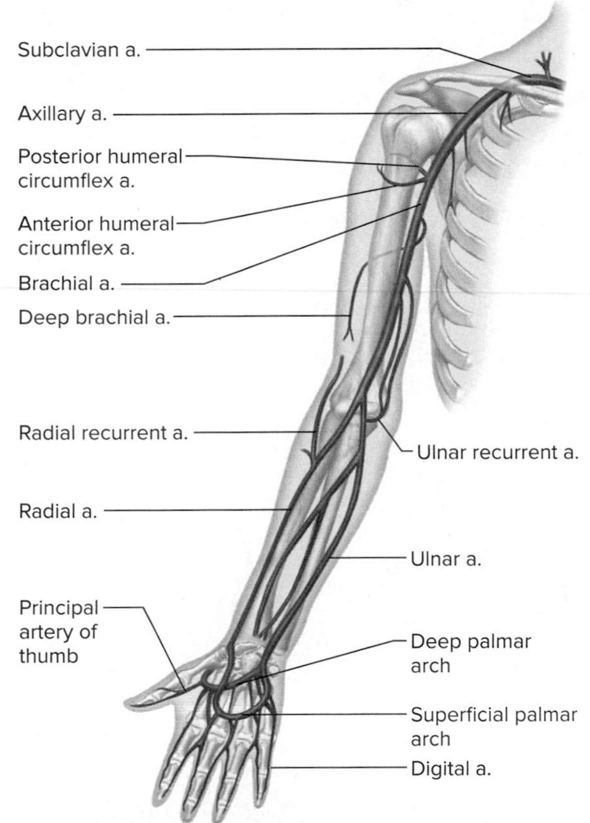

Figure 13.30 The major arteries to the shoulder and upper limb. (**a.** stands for **artery.**) APR

PRACTICE FIGURE 13.30

Blood from the brachial artery flows into which artery (arteries) in the forearm?

Answer can be found in Appendix E.

The **ulnar artery** leads downward on the ulnar side of the forearm to the wrist. Some of its branches supply the elbow joint, and some supply blood to muscles in the forearm.

The **radial artery** extends along the radial side of the forearm to the wrist, supplying the lateral muscles of the forearm. As the radial artery nears the wrist, it comes close to the surface and provides a convenient vessel for taking the pulse (radial pulse).

At the wrist, the branches of the ulnar and radial arteries join to form a network of vessels. Arteries arising from this network supply blood to the hand.

Arteries to the Thoracic and Abdominal Walls

Blood reaches the thoracic wall through several vessels. The **internal thoracic artery,** a branch of the subclavian artery, gives off *anterior intercostal* (in″ter-kos′tal) *arteries* that supply the intercostal muscles and mammary glands. The *posterior intercostal arteries* arise from the thoracic aorta and enter the intercostal spaces. They supply the intercostal muscles, the vertebrae, the spinal cord, and the deep muscles of the back.

Branches of the *internal thoracic* and *external iliac arteries* provide blood to the anterior abdominal wall. Paired vessels originating from the abdominal aorta, including the *phrenic* and *lumbar arteries,* supply blood to structures in the posterior and lateral abdominal wall.

Arteries to the Pelvis and Lower Limb

The abdominal aorta divides to form the **common iliac** (il′ē-ak) **arteries** at the level of the pelvic brim (fig. 13.31). Each common iliac artery divides into an internal and an external branch. The **internal iliac artery** gives off many branches to pelvic muscles and visceral structures, as well as to the gluteal muscles and the external reproductive organs. The **external iliac artery** provides the main blood supply to the lower limbs. It passes downward along the brim of the pelvis and branches to supply the muscles and skin in the lower abdominal wall. Midway between the pubic symphysis and the anterior superior iliac spine, the external iliac artery becomes the femoral artery.

The **femoral** (fem′or-al) **artery,** which approaches the anterior surface of the upper thigh, gives off branches to muscles and superficial tissues of the thigh. These

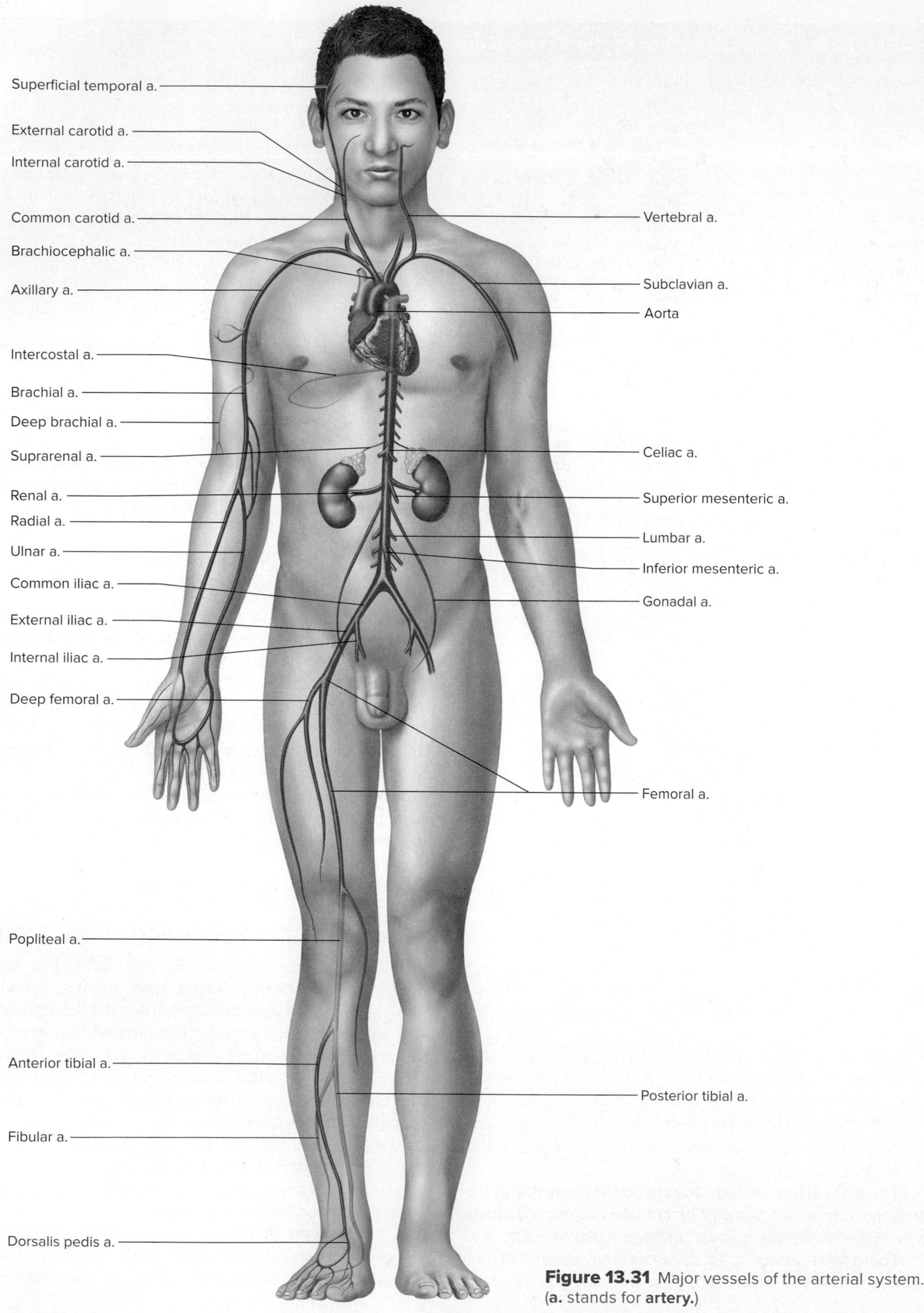

Figure 13.31 Major vessels of the arterial system. (**a.** stands for **artery.**)

branches also supply the skin of the groin and the lower abdominal wall.

As the femoral artery passes behind the medial distal femur and reaches the proximal border of the space behind the knee, it becomes the **popliteal** (pop-lit′ē-al) **artery.** Branches of this artery supply blood to the knee joint and to certain muscles in the thigh and calf. The popliteal artery diverges into the anterior and posterior tibial arteries.

The **anterior tibial artery** passes downward between the tibia and fibula, giving off branches to the skin and muscles in the anterior and lateral regions of the leg. This vessel continues into the foot as the *dorsalis pedis artery* (dorsal pedis artery), which supplies blood to the foot. The **posterior tibial artery,** the larger of the two popliteal branches, descends beneath the calf muscles, giving off branches to the skin, muscles, and other tissues of the leg along the way to the foot.

PRACTICE 13.6

1. Name the parts of the aorta.
2. Name the vessels that arise from the aortic arch.
3. Name the branches of the thoracic and abdominal aorta.
4. Which vessels supply blood to the head? To the upper limb? To the abdominal wall? To the lower limb?

13.7 | Venous System

LEARN

1. Identify and locate the major veins.

Venous circulation returns blood to the heart after gases, nutrients, and wastes are exchanged between the blood and body cells.

Characteristics of Venous Pathways

The vessels of the venous system originate with the merging of the capillaries into venules. Venules then merge into small veins, and small veins meet to form larger ones. Unlike the arterial pathways, however, the vessels of the venous system are difficult to follow because they commonly connect in irregular networks. Many unnamed tributaries may join to form a large vein.

The pathways of larger veins are less variable than those of smaller veins. These veins typically parallel the courses of named arteries, and many bear the same names as their arterial counterparts. For example, the renal vein parallels the renal artery, and the common iliac vein accompanies the common iliac artery.

The veins that carry blood from the lungs and myocardium back to the heart have been described. The veins from all the other parts of the body converge into two major vessels, the **superior** and **inferior venae cavae,** which lead to the right atrium.

Veins from the Brain, Head, and Neck

The **external jugular** (jug′ū-lar) **veins** drain blood from the face, scalp, and superficial regions of the neck. These vessels descend on either side of the neck and empty into the *right* and *left subclavian veins* (fig. 13.32).

The **internal jugular veins,** which are somewhat larger than the external jugular veins, arise from numerous veins and venous sinuses of the brain and from deep veins in parts of the face and neck. They descend through the neck and join the subclavian veins. These unions of the internal jugular and subclavian veins form large **brachiocephalic veins** on each side. These vessels then merge and give rise to the superior vena cava, which enters the right atrium.

Veins from the Upper Limb and Shoulder

A set of deep veins and a set of superficial veins drain the upper limb. The deep veins generally parallel the arteries in each region and have similar names. Deep venous drainage

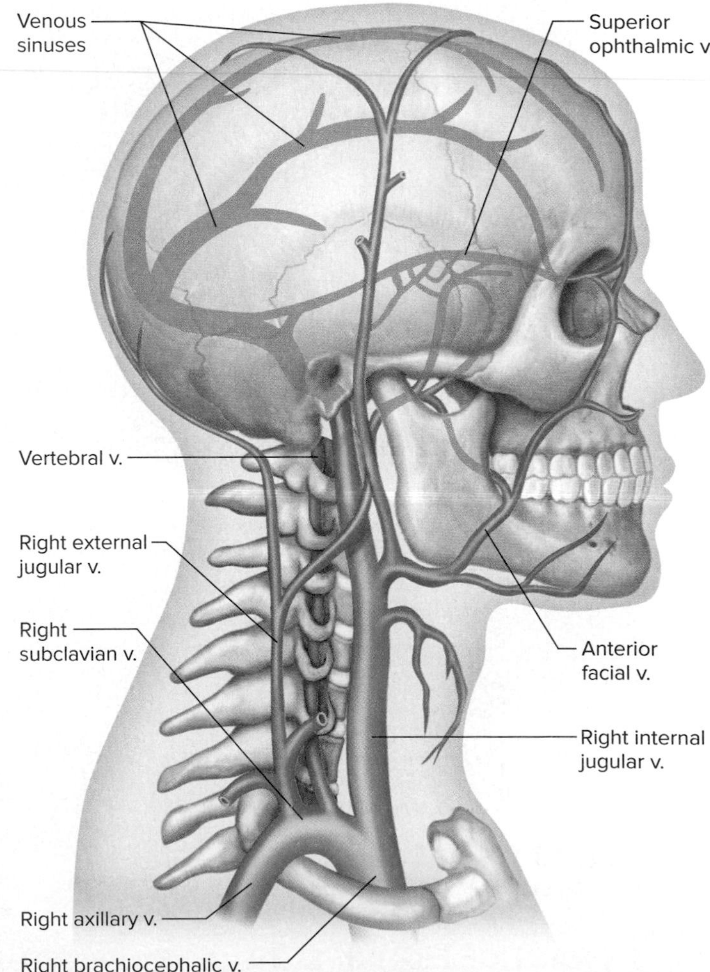

Figure 13.32 The major veins of the brain, head, and neck. Note that the clavicle has been removed. (**v.** stands for **vein.**) APR

of the upper limb begins in the digital veins that drain into pairs of **radial veins** and pairs of **ulnar veins,** which empty into a pair of **brachial veins.** The superficial veins connect in complex networks just beneath the skin. They also communicate with the deep vessels of the upper limb, providing many alternate pathways through which blood can leave the tissues (fig. 13.33). The main vessels of the superficial network are the basilic and cephalic veins.

The **basilic** (bah-sil′ik) **vein** ascends medially from the forearm to the middle of the arm, where it penetrates deeply and joins the **brachial vein.** The basilic and brachial veins merge, forming the **axillary vein.**

The **cephalic** (sĕ-fal′ik) **vein** courses upward laterally from the hand to the shoulder. In the shoulder, it pierces the tissues and empties into the axillary vein. Beyond the axilla, the axillary vein becomes the subclavian vein.

In the bend of the elbow, a *median cubital vein* ascends from the cephalic vein on the lateral side of the forearm to the basilic vein on the medial side. This large vein is usually visible just beneath the skin. It is often used as a site for *venipuncture,* when it is necessary to remove a blood sample or to add fluids to blood.

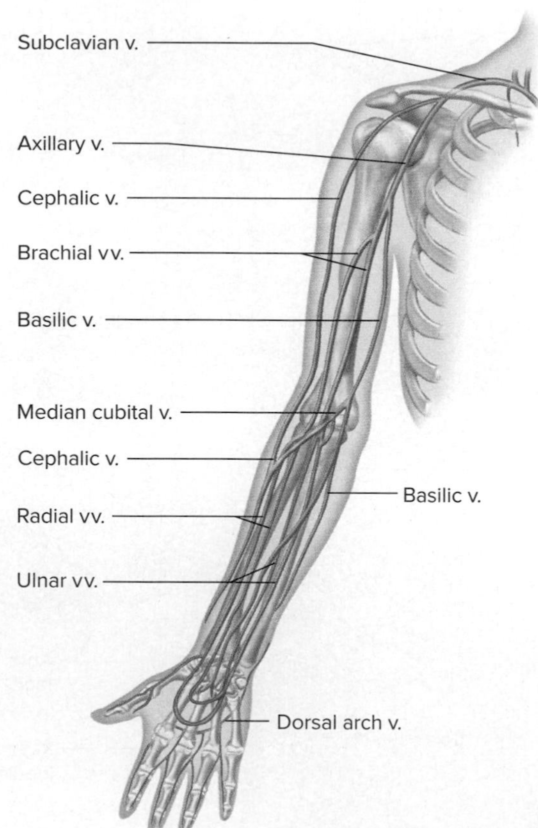

Figure 13.33 The major veins of the upper limb and shoulder. (**v.** stands for **vein, vv.** stands for **veins.**) APR

PRACTICE FIGURE 13.33

Blood from the brachial vein and basilic vein drains into which vein(s)?

Answer can be found in Appendix E.

Veins from the Abdominal and Thoracic Walls

Tributaries of the brachiocephalic and azygos veins drain the abdominal and thoracic walls. For example, the **brachiocephalic vein** receives blood from the *internal thoracic vein,* which generally drains the tissues the internal thoracic artery supplies. Some *intercostal veins* also empty into the brachiocephalic vein.

The **azygos** (az′ĭ-gos) **vein** originates in the dorsal abdominal wall and ascends through the mediastinum on the right side of the vertebral column to join the superior vena cava. It drains most of the muscular tissue in the abdominal and thoracic walls.

Tributaries of the azygos vein include the *posterior intercostal veins* on the right side, which drain the intercostal spaces, and the *superior* and *inferior hemiazygos veins,* which receive blood from the posterior intercostal veins on the left. The right and left *ascending lumbar veins,* with tributaries that include vessels from the lumbar and sacral regions, also connect to the azygos system.

Veins from the Abdominal Viscera

Most veins transport blood directly to the atria of the heart. Veins that drain the abdominal viscera are exceptions (fig. 13.34). They originate in the capillary networks of the stomach, intestines, pancreas, and spleen and carry blood from these organs through a **hepatic** (hĭ-păt′ĭk) **portal vein** to the liver. This venous pathway is called the **hepatic portal system.**

Tributaries of the hepatic portal vein include:

1. right and left *gastric veins* from the stomach.
2. *superior mesenteric vein* from the small intestine, ascending colon, and transverse colon.
3. *splenic vein* from a convergence of several veins draining the spleen, the pancreas, and part of the stomach. Its largest tributary, the *inferior mesenteric vein,* brings blood upward from the descending colon, sigmoid colon, and rectum.

About 80% of the blood flowing to the liver in the hepatic portal system comes from capillaries in the stomach and intestines. It is oxygen-poor but nutrient-rich. As discussed in section 15.8, Liver and Gallbladder, the liver handles these nutrients in a variety of ways. It regulates blood glucose concentration by joining (polymerizing) excess glucose molecules into glycogen for storage, or by breaking down glycogen into glucose when blood glucose concentration drops below normal. The liver helps regulate blood concentrations of recently absorbed amino acids and lipids by modifying them into forms that the cells can use, by oxidizing them, or by changing them into storage forms. The liver also stores certain vitamins and detoxifies harmful

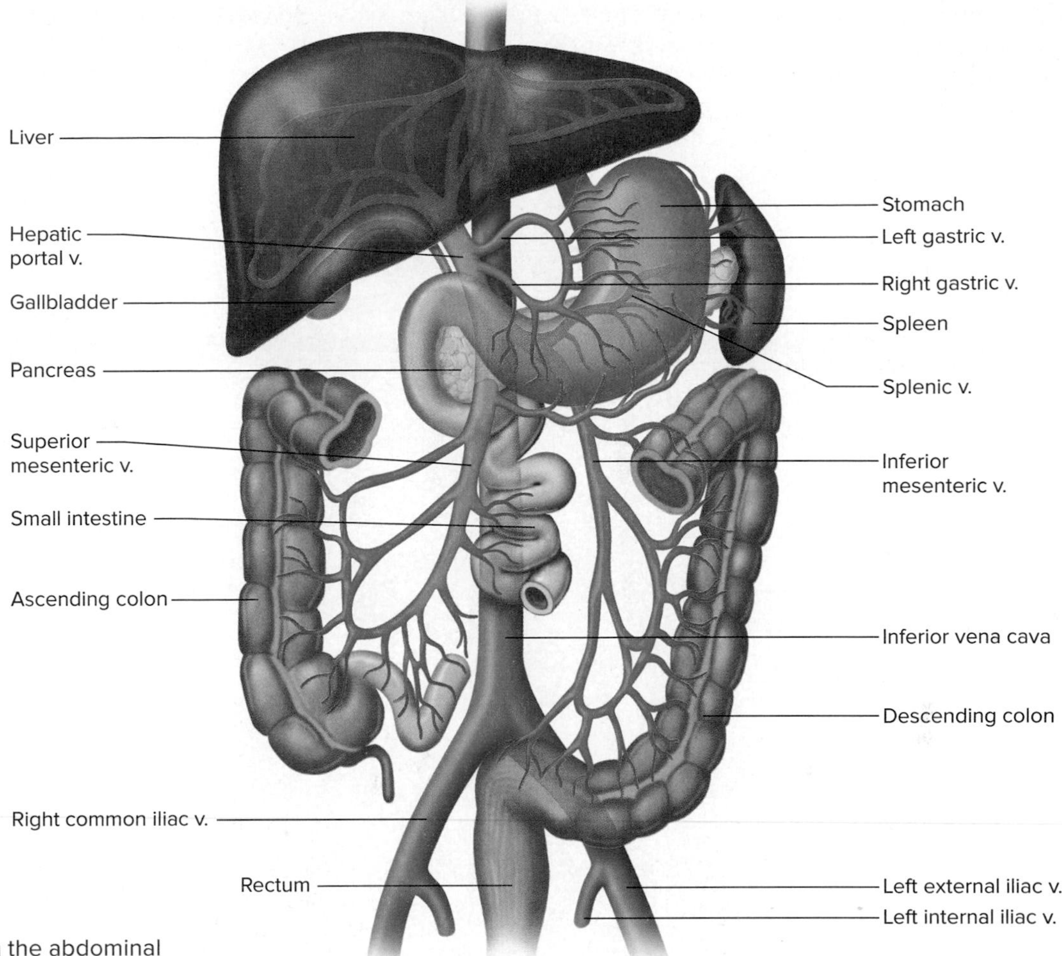

Figure 13.34 Veins that drain the abdominal viscera. (**v.** stands for **vein.**) APR

substances. Blood in the hepatic portal vein nearly always contains bacteria that have entered through intestinal capillaries. Large *Kupffer cells* lining small vessels in the liver called hepatic sinusoids phagocytize these microorganisms, removing them from portal blood before it leaves the liver.

After passing through the hepatic sinusoids of the liver, the blood in the hepatic portal system travels through a series of merging vessels into **hepatic veins.** These veins empty into the inferior vena cava, returning the blood to the general circulation.

Veins from the Lower Limb and Pelvis

Veins that drain blood from the lower limb can be divided into deep and superficial groups, as in the upper limb (fig. 13.35). The deep veins of the leg, such as the *anterior* and *posterior tibial veins,* are named for the arteries they accompany. At the level of the knee, these vessels form a single trunk, the **popliteal vein.** This vein continues upward through the thigh as the **femoral vein,** which in turn becomes the **external iliac vein.**

The superficial veins of the foot, leg, and thigh connect to form a complex network beneath the skin. These vessels drain into two major trunks—the small and great saphenous veins. The **small saphenous** (sah-fe′nus) **vein** ascends along the back of the calf, behind the knee, and joins the popliteal vein. The **great saphenous vein,** the longest vein in the body, ascends in front of the medial malleolus and extends upward along the medial side of the leg and thigh. In the thigh, it penetrates deeply and joins the femoral vein. Near its termination, the great saphenous vein receives tributaries from a number of vessels that drain the upper thigh, groin, and lower abdominal wall.

The saphenous veins communicate freely with each other and extensively with the deep veins of the leg and thigh. Blood can thus return to the heart from the lower extremities by several routes.

In the pelvic region, vessels leading to the **internal iliac vein** transport blood away from the organs of the reproductive, urinary, and digestive systems. The internal iliac veins unite with the right and left external iliac veins to form the **common iliac veins.** These veins, in turn, merge to produce the inferior vena cava.

PRACTICE 13.7

1. Name the veins that return blood to the right atrium.
2. Which major veins drain blood from the head? From the upper limbs? From the abdominal viscera? From the lower limbs?

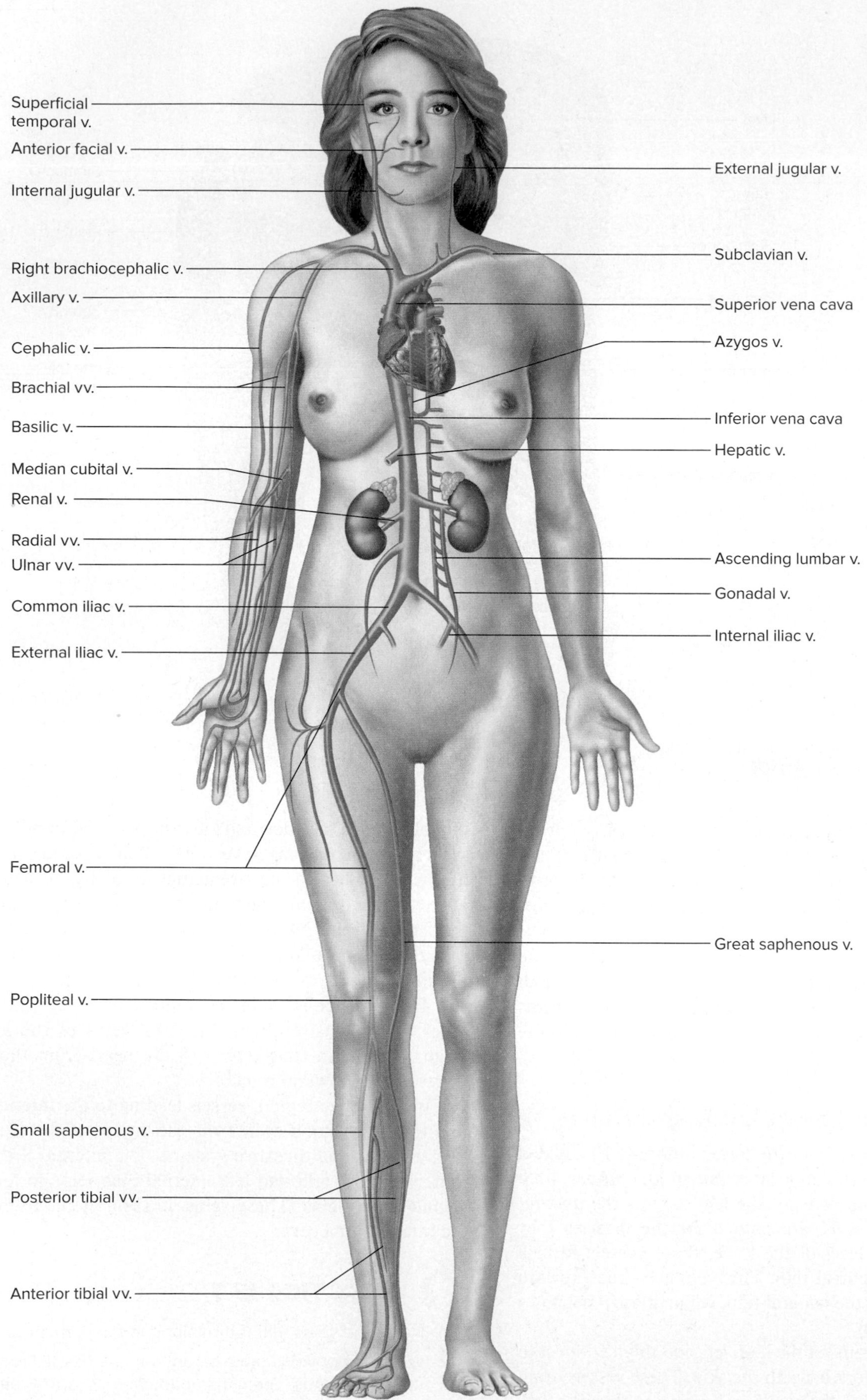

Figure 13.35 Major vessels of the venous system. (**v.** stands for **vein, vv.** stands for **veins.**)

Cardiovascular System

INTEGUMENTARY SYSTEM

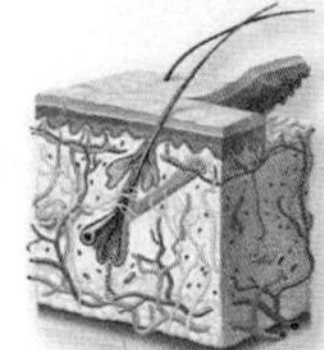

Changes in skin blood flow are important in temperature control.

SKELETAL SYSTEM

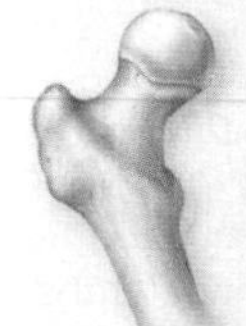

Bones help control plasma levels of calcium ions, which influence heart action.

MUSCULAR SYSTEM

Blood flow increases to exercising skeletal muscle, delivering oxygen and nutrients and removing wastes. Muscle actions help the blood circulate.

NERVOUS SYSTEM

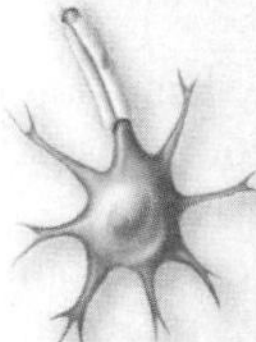

The brain is especially dependent on blood flow for survival. The nervous system helps control blood flow and blood pressure.

ENDOCRINE SYSTEM

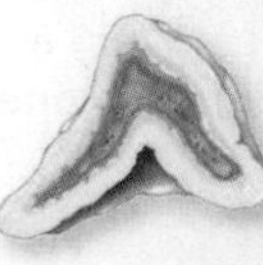

Hormones are carried in the bloodstream. Some hormones directly affect the heart and blood vessels.

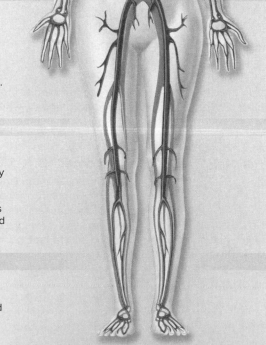

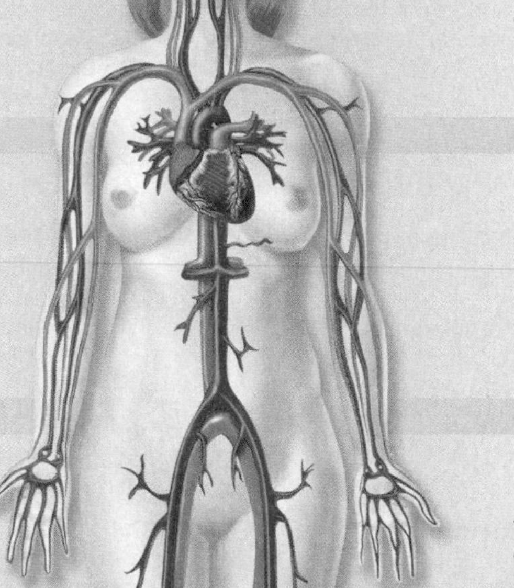

The heart pumps blood through as many as 60,000 miles of blood vessels, delivering nutrients to, and removing wastes from, all body cells.

LYMPHATIC SYSTEM

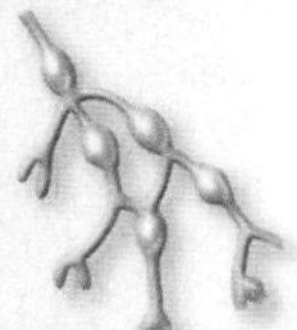

The lymphatic system returns tissue fluids to the bloodstream.

DIGESTIVE SYSTEM

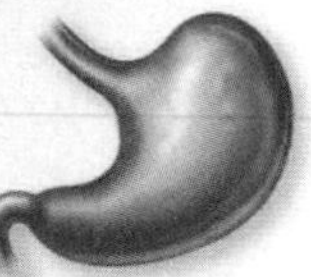

The digestive system breaks down nutrients into forms readily absorbed by the bloodstream.

RESPIRATORY SYSTEM

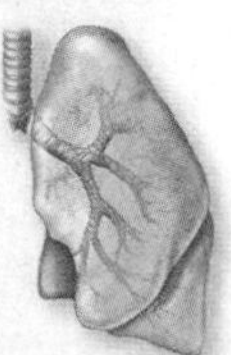

The respiratory system oxygenates the blood and removes carbon dioxide. Respiratory movements help the blood circulate.

URINARY SYSTEM

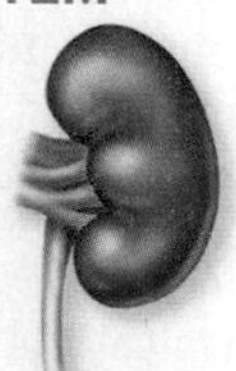

The kidneys clear the blood of wastes. The kidneys help control blood pressure and blood volume.

REPRODUCTIVE SYSTEM

Blood pressure is important in normal function of the sex organs.

ASSESS

CHAPTER ASSESSMENTS

13.1 Introduction

1. The cardiovascular system includes ______.
   a. the heart
   b. the arteries and veins
   c. the capillaries
   d. all of the above

13.2 Structure of the Heart

2. Describe the pericardium.
3. Compare the layers of the heart wall.
4. Draw a heart and label the chambers and valves.
5. Blood flows through the vena cavae and coronary sinus into the right atrium through the ______ to the right ventricle, through the pulmonary valve to the pulmonary trunk into the right and left ______ to the lungs, then leaves the lungs through the pulmonary veins and flows into the ______, through the mitral valve to the ______, and through the ______ to the aorta.
6. List the vessels through which blood flows from the aorta to the myocardium and back to the right atrium.

13.3 Heart Actions

7. Distinguish between the roles of the SA node and the AV node.
8. Explain how the cardiac conduction system controls the cardiac cycle.
9. Describe and explain the normal ECG pattern.
10. Describe the pressure changes in the atria and ventricles during a cardiac cycle.
11. Discuss how the nervous system regulates the cardiac cycle.

13.4 Blood Vessels

12. Distinguish between an artery and an arteriole.
13. Explain control of vasoconstriction and vasodilation.
14. Describe the structure and function of a capillary.
15. Relate how diffusion functions in the exchange of substances between the blood and tissues.
16. Explain why water and dissolved substances leave the arteriolar end of a capillary and enter the venular end.
17. Distinguish between a vein and a venule.

13.5 Blood Pressure

18. Arterial blood pressure peaks when the ventricles contract. This maximum pressure achieved is called ______.
19. Name several factors that influence blood pressure, and explain how each produces its effect.
20. Describe the control of blood pressure.
21. Explain how skeletal muscle contraction, breathing, and venoconstriction promote the flow of venous blood.
22. Distinguish between the pulmonary and systemic circuits of the cardiovascular system.

13.6–13.7 Arterial System–Venous System

23. Describe the aorta, and name its principal branches.
24. Discuss the relationship between the major venous pathways and the major arterial pathways to the head, upper limbs, abdominal viscera, and lower limbs.

ASSESS

INTEGRATIVE ASSESSMENTS/CRITICAL THINKING

Outcomes 3.3, 12.3, 13.4

1. How would low plasma proteins affect the net force of exchange at the venular end of the capillary? What are possible consequences?

Outcomes 5.5, 9.16, 13.2, 13.3

2. What structures and properties should an artificial heart have?

Outcomes 12.2, 13.2

3. Karen was noticeably fatigued and her capacity for physical exertion was reduced. Her physician suspected that Karen had an incompetent valve, possibly caused by a congenital defect or by valvular stenosis from repeated bacterial infections. Given this diagnosis, explain her symptoms.

Outcomes 13.2, 13.4, 13.6, 13.7, 13.8

4. If a patient develops a blood clot in the femoral vein of the left lower limb and a portion of the clot breaks loose, where is the blood flow likely to carry the embolus? What symptoms are likely?

Outcomes 13.2, 13.7

5. If a cardiologist inserts a catheter into a patient's right femoral artery, which arteries will the tube have to pass through in order to reach the entrance to the left coronary artery?

Outcomes 13.4, 13.5, 13.8

6. Cirrhosis of the liver, a disease commonly associated with alcoholism, obstructs blood flow through hepatic blood vessels. As a result, blood backs up and capillary pressure greatly increases in organs drained by the hepatic portal system. What effects might this increasing capillary pressure produce, and which organs would it affect?

Outcomes 13.6, 13.7

7. The liver receives blood from two sources. What are these two sources, and how do they differ from one another?

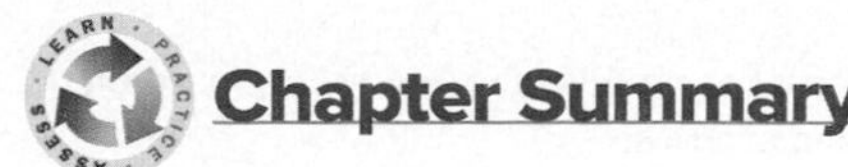

Chapter Summary

13.1 Introduction

*The **cardiovascular system,** consisting of the heart and blood vessels, provides oxygen and nutrients to and removes wastes from body cells.*

13.2 Structure of the Heart

1. Size and location of the heart
   a. The heart is about 14 centimeters long and 9 centimeters wide.
   b. It is located within the mediastinum, superior to the diaphragm.
2. Coverings of the heart
   a. A layered pericardium encloses the heart.
   b. The pericardial cavity is a space between the parietal and visceral layers of the **pericardium.**
3. Wall of the heart
   The wall of the heart has three layers—an **epicardium,** a **myocardium,** and an **endocardium.**
4. Heart chambers and valves
   a. The heart is divided into two **atria** and two **ventricles.**
   b. Right chambers and valves
      (1) The right atrium receives blood from the venae cavae and coronary sinus.
      (2) The **tricuspid valve** separates the right atrium from the right ventricle.
      (3) The right ventricle pumps blood into the pulmonary trunk.
      (4) A **pulmonary valve** guards the base of the pulmonary trunk.
   c. Left chambers and valves
      (1) The left atrium receives blood from the pulmonary veins.
      (2) The **mitral valve** separates the left atrium from the left ventricle.
      (3) An **aortic valve** guards the base of the aorta.
5. Skeleton of the heart
   The skeleton of the heart consists of masses of dense connective tissues in the septum between the ventricles, and fibrous rings that enclose the bases of the pulmonary trunk and aorta.
6. Blood flow through the heart
   a. Blood low in oxygen and high in carbon dioxide enters the right side of the heart and is pumped into the pulmonary circulation.
   b. After blood is oxygenated in the lungs and some carbon dioxide is removed, it returns to the left side of the heart through the pulmonary veins.
   c. From the left ventricle, the blood moves into the aorta.
7. Blood supply to the heart
   a. The **coronary arteries** supply blood to the myocardium.
   b. Blood returns to the right atrium through the **cardiac veins** and **coronary sinus.**

13.3 Heart Actions

1. Cardiac muscle cells
   a. Cardiac muscle cells connect to form a **functional syncytium.**
   b. If any part of the syncytium is stimulated, the whole structure contracts as a unit.
2. Cardiac conduction system
   a. The **cardiac conduction system** initiates and conducts impulses throughout the myocardium.
   b. Impulses from the **SA node** pass slowly to the **AV node;** impulses are conducted rapidly along the **AV bundle** and **Purkinje fibers.**
3. **Electrocardiogram (ECG)**
   a. An ECG records electrical changes in the myocardium during a cardiac cycle.
   b. The pattern contains several waves.
      (1) The P wave represents atrial depolarization.
      (2) The QRS complex represents ventricular depolarization.
      (3) The T wave represents ventricular repolarization.
4. Heart sounds
   Heart sounds are due to the vibrations produced when the valves close.
5. Pressure and volume changes during a cardiac cycle
   a. The atria contract (atrial **systole**) while the ventricles relax (ventricular **diastole**). The ventricles contract (ventricular systole) while the atria relax (atrial diastole).
   b. Pressure within the chambers rises and falls in repeated **cardiac cycles.**
6. Heart rate and regulation of the cardiac cycle
   a. Physical exercise, body temperature, and the concentration of various ions affect heart rate.
   b. Branches of sympathetic and parasympathetic nerve fibers innervate the SA and AV nodes.
   c. The cardiac control center in the medulla oblongata regulates autonomic impulses to the heart.

13.4 Blood Vessels

Blood vessels form a closed circuit of tubes that transport blood from the heart to body cells and back again.

1. **Arteries and arterioles**
   a. Arteries are adapted to transport blood under relatively high pressure away from the heart.
   b. The walls of arteries and arterioles consist of layers of endothelium, smooth muscle, and connective tissue.
   c. Autonomic fibers innervate smooth muscle in vessel walls, causing **vasoconstriction** or **vasodilation.**
2. **Capillaries**
   a. Capillaries connect arterioles and venules.
   b. The capillary wall is a single layer of cells that forms a semipermeable membrane.
   c. Openings in capillary walls, where endothelial cells overlap, vary in size from tissue to tissue.
   d. Precapillary sphincters regulate capillary blood flow.
   e. Capillary blood and tissue fluid exchange gases, nutrients, and metabolic by-products.
      (1) Diffusion provides the most important means of transport.
      (2) Filtration, due to the hydrostatic pressure of blood, causes a net outward movement of fluid at the arteriolar end of a capillary.
      (3) Osmosis due to colloid osmotic pressure causes a net inward movement of fluid at the venular end of a capillary.

3. **Venules and veins**
   a. Venules continue from capillaries and merge to form veins.
   b. Veins transport blood to the heart.
   c. Venous walls are similar to arterial walls, but are thinner and contain less smooth muscle and elastic tissue.

13.5 Blood Pressure

Blood pressure is the force blood exerts against the inner walls of blood vessels.

1. Arterial blood pressure
   a. Arterial blood pressure rises and falls with the phases of the cardiac cycle.
   b. **Systolic pressure** occurs when the ventricle contracts; **diastolic pressure** occurs when the ventricle relaxes.
2. Factors that influence arterial blood pressure
   Arterial blood pressure increases as **cardiac output, blood volume, peripheral resistance,** or blood **viscosity** increases.
3. Control of blood pressure
   a. Blood pressure is controlled in part by the mechanisms that regulate cardiac output and peripheral resistance.
   b. The more blood that enters the heart, the stronger the ventricular contraction, the greater the stroke volume, and the greater the cardiac output.
   c. The baroreceptor reflexes involving the cardiac control center of the medulla oblongata regulate heart rate.
4. Venous blood flow
   a. Venous blood flow depends on skeletal muscle contraction, breathing movements, and venoconstriction.
   b. Many veins contain flaplike valves that prevent blood from backing up.

13.6 Arterial System

1. Principal branches of the aorta
   a. The **aorta** is the largest artery with respect to diameter.
   b. The branches of the ascending aorta include the right and left coronary arteries.
   c. The branches of the aortic arch include the **brachiocephalic,** left **common carotid,** and left **subclavian arteries.**
   d. The branches of the descending aorta include the **thoracic aorta** and **abdominal aorta** and their respective branches.
   e. The abdominal aorta diverges into the right and left **common iliac arteries.**
2. Arteries to the neck, head, and brain
   These include branches of the subclavian and common carotid arteries.
3. Arteries to the shoulder and upper limb
   a. The subclavian artery passes into the upper limb, and becomes the **axillary artery** and **brachial artery.**
   b. Branches of the brachial artery include the **ulnar arteries** and **radial arteries.**
4. Arteries to the thoracic and abdominal walls
   a. Branches of the subclavian artery and thoracic aorta supply the thoracic wall.
   b. Branches of the abdominal aorta and other arteries supply the abdominal wall.
5. Arteries to the pelvis and lower limb
   The common iliac arteries supply the pelvic organs, gluteal region, and lower limbs.

13.7 Venous System

1. Characteristics of venous pathways
   a. Veins return blood to the heart.
   b. Larger veins usually parallel the paths of major arteries.
2. Veins from the brain, head, and neck
   a. **Jugular veins** drain these regions.
   b. Jugular veins unite with subclavian veins to form the **brachiocephalic veins.**
3. Veins from the upper limb and shoulder
   a. Sets of superficial and deep veins drain these regions.
   b. Deep veins parallel arteries with similar names.
4. Veins from the abdominal and thoracic walls
   Tributaries of the brachiocephalic and **azygos veins** drain these walls.
5. Veins from the abdominal viscera
   a. Blood from the abdominal viscera enters the **hepatic portal system** and is transported to the liver.
   b. From the liver, hepatic veins transport blood to the inferior vena cava.
6. Veins from the lower limb and pelvis
   a. Sets of deep and superficial veins drain these regions.
   b. The deep veins include the tibial veins, and the superficial veins include the **saphenous veins.**

Lymphatic System and Immunity

Peculiarities of peanuts, combined with our fondness for them, sets the stage for allergy, a misdirected immune reaction. anopdesignstock/iStock/Getty Images Plus

LEARN • PRACTICE • ASSESS

Peanut allergy. The young woman went to the emergency department for sudden onset of difficulty in breathing. She was also flushed and had vomited. An astute medical student, taking a quick history from the woman's roommates, discovered she had just eaten cookies from a vending machine in their dorm. Suspecting that the cookies may have contained peanuts, the medical student alerted the attending physician, who treated the woman for suspected peanut allergy—giving oxygen, an antihistamine, a steroid drug, and epinephrine. She recovered.

Peanut allergy is common and on the rise in certain countries. In the United States, about 400,000 school-aged children are allergic to peanuts. Certain glycoproteins in peanuts are allergens, causing the misdirected immune response that is an allergy. These glycoproteins are highly concentrated in the peanut and are resistant to digestion. When eaten, they disturb the intestinal lining in such a way that they enter the circulation rapidly.

Another reason peanut allergy is common in the United States is that we eat many peanuts. Virtually everyone has eaten a peanut by two years of age, usually in peanut butter. This is sufficient exposure to set the stage for a later allergy in genetically predisposed individuals. The average age of the first allergic reaction to peanuts—only fourteen months—suggests that the initial exposure necessary to "prime" the immune system for future allergic response may happen through breast milk or in the uterus.

The dry roasting of peanuts in the United States may make the three glycoproteins that evoke the allergic response more active. In China, where peanuts are equally popular but are eaten boiled or fried, peanut allergy is rare. However, children of Chinese immigrants in the United States have the same incidence of peanut allergy as other children in the United States, suggesting that the method of preparation may contribute to allergenicity.

In an experimental strategy called immunotherapy, an individual allergic to peanuts eats a tiny amount each day for at least two years. Most people then become "desensitized," able to eat up to ten peanuts a day safely.

LEARNING OUTLINE

After studying this chapter, you should be able to complete the "Learning Outcomes" that follow the major headings throughout the chapter.

14.1 Introduction
14.2 Lymphatic Pathways
14.3 Tissue Fluid and Lymph
14.4 Lymphatic Tissues and Organs
14.5 Body Defenses Against Infection
14.6 Innate (Nonspecific) Defenses
14.7 Immunity: Adaptive (Specific) Defenses

AIDS TO UNDERSTANDING WORDS

-gen [be produced] aller*gen:* substance that evokes an allergic response.

humor- [fluid] *humor*al immunity: immunity resulting from antibodies in body fluids.

immun- [free] *immun*ity: resistance to (freedom from) a specific disease.

inflamm- [set on fire] *inflamm*ation: localized redness, heat, swelling, and pain in tissues.

nod- [knot] *nod*ule: small mass of lymphocytes surrounded by connective tissue.

patho- [disease] *patho*gen: disease-causing agent.

(Appendix A has a complete list of Aids to Understanding Words.)

14.1 | Introduction

LEARN

1. Describe the general functions of the lymphatic system.

Often overlooked as one of the body systems, the **lymphatic** (lim-fat′ik) **system** is actually a secondary circulatory system. In this capacity, its function is to pick up excess tissue fluid. During the exchanges of gases, nutrients, and wastes with all body cells, much of the fluid from the blood ends up in the spaces between the cells. In these spaces, it is called interstitial fluid. To satisfy homeostatic balance and to maintain proper blood volume, the lymphatic vessels return the interstitial fluid to the cardiovascular system (fig. 14.1). Without this function, fluid could build up in tissues, causing edema.

The lymphatic vessels have enlarged regions throughout the body known as lymph *nodes*. These nodes contain masses of lymphocytes (B cells and T cells). For this reason, the lymphatic system also plays a role in immunity and defense.

Special lymphatic capillaries associated with the digestive tract, *lacteals,* are instrumental in fat absorption. This is discussed in chapter 15.

PRACTICE 14.1

Answers to the Practice questions can be found in the eBook.

1. What are the general functions of the lymphatic system?

14.2 | Lymphatic Pathways

LEARN

1. Identify the locations of the major lymphatic pathways.

The **lymphatic pathways** begin as lymphatic capillaries. These tiny tubes merge to form larger lymphatic vessels, which in turn lead to even larger lymphatic trunks that unite with the veins in the thorax.

CAREER CORNER
Public Health Nurse

A hurricane unexpectedly veers inland and churns up rivers, flooding several towns. Residents are evacuated. When the residents return weeks later, public health nurses go door-to-door, warning people about the dangers of exposure to respiratory irritants such as mold and assisting them with getting help.

Public health nurses work in a variety of settings. They vaccinate children in schools, talk to senior citizens about coping with chronic illness or physical limitations, educate parents of preschoolers about avoiding or treating head lice, and conduct blood pressure screenings at public places. These health-care professionals are particularly valuable in underserved communities, where they educate people and help them to access health-care services.

A public health nurse is a registered nurse with specialized training in community-based care. Training includes courses in public policy, health administration, and public health. It is very helpful for a public health nurse to speak the primary languages of the communities served (for instance, to be bilingual in English and Spanish). Volunteer work at the community level is a good way to prepare for this career. A public health nurse must be able to work well with groups. Employment requires passing a national licensing exam for nursing.

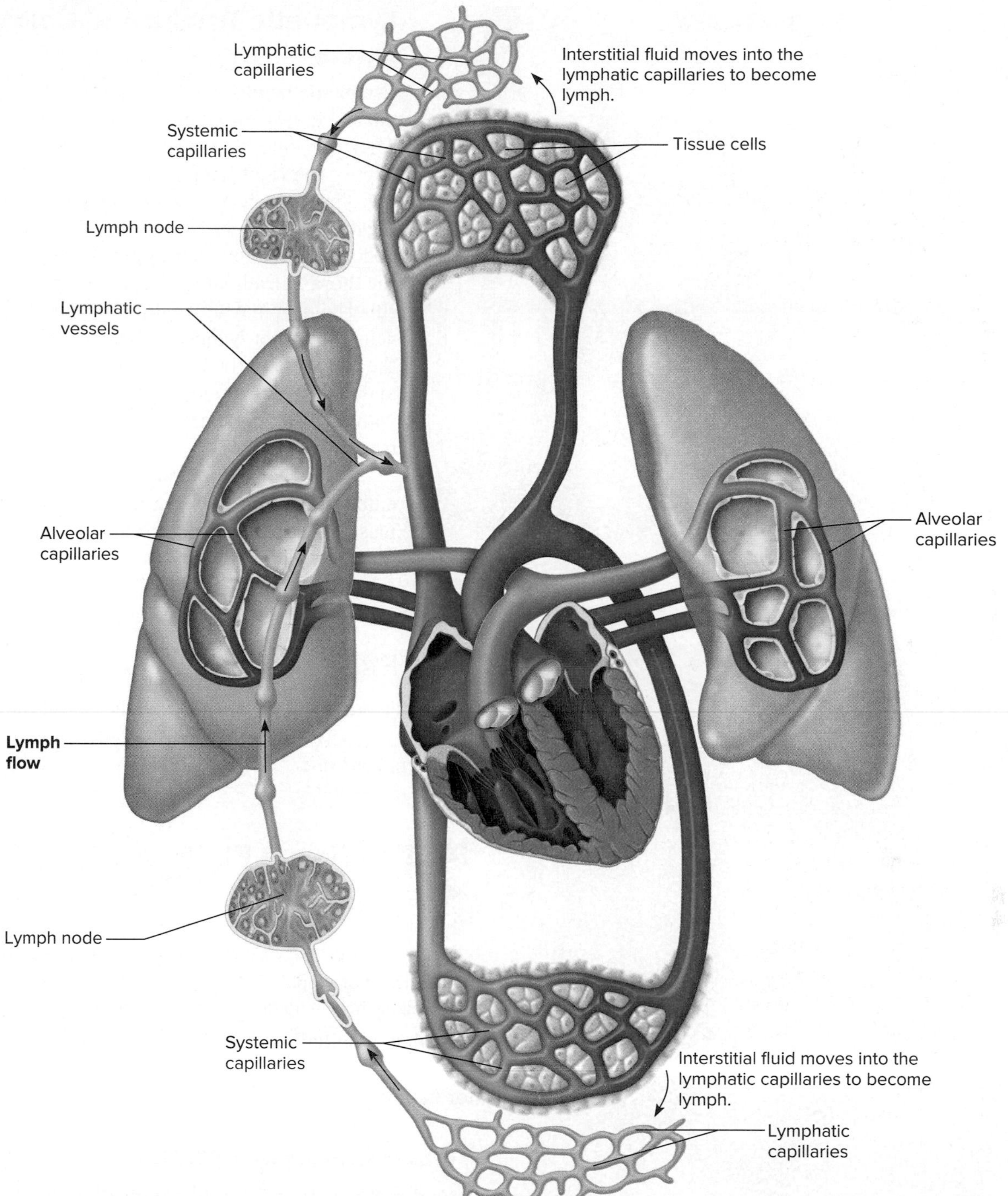

Figure 14.1 Schematic representation of lymphatic vessels transporting fluid from interstitial spaces to the bloodstream. Depending on its origin, lymph enters the right or left subclavian vein. APR

Lymphatic Capillaries

Lymphatic capillaries are microscopic, closed-ended tubes (fig. 14.2). They extend into interstitial spaces, forming complex networks that parallel the networks of the blood capillaries. Lymphatic capillaries are nearly everywhere there are blood capillaries. The walls of lymphatic capillaries, like those of blood capillaries, are formed from a single layer of squamous epithelial cells. These thin walls allow tissue fluid (interstitial fluid) to enter lymphatic capillaries. Once inside lymphatic capillaries, the fluid is called **lymph** (limf).

Lymphatic Vessels

The walls of **lymphatic vessels** are similar to those of veins, but are thinner. Like some peripheral veins, lymphatic vessels have valves that help prevent backflow of lymph (fig. 14.3).

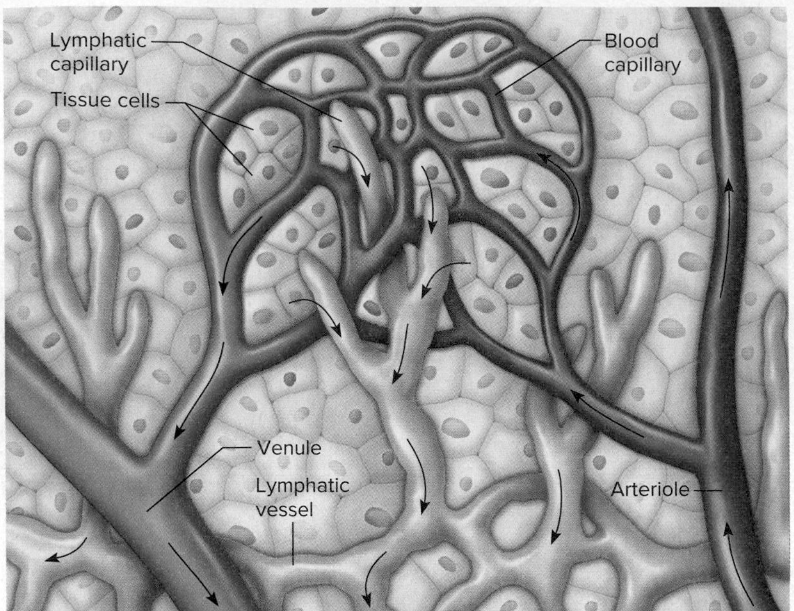

Figure 14.2 Lymphatic capillaries are microscopic, closed-ended tubes that originate in the interstitial spaces of most tissues.

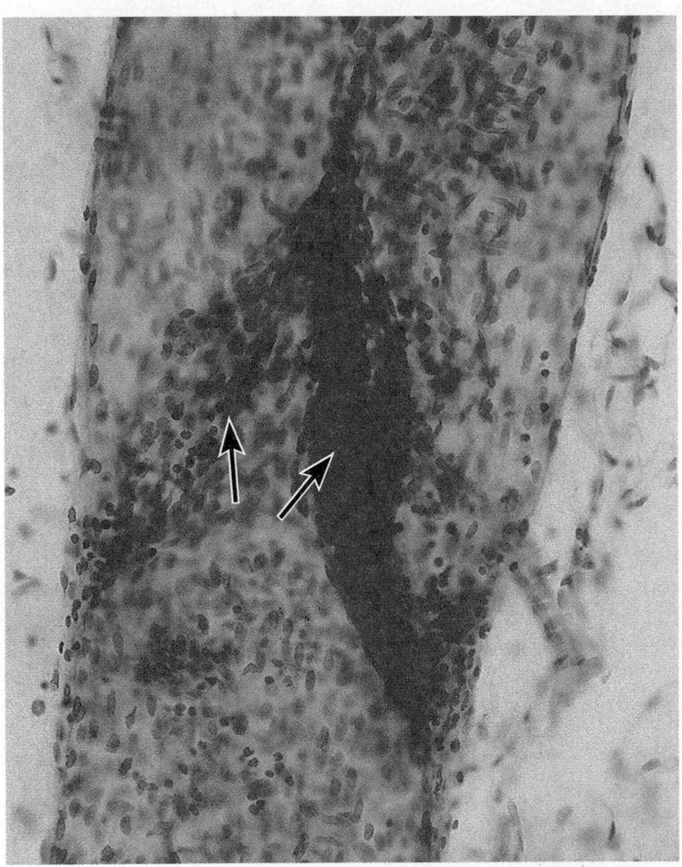

Figure 14.3 Light micrograph of the valve (arrows) within a lymphatic vessel (60x). Dennis Strete/McGraw-Hill Education

The larger lymphatic vessels lead to specialized organs called **lymph nodes** (limf nōdz). After leaving the nodes, the vessels merge to form still larger lymphatic trunks.

Lymphatic Trunks and Collecting Ducts

Lymphatic trunks, which drain lymph from the lymphatic vessels, are named for the regions they serve. They join one of two **collecting ducts**—the thoracic duct or the right lymphatic duct (fig. 14.4).

The **thoracic duct** is the wider and longer collecting duct. It receives lymph from the lower limbs and abdominal regions, left upper limb, and left side of the thorax, head, and neck, and empties into the left subclavian vein near its junction with the left jugular vein. The **right lymphatic duct** receives lymph from the right side of the head and neck, right upper limb, and right thorax, and empties into the right subclavian vein near its junction with the right jugular vein (see fig. 14.4*a*).

After leaving the two collecting ducts, lymph enters the venous system and becomes part of the plasma just before blood returns to the right atrium. Figure 14.5 summarizes the typical lymphatic pathway.

PRACTICE 14.2

1. Compare the structure of lymphatic vessels and veins.
2. Distinguish between the thoracic duct and the right lymphatic duct.

14.3 | Tissue Fluid and Lymph

LEARN

1. Describe how tissue fluid and lymph form, and explain the function of lymph.
2. Explain how lymphatic circulation is maintained.

Lymph is essentially tissue fluid that has entered a lymphatic capillary. Thus, lymph formation depends upon tissue fluid formation.

Tissue Fluid Formation

Recall from section 13.4, Blood Vessels, that tissue fluid originates from blood plasma. Tissue fluid is composed of water and dissolved substances that leave blood capillaries by filtration as capillary blood pressure forces water and small molecules from the plasma. The resulting fluid is very similar in composition to the blood plasma (including nutrients, gases, and hormones), with the important exception of the plasma proteins, most of which are too large to pass through the blood capillary walls. The osmotic effect of the plasma proteins (called the *plasma colloid osmotic pressure*) helps draw fluid back into the blood capillaries by osmosis. Some of the smaller plasma proteins do pass through the blood capillary walls, and do not contribute to the plasma colloid osmotic pressure.

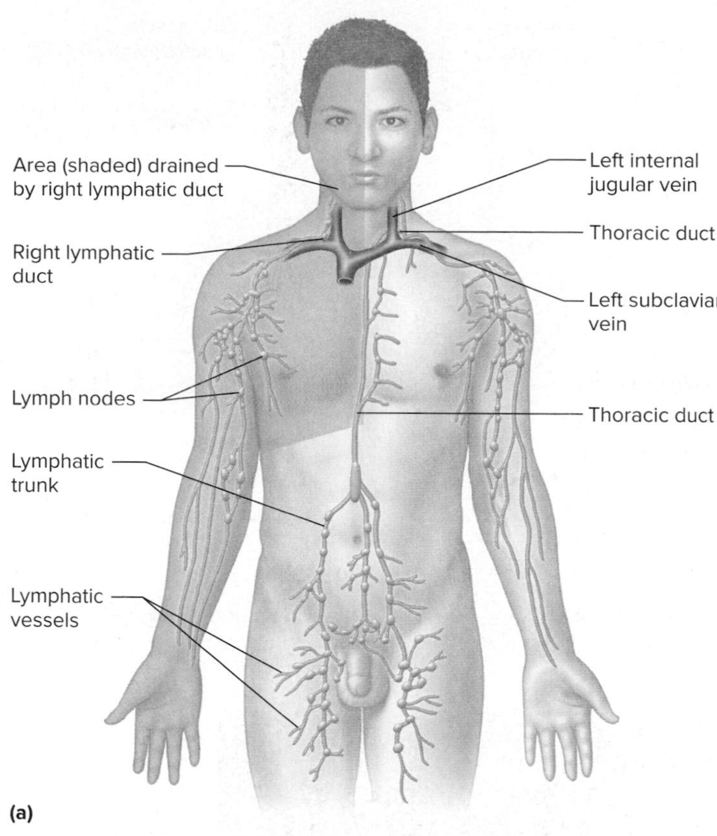

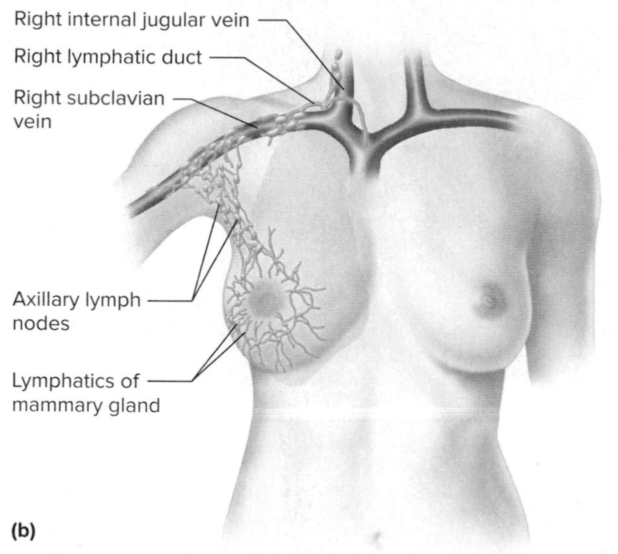

Figure 14.4 Lymphatic pathways. **(a)** The right lymphatic duct drains lymph from the upper right side of the body, whereas the thoracic duct drains lymph from the rest of the body. **(b)** Lymph drainage of the right breast illustrates a localized function of the lymphatic system. Surgery to treat breast cancer can disrupt this drainage, causing painful swelling (edema) in the upper limb of the treated side. APR

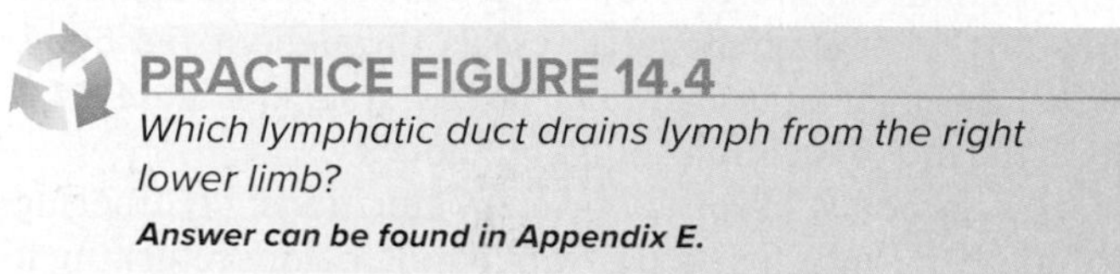

PRACTICE FIGURE 14.4

Which lymphatic duct drains lymph from the right lower limb?

Answer can be found in Appendix E.

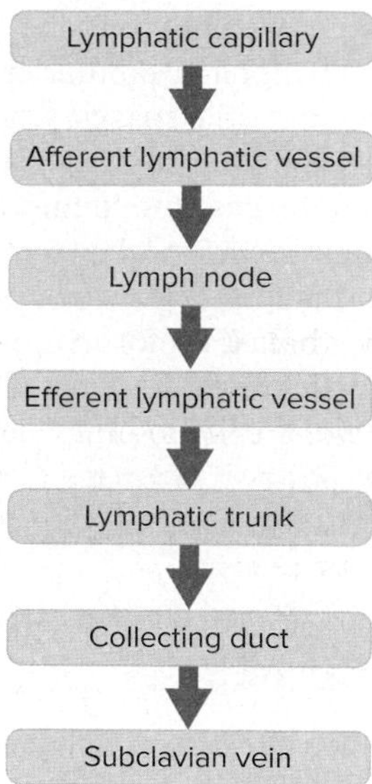

Figure 14.5 The lymphatic pathway. This pathway occurs on both the right and left sides of the body.

Lymph Formation and Function

Filtration from the plasma normally exceeds reabsorption, leading to the net formation of tissue fluid. This accumulation of tissue fluid increases the tissue fluid hydrostatic pressure, which moves tissue fluid into lymphatic capillaries, forming lymph (see fig. 14.2). Lymph returns to the bloodstream most of the small proteins that passed through the blood capillary walls. At the same time, lymph transports foreign particles, such as bacteria and viruses, to lymph nodes.

Lymph Movement

The hydrostatic pressure of tissue fluid drives lymph into lymphatic capillaries. However, muscular activity largely influences the movement of lymph through the lymphatic vessels. Lymph within lymphatic vessels, like venous blood, is under relatively low hydrostatic pressure and may not flow readily through the lymphatic vessels without help from contraction of skeletal muscles in the limbs, contraction of the smooth muscle in the walls of the larger lymphatic trunks, and pressure changes associated with breathing.

Contracting skeletal muscles compress lymphatic vessels. This squeezing action moves the lymph inside lymphatic vessels. Valves in these vessels prevent backflow, so lymph can only move toward a collecting duct. Additionally, smooth muscle in the walls of larger lymphatic trunks contracts rhythmically and compresses the lymph inside, forcing the fluid onward.

Breathing aids lymph circulation by creating a relatively low pressure in the thoracic cavity during inhalation. At the same time, the contracting diaphragm increases the pressure in the abdominal cavity. Consequently, lymph is squeezed out of the abdominal vessels and forced into the thoracic vessels. Once again, valves in lymphatic vessels prevent lymph backflow.

The continuous movement of fluid from interstitial spaces into blood and lymphatic capillaries stabilizes the volume of fluid in these interstitial spaces. Conditions that interfere with lymph movement cause tissue fluid to accumulate within the interstitial spaces, producing **edema** (ĕ-dēmah), or swelling. Edema may develop when surgery removes lymphatic tissue, preventing lymph flow. For example, a surgeon removing a cancerous breast tumor may also remove nearby axillary lymph nodes to prevent associated lymphatic vessels from transporting cancer cells to other sites. Removing this lymphatic tissue can obstruct drainage from the upper limb, causing edema (see fig. 14.4*b*).

PRACTICE 14.3

1. What is the relationship between tissue fluid and lymph?
2. How do plasma proteins in blood capillaries affect lymph formation?
3. What are the major functions of lymph?
4. What factors promote lymph flow?
5. What is the consequence of lymphatic obstruction?

14.4 | Lymphatic Tissues and Organs

LEARN

1. Describe a lymph node and its major functions.
2. Discuss the location and function of the thymus and the spleen.

Lymphatic tissue contains lymphocytes, macrophages, and other cells. The unencapsulated diffuse lymphatic tissue associated with the digestive, respiratory, urinary, and reproductive tracts is called the **mucosa-associated lymphoid tissue** (**MALT**). Compact masses of lymphatic tissue compose the tonsils and appendix (see sections 15.3, Mouth, and 15.10, Large Intestine). MALT aggregates of lymphatic tissue, called *Peyer's patches,* are scattered throughout the mucosal lining of the distal portion of the small intestine. The lymphatic organs, including the lymph nodes, thymus, and spleen, are encapsulated lymphatic tissue. A *capsule* of connective tissue with many fibers encloses each organ.

Lymph Nodes

Lymph nodes vary in size and shape, but are usually less than 2.5 centimeters long and somewhat bean-shaped (figs. 14.6 and 14.7). Blood vessels join a lymph node through the indented region of the lymph node, called the **hilum.** The lymphatic vessels leading to a lymph node (afferent vessels) enter separately at various points on its convex surface, but the lymphatic vessels leaving the lymph node (efferent vessels) exit from the hilum.

A *capsule* of connective tissue encloses each lymph node and subdivides it into compartments. Lymph nodes contain large numbers of lymphocytes (B cells and T cells) and macrophages that fight invading microorganisms. Masses of B cells and macrophages in the cortex are contained within **lymphatic nodules,** also called **lymphatic follicles,** the functional units of the lymph node. The spaces within a node, called **lymphatic sinuses,** provide a complex network of chambers and channels through which lymph circulates.

Lymph nodes are generally in groups or chains along the paths of the larger lymphatic vessels throughout the body, but are absent in the central nervous system. Figure 14.8 shows the major locations of lymph nodes.

Lymph nodes have two primary functions: (1) filtering potentially harmful particles from lymph before returning it

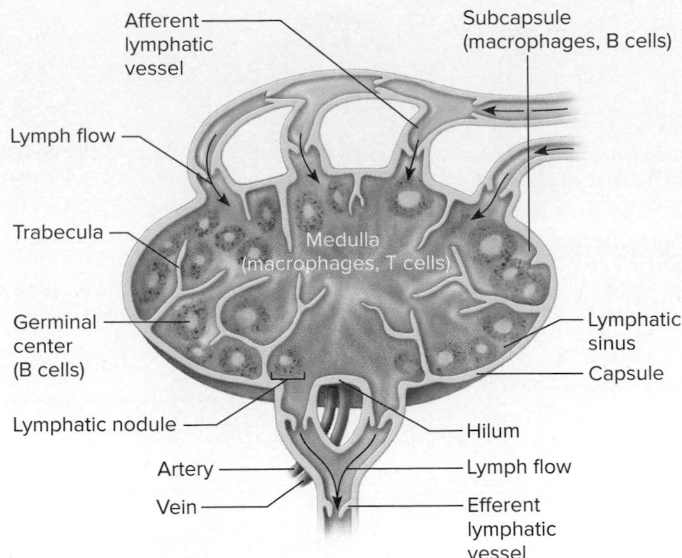

Figure 14.6 A section through a lymph node. APR

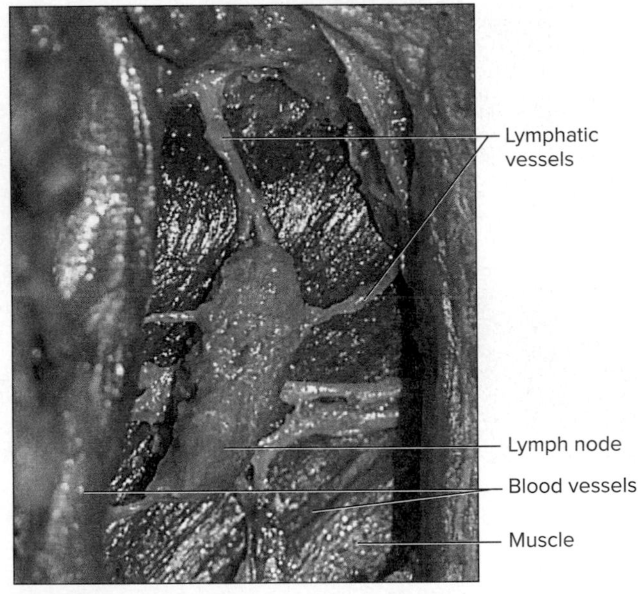

Figure 14.7 Lymph enters and leaves a lymph node through lymphatic vessels. Dr. Kent M. Van De Graaff

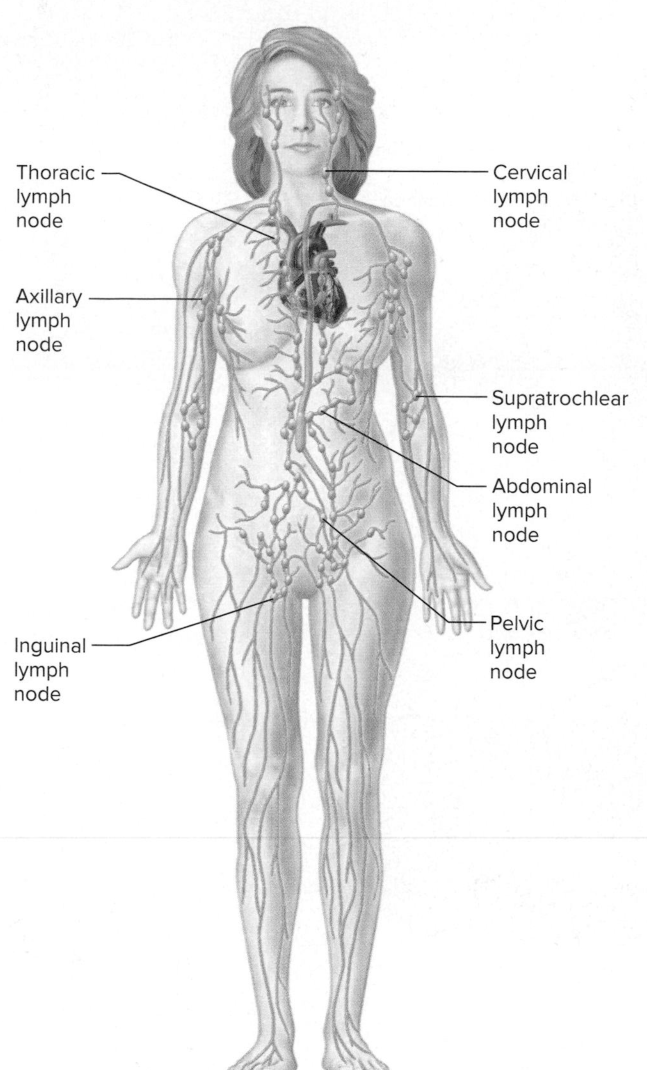

Figure 14.8 Locations of major lymph nodes.

to the bloodstream, and (2) monitoring body fluids (immune surveillance), a function performed by lymphocytes and macrophages. Along with red bone marrow, the lymph nodes are centers for lymphocyte production. Lymphocytes attack viruses, bacteria, and other parasitic cells that are brought to the lymph nodes by lymph in the lymphatic vessels. Macrophages in the lymph nodes engulf and destroy foreign substances, damaged cells, and cellular debris.

Superficial lymphatic vessels inflamed by bacterial infection appear as red streaks beneath the skin, a condition called *lymphangitis.* Inflammation of the lymph nodes, called *lymphadenitis,* often follows. In *lymphadenopathy,* affected lymph nodes enlarge and may be quite painful.

PRACTICE 14.4

1. What is the size and shape of a lymph node?
2. Where are lymph nodes located and what are their functions?

Thymus

The **thymus** (thī′mus) is a soft, bilobed gland enclosed in a connective tissue capsule and located anterior to the aorta. It is posterior to the upper part of the sternum (fig. 14.9*a*). The thymus is usually proportionately larger during infancy and early childhood, but shrinks after puberty and may be quite small in an adult. In elderly people, adipose and connective tissues replace lymphatic tissue in the thymus. In a person aged seventy, the thymus is one-tenth the size it was in that person when he or she was age ten.

Connective tissues extend inward from the surface of the thymus, subdividing it into *lobules* (fig. 14.9*b*). The lobules house many lymphocytes. Most of these cells (thymocytes) are inactive; however, some mature into *T lymphocytes* (T cells), which leave the thymus and provide immunity. Epithelial cells in the thymus secrete hormones called thymosins, which stimulate maturation of T lymphocytes.

Spleen

The **spleen,** the largest lymphatic organ, is in the upper left portion of the abdominal cavity, just inferior to the diaphragm. It is posterior and lateral to the stomach (fig. 14.9*a*). The spleen resembles a large lymph node and is subdivided into lobules. However, unlike the lymphatic sinuses of a lymph node, the spaces in the spleen, called venous sinuses, are filled with blood instead of lymph.

The tissues within splenic lobules are of two types (fig. 14.10). The *white pulp* is distributed throughout the spleen in tiny islands. This tissue is composed of splenic nodules, which are similar to the lymphatic nodules in lymph nodes and are packed with lymphocytes. The *red pulp,* which fills the remaining spaces of the lobules, surrounds the venous sinuses. This pulp contains numerous red blood cells, which impart its color, plus many lymphocytes and macrophages.

Blood capillaries in the red pulp are quite permeable. Red blood cells can squeeze through the pores in these capillary walls and enter the venous sinuses. The older, more fragile red blood cells may rupture during this passage, and the resulting cellular debris is removed by phagocytic macrophages in the venous sinuses. These macrophages also engulf and destroy foreign particles, such as bacteria, that may be carried in the blood as it flows through the venous sinuses. Thus, the spleen filters blood much as the lymph nodes filter lymph.

PRACTICE 14.4

3. Why are the thymus and the spleen considered organs of the lymphatic system?
4. What are the major functions of the thymus and the spleen?

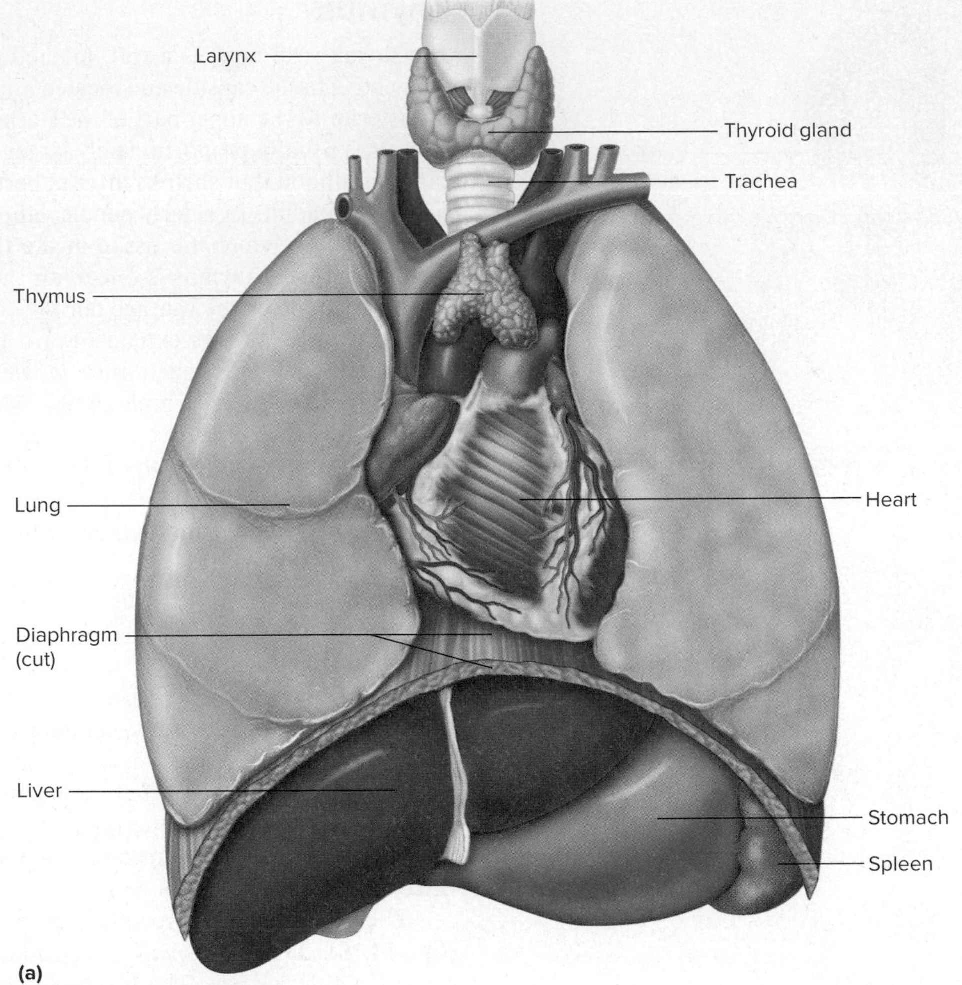

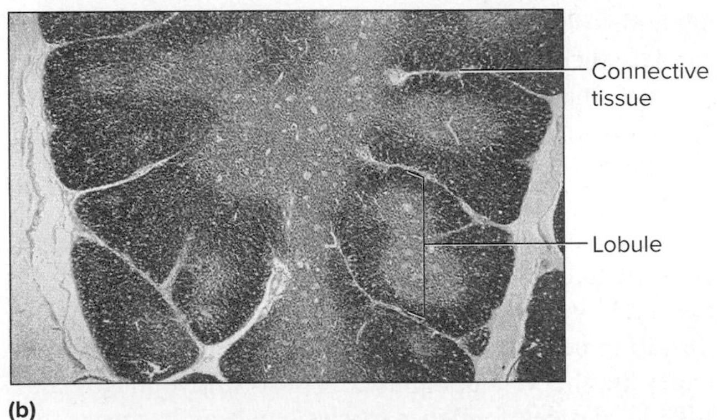

Figure 14.9 Thymus and spleen. **(a)** The thymus is bilobed, located between the lungs and superior to the heart. The spleen is located inferior to the diaphragm and posterior and lateral to the stomach. **(b)** A section through the thymus (15x). Note how the thymus is subdivided into lobules. APR

(b): Dennis Strete/McGraw-Hill Education

14.5 Body Defenses Against Infection

LEARN

1. Distinguish between innate (nonspecific) and adaptive (specific) defenses.

The presence and multiplication of a disease-causing agent, or **pathogen** (path'o-jen), if unchecked, may cause an **infection.** Pathogens include viruses, bacteria, fungi, and protozoans.

A virus is just a nucleic acid (DNA or RNA) inside a coat of proteins, and perhaps glycoproteins. An example of a DNA virus is herpes virus. Examples of diseases caused by RNA viruses are influenza, polio, and measles. Viruses are simpler than cells, and reproduce using the organelles of the cells they infect to synthesize their own proteins. Bacteria are single, simple cells and cause many common infections, such as "staph" and "strep." Fungi may be single-celled or multi-celled. Yeast infections, ringworm, and athlete's foot are fungal infections. Protozoans are single, complex cells that may cause diseases, including malaria.

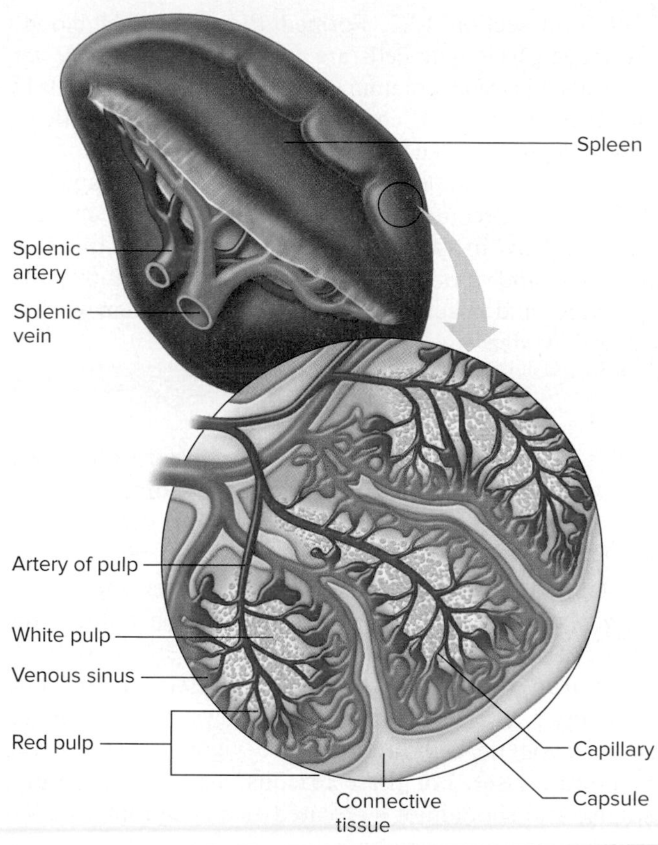

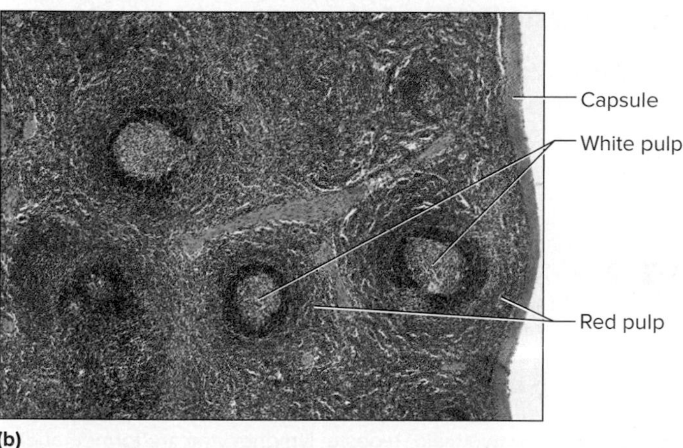

Figure 14.10 Spleen. **(a)** The spleen resembles a large lymph node. **(b)** Light micrograph of the spleen (40x). APR
(b): Al Telser/McGraw-Hill Education

The human body can prevent the entry of pathogens or destroy them if they enter. Some mechanisms are general in that they protect against many types of pathogens, providing **innate (nonspecific) defense.** These mechanisms include species resistance, mechanical barriers, inflammation, chemical barriers (enzyme action, interferon, and complement), natural killer cells, phagocytosis, and fever. Other defense mechanisms are very precise, targeting specific pathogens and providing **adaptive (specific) defense,** or **immunity** (immunity = free form or resistant to). Specialized lymphocytes that recognize foreign molecules (nonself antigens) in the body act against the foreign molecules in several ways, including the production of cytokines and antibodies. Innate and adaptive defense mechanisms work together to protect the body against infection. The innate defenses respond quite rapidly, while adaptive defenses develop more slowly.

PRACTICE 14.5

1. What may cause an infection?
2. In general, how do innate and adaptive defenses differ?

14.6 | Innate (Nonspecific) Defenses

LEARN

1. List seven innate body defense mechanisms, and describe the action of each mechanism.

Species Resistance

Species resistance refers to the fact that a given type of organism, or *species* (such as the human species, *Homo sapiens*), may be resistant to infectious diseases that affect other species. The basis of species resistance is that the cells of a resistant species do not have receptors for the pathogen, or the resistant species' tissues do not provide the temperature or chemical environment that a particular pathogen requires. For example, humans contract measles, mumps, gonorrhea, and syphilis, but other animal species do not.

Mechanical Barriers

The skin and mucous membranes lining the passageways of the respiratory, digestive, urinary, and reproductive systems create **mechanical barriers** that prevent the entrance of some infectious agents. As long as the skin and mucous membranes remain intact, many pathogens are unable to penetrate them. Hair traps infectious agents associated with the skin and mucous membranes, and sweat and mucus rinse away microorganisms. These mechanical barriers provide a *first line of defense.* The other innate defenses discussed in this section are part of the *second line of defense.*

Inflammation

Inflammation is a tissue response to injury or infection, producing localized redness, swelling, heat, and pain. The redness is a result of blood vessel dilation that increases blood flow and volume in the affected tissues. This effect, coupled with an increase in the permeability of nearby capillaries and subsequent leakage of protein-rich fluid into tissue spaces, swells tissues (edema). The heat comes as blood enters from deeper body parts, which are warmer than the surface. Pain results from stimulation of nearby pain receptors.

Infected cells release chemicals that attract white blood cells to sites of inflammation. Here the white blood cells phagocytize pathogens. Local heat speeds up phagocytic activity. In bacterial infections, the resulting mass of white blood cells, bacterial cells, and damaged tissue may form a thick fluid called **pus.**

Fluids that leak out of the capillaries, called *exudates,* also collect in inflamed tissues. These fluids contain fibrinogen and other blood-clotting factors. Clotting forms a network of fibrin threads in the affected region. Later, fibroblasts may arrive and secrete matrix components until the area is enclosed in a connective tissue sac. This walling off of the infected area helps inhibit the spread of pathogens and toxins to adjacent tissues.

Chemical Barriers

Enzymes in body fluids provide a **chemical barrier** to pathogens. Gastric juice, for example, contains the protein-splitting enzyme pepsin and has a low pH due to the presence of hydrochloric acid (HCl) (see section 15.6, Stomach). The combined effect of pepsin and HCl kills many pathogens that enter the stomach. Similarly, tears contain the enzyme lysozyme, which destroys certain bacteria on the eyes. The accumulation of salt from perspiration kills certain bacteria on the skin.

Lymphocytes and fibroblasts produce hormonelike peptides called **interferons** in response to viruses or tumor cells. Once released from the virus-infected cell, interferon binds to receptors on uninfected cells, stimulating them to synthesize proteins that block replication of a variety of viruses. Thus, interferon's effect is nonspecific. Interferons also stimulate phagocytosis and enhance the activity of other cells that help resist infections and the growth of tumors.

The **complement system** is a group of plasma proteins. Part of the immune system, it enhances, or "complements," the action of antibodies and phagocytes. Activation of the complement system also stimulates the inflammation process.

Natural Killer (NK) Cells

Natural killer (NK) cells are a small population of lymphocytes. They are different from the lymphocytes (T and B cells) that provide adaptive (specific) defense mechanisms discussed later in this chapter. NK cells defend the body against various viruses and cancer cells by secreting cytolytic ("cell-cutting") substances called **perforins** that lyse the cell membrane, destroying the infected cell. NK cells also secrete chemicals that enhance inflammation.

Phagocytosis

Phagocytosis removes foreign particles from the lymph as it moves from the interstitial spaces to the bloodstream. Phagocytes in the blood vessels and in the tissues of the spleen, liver, or bone marrow remove particles that reach the blood. Recall from section 12.2, Formed Elements, that blood's most active phagocytic cells are *neutrophils* and *monocytes.* Chemicals released from injured tissues attract these cells by *chemotaxis.* Neutrophils engulf and digest smaller particles; monocytes phagocytize larger ones.

Monocytes that leave the bloodstream by diapedesis (see fig. 12.8) become *macrophages.* These large cells may be *free,* or *fixed* in various tissues. The fixed macrophages can divide and produce new macrophages. Neutrophils, monocytes, and macrophages constitute the **mononuclear phagocytic system** (reticuloendothelial system).

Fever

Fever is body temperature elevated above an individual's normal temperature due to an elevated setpoint. It is part of the innate defense because, as a result of the fever, the body becomes inhospitable to certain pathogens. Higher body temperature inhibits certain pathogens and causes the liver and spleen to sequester iron, which reduces the level of iron in the blood. Because bacteria and fungi require iron for normal metabolism, their growth and reproduction in a fever-ridden body slows and may cease. Also, phagocytic cells attack more vigorously when the temperature rises. For these reasons, a fever is a natural response and should not be treated as a symptom unless it lasts too long or gets too high.

PRACTICE 14.6

1. What is considered a first line of defense? Provide examples.
2. Explain seven innate (nonspecific) defense mechanisms.

14.7 Immunity: Adaptive (Specific) Defenses

LEARN

1. Explain how two major types of lymphocytes are formed and activated, and how they function in immune mechanisms.
2. Discuss the actions of the five types of antibodies.
3. Distinguish between primary and secondary immune responses.
4. Distinguish between active and passive immunity.
5. Explain how allergic reactions, tissue rejection reactions, and autoimmunity arise from immune mechanisms.

The *third line of defense,* **immunity,** is resistance to specific pathogens or to their toxins or metabolic by-products. Lymphocytes and macrophages that recognize and remember specific foreign molecules carry out adaptive immune responses, which include the *cellular immune response* and the *humoral immune response.* **Figure 14.11** shows a schematic representation of body defenses against pathogens.

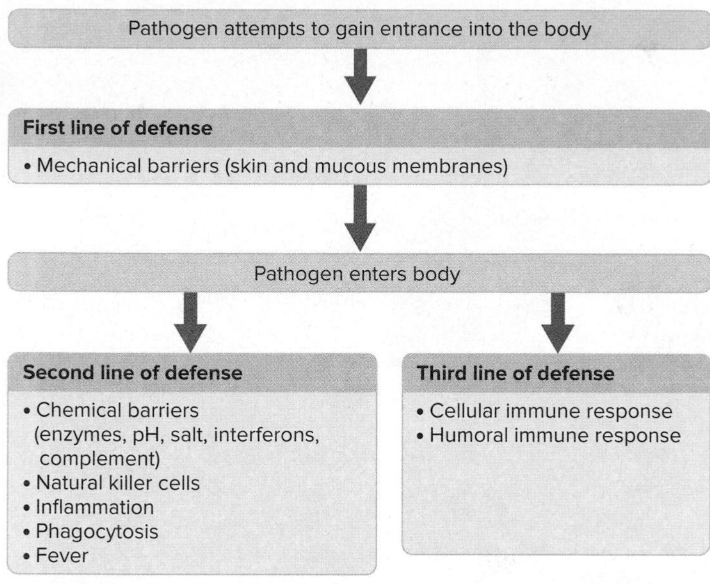

Pathogen attempts to gain entrance into the body
First line of defense
• Mechanical barriers (skin and mucous membranes)
Pathogen enters body
Second line of defense
• Chemical barriers (enzymes, pH, salt, interferons, complement)
• Natural killer cells
• Inflammation
• Phagocytosis
• Fever
Third line of defense
• Cellular immune response
• Humoral immune response

Figure 14.11 Body defenses against pathogens.

Antigens

Antigens (an'tĭ-jenz) are proteins, polysaccharides, glycoproteins, or glycolipids that can elicit an immune response. Before birth, cells inventory the antigens in the body, learning to identify these as "self." The immune response is to "nonself," or foreign, antigens, but not normally to self antigens. Receptors on lymphocyte surfaces enable these cells to recognize nonself antigens.

The antigens most effective in eliciting an immune response are large and complex, with few repeating parts. A smaller molecule that cannot by itself stimulate an immune response may combine with a larger molecule, which makes it detectable. Such a small molecule is called a **hapten** (hap'ten). Stimulated lymphocytes react either to the hapten or to the larger molecule of the combination. Hapten molecules are in drugs such as penicillin, in household and industrial chemicals, in dust particles, and in animal dander.

Lymphocyte Origins

Lymphocyte production begins during fetal development and continues throughout life, with red bone marrow releasing unspecialized precursors to lymphocytes into the circulation. About half of these cells reach the thymus, where they specialize into **T lymphocytes,** or **T cells.** After leaving the thymus, some of these T cells constitute 70% to 80% of the circulating lymphocytes in blood. Other T cells reside in lymphatic organs and are particularly abundant in the lymph nodes, thoracic duct, and white pulp of the spleen.

Other lymphocytes remain in the red bone marrow until they differentiate into **B lymphocytes,** or **B cells.** The blood distributes B cells, which constitute 20% to 30% of circulating lymphocytes (fig. 14.12). B cells settle in lymphatic organs along with T cells and are abundant in the lymph nodes, spleen, bone marrow, and intestinal lining. Figure 14.13 illustrates B cell and T cell production. Table 14.1 compares the characteristics of T cells and B cells.

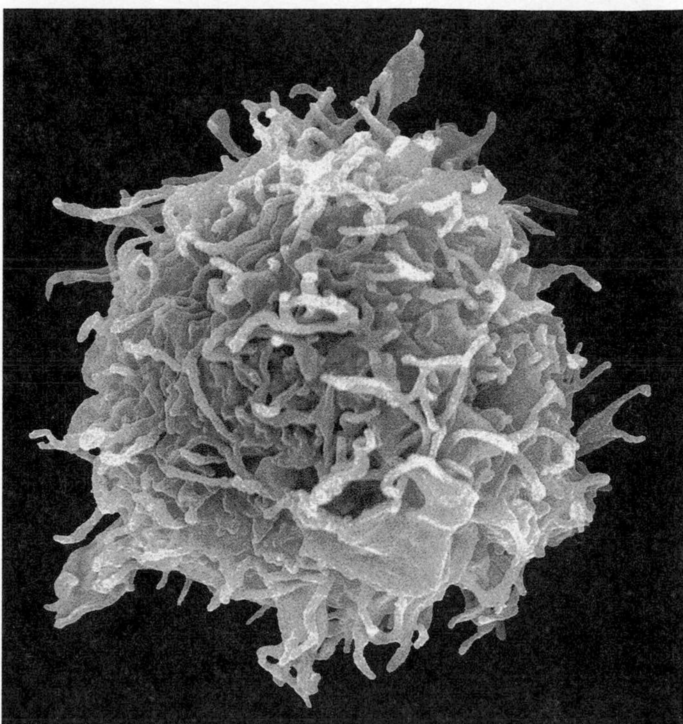

Figure 14.12 Falsely colored scanning electron micrograph of a B cell (6,000x). Science Photo Library/Getty Images

PRACTICE 14.7

1. What is immunity?
2. What is the difference between an antigen and a hapten?
3. How do T cells and B cells originate?

T Cells and the Cellular Immune Response

A lymphocyte must be activated before it can respond to an antigen. T cell activation requires that processed fragments of the antigen be attached to the surface of another type of cell, called an **antigen-presenting cell** (accessory cell). Macrophages, B cells, and several other cell types can be antigen-presenting cells.

T cell activation may occur when a macrophage phagocytizes a bacterium and digests it within a *phagolysosome* formed by the fusion of the vesicle containing the bacterium (phagosome) and a lysosome. Some of the resulting bacterial antigens are then displayed on the macrophage's cell membrane near certain protein molecules that are part of a

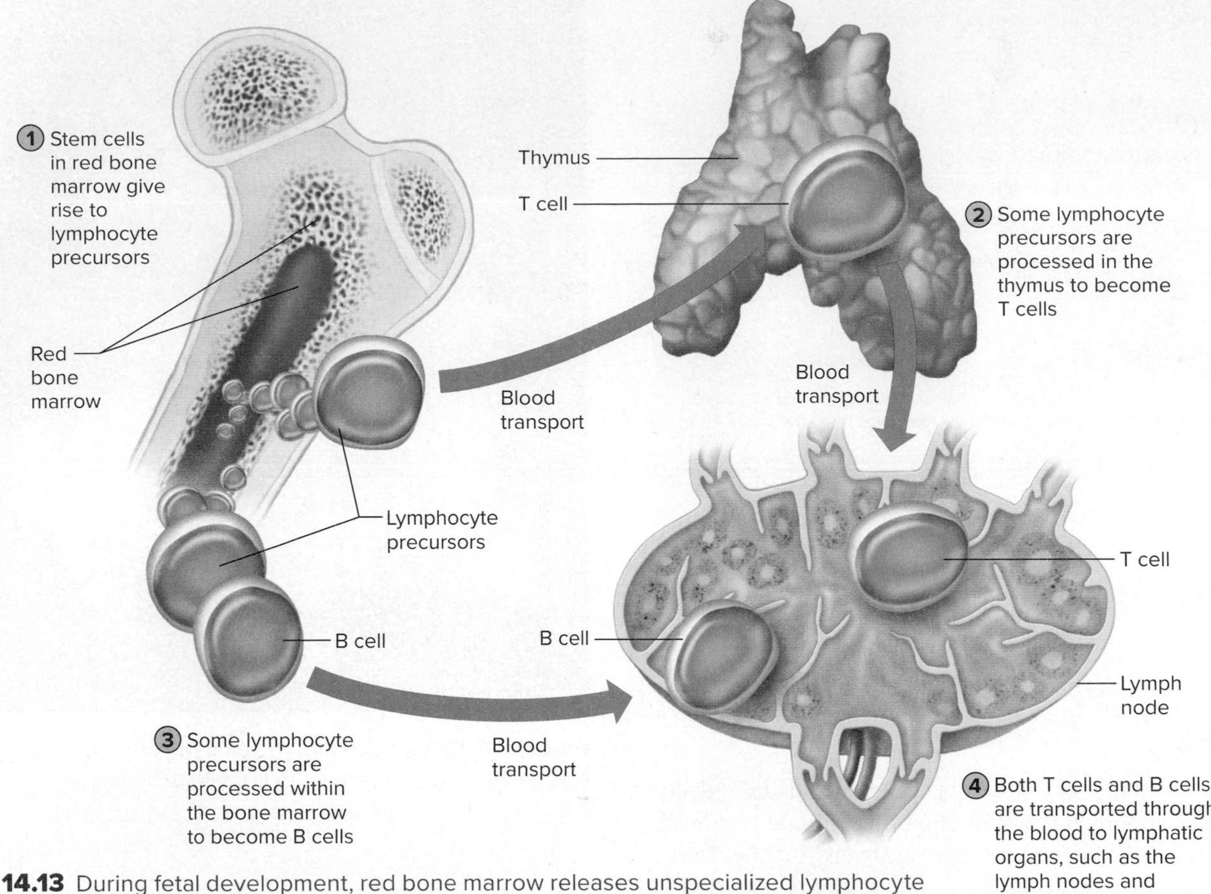

Figure 14.13 During fetal development, red bone marrow releases unspecialized lymphocyte precursors, which after processing specialize as T cells (T lymphocytes) or B cells (B lymphocytes). Lymphocyte production continues throughout life, with most T cells arising before puberty.

TABLE 14.1 A Comparison of T Cells and B Cells APR

Characteristic	T Cells	B Cells
Origin of undifferentiated cell	Red bone marrow	Red bone marrow
Site of differentiation	Thymus	Red bone marrow
Primary locations	Lymphatic tissues, 70–80% of the circulating lymphocytes in the blood	Lymphatic tissues, 20–30% of the circulating lymphocytes in the blood
Primary functions	Provides cellular immune response in which T cells interact directly with the antigens or antigen-bearing agents, to destroy them	Provides humoral immune response in which B cells interact indirectly, producing antibodies that destroy the antigens or antigen-bearing agents

group of proteins called the *major histocompatibility complex* (*MHC*). MHC antigens help T cells recognize that a newly displayed antigen is foreign (nonself).

Activated T cells interact directly with antigen-bearing cells. Such cell-to-cell contact is called the **cellular immune response,** or cell-mediated immunity. T cells (and some macrophages) also synthesize and secrete polypeptides called *cytokines* that enhance certain cellular responses to antigens. For example, *interleukin-1* and *interleukin-2* stimulate the synthesis of several other cytokines from other T cells. Additionally, interleukin-1 helps activate T cells, whereas interleukin-2 causes T cells to proliferate. This proliferation increases the number of T cells in a **clone** (klōn), which is a group of genetically identical cells that descend from a single, original cell. Other cytokines, called *colony-stimulating factors* (*CSFs*), stimulate leukocyte production in red bone marrow and activate macrophages. T cells may also secrete toxins that kill their antigen-bearing target cells, growth-inhibiting factors that prevent target cell growth, or interferon that inhibits the proliferation of viruses and tumor cells. Several types of T cells have distinct functions.

A specialized type of T cell, called a *helper T cell,* is activated when its antigen receptor combines with a displayed foreign antigen (fig. 14.14). Once activated, the helper T cell

③ Macrophage displays antigen on its surface
Phagolysosome
Displayed antigen
Cytotoxic T cell
② Macrophage digests antigen-bearing agent
B cells
① Macrophage phagocytizes antigen-bearing agent
Cytotoxic T cell contacts displayed antigen
④
Helper T cell
Helper T cell contacts displayed antigen and proliferates
⑦ B cell combines with antigen
Cytokines
Proliferation
Interleukin
Antigen receptor
Antigen
Helper T cells
Cytotoxic T cell
⑤
Activated helper T cell interacts with B cell (which has combined with an identical antigen) and releases cytokines, which activate the B cell
Activated helper T cell interacts with cytotoxic T cell (which has combined with an identical antigen) and releases interleukin-2, which activates the cytotoxic T cell
⑥ Proliferation and differentiation
⑧ Activated B cell
Memory T cell
Cytotoxic T cell

Figure 14.14 T and B cell activation. Antigen-presenting cells activate T cells (① – ⑥) and B cells (⑦ – ⑧), triggering an immune response. APR

proliferates and the resulting cells stimulate B cells to produce antibodies that are specific for the displayed antigen.

Another type of T cell is a *cytotoxic T cell,* which recognizes and combines with nonself antigens that cancerous cells or virally infected cells display on their surfaces near certain MHC proteins. Cytokines from helper T cells activate the cytotoxic T cell. Next, the cytotoxic T cell proliferates. Cytotoxic T cells then bind to the surfaces of antigen-bearing cells, where they release *perforin* protein that cuts porelike openings in the cell membrane, destroying these cells. In this way, cytotoxic T cells continually monitor the body's cells, recognizing and eliminating tumor cells and cells infected with viruses. Cytotoxic T cells provide much of the body's defense against HIV infection, discussed in Clinical Application 14.1.

Some cytotoxic T cells do not respond to a nonself antigen on first exposure, but remain as *memory T cells* that provide for future immune protection. Upon subsequent exposure to the same antigen, these memory cells immediately divide to yield more cytotoxic T cells and helper T cells, often before symptoms arise.

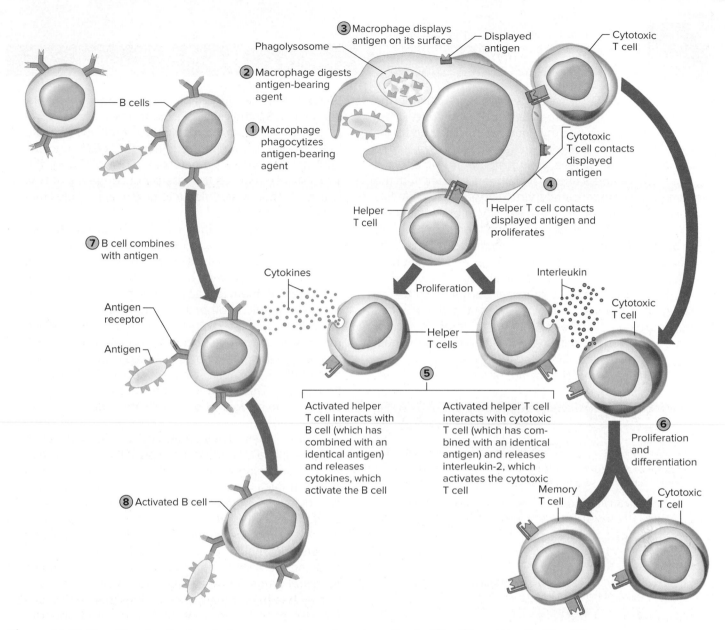

PRACTICE 14.7

4. How do T cells become activated?
5. What are some functions of cytokines?
6. Name three types of T cells.
7. How do cytotoxic T cells destroy cells bearing foreign antigens?

CLINICAL APPLICATION 14.1

Immunity Breakdown: HIV/AIDS

In the early 1980s, physicians from large cities in the United States began reporting cases of rare infections in otherwise healthy young men. The conditions were opportunistic, taking advantage of a weakened immune system. These young men were among the first in the United States to have been infected with the human immunodeficiency virus (HIV). Table 14A lists how HIV is, and isn't, spread. The HIV infection typically remains silent for several years and then may progress to *acquired immune deficiency syndrome* (*AIDS*), which starts with recurrent fever, weakness, and weight loss. Then, the opportunistic infections begin.

First, HIV crosses a mucosal barrier, such as the lining of the anus or vagina. Then, the virus enters macrophages. In these cells, and later in helper T cells, the virus adheres to two receptors on the host cell surface, called CD4 and CCR5. Once the virus enters the cell, a viral enzyme, reverse transcriptase, builds a DNA strand complementary to the viral RNA sequence. Synthesis of a second DNA strand complementary to the first results in a viral DNA double helix, which is transported into the cell's nucleus and inserted into a chromosome. The cell uses the viral DNA sequences to mass-produce pieces of HIV, which are then assembled into new viral particles that burst from the cell.

TABLE 14A	HIV Transmission
How HIV Is Transmitted	
Sexual contact, particularly anal intercourse, but also vaginal intercourse and oral sex	
Contaminated needles (intravenous drug use, injection of anabolic steroids, accidental needle stick in medical setting)	
During birth from infected mother	
Breast milk from infected mother	
Receiving infected blood or other tissue (precautions usually prevent this)	
How HIV Is Not Transmitted	
Casual contact (social kissing, hugging, handshakes)	
Objects (toilet seats, deodorant sticks, doorknobs)	
Mosquitoes	
Sneezing and coughing	
Sharing food	
Swimming in the same water	
Donating blood (donated blood is screened)	

Once infected helper T cells start to rapidly die, bacterial infections begin because B cells aren't activated to produce antibodies. Much later in infection, HIV variants arise that bind receptors on cytotoxic T cells, killing them too. Loss of these cells renders the body vulnerable to other infections and to cancers.

HIV replicates quickly and alters its surface features in ways that evade recognition and attack by the immune response. The virus is especially prone to mutations because it cannot repair DNA replication errors. The immune response cannot keep up, with antibodies against one viral variant being useless against another.

Within days of infection, HIV variants can arise that resist the drugs used to treat HIV infection and AIDS. Combining drugs that act at different steps in infection minimizes the number of viruses (viral load) and delays symptom onset and progression. These drugs block HIV from binding to cells, fusing with the cell membrane, entering the cell, replicating viral genetic material, or processing viral proteins to functional sizes. Many people with HIV infection who use retroviral drugs can keep their viral levels low enough to stay healthy for many years. More than 200 drugs are also available to treat AIDS-associated opportunistic infections and cancers.

Developing a vaccine against HIV has been challenging, because of the great variability of HIV and the difficulty of testing the vaccine. A live vaccine is too dangerous; and a killed vaccine does not generate a sufficient immune response. HIV infection may be prevented other ways, including education about reducing risk factors, and giving infected pregnant women antiretroviral drugs to prevent transmission of the virus to offspring.

B Cells and the Humoral Immune Response

When a B cell encounters an antigen whose molecular shape fits the shape of the B cell's antigen receptors, it becomes activated. In response to the receptor-antigen combination, the B cell divides repeatedly, expanding its clone. However, most of the time B cell activation requires T cell "help."

When an activated helper T cell encounters a B cell already combined with a foreign antigen identical to the one that activated the helper T cell, the helper T cell releases certain cytokines. These cytokines stimulate the B cell to proliferate, increasing the number of cells in its clone of antibody-producing cells (fig. 14.15). The cytokines also attract macrophages and leukocytes into inflamed tissues and help keep them there.

Some members of the activated B cell's clone differentiate further into **plasma cells,** which can produce and secrete large globular proteins called **antibodies** (an'tĭ-bod"ēz), also called **immunoglobulins** (im"u-nō-glob'u-linz), at the rate of 2,000 molecules per second. Antibodies are similar in structure to the antigen-receptor molecules on the original B cell's surface. Body fluids carry antibodies, which then react in various ways to destroy specific antigens or antigen-bearing particles. This antibody-mediated immune response is called the **humoral immune response** ("humoral" refers to body fluids).

An individual's B cells can produce more than 1,000,000,000 different antibodies, each reacting against a specific antigen. The enormity and diversity of the antibody response defends against many pathogens. Table 14.2 summarizes the steps leading to antibody production as a result of B cell and T cell activities.

Other members of the activated B cell's clone differentiate further into *memory B cells.* Like memory T cells, these memory B cells respond rapidly to subsequent exposure to a specific antigen.

Types of Antibodies

Antibodies (immunoglobulins) are soluble, globular proteins that constitute the *gamma globulin* fraction of plasma

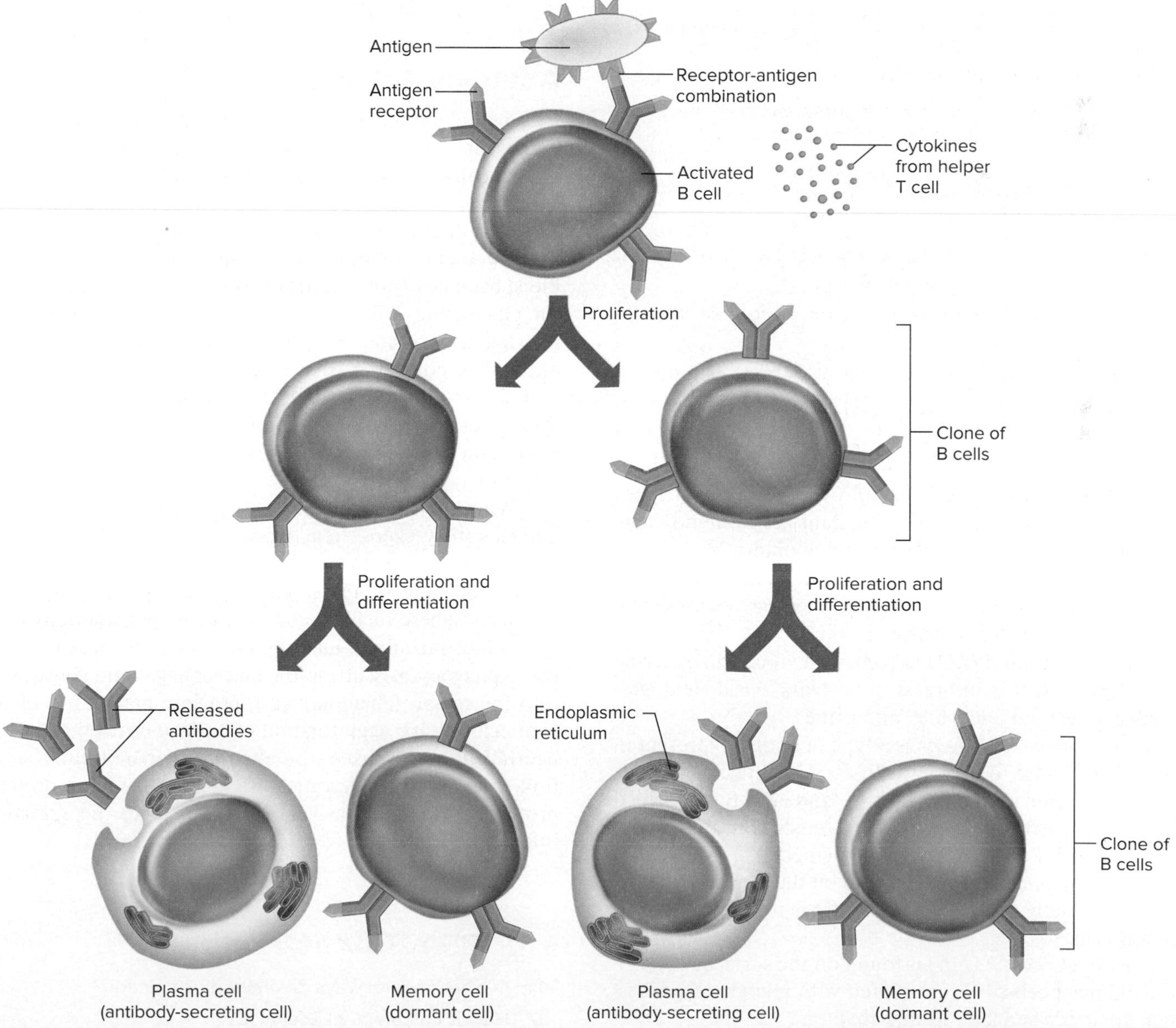

Figure 14.15 An activated B cell proliferates after stimulation by cytokines released from helper T cells. The B cell's clone enlarges. Some cells of the clone give rise to antibody-secreting plasma cells and others to dormant memory cells.

TABLE 14.2 Steps in Antibody Production

B Cell Activities
1. Antigen-bearing agents enter tissues.
2. B cell encounters an antigen that fits its antigen receptors.
3. Either alone or more often in conjunction with helper T cells, the B cell is activated. The B cell proliferates, enlarging its clone.
4. Some of the newly formed B cells differentiate further to become plasma cells.
5. Plasma cells synthesize and secrete antibodies whose molecular structure is similar to the activated B cell's antigen receptors.
T Cell Activities
1. Antigen-bearing agents enter tissues.
2. An accessory cell, such as a macrophage, phagocytizes the antigen-bearing agent, and the macrophage's lysosomes digest the agent.
3. Antigens from the digested antigen-bearing agents are displayed on the membrane of the accessory cell.
4. Helper T cell becomes activated when it encounters a displayed antigen that fits its antigen receptors.
5. Activated helper T cell releases cytokines when it encounters a B cell that has previously combined with an identical antigen-bearing agent.
6. Cytokines stimulate the B cell to proliferate, enlarging its clone.
7. Some of the newly formed B cells give rise to cells that differentiate into antibody-secreting plasma cells.

proteins (see section 12.3, Plasma). Of the five major types of immunoglobulins, the most abundant are immunoglobulin G, immunoglobulin A, and immunoglobulin M.

Immunoglobulin G (IgG) is in plasma and tissue fluids and is particularly effective against bacteria, viruses, and toxins. It also activates *complement.*

Immunoglobulin A (IgA) is commonly found in exocrine gland secretions. It is in breast milk, tears, nasal fluid, gastric juice, intestinal juice, bile, and urine.

Immunoglobulin M (IgM) is a type of antibody present in plasma in response to contact with certain antigens in foods or bacteria. The antibodies anti-A and anti-B, described in section 12.5, Blood Groups and Transfusions, are examples of IgM. IgM also activates complement.

Immunoglobulin D (IgD) is found on the surfaces of most B cells, especially those of infants. IgD is important in activating B cells.

Immunoglobulin E (IgE) is found on the surfaces of basophils and mast cells. It is associated with allergic responses, which are described later in this chapter.

A newborn does not yet have its own antibodies, but does retain for a while IgG that passed through the placenta from the mother. These maternal antibodies protect the infant against some illnesses to which the mother is immune. As the maternal antibody supply falls, the infant begins to manufacture its own antibodies. The newborn also receives IgA from colostrum, a substance secreted from the mother's breasts for the first few days after birth, and then from breast milk. Antibodies in colostrum protect against certain digestive and respiratory infections.

PRACTICE 14.7

8. How are B cells activated?
9. How does the antibody response protect against diverse infections?
10. Which immunoglobulins are most abundant, and how do they differ from each other?

Antibody Actions

In general, antibodies react to antigens in three ways. Antibodies directly attack antigens, activate complement, or stimulate localized changes (inflammation) that help prevent the spread of pathogens or cells bearing foreign antigens.

In a direct attack, antibodies combine with antigens, causing them to clump (agglutination) or to form insoluble substances (precipitation). Such actions make it easier for phagocytic cells to recognize and engulf the antigen-bearing agents and eliminate them. In other instances, antibodies cover the toxic portions of antigen molecules and neutralize their effects (neutralization). However, under normal conditions, direct antibody attack is not as important as complement activation in protecting against infection.

When certain IgG or IgM antibodies combine with antigens, they expose reactive sites on antibody molecules. This triggers a series of reactions, leading to activation of the complement proteins, which in turn produce a variety of effects. These include: coating the antigen-antibody complexes (opsonization), making the complexes more susceptible to phagocytosis; attracting macrophages and neutrophils into the region (chemotaxis); rupturing membranes of foreign cells (lysis); agglutination of antigen-bearing cells; and neutralization of viruses by altering their molecular structure, making them harmless (fig. 14.16). Other proteins promote inflammation, which helps prevent the spread of infectious agents.

PRACTICE 14.7

11. In what general ways do antibodies function?
12. How is complement activated?
13. What are the effects of complement activation?

Immune Responses

Activation of B cells or T cells after they first encounter the antigens for which they are specialized to react constitutes a **primary immune response.** During such a response, plasma cells release antibodies (IgM, followed by IgG) into the lymph. The antibodies are transported to the blood and then throughout the body, where they help destroy antigen-bearing agents. Production and release of antibodies continues for several weeks.

Following a primary immune response, some of the B cells produced during proliferation of the clone remain dormant as memory cells (see fig. 14.15). If the same antigen is encountered again, the clones of these memory cells enlarge, and they can respond rapidly by producing IgG to the antigen to which they were previously sensitized. These memory B cells, along with the memory cytotoxic T cells, produce a **secondary immune response.**

As a result of a primary immune response, detectable concentrations of antibodies usually appear in the blood plasma five to ten days after exposure to antigens. If the same type of antigen is encountered later, a secondary immune response may produce the same antibodies within a day or two (fig. 14.17). Although newly formed antibodies may persist in the body for only a few months or years, memory cells live much longer.

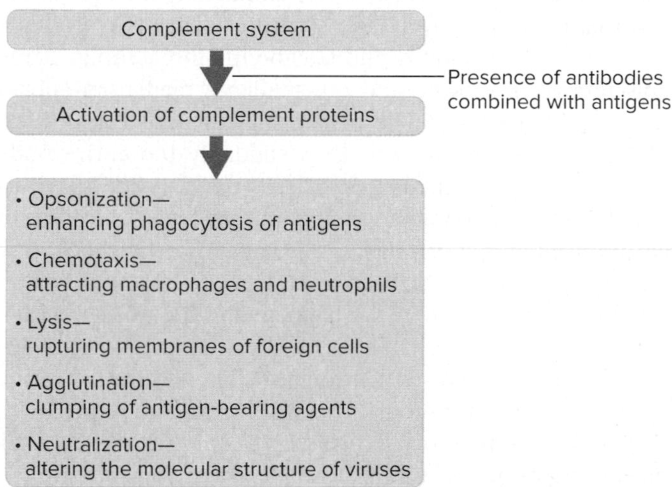

Figure 14.16 Actions of the complement system.

Practical Classification of Immunity

Adaptive, also known as acquired, immunity can arise in response to natural events or be induced artificially by injection. Both naturally and artificially acquired immunities can be either active or passive. Active immunity results when the person produces an immune response (including memory cells) to the antigen; it is long-lasting. Passive immunity occurs when a person receives antibodies produced by another individual. Because the person does not produce an immune response, passive immunity is short-term. That is, the individual will be susceptible upon exposure to the antigen at some later date.

Naturally acquired active immunity occurs when a person exposed to a pathogen develops a disease. Resistance to that pathogen is the result of a primary immune response.

A **vaccine** (vak′sēn) is a preparation that produces *artificially acquired active immunity.* A vaccine might consist of bacteria or viruses that have been killed or weakened so that they cannot cause a serious infection, or only molecules unique to the pathogens. A vaccine might also be a toxoid, which is a toxin from an infectious organism that has been chemically altered to destroy its dangerous effects. Whatever its composition, a vaccine includes the antigens that

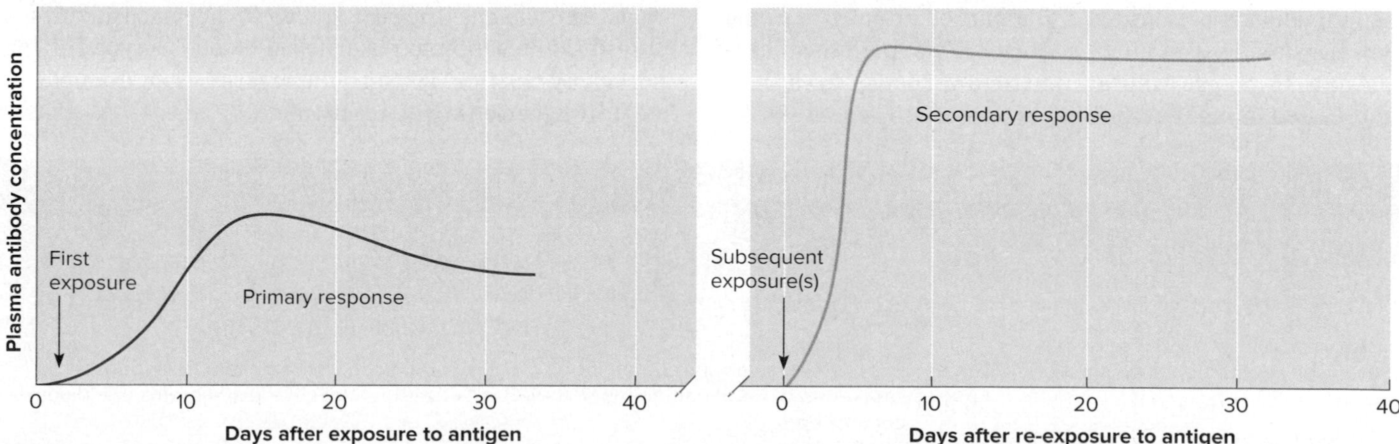

Figure 14.17 A primary immune response occurs after the first exposure to an antigen. A secondary immune response occurs after subsequent exposure(s) to the same antigen. The break in the timeline represents a time lapse between exposure(s) to the antigen.

PRACTICE FIGURE 14.17

Which immune response produces antibodies to a specific antigen more rapidly, primary or secondary?

Answer can be found in Appendix E.

stimulate a primary immune response, but does not produce symptoms of disease and the associated infections.

Specific vaccines stimulate active immunity against a variety of diseases, including typhoid fever, cholera, whooping cough, diphtheria, tetanus, polio, chickenpox, measles (rubeola), German measles (rubella), mumps, influenza, hepatitis A, hepatitis B, and bacterial pneumonia. A vaccine has eliminated naturally acquired smallpox from the world.

A person who has been exposed to infection may require protection against a pathogen before active immunity has had the time to develop. An injection of antiserum (antibody-rich serum) may help. Antibodies may be obtained by separating gamma globulins from the plasma of persons who have already developed immunity against the particular disease. Injection of either antiserum or gamma globulins provides *artificially acquired passive immunity.*

During pregnancy, certain antibodies (IgG) pass from the maternal blood into the fetal bloodstream. As a result, the fetus acquires limited immunity against pathogens that the pregnant woman has developed active immunities against. The fetus thus has *naturally acquired passive immunity,* which may last for six months to a year after birth. Table 14.3 summarizes the types of acquired immunity.

PRACTICE 14.7

14. Distinguish between a primary and a secondary immune response.
15. Distinguish between active and passive immunity.

Hypersensitivity APR

A hypersensitivity reaction is an exaggerated immune response to a nonharmful antigen. In all hypersensitivities, the individual is presensitized to a particular antigen. Some hypersensitivities can affect almost anyone, but others happen only to people with an inherited tendency toward an exaggerated immune response.

A type I hypersensitivity (*immediate-reaction hypersensitivity*) is commonly called an **allergy,** and the antigens that trigger allergic responses are called **allergens** (al′er-jenz). In a type I hypersensitivity, a person becomes sensitized by producing IgE antibodies in response to a certain allergen. In the initial exposure, IgE attaches to the cell membranes of widely distributed mast cells and basophils. When a subsequent exposure to the same allergen occurs, these cells release allergy mediators such as *histamine, prostaglandin* D_2, and *leukotrienes.* This subsequent reaction occurs within seconds, therefore the term "immediate-reaction."

The effect of an allergy depends on how widespread the response is. Allergy mediators dilate arterioles and increase vascular permeability, causing edema. Allergy mediators contract bronchial and intestinal smooth muscle and increase mucus production. The overall result is an inflammation reaction that is responsible for the symptoms of the allergy, such as occurs in hives, hay fever, asthma, eczema, or gastric disturbances. Anaphylaxis is a severe form of immediate-reaction hypersensitivity and may be life-threatening. Large amounts of histamine and other allergy mediators spread throughout the body. The person may at first feel an inexplicable apprehension, and then suddenly the entire body itches and breaks out in red hives. Vomiting and diarrhea may follow, and systemic vasodilation may lead to a form of cardiovascular system failure called anaphylactic shock. The face, tongue, and larynx may swell from edema, and breathing may become difficult, as happened to the woman with a peanut allergy in the chapter-opening vignette. Anaphylactic shock most often results from an allergy to the antibiotic penicillin or from insect stings. With prompt medical attention and avoidance of allergens by people who know they have allergies, fewer than 100 people a year die from it.

Hypersensitivities that involve IgG (sometimes IgM) antibodies and take one to three hours or more to develop include type II (*antibody-dependent cytotoxic reactions*) and type III (*immune complex reactions*) hypersensitivities. A transfusion reaction to mismatched blood is a type II hypersensitivity reaction. Rheumatoid arthritis is an example of a type III hypersensitivity reaction.

TABLE 14.3 Practical Classification of Immunity

Type	Mechanism	Result
Naturally acquired active immunity	Exposure to live pathogens	Stimulation of an immune response with symptoms of a disease
Artificially acquired active immunity	Exposure to a vaccine containing weakened or dead pathogens or their components	Stimulation of an immune response without the severe symptoms of a disease
Naturally acquired passive immunity	Antibodies passed to fetus from pregnant woman with active immunity or to newborn through colostrum or breast milk from a woman with active immunity	Short-term immunity for a newborn without stimulating an immune response
Artificially acquired passive immunity	Injection of antiserum or gamma globulins	Short-term immunity without stimulating an immune response

A type IV hypersensitivity (*delayed-reaction hypersensitivity*) may affect anyone. It results from repeated exposure of the skin to certain chemicals–commonly, household or industrial chemicals or some cosmetics. Eventually the foreign substance activates T cells, many of which collect in the skin. The T cells and the macrophages they attract release chemical factors, which in turn cause eruptions and inflammation of the skin (dermatitis). This reaction is called *delayed* because it usually takes about 48 hours to begin. An example is seen in how some people react to poison ivy.

Transplantation and Tissue Rejection

Transplantation of tissues or an organ, such as the skin, kidney, heart, or liver, from one person to another can replace a nonfunctional, damaged, or lost body part. However, the recipient's immune response may recognize the donor's cell surfaces as foreign and attempt to destroy the transplanted tissue, causing a **tissue rejection reaction.**

Tissue rejection resembles the cellular immune response against a foreign antigen. The greater the antigenic difference between the cell surface molecules (MHC antigens, discussed earlier in this section) of the recipient tissues and the donor tissues, the more rapid and severe the rejection reaction. Matching donor and recipient tissues can minimize the rejection reaction.

Immunosuppressive drugs are used to reduce rejection of transplanted tissues. These drugs interfere with the recipient's immune response. Use of these drugs may make the recipient more susceptible to infection by suppressing the formation of antibodies or production of T cells, thereby dampening the humoral and cellular immune responses.

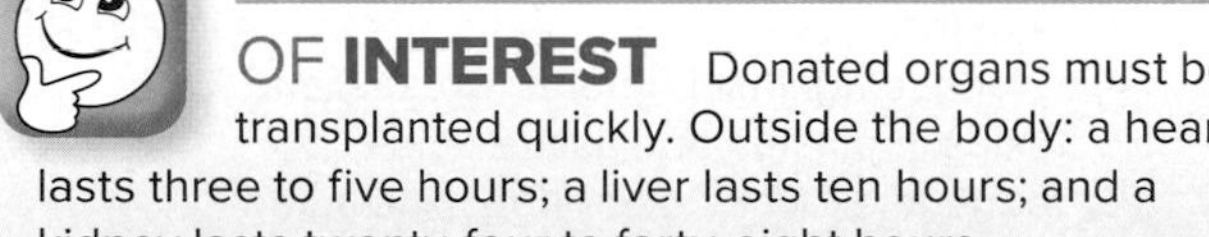

OF **INTEREST** Donated organs must be transplanted quickly. Outside the body: a heart lasts three to five hours; a liver lasts ten hours; and a kidney lasts twenty-four to forty-eight hours.

Autoimmunity

The immune response can turn against the body itself. It may become unable to distinguish a particular self antigen from a nonself antigen, producing **autoantibodies** and cytotoxic T cells that attack and damage the body's tissues and organs. This reaction against self is called **autoimmunity.** The specific nature of an autoimmune disorder reflects the cell types that are the target of the immune attack. In type 1 (insulin-dependent) diabetes mellitus, the target is beta cells in the pancreas. The tissues within the joints are targeted in rheumatoid arthritis. In systemic lupus erythematosus, the target is DNA and proteins associated with it in the cell nuclei. About 5% of the population has an autoimmune disorder.

Why might the immune response attack body tissues? Perhaps a virus, while replicating inside a human cell, takes proteins from the host cell's surface and incorporates them onto its own surface. When the immune response "learns" the surface of the virus in order to destroy it, it also learns to attack the human cells that normally bear those particular proteins. Another explanation of autoimmunity is that somehow T cells never learn to distinguish self from nonself. A third possible route of autoimmunity is when a nonself antigen coincidentally resembles a self antigen. Clinical Application 14.2 discusses how some disorders thought to be autoimmune may be the result of lingering fetal cells.

PRACTICE 14.7

16. How are allergic reactions and immune reactions similar yet different?
17. How does a tissue rejection reaction involve an immune response?
18. How is autoimmunity an abnormal functioning of the immune response?

CLINICAL APPLICATION 14.2

Persisting Fetal Cells and Autoimmunity

Some disorders thought to be autoimmune may have a stranger cause—fetal cells persisting in a woman's circulation, for decades. In response to an as yet unknown trigger, the fetal cells, perhaps "hiding" in a tissue such as skin, emerge and stimulate antibody production. The resulting antibodies and symptoms appear to be an autoimmune disorder. This situation, in which persisting fetal cells provoke an immune response, can be seen in a disorder called scleroderma, which means "hard skin."

Scleroderma, which is four times more common in women, typically begins between ages forty-five and fifty-five. It is described as "the body turning to stone." Symptoms include fatigue, swollen joints, stiff fingers, and a masklike face. The hardening may affect blood

—*Continued next page*

Continued—

vessels, the lungs, and the esophagus. Clues that scleroderma is a delayed response to persisting fetal cells include the following observations:

- It is much more common among women past their reproductive years.
- Symptoms resemble those of graft-versus-host disease (GVHD), in which transplanted tissue produces chemicals that destroy the recipient's tissues.
- Mothers who have scleroderma and their sons have cell surfaces more similar than those of unaffected mothers and their sons. Perhaps the similarity of cell surfaces enables the fetal cells to escape destruction by the woman's immune system. Female fetal cells probably have the same effect, but they are not observable because the method used to detect fetal cells identifies a DNA sequence unique to the Y chromosome (which is only in males).

ASSESS

CHAPTER ASSESSMENTS

14.1 Introduction

1. Explain the functions of the lymphatic system.

14.2 Lymphatic Pathways

2. Trace the general pathway of lymph from the interstitial spaces to the bloodstream.

14.3 Tissue Fluid and Lymph

3. Tissue fluid forms as a result of filtration from blood capillaries exceeding ______, whereas lymph forms due to increasing ______ ______ in the tissue fluid.
4. Describe two functions of lymph.
5. Explain why physical exercise promotes lymphatic circulation.

14.4 Lymphatic Tissues and Organs

6. Draw a lymph node and label its parts.
7. Explain the functions of a lymph node.
8. Indicate the locations of the thymus and spleen.
9. Compare and contrast the functions of the thymus and spleen.

14.5 Body Defenses Against Infection

10. Defense mechanisms that prevent the entry of many types of pathogens and destroy them if they enter provide ______ (nonspecific) defense. Precise mechanisms targeting specific pathogens provide ______ (specific) defense.

14.6 Innate (Nonspecific) Defenses

11. Define *species resistance*.
12. Identify the barriers that provide the body's first line of defense against infectious agents.
13. List the major effects of inflammation, and explain why each occurs.
14. Describe how enzymatic actions function as defense mechanisms.
15. Define *interferon* and explain its action.
16. ______ is a group of plasma proteins that when activated stimulate inflammation, attract phagocytes, and enhance phagocytosis.
17. ______ ______ ______ are specialized lymphocytes that secrete perforins to lyse cell membranes of virus-infected cells.
18. Identify the major phagocytic cells in blood and other tissues.
19. Discuss how a low-grade fever of short duration may be a natural response to infection.

14.7 Immunity: Adaptive (Specific) Defenses

20. Review the origin of T cells and B cells.
21. Explain the cellular immune response including the activation of T cells.
22. List three types of T cells and describe the function of each in the immune response.
23. Explain the humoral immune response, including the activation of B cells.
24. Explain the function of plasma cells.
25. Match the major types of antibodies with their function and/or where each is found.

(1) associated with allergic reactions	A. IgA
(2) important in B cell activation, on surfaces of most B cells	B. IgM
(3) activates complement, anti-A and anti-B in blood	C. IgG
(4) effective against bacteria, viruses, and toxins in plasma and tissue fluids	D. IgD
(5) found in exocrine secretions, including breast milk	E. IgE

26. Describe three ways in which an antibody's direct attack on an antigen helps remove that antigen.
27. List the various effects of complement activation.
28. Contrast a primary and a secondary immune response.
29. Match the practical classification of immunity with its example.

(1) naturally acquired active immunity	A. a breast-fed newborn
(2) artificially acquired active immunity	B. gamma globulin injection
(3) naturally acquired passive immunity	C. vaccination
(4) artificially acquired passive immunity	D. measles infection

30. Describe how an immediate-reaction hypersensitivity (allergy) response may occur.
31. List the major events leading to a delayed-reaction hypersensitivity response.
32. Explain the relationship between tissue rejection and an immune response.
33. Explain the relationship between autoimmunity and an immune response.

Lymphatic System

INTEGUMENTARY SYSTEM

The skin is a first line of defense against infection.

SKELETAL SYSTEM

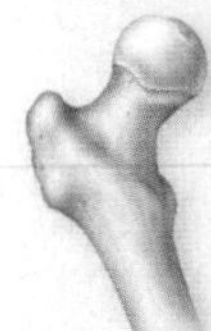

Cells of the immune system originate in the bone marrow.

MUSCULAR SYSTEM

Muscle action helps pump lymph through the lymphatic vessels.

NERVOUS SYSTEM

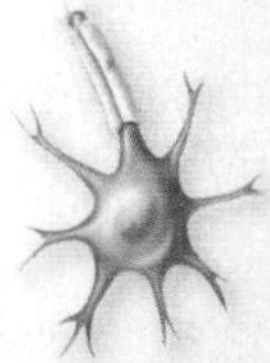

Stress may impair the immune response.

ENDOCRINE SYSTEM

Hormones stimulate lymphocyte production.

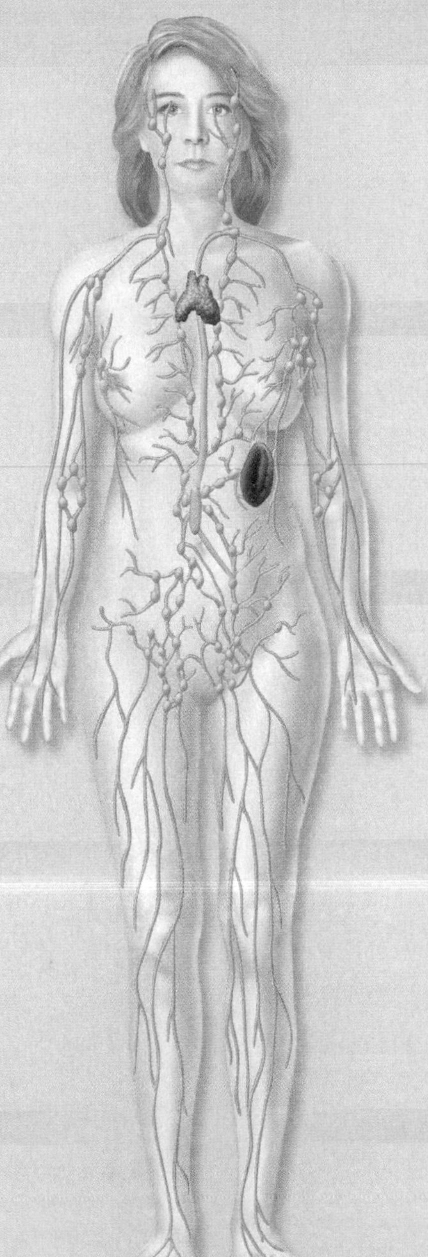

The lymphatic system is an important link between tissue fluid and the plasma; it also plays a major role in the body's response to infection.

CARDIOVASCULAR SYSTEM

The lymphatic system returns tissue fluid to the bloodstream. Lymph originates as tissue fluid, formed by the action of blood pressure.

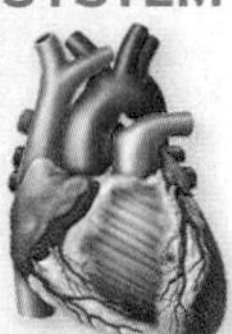

DIGESTIVE SYSTEM

Lymph plays a major role in the absorption of fats.

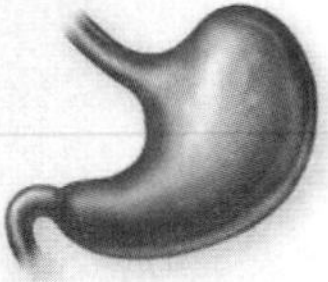

RESPIRATORY SYSTEM

Cells of the immune system patrol the respiratory system to defend against infection.

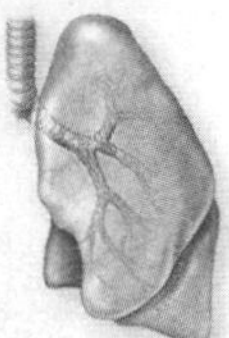

URINARY SYSTEM

The kidneys control the volume of extracellular fluid, including lymph.

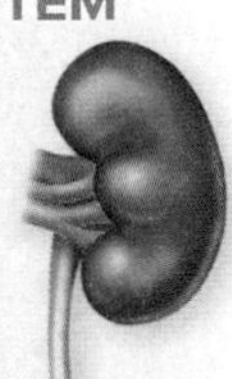

REPRODUCTIVE SYSTEM

Special mechanisms inhibit the female immune system in its attack of sperm as foreign invaders.

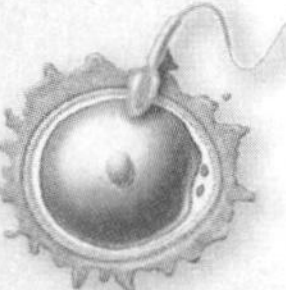

ASSESS

INTEGRATIVE ASSESSMENTS/CRITICAL THINKING

Outcomes 3.3, 13.4, 14.3, 14.4

1. During or before surgery to remove a tumor, a sentinel lymph node biopsy may be done. Biopsy results determine the number of lymph nodes that should be removed. Explain the influence of lymph node removal on exchanges in blood capillaries.

Outcomes 5.3, 12.2, 14.6, 14.7

2. Macrophages are part of our innate (nonspecific) defenses, and without them, T cell activation may be impeded. What is the connection between innate and adaptive defenses in the role that macrophages play?

Outcomes 6.2, 14.2, 14.3

3. Why is injecting a substance into the skin similar to injecting it into the lymphatic system?

Outcomes 13.5, 13.7, 13.8, 14.2

4. Lymphatic vessels deliver lymph to blood circulation through subclavian veins. Why are subclavian veins more suitable for this than subclavian arteries?

Outcomes 14.2, 14.3, 14.4

5. How can the removal of enlarged lymph nodes for microscopic examination aid in diagnosing certain diseases?

Outcome 14.7

6. The immune response is specific, diverse, and has memory. Give examples of each of these characteristics.
7. Some parents keep their preschoolers away from other children to prevent them from catching illnesses. How might these well-meaning parents actually be harming their children?
8. Why does vaccination provide long-lasting protection against a disease, while gamma globulin (IgG) provides only short-term protection?
9. Why is a transplant consisting of fetal tissue less likely to provoke an immune rejection response than tissue from an adult?

Chapter Summary

14.1 Introduction

*The **lymphatic system** is closely associated with the cardiovascular system. It transports excess tissue fluid to the bloodstream, absorbs fats, and helps defend the body against disease-causing agents.*

14.2 Lymphatic Pathways

1. Lymphatic capillaries
   a. **Lymphatic capillaries** are microscopic, closed-ended tubes that extend into interstitial spaces.
   b. They receive tissue fluid through their thin walls, and once inside the lymphatic capillaries the fluid is **lymph.**
2. Lymphatic vessels
   a. **Lymphatic vessels** are formed by the merging of lymphatic capillaries.
   b. Lymphatic vessels have walls similar to those of veins, only thinner, and possess valves that prevent backflow of lymph.
   c. The larger lymphatic vessels lead to **lymph nodes** and then merge into lymphatic trunks.
3. Lymphatic trunks and collecting ducts
   a. Lymphatic trunks lead to two **collecting ducts**—the **thoracic duct** and the **right lymphatic duct.**
   b. Collecting ducts empty into the subclavian veins.

14.3 Tissue Fluid and Lymph

1. Tissue fluid formation
   a. Tissue fluid originates from plasma and includes water and dissolved substances that have passed through the blood capillary wall.
   b. It generally lacks large proteins, but some smaller proteins are filtered out of blood capillaries into interstitial spaces.
   c. As the protein concentration of tissue fluid increases, colloid osmotic pressure increases.
2. Lymph formation and function
   a. Increasing hydrostatic pressure within interstitial spaces forces some tissue fluid into lymphatic capillaries, and this fluid becomes lymph.
   b. Lymph returns small protein molecules to the bloodstream and transports foreign particles to lymph nodes.
3. Lymph Movement
   a. Lymph is under relatively low hydrostatic pressure and may not flow readily without external aid.
   b. Lymph is moved by the contraction of skeletal muscles, contraction of smooth muscle in the walls of large lymphatic trunks, and low pressure in the thorax created by breathing movements.

14.4 Lymphatic Tissues and Organs

Unencapsulated lymphatic tissue includes the diffuse MALT (mucosa-associated lymphoid tissue), tonsils, and appendix. The lymph nodes, thymus, and spleen are encapsulated lymphatic organs.

1. Lymph nodes
   a. Lymph nodes are subdivided into **lymphatic nodules.**
   b. Lymphatic nodules contain masses of lymphocytes and macrophages.

   c. Lymph nodes aggregate in groups or chains along the paths of larger lymphatic vessels.
   d. Lymph nodes filter potentially harmful foreign particles from lymph.
   e. Lymph nodes are centers for the production of lymphocytes, and they also contain phagocytic cells.
2. Thymus
   a. The **thymus,** located anterior to the aorta and posterior to the upper part of the sternum, is composed of lymphatic tissue subdivided into lobules.
   b. The thymus slowly shrinks after puberty.
   c. Some lymphocytes mature in the thymus and provide immunity.
3. Spleen
   a. The **spleen,** just inferior to the diaphragm and posterior and lateral to the stomach, resembles a large lymph node subdivided into lobules.
   b. Spaces within splenic lobules are filled with blood.
   c. The spleen contains many macrophages, which filter foreign particles and damaged red blood cells from blood.

14.5 Body Defenses Against Infection

*The body has **innate (nonspecific) defenses** and **adaptive (specific) defenses** against **infection.***

14.6 Innate (Nonspecific) Defenses

1. **Species resistance**
   Each species is resistant to certain diseases that may affect other species.
2. **Mechanical barriers**
   a. Mechanical barriers include the skin and mucous membranes, which block the entrance of some pathogens.
   b. Hair traps infectious agents; fluids such as tears, sweat, saliva, mucus, and urine wash away microorganisms before they can firmly attach.
3. **Inflammation**
   a. Inflammation is a tissue response to injury or infection, and includes localized redness, swelling, heat, and pain.
   b. Chemicals released by damaged tissues attract white blood cells to the site.
   c. Connective tissue may form a sac around injured tissue and thus block the spread of pathogens.
4. **Chemical barriers**
   a. Enzymes in gastric juice and tears kill some pathogens.
   b. **Interferons** stimulate uninfected cells to synthesize antiviral proteins that stimulate phagocytosis, block proliferation of viruses, and enhance the activity of cells that help resist infections and stifle tumor growth.
   c. Activation of **complement** proteins in plasma stimulates inflammation, attracts phagocytes, and enhances phagocytosis.
5. **Natural killer (NK) cells**
   Natural killer cells secrete perforins, which destroy cancer cells and cells infected with viruses.
6. **Phagocytosis**
   a. The most active phagocytes in blood are neutrophils and monocytes. Monocytes give rise to macrophages, which may be free or fixed in various tissues.
   b. Phagocytic cells are associated with the linings of blood vessels, and are in red bone marrow, the liver, spleen, lungs, and lymph nodes.
   c. Phagocytes remove foreign particles from tissues and body fluids.
7. **Fever**
   Elevated body temperature, with the resulting decrease in blood iron level and increase in phagocytic activity hampering infection.

14.7 Immunity: Adaptive (Specific) Defenses

1. **Antigens**
   a. During fetal development, body cells inventory "self" proteins and other large molecules.
   b. After inventory, lymphocytes develop receptors that allow them to differentiate between nonself (foreign) and self antigens.
   c. **Haptens** are small molecules that can combine with larger ones, becoming antigenic.
2. Lymphocyte origins
   a. Lymphocytes originate in red bone marrow and are released into the blood.
   b. Some reach the thymus, where they mature into **T cells.**
   c. Others, the **B cells,** mature in red bone marrow.
   d. Both T cells and B cells reside in lymphatic tissues and organs.
3. T cells and the cellular immune response
   a. T cells are activated when an **antigen-presenting cell** displays a foreign antigen.
   b. When a macrophage acts as an antigen-presenting cell, it phagocytizes an antigen-bearing agent, digests the agent, and displays the resulting antigens on its cell membrane in association with certain MHC proteins.
   c. T cells respond to antigens through cell-to-cell contact (**cellular immune response**).
   d. T cells secrete cytokines that enhance cellular responses to antigens, and stimulate proliferation of a T cell to enlarge its clone.
   e. T cells may also secrete substances that are toxic to their target cells.
   f. A helper T cell becomes activated when it encounters displayed antigens for which it is specialized to react.
   g. Once activated, helper T cells stimulate B cells to produce antibodies.
   h. Cytotoxic T cells recognize foreign antigens on tumor cells, as well as cells whose surfaces indicate that they are infected by viruses, and then release perforin to destroy these cells.
   i. Memory T cells allow for an immediate response to a second and subsequent exposure of the same antigen.
4. B cells and the humoral immune response
   a. Sometimes a B cell is activated when it encounters an antigen that fits its antigen receptors. More often, a B cell is activated when stimulated by a helper T cell.
   b. An activated B cell proliferates (especially when stimulated by a T cell), enlarging its clone.

c. Some activated B cells differentiate into antibody-producing **plasma cells.**
d. **Antibodies** react against the antigen-bearing agent that stimulated their production (**humoral immune response**).
e. Other activated B cells differentiate further into memory B cells.
f. An individual's diverse B cells defend against many pathogens.

5. Types of antibodies
   a. Antibodies are soluble proteins called immunoglobulins.
   b. The five major types of immunoglobulins are IgG, IgA, IgM, IgD, and IgE.
6. Antibody actions
   a. Antibodies directly attach to antigens, activate complement, or stimulate local tissue changes that are unfavorable to antigen-bearing agents.
   b. Direct attachment results in agglutination, precipitation, or neutralization.
   c. Activated complement proteins attract phagocytes, alter cells so that they become more susceptible to phagocytosis, and rupture foreign cell membranes (lysis).
7. Immune responses
   a. The first reaction to an antigen is called a **primary immune response.**
      (1) During this response, antibodies are produced for several weeks.
      (2) Some T cells and B cells remain dormant as memory cells.
   b. A **secondary immune response** occurs as a result of memory cells rapidly responding to subsequent exposure to an antigen.
8. Practical classification of immunity
   a. Naturally acquired immunity arises in the course of natural events, whereas artificially acquired immunity is the consequence of a medical procedure.
   b. Active immunity lasts much longer than passive immunity.
   c. A person who encounters a pathogen and has a primary immune response develops naturally acquired active immunity.
   d. A person who receives a **vaccine** containing a dead or weakened pathogen, or part of one, develops artificially acquired active immunity.
   e. A person who receives an injection of antiserum or gamma globulins has artificially acquired passive immunity.
   f. When antibodies pass through a placental membrane from a pregnant woman to her fetus, the fetus develops naturally acquired passive immunity.
9. Hypersensitivity
   a. Hypersensitivity reactions are excessive and misdirected immune responses that may damage tissue.
   b. Type I hypersensitivity (immediate-reaction hypersensitivity) is an inborn ability to overproduce IgE in response to an **allergen.**
      (1) Allergic reactions result from mast cells bursting and releasing allergy mediators such as histamine.
      (2) The released chemicals cause allergy symptoms such as hives, hay fever, asthma, eczema, or gastric disturbances.
   c. Type II hypersensitivity reactions (antibody-dependent cytotoxic reactions) occur when blood transfusions are mismatched.
   d. Rheumatoid arthritis is an example of a Type III hypersensitivity reaction (immune-complex reaction).
   e. Type IV hypersensitivity (delayed-reaction hypersensitivity), which can occur in anyone and inflame the skin, results from repeated exposure to household or industrial chemicals or some cosmetics.
10. Transplantation and tissue rejection
    a. A transplant recipient's immune response may react against the donated tissue in a tissue rejection reaction.
    b. Matching donor and recipient tissues and using immunosuppressive drugs can minimize a **tissue rejection reaction.**
    c. Immunosuppressive drugs may increase susceptibility to infection.
11. **Autoimmunity**
    a. In autoimmune disorders, the immune response manufactures **autoantibodies** that attack a person's own body tissues.
    b. Autoimmune disorders may result from a previous viral infection, faulty T cell development, or reaction to a nonself antigen that resembles a self antigen.

UNIT 5 ABSORPTION AND EXCRETION

Digestive System and Nutrition

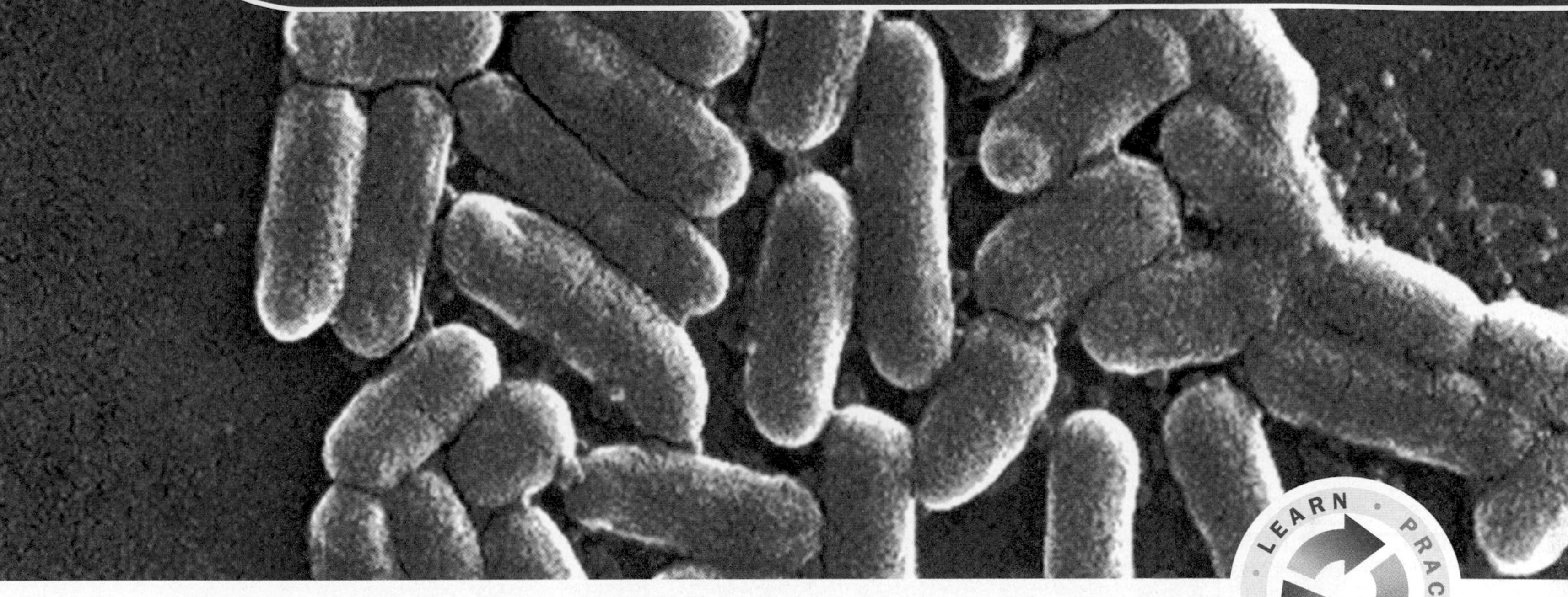

Several million microorganisms are normal residents of our digestive tracts. Escherichia coli, *pictured here (6,800x), produce vitamin K and, if present in low numbers, will not cause diarrhea.* Source: CDC/Janice Haney Carr

LEARN • PRACTICE • ASSESS •

The gut microbiome. Not all of the cells in an adult body are human; about half are microorganisms traditionally called microflora, but more recently called the microbiome. The "human oral microbiome," for example, includes more than 600 species that can live in the mouth. Each person has about 200 of these oral bacterial types. The other end of the digestive tract houses the "distal gut microbiome," which includes more than 6,800 species.

Researchers tracked the formation and changing nature of the human gut microbiome by classifying microbial DNA in a year's worth of stool collected daily from the soiled diapers of a number of babies. Bacteria in the stool varied greatly from baby to baby at the onset, but by the babies' first birthdays, the gut communities were more alike and more closely resembled the microbial communities in adults.

The microorganisms that live in our large intestines are crucial to our health. They produce more than eighty types of enzymes that digest plant polysaccharides that our bodies cannot break down, and help process certain sugars. Our "gut" residents also synthesize vitamins and amino acids, and break down certain toxins and drugs.

We can use knowledge of our gut microbiome to improve health, because illness can alter the bacterial populations within us. An approach called probiotics adds bacteria to foods to prevent certain infections. For example, certain *Lactobacillus* strains added to yogurt help protect against *Salmonella* food-borne infection.

A procedure called fecal microbiota transplantation reconstitutes a healthy gut microbiome in people who have recurrent infection from *Clostridium difficile,* which causes severe diarrhea. As public stool banks are screening donations and providing them to physicians, clinical trials are evaluating how to best deliver the material: fresh or frozen, and by enema, nasogastric tube, or capsule. Analysis of the bacterial genomes in the stool of treated patients indicates that fecal transplant treats diarrhea.

Anatomy & Physiology Revealed 4.0

Module 12 Digestive System

LEARNING OUTLINE

After studying this chapter, you should be able to complete the "Learning Outcomes" that follow the major headings throughout the chapter.

AIDS TO UNDERSTANDING WORDS

aliment- [food] *alimentary* canal: tubelike part of the digestive system.

chym- [juice] *chyme*: semifluid paste of food particles and gastric juice formed in the stomach.

decidu- [falling off] *deciduous* teeth: teeth shed during childhood.

gastr- [stomach] *gastric* gland: part of the stomach that secretes gastric juice.

hepat- [liver] *hepatic* duct: duct that carries bile from the liver to the bile duct.

lingu- [tongue] *lingual* tonsil: mass of lymphatic tissue at the root of the tongue.

nutri- [nourish] *nutrient*: substance needed to nourish cells.

peri- [around] *peristalsis*: wavelike ring of contraction that moves material along the alimentary canal.

pyl- [gatekeeper] *pyloric* sphincter: muscle that serves as a valve between the stomach and the small intestine.

vill- [hairy] *villi*: tiny projections of mucous membrane in the small intestine.

(Appendix A has a complete list of Aids to Understanding Words.)

15.1 | Introduction

LEARN

1. Describe the general functions of the digestive system.
2. Name the major organs of the digestive system.

Whether your favorite food is pizza, a hamburger, fries, or a salad, **digestion** (di-jest′yun) is the same. It is the process of breaking down all foods into usable nutrients for absorption into the bloodstream.

Mechanical digestion breaks large pieces of food into smaller ones without altering their chemical composition. *Chemical digestion* breaks down larger nutrient molecules into simpler chemicals, allowing them to be absorbed. These include amino acids from proteins, fatty acids from lipids, and glucose and other monosaccharides from carbohydrates. Vitamins provide assistance with chemical digestion, and minerals are vital for many physiological processes.

The digestive system consists of the digestive tract, or **alimentary canal** (al″i-men′tar-ē kah-nal′), and several **accessory organs** that aid in chemical digestion. The alimentary canal is a continuous tube that (from beginning to end) includes the mouth, pharynx, esophagus, stomach, small intestine, large intestine, rectum, and anus. The accessory organs include the salivary glands, liver, gallbladder, and pancreas (fig. 15.1; see reference plates 4, 5, and 6).

PRACTICE 15.1

Answers to the Practice questions can be found in the eBook.

1. What are the general functions of the digestive system?
2. Which organs constitute the digestive system?

CAREER CORNER
Registered Dietitian

In preparation for weight-loss surgery, the young woman must lose 10% of her body weight. A registered dietitian (RD) assists by discussing the types of foods that an obese person with type 2 diabetes mellitus should and shouldn't eat. The RD uses plastic models of foods to demonstrate the makeup of appropriate meals and portion sizes, and then works with the patient to develop a meal plan.

Dietitians work in diverse settings. These include businesses, hotels, food service corporations, community agencies, schools, senior centers, health-care facilities, prisons, and restaurants. Registered dietitians can also have careers in research or teaching other health-care professionals, or open their own businesses. Educating the public about healthful food choices can be a major part of the job.

Training to become a registered dietitian includes earning a bachelor's degree in an accredited program and completing a supervised practice program typically lasting 6 to 12 months. The individual must then pass a national exam.

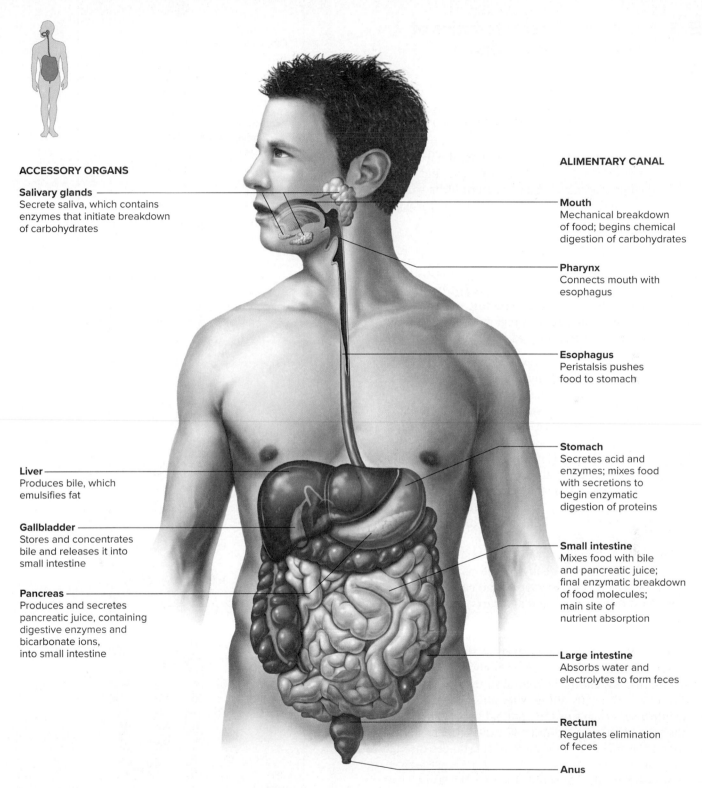

ACCESSORY ORGANS

Salivary glands
Secrete saliva, which contains enzymes that initiate breakdown of carbohydrates

Liver
Produces bile, which emulsifies fat

Gallbladder
Stores and concentrates bile and releases it into small intestine

Pancreas
Produces and secretes pancreatic juice, containing digestive enzymes and bicarbonate ions, into small intestine

ALIMENTARY CANAL

Mouth
Mechanical breakdown of food; begins chemical digestion of carbohydrates

Pharynx
Connects mouth with esophagus

Esophagus
Peristalsis pushes food to stomach

Stomach
Secretes acid and enzymes; mixes food with secretions to begin enzymatic digestion of proteins

Small intestine
Mixes food with bile and pancreatic juice; final enzymatic breakdown of food molecules; main site of nutrient absorption

Large intestine
Absorbs water and electrolytes to form feces

Rectum
Regulates elimination of feces

Anus

Figure 15.1 Organs of the digestive system. 

15.2 | General Characteristics of the Alimentary Canal

LEARN

1. Describe the structure of the wall of the alimentary canal.
2. Explain how the contents of the alimentary canal are mixed and moved.

The alimentary canal is a muscular tube about 8 meters long that passes through the body's thoracic and abdominopelvic cavities (fig. 15.2). The structure of its wall, how it moves food, and its innervation are similar throughout its length.

Structure of the Wall

The wall of the alimentary canal consists of four distinct layers that are developed to different degrees from region to region. Although the four-layered structure persists throughout the alimentary canal, certain regions are specialized for particular functions. Beginning with the innermost tissues, these layers are (fig. 15.3):

1. **Mucosa** (mū-kō′sah), or **mucous membrane** (mū′kus mem′brān): Surface epithelium, underlying connective tissue, and a small amount of smooth muscle form this layer. In some regions, the mucosa is folded, with tiny projections that extend into the passageway, or **lumen** (lū′men), of the digestive tube. The folds increase the absorptive surface area. The mucosa also has glands that secrete mucus and digestive enzymes. The mucosa protects the tissues beneath it, secretes into the lumen, and absorbs substances from the diet.
2. **Submucosa** (sub″mū-kō′sah): The submucosa contains considerable loose connective tissue, as well as glands, blood vessels, lymphatic vessels, and nerves. Its vessels nourish surrounding tissues and carry away absorbed materials.
3. **Muscularis**: This layer, which provides movements of the tube, consists of two layers of smooth muscle tissue. The cells of the inner layer encircle the tube. When this circular layer contracts, the tube's diameter decreases. The cells of the outer muscular layer run lengthwise. When this longitudinal layer contracts, the tube shortens. Coordinated contractions of both muscle layers cause movement of substances through the tube.
4. **Serosa** (sĕ′rō-sah), or serous layer (se′rus lā′er): The layer of epithelium on the outside of the tube with the connective tissue beneath it compose the serous layer. This is also called the *visceral peritoneum* (see section 1.6, Organization of the Human Body). The cells of the serosa protect underlying tissues and secrete serous fluid, which moistens and lubricates the tube's outer surface. As a result, organs within the abdominal cavity slide freely against one another.

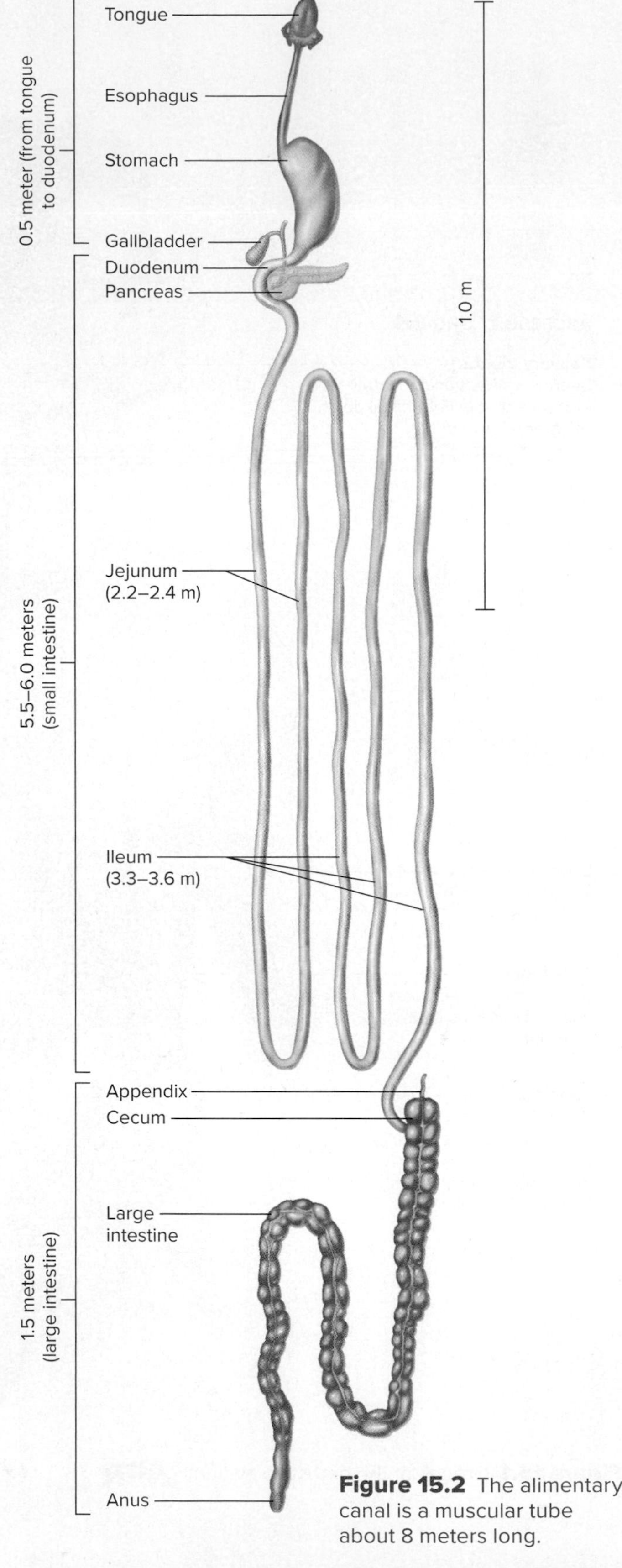

Figure 15.2 The alimentary canal is a muscular tube about 8 meters long.

PRACTICE FIGURE 15.2

What is the distance from the tongue to the duodenum in English units (inches)?

Answer can be found in Appendix E.

Artery
Vein
Mesentery
Serosa
Circular fold
Mucosa
Submucosa
Longitudinal muscle
Circular muscle
Microvilli
Goblet cell
Nucleus
Epithelium
Lacteal
Villi
Mucosa
Simple columnar epithelium
Capillaries
Lacteal
Lymph nodule
Intestinal gland
Submucosa
Mucous gland in submucosa
Circular muscle
Muscularis
Longitudinal muscle
Nerve plexuses
Serosa

Figure 15.3 The wall of the small intestine, as in other portions of the alimentary canal, consists of four layers: the inner mucosa, the submucosa, the muscularis, and the outer serosa. (Recall from chapter 14 that a lacteal is a specialized lymphatic capillary that transports absorbed fats.)

Movements of the Tube

The motor functions of the alimentary canal are of two basic types—*mixing movements* and *propelling movements.* Mixing occurs when smooth muscle in small segments of the tube contracts rhythmically (fig. 15.4*a*). For example, when the stomach is full, waves of muscular contractions move along its walls from one end to the other. These waves mix food with digestive juices that the mucosa secretes. In the small intestine, a process called **segmentation** aids mixing by alternately contracting and relaxing the smooth muscle in nonadjacent segments of the organ. Because segmentation follows a back-and-forth pattern, materials are not moved along the tract in one direction (fig. 15.4*b*).

Propelling movements include a wavelike motion called **peristalsis** (per″ĭ-stal′sis). In peristalsis, a ring of contraction occurs in the wall of the tube. At the same time, the muscular wall just ahead of the ring relaxes. As the peristaltic wave moves along the tube, it pushes the tubular contents ahead of it (fig. 15.4*c*).

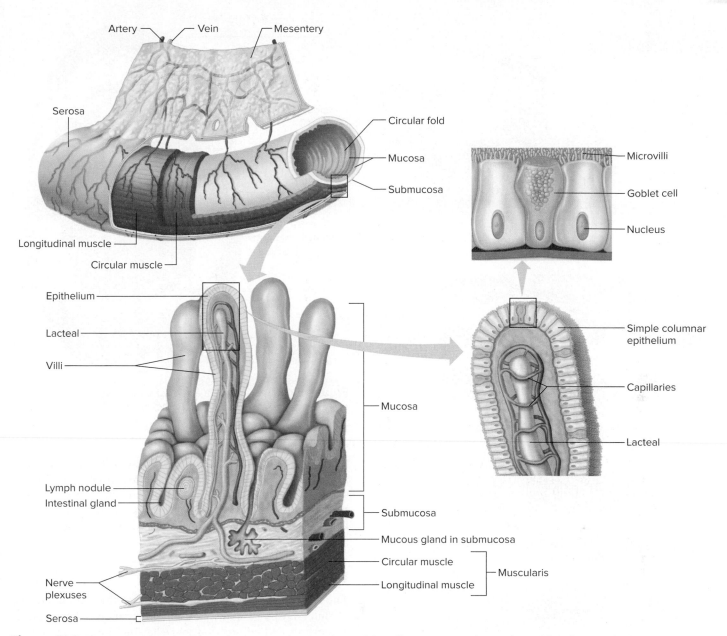

PRACTICE 15.2

1. Describe the wall of the alimentary canal.
2. Name the two basic types of movements in the alimentary canal.

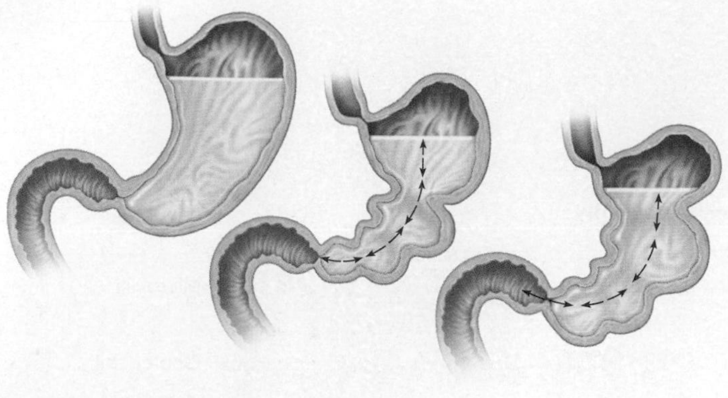

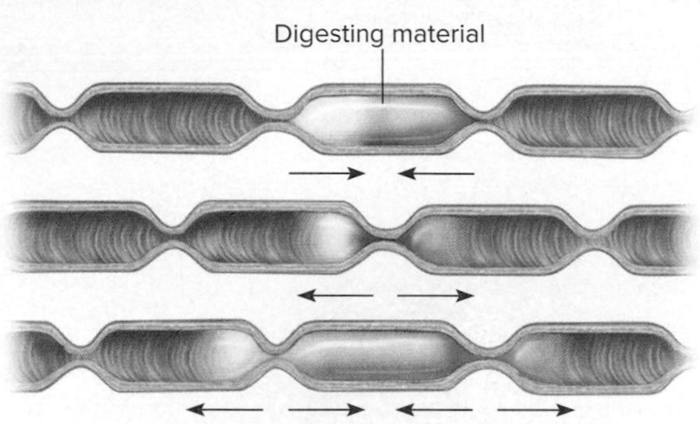

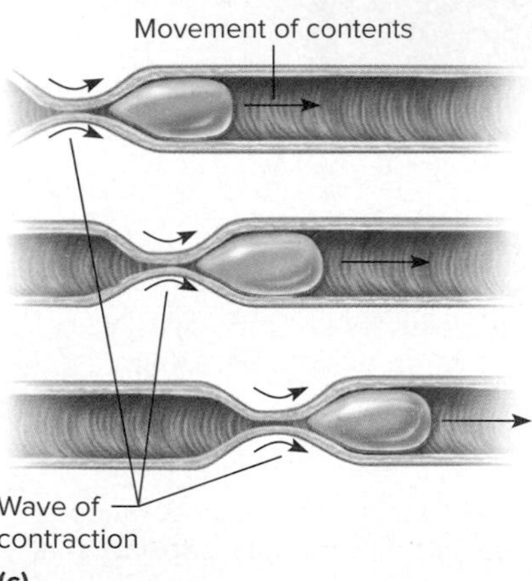

Figure 15.4 Movements through the alimentary canal. (**a**) Mixing movements occur when small segments of the muscular wall of the stomach rhythmically contract. (**b**) Segmentation mixes contents of the small intestine. (**c**) Peristaltic waves move the contents along the canal.

15.3 | Mouth

LEARN

1. Describe the functions of the structures associated with the mouth.
2. Describe how different types of teeth are adapted for different functions, and list the parts of a tooth.

The **mouth** receives food and begins digestion by mechanically breaking up solid particles into smaller pieces and mixing them with saliva. This chewing action is called *mastication* (mas″tĭ-kā′shun). The lips, cheeks, tongue, and palate surround the mouth, which includes a chamber between the palate and tongue called the *oral cavity,* as well as a narrow space between the teeth, cheeks, and lips called the *vestibule* (fig. 15.5).

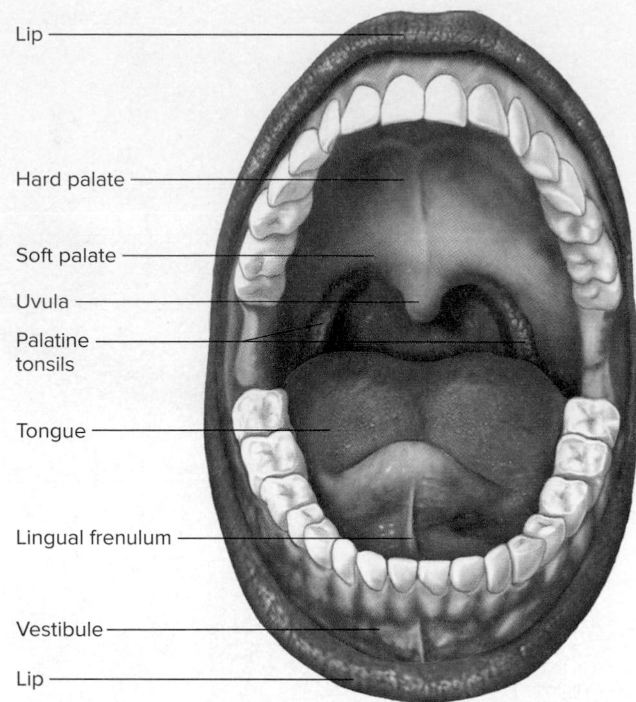

Figure 15.5 The mouth is adapted for ingesting food and beginning digestion, both mechanically and chemically.

Cheeks and Lips

The **cheeks,** forming the lateral walls of the mouth, consist of outer layers of skin, pads of subcutaneous fat, muscles associated with expression and chewing, and inner linings of moist, stratified squamous epithelium. The **lips** are highly mobile structures that surround the mouth opening. They contain skeletal muscles and sensory receptors useful in judging the temperature and texture of foods. Blood vessels near lip surfaces impart a reddish color.

Tongue

The **tongue** nearly fills the oral cavity when the mouth is closed. Mucous membrane covers the tongue, and a membranous fold called the **lingual frenulum** (ling′gwahl fren′ū-lum) connects the midline of the tongue to the floor of the mouth.

The *body* of the tongue is mostly skeletal muscle. Muscular action mixes food particles with saliva during chewing and moves food toward the pharynx during swallowing. The tongue also helps position food for chewing. Rough projections called **papillae** on the tongue surface provide friction, which helps move food. These papillae also bear taste buds (see section 10.6, Sense of Taste).

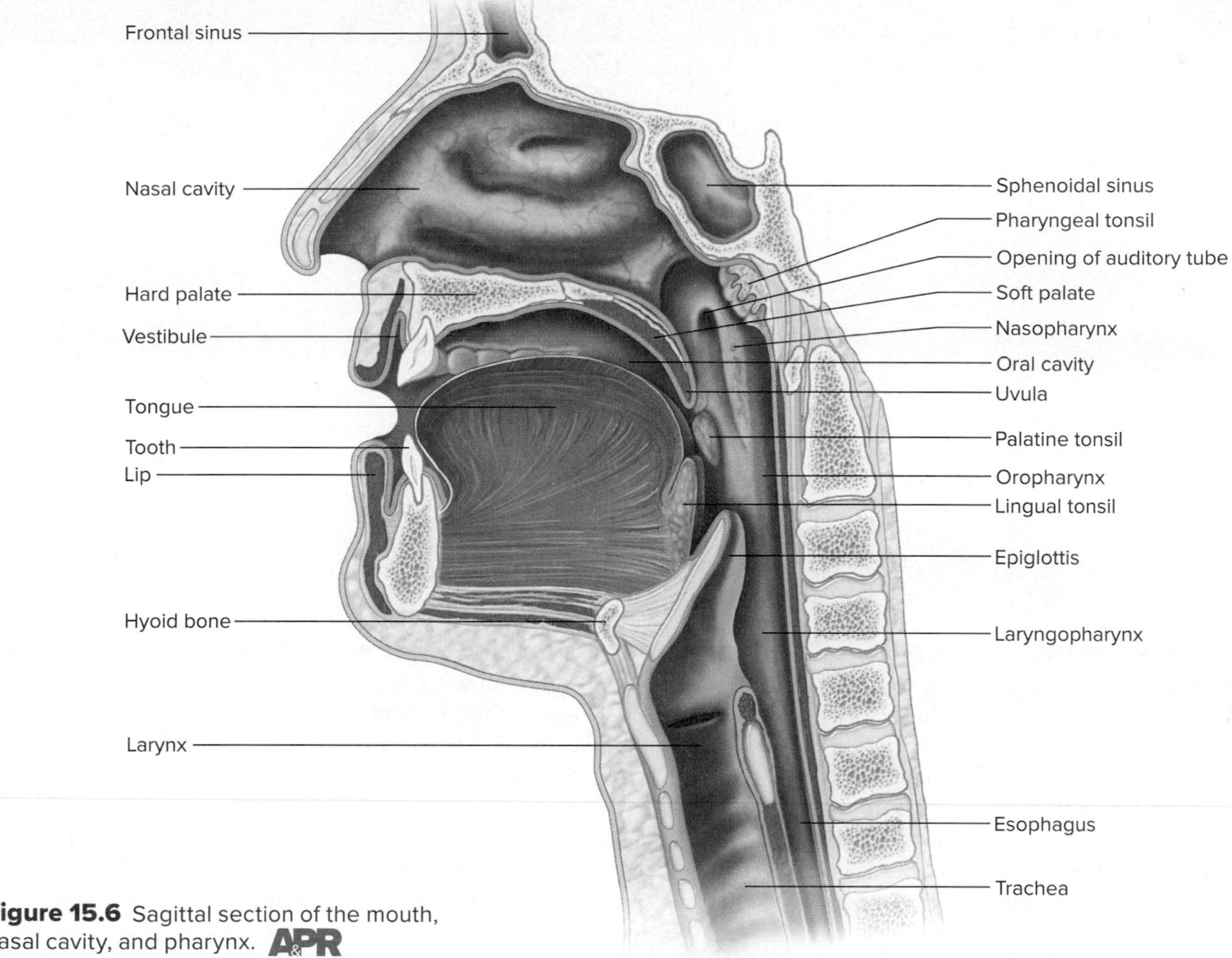

Figure 15.6 Sagittal section of the mouth, nasal cavity, and pharynx. APR

The posterior region, or *root,* of the tongue is anchored to the hyoid bone. It is covered with rounded masses of lymphatic tissue called **lingual tonsils** (ton'silz) (fig. 15.6).

Palate

The **palate** (pal'at) forms the roof of the oral cavity and consists of a bony anterior part (*hard palate*) and a muscular posterior part (*soft palate*). A muscular arch of the soft palate extends posteriorly and downward as a cone-shaped projection called the **uvula** (ū'vu-lah).

In the back of the mouth, on either side of the tongue and closely associated with the palate, are masses of lymphatic tissue called **palatine** (pal'ah-tīn) **tonsils** (see figs. 15.5 and 15.6). These structures lie beneath the epithelial lining of the mouth. Like other lymphatic tissues, the palatine tonsils help protect the body against infection.

Other masses of lymphatic tissue, called **pharyngeal** (fah-rin'jē-al) **tonsils,** or *adenoids,* are on the posterior wall of the pharynx, above the border of the soft palate (fig. 15.6). Enlarged pharyngeal tonsils that block the passage between the nasal cavity and the pharynx may be surgically removed.

The palatine tonsils are common sites of infection, and become inflamed in *tonsillitis.* Infected tonsils may swell so greatly that they block the passageways of the pharynx and interfere with breathing and swallowing. When tonsillitis recurs and does not respond to antibiotic treatment, the tonsils may be surgically removed. Such tonsillectomies are done less often today than they were a generation ago because the tonsils' role in immunity is now recognized.

PRACTICE 15.3

1. How does the tongue function as part of the digestive system?
2. Where are the tonsils located?

Teeth

The shape, size, and number of the teeth in an animal's mouth is called its **dentition** and is an indicator of what it eats. A human dentition reveals generalized teeth for an omnivorous diet, eating all types of food (table 15.1). Two different sets of **teeth** form during development. The first set, the *primary teeth* (deciduous teeth), usually erupt through the gums (gingiva) at regular intervals between the ages

TABLE 15.1 Primary and Secondary Teeth

Type	Number of Primary Teeth	Number of Secondary Teeth	Function
Incisor			Bite off pieces of food
Central	4	4	
Lateral	4	4	
Canine (cuspid)	4	4	Grasp and tear food
Premolar (bicuspid)			Grind food particles
First	0	4	
Second	0	4	
Molar			Grind food particles
First	4	4	
Second	4	4	
Third	0	4	
Total	20	32	

of six months and two to four years (fig. 15.7). There are twenty primary teeth—ten in each jaw.

The primary teeth are usually shed in the order in which they erupted. After their roots are resorbed, the *secondary teeth* (permanent teeth) push the primary teeth out of their sockets. This secondary set consists of thirty-two teeth—sixteen in each jaw (fig. 15.8). The secondary teeth usually begin to erupt at six years of age, but the set may not be complete until the third molars (wisdom teeth) emerge between seventeen and twenty-five years of age.

Each tooth consists of two main parts—the *crown,* which projects beyond the gum, and the *root,* which is anchored to the alveolar process of the jaw. These structures meet at the *neck* of the tooth.

Glossy white **enamel** covers the crown. Enamel mainly consists of calcium salts and is the hardest substance in the body. Enamel damaged by abrasive action or injury is not replaced. Enamel also tends to wear away with age. Clinical Application 15.1 discusses the destruction of tooth enamel.

The bulk of a tooth beneath the enamel is **dentin,** a substance much like bone but harder. Dentin surrounds the tooth's central cavity (pulp cavity), which contains blood

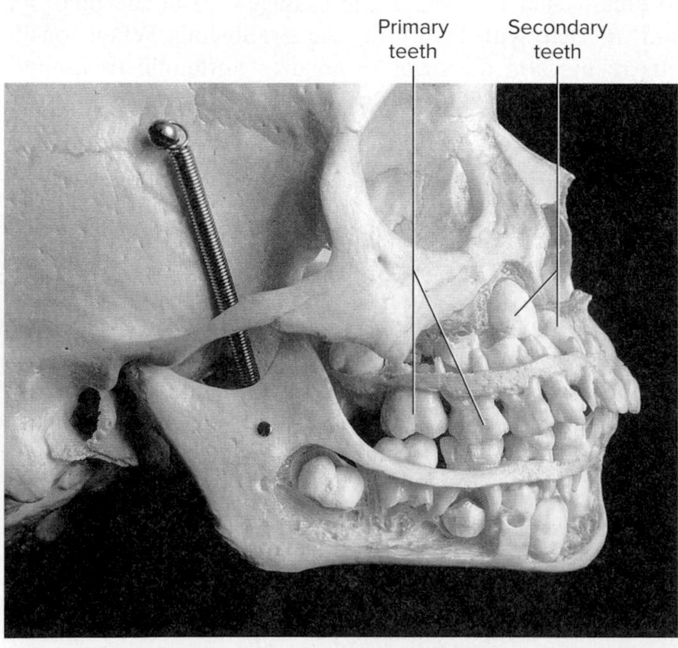

Figure 15.7 This partially dissected child's skull reveals primary and developing secondary teeth in the maxilla and mandible. Rebecca Gray/Don Kincaid/McGraw-Hill Education

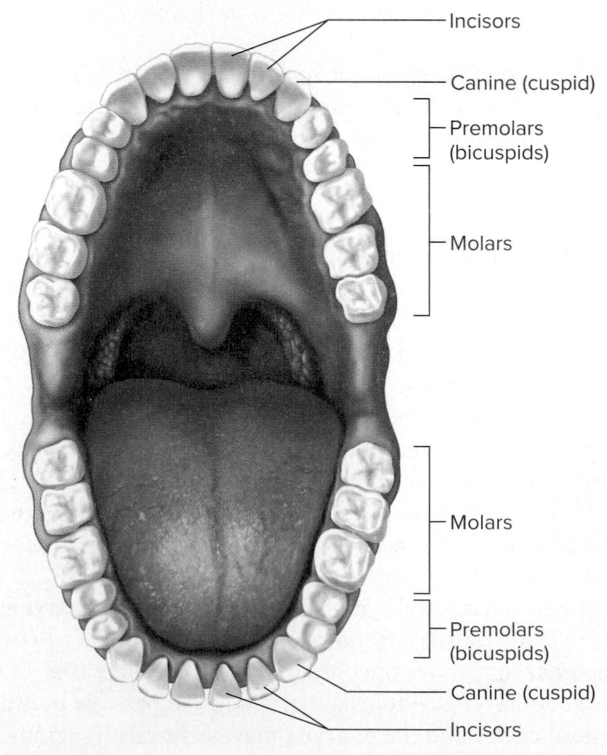

Figure 15.8 The secondary teeth of the upper and lower jaws. APR

CLINICAL APPLICATION 15.1
Dental Caries

Sticky foods, such as caramel, lodge between the teeth and in the crevices of molars, feeding bacteria such as *Actinomyces, Streptococcus mutans,* and *Lactobacillus.* These microorganisms metabolize carbohydrates in the food, producing acid by-products that destroy tooth enamel and dentin. The bacteria also produce sticky substances that hold them in place.

If a person eats a candy bar but does not brush the teeth soon afterward, the acid-forming bacteria may cause tooth decay, creating a condition called *dental caries.* Unless a dentist cleans and fills the resulting cavity that forms where enamel is destroyed, the damage will spread to the underlying dentin.

According to the American Dental Association, the following may help prevent dental caries:

1. Brush and floss teeth regularly.
2. Have regular dental exams and cleanings.
3. Talk with the dentist about receiving a fluoride treatment. Fluoride is added to the water supply in many communities. Fluoride is incorporated into the enamel's chemical structure, strengthening it.
4. The dentist may apply a sealant to children's and adolescents' teeth where crevices might hold onto decay-causing bacteria. The sealant is a coating that keeps acids from eating away at tooth enamel.

vessels, nerves, and connective tissue, collectively called *pulp.* Blood vessels and nerves reach this cavity through tubular *root canals,* which extend into the root.

A thin layer of bonelike material called **cementum,** surrounded by a **periodontal ligament,** encloses the root. This ligament contains blood vessels and nerves, as well as bundles of thick collagen fibers that pass between the cementum and the bone of the alveolar process, firmly attaching the tooth to the jaw (fig. 15.9).

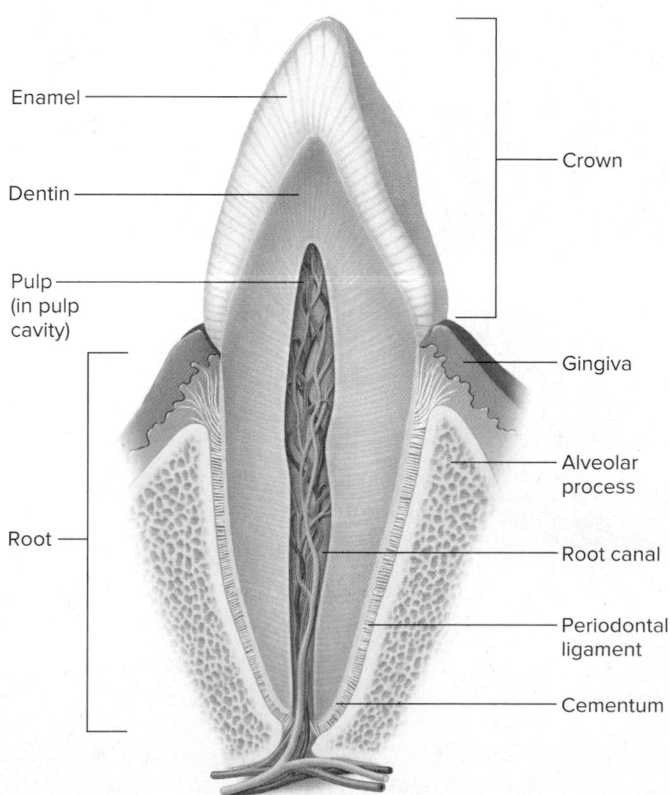

Figure 15.9 A section of a tooth.

Teeth begin mechanical digestion by breaking pieces of food into smaller pieces. This action increases the surface area of food particles, allowing more digestive enzyme access to the food molecules.

PRACTICE 15.3

3. How do primary teeth differ from secondary teeth?
4. Describe the structure of a tooth.
5. Explain how a tooth is attached to the bone of the jaw.

15.4 | Salivary Glands

LEARN

1. Locate the salivary glands and describe their secretions.
2. Identify the function of salivary amylase.
3. Describe how saliva secretion is controlled.

The **salivary** (sal'ī-ver-e) **glands** secrete *saliva.* Saliva is mostly water but also contains mucus, electrolytes, and enzymes. It functions to moisten and dissolve food particles so they can be tasted, and begins the chemical digestion of carbohydrates. Saliva also helps cleanse the mouth and teeth through the action of an antibacterial enzyme, lysozyme.

Salivary Secretions

A salivary gland has two types of secretory cells—*serous cells* and *mucous cells.* Proportions of these cells vary in the different types of salivary glands. Serous cells produce a watery fluid that contains the digestive enzyme **salivary amylase** (am'i-lās). This enzyme splits starch molecules into disaccharides—the first step in the chemical digestion of carbohydrates. Mucous cells secrete a thick liquid called

mucus, which binds food particles and acts as a lubricant during swallowing.

When a person sees, smells, tastes, or even thinks about appealing food, parasympathetic impulses elicit the secretion of a large volume of watery saliva. Conversely, food that looks, smells, or tastes unpleasant inhibits parasympathetic activity and less saliva is produced. Swallowing may become difficult.

Major Salivary Glands

Three pairs of major salivary glands—the parotid, submandibular, and sublingual glands—and many minor ones are associated with the mucous membranes of the tongue, palate, and cheeks (fig. 15.10). The **parotid glands** (pah-rot′id glandz) are the largest of the major salivary glands. Each parotid gland lies anterior and somewhat inferior to each ear, between the skin of the cheek and the masseter muscle. The secretory cells of the parotid glands are primarily serous cells. These glands secrete a clear, watery fluid that is rich in salivary amylase.

The **submandibular** (sub″man-dib′u-lar) **glands** are located in the floor of the mouth on the inside surface of the lower jaw. The secretory cells of these glands are about equally serous and mucous. Consequently, the submandibular glands secrete a more viscous fluid than the parotid glands.

The **sublingual** (sub-ling′gwal) **glands,** the smallest of the major salivary glands, are in the floor of the mouth inferior to the tongue. Their secretory cells are primarily the mucous type, making their secretions thick and stringy.

PRACTICE 15.4

1. What is the function of saliva?
2. What stimulates salivary glands to secrete saliva?
3. Where are the major salivary glands located?

15.5 | Pharynx and Esophagus

LEARN

1. Locate the pharynx and esophagus, and describe their general functions.
2. Describe the mechanism of swallowing.

The **pharynx** (far inks), commonly called the "throat," is a cavity posterior to the mouth. The tubular esophagus extends from the pharynx to the stomach (figs. 15.1 and 15.6). The pharynx and the esophagus do not digest food, but both are important passageways. Their muscular walls function in swallowing.

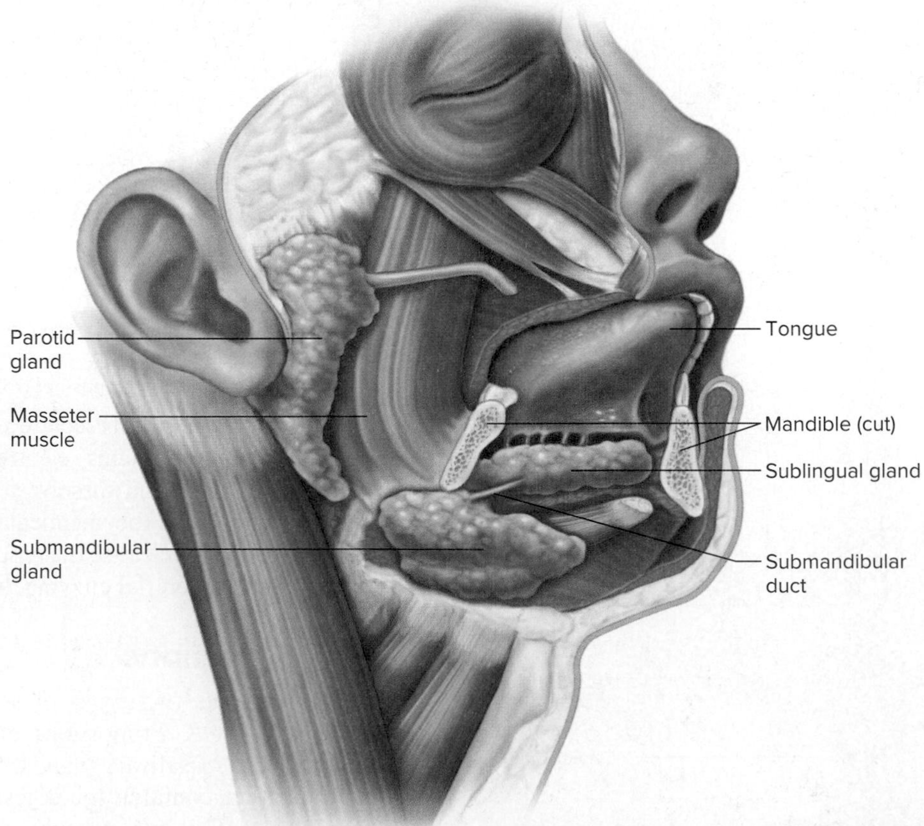

Figure 15.10 Locations of the major salivary glands. 

Structure of the Pharynx

The pharynx connects the nasal and oral cavities with the larynx and esophagus (see fig. 15.6). It has three parts:

1. The **nasopharynx** (nā″zō-far′inks) communicates with the nasal cavity and provides a passageway for air during breathing. The auditory tubes, which connect the pharynx with the middle ears, open through the walls of the nasopharynx.
2. The **oropharynx** (o″rō-far′inks) is posterior to the soft palate and inferior to the nasopharynx. It is a passageway for food moving downward from the mouth and for air moving to and from the nasal cavity.
3. The **laryngopharynx** (lah-ring″gō-far′inks), posterior to the larynx and inferior to the oropharynx, is a passageway to the esophagus.

Swallowing Mechanism

Swallowing has three stages. In the first stage, which is voluntary, food is chewed and mixed with saliva. Then, the tongue rolls this mixture into a mass, or **bolus,** and forces it into the oropharynx.

The second stage of swallowing begins as food reaches the oropharynx and stimulates sensory receptors around the pharyngeal opening. This triggers the swallowing reflex, which includes the following actions:

- The soft palate (including the uvula) rises, preventing food from entering the nasal cavity.
- The hyoid bone and the larynx are elevated. A flaplike structure attached to the larynx, called the *epiglottis* (ep″ĭ-glot′is), closes off the top of the larynx so that food is less likely to enter the trachea. Breathing is temporarily suspended and the tongue is pressed against the soft palate, sealing off the oral cavity from the nasopharynx to prevent choking.
- The longitudinal muscles in the pharyngeal wall contract, pulling the pharynx upward toward the food.
- Muscles in the laryngopharynx relax, opening the esophagus.
- A peristaltic wave begins in the pharyngeal muscles and forces food into the esophagus.

Then, during the third stage of swallowing, peristalsis transports the food in the esophagus to the stomach.

Esophagus

The **esophagus** (ĕ-sof′ah-gus) is a straight, somewhat "rubbery" collapsible tube about 25 centimeters long. Lying behind the trachea, it serves as a passageway for food from the pharynx to the stomach (see figs. 15.1 and 15.6). It penetrates the diaphragm through an opening, the *esophageal hiatus* (ĕ-sof″ah-jē′al hi-a′tus), and is continuous with the stomach.

Mucous glands in the esophagus provide lubrication to ease the movement of food. Some of the smooth muscle just superior to the point where the esophagus joins the stomach has increased muscle tone, forming the **lower esophageal sphincter** (loh′er ĕ-sof″ah-jē′al sfingk′ter), or cardiac sphincter (fig. 15.11). This sphincter regulates passage of food into the stomach and also prevents regurgitation of food back into the esophagus.

PRACTICE 15.5

1. Describe the regions of the pharynx.
2. List the major events of swallowing.
3. What is the function of the esophagus?

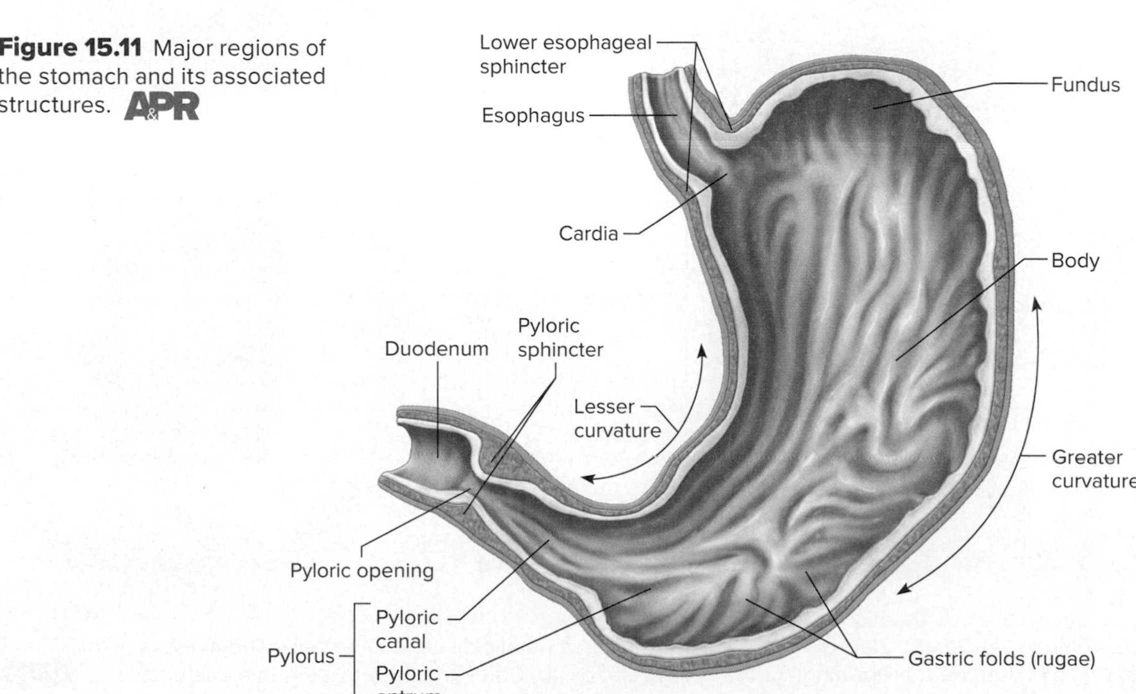

Figure 15.11 Major regions of the stomach and its associated structures. APR

15.6 | Stomach

LEARN

1. Locate the stomach and explain its functions.
2. Describe the secretions of the stomach.
3. Describe how stomach secretion is regulated.

The **stomach** is a J-shaped, pouchlike organ that hangs inferior to the diaphragm in the upper left portion of the abdominal cavity and has a capacity of about 1 liter or more (figs. 15.1 and 15.11; see reference plates 4 and 5). Thick folds (*rugae*) of mucosal and submucosal layers mark the stomach's inner lining and disappear when the stomach wall is distended. The stomach receives food from the esophagus, mixes the food with gastric juice, initiates protein digestion, carries on limited absorption, and moves food into the small intestine.

Parts of the Stomach

The stomach is divided into the cardia, fundus, body, and pylorus (fig. 15.11). The *cardia* is a small area near the esophageal opening. The *fundus,* which balloons superior to the cardia, is a temporary storage area. The dilated *body region,* which is the main part of the stomach, lies between the fundus and pylorus. The *pylorus* is the distal portion of the stomach where it approaches the small intestine. The **pyloric canal** is a narrowing of the pylorus as it approaches the small intestine. At the end of the pyloric canal, the muscular wall thickens, forming a powerful circular muscle, the **pyloric sphincter.** This muscle is a valve that controls gastric emptying.

Gastric Secretions

The mucous membrane that forms the inner lining of the stomach is thick. Its surface is studded with many small openings called *gastric pits* located at the ends of tubular **gastric glands** (gas'trik glandz) (fig. 15.12*a*).

Gastric glands generally contain three types of secretory cells. **Mucous cells,** in the necks of the glands near the openings of the gastric pits, secrete mucus. Chief cells and parietal cells are in the deeper parts of the glands. The **chief cells** secrete digestive enzymes, and the **parietal cells** release a solution containing hydrochloric acid HCl. The products of the mucous cells, chief cells, and parietal cells together form **gastric juice.**

Pepsin (pep'sin) is by far the most important digestive enzyme in gastric juice. The chief cells secrete pepsin in the form of an inactive enzyme precursor called **pepsinogen** (pep-sin'o-jen). When pepsinogen contacts hydrochloric acid from the parietal cells, it breaks down rapidly, forming pepsin (fig. 15.12*b*). Pepsin begins the digestion of nearly all types of dietary protein into polypeptides.

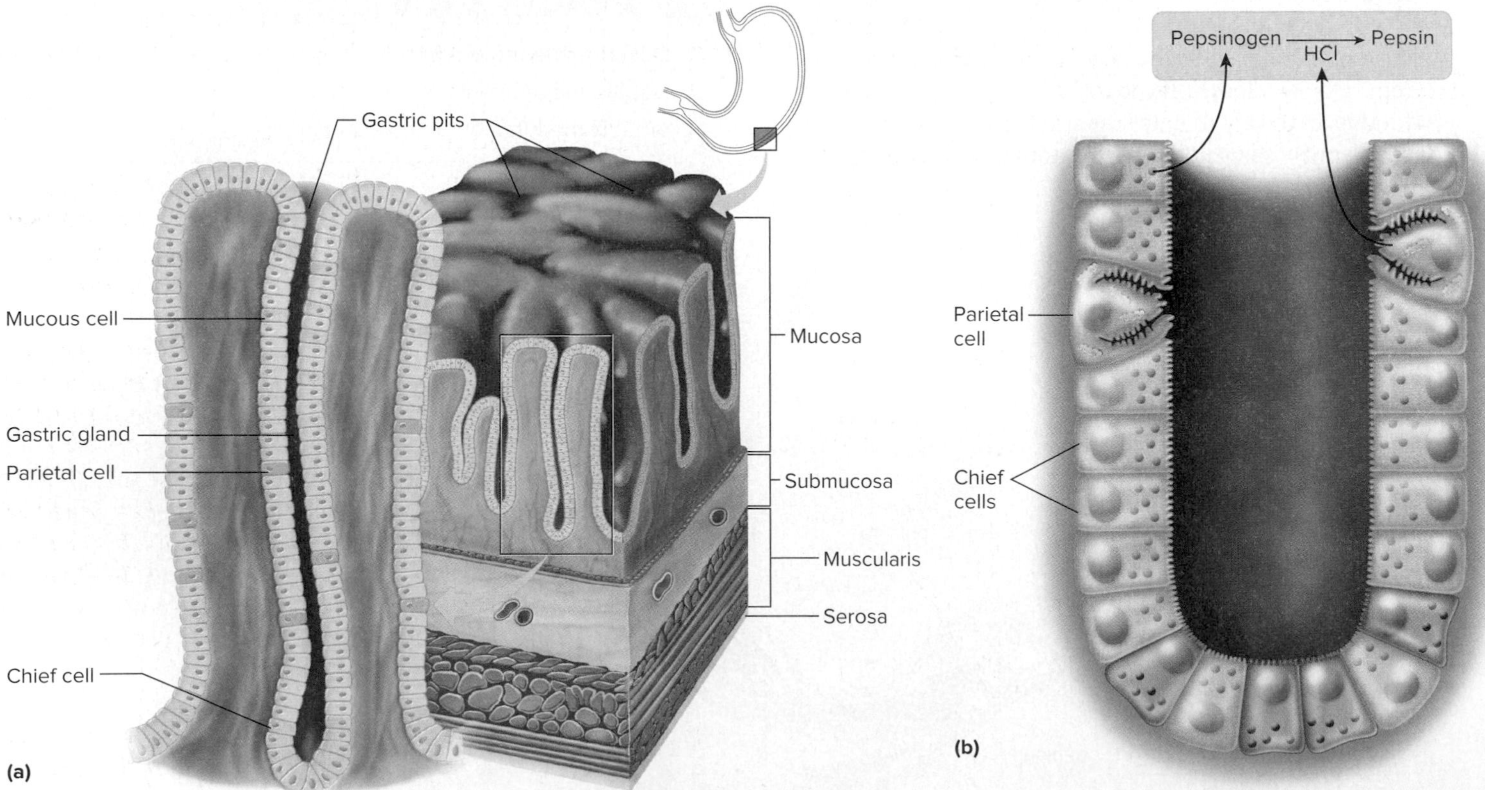

Figure 15.12 Lining of the stomach. (**a**) Gastric glands include mucous cells, parietal cells, and chief cells. The mucosa of the stomach is studded with gastric pits that are the openings of the gastric glands. (**b**) Chief and parietal cells in a gastric gland. APR

(b): Al Telser,/McGraw-Hill Education

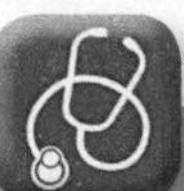

CLINICAL APPLICATION 15.2

Hiatal Hernia and Ulcers

In a hiatal hernia, part of the stomach protrudes through a weakened area of the diaphragm, through the esophageal hiatus, and into the thorax. Regurgitation (reflux) of gastric juice into the esophagus becomes more likely. This may inflame the esophageal mucosa, causing a burning sensation (heartburn), difficulty in swallowing, or ulceration and blood loss. In response to the destructive action of gastric juice, columnar epithelium may replace the squamous epithelium that normally lines the esophagus (see section 5.2, Epithelial Tissues). This condition, called Barrett's esophagus, increases the risk of developing esophageal cancer.

An ulcer is an open sore in the skin or a mucous membrane resulting from localized tissue breakdown. Gastric ulcers form in the stomach, and duodenal ulcers form in the region of the small intestine nearest the stomach.

For many years, gastric and duodenal ulcers were attributed to stress and treated with medications to decrease stomach acid secretion. In 1982, two Australian researchers boldly suggested that stomach infection by the bacterium *Helicobacter pylori* causes gastric ulcers. When the medical community did not believe them, one of the researchers swallowed a solution of some bacteria, calling it "swamp water," to demonstrate the effect. The researcher soon developed gastritis (inflammation of the stomach lining). Today, a short course of antibiotics, often combined with drugs that decrease stomach acidity, can be used to treat many gastric ulcers.

This enzyme is most active in an acidic environment, which is provided by the hydrochloric acid in gastric juice. Clinical Application 15.2 discusses the effect of gastric juice in conditions of a hiatal hernia and gastric and duodenal ulcers.

The mucous cells of the gastric glands (*mucous neck cells*) and the mucous cells associated with the stomach's inner surface release a viscous, alkaline secretion that coats the inside of the stomach wall. This coating normally prevents the stomach from digesting itself.

Another component of gastric juice is **intrinsic factor** (in-trin'sik fak'tor), which the parietal cells secrete. Intrinsic factor is necessary for the absorption of vitamin B_{12} in the small intestine. Table 15.2 summarizes the major components of gastric juice. The 40 million cells that line the stomach's interior can secrete 2 to 3 quarts (about 2 to 3 liters) of gastric juice per day.

PRACTICE 15.6

1. What are the secretions of the chief cells and parietal cells?
2. Which is the most important digestive enzyme in gastric juice?
3. Why doesn't the stomach digest itself?

Regulation of Gastric Secretions

Gastric juice is produced continuously, but the rate varies considerably and is controlled both neurally and hormonally. When a person tastes, smells, or even sees appetizing food, or when food enters the stomach, parasympathetic impulses on the vagus nerves stimulate the release of the neurotransmitter acetylcholine (ACh). This ACh stimulates gastric glands to secrete abundant gastric juice, which is rich

TABLE 15.2 Major Components of Gastric Juice APR

Component	Source	Function
Pepsinogen	Chief cells of the gastric glands	Inactive form of pepsin
Pepsin	Formed from pepsinogen in the presence of hydrochloric acid	A protein-splitting enzyme that digests nearly all types of dietary protein into polypeptides
Hydrochloric acid	Parietal cells of the gastric glands	Provides the acid environment needed for the production and action of pepsin
Mucus	Mucous cells	Provides a viscous, alkaline protective layer on the stomach's inner surface
Intrinsic factor	Parietal cells of the gastric glands	Necessary for vitamin B_{12} absorption in the small intestine

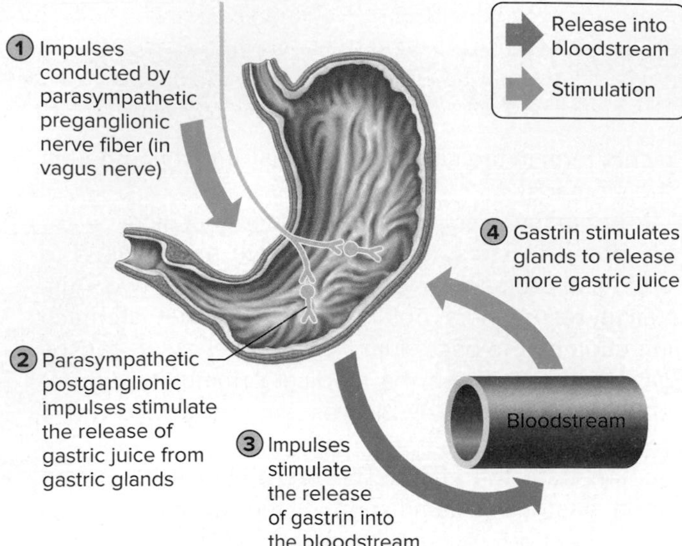

Release into bloodstream
Stimulation
1 Impulses conducted by parasympathetic preganglionic nerve fiber (in vagus nerve)
4 Gastrin stimulates glands to release more gastric juice
2 Parasympathetic postganglionic impulses stimulate the release of gastric juice from gastric glands
Bloodstream
3 Impulses stimulate the release of gastrin into the bloodstream

Figure 15.13 The secretion of gastric juice is regulated in part by parasympathetic impulses that stimulate the release of gastric juice and gastrin. APR

in hydrochloric acid and pepsinogen. These parasympathetic impulses also stimulate certain stomach cells to release the peptide hormone **gastrin** (gas′trin), which increases the secretory activity of gastric glands (fig. 15.13). Gastrin stimulates cell division in the mucosa of the stomach and intestines, which replaces mucosal cells damaged by normal stomach function, disease, or medical treatments.

As food moves into the small intestine, acid triggers sympathetic impulses that inhibit gastric juice secretion. At the same time, proteins and fats in this region of the intestine cause the intestinal wall to release the peptide hormone **cholecystokinin** (kō″lē-sis″tō-kī′nin). This hormonal action decreases gastric motility as the small intestine fills with food.

Gastric Absorption

Gastric enzymes begin breaking down proteins, but the stomach wall is not well adapted to absorb digestive products. The stomach absorbs only small volumes of water and certain salts, as well as certain lipid-soluble drugs. Alcohol, which is not a nutrient, is absorbed both in the small intestine and in the stomach.

PRACTICE 15.6

4. What controls gastric juice secretion?
5. What is the function of cholecystokinin?
6. Which substances can the stomach absorb?

Mixing and Emptying Actions

Following a meal, the mixing movements of the stomach wall aid in producing a semifluid paste of food particles and gastric juice called **chyme** (kīm). Peristaltic waves push the chyme toward the pylorus of the stomach. As chyme accumulates near the pyloric sphincter, the sphincter begins to relax. Stomach contractions push chyme a little at a time into the small intestine.

The rate at which the stomach empties depends on the fluidity of the chyme and the type of food present. Liquids usually pass through the stomach rapidly, but solids remain until they are well mixed with gastric juice. Fatty foods may remain in the stomach from three to six hours; foods high in proteins move through more quickly; carbohydrates usually pass through faster than either fats or proteins.

As chyme enters the duodenum (the proximal portion of the small intestine), accessory organs—the pancreas, liver, and gallbladder—add their secretions.

Vomiting

Vomiting is the process by which the contents of the stomach and intestines are emptied in the reverse of the normal direction. Usually, this is by a rather forceful ejection through the mouth. Emotional distress, pain, sensory overload, distension in the stomach, or irritation from toxins such as alcohol can trigger vomiting. This is controlled by a complex reflex where sensory impulses travel from the site of stimulation to the *vomiting center* in the medulla oblongata, followed by a motor response. These include taking a deep breath, raising the soft palate and thus closing the nasal cavity, closing the opening to the trachea (glottis), relaxing the lower esophageal sphincter, contracting the diaphragm so it presses downward over the stomach, and contracting the abdominal wall muscles to increase pressure inside the abdominal cavity. As a result, the stomach is squeezed from all sides, forcing its contents upward and out via the open pathway through the esophagus, pharynx, and mouth.

PRACTICE 15.6

7. How is chyme produced?
8. What factors influence how quickly chyme leaves the stomach?
9. What is the result of the vomiting center's motor response?

15.7 | Pancreas

LEARN

1. Locate the pancreas and describe its secretions.
2. Identify the function of each enzyme secreted by the pancreas.
3. Describe how pancreatic secretions are regulated.

The **pancreas** was discussed as an endocrine gland in chapter 11 (see section 11.8, Pancreas). It also has an exocrine function—secretion of a digestive fluid called **pancreatic juice** (pan″krē-at′ik joos).

Structure of the Pancreas

The pancreas is closely associated with the small intestine. It extends horizontally across the posterior abdominal wall, with its head in the C-shaped curve of the duodenum and its tail against the spleen (figs. 15.1 and 15.14).

The cells that produce pancreatic juice, called *pancreatic acinar* (a'sĭ-nar) *cells,* make up the bulk of the pancreas. These cells cluster around tiny tubes into which they release their secretions. The smaller tubes unite to form larger ones, which join a *pancreatic duct* that extends the length of the pancreas. The pancreatic duct usually connects with the duodenum at the same place where the bile duct from the liver and gallbladder joins the duodenum, although divisions of the pancreatic duct and connections to other parts of the duodenum may be present (fig. 15.14). A *hepatopancreatic sphincter* controls the movement of pancreatic juice into the duodenum.

Pancreatic Juice

Pancreatic juice contains enzymes that digest carbohydrates, fats, nucleic acids, and proteins. The carbohydrate-digesting enzyme **pancreatic amylase** splits molecules of starch into disaccharides. The fat-digesting enzyme **pancreatic lipase** breaks triglyceride molecules into fatty acids and glycerol. Two types of **nucleases** break down nucleic acid molecules into nucleotides.

The protein-splitting (proteolytic) pancreatic enzymes are **trypsin, chymotrypsin,** and **carboxypeptidase** (kar-bok"sē-pep'tĭ-dā-s). These enzymes split the bonds between particular combinations of amino acids in proteins. No single enzyme can split all possible amino acid combinations, so several enzymes are necessary to completely digest protein molecules.

Painful *acute pancreatitis* results when pancreatic enzymes become active before they are secreted, and digest parts of the pancreas. Blockage of the release of pancreatic juice by gallstones can cause pancreatitis, as can alcoholism, certain infections, traumatic injuries, and some medications.

Bicarbonate ions make the pancreatic juice alkaline, providing a favorable environment for the actions of the digestive enzymes and helping neutralize acidic chyme as

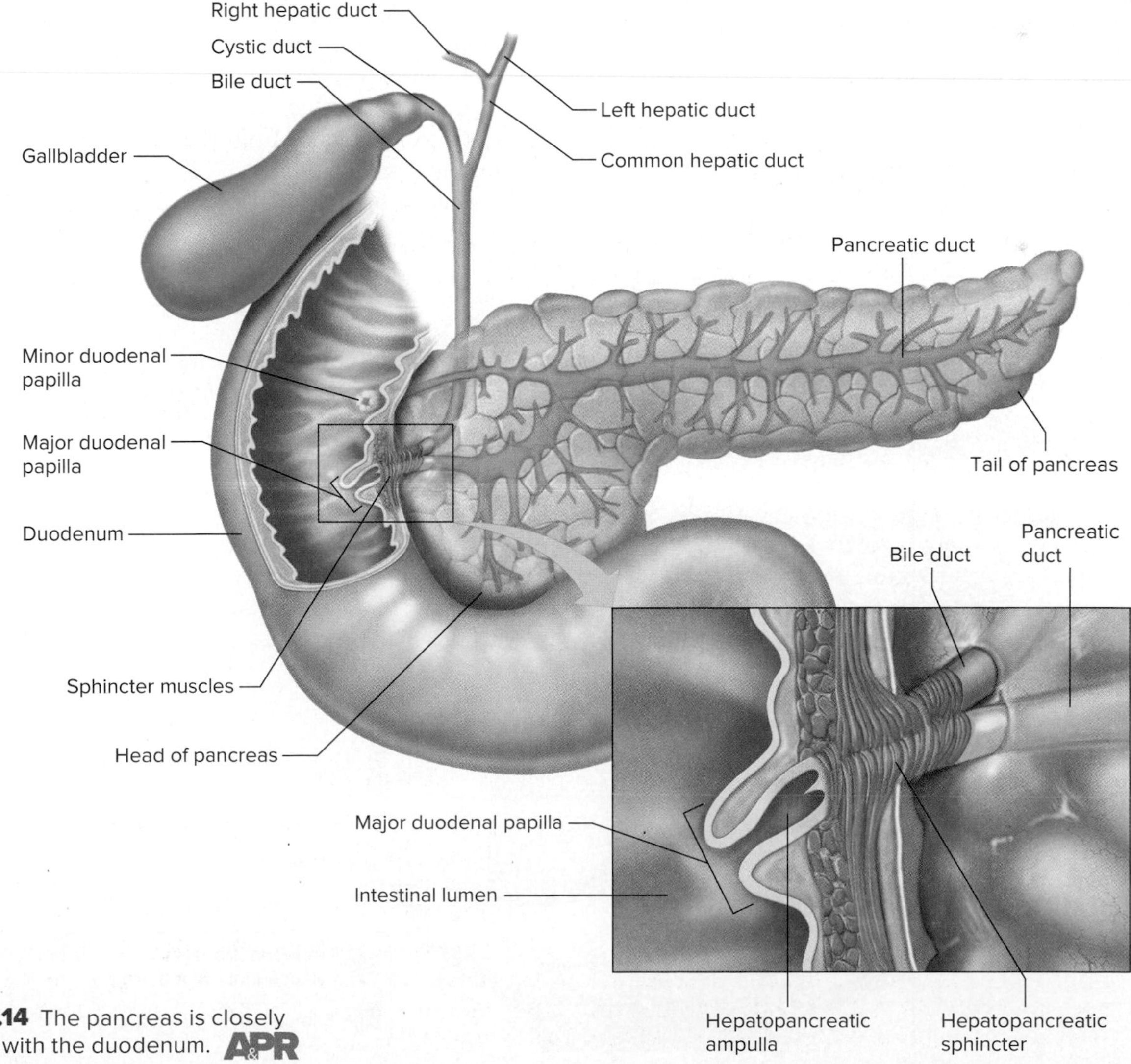

Figure 15.14 The pancreas is closely associated with the duodenum. APR

it moves from the stomach into the small intestine. The less acidic environment in the small intestine also blocks the action of pepsin, which might otherwise damage the duodenal wall.

Regulation of Pancreatic Secretion

The nervous and endocrine systems regulate release of pancreatic juice, as they do gastric and small intestinal secretions. For example, when parasympathetic impulses stimulate gastric juice secretion, other parasympathetic impulses stimulate the pancreas to release digestive enzymes. Also, as acidic chyme enters the duodenum, the duodenal mucous membrane releases the peptide hormone **secretin** (se-krē′tin) into the bloodstream (fig. 15.15). This hormone stimulates secretion of pancreatic juice that has a high concentration of bicarbonate ions.

Proteins and fats in chyme in the duodenum also stimulate the intestinal wall to release *cholecystokinin.* Like secretin, cholecystokinin travels via the bloodstream to the pancreas. Pancreatic juice secreted in response to cholecystokinin has a high concentration of digestive enzymes.

PRACTICE 15.7

1. List the enzymes in pancreatic juice.
2. What are the functions of the enzymes in pancreatic juice?
3. What regulates secretion of pancreatic juice?

15.8 | Liver and Gallbladder

LEARN

1. Locate the liver and describe its structure.
2. Explain the various liver functions.
3. Locate the gallbladder and describe the release of bile.

The **liver** is in the right upper quadrant of the abdominal cavity, just below the diaphragm. Weighing about 3 pounds, it is the heaviest internal organ, and is partially surrounded by the ribs (see fig. 15.1 and reference plate 4). The reddish-brown liver is unique in that it has a double supply of blood: The hepatic artery brings oxygen-rich blood from the aorta, and the hepatic portal vein brings nutrient-rich blood from the digestive tract (figs. 15.16, 15.17, and 15.18).

Liver Structure

A fibrous capsule encloses the liver, and ligaments divide the organ into a large *right lobe* and a smaller *left lobe* (fig. 15.16). The liver also has two minor lobes, the *quadrate lobe* and the *caudate lobe.* Each lobe is separated into many microscopic **hepatic lobules** (hĕ-pat′ik lob′ulz), the liver's functional units (fig. 15.17). A lobule consists of many hepatic cells radiating outward from a *central vein.* Blood-filled channels called **hepatic sinusoids** separate platelike groups of these cells from each other. Associated with every lobule is a hepatic triad consisting of branches from the hepatic artery, the hepatic portal vein, and the bile duct. Blood flows toward the central vein, whereas bile flows out of the liver (figs. 15.17 and 15.18).

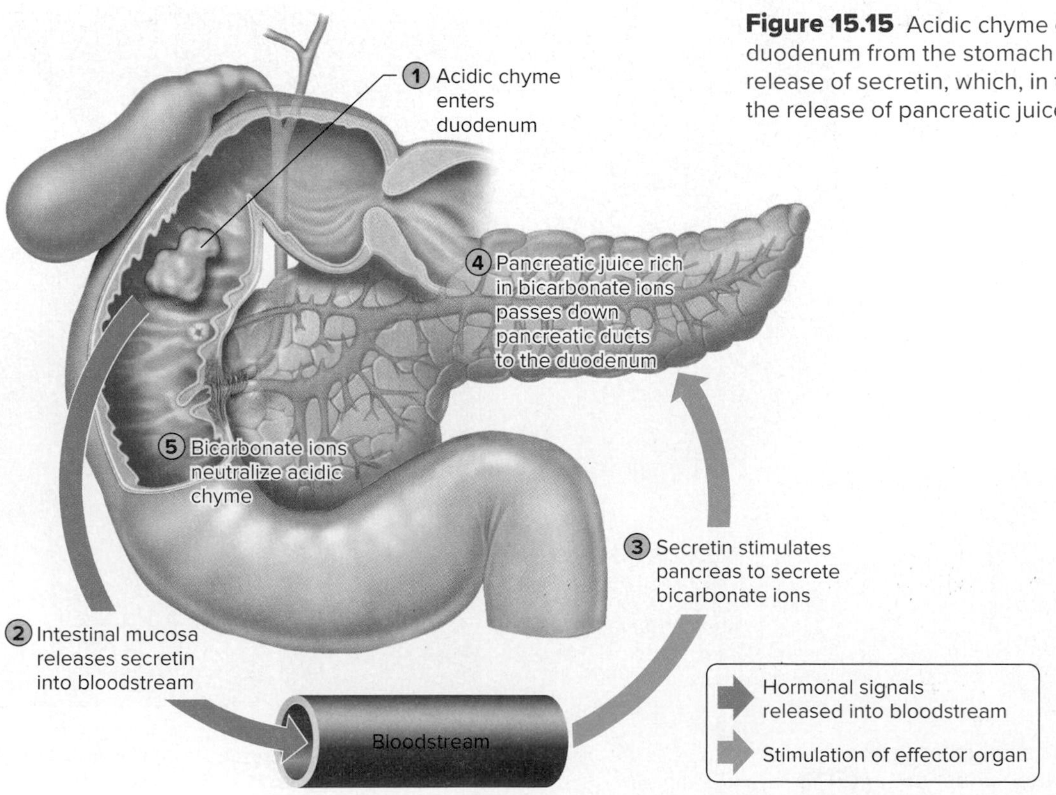

1 Acidic chyme enters duodenum
4 Pancreatic juice rich in bicarbonate ions passes down pancreatic ducts to the duodenum
5 Bicarbonate ions neutralize acidic chyme
3 Secretin stimulates pancreas to secrete bicarbonate ions
2 Intestinal mucosa releases secretin into bloodstream
Bloodstream
Hormonal signals released into bloodstream
Stimulation of effector organ

Figure 15.15 Acidic chyme entering the duodenum from the stomach stimulates the release of secretin, which, in turn, stimulates the release of pancreatic juice.

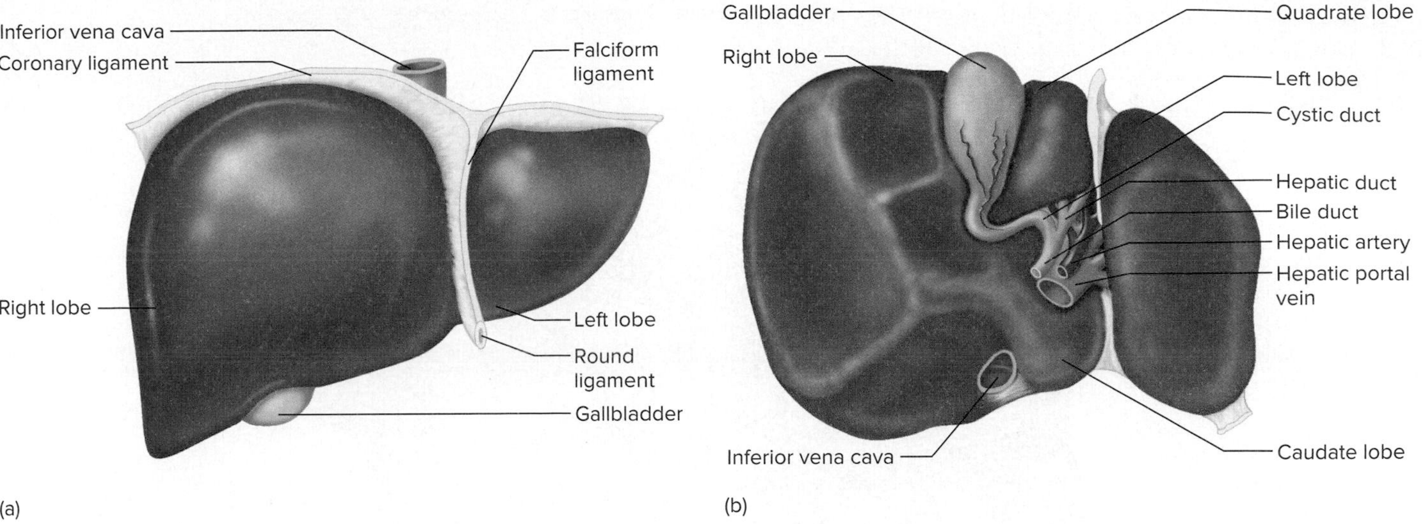

Figure 15.16 Lobes of the liver, viewed (**a**) anteriorly and (**b**) inferiorly. APR

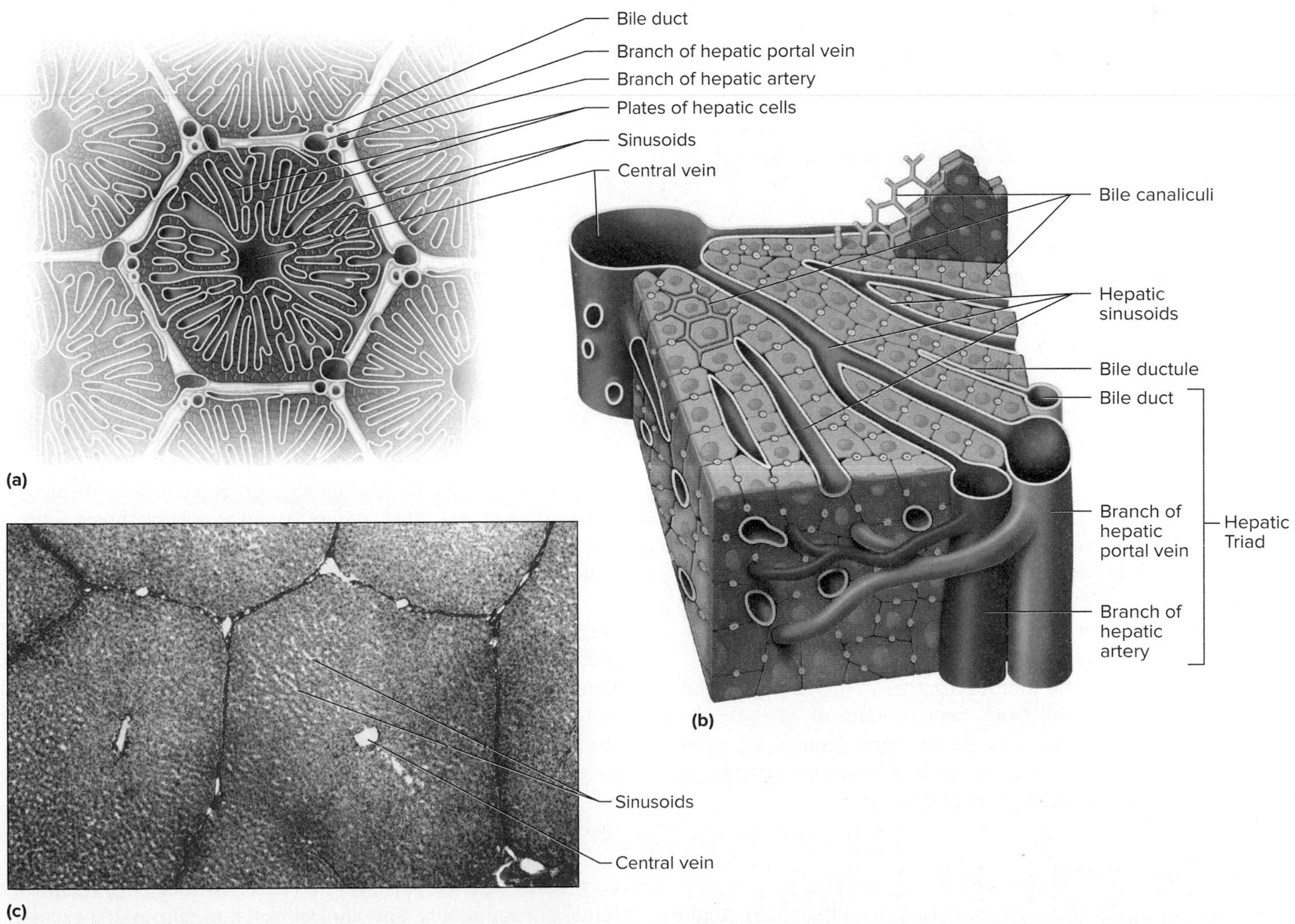

Figure 15.17 Hepatic lobule. (**a**) Cross section of a hepatic lobule. (**b**) Enlarged longitudinal section of a hepatic lobule. (**c**) Light micrograph of hepatic lobules in cross section (17x). APR (c): Victor P. Eroschenko

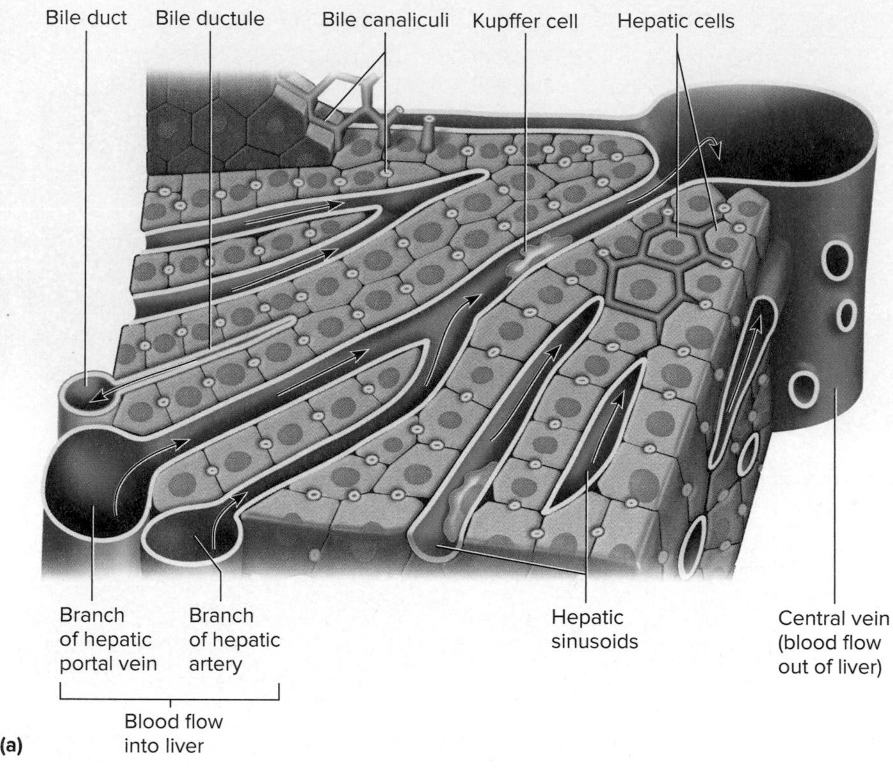

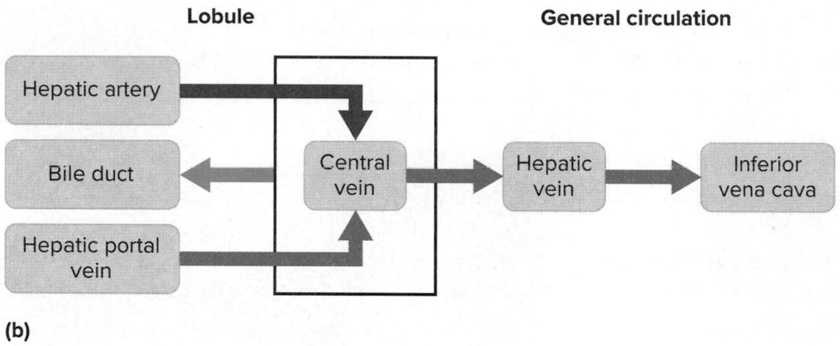

Figure 15.18 The paths of blood and bile within a hepatic lobule. APR

Large phagocytic macrophages called **Kupffer cells** (koop'fer selz) are fixed to the inner linings of the hepatic sinusoids. They remove bacteria or other foreign particles that enter the blood through the intestinal wall, and are brought to the liver via the hepatic portal vein. Blood passes from these sinusoids into the central veins of the hepatic lobules and exits the liver via the hepatic veins.

Within the hepatic lobules are many fine *bile canaliculi,* which carry secretions from hepatic cells to *bile ductules* (fig. 15.18). The ductules of neighboring lobules converge to ultimately form the **hepatic ducts.** These ducts merge, in turn, to form the **common hepatic duct.**

Liver Functions

The liver removes toxic substances from blood that might have been ingested, such as alcohol and certain drugs. This detoxification function is one of the primary reasons that blood from the digestive tract is sent to the liver before it flows into the general circulation. Long-term drug or alcohol abuse can result in a condition known as cirrhosis. This occurs when the liver cells have been overtaxed with detoxification efforts. The cells die and are replaced with scar tissue.

The liver carries on many other important metabolic activities. Recall from section 11.8, Pancreas, that the liver plays a key role in carbohydrate metabolism by helping maintain concentration of blood glucose within the normal range. Liver cells responding to the hormone insulin lower the blood glucose level by polymerizing glucose to glycogen. Liver cells responding to the hormone glucagon raise the blood glucose level by breaking down glycogen to glucose or by converting noncarbohydrates into glucose.

The liver's effects on lipid metabolism include oxidizing fatty acids at an especially high rate; synthesizing lipoproteins, phospholipids, and cholesterol; and converting excess portions of carbohydrate molecules into fat molecules. The blood transports fats synthesized in the liver to adipose tissue for storage.

TABLE 15.3 Major Functions of the Liver

General Function	Specific Function
Detoxification	Removes toxins from blood
Carbohydrate metabolism	Polymerizes glucose to glycogen; breaks down glycogen to glucose; converts noncarbohydrates to glucose
Lipid metabolism	Oxidizes fatty acids; synthesizes lipoproteins, phospholipids, and cholesterol; converts excess portions of carbohydrate molecules into fats
Protein metabolism	Deaminates amino acids; forms urea; synthesizes plasma proteins; converts certain amino acids into other amino acids
Storage	Stores glycogen, iron, and vitamins A, D, and B_{12}
Blood filtering	Removes damaged red blood cells and foreign substances by phagocytosis
Secretion	Produces and secretes bile

Other liver functions concern protein metabolism. They include deaminating amino acids; forming urea (see section 15.11, Nutrition and Nutrients); synthesizing plasma proteins such as clotting factors (see section 12.3, Plasma); and converting certain amino acids into other amino acids.

The liver also stores many substances, including glycogen, iron, and vitamins A, D, and B_{12}. In addition, macrophages in the liver help destroy damaged red blood cells (see section 12.2, Formed Elements) and phagocytize foreign antigens.

A liver function important to digestion is bile secretion. Table 15.3 summarizes the major functions of the liver. Clinical Application 15.3 discusses viral infections of the liver.

OF INTEREST The liver is unlike most organs in that it can regenerate. Up to 75% of a liver can be destroyed and the organ can still regenerate and recover. For this reason, people can donate parts of their livers to people with liver failure, if the tissues of donor and recipient are compatible.

PRACTICE 15.8

1. Locate the liver.
2. Describe a hepatic lobule.
3. Review liver functions.

Composition of Bile

Bile (bīl) is a yellowish-green liquid continuously secreted from hepatic cells. In addition to water, bile contains *bile salts, bile pigments* (bilirubin and biliverdin), *cholesterol,* and *electrolytes.* Of these, bile salts are the most abundant and are the only bile components that have a digestive function. Bile pigments are breakdown products of hemoglobin from red blood cells and are normally secreted in the bile (see section 12.2, Formed Elements).

Jaundice, a yellowing of the skin and mucous membranes due to accumulation of bile pigment, has several causes. In *obstructive jaundice,* bile ducts are blocked, perhaps by gallstones or tumors. In *hepatocellular jaundice,* the liver is diseased, as in cirrhosis or hepatitis. In *hemolytic jaundice,* red blood cells are destroyed too rapidly, as happens with an incompatible blood transfusion or a blood infection.

Gallbladder

The **gallbladder** (gawl′blad-er) is a pear-shaped sac in a depression on the liver's inferior surface. The gallbladder is lined with epithelial cells and has a strong layer of smooth muscle in its wall. The gallbladder stores bile between meals, reabsorbs water to concentrate bile, and contracts to release bile into the small intestine. It connects to the **cystic duct** (sis′tik dukt), which in turn joins the common hepatic duct (figs. 15.1 and 15.19).

The common hepatic duct and cystic duct join to form the **bile duct** (common bile duct). It leads to the duodenum where the *hepatopancreatic sphincter* guards its exit (figs. 15.14 and 15.19). Because this sphincter normally remains contracted, bile collects in the bile duct. It backs up into the cystic duct and flows into the gallbladder, where it is stored.

Cholesterol in bile may precipitate under certain conditions and form crystals called *gallstones* (fig. 15.20). Gallstones in the bile duct may block bile flow into the small intestine and cause considerable pain. A surgical procedure called a *cholecystectomy* can remove the gallbladder when gallstones are obstructive. The surgery can often be done with a laparoscope (small, lit probe) on an outpatient basis.

Regulation of Bile Release

Normally bile does not enter the duodenum until *cholecystokinin* stimulates the gallbladder to contract. The intestinal mucosa releases this hormone in response to proteins and fats in the small intestine. (Recall its action to stimulate pancreatic enzyme secretion, section 15.7, Pancreas.) The hepatopancreatic sphincter usually remains contracted until a peristaltic wave in the duodenal wall approaches it. Then the sphincter relaxes, and bile is squirted into the duodenum (see fig. 15.19). Table 15.4 summarizes the hormones that help control digestion.

Functions of Bile Salts

Bile salts aid digestive enzymes. Bile salts affect *fat globules* (clumped molecules of fats) much like a soap or detergent would affect them. That is, bile salts break fat globules into smaller droplets that are more soluble in water. This action, called **emulsification** (e-mul″sĭ-fĭ-kā′shun), greatly increases the total surface area of the fatty substance. The resulting fat droplets disperse in water. Fat-splitting enzymes (lipases) can then digest the fat molecules more effectively.

CLINICAL APPLICATION 15.3

Hepatitis

Hepatitis is an inflammation of the liver. About half a million people develop hepatitis in the United States each year, and 6,000 die of the disease. Hepatitis has several causes, but the various types have similar symptoms.

Acute hepatitis may at first resemble the flu, producing mild headache, low fever, fatigue, lack of appetite, nausea and vomiting, and sometimes stiff joints. By the end of the first week, more distinctive symptoms arise: a rash, pain in the upper right quadrant of the abdomen, dark and foamy urine, and pale feces. The skin and sclera of the eyes turn yellow due to accumulating bile pigments (jaundice). Great fatigue may continue for two or three weeks, and then gradually the person begins to feel better. This is hepatitis in its most common, least dangerous acute guise. In a rare acute form called *fulminant hepatitis,* symptoms are sudden and severe, along with altered behavior and personality. Medical attention is necessary to prevent kidney or liver failure, or coma.

Chronic hepatitis is a disease that persists for more than six months. As many as 300 million people worldwide are hepatitis carriers. They do not have symptoms but can infect others. Five percent of carriers eventually develop liver cancer.

Only rarely does hepatitis result from alcoholism, autoimmunity, or the use of certain drugs. Usually, one of several types of viruses causes hepatitis. Viral types are distinguished by the route of infection, surface features, and whether the viral genetic material is DNA or RNA. Hepatitis B virus has DNA; the others have RNA. The viral types are classified as follows:

Hepatitis A spreads by contact with food or objects contaminated with virus-containing feces, including diapers. The course of hepatitis A is short and mild.

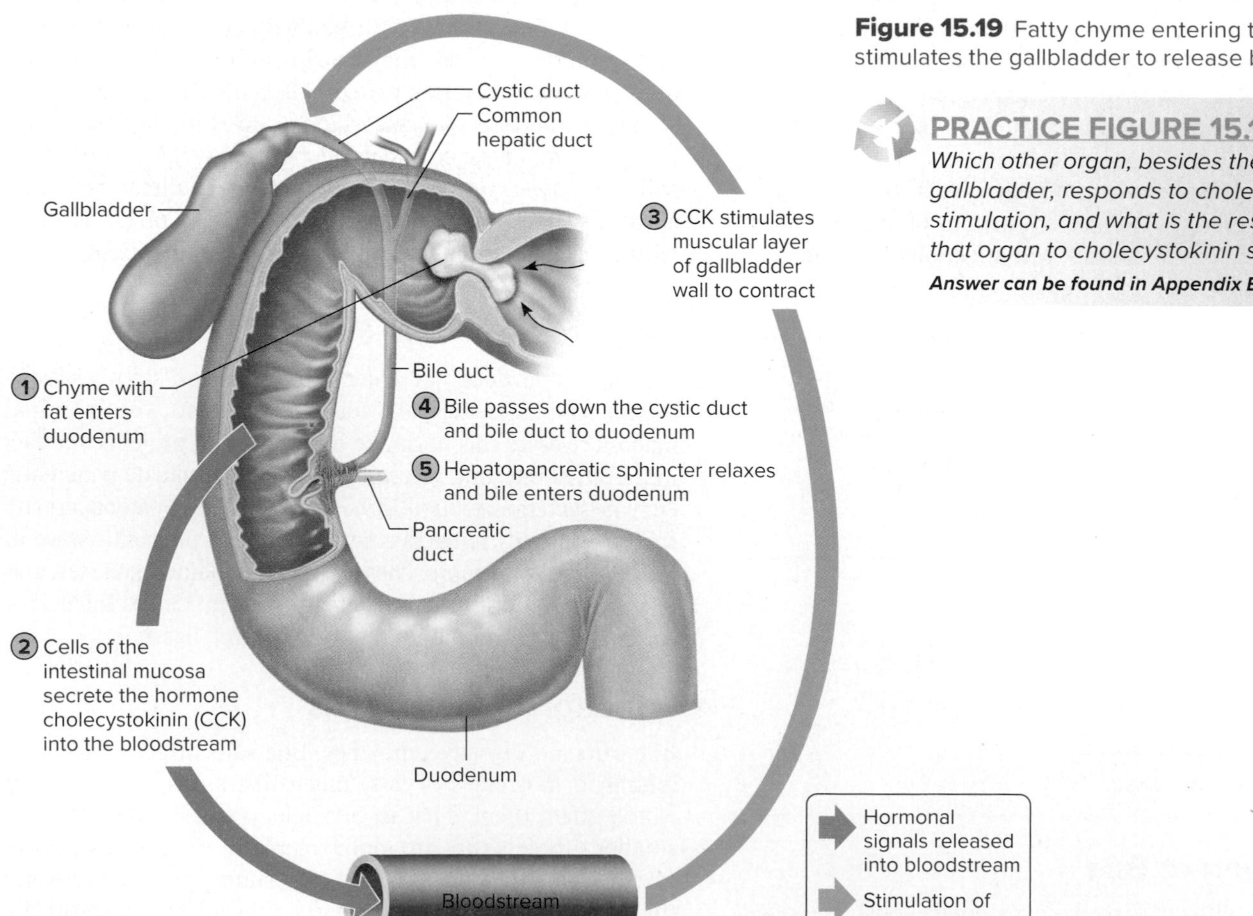

Figure 15.19 Fatty chyme entering the duodenum stimulates the gallbladder to release bile.

PRACTICE FIGURE 15.19

Which other organ, besides the gallbladder, responds to cholecystokinin stimulation, and what is the response of that organ to cholecystokinin stimulation?

Answer can be found in Appendix E.

Continued—

Hepatitis B spreads by contact with virus-containing body fluids, such as blood, saliva, or semen. It may be transmitted by blood transfusions, hypodermic needles, or sexual activity.

Hepatitis C accounts for about half of all known cases of hepatitis. This virus is primarily transmitted in blood—by sharing razors or needles, from pregnant woman to fetus, or through blood transfusions or use of blood products. As many as 60% of individuals infected with the hepatitis C virus suffer chronic symptoms.

Hepatitis D infection occurs in people already infected with the hepatitis B virus. It is bloodborne and associated with blood transfusions and intravenous drug use. About 20% of individuals infected with this virus die from the infection.

Hepatitis E virus is usually transmitted in water contaminated with feces. It most often affects visitors to developing nations.

Hepatitis G infection is rare but accounts for a significant percentage of cases of fulminant hepatitis. In people with healthy immune systems, the virus produces symptoms so mild that they may not be noticed.

Antibiotic drugs, which are effective against bacteria, are not helpful against viral hepatitis. For some types of hepatitis, the person must just wait out the symptoms. The clinical picture is changing, however, for many of the approximately 2.7 million people in the United States who have chronic hepatitis C. Treatment with several new drugs, used in combination to block viral activity in several ways, is now simpler, faster, and has fewer side effects than older drugs such as interferon and ribavirin.

Bile salts also enhance absorption of fatty acids, cholesterol, and the fat-soluble vitamins A, D, E, and K. Low levels of bile salts result in poor lipid absorption and vitamin deficiencies.

PRACTICE 15.8

4. Explain how bile forms.
5. Describe the function of the gallbladder.
6. How is secretion of bile regulated?
7. How do bile salts function in digestion?

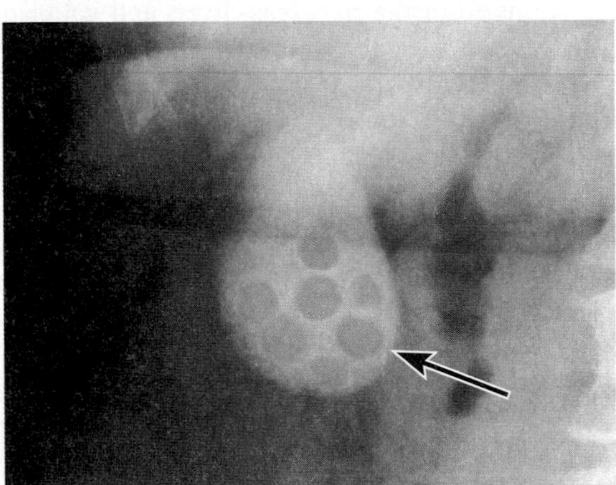

Figure 15.20 Falsely colored radiograph of a gallbladder that contains gallstones (arrow). Southern Illinois University/Science Source

15.9 | Small Intestine

LEARN

1. Locate the small intestine and describe its structure.
2. Identify the function of each enzyme secreted by the small intestine.
3. Describe how small intestinal secretions are regulated.
4. Explain how the products of digestion are absorbed in the small intestine.

The **small intestine** is a tubular organ that extends from the stomach to the beginning of the large intestine. With its many loops and coils, the small intestine fills much of the abdominal cavity (see fig. 15.1 and reference plates 4 and 5).

TABLE 15.4 Hormones of the Digestive Tract

Hormone	Source	Function
Gastrin	Gastric cells, in response to food	Increases secretory activity of gastric glands
Cholecystokinin	Intestinal wall cells, in response to proteins and fats in the small intestine	Decreases secretory activity of gastric glands and inhibits gastric motility; stimulates pancreas to secrete fluid with a high digestive enzyme concentration; stimulates gallbladder to contract and release bile
Secretin	Cells in the duodenal wall, in response to acidic chyme entering the small intestine	Stimulates pancreas to secrete fluid with a high bicarbonate ion concentration

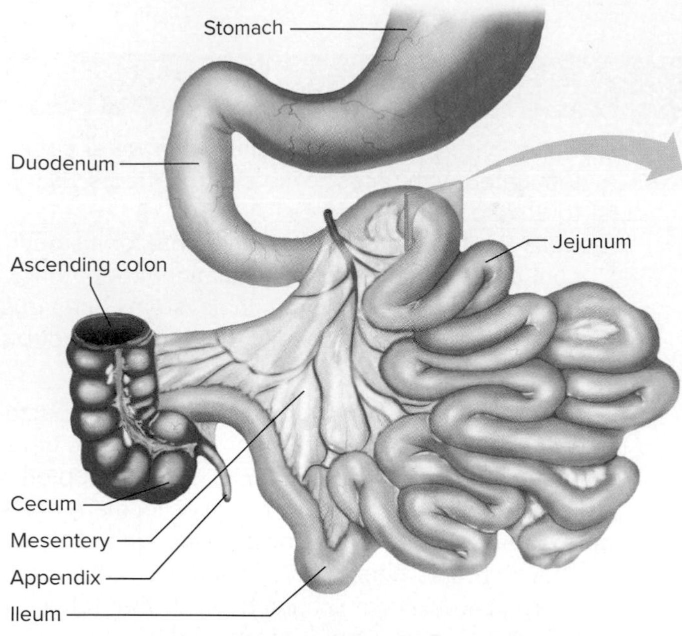

Figure 15.21 The three parts of the small intestine are the duodenum, the jejunum, and the ileum. APR

The small intestine receives chyme from the stomach and secretions from the pancreas, liver, and gallbladder. It completes digestion of the nutrients in chyme, absorbs the products of digestion, and transports the residue to the large intestine.

Parts of the Small Intestine

The small intestine consists of three parts: the duodenum, the jejunum, and the ileum (figs. 15.21 and 15.22). The **duodenum** (dū″o-de′num), about 25 centimeters long and 5 centimeters in diameter, lies posterior to the parietal peritoneum and is the most fixed portion of the small intestine. It follows a C-shaped path as it passes anterior to the right kidney and the upper three lumbar vertebrae.

The remainder of the small intestine is mobile and lies free in the peritoneal cavity. The proximal two-fifths of this portion of the small intestine is the **jejunum** (jĕ-joo′num), and the remainder is the **ileum** (il′ē-um). The jejunum and ileum are not easily distinguished as separate parts; however, the diameter of the jejunum is typically greater than that of the ileum, and its wall is thicker, more vascular, and more active.

A double-layered fold of peritoneal membrane called **mesentery** (mes′en-ter″ē) suspends the jejunum and ileum from the posterior abdominal wall (figs. 15.21, 15.23 and reference plate 5). The mesentery, now considered a full-fledged organ, supports the blood vessels, nerves, and lymphatic vessels that supply the intestinal wall.

A filmy, double fold of peritoneal membrane called the *greater omentum* drapes like an apron from the stomach over the transverse colon and the folds of the small intestine (fig. 15.23 and reference plate 3). If the wall of the alimentary canal becomes infected, cells from the omentum may adhere to the inflamed region, helping to wall off the area. This action prevents spread of the infection to the peritoneal cavity.

Structure of the Small Intestinal Wall

The inner surface of the small intestine throughout its length appears velvety due to many tiny projections of mucous membrane called **intestinal villi** (vil′ī) (figs. 15.24 and 15.25; see fig. 15.3). These structures are most numerous in the duodenum and the proximal jejunum. They project into the lumen of the alimentary canal, contacting the intestinal contents. Villi greatly increase the surface area of the intestinal lining, aiding the absorption of digestive products.

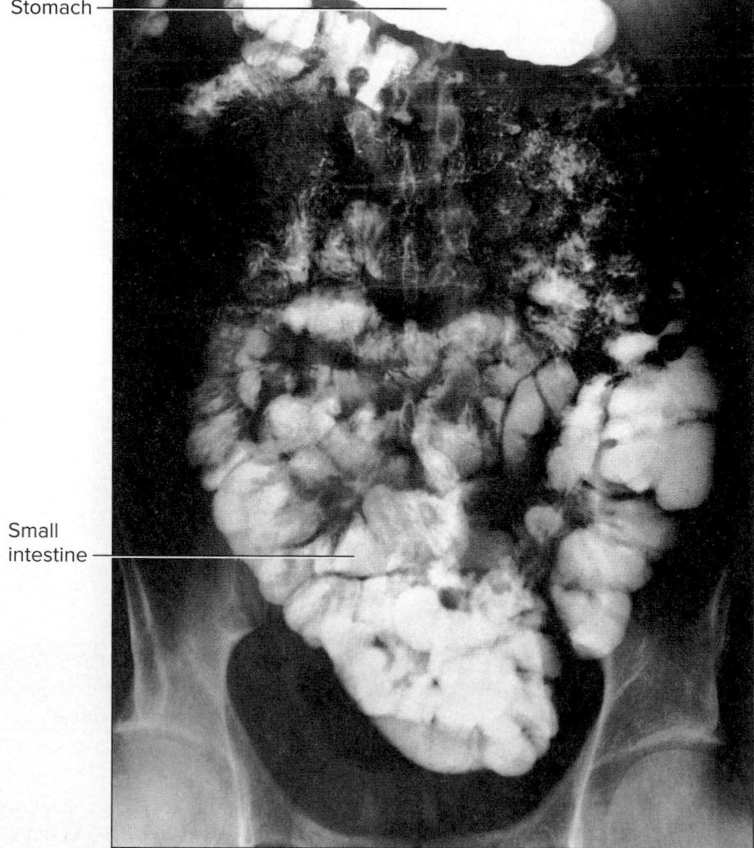

Figure 15.22 Radiograph showing a normal small intestine containing a radiopaque substance that the patient ingested. APR tbradford/Getty Images

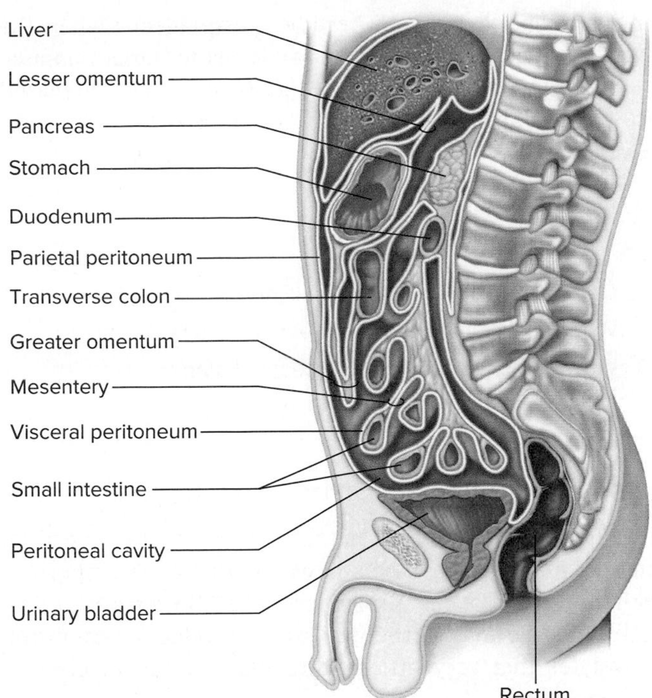

Figure 15.23 Mesentery formed by folds of the peritoneal membrane suspends portions of the small intestine from the posterior abdominal wall. APR

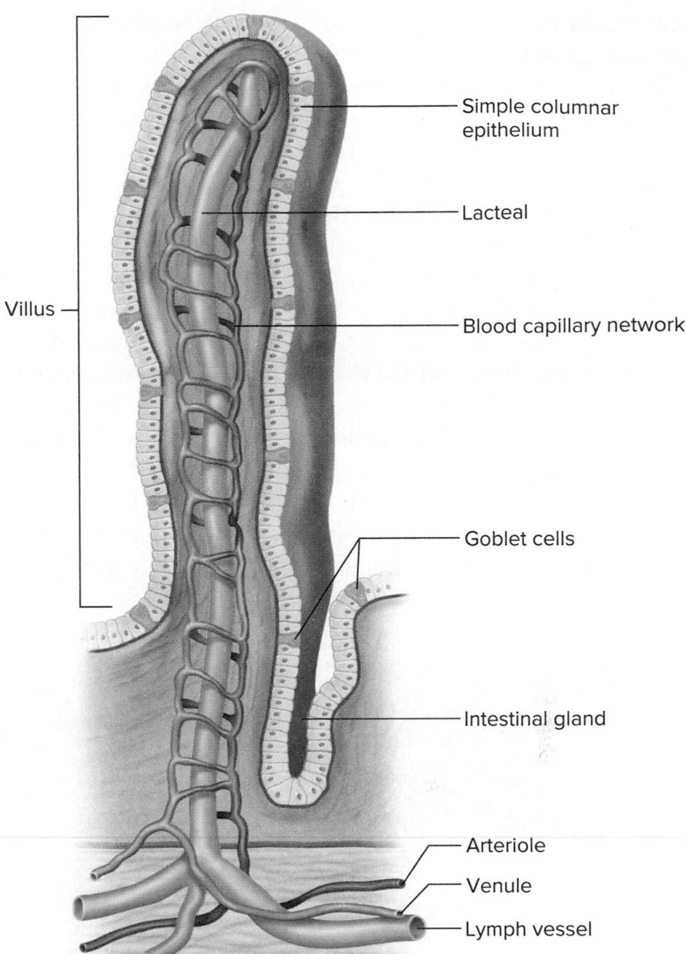

Figure 15.24 Structure of a single intestinal villus. The elongated shapes of intestinal villi dramatically increase the absorptive surface area of the small intestine.

Each villus consists of a layer of simple columnar epithelium and a core of connective tissue containing blood capillaries, a lymphatic capillary called a **lacteal,** and nerve fibers. Blood capillaries and lacteals carry away absorbed nutrients, and nerve fibers conduct impulses to stimulate or inhibit villus activities. Between the bases of adjacent villi are tubular **intestinal glands** that extend downward into the mucous membrane (figs. 15.24 and 15.25; see fig. 15.3).

The epithelial cells that form the lining of the small intestine are continually replaced. New cells form in the intestinal glands by mitosis and migrate outward onto the villus surface. When the migrating cells reach the tip of the villus, they are shed. As a result, nearly one-quarter of the bulk of feces consists of dead epithelial cells from the small intestine. This cellular turnover renews the small intestine's epithelial lining every three to six days.

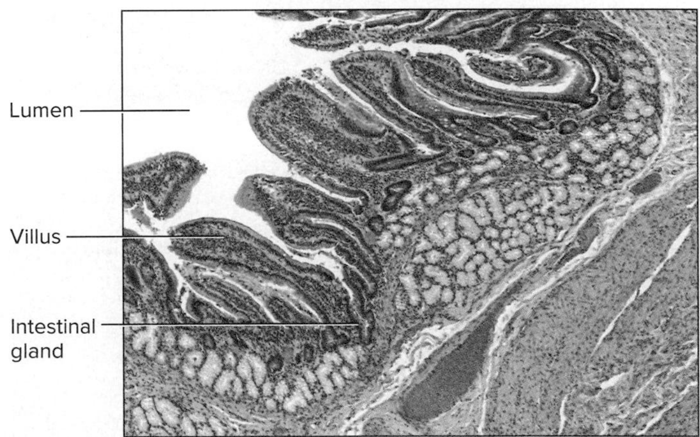

Figure 15.25 Light micrograph of intestinal villi from the wall of the duodenum (50x). APR Al Telser/McGraw-Hill Education

Secretions of the Small Intestine

Mucus-secreting **goblet cells** are abundant throughout the mucosa of the small intestine. In addition, many specialized *mucus-secreting glands* in the submucosa in the proximal portion of the duodenum secrete a thick, alkaline mucus in response to mechanical and chemical stimuli.

Distension of the intestinal wall activates the nerve plexuses within the wall and stimulates parasympathetic reflexes that also trigger release of small intestinal secretions.

The intestinal glands at the bases of the villi secrete large volumes of a watery fluid, which brings digestive products to the villi. The fluid has a nearly neutral pH (6.5–7.5), and it does not carry digestive enzymes. On their luminal surfaces, the epithelial cells of the intestinal mucosa have digestive enzymes embedded in the membranes of their microvilli. These enzymes break down food molecules just before absorption takes place. The enzymes include **peptidases,** which split peptides into their constituent amino acids; **sucrase, maltase,** and **lactase,** which split the disaccharides sucrose, maltose, and lactose into the monosaccharides glucose, fructose, and galactose; and **intestinal lipase,** which splits fats into fatty acids and glycerol. Table 15.5 summarizes the sources and actions of the major digestive enzymes.

Adults who have *lactose intolerance* do not produce sufficient lactase to adequately digest lactose, or milk sugar. Undigested lactose increases the osmotic pressure of the intestinal contents and draws water into the intestines. At the same time, intestinal bacteria metabolize undigested sugar, producing organic acids and gases. The overall result is bloating, intestinal cramps, and diarrhea.

Genetic evidence suggests that lactose intolerance may be the "normal" condition, and that the ability to digest lactose is the result of a mutation that occurred recently in our evolutionary past and became advantageous when the advent of agriculture brought dairy foods to human populations. The trait of the ability to digest lactose has increased with the increased use of dairy foods at least three times in history, in different populations.

PRACTICE 15.9

1. Describe the parts of the small intestine.
2. What is the function of an intestinal villus?
3. What is the function of the intestinal glands?
4. List the intestinal digestive enzymes.

TABLE 15.5 Summary of the Major Digestive Enzymes

Enzyme	Source	Digestive Action
Salivary amylase	Salivary glands	Begins carbohydrate digestion by breaking down starch to disaccharides
Pepsin	Gastric chief cells	Begins protein digestion
Pancreatic amylase	Pancreas	Breaks down starch into disaccharides
Pancreatic lipase	Pancreas	Breaks down fats into fatty acids and glycerol
Proteolytic enzymes (a) Trypsin (b) Chymotrypsin (c) Carboxypeptidase	Pancreas	Break down proteins or partially digested proteins into peptides
Nucleases	Pancreas	Break down nucleic acids into nucleotides
Peptidase	Intestinal mucosal cells	Breaks down peptides into amino acids
Sucrase, maltase, lactase	Intestinal mucosal cells	Break down disaccharides into monosaccharides
Intestinal lipase	Intestinal mucosal cells	Breaks down fats into fatty acids and glycerol
Enterokinase	Intestinal mucosal cells	Converts trypsinogen into trypsin

Absorption in the Small Intestine

Villi greatly increase the surface area of the intestinal mucosa, making the small intestine the most important absorbing organ of the alimentary canal. So effective is the small intestine in absorbing digestive products, water, and electrolytes that very little absorbable material reaches its distal end.

Carbohydrate digestion begins in the mouth with the activity of salivary amylase, and completes in the small intestine as enzymes from the pancreas and the intestinal mucosa break down their substrates. Villi absorb the resulting monosaccharides via facilitated diffusion or active transport (see section 3.3, Movements Into and Out of the Cell). The resulting simple sugars then enter blood capillaries.

Protein digestion begins in the stomach as a result of pepsin activity, and completes in the small intestine as enzymes from the pancreas and the intestinal mucosa break down large protein molecules into amino acids. These products of protein digestion are then actively transported into the villi and carried away by the blood.

Triglyceride molecules are digested almost entirely by enzymes from the pancreas and intestinal mucosa. The resulting fatty acids and glycerol molecules diffuse into villi epithelial cells (fig. 15.26*[1]*). The endoplasmic reticula of the cells use the fatty acids to resynthesize triglyceride molecules similar to those previously digested (fig. 15.26*[2]*). These triglycerides are encased in protein to form **chylomicrons** (fig. 15.26*[3]*), which make their way to the lacteals of the villi (fig. 15.26*[4]*). Lymph in the lacteals and other lymphatic vessels carries chylomicrons to the bloodstream, as discussed in section 14.1, Introduction (fig. 15.26*[5]*). Some fatty acids with very short carbon chains may be absorbed directly into the blood capillary of a villus without being changed back into triglyceride molecules.

Chylomicrons transport dietary triglycerides to muscle and adipose cells. Similarly, VLDL (very low-density lipoprotein, with a high concentration of triglycerides) particles, produced in the liver, transport triglycerides synthesized from excess dietary carbohydrates. As VLDL particles reach adipose cells, an enzyme, *lipoprotein lipase,* catalyzes reactions that unload their triglycerides, converting the VLDL to LDL (low-density lipoproteins). Because most of

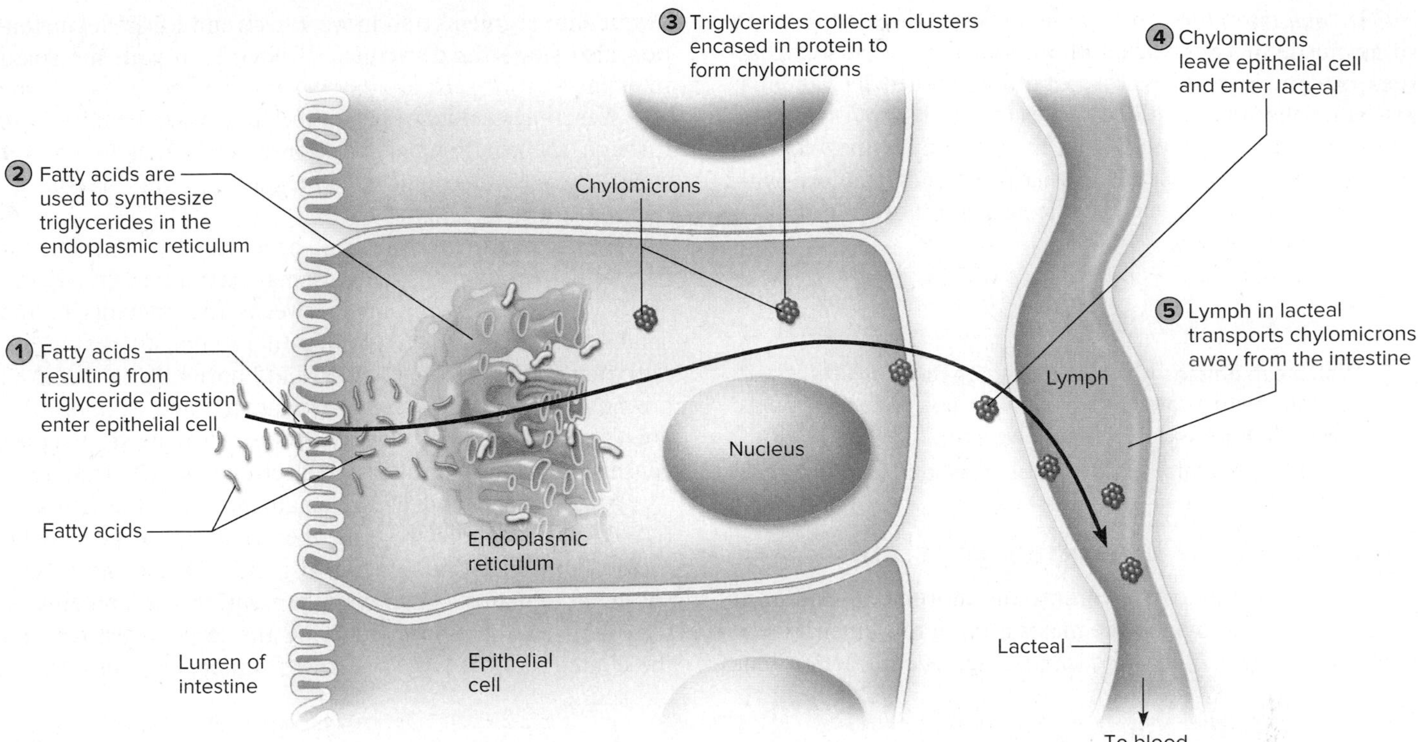

Figure 15.26 Triglyceride absorption involves several steps. (The microvilli on the simple columnar epithelial cells further increase the surface area of the cell membranes for greater absorption.)

the triglycerides have been removed, LDL particles have a higher cholesterol content than do the original VLDL particles. Cells in the peripheral tissues remove LDL from plasma by receptor-mediated endocytosis, thus obtaining a supply of cholesterol (see section 3.3, Movements Into and Out of the Cell).

While LDL delivers cholesterol to tissues, HDL (high-density lipoprotein, with a high concentration of protein and a low concentration of lipids) removes cholesterol from tissues. The liver produces the basic HDL framework and secretes the HDL particles into the bloodstream. As they circulate, the HDL particles pick up cholesterol from peripheral tissues and return to the liver, where they enter liver cells by receptor-mediated endocytosis. The liver secretes the cholesterol it obtains in this manner into bile or uses it to synthesize bile salts.

The small intestine reabsorbs much of the cholesterol and bile salts in bile, which are then transported back to the liver, and the secretion-reabsorption cycle repeats. During each cycle, some of the cholesterol and bile salts escape reabsorption, reach the large intestine, and are excreted as part of the feces.

The intestinal villi absorb electrolytes by diffusion and active transport and water by osmosis, in addition to absorbing the products of carbohydrate, protein, and fat digestion. Table 15.6 summarizes the intestinal absorption process.

TABLE 15.6 Intestinal Absorption of Nutrients

Nutrient	Absorption Mechanism	Means of Circulation
Monosaccharides	Facilitated diffusion and active transport	Blood in capillaries
Amino acids	Active transport	Blood in capillaries
Fatty acids and glycerol	Facilitated diffusion of glycerol; diffusion of fatty acids into cells or blood capillaries	
	(a) Most fatty acids are resynthesized into triglycerides and incorporated in chylomicrons	Lymph in lacteals
	(b) Some fatty acids with relatively short carbon chains are absorbed without being changed back into triglycerides	Blood in capillaries
Electrolytes	Diffusion and active transport	Blood in capillaries
Water	Osmosis	Blood in capillaries

In *malabsorption,* the small intestine does not absorb some nutrients. Symptoms of malabsorption include diarrhea, weight loss, weakness, vitamin deficiencies, anemia, and bone demineralization. Causes of malabsorption include surgical removal of a portion of the small intestine, obstruction of lymphatic vessels due to a tumor, interference with the production and release of bile as a result of liver disease, or enzyme deficiency.

PRACTICE 15.9

5. Which substances resulting from digestion of carbohydrate, protein, and fat molecules does the small intestine absorb?
6. Describe how fatty acids are absorbed and transported.

Movements of the Small Intestine

The small intestine carries on mixing movements and peristalsis, like the stomach. The major mixing movement is segmentation, in which periodic small, ringlike contractions cut chyme into segments and move it back and forth. Segmentation also slows the movement of chyme through the small intestine.

Weak peristaltic waves propel chyme short distances through the small intestine. Consequently, chyme moves slowly through the small intestine, taking from three to ten hours to travel its length.

If the small intestine wall becomes overdistended or irritated, a strong *peristaltic rush* may pass along the organ's entire length. This movement sweeps the contents of the small intestine into the large intestine so quickly that water, nutrients, and electrolytes that would normally be absorbed are not. The result is *diarrhea,* characterized by more frequent defecation and watery stools. Prolonged diarrhea causes imbalances in water and electrolyte concentrations.

At the distal end of the small intestine, the **ileocecal** (il″ē-ō-sē′kal) **sphincter** joins the small intestine's ileum to the large intestine's cecum (fig. 15.27). Normally, this sphincter remains constricted, preventing the contents of the small intestine from entering the large intestine, and the contents of the large intestine from backing up into the

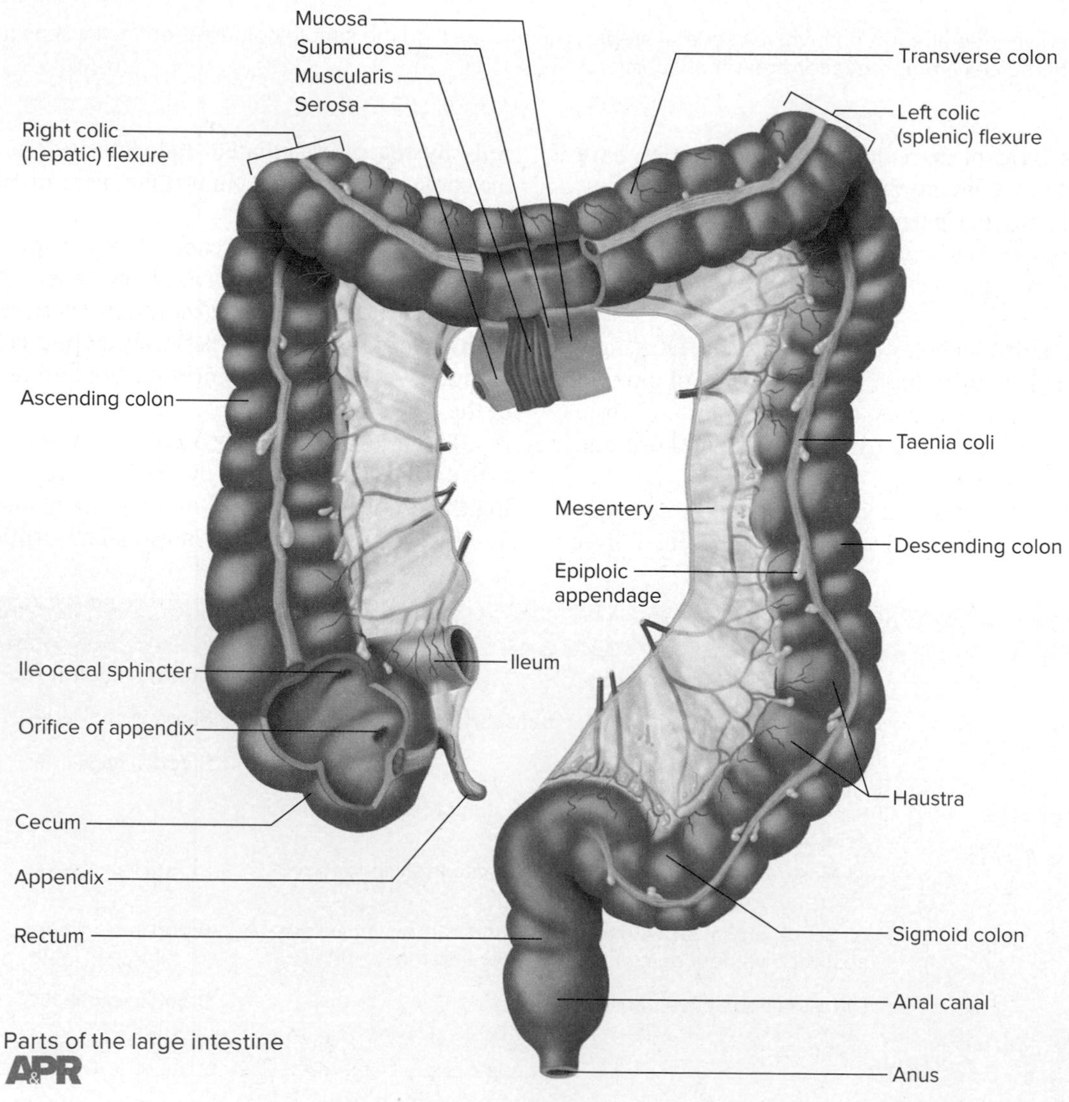

Figure 15.27 Parts of the large intestine (anterior view). APR

ileum. However, after a meal, a gastroileal reflex increases peristalsis in the ileum and relaxes the sphincter, forcing some of the contents of the small intestine into the cecum.

PRACTICE 15.9

7. Describe the movements of the small intestine.
8. What is a peristaltic rush?

15.10 | Large Intestine

LEARN

1. Locate the large intestine and describe its structure.
2. Identify the functions of the large intestine.
3. Describe the mechanism of defecation.

The **large intestine** is so named because its diameter is greater than that of the small intestine. This part of the alimentary canal is about 1.5 meters long. It begins in the lower right side of the abdominal cavity, where the ileum joins the cecum. From there, the large intestine ascends on the right side, crosses obliquely to the left, and descends into the pelvis. At its distal end, it opens to the outside of the body as the anus (see fig. 15.1).

The large intestine absorbs water and electrolytes from chyme remaining in the alimentary canal. It also forms and stores feces.

Parts of the Large Intestine

The large intestine consists of the cecum, colon, rectum, and anal canal (figs. 15.27 and 15.28; see reference plates 4 and 5). The **cecum,** at the beginning of the large intestine, is a dilated, pouchlike structure that hangs slightly below the ileocecal opening. Projecting downward from it is a closed ended, narrow tube containing lymphatic tissue called the **appendix.** The lymph nodules in the human appendix function in the immune response.

The **colon** is divided into four parts—the ascending, transverse, descending, and sigmoid colons. The **ascending colon** begins at the cecum and continues upward against the posterior abdominal wall to a point just inferior to the liver. There, it turns sharply to the left and becomes the **transverse colon,** the longest and most movable part of the large intestine. It is suspended by a fold of peritoneum and sags in the middle below the stomach. As the transverse colon approaches the spleen, it turns abruptly downward and becomes the **descending colon.** At the brim of the pelvis, the descending colon makes an S-shaped curve called the **sigmoid colon** and then becomes the rectum.

The **rectum** lies next to the sacrum and generally follows its curvature. The peritoneum firmly attaches the rectum to the sacrum. The rectum ends about 5 centimeters below the tip of the coccyx, where it becomes the **anal canal** (see fig. 15.27).

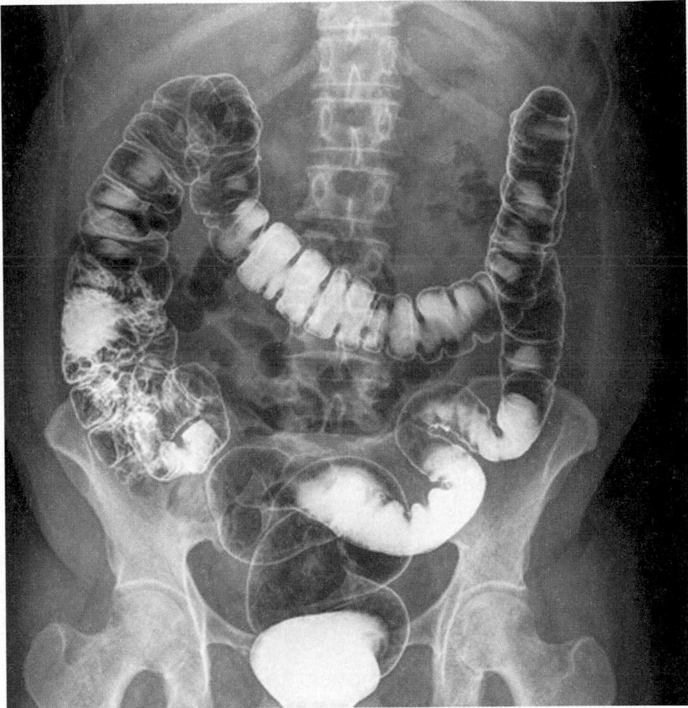

Figure 15.28 Radiograph of the large intestine containing a radiopaque substance that the patient ingested. akesak/iStock/Getty Images Plus

The last 2.5 to 4.0 centimeters of the large intestine form the anal canal (fig. 15.29). The mucous membrane in the canal is folded into six to eight longitudinal *anal columns.* At its distal end, the canal opens to the outside as the **anus.** Two sphincter muscles guard the anus—an *internal anal sphincter muscle,* composed of smooth muscle under involuntary control, and an *external anal sphincter muscle,* composed of skeletal muscle

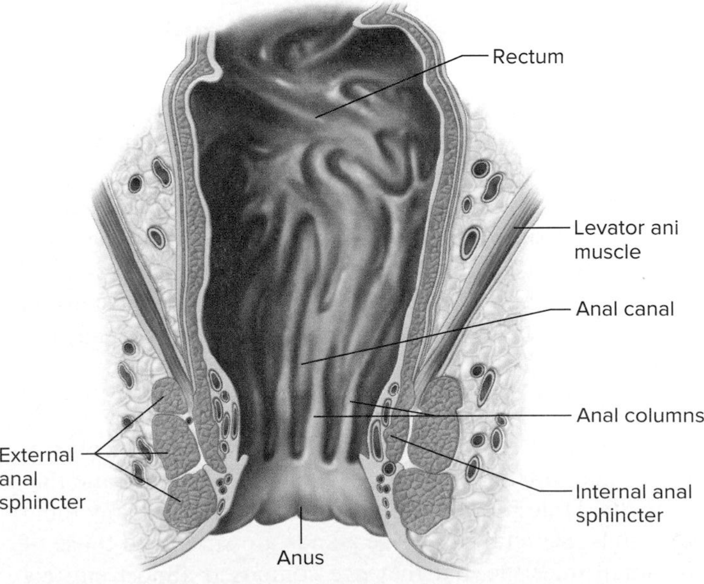

Figure 15.29 The rectum and the anal canal are at the distal end of the alimentary canal.

CLINICAL APPLICATION 15.4
Appendicitis, Hemorrhoids, and Colorectal Cancer

In appendicitis, the appendix becomes inflamed. Surgery is often required to remove the appendix before it ruptures. If it does rupture, this may allow contents of the large intestine, including infectious organisms, to enter the abdominal cavity and cause a serious inflammation of the peritoneum called peritonitis.

Hemorrhoids are enlarged and inflamed branches of the rectal vein in the anal columns that cause intense itching, sharp pain, and sometimes bright red bleeding. The hemorrhoids may be internal or bulge out of the anus (external). Causes of hemorrhoids include anything that puts prolonged pressure on the delicate rectal tissue, including obesity, pregnancy, constipation, diarrhea, and liver disease.

Fiberoptic colonoscopy is a test commonly performed on people over age 50, when the risk of colorectal cancer increases. Under sedation, a flexible lit tube is inserted into the rectum, and polyps and tumors are identified and removed. Computed tomographic colonography (popularly called a virtual colonoscopy) requires the same preparatory bowel cleansing, but does not require sedation and is faster. However, if a lesion is detected, the more invasive fiberoptic colonoscopy must be used to remove the suspicious tissue, which is then examined for the presence of cancer cells.

In a less invasive test recommended for people under age 50 at average risk for colorectal cancer, a patient sends a stool sample to a testing company. The stool sample is checked for blood as well as mutations, known to cause colorectal cancer, in the patient's cells within the stool sample. The company reports back to the physician. The test is available by prescription and does not replace colonoscopy in high-risk individuals.

under voluntary control. Clinical Application 15.4 discusses inflammation of the appendix, enlargement and inflamed rectal vein, and methods to screen for colorectal cancer.

PRACTICE 15.10

1. What is the general function of the large intestine?
2. Describe the parts of the large intestine.

Structure of the Large Intestinal Wall

The wall of the large intestine is composed of the same types of tissues as other parts of the alimentary canal but also has unique features. The large intestinal wall does not have the villi characteristic of the small intestine. The layer of longitudinal smooth muscle is not uniformly distributed throughout the large intestinal wall. Instead, the smooth muscle is mostly in three distinct bands (teniae coli) that extend the entire length of the colon (see fig. 15.27). These bands exert tension lengthwise on the wall, creating a series of pouches (haustra).

Functions of the Large Intestine

The large intestine has little digestive function, in contrast to the small intestine. However, the mucous membrane that forms the large intestine's inner lining contains many tubular glands. Structurally, these glands are similar to those of the small intestine, but they are composed almost entirely of goblet cells (fig. 15.30). Consequently, mucus is the large intestine's only significant secretion.

Mucus secreted into the large intestine protects the intestinal wall against the abrasive action of the materials passing through it. Mucus also binds particles of fecal matter, and its alkalinity helps control the pH of the large intestinal contents.

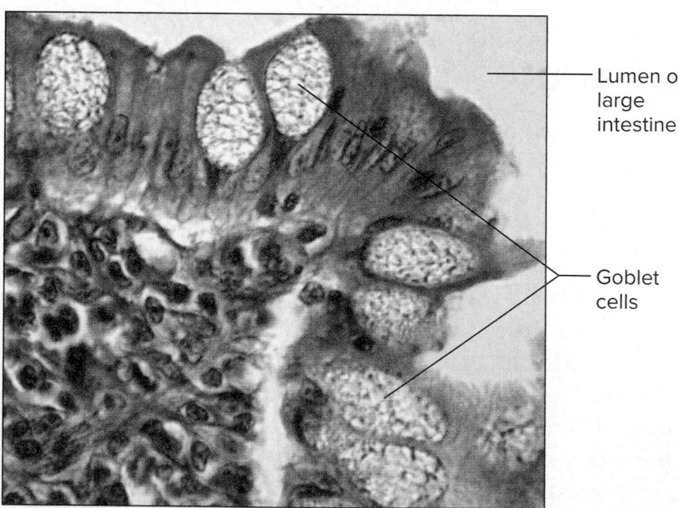

Figure 15.30 Light micrograph of the large intestinal mucosa (560x). APR Ed Reschke

PRACTICE FIGURE 15.30

Note the many goblet cells in the mucosa of the large intestine. Why are there so many more of these cells in the large intestinal wall than in the small intestinal wall?

Answer can be found in Appendix E.

Chyme entering the large intestine contains materials that the small intestine did not digest or absorb. It also contains water, electrolytes, mucus, and bacteria. The large intestine normally absorbs water (although most water is absorbed in the small intestine) and electrolytes in the proximal half of the tube. Substances that remain in the tube become feces and are stored in the distal part of the large intestine.

The many bacteria that normally inhabit the large intestine, called *intestinal flora,* break down some of the molecules that escape the actions of human digestive enzymes. For instance, cellulose, a complex carbohydrate in food of plant origin, passes through the alimentary canal almost unchanged, but colon bacteria can break down cellulose and use it as an energy source. These bacteria, in turn, synthesize certain vitamins, such as K, B_{12}, thiamine, and riboflavin, which the intestinal mucosa absorbs. Bacterial actions in the large intestine may produce intestinal gas (flatus).

PRACTICE 15.10

3. How does the structure of the large intestine differ from that of the small intestine?
4. Which substances does the large intestine absorb?

Movements of the Large Intestine

The movements of the large intestine—mixing and peristalsis—are similar to those of the small intestine, although usually slower. Also, peristaltic waves of the large intestine happen only two or three times each day. These waves produce *mass movements* in which a large section of the intestinal wall constricts vigorously, forcing the intestinal contents toward the rectum. Typically, mass movements follow a meal as a result of the gastrocolic reflex initiated in the small intestine. Irritations of the intestinal mucosa also can trigger such movements. For instance, a person with an inflamed colon (colitis) may experience frequent mass movements. Clinical Application 15.5 discusses inflammatory bowel disease.

A person can usually initiate a *defecation reflex* by holding a deep breath and contracting the abdominal wall muscles. This action increases internal abdominal pressure and forces feces into the rectum. As the rectum fills, its wall distends, triggering the defecation reflex that stimulates peristaltic waves in the descending colon. The internal anal sphincter relaxes. At the same time, other reflexes involving the sacral region of the spinal cord strengthen the peristaltic waves, lower the diaphragm, close the glottis, and contract the abdominal wall muscles. These actions further increase internal abdominal pressure and squeeze the rectum. The

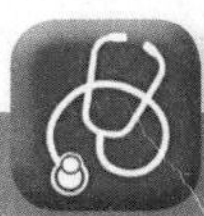

CLINICAL APPLICATION 15.5
Inflammatory Bowel Disease

Inflammatory bowel disease (IBD) is a group of disorders that affect about a million people in the United States. The most common ages of onset are between ten and thirty years, and between fifty and sixty years. IBD includes ulcerative colitis and Crohn's disease. These disorders differ by the site and extent of inflammation and ulceration of the intestines. Both produce abdominal cramps and diarrhea. (Irritable bowel syndrome [IBS] is a much less severe condition of cramping and diarrhea, but not inflammation. In IBS, unlike IBD, the large intestine is not abnormal or damaged.)

IBD is usually the result of an inflammatory response, such as to infection, that does not stop. About 10% to 20% of people with IBD have relatives with the condition. Mutations in several genes cause severe IBD in young children. The genes normally encode receptors for cytokines that halt the inflammatory response. When mutations impair the receptors so that inflammation continues, painfully swollen intestines result.

Ulcerative colitis affects the mucosa and submucosa of the distal large intestine and the rectum. In about 25% of cases, the disease extends no farther than the rectum. Bloody diarrhea and cramps may last for days or weeks, and vary in frequency. The severe diarrhea leads to weight loss and electrolyte imbalances and may affect other organs, including the skin, eyes, and liver. The inflamed and ulcerous tissue is continuous. Severe ulcerative colitis is also associated with increased risk of developing colon cancer.

Crohn's disease is more extensive than ulcerative colitis, infiltrating the small and large intestines and penetrating all tissue layers. In contrast to the uniformity of ulcerative colitis, affected portions of intestine in Crohn's disease are interspersed with unaffected areas, producing a "cobblestone" effect after many years. Rarely, the disease affects more proximal structures of the gastrointestinal tract. The diarrhea is often not bloody, and complications such as cancer are rare.

Several classes of drugs are used to control IBD. They affect different parts of the gastrointestinal tract and/or different parts of the immune response that contribute to the excess inflammation. Surgery may be necessary if drug therapy is ineffective or if cancer develops. For at least one very severe type of IBD that is inherited and primarily affects the immune system, a bone marrow or stem cell transplant can be lifesaving (discussed in Clinical Application 4.1).

external anal sphincter is signaled to relax, and the feces are forced to the outside. Contracting the external anal sphincter allows voluntary inhibition of defecation.

Feces

Feces (fē′sēz) include materials not digested or absorbed, plus water, electrolytes, mucus, shed intestinal cells, and bacteria. Usually, feces are about 75% water, and their color derives from bile pigments altered by bacterial action. Feces' pungent odor results from a variety of compounds that bacteria produce.

PRACTICE 15.10

5. How does peristalsis in the large intestine differ from peristalsis in the small intestine?
6. List the major events of defecation.
7. Describe the composition of feces.

15.11 | Nutrition and Nutrients

LEARN

1. List the major dietary sources of carbohydrates, lipids, and proteins.
2. Describe how cells use carbohydrates, lipids, and proteins.
3. Identify the functions of each fat-soluble and water-soluble vitamin.
4. Identify the functions of each major mineral and trace element.
5. Describe an adequate diet.

Nutrition is the study of nutrients and how the body utilizes them. **Nutrients** are chemicals supplied from the environment that an organism requires for survival, and include carbohydrates, lipids, proteins, vitamins, minerals, and water (see section 2.6, Chemical Constituents of Cells, and section 4.4, Energy for Metabolic Reactions). Carbohydrates, lipids, and proteins are called **macronutrients** because they are required in large amounts. They provide energy as well as other specific functions. Vitamins and minerals are required in much smaller amounts and are therefore called **micronutrients.** They do not directly provide energy, but make possible the biochemical reactions that extract energy from macronutrient molecules.

Macronutrients provide potential energy that can be expressed in calories, which are units of heat. A **calorie** (kal′ō-rē) is the amount of heat required to raise the temperature of a gram of water by 1° Celsius. The calorie used to measure food energy is 1,000 times greater. This larger calorie (Cal) is technically a kilocalorie, but nutritional studies commonly refer to it simply as a calorie. As a result of cellular oxidation, 1 gram of carbohydrate or 1 gram of protein yields, on average, about 4.1 calories, and 1 gram of fat yields about 9.5 calories (more than twice as much chemical energy as carbohydrates or proteins).

Foods provide nutrients, and digestion breaks down nutrient molecules to sizes that can be absorbed and transported in the bloodstream. Nutrients that human cells cannot synthesize, such as certain amino acids, are called **essential nutrients.**

PRACTICE 15.11

1. Identify and distinguish between macronutrients and micronutrients.
2. How is food energy measured?

Carbohydrates

Carbohydrates are organic compounds that include the sugars and starches. The energy held in their chemical bonds is used primarily to power cellular processes.

Carbohydrate Sources

Carbohydrates are ingested in a variety of forms, including starch from grains and vegetables; **glycogen** from meats; disaccharides from milk sugar, cane sugar, beet sugar, and molasses; and monosaccharides from honey and fruits (see section 2.6, Chemical Constituents of Cells). Digestion breaks down carbohydrates into monosaccharides, which are small enough to be absorbed into the bloodstream.

Cellulose is a complex plant carbohydrate that is abundant in food—it gives celery its crunch and lettuce its crispness. Humans cannot digest cellulose, so the portion of it that is not broken down by intestinal flora passes through the alimentary canal largely unchanged. In this way, cellulose provides bulk (also called fiber or roughage) against which the muscular wall of the digestive system can push, easing the movement of intestinal contents.

Carbohydrate Use

The monosaccharides absorbed from the digestive tract include *fructose, galactose,* and *glucose.* Liver enzymes catalyze reactions that convert fructose and galactose into **glucose,** which is the carbohydrate form most commonly oxidized for cellular fuel (see section 4.4, Energy for Metabolic Reactions).

Many cells obtain energy by oxidizing fatty acids when glucose levels are low. However, some cells, such as neurons, normally require a continuous supply of glucose for survival. Even a temporary decrease in the glucose supply may seriously impair nervous system function. Consequently, the body requires a minimum amount of carbohydrates. If foods do not provide an adequate carbohydrate supply, the liver may convert some noncarbohydrates, such as amino acids from proteins, into glucose. The requirement for glucose has physiological priority over the requirement to synthesize proteins from available amino acids.

Some excess glucose is polymerized to form *glycogen,* which is stored in the liver and muscles. When glucose is required to supply energy, it can be mobilized rapidly by breaking down glycogen. However, the body can store only a certain amount of glycogen, so excess glucose beyond that converted to glycogen is usually converted into triglycerides (fats), which are stored in adipose tissue. To obtain energy, the body first metabolizes glucose, then glycogen stores, and finally fats and proteins.

Cells use carbohydrates as starting materials for synthesizing such vital biochemicals as the five-carbon sugars *ribose* and *deoxyribose.* These sugars are required for production of the nucleic acids RNA and DNA. Carbohydrates are also required to synthesize the disaccharide *lactose* (milk sugar) when the mammary glands are actively producing milk.

Carbohydrate Requirements

Carbohydrates provide the primary fuel source for cellular processes, so the need for carbohydrates varies with individual energy expenditure. Physically active individuals require more fuel than those who are sedentary. The minimal requirement for carbohydrates in the human diet is unknown. It is estimated, however, that an intake of at least 125 to 175 grams daily is necessary to avoid protein breakdown and to avoid metabolic disorders resulting from excess fat use.

PRACTICE 15.11

3. List several common sources of carbohydrates.
4. Explain the importance of cellulose in the diet.
5. Explain why the requirement for glucose has priority over protein synthesis.
6. Why do daily requirements for carbohydrates vary from person to person?

Lipids

Lipids are organic compounds that include fats, oils, phospholipids, and cholesterol. They supply energy for cellular processes and help build structures, such as cell membranes. The most common dietary lipids are the fats called **triglycerides.** Recall from section 2.6, Chemical Constituents of Cells, that a triglyceride molecule consists of a glycerol and three fatty acids.

Lipid Sources

Triglycerides are found in plant- and animal-based foods. Saturated fats are mainly found in foods of animal origin, such as meats, eggs, milk, and lard, as well as in palm and coconut oils. Unsaturated fats are in seeds, nuts, and plant oils. Monounsaturated fats (in which fatty acids contain one double bond), such as those in olive, peanut, and canola oils, are the healthiest. Saturated fats in excess are a risk factor for cardiovascular disease.

Cholesterol is abundant in liver and egg yolk and, to a lesser extent, in whole milk, butter, cheese, and meats. It is not present in foods of plant origin.

Lipid Use

The lipids in foods are phospholipids, cholesterol, or triglycerides. Lipids provide a variety of physiological functions; however, triglycerides mainly supply energy. Before a triglyceride molecule can release energy, it must undergo hydrolysis (section 4.2, Metabolic Reactions) as part of digestion, releasing the constituent fatty acids and glycerol (see fig. 2.14 and fig. 4.3). After being absorbed, these products are carried in lymph to the blood, then on to tissues. Figure 15.31 shows that some of the fatty acid portions can react to form molecules of acetyl coenzyme A by a series of reactions called **beta oxidation** (bā'tah ok"sĭ-dā'shun). Excess acetyl coenzyme A can be converted into compounds called **ketone bodies,** such as acetone, which later may be changed back to acetyl coenzyme A. In either case, the resulting acetyl coenzyme A can be oxidized in the citric acid cycle. The glycerol parts of the triglyceride molecules can also enter metabolic pathways leading to the citric acid cycle, or they can be used to synthesize

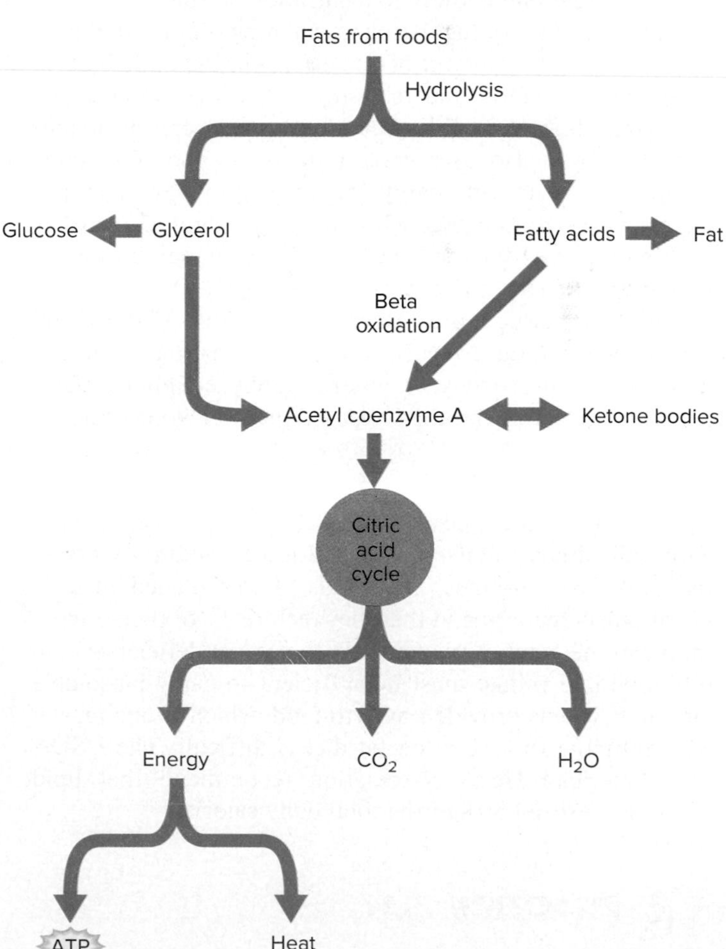

Figure 15.31 The body digests fats from foods into glycerol and fatty acids, which may enter catabolic pathways and provide energy.

glucose. Fatty acid molecules released from fat hydrolysis can combine to form fat molecules and be stored in adipose tissue.

The liver can convert fatty acids from one form to another, but it cannot synthesize certain fatty acids, called **essential fatty acids,** which must be obtained in the diet. *Linoleic acid,* for example, is an essential fatty acid required for phospholipid synthesis, which in turn is necessary for constructing cell membranes and myelin sheaths, and transporting circulating lipids. Good sources of linoleic acid include corn, cottonseed, and soy oils. Another essential fatty acid is *linolenic acid.*

The liver regulates circulating lipids by using free fatty acids to synthesize triglycerides, phospholipids, and lipoproteins that may then be released into the bloodstream. Lipoproteins are classified on the basis of their densities, which reflect their composition. As the proportion of lipids in a lipoprotein increases, the density of the particle decreases because lipids are less dense than proteins. Conversely, as the proportion of lipids decreases, the density increases. *Very-low-density lipoproteins* (*VLDLs*) have a relatively high concentration of triglycerides. *Low-density lipoproteins* (*LDLs*) have a relatively high concentration of cholesterol and are the major cholesterol-carrying lipoproteins. *High-density lipoproteins* (*HDLs*) have a relatively high concentration of protein and a lower concentration of lipids.

In addition to regulating circulating lipids, the liver controls the total amount of cholesterol in the body by synthesizing cholesterol and releasing it into the blood, or by removing cholesterol from the blood and excreting it into bile. The liver also uses cholesterol to produce bile salts. Cholesterol is not an energy source. It provides structural material for cell and organelle membranes and it furnishes starting material for the synthesis of certain sex hormones and adrenal cortex hormones.

Adipose tissue stores excess triglycerides. If the blood lipid concentration drops (in response to fasting, for example), some of these triglycerides are hydrolyzed into free fatty acids and glycerol, and then released into the bloodstream.

Lipid Requirements

The amounts and types of lipids required for health vary with individuals' habits and goals. Linoleic acid is an essential fatty acid. To meet the needs of formula-fed infants, nutritionists recommend that they receive 3% of their energy intake in the form of linoleic acid to prevent deficiency conditions. Lipid intake must be sufficient to carry fat-soluble vitamins. Lipids provide flavor to food, which is one reason why adhering to a very low-fat diet is difficult. The USDA and American Heart Association recommend that lipid intake not exceed 30% of the total daily calories.

PRACTICE 15.11

7. Which fatty acids are essential nutrients?
8. What is the liver's role in the use of lipids?
9. What are the functions of cholesterol?

Proteins

Proteins in the body are composed of twenty types of amino acids. Proteins have a wide variety of functions.

Protein Sources

Foods rich in proteins include meats, seafood, poultry, cheese, nuts, milk, eggs, and cereals. Legumes, including beans and peas, contain less protein. The cells of an adult can synthesize twelve of the twenty required amino acids, and the cells of a child can produce ten types. Amino acids that the body can synthesize are termed nonessential; those that the body cannot synthesize and must obtain from the diet are **essential amino acids.** Table 15.7 lists the amino acids in foods and indicates those that are essential.

All twenty types of amino acids must be in the body at the same time to provide the raw materials for growth and tissue repair. Therefore, if just one type of essential amino acid is missing from the diet, normal protein synthesis cannot take place.

Proteins are classified as complete or incomplete on the basis of the amino acids they provide. **Complete proteins** (also called high-quality proteins), such as those in milk, meats, and eggs, have adequate amounts of all of the essential amino acids. **Incomplete proteins** (also called low-quality proteins) lack one or more of the essential amino acids. Many plant proteins have too little of one or more essential amino acids to provide adequate nutrition for a person. For example, *zein* in corn has too little of the essential amino acids tryptophan and lysine. Zein is unable by itself to maintain human tissues or to support normal growth and development. However, combining appropriate plant foods can supply an adequate diversity of dietary amino acids. Beans and rice provide an example. Beans are low in methionine

TABLE 15.7 Amino Acids in Foods

Alanine	Leucine (e)
Arginine (ch)	Lysine (e)
Asparagine	Methionine (e)
Aspartic acid	Phenylalanine (e)
Cysteine	Proline
Glutamic acid	Serine
Glutamine	Threonine (e)
Glycine	Tryptophan (e)
Histidine (ch)	Tyrosine
Isoleucine (e)	Valine (e)

Eight essential amino acids (e) cannot be synthesized by human cells and must be provided in the diet. Two additional amino acids (ch) are essential in growing children.

but have enough lysine. Rice lacks lysine but has enough methionine. A meal of beans and rice provides enough of both types of amino acids.

Protein Use

When dietary proteins are digested, the resulting amino acids are absorbed and transported by the blood to the cells, which use the amino acids to synthesize proteins. These new proteins include enzymes that control the rates of metabolic reactions, clotting factors, the keratin of skin and hair, elastin and collagen of connective tissue, plasma proteins that regulate water balance, the muscle components actin and myosin, certain hormones, and the antibodies that protect against infection.

Proteins may also supply energy after digestion breaks them down into amino acids. The liberated amino acids are transported to the liver, where they undergo **deamination,** losing their nitrogen-containing ($-NH_2$) groups (see fig. 2.17). These $-NH_2$ groups subsequently react to form the waste *urea* (ū-rē′ah), which is excreted in urine.

Depending upon the particular amino acids involved, the remaining deaminated parts are decomposed in one of several pathways (fig. 15.32). Some of these pathways lead to formation of acetyl coenzyme A, and others lead more directly to the steps of the citric acid cycle. Most of the energy released from the cycle is captured in ATP molecules. If energy is not required immediately, the deaminated parts of the amino acids may react to form glucose or fat molecules in other metabolic pathways.

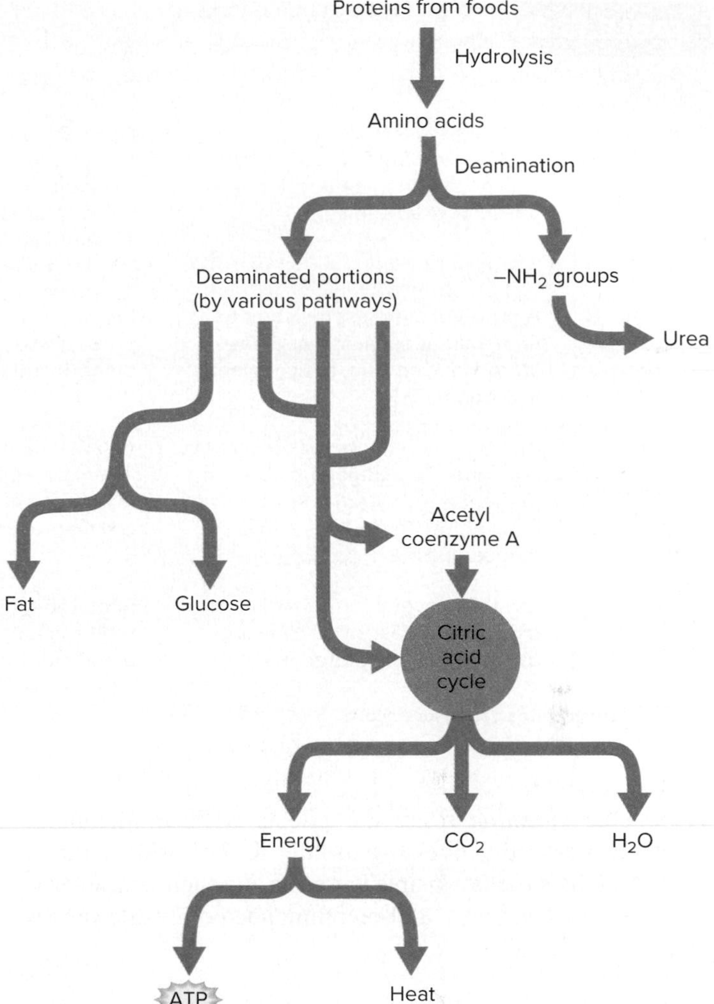

Figure 15.32 The body digests proteins from foods into amino acids, but must deaminate these smaller molecules before they can be used as energy sources.

Protein Requirements

Proteins may supply essential amino acids. They also provide nitrogen and other elements for the synthesis of nonessential amino acids and certain nonprotein nitrogenous substances. The amount of dietary protein individuals require varies according to body size, metabolic rate, and other factors, such as activity level. Bodybuilders, for example, require more protein to help heal small muscle tears that result from weight lifting.

For an average adult, nutritionists recommend a daily protein intake of about 0.8 gram per kilogram of body weight. Another way to estimate desirable protein intake is to divide weight in pounds by two. Most people should consume 60 to 150 grams of protein a day. For a pregnant woman, the recommendation adds 30 grams of protein per day. Similarly, a nursing mother requires an additional 20 grams of protein per day to maintain milk production.

PRACTICE 15.11

10. Which foods are rich sources of proteins?
11. Why are some amino acids called essential?
12. Distinguish between complete and incomplete proteins.
13. List some proteins synthesized in the body.
14. How does dietary protein provide energy?

Vitamins

Vitamins are organic compounds required in small amounts for normal metabolism, that cells cannot synthesize in adequate amounts. Therefore, they are essential nutrients that must come from foods.

Vitamins are classified on the basis of solubility. Some are soluble in fats and others are soluble in water. *Fat-soluble vitamins* are vitamins A, D, E, and K; *water-soluble vitamins* are the B vitamins and vitamin C.

Fat-Soluble Vitamins

Fat-soluble vitamins dissolve in fats, and therefore they associate with lipids and respond to the same factors that affect lipid absorption. For example, bile salts in the intestine promote absorption of these vitamins. Fat-soluble vitamins accumulate in various tissues, which is why excess intake can lead to overdose conditions. For example, too much beta carotene, a vitamin A precursor, can tinge the skin orange.

TABLE 15.8 Fat-Soluble Vitamins

Vitamin	Characteristics	Functions	Sources and RDA* for Adults
Vitamin A	Exists in several forms; synthesized from carotenes; stored in liver; stable in heat, acids, and bases; unstable in light	An antioxidant necessary for synthesis of visual pigments, mucoproteins, and mucopolysaccharides; for normal development of bones and teeth; and for maintenance of epithelial cells	Liver, fish, whole milk, butter, eggs, leafy green vegetables, yellow and orange vegetables and fruits; RDA = 2,300–3,000 IU†
Vitamin D	A group of steroids; resistant to heat, oxidation, acids, and bases; stored in liver, skin, brain, spleen, and bones	Promotes absorption of calcium and phosphorus; promotes development of teeth and bones	Produced in skin exposed to ultraviolet light; in milk, egg yolk, fish liver oils, fortified foods; RDA = 600 IU
Vitamin E	A group of compounds; resistant to heat and visible light; unstable in presence of oxygen and ultraviolet light; stored in muscles and adipose tissue	An antioxidant; prevents oxidation of vitamin A and polyunsaturated fatty acids; may help maintain stability of cell membranes	Oils from cereal seeds, salad oils, margarine, shortenings, fruits, nuts, and vegetables; RDA = 22.5 IU
Vitamin K	Exists in several forms; resistant to heat, but destroyed by acids, bases, and light; stored in liver	Required for synthesis of prothrombin, which functions in blood clotting	Leafy green vegetables, egg yolk, pork liver, soy oil, tomatoes, cauliflower; RDA = 90 μg

*RDA = recommended daily allowance.
†IU = international unit.

Fat-soluble vitamins resist the effects of heat; therefore, cooking and food processing usually do not destroy them. Table 15.8 lists the fat-soluble vitamins and their characteristics, functions, sources, and recommended daily allowances (RDAs) for adults.

Water-Soluble Vitamins

The water-soluble vitamins include the B vitamins and vitamin C. The **B vitamins** are several compounds that are essential for normal cellular metabolism. They help oxidize carbohydrates, lipids, and proteins. The B vitamins are usually in the same foods, so they are called the *vitamin B complex.* Members of this group differ chemically and functionally. Cooking and food processing destroy some of them.

Vitamin C (ascorbic acid) is one of the least stable vitamins and is fairly widespread in plant foods. It is necessary for collagen production, the activation of folacin, and the metabolism of certain amino acids. Vitamin C also promotes iron absorption and synthesis of certain hormones from cholesterol. Table 15.9 lists the water-soluble vitamins and their characteristics, functions, sources, and RDAs for adults.

OF INTEREST Sailors on English ships ate limes to protect them from scurvy (vitamin C deficiency). American ships carried cranberries for the same purpose.

PRACTICE 15.11

15. What are vitamins?
16. How are vitamins classified?
17. How do bile salts affect the absorption of fat-soluble vitamins?
18. List the fat-soluble and water-soluble vitamins.

Minerals

Dietary **minerals** are inorganic elements essential in human metabolism. Plants usually extract these elements from soil, and humans obtain them from plant foods or from animals that have eaten plants.

Characteristics of Minerals

Minerals contribute about 4% of body weight and are most concentrated in the bones and teeth. Many minerals are incorporated into organic molecules. For example, phosphorus is found in phospholipids, iron in hemoglobin, and iodine in thyroxine. However, some minerals are part of inorganic compounds, such as the calcium phosphate of bone. Other minerals are free ions, such as sodium, chloride, and calcium ions in blood.

Minerals are parts of the structural materials of all body cells. They also constitute portions of enzyme molecules, contribute to the osmotic pressure of body fluids, and play vital roles in impulse conduction in neurons, muscle fiber contraction, blood coagulation, and maintenance of the pH of body fluids.

Major Minerals

Calcium and *phosphorus* account for nearly 75% by weight of the mineral elements in the body. Therefore, they are

TABLE 15.9 Water-Soluble Vitamins

Vitamin	Characteristics	Functions	Sources and RDA* for Adults
Thiamine (vitamin B_1)	Destroyed by heat and oxygen, especially in alkaline environment	Part of coenzyme required for oxidation of carbohydrates; coenzyme required for ribose synthesis	Lean meats, liver, eggs, whole-grain cereals, leafy green vegetables, legumes; RDA = 1.2 mg
Riboflavin (vitamin B_2)	Stable to heat, acids, and oxidation; destroyed by bases and ultraviolet light	Part of enzymes and coenzymes required for oxidation of glucose and fatty acids and for cellular growth	Meats, dairy products, leafy green vegetables, whole-grain cereals; RDA = 1.3 mg
Niacin (nicotinic acid) (vitamin B_3)	Stable to heat, acids, and bases; converted to niacinamide by cells; synthesized from tryptophan	Part of coenzymes required for oxidation of glucose and synthesis of proteins, fats, and nucleic acids	Liver, lean meats, peanut butter, legumes; RDA = 14–16 mg
Pantothenic acid (vitamin B_5)	Destroyed by heat, acids, and bases	Part of coenzyme A required for oxidation of carbohydrates and fats	Meats, whole-grain cereals, legumes, milk, fruits, vegetables; RDA = 5 mg
Vitamin B_6	Group of three compounds; stable to heat and acids; destroyed by oxidation, bases, and ultraviolet light	Coenzyme required for synthesis of proteins and certain amino acids, for conversion of tryptophan to niacin, for production of antibodies, and for nucleic acid synthesis	Liver, meats, bananas, avocados, beans, peanuts, whole-grain cereals, egg yolk; RDA = 1.3–1.7 mg
Biotin (vitamin B_7)	Stable to heat, acids, and light; destroyed by oxidation and bases	Coenzyme required for metabolism of amino acids and fatty acids, and for nucleic acid synthesis	Liver, egg yolk, nuts, legumes, mushrooms; RDA = 0.3 mg
Folacin (folic acid) (vitamin B_9)	Occurs in several forms; destroyed by oxidation in an acid environment or by heat in an alkaline environment; stored in liver, where it is converted into folinic acid	Coenzyme required for metabolism of certain amino acids and for DNA synthesis; promotes production of normal red blood cells	Liver, leafy green vegetables, whole-grain cereals, legumes; RDA = 0.4 mg
Cyanocobalamin (vitamin B_{12})	Complex, cobalt-containing compound; stable to heat; inactivated by light, strong acids, and strong bases; absorption regulated by intrinsic factor from gastric glands; stored in liver	Part of a coenzyme required for synthesis of nucleic acids and for metabolism of carbohydrates; plays role in myelin synthesis; needed for normal red blood cell production	Liver, meats, milk, cheese, eggs; RDA = 2.4 μg
Ascorbic acid (vitamin C)	Chemically similar to monosaccharides; stable in acids but destroyed by oxidation, heat, light, and bases	Required for collagen production, conversion of folacin to folinic acid, and metabolism of certain amino acids; promotes absorption of iron and synthesis of hormones from cholesterol	Citrus fruits, tomatoes, leafy green vegetables; RDA = 75–90 mg

termed **major minerals.** Other major minerals, each of which accounts for 0.05% or more of the body weight, include potassium, sulfur, sodium, chlorine, and magnesium. Table 15.10 lists the distribution, functions, sources, and adult RDAs of major minerals.

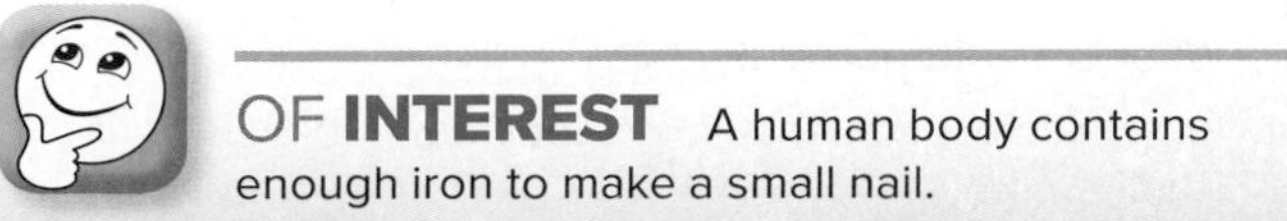

OF INTEREST A human body contains enough iron to make a small nail.

Trace Elements

Trace elements are essential minerals found in minute amounts, each making up less than 0.005% of adult body weight. They include iron, manganese, copper, iodine, cobalt, zinc, fluorine, selenium, and chromium. Table 15.11 lists the distribution, functions, sources, and adult RDAs of the trace elements.

PRACTICE 15.11

19. What are minerals?
20. What are the major functions of minerals?
21. Distinguish between a major mineral and a trace element.
22. Name the major minerals and trace elements.

TABLE 15.10 Major Minerals

Mineral	Distribution	Functions	Sources and RDA* for Adults
Calcium (Ca)	Mostly in the inorganic salts of bones and teeth	Structure of bones and teeth; essential for neurotransmitter release, muscle fiber contraction, and blood coagulation; increases permeability of cell membranes; activates certain enzymes	Milk, milk products, leafy green vegetables; RDA = 1,000–1,200 mg
Phosphorus (P)	Mostly in the inorganic salts of bones and teeth	Structure of bones and teeth; component in nearly all metabolic reactions; in nucleic acids, many proteins, some enzymes, and some vitamins; in cell membrane, ATP, and phosphates of body fluids	Meats, cheese, nuts, whole-grain cereals, milk, legumes; RDA = 700 mg
Potassium (K)	Widely distributed; tends to be concentrated inside cells	Helps maintain intracellular osmotic pressure and regulate pH; required for impulse conduction in neurons	Avocados, dried apricots, meats, peanut butter, potatoes, bananas; RDA = 4,700 mg
Sulfur (S)	Widely distributed; abundant in skin, hair, and nails	Essential part of certain amino acids, thiamine, insulin, biotin, and mucopolysaccharides	Meats, milk, eggs, legumes; No RDA established
Sodium (Na)	Widely distributed; mostly in extracellular fluids and bound to inorganic salts of bone	Helps maintain osmotic pressure of extracellular fluids; regulates water movement; plays a role in impulse conduction in neurons; regulates pH and transport of substances across cell membranes	Table salt, cured ham, sauerkraut, cheese; RDA = 2,300 mg
Chlorine (Cl)	Closely associated with sodium (as chloride); most highly concentrated in cerebrospinal fluid and gastric juice	Helps maintain osmotic pressure of extracellular fluids; regulates pH; maintains electrolyte balance; forms hydrochloric acid; aids transport of carbon dioxide by red blood cells	Same as for sodium; No RDA established
Magnesium (Mg)	Abundant in bones	Required in metabolic reactions in mitochondria that produce ATP; plays a role in the breakdown of ATP to ADP	Milk, dairy products, legumes, nuts, leafy green vegetables; RDA = 320–420 mg

*RDA = recommended daily allowance.

Adequate Diets

An *adequate diet* provides sufficient energy (calories), essential fatty acids, essential amino acids, vitamins, and minerals to support optimal growth and to maintain and repair body tissues. Individual requirements for nutrients vary greatly with age, sex, growth rate, physical activity, and level of stress, as well as with genetic and environmental factors. Therefore, designing a diet that is adequate for everyone is impossible. In the past, diagrams called food pyramids were used to organize foods according to suggested relative amounts. Figure 15.33 depicts a replacement design, *MyPlate,* more recently developed by the U.S. Department of Agriculture.

Malnutrition (mal″nu-trish′un) is poor nutrition that results from a lack of essential nutrients or an inability to utilize them. It may result from *undernutrition* and produce the symptoms of deficiency diseases, or it may be due to *overnutrition* arising from excess nutrient intake.

A variety of factors can lead to malnutrition. For example, a deficiency condition may stem from lack of availability or the poor quality of food. On the other hand, malnutrition may result from overeating or from taking too many vitamin supplements.

A measurement called the *body mass index* (*BMI*) is used to determine whether a person is of adequate weight, overweight, or obese. To calculate BMI, divide body weight in kilograms (1 kilogram equals 2.2 pounds) by body height in meters squared (1 foot equals about 0.3 meter). Figure 15.34 interprets the BMI. Clinical Application 15.6 discusses the effects of undereating and overeating.

PRACTICE 15.11

23. What is an adequate diet?
24. Which factors influence individual nutrient requirements?
25. What causes malnutrition?

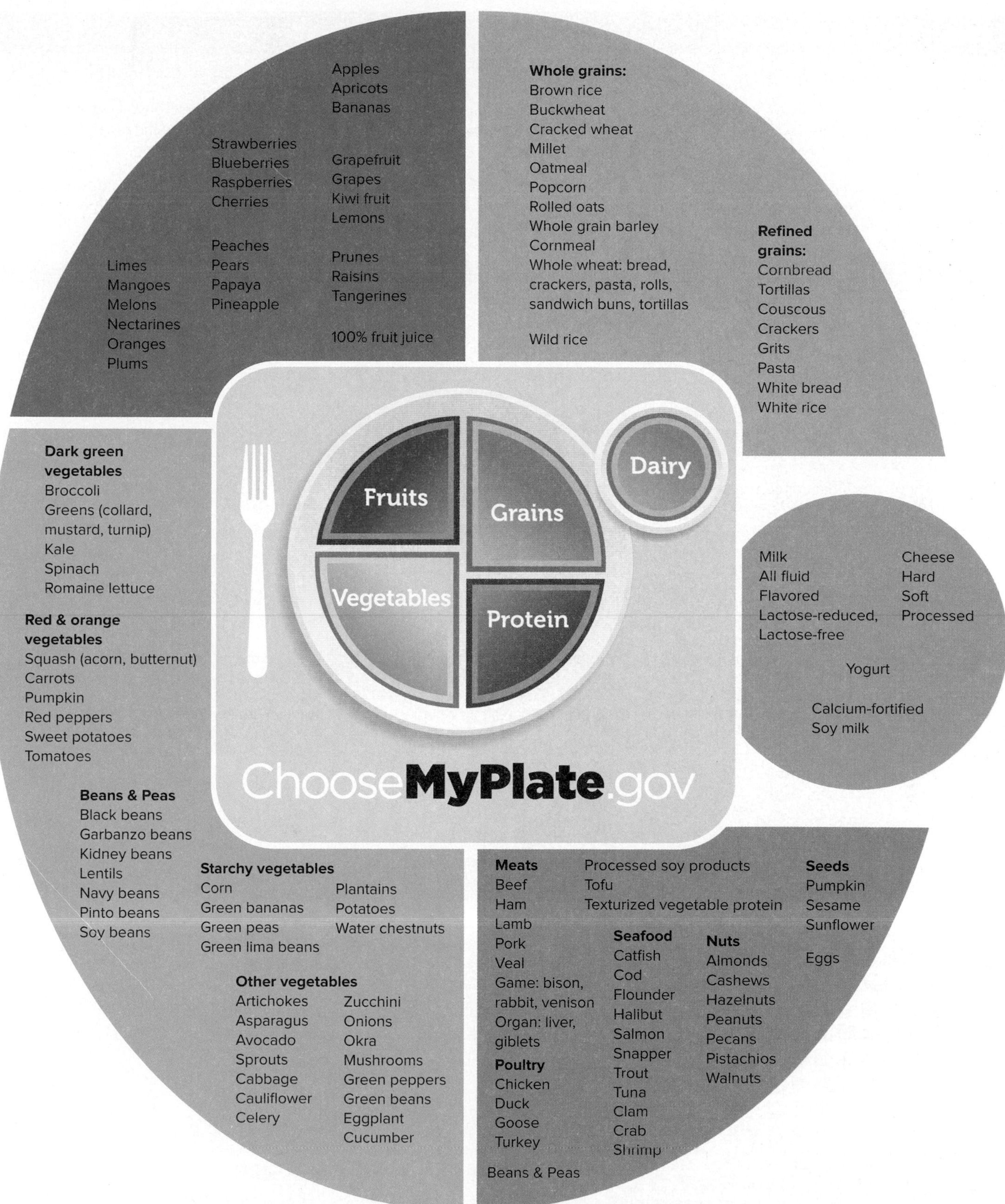
Apples
Apricots
Bananas
Strawberries
Blueberries
Raspberries
Cherries
Grapefruit
Grapes
Kiwi fruit
Lemons
Peaches
Pears
Papaya
Pineapple
Prunes
Raisins
Tangerines
Limes
Mangoes
Melons
Nectarines
Oranges
Plums
100% fruit juice
Whole grains:
Brown rice
Buckwheat
Cracked wheat
Millet
Oatmeal
Popcorn
Rolled oats
Whole grain barley
Cornmeal
Whole wheat: bread, crackers, pasta, rolls, sandwich buns, tortillas
Wild rice
Refined grains:
Cornbread
Tortillas
Couscous
Crackers
Grits
Pasta
White bread
White rice
Dairy
Fruits
Grains
Vegetables
Protein
ChooseMyPlate.gov
Milk
All fluid
Flavored
Lactose-reduced,
Lactose-free
Cheese
Hard
Soft
Processed
Yogurt
Calcium-fortified
Soy milk
Dark green vegetables
Broccoli
Greens (collard, mustard, turnip)
Kale
Spinach
Romaine lettuce
Red & orange vegetables
Squash (acorn, butternut)
Carrots
Pumpkin
Red peppers
Sweet potatoes
Tomatoes
Beans & Peas
Black beans
Garbanzo beans
Kidney beans
Lentils
Navy beans
Pinto beans
Soy beans
Starchy vegetables
Corn
Green bananas
Green peas
Green lima beans
Plantains
Potatoes
Water chestnuts
Other vegetables
Artichokes
Asparagus
Avocado
Sprouts
Cabbage
Cauliflower
Celery
Zucchini
Onions
Okra
Mushrooms
Green peppers
Green beans
Eggplant
Cucumber
Meats
Beef
Ham
Lamb
Pork
Veal
Game: bison, rabbit, venison
Organ: liver, giblets
Poultry
Chicken
Duck
Goose
Turkey
Processed soy products
Tofu
Texturized vegetable protein
Seafood
Catfish
Cod
Flounder
Halibut
Salmon
Snapper
Trout
Tuna
Clam
Crab
Shrimp
Nuts
Almonds
Cashews
Hazelnuts
Peanuts
Pecans
Pistachios
Walnuts
Seeds
Pumpkin
Sesame
Sunflower
Eggs
Beans & Peas

Figure 15.33 MyPlate, developed by the U.S. Department of Agriculture, depicts the foods and appropriate proportions that should make up a healthy diet. U.S. Department of Agriculture

TABLE 15.11 Trace Elements

Trace Element	Distribution	Functions	Sources and RDA* for Adults
Iron (Fe)	Primarily in blood; stored in liver, spleen, and bone marrow	Part of hemoglobin molecule; assists in vitamin A synthesis; incorporated into a number of enzymes	Liver, lean meats, dried apricots, raisins, enriched whole-grain cereals, legumes, molasses; RDA = 8 mg
Manganese (Mn)	Most concentrated in liver, kidneys, and pancreas	Part of enzymes required for fatty acids and cholesterol synthesis, urea formation, and normal functioning of the nervous system	Nuts, legumes, whole-grain cereals, leafy green vegetables, fruits; RDA = 1.8–2.3 mg
Copper (Cu)	Most highly concentrated in liver, heart, and brain	Essential for hemoglobin synthesis, bone development, melanin production, and myelin formation	Liver, oysters, crabmeat, nuts, whole-grain cereals, legumes; RDA = 0.9 mg
Iodine (I)	Concentrated in thyroid gland	Essential component for synthesis of thyroid hormones	Food content varies with soil content in different geographic regions; iodized table salt; RDA = 0.15 mg
Cobalt (Co)	Widely distributed	Component of cyanocobalamin; required for synthesis of several enzymes	Liver, lean meats, milk; No RDA established
Zinc (Zn)	Most concentrated in liver, kidneys, and brain	Component of enzymes involved in digestion, respiration, bone metabolism, liver metabolism; necessary for normal wound healing and maintaining skin integrity	Meats, cereals, legumes, nuts, vegetables; RDA = 8–11 mg
Fluorine (F)	Primarily in bones and teeth	Component of tooth enamel	Fluoridated water; RDA = 3–4 mg
Selenium (Se)	Concentrated in liver and kidneys	Component of certain enzymes	Lean meats, onions, cereals; RDA = 0.055 mg
Chromium (Cr)	Widely distributed	Essential for use of carbohydrates	Liver, lean meats, yeast; RDA = 20–30 mg

*RDA = recommended daily allowance.

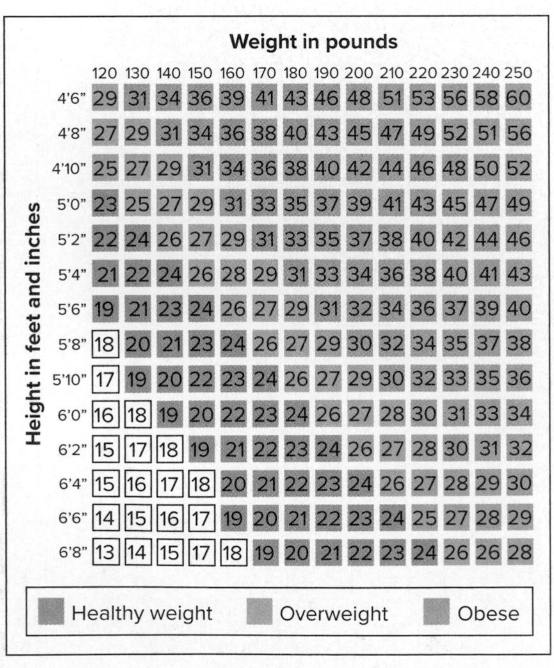

Weight in pounds

Height in feet and inches	120	130	140	150	160	170	180	190	200	210	220	230	240	250
4'6"	29	31	34	36	39	41	43	46	48	51	53	56	58	60
4'8"	27	29	31	34	36	38	40	43	45	47	49	52	51	56
4'10"	25	27	29	31	34	36	38	40	42	44	46	48	50	52
5'0"	23	25	27	29	31	33	35	37	39	41	43	45	47	49
5'2"	22	24	26	27	29	31	33	35	37	38	40	42	44	46
5'4"	21	22	24	26	28	29	31	33	34	36	38	40	41	43
5'6"	19	21	23	24	26	27	29	31	32	34	36	37	39	40
5'8"	18	20	21	23	24	26	27	29	30	32	34	35	37	38
5'10"	17	19	20	22	23	24	26	27	29	30	32	33	35	36
6'0"	16	18	19	20	22	23	24	26	27	28	30	31	33	34
6'2"	15	17	18	19	21	22	23	24	26	27	28	30	31	32
6'4"	15	16	17	18	20	21	22	23	24	26	27	28	29	30
6'6"	14	15	16	17	19	20	21	22	23	24	25	27	28	29
6'8"	13	14	15	17	18	19	20	21	22	23	24	26	26	28

Healthy weight | Overweight | Obese

Figure 15.34 BMI can be calculated. In this chart, the calculations have been done. The uncolored squares indicate lower than healthy weight according to this index. Source: Developed by the National Center for Health Statistics in collaboration with the National Center for Chronic Disease Prevention and Health Promotion

CLINICAL APPLICATION 15.6

Eating Extremes: Undereating and Overeating

It is difficult to determine a desirable body weight. The body mass index (BMI) is used to classify a person as underweight, normal weight, overweight, or obese.

Being thin is not a disease, but an abnormal eating behavior may be a sign of illness. Anorexia nervosa is self-imposed starvation. The sufferer perceives herself or himself as overweight and eats barely enough to survive, losing as much as 25% of her or his body weight. Anorexia leads to low blood pressure, slowed or irregular heartbeat, constipation, and constantly feeling chilly. In the female, menstruation may stop as the body fat level plunges. Hair becomes brittle, the skin dries out, and soft, pale, fine body hair called *lanugo,* normally seen only on a fetus, grows to preserve body heat. A person with anorexia may be hospitalized so that intravenous feeding can prevent sudden death from heart failure due to an electrolyte imbalance. Psychotherapy and nutritional counseling may help identify and remedy the underlying cause of the abnormal eating behavior. Despite these interventions, 15% to 21% of people with anorexia die from the disease.

In bulimia, a person eats large amounts of food and then gets rid of the thousands of extra calories by vomiting, taking laxatives, or exercising frantically. The binge-and-purge cycle is very hard to break, even with psychotherapy and nutritional counseling.

Because body weight reflects energy balance, excess food means, ultimately, excess weight. Being overweight or obese raises the risk of developing hypertension, diabetes, stroke, gallstones, sleep apnea, and certain cancers. The body strains to support the extra weight—miles of blood vessels are required to nourish additional body mass.

Usually, being overweight stems from overeating and inactivity. Weight loss requires eating less and exercising more. The National Heart, Lung, and Blood Institute recommends that overweight and obese individuals aim to lose 10% of their body weight over a six-month period. For people with BMIs from 27 to 35, that means a decrease of 300 to 500 calories per day, to lose 1/2 to 1 pound of body weight per week. For people with BMIs exceeding 35, a decrease of up to 500 to 1,000 calories per day will translate to a loss of 1 to 2 pounds of body weight per week.

For people with BMIs above 40, or above 35 in addition to an obesity-related disorder, bariatric surgery can lead to great weight loss. Three types of procedures are done. In laparoscopic adjustable gastric banding, a silicone band ties off part of the stomach, limiting its capacity to hold food. The band can be inflated or deflated in a doctor's office by adding or removing saline. The band may need to be removed if it slips out of place, erodes the stomach lining, the body rejects it, or the port to add or remove fluid becomes displaced. In the second type of bariatric surgery, sleeve gastrectomy, a surgeon removes about 75% of the stomach, leaving a banana-shaped "sleeve." The sleeve procedure is irreversible. The third type of bariatric surgery is gastric bypass, in which part of the stomach is stapled shut, forming a pouch that is surgically connected to the jejunum, bypassing the duodenum. Gastric bypass surgery is not reversible.

All bariatric surgeries lead to decreased hunger, very reduced food intake, and some decrease in the absorption of nutrients. A special diet, liquid at first, must be followed. Many patients who have had bariatric surgery report improvement in or disappearance of type 2 diabetes, back pain, arthritis, varicose veins, sleep apnea, and hypertension. However, people can regain the weight and lose the benefits if they fail to adhere to the dietary restrictions and fail to exercise regularly.

ASSESS

CHAPTER ASSESSMENTS

15.1 Introduction

1. Functions of the digestive system include __________ .
   a. mechanical breakdown of foods
   b. chemical breakdown of foods
   c. breaking large pieces into smaller ones without altering their chemical composition
   d. all of these
2. List the major parts of the alimentary canal, then separately list the accessory organs of the digestive system.

15.2 General Characteristics of the Alimentary Canal

3. Contrast the composition of the layers of the wall of the alimentary canal.
4. Distinguish between mixing and propelling movements.

15.3 Mouth

5. Discuss the functions of the mouth and its parts.
6. Distinguish between primary and secondary teeth.
7. The teeth that are best adapted for grasping and tearing food are the ______________________.
   a. incisors
   b. canines
   c. premolars
   d. molars
8. Describe the structure of a tooth.

15.4–15.10 Salivary Glands–Large Intestine

9. Match the organ or gland with the enzyme(s) it secretes. Enzymes may be used more than once. An organ or gland may secrete more than one enzyme.

(1) salivary glands (serous cells)	A. peptidase
(2) stomach (chief cells)	B. amylase
(3) pancreas (acinar cells)	C. nuclease
(4) small intestine	D. lipase
	E. pepsinogen
	F. trypsin, chymotrypsin, carboxypeptidase
	G. sucrase, maltase, lactase

10. Match the enzyme(s) with its (their) function(s).

(1) peptidase	A. begins protein digestion
(2) amylase	B. breaks fats into fatty acids and glycerol
(3) nuclease	C. breaks down proteins into peptides
(4) lipase	D. breaks down starch into disaccharides
(5) pepsin	E. breaks down peptides into amino acids
(6) trypsin, chymotrypsin, carboxypeptidase	F. breaks down nucleic acids into nucleotides
(7) sucrase, maltase, lactase	G. breaks down disaccharides into monosaccharides

11. List the steps in swallowing.
12. Explain the stimulus for, and response of, the parasympathetic nervous system in digestion.
13. Explain how hormones control the secretions and/or release of secretions from the stomach, pancreas, and gallbladder.
14. Discuss absorption of amino acids, monosaccharides, glycerol, fatty acids, electrolytes, and water from substances in the small and large intestines.
15. List the steps in defecating.

15.11 Nutrition and Nutrients

16. Identify dietary sources of carbohydrates, lipids, and proteins.
17. Explain how cells use carbohydrates, lipids, and proteins for the normal functioning of the body.
18. Match the vitamins with their general functions, and indicate if the vitamin is fat-soluble or water-soluble. Functions may be used more than once, and more than one function may be applied to a vitamin.

(1) vitamin A	A. part of coenzyme A in oxidation of carbohydrates
(2) vitamin B_1 (thiamine)	B. coenzyme in ribose synthesis
(3) vitamin B_2 (riboflavin)	C. necessary for synthesis of visual pigments
(4) vitamin B_3 (niacin)	D. required for synthesis of prothrombin
(5) vitamin B_5 (pantothenic acid)	E. required to produce collagen
(6) vitamin B_6	F. required to synthesize nucleic acids
(7) vitamin B_7 (biotin)	G. promotes normal red blood cell production
(8) vitamin B_9 (folacin)	H. plays a role in myelin synthesis
(9) vitamin B_{12} (cyanocobalamin)	I. antioxidant, helps stabilize cell membranes
(10) vitamin C (ascorbic acid)	J. promotes development of teeth and bones
(11) vitamin D	K. required to produce antibodies
(12) vitamin E	L. required for oxidation of glucose
(13) vitamin K	M. part of coenzymes to synthesize proteins and fats

19. Match the minerals with their functions, and indicate whether each is a major mineral or a trace element required for nutrition. (The functions of the minerals may be used more than once.)

(1) calcium	A. essential for the use of glucose
(2) chlorine	B. component of certain enzymes
(3) chromium	C. component of tooth enamel
(4) cobalt	D. component of teeth and bones
(5) copper	E. helps maintain intracellular osmotic pressure
(6) fluorine	F. essential part of certain amino acids
(7) iodine	G. helps maintain extracellular fluid osmotic pressure
(8) iron	H. necessary for normal wound healing
(9) magnesium	I. component of cyanocobalamin
(10) manganese	J. essential for synthesis of thyroid hormones
(11) phosphorus	K. required in metabolic reactions associated with ATP production
(12) potassium	L. component of hemoglobin molecules
(13) selenium	M. essential for hemoglobin synthesis and melanin production
(14) sodium	N. required for cholesterol synthesis and urea formation
(15) sulfur	
(16) zinc	

20. Define *adequate diet.*

Digestive System

INTEGUMENTARY SYSTEM

Vitamin D activated in the skin plays a role in absorption of calcium from the digestive tract.

SKELETAL SYSTEM

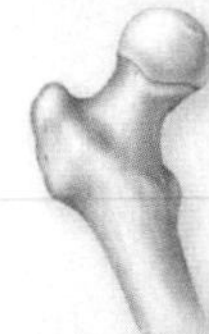

Bones are important in mastication. Calcium absorption is necessary to maintain bone matrix.

MUSCULAR SYSTEM

Muscles are important in mastication, swallowing, and the mixing and moving of digestion products through the gastrointestinal tract.

NERVOUS SYSTEM

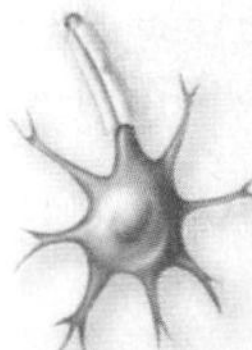

The nervous system can influence digestive system activity.

ENDOCRINE SYSTEM

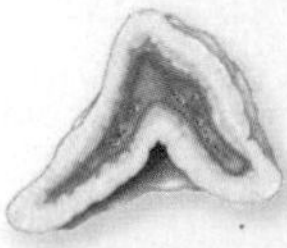

Hormones can influence digestive system activity.

CARDIOVASCULAR SYSTEM

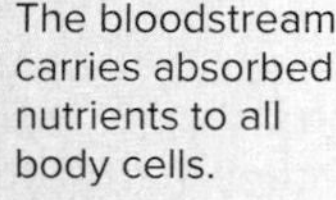

The bloodstream carries absorbed nutrients to all body cells.

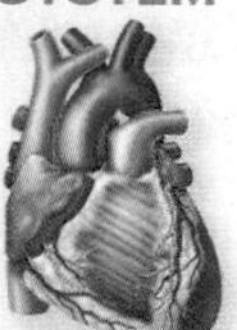

LYMPHATIC SYSTEM

The lymphatic system plays a major role in the absorption of fats.

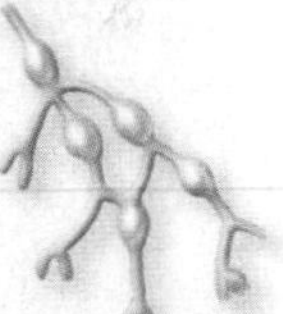

RESPIRATORY SYSTEM

The digestive system and the respiratory system share anatomical structures.

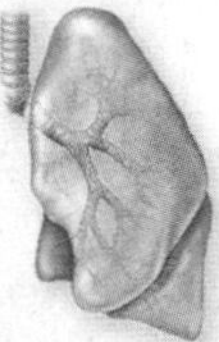

URINARY SYSTEM

The kidneys and liver work together to activate vitamin D.

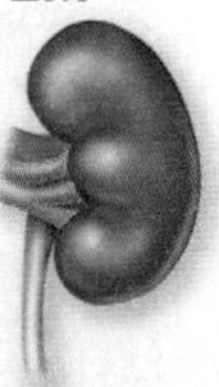

REPRODUCTIVE SYSTEM

In a woman, nutrition is essential for conception and normal development of an embryo and fetus.

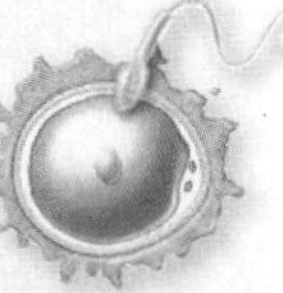

The digestive system ingests, digests, and absorbs nutrients for use by all body cells.

ASSESS

INTEGRATIVE ASSESSMENTS/CRITICAL THINKING

Outcomes 11.8, 15.8, 15.11

1. Why does blood sugar concentration remain stable in a person whose diet is low in carbohydrates?

Outcomes 13.8, 15.11

2. Charlotte consumed a Bloody Mary, bacon, eggs, and toast for breakfast. After digestion, which blood vessels receive the absorbed substances from these foods, and to where are they transported? What might the hepatocytes (liver cells) do with the substances they receive?

Outcomes 14.4, 15.9, 15.10

3. Given the location of the ileocecal valve and the abundant microbiome of bacteria species in the cecum and colon of the large intestine, what is the importance of having *Peyer's patches* in the mucosal lining of the ileum?

Outcomes 15.4, 15.7, 15.8, 15.9

4. The term *accessory organs* of the digestive system is used for the salivary glands, liver, gallbladder, and pancreas. Why are these organs categorized in this way? Make an argument as to which of these organs can be removed from the body without a huge impact on overall health?

Outcomes 15.6, 15.11

5. How would removal of 95% of the stomach (subtotal gastrectomy) to treat severe ulcers or cancer affect digestion and absorption? How would the patients have to alter their eating habits? Why? Do you think that people should have this type of surgery to treat life-threatening obesity?

Outcomes 15.7, 15.8

6. Why might a person with inflammation of the gallbladder (cholecystitis) also develop inflammation of the pancreas (pancreatitis)?

Outcome 15.11

7. Examine the label information on the packages of a variety of breakfast cereals. Which types of cereals provide the best sources of carbohydrates, lipids, proteins, vitamins, and minerals? Which major nutrients are lacking in these cereals?

Chapter Summary

15.1 Introduction

Digestion mechanically and chemically breaks down foods and absorbs the products. The digestive system consists of an alimentary canal and several accessory organs.

15.2 General Characteristics of the Alimentary Canal

The alimentary canal is a muscular tube.

1. Structure of the wall
   The wall consists of four layers—the **mucosa, submucosa, muscularis,** and **serosa.**
2. Movements of the tube
   Motor functions include mixing and propelling movements.

15.3 Mouth

*The **mouth** receives food and begins digestion.*

1. Cheeks and lips
   a. **Cheeks** consist of outer layers of skin, pads of fat, muscles associated with expression and chewing, and inner linings of epithelium.
   b. **Lips** are highly mobile and have sensory receptors.
2. Tongue
   a. The **tongue's** rough surface handles food and has taste buds.
   b. **Lingual tonsils** are on the root of the tongue.
3. Palate
   a. The **palate** includes hard and soft portions.
   b. **Palatine tonsils** are located on either side of the tongue in the back of the mouth.
4. Teeth
   a. There are two sets of **teeth,** twenty primary and thirty-two secondary teeth.
   b. Teeth begin mechanical digestion by breaking food into smaller pieces, increasing the surface area exposed to digestive actions.
   c. Each tooth consists of a crown and root, and is composed of enamel, dentin, pulp, nerves, and blood vessels.
   d. A periodontal ligament attaches a tooth to the alveolar process.

15.4 Salivary Glands

Salivary glands secrete saliva, which moistens food, helps bind food particles, begins chemical digestion of carbohydrates, makes taste possible, and helps cleanse the mouth.

1. Salivary secretions
   Salivary glands include serous cells that secrete **salivary amylase** and mucous cells that secrete **mucus.**
2. Major salivary glands
   a. The **parotid glands** secrete saliva rich in amylase.
   b. The **submandibular glands** produce viscous saliva.
   c. The **sublingual glands** primarily secrete mucus.

15.5 Pharynx and Esophagus

The pharynx and esophagus are important passageways.

1. Structure of the pharynx
   The **pharynx** is divided into a **nasopharynx, oropharynx,** and **laryngopharynx.**
2. Swallowing mechanism
   Swallowing occurs in three stages:
   a. Food is mixed with saliva and forced into the oropharynx.
   b. Involuntary reflex actions move the food into the esophagus.
   c. Peristalsis transports food to the stomach.
3. Esophagus
   a. The **esophagus** passes through the diaphragm and joins the stomach.
   b. Circular muscle fibers at the distal end of the esophagus help prevent regurgitation of food from the stomach.

15.6 Stomach

*The **stomach** receives food, mixes it with gastric juice, carries on a limited amount of absorption, and moves food into the small intestine.*

1. Parts of the stomach
   a. The stomach is divided into cardia, fundus, body, and pylorus.
   b. The **pyloric sphincter** is a valve between the stomach and small intestine.
2. Gastric secretions
   a. **Gastric glands** secrete **gastric juice.**
   b. Gastric juice contains **pepsin** (begins chemical digestion of proteins), hydrochloric acid, and **intrinsic factor.**
3. Regulation of gastric secretions
   a. Parasympathetic impulses and the hormone **gastrin** enhance gastric secretion.
   b. Food in the small intestine reflexively inhibits gastric secretions.
4. Gastric absorption
   The stomach wall may absorb a few substances, such as water and other small molecules.
5. Mixing and emptying actions
   a. Mixing movements help produce **chyme.** Peristaltic waves move chyme into the pylorus.
   b. The muscular wall of the pylorus regulates chyme movement into the small intestine.
   c. The rate of emptying depends on the fluidity of chyme and the type of food present.
6. Vomiting
   Irritants can cause food to be expelled through the mouth.

15.7 Pancreas

1. Structure of the pancreas
   a. The **pancreas** produces pancreatic juice that is secreted into a pancreatic duct.
   b. The pancreatic duct leads to the duodenum.
2. Pancreatic juice
   a. Pancreatic juice contains enzymes that can break down carbohydrates, fats, nucleic acids, and proteins.
   b. Pancreatic juice has a high bicarbonate ion concentration that helps neutralize chyme and causes intestinal contents to be alkaline.
3. Hormones regulate pancreatic secretion
   a. **Secretin** stimulates the release of pancreatic juice with a high bicarbonate ion concentration.
   b. Cholecystokinin stimulates the release of pancreatic juice with a high concentration of digestive enzymes.

15.8 Liver and Gallbladder

1. Liver structure
   a. The lobes of the **liver** consist of **hepatic lobules,** the functional units of the organ.
   b. Bile canals carry bile from hepatic lobules to **hepatic ducts** that unite to form the **common hepatic duct.**
2. Liver functions
   a. The liver metabolizes carbohydrates, lipids, and proteins; stores some substances; and removes toxic substances from the blood (detoxifies)
   b. Bile is the only liver secretion that directly affects digestion.
3. Composition of bile
   a. **Bile** contains bile salts, bile pigments, cholesterol, and electrolytes.
   b. Only the bile salts have digestive functions.
4. Gallbladder
   a. The **gallbladder** stores bile between meals.
   b. A sphincter muscle controls release of bile from the bile duct.
5. Regulation of bile release
   a. Cholecystokinin from the small intestine stimulates bile release.
   b. The sphincter muscle at the base of the bile duct relaxes as a peristaltic wave in the duodenal wall approaches.
6. Functions of bile salts
   Bile salts emulsify fats and aid in the absorption of fatty acids, cholesterol, and certain vitamins.

15.9 Small Intestine

The small intestine receives chyme from the stomach and secretions from the pancreas and liver, completes nutrient digestion, absorbs the products of digestion, and transports the residues to the large intestine.

1. Parts of the small intestine
   The **small intestine** consists of the **duodenum, jejunum,** and **ileum.**
2. Structure of the small intestinal wall
   a. The wall is lined with **villi** that greatly increase the surface area of the intestinal lining, aiding the absorption of digestive products.
   b. **Intestinal glands** are located between the villi.
3. Secretions of the small intestine
   a. Secretions include mucus and digestive enzymes.
   b. Digestive enzymes embedded in the surfaces of microvilli break down molecules of sugars, proteins, and fats into simpler forms.
4. Absorption in the small intestine
   a. Blood capillaries in the villi absorb monosaccharides and amino acids.

   b. Blood capillaries in the villi also absorb water and electrolytes.
   c. Fat molecules with longer chains of carbon atoms enter the lacteals of the villi.
   d. Fatty acids with relatively short carbon chains enter blood capillaries of the villi.
5. Movements of the small intestine
   a. Movements include mixing by segmentation and peristalsis.
   b. The **ileocecal sphincter** controls movement of the intestinal contents from the small intestine into the large intestine.

15.10 Large Intestine

The large intestine reabsorbs water and electrolytes, and forms and stores feces.

1. Parts of the large intestine
   a. The large intestine consists of the **cecum, colon, rectum,** and **anal canal.**
   b. The colon is divided into **ascending, transverse, descending,** and **sigmoid** portions.
2. Structure of the large intestinal wall
   a. The large intestinal wall resembles the wall in other parts of the alimentary canal.
   b. The large intestinal wall has a unique layer of longitudinal muscle fibers arranged in distinct bands.
3. Functions of the large intestine
   a. The large intestine has little digestive function.
   b. It secretes mucus.
   c. The large intestine absorbs water and electrolytes.
   d. The large intestine forms and stores feces.
4. Movements of the large intestine
   a. Movements are similar to those in the small intestine.
   b. Mass movements occur two to three times each day.
   c. A reflex stimulates defecation.
5. Feces
   a. **Feces** consist of water, undigested material, electrolytes, mucus, shed intestinal cells, and bacteria.
   b. The color of feces is due to bile pigments that have been altered by bacterial actions.

15.11 Nutrition and Nutrients

Nutrition is the study of nutrients and how the body utilizes them. The macronutrients (carbohydrates, lipids, and proteins) are required in large amounts. The micronutrients (vitamins and minerals) are required in smaller amounts. Calories measure potential energy in foods.

1. **Carbohydrates**
   a. Carbohydrate sources
      (1) Starch, glycogen, disaccharides, and monosaccharides are carbohydrates.
      (2) Cellulose is a polysaccharide that human enzymes cannot digest.
   b. Carbohydrate use
      (1) Oxidation releases energy from glucose.
      (2) Excess glucose is stored as glycogen or combined to produce fat.
      (3) Carbohydrates supply energy and are also part of nucleic acids and milk.
   c. Carbohydrate requirements
      (1) Humans survive with a wide range of carbohydrate intakes.
      (2) Excess carbohydrates may lead to weight gain.
2. **Lipids**
   a. Lipid sources
      (1) Foods of plant and animal origin provide triglycerides.
      (2) Foods of animal origin provide dietary cholesterol.
   b. Lipid use
      (1) The liver and adipose tissue control triglyceride metabolism.
      (2) Linoleic acid and linolenic acid are **essential fatty acids.**
      (3) Lipids supply energy and are used to build cell and organelle membranes and steroid hormones.
   c. Lipid requirements
      (1) The amounts and types of lipids needed for health are unknown.
      (2) Fat intake must be sufficient to carry fat-soluble vitamins.
3. **Proteins**
   a. Protein sources
      (1) Proteins are mainly obtained from meats, dairy products, cereals, and legumes.
      (2) **Complete proteins** contain adequate amounts of all the **essential amino acids.**
      (3) **Incomplete proteins** lack adequate amounts of one or more essential amino acids.
   b. Protein use
      Proteins serve as structural materials, function as enzymes, and provide energy.
   c. Protein requirements
      Proteins and amino acids must supply essential amino acids and nitrogen for the synthesis of nitrogen-containing molecules.
4. **Vitamins**
   a. Fat-soluble vitamins
      (1) These include vitamins A, D, E, and K.
      (2) They are carried in lipids and are influenced by the same factors that affect lipid absorption.
      (3) They resist the effects of heat; thus, they are not destroyed by cooking or food processing.
   b. Water-soluble vitamins
      (1) This group includes the **B vitamins** and **vitamin C.**
      (2) B vitamins make up a group (the vitamin B complex) and oxidize carbohydrates, lipids, and proteins.
      (3) Cooking or processing food destroys some water-soluble vitamins.
5. **Minerals**
   a. Characteristics of minerals
      (1) Most minerals are in the bones and teeth.
      (2) Minerals are usually incorporated into organic molecules; some occur in inorganic compounds or as free ions.

(3) They serve as structural materials, function in enzymes, and play vital roles in metabolism.

b. **Major minerals** include calcium, phosphorus, potassium, sulfur, sodium, chlorine, and magnesium.

c. **Trace elements** include iron, manganese, copper, iodine, cobalt, zinc, fluorine, selenium, and chromium.

6. Adequate diets
   a. An adequate diet provides sufficient energy and essential nutrients to support optimal growth, maintenance, and repair of tissues.
   b. Individual requirements vary so greatly that designing a diet that is adequate for everyone is not possible. MyPlate can help to personalize diets.
   c. **Malnutrition** may result from undernutrition or overnutrition.

16 Respiratory System

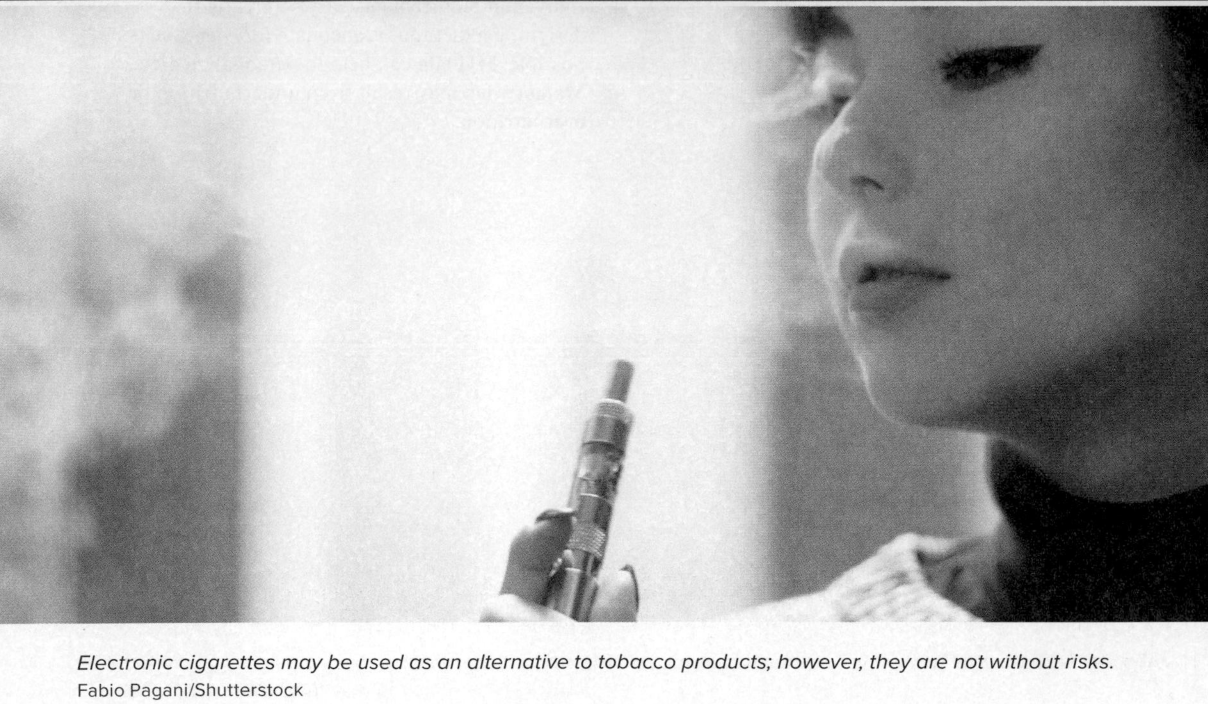

LEARN • PRACTICE • ASSESS

Electronic cigarettes may be used as an alternative to tobacco products; however, they are not without risks.
Fabio Pagani/Shutterstock

Her parents have smoked cigarettes longer than Maria, age 24, has been alive. She has heard of the dangers of secondhand and thirdhand smoke, and that exposure can be as dangerous as actual smoking. This awareness brought up concern for her health, as well as her parents' health.

Smoke contains more than 7,000 chemicals, many of which are irritants, carcinogens, mutagens, and systemic or developmental toxins. More than seventy carcinogens are in tobacco smoke. Toxins include benzene, formaldehyde, vinyl chloride, ammonia, arsenic, and cyanide.

Secondhand smoke comes from lit cigarettes, cigars, or pipes, and from what smokers exhale. Researchers assess exposure to secondhand smoke by measuring a breakdown product of nicotine, called cotinine, in samples of saliva, urine, or blood. Even short exposures to smoke are dangerous. A half hour of breathing someone else's smoke can activate platelets, damage endothelium, and decrease coronary artery blood flow—all changes that set the stage for cardiovascular disease. In the United States, exposure to secondhand smoke causes about 34,000 deaths from heart disease each year, and approximately 7,300 deaths from lung cancer. About 2.5 million U.S. nonsmokers have died from the effects of exposure to secondhand smoke since studies began in 1964.

Thirdhand smoke is the residual effect of the nicotine and other chemicals left on surfaces, like dust, clothing, flooring, and upholstery, as Maria would find in her parent's home and cars. Thirdhand smoke is a relatively new finding, but one that should not be overlooked. Babies and toddlers, in particular, touch these contaminated surfaces and put objects in their mouths, increasing their exposure. Respiratory complications, such as coughing, infection, and asthma, may result.

Maria thought that electronic cigarettes (e-cigarettes) may be a healthier alternative for her parents. When using an e-cigarette, a liquid is loaded and electricity is applied to heat the liquid until it vaporizes. Chemical reactions occur during heating, producing new toxic chemicals, like the cancer-causing agent, formaldehyde. The e-cigarette vapor is inhaled as a chemical-filled aerosol in what is called "vaping." Although studies have shown that smokers who switch to e-cigarettes are more likely to quit smoking than smokers who try to quit using nicotine patches or similar products, what they found is the tendency to become hooked on e-cigarettes. However, if these people gradually lower the dose of nicotine in these vaping devices, it may become easier for them to quit vaping.

In 2016, the Surgeon General concluded that secondhand emissions of e-cigarettes contain "nicotine; ultrafine particles; flavorings such as diacetyl, a chemical linked to serious lung disease; volatile organic compounds such as benzene, which is found in car exhaust; and heavy metals, such as nickel, tin, and lead" that can be inhaled deeply into the lungs. A huge concern is that many e-liquids are flavored to be attractive to children. In children, teenagers, and young adults whose brains are still developing, exposure to nicotine may impair brain development and predispose them to drug addiction, because nicotine is highly addictive. Studies are still being done to confirm the long-term health consequences from using these products.

Module 11 Respiratory System

LEARNING OUTLINE

After studying this chapter, you should be able to complete the "Learning Outcomes" that follow the major headings throughout the chapter.

AIDS TO UNDERSTANDING WORDS

alveol- [small cavity] *alveol*us: microscopic air sac within a lung.

bronch- [windpipe] *bronch*us: branch of the trachea.

cric- [ring] *cric*oid cartilage: ring-shaped mass of cartilage at the base of the larynx.

epi- [upon] *epi*glottis: flaplike structure that partially covers the opening into the larynx during swallowing.

hemo- [blood] *hemo*globin: pigment in red blood cells that transports oxygen and carbon dioxide.

(Appendix A has a complete list of Aids to Understanding Words.)

16.1 | Introduction

LEARN

1. Identify the general functions of the respiratory system.

If asked what **respiration** means, you would most likely say breathing. After all, a respiratory therapist will aid patients with breathing problems. This is part of the truth, but not all of it. The term *respiration* actually refers to how an organism makes energy. Oxygen is essential for our energy production. Carbon dioxide is a byproduct of this process and is expelled. Therefore, the **respiratory system** is custom built to obtain oxygen and to remove carbon dioxide. It includes tubes that transport air into and out of the lungs, as well as microscopic air sacs where the gases are exchanged.

Although it is a continual process, or cycle, we recognize several stages of respiration that require an intimate relationship with the cardiovascular system. **External respiration** includes *ventilation*, or breathing. This allows for an exchange of gases between the capillaries in the lungs and the air. **Internal respiration** is the transport of these gases in the blood for exchange between all body cells. **Cellular respiration** occurs in cells in the mitochondria, where the oxygen is actually used for energy production. The carbon dioxide given off in this process is then transported back to the lungs.

The respiratory system also removes particles from incoming air, helps control the temperature and water content of the air, produces vocal sounds, and participates in the sense of smell and the regulation of blood pH.

PRACTICE 16.1

Answers to the Practice questions can be found in the eBook.

1. What is respiration?
2. During what stage of respiration is oxygen used and carbon dioxide produced?

CAREER CORNER
Respiratory Therapist

The teachers welcome an education session from the local hospital's respiratory therapist, who has taken time away from her patient responsibilities to talk to the class. Several children in the elementary school have asthma, and the teachers have requested this education session. After the visit, the teachers feel more confident that they can help both the students who have asthma and their classmates, to better understand what is happening.

Respiratory therapists assess patients who are experiencing breathing problems and treat them, under a physician's supervision. They work in diverse places and under a variety of situations. A respiratory therapist might help patients breathe on their own following anesthesia, instruct a patient newly diagnosed with emphysema on how to use supplemental oxygen, check on how hospitalized patients are breathing, or teach parents of a young child with cystic fibrosis how to apply pressure to the chest to shake free the thick mucus that builds up. They assist patients in using specialized equipment, such as that used to treat sleep apnea or a home ventilator for a patient with a spinal cord injury. The overall goal is to help patients be as independent as possible.

A respiratory therapist applies knowledge of how the heart and lungs function. Educational requirements include at least an associate's degree and certification, plus continuing education.

16.2 Organs and Structures of the Respiratory System

LEARN

1. Locate the organs and associated structures of the respiratory system.
2. Describe the functions of each organ of the respiratory system.

The respiratory system can be divided into two parts, or tracts. The *upper respiratory tract* includes the nose, nasal cavity, paranasal sinuses, pharynx, and larynx. The *lower respiratory tract* includes the trachea, bronchial tree, and lungs (fig. 16.1; see reference plates 3, 4, 5, and 6).

Nose

Bone and cartilage internally support the facial structure called the **nose.** Its two *nostrils* are openings through which air can enter and leave the nasal cavity. Many internal hairs guard the nostrils, preventing entry of large particles carried in the air.

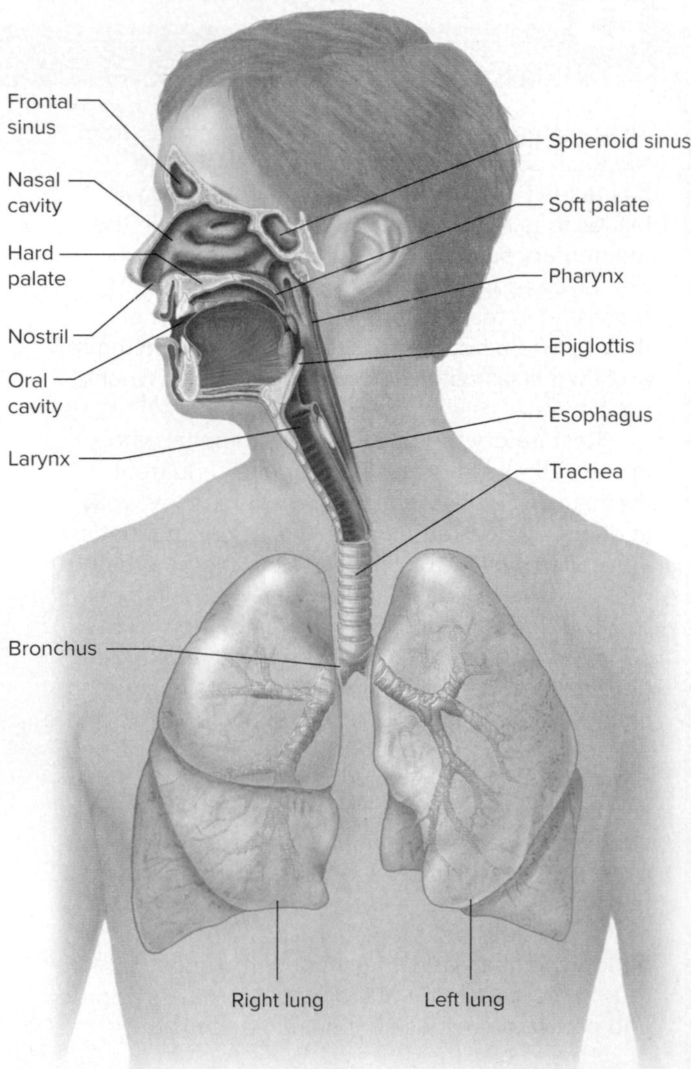

Figure 16.1 Organs and associated structures of the respiratory system.

Nasal Cavity

The **nasal cavity** is a hollow space behind the nose (fig. 16.1). The **nasal septum,** composed of bone and cartilage, divides the nasal cavity into right and left parts. The nasal septum is usually straight at birth, but it can bend as the result of a birth injury. With age, the septum bends toward one side or the other. If such a deviated septum is severe, it may obstruct the nasal cavity, making breathing difficult.

Nasal conchae are bones and bone processes that curl out from the lateral walls of the nasal cavity on each side, dividing the cavity into passageways (fig. 16.2). Nasal conchae support the mucous membrane that lines the nasal cavity. The conchae also help increase the mucous membrane's surface area.

The mucous membrane has pseudostratified ciliated epithelium that is rich in mucus-secreting goblet cells (see section 5.2, Epithelial Tissues). It also includes an extensive network of blood vessels. As air passes over the mucous membrane, heat leaves the blood and warms the air, adjusting the air's temperature to that of the body. In addition, incoming air is moistened as water evaporates from the mucous membrane. The sticky mucus that the mucous membrane secretes entraps dust and other small particles entering with the air.

As the cilia of the epithelial lining move, they push a thin layer of mucus and entrapped particles toward the pharynx, where the mucus is swallowed (fig. 16.3). In the stomach, gastric juice destroys microorganisms in the mucus.

OF INTEREST A spore of the bacterium that causes anthrax is only half a micrometer wide. When spores are combined with powder to create a "bioweapon," they are still small enough to bypass the hairs guarding the nostrils and the sticky mucous membranes, reaching the lungs, where they can cause inhalation anthrax. The bacteria release a toxin that causes death.

Paranasal Sinuses

Recall from section 7.6, Skull, that the **paranasal sinuses** are air-filled spaces within the frontal, ethmoid, sphenoid, and maxillary bones of the skull that open into the nasal cavity. Mucous membranes line the sinuses and are continuous with the lining of the nasal cavity. The paranasal sinuses reduce the weight of the skull and are resonant chambers that affect the quality of the voice. The frontal and sphenoid sinuses are shown in figures 16.1 and 16.2. A painful sinus headache can result from blocked drainage caused by an infection or allergic reaction.

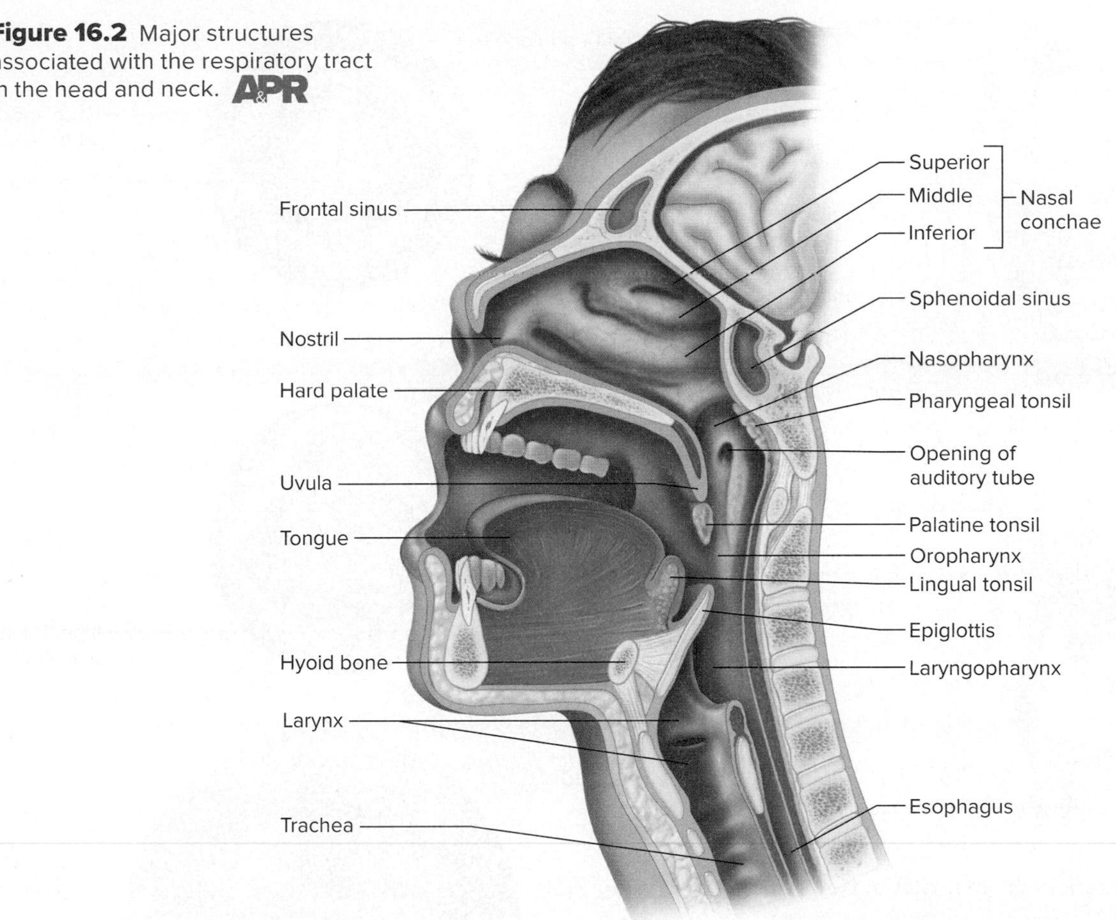

Figure 16.2 Major structures associated with the respiratory tract in the head and neck. APR

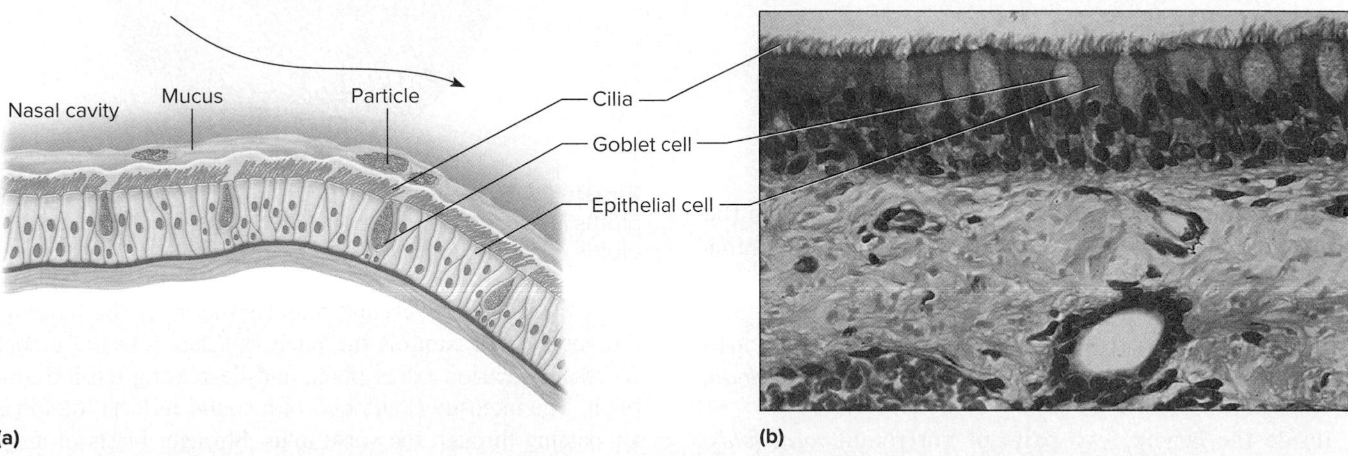

Figure 16.3 Mucus movement in the respiratory tract. **(a)** Cilia move mucus and trapped particles from the nasal cavity to the pharynx. **(b)** Micrograph of pseudostratified ciliated epithelium in the respiratory tract (275x). APR (b): Biophoto Associates/Science Source

PRACTICE 16.2

1. Which organs constitute the respiratory system?
2. What are the functions of the mucous membrane that lines the nasal cavity?
3. Where are the paranasal sinuses?
4. What are the functions of the paranasal sinuses?

Pharynx

Commonly called the throat, the **pharynx** is the space behind the oral cavity, the nasal cavity, and the epiglottis. Behind the larynx it transitions into the esophagus (see fig. 16.1). It is a passageway for food moving from the oral cavity to the esophagus and for air passing between the nasal cavity and the larynx. The pharynx also helps produce the sounds of speech. Section 15.5, Pharynx and Esophagus, describes the

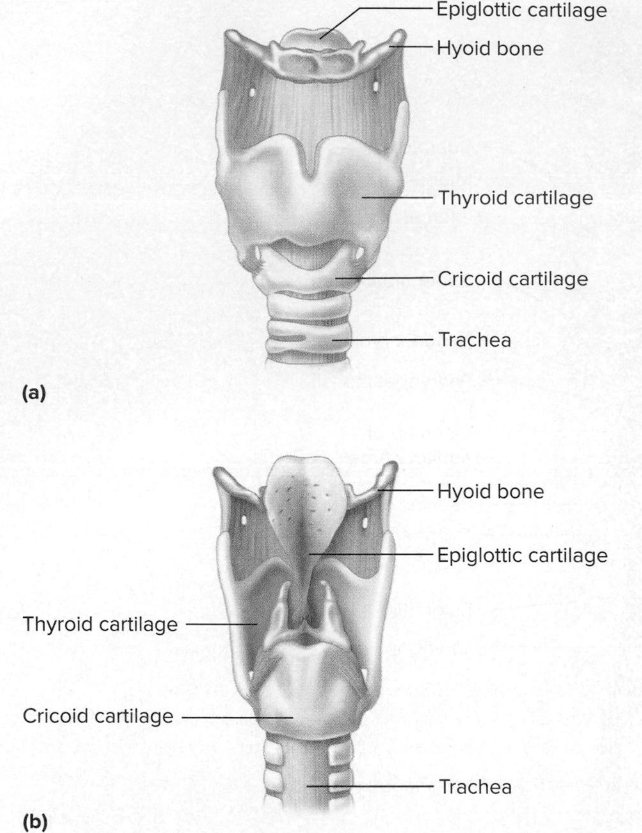

Figure 16.4 Larynx. (**a**) Anterior and (**b**) posterior views of the larynx. APR

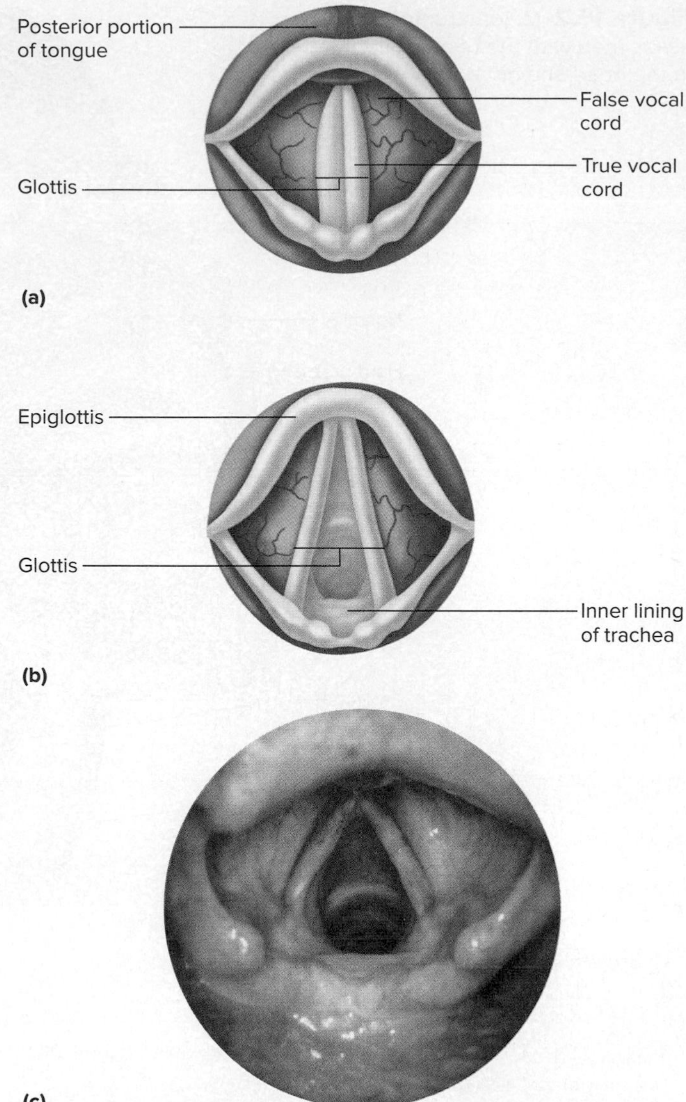

Figure 16.5 The vocal cords as viewed from above with the glottis (**a**) closed and (**b**) open. (**c**) Photograph of the open glottis and vocal folds. (c): CNRI/Science Source

subdivisions of the pharynx—the nasopharynx, oropharynx, and laryngopharynx, which are shown in figure 16.2.

Larynx

The **larynx** (lar′inks), or voice box, is an enlargement in the airway at the top of the trachea, where the epiglottis ensures that only air enters. The larynx also houses the **vocal cords.**

The larynx is composed of a framework of muscles and cartilages bound by elastic tissue. The largest of the cartilages are the *thyroid* ("Adam's apple"), *cricoid,* and *epiglottic cartilages* (fig. 16.4).

Inside the larynx, two pairs of horizontal *vocal folds,* composed of muscle tissue and connective tissue with a covering of mucous membrane, extend inward from the lateral walls. The upper folds are called *false vocal cords* because they do not produce sounds (fig. 16.5*a*). The muscle fibers within these folds help close the airway during swallowing.

The lower folds have muscle fibers and strong but elastic connective tissue. These are the *true vocal cords.* The true vocal cords and the opening between them form the **glottis** (glot′is) (figs. 16.5*b* and 16.5*c*). During normal breathing, the vocal cords are relaxed and the glottis is a triangular slit. Air forced through the glottis causes the true vocal cords to vibrate, which produces sound (fig. 16.5).

Contracting or relaxing muscles that alter the tension on the vocal cords controls the pitch (musical tone) of a sound. Increasing tension raises pitch, and decreasing tension lowers pitch. The intensity (loudness) of a sound reflects the force of air passing through the vocal folds. Stronger blasts of air produce louder sound; weaker movements of air produce softer sound. Changing the shapes of the pharynx and oral cavity and using the tongue and lips transform the sound into words. Nerve injury and inflammation due to infection or irritants may affect laryngeal health, as discussed in Clinical Application 16.1.

The epiglottic cartilage is the central part of a flaplike structure called the **epiglottis.** This structure usually stands upright and allows air to enter the larynx. During swallowing, the larynx rises and the tongue presses the epiglottis downward, covering the opening into the larynx. In addition, contraction of the false vocal cords closes off the larynx. These actions help prevent foods and liquids from entering the air passages (see section 15.5, Pharynx and Esophagus).

CLINICAL APPLICATION 16.1
Laryngeal Health

Damage to the nerves (recurrent laryngeal nerves) that supply the laryngeal muscles can alter the quality of a person's voice. These nerves pass through the neck as parts of the vagus nerves, and trauma or surgery to the neck or thorax can injure them. Nodules or other growths on the margins of the vocal folds that interfere with the free flow of air can also cause vocal problems. Sometimes surgery may be necessary to remove such lesions.

Laryngitis—which causes hoarseness or lack of voice—occurs when the mucous membrane of the larynx becomes inflamed and swollen, due to an infection or an irritation from inhaled vapors, and prevents the vocal cords from vibrating freely. Laryngitis is usually mild, but may be dangerous if swollen tissues obstruct the airway and interfere with breathing. In such cases, a clinician may insert a tube (endotracheal tube) into the trachea through the mouth to restore the passageway until the inflammation subsides.

PRACTICE 16.2

5. Describe the structure of the larynx.
6. How do the vocal cords produce sounds?
7. What is the function of the glottis? The epiglottis?

Trachea

The **trachea** (trā'kē-ah), or windpipe, is a flexible cylindrical tube about 2.5 centimeters in diameter and 12.5 centimeters in length (fig. 16.6). It extends downward in front of the esophagus to about the level of the fifth thoracic vertebra, past the top fourth of the lungs. The carina is the spot where the trachea splits into the right and left bronchi.

A ciliated mucous membrane with many goblet cells lines the trachea's inner wall. This membrane filters incoming air and moves entrapped particles upward into the pharynx, where the mucus can be swallowed. Clinical Application 16.2, "Cystic Fibrosis," discusses what happens when the mucus formed is extremely thick.

Within the tracheal wall are about twenty C-shaped pieces of hyaline cartilage, one above the other. The open ends of these incomplete rings are directed posteriorly, and smooth muscle and connective tissues fill the gaps between the ends. These cartilaginous rings prevent the trachea from collapsing and blocking the airway. The soft tissues that complete the rings in the back allow the nearby esophagus to expand as food moves through it to the stomach.

Bronchial Tree

The **bronchial tree** (brong'kē-al trē) consists of branched airways leading from the trachea to the microscopic air sacs in the lungs (fig. 16.7). Its branches begin with the right and left **main (primary) bronchi,** which arise from the trachea at the level of the fifth thoracic vertebra. Each **bronchus** enters its respective lung.

A short distance from its origin, each main bronchus divides into *lobar* (secondary) bronchi. The lobar bronchi branch into *segmental* (tertiary) bronchi, and then into increasingly finer tubes. Among these smaller tubes are **bronchioles** that continue to branch, giving rise to *terminal bronchioles, respiratory bronchioles,* and finally to very thin

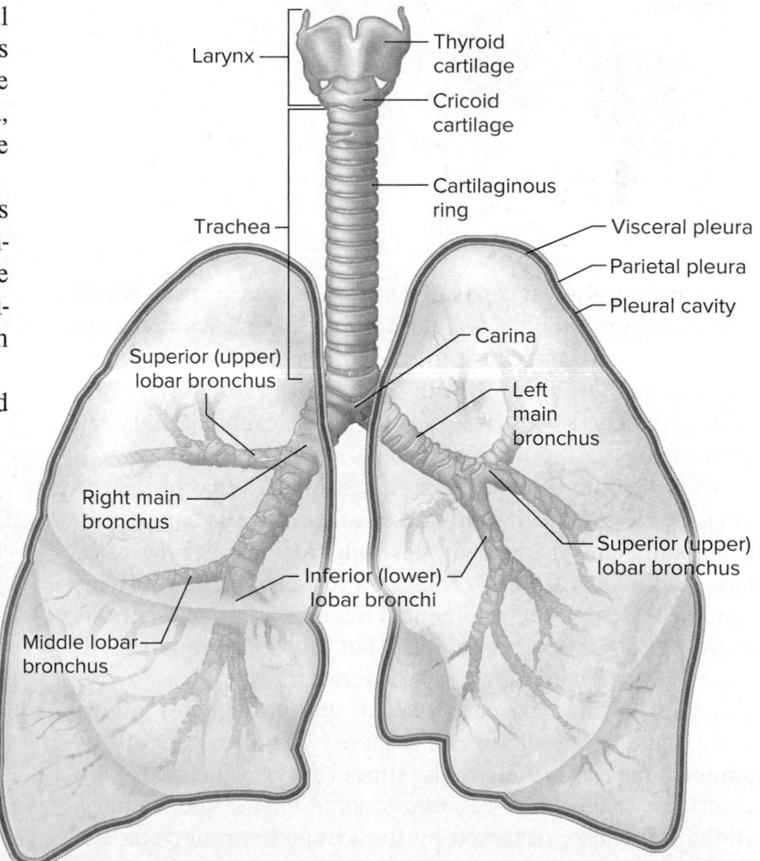

Figure 16.6 The trachea conducts air between the larynx and the bronchi.

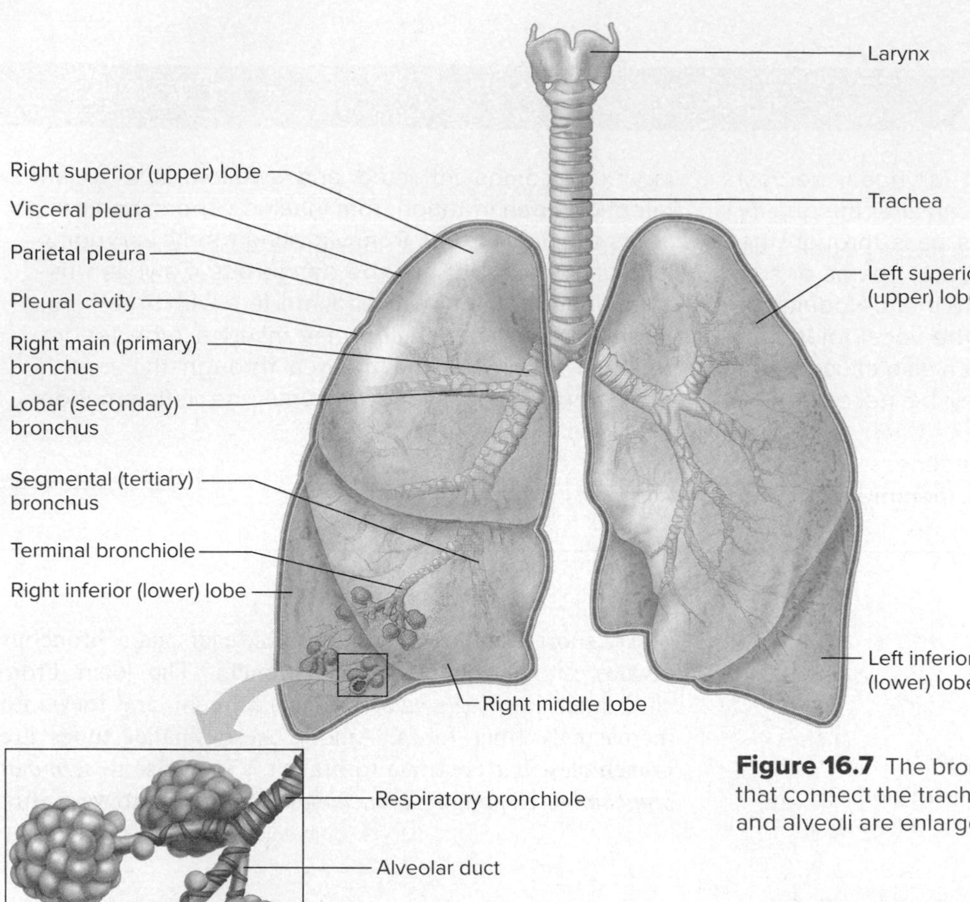

Figure 16.7 The bronchial tree consists of the passageways that connect the trachea and the alveoli. The alveolar ducts and alveoli are enlarged to show their locations. APR

tubes called **alveolar ducts.** These ducts lead to thin-walled outpouchings called **alveolar sacs.** Alveolar sacs are composed of microscopic air sacs called **alveoli** (al-vē′o-lī; singular, alveolus), which lie within capillary networks (figs. 16.8 and 16.9). The alveoli are the sites of gas exchange between the inhaled air and the bloodstream.

The structure of a bronchus is similar to that of the trachea, but the tubes that branch from it have less cartilage in their walls, and the bronchioles lack cartilage. As the cartilage diminishes, a layer of smooth muscle surrounding the tube becomes more prominent. This muscular layer persists even in the smallest bronchioles, but only a few muscle cells are associated with the alveolar ducts.

The absence of cartilage in the bronchioles allows their diameters to change in response to contraction of the smooth muscle in their walls, similar to what happens with arterioles of the cardiovascular system. Part of the "fight-or-flight" response, triggered by the sympathetic nervous system, is **bronchodilation,** in which the smooth muscle relaxes and the airways become wider and allow more airflow. The opposite, **bronchoconstriction,** occurs when the smooth muscle contracts and it becomes difficult to move air in and out of the lungs. Bronchoconstriction can occur with allergies. Asthma is an extreme example of bronchoconstriction.

The mucous membranes of the bronchial tree continue to filter the incoming air, and the many branches of the tree distribute the air to alveoli throughout the lungs. The alveoli, in turn, provide a large surface area of thin simple squamous epithelial cells through which gases are easily exchanged. Oxygen diffuses from the alveoli into the blood in nearby capillaries, and carbon dioxide diffuses from the blood into the alveoli (fig. 16.10).

OF INTEREST Combined, two adult lungs have about 300 million alveoli, providing a total surface area nearly half the size of a tennis court.

PRACTICE 16.2

8. What is the function of the cartilaginous rings in the tracheal wall?
9. Describe the bronchial tree.
10. Predict the direction of diffusion of gases between alveoli and alveolar capillaries.

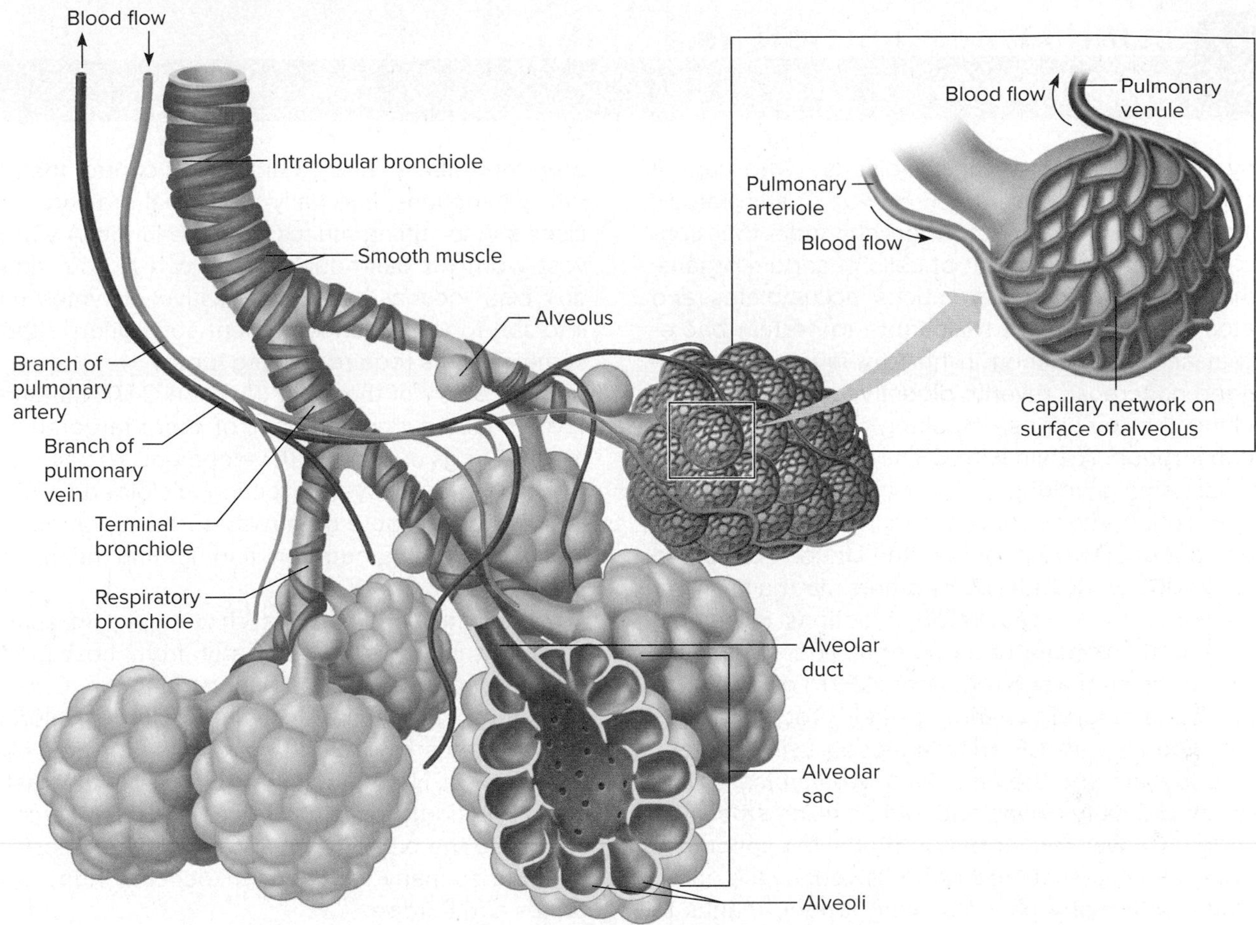

Figure 16.8 The respiratory tubes end in tiny alveoli, each of which is surrounded by a capillary network. APR

PRACTICE FIGURE 16.8

Why is the pulmonary artery colored blue in this figure?

Answer can be found in Appendix E.

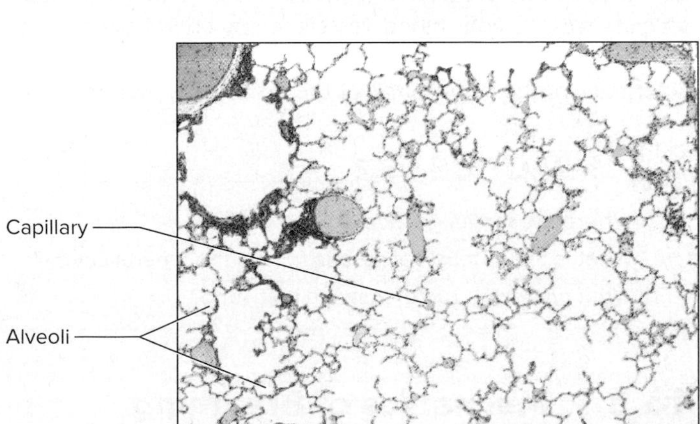

Figure 16.9 Light micrograph of alveoli (250x). APR

Al Telser/McGraw-Hill Education

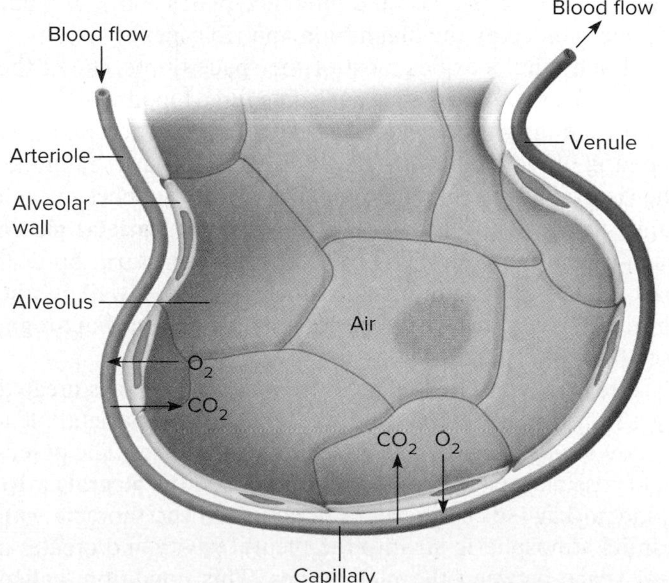

Figure 16.10 Oxygen (O_2) diffuses from air within the alveolus into the capillary, while carbon dioxide (CO_2) diffuses from blood within the capillary into the alveolus. APR

CLINICAL APPLICATION 16.2
Cystic Fibrosis

Many young children with this disease who cannot pronounce its name call it "65 roses." Cystic fibrosis (CF) is an inherited defect in ion channels that control chloride movement out of cells in certain organs. In the lungs, thick, sticky mucus accumulates and creates an environment hospitable to certain bacteria that are not common in healthy lungs. A mucus-clogged pancreas prevents digestive secretions from reaching the intestines, impairing nutrient digestion and absorption. A child with CF has trouble breathing and maintaining weight.

CF is inherited from two carrier parents, and affects about 30,000 people in the United States and about 70,000 worldwide. Many others may have milder cases, with recurrent respiratory infections. More than 2,000 mutations have been recognized in the cystic fibrosis transmembrane regulator (*CFTR*) gene, which encodes the chloride channel protein. Today, fetuses and newborns with CF are diagnosed using genetic tests, but years ago the first signs were typically "failure to thrive," salty sweat, and foul-smelling stools.

When CF was recognized in 1938, life expectancy was only five years, but today median survival is about age fifty, with many patients living longer, thanks to drug treatments. Inhaled antibiotics control the respiratory infections, and daily "bronchial drainage" exercises shake stifling mucus from the lungs. A vibrating vest worn for half-hour periods two to four times a day also loosens mucus. Digestive enzymes mixed into soft foods enhance nutrient absorption, although some patients require feeding tubes.

Discovery of the most common *CFTR* mutation in 1989 enabled development of more-targeted treatments. Drugs are now in development. The new drugs work in various ways: correcting misfolded *CFTR* protein, restoring liquid on airway surfaces, breaking up mucus, improving nutrition, and fighting inflammation and infection.

Life with severe CF is challenging. In summertime, a child must avoid water from hoses, which harbor lung-loving *Pseudomonas* bacteria. Cookouts spew lung-irritating particulates. Too much chlorine in pools irritates lungs, whereas too little invites bacterial infection. New infections arise, too. In the past few years, multidrug-resistant *Mycobacterium abscessus,* related to the pathogen that causes tuberculosis, has affected as many as 10% of CF patients in the United States and Europe.

Lungs

The **lungs** are soft, spongy, cone-shaped organs in the thoracic cavity (see fig. 16.1 and reference plates 3 and 4). They are surrounded by the diaphragm and rib cage.

Each lung occupies most of the space on its side of the thoracic cavity. A bronchus and some large blood vessels suspend each lung in the cavity. These tubular structures enter the lung on its medial surface. A layer of serous membrane, the **visceral pleura** (vis′er-al ploo′rah), firmly attaches to each lung surface and folds back to become the **parietal pleura** (pah-rī′ĕ-tal ploo′rah). The parietal pleura, in turn, borders part of the mediastinum and lines the inner wall of the thoracic cavity and the superior surface of the diaphragm (see figs. 16.7 and 16.11).

Held together by low pressure and wet surfaces created by a thin film of serous fluid, the visceral and parietal pleurae are almost entirely in contact with each other. The potential (possible) space between them is called the **pleural cavity** (ploo′ral kav′ĭ-tē). However, a puncture in the thoracic wall admits atmospheric air into the pleural cavity and creates a real space between the membranes. This condition, called *pneumothorax,* may collapse the lung on the affected side because of the lung's elasticity. A collapsed lung is called *atelectasis.*

The right lung is larger than the left one and is divided into three lobes. The left lung has two lobes (see figs. 16.1 and 16.7). A lobar bronchus supplies each lobe of each lung. A lobe also has connections to blood and lymphatic vessels and lies within connective tissues. Thus, a lung includes air passages, alveoli, blood vessels, connective tissues, lymphatic vessels, and nerves. Table 16.1 summarizes the characteristics of the major parts of the respiratory system.

PRACTICE 16.2

11. Where are the lungs located?
12. What is the function of serous fluid in the pleural cavity?
13. What types of structures make up a lung?

16.3 | Mechanics of Breathing

LEARN

1. Explain the mechanisms of inspiration and expiration.
2. Describe each of the respiratory volumes and capacities.

Breathing, or ventilation, is the movement of air into and out of the lungs. This allows for gas exchange in the respiratory

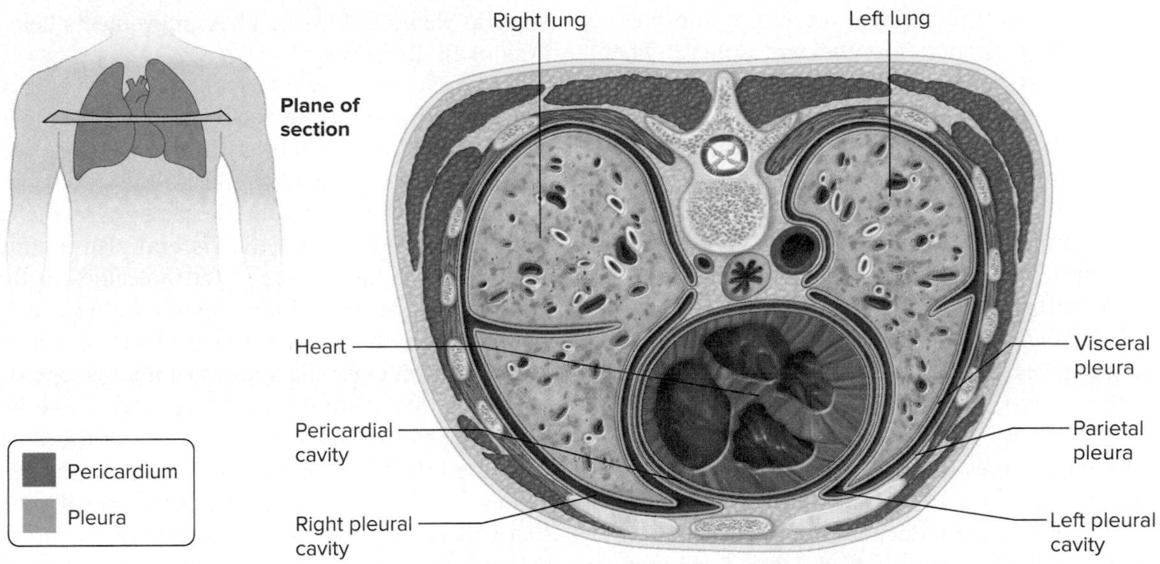

Figure 16.11 The potential spaces between the pleural membranes, called the left and right pleural cavities, are shown here as actual spaces.

TABLE 16.1 Parts of the Respiratory System

Part	Description	Function
Nose	Part of face centered above mouth, in and below space between eyes	Nostrils provide entrance to nasal cavity; internal hairs begin to filter incoming air
Nasal cavity	Hollow space behind nose	Conducts air to pharynx; mucous lining filters, warms, and moistens incoming air
Paranasal sinuses	Hollow spaces in certain skull bones	Reduce weight of skull; serve as resonant chambers
Pharynx	Chamber behind nasal cavity, oral cavity, and larynx	Passageway for air moving from nasal cavity to larynx and for food moving from oral cavity to esophagus
Larynx	Enlargement at top of trachea	Passageway for air; prevents foreign objects from entering trachea; houses vocal cords
Trachea	Flexible tube that connects larynx with bronchial tree	Passageway for air; mucous lining continues to filter particles from incoming air
Bronchial tree	Branched tubes that lead from trachea to alveoli	Conducts air from trachea to alveoli; mucous lining continues to filter incoming air
Lungs	Soft, cone-shaped organs that occupy a large portion of the thoracic cavity	Contain air passages, alveoli, blood vessels, connective tissues, lymphatic vessels, and nerves

membrane found in the alveoli. The actions providing these air movements are termed **inspiration** (in″spĭ-rā′shun), or inhalation, and **expiration** (ek″spĭ-rā′shun), or exhalation.

Inspiration

Just as blood flows from high pressure to low pressure, so does a mixture of gases, such as ordinary air. **Atmospheric pressure,** the pressure of the air around us, provides the force that moves air into the lungs. At sea level, this pressure is sufficient to support a column of mercury about 760 millimeters (mm) high in a tube. Thus, normal air pressure at sea level is equal to about 760 mm of mercury (Hg).

Air pressure is exerted on all surfaces in contact with the air. Because the airways are open to the outside, the airways and alveoli are subjected to outside air pressure. This is easiest to envision at the end of a normal, resting expiration, when no air is moving in or out. At this point, the pressures on the inside of the airways and alveoli and on the outside of the thoracic wall are about the same. (This is also true at the end of an inspiration, before the expiration begins.)

Pressure and volume are related in an opposite (inverse) way. For example, pulling back on the plunger of a syringe increases the volume inside the barrel, which decreases the air pressure inside. Atmospheric pressure then pushes outside air into the syringe. In contrast, pushing on the plunger reduces the volume inside the syringe, increasing the pressure inside and forcing air out into the atmosphere. The movement of air into and out of the lungs occurs in much the same way.

If the pressure inside the lungs decreases, atmospheric pressure will push outside air into the airways and alveoli. This is what happens during normal, resting inspiration. Impulses conducted on the phrenic nerves, which are associated with the cervical plexuses (see section 9.15, Peripheral Nervous System), stimulate skeletal muscle fibers in the dome-shaped *diaphragm* below the lungs to contract. The diaphragm moves downward, the thoracic cavity enlarges, and the pressure in the alveoli falls to about 2 mm Hg below that of atmospheric pressure. In response, atmospheric pressure forces air through the airways into the alveoli (fig. 16.12).

If a person needs to take a deeper-than-normal breath, while the diaphragm is contracting and moving downward, the *external (inspiratory) intercostal muscles* between the ribs may be stimulated to contract. Additional muscles, such as the *pectoralis minors,* the *sternocleidomastoids,* and the *scalenes,* can also pull the thoracic cage farther upward and outward. These actions enlarge the thoracic cavity even more. As a result, the alveolar pressure is reduced further, and atmospheric pressure forces even more air into the alveoli (fig. 16.13).

Lung expansion depends on movements of the pleural membranes in response to movements of the diaphragm and chest wall. When the external intercostal muscles move the thoracic wall upward and outward, and the diaphragm moves downward, the parietal pleura moves too, and the visceral pleura follows it. These movements help expand the lung in all directions.

Two factors ensure that the pleural membranes always move as a unit. First, any tendency for the pleural membranes to pull away from each other decreases pressure in the pleural cavity, holding them together. Second, only a thin film of serous fluid separates the parietal pleura on the inner wall of the thoracic cavity from the visceral pleura attached to the surface of the lungs. The water molecules in this fluid are attracted to the pleural membranes and to each other, helping to adhere the moist surfaces of the pleural membranes, much as a wet coverslip sticks to a microscope slide.

The moist pleural membranes play a role in expanding the lungs, but the moist inner surfaces of the alveoli have the opposite effect. In the alveoli, the attraction of water molecules to each other creates a force called **surface tension** that makes it difficult to inflate the alveoli and may actually collapse them. Certain alveolar cells, however, synthesize a mixture of lipids and proteins called **surfactant** (ser-fak′tant). It is secreted continuously into alveolar air spaces and reduces the alveoli's tendency to collapse, especially when lung volumes are low. Surfactant makes it easier for inspiratory efforts to expand the alveoli.

Surfactant is particularly important in the minutes after birth, when the newborn's lungs fully inflate for the first time. Premature infants may suffer respiratory distress syndrome if they do not produce sufficient surfactant. To help many of these newborns survive, physicians introduce synthetic surfactant into the tiny lungs through an endotracheal tube. A ventilator machine especially geared to an infant's size assists breathing.

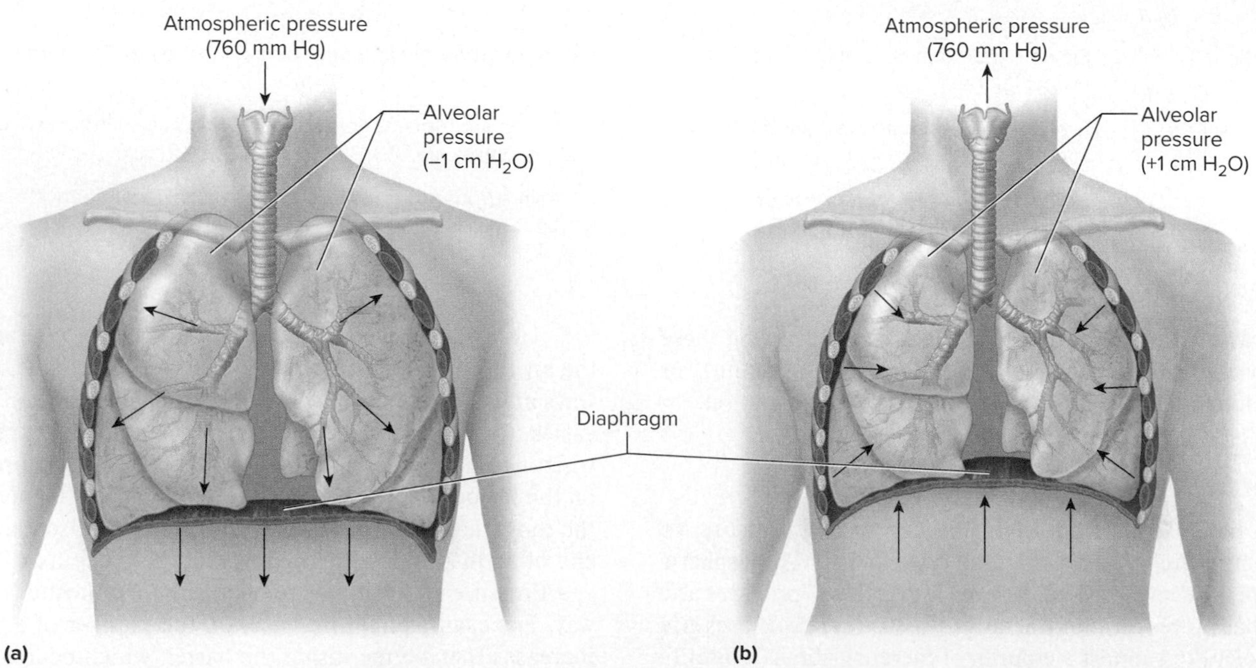

Figure 16.12 Mechanisms of breathing. **(a)** During inspiration, the diaphragm contracts, reducing aveolar pressure and allowing air to flow in. **(b)** When the diaphragm relaxes, air flows out due to the increase in alveolar pressure. APR

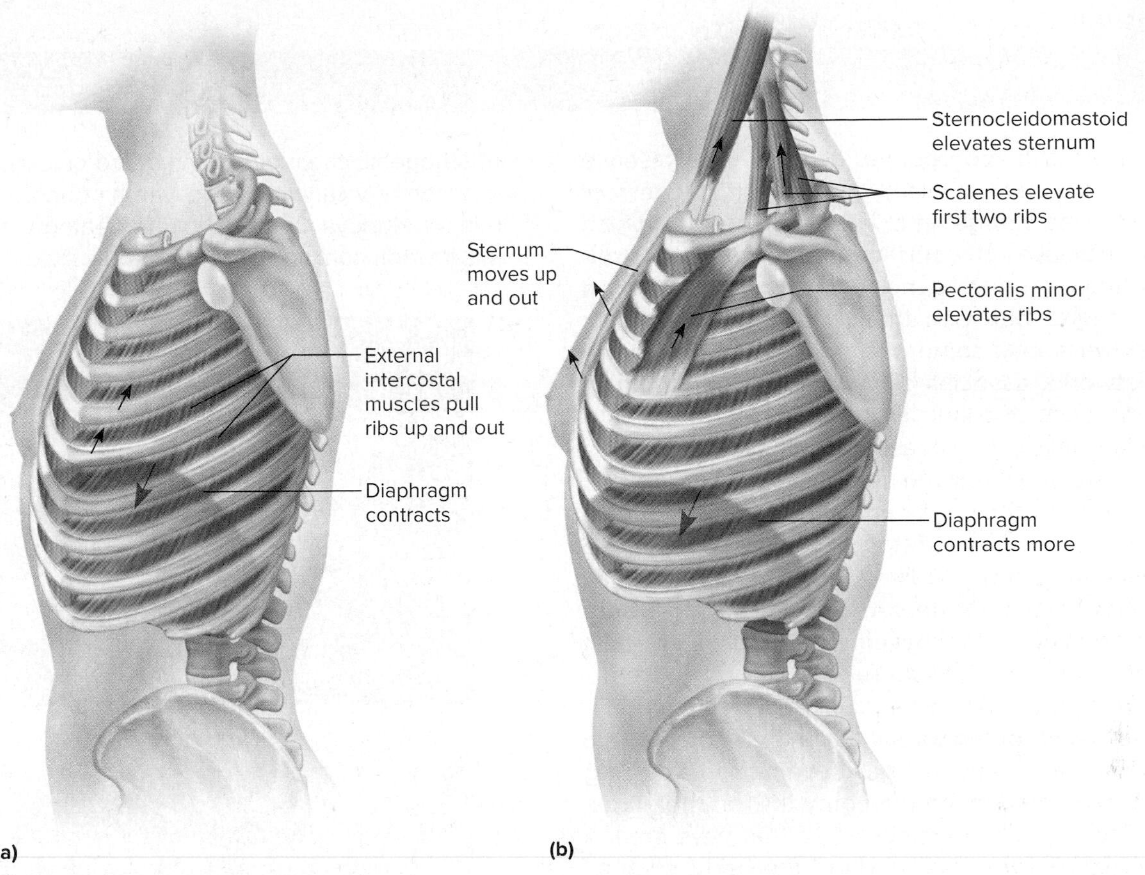
Sternum moves up and out
External intercostal muscles pull ribs up and out
Diaphragm contracts
Sternocleidomastoid elevates sternum
Scalenes elevate first two ribs
Pectoralis minor elevates ribs
Diaphragm contracts more
(a)
(b)

Figure 16.13 Maximal inspiration. **(a)** Shape of the thorax at the end of normal inspiration. **(b)** Shape of the thorax at the end of maximal inspiration, enlarged further by contraction of the sternocleidomastoid, pectoralis minor, and scalene muscles.

OF **INTEREST** The first breath is the toughest. A newborn must use twenty times the energy to take the first breath as for subsequent breaths.

Expiration

The forces for resting expiration come from the *elastic recoil* of tissues and from surface tension. The lungs contain considerable elastic tissue, which stretches with lung expansion during inspiration. As the diaphragm and external intercostal muscles relax following inspiration, the elastic tissues cause the lungs to recoil and return to their original shapes. This pulls the visceral pleural membrane inward, and the parietal pleura and chest wall follow. Also, during inspiration the diaphragm compresses the abdominal organs beneath it. When the diaphragm relaxes, the abdominal organs spring back into their previous shapes, pushing the diaphragm upward (figs. 16.12 and 16.14*a*). At the same time, the surface tension that develops on the moist surfaces of the alveolar linings decreases the diameters of the alveoli. Together, these factors increase alveolar pressure about 1 mm Hg above atmospheric pressure, so that the air inside the lungs is forced out through respiratory passages with no muscle action. Thus, normal resting expiration is a passive process.

If a person needs to exhale more air than normal, the *internal (expiratory) intercostal muscles* can contract (fig. 16.14*b*). These muscles pull the ribs and sternum downward and inward, increasing the air pressure in the lungs to force more air out. Also, the abdominal wall muscles, including the *external* and *internal obliques, transversus abdominis,* and *rectus abdominis,* can squeeze the abdominal organs inward (see fig. 8.20). In this way, the abdominal wall muscles can increase pressure in the abdominal cavity and force the diaphragm still higher against the lungs. These actions force additional air out of the lungs. Clinical Application 16.3 discusses problems in breathing associated with two common diseases: emphysema and lung cancer.

PRACTICE 16.3

1. Describe the events in inspiration.
2. How does expansion of the chest wall expand the lungs during inspiration?
3. Which forces cause normal expiration?

CLINICAL APPLICATION 16.3

Emphysema and Lung Cancer

Emphysema is a progressive, degenerative disease that destroys alveolar walls. As a result, clusters of small air sacs merge into larger chambers, which greatly decreases the surface area available for diffusion, thereby reducing the volume of gases that can be exchanged between the alveoli and the blood. Alveolar walls lose some of their elasticity, and capillary networks associated with the alveoli diminish (fig. 16A). Loss of tissue elasticity in the lungs contributes to making it increasingly difficult for a person with emphysema to exhale.

Emphysema may develop in response to prolonged exposure to respiratory irritants, such as those in tobacco smoke and polluted air. The disease may also result from an inherited enzyme (alpha-1 antitrypsin) deficiency. Emphysema is a type of chronic obstructive pulmonary disease (COPD).

Lung cancer, like other cancers, is the uncontrolled division of abnormal cells that rob normal cells of nutrients and oxygen, eventually crowding them out. Some cancerous growths in the lungs result secondarily from cancer cells that have spread (metastasized) from other parts of the body, such as the breasts, intestines, liver, or kidneys. Cancers that begin in the lungs are called *primary pulmonary cancers.* These may arise from epithelia, connective tissue, or blood cells. The most common form originates from epithelium in a bronchiole (fig. 16B) and is called *bronchogenic carcinoma.* This type of cancer is a response to irritation, such as prolonged exposure to tobacco smoke. Susceptibility to primary pulmonary cancers may be inherited.

Cancer cells divide to form tumor masses that obstruct air passages and reduce gas exchange. Bronchogenic carcinoma can spread quickly, establishing secondary cancers in the lymph nodes, liver, bones, brain, or kidneys. Lung cancer is treated with surgery, ionizing radiation, and drugs, but the survival rate is low.

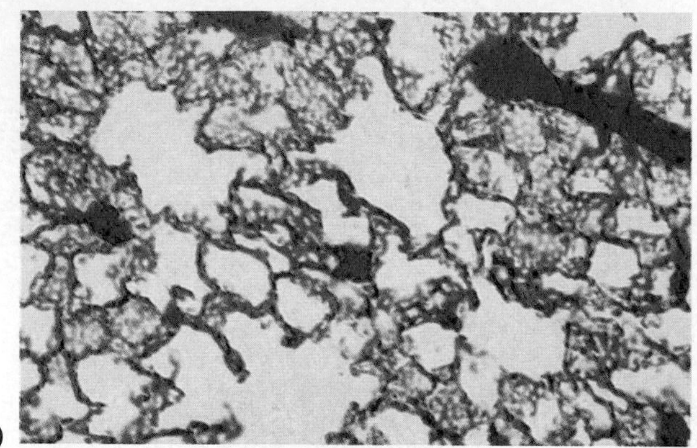

(a)

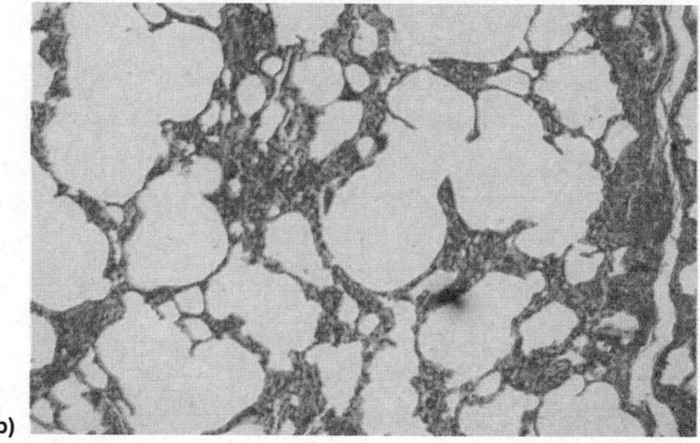

(b)

Figure 16A Comparison of lung tissues. **(a)** Normal lung tissue (100x). **(b)** As emphysema develops, alveoli merge, forming larger chambers (100x). (a, b): Victor B. Eichler, Ph.D.

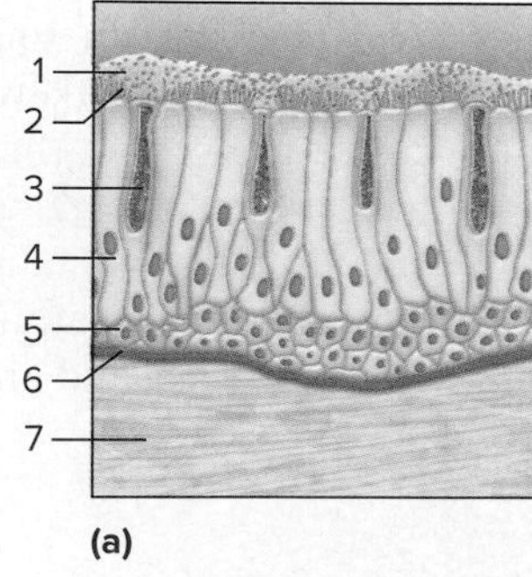

(a)

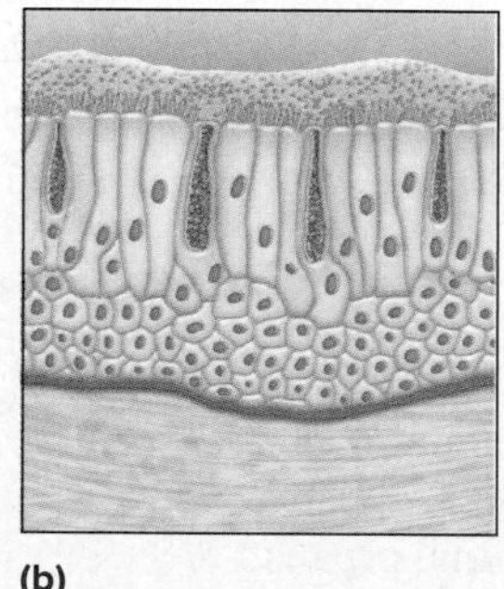

(b)

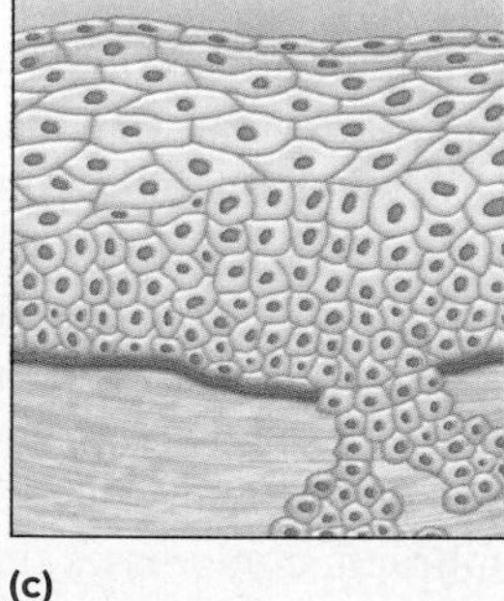

(c)

Figure 16B About 95% of lung cancers start in the lining (epithelium) of a bronchiole. **(a)** The normal lining shows (**4**) columnar cells with (**2**) hairlike cilia, (**3**) goblet cells that secrete (**1**) mucus, and (**5**) basal cells from which new columnar cells arise. (**6**) A basement membrane separates the epithelial cells from (**7**) the underlying connective tissue. **(b)** In the first stage of lung cancer, the basal cells divide repeatedly. The goblet cells secrete excess mucus, and the cilia are less efficient in moving the heavy mucus secretion. **(c)** Continued division of basal cells displaces the columnar and goblet cells. The basal cells penetrate the basement membrane and invade the deeper connective tissue.

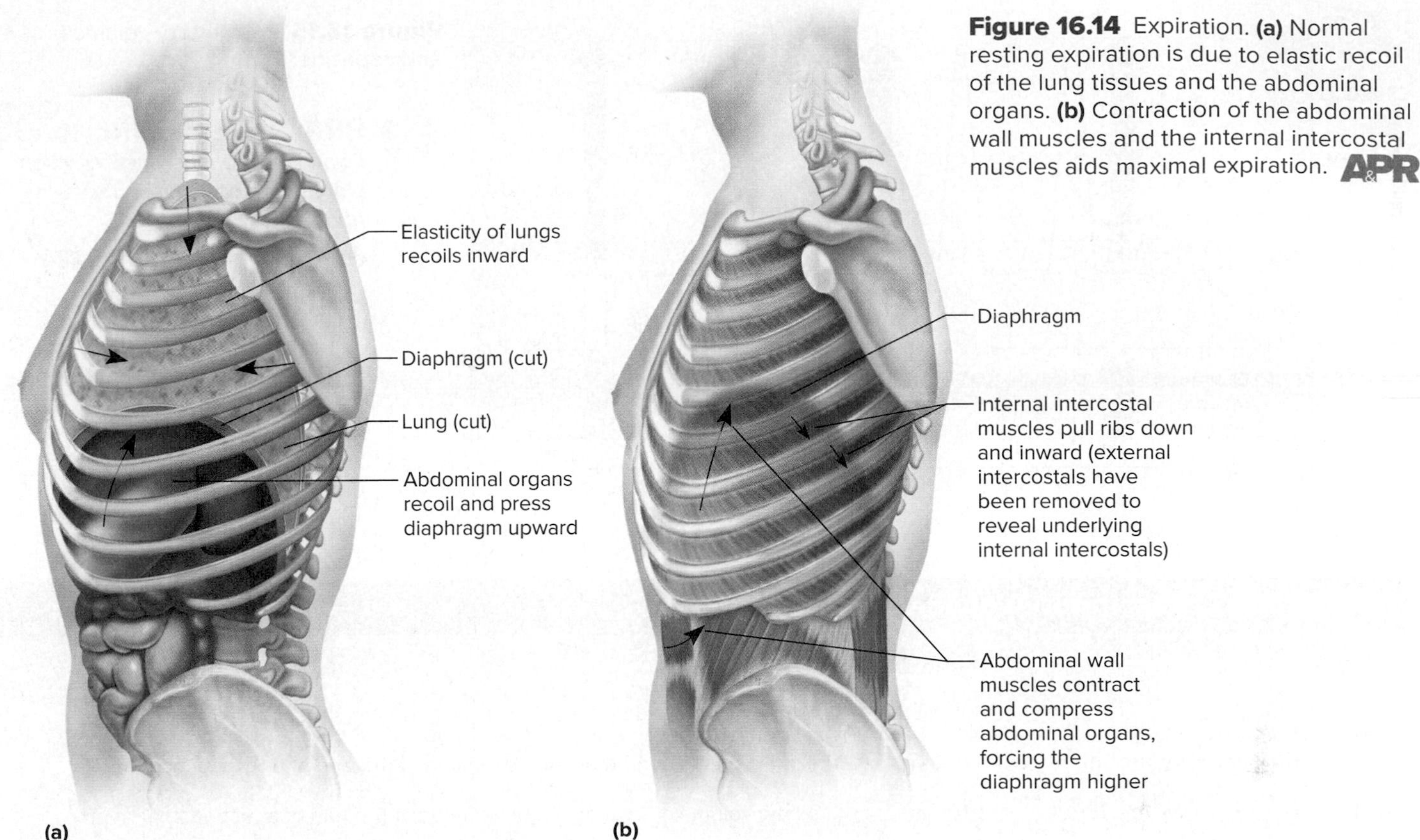
Elasticity of lungs recoils inward
Diaphragm (cut)
Lung (cut)
Abdominal organs recoil and press diaphragm upward
(a)
Diaphragm
Internal intercostal muscles pull ribs down and inward (external intercostals have been removed to reveal underlying internal intercostals)
Abdominal wall muscles contract and compress abdominal organs, forcing the diaphragm higher
(b)

Figure 16.14 Expiration. **(a)** Normal resting expiration is due to elastic recoil of the lung tissues and the abdominal organs. **(b)** Contraction of the abdominal wall muscles and the internal intercostal muscles aids maximal expiration. APR

Respiratory Volumes and Capacities

Spirometry is a procedure that measures air volumes. Using an instrument called a *spirometer,* three distinct **respiratory volumes** can be measured. Such measurements are used to help evaluate the courses of emphysema, pneumonia, and lung cancer, conditions in which functional lung tissue is lost. Spirometry may also be used to track the progress of diseases, such as bronchial asthma, that obstruct air passages.

A *respiratory cycle* is an inspiration plus the following expiration. The **tidal volume** is the volume of air that moves, in then out, during a single respiratory cycle. About 500 milliliters (mL) of air enter during a normal, resting inspiration. Approximately the same volume leaves during a normal, resting expiration. Thus, the **resting tidal volume** is about 500 mL (fig. 16.15).

During forced maximal inspiration, air in addition to the resting tidal volume enters the lungs. This extra volume is the **inspiratory reserve volume** (complemental air). It equals about 3,000 mL.

During forced maximal expiration, the lungs can expel up to about 1,100 mL of air beyond the resting tidal volume. This volume is called the **expiratory reserve volume** (supplemental air).

Even after the most forceful expiration, about 1,200 mL of air remains in the lungs. This is called the **residual volume** and can only be measured using special gas dilution techniques.

As figure 16.15 depicts, each resting inspiration adds about 500 mL of air to about 2,400 mL of air already in the lungs. Normally, this newly inhaled air mixes completely with air already in the lungs. This mixing prevents the oxygen and carbon dioxide concentrations in the alveoli from fluctuating greatly with each respiratory cycle.

Four **respiratory capacities** (re-spi′rah-to″rē kah-pas′ĭ-tēz) can be determined by combining two or more of the respiratory volumes. Combining the inspiratory reserve volume (3,000 mL) with the tidal volume (500 mL) and the expiratory reserve volume (1,100 mL) gives the **vital capacity** (4,600 mL). This is the maximum volume of air a person can exhale after taking the deepest breath possible.

The tidal volume (500 mL) plus the inspiratory reserve volume (3,000 mL) gives the **inspiratory capacity** (3,500 mL). This is the maximum volume of air a person can inhale following a resting expiration. Similarly, the expiratory reserve volume (1,100 mL) plus the residual volume (1,200 mL) equals the **functional residual capacity** (2,300 mL). This is the volume of air that remains in the lungs following a resting expiration.

The vital capacity plus the residual volume equals the **total lung capacity** (5,800 mL). Total lung capacity varies with age, sex, and body size. Table 16.2 summarizes the respiratory air volumes and capacities.

Some of the air that enters the respiratory tract during breathing does not reach the alveoli. This volume (about 150 mL) remains in the passageways of the trachea, bronchi, and bronchioles. Because gas is not exchanged through the walls of these passages, this air is said to occupy *anatomic dead space.*

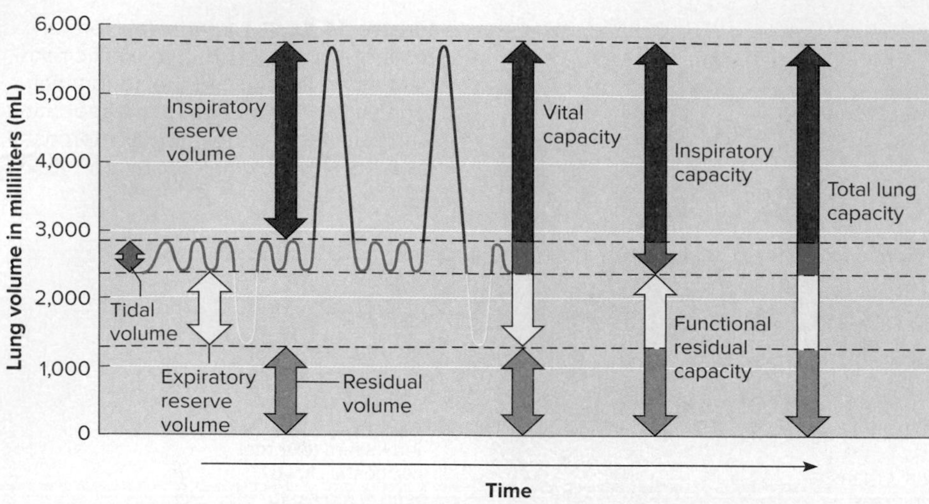

Figure 16.15 Respiratory volumes and capacities.

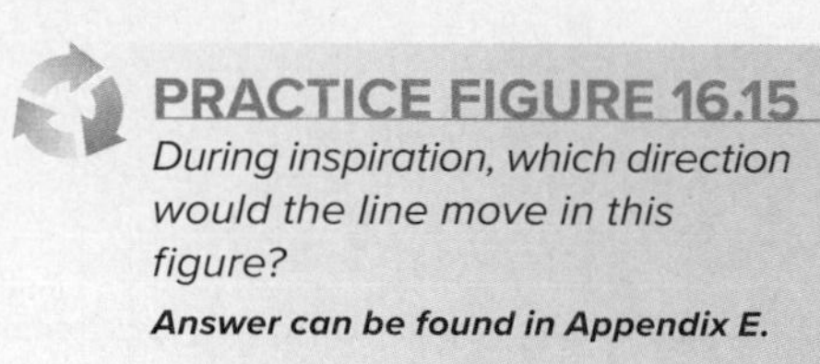

PRACTICE FIGURE 16.15

During inspiration, which direction would the line move in this figure?

Answer can be found in Appendix E.

TABLE 16.2 Respiratory Volumes and Capacities

Name	Volume*	Description
Tidal volume (TV)	500 mL	Volume of air moved in or out of the lungs during a respiratory cycle
Inspiratory reserve volume (IRV)	3,000 mL	Maximal volume of air that can be inhaled at the end of a resting inspiration
Expiratory reserve volume (ERV)	1,100 mL	Maximal volume of air that can be exhaled at the end of a resting expiration
Residual volume (RV)	1,200 mL	Volume of air that remains in the lungs even after a maximal expiration
Vital capacity (VC)	4,600 mL	Maximum volume of air that can be exhaled after taking the deepest breath possible: VC = TV + IRV + ERV
Inspiratory capacity (IC)	3,500 mL	Maximum volume of air that can be inhaled following exhalation of resting tidal volume: IC = TV + IRV
Functional residual capacity (FRC)	2,300 mL	Volume of air that remains in the lungs following exhalation of resting tidal volume: FRC = ERV + RV
Total lung capacity (TLC)	5,800 mL	Total volume of air that the lungs can hold: TLC = VC + RV

*Values are typical for a tall, young adult.

Nonrespiratory Movements

Air movements other than breathing are called *nonrespiratory movements.* They are used to clear air passages, as in coughing and sneezing, or to express emotion, as in laughing and crying.

Nonrespiratory movements usually result from *reflexes,* although sometimes they are initiated voluntarily. A *cough,* for example, can be produced through conscious effort or may be triggered by a foreign object in an air passage.

Coughing involves taking a deep breath, closing the glottis, and forcing air upward from the lungs against the closure. Then the glottis is suddenly opened, and a blast of air is forced upward from the lower respiratory tract. Usually, this rapid rush of air removes the substance that triggered the reflex.

A *sneeze* is much like a cough, but it clears the upper respiratory passages rather than the lower ones. This reflex is usually initiated by a mild irritation in the lining of the nasal cavity, and in response, a blast of air is forced up through the glottis. The air is directed into the nasal passages by depressing the uvula, closing the opening between the pharynx and the oral cavity. A sneeze can propel a particle out of the nose at 200 miles per hour.

Laughing involves taking a breath and releasing it in a series of short expirations. *Crying* consists of very similar movements. It may be necessary to note a person's facial expression to distinguish laughing from crying.

A *hiccup* is caused by sudden inspiration due to a spasmodic contraction of the diaphragm while the glottis is closed. Air striking the vocal folds generates the sound of the hiccup. The function of hiccups, if there is one, is unknown.

Yawning is familiar to everyone, yet its significance and how it is contagious remain poorly understood. Evidence points away from a role in increasing oxygen intake, as had long been thought.

PRACTICE 16.3

4. What is tidal volume?
5. Distinguish between inspiratory and expiratory reserve volumes.
6. How is vital capacity determined?
7. How is total lung capacity calculated?

16.4 | Control of Breathing

LEARN

1. Locate the respiratory areas in the brainstem and explain how they control breathing.
2. Discuss how various factors affect the respiratory areas.

Normal breathing is a rhythmic, involuntary act that continues even when a person is unconscious. The respiratory muscles, however, are also under voluntary control. (Take a deep breath and consider this!)

Respiratory Areas

Groups of neurons in the brainstem form the **respiratory areas,** which control both inspiration and expiration. The components of the respiratory areas are widely distributed throughout the pons and medulla oblongata. Two parts of the respiratory areas are of special interest: the respiratory center of the medulla and the respiratory group of the pons (fig. 16.16).

The **medullary respiratory center** includes two bilateral groups of neurons that extend throughout the length of the medulla oblongata. They are called the ventral respiratory group and the dorsal respiratory group.

Current evidence suggests that the basic rhythm of breathing arises from the *ventral respiratory group,* which stimulates the inspiratory muscles. The *dorsal respiratory group* helps process sensory information regarding the respiratory system and plays a role in certain cardiopulmonary reflexes that affect respiratory rhythm and depth. The dorsal respiratory group also stimulates the inspiratory muscles directly, primarily the diaphragm (fig. 16.16).

Neurons in another part of the brainstem, the pons, form the *pontine respiratory group* (formerly called the *pneumotaxic center*). They may contribute to the rhythm of breathing by limiting inspiration, thus affecting both respiratory rate and depth.

PRACTICE 16.4

1. Where are the respiratory areas?
2. Describe how the respiratory areas maintain a normal breathing pattern.
3. How might the breathing pattern change?

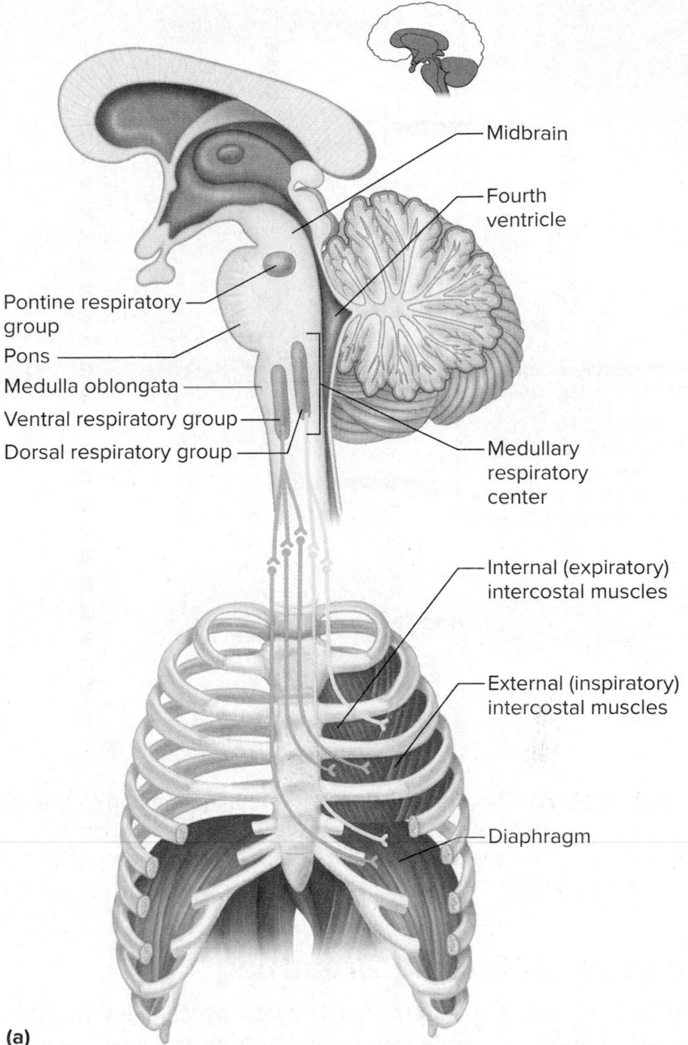

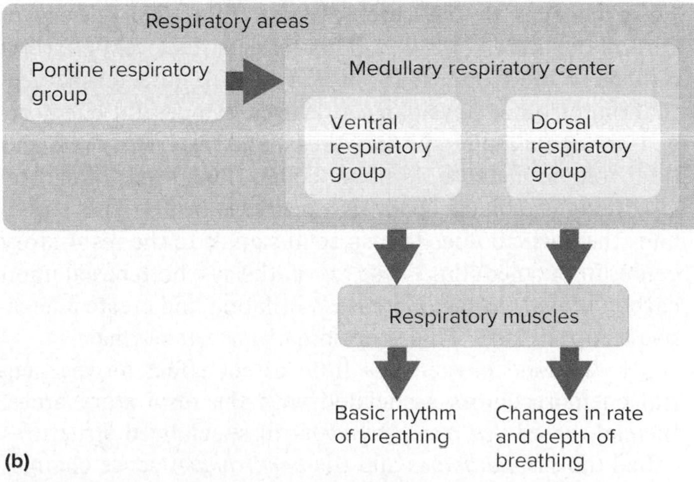

Figure 16.16 **(a)** The respiratory areas are located in the pons and the medulla oblongata. **(b)** The medullary respiratory center and the pontine respiratory group control breathing.

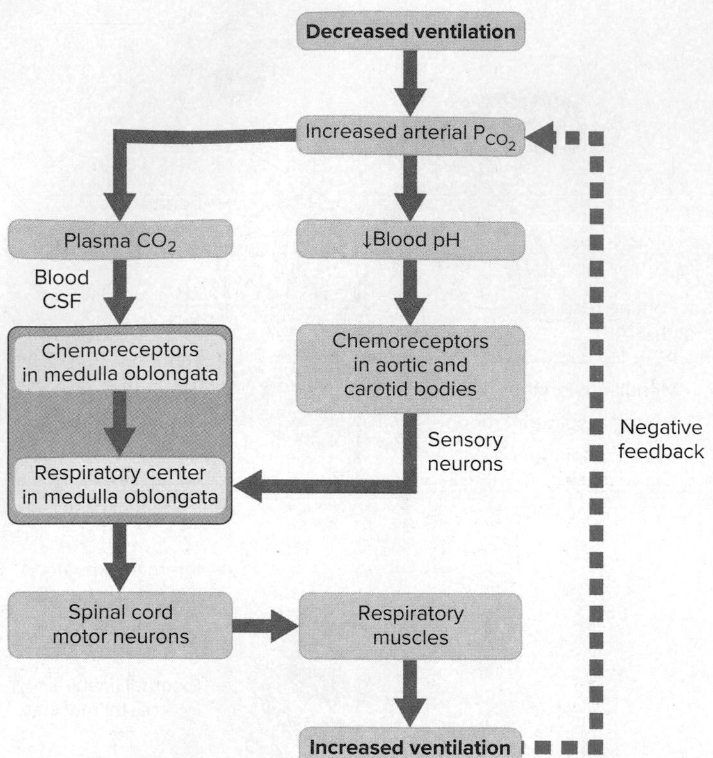

Figure 16.17 Chemoreceptors detect an increase in CO_2 and a decrease in pH of the blood and CSF when ventilation is decreased. This results in stimulation of respiratory muscles to increase ventilation.

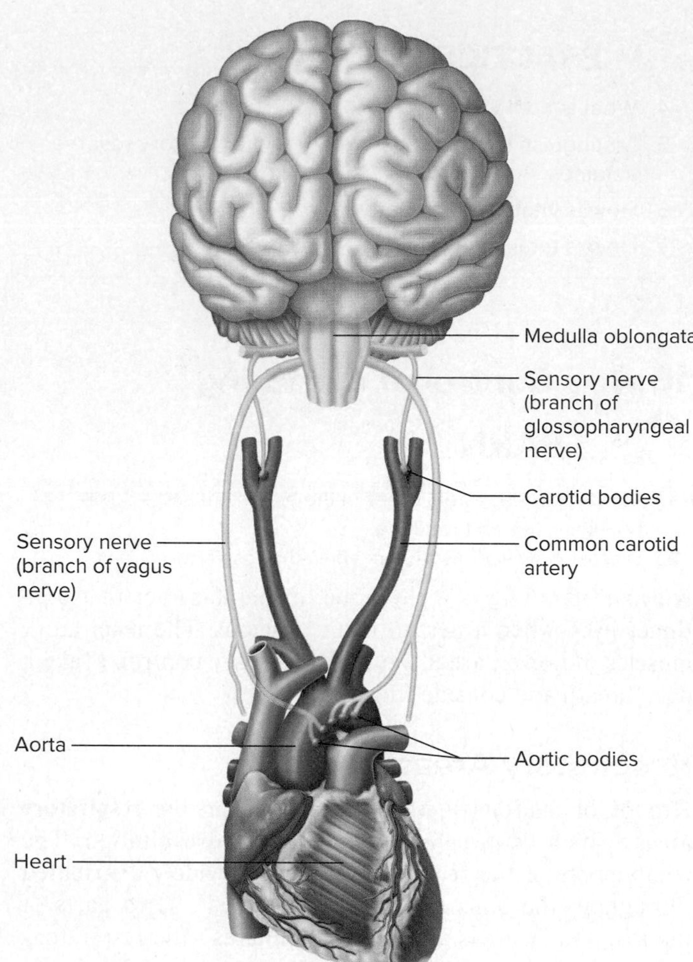

Figure 16.18 Decreased blood oxygen concentration stimulates peripheral chemoreceptors in the carotid and aortic bodies.

Factors Affecting Breathing

Breathing rate and depth are affected by changes in body fluid chemistry, the degree to which lung tissues stretch, a person's emotional state, and level of physical activity. *Chemoreceptors* in the ventral part of the medulla oblongata sense changes in the cerebrospinal fluid (CSF) levels of carbon dioxide (fig. 16.17). This reflects the level of carbon dioxide in the blood. If the level rises, from decreased ventilation, or other physiological causes, the central chemoreceptors signal the respiratory areas, and respiratory rate and tidal volume increase. Simultaneously, high levels of carbon dioxide in the blood result in a decrease in pH. This stimulates the aortic bodies to also send signals to the respiratory center in the medulla. These two pathways, both based upon carbon dioxide levels, increase ventilation and create a negative feedback loop. This keeps breathing rate normal.

Low blood oxygen has little direct effect on the central chemoreceptors associated with the respiratory areas. Instead, *peripheral chemoreceptors* in specialized structures called the *carotid bodies* and the *aortic bodies* sense changes in blood oxygen levels. Peripheral chemoreceptors are in the walls of certain large arteries (the carotid arteries and the aorta) in the neck and thorax (fig. 16.18). When decreased blood oxygen stimulates these peripheral receptors, they send impulses to the respiratory center, and breathing rate and tidal volume rise. However, blood oxygen levels must be about 50% of normal to trigger this mechanism. Thus, oxygen plays only a minor role in the control of normal respiration. For this reason, carbon dioxide, having greater control over respiration, is known as the "limiting" blood gas.

An *inflation reflex* helps regulate the depth of breathing. This reflex occurs when stretched lung tissues stimulate stretch receptors in the visceral pleura, bronchioles, and alveoli. The sensory impulses of this reflex travel via the vagus nerves to the brainstem respiratory areas and shorten the duration of inspiratory movements. This action prevents overinflation of the lungs during forceful breathing.

Emotional upset can alter the normal breathing pattern. Fear and pain typically increase the breathing rate. Clinical Application 16.4 describes the changes in breathing that occur during exercise.

A person can voluntarily stop breathing for a very short time because the respiratory muscles are voluntary. If breathing stops, blood levels of carbon dioxide rise, and oxygen levels fall. These changes (primarily the increased carbon dioxide) stimulate the chemoreceptors, and soon the urge to inhale overpowers the desire to hold the breath.

A person can increase breath-holding time by breathing rapidly and deeply in advance, exhaling more carbon dioxide

CLINICAL APPLICATION 16.4

Exercise and Breathing

Moderate to heavy physical exercise greatly increases the volume of oxygen the skeletal muscles use. For example, a young man at rest utilizes about 250 mL of oxygen per minute, but maximal exercise may require 3,600 mL of oxygen per minute.

As oxygen utilization increases, the volume of carbon dioxide produced also increases. Because decreased blood oxygen and increased blood carbon dioxide concentrations stimulate the respiratory areas, exercise would be expected to increase breathing rate. Studies reveal, however, that blood oxygen and carbon dioxide concentrations usually do not change significantly during exercise.

Sensory structures called *proprioceptors* that are associated with muscles and joints cause much of the increased breathing rate during vigorous exercise. Muscular movements stimulate proprioceptors, triggering a *joint reflex* that sends impulses to the respiratory center, increasing breathing rate.

When breathing rate increases during exercise, increased blood flow is also required to power skeletal muscles. Thus, physical exercise taxes both the cardiovascular and respiratory systems. If either of these systems does not keep pace with cellular demands, the person feels short of breath. This feeling may reflect an inability of the heart and blood vessels to move enough blood between the lungs and cells, or the respiratory system's inability to provide enough air.

than normal. This action, called **hyperventilation** (hī″per-ven″tĭ-lā′shun), lowers the blood carbon dioxide level below normal. Following hyperventilation, it takes longer for carbon dioxide to rise to the level that stimulates the respiratory areas. (*Note:* Prolonging breath-holding in this way can cause abnormally low blood oxygen levels. Use of hyperventilation to help hold the breath while swimming is potentially dangerous because the person may lose consciousness underwater and drown.)

A person who is emotionally upset may hyperventilate, become dizzy, and lose consciousness. This condition is due to a lowered carbon dioxide concentration followed by a rise in pH (alkalosis), a localized vasoconstriction of cerebral arterioles, and resulting decreased blood flow to nearby brain cells. Interference with the oxygen supply to the brain may cause fainting.

PRACTICE 16.4

4. Which factors affect breathing?
5. What is the relationship of chemoreceptors, carbon dioxide, pH, and ventilation rate?
6. Describe the inflation reflex.
7. How does hyperventilation decrease the respiratory rate?

16.5 | Alveolar Gas Exchanges

LEARN

1. Describe the structure and function of the respiratory membrane.
2. Explain how air and blood exchange gases.

The parts of the respiratory system discussed so far conduct air in and out of air passages. The alveoli carry on the vital process of exchanging gases between the air and the blood.

Alveoli

Alveoli are microscopic cavities that resemble a cluster of grapes at the end of the alveolar ducts (see fig. 16.8). Each alveolus consists of a tiny space within a thin wall that separates it from adjacent alveoli.

Respiratory Membrane

The wall of an alveolus is primarily simple squamous epithelium, along with surfactant-producing cells. In close association with an alveolus is a dense network of capillaries, which also have walls of simple squamous epithelium. The fused basement membranes of these epithelial layers join the alveolar and capillary walls. Thus, two thicknesses of epithelial cells and a layer of fused basement membranes separate the air in an alveolus from the blood in a capillary. These layers constitute the **respiratory membrane** (re-spi′rah-to″rē mem′brān) across which blood and alveolar air exchange gases (fig. 16.19).

Diffusion Across the Respiratory Membrane

Ordinary air is about 78% nitrogen, 21% oxygen, and 0.04% carbon dioxide measured by volume. Air also has traces of other gases that have little or no physiological importance.

OF **INTEREST** If all of the capillaries that surround the alveoli were unwound and laid end to end, they would extend for about 600 miles.

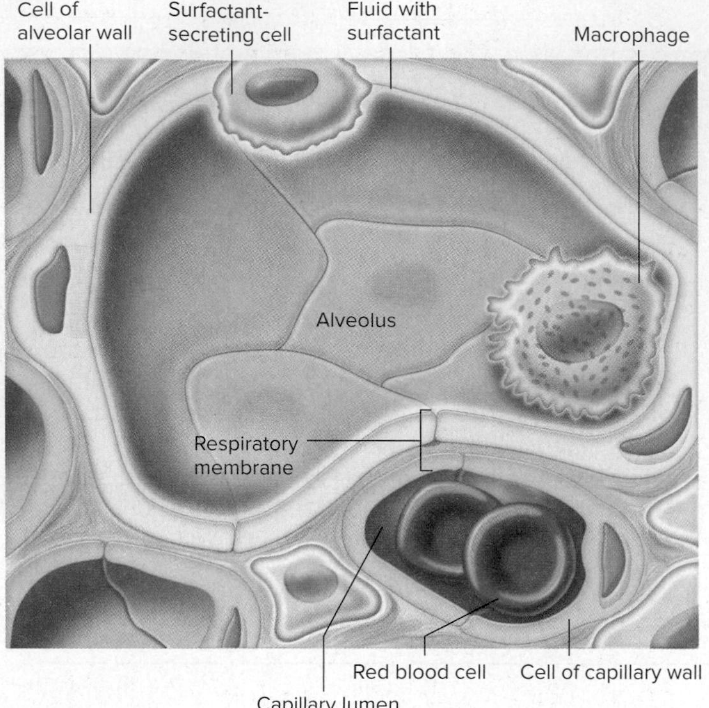

Figure 16.19 The respiratory membrane consists of the wall of the alveolus, the wall of the capillary, and their basement membranes.

In a mixture of gases such as air, each gas accounts for a portion of the total pressure the mixture produces. The amount of pressure each gas contributes is called the **partial pressure** of that gas and is proportional to its concentration. For example, because air is 21% oxygen, oxygen accounts for 21% of the atmospheric pressure (21% of 760 mm Hg), or 160 mm Hg. Thus, the partial pressure of oxygen, symbolized Po_2, in air is 160 mm Hg. Similarly, the partial pressure of carbon dioxide (Pco_2) in air is 0.3 mm Hg.

Gas molecules may enter, or dissolve in, a liquid. Air pressure is what causes the gas molecules to dissolve. This happens when carbon dioxide is added to a carbonated beverage (under artificially high pressure), or when inspired gases dissolve in the blood in the alveolar capillaries.

Recall from section 3.3, Movements Into and Out of the Cell, that molecules diffuse from regions where they are in higher concentration toward regions where they are in lower concentration. The same principle applies to diffusion of dissolved gases, but for dissolved gases, it is more useful to think of diffusion from regions of higher partial pressure toward regions of lower partial pressure. Each gas diffuses between blood and its surroundings from areas of higher partial pressure to areas of lower partial pressure until the partial pressures in the two regions reach equilibrium.

In a mixture of gases, such as room air, each gas will dissolve in a liquid according to its partial pressure gradient. Furthermore, when equilibrium is reached, the partial pressure of the dissolved gas is said to be the same as the partial pressure of that air to which it was exposed. For example, the partial pressure of oxygen dissolved in a glass of water that has been sitting on your table for a while must be close to 160 mm Hg, the same as in the air around it.

The Pco_2 in capillary blood is 45 mm Hg, but the Pco_2 in alveolar air is 40 mm Hg. Because of the difference in these partial pressures, carbon dioxide diffuses from blood, where its partial pressure is higher, across the respiratory membrane and into alveolar air (fig. 16.20). When blood leaves the alveolar capillaries, its Pco_2 is 40 mm Hg, which is the same as the Pco_2 of alveolar air with which it has reached equilibrium.

Similarly, the Po_2 of alveolar capillary blood is 40 mm Hg, but that of alveolar air is 104 mm Hg. Thus, oxygen diffuses from alveolar air into the blood, and the blood leaves the alveolar capillaries with a Po_2 of 104 mm Hg. (Because of the relatively large volume of air always in the lungs, as long as breathing continues, alveolar Po_2 stays relatively constant at 104 mm Hg. It is lower than the 160 mm Hg of inspired air primarily because of the diffusion of oxygen from the alveoli into the blood.)

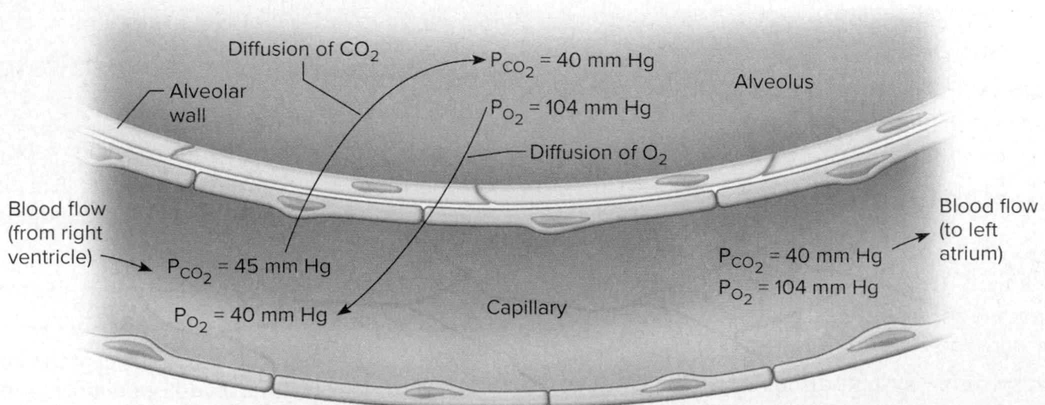

Figure 16.20 Gases are exchanged between alveolar air and alveolar capillary blood because of differences in partial pressures.

OF INTEREST The respiratory membrane is normally so thin that certain soluble chemicals other than carbon dioxide may diffuse into alveolar air and be exhaled. This is why breath analysis can reflect alcohol in the blood of a person who has been drinking, or reveal acetone on the breath of a person who has untreated diabetes mellitus. Breath analysis may also detect substances associated with kidney failure, certain digestive disturbances, and liver disease. A breath analysis is also called a "breathprint" because studies show it is as unique to an individual as a fingerprint.

PRACTICE 16.5

1. Describe the structure of the respiratory membrane.
2. What is the partial pressure of a gas?
3. Which force moves oxygen and carbon dioxide across the respiratory membrane?

16.6 | Gas Transport

LEARN

1. List the ways blood transports oxygen and carbon dioxide.

Blood transports oxygen and carbon dioxide between the lungs and the cells. As these gases enter blood, they dissolve in the liquid portion (plasma) or combine chemically with other blood components.

Oxygen Transport

Only a small percentage of the oxygen entering the blood in the alveolar capillaries is transported dissolved in the plasma. Almost all of the oxygen (over 98%) that blood transports binds the iron-containing protein **hemoglobin** (hē″mo-glō′bin) in red blood cells.

Dissolved Oxygen

If a person is breathing normal air, alveolar P_{O_2} is always higher than the P_{O_2} of the blood entering the alveolar capillaries. Thus oxygen leaves the alveoli, crosses the respiratory membrane, and enters the plasma by diffusion. The amount of oxygen that dissolves in plasma is determined by its partial pressure. When breathing air, only about 1% to 2% of the oxygen that enters the blood is transported dissolved in the plasma.

Oxyhemoglobin

Most of the oxygen entering the blood in the alveolar capillaries enters red cells and combines rapidly with the iron atoms of hemoglobin, forming **oxyhemoglobin** (ok″sĭ-hē″mo-glō′bin). Each hemoglobin molecule can bind up to four oxygen molecules (fig. 16.21). At the normal alveolar P_{O_2} of 104 mm Hg, all of the oxygen-binding sites on hemoglobin are occupied by molecules of oxygen.

Although blood leaves the alveolar capillaries with a P_{O_2} of 104 mm Hg, it enters the systemic capillaries with a P_{O_2} of 95 mm Hg. A number of factors can contribute to this difference in partial pressure. For example, some oxygen-poor systemic venous blood draining the bronchi and bronchioles mixes with pulmonary venous blood before returning to the heart. As a result, the P_{O_2} of left atrial, left ventricular, and systemic arterial blood drops to 95 mm Hg. However, even at the systemic arterial P_{O_2} of 95 mm Hg, hemoglobin is in the form of oxyhemoglobin, so systemic arterial blood is oxygen-rich.

The chemical bonds between oxygen and hemoglobin molecules can break. In the systemic circuit, where the tissue P_{O_2} is 40 mm Hg, oxyhemoglobin molecules release oxygen, which diffuses into nearby cells that have depleted their oxygen supplies in the reactions of cellular respiration (fig. 16.21). With the addition of oxygen, tissue P_{O_2} remains at 40 mm Hg even though cellular respiration continues to consume it.

Notice that oxygen-poor systemic venous blood still has 75% of its oxygen binding sites carrying oxygen (fig. 16.21). In other words, as the blood circulates through the systemic circuit, it gives up only 25% of the oxygen it is carrying. This ensures a temporary reserve supply of oxygen at times when inspired oxygen might not be available (during breath-holding, for example).

Several other factors affect how much oxygen oxyhemoglobin releases. More oxygen is released as the blood level of carbon dioxide increases, as blood becomes more acidic, or as blood temperature increases. This explains why more oxygen is released from oxyhemoglobin to skeletal muscles during physical exercise. The increased muscular activity and oxygen utilization increase carbon dioxide concentration, decrease pH, and raise temperature. Less active cells receive proportionately less oxygen.

A deficiency of O_2 reaching the tissues is called **hypoxia.** Examples include decreased arterial P_{O_2} (*hypoxemia*), diminished ability of the blood to transport O_2 (*anemic hypoxia*), and inadequate blood flow (*ischemic hypoxia*). *Histotoxic hypoxia* is a special case in which sufficient oxygen reaches the tissues, but cells are unable to use the oxygen because the pathways of aerobic respiration are blocked. An example is cyanide poisoning.

PRACTICE 16.6

1. How is oxygen transported from the lungs to cells?
2. What stimulates blood to release oxygen to tissues?

Carbon Dioxide Transport

Blood flowing through systemic capillaries gains carbon dioxide because the tissues have a relatively high P_{CO_2}. Blood transports carbon dioxide to the lungs in one of three forms: as carbon dioxide dissolved in plasma, as part of a compound formed by bonding to hemoglobin, or as a bicarbonate ion (fig. 16.22*a*).

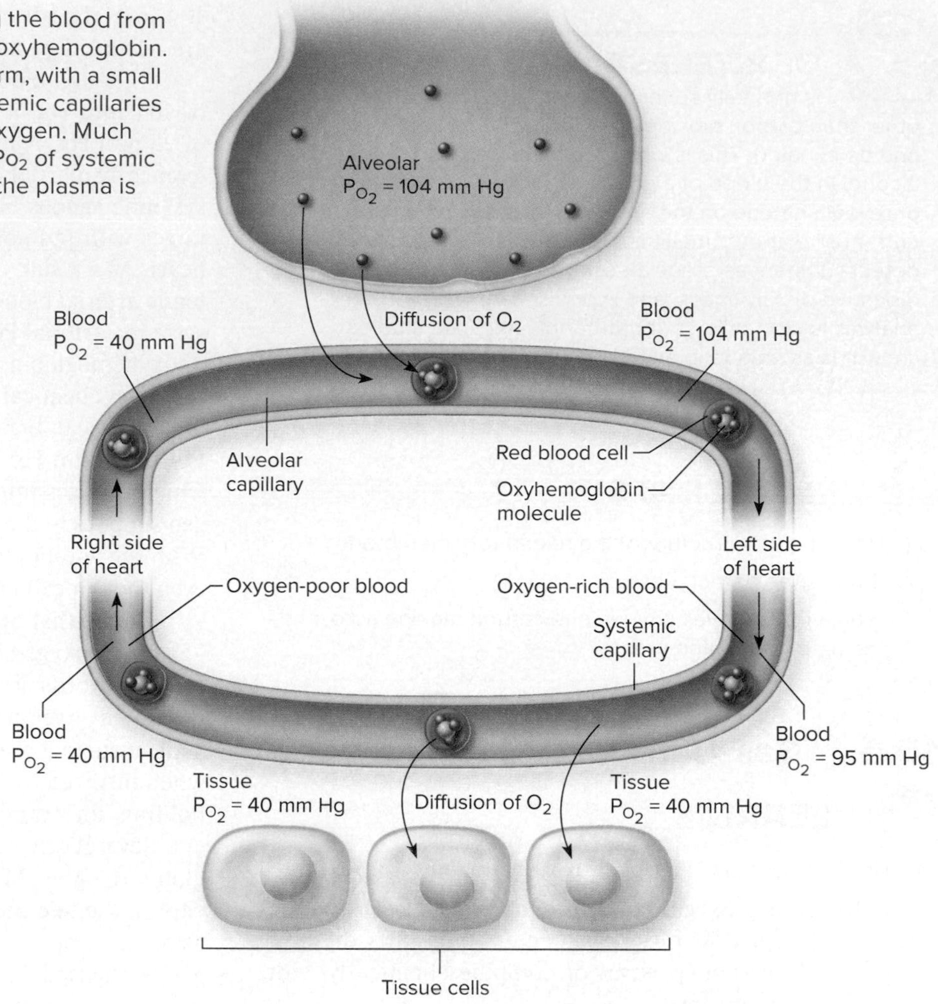

Figure 16.21 Oxygen molecules, entering the blood from the alveolus, bond to hemoglobin, forming oxyhemoglobin. Blood transports oxygen primarily in this form, with a small fraction dissolved in the plasma. In the systemic capillaries near body cells, oxyhemoglobin releases oxygen. Much oxygen is still bound to hemoglobin at the P_{O_2} of systemic venous blood. (Note: Oxygen dissolved in the plasma is not shown.) APR

Dissolved CO_2

The amount of carbon dioxide that dissolves in plasma is determined by its partial pressure. The higher the P_{CO_2} of the tissues, the more carbon dioxide will go into solution. However, only about 7% of the carbon dioxide that enters the blood dissolves in the plasma.

Carbaminohemoglobin

Unlike oxygen, which binds to the iron atoms (part of the "heme" portion) of hemoglobin molecules, carbon dioxide bonds with the amino groups ($-NH_2$) of the "globin" or protein portions of these molecules. Consequently, oxygen and carbon dioxide do not compete for binding sites, and a hemoglobin molecule can transport both gases at the same time.

Carbon dioxide bonds loosely with hemoglobin amino groups. About 23% of the carbon dioxide that enters the blood reacts with hemoglobin to form **carbaminohemoglobin** (kar-bam″ĭ-nō-hē″mo-glō′bin). This molecule decomposes readily at the low P_{CO_2} near the alveoli, releasing its carbon dioxide. The released carbon dioxide leaves the red blood cells, crosses the respiratory membrane, and enters the alveoli by diffusion (fig. 16.22*b*).

Bicarbonate Ions

The most important carbon dioxide transport mechanism forms bicarbonate ions (HCO_3^-). Carbon dioxide entering the blood reacts with water to form carbonic acid (H_2CO_3):

$$CO_2 + H_2O \longrightarrow H_2CO_3$$

This reaction occurs slowly in plasma, but much of the carbon dioxide diffuses into red blood cells. These cells have the enzyme **carbonic anhydrase** (kar-bon′ik an-hī′drās), which greatly speeds the reaction between carbon dioxide and water. The resulting carbonic acid then dissociates, releasing hydrogen ions (H^+) and bicarbonate ions (HCO_3^-):

$$H_2CO_3 \longrightarrow H^+ + HCO_3^-$$

The hemoglobin that has given up oxygen is an excellent buffer and can bind hydrogen ions inside the red blood cells. As blood passes through the systemic capillaries, most of the hydrogen ions bind to hemoglobin molecules. In contrast, the bicarbonate ions diffuse out of red blood cells and enter the plasma. Nearly 70% of the carbon dioxide that enters the blood is transported in this way.

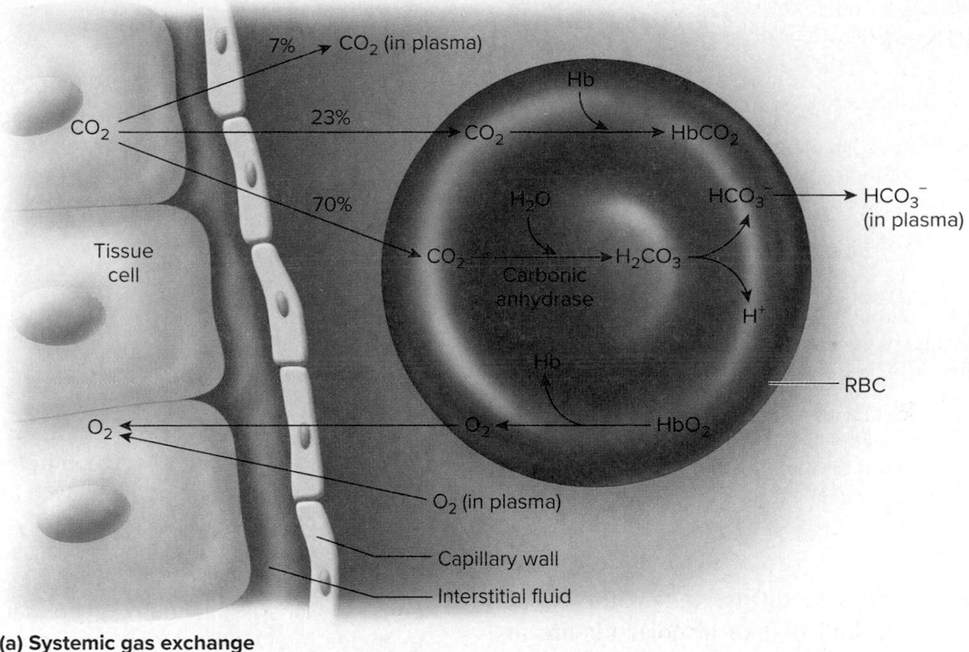

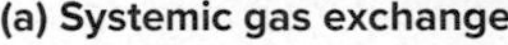
(a) Systemic gas exchange

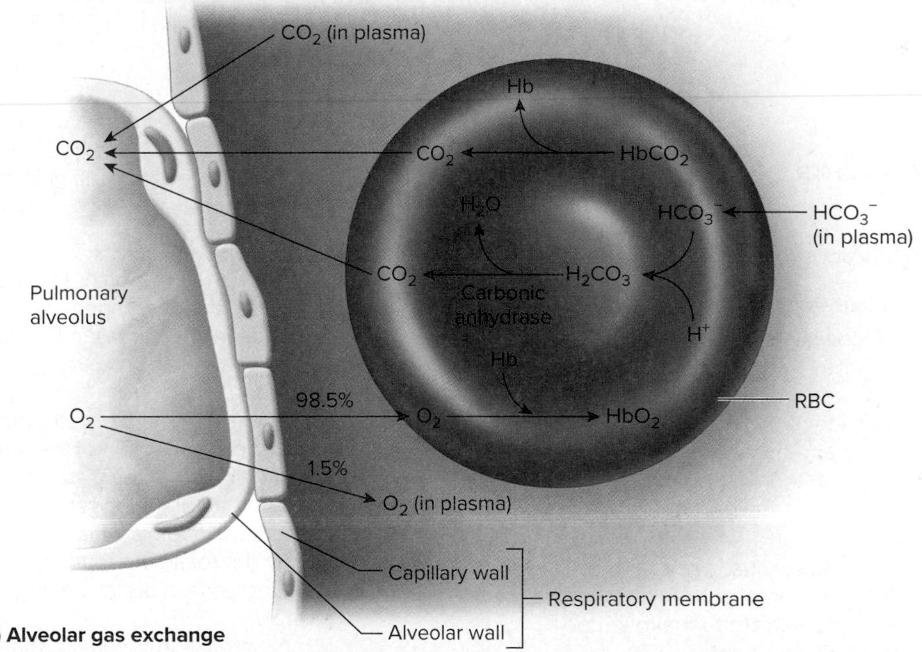

(b) Alveolar gas exchange

Figure 16.22 Transport and exchange of CO_2 and O_2. **(a)** Systemic gas exchange—O_2 diffuses into the tissue cells from the blood capillary, as CO_2 diffuses from the cells into the blood capillary to be delivered to the lungs, primarily in the form of bicarbonate ion (HCO_3^-). **(b)** Alveolar gas exchange—CO_2 is released from the blood capillary and diffuses into the alveoli to be exhaled, as inhaled O_2 diffuses from the alveoli into the blood, where it is primarily carried attached to hemoglobin (Hb) molecules in the red blood cells.

These reactions of CO_2 transport are reversible. If a person is breathing normal air, alveolar Pco_2 is always lower than the Pco_2 of the blood entering the pulmonary capillaries. Thus the dissolved carbon dioxide picked up in the systemic capillaries leaves the plasma, crosses the respiratory membrane, and enters the alveoli by diffusion (fig. 16.22*b*). At the same time, hydrogen ions and bicarbonate ions in red blood cells recombine to form carbonic acid, and under the influence of carbonic anhydrase, the carbonic acid quickly breaks down to yield carbon dioxide and water:

$$H^+ + HCO_3^- \longrightarrow H_2CO_3 \longrightarrow CO_2 + H_2O$$

The newly formed carbon dioxide crosses the respiratory membrane and enters the alveoli by diffusion (fig. 16.22*b*).

TABLE 16.3 Gases Transported in Blood

Gas	Reaction Involved	Substance Transported
Oxygen	1% to 2% dissolves in plasma; 98% to 99% combines with iron atoms of hemoglobin molecules	Oxyhemoglobin
Carbon dioxide	About 7% dissolves in plasma	Carbon dioxide
	About 23% combines with amino groups of hemoglobin molecules	Carbamino-hemoglobin
	About 70% reacts with water to form carbonic acid; the carbonic acid then dissociates to release hydrogen ions and bicarbonate ions	Bicarbonate ions

Carbon dioxide continues to diffuse out of the blood until the P_{CO_2} of the blood and that of alveolar air are in equilibrium. As long as breathing continues, the P_{CO_2} of alveolar air normally stays relatively constant at 40 mm Hg. Table 16.3 summarizes the transport of blood gases.

PRACTICE 16.6

3. Describe three forms in which blood can transport carbon dioxide from cells to the lungs.
4. How can hemoglobin carry oxygen and carbon dioxide at the same time?
5. How is carbon dioxide released from blood into the lungs?

OF INTEREST *Carbon monoxide* (CO) is a toxic gas. CO outcompetes oxygen for the binding sites on hemoglobin with the ability to bind about 200 times more tightly than oxygen. Significant exposure can result in CO poisoning, starving tissues of oxygen causing chest pain, shortness of breath, fatigue, confusion, an irregular pulse, abnormal heart rhythm, and even death. CO is produced in gasoline engines and some stoves as a result of incomplete combustion of fuels. CO is also a component of tobacco smoke.

ASSESS

CHAPTER ASSESSMENTS

16.1 Introduction

1. List the general functions of the respiratory system.

16.2 Organs and Structures of the Respiratory System

2. Which one of the following is the beginning of the lower respiratory tract?
   a. trachea b. nasal cavity
   c. pharynx d. larynx
3. Explain how the nose and nasal cavity filter the incoming air.
4. Identify the locations of the major paranasal sinuses.
5. Match the following structures with their descriptions:
   (1) true vocal cords A. serous membrane on lungs
   (2) false vocal cords B. contains the vocal cords
   (3) larynx C. vibrate to make sound
   (4) visceral pleura D. air sacs
   (5) alveoli E. muscular folds
6. Name and describe the locations of the larger cartilages of the larynx.
7. What are the structural characteristics and functions of the C-shaped rings of the trachea?

16.3 Mechanics of Breathing

8. Explain how inspiration and expiration depend on pressure changes.
9. Compare the muscles used in a resting inspiration with those in a forced inspiration.
10. Define *surface tension* and explain how it aids breathing.
11. Define *surfactant* and explain its function.
12. Compare the muscles used (if any) in a resting expiration with those in a forced expiration.
13. Distinguish between the vital capacity and the total lung capacity.

16.4 Control of Breathing

14. Describe the location of the respiratory areas and name the major components.
15. Chemosensitive areas in the medulla oblongata are most sensitive to levels of __________.
    a. nitrogen b. oxygen
    c. carbon dioxide d. sodium
16. Describe the function of the chemoreceptors in the carotid and aortic bodies.
17. Describe the inflation reflex.
18. Hyperventilation is which one of the following?
    a. any decrease in breathing
    b. a decrease in breathing that brings in oxygen too slowly
    c. an increase in breathing that eliminates carbon dioxide too quickly
    d. any increase in breathing

16.5 Alveolar Gas Exchanges

19. Define *respiratory membrane* and indicate its function.
20. Explain the relationship between the partial pressure of a gas and diffusion of that gas.
21. Summarize the exchange of oxygen and carbon dioxide across the respiratory membrane.

16.6 Gas Transport

22. Identify how blood transports oxygen.
23. List three factors that increase the release of oxygen from hemoglobin.
24. Identify the three ways blood transports carbon dioxide.

Respiratory System

INTEGUMENTARY SYSTEM

Stimulation of skin receptors may alter respiratory rate.

CARDIOVASCULAR SYSTEM

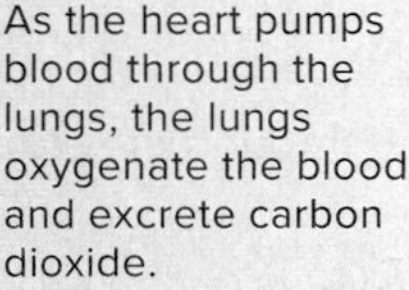

As the heart pumps blood through the lungs, the lungs oxygenate the blood and excrete carbon dioxide.

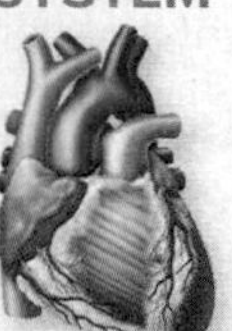

SKELETAL SYSTEM

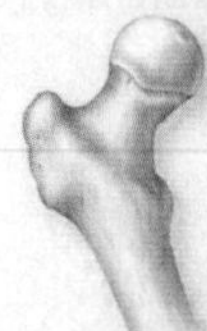

Bones provide attachments for muscles involved in breathing.

LYMPHATIC SYSTEM

Cells of the immune system patrol the lungs and defend against infection.

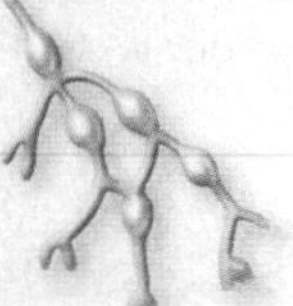

MUSCULAR SYSTEM

The respiratory system eliminates carbon dioxide produced by exercising muscles.

DIGESTIVE SYSTEM

The digestive system and respiratory system share openings to the outside.

NERVOUS SYSTEM

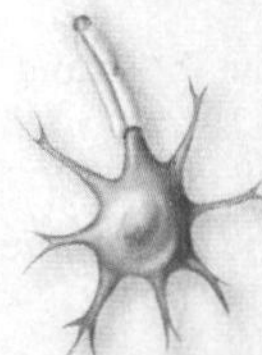

The brain controls the respiratory system.

URINARY SYSTEM

The kidneys and the respiratory system work together to maintain blood pH. The kidneys compensate for water lost through breathing.

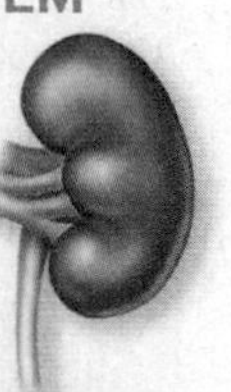

ENDOCRINE SYSTEM

Hormone-like substances control the production of red blood cells that transport oxygen and carbon dioxide.

REPRODUCTIVE SYSTEM

Respiration increases during sexual activity. Fetal gas exchange begins before birth.

The respiratory system provides oxygen for the internal environment and excretes carbon dioxide.

ASSESS

INTEGRATIVE ASSESSMENTS/CRITICAL THINKING

Outcomes 3.2, 16.2

1. Smoking causes paralysis and eventual destruction of cilia. Sherry is a chronic smoker who is often coughing up mucus. How would you explain to Sherry why she has this cough?

Outcomes 3.3, 13.2, 13.4, 16.5

2. When the left ventricle of the heart is not able to pump blood efficiently, an imbalance occurs, causing congestion in the pulmonary veins that leads through the lungs, resulting in pulmonary edema. Explain why edema occurs and its effect on the respiratory membranes. What might a patient with this condition be feeling, and why?

Outcome 16.2

3. Why does breathing through the mouth dry out the throat?

Outcomes 16.2, 16.3

4. It is below 0°F outside, but the dedicated runner bundles up and hits the road anyway. "You're crazy," shouts a neighbor. "Your lungs will freeze." Why is the well-meaning neighbor wrong?

Outcomes 16.3, 16.4, 16.5

5. When a woman is very close to delivering a baby, she may hyperventilate. Breathing into a paper bag regulates her breathing. How does this action return her breathing to normal?

Outcomes 16.3, 16.6

6. Ten-year-old Matt was bicycling with his mom and brothers while on vacation in central Oregon when Matt showed signs of breathing difficulties as he was climbing a large hill. Matt was clearly scared. When they returned from vacation, Matt's mom took him to see his pediatrician, where Matt's vital capacity (VC) was measured. Why was his VC measured, what do you predict were the results of Matt's VC, and why? Would Matt's breathing difficulties cause his blood pH to increase or decrease?

Outcomes 16.3, 16.5, 16.6

7. Why were the finishing times of endurance events rather slow at the 1968 Olympics, held in 2,200-meter-high Mexico City?

Outcomes 16.4, 16.5

8. If a person has stopped breathing and is receiving pulmonary resuscitation, would it be better to administer pure oxygen or a mixture of oxygen and carbon dioxide? Why?

Chapter Summary

16.1 Introduction

Obtaining oxygen and removing carbon dioxide are the primary functions of the ***respiratory system.*** *It includes tubes that transport air into and out of the lungs, as well as microscopic air sacs where gases are exchanged. The entire process of gas exchange between the atmosphere and cells is called* ***respiration.***

16.2 Organs and Structures of the Respiratory System

The respiratory system can be divided into two parts or tracts. The upper respiratory tract includes the nose, nasal cavity, paranasal sinuses, pharynx, and larynx; the lower respiratory tract includes the trachea, bronchial tree, and lungs.

1. **Nose**
   a. Bone and cartilage support the nose.
   b. The nostrils are openings for air.
2. **Nasal cavity**
   a. **Nasal conchae** divide the nasal cavity into passageways and help increase the surface area of the mucous membrane.
   b. The mucous membrane filters, warms, and moistens incoming air.
   c. Ciliary action carries particles trapped in mucus to the pharynx, where they are swallowed.
3. **Paranasal sinuses**
   a. The paranasal sinuses are spaces in the bones of the skull that open into the nasal cavity.
   b. Mucous membrane lines the sinuses.
4. **Pharynx**
   a. The pharynx is behind the nasal cavity, oral cavity, and larynx.
   b. It is a passageway for air and food.
5. **Larynx**
   a. The larynx conducts air and helps prevent foreign objects from entering the trachea.
   b. It is composed of muscles and cartilages and is lined with mucous membrane.
   c. The larynx contains the **vocal cords,** which vibrate from side to side and produce sounds when air passes between them.
   d. The **glottis** and **epiglottis** help prevent foods and liquids from entering the trachea.
6. **Trachea**
   a. The trachea extends into the thoracic cavity anterior to the esophagus.
   b. It divides into right and left main bronchi.
7. **Bronchial tree**
   a. The bronchial tree consists of branched air passages that lead from the trachea to the air sacs.
   b. **Alveoli** are at the distal ends of the narrowest tubes, the alveolar ducts.
8. **Lungs**
   a. The mediastinum separates the left and right lungs, and the diaphragm and thoracic cage enclose them.
   b. The **visceral pleura** attaches to the surface of the lungs. The **parietal pleura** lines the thoracic cavity.
   c. Each lobe of the lungs is composed of alveoli, blood vessels, and supporting tissues.

16.3 Mechanics of Breathing

Changes in the size of the thoracic cavity accompany inspiration and expiration.

1. **Inspiration**
   a. **Atmospheric pressure** provides the force that moves air into the lungs.
   b. Inspiration occurs when the pressure inside alveoli decreases.
   c. Pressure within alveoli decreases when the diaphragm moves downward and the thoracic cage moves upward and outward.
   d. The moist pleural membranes aid lung expansion.
2. **Expiration**
   a. Elastic recoil of tissues and **surface tension** within alveoli provide the forces of resting expiration.
   b. Thoracic and abdominal wall muscles aid forced expiration.
3. **Respiratory volumes** and **capacities**
   a. One inspiration followed by one expiration is a respiratory cycle.
   b. The amount of air that moves in (or out) during a single respiratory cycle is the **tidal volume.**
   c. Additional air that can be inhaled is the **inspiratory reserve volume.** Additional air that can be exhaled is the **expiratory reserve volume.**
   d. **Residual volume** remains in the lungs after a maximal expiration.
   e. The **vital capacity** is the maximum amount of air a person can exhale after taking the deepest breath possible.
   f. The **inspiratory capacity** is the maximum volume of air a person can inhale following exhalation of the tidal volume.
   g. The **functional residual capacity** is the volume of air that remains in the lungs after a person exhales the tidal volume.
   h. The **total lung capacity** equals the vital capacity plus the residual volume.
4. **Nonrespiratory Movements**
   a. Nonrespiratory movements usually result from reflexes.
   b. Movements include coughing, sneezing, laughing, crying, hiccupping, and yawning.

16.4 Control of Breathing

Normal breathing is rhythmic and involuntary.

1. **Respiratory areas**
   a. The respiratory areas are in the brainstem and include parts of the medulla oblongata and pons.
   b. The **medullary respiratory center** includes two groups of neurons.
      (1) The ventral respiratory group gives rise to the basic rhythm of breathing.
      (2) The dorsal respiratory group stimulates inspiratory muscles.
   c. The pontine respiratory group may contribute to the rhythm of breathing by limiting inspiration.
2. Factors affecting breathing
   a. Chemicals, stretching of lung tissues, emotional state, and exercise affect breathing.
   b. Chemosensitive areas (central chemoreceptors) are associated with the respiratory center.
      (1) CSF levels of carbon dioxide and hydrogen ions (which reflect blood levels) influence the central chemoreceptors.
      (2) Stimulation of these receptors increases breathing rate and tidal volume.
   c. Peripheral chemoreceptors are in the walls of certain large arteries.
      (1) These chemoreceptors sense low oxygen levels.
      (2) When oxygen levels are low, breathing rate increases.
   d. Overstretching lung tissues triggers an inflation reflex.
      (1) This reflex shortens the duration of inspiratory movements.
      (2) The inflation reflex prevents overinflation of the lungs during forceful breathing.
   e. **Hyperventilation** decreases blood carbon dioxide levels, but *this can be dangerous when done before swimming underwater.*

16.5 Alveolar Gas Exchanges

Gas exchange between air and blood occurs in alveoli.

1. Alveoli
   Alveoli are tiny air sacs clustered at the distal ends of alveolar ducts.
2. **Respiratory membrane**
   a. This membrane consists of alveolar and capillary walls and their basement membranes.
   b. Blood and alveolar air exchange gases across this membrane.
3. Diffusion across the respiratory membrane
   a. The **partial pressure** of a gas is proportional to the concentration of that gas in a mixture or the concentration dissolved in a liquid.
   b. Gases diffuse from regions of higher partial pressure toward regions of lower partial pressure.
   c. Carbon dioxide diffuses from blood into alveolar air. Oxygen diffuses from alveolar air into blood.

16.6 Gas Transport

Blood transports gases between the lungs and cells.

1. Oxygen transport
   a. Blood mainly transports oxygen in combination with **hemoglobin** molecules.
   b. The resulting **oxyhemoglobin** releases its oxygen in regions where the Po_2 is low.
   c. More oxygen is released as the plasma Pco_2 increases, as blood becomes more acidic, and as blood temperature increases.
2. Carbon dioxide transport
   a. Carbon dioxide may be carried dissolved in plasma, bound to hemoglobin, or as a bicarbonate ion.
   b. Most carbon dioxide is transported in the form of bicarbonate ions.
   c. The enzyme **carbonic anhydrase** speeds the reaction between carbon dioxide and water to form carbonic acid.
   d. Carbonic acid dissociates to release hydrogen ions and bicarbonate ions.

17 Urinary System

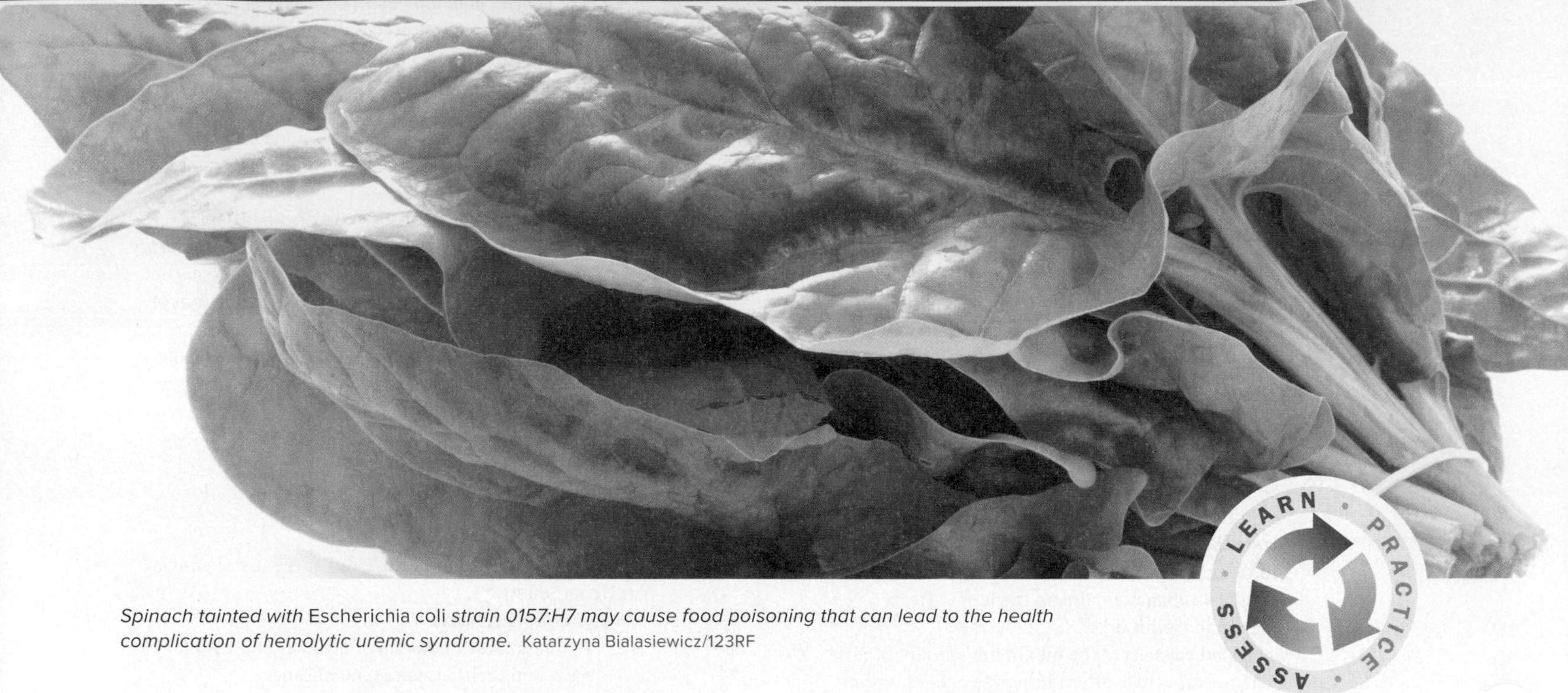

Spinach tainted with Escherichia coli *strain 0157:H7 may cause food poisoning that can lead to the health complication of hemolytic uremic syndrome.* Katarzyna Bialasiewicz/123RF

Each year, about 6,500 people in the United States, most of them young children, develop bloody diarrhea and sharp abdominal pain after eating food or drinking water tainted with a bacterial toxin. Because the condition can progress to kidney damage, rapidly identifying the source of an outbreak is important.

Escherichia coli (*E. coli*) strain 0157:H7 is a common bacterium that can cause severe illness when it produces a poison called shigatoxin. Many affected individuals seek care at emergency departments because the pain and diarrhea can be severe. For 5% to 10% of them, the condition worsens to hemolytic uremic syndrome (HUS). This complication develops as the bloodstream transports the toxin to the kidneys, where the toxin destroys the microscopic capillaries that normally prevent proteins and blood cells from being excreted. With the capillaries compromised, proteins and blood cells, as well as damaged kidney cells, appear in the urine. HUS is the main cause of acute kidney failure in children. Blood forms clots around the sites of the damaged capillaries. Usually after weeks of hospitalization the person recovers, but in some cases, the kidney damage may be permanent. For a few people, HUS is deadly.

E. coli outbreaks have occurred in a variety of circumstances involving exposure to bacteria-laden manure—such as eating undercooked beef, drinking unpasteurized apple cider, or picking up bacteria from petting farm animals. Epidemiologists used DNA tests to identify raw spinach in a smoothie that sickened a toddler. They then traced the vegetable, associated with other sick people in twenty-two states, to a single facility in California where the vegetable was contaminated with runoff from a nearby cattle ranch. In another instance, epidemiologists traced the source of infection in twenty-six people who developed HUS in Germany to contaminated sprouts served at a single restaurant.

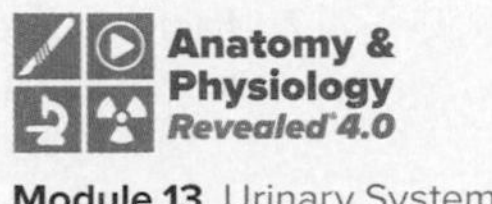

Module 13 Urinary System

LEARNING OUTLINE

After studying this chapter, you should be able to complete the "Learning Outcomes" that follow the major headings throughout the chapter.

AIDS TO UNDERSTANDING WORDS

calyc- [small cup] major *calyc*es: cuplike divisions of the renal pelvis.

cort- [covering] renal *cort*ex: shell of tissues surrounding the inner kidney.

detrus- [to force away] *detrus*or muscle: muscle within the bladder wall that expels urine.

glom- [little ball] *glom*erulus: cluster of capillaries within a renal corpuscle.

mict- [to pass urine] *mict*urition: process of expelling urine from the urinary bladder.

nephr- [pertaining to the kidney] *nephr*on: functional unit of a kidney.

papill- [nipple] renal *papill*ae: small elevations that project into a renal calyx.

trigon- [triangle] *trigon*e: triangular area on the internal floor of the urinary bladder.

(Appendix A has a complete list of Aids to Understanding Words.)

17.1 | Introduction

LEARN

1. List the general functions of the organs of the urinary system.

The urinary system consists of a pair of kidneys, a pair of tubular ureters, a saclike urinary bladder, and a tubular urethra. Acting as major contributors to homeostasis, the kidneys are vital excretory organs. They filter wastes and water to form urine and are instrumental in maintaining electrolyte balance and pH. By secreting and responding to hormones, the kidneys help with the maintenance of blood pressure and the production of red blood cells. In addition, the last step in a biochemical pathway to produce vitamin D occurs in the kidneys.

The ureters transport urine from the kidneys to the urinary bladder. The urinary bladder serves as a urine reservoir. From the bladder, urine reaches the outside of the body through the urethra. **Figure 17.1*a*** shows these organs.

PRACTICE 17.1

Answers to the Practice questions can be found in the eBook.

1. Describe the general functions of the organs of the urinary system.

17.2 | Kidneys

LEARN

1. Describe the locations and structure of the kidneys.
2. Provide the functions of the kidneys.
3. Trace the pathway of blood through the major vessels in a kidney.
4. Describe a nephron, and explain the functions of its major parts.

Location of the Kidneys

The kidneys lie on either side of the vertebral column in a depression high on the posterior wall of the abdominal cavity. The upper and lower borders of the kidneys are generally at the levels of the twelfth thoracic and third lumbar vertebrae, respectively, so they receive some protection from the lower rib cage (see fig. 17.1*a*). In most individuals, the right kidney is 1.5 to 2.0 centimeters inferior to the left kidney, providing room for the liver.

The kidneys are positioned **retroperitoneally** (ret″rō-per″ĭ-to-nē′alē), which means they are behind the parietal peritoneum and against the deep muscles of the back. Each kidney is enclosed and protected by a tough, fibrous capsule covered by a mass of fat, and an outermost fascia composed of dense fibrous connective tissue. These tissues help hold the kidneys in position (**fig. 17.1*b***).

Kidney Structure

A kidney is a reddish-brown, bean-shaped organ with a smooth surface. An adult kidney is about 12 centimeters long, 6 centimeters wide, and 3 centimeters thick. The lateral surface of each kidney is convex, while its medial side is deeply concave. The medial depression leads into a hollow chamber called the **renal sinus,** and the entrance to this sinus is the *renal hilum* where ureters, blood vessels, lymphatic vessels, and nerves enter or exit the kidney (see fig. 17.1).

When observing a frontal section of a kidney, there are two distinct regions: an inner medulla and an outer cortex. The **renal medulla** (rē′nal mĕ-dul′ah) is composed of conical masses of tissue called *renal pyramids* that look striated in appearance. The broad base of each pyramid faces the cortex, while the pointed apex points toward the inner part of the kidney. The **renal cortex** (rē′nal kor′teks), which appears

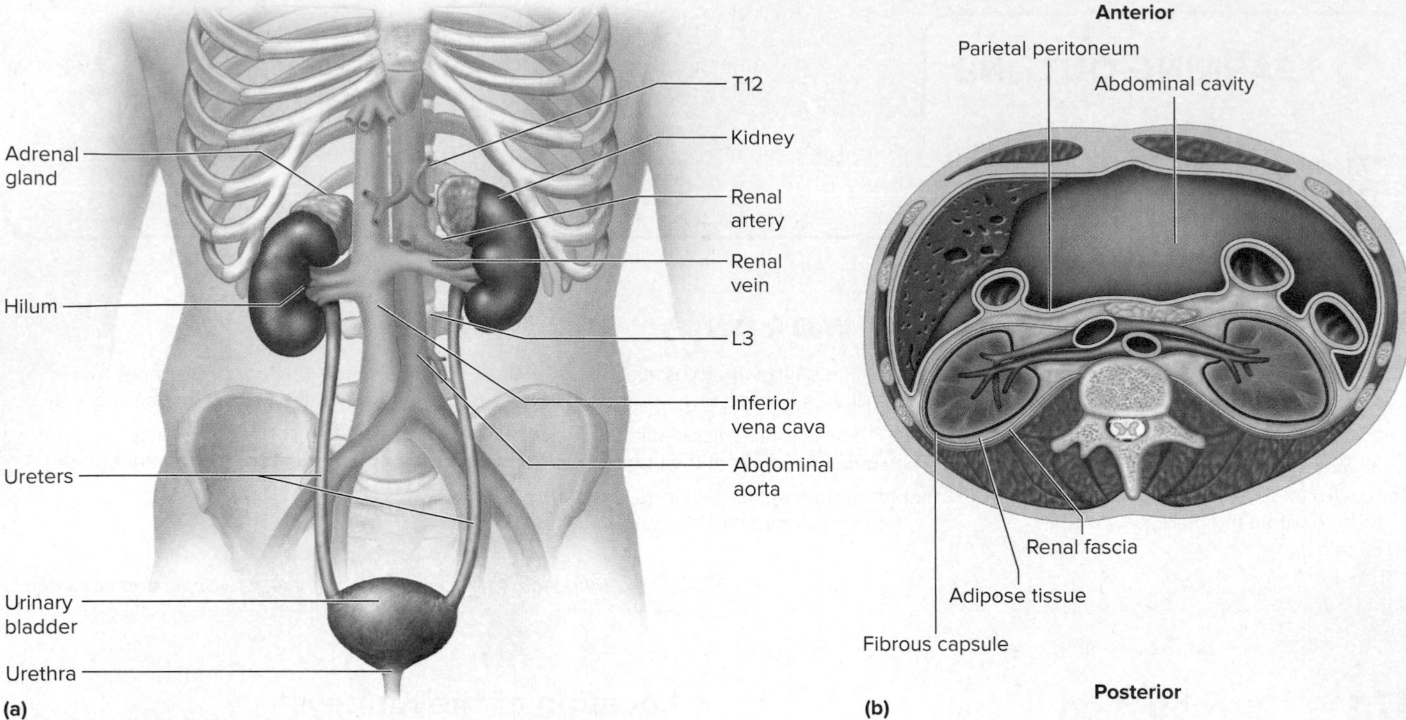

Figure 17.1 The urinary system. **(a)** Anterior view showing the kidneys, ureters, urinary bladder, and urethra. Note the relationship of these structures to blood vessels and vertebrae. **(b)** Transverse section showing the retroperitoneal position of the kidneys and surrounding supportive tissues. APR

granular, forms a shell around the renal medulla and dips into the medulla between the renal pyramids, forming *renal columns*. The granular appearance of the cortex and the striations of the medulla are due to the organization of the tiny renal corpuscles and renal tubules associated with the nephrons (nef'ronz), the functional units of the kidneys that produce urine (fig. 17.2).

Inside the renal sinus and superior to the ureter is a funnel-shaped sac called the **renal pelvis** (rē'nal pel'vis) with extending branches called *major calyces* (kal-uh-seez; singular, *calyx*) that further branch into several *minor calyces.* At the apex of the renal pyramids is a projection, called a *papilla,* with tiny openings that allow urine to drip from the collecting ducts into the minor calyces (fig. 17.2*a*). The urine flow continues into the major calyces, the renal pelvis, the ureter, the urinary bladder, the urethra, and out the body.

PRACTICE 17.2

1. Where are the kidneys located?
2. Describe kidney structure.
3. Beginning at the renal papilla, what is the direction of urine flow out the body?

Kidney Functions

The primary function of the kidneys is to help maintain homeostasis by regulating the composition (including pH) and the volume of the extracellular fluid. They accomplish

CAREER CORNER
Dialysis Technician

The eighty-three-year-old woman's kidneys have been slowly failing for several years. Now, with 90% of the tiny subunits (nephrons) of her kidney damaged, she requires hemodialysis to cleanse her blood of toxins and wastes and remove excess fluid. A nephrologist has prescribed dialysis three times a week, for three to four hours per session, to treat her chronic kidney disease.

A dialysis technician performs the tasks of hemodialysis, including:

- Setting up, cleaning, sterilizing, and using the dialysis machine.
- Preparing dialysis solutions.
- Monitoring and documenting vital signs.
- Preparing and administering medication.

Dialysis technicians work in hospitals and dialysis centers, as well as assist patients performing peritoneal dialysis in their homes. Training programs typically take two semesters, leading to a written exam for certification and national licensure. Students study anatomy and physiology, nutrition, transplantation medicine, pharmacology, health information management, medical terminology, and clinical skills related to performing dialysis.

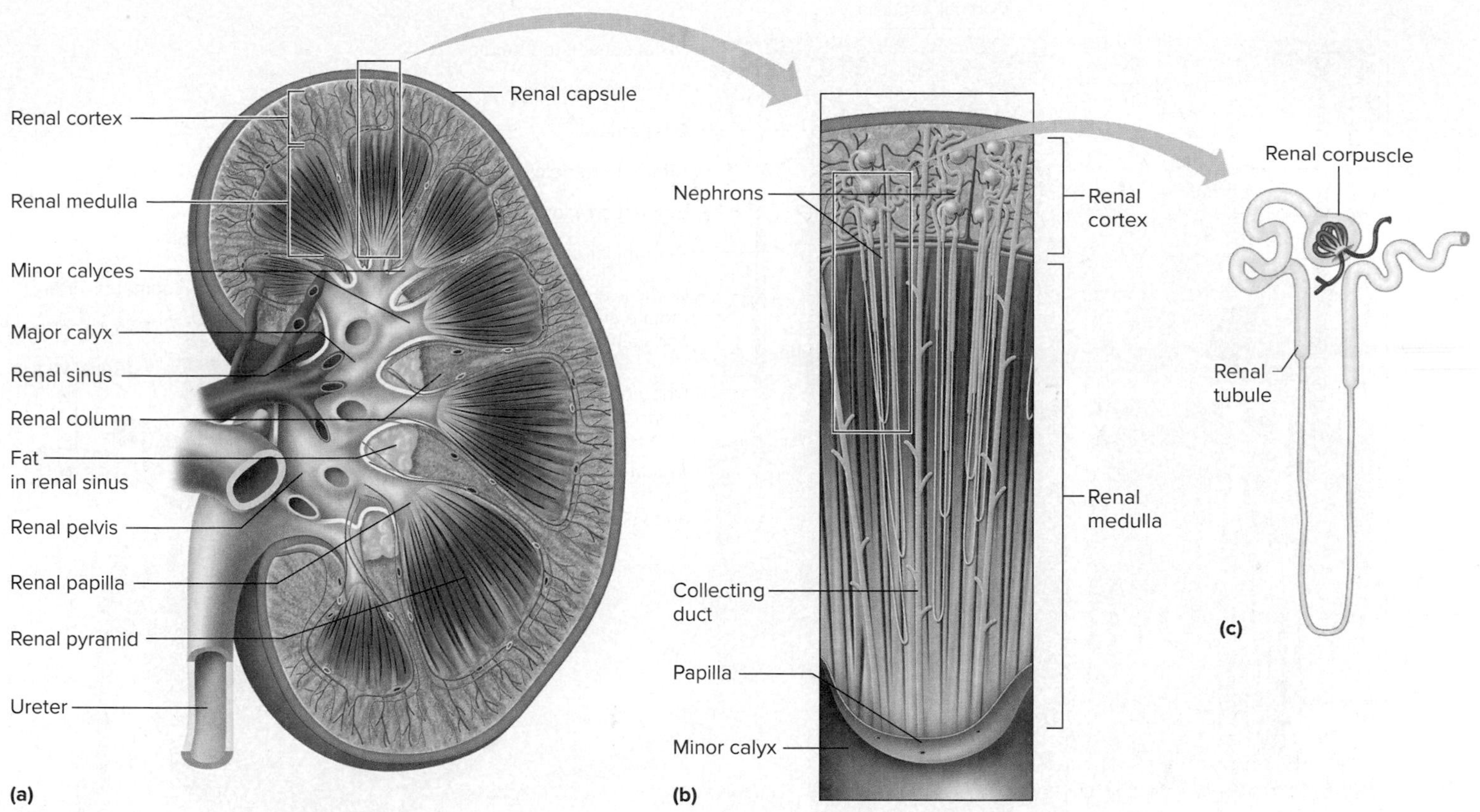

Figure 17.2 The kidney. (**a**) Longitudinal section of a kidney. (**b**) Part of a renal pyramid containing nephrons. (**c**) A single nephron. APR

this by removing metabolic wastes from the blood and combining the wastes with excess water and electrolytes to form **urine,** which they then excrete.

The kidneys have several other important functions:

- Secreting the hormone *erythropoietin* (see section 12.2, Formed Elements), which targets red bone marrow to control the rate of red blood cell production.
- Playing a role in forming the active form of vitamin D, called *calcitriol,* that acts to raise blood calcium levels when levels fall too low.
- Helping to maintain blood volume and blood pressure by secreting the enzyme *renin.*

Renal Blood Supply

The **renal arteries,** which arise from the abdominal aorta, supply blood to the kidneys. These arteries transport a large volume of blood. When a person is at rest, the renal arteries usually carry about one-fourth of the total cardiac output into the kidneys. Because the blood exchanges substances with the interstitial fluid, this large blood flow enables the kidneys to continuously process and cleanse the body fluids.

A renal artery enters a kidney through the hilum and divides into *segmental arteries* that branch into several *interlobar arteries,* which pass between the renal pyramids. At the junction between the medulla and the cortex, the interlobar arteries branch, forming a series of incomplete arches, the *arcuate arteries,* which in turn give rise to *cortical radiate arteries* that project up into the cortex. The final branches of the cortical radiate arteries, called **afferent arterioles** (af′er-ent ar-te′rē-ōlz), lead to the nephrons. Each afferent arteriole delivers blood to a mass of fenestrated blood capillaries, called the **glomerulus** (figs. 17.3, 17.4, and 17.5).

Blood that passes through the glomerulus, and is not filtered, will enter the **efferent arteriole** (ef′er-ent ar-te′rē-ōl). (Note that this is different from the typical circulatory route where blood enters a venule.) The efferent arterioles most often lead to low-pressure, porous *peritubular capillaries* surrounding the renal tubules of the nephron in the renal cortex. The peritubular capillaries deliver blood into the venous system. A small amount of blood from the efferent arterioles descends into small blood vessels called the *vasa recta* deep within the renal medulla, which are important in helping the body conserve water (figs. 17.3, 17.4, and 17.5).

Venous blood returns from the nephrons through a series of vessels that correspond generally to arterial pathways, although there are no segmental veins. The **renal vein** then joins the inferior vena cava as it courses through the abdominal cavity (see figs. 17.1 and 17.4).

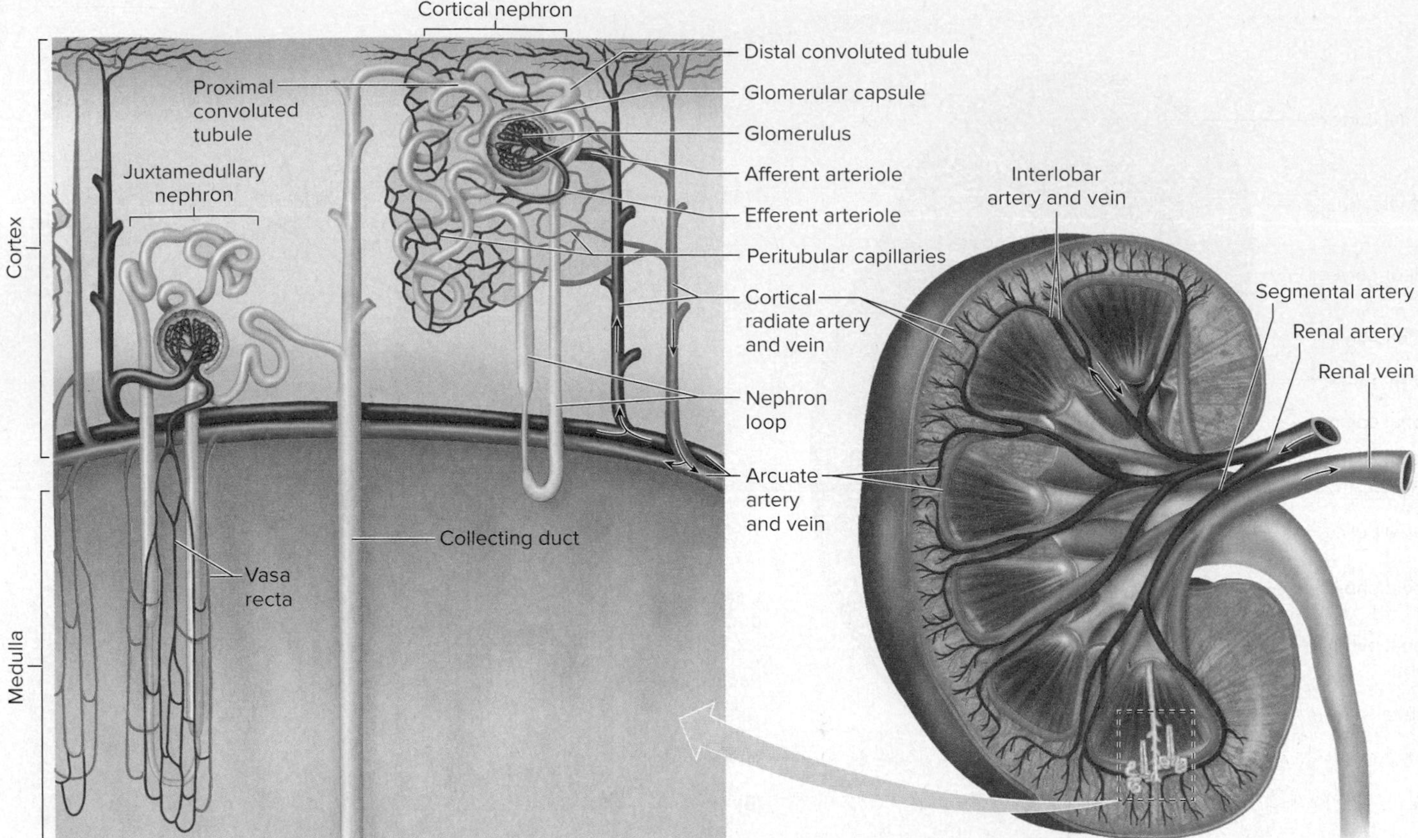

Figure 17.3 Major renal blood vessels, including a simplified representation of blood vessels associated with a typical cortical nephron and juxtamedullary nephron. APR

Figure 17.4 Flowchart of renal blood supply. Orange arrows indicate arterial supply to the nephrons; purple indicate the circulatory route of the connected capillary beds; and blue indicate the venous drainage of the kidney.

(The pathway through the vasa recta [instead of peritubular capillaries] applies only to the juxtamedullary nephrons.)

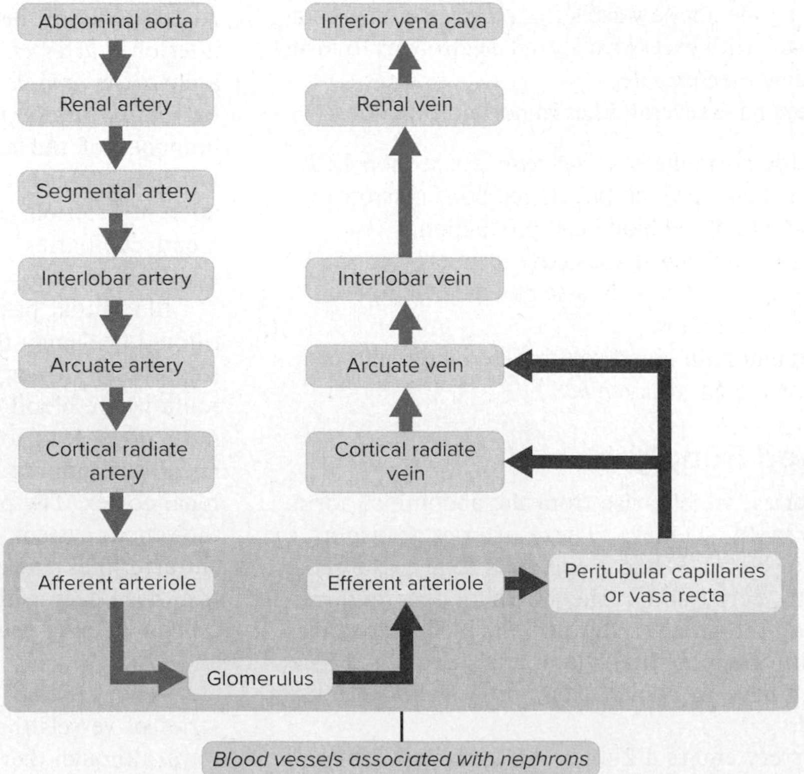

Note: Representation of renal blood flow, including nephron-associated blood vessels.

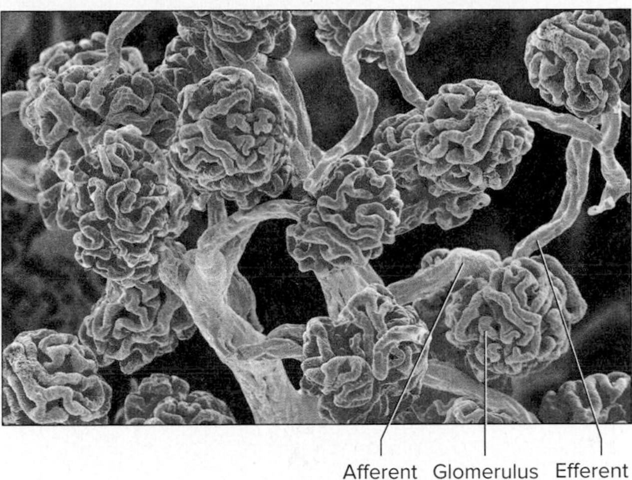

Figure 17.5 Scanning electron micrograph of a cast of the renal blood vessels associated with glomeruli (200x). From *Tissues and Organs: A Text-Atlas of Scanning Electron Microscopy,* by R. G. Kessel and R. H. Kardon, © 1979 W. H. Freeman and Company, all rights reserved.
©Susumu Nishinaga/Science Source

PRACTICE FIGURE 17.5

Which arteriole has a larger diameter, and what is the functional significance?

Answer can be found in Appendix E.

Nephrons

Nephron Structure

The functional units of the kidneys are the nephrons, meaning they perform the functions of the organ as a whole. A kidney contains about one million nephrons. Each nephron consists of a **renal corpuscle** (rē'nal kor'pusl), located in the renal cortex, and a **renal tubule** (rē'nal too'būl), with portions in both the cortex and medulla (see fig. 17.2*c*).

A renal corpuscle is composed of a tangled cluster of blood capillaries called a **glomerulus** (glo-mer'u-lus). Glomerular capillaries filter fluid and some solutes, collectively called filtrate, which is the first step in urine formation. At the beginning (proximal end) of the renal tubules is a thin-walled, saclike structure called a **glomerular capsule** (glo-mer'u-lar kap'sūl), also a component of the renal corpuscle. This capsule surrounds the glomerulus and receives the filtrate from the glomerulus (figs. 17.6*a* and 17.7). The renal tubule leads away from the glomerular capsule. The renal tubules from multiple nephrons connect to a collecting duct that terminates at the apex of the renal pyramid (fig. 17.6*b*). In sequence from the glomerular capsule, the renal tubule consists of the longer coiled *proximal convoluted tubule,* the hairpin *nephron loop,* and the smaller coiled *distal convoluted tubule* of the nephron, as well as the *collecting duct* (not considered part of a nephron) that collects fluid from many nephrons and empties the fluid as urine into a minor calyx through the opening in the renal papilla (figs. 17.2, 17.6, and 17.7). Figure 17.8 summarizes the urinary system structures and the sequence in which they function.

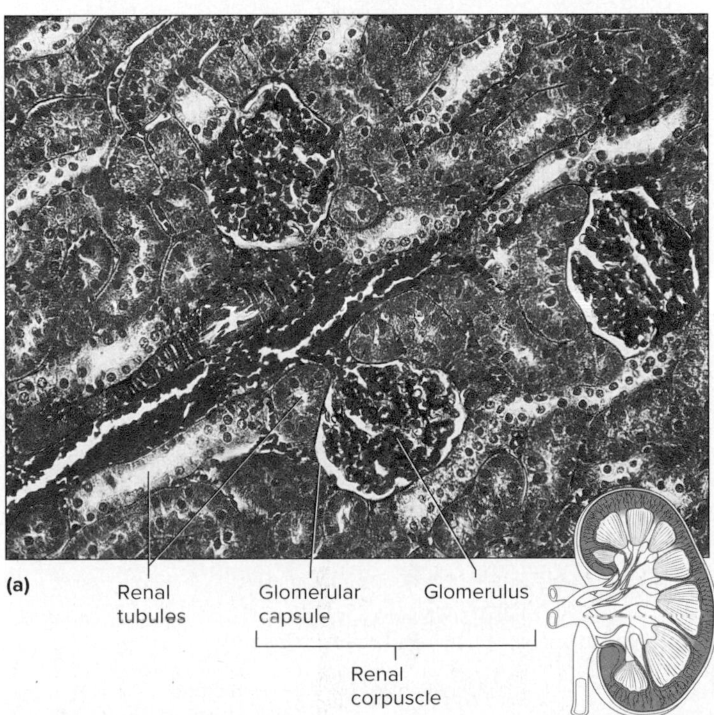

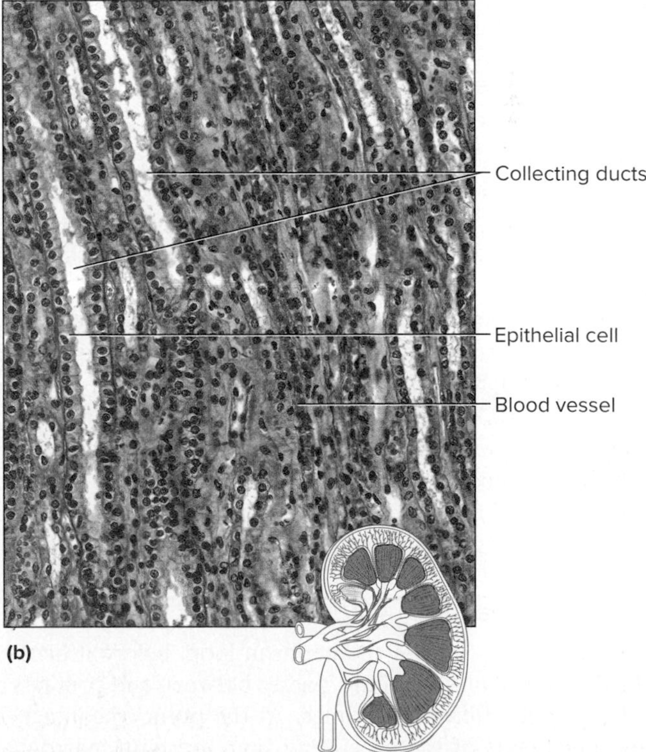

Figure 17.6 Microscopic view of the kidney. (**a**) Light micrograph of a section of the human renal cortex (220x). (**b**) Light micrograph of a section of the renal medulla (80x). APR (a): ©Manfred Kage/Science Source; (b): ©McGraw-Hill Education/Al Telser, photographer

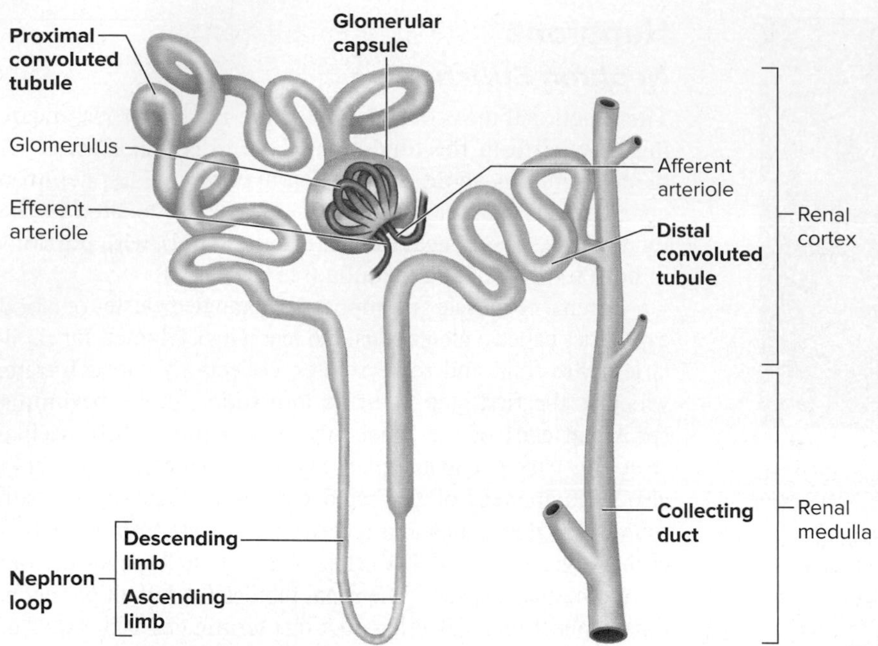

Figure 17.7 A general structure of a nephron with most of the associated blood supply removed. APR

PRACTICE 17.2

4. Explain the general functions of the kidneys.
5. Trace the renal blood supply, beginning at the abdominal aorta and ending at the inferior vena cava.
6. Name the parts of a nephron.

Types of Nephrons

There are two distinct types of nephrons—cortical nephrons and juxtamedullary nephrons. The most abundant are the **cortical nephrons** that reside almost entirely in the renal cortex. About 15% of the nephrons are the **juxtamedullary nephrons** whose renal corpuscles reside deep in the cortex and whose nephron loops project deep into the medulla. Reabsorbing most of what gets filtered, the peritubular capillaries surround the renal tubules of cortical nephrons, primarily. Surrounding the nephron loops of the juxtamedullary nephrons are the vasa recta, which are important in transporting water and solutes within the renal medulla to dilute or concentrate urine based upon the needs of the body (fig. 17.3). Both of these nephron types are innervated by renal nerves, most of which are sympathetic fibers.

Juxtaglomerular Apparatus

The ascending limb of the nephron loop, before it becomes the distal convoluted tubule, passes between and contacts the afferent and efferent arterioles. At the point of contact, the epithelial cells of the ascending limb are quite narrow and densely packed, forming a structure called the *macula densa.* These cells monitor the NaCl composition of filtrate before it enters the distal convoluted tubule.

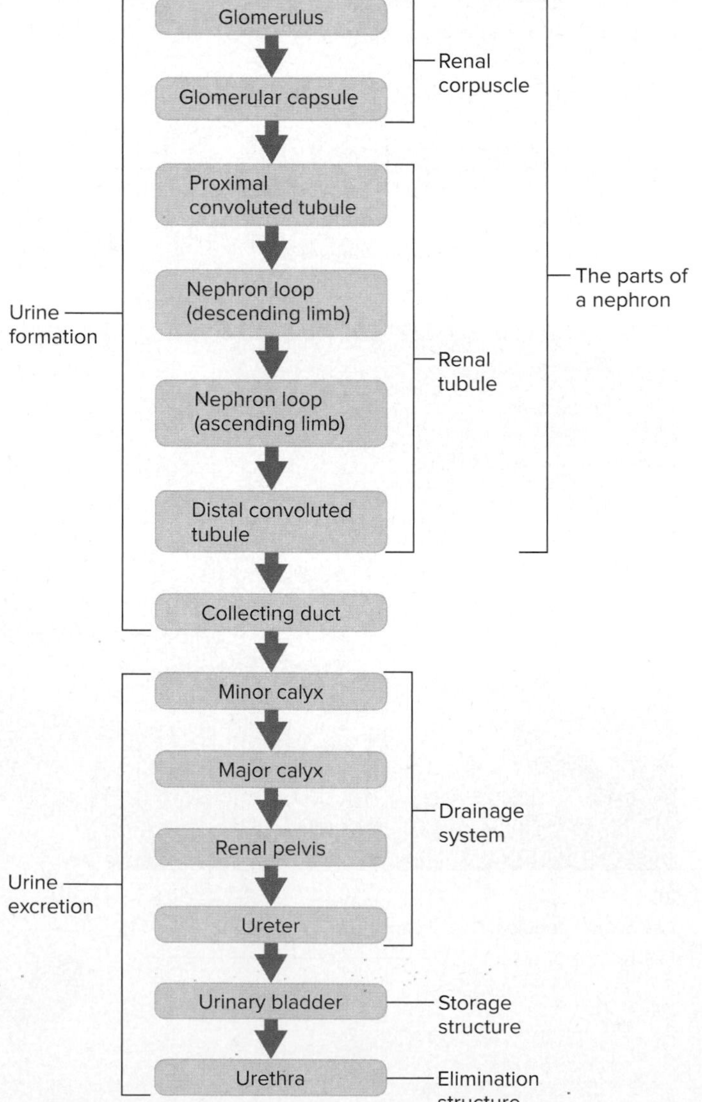

Figure 17.8 Structures associated with urine formation and excretion.

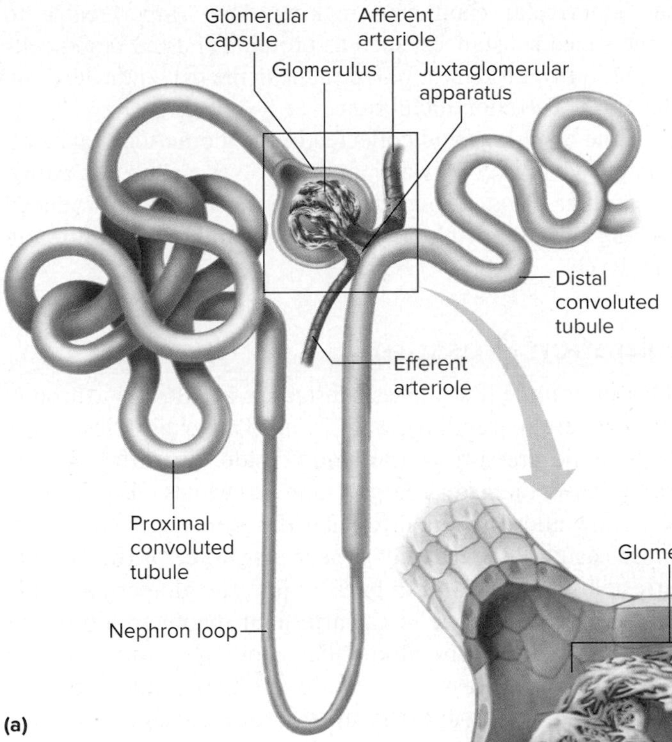

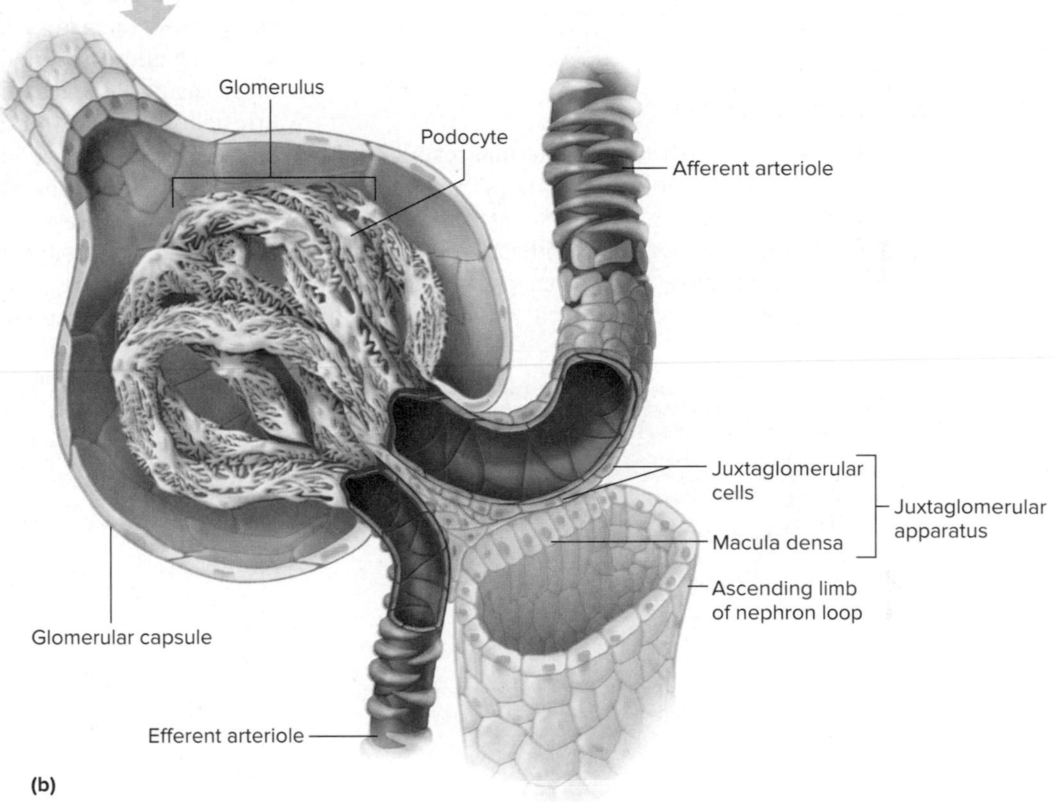

Figure 17.9 Juxtaglomerular apparatus. **(a)** Location of the renal corpuscle, renal tubules, and juxtaglomerular apparatus within a nephron. **(b)** Enlargement of a section of the juxtaglomerular apparatus, which consists of the macula densa and the juxtaglomerular cells.

In the wall of the afferent arteriole near its attachment to the glomerulus are large smooth muscle cells called *juxtaglomerular (JG) cells* that sense afferent arteriole blood pressure. The JG cells and cells of the macula densa together constitute the **juxtaglomerular apparatus** (juks″tah-glo-mer′u-lar ap″ah-ra′tus) (fig. 17.9). Its role in the control of renin secretion is described in section 17.3, Urine Formation.

PRACTICE 17.2

7. Identify the two types of nephrons and their associated blood vessels.
8. Which structures form the juxtaglomerular apparatus?

17.3 | Urine Formation

LEARN

1. Explain how glomerular filtrate is produced, and describe its composition.
2. Explain the factors that affect the rate of glomerular filtration and how this rate is regulated.
3. Discuss the role of tubular reabsorption in urine formation.
4. Define *tubular secretion,* and explain its role in urine formation.

The formation of urine involves three processes: *glomerular filtration, tubular reabsorption,* and *tubular secretion.* Recall from section 13.4, Blood Vessels, that the force of blood pressure at the arterial end promotes filtration in capillaries throughout the body, and most of this filtered fluid is reabsorbed into the bloodstream at the venular end by the colloid osmotic pressure of the plasma (fig. 17.10*a*). Filtration and reabsorption also occur in nephrons; however, these processes involve two capillary beds: the glomerulus and the peritubular capillaries. The glomerular capillaries are specialized to filter, while the peritubular capillaries reabsorb or secrete substances (fig. 17.10*b*). The final product of these three processes is *urine.* The following relationship determines the amount of any given substance excreted in the urine:

Amount *filtered* at the glomerulus
− Amount *reabsorbed* by the renal tubule (and collecting duct)
+ Amount *secreted* by the renal tubule
= Amount excreted in the urine

Glomerular Filtration

Urine formation begins when the glomerular capillaries filter water and certain dissolved substances from blood plasma into glomerular capsules, a process called **glomerular filtration** (fig. 17.11*a*). Glomerular capillaries have tiny openings (*fenestrae*) that make them more permeable than a typical capillary; however, cells called *podocytes* cover the glomerular capillaries making them impermeable to larger-sized substances, such as proteins and red blood cells (fig. 17.11*b*). Protein or blood in the urine may indicate damage to the filtration membrane.

The resulting fluid collected in the glomerular capsules, composed mostly of water and dissolved solutes, is called **glomerular filtrate.** This filtrate will continue to be produced as long as systemic blood pressure is maintained within normal levels.

Filtration Pressure

The main force that moves substances by filtration through the glomerular capillary wall, as in other capillaries, is the hydrostatic pressure of the blood inside. The afferent arterioles have diameters larger than arterioles elsewhere in the body, allowing blood to enter the glomerular capillaries more easily. The relatively greater resistance of the efferent arterioles causes blood to back up into the glomerular capillaries. This difference in the arteriolar diameters raises the blood pressure in the glomerular capillaries, thus favoring filtration. In contrast, the colloid osmotic pressure of plasma in the glomerulus and the hydrostatic pressure inside the glomerular capsule oppose this movement. An increase in either of these pressures reduces filtration.

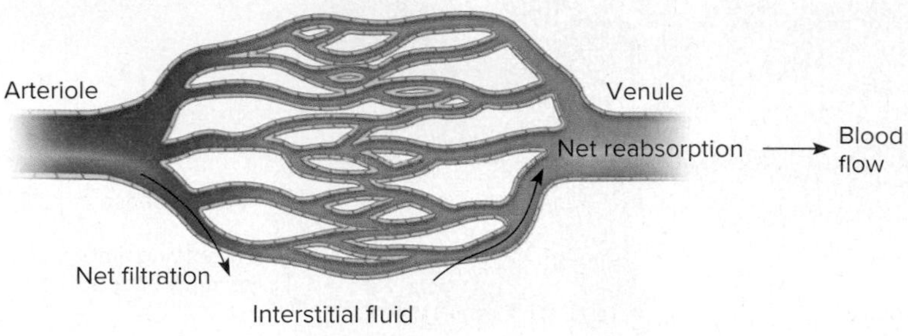

(a) In most systemic capillaries, filtration predominates at the arteriolar end and osmotic reabsorption predominates at the venular end.

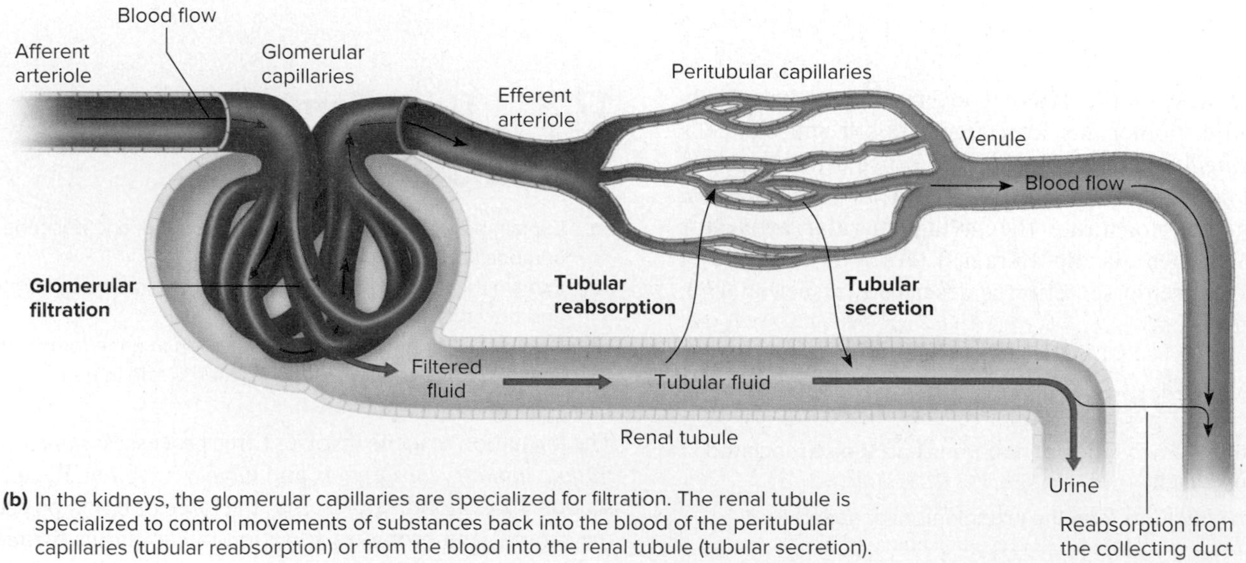

(b) In the kidneys, the glomerular capillaries are specialized for filtration. The renal tubule is specialized to control movements of substances back into the blood of the peritubular capillaries (tubular reabsorption) or from the blood into the renal tubule (tubular secretion).

Figure 17.10 A comparison of most capillary beds shown in **(a)** systemic circulation, and **(b)** two capillary beds of the nephron.

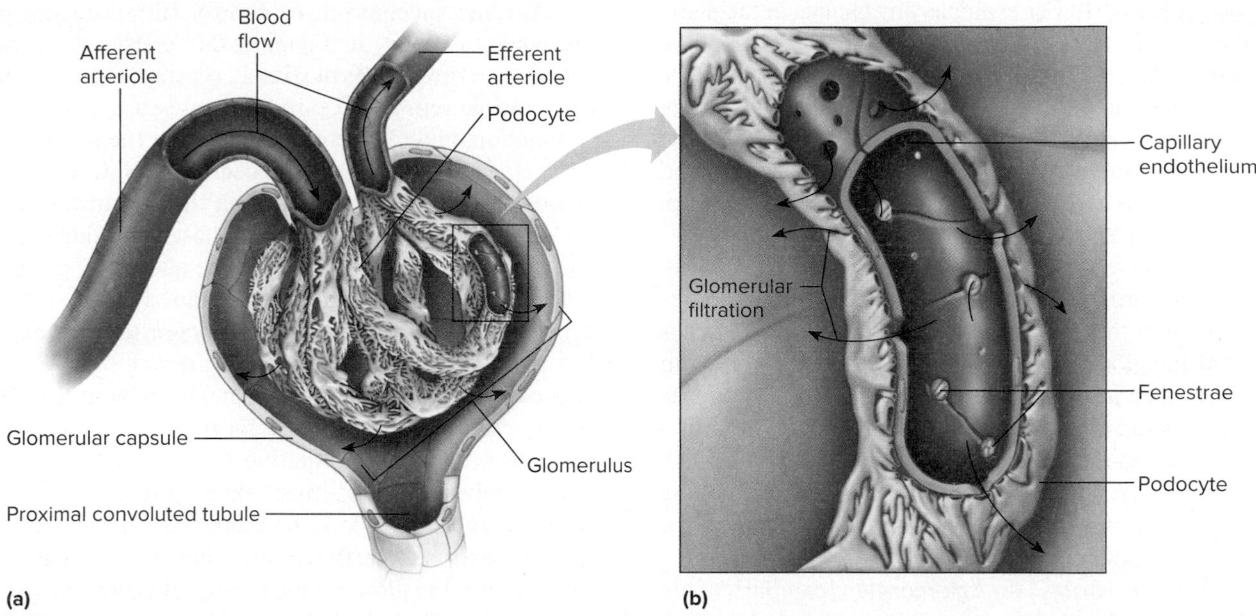

Figure 17.11 Glomerular filtration. (**a**) The first step in urine formation is filtration of substances out of glomerular capillaries and into the glomerular capsule. (**b**) Glomerular filtrate passes through the fenestrae of the capillary endothelium. (Part of the capillary and podocytes have been removed to reveal the fenestrae.)

PRACTICE FIGURE 17.11

Vasoconstriction of which arteriole in part (a) of this figure would decrease glomerular filtration?

Answer can be found in Appendix E.

The result of these forces is the **net filtration pressure.** Normally, net filtration pressure is positive throughout the glomerular capillaries, causing filtration to occur (fig. 17.12).

If arterial blood pressure drops greatly, as can occur during *shock,* glomerular hydrostatic pressure may fall below the level required for filtration. At the same time, epithelial cells of the renal tubules may not receive sufficient nutrients to maintain their high metabolic rates. As a result, tubular cells may die (tubular necrosis), impairing renal functions. Such changes can cause renal failure.

Glomerular Filtration Rate

At rest, the kidneys receive about 25% of the cardiac output, and about 20% of the blood plasma is filtered as it flows through the glomerular capillaries. This means that in an average adult, the **glomerular filtration rate (GFR)** for the nephrons of both kidneys is about 120–125 milliliters per minute, or approximately 180 liters (nearly 45 gallons) in 24 hours. Of course, only a small fraction is excreted as urine. Instead, most of the fluid that passes through the renal tubules is reabsorbed and reenters the plasma.

The GFR, the most commonly measured index of kidney function, is directly proportional to net filtration pressure. If the net filtration pressure increases, GFR goes up, and if net filtration pressure decreases, GFR goes down. Consequently, factors that affect glomerular hydrostatic pressure, glomerular plasma osmotic pressure, or hydrostatic pressure in the glomerular

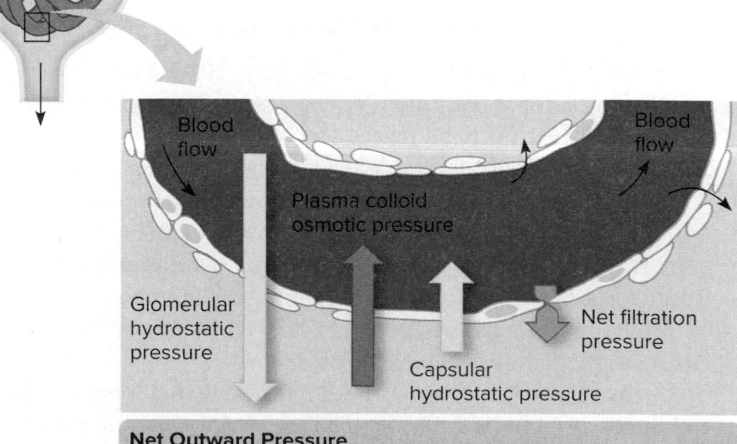

Net Outward Pressure	
Outward force, glomerular hydrostatic pressure	= +60 mm
Inward force of plasma colloid osmotic pressure	= −32 mm
Inward force of capsular hydrostatic pressure	= −18 mm
Net filtration pressure	= +10 mm

Figure 17.12 Normally, the glomerular net filtration pressure is positive, causing filtration. The forces involved include the hydrostatic and osmotic pressure of the plasma and the hydrostatic pressure of the fluid in the glomerular capsule.

capsule, also affect GFR. For example, any change in the diameters of the afferent and efferent arterioles changes net filtration pressure, also altering the GFR. If the afferent arteriole (through which blood enters the glomerulus) constricts, filtration pressure will decrease. In contrast, if the efferent arteriole (through which blood leaves the glomerulus) constricts, blood backs up into the glomerulus and filtration pressure rises. Vasodilation of these vessels produces opposite effects.

In capillaries, the plasma colloid osmotic pressure that attracts water inward (see section 13.4, Blood Vessels) opposes the blood pressure that forces water and dissolved substances outward. During glomerular filtration, proteins remaining in the plasma raise colloid osmotic pressure within the glomerular capillary. As the colloid osmotic pressure rises, net filtration pressure decreases, but the filtration pressure is normally still sufficient to maintain filtration. Conditions that decrease plasma colloid osmotic pressure, such as a decrease in plasma protein concentration, increase the filtration rate.

In *glomerulonephritis,* the glomerular capillaries are inflamed and become more permeable to proteins, which appear in the glomerular filtrate and in urine (proteinuria). At the same time, the protein concentration in blood plasma decreases (hypoproteinemia), and this decreases plasma colloid osmotic pressure. As a result, less tissue fluid moves into the capillaries throughout the body, and tissue fluid accumulates (edema).

PRACTICE 17.3

1. Which processes form urine?
2. Which forces affect net filtration pressure?
3. Which factors influence the rate of glomerular filtration?

Regulation of Filtration Rate

Controlling glomerular filtration rate (GFR) is vital in achieving proper filtration of wastes and composition of body fluids. If GFR increases too much, fluids move faster through the renal tubules, and urine output rises as it takes usable substances with it. If GFR substantially decreases, fluids move too slowly through the renal tubules, reducing urine output as more fluids and nitrogenous wastes are reabsorbed rather than eliminated. To maintain homeostasis and keep the GFR relatively constant, the GFR may be adjusted by changing glomerular blood pressure.

Sympathetic nervous system reflexes that respond to changes in blood pressure and blood volume can stimulate vasoconstriction of the afferent arterioles, and to a lesser degree, the efferent arterioles. Afferent arteriole constriction helps maintain peripheral resistance and systemic blood pressure, but lowers filtration pressure. Simultaneous contraction of the efferent arteriole counteracts this effect on filtration pressure. Thus, systemic pressure is protected and GFR remains relatively constant. Conversely, vasodilation of afferent arterioles increases the glomerular filtration rate to counter increased blood volume or blood pressure.

Another mechanism to control filtration rate involves the enzyme **renin.** Juxtaglomerular cells secrete renin in response to three types of stimuli: (1) special cells in the afferent arteriole sense a drop in blood pressure; (2) sympathetic stimulation; and (3) the macula densa of the juxtaglomerular apparatus (see fig. 17.9) senses decreased NaCl concentration of slowly moving filtrate reaching the end of the ascending limb of the nephron loop. (All of these conditions are consistent with a decrease in blood volume and blood pressure, and subsequent low GFR.) Once in the bloodstream, renin reacts with the plasma protein *angiotensinogen* to form *angiotensin I.* A second enzyme (*angiotensin-converting enzyme,* or ACE), located primarily in the lungs, quickly converts angiotensin I to the active hormone *angiotensin II.*

Angiotensin II, circulating in the bloodstream, carries out a number of actions that help maintain sodium balance, water balance, and blood pressure (fig. 17.13). Angiotensin II vasoconstricts the efferent arteriole, which causes blood to back up into the glomerulus, raising filtration pressure. This important action helps minimize the decrease in GFR when systemic blood pressure is low. Angiotensin II has a major

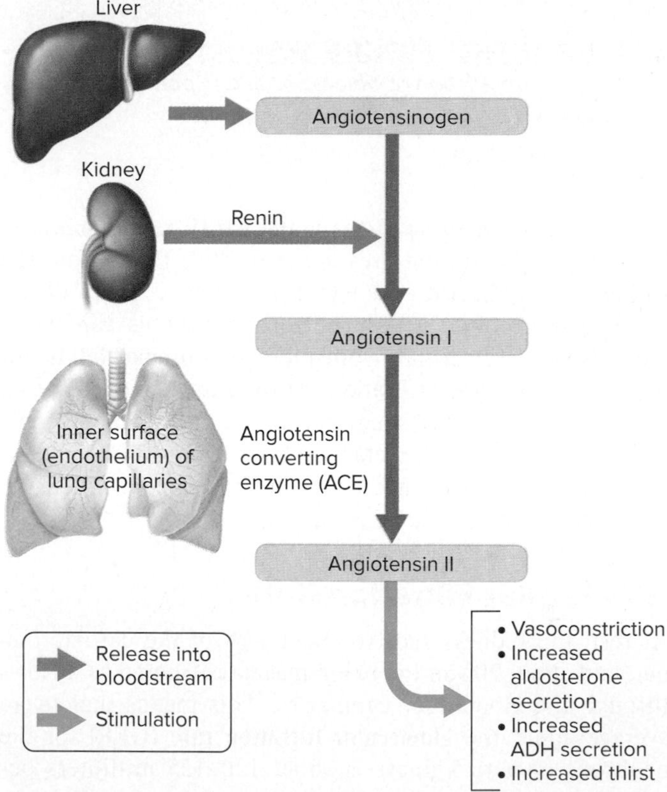

Figure 17.13 The formation of angiotensin II involves several organs and results in multiple actions that conserve sodium and water.

PRACTICE FIGURE 17.13

Would renin inhibitor *medication be given to a patient who has low blood pressure (hypotension) or high blood pressure (hypertension)?*

Answer can be found in Appendix E.

effect on the kidneys by stimulating secretion of the adrenal hormone aldosterone, which stimulates tubular reabsorption of sodium. Water follows sodium as it is reabsorbed, resulting in decreased urine output. Angiotensin II also stimulates the posterior pituitary to secrete antidiuretic hormone (ADH) that acts to retain water, and it stimulates the thirst center in the hypothalamus to encourage water ingestion. Increased water retention and intake, and reduction in water loss, helps raise blood pressure, and consequently increases GFR back to normal levels.

The heart secretes a hormone, *atrial natriuretic peptide* (ANP), when blood volume and pressure increases. ANP increases sodium and water excretion by a number of mechanisms, including increasing the glomerular filtration rate.

Elevated blood pressure (hypertension) is sometimes associated with excessive release of renin, followed by increased formation of the vasoconstrictor angiotensin II. Some patients with this form of high blood pressure may take a drug called an *angiotensin-converting enzyme inhibitor.* These "ACE inhibitors" prevent the formation of angiotensin II by inhibiting the action of the enzyme that converts angiotensin I into angiotensin II.

PRACTICE 17.3

4. What is the function of the macula densa?
5. How does renin help regulate glomerular filtration rate when blood pressure drops?

Tubular Reabsorption

There are substances that were nonselectively filtered through the glomerulus into the glomerular capsule and renal tubules that are useful to the body, such as water, electrolytes, amino acids, and glucose, so they should be returned to the blood rather than eliminated in urine. The process of selectively moving substances from the tubular fluid back into the blood within the peritubular capillaries is called **tubular reabsorption.** In this process, substances are transported out of the tubular fluid, through the epithelium of the renal tubule, and into the interstitial fluid, where they then diffuse into the peritubular capillaries (fig. 17.14). Thus, tubular reabsorption returns substances to the internal environment.

A number of factors contribute to tubular reabsorption by enhancing the rate of fluid movement from the interstitial fluid into the peritubular capillaries. Peritubular capillary blood is under relatively low pressure because it has already passed through two arterioles. Also, the walls of the peritubular capillaries are more permeable than other capillaries. Finally, because so much fluid is lost through glomerular filtration, the plasma protein concentration, and therefore, the colloid osmotic pressure, in the peritubular capillaries is relatively high.

Movement across the tubule cell membranes occurs passively and actively. Water moves passively by osmosis and may carry dissolved solutes with it. Substances moving down a

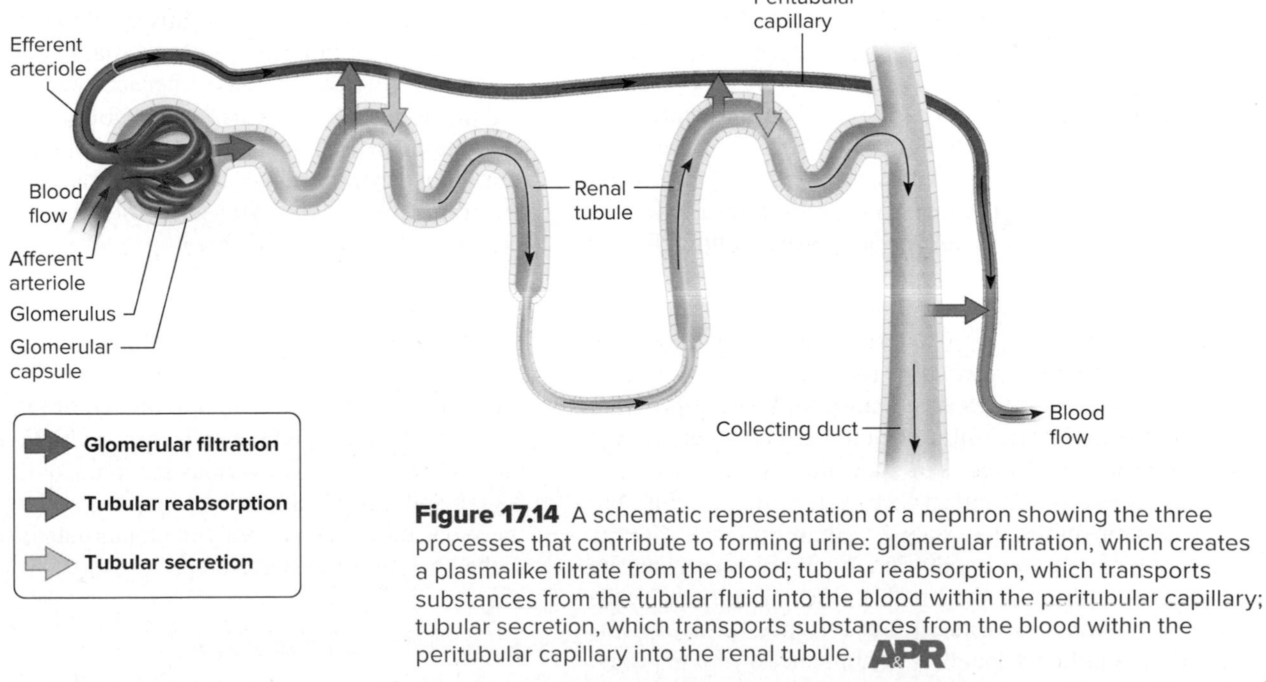

Figure 17.14 A schematic representation of a nephron showing the three processes that contribute to forming urine: glomerular filtration, which creates a plasmalike filtrate from the blood; tubular reabsorption, which transports substances from the tubular fluid into the blood within the peritubular capillary; tubular secretion, which transports substances from the blood within the peritubular capillary into the renal tubule. APR

PRACTICE FIGURE 17.14

Which of the three processes (glomerular filtration, tubular reabsorption, tubular secretion), if increased for a substance, would reduce urinary excretion of that substance?

Answer can be found in Appendix E.

concentration gradient across a cell membrane must be lipid-soluble, or there must be a specific protein carrier or channel for that substance. Active transport, requiring ATP, uses membrane carriers to selectively move substances against their concentration gradients. It makes sense that there are many carriers for those substances that the body needs to keep, while little to no carriers for those substances that are useless.

Most tubular reabsorption takes place in the proximal convoluted tubule. The epithelial cells here have many microscopic projections called *microvilli* that form a "brush border" on their free surfaces. These tiny extensions greatly increase the surface area exposed to glomerular filtrate and enhance reabsorption.

Segments of the renal tubule are adapted to reabsorb specific substances, using particular modes of transport that include active transport, diffusion, and osmosis. Active transport, for example, reabsorbs sodium ions and glucose through the walls of the proximal convoluted tubule. Water is then reabsorbed by osmosis. However, parts of the distal convoluted tubule and the collecting duct may be almost impermeable to water. This is a characteristic important in regulating urine concentration and volume, described later in this section.

Active transport utilizes membrane protein carriers that transport molecules across the membrane, release them, and then repeat the process (see section 3.3, Movements Into and Out of the Cell). This mechanism of transport has a *limited transport capacity,* so only a certain number of molecules can be transported in a given time, called **transport maximum** (T_m). For example, glucose carrier proteins are usually able to transport all of the glucose in glomerular filtrate; however, when blood glucose concentration exceeds a critical level, called the *renal plasma threshold,* where all carriers are saturated, more glucose molecules will be in the filtrate than can be reabsorbed. Consequently, the remaining glucose in the tubular fluid will be excreted in urine. This explains why the elevated blood glucose (hyperglycemia) of uncontrolled diabetes mellitus results in glucose in the urine (glucosuria). Because water follows higher solute concentration, more water will be eliminated with the glucose, a process called *osmotic diuresis.*

Amino acids enter the glomerular filtrate and are reabsorbed in the proximal convoluted tubule. Three active transport mechanisms reabsorb different groups of amino acids whose members have similar structures. Normally, only a trace of amino acids remains in the urine.

The glomerular filtrate is nearly free of proteins, but smaller protein molecules, such as albumins, may squeeze through the glomerular capillaries. They are then taken up by endocytosis through the brush border of epithelial cells lining the proximal convoluted tubule. Once these proteins are inside an epithelial cell, they are broken down into amino acids, which then move into the blood of the peritubular capillary.

The epithelium of the proximal convoluted tubule reabsorbs other substances, including lactic, citric, uric, and ascorbic (vitamin C) acids; and phosphate, sulfate, calcium, potassium, and sodium ions. Active transport mechanisms with limited transport capacities reabsorb these chemicals. These substances usually do not appear in urine until the glomerular filtrate concentration exceeds a particular substance's threshold.

Sodium and Water Reabsorption

Approximately 70% of sodium ions are reabsorbed by active transport in the proximal convoluted tubule. The reabsorption of sodium (Na^+) by primary active transport provides the energy and the mechanism to allow the reabsorption of most nutrients, such as glucose, amino acids, some ions, and vitamins, as well as water. The gradient created by the Na^+-K^+ pumps allows these substances to be reabsorbed by secondary active transport, so as Na^+ diffuses down its concentration gradient ("downhill"), these substances are cotransported, or carried by facilitated diffusion. For example, glucose is cotransported with sodium using carrier proteins called sodium-glucose transporters (SGLTs).

Sodium reabsorption creates an electrical gradient and an osmotic gradient that drives the reabsorption of other solutes and water. When the positively charged sodium ions (Na^+) are moved through the tubular wall, negatively charged ions, including chloride ions (Cl^-), accompany them. This movement of negatively charged ions is due to the electrochemical attraction between particles of opposite electrical charge. Although dependent on the active transport of sodium, it is considered a passive process because it does not require direct expenditure of cellular energy (ATP).

As more sodium ions are reabsorbed into the peritubular capillaries along with negatively charged ions, the concentration of sodium in the peritubular capillaries might be expected to increase; however, because water moves through cell membranes by osmosis from regions of lesser solute concentration (hypotonic) toward regions of greater solute concentration (hypertonic), water is also reabsorbed as it follows the reabsorbed solute. Movement of solutes and water into the peritubular capillary greatly reduces the fluid volume within the renal tubule. By the end of the proximal convoluted tubule, the tubular fluid is in osmotic equilibrium with blood plasma, and is therefore isotonic (fig. 17.15).

Active transport continues to reabsorb sodium ions as the tubular fluid moves through the nephron loop, the distal convoluted tubule, and the collecting duct. Water is absorbed passively by osmosis in various segments of the renal tubule. As a result, almost all the sodium ions and water that enter the renal tubule as part of the glomerular filtrate are reabsorbed before urine is excreted.

PRACTICE 17.3

6. Which chemicals are normally present in the glomerular filtrate but not in urine?
7. Which mechanisms reabsorb solutes from the glomerular filtrate?
8. Describe the role of sodium reabsorption in urine formation.

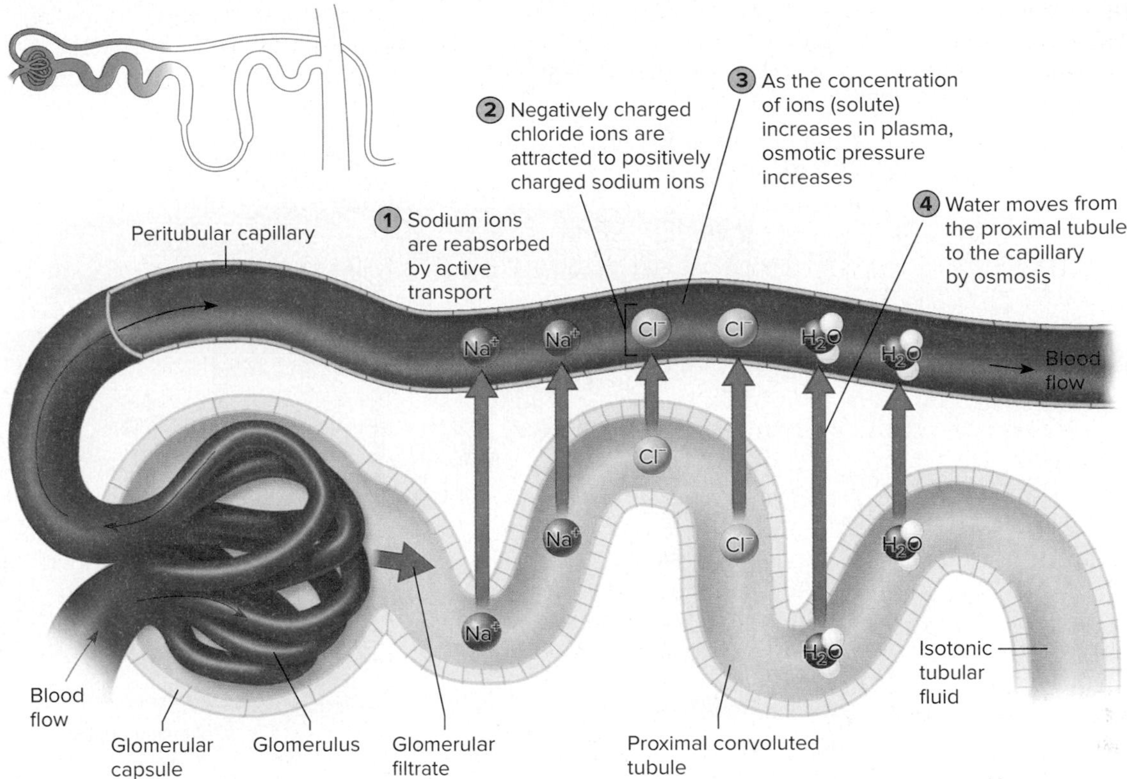

Figure 17.15 In the proximal portion of the renal tubule, osmosis reabsorbs water in response to reabsorption of sodium and chloride ions.

Tubular Secretion

Tubular secretion is essentially the reverse of tubular reabsorption. About 20% of the plasma flowing through the kidneys is filtered in the glomeruli, and approximately 80% of the plasma escapes filtration and continues on through the peritubular capillaries; however, there are substances that still need to be cleared from the body. In tubular secretion, certain substances move from the blood plasma in the peritubular capillary into the tubular fluid. As a result, the amount of a particular chemical excreted in the urine may exceed the amount filtered from the plasma in the glomerulus (see fig. 17.14).

Active transport mechanisms similar to those that function in tubular reabsorption secrete some substances. For example, the epithelium of the proximal convoluted tubule actively secretes certain organic compounds, including histamine and the drug penicillin, into the tubular fluid.

Hydrogen ions (H^+) are also actively secreted throughout the entire renal tubule. As sodium ions move into the tubular cell to be reabsorbed, hydrogen ions are transported in the opposite direction into the tubular fluid to be excreted in urine. Secretion of hydrogen ions is important in regulating the pH of body fluids, as section 18.5, Acid-Base Balance, explains.

Most potassium ions (K^+) present in the glomerular filtrate are actively reabsorbed from the tubular fluid in the proximal convoluted tubule, but some may be secreted in the distal convoluted tubule and collecting duct. During this process, active reabsorption of sodium ions from the tubular fluid results in a negative electrical charge within the tubule. Because positively charged potassium ions (K^+) are attracted to negatively charged regions, these ions move passively through the tubular epithelium and enter the tubular fluid (fig. 17.16). Potassium ions are also secreted by active processes.

To summarize, urine forms as a result of the following:

- Glomerular filtration of materials from blood plasma into the glomerular capsule.
- Reabsorption of substances from the tubular fluid into the blood plasma, including glucose; water; creatine; amino acids; lactic, citric, and uric acids; and phosphate, sulfate, calcium, potassium, and sodium ions.
- Secretion of substances from the blood plasma into the tubular fluid, including penicillin, histamine, phenobarbital, hydrogen ions, ammonia, and potassium ions.

PRACTICE 17.3

9. Define *tubular secretion*.
10. Which substances are actively secreted?
11. How does sodium reabsorption affect potassium secretion?

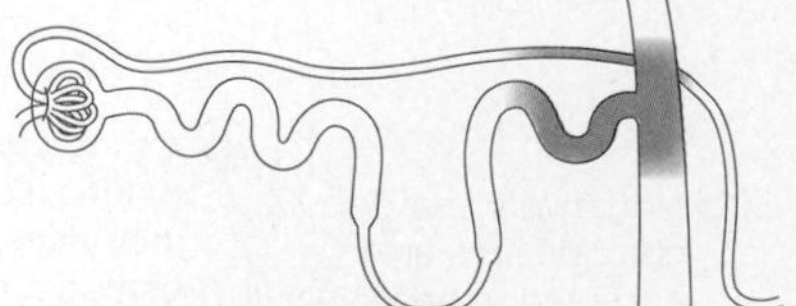

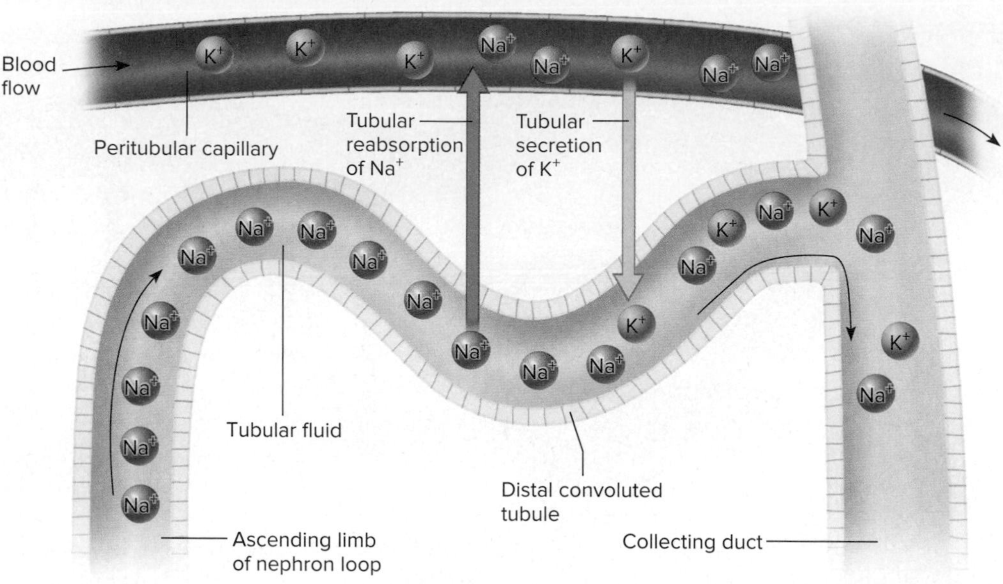

Figure 17.16 In the distal convoluted tubule, potassium ions may be passively secreted in response to the active reabsorption of sodium ions.

Regulation of Urine Concentration and Volume

The hormones aldosterone and ADH (antidiuretic hormone) may stimulate additional reabsorption of sodium and water, respectively. Atrial natriuretic peptide (ANP) works in opposition to aldosterone and ADH. The changes in sodium and water excretion in response to these hormones are the final adjustments the kidney makes to maintain a constant internal environment.

Aldosterone, the "salt-retaining hormone," is secreted by cells of the adrenal gland when blood sodium concentration falls or if blood potassium concentration rises (see section 11.7, Adrenal Glands). In response, aldosterone stimulates the distal convoluted tubule and collecting duct to reabsorb sodium and secrete potassium. As sodium is reabsorbed, water and chloride ions follow the sodium ions, which reduces urine output. Angiotensin II is an important stimulator of aldosterone secretion in response to a decrease in blood pressure (see fig. 17.13).

Neurons in the hypothalamus produce ADH, which the posterior pituitary releases in response to a decreasing water concentration in blood, an increase in blood osmolarity, or a decrease in blood volume. When ADH reaches the kidney, it increases the water permeability of the epithelial linings of the distal convoluted tubule and collecting duct, and water moves rapidly out of these segments by osmosis—that is, water is reabsorbed. Urine volume falls, and soluble wastes and other substances become more concentrated, which minimizes loss of water from the body fluids when dehydration is likely.

If the body fluids have excess water, ADH secretion decreases. As blood levels of ADH drop, the epithelial linings of the distal segment and collecting duct become less permeable to water, less water is reabsorbed, and urine is more dilute, excreting the excess water. Table 17.1 summarizes the role of ADH in urine production.

TABLE 17.1 Role of ADH in Regulating Urine Concentration and Volume

	Body Fluids Too Concentrated	Body Fluids Too Dilute
1.	Concentration of water in blood decreases.	Concentration of water in blood increases.
2.	Increase in osmotic pressure of body fluids is sensed by osmoreceptors in hypothalamus of the brain.	Decrease in osmotic pressure of body fluids is sensed by osmoreceptors in hypothalamus of the brain.
3.	Hypothalamus signals posterior pituitary gland to increase secretion of ADH.	ADH secretion by posterior pituitary gland is reduced.
4.	More ADH reaches kidneys via the bloodstream.	Less ADH reaches kidneys via the bloodstream.
5.	ADH causes distal convoluted tubules and collecting ducts to increase water reabsorption by osmosis.	Distal convoluted tubules and collecting ducts decrease water reabsorption.
6.	Urine volume decreases, urine becomes concentrated, and water is conserved.	Urine volume increases, urine becomes dilute, and excess water is excreted.

In response to high blood pressure entering the heart, natriuretic peptides are secreted and act to reduce blood volume and pressure by dilating the afferent arteriole and constricting the efferent arteriole, which increases the GFR, and to inhibit renin, aldosterone, and ADH secretion.

Urea and Uric Acid Excretion

Urea (yoo-rē′-ah) is a by-product of amino acid catabolism. Consequently, its plasma concentration reflects the amount of protein in the diet. Urea enters the renal tubule by filtration. About 40% to 60% of it is reabsorbed, and the remainder, as well as any urea that is secreted, is excreted in the urine.

Uric acid (yoo′rik as′id) is a product of the catabolism of certain organic bases in nucleic acids. Active transport reabsorbs all the uric acid normally present in glomerular filtrate, but a small amount is secreted into the renal tubule and is excreted in urine. Table 17.2 summarizes some specific functions of the nephron segments and the collecting duct.

Excess uric acid may precipitate in the plasma, ultimately deposited as crystals in joints. When this results in inflammation and associated pain, it is known as gout. It is particularly prevalent in the digits, especially in the great toe. Gout is treated with drugs that inhibit uric acid reabsorption or block its production in a biosynthetic pathway. Limiting foods rich in uric acid, such as organ meats and seafood, and drinking more water to dilute the urine, can help. Susceptibility to gout is inherited, but an attack may not occur until the person eats the offending foods.

PRACTICE 17.3

12. What stimulates the adrenal glands to secrete aldosterone?
13. How does the hypothalamus regulate urine concentration and volume?
14. What produces urea and uric acid?

Urine Composition

Urine composition reflects the volumes of water and amounts of solutes that the kidneys must eliminate from the body or retain in the internal environment to maintain homeostasis. The kidney is able to accomplish this because it handles each of the substances discussed above independently. Urinary excretion of some substances may increase, while that of other substances may decrease. Urine composition varies considerably from time to time because of variations in dietary intake and physical activity. Urine, which is about 95% water, usually contains urea and uric acid. It may also have a trace of amino acids and a variety of other electrolytes.

Several factors contribute to the volume of urine produced, which is usually between 0.6 and 2.5 liters per day. The volume depends largely on fluid intake, and whether the intake is sufficient to balance fluid lost through sweating and exhaling. Water loss through these routes can be affected by the environmental temperature, a person's body temperature, and relative humidity of the surrounding air.

TABLE 17.2 Functions of Nephron Components

Part			Function
Nephron			
	Renal corpuscle		
		Glomerulus	Filtration of water and dissolved substances from plasma
		Glomerular capsule	Recipient of glomerular filtrate
	Renal tubule		
		Proximal convoluted tubule	Reabsorption of glucose; amino acids; lactic, uric, citric, and ascorbic acids; phosphate, sulfate, calcium, potassium, and sodium ions by active transport
			Reabsorption of water by osmosis
			Reabsorption of chloride ions and other negatively charged ions by electrochemical attraction
			Active secretion of substances such as penicillin, histamine, creatinine, and hydrogen ions
		Descending limb of nephron loop	Reabsorption of water by osmosis
		Ascending limb of nephron loop	Reabsorption of sodium, potassium, and chloride ions by active transport
		Distal convoluted tubule	Reabsorption of sodium ions by active transport
			Reabsorption of water by osmosis
			Secretion of hydrogen and potassium ions both actively and passively
Collecting duct			Reabsorption of water by osmosis (Although the collecting duct is not anatomically part of the nephron, it is included here because of its functional importance in urine formation.)

TABLE 17.3 Relative Concentrations of Substances in the Plasma, Glomerular Filtrate, and Urine

Substance	Plasma (mEq/L)	Glomerular Filtrate (mEq/L)	Urine (mEq/L)
Sodium (Na^+)	142	142	128
Potassium (K^+)	5	5	60
Calcium (Ca^{+2})	4	4	5
Magnesium (Mg^{+2})	3	3	15
Chloride (Cl^-)	103	103	134
Bicarbonate (HCO_3^-)	27	27	14
Sulfate (SO_4^{-2})	1	1	33
Phosphate (PO_4^{-3})	4	4	40

Substance	Plasma (mEq/L)	Glomerular Filtrate (mEq/L)	Urine (mEq/L)
Glucose	100	100	0
Urea	26	26	1,820
Uric acid	4	4	53

Note: mEq/L = milliequivalents per liter.

A person's emotional condition can also influence urine volume. Anxiety can increase water loss through increased breathing rate, and stress can stimulate ADH secretion, reducing urine volume. Urine output of 50 to 60 milliliters per hour is normal; output of less than 30 milliliters per hour may indicate kidney disease.

The concentration of a substance in the urine compared to its concentration in the plasma reflects the combination of glomerular filtration, tubular reabsorption, and tubular secretion characteristic of that substance, as well as the amount of water reabsorbed. Table 17.3 shows the relative concentrations of some substances in plasma, glomerular filtrate, and urine.

Glucose, proteins, ketones, and blood cells are not typically in urine. When in urine, these components may reflect certain normal circumstances, or disease. For example, glucose in urine may follow a large intake of carbohydrates, precede giving birth, or be due to diabetes mellitus (see Clinical Application 11.3). Proteins may appear following vigorous physical exercise, and ketones may appear after a prolonged fast. Sections 18.3, Water Balance, 18.4, Electrolyte Balance, and 18.5, Acid-Base Balance discuss urine volume and composition further.

PRACTICE 17.3

15. List the normal constituents of urine.
16. Which factors affect urine volume?

17.4 | Urine Elimination

LEARN

1. Describe the structure of the ureters, urinary bladder, and urethra.
2. Explain the process and control of micturition.

After urine forms in the nephrons and collecting ducts, it passes through openings in the renal papillae and enters the calyces of a kidney (see fig. 17.2). From there, it passes through the renal pelvis, and a ureter conveys it to the urinary bladder (see figs. 17.1, 17.8, and reference plate 6). The urethra passes urine to the outside.

Ureters

Each **ureter** (yoo-rē′ter) is a tube about 25 centimeters long that begins with the funnel-shaped renal pelvis. It descends behind the parietal peritoneum and runs parallel to the vertebral column. In the pelvic cavity, each ureter courses forward and medially, joining the posterior portion of the urinary bladder from underneath.

The ureter wall has three layers. The inner layer, or *mucosa,* is composed of transitional epithelium and is continuous with the linings of the renal tubules and the urinary bladder. The middle layer, or *muscularis,* consists largely of smooth muscle. The outer layer, or *adventitia,* is composed of a fibrous connective tissue and binds to nearby tissues (fig. 17.17).

Movements of the muscular walls of the ureters propel the urine. Muscular peristaltic waves, originating in the renal pelvis, force urine along the length of the ureter. When a peristaltic wave reaches the urinary bladder, a jet of urine spurts through a fold of mucous membrane that surrounds the opening, and the urine enters. Because of the angle at which the ureters enter the bladder, the bladder wall acts as a valve, allowing urine to enter the bladder from the ureter but preventing it from flowing backward as the bladder fills. Clinical Application 17.1 discusses kidney stones, which cause sharp pain when they enter a ureter.

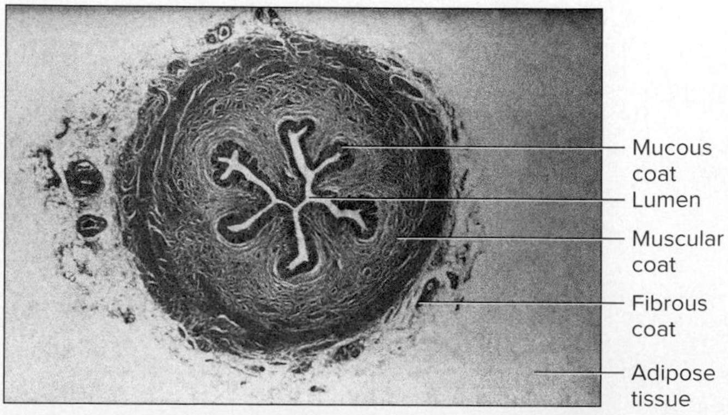

Figure 17.17 Cross section of a ureter (75x). APR

PRACTICE 17.4

1. Describe the structure of a ureter.
2. How is urine moved from the renal pelvis to the urinary bladder?
3. What prevents urine from backing up from the urinary bladder into the ureters?

Urinary Bladder

The **urinary bladder** is a hollow, distensible, muscular organ that stores urine and forces it into the urethra (see fig. 17.1 and reference plate 6). The bladder is in the pelvic cavity, behind the pubic symphysis and beneath the parietal peritoneum.

The pressure of surrounding organs alters the bladder's somewhat spherical shape. When empty, the inner wall of the bladder forms many folds called *rugae,* but as the bladder fills with urine, the wall becomes smoother. At the same time, the superior surface of the bladder expands upward into a dome.

The internal floor of the bladder includes a triangular area called the *trigone,* which has an opening at each of its three angles (**fig. 17.18*a***). Posteriorly, at the base of the trigone, the openings are those of the ureters. Anteriorly and inferiorly, at the apex of the trigone, a short, funnel-shaped extension called the *neck* of the bladder contains the opening into the urethra.

The wall of the urinary bladder has four layers. The inner layer, or *mucosa,* consists of transitional epithelium. The thickness of this tissue changes as the bladder expands and contracts. During distension, the tissue may be only two or three cells thick; during contraction, it may be five or six cells thick (see fig. 5.9).

The second layer of the bladder wall is the *submucosa.* It consists of connective tissue and has many elastic fibers. The third layer of the bladder wall, or *muscularis,* is composed primarily of coarse bundles of smooth muscle cells. These bundles are interlaced in all directions and at all depths, and together they comprise the **detrusor muscle** (dī-troo′zer mus′l). The part of the detrusor muscle that surrounds the neck of the bladder has increased muscle tone, forming the *internal urethral sphincter.* Sustained contraction of this sphincter prevents the bladder from emptying until pressure in the bladder increases to a certain level. The detrusor muscle is innervated with parasympathetic nerve fibers that function in the micturition (urination) reflex, discussed in the next section.

The outer layer of the bladder wall, or *serosa,* consists of the parietal peritoneum. This layer is only on the bladder's upper surface. Elsewhere, the outer coat is connective tissue (**fig. 17.18*b***).

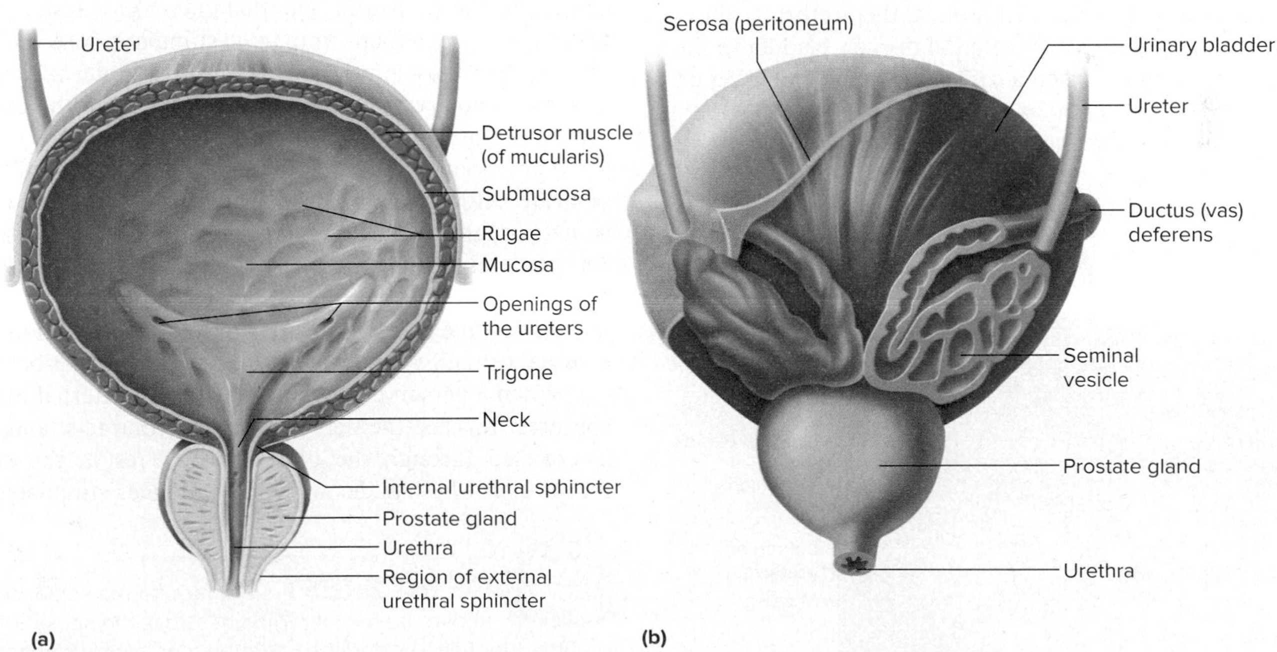

Figure 17.18 A urinary bladder in a male. **(a)** Longitudinal section. **(b)** Posterior view. APR

PRACTICE FIGURE 17.18

How would prostatitis (inflammation of the prostate gland) affect urine flow in a male?

Answer can be found in Appendix E.

PRACTICE 17.4

4. Describe the trigone of the urinary bladder.
5. Describe the structure of the bladder wall.
6. What are the nerve fibers that innervate the detrusor muscle?

Urethra

The **urethra** (yoo-rē'thrah) is a tube that conveys urine from the urinary bladder to the outside (see fig. 17.1 and reference plate 7). The innermost wall of the urethra is composed of a *mucosa* that consists of transitional epithelium nearest the urinary bladder, but then it progresses to pseudostratified epithelium, and eventually becomes stratified squamous epithelium near the opening to outside the body, called the *external urethral orifice.* The thick layer of smooth muscle tissue that makes up the *muscularis* has cells that are generally directed longitudinally. Near the neck of the bladder, the detrusor muscle thickens to form an involuntary sphincter composed of smooth muscle, called the *internal urethral sphincter,* while a voluntary sphincter composed of skeletal muscle surrounds the urethra about 3 centimeters down from the urinary bladder, and is called the *external urethral sphincter.* The urethral wall also has abundant mucous glands, called *urethral glands,* which secrete mucus into the urethral canal (fig. 17.19).

In a female, the urethra is about 4 centimeters long. Its external urethral orifice is anterior to the vaginal opening and posterior to the clitoris. In a male, the urethra is about 20 centimeters long, extends from the urinary bladder to the tip of the penis, and is given a different name depending on the region: *prostatic urethra, membranous urethra,* and *spongy* (or *penile*) *urethra.* The male urethra functions as part of both the urinary system and the reproductive system, so it carries both urine and semen.

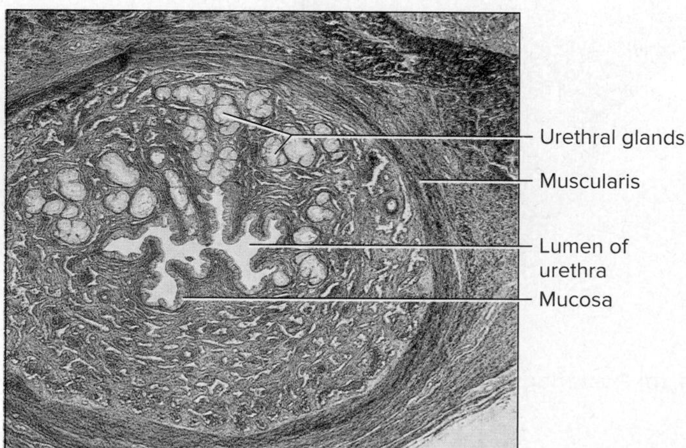

Figure 17.19 Cross section through the urethra (10x).

Inflammation of the urinary bladder, called *cystitis,* is more common in women than in men because the female urethral pathway is shorter, making it easier for bacteria to enter from the outside. Infectious agents, such as bacteria, may ascend from the urinary bladder into the ureters because the linings of these structures are continuous. Inflammation of the ureter is called *ureteritis.* In either case, a urinary tract infection (UTI) causes urination that is frequent, painful, scant, and may be bloody, with pelvic pain. Pathogenic bacteria can be easily killed with antibiotic drugs, or possibly prevented by ingesting cranberry juice that contains compounds that prevent bacteria, such as *Escherichia coli,* from attaching to cells lining the urinary tract.

Micturition APR

Micturition (mik"chuh-rish'un), or urination, is the process of expelling urine from the urinary bladder. Micturition involves reflex contraction of the detrusor muscle and reflex relaxation of the internal urethral sphincter. It also requires voluntary (conscious) relaxation of the external urethral sphincter, which is part of the urogenital diaphragm described in section 8.8, Major Skeletal Muscles.

The kidneys produce urine constantly. Spinal reflexes keep the internal and external urethral sphincters contracted as the bladder fills, and keep the detrusor muscle relaxed, so the bladder can expand.

Continued filling of the bladder stimulates stretch receptors in the bladder wall, eventually triggering the *micturition reflex.* This is a spinal reflex, but receives input from neural centers in the pons and hypothalamus. As a result of this reflex, parasympathetic impulses stimulate contraction of the detrusor muscle and relaxation of the internal urethral sphincter. However, conscious effort may keep the external urethral sphincter closed, preventing urination.

The urinary bladder may hold as much as 600 milliliters of urine before stimulating pain receptors, but the urge to urinate usually begins when it contains about 150 milliliters. As urine volume increases to 300 milliliters or more, the sensation of fullness intensifies and contractions of the detrusor muscle become more powerful. Conscious contraction of the external urinary sphincter may still prevent urination.

When a person decides to urinate, the external urethral sphincter relaxes, the detrusor muscle contracts, and urine is excreted through the urethra. As a result, the stretch receptors in the bladder wall are no longer stimulated, the

OF **INTEREST** Damage to the spinal cord above the sacral region destroys the sensation of bladder fullness and the voluntary control of urination. However, if the micturition reflex center and its sensory and motor fibers are uninjured, micturition may continue to occur by reflex. In this case, the urinary bladder collects urine until its walls stretch enough to trigger a micturition reflex, and the detrusor muscle contracts in response. This condition is called an *automatic bladder.*

CLINICAL APPLICATION 17.1
Kidney Stones

Kidney stones, which are usually composed of uric acid, calcium oxalate, calcium phosphate, or magnesium phosphate, can form in the collecting ducts and renal pelvis (fig. 17A). Such a stone passing into a ureter causes sudden, severe pain that begins in the region of the kidney and radiates into the abdomen, pelvis, and lower limbs. It may also cause nausea and vomiting, and blood in the urine.

An untreated stone may cause pressure in the ureter to rise. As a result, pressure may back up into renal tubules and raise the hydrostatic pressure in glomerular capsules. Any increase in capsular pressure opposes glomerular filtration, and the GFR may decrease significantly.

About 60% of kidney stones pass from the body on their own. The other 40% of stones were at one time removed surgically but are now shattered with intense sound waves. In this far less invasive procedure, called *extracorporeal shock-wave lithotripsy (ESWL),* the patient is placed in a stainless steel tub filled with water. A spark-gap electrode produces underwater shock waves, and a reflector concentrates and focuses the shock-wave energy on the stones. They shatter, and the resulting sandlike fragments then leave in urine.

The tendency to form kidney stones is inherited, particularly the stones that contain calcium, which account for more than half of all cases. Eating calcium-rich foods does not increase the risk, but taking calcium supplements can. People who have calcium oxalate stones can reduce the risk of recurrence by avoiding specific foods: chocolate, coffee, wheat bran, cola, strawberries, spinach, nuts, and tea. Other causes of kidney stones include excess vitamin D, blockage of the urinary tract, or a complication of a urinary tract infection.

It is very helpful for a physician to analyze the composition of the stones, because certain drugs may prevent recurrence. Stones can be collected during surgery, or in some cases by the patient using a special collection device.

Figure 17A This kidney stone is less than 4 mm in diameter, but it is large enough to cause severe pain.

detrusor muscle relaxes, and the urinary bladder begins to fill with urine again.

When a person is unable to voluntarily control the external urethral sphincter, it is called *incontinence.* This is normal in infants and toddlers who have not yet gained control of this sphincter. There are a variety of causes of incontinence in adults; for example, in some women, it may occur as a result of childbirth that weakens muscles needed for bladder control.

PRACTICE 17.4

7. Describe the structure of the urethra.
8. Describe micturition.
9. How is it possible to consciously inhibit the micturition reflex?

ASSESS

CHAPTER ASSESSMENTS

17.1 Introduction

1. Name and identify the general functions of the organs of the urinary system.

17.2 Kidneys

2. Explain why the kidneys are said to be retroperitoneal.
3. Describe the external and internal structure of a kidney.
4. Identify the functions of the kidneys.
5. List in correct order the vessels through which blood passes as it travels from the renal artery to the renal vein.
6. Distinguish between a renal corpuscle and a renal tubule.
7. Name in correct order the parts of the nephron through which fluid passes from the glomerulus to the collecting duct.
8. Compare the structure and function of a cortical nephron and a juxtamedullary nephron.
9. Describe the location and structure of the juxtaglomerular apparatus.

17.3 Urine Formation

10. Which one of the following is abundant in blood plasma, but only in small amounts in glomerular filtrate?
    a. sodium ions
    b. water
    c. glucose
    d. protein
11. Define *net filtration pressure.*
12. Explain how the diameters of the afferent and efferent arterioles affect the rate of glomerular filtration.
13. Explain how changes in the osmotic pressure of blood plasma affect the glomerular filtration rate.
14. Explain how the hydrostatic pressure of a glomerular capsule affects the rate of glomerular filtration.
15. Describe two mechanisms by which the body regulates filtration rate.
16. Discuss how tubular reabsorption is selective.
17. Explain how the peritubular capillary is adapted for reabsorption.
18. Explain how epithelial cells of the proximal convoluted tubule are adapted for reabsorption.
19. Explain why active transport mechanisms have limited transport capacities.
20. Define *renal plasma threshold.*
21. Explain how amino acids and proteins are reabsorbed.
22. Describe the effect of sodium reabsorption on the reabsorption of negatively charged ions.
23. Explain how sodium reabsorption affects water reabsorption.
24. Explain how potassium ions may be secreted passively.
25. The major action of ADH in the kidneys is to:
    a. increase water absorption by the proximal convoluted tubule.
    b. increase glomerular filtration rate.
    c. increase water reabsorption by the collecting duct.
    d. increase potassium's excretion.
26. Compare the processes that reabsorb urea and uric acid.
27. List the common constituents of urine and their sources.
28. Identify some of the factors that affect the volume of urine produced daily.

17.4 Urine Elimination

29. Describe the structure and function of a ureter.
30. Explain how the muscular wall of the ureter helps move urine.
31. Describe the structure and location of the urinary bladder.
32. Define *detrusor muscle.*
33. Compare the female and male urethra.
34. Distinguish between the internal and external urethral sphincters.
35. Describe the micturition reflex.
36. Which of the following involves skeletal muscle?
    a. contraction of the internal urethral sphincter
    b. contraction of the external urethral sphincter
    c. ureteral peristalsis
    d. detrusor muscle contraction

ASSESS

INTEGRATIVE ASSESSMENTS/CRITICAL THINKING

Outcomes 3.3, 17.1, 17.2, 17.3

1. Imagine you are adrift at sea. Why will you dehydrate more quickly if you drink seawater instead of fresh water to quench your thirst?
2. Why are people following high-protein diets advised to drink large volumes of water?

Outcomes 13.5, 17.2, 17.3

3. If blood pressure drops in a patient in shock as a result of a severe injury, how would you expect urine volume to change? Why?

Outcomes 13.5, 17.3

4. Judy was diagnosed with hypertension. Explain why her cardiologist prescribed an ACE inhibitor medication. Would you expect Judy's urine output to increase or decrease while on this medication?

Outcomes 13.7, 17.2

5. Ryan was asked by his teacher to identify a portal system found in the kidneys. He remembered the portal system of the veins in the abdominal viscera that he had learned about in an earlier chapter, so he applied this knowledge to help him answer the teacher's question. What was Ryan's answer?

Urinary System

INTEGUMENTARY SYSTEM

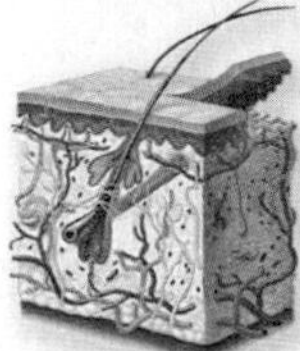

The urinary system compensates for water loss due to sweating. The kidneys and skin both play a role in vitamin D production.

SKELETAL SYSTEM

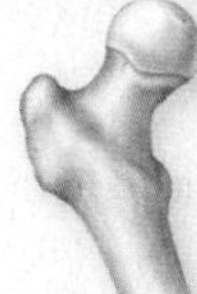

The kidneys and bone tissue work together to control plasma calcium levels.

MUSCULAR SYSTEM

Muscle tissue provides voluntary control of urine elimination from the bladder.

NERVOUS SYSTEM

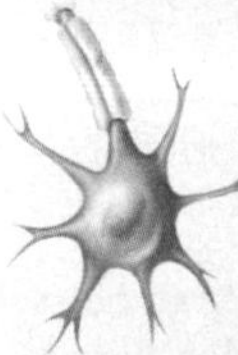

The nervous system influences urine production and elimination.

ENDOCRINE SYSTEM

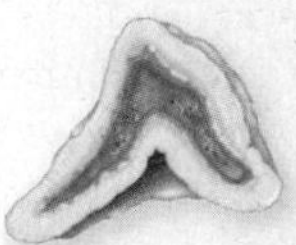

The endocrine system influences urine production.

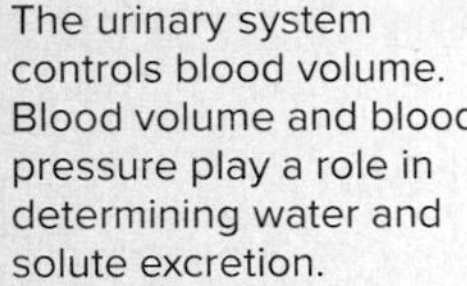

The urinary system controls the composition of the internal environment.

CARDIOVASCULAR SYSTEM

The urinary system controls blood volume. Blood volume and blood pressure play a role in determining water and solute excretion.

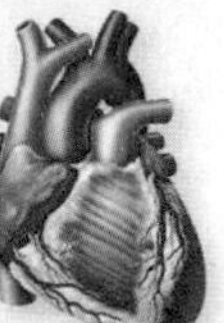

LYMPHATIC SYSTEM

The kidneys control extracellular fluid (including lymph) volume and composition.

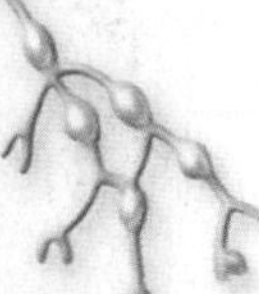

DIGESTIVE SYSTEM

The kidneys compensate for fluids lost by the digestive system.

RESPIRATORY SYSTEM

The kidneys and the lungs work together to control the pH of the internal environment.

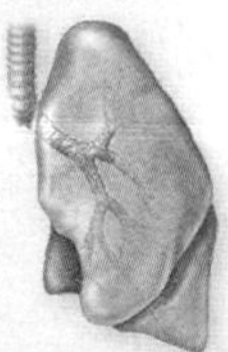

REPRODUCTIVE SYSTEM

The urinary system in males shares organs with the reproductive system. The kidneys compensate for fluids lost from the male and female reproductive systems.

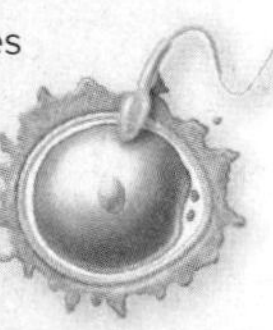

Outcomes 15.11, 17.3

6. Kris decided to try a ketogenic diet of very low carbohydrates and large quantities of fats and proteins, because she heard it would help her lose weight. Explain why a routine urinalysis during her annual physical showed that she had ketonuria (ketone bodies in her urine).

Outcomes 17.2, 17.3

7. Why may protein in the urine be a sign of kidney damage? What structures in the kidney are probably affected?
8. An infant is born with narrowed renal arteries. What effect will this condition have on urine volume?

Outcomes 17.2, 17.4

9. Why do urinary tract infections frequently accompany sexually transmitted diseases?

Chapter Summary

17.1 Introduction

The urinary system consists of the kidneys, ureters, urinary bladder, and urethra.

17.2 Kidneys

1. Location of the **kidneys**
   a. The kidneys are high on the posterior wall of the abdominal cavity.
   b. They are **retroperitoneal** (behind the parietal peritoneum).
2. Kidney structure
   a. A kidney has a hollow **renal sinus.**
   b. Each kidney is divided into a **renal medulla** and a **renal cortex.**
   c. Urine flows from collecting ducts through openings in renal papillae that project into the minor calyces, and continues to flow into major calyces and the **renal pelvis** before entering the ureter.
   d. The functional units of the kidney are the **nephrons.**
3. Kidney functions
   a. The kidneys maintain homeostasis by removing metabolic wastes from blood and excreting them in **urine.**
   b. They also help regulate red blood cell production; blood volume and blood pressure; and the volume, composition, and pH of body fluids.
4. Renal blood supply
   a. Arterial blood flows through the **renal artery,** segmental arteries, interlobar arteries, arcuate arteries, cortical radiate arteries, and **afferent arterioles** to the **glomerulus** of the nephrons.
   b. Unfiltered blood leaves the glomerulus through the **efferent arteriole,** which gives rise to **peritubular capillaries,** or **vasa recta,** that surround the renal tubule.
   c. Venous blood returns through a series of vessels that correspond to the arterial pathways (except segmental arteries), leading to the **renal vein.**
5. Nephrons
   a. Nephron structure
      (1) A nephron is the functional unit of the kidney.
      (2) It consists of a **renal corpuscle** and a **renal tubule.**
         (a) The corpuscle consists of a **glomerulus** and a **glomerular capsule.**
         (b) Segments of the renal tubule include the proximal convoluted tubule, nephron loop (descending and ascending limbs), and distal convoluted tubule, which empties into a collecting duct.
      (3) The collecting duct (technically not part of a nephron) empties urine into the minor calyx.
   b. Types of nephrons
      (1) The **cortical nephron** is the most abundant type and is almost entirely located in the renal cortex, with peritubular capillaries surrounding the renal tubule.
      (2) The **juxtamedullary nephron** has its renal corpuscle deep in the cortex with a long nephron loop extending deep into the medulla, surrounded by vasa recta.
   c. Juxtaglomerular apparatus
      (1) The juxtaglomerular apparatus is at the point of contact between the last portion of the ascending limb of the nephron loop and the afferent and efferent arterioles.
      (2) It consists of the macula densa and juxtaglomerular cells.

17.3 Urine Formation

Nephrons remove wastes from blood and regulate water and electrolyte concentrations. Urine is the end product.

1. **Glomerular filtration**
   a. Urine formation begins when water and dissolved materials filter out of glomerular capillaries.
   b. The glomerular capsule receives the resulting **glomerular filtrate** (filtered fluid).
   c. Glomerular capillaries are much more permeable than the capillaries in other tissues.
   d. The composition of the filtered fluid is similar to that of tissue fluid.
   e. The filtered fluid becomes tubular fluid as it moves through the renal tubule and is modified by the processes of **tubular reabsorption** and **tubular secretion.**
   f. The final product of glomerular filtration, tubular reabsorption, and tubular secretion is **urine.**
2. Filtration pressure
   a. Filtration is due mainly to hydrostatic pressure inside glomerular capillaries.

   b. The osmotic pressure of plasma in the glomerular capillaries and the hydrostatic pressure in the glomerular capsule also affect filtration.
   c. **Net filtration pressure** is the net force moving material out of the glomerular capillaries and into the glomerular capsule.
3. **Glomerular filtration rate**
   a. Rate of filtration varies with filtration pressure.
   b. Filtration pressure changes with the diameters of the afferent and efferent arterioles.
   c. As colloid osmotic pressure in the glomerulus increases, filtration rate decreases.
   d. As hydrostatic pressure in a glomerular capsule increases, filtration rate decreases.
   e. The kidneys produce about 120 to 125 milliliters of glomerular filtrate per minute, most of which is reabsorbed.
4. Regulation of filtration rate
   a. Glomerular filtration rate (GFR) remains relatively constant, but may increase or decrease as required.
   b. Increased activity of the sympathetic nervous system can maintain blood pressure with minimal change in GFR.
   c. When the macula densa senses decreased NaCl concentration in the last part of the ascending limb of the nephron loop, it causes juxtaglomerular cells to release **renin.**
   d. Renin release triggers a series of changes leading to angiotensin II formation, which causes vasoconstriction of afferent and efferent arterioles, minimizing any decrease in glomerular filtration rate, and also stimulates aldosterone secretion, resulting in increased tubular sodium reabsorption.
5. Tubular reabsorption
   a. Substances are selectively reabsorbed from glomerular filtrate.
   b. The peritubular capillary's permeability adapts it for reabsorption.
   c. Most reabsorption occurs in the proximal tubule, where epithelial cells have microvilli.
   d. Different modes of transport reabsorb various substances, in particular segments of the renal tubule. Examples include:
      (1) Active transport reabsorbs glucose and amino acids.
      (2) Osmosis reabsorbs water.
   e. Active transport mechanisms have limited transport capacities.
6. Sodium and water reabsorption
   a. Active transport reabsorbs sodium ions.
   b. As positively charged sodium ions move out of the filtrate, negatively charged chloride ions follow them.
   c. Water is passively reabsorbed by osmosis.
   d. Substances that remain in the filtrate are concentrated as water is reabsorbed.
7. Tubular secretion
   a. Secretion transports substances from plasma in the peritubular capillaries to the renal tubular fluid.
   b. Various organic compounds are secreted actively.
   c. Potassium and hydrogen ions are both secreted actively. Potassium ions are also secreted passively.
8. Regulation of urine concentration and volume
   a. Most sodium is reabsorbed before urine is excreted, assisted by aldosterone in the distal convoluted tubule and collecting duct.
   b. Antidiuretic hormone increases the permeability of the distal convoluted tubule and collecting duct, promoting water reabsorption.
   c. Atrial naturetic peptide opposes actions of renin, aldosterone, and ADH.
9. **Urea** and **uric acid** excretion
   a. Diffusion passively reabsorbs about 40% to 60% of the urea.
   b. Active transport reabsorbs uric acid. Some uric acid is secreted into the renal tubule.
10. Urine composition
   a. Urine is about 95% water, and it also usually contains urea and uric acid.
   b. Urine contains varying amounts of electrolytes and may contain a trace of amino acids.
   c. Urine volume varies with fluid intake, certain environmental factors, a person's emotional state, and body temperature.

17.4 Urine Elimination

1. **Ureters**
   a. The ureter extends from the kidney to the urinary bladder.
   b. Peristaltic waves in the ureter force urine to the urinary bladder.
2. **Urinary bladder**
   a. The urinary bladder stores urine and forces it through the urethra during micturition.
   b. The openings for the ureters and urethra are at the three angles of the trigone.
   c. A portion of the **detrusor** muscle forms an internal urethral sphincter.
3. **Urethra**
   The urethra conveys urine from the urinary bladder to the outside.
4. **Micturition**
   a. Micturition is the expulsion of urine.
   b. Micturition results from contraction of the detrusor muscle and relaxation of the external urethral sphincter.
   c. Micturition reflex
      (1) Distension stimulates stretch receptors in the bladder wall.
      (2) The micturition reflex center in the spinal cord sends parasympathetic motor impulses to the detrusor muscle.
      (3) As the bladder fills, its internal pressure increases, forcing the internal urethral sphincter open.
      (4) A second reflex relaxes the external urethral sphincter unless voluntary control maintains its contraction.
      (5) Nerve centers in the cerebral cortex and brainstem aid control of urination.

18 Water, Electrolyte, and Acid-Base Balance

An extremely high environmental temperature can challenge the body's ability to maintain a normal body temperature. ©Tetra Images/Getty Images

Heatstroke can kill. Heatstroke, a form of hyperthermia, is a result of exposure to extreme environmental heat combined with a failure of normal body temperature control. It can be quickly fatal. It occurs when the body is exposed to a heat index (heat considering humidity) of more than 105°F and body temperature reaches 104°F. Under these conditions, evaporation of sweat becomes less efficient at cooling the body, and organs begin to fail.

The symptoms of heatstroke happen in a sequence. First come headache, dizziness, and exhaustion. Sweating is profuse, then stops, and the skin becomes dry, hot, and red. Respiratory rate rises, and the pulse may race up to 180 beats per minute. If the person isn't cooled with drinking fluids, water applied to the skin, fanning, and removal of clothing, neurological symptoms may begin. These include disorientation, hallucinations, and odd behavior. Kidney failure and/or heart arrhythmia may prove fatal.

During heat waves, the very young and the very old are more susceptible to heatstroke than others because their body temperature control mechanisms may be compromised. However, heatstroke also affects two groups of young, otherwise healthy individuals—athletes who work out in extreme heat and soldiers deployed to hot climates.

Military researchers are conducting experiments to better understand the conditions under which heatstroke occurs. They are looking for biomarkers (see Clinical Application 2.2) to identify particularly susceptible individuals and to determine when it is safe for a soldier who has suffered a heat injury to return to battle.

Females and males differ somewhat in their responses to extreme, sustained heat. Women begin to sweat at higher environmental temperatures than do men, and women's sweat is effective for a longer proportion of total sweating time. Women are more susceptible to heatstroke during the second half of the menstrual cycle, when core body temperature is higher. One study showed that women are more susceptible to developing heat intolerance than men. (Heat intolerance is maximum heart rate exceeding 150 beats per minute or a core body temperature greater than 101.3°F when exposed to sustained environmental heat. It can precede heatstroke or continue after it.) United States Army investigators compared heat intolerance in 55 men and 20 women. The participants took a standardized heat tolerance test (treadmill walking at 5 kilometers per hour at a 2% grade for 2 hours at 104°F and 40% relative humidity). The women had a 3.7-fold greater likelihood of suffering heat intolerance than men.

Module 13 Urinary System

LEARNING OUTLINE

After studying this chapter, you should be able to complete the "Learning Outcomes" that follow the major headings throughout the chapter.

AIDS TO UNDERSTANDING WORDS

de- [separation from] *de*hydration: removal of water from the cells or body fluids.

extra- [outside] *extra*cellular fluid: fluid outside of the body cells.

im- [not] *im*balance: condition in which factors are not in equilibrium.

intra- [within] *intra*cellular fluid: fluid in body cells.

neutr- [neither one nor the other] *neutr*al solution: solution that is neither acidic nor basic.

(Appendix A has a complete list of Aids to Understanding Words.)

18.1 | Introduction

LEARN

1. Explain water and electrolyte balance.

Water and electrolytes have a unique and special relationship within our bodies. This is because electrolytes, or ions, are dissolved in the water in all body fluids. Recall from chapter 3 that water moves into and out of cells based upon the concentration of the dissolved electrolytes called solutes. Some of the most important electrolytes include Na^+, K^+, Ca^{+2}, H^+, OH^-, Cl^-, and Mg^+.

For homeostasis, water and electrolyte balance must be maintained; that is, the quantities entering the body must equal those leaving it. Therefore, the body requires mechanisms to (1) replace lost water and electrolytes, and (2) excrete any excess water and electrolytes. These mechanisms include eating, drinking, urinating, and perspiring.

Water can dissociate into the H^+ and OH^- ions. Along with other various ions, they play a major role in acid-base balance.

PRACTICE 18.1

Answers to the Practice questions can be found in the eBook.

1. How are water and electrolyte balance interdependent?

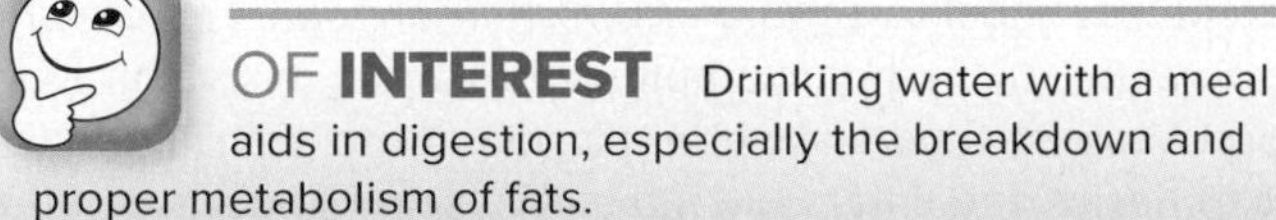

OF **INTEREST** Drinking water with a meal aids in digestion, especially the breakdown and proper metabolism of fats.

18.2 | Distribution of Body Fluids

LEARN

1. Describe how body fluids are distributed in compartments.

Body fluids are not uniformly distributed. Instead, they occupy regions, or *compartments*, of different volumes that contain fluids of varying compositions. The movement of water and electrolytes between these compartments is regulated to stabilize both the distribution and the composition of body fluids.

CAREER CORNER
Medical Laboratory Technician

A story lies behind every patient sample, and that's what the medical laboratory technician loves about her job. She analyzes specimens of blood, urine, saliva, and other body fluids or tissues to assist physicians in diagnosing disease. The urine sample before her now has tested positive for protein, an indication that the kidneys are not filtering normally. This is consistent with depressed levels of the protein albumin measured in the patient's blood, and with the accumulation of fluid in the patient's tissues, called edema.

A medical laboratory technician follows a patient sample as it is analyzed, conducting automated tests as well as observing cells under a microscope. Medical laboratory technicians work in hospital labs, physicians' offices, diagnostic laboratories, forensics laboratories, and pharmaceutical and biotechnology companies.

Training varies. Many medical laboratory technicians earn associate's degrees. Hospitals, technical programs, and the military also provide training. Education must include a clinical component, and the medical laboratory technician must pass a national certification exam.

Fluid Compartments

The body of an average adult female is about 52% water by weight and that of an average male is about 63% water by weight. The reason for this difference is that females generally have more adipose tissue, which contains little water. Males generally have proportionately more muscle tissue, which contains a great deal of water. Water in the adult human body (about 40 liters), with its dissolved electrolytes, is distributed into two major compartments: an intracellular fluid compartment and an extracellular fluid compartment.

The **intracellular** (in"trah-sel'u-lar) **fluid compartment** includes all the water and electrolytes that cell membranes enclose. In other words, intracellular fluid is the fluid inside cells. In an adult, it accounts for about 63% by volume of total body water.

The **extracellular** (ek"strah-sel'u-lar) **fluid compartment** includes all the fluid outside of cells—in tissue spaces (interstitial fluid), blood vessels (plasma), and lymphatic vessels (lymph). **Transcellular fluid,** a type of extracellular fluid, is found in cavities separated from other extracellular fluids by epithelial or connective tissue membranes. Transcellular fluid includes cerebrospinal fluid of the central nervous system, aqueous and vitreous humors of the eyes, synovial fluid of the joints, and serous fluid in the body cavities. The fluids of the extracellular compartment constitute about 37% by volume of total body water (fig. 18.1).

Body Fluid Composition

Extracellular fluids generally are similar in composition, including high concentrations of sodium, chloride, calcium, and bicarbonate ions. The blood plasma portion of extracellular fluid has considerably more protein than does either interstitial fluid or lymph.

Intracellular fluid has high concentrations of potassium, phosphate, and magnesium ions. It also includes a greater concentration of sulfate ions than extracellular fluid and a greater concentration of proteins than plasma. Figure 18.2 shows these relative concentrations.

PRACTICE 18.2

1. Describe the normal distribution of water in the body.
2. Which electrolytes are in higher concentrations in extracellular fluids? In intracellular fluid?
3. How does the protein concentration vary among body fluids?

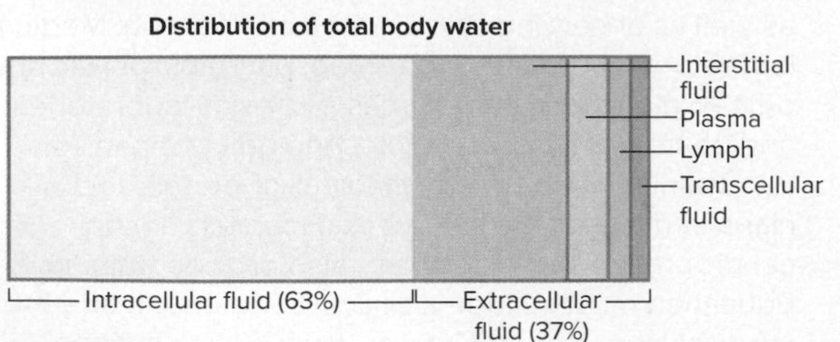

Figure 18.1 Approximately two-thirds of the water in the body is inside cells.

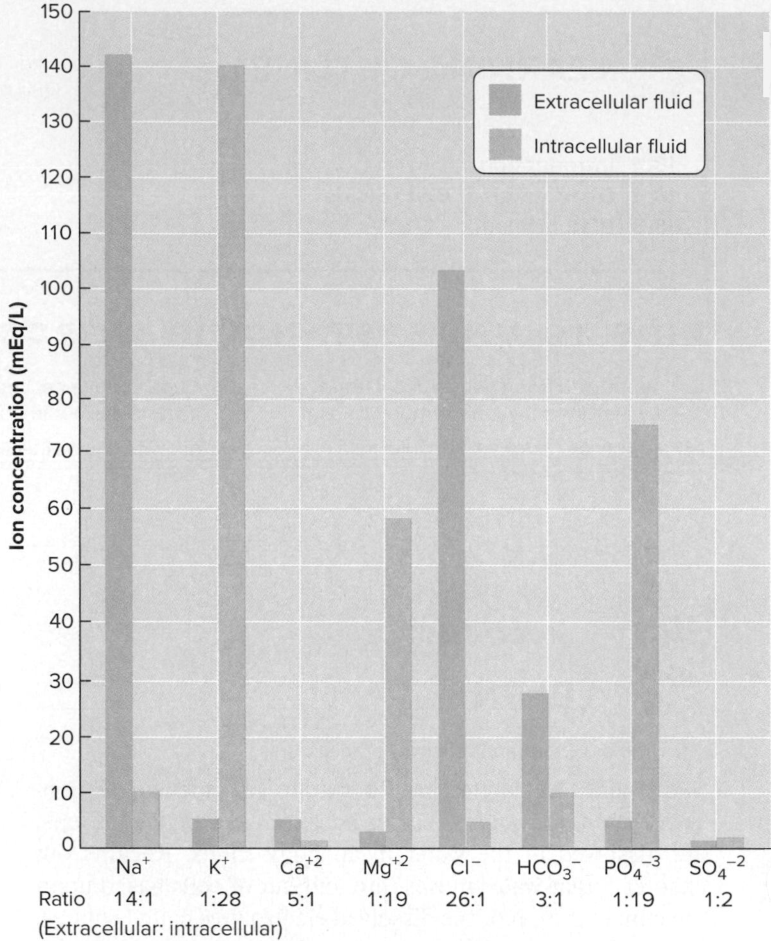

Figure 18.2 Extracellular fluids have relatively high concentrations of sodium (Na^+), calcium (Ca^{+2}), chloride (Cl^-), and bicarbonate (HCO_3^-) ions. Intracellular fluid has relatively high concentrations of potassium (K^+), magnesium (Mg^{+2}), phosphate (PO_4^{-3}), and sulfate (SO_4^{-2}) ions.

PRACTICE FIGURE 18.2

According to the graph, which positively charged intracellular fluid ion is in the highest concentration?

Answer can be found in Appendix E.

Movement of Fluid Between Compartments

Two major factors contribute to the movement of fluid from one compartment to another: *hydrostatic pressure* and *osmotic pressure* (fig. 18.3). For example, as explained in section 13.4, Blood Vessels, fluid leaves the plasma at the arteriolar ends of capillaries and enters the interstitial spaces because of the net outward force of hydrostatic pressure (blood pressure). Fluid returns to the plasma from the interstitial spaces at the venular ends of capillaries because of the net inward force of *colloid osmotic pressure* due to the plasma proteins. Likewise, as mentioned in section 14.3, Tissue Fluid and Lymph, fluid leaves the interstitial spaces and enters the lymph capillaries

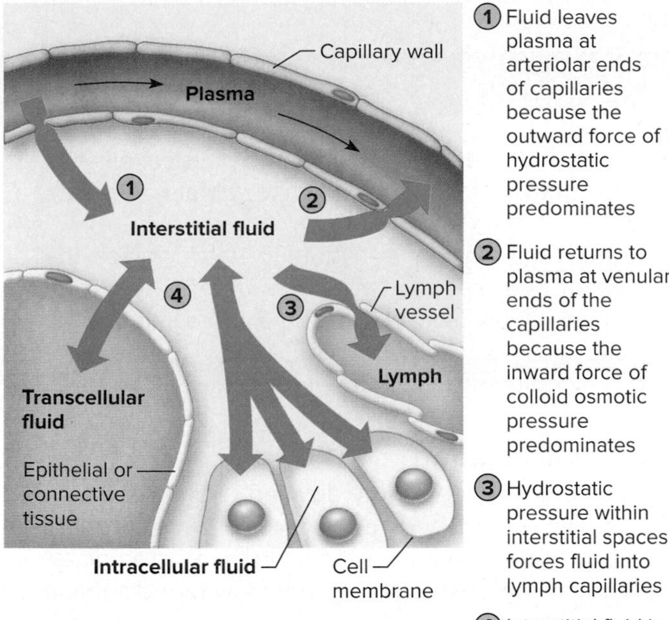

Figure 18.3 Net movements of fluids between compartments result from differences in hydrostatic and osmotic pressures.

due to the hydrostatic pressure of the interstitial fluid. The circulation of lymph returns interstitial fluid to the plasma.

Hydrostatic pressure in the cells and surrounding interstitial fluid is ordinarily equal and remains stable. Therefore, any net fluid movement is likely to be the result of changes in osmotic pressure.

The total solute concentration in extracellular and intracellular fluids is normally equal. However, a decrease in extracellular sodium ion concentration causes a net movement of water from the extracellular compartment into the intracellular compartment by osmosis. The cells swell. Conversely, if the extracellular sodium ion concentration increases, cells shrink as they lose water by osmosis (see section 3.3, Movements Into and Out of the Cell).

PRACTICE 18.2

4. Which factors control the movement of water and electrolytes from one fluid compartment to another?
5. How does the sodium ion concentration in body fluids affect the net movement of water between the compartments?

18.3 | Water Balance

LEARN

1. List the routes by which water enters and leaves the body.
2. Explain how water intake and output are regulated.

Homeostasis requires control of both water intake and water output. **Water balance** exists when water intake equals water output. Clinical Application 18.1 discusses water balance disorders, including dehydration, edema, and water intoxication.

Water Intake

The volume of water gained each day varies among individuals. An average adult living in a moderate environment takes in about 2,500 milliliters. Probably 60% is obtained from drinking water or beverages, and another 30% comes from moist foods. The remaining 10% is a by-product of the oxidative metabolism of nutrients (see section 4.4, Energy for Metabolic Reactions), called **metabolic water** (fig. 18.4*a*).

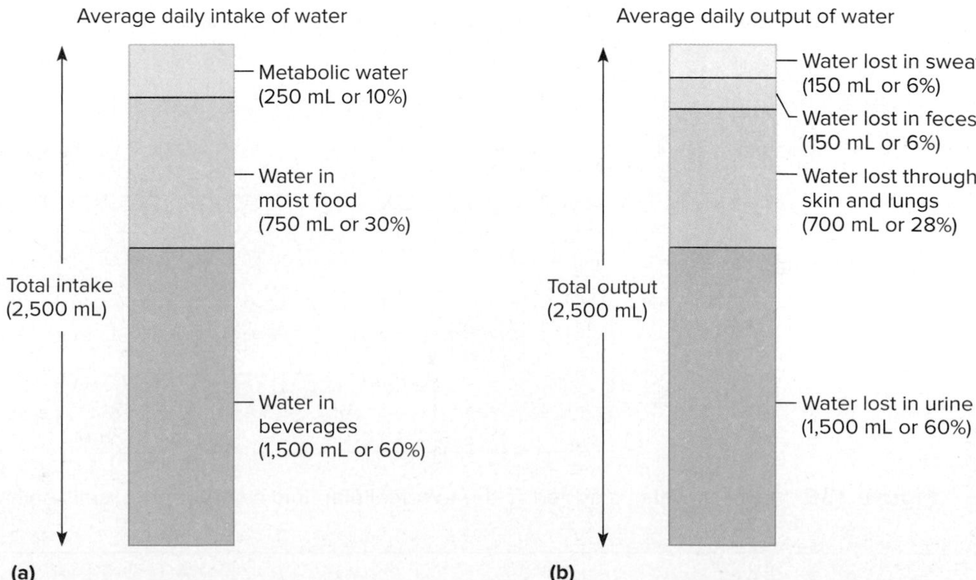

Figure 18.4 Water balance. **(a)** Major sources of body water. **(b)** Routes by which the body loses water.

CLINICAL APPLICATION 18.1

Water Balance Disorders

Dehydration, water intoxication, and edema are among the more common disorders that involve a water imbalance in body fluids.

Dehydration

In *dehydration*, water output exceeds water intake. Dehydration may develop following excessive sweating or as a result of prolonged water deprivation accompanied by continued water output. The extracellular fluid becomes more concentrated, and water leaves cells by osmosis (fig. 18A). Dehydration may also accompany prolonged vomiting or diarrhea that depletes body fluids.

During dehydration, the skin and mucous membranes of the mouth feel dry, and body weight drops. Severe hyperthermia may develop as the body's temperature-regulating mechanism falters due to lack of water for sweat production.

Infants are more likely to become dehydrated because their kidneys are less efficient at conserving water than those of adults. Elderly people are also especially susceptible to developing water imbalances because the sensitivity of their thirst mechanism decreases with age, and physical disabilities may make it difficult for them to obtain adequate fluids.

The treatment for dehydration is to replace the lost water and electrolytes. If only water is replaced, the extracellular fluid will become more dilute than normal, causing cells to swell (fig. 18B). This may produce a condition called water intoxication.

Water Intoxication

Until recently, runners were advised to drink as much fluid as they could, particularly in long events. But the death of a young woman running in the Boston marathon, following brain swelling, from low blood sodium (*hyponatremia*) due to excessive water intake leading to *water intoxication*, inspired further study and a reevaluation of this advice. Researchers from Harvard Medical School studied 488 runners from the marathon and found that 13% of them had developed hyponatremia. The tendency to develop the condition was associated with longer race time and high or low body mass index.

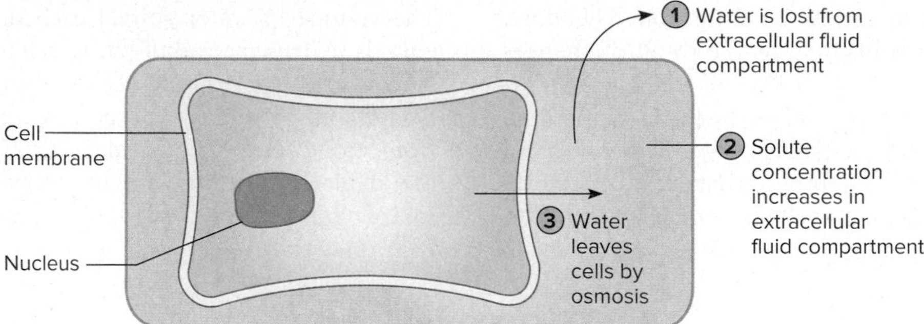
1 Water is lost from extracellular fluid compartment
Cell membrane
2 Solute concentration increases in extracellular fluid compartment
3 Water leaves cells by osmosis
Nucleus

Figure 18A If excess extracellular fluids are lost, cells dehydrate by osmosis.

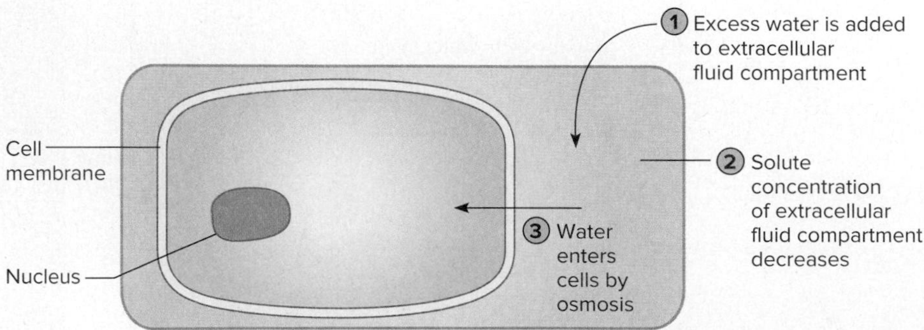
1 Excess water is added to extracellular fluid compartment
Cell membrane
2 Solute concentration of extracellular fluid compartment decreases
3 Water enters cells by osmosis
Nucleus

Figure 18B If excess water is added to the extracellular fluid compartment, cells gain water by osmosis.

In recognition of the possibility of hyponatremia, USA Track and Field, the national governing body for the sport, advises runners how to determine how much water to drink during a one-hour training run. The goal is to replace exactly what is lost. Hyponatremia in the general population is rare. A person must drink several liters of fluid to be at risk. Symptoms are similar to those of heatstroke, plus exhaustion, headache, diarrhea, nausea, and vomiting.

Edema

Edema is an abnormal accumulation of extracellular fluid in the interstitial spaces. Body parts swell. Causes of edema include a decrease in the plasma protein concentration (*hypoproteinemia*), obstructions in lymphatic vessels, increased venous pressure, and increased capillary permeability.

Hypoproteinemia may result from failure of the liver to synthesize plasma proteins; kidney disease that damages glomerular capillaries, allowing proteins to enter the urine; or starvation, in which amino acid intake is insufficient to support synthesis of plasma proteins. In each of these instances, the plasma protein concentration is decreased, which decreases plasma colloid osmotic pressure, reducing the normal return of tissue fluid to the venular ends of capillaries. Consequently, tissue fluid accumulates in the interstitial spaces.

Edema may also result from lymphatic obstructions due to surgery or parasitic infections of lymphatic vessels, as discussed in section 14.3, Tissue Fluid and Lymph. Back pressure develops in the lymphatic vessels, which interferes with the normal movement of tissue fluid into them. At the same time, proteins that are usually removed by the lymphatic circulation accumulate in interstitial spaces, raising the osmotic pressure of interstitial fluid. This effect draws still more fluid into the interstitial spaces.

If blood outflow from the liver into the inferior vena cava is blocked, the venous pressure in the liver and portal blood vessels greatly increases. As a result, fluid with a high protein concentration is exuded from the surfaces of the liver and intestine into the peritoneal cavity. The protein buildup elevates the osmotic pressure of the abdominal fluid, which in turn draws more water into the peritoneal cavity by osmosis, causing the painful condition *ascites*.

Edema may also result from increased capillary permeability accompanying inflammation. Recall that inflammation is a response to tissue damage and usually releases chemicals such as histamine from damaged cells. Histamine causes vasodilation and increased capillary permeability. As a result, excess fluid leaks out of the capillary and enters the interstitial spaces. Table 18A summarizes the factors that result in edema.

TABLE 18A Factors Associated with Edema

Factor	Cause	Effect
Low plasma protein concentration	Liver disease and failure to synthesize proteins; kidney disease and loss of proteins in urine; lack of proteins in diet due to starvation	Plasma colloid osmotic pressure decreases; less fluid reabsorbed at venular ends of capillaries by osmosis
Obstruction of lymphatic vessels	Surgical removal of portions of lymphatic pathways; certain parasitic infections	Back pressure in lymph vessels interferes with movement of fluid from interstitial spaces into lymph capillaries
Increased venous pressure	Venous obstructions or faulty venous valves	Back pressure in veins increases capillary filtration
Inflammation	Tissue damage	Vasodilation and increased capillary permeability lead to increased filtration

The primary regulator of water intake is thirst. The intense feeling of thirst derives from the osmotic pressure of extracellular fluids and both neural and hormonal input to the brain. As the body loses water, the osmotic pressure of extracellular fluids increases. Such a change stimulates *osmoreceptors* in the hypothalamus, and in response, the hypothalamus causes the person to feel thirsty and to seek water.

Thirst is a homeostatic mechanism, normally triggered when total body water decreases by as little as 1%. The act of drinking and the resulting distension of the stomach wall trigger impulses that inhibit the thirst mechanism. In this way, drinking stops even before the swallowed water is absorbed, preventing the person from drinking more than is required to replace the volume lost, avoiding development of an imbalance.

PRACTICE 18.3

1. What is water balance?
2. Where is the thirst center?
3. What stimulates fluid intake? What inhibits it?

Water Output

Water normally enters the body only through the mouth, but it can be lost by a variety of routes. These include obvious losses in urine, feces, and sweat (sensible perspiration), as well as evaporation of water from the skin (insensible perspiration) and from the lungs during breathing.

If an average adult takes in 2,500 milliliters of water each day, then 2,500 milliliters must be eliminated to maintain water balance. Of this volume, perhaps 60% is lost in urine, 6% in feces, and 6% in sweat. About 28% is lost by evaporation from the skin and lungs (fig. 18.4*b*). These percentages vary with such environmental factors as temperature and relative humidity and with physical exercise.

The primary means of regulating water output is urine production. The renal distal convoluted tubules of the nephrons and collecting ducts are the effectors of the mechanism that regulates urine volume. The epithelial linings in these structures remain relatively impermeable to water unless antidiuretic hormone (ADH) is present. ADH increases the permeability of the distal convoluted tubule and collecting duct, thereby increasing water reabsorption and reducing urine production. In the absence of ADH, less water is reabsorbed and more urine is produced (see section 17.3, Urine Formation).

Diuretics are chemicals that promote urine production. They act in different ways. Alcohol and certain narcotic drugs promote urine formation by inhibiting ADH release. A drug called mannitol is an osmotic diuretic. It is filtered by the kidneys, but is not reabsorbed, so it draws water into the renal tubules by osmosis and more water is lost in the urine. Mannitol is used in some patients to increase urinary excretion of toxins. It is also used to sweeten chewing gum, but not in sufficient amounts to have a diuretic effect.

PRACTICE 18.3

4. By what routes does the body lose water?
5. What is the primary regulator of water loss?

18.4 | Electrolyte Balance

LEARN

1. List the routes by which electrolytes enter and leave the body.
2. Explain how electrolyte intake and output are regulated.

Electrolyte balance exists when the quantities of electrolytes the body gains equal those lost. The homeostatic goal is to keep the associated ions in appropriate concentrations within the plasma and the interstitial fluid.

Electrolyte Intake

The electrolytes of greatest importance to cellular functions dissociate to release sodium, potassium, calcium, magnesium, chloride, sulfate, phosphate, bicarbonate, and hydrogen ions. These electrolytes are primarily obtained from foods, but they may also be found in drinking water and other beverages. In addition, some electrolytes are byproducts of metabolic reactions.

Ordinarily, a person obtains sufficient electrolytes by responding to hunger and thirst. However, a severe electrolyte deficiency may cause *salt craving*, which is a strong desire to eat salty foods.

Electrolyte Output

The body loses some electrolytes by perspiring, with more lost in sweat on warmer days and during strenuous exercise. Varying amounts of electrolytes are lost in the feces. The greatest electrolyte output occurs as a result of kidney function and urine production. The kidneys alter electrolyte output to maintain the proper composition of body fluids, thereby promoting homeostasis.

PRACTICE 18.4

1. Which electrolytes are most important to cellular functions?
2. Which mechanisms ordinarily regulate electrolyte intake?
3. By what routes does the body lose electrolytes?

Precise concentrations of positively charged ions, such as sodium (Na^+), potassium (K^+), and calcium (Ca^{+2}), are required for impulse conduction along an axon, muscle fiber contraction, and maintenance of cell membrane potential. Sodium ions account for nearly 90% of positively charged ions in extracellular fluids. The kidneys and the hormone aldosterone regulate these ions. Aldosterone, which the adrenal cortex secretes, increases sodium ion reabsorption in the distal convoluted tubules of the kidneys' nephrons and in the collecting ducts.

Aldosterone also regulates potassium ion concentration. A rising potassium ion concentration directly stimulates cells of the adrenal cortex to secrete aldosterone. This hormone enhances tubular secretion of potassium ions at the same time it causes tubular reabsorption of sodium ions (fig. 18.5). Clinical Application 18.2 discusses conditions resulting from an imbalance of sodium and potassium ions.

Recall from section 11.6, Parathyroid Glands, that the calcium ion concentration dropping below normal directly stimulates the parathyroid glands to secrete parathyroid hormone. This hormone returns the concentration of calcium in extracellular fluids toward normal.

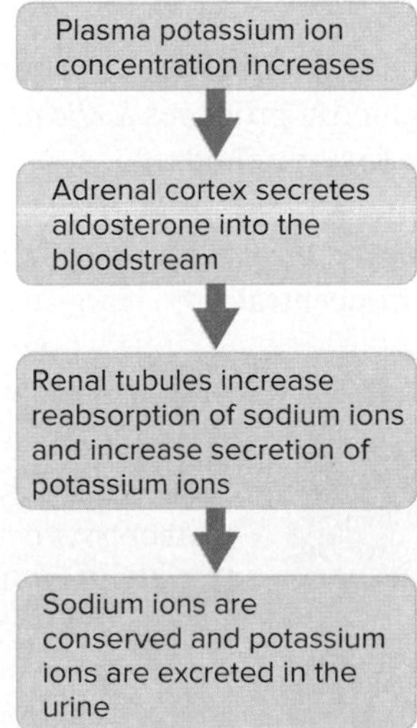

Figure 18.5 If the potassium ion concentration increases, the kidneys conserve sodium ions and excrete potassium ions.

Generally, the regulatory mechanisms that control positively charged ions secondarily control the concentrations of negatively charged ions. For example, chloride ions (Cl^-), the most abundant negatively charged ions in extracellular fluids, are passively reabsorbed in response to the active tubular reabsorption of sodium ions. That is, the negatively charged chloride ions are electrically attracted to positively charged sodium ions and accompany them as they are reabsorbed (see section 17.3, Urine Formation). Water is then reabsorbed by osmosis.

Active transport mechanisms with limited transport capacities partially regulate some negatively charged ions, such as phosphate ions (PO_4^{-3}) and sulfate ions (SO_4^{-2}). Therefore, if the extracellular phosphate ion concentration is low, renal tubules reabsorb phosphate ions. On the other hand, if the renal plasma threshold is exceeded, excess phosphate is excreted in urine (see section 17.3, Urine Formation).

PRACTICE 18.4

4. How does aldosterone regulate the sodium and potassium ion concentrations?
5. How is calcium regulated?
6. What mechanism regulates the concentrations of most negatively charged ions?

18.5 | Acid-Base Balance

LEARN

1. List the major sources of hydrogen ions in the body.
2. Distinguish between strong and weak acids and bases.
3. Explain how chemical buffer systems, the respiratory center, and the kidneys keep the pH of body fluids relatively constant.

Electrolytes that release hydrogen ions are called **acids,** and electrolytes that release ions that combine with hydrogen ions are called **bases,** as section 2.5, Acids and Bases, discussed. Maintenance of homeostasis depends on balancing the concentrations of acids and bases in body fluids.

Sources of Hydrogen Ions

Most of the hydrogen ions in body fluids originate as byproducts of metabolic processes, although the digestive tract may directly absorb some hydrogen ions. The major metabolic sources of hydrogen ions include the following (fig. 18.6):

CLINICAL APPLICATION 18.2

Sodium and Potassium Imbalances

Extracellular fluids usually have high sodium ion concentrations, and intracellular fluid usually has a high potassium ion concentration. Renal regulation of sodium is closely related to that of potassium, because active reabsorption of sodium (under the influence of aldosterone) is accompanied by tubular secretion (and excretion) of potassium. Therefore, conditions resulting from sodium ion imbalance often also involve potassium ion imbalance.

Such disorders include:

1. *Low blood sodium concentration (hyponatremia)* Possible causes of sodium deficiencies include prolonged sweating, vomiting, or diarrhea; renal disease in which sodium is inadequately reabsorbed; adrenal cortex disorders in which aldosterone secretion is insufficient to promote sodium reabsorption (Addison disease); and drinking too much water. One possible effect of hyponatremia is the development of hypotonic extracellular fluid that promotes water movement into cells by osmosis, producing symptoms of water intoxication.
2. *High blood sodium concentration (hypernatremia)* Possible causes of the elevated sodium ion concentration include excessive water loss by evaporation (despite decreased sweating, as may occur during high fever), or increased water loss accompanying diabetes insipidus. In one form of diabetes insipidus, the secretion of antidiuretic hormone (ADH) is insufficient for the renal tubules and collecting ducts to conserve water. Hypernatremia may disturb the central nervous system, causing confusion, stupor, and coma.
3. *Low blood potassium concentration (hypokalemia)* Possible causes of potassium deficiency include the release of excess aldosterone by the adrenal cortex (Cushing syndrome), which increases renal excretion of potassium; use of diuretic drugs that promote potassium excretion; kidney disease; and prolonged vomiting or diarrhea. Possible effects of hypokalemia include muscular weakness or paralysis, respiratory difficulty, and severe cardiac disturbances, such as atrial or ventricular arrhythmia.
4. *High blood potassium concentration (hyperkalemia)* Possible causes of the elevated potassium ion concentration include renal disease, which decreases potassium excretion; use of drugs that promote renal conservation of potassium; the release of insufficient aldosterone by the adrenal cortex (Addison disease); or a shift of potassium from the intracellular to the extracellular fluid, a change that accompanies an increase in the plasma hydrogen ion concentration (acidosis). Possible effects of hyperkalemia include paralysis of the skeletal muscles and severe cardiac disturbances, such as cardiac arrest.

- **Aerobic respiration of glucose** This process produces carbon dioxide and water. Carbon dioxide diffuses out of the cells and reacts with the water in the extracellular fluids to form *carbonic acid,* which then ionizes to release hydrogen ions and bicarbonate ions:

$$H_2CO_3 \rightarrow H^+ + HCO_3^-$$

- **Anaerobic respiration of glucose** Anaerobically metabolized glucose produces *lactic acid,* which adds hydrogen ions to body fluids.
- **Incomplete oxidation of fatty acids** This process produces *acidic ketone bodies,* which increase the hydrogen ion concentration.

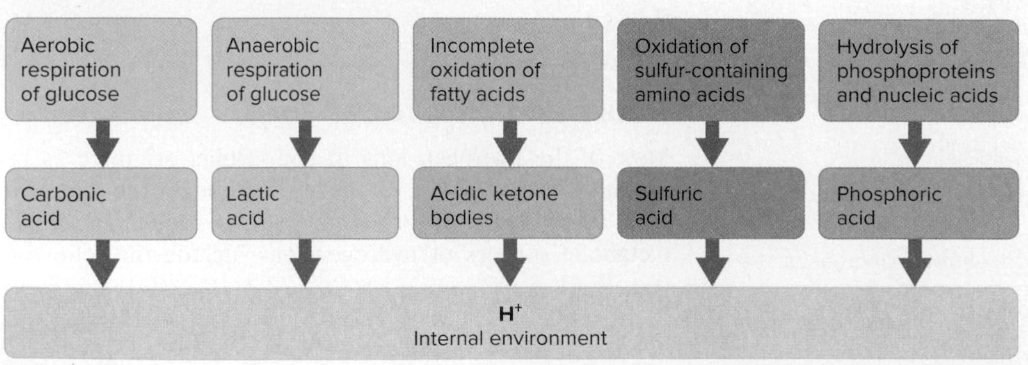

Figure 18.6 Some of the metabolic processes that provide hydrogen ions.

- **Oxidation of amino acids containing sulfur** This process yields *sulfuric acid* (H_2SO_4), which ionizes to release hydrogen ions.
- **Hydrolysis of phosphoproteins and nucleic acids** Phosphoproteins and nucleic acids contain phosphorus. Their oxidation produces *phosphoric acid* (H_3PO_4), which ionizes to release hydrogen ions.

The acids resulting from metabolism vary in strength. Therefore, their effects on the hydrogen ion concentration of body fluids vary.

PRACTICE 18.5

1. Distinguish between an acid and a base.
2. What are the major sources of hydrogen ions in the body?

Strengths of Acids and Bases

Acids that ionize more completely are *strong acids*, and those that ionize less completely are *weak acids*. For example, the hydrochloric acid (HCl) of gastric juice is a strong acid, but the carbonic acid (H_2CO_3) produced when carbon dioxide reacts with water is weak.

Bases release ions, such as hydroxide ions (OH^-), which can combine with hydrogen ions to form water (H_2O), thereby lowering their concentration. Sodium hydroxide (NaOH), which releases hydroxide ions, and sodium bicarbonate ($NaHCO_3$), which releases bicarbonate ions (HCO_3^-), are bases. *Strong bases* dissociate to release more OH^- or its equivalent than do *weak bases*. Often, the negative ions themselves are called bases. For example, HCO_3^- acting as a base combines with H^+ from the strong acid HCl to form the weak acid carbonic acid (H_2CO_3).

Regulation of Hydrogen Ion Concentration

Chemical buffer systems, the respiratory center in the brainstem, and the nephrons in the kidneys regulate hydrogen ion concentration in body fluids. The pH scale (see fig. 2.11) is used to express the hydrogen ion concentration.

Chemical Buffer Systems

Chemical buffer systems, in all body fluids, consist of chemicals that combine with excess acids or bases. More specifically, buffering systems help control the levels of H^+ and OH^- ions in an effort to maintain pH levels. The three most important chemical buffer systems in the body fluids are:

1. **Bicarbonate buffer system** The bicarbonate buffer system, which is present in both intracellular and extracellular fluids, uses the bicarbonate ion (HCO_3^-), acting as a weak base, and carbonic acid (H_2CO_3), acting as a weak acid. In the presence of excess hydrogen ions, bicarbonate ions combine with hydrogen ions to form carbonic acid, thereby minimizing any increase in the hydrogen ion concentration of the body fluids:

$$H^+ + HCO_3^- \rightarrow H_2CO_3$$

On the other hand, if conditions are basic or alkaline, carbonic acid dissociates to release bicarbonate ions and hydrogen ions:

$$H_2CO_3 \rightarrow H^+ + HCO_3^-$$

Although this reaction releases bicarbonate ions, it is the increase of free hydrogen ions at equilibrium that minimizes the shift toward a more alkaline pH.

2. **Phosphate buffer system** The phosphate buffer system also operates in both intracellular and extracellular body fluids. However, it is particularly important in the control of hydrogen ion concentrations in the ICF, tubular fluid of the nephrons, and in urine, where phosphate concentrations are higher. This buffer system consists of two phosphate ions, monohydrogen phosphate (HPO_4^{-2}), acting as a weak base, and dihydrogen phosphate ($H_2PO_4^-$), acting as a weak acid. In the presence of excess hydrogen ions, monohydrogen phosphate ions combine with hydrogen ions to form dihydrogen phosphate, thereby minimizing the increase in the hydrogen ion concentration of body fluids:

$$H^+ + HPO_4^{-2} \rightarrow H_2PO_4^-$$

On the other hand, if conditions are basic or alkaline, dihydrogen phosphate dissociates to release monohydrogen phosphate and hydrogen ions:

$$H_2PO_4^{-2} \rightarrow H^+ + H_2PO_4^{-2}$$

3. **Protein buffer system** The protein buffer system consists of the plasma proteins, such as albumins, and certain proteins in cells, including the hemoglobin of red blood cells. As described in section 2.6, Chemical Constituents of Cells, proteins are chains of amino acids. Some of these amino acids have freely exposed amino groups ($-NH_2$). If the solution pH falls (the hydrogen ion concentration rises), these amino groups can accept hydrogen ions:

$$-NH_2 + H^+ \rightarrow -NH_3^+$$

Some amino acids of a protein also have freely exposed *carboxyl groups* (–COOH). If the solution pH rises (the hydrogen ion concentration falls), these carboxyl groups can ionize, releasing hydrogen ions:

$$-COOH \rightarrow -COO^- + H^+$$

TABLE 18.1 Chemical Buffer Systems

Buffer System	Constituents	Actions
Bicarbonate system	Bicarbonate ion (HCO_3^-) Carbonic acid (H_2CO_3)	Combines with a hydrogen ion in the presence of excess acid Releases a hydrogen ion in the presence of excess base
Phosphate system	Monohydrogen phosphate ion (HPO_4^{-2}) Dihydrogen phosphate ion ($H_2PO_4^-$)	Combines with a hydrogen ion in the presence of excess acid Releases a hydrogen ion in the presence of excess base
Protein system (and amino acids)	$-NH_2$ group of an amino acid or protein $-COOH$ group of an amino acid or protein	Combines with a hydrogen ion in the presence of excess acid Releases a hydrogen ion in the presence of excess base

Therefore, protein molecules can function as bases by accepting hydrogen ions into their amino groups or as acids by releasing hydrogen ions from their carboxyl groups. This special property allows protein molecules to operate as an acid-base buffer system, minimizing changes in pH.

Table 18.1 summarizes the actions of the three major chemical buffer systems.

PRACTICE 18.5

3. What is the difference between a strong acid or base and a weak acid or base?
4. How does a chemical buffer system help regulate the pH of body fluids?
5. List the major chemical buffer systems of the body.

Respiratory Excretion of Carbon Dioxide

The respiratory center in the brainstem helps regulate the hydrogen ion concentrations in the body fluids by controlling the rate and depth of breathing (see section 16.4, Control of Breathing). Figure 18.7 traces this process. Specifically, if body cells increase their production of carbon dioxide, as occurs during periods of physical exertion, carbonic acid production increases. As the carbonic acid dissociates, the concentration of hydrogen ions increases, and the pH of the internal environment drops. Such an increasing concentration of carbon dioxide in the central nervous system and the subsequent increase in the hydrogen ion concentration in the cerebrospinal fluid stimulate chemosensitive areas in the respiratory center.

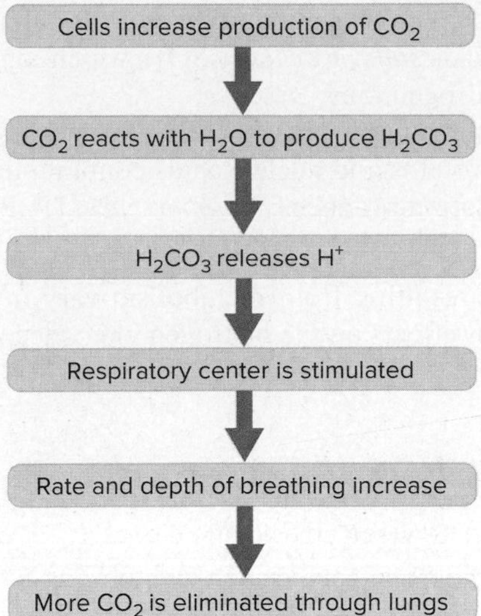

Figure 18.7 An increase in carbon dioxide production leads to an increase in carbon dioxide elimination.

In response to stimulation, the respiratory center increases the depth and rate of breathing, so that the lungs excrete more carbon dioxide. The hydrogen ion concentration in body fluids returns toward normal because the released carbon dioxide comes from carbonic acid:

$$H_2CO_3 \rightarrow CO_2 + H_2O$$

Conversely, if body cells are less active, concentrations of carbon dioxide and hydrogen ions in body fluids remain low. As a result, breathing rate and depth stay closer to resting levels.

Renal Excretion of Hydrogen Ions

Nephrons help regulate the hydrogen ion concentration of body fluids by excreting hydrogen ions in urine. Recall from section 17.3, Urine Formation, that epithelial cells lining certain segments of the renal tubules secrete hydrogen ions into the tubular fluid.

Time Course of Hydrogen Ion Regulation

The various regulators of the hydrogen ion concentration operate at different rates. Chemical buffers can convert strong acids or bases into weak acids or bases almost immediately. For this reason, these chemical buffer systems are called the body's *first line of defense* against shifts in pH.

Physiological buffer systems, such as the respiratory and renal mechanisms, function more slowly and constitute the *second line of defense* against shifts in pH. The respiratory mechanism may require several minutes to begin resisting a change in pH, and the renal mechanism may require one to three days to regulate a changing hydrogen ion concentration. Figure 18.8 compares the actions of chemical buffers and physiological buffers.

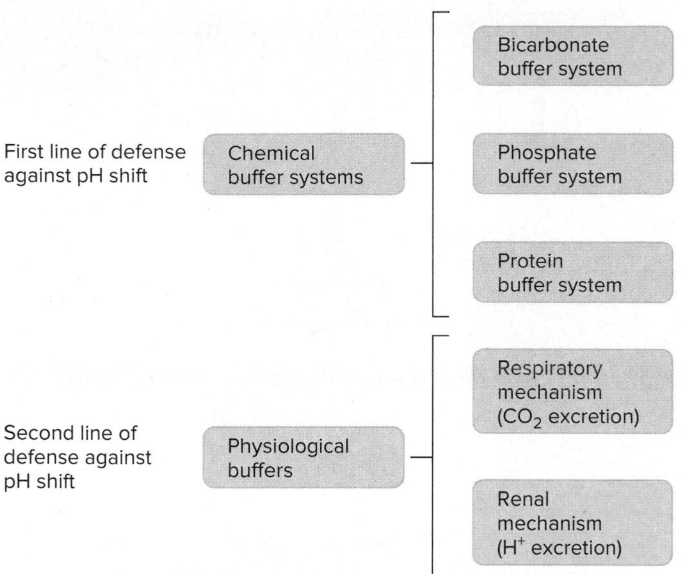

Figure 18.8 Chemical buffers act rapidly, while physiological buffers may require several minutes to several days to begin resisting a change in pH.

PRACTICE FIGURE 18.8

How does respiratory excretion of CO_2 buffer the pH of body fluids?

Answer can be found in Appendix E.

PRACTICE 18.5

6. How does the respiratory system help regulate acid-base balance?
7. How do the kidneys respond to excess hydrogen ions?
8. How do the reaction rates of chemical and physiological buffer systems differ?

18.6 | Acid-Base Imbalances

LEARN

1. Describe the causes and consequences of an increase or a decrease in body fluid pH.

Chemical and physiological buffer systems ordinarily maintain the hydrogen ion concentration of body fluids within very narrow pH ranges. Abnormal conditions may disturb the acid-base balance. For example, the pH of arterial blood is normally 7.35 to 7.45. A pH value below 7.35 produces **acidosis**. A pH above 7.45 produces **alkalosis**. Such shifts in the pH of body fluids can be life-threatening. A person usually cannot survive if the pH drops to 6.8 or rises to 8.0 for more than a few hours (fig. 18.9). Neurons are particularly sensitive to changes in the pH of body fluids. If the interstitial fluid becomes more alkaline than normal, neurons become more excitable and seizures may result. Conversely, acidic conditions depress neuron activity, and may reduce the level of consciousness.

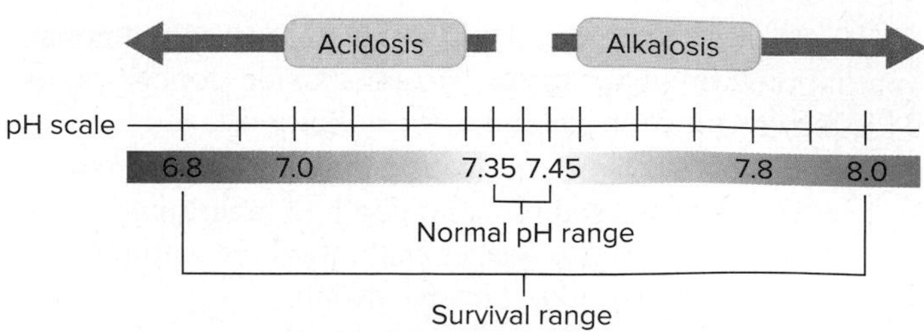

Figure 18.9 If the pH of arterial blood drops to 6.8 or rises to 8.0 for more than a few hours, the person usually cannot survive.

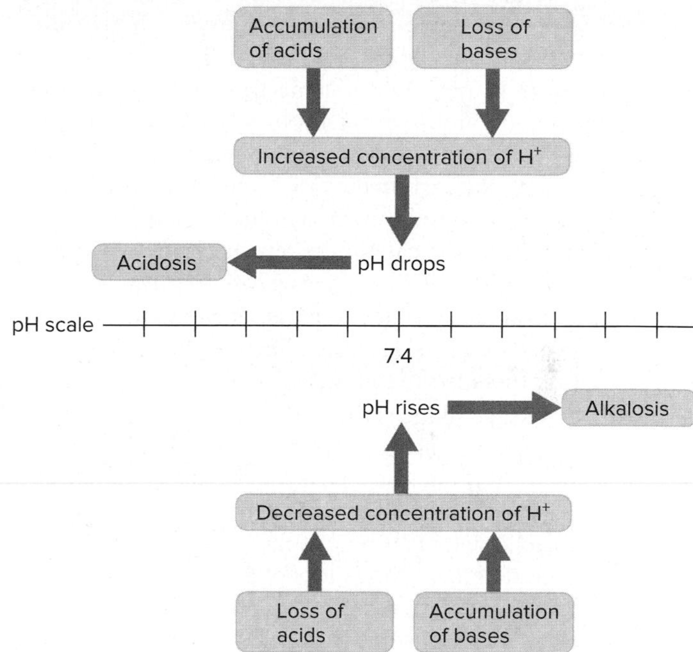

Figure 18.10 Acidosis results from accumulation of acids or loss of bases. Alkalosis results from loss of acids or accumulation of bases.

Acidosis results from an accumulation of acids or a loss of bases, both of which cause abnormal increases in the hydrogen ion concentrations of body fluids. Conversely, alkalosis results from a loss of acids or an accumulation of bases accompanied by a decrease in the hydrogen ion concentrations (fig. 18.10).

Acidosis

The symptoms of acidosis result from depression of central nervous system function. They include drowsiness, disorientation, stupor, and cyanosis.

The two major types of acidosis are *respiratory acidosis* and *metabolic acidosis*. Factors that increase carbon dioxide levels, which increases the concentration of carbonic acid (the respiratory acid), cause respiratory acidosis. Metabolic acidosis is due to an abnormal accumulation of any other acids in the body fluids or to a loss of bases, including bicarbonate ions.

Respiratory acidosis may be due to hindered pulmonary ventilation, which increases carbon dioxide concentration. This may result from the following conditions:

- Injury to the respiratory center of the brainstem that results in decreased rate and depth of breathing.
- Obstruction in air passages that interferes with air movement into and out of the alveoli.
- Diseases that decrease gas exchange, such as pneumonia or emphysema.

Figure 18.11 summarizes the factors that can lead to respiratory acidosis. Any of these conditions can increase the level of carbonic acid and hydrogen ions in body fluids, lowering pH. Chemical buffers, such as hemoglobin, may resist this shift in pH. At the same time, increasing levels of carbon dioxide and hydrogen ions stimulate the respiratory center, increasing the breathing rate and depth and thereby lowering the carbon dioxide levels. Also, the kidneys may begin to excrete more hydrogen ions. Eventually, these chemical and physiological buffers return the pH of the body fluids to normal. The acidosis is thus *compensated*.

Metabolic acidosis is due to either accumulation of nonrespiratory acids or loss of bases. Factors that may lead to this condition include the following:

- Kidney disease reduces the ability of the kidneys to excrete acids produced in metabolism (uremic acidosis).
- Prolonged vomiting loses the acidic stomach contents and alkaline contents of the small intestine. (Losing only the stomach contents produces metabolic alkalosis.)
- Prolonged diarrhea causes loss of alkaline intestinal secretions (especially in infants).
- In diabetes mellitus, some fatty acids react to produce ketone bodies, such as *acetoacetic acid*, *beta-hydroxybutyric acid*, and *acetone*. Normally, these molecules are scarce, and cells oxidize them as energy sources. However, if fats are used at an abnormally high rate, as may occur in diabetes mellitus, ketone bodies may accumulate faster than they can be oxidized and, as a result, spill over into the urine (ketonuria). In addition, the lungs may release acetone, which is volatile and imparts a fruity odor to the breath. More seriously, the accumulation of acetoacetic acid and beta-hydroxybutyric acid in the blood may lower pH (ketonemic acidosis).

Figure 18.12 summarizes the factors leading to metabolic acidosis. In each case, pH is lowered. Countering this lower pH are chemical buffer systems, which accept excess hydrogen ions; the respiratory center, which increases breathing rate and depth to eliminate more CO_2; and the kidneys, which excrete more hydrogen ions.

Alkalosis

The symptoms of alkalosis include light-headedness, agitation, dizziness, and tingling sensations. In severe cases, impulses may be triggered spontaneously on motor neurons, and muscles may respond with tetanic contractions (see section 8.4, Muscular Responses).

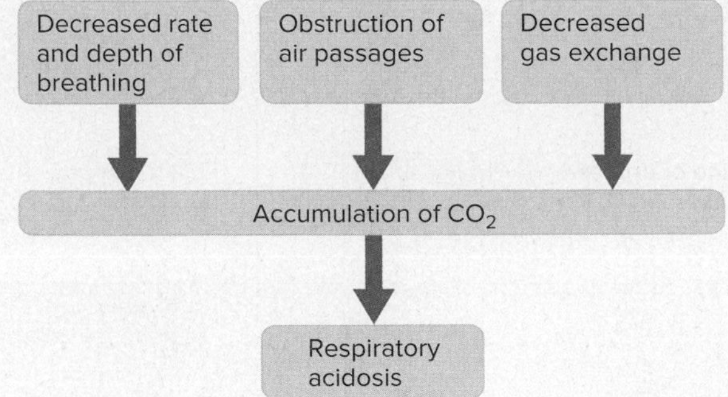

Figure 18.11 Some of the factors that lead to respiratory acidosis.

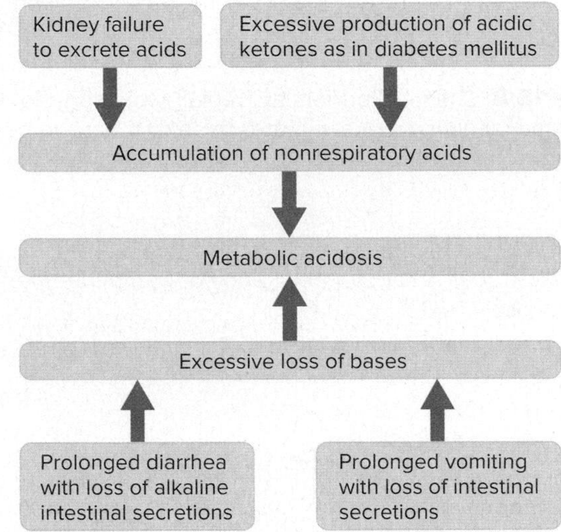

Figure 18.12 Some of the factors that lead to metabolic acidosis.

The two major types of alkalosis are *respiratory alkalosis* and *metabolic alkalosis*. Respiratory alkalosis results from excessive loss of carbon dioxide and consequent loss of carbonic acid. Metabolic alkalosis is due to excessive loss of hydrogen ions or gain of bases.

Respiratory alkalosis develops as a result of **hyperventilation** (described in section 16.4, Control of Breathing), in which too much carbon dioxide is lost, decreasing carbonic acid and hydrogen ion concentrations. Hyperventilation may happen during periods of anxiety or may accompany fever or poisoning from salicylates, such as aspirin. At high altitudes, hyperventilation may be a response to low oxygen partial pressure. Musicians may hyperventilate to provide the large volume of air needed to play sustained passages on wind instruments. In each case, rapid, deep breathing depletes carbon dioxide, and the pH of body fluids increases. Figure 18.13 illustrates the factors leading to respiratory alkalosis.

Chemical buffers, such as hemoglobin, that release hydrogen ions resist the increase in pH. The lower levels

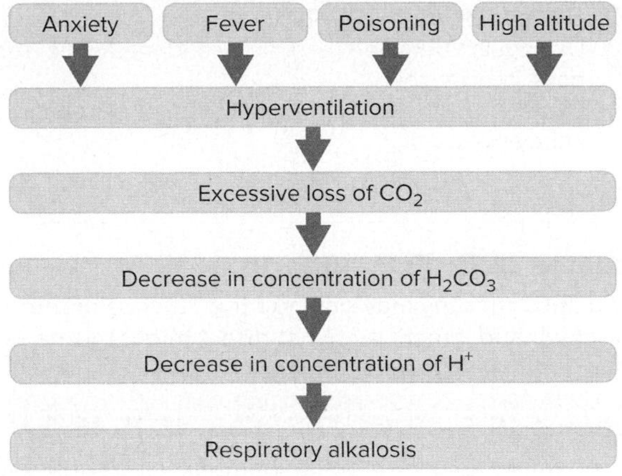

Figure 18.13 Some of the factors that lead to respiratory alkalosis.

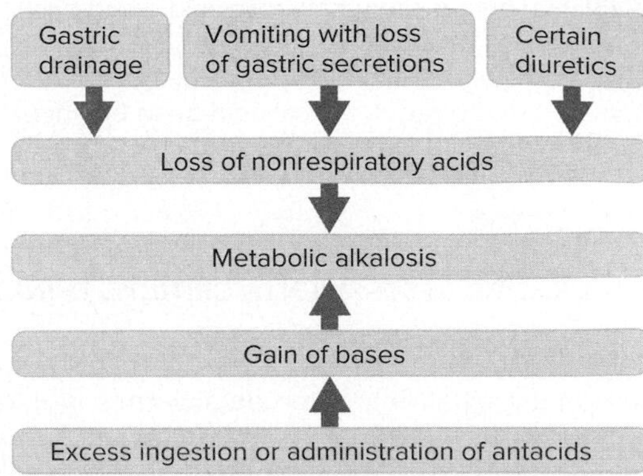

Figure 18.14 Some of the factors that lead to metabolic alkalosis.

of carbon dioxide and hydrogen ions stimulate the respiratory center to a lesser degree. This inhibits hyperventilation, thereby reducing further carbon dioxide loss. At the same time, the kidneys decrease their secretion of hydrogen ions, and the urine becomes alkaline as bases are excreted.

Metabolic alkalosis results from a loss of hydrogen ions or from a gain in bases, both accompanied by a rise in the pH of the blood (alkalemia). This condition may occur following gastric drainage (lavage), vomiting in which only the stomach contents are lost, or the use of certain diuretic drugs. Gastric juice is acidic, so its loss leaves body fluids more basic. Metabolic alkalosis may also develop as a result of ingesting too much antacid, such as sodium bicarbonate to relieve the symptoms of indigestion. Compensation for metabolic alkalosis includes a decrease in the breathing rate and depth, which in turn results in an increased concentration of carbon dioxide in the blood. Figure 18.14 illustrates the factors leading to metabolic alkalosis.

PRACTICE 18.6

1. What is the difference between a respiratory acid-base disturbance and a metabolic acid-base disturbance?
2. How do the symptoms of acidosis compare with those of alkalosis?

ASSESS

CHAPTER ASSESSMENTS

18.1 Introduction

1. Explain how water balance and electrolyte balance are interdependent.

18.2 Distribution of Body Fluids

2. Water and electrolytes enclosed by cell membranes constitute the ________.
   a. transcellular fluid
   b. intracellular fluid
   c. extracellular fluid
   d. lymph
3. Explain how the fluids in the compartments differ in composition.
4. Describe how fluid movements between the compartments are regulated.

18.3 Water Balance

5. Prepare a list of sources of normal water gain and loss to illustrate how the input of water equals the output of water.
6. Define *metabolic water.*
7. Explain how water intake is regulated.
8. Explain how the kidneys regulate water output.

18.4 Electrolyte Balance

9. Electrolytes in body fluids of importance to cellular functions include ________.
   a. sodium
   b. potassium
   c. calcium
   d. all of the above
10. Explain how electrolyte intake is regulated.
11. List the routes by which electrolytes leave the body.
12. Explain how the adrenal cortex functions to regulate electrolyte balance.
13. Describe the role of the parathyroid glands in regulating electrolyte balance.

18.5 Acid-Base Balance

14. List five sources of hydrogen ions in body fluids, and name an acid that originates from each source.
15. ________ ionize more completely. An example is hydrochloric acid.
16. ________ dissociate to release fewer hydroxide ions.
17. Explain how the bicarbonate and phosphate buffer systems resist pH changes.

18. Explain why a protein has both acidic and basic properties.
19. Explain how the respiratory system and the kidneys function in the regulation of acid-base balance.

18.6 Acid-Base Imbalances

20. Distinguish between respiratory and metabolic acid-base imbalances.
21. Explain how the body compensates for acid-base imbalances.

ASSESS

INTEGRATIVE ASSESSMENTS/CRITICAL THINKING

Outcomes 3.3, 18.3

1. While running a half-marathon on a warm spring day, Tina chose not to drink fluids along the route because she did not want to disrupt her event time, so she just kept on running without hydrating. After the race, Tina quickly consumed a large quantity of water because she was so thirsty. It didn't take long before her head ached, and she felt weak and nauseous. What was the likely cause of her symptoms?

Outcomes 11.6, 18.4

2. Grandma was feeling disoriented and had a lack of appetite, so her son took her to see a doctor. It turns out that Grandma was excessively consuming Tums ($CaCO_3$), Alka-Seltzer ($NaHCO_3$), and milk daily to help soothe her ulcer symptoms. Blood test results showed that her blood calcium levels were elevated. What was Grandma's acid-base imbalance as a result of consuming these products? Grandma's physician expected her parathyroid hormone (PTH) levels would be decreased. Why?

Outcomes 13.2, 13.4, 13.5, 14.3, 18.2

3. If the right ventricle of a patient's heart is failing, increasing the systemic venous pressure, what changes might occur in the patient's extracellular fluid compartments?

Outcomes 15.2, 15.6, 15.9, 18.4, 18.6

4. Radiation therapy may damage the mucosa of the stomach and intestines. What effect might this have on the patient's electrolyte balance?

Outcomes 15.9, 15.10, 17.2, 17.3, 18.5, 18.6

5. After eating an undercooked hamburger, a twenty-five-year-old male developed diarrhea due to infection with a strain of *Escherichia coli* that produces a shigatoxin. How would this affect his blood pH, urine pH, and respiratory rate?

Outcomes 15.11, 17.3, 18.5, 18.6

6. A distraught, grieving widow had not been eating much food for days. Her breath had a fruity odor. Explain how her reduced caloric intake will affect the pH of her urine.

Outcomes 16.4, 16.6, 18.5, 18.6

7. A student hyperventilates and is disoriented just before an exam. Is this student likely to be experiencing acidosis or alkalosis? How will the body compensate in an effort to maintain homeostasis?

Chapter Summary

18.1 Introduction

Maintenance of water and electrolyte balance requires that the quantities of these substances entering the body equal the quantities leaving it. Altering the water balance affects the electrolyte balance.

18.2 Distribution of Body Fluids

1. Fluid compartments
   a. The **intracellular fluid compartment** includes the fluids and electrolytes cell membranes enclose.
   b. The **extracellular fluid compartment** includes all the fluids and electrolytes outside cell membranes.
2. Body fluid composition
   a. Extracellular fluids have high concentrations of sodium, chloride, calcium, and bicarbonate ions. Plasma contains more protein than does either interstitial fluid or lymph.
   b. Intracellular fluid contains high concentrations of potassium, magnesium, and phosphate ions. It also has a greater concentration of sulfate ions than extracellular fluid and a greater concentration of proteins than plasma.
3. Movement of fluid between compartments
   a. Hydrostatic and osmotic pressure regulate fluid movements.
      (1) Hydrostatic pressure forces fluid out of plasma, and colloid osmotic pressure returns fluid to plasma.
      (2) Hydrostatic pressure drives fluid into lymph vessels.
      (3) Osmotic pressure regulates fluid movement in and out of blood vessels and cells.
   b. Sodium ion concentrations are especially important in regulating fluid movement.

18.3 Water Balance

1. Water intake
   a. Most water comes from consuming liquids or moist foods.
   b. Oxidative metabolism produces some water.
   c. Thirst is the primary regulator of water intake.

d. Drinking and the resulting stomach distension inhibit thirst.
2. Water output
 a. Water is lost in urine, feces, and sweat, and by evaporation from the skin and lungs.
 b. The distal convoluted tubules of the nephrons and the collecting ducts are the effectors of the control system that regulate water output.

18.4 Electrolyte Balance

1. Electrolyte intake
 a. The electrolytes of greatest importance to cellular functions in body fluids dissociate to release ions of sodium, potassium, calcium, magnesium, chloride, sulfate, phosphate, bicarbonate, and hydrogen.
 b. These ions are obtained in foods and beverages or as by-products of metabolic processes.
 c. Food and drink usually provide sufficient electrolytes.
 d. A severe electrolyte deficiency may produce a salt craving.
2. Electrolyte output
 a. Electrolytes are lost through perspiration, feces, and urine.
 b. Quantities lost vary with temperature and physical exercise.
 c. Most electrolytes are lost as a result of kidney function.
 d. Concentrations of sodium, potassium, and calcium ions in body fluids are particularly important.
 e. The adrenal cortex secretes aldosterone to regulate sodium and potassium ions.
 f. Parathyroid hormone regulates calcium ion concentration in body fluids.
 g. The mechanisms that control positively charged ions secondarily regulate negatively charged ions.

18.5 Acid-Base Balance

***Acids** are electrolytes that release hydrogen ions. **Bases** release ions that combine with hydrogen ions. Body fluid pH must remain within a certain range.*

1. Sources of hydrogen ions
 a. **Aerobic respiration of glucose** produces carbonic acid.
 b. **Anaerobic respiration of glucose** produces lactic acid.
 c. **Incomplete oxidation of fatty acids** releases acidic ketone bodies.
 d. **Oxidation of amino acids containing sulfur** produces sulfuric acid.
 e. **Hydrolysis of phosphoproteins and nucleic acids** produces phosphoric acid.
2. Strengths of acids and bases
 a. Acids vary in the extent to which they ionize to release ions.
  (1) Strong acids, such as hydrochloric acid, ionize more completely.
  (2) Weak acids, such as carbonic acid, ionize less completely.
 b. Bases also vary in strength.
3. Regulation of hydrogen ion concentration
 a. **Chemical buffer systems**
  (1) Buffer systems convert strong acids into weaker acids or strong bases into weaker bases.
  (2) They include the **bicarbonate buffer system, phosphate buffer system,** and **protein buffer system.**
  (3) Buffer systems minimize pH changes.
 b. The respiratory center controls the rate and depth of breathing to regulate pH.
 c. The kidneys excrete hydrogen ions to regulate pH.
 d. Chemical buffers act more rapidly. Physiological buffers act more slowly.

18.6 Acid-Base Imbalances

1. **Acidosis**
 a. Respiratory acidosis results from increased levels of carbon dioxide and carbonic acid.
 b. Metabolic acidosis results from accumulation of nonrespiratory acids or loss of bases.
2. **Alkalosis**
 a. Respiratory alkalosis results from loss of carbon dioxide and carbonic acid.
 b. Metabolic alkalosis results from loss of hydrogen ions or gain of bases.

19

UNIT 6 THE HUMAN LIFE CYCLE

Reproductive Systems

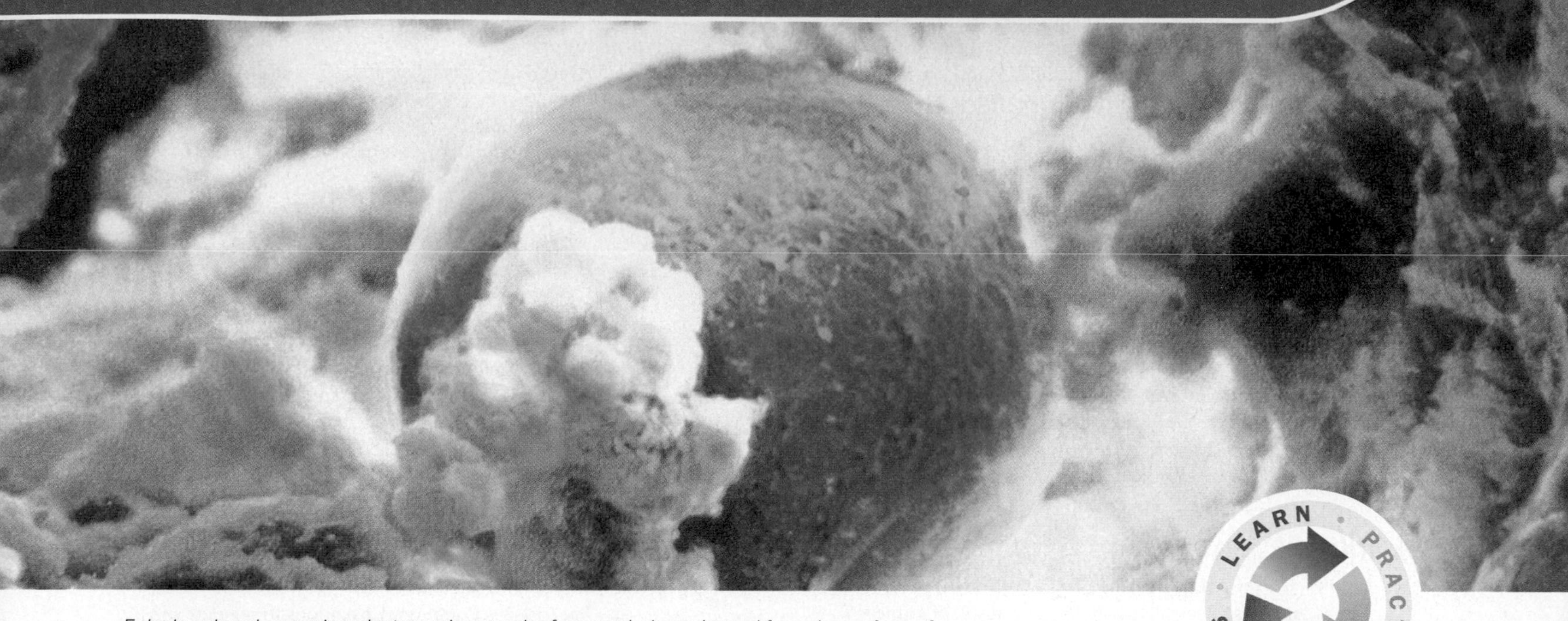

Falsely colored, scanning electron micrograph of an egg being released from the surface of an ovary.
Profs. P.M. Motta & J. Van Blerkom/Science Source

"Selling eggs." The ad in the student newspaper seemed too good to be true—the fee for donating a few eggs would pay nearly a semester's tuition. Intrigued, the young woman submitted a health history, had a checkup, and a month later received a call. A young couple, Linda and Tom, who were struggling with infertility sought an egg donor. They'd chosen Sherrie because, with her strawberry-blond hair, she looked like Linda, whose cancer had left her unable to conceive. The donor eggs would be fertilized in a laboratory dish (*in vitro*) with sperm from Linda's partner, and then implanted in Linda's uterus.

For two weeks, Sherrie injected herself in the thigh with a drug that acts like gonadotropin-releasing hormone, suppressing release of an egg from an ovary (ovulation). When daily hormone checks indicated that her endocrine system was in sync with Linda's, Sherrie began giving herself shots twice a day at the back of the hip. This second drug mimicked a follicle-stimulating hormone, and it caused several ovarian follicles, the structures that contain developing eggs, to mature. Finally, injections of luteinizing hormone brought the eggs close to being released. Then, at a health-care facility, Sherrie received pain medication and light sedation. A needle inserted through her vaginal wall retrieved a dozen mature eggs from the surface of her ovary.

Two *in vitro* fertilized ova divided a few times, then the embryos were implanted into Linda's uterus. The rest were frozen for possible later use. The preparation and procedure weren't too painful. Sherrie had felt a dull aching the last day before, and experienced bloating for a few days after the egg retrieval, but she did not experience bleeding, infection, cramping, or mood swings.

Module 14 Reproductive Systems

LEARNING OUTLINE

After studying this chapter, you should be able to complete the "Learning Outcomes" that follow the major headings throughout the chapter.

AIDS TO UNDERSTANDING WORDS

andr- [man] *andr*ogens: male sex hormones.

ejacul- [to shoot forth] *ejacul*ation: expulsion of semen from the male reproductive tract.

fimb- [fringe] *fimb*riae: irregular extensions on the margin of the infundibulum of the uterine tube.

follic- [small bag] *follic*le: ovarian structure that contains an egg.

-genesis [origin] spermato*genesis:* formation of sperm cells.

labi- [lip] *labi*a minora: flattened, longitudinal folds that extend along the margins of the female vestibule.

mens- [month] *mens*es: monthly flow of blood from the female reproductive tract.

mons- [an eminence] *mons* pubis: rounded elevation of fatty tissue overlying the pubic symphysis in a female.

puber- [adult] *puber*ty: the time in life when a person becomes able to reproduce.

(*Appendix A has a complete list of Aids to Understanding Words.*)

19.1 | Introduction

LEARN

1. State the general functions of the male and female reproductive systems.

The reproductive system is the one organ system a person can live without. It only serves to develop and sustain organs essential for producing offspring. These organs fully develop during puberty and can function well into old age in males. Females have a smaller window for reproductive viability, but invest much more into the process.

The **gonads** are structures that produce hormones and **gametes,** or sex cells. Male gonads are the testes (testicles) that produce sperm cells. Ovaries, the gonads in females, produce **ova,** or eggs.

The male's sole reproductive contribution is to serve as a "delivery system" for sperm to fertilize eggs. A female reproductive system houses, protects, and nourishes the developing embryo until birth. For this reason, it is much more anatomically and physiologically complex than that of a male.

PRACTICE 19.1

Answers to the Practice questions can be found in the eBook.

1. What are the male and the female sex cells called?

19.2 | Organs of the Male Reproductive System

LEARN

1. Describe the structure and function(s) of each part of the male reproductive system.
2. Describe semen production and exit from the body.

The **primary sex organs** (gonads) of the male reproductive system are the two testes, in which sperm cells develop and the male sex hormones are synthesized. The *accessory sex organs* (secondary sex organs) of the male reproductive system are the internal and external reproductive organs (fig. 19.1; reference plates 3 and 4).

Testes

The **testes** (tes′tēz; sing., *testis*) are ovoid structures about 5 centimeters in length and 3 centimeters in diameter. Each testis is suspended from the trunk by a spermatic cord that contains blood vessels, nerves, and the ductus deferens. Both testes are within the cavity of the saclike *scrotum.*

A tough, white, fibrous capsule, called the **tunica albuginea,** encloses each testis. Along the capsule's posterior border, the connective tissue thickens and extends into the testis, forming thin septa that divide the testis into about 250 *lobules* (fig. 19.2*a*).

A lobule contains one to four highly coiled, convoluted **seminiferous tubules** (se″mĭ-nif′er-us too′būlz), each approximately 70 centimeters long uncoiled. These tubules course posteriorly and unite to form a complex network of channels that give rise to several ducts that join a tube called the *epididymis.* The epididymis is coiled on the outer surface of the testis and continues to become the *ductus deferens* (fig. 19.2*a*).

The epithelial cells of the seminiferous tubules can give rise to *testicular cancer,* a common cancer in young men. In most cases, the first sign is a painless testis enlargement or a small lump or area of hardness on the testis. If a biopsy (tissue sample) reveals cancer cells, surgery is performed to remove the affected testis (orchiectomy). Radiation and/or chemotherapy often prevents the cancer from recurring.

A specialized stratified epithelium with **spermatogenic** (sper″mah-to-jen′ik) **cells** (germ cells), which give rise to sperm cells, lines the seminiferous tubules. Other specialized cells, called **interstitial cells** (cells of Leydig), lie in the spaces between the seminiferous tubules (fig. 19.2*b,c*). Interstitial cells produce and secrete male sex hormones.

PRACTICE 19.2

1. Describe the structure of a testis.
2. Where in the testes are the sperm cells produced?
3. Which cells produce male sex hormones?

Male Internal Accessory Reproductive Organs

The internal accessory organs of the male reproductive system are specialized to nurture and transport sperm cells. Each testis has an associated epididymis, ductus deferens, and seminal vesicle. Other supportive structures include the prostate gland and the two bulbourethral glands.

Epididymis

The **epididymis** is a tightly coiled tube about 6 meters long (see figs. 19.1 and 19.2). Each epididymis is connected to ducts within a testis. It emerges from the top of the testis, descends along the posterior surface of the testis, and then courses upward to become the ductus deferens.

When sperm cells reach the epididymis, they are non-motile. As rhythmic peristaltic contractions help move these cells through the epididymis, the cells mature. Following this aging process, the sperm cells can move independently and fertilize oocytes. However, sperm cells usually do not move independently until after ejaculation.

Ductus Deferens

The **ductus deferens,** also called the *vas deferens,* is a muscular tube about 45 centimeters long (see fig. 19.1). Each ductus deferens passes upward along the medial side of a testis and through a passage in the lower abdominal wall (inguinal canal), enters the pelvic cavity, and ends behind the urinary bladder. Just outside the prostate gland, the ductus deferens unites with the duct of a seminal vesicle to form an **ejaculatory duct,** which passes through the prostate gland and empties into the urethra.

Seminal Vesicles

The **seminal vesicles** are convoluted, saclike structures about 5 centimeters long. Each attaches to a ductus deferens on the posterior surface and near the base of the urinary bladder (see fig. 19.1). The glandular tissue lining the inner wall of a seminal vesicle secretes a slightly alkaline fluid. This fluid helps regulate the pH of the tubular contents as sperm cells travel to the outside. Additionally, seminal vesicle fluid neutralizes the acidic secretions of the vagina, helping to sustain sperm cells that enter the female reproductive tract. Seminal vesicle secretions also include *fructose,* a monosaccharide that provides energy to sperm cells, and **prostaglandins** (see section 11.2, Hormone Action), which stimulate muscular contractions within the female reproductive organs, aiding the movement of sperm cells toward the oocyte.

PRACTICE 19.2

4. Describe the structure of the epididymis.
5. Trace the path of the ductus deferens.
6. What is the function of a seminal vesicle?

CAREER CORNER

Nurse-Midwife

A couple chooses a certified nurse-midwife (CNM) to deliver their third child because they prefer a home birth. Their first two children had been delivered in a hospital, but they are healthy and the births were uncomplicated.

A certified nurse-midwife (CNM) is a health-care professional who meets with the patient throughout the pregnancy, for at least an hour each time, and is present for the entire labor and birth. Working in the patient's home, at a birthing center, or at a hospital, the nurse-midwife uses soothing words and gentle massage to ease the baby into the world, placing the newborn on the mother's chest while completing the initial physical examination of the child.

A CNM can also provide primary care, gynecologic care, family planning, treatment for sexually transmitted infections, and newborn care for the first month of life. She or he can do physical exams, prescribe medications, and order and interpret medical tests to provide a diagnosis. A nurse-midwife is a registered nurse who has a graduate degree in midwifery from an accredited program, and has American Midwifery Certification Board approval.

In the United States, about 1.5% of babies are born at home, according to the Centers for Disease Control and Prevention National Center for Health Statistics. Because babies born at home face higher risks of complications, patients are carefully screened and plans are made so that a physician can arrive or the patient can be transported to a hospital within minutes. Risk factors that rule out a home birth include evidence of twins or breech presentation (buttocks first), previous cesarean section delivery, diabetes, high blood pressure, and being past the due date. But emergencies, such as the placenta detaching or the umbilical cord preceding the baby, are unpredictable. An important part of the CNM's job is to thoroughly evaluate the risks for patients to help them choose the safest site for delivery.

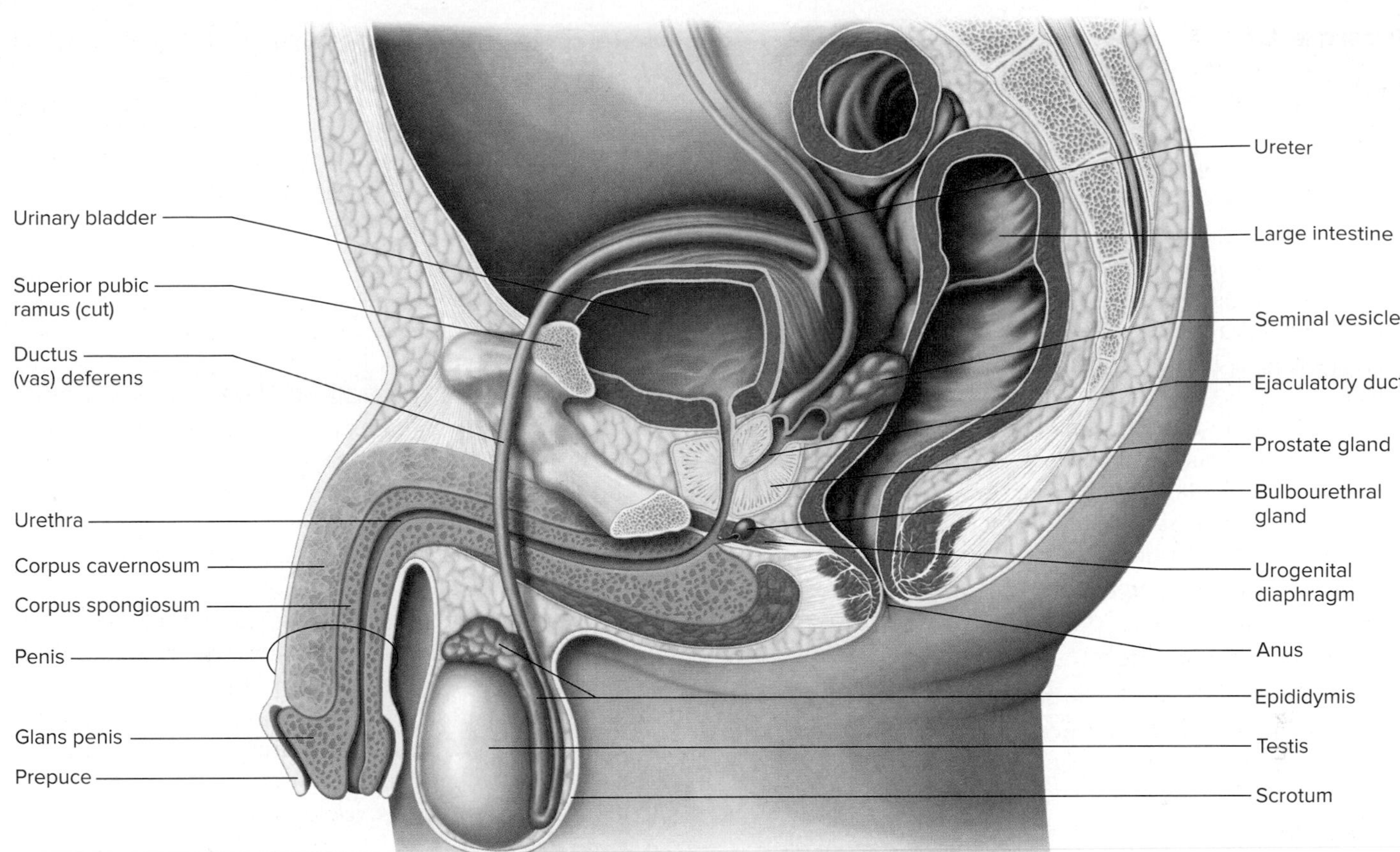

Figure 19.1 Male reproductive organs (sagittal view). The paired testes are the primary sex organs, and the other reproductive structures, both internal and external, are accessory sex organs. APR

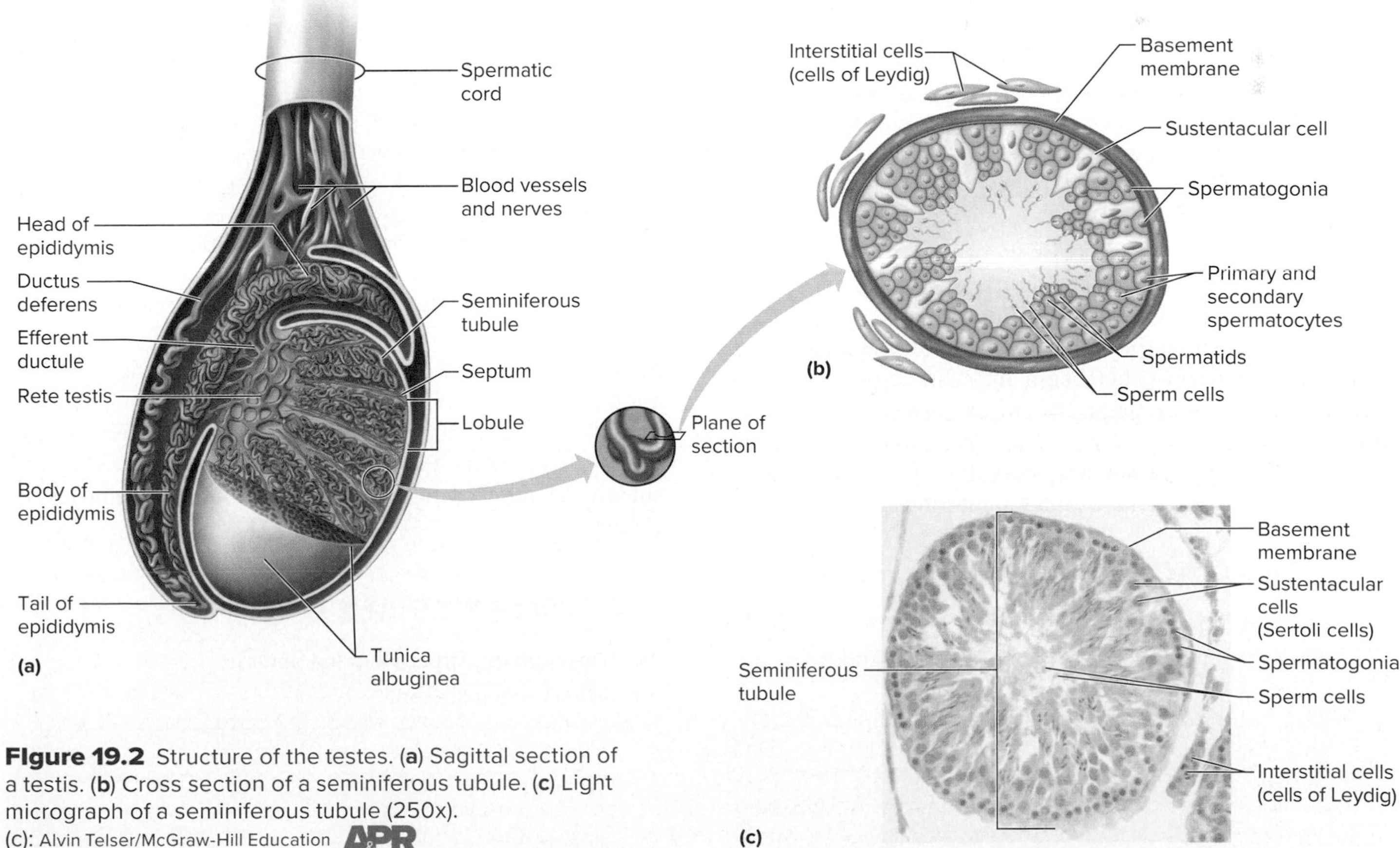

Figure 19.2 Structure of the testes. (**a**) Sagittal section of a testis. (**b**) Cross section of a seminiferous tubule. (**c**) Light micrograph of a seminiferous tubule (250x).
(c): Alvin Telser/McGraw-Hill Education APR

Prostate Gland

The **prostate** (pros′tāt) **gland** is a chestnut-shaped structure about 4 centimeters across and 3 centimeters thick that surrounds the proximal part of the urethra, just inferior to the urinary bladder (see fig. 19.1). It is enclosed in connective tissue and composed of many branched tubular glands whose ducts open into the urethra.

The prostate gland secretes a thin, milky fluid. This secretion includes citrate, which is a nutrient for the sperm, and prostate-specific antigen (PSA), which is an enzyme that helps liquefy semen. Clinical Application 19.1 discusses prostate cancer.

The prostate gland is small in boys, begins to grow in early adolescence, and reaches adult size several years later. Usually, the gland does not grow again until age fifty, when in about half of all men it enlarges enough to press on the urethra. The man feels the need to urinate frequently because the bladder cannot empty completely.

Bulbourethral Glands

The two **bulbourethral** (bul″bō-yoo-rē′thral) **glands** (Cowper's glands) are each about a centimeter in diameter and are inferior to the prostate gland (see fig. 19.1). Bulbourethral glands are composed of many tubes whose epithelial linings secrete a mucuslike fluid in response to sexual stimulation. This fluid lubricates the end of the penis in preparation for sexual intercourse. However, most of the lubricating fluid for sexual intercourse is secreted by glands of the female reproductive tract.

Semen

Semen (sē′men) is the fluid the male urethra conveys to the outside during ejaculation. It consists of sperm cells from the testes and secretions of the seminal vesicles, prostate gland, and bulbourethral glands. Semen is slightly alkaline (pH about 7.5), and it includes prostaglandins and nutrients.

The volume of semen released at one time varies from 2 to 5 milliliters. The average number of sperm cells in the fluid is about 120 million per milliliter.

Sperm cells remain nonmotile while in the ducts of the testis and epididymis, but begin to swim as they mix with accessory gland secretions. Sperm cells cannot fertilize an oocyte until they enter the female reproductive tract. Here, they undergo **capacitation,** a process that results in a weakening of the sperm cells' acrosomal membranes.

PRACTICE 19.2

7. Where is the prostate gland located?
8. What is the function of the bulbourethral glands?
9. What are the components of semen?

Male External Accessory Reproductive Organs

The external accessory organs of the male reproductive system are the scrotum, which encloses two testes, and the penis. The urethra passes through the penis.

Scrotum

The **scrotum** is a pouch of skin and subcutaneous tissue that hangs from the lower abdominal region posterior to the penis (see fig. 19.1). A medial septum divides the scrotum into two chambers, each of which encloses a testis. Each chamber also contains a serous membrane that covers the testis. The serous membrane secretes serous fluid, which reduces friction as the testis moves within the scrotum. The scrotum protects and helps regulate the temperature of the testes. These factors are important to sex cell production.

Exposure to cold stimulates the smooth muscle cells in the wall of the scrotum to contract, moving the testes closer to the pelvic cavity where they can absorb heat. Exposure to warmth stimulates the smooth muscle cells to relax and the scrotum to hang loosely, providing an environment 3°C (about 5°F) below body temperature, which is important to sperm production and survival.

Penis

The **penis** is a cylindrical organ that conveys urine and semen through the urethra to the outside (see fig. 19.1). During erection, it enlarges and stiffens, enabling it to be inserted into the vagina during sexual intercourse.

The *body,* or shaft, of the penis has three columns of erectile tissue—a pair of dorsally located *corpora cavernosa* and a single, ventral *corpus spongiosum.* A tough capsule of dense connective tissue surrounds each column. Skin, a thin layer of subcutaneous tissue, and a layer of connective tissue enclose the penis.

The corpus spongiosum, through which the urethra extends, enlarges at its distal end to form a sensitive, cone-shaped **glans penis.** The glans covers the ends of the corpora cavernosa and bears the urethral opening (external urethral orifice). The skin of the glans is very thin and hairless, and contains sensory receptors for sexual stimulation. A loose fold of skin called the *prepuce* (foreskin) originates just posterior to the glans and extends anteriorly to cover it as a sheath. A surgical procedure called *circumcision* removes the prepuce.

PRACTICE 19.2

10. Describe the structure of the penis.
11. What is circumcision?

CLINICAL APPLICATION 19.1

Prostate Cancer

Each year in the United States, nearly 240,000 men receive a diagnosis of prostate cancer and about 30,000 men die of the disease. The diagnostic process typically begins with a rectal exam, in which a health-care practitioner feels an enlargement of the prostate gland, and a blood test to detect elevated levels of a biomarker called prostate-specific antigen (PSA). Normally, secretory epithelium in the prostate gland releases PSA, which liquefies the ejaculate. When cancer cells accumulate, more PSA is produced, and it enters capillaries in the prostate.

Health-care organizations' recommendations for PSA screening change often, and range from screening all men to screening none, although any man with symptoms of frequent or slowed urination should be tested. The controversy is that the value of saving lives following screening must be balanced against the risk that high levels of PSA in the absence of cancer can lead to unnecessary biopsies.

If the physician feels an enlargement of the patient's prostate, or if the PSA level remains high or rises on tests repeated a few months later, the next step is a biopsy procedure to sample cells from several sites in the gland. A cancer detected with a biopsy is assigned a two-digit number, called a Gleason score, which indicates how specialized the cancer cells are. The less specialized the cancer cells, the more aggressive the disease. Imaging technologies can be used to assess whether the cancer has spread beyond the prostate capsule.

Treatment of prostate cancer may be necessary if the tumor has a high Gleason score or fits the genetic profile of being likely to spread. Treatments include radiation, hormones, and surgery to remove the prostate gland. Adverse effects of treatment include urinary incontinence and erectile dysfunction. For many men, active surveillance to regularly monitor the disease is sufficient. This means PSA tests twice a year and a biopsy every one to two years, with treatment if the condition worsens.

Erection, Orgasm, and Ejaculation

During sexual stimulation, parasympathetic impulses from the sacral part of the spinal cord cause the release of the neurotransmitter nitric oxide (NO), which dilates the arteries leading into the penis, increasing blood flow into erectile tissues. At the same time, the increasing pressure of arterial blood entering the vascular spaces of erectile tissue compresses the veins of the penis, reducing the flow of venous blood away from the organ. Consequently, blood accumulates in the erectile tissues, and the penis swells and elongates, producing an **erection.**

The culmination of sexual stimulation is **orgasm** (or′gazm), a pleasurable feeling of physiological and psychological release. Orgasm in the male is accompanied by emission and ejaculation.

Emission (e-mish′un) is the movement of sperm cells from the testes and secretions from the prostate gland and seminal vesicles into the urethra, where they mix to form semen. Emission is a response to sympathetic impulses from the spinal cord, which stimulate peristaltic contractions in smooth muscle in the walls of the testicular ducts, epididymides, ductus deferentia, and ejaculatory ducts. Other sympathetic impulses stimulate rhythmic contractions of the seminal vesicles and prostate gland.

As the urethra fills with semen, sensory impulses pass into the sacral part of the spinal cord. In response, motor impulses are conducted from the spinal cord to certain skeletal muscles at the base of the penile erectile columns, rhythmically contracting them. This increases the pressure in the erectile tissues and aids in forcing the semen through the urethra to the outside, a process called **ejaculation** (e-jak″u-lā′shun). At the time of ejaculation, the posterior pituitary gland releases a burst of oxytocin, which stimulates contractions of the epididymides, seminiferous tubules, and prostate gland, aiding the movement of sperm.

The sequence of events during emission and ejaculation is coordinated so that the fluid from the bulbourethral glands is expelled first. This is followed by the release of fluid from the prostate gland, the passage of sperm cells, and finally the ejection of fluid from the seminal vesicles into the urethra. Ejaculation forcefully expels the semen from the body.

Spontaneous emission and ejaculation are common in sleeping adolescent males. Changes in hormonal concentrations that accompany adolescent development and sexual maturation cause these events.

Immediately after ejaculation, sympathetic impulses constrict the arteries that supply the erectile tissue, reducing

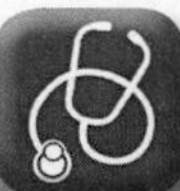

CLINICAL APPLICATION 19.2
Male Infertility

Male infertility has several causes. If, during fetal development, the testes do not descend into the scrotum, the higher temperature of the abdominal cavity or inguinal canal hinders development of sperm cells in the seminiferous tubules. Certain diseases, such as mumps, may inflame the testes (orchitis), impairing fertility.

The quality and the quantity of sperm cells are essential factors in the ability of a man to father a child. If a sperm head is misshapen, if the acrosome is too tough to burst and release enzymes, or if too few sperm cells reach the well-protected oocyte, fertilization may not happen.

Computer-aided semen analysis (CASA) is a technique used to evaluate a man's fertility. For this analysis, a man abstains from intercourse for two to three days and then provides a sperm sample. The man also provides information about his reproductive history and possible exposure to toxins. The CASA system captures images digitally and analyzes and integrates information on sperm cell density, motility, and morphology. The result is a "spermiogram." Figure 19A shows a CASA of normal sperm cells, depicting different swimming patterns as they travel. Table 19A lists the components of a semen analysis.

Devices are being developed that will enable a man to estimate his sperm count at home. They indicate whether a man's sperm count is above or below the World Health Organization's designation of 20 million sperm per milliliter of ejaculate as the lower limit for normal fertility.

TABLE 19A Ranges of Semen and Sperm Characteristics in Healthy Men

Characteristic	Range
Volume/ejaculate	1.5–5.0 mL
Number of sperm	20–150 million/mL
Concentration of sperm	12–16 million/mL
Sperm vitality	55–63%
Sperm motility	38–42%
Morphologically normal	3–4%

mL = milliliter

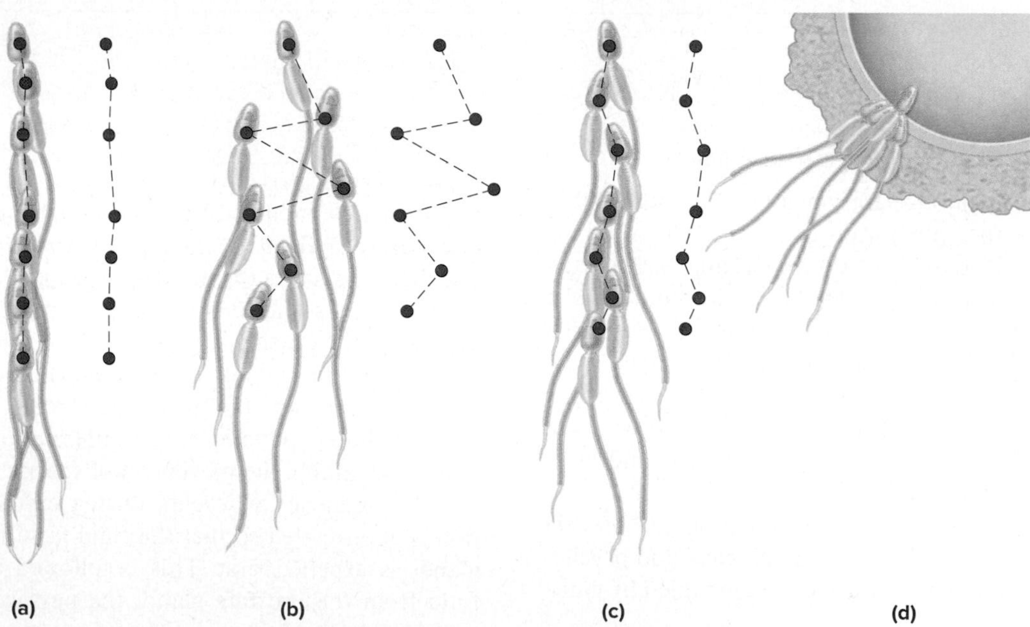

Figure 19A A computer tracks sperm cell movements. In semen, sperm cells swim in a straight line (**a**), but as they are activated by biochemicals in the woman's body, their trajectories widen (**b**). The sperm cells in (**c**) are in the mucus of a woman's cervix, and the sperm cells in (**d**) are attempting to digest through the structures surrounding an oocyte.

inflow of blood. Smooth muscle in the walls of the vascular spaces partially contracts, and the veins of the penis carry excess blood out of these spaces. The penis gradually returns to its flaccid state. Table 19.1 summarizes the functions of the male reproductive organs. Clinical Application 19.2 discusses male infertility.

PRACTICE 19.2

12. What controls blood flow into penile erectile tissues?
13. Distinguish among orgasm, emission, and ejaculation.
14. Review the events associated with emission and ejaculation.

TABLE 19.1 Functions of the Male Reproductive Organs

Organ	Function
Testis	
Seminiferous tubules	Produce sperm cells
Interstitial cells	Produce and secrete male sex hormones
Epididymis	Promotes sperm cell maturation; stores sperm cells; conveys sperm cells to ductus deferens
Ductus deferens	Conveys sperm cells to ejaculatory duct
Seminal vesicle	Secretes an alkaline fluid containing nutrients and prostaglandins that helps regulate pH of semen
Prostate gland	Secretes a fluid that contains citrate, a nutrient for sperm
Bulbourethral gland	Secretes fluid that lubricates end of penis
Scrotum	Encloses, protects, and regulates temperature of testes
Penis	Conveys urine and semen to outside of body; inserted into vagina during sexual intercourse; the glans penis is richly supplied with sensory nerve endings associated with feelings of pleasure during sexual stimulation

19.3 | Spermatogenesis

LEARN

1. Outline the process of spermatogenesis.

The formation of sperm cells is called spermatogenesis. Meiosis is the process that ensures that sperm cells will have only 23 chromosomes.

Formation of Sperm Cells

The epithelium of the seminiferous tubules consists of *sustentacular cells* (Sertoli cells) and spermatogenic cells. Sustentacular cells support, nourish, and regulate the spermatogenic cells.

In the male embryo, the undifferentiated spermatogenic cells are called **spermatogonia** (sper″mah-to-go′ne-ah). Each spermatogonium contains 46 chromosomes (23 pairs) in its nucleus, the usual number for human body cells. Beginning during embryonic development, hormones stimulate spermatogonia to undergo mitosis (see section 3.4, The Cell Cycle). Each cell division gives rise to two new cells, one of which (type A) maintains the supply of undifferentiated cells, the other of which (type B) differentiates, becoming a *primary spermatocyte.* Sperm cell production, or **spermatogenesis** (sper″mah-to-jen′ĕ-sis), pauses at this stage.

At puberty, mitosis resumes, and new spermatogonia form. Testosterone secretion increases and the primary spermatocytes then reproduce by a special type of cell division called **meiosis** (mī-ō′sis) (fig. 19.3). Meiosis includes two successive divisions, called the *first* and *second meiotic divisions.*

Half of an individual's 46 chromosomes are inherited from the mother (23), and half from the father (23). The 23 pairs of corresponding chromosomes, called homologous pairs, are the same, gene for gene. They may not be identical, however, because a gene may have variants, and a given chromosome that comes from the mother may carry a different variant of a gene than that chromosome from the father. Before meiosis I, each homologous chromosome is replicated, so it consists of two identical DNA strands called **chromatids.** The chromatids of a replicated chromosome attach at regions called **centromeres.**

The first meiotic division (meiosis I) separates homologous chromosome pairs. Thus, each of the secondary spermatocytes that undergoes the second meiotic division (meiosis II) begins with one member of each homologous pair, a condition termed **haploid** (hap′loyd). This second division separates the chromatids, producing cells that are still haploid, but whose chromosomes are no longer in the replicated, double-stranded (two chromatids) form. After meiosis II, each of the chromatids is an independent, single-stranded chromosome (fig. 19.3).

The halving of chromosome number is accomplished as each primary spermatocyte divides to form two secondary spermatocytes. Each of these cells, in turn, divides to form two **spermatids,** which mature into sperm cells (fig. 19.4). Consequently, for the primary spermatocyte that undergoes meiosis, four sperm cells form, with 23 chromosomes each. Spermatogenesis occurs continually in a male, starting at puberty. The resulting sperm cells collect in the lumen of each seminiferous tubule, and then pass to the epididymis, where they accumulate and mature.

Structure of a Sperm Cell

A mature sperm cell is a tiny, tadpole-shaped structure about 0.06 millimeters long. It consists of a flattened head, a cylindrical midpiece (body), and an elongated tail (fig. 19.5; see fig. 3.9*b*).

The oval *head* of a sperm cell is primarily composed of a nucleus and contains highly compacted chromatin consisting of 23 chromosomes. A small protrusion of its head, called the *acrosome,* contains enzymes that aid the sperm cell in penetrating the layers surrounding the oocyte during fertilization. One of the enzymes on the sperm cell membrane contributes to entering the oocyte. (Section 20.2, Fertilization, describes this process.)

The *midpiece* of a sperm cell has a central, filamentous core and many mitochondria organized in a spiral. The *tail* (flagellum) consists of several microtubules enclosed in an extension of the cell membrane. The mitochondria provide ATP for the tail's lashing movement that propels the sperm cell through fluid.

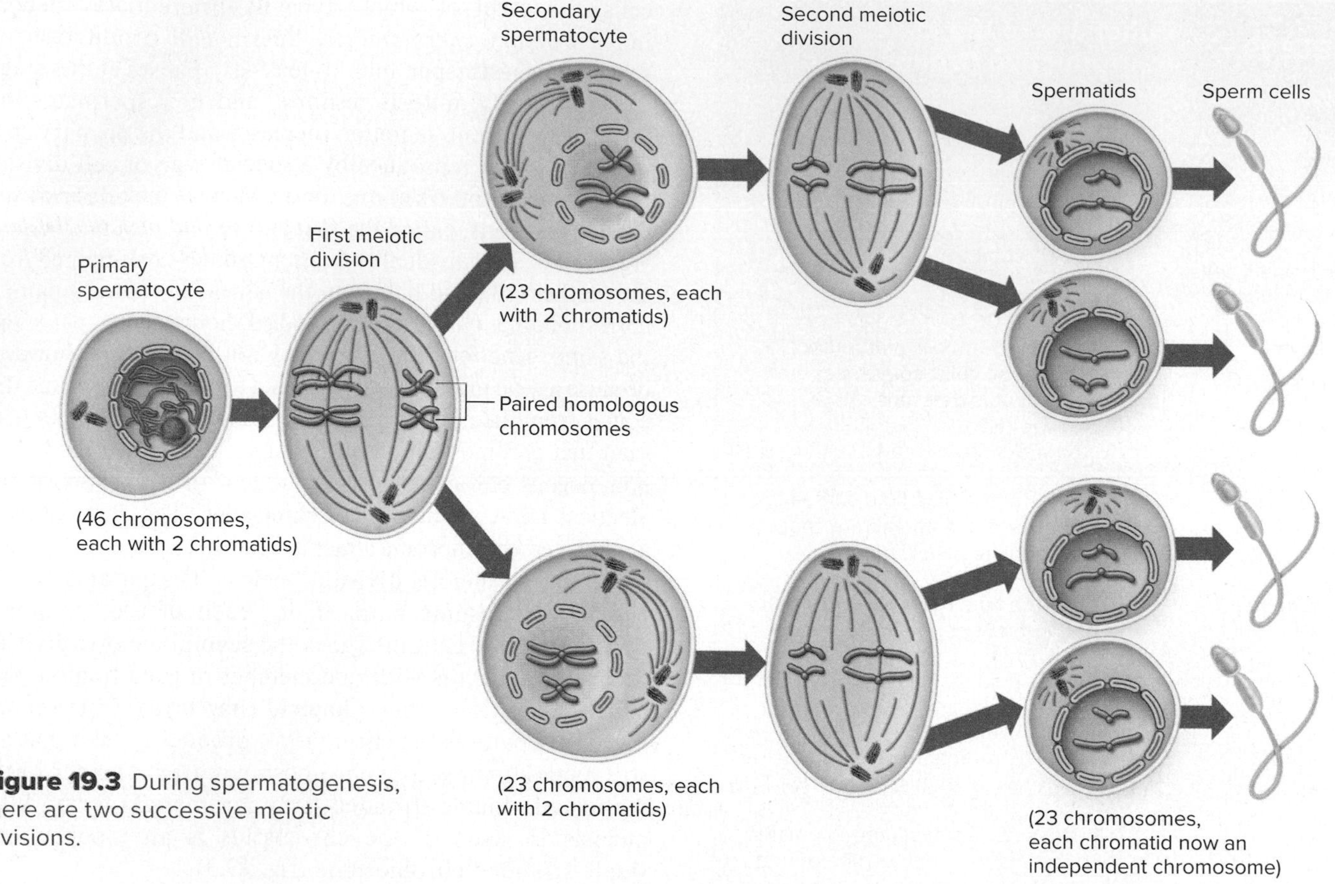

Figure 19.3 During spermatogenesis, there are two successive meiotic divisions.

PRACTICE FIGURE 19.3

Why is it important that a sperm possess only 23 chromosomes?

Answer can be found in Appendix E.

PRACTICE 19.3

1. Explain the function of sustentacular cells in the seminiferous tubules.
2. Review the events of spermatogenesis.
3. Describe the structure of a sperm cell.

19.4 Hormonal Control of Male Reproductive Functions

LEARN

1. Explain how hormones control the activities of the male reproductive organs and the development of male secondary sex characteristics.

The hypothalamus, anterior pituitary gland, and testes secrete hormones that control male reproductive functions. These hormones initiate and maintain sperm cell production, and oversee the development and maintenance of male secondary sex characteristics, which are special features associated with the adult male body.

Hypothalamic and Pituitary Hormones

The male body before ten years of age is reproductively immature, with undifferentiated spermatogenic cells. Then, a series of changes leads to development of a reproductively functional adult. The hypothalamus controls many of these changes.

Recall from section 11.4, Pituitary Gland, that the hypothalamus secretes gonadotropin-releasing hormone (GnRH), which enters the blood vessels leading to the anterior pituitary gland. In response, the anterior pituitary secretes the **gonadotropins** (go-nad″o-trōp′inz) called *luteinizing hormone (LH)* and *follicle-stimulating hormone (FSH).* LH, which in males has been referred to as interstitial cell stimulating hormone (ICSH), promotes development of interstitial cells of the testes, and they, in turn, secrete male sex hormones. FSH stimulates the sustentacular cells of the seminiferous tubules to respond to the effects of the male sex hormone *testosterone.* Then, in the presence of FSH and testosterone, the sustentacular cells stimulate spermatogenic cells to undergo spermatogenesis, giving rise to sperm cells (fig. 19.6). The sustentacular cells also secrete a hormone called *inhibin,* which inhibits the anterior pituitary gland by negative feedback. This action prevents oversecretion of FSH.

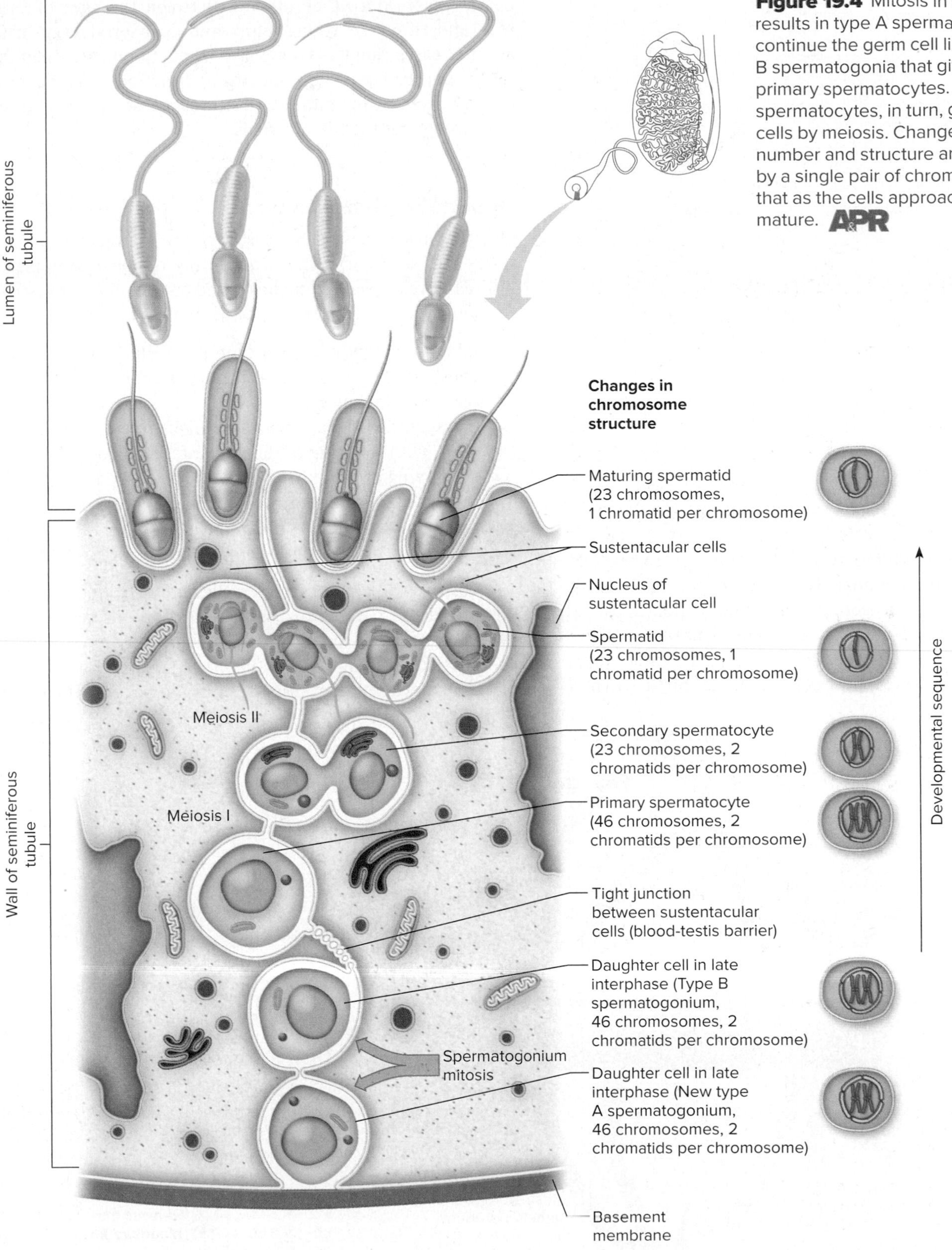

Figure 19.4 Mitosis in spermatogonia results in type A spermatogonia that continue the germ cell line and type B spermatogonia that give rise to primary spermatocytes. The primary spermatocytes, in turn, give rise to sperm cells by meiosis. Changes in chromosome number and structure are represented by a single pair of chromosomes. Note that as the cells approach the lumen they mature. APR

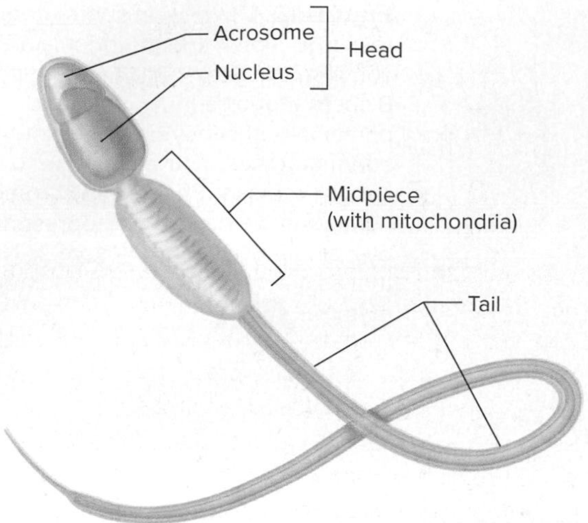

Figure 19.5 Parts of a mature sperm cell.

Male Sex Hormones

Male sex hormones are termed **androgens** (an′dro-jenz). Interstitial cells of the testes produce most of them, but small amounts are synthesized in the adrenal cortex (see section 11.7, Adrenal Glands). **Testosterone** (tes-tos′tĕ-rōn) is the most important androgen. It is secreted by the testes and transported in the blood, loosely attached to plasma proteins.

Testosterone secretion begins during fetal development and continues for several weeks following birth; then it nearly ceases during childhood. Between the ages of thirteen and fifteen, a young man's androgen production usually increases rapidly. This phase in development, when an individual becomes reproductively functional, is **puberty** (pyoo′ber-tē). After puberty, testosterone secretion continues throughout the life of a male.

Actions of Testosterone

During puberty, testosterone stimulates enlargement of the testes and accessory organs of the reproductive system, as well as development of male secondary sex characteristics. Secondary sex characteristics in the male include:

- increased growth of body hair, particularly on the face, chest, axillary region, and pubic region; growth of hair on the scalp may slow
- enlargement of the larynx and thickening of the vocal folds, with lowering of the pitch of the voice
- thickening of the skin
- increased muscular growth, broadening of the shoulders, and narrowing of the waist
- thickening and strengthening of the bones

Testosterone also increases the rate of cellular metabolism and red blood cell production. For this reason, the average number of red blood cells in a microliter of blood is usually greater in males than in females. Testosterone stimulates sexual activity by affecting certain parts of the brain.

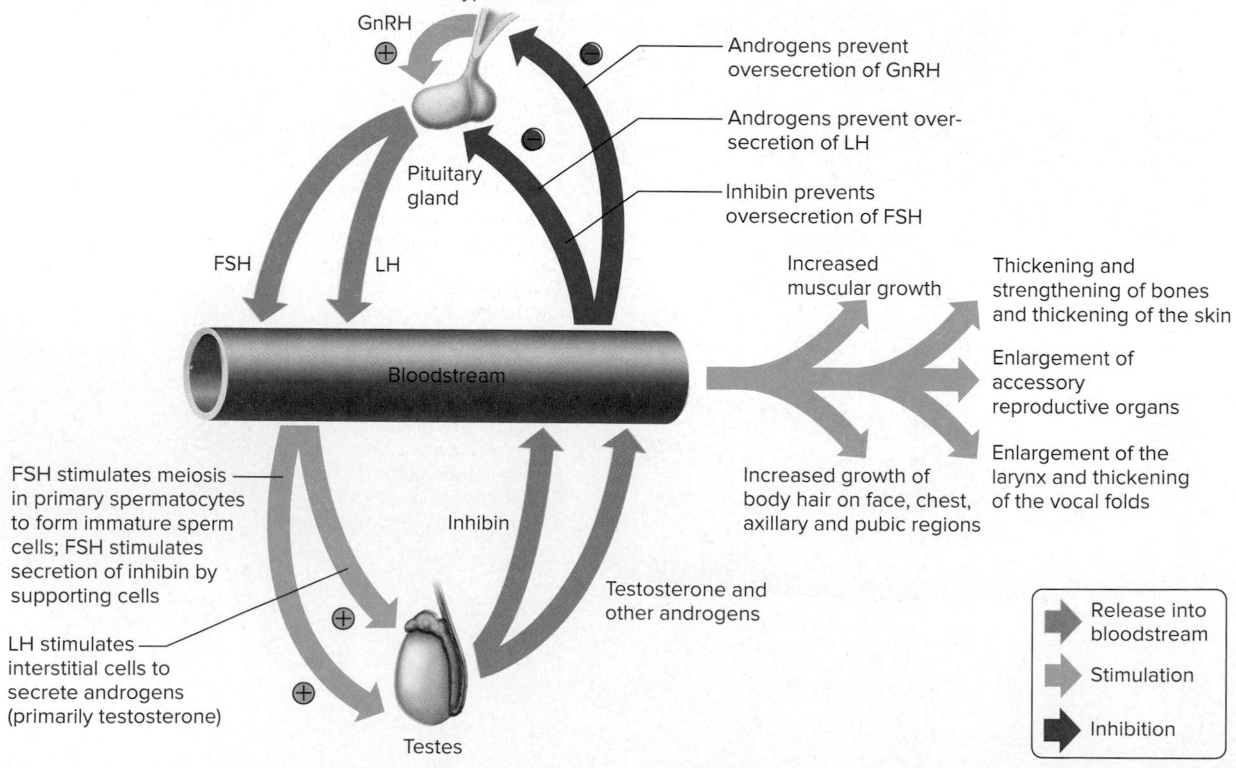

Figure 19.6 The hypothalamus controls maturation of sperm cells and development of male secondary sex characteristics. Negative feedback among the hypothalamus, the anterior lobe of the pituitary gland, and the testes controls the concentration of testosterone in the male body.

Regulation of Male Sex Hormones

The extent to which male secondary sex characteristics develop is directly related to the amount of testosterone that interstitial cells secrete. The hypothalamus regulates testosterone output through negative feedback (fig. 19.6).

An increasing blood testosterone concentration inhibits the hypothalamus, and hypothalamic stimulation of the anterior pituitary gland by GnRH decreases. As the pituitary gland's secretion of LH falls in response, testosterone release from the interstitial cells decreases.

As the blood testosterone concentration drops, the hypothalamus becomes less inhibited, and it once again stimulates the anterior pituitary gland to release LH. Increasing LH secretion then causes interstitial cells to release more testosterone, and the blood testosterone concentration increases.

Testosterone level decreases somewhat during and after the *male climacteric,* which is a decline in sexual function associated with aging. At any given age, the testosterone concentration in the male body is regulated to remain relatively constant.

PRACTICE 19.4

1. Which hormone initiates the changes associated with male sexual maturity?
2. Describe several male secondary sex characteristics.
3. List the functions of testosterone.
4. Explain how the secretion of male sex hormones is regulated.

19.5 | Organs of the Female Reproductive System

LEARN

1. Describe the structure and function(s) of each part of the female reproductive system.

The organs of the female reproductive system produce and maintain the female sex cells, the oocytes; transport these cells to the site of fertilization; provide a favorable environment for a developing offspring; move the offspring to the outside; and produce female sex hormones. The *primary sex organs* (gonads) of this system are the two ovaries, which produce the female sex cells and sex hormones. The *accessory sex organs* (secondary sex organs) of the female reproductive system are the internal and external reproductive organs (fig. 19.7; reference plates 5 and 6).

Ovaries

The two **ovaries** are solid, ovoid structures, each about 3.5 centimeters long, 2 centimeters wide, and 1 centimeter thick. The ovaries lie in shallow depressions in the lateral wall of the pelvic cavity (fig. 19.7).

Ovary Structure

Ovarian tissues are divided into two indistinct regions—an inner *medulla* and an outer *cortex.* The ovarian medulla is mostly composed of loose connective tissue and contains

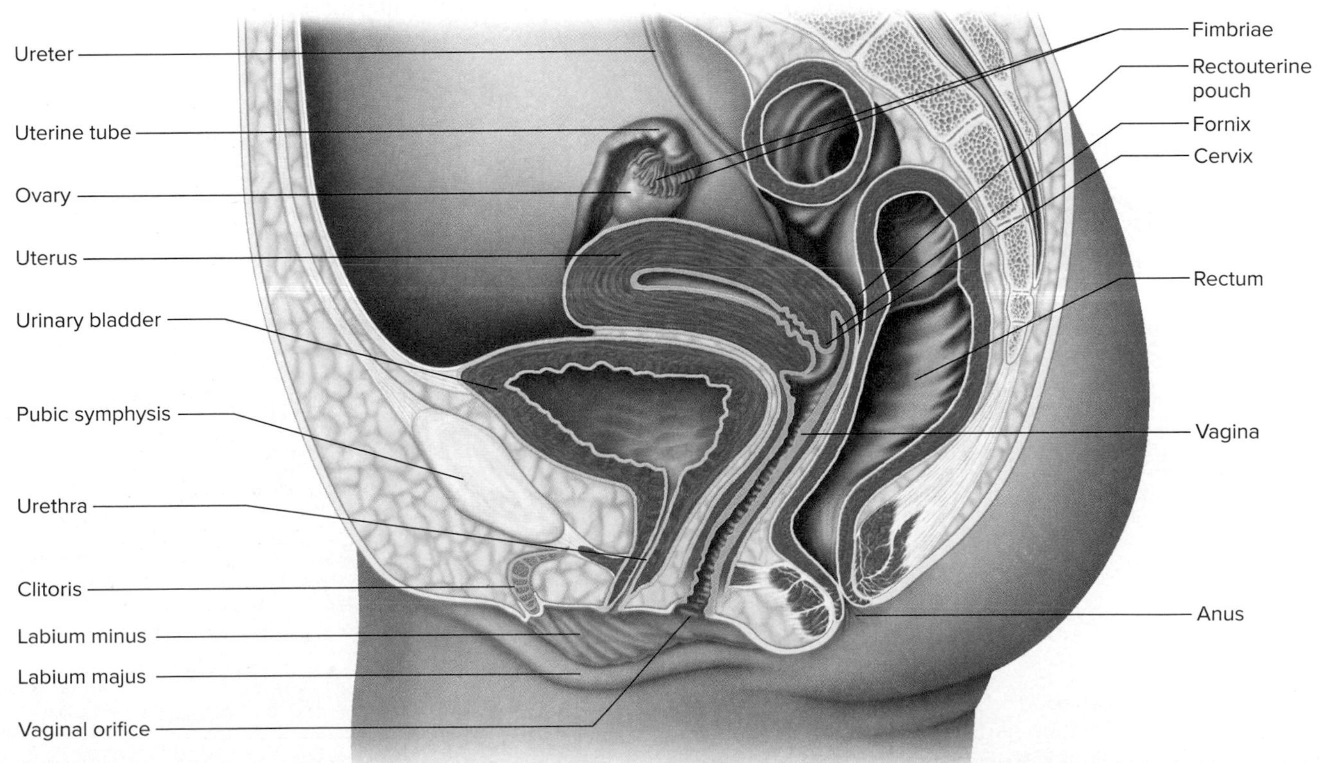

Figure 19.7 Female reproductive organs (sagittal view). The paired ovaries are the primary sex organs, and the other reproductive structures, both internal and external, are accessory sex organs. APR

many blood vessels, lymphatic vessels, and nerve fibers. The ovarian cortex consists of more compact tissue and has a granular appearance due to tiny masses of cells called *ovarian follicles.*

A layer of cuboidal epithelium covers the ovary's free surface. Just beneath this epithelium is a layer of dense connective tissue.

PRACTICE 19.5

1. What are the primary sex organs of the female?
2. Describe the structure of an ovary.

Primordial Follicles

During prenatal development of a female, small groups of cells in the outer region of the ovarian cortex form several million **primordial follicles.** Each follicle consists of a single, large cell, called a *primary oocyte,* surrounded by epithelial cells called *follicular cells.*

Early in development, the primary oocytes begin to undergo meiosis, but the process soon halts and does not continue until the individual reaches puberty. Once the primordial follicles appear, no new ones form. Instead, the number of primary oocytes in the ovary steadily declines as many of the primary oocytes degenerate. Of the six to seven million primary oocytes that formed in the embryonic ovary, only a million or so remain at birth, and perhaps 300,000 are present at puberty. Of these, probably fewer than 300 to 400 oocytes will be released from the ovary during a female's reproductive life.

PRACTICE 19.5

3. Describe a primordial follicle and the cell it contains.
4. Approximately how many primary oocytes are present at birth? At puberty?

Female Internal Accessory Reproductive Organs

The internal accessory organs of the female reproductive system include a pair of uterine tubes, a uterus, and a vagina.

Uterine Tubes

The **uterine tubes** (fallopian tubes or oviducts) open near the ovaries (fig. 19.8). Each tube is about 10 centimeters long and 0.7 centimeters in diameter, passes medially to the uterus, penetrates its wall, and opens into the uterine cavity.

Near each ovary, a uterine tube expands, forming a funnel-shaped *infundibulum* (in″fun-dib′u-lum), which partially encircles the ovary. Fingerlike extensions called *fimbriae* (fim′brē) fringe the infundibulum margin. Although the infundibulum generally does not touch the ovary, one of the larger fimbriae connects directly to the ovary.

Simple columnar epithelial cells, some *ciliated,* line the uterine tube. The epithelium secretes mucus, and the cilia beat toward the uterus. These actions help draw the secondary oocyte and expelled follicular fluid into the infundibulum following ovulation. Ciliary action and peristaltic contractions of the uterine tube's smooth muscle layer aid

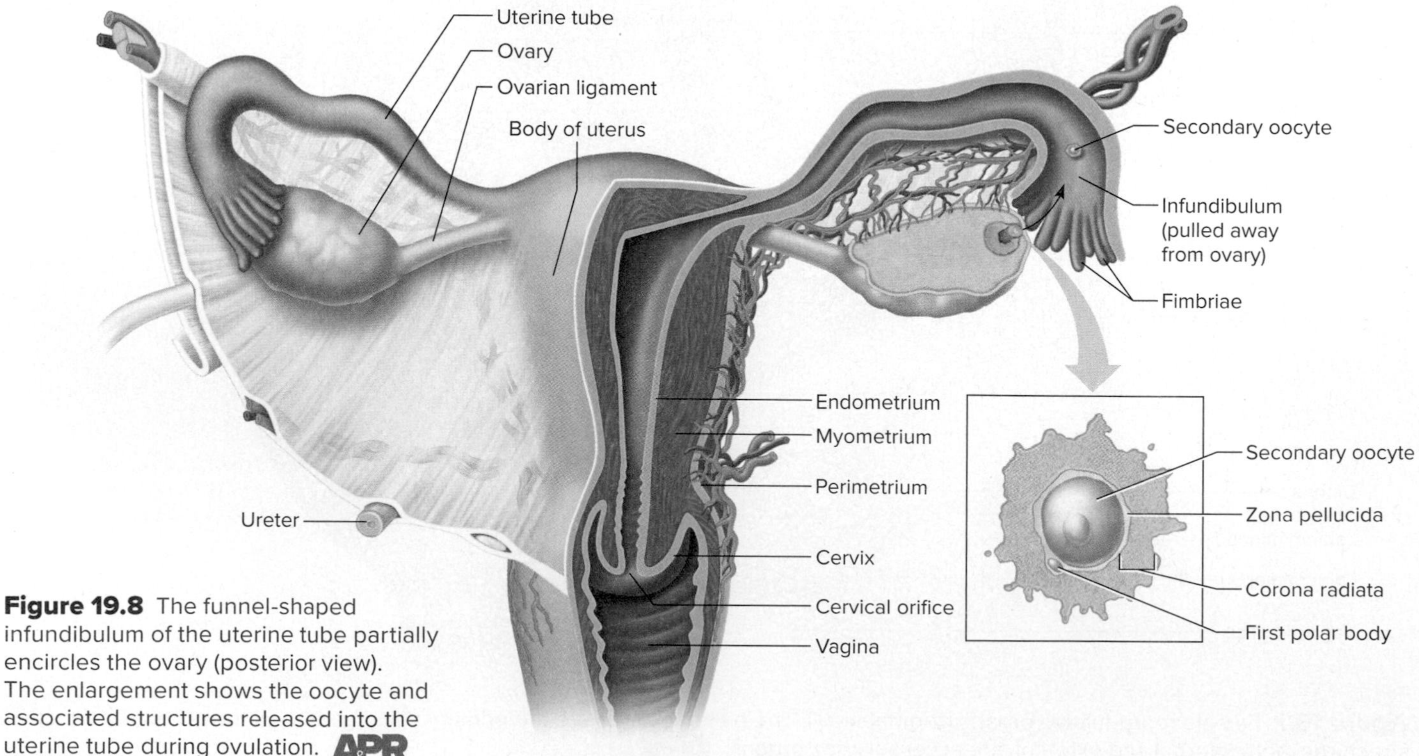

Figure 19.8 The funnel-shaped infundibulum of the uterine tube partially encircles the ovary (posterior view). The enlargement shows the oocyte and associated structures released into the uterine tube during ovulation. APR

transport of the oocyte down the uterine tube. Fertilization usually occurs in the uterine tube.

Uterus

The **uterus** is a hollow, muscular organ shaped somewhat like an inverted pear. The uterus receives the embryo that develops from an oocyte fertilized in the uterine tube, and sustains its development.

The size of the uterus changes greatly during pregnancy. In its nonpregnant, adult state, the uterus is about 7 centimeters long, 5 centimeters wide (at its broadest point), and 2.5 centimeters in diameter. The uterus is located medially in the anterior part of the pelvic cavity, superior to the vagina, and usually bends forward over the urinary bladder.

The upper two-thirds, or *body,* of the uterus has a dome-shaped top called the *fundus* (fig. 19.8). The uterine tubes connect at the upper lateral edges of the uterus. The lower third of the uterus is called the **cervix.** This tubular part extends downward into the upper part of the vagina. The cervix surrounds the opening called the *cervical orifice,* through which the uterus opens to the vagina.

The uterine wall is thick and has three layers (figs. 19.8 and 19.9). The **endometrium** (en″dō-mē′trē-um), the inner mucosal layer, is covered with simple columnar epithelium and contains abundant tubular glands surrounded by loose connective tissue. The **myometrium** (mī″ō-mē′trē-um), a thick, middle, muscular layer, consists largely of bundles of smooth muscle cells. During the monthly female menstrual cycles and during pregnancy, the endometrium and myometrium change extensively. The **perimetrium** (per-ĭ-mē′trē-um) consists of an outer serosal layer, which covers the body of the uterus and part of the cervix.

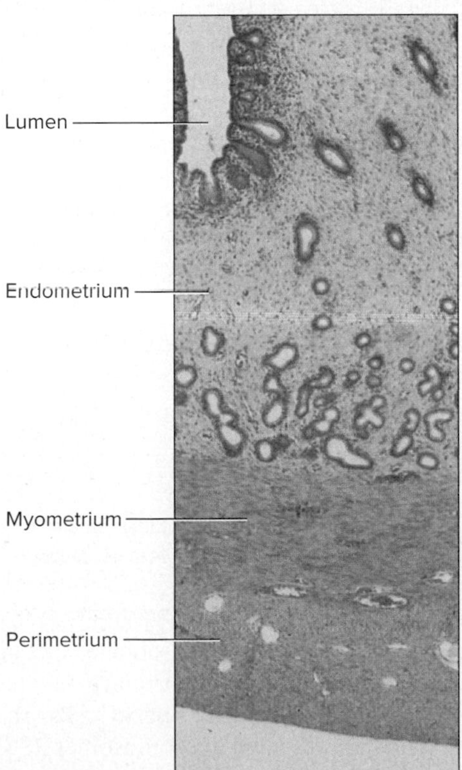

Figure 19.9 Light micrograph of the uterine wall (10x). APR Carol D. Jacobson, Ph.D., Department of Veterinary Anatomy, Iowa State University/McGraw-Hill Education

Vagina

The **vagina** is a fibromuscular tube, about 9 centimeters long, extending from the uterus to the outside of the body (see fig. 19.7). It conveys uterine secretions, receives the erect penis during sexual intercourse, and provides a passageway for the offspring to exit the body during birth.

The vagina extends upward and back into the pelvic cavity. It is posterior to the urinary bladder and urethra, anterior to the rectum, and attached to these structures by connective tissues.

A thin membrane of connective tissue and stratified squamous epithelium called the **hymen** partially covers the *vaginal orifice.* A central opening of varying size allows uterine and vaginal secretions to pass to the outside.

The vaginal wall has three layers. The inner *mucosal layer* is stratified squamous epithelium. This layer lacks mucous glands; the mucus in the lumen of the vagina comes from uterine glands and from vestibular glands at the mouth of the vagina.

The middle *muscular layer* consists of smooth muscle in an outer circular layer and an inner longitudinal layer. The smooth muscle can stretch to accommodate the penis during sexual intercourse and a child during childbirth.

The outer *fibrous layer* consists of dense connective tissue interlaced with elastic fibers. It attaches the vagina to surrounding organs.

PRACTICE 19.5

5. How is a secondary oocyte moved along a uterine tube?
6. Describe the structure of the uterus.
7. Describe the structure of the vagina.

Female External Accessory Reproductive Organs

The external accessory organs of the female reproductive system include the labia majora, labia minora, clitoris, and the vestibular glands (see fig. 19.7). These structures that surround the openings of the urethra and vagina compose the **vulva.**

Labia Majora

The **labia majora** (singular, *labium majus*) enclose and protect the other external reproductive organs. The labia majora correspond to the scrotum of the male and are composed of rounded folds of adipose tissue and a thin layer of smooth muscle, covered by skin.

The labia majora lie close together. A cleft that includes the urethral and vaginal openings separates the labia longitudinally. At their anterior ends, the labia merge to form a medial, rounded elevation of adipose tissue called the *mons pubis,* which overlies the pubic symphysis (see fig. 7.26). The labia taper at their posterior ends and merge near the anus.

Labia Minora

The **labia minora** (singular, *labium minus*) are flattened, longitudinal folds between the labia majora (see fig. 19.7). They are composed of connective tissue richly supplied with blood vessels, giving a pinkish appearance. Posteriorly, the labia minora merge with the labia majora, while anteriorly, they converge to form a hoodlike covering around the clitoris.

Clitoris

The **clitoris** (klit′o-ris) is visible as a small projection at the anterior end of the vulva between the labia minora (see fig. 19.7). In most women, it is about 2 centimeters long and 0.5 centimeters in diameter, most of which is embedded in surrounding tissues. The clitoris corresponds to the penis in males and has a similar structure. It is composed of two columns of erectile tissue called *corpora cavernosa.* At its anterior end, a small mass of erectile tissue forms a glans, which is richly supplied with sensory nerve fibers.

Vestibule

The labia minora enclose a space called the **vestibule.** The vagina opens into the posterior portion of the vestibule, and the urethra opens in the midline, just anterior to the vagina and about 2.5 centimeters posterior to the glans of the clitoris.

A pair of **vestibular glands,** corresponding to the bulbourethral glands in the male, lie one on either side of the vaginal opening. Beneath the mucosa of the vestibule on either side is a mass of vascular erectile tissue called the *vestibular bulb.*

PRACTICE 19.5

8. What is the male counterpart of the labia majora? Of the clitoris?
9. Which structures are within the vestibule?

Erection, Lubrication, and Orgasm

Erectile tissues in the clitoris and around the vaginal entrance respond to sexual stimulation. Following such stimulation, parasympathetic impulses from the sacral portion of the spinal cord cause the release of the neurotransmitter nitric oxide (NO), which dilates the arteries associated with the erectile tissues. As a result, blood inflow increases, erectile tissues swell, and the vagina expands and elongates.

If sexual stimulation is sufficiently intense, parasympathetic impulses stimulate the vestibular glands to secrete mucus into the vestibule. This secretion moistens and lubricates the tissues surrounding the vestibule and the lower end of the vagina, facilitating insertion of the penis into the vagina.

The clitoris is abundantly supplied with sensory nerve fibers, which are especially sensitive to local stimulation. Such stimulation culminates in orgasm.

Just prior to orgasm, the tissues of the outer third of the vagina engorge with blood and swell. This increases the friction on the penis during intercourse. Orgasm initiates a series of reflexes involving the sacral and lumbar parts of the spinal cord. In response to these reflexes, the muscles of the perineum and the walls of the uterus and uterine tubes contract rhythmically. These contractions help transport sperm cells through the female reproductive tract toward the upper ends of the uterine tubes. Table 19.2 summarizes the functions of the female reproductive organs.

TABLE 19.2 Functions of the Female Reproductive Organs APR

Organ	Function
Ovary	Produces oocytes and female sex hormones
Uterine tube	Conveys secondary oocyte toward uterus; site of fertilization; conveys developing embryo to uterus
Uterus	Protects and sustains embryo during pregnancy
Vagina	Conveys uterine secretions to outside of body; receives erect penis during sexual intercourse; provides a passageway for offspring during birth process
Labia majora	Enclose and protect other external reproductive organs
Labia minora	Form margins of vestibule; protect openings of vagina and urethra
Clitoris	Produces feelings of pleasure during sexual stimulation due to abundant sensory nerve endings in glans
Vestibule	Space between labia minora that contains vaginal and urethral openings
Vestibular glands	Secrete fluid that moistens and lubricates vestibule

PRACTICE 19.5

10. What events result from parasympathetic stimulation of the female reproductive organs?
11. What changes occur in the vagina just prior to and during orgasm?

19.6 | Oogenesis and the Ovarian Cycle

LEARN

1. Outline the process of oogenesis.
2. Explain how hormones control the activities of the female reproductive organs and the development of female secondary sex characteristics.
3. Describe the major events during a female reproductive cycle.

Oogenesis (ō″uh-jen′ĕ-sis) is the process of oocyte, or egg cell formation. Beginning at puberty, some primary oocytes are stimulated to continue meiosis. Like sperm cells, the resulting cells have one-half as many chromosomes (23) in their nuclei as their parent cells. Unlike a primary spermatocyte, when a primary oocyte divides, the cytoplasm is distributed unequally. One of the resulting cells, called a *secondary oocyte,* is large, and the other, called the first **polar body,** is small (fig. 19.10). If fertilization occurs, the

① Oogonia give rise to oocytes. Before birth, oogonia multiply by mitosis. During development of the fetus, many oogonia begin meiosis, but stop in prophase I and are now called primary oocytes. They remain in this state until puberty.

② Before birth, the primary oocytes become surrounded by a single layer of granulosa cells, creating a primordial follicle. These are present until puberty.

③ After puberty, primordial follicles develop into primary follicles when the granulosa cells enlarge and increase in number.

④ Secondary follicles form when fluid-filled vesicles develop and theca cells arise on the outside of the follicle.

⑤ Mature follicles form when the vesicles create a single antrum.

⑥ Just before ovulation, the primary oocyte completes meiosis I, creating a secondary oocyte and a nonviable polar body.

⑦ The secondary oocyte begins meiosis II, but stops at metaphase II.

⑧ During ovulation, the secondary oocyte is released from the ovary.

⑨ The secondary oocyte completes meiosis II only if it is fertilized by a sperm cell. The completion of meiosis II forms an oocyte and a second polar body. Fertilization is complete when the oocyte nucleus and the sperm cell nucleus unite, creating a zygote.

⑩ Following ovulation, the granulosa cells divide rapidly and enlarge to form the corpus luteum.

⑪ The corpus luteum degenerates to form a scar, or corpus albicans.

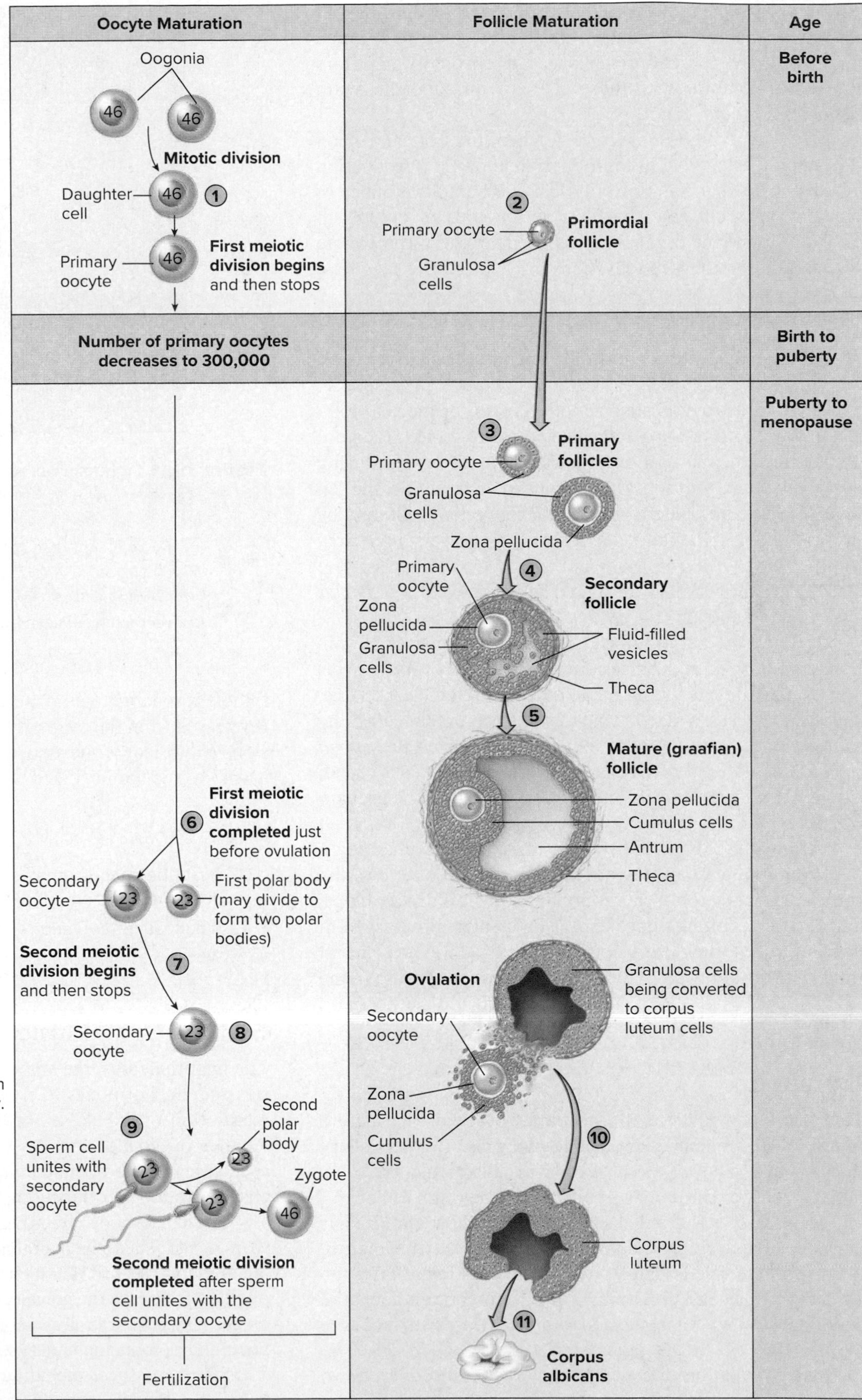

Figure 19.10 During oogenesis, a secondary oocyte results from meiosis of a primary oocyte. If fertilized, the egg cell undergoes a second meiotic division. This division generates a second polar body and a zygote. (Note: The second meiotic division may not occur in the secondary oocyte if it is not fertilized.)

secondary oocyte undergoes the second meiotic division and divides unequally to produce a tiny *second polar body* and a large *ovum*. At the end of meiosis, the chromosomes from the sperm combine with those of the ovum, and the ovum becomes a **zygote** (zī'gōt).

The polar bodies have no further function and soon degenerate. Their role in reproduction allows for production of an oocyte that has not only just the haploid number of chromosomes but also the massive amounts of cytoplasm and the abundant organelles required to carry the zygote through the first few cell divisions.

Follicle Maturation

At puberty, the anterior pituitary gland secretes increased amounts of FSH, and the ovaries enlarge in response. Throughout a woman's reproductive years, primordial follicles mature into **primary follicles** (prī'ma-rē fol'ĭ-klz). Figure 19.10 traces the maturation of a follicle in an ovary. During early follicle maturation, the primary oocyte within the follicle enlarges, and surrounding **follicular cells** proliferate by mitosis. These follicular cells organize into layers, and eventually a cavity (*antrum*) appears in the cellular mass. The structure is now called an **antral follicle.** A clear *follicular fluid* fills the antrum (fig. 19.10).

In time, the antral follicle reaches a diameter of 10 millimeters or more and bulges outward on the ovary surface. The primary oocyte within the mature antral follicle is a large, spherical cell, surrounded by a layer of glycoprotein called the *zona pellucida,* with layers of follicular cells called the *corona radiata* (fig. 19.11). Processes from these follicular cells extend through the zona pellucida and supply nutrients to the primary oocyte.

Although as many as twenty primary follicles may begin maturing at any one time, one follicle usually outgrows the others. Typically, only the dominant follicle fully develops, and the other follicles degenerate. This entire process, from a primordial follicle to a dominant, fully developed antral follicle, takes almost 300 days. The process is ongoing in the ovaries, such that a new dominant antral follicle becomes ready for ovulation approximately every 28 days.

Ovulation

Just prior to ovulation, the primary oocyte in the mature antral follicle completes meiosis I, giving rise to a secondary oocyte and a first polar body. A process called **ovulation** (o″vu-lā'shun) releases these cells from the mature antral follicle.

Release of LH from the anterior pituitary gland plays a role in triggering ovulation, during which the mature antral follicle swells and its wall weakens. Eventually the wall ruptures, and follicular fluid, accompanied by the secondary oocyte, is released from the ovary's surface (see fig. 19.10).

After ovulation, the secondary oocyte and one or two layers of follicular cells surrounding it are normally propelled to the opening of a nearby uterine tube (fig. 19.8). Compared to the overall size of the ovary, the mature antral

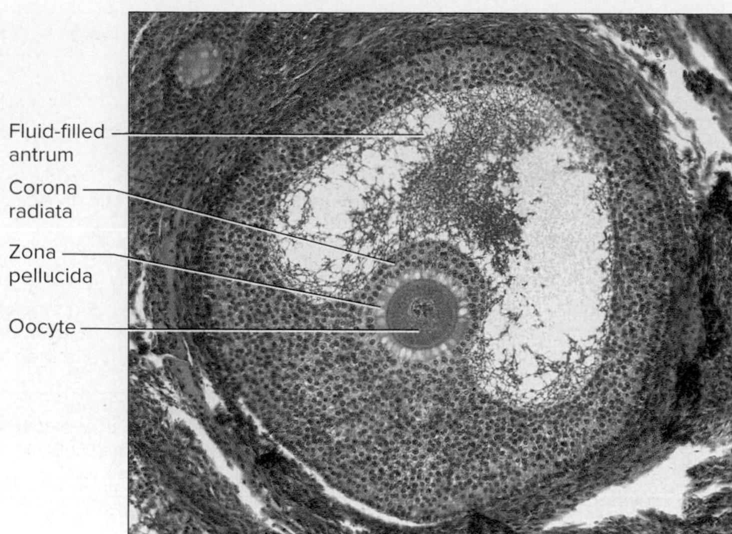

Figure 19.11 Light micrograph of a mature antral follicle (250x). APR Al Telser/McGraw-Hill Education

PRACTICE FIGURE 19.11

Which structure is formed by the follicular cells attached to and surrounding the oocyte?

Answer can be found in Appendix E.

follicle is so large that no matter where the secondary oocyte is released from the ovary, it will contact the branched extensions (fimbriae) of the associated uterine tube. If the secondary oocyte is not fertilized within hours, it degenerates.

PRACTICE 19.6

1. Describe changes that occur in a follicle and its oocyte during maturation.
2. What causes ovulation?
3. What happens to an oocyte following ovulation?

Female Sex Hormones

The hypothalamus, the anterior pituitary gland, and the ovaries secrete hormones. Certain of these hormones control maturation of female sex cells, the development and maintenance of female secondary sex characteristics (which are special features associated with the adult female body), and changes that occur during the menstrual cycle.

The female body is reproductively immature until about ten years of age. Then, the hypothalamus begins to secrete increasing amounts of GnRH, which, in turn, stimulates the anterior pituitary to release the gonadotropins FSH and LH. These hormones play primary roles in controlling female sex cell maturation and in producing female sex hormones (fig. 19.12).

Several tissues, including the ovaries, the adrenal cortices, and the placenta (during pregnancy), secrete female sex hormones belonging to two major groups—**estrogens** (es'tro-jenz) and **progesterone** (prō-jes'tĭ-rōn). *Estradiol* is the

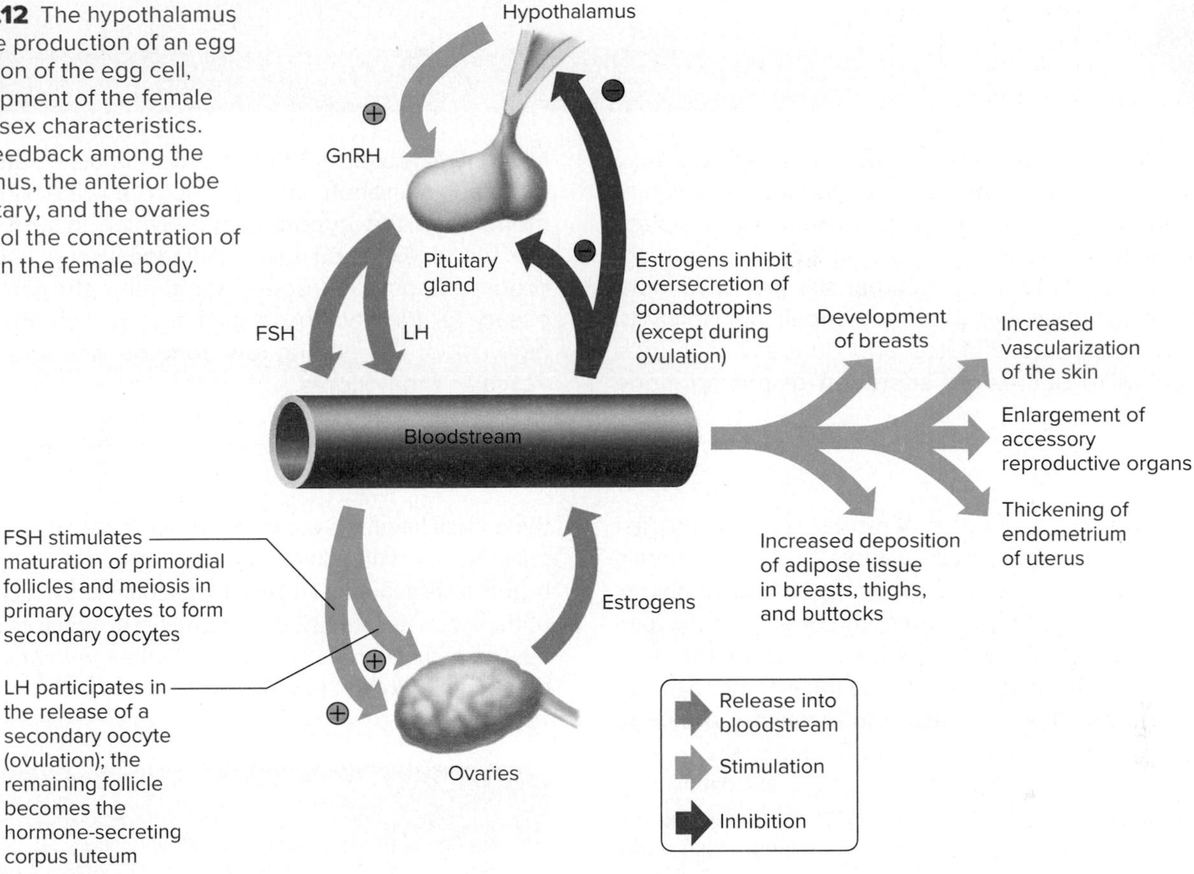
Hypothalamus
GnRH
Pituitary gland
FSH
LH
Estrogens inhibit oversecretion of gonadotropins (except during ovulation)
Development of breasts
Increased vascularization of the skin
Enlargement of accessory reproductive organs
Thickening of endometrium of uterus
Bloodstream
Increased deposition of adipose tissue in breasts, thighs, and buttocks
Estrogens
FSH stimulates maturation of primordial follicles and meiosis in primary oocytes to form secondary oocytes
LH participates in the release of a secondary oocyte (ovulation); the remaining follicle becomes the hormone-secreting corpus luteum
Ovaries
Release into bloodstream
Stimulation
Inhibition

Figure 19.12 The hypothalamus controls the production of an egg cell, ovulation of the egg cell, and development of the female secondary sex characteristics. Negative feedback among the hypothalamus, the anterior lobe of the pituitary, and the ovaries helps control the concentration of estrogens in the female body.

most abundant of the estrogens, which also include *estrone* and *estriol.*

The ovaries are the primary source of estrogens (in a nonpregnant female). At puberty, under the influence of the anterior pituitary, the ovaries secrete increasing amounts of estrogens. Estrogens stimulate enlargement of reproductive organs, including the vagina, uterus, uterine tubes, ovaries, and external reproductive structures. Estrogens also develop and maintain the female secondary sex characteristics, which include (fig. 19.12):

- development of the breasts and the ductile system of the mammary glands in the breasts
- increased deposition of adipose tissue in the subcutaneous layer generally and in the breasts, thighs, and buttocks particularly
- increased vascularization of the skin

The ovaries are also the primary source of progesterone (in a nonpregnant female). This hormone promotes changes in the uterus during the female menstrual cycle, affects the mammary glands, and helps regulate the secretion of gonadotropins from the anterior pituitary gland.

Androgens (male sex hormone) produce certain other changes in females at puberty. For example, increased hair growth in the pubic and axillary regions is due to androgen secreted by the adrenal cortices. Conversely, development of the female skeletal configuration, which includes narrow shoulders and broad hips, is a response to a low androgen concentration.

PRACTICE 19.6

4. What stimulates sexual maturation in a female?
5. What is the function of estrogens?
6. What is the function of androgen in a female?

Menstrual Cycle

The menstrual cycle is characterized by regular, recurring changes in the endometrium, which culminate in menstrual bleeding (menses). Such cycles usually begin around age thirteen and continue into the early fifties, then cease. Clinical Application 19.3 discusses how low body fat may cause disruption in a woman's menstrual cycle.

A female's first menstrual cycle, called **menarche** (mě-nar′kē), occurs after the ovaries and other organs of the reproductive control system mature and begin responding to certain hormones. Then, the hypothalamic secretion of GnRH stimulates the anterior pituitary to release threshold levels of FSH and LH. FSH stimulates the final maturation of an ovarian follicle. The follicular cells produce increasing amounts of estrogens and some progesterone. LH stimulates certain ovarian cells to secrete precursor molecules (such as testosterone), also used to produce estrogens. In a young female, estrogens stimulate the development of secondary sex characteristics. Estrogens secreted during subsequent menstrual cycles continue the development and maintenance of these characteristics.

CLINICAL APPLICATION 19.3
Disruption of Menstrual Cycle

Elite female athletes and professional dancers may have disturbed menstrual cycles, ranging from diminished menstrual flow (oligomenorrhea) to complete stoppage (amenorrhea). The more active an athlete or dancer, the more likely it is that she will have menstrual irregularities, and this may impair her ability to conceive. Trim elite athletes and dancers have little fat, leading to decreased secretion of the hormone leptin. Decreased leptin secretion is associated with lowered secretion of gonadotropin-releasing hormone from the hypothalamus, which in turn lowers the blood estrogen level. Adipose tissue itself also produces some estrogen. Adequate estrogen is necessary for fertility. These girls and women are also at high risk of developing low bone density and cardiovascular impairments.

The increasing concentration of estrogens during the first week or so of a menstrual cycle changes the uterine lining, thickening the glandular endometrium (proliferative phase). Meanwhile, the follicle fully matures, and by around the fourteenth day of the cycle, the antral follicle appears on the ovary surface as a blisterlike bulge. Within the follicle, the follicular cells, which surround and connect the secondary oocyte to the inner wall, loosen. Follicular fluid accumulates.

While the follicle matures, it secretes estrogens that inhibit the release of LH from the anterior pituitary gland but allow LH to be stored in the gland. Estrogens also make anterior pituitary cells more sensitive to the action of GnRH, which is released from the hypothalamus in rhythmic pulses about ninety minutes apart.

Near the fourteenth day of follicular development, the anterior pituitary cells finally respond to the pulses of GnRH and release the stored LH. The resulting surge in LH concentration lasts about thirty-six hours. In response to the LH, the primary oocyte completes meiosis I. The LH also acts with FSH, inducing complex interactions with prostaglandins, progesterone, plasmin, and proteolytic enzymes, leading to the weakening and rupturing of the bulging follicular wall. This event sends the secondary oocyte and follicular fluid out of the ovary (ovulation).

Following ovulation, the space containing the follicular fluid fills with blood, which soon clots. Under the influence of LH, the remnants of the follicle within the ovary form a temporary glandular structure in the ovary called a **corpus luteum** (kor'pus loot'ē-um) ("yellow body") (see fig. 19.10).

Follicular cells secrete some progesterone during the first part of the menstrual cycle. During the second half of the cycle, cells of the corpus luteum secrete abundant progesterone and estrogens. Consequently, as a corpus luteum forms, the blood progesterone concentration sharply increases.

Progesterone causes the endometrium to become more vascular and glandular. It also stimulates the uterine glands to secrete more glycogen and lipids (secretory phase). The endometrial tissues fill with fluids containing nutrients and electrolytes, which provide a favorable environment for an embryo to develop.

High levels of estrogens and progesterone inhibit the anterior pituitary gland's release of LH and FSH. Consequently, no other follicles are stimulated to complete development when the corpus luteum is active. However, if the secondary oocyte released at ovulation is not fertilized, the corpus luteum begins to degenerate (regress) on about the twenty-fourth day of the cycle. Eventually, connective tissue replaces it. The remnant of such a corpus luteum is called a *corpus albicans,* and is eventually absorbed (see fig. 19.10).

TABLE 19.3 Major Events in a Menstrual Cycle
1. The anterior pituitary gland secretes follicle-stimulating hormone (FSH) and luteinizing hormone (LH).
2. FSH stimulates maturation of a dominant follicle.
3. Follicular cells produce and secrete estrogens. a. Estrogens maintain secondary sex characteristics. b. Estrogens cause the endometrium to thicken.
4. The anterior pituitary releases a surge of LH, which leads to ovulation.
5. Follicular cells become corpus luteum cells, which secrete estrogens and progesterone. a. Estrogens continue to stimulate uterine wall development. b. Progesterone stimulates the endometrium to become more glandular and vascular. c. Estrogens and progesterone inhibit the secretion of FSH and LH from the anterior pituitary gland.
6. If the secondary oocyte is not fertilized, the corpus luteum degenerates and no longer secretes estrogens and progesterone.
7. As the concentrations of estrogens and progesterone decline, blood vessels in the endometrium constrict.
8. The uterine lining disintegrates and sloughs off, producing a menstrual flow.
9. The anterior pituitary gland is no longer inhibited and again secretes FSH and LH.
10. The menstrual cycle repeats.

When the corpus luteum ceases to function, concentrations of estrogens and progesterone rapidly decline, and in response, blood vessels in the endometrium constrict. This reduces the supply of oxygen and nutrients to the thickened endometrium (stratum functionalis and stratum basalis), and most of these lining tissues soon disintegrate and slough off. At the same time, blood leaves damaged capillaries, creating a flow of blood and cellular debris that passes through the vagina as the *menstrual flow* (menses). This flow usually begins about the twenty-eighth day of the cycle and continues for three to five days, while the concentrations of estrogens are relatively low. The beginning of the menstrual flow marks the end of a menstrual cycle and the beginning of the next cycle as a new developing antral follicle becomes available. This cycle is summarized in table 19.3 and diagrammed in figure 19.13.

Low blood concentrations of estrogens and progesterone at the beginning of the menstrual cycle mean that the hypothalamus and anterior pituitary gland are no longer inhibited. Consequently, FSH and LH concentrations soon

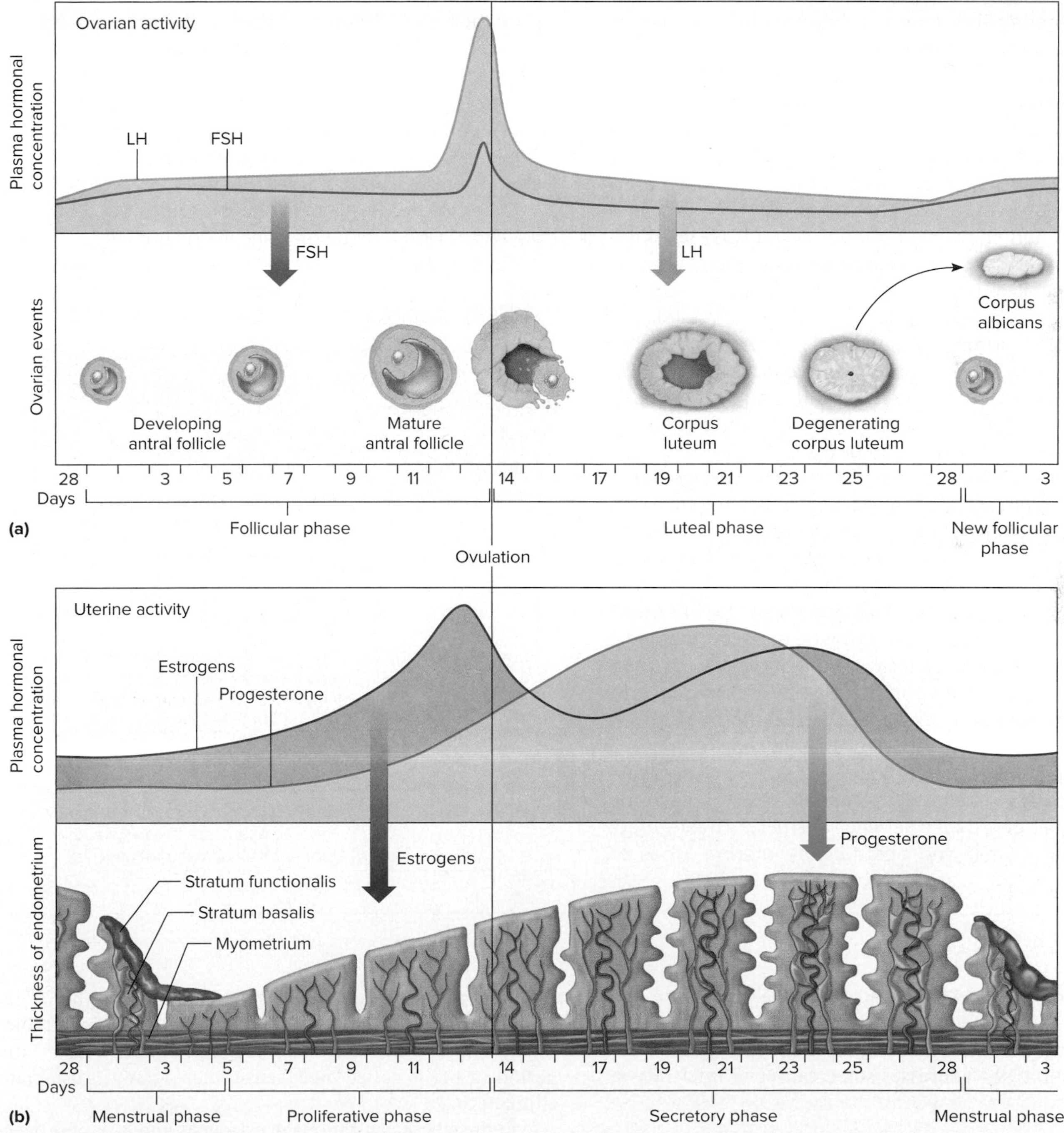

Figure 19.13 Major events in the female menstrual cycle. (**a**) Plasma hormonal concentrations of FSH and LH affect follicle maturation in the ovaries. (**b**) Plasma hormonal concentrations of estrogen and progesterone influence changes in the uterine lining. APR

CLINICAL APPLICATION 19.4
Female Infertility

Infertility is the inability to conceive after a year of trying. Males and females contribute equally to infertility, with 25% of the cases due to more than one factor, according to the American Society for Reproductive Medicine.

A common cause of female infertility is insufficient secretion (hyposecretion) of gonadotropic hormones by the anterior pituitary, preventing ovulation (anovulation). Testing the urine for *pregnanediol,* a product of progesterone metabolism, can detect an anovulatory ovarian cycle. Because progesterone concentration normally rises following ovulation, no increase in pregnanediol in the urine during the latter part of the ovarian cycle suggests ovulation has not occurred.

Fertility specialists may treat anovulation due to hyposecretion of gonadotropic hormones by administering human chorionic gonadotropin (hCG) obtained from human placentas. Another ovulation-stimulating biochemical, human menopausal gonadotropin (hMG), contains luteinizing hormone (LH) and follicle-stimulating hormone (FSH) and is obtained from the urine of postmenopausal women. However, either hCG or hMG may overstimulate the ovaries and cause many follicles to release secondary oocytes simultaneously, which may result in multiple births.

Another cause of female infertility is *endometriosis,* in which small pieces of the uterine lining (endometrium) move up through the uterine tubes during menstruation and attach somewhere in the abdominal cavity. Here, the tissue changes as it would in the uterine lining during the ovarian cycle. The misplaced tissue breaks down at the end of the cycle but cannot be expelled. Instead it remains in the abdominal cavity, irritating the lining (peritoneum) and causing considerable pain. This tissue also stimulates formation of fibrous tissue (fibrosis), which may encase the ovary and prevent ovulation or obstruct the uterine tubes. Conception may become impossible.

Sexually transmitted infections (STIs), such as gonorrhea, can cause female infertility. These infections can inflame and obstruct the uterine tubes or stimulate production of viscous mucus that can plug the cervix and prevent sperm entry.

Women become infertile if their ovaries are removed, which may be part of cancer treatment, or are damaged by cancer treatments such as chemotherapy and radiation. To make future pregnancies possible, these women can have strips of ovarian tissue removed before their cancer treatment begins. The strips are frozen and stored, then thawed and implanted under the skin of the forearm or abdomen or in the pelvic cavity near the ovaries, such that secondary oocytes can enter the uterine tubes. Another approach is to freeze oocytes in liquid nitrogen at −30°C to −40°C, first protecting them with a chemical that removes water and prevents ice crystal formation. Thousands of babies have been born following *in vitro* fertilization using frozen oocytes.

Finding the right treatment for a particular patient requires determining the infertility's cause. Table 19B describes diagnostic tests for female infertility.

TABLE 19B Tests to Assess Female Infertility

Test	What It Checks
Hormone levels	If ovulation occurs
Ultrasound	Placement and appearance of reproductive organs and structures
Postcoital test	Cervix examined soon after unprotected intercourse to see if mucus is thin enough to allow sperm through
Endometrial biopsy	Small piece of uterine lining sampled and viewed under microscope to see if it can support an embryo
Hysterosalpingogram	Dye injected into uterine tube and followed with scanner to show if tube is clear or blocked
Laparoscopy	Small, lit optical device inserted near navel to detect scar tissue blocking tubes, which ultrasound may miss

increase, stimulating a new antral follicle to mature. As this follicle secretes estrogens, the uterine lining undergoes repair, and the endometrium begins to thicken again. Clinical Application 19.4 addresses some causes of infertility in the female.

Menopause

After puberty, menstrual cycles continue at regular intervals into the late forties or early fifties, when the ovaries start to produce less estrogen and progesterone. This results in the menstrual cycles becoming less predictable. Then, within a few months or years, the cycles cease. This period in life is called **menopause** (men'o-pawz), or female climacteric.

Reduced concentrations of estrogens and lack of progesterone may change the female secondary sex characteristics. The breasts, vagina, uterus, and uterine tubes may shrink, and the pubic and axillary hair may thin.

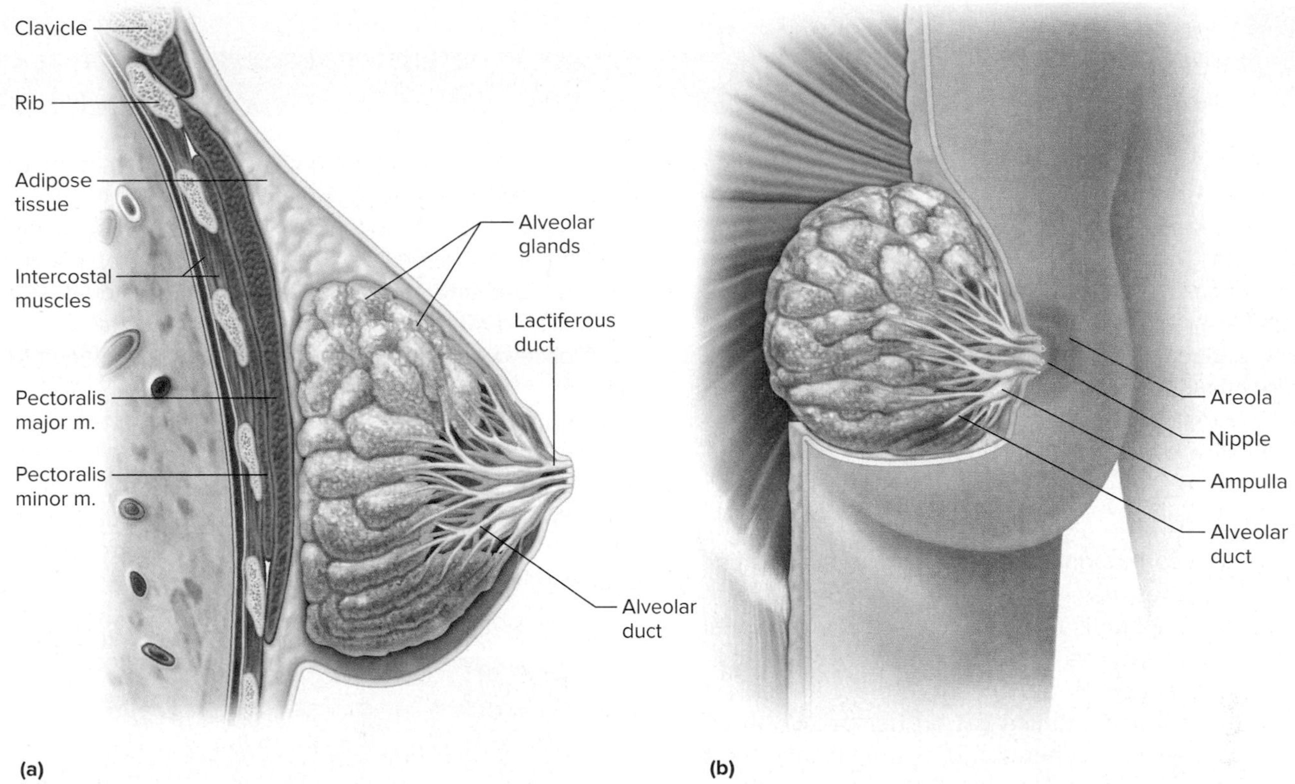

Figure 19.14 Structure of the female breast and mammary glands. (**a**) Sagittal section. (**b**) Anterior view. (**m.** stands for **muscle.**) 

PRACTICE 19.6

7. Trace the events of the menstrual cycle.
8. What causes menstrual flow?
9. What are some changes that may occur at menopause?

19.7 | Mammary Glands

LEARN

1. Review the structure of the mammary glands.

The **mammary glands** are accessory organs of the female reproductive system specialized to secrete milk following pregnancy. They are in the subcutaneous tissue of the anterior thorax within elevations called *breasts* (fig. 19.14*a*; reference plate 1). The breasts overlie the *pectoralis major* muscles and extend from the second to the sixth ribs and from the sternum to the axillae.

A *nipple* is located near the tip of each breast at about the level of the fourth intercostal space. A circular area of pigmented skin, called the *areola,* surrounds each nipple (fig. 19.14*b*).

A mammary gland is composed of fifteen to twenty lobes. Each lobe contains glands (alveolar glands), drained by alveolar ducts, which drain into a lactiferous duct that leads to the nipple and opens to the outside (fig. 19.14). Adipose and dense connective tissues separate the lobes. These tissues also support the glands and attach them to the fascia of the underlying pectoral muscles. Other connective tissue, which forms dense strands called *suspensory ligaments,* extends inward from the dermis of the breast to the fascia, helping support the breast.

The mammary glands of males and females are similar. As children reach puberty, the glands in males do not develop, whereas ovarian hormones stimulate development of the glands in females. The alveolar glands and ducts enlarge, and fat is deposited around and within the breasts. Section 20.3, Pregnancy and the Prenatal Period, describes the hormonal mechanism that stimulates mammary glands to produce and secrete milk. Clinical Application 19.5 discusses diagnosis, treatment, and prevention of breast cancer.

PRACTICE 19.7

1. Describe the structure of a mammary gland.
2. How does ovarian hormone secretion change the mammary glands?

19.8 | Birth Control

LEARN

1. Describe several methods of birth control, including the relative effectiveness of each method.

Birth control is the voluntary regulation of the number of offspring produced and the time they are conceived. This control requires a method of **contraception** (kon″trah-sep′shun)

CLINICAL APPLICATION 19.5

Treating Breast Cancer

According to the National Cancer Institute, the lifetime risk of a woman developing breast cancer is about one in eight. However, as table 19C illustrates, this is not the same as one in eight women at any point in time having breast cancer. Less than 1% of breast cancer cases are in men. Several types of breast cancer have been traditionally distinguished by the types of cells affected and their locations, but recent genetic testing has revealed mutations that occur in the cancer cells that increase susceptibility to or cause the disease.

Warning Signs

Changes that could signal breast cancer include a small area of thickened tissue; a dimple; a change in contour; or a nipple that is flattened, points in an unusual direction, or produces a discharge. A woman can note these changes by performing a monthly "breast self-exam," in which she lies flat on her back with an arm raised behind her head and systematically feels all parts of each breast. But sometimes breast cancer gives no warning—fatigue and feeling ill may not occur until the disease has spread beyond the breast.

After finding a lump, the next step is a physical exam, where a health-care provider palpates the breast and does a mammogram, which is an X-ray scan that can pinpoint the location and approximate extent of abnormal tissue (fig. 19B). A mammogram also indicates the proportions of tissues making up the breasts. High "breast density" refers to an increased proportion of connective tissue and glandular tissue compared to fat, and is associated with a greater risk of developing breast cancer. Dense breasts may make mammograms difficult to interpret.

An ultrasound scan can distinguish between a cyst (a fluid-filled sac of glandular tissue) and a tumor (a solid mass). If an area is suspicious, a thin needle is used to take a biopsy (sample) of the tissue and cells, which is then scrutinized under a microscope for the characteristics of cancer.

TABLE 19C Breast Cancer Risk

By Age	Odds	By Age	Odds
25	1 in 19,608	60	1 in 24
30	1 in 2,525	65	1 in 17
35	1 in 622	70	1 in 14
40	1 in 217	75	1 in 11
45	1 in 93	80	1 in 10
50	1 in 50	85	1 in 9
55	1 in 33	95 or older	1 in 8

Source: Southern Illinois University School of Medicine.

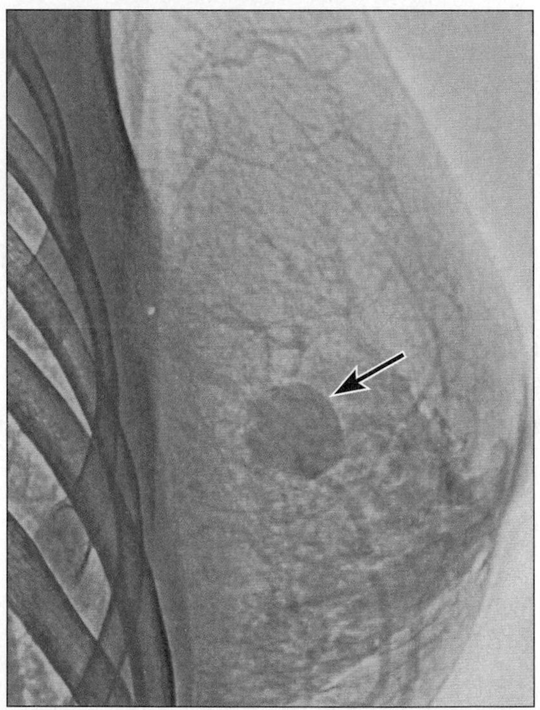

Figure 19.B Mammogram of a breast with a tumor (arrow). APR Source: Southern Illinois University, School of Medicine

designed to avoid fertilization of an egg cell following sexual intercourse (coitus) or to prevent implantation of an embryo. The several methods of contraception have varying degrees of effectiveness.

Coitus Interruptus

Coitus interruptus is the practice of withdrawing the penis from the vagina before ejaculation, preventing entry of sperm cells into the female reproductive tract. This method can still result in pregnancy because a male may find it difficult to withdraw just prior to ejaculation. Also, some semen containing sperm cells may reach the vagina before ejaculation occurs.

Rhythm Method

The *rhythm method* (also called timed coitus or natural family planning) requires abstinence from sexual intercourse two days before and one day after ovulation. The rhythm

Further tests can identify estrogen and progesterone receptors on cancer cells, providing information used to guide treatment choices. A very aggressive form of breast cancer is called triple negative, because affected cells do not have receptors (for estrogen, progesterone, and epidermal growth factor) that many drugs target. Triple negative breast cancer tends to start earlier, spread faster, and recur more often than other types. It is more prevalent among women of color.

Surgery, Radiation, and Drugs

If biopsied breast cells are cancerous, treatment usually begins with surgery. A lumpectomy removes a small tumor and some surrounding tissue; a simple mastectomy removes a breast; and a modified radical mastectomy removes the breast and usually one or two lymph nodes, but preserves the pectoral muscles. If cancer cells are detected in the lymph nodes, further surgery is performed.

Many breast cancers are then treated with radiation and combinations of drugs, plus sometimes newer drugs that are targeted to certain types of breast cancer. Standard drugs kill all rapidly dividing cells; those used for breast cancer include fluorouracil, doxorubicin, cyclophosphamide, methotrexate, and paclitaxol. Protocols that provide more frequent, lower doses can temper some of the side effects of these powerful drugs.

Newer treatments for breast cancer are easier to tolerate and can be extremely effective. Three types of drugs keep signals (estrogen and growth factors) from stimulating cancer cells to divide:

- Selective estrogen receptor modulators (SERMs), such as tamoxifen and raloxifene, block estrogen receptors. About half of people with breast cancer have receptors for estrogen on their cancer cells and can benefit from these drugs.
- Aromatase inhibitors block an enzyme required for tissues other than those of the ovaries to synthesize estrogens. These drugs are used in women who are past menopause, whose ovaries no longer synthesize estrogen. They are prescribed after a five-year course of a SERM.
- Herceptin (trastuzumab) can help people whose cancer cells bear too many receptors that bind epidermal growth factor. It is a monoclonal antibody, based on an immune system protein. Herceptin blocks the growth factor from signaling cell division. This drug treats a particularly aggressive form of the disease that strikes younger women.

Prevention Strategies

How often to have mammography depends on an individual's family history of breast and other cancers. The American College of Obstetrics and Gynecology advises physicians to offer mammography annually to female patients after age 40; however, the U.S. Preventive Services Task Force recommends every other year for women aged 50 through 74, but starting at age 40 for those with a parent, child, or sibling who has breast cancer. Mammography is generally not offered to women 75 or older. Although a mammogram can detect a tumor up to two years before it can be felt, it can also miss some tumors. Thus, the breast self-exam is also important in early detection.

Genetic tests can identify women who have inherited certain variants of genes—such as *BRCA1, BRCA2, TP53,* and *HER-2/neu*—that place them at high risk for developing breast cancer. Some of these women have their breasts removed to prevent the disease. Gene expression profiling (determining which genes are turned on or off in cancer cells) can determine which drugs are most likely to help particular patients, and estimate the risk of recurrence after surgery. This type of information may help patients and their health-care providers decide on treatments that follow surgery.

Only 5% to 10% of all breast cancers arise from an inherited tendency. Much research seeks to identify the environmental triggers that contribute to causing the majority of cases.

---

method results in a relatively high rate of pregnancy because accurately identifying infertile ("safe") times to have intercourse is difficult. Another disadvantage of the rhythm method is that it requires adherence to a particular pattern of behavior and restricts spontaneity in sexual activity.

Mechanical Barriers

Mechanical barrier contraceptives prevent sperm cells from entering the female reproductive tract during sexual intercourse. The *male condom* is a thin latex or polyurethane or natural membrane sheath placed over the erect penis before intercourse to prevent semen from entering the vagina upon ejaculation (fig. 19.15*a*). A *female condom* resembles a small plastic bag. A woman inserts it into her vagina prior to intercourse. The device blocks sperm cells from reaching the cervix (fig. 19.15*a*).

Some men feel that a condom decreases the sensitivity of the penis during intercourse, and condom use may

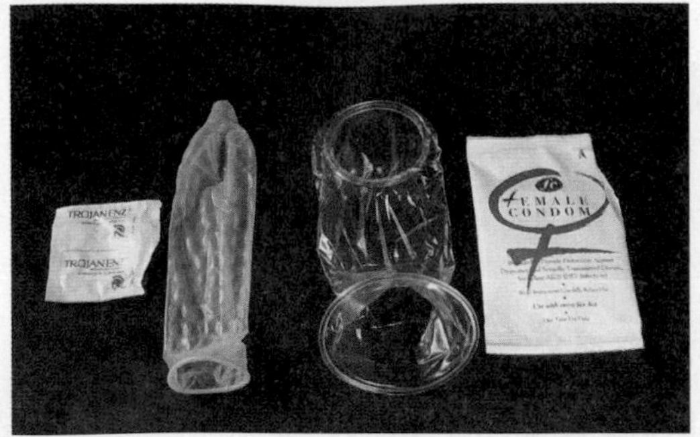

(a)

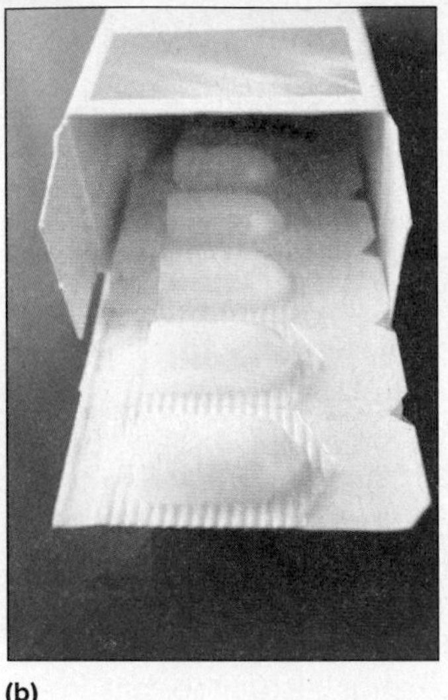

(b)

(c)

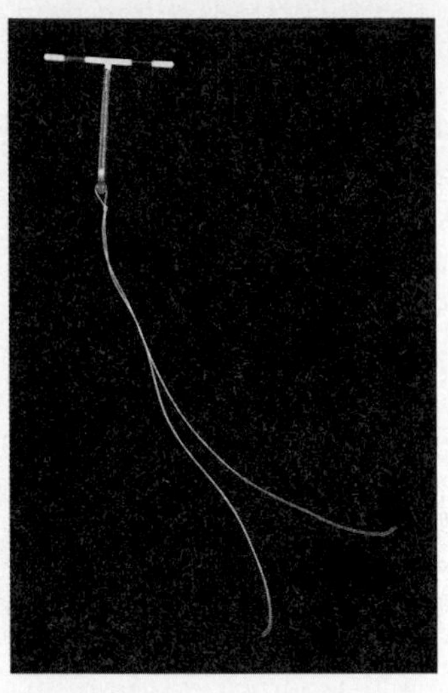

(d)

Figure 19.15 Devices and substances used for birth control include (**a**) male and female condoms; (**b**) spermicide in film, sponge, suppositories, and gel; (**c**) oral contraceptives; and (**d**) a copper IUD. (a, d): Jill Braaten/McGraw-Hill Education; (b): Image Point Fr/Shutterstock; (c): Don Farrall/Getty Images

PRACTICE FIGURE 19.15

Which of these methods of birth control use(s) hormones to prevent pregnancy?

Answer can be found in Appendix E.

interrupt spontaneity. However, condoms are inexpensive and may also protect the user from contracting or spreading some sexually transmitted infections.

Chemical Barriers

Chemical barrier contraceptives include creams, foams, and jellies with spermicidal properties (fig. 19.15*b*). These chemicals create an unfavorable environment in the vagina for sperm cells.

Chemical barrier contraceptives are fairly easy to use but have a high failure rate when used alone. They are more effective when used with a condom. The contraceptive sponge contains a spermicide and blocks entry of sperm into the uterus.

Combined Hormone Contraceptives

Combined hormone contraceptives deliver estrogen and progestin to prevent pregnancy. Various methods are used to administer the hormones, but all work on the same principle with about the same efficacy, although the amounts of the component hormones may vary. One such method is a small flexible chemical ring inserted deep into the vagina once a month, remaining in place three out of four weeks. A plastic patch impregnated with the hormones may be applied to the skin on the buttocks, stomach, arm, or upper torso once a week for three out of four weeks. The most commonly used method to deliver the hormones is orally, in pill form (fig. 19.15*c*).

Combined hormone contraceptives contain synthetic estrogen-like and progesterone-like chemicals. These drugs disrupt the normal pattern of gonadotropin (FSH and LH) secretion, preventing follicle maturation and the LH surge that leads to ovulation. They also thicken cervical mucus to prevent the sperm from joining an oocyte. Most combined hormone contraceptives cause light monthly bleeding. A new type of pill causes only four bleeding periods a year.

If used correctly, combined hormone contraceptives prevent pregnancy nearly 100% of the time. However, they may cause nausea, retention of body fluids, increased skin pigmentation, and breast tenderness. Some women, particularly those over thirty-five years of age who smoke, may develop intravascular blood clots, liver disorders, or high blood pressure when using certain types of these contraceptives.

Other Hormone Contraceptives

An intramuscular injection of medroxyprogesterone acetate protects against pregnancy for three months by preventing follicle maturation and release of a secondary oocyte. It also alters the cervix to prevent the sperm from joining an oocyte. Medroxyprogesterone acetate is long-acting; it takes ten to eighteen months after the last injection for the effects to wear off. Use of this drug requires a doctor's care because of potential side effects and risks.

An implantable rod containing progestin may be inserted under the skin of the arm. This hormone insert also prevents follicle maturation and ovulation for up to three years.

Intrauterine Devices

An *intrauterine device (IUD)* is a small, solid object that a physician places in the uterine cavity (fig. 19.15*d*). A copper-containing IUD may be effective for up to ten years. The copper wire causes an inflammatory reaction that is usually toxic to sperm and egg cells, ultimately preventing pregnancy. A levonorgestrel-releasing IUD, combining hormone contraception with an IUD, is effective in preventing pregnancy for up to five years. It thickens the cervical mucus so the sperm has trouble entering the uterus, creates a hostile environment for the sperm, and inhibits endometrial growth so the uterus is not conducive to implantation if the oocyte is fertilized.

An IUD may be spontaneously expelled from the uterus, or produce abdominal pain or excessive menstrual bleeding. It may also harm the uterus or produce other serious health problems.

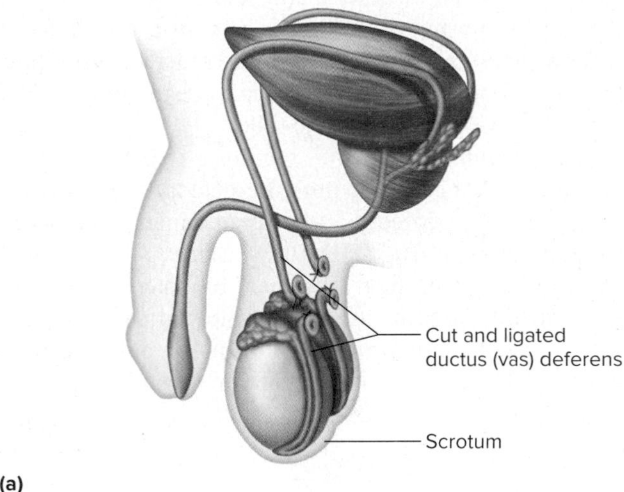

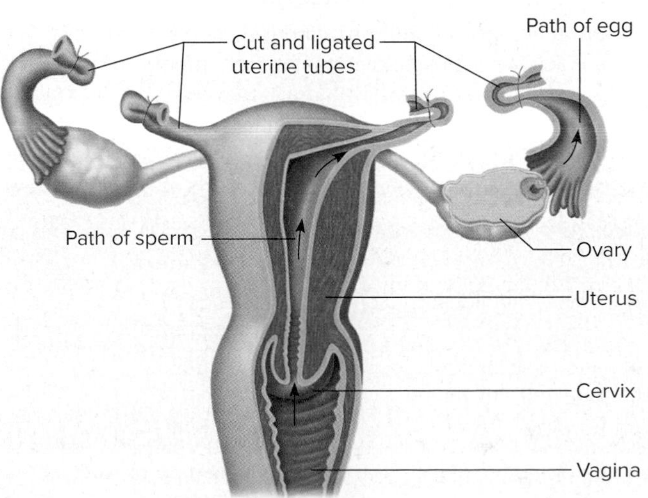

Figure 19.16 Surgical methods of birth control. (**a**) In a vasectomy, each ductus (vas) deferens is cut and ligated. (**b**) In a tubal ligation, each uterine tube is cut and ligated. The egg released at ovulation consists of the secondary oocyte and its surrounding layers.

Sterilization

Surgical methods of contraception sterilize the male or female. In the male, a physician removes a small section of each ductus (vas) deferens near the epididymis and ties (ligates) the cut ends of the ducts (fig. 19.16*a*). This is a *vasectomy,* an operation that produces few side effects, although it may cause some pain for a week or two.

After a vasectomy, sperm cells cannot leave the epididymis; thus, they are excluded from the semen. However, sperm cells may already be present in the ducts distal to the cuts. Consequently, the sperm count may not reach zero for several weeks.

The corresponding procedure in the female is *tubal ligation* (fig. 19.16*b*). The uterine tubes are cut and ligated (tied) so that sperm cells cannot reach an egg. In a nonsurgical method of sterilization in the female, a physician inserts a tube through the vagina and uterus into the uterine tubes that delivers a tiny spring-like device. Usually within three months, scar tissue blocks the uterine tubes so sperm cannot reach the egg.

Sterilization procedures do not change hormonal concentrations or sex drives. These procedures are the most reliable forms of contraception. Reversing them requires microsurgery.

PRACTICE 19.8

1. What factors make the rhythm method less reliable than some other methods of contraception?
2. Describe two methods of contraception that use mechanical barriers.
3. How do combined hormone contraceptives prevent pregnancy?
4. How does an IUD prevent pregnancy?

19.9 | Sexually Transmitted Infections

LEARN

1. List the general symptoms of sexually transmitted infections.

The term **sexually transmitted infection (STI)** is replacing the term *sexually transmitted disease (STD)* because a person can be infected with a pathogen and transmit the pathogen to others but not develop the symptoms of the disease. By the time symptoms appear, it is often too late to prevent

CLINICAL APPLICATION 19.6
Pap Smears

Carcinoma in situ (CIS) is a condition in which the cells that line the cervix develop some of the characteristics of cancer cells. CIS is considered to be "precancerous." A *Pap (Papanicolaou) smear test* identifies these cellular changes. A Pap sample is often also used to test for the presence of certain strains of the sexually transmitted human papillomavirus (HPV), which can cause cervical cancer. Government public health guidelines recommend PAP smears every three years beginning at age twenty-one or when a woman becomes sexually active, until age twenty-nine. From then until age 65, women should be tested every five years. PAP smear testing can stop after age 65 if the most recent three tests were normal.

complications or spread of the infection to sexual partners. Many diseases associated with STIs have similar symptoms, and some of the symptoms are also seen in diseases or allergies not sexually related, so it is wise to consult a physician if one or a combination of these symptoms appears:

- burning sensation during urination
- pain in the lower abdomen
- fever or swollen glands in the neck
- discharge from the vagina or penis
- pain, itching, or inflammation in the genital or anal area
- pain during intercourse
- sores, blisters, bumps, or a rash anywhere on the body, particularly the mouth or genitals
- itchy, runny eyes

Table 19.4 describes some prevalent diseases associated with sexually transmitted infections. Clinical Application 19.6 discusses the use of a Pap smear to detect certain sexually transmitted viruses and other disease. One possible complication of the associated diseases gonorrhea and chlamydia is **pelvic inflammatory disease (PID),** in which bacteria enter the vagina and spread throughout the reproductive organs. The disease begins with intermittent cramps, followed by sudden fever, chills, weakness, and severe cramps. Hospitalization and intravenous antibiotics can stop the infection. The uterus and uterine tubes are often scarred, resulting in infertility and increased risk of ectopic pregnancy, in which the embryo develops in a uterine tube.

Acquired immune deficiency syndrome (AIDS) is a steady deterioration of the body's immune defenses in which the body is overrun by infection and often cancer. The human immunodeficiency virus (HIV) that causes AIDS is transmitted in body fluids such as semen, blood, and milk. It is most frequently passed during unprotected intercourse or by using a needle containing contaminated blood. Clinical Application 14.1 explores HIV infection further.

PRACTICE 19.9

1. Why is the term *sexually transmitted infection* replacing the term *sexually transmitted disease*?
2. What are some common symptoms of diseases associated with sexually transmitted infections?

TABLE 19.4 Some Diseases Associated with Sexually Transmitted Infections

Associated Disease	Cause	Symptoms	Treatment
Acquired immune deficiency syndrome (AIDS)	Human immunodeficiency virus (HIV)	Fever, weakness, infections, cancer	Drugs to treat or delay symptoms
Chlamydia infection	*Chlamydia trachomatis* bacteria	Painful urination and intercourse, mucous discharge from penis or vagina	Antibiotics
Genital herpes	Herpes simplex 2 virus (HSV2)	Genital sores, fever	Antiviral drug (acyclovir)
Genital warts	Human papilloma virus (HPV)	Warts on genitals	Chemical or surgical removal
Gonorrhea	*Neisseria gonorrhoeae* bacteria	In women, usually none; in men, painful urination	Antibiotics
Syphilis	*Treponema pallidum* bacteria	Initial chancre usually on genitals or mouth; rash six months later; several years with no symptoms as infection spreads; finally damage to heart, liver, nerves, brain	Antibiotics

Reproductive Systems

INTEGUMENTARY SYSTEM

Skin sensory receptors play a role in sexual pleasure.

SKELETAL SYSTEM

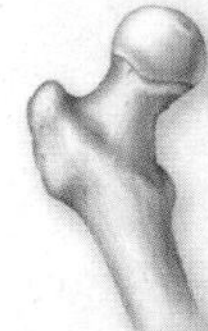

Bones can be a temporary source of calcium during lactation.

MUSCULAR SYSTEM

Skeletal, cardiac, and smooth muscles all play a role in reproductive processes and sexual activity.

NERVOUS SYSTEM

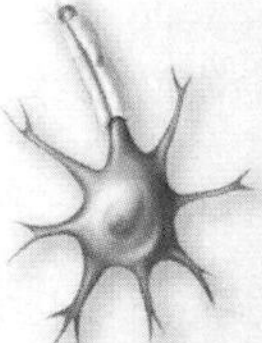

The nervous system plays a major role in sexual activity and sexual pleasure.

ENDOCRINE SYSTEM

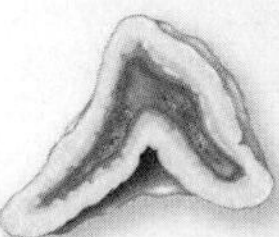

Hormones control the production of eggs in the female and sperm in the male.

CARDIOVASCULAR SYSTEM

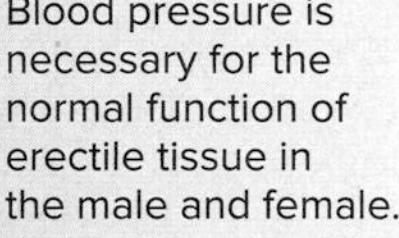

Blood pressure is necessary for the normal function of erectile tissue in the male and female.

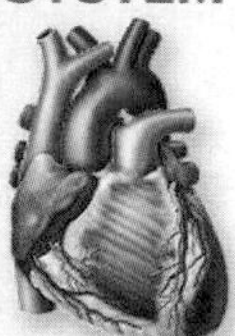

LYMPHATIC SYSTEM

Special mechanisms inhibit the female immune system from attacking sperm as foreign invaders.

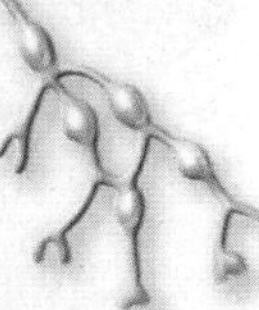

DIGESTIVE SYSTEM

Proper nutrition is essential for the formation of normal gametes.

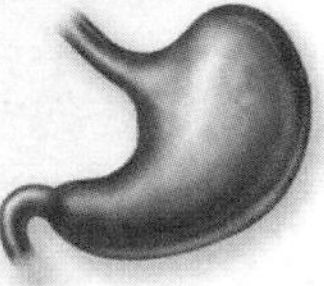

RESPIRATORY SYSTEM

Breathing provides oxygen that assists in the production of ATP needed for egg and sperm development.

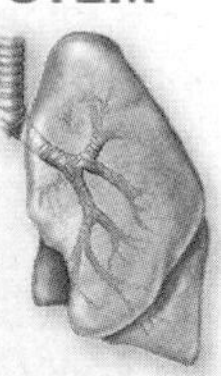

URINARY SYSTEM

Male urinary and reproductive systems share structures. Kidneys help compensate for fluid loss from the reproductive systems.

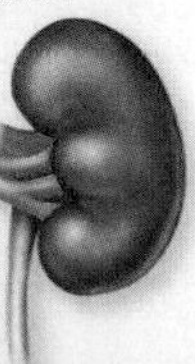

Gamete production, fertilization, embryonic and fetal development, and childbirth are essential for survival of the species.

ASSESS

CHAPTER ASSESSMENTS

19.1 Introduction

1. General functions of the male and female reproductive systems include ________.
   a. producing sex cells
   b. transporting sex cells to sites of fertilization
   c. secreting hormones
   d. all of the above

19.2 Organs of the Male Reproductive System

2. List the organs (both primary and accessory) of the male reproductive system, and explain how each organ's structure affects the organ's function.

19.3 Spermatogenesis

3. Trace the path of sperm cells from their site of formation to the outside of the body. Indicate composition and when and where secretions are added to produce semen.
4. Distinguish between emission and ejaculation.
5. List the major steps in spermatogenesis.

19.4 Hormonal Control of Male Reproductive Functions

6. Describe the role of gonadotropin-releasing hormone (GnRH) in the control of male reproductive functions.
7. Discuss the actions of testosterone.

19.5 Organs of the Female Reproductive System

8. List the organs (both primary and accessory) of the female reproductive system, and explain how each organ's structure affects the organ's function.

19.6 Oogenesis and the Ovarian Cycle

9. List the major steps in oogenesis.
10. Describe how a follicle matures.
11. Define *ovulation*.
12. Describe the role of gonadotropin-releasing hormone (GnRH) in the control of female reproductive functions.
13. Discuss the actions of estrogens.
14. Summarize the major events in the menstrual cycle.

19.7 Mammary Glands

15. Describe the structure of a mammary gland.

19.8 Birth Control

16. Match the birth control method with its description.

(1) withdrawal	a. kills sperm
(2) rhythm method	b. keeps sperm out of vagina or from entering cervix
(3) condom	c. increases thickness of cervical mucus to inhibit sperm from entering the uterus
(4) spermicide	d. no intercourse during fertile times
(5) estrogen and progesterone pills	e. penis removed from vagina before ejaculation
(6) IUD	f. sperm cells never reach penis
(7) vasectomy	g. prevent follicle maturation and ovulation
(8) tubal ligation	h. oocytes never reach uterus

19.9 Sexually Transmitted Infections

17. Common symptoms of diseases associated with sexually transmitted infections include ________.
    a. a burning sensation during urination
    b. discharge from vagina or penis
    c. sores, blisters, or rash on genitals
    d. all of the above
18. If left untreated, a complication of the diseases gonorrhea and chlamydia is ________.

ASSESS

INTEGRATIVE ASSESSMENTS/CRITICAL THINKING

Outcomes 7.4, 11.4, 11.5, 19.4

1. Joshua, a seven-year-old, had noticeable pubic hair, underarm hair growth, and adult body odor. His mom took him to see an endocrinologist who requested that Joshua get an X-ray of his hand. What was the purpose of the hand X-ray? Josh was diagnosed with precocious puberty (early puberty). The next step was to seek the source of the problem. What organs of the body might be tested, and why?

Outcomes 11.4, 11.7, 19.2, 19.3, 19.4, 19.5, 19.7, 19.8

2. Understanding the causes of infertility can be valuable in developing new birth control methods. Cite a type of contraceptive that might be used based on each of the following causes of infertility:
(a) failure to ovulate due to a hormonal imbalance
(b) a large fibroid tumor that disturbs the uterine lining
(c) endometrial tissue blocking uterine tubes
(d) low sperm count (too few sperm per ejaculate)

Outcomes 11.5, 11.8, 19.2, 19.3, 19.4, 19.5

3. What changes, if any, would a male who has had one testis removed experience? A female who has had one ovary removed?

Outcomes 11.4, 11.9, 19.5, 19.6

4. Michelle, age forty, had a hysterectomy and an oophorectomy (ovariectomy) as suggested by her genetics specialist and primary care physician based upon results of genetic testing that showed a high risk for endometrial and ovarian cancers. Post-surgery, explain why Michelle was experiencing periodic chills, hot flashes, vaginal dryness, and moodiness.

Outcomes 19.2, 19.3

5. Some men are unable to become fathers because their spermatids do not mature into sperm. Injection of their spermatids into their partners' secondary oocytes sometimes results in conception. Men have fathered healthy babies this way. Why would this procedure work with spermatids, but not with primary spermatocytes?

Outcome 19.6

6. Sometimes a sperm cell fertilizes a polar body rather than a secondary oocyte. An embryo does not develop, and the fertilized polar body degenerates. Why is a polar body unable to support development of an embryo?

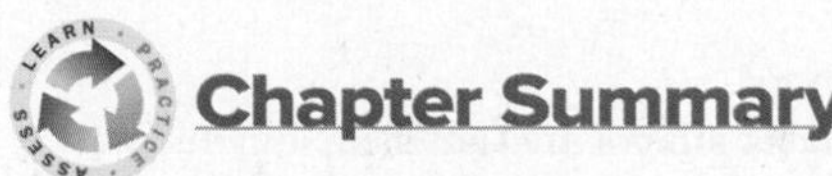

Chapter Summary

19.1 Introduction

Reproductive organs produce sex cells and sex hormones, nurture these cells, or transport them.

19.2 Organs of the Male Reproductive System

The primary male sex organs are the ***testes,*** *which produce sperm cells and male sex hormones. Accessory reproductive organs are internal and external.*

1. Testes
   a. Structure of the testes
      (1) The testes are separated by connective tissue and filled with seminiferous tubules.
      (2) The **seminiferous tubules** contain **spermatogenic cells** that give rise to sperm cells.
      (3) The **interstitial cells** produce male sex hormones.
2. Male internal accessory reproductive organs
   a. Epididymis
      (1) Each **epididymis** is a tightly coiled tube that leads into a ductus deferens.
      (2) They store and nourish immature sperm cells and promote their maturation.
   b. Ductus deferens
      (1) Each **ductus deferens** is a muscular tube.
      (2) They pass through the inguinal canal, enter the pelvic cavity, course medially, and end behind the urinary bladder.
      (3) They fuse with the ducts from seminal vesicles to form the **ejaculatory ducts.**
   c. Seminal vesicles
      (1) Each **seminal vesicle** is a saclike structure attached to a ductus deferens.
      (2) They secrete an alkaline fluid, to help regulate the pH of semen, that contains fructose and prostaglandins.
   d. Prostate gland
      (1) The **prostate gland** surrounds the urethra just inferior to the urinary bladder.
      (2) It secretes a thin, milky fluid containing citrate and prostate-specific antigen.
   e. Bulbourethral glands
      (1) The **bulbourethral glands** are two small structures inferior to the prostate gland.
      (2) They secrete a fluid that lubricates the penis in preparation for sexual intercourse.
   f. Semen
      (1) **Semen** consists of sperm cells and secretions of the seminal vesicles, prostate gland, and bulbourethral glands.
      (2) This fluid is slightly alkaline and contains nutrients and prostaglandins.
      (3) Sperm cells in semen begin to swim, but are unable to fertilize oocytes until they are activated in the female reproductive tract.
3. Male external accessory reproductive organs
   a. Scrotum
   The **scrotum** is a pouch of skin and subcutaneous tissue that encloses the testes for protection and temperature regulation.
   b. Penis
      (1) The **penis** becomes erect for insertion into the vagina during sexual intercourse.
      (2) Its body is composed of three columns of erectile tissue.
4. Erection, orgasm, and ejaculation
   a. During **erection,** vascular spaces in the erectile tissue become engorged with blood.
   b. **Orgasm** is the culmination of sexual stimulation and is accompanied by **emission** and **ejaculation.**
   c. Semen moves along the reproductive tract as smooth muscle in the walls of the tubular structures contracts by reflex.

19.3 Spermatogenesis

The formation of sperm cells.

a. The epithelium lining the seminiferous tubules includes sustentacular cells and spermatogenic cells.
   (1) Sustentacular cells support and nourish spermatogenic cells.
   (2) Spermatogenic cells (**spermatogonia**) give rise to sperm cells.
b. **Spermatogenesis** produces four sperm cells from each primary spermatocyte.
c. **Meiosis** reduces the number of chromosomes in sperm cells by one-half (from 46 to 23).
d. Structure of a sperm cell
A sperm cell consists of a head, midpiece, and tail.

19.4 Hormonal Control of Male Reproductive Functions

1. Hypothalamic and pituitary hormones
   a. The male body remains reproductively immature until the hypothalamus releases gonadotropin-releasing hormone (GnRH), which stimulates the anterior pituitary gland to release **gonadotropins.**
   b. Follicle-stimulating hormone (FSH) stimulates spermatogenesis.
   c. Luteinizing hormone (LH), sometimes known in males as interstitial cell stimulating hormone (ICSH), stimulates interstitial cells to produce male sex hormones.
2. Male sex hormones
   a. Male sex hormones are called **androgens,** with **testosterone** the most important.
   b. Androgen production increases rapidly at **puberty.**
   c. Actions of testosterone
      (1) Testosterone stimulates development of the male reproductive organs.
      (2) It also develops and maintains male secondary sex characteristics.

3. Regulation of male sex hormones
   a. A negative feedback mechanism regulates testosterone concentration.
      (1) A rising testosterone concentration inhibits the hypothalamus and reduces the anterior pituitary gland's secretion of gonadotropins.
      (2) As testosterone concentration falls, the hypothalamus signals the anterior pituitary gland to secrete gonadotropins.
   b. The testosterone concentration remains relatively stable from day to day.

19.5 Organs of the Female Reproductive System

*The primary female sex organs are the **ovaries,** which produce female sex cells and sex hormones. Accessory reproductive organs are internal and external.*

1. Ovaries
   a. Ovary structure
      (1) Each ovary is subdivided into a medulla and a cortex.
      (2) The medulla is composed of connective tissue, blood vessels, lymphatic vessels, and nerves.
      (3) The cortex contains ovarian follicles and is covered by cuboidal epithelium.
   b. Primordial follicles
      (1) During prenatal development, groups of cells in the ovarian cortex form millions of **primordial follicles.**
      (2) Each primordial follicle contains a primary oocyte and a layer of follicular cells.
      (3) The primary oocytes begin meiosis, but the process halts until puberty.
      (4) The number of primary oocytes steadily declines throughout a female's life.
2. Female internal accessory reproductive organs
   a. Uterine tubes
      (1) The end of each **uterine tube** expands, and its margin bears irregular extensions.
      (2) Ciliated cells that line the tube and peristaltic contractions in the wall of the tube help transport the secondary oocyte down the uterine tube. Fertilization of the oocyte occurs here.
   b. Uterus
      (1) The **uterus** receives the embryo and sustains it during development.
      (2) The uterine wall includes the **endometrium, myometrium,** and **perimetrium.**
   c. Vagina
      (1) The **vagina** receives the erect penis, conveys uterine secretions to the outside, and provides an open channel for the fetus during birth.
      (2) Its wall consists of a mucosal layer, muscular layer, and fibrous layer.
3. Female external accessory reproductive organs
   a. Labia majora
      (1) The **labia majora** are rounded folds of adipose tissue and skin.
      (2) The anterior ends form a rounded elevation over the pubic symphysis, called the mons pubis.
   b. Labia minora
      (1) The **labia minora** are flattened, longitudinal folds between the labia majora.
      (2) They are well supplied with blood vessels.
   c. Clitoris
      (1) The **clitoris** is a small projection at the anterior end of the vulva. It corresponds to the male penis.
      (2) It is composed of two columns of erectile tissue.
   d. Vestibule
      (1) The **vestibule** is the space between the labia minora.
      (2) The **vestibular glands** secrete mucus into the vestibule during sexual stimulation.
4. Erection, lubrication, and orgasm
   a. During periods of sexual stimulation, the erectile tissues of the clitoris and vestibular bulbs become engorged with blood and swell.
   b. The vestibular glands secrete mucus into the vestibule and vagina.
   c. During orgasm, the muscles of the perineum, uterine wall, and uterine tubes contract rhythmically.

19.6 Oogenesis and the Ovarian Cycle

The hypothalamus, anterior pituitary gland, and ovaries secrete hormones that control sex cell maturation, the development and maintenance of female secondary sex characteristics, and changes that occur during the menstrual cycle.

1. Oogenesis
   a. Beginning at puberty, some primary oocytes are stimulated to continue meiosis.
   b. When a primary oocyte undergoes meiosis, it gives rise to a secondary oocyte in which the original chromosome number is reduced by one-half (from 46 to 23).
   c. If fertilization occurs, the secondary oocyte undergoes a second meiotic division, which produces an ovum.
   d. The chromosomes of the sperm combine with the chromosomes of the ovum, producing a **zygote**.
2. Follicle maturation
   a. At puberty, FSH initiates follicle maturation.
   b. During maturation of the follicle, the follicular cells multiply, and a fluid-filled cavity (antrum) forms.
   c. Usually, only one dominant follicle fully develops in preparation for ovulation every 28 days.
   d. Just prior to ovulation, the primary oocyte completes meiosis to give rise to a secondary oocyte and first polar body.
3. Ovulation
   a. **Ovulation** is the release of a secondary oocyte (within its surrounding one to two layers of follicular cells) from an ovary.
   b. A rupturing follicle releases the secondary oocyte.
   c. After ovulation, the secondary oocyte is drawn into the opening of the uterine tube.
4. Female sex hormones
   a. A female body remains reproductively immature until about ten years of age, when gonadotropin secretion increases.

b. The most important female sex hormones are **estrogens** and **progesterone.**
   (1) Estrogens develop and maintain most of the female secondary sex characteristics.
   (2) Progesterone prepares the uterus for pregnancy.

5. Menstrual cycle
   a. FSH from the anterior pituitary gland initiates a menstrual cycle by stimulating follicle maturation.
   b. Maturing follicular cells secrete estrogens, which maintain the secondary sex characteristics and thicken the uterine lining.
   c. Secretion of a relatively large amount of LH (LH "surge") by the anterior pituitary gland leads to ovulation. The LH surge causes the primary oocyte to complete meiosis, giving rise to a secondary oocyte and first polar body.
   d. After ovulation, follicle remnants form the **corpus luteum.**
      (1) The corpus luteum secretes estrogens and progesterone, which causes the endometrium to become more vascular and glandular.
      (2) If a secondary oocyte is not fertilized, the corpus luteum begins to degenerate.
      (3) As concentrations of estrogens and progesterone decline, the uterine lining disintegrates, causing menstrual flow.
   e. During this cycle, estrogens and progesterone inhibit the release of LH and FSH (except for the large increase at mid-cycle). As concentrations of estrogens and progesterone decline, the anterior pituitary gland secretes FSH and LH again, stimulating the next menstrual cycle.
6. Menopause
   a. **Menopause** is cessation of the menstrual cycles. Less estrogen and progesterone are produced.
   b. Reduced concentrations of estrogens and lack of progesterone may cause regressive changes in female secondary sex characteristics.

19.7 Mammary Glands

1. The **mammary glands** are in the subcutaneous tissue of the anterior thorax within the breasts.
2. They are composed of lobes that contain glands and ducts.
3. Dense connective and adipose tissues separate the lobes.
4. Ducts connect the mammary glands to the nipple.
5. Ovarian hormones stimulate female breast development.
   a. Alveolar glands and ducts enlarge.
   b. Fat is deposited around and within the breasts.

19.8 Birth Control

Birth control is voluntary regulation of how many offspring are produced and when they are conceived. It usually involves some method of ***contraception.***

1. Coitus interruptus is withdrawal of the penis from the vagina before ejaculation.
2. The rhythm method is abstinence from sexual intercourse for several days before and after ovulation.
3. Mechanical barriers
   Males and females can use condoms.
4. Chemical barriers
   Spermicidal creams, foams, and jellies provide an unfavorable environment in the vagina for sperm survival.
5. Combined hormone contraceptives
   a. A flexible ring inserted deep into the vagina, a plastic patch, or a pill can deliver estrogen and progestin to prevent pregnancy.
   b. They disrupt a female's normal pattern of gonadotropin secretion, which prevents follicle maturation and ovulation, and they thicken cervical mucus to prevent the sperm from joining the oocyte.
6. Other hormone contraceptives
   a. Intramuscular injection with medroxyprogesterone every three months acts similarly to oral contraceptives to prevent pregnancy.
   b. An implantable rod containing progestin may prevent pregnancy for up to three years.
7. Intrauterine devices (IUDs)
   An IUD is a solid object inserted in the uterine cavity that prevents pregnancy by thickening cervical mucus to prevent the sperm from entering the uterus, create an inhospitable environment for the sperm, and inhibit endometrial growth.
8. Sterilization
   a. Vasectomies in males and tubal ligations in females are surgical sterilization procedures.
   b. A nonsurgical method involves insertion of a tiny coil into each uterine tube, causing scarring that blocks the sperm from joining the egg.

19.9 Sexually Transmitted Infections

1. **Sexually transmitted infections** are passed during sexual contact and may go undetected for years.
2. Many of the diseases associated with sexually transmitted infections share similar symptoms.

20 Pregnancy, Growth, Development, and Genetics

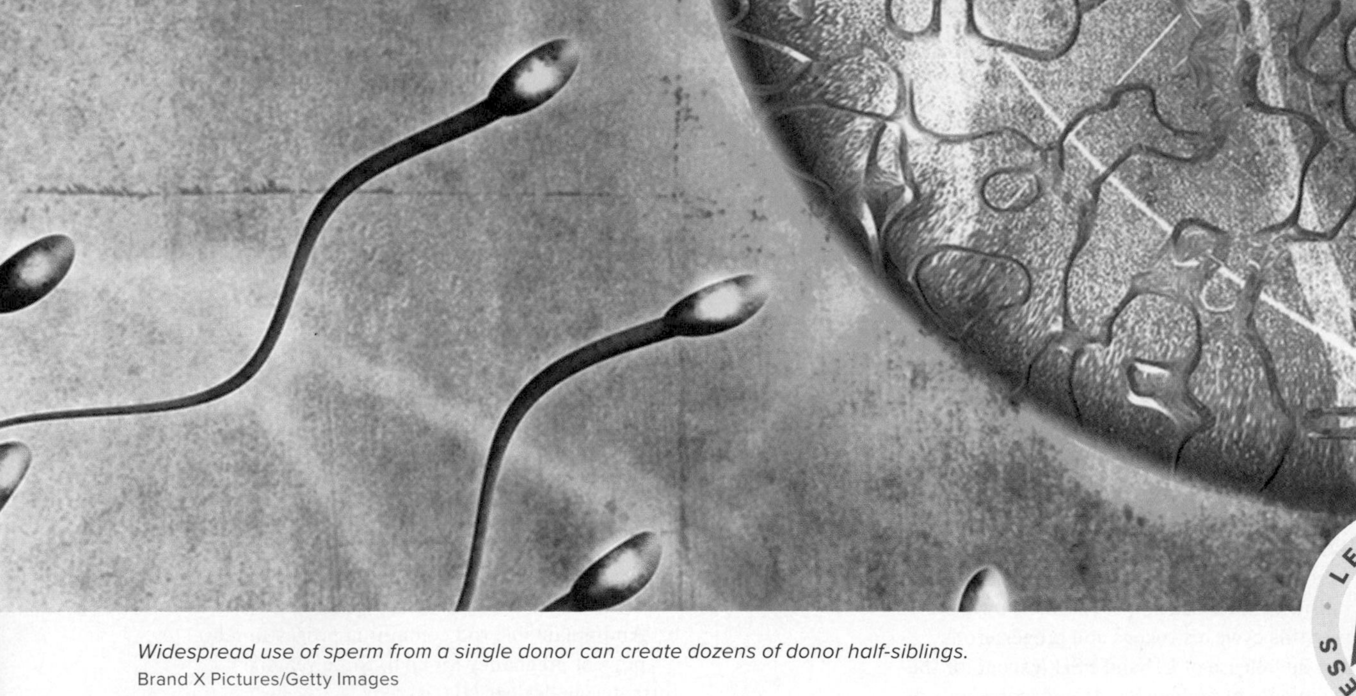

Widespread use of sperm from a single donor can create dozens of donor half-siblings.
Brand X Pictures/Getty Images

Sperm donation can create many donor half-siblings. Before the age of the Internet, it was highly unusual for children conceived with the same donated sperm to meet each other. Today, half-siblings can find each other and are even meeting, thanks to the Donor Sibling Registry. This and other websites enable parents who used donated sperm to learn of other families that used the same sperm, using identifying numbers that the sperm banks assigned. Several of these half-sibling groups number up to 50, with one including 150 children who share their father, but not their mothers.

Some families who discover many half-siblings react positively, even holding social events where everyone marvels at the resemblances among the children. But others seek legislation limiting the number of children a single sperm donor can father. Because mothers are not required to report births using sperm donors to the facilities that provided the sperm, the true number of children born each year using donor sperm is unknown.

Use of sperm from one donor to inseminate many women can have negative biological consequences. Two half-siblings might wish to start a family together, not knowing they are related. This increases the likelihood that they carry the same mutations inherited from a shared ancestor, compared to unrelated individuals. The gene pool (a population measure) is affected when one individual contributes disproportionately to the next generation, increasing the prevalence of mutations that cause rare inherited diseases, because a man would not ordinarily father 50 to 100 children.

Many people who find out that their children have dozens of half-siblings are upset, as are the men who donated the sperm. The American Society for Reproductive Medicine advises that one man's donated sperm be used for no more than 25 births for every 800,000 births.

Module 14 Reproductive Systems

LEARNING OUTLINE

After studying this chapter, you should be able to complete the "Learning Outcomes" that follow the major headings throughout the chapter.

AIDS TO UNDERSTANDING WORDS

allant- [sausage] *allantois*: tubelike structure extending from the yolk sac into the connecting stalk of an embryo.

chorio- [skin] *chorion*: outermost embryonic membrane.

cleav- [to divide] *cleavage*: period of development when a zygote divides, producing increasingly smaller cells.

hetero- [other, different] *heterozygous*: when the two copies of a gene variant in an individual are of different DNA sequence.

hom- [same, common] *homozygous*: when the two copies of a gene variant in an individual are identical in the DNA sequence.

lacun- [pool] *lacuna*: space between the chorionic villi that fills with maternal blood.

morul- [mulberry] *morula*: embryonic structure consisting of a solid ball of about sixteen cells that resembles a mulberry.

nat- [to be born] *prenatal*: before birth.

troph- [well fed] *trophoblast*: cellular layer that surrounds the inner cell mass and helps nourish it.

umbil- [navel] *umbilical cord*: structure attached to the fetal navel (umbilicus) that connects the fetus to the placenta.

(Appendix A has a complete list of Aids to Understanding Words.)

20.1 | Introduction

LEARN

1. Distinguish between growth and development.
2. Distinguish between prenatal and postnatal.

When a sperm cell joins a secondary oocyte, a zygote is formed, and the journey of development begins. Following thirty-eight weeks of cell division, growth, and specialization into distinctive tissues and organs, a new human being enters the world.

Humans grow, develop, and age. **Growth** is an increase in size. In humans and other multicellular organisms, growth entails an increase in cell numbers as a result of mitosis, followed by enlargement of the newly formed cells.

Development, which includes growth and aging, is the continuous process by which an individual changes during their life span. This span includes a **prenatal** (prē-nā′tal) **period,** which begins with fertilization and ends at birth, and a **postnatal** (pōst-nā′tal) **period,** which begins at birth and ends with death.

PRACTICE 20.1

Answers to the Practice questions can be found in the eBook.

1. Distinguish between growth and development.
2. When is the beginning and ending of the postnatal period?

20.2 | Fertilization

LEARN

1. Describe fertilization.

The union of a secondary oocyte and a sperm cell is called **fertilization** (fer″tĭ-lĭ-zā′shun), or conception. Fertilization takes place in a uterine (fallopian) tube.

Transport of Sex Cells

A female of reproductive age usually ovulates a secondary oocyte and its surrounding cells each month. The released secondary oocyte then usually enters a uterine tube.

Several functions of the male and female reproductive systems assist sperm in reaching the secondary oocyte. During sexual intercourse, the male deposits semen containing sperm in the vagina near the cervix. The sperm must then move upward through the uterus and uterine tube. Prostaglandins in the semen stimulate lashing of sperm tails, and muscular contractions within the walls of the uterus and uterine tube aid the sperm cells' journey. Also, under the influence of high concentrations of estrogens during the first part of the menstrual cycle, the uterus and cervix secrete a watery fluid that promotes sperm transport and survival. Conversely, during the latter part of the cycle, when the progesterone concentration is high, the female reproductive tract secretes a viscous fluid that hampers sperm transport and survival.

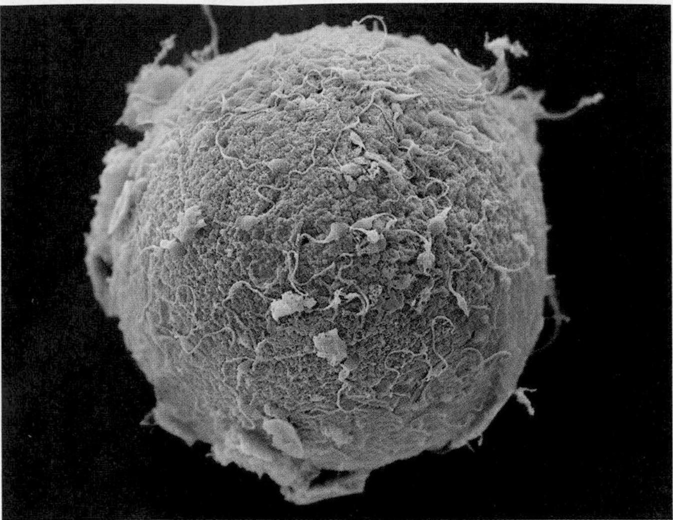

Figure 20.1 Scanning electron micrograph of sperm cells on the surface of a secondary oocyte (650x). Only one sperm cell usually fertilizes a secondary oocyte. Eye of Science/Science Source

Sperm reach the upper part of the uterine tube in less than an hour following sexual intercourse. Many sperm cells may reach a secondary oocyte, but usually only one sperm cell fertilizes it (fig. 20.1; see fig. 20.4*a*).

Sperm Cell Joins Secondary Oocyte

Sperm cells first invade the corona radiata, which is a layer that consists of follicular cells surrounding the secondary oocyte. An enzyme associated with the sperm cell membrane (hyaluronidase) aids penetration of the sperm head by digesting proteins in the corona radiata (fig. 20.2). The sperm next bind to the *zona pellucida,* which is a membrane rich in glycoproteins that is the layer closest to the secondary oocyte's cell membrane. As sperm bind to a specific class of protein on the zona pellucida, the acrosomes release their enzymes by exocytosis, and those enzymes digest the material of the zona pellucida. The sperm now have direct access to the secondary oocyte.

Hundreds of sperm take part in removing these barriers, but usually only one sperm, the first to reach the secondary oocyte's cell membrane, will fertilize the secondary oocyte. The head portion of that first sperm cell fuses with the secondary oocyte's cell membrane, and the sperm nucleus enters the secondary oocyte, leaving the mitochondria-rich middle section and tail outside. Sperm entry triggers lysosome-like vesicles just beneath the secondary oocyte's cell membrane to release enzymes that harden the zona pellucida. This reduces the chance that other sperm cells will penetrate.

The sperm cell nucleus enters the secondary oocyte's cytoplasm and swells. This triggers completion of meiosis II in the secondary oocyte, resulting in the formation of a large cell, the mature **egg** (or ovum) whose nucleus contains the female's genetic contribution (DNA), and a tiny second polar body, which is later expelled. The approaching nuclei from the two sex cells are called pronuclei (fig. 20.3). When the pronuclei unite, their nuclear membranes disassemble, and their chromosomes mingle, completing fertilization. This cell, called a **zygote** (zī′gōt), is the first cell of the future offspring.

Each sex cell provides 23 chromosomes, so the product of fertilization is a cell with 46 chromosomes—the usual number in a human body cell (somatic cell). Table 20.1 describes assisted reproductive technologies used to achieve fertilization.

PRACTICE 20.2

1. What factors aid the movements of the egg and sperm cells through the female reproductive tract?
2. Where in the female reproductive system does fertilization take place?
3. List the events of fertilization.

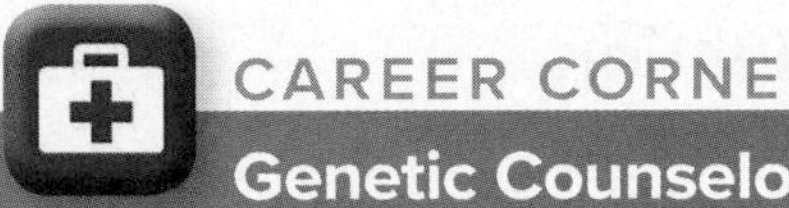

CAREER CORNER
Genetic Counselor

The pregnant patient is anxious after receiving a phone call reporting the test result that she is a carrier for cystic fibrosis. She soon meets with a genetic counselor, who explains how the disease is inherited and describes testing options. The next step is to test the father-to-be, and when he is found not to have a mutation, the couple relaxes. Had he also carried a mutation, the counselor explained, the child and any future children would have a 25% chance each of inheriting the disease, which could have been detected with further testing.

A genetic counselor is a health-care professional trained in genetics, psychology, statistics, and counseling techniques to provide guidance to patients taking genetic tests or with inherited diseases in their families. A certified genetic counselor has a master's degree in the field, but nurses, social workers, physicians, and PhD geneticists can also provide genetic counseling. The counselor explains the inheritance patterns, assesses risks for specific individuals in a family, recommends tests, and interprets the results.

Genetic counselors work in medical centers, in physicians' offices, in research laboratories, and at companies that provide genetic tests. They may specialize in prenatal care or work in clinics for patients with specific diseases, such as familial cancers, hereditary forms of blindness, or inherited blood disorders.

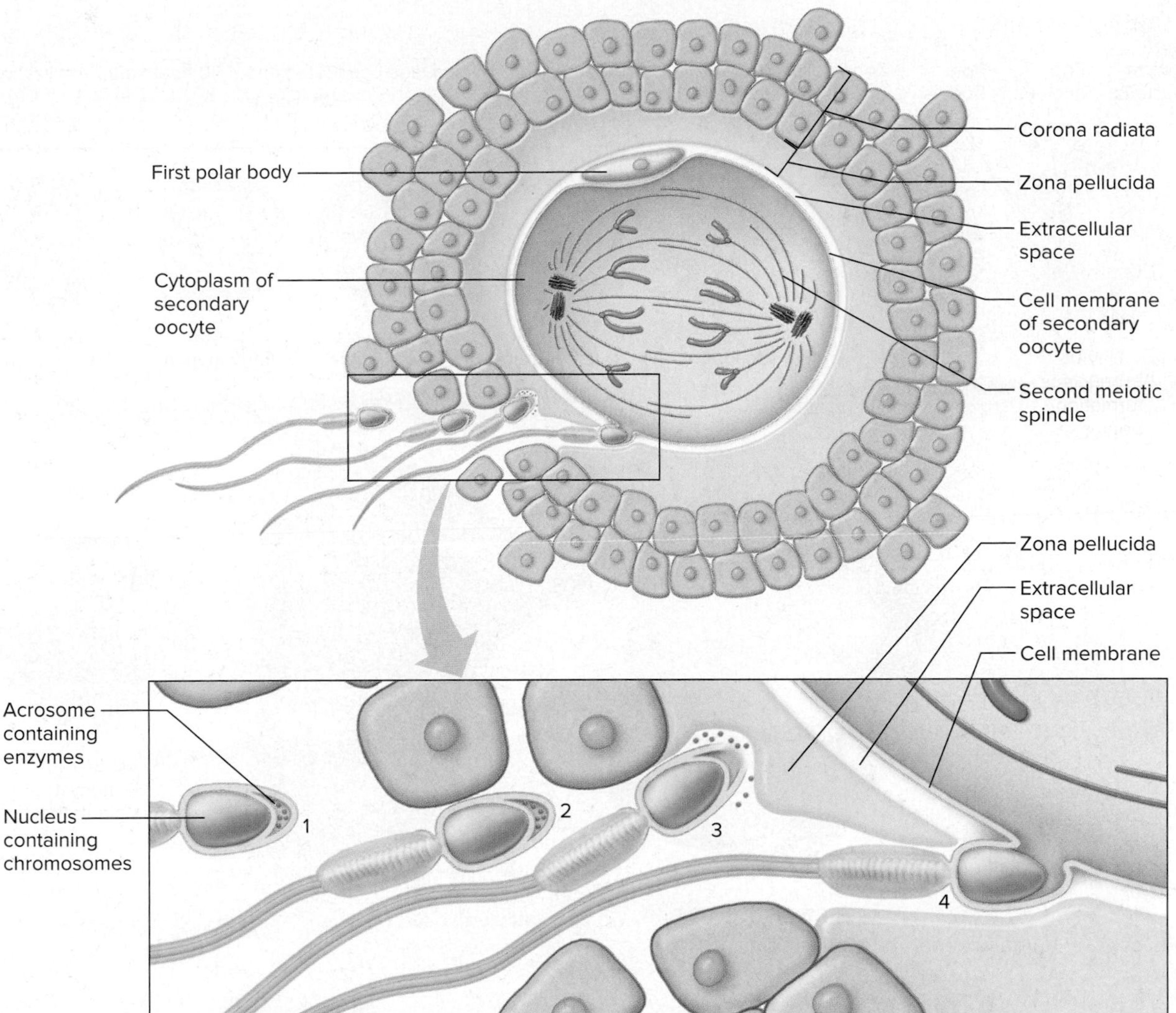

Figure 20.2 Steps in fertilization: (**1**) The sperm reaches the corona radiata surrounding the secondary oocyte. (**2**) An enzyme on the sperm surface digests a path through the corona radiata. (**3**) The sperm binds to proteins in the zona pellucida, triggering the release of digestive enzymes from the acrosome by exocytosis. (**4**) Once the sperm is through the zona pellucida, the sperm head fuses with the secondary oocyte's cell membrane, and the sperm nucleus enters the egg. Although usually only a single sperm can fertilize the secondary oocyte, hundreds of sperm are necessary to break down these barriers encountered along the way.

PRACTICE FIGURE 20.2

How many chromosomes are contained in the secondary oocyte prior to fertilization?

Answer can be found in Appendix E.

20.3 | Pregnancy and the Prenatal Period

LEARN

1. List and provide details of the major events of cleavage.
2. Describe implantation.
3. Describe the extraembryonic membranes, as well as the formation and function of the placenta.
4. List the structures produced by each of the primary germ layers.
5. Describe the major events of the fetal stage of development.
6. Trace the path of blood through the fetal cardiovascular system.
7. Discuss the hormonal and other changes in the maternal body during pregnancy.
8. Explain the role of hormones in the birth process and milk production.

Pregnancy is the presence of a developing offspring in the uterus. Pregnancy consists of three periods called trimesters, each about three months long.

The prenatal period of development of the offspring usually lasts for thirty-eight weeks from conception. It can be divided into an embryonic stage and a fetal stage.

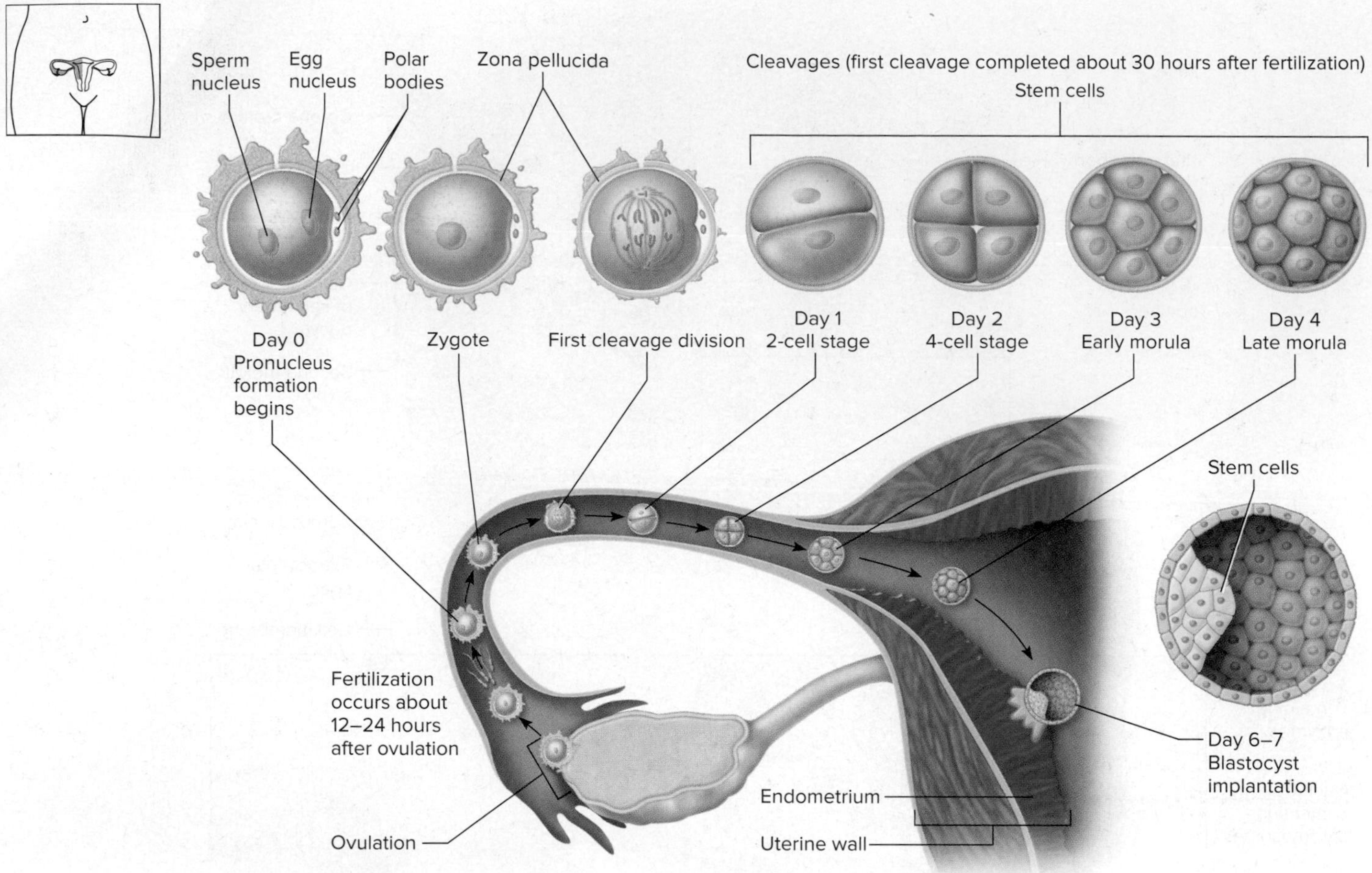

Figure 20.3 Stages of early human prenatal development. APR

TABLE 20.1 Assisted Reproductive Technologies

Technology	Condition It Treats	Procedure
Intrauterine insemination	Male infertility—lack of sperm cells or low sperm count	Donated sperm cells are placed into the cervix or body of the uterus.
Surrogate mother	Female infertility—a woman has healthy ovaries, but lacks a uterus	An egg fertilized *in vitro* is implanted in a woman other than its donor. The surrogate, or "gestational mother," gives the newborn to the "genetic mother" and her partner, the sperm donor.
Gamete intrafallopian transfer (GIFT)	Female infertility—bypasses blocked uterine tube	Eggs are removed from a woman's ovary, then placed along with donated sperm cells into a uterine tube at a site past any obstruction.
Zygote intrafallopian transfer (ZIFT)	Female infertility—bypasses blocked uterine tube	An egg fertilized *in vitro* is placed in a uterine tube past the blockage. It travels to the uterus on its own.
Embryo donation	Female infertility—a woman has nonfunctional ovaries, but a healthy uterus	A woman undergoes intrauterine insemination with sperm cells from a man whose partner cannot ovulate. If the woman conceives, the embryo is flushed from her uterus and implanted in the uterus of the sperm donor's partner.
Intracytoplasmic sperm injection	Low sperm count; sperm that cannot mature past spermatid stage; men with paralysis who cannot ejaculate	Sperm are injected into secondary oocytes *in vitro*.

Embryonic Stage

The **embryonic stage** extends from fertilization through the eighth week of prenatal development. During this time, the placenta forms, the main internal organs develop, and the external body structures appear.

Period of Cleavage

About thirty hours after the zygote forms, it undergoes *mitosis,* giving rise to two new cells (fig. 20.4*b*). These cells, in turn, divide into four cells, which divide into eight cells, and so forth. The divisions occur rapidly, with little time for the cells to grow. As a result, each division yields smaller cells (blastomeres). This phase of early rapid cell division is termed **cleavage** (klēv′ij) (see fig. 20.3).

During cleavage, the tiny mass of cells moves through the uterine tube to the uterine cavity. The trip takes about three days, and by then the structure consists of a solid ball, called a *morula* (mor′oo-lah), of about sixteen cells (figs. 20.3 and 20.4*c*).

The morula remains unattached within the uterine cavity for about three days. During this stage, the zona pellucida of the original secondary oocyte degenerates. As cell division continues and cell number is increasing, the morula hollows out, forming a **blastocyst,** which begins to attach to the endometrium. By the end of the first week of development, the blastocyst superficially attaches to the endometrium (fig. 20.5). Up until this point, the cells that will become the developing offspring are pluripotent stem cells (see fig. 20.3), which means they can give rise to several specialized types of cells, as well as yield additional stem cells.

At about the time of attachment, certain cells on the inner face of the blastocyst organize into a group, called the *inner cell mass* (or embryoblast), that eventually gives rise to the **embryo proper,** the body of the developing offspring. The cells that form the wall of the blastocyst make up the *trophoblast,* which develops into structures that assist the development of the **embryo** (em′brē-ō).

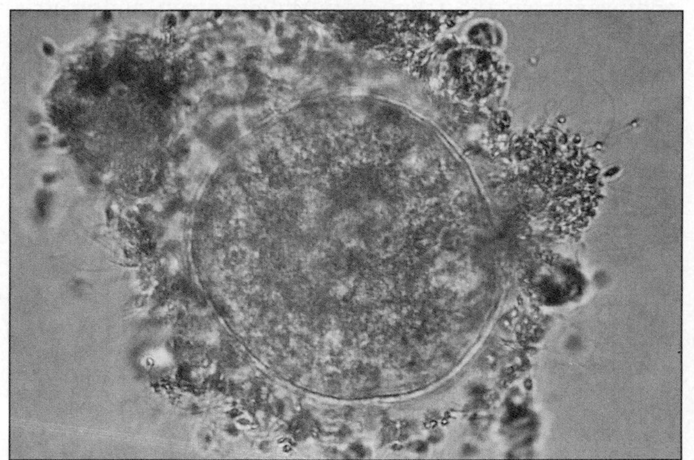

(a)

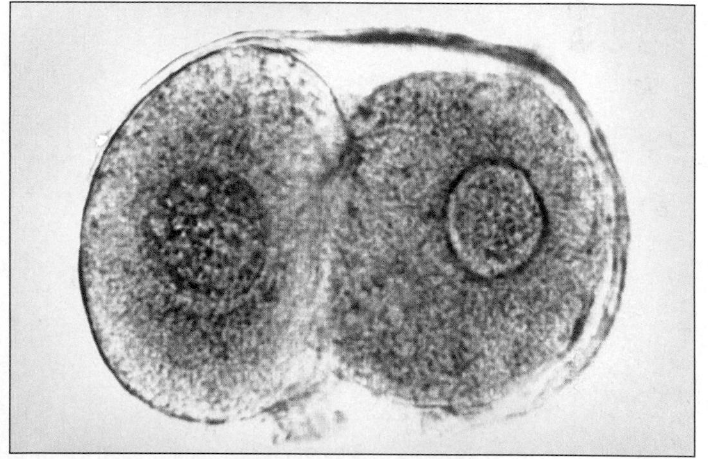

(b)

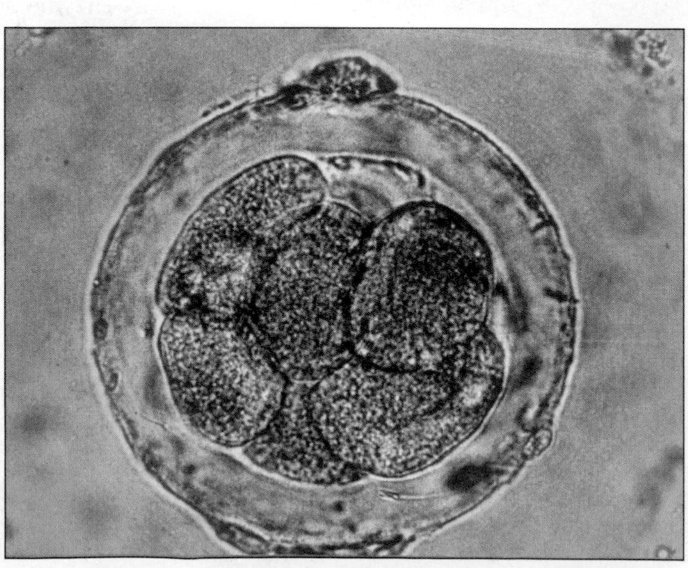

(c)

Figure 20.4 Light micrographs of (**a**) a human secondary oocyte surrounded by follicular cells and sperm cells (250x), (**b**) the two-cell stage (600x), and (**c**) a morula (500x). (a): Alexander Tsiaras/Science Source; (b): Omikron/Science Source; (c): Petit Format/Nestle/Science Source

The cells of the trophoblast begin to produce tiny, finger-like extensions (microvilli) that grow into the endometrium. At the same time, growth of the endometrium envelopes the blastocyst until it is completely embedded in the uterine wall. This results in **implantation** (im′plan-tā′shun) into the uterine lining.

If a blastocyst implants in tissues outside the uterus, such as those of a uterine tube, an ovary, the cervix, or an organ in the abdominal cavity, the result is an *ectopic pregnancy.* A blastocyst implanted in the uterine tube is a *tubal pregnancy.* The tube usually ruptures as the developing embryo enlarges, resulting in severe pain and heavy vaginal bleeding. Treatment is prompt surgical removal of the embryo and repair or removal of the damaged uterine tube.

The trophoblast secretes the hormone **human chorionic gonadotropin (hCG),** which maintains the corpus luteum during the early stages of pregnancy and keeps the immune system from rejecting the blastocyst. This hormone also stimulates synthesis of other hormones from the developing **placenta** (plah-sen′tah) (see fig. 20.5). The placenta is a vascular structure, formed by the cells surrounding the embryo and cells of the endometrium, that anchors the embryo to the uterine wall and exchanges nutrients, gases, and wastes between the maternal blood and the embryo's blood.

If a pregnant woman repeatedly ingests an addictive substance that crosses the placenta, her newborn may suffer from withdrawal symptoms when amounts of the chemical the fetus was accustomed to receiving suddenly plummet after birth. Newborn addiction can occur with certain drugs of abuse, such as heroin, and with certain prescription drugs used to treat anxiety.

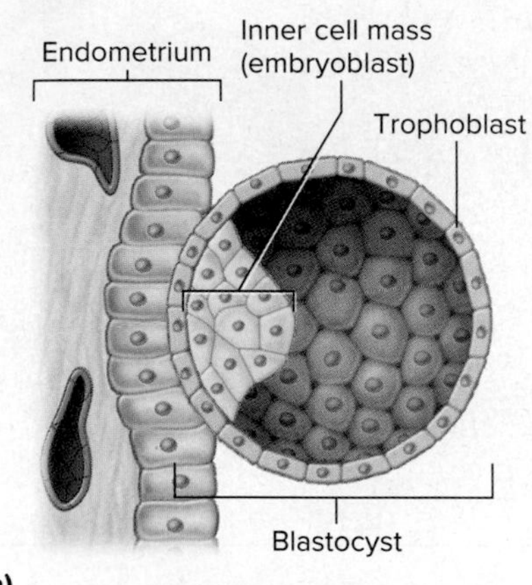

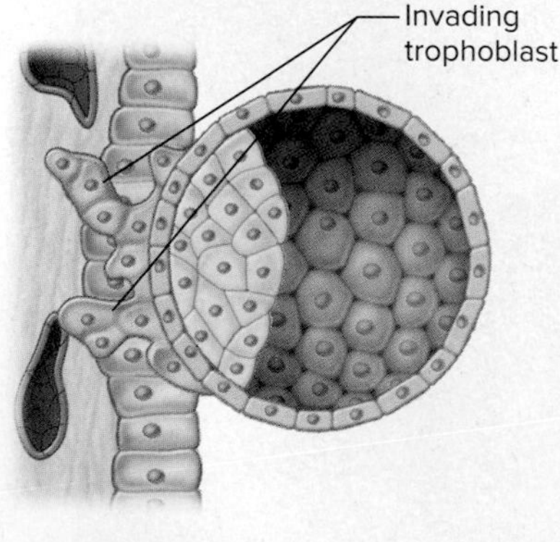

Figure 20.5 At about the sixth day of prenatal development, the blastocyst (**a**) attaches to the uterine wall and (**b**) begins to implant. The trophoblast, which will help form the placenta, secretes hCG, a hormone that maintains the pregnancy. APR

PRACTICE 20.3

1. What is cleavage?
2. What is implantation?
3. What is the function of hCG?

Extraembryonic Membrane Formation and Placentation

As the embryo implants in the uterus, proteolytic enzymes from the trophoblast break down endometrial tissue, providing nutrients for the developing embryo. A second layer of cells begins to line the trophoblast, and together those two layers form a structure called the **chorion** (ko′rē-on), the outermost embryonic membrane. Soon slender projections, including the new cell layer, grow out from the trophoblast, making their way into the surrounding endometrium by eroding it with their secretions of proteolytic enzymes. These projections become increasingly complex and form the highly branched **chorionic villi,** which are well established by the end of the fourth week (fig. 20.6).

Continued secretion of proteolytic enzymes forms irregular spaces called **lacunae** in the endometrium around and between the chorionic villi. These spaces fill with maternal blood that escapes from endometrial blood vessels eroded by enzyme action. At the same time, embryonic blood vessels extend through the *connecting stalk,* which attaches the embryo to the developing placenta. They establish capillary networks in the developing chorionic villi. These embryonic vessels allow nutrient exchange with the blood in the lacunae, meeting the increased nutrient demands of the growing embryo.

While the placenta is forming from the chorion, a second membrane, called the **amnion** (am′nē-on), develops around the embryo (fig. 20.7). It appears during the second week. Its margin is attached around the edge of the inner cell mass or **embryonic disc.** Fluid, called **amniotic fluid,** fills the space between the amnion and the embryonic disc. The amniotic fluid allows the embryo to grow freely without compression by surrounding tissues. The amniotic fluid also protects the embryo from jarring movements of the woman's body, and helps to maintain a stable temperature for embryonic and fetal development.

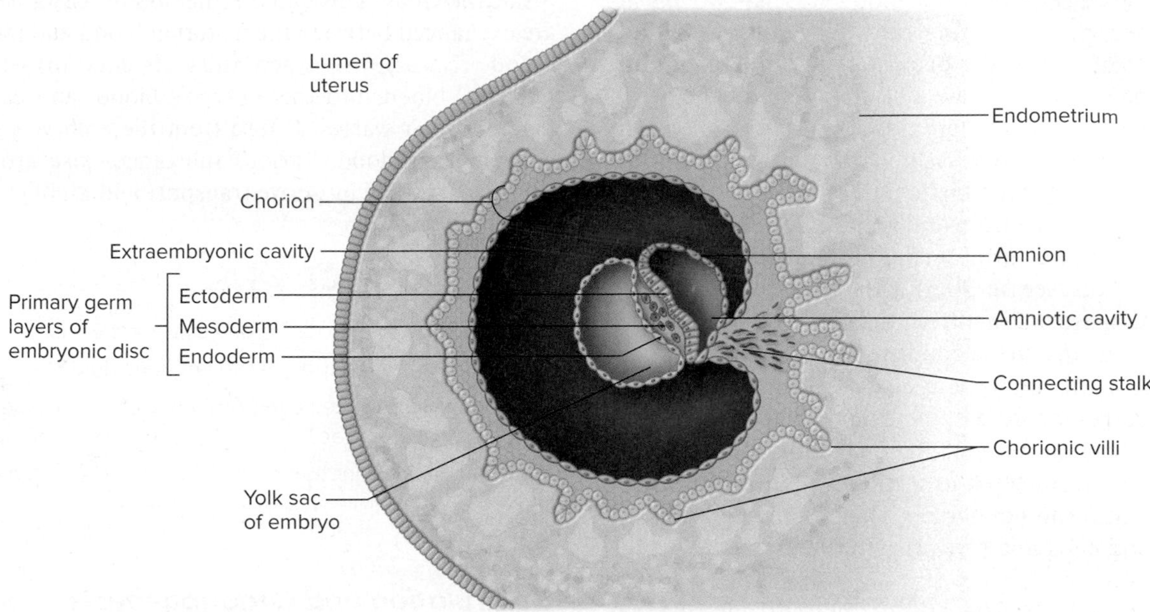

Figure 20.6 Early in the embryonic stage of development, the extraembryonic membranes form.

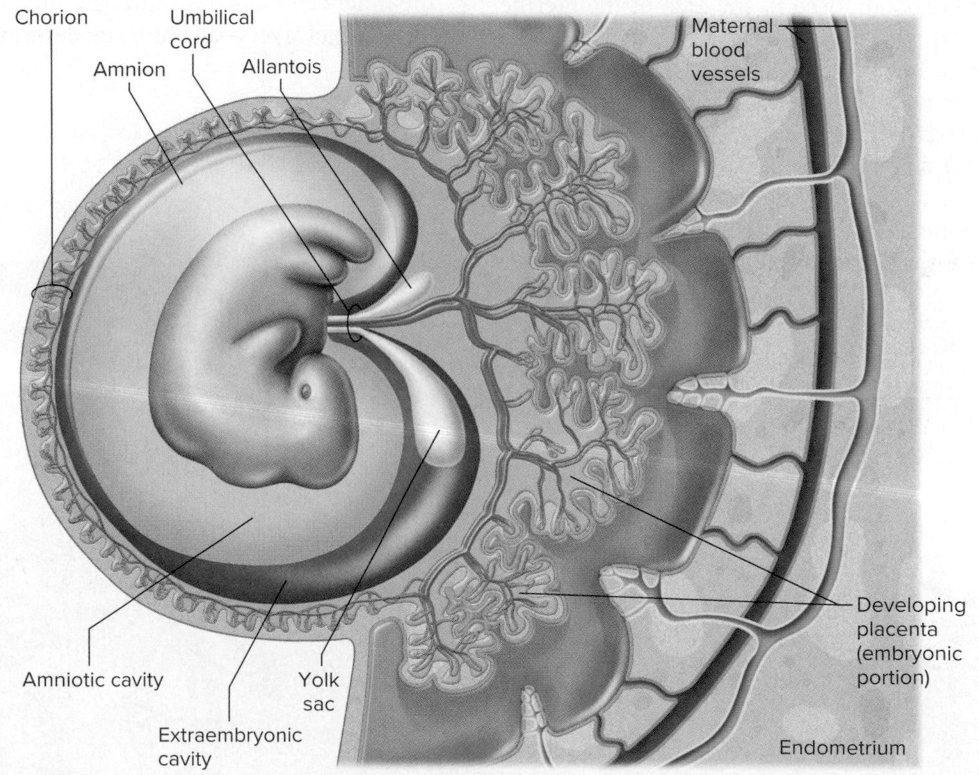

Figure 20.7 As the amnion develops, it surrounds the embryo, and the umbilical cord begins to form from structures in the connecting stalk.

The margins of the amnion fold in such a way as to enclose the embryo in the amnion and amniotic fluid. The amnion envelops the tissues on the underside of the embryo, particularly the connecting stalk, by which the embryo is attached to the chorion and the developing placenta. In this manner, the **umbilical cord** (um-bil′ĭ-kal kord) forms.

The umbilical cord originates at the umbilicus of the embryo and inserts into the center of the placenta (fig. 20.7). It suspends the embryo in the *amniotic cavity.*

The umbilical cord contains three blood vessels—two *umbilical arteries* and one *umbilical vein*—that transport blood between the embryo and the placenta (fig. 20.8). On rare occasions, a newborn will have only one umbilical artery. Because this condition is often associated with other cardiovascular disorders, the vessels in the severed cord are routinely counted following birth.

In addition to the chorion and amnion, two other extraembryonic membranes form during development—the yolk sac and the allantois (see fig. 20.7). ("Extraembryonic" refers to structures that are distinct from the embryo proper.) The **yolk sac** forms during the second week and is attached to the underside of the embryonic disc. It forms blood cells in the early stages of development and gives rise to the cells that later become sex cells. The **allantois** (ah-lan′tō-is) forms during the third week as a tube extending from the early yolk sac into the connecting stalk of the embryo. It, too, forms blood cells and gives rise to the umbilical blood vessels.

The disc-shaped area where the chorion still contacts the uterine wall develops into the placenta. The embryonic portion of the placenta is composed of the chorion and its villi; the maternal portion is composed of the area of the uterine wall (decidua basalis) where the villi attach (fig. 20.8). The fully formed placenta is a reddish-brown disc about 22 centimeters in diameter and 2.0 to 2.5 centimeters thick, and weighs about 0.5 kilogram.

A thin **placental membrane** separates embryonic blood in the capillary of a chorionic villus from maternal blood in a lacuna. This membrane is composed of the epithelium of the chorionic villus and the endothelium of the capillary inside the villus. Through this membrane, certain substances are exchanged between the maternal blood and the embryo's blood (fig. 20.9). Oxygen and nutrients diffuse from the maternal blood into the embryo's blood, and carbon dioxide and other wastes diffuse from the embryo's blood into the maternal blood. Various substances also cross the placental membrane by active transport and pinocytosis.

PRACTICE 20.3

4. Describe the structure of a chorionic villus.
5. What is the function of the amniotic fluid?
6. What types of cells and other structures are derived from the yolk sac?
7. What is the function of the placental membrane?

Umbilical cord
Umbilical arteries
Umbilical vein
Embryonic blood vessels
Lacuna
Villi (embryonic portion of placenta)
Decidua basalis (maternal portion of placenta)
Maternal blood vessels
Chorion
Endometrium
Placenta
Myometrium

Figure 20.8 The placenta consists of an embryonic portion and a maternal portion. Note that the umbilical vessels are named with respect to the embryo. Umbilical arteries transport oxygen-poor blood away from the embryo, and the umbilical vein returns oxygen-rich blood to the embryo.

Gastrulation and Organogenesis

Gastrulation is the movement of cells within the embryonic disc to form multiple layers. By the end of the second week, the inner cell mass has flattened into an embryonic disc with two distinct layers—an outer **ectoderm** and an inner **endoderm.**

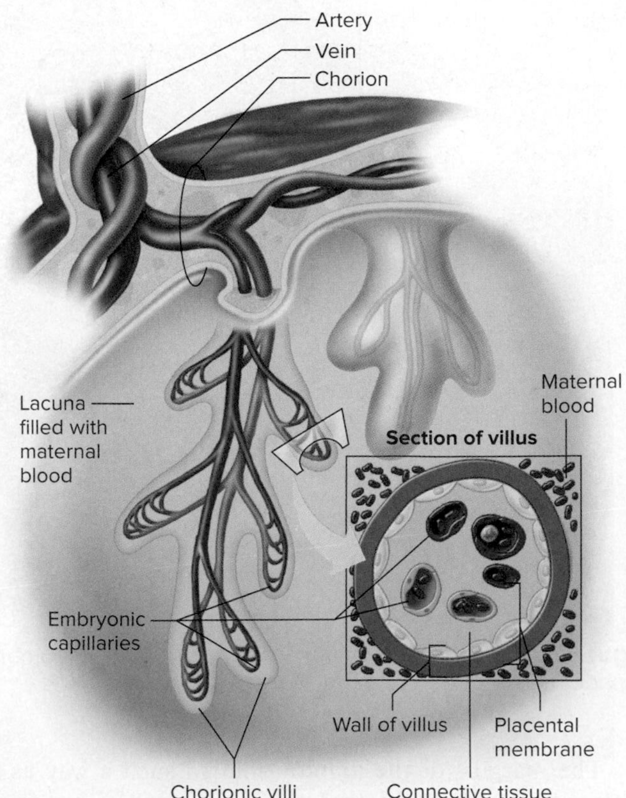

Figure 20.9 The placental membrane consists of the endothelium of an embryonic capillary and the epithelial wall of a chorionic villus, as illustrated in the section of the villus (lower part of the figure).

TABLE 20.2 Stages and Events of Early Human Prenatal Development

Stage	Time Period	Principal Events
Zygote	12 to 24 hours following ovulation	Egg is fertilized, meiosis is completed; zygote has 46 chromosomes and is genetically distinct
Cleavage	30 hours to third day	Mitosis increases cell number
Morula	Third to fourth day	Consists of a solid ball of cells
Blastocyst	Fifth day through second week	Consists of a hollow ball consisting of the trophoblast (outside), which implants and helps form the placenta, and inner cell mass, which flattens to form the embryonic disc
Gastrula	End of second week	Primary germ layers form

A short time later, the ectoderm and endoderm fold, and a third layer of cells, the **mesoderm,** forms between them. All organs form from these three cell layers, called the **primary germ layers**, in a process called **organogenesis** (see fig. 20.6). The two-week embryo, with its three primary germ layers, is called a **gastrula** (gas′troo-lah). Table 20.2 summarizes the stages of early human prenatal development.

Gastrulation is an important process in prenatal development because each cell's fate is determined by which layer it is in. Ectodermal cells give rise to the nervous system, parts of special sensory organs, the epidermis, hair, nails, glands of the skin, and linings of the mouth and anal canal. Mesodermal cells form all types of muscle tissue, bone tissue, bone marrow, blood, blood vessels, lymphatic vessels, internal reproductive organs, kidneys, and the epithelial linings of the body cavities. Endodermal cells produce the epithelial linings of the digestive tract, respiratory tract, urinary bladder, and urethra.

During the fourth week of development, part of the flat embryonic disc becomes cylindrical to form the neural tube, which will become the central nervous system. By the end of week four, the head and jaws appear, the heart beats and forces blood through the blood vessels, and tiny buds form, which will give rise to the upper and lower limbs (fig. 20.10).

During the fifth through seventh weeks, as figure 20.10 shows, the head grows rapidly and becomes rounded and erect. The face develops eyes, nose, and mouth. The upper and lower limbs elongate, and fingers and toes form. By the end of the seventh week, all the main internal organs are established. By the beginning of the eighth week, the embryo is about 25 millimeters long and weighs less than a gram (fig. 20.11).

Until about the end of the eighth week, the chorionic villi cover the entire surface of the former trophoblast. However, as the embryo and the surrounding chorion enlarge, only villi that contact the endometrium endure. The others degenerate, and the areas of the chorion where they were attached become smooth. The region of the chorion still in contact with the uterine wall is restricted to the placenta.

The embryonic stage concludes at the end of the eighth week. It is the most critical period of development, when all the essential external and internal body parts form. Factors that cause congenital malformations by affecting an embryo are called **teratogens.** Such agents include drugs, viruses, radiation, and even large amounts of otherwise healthful substances, such as fat-soluble vitamins.

The specific nature of a birth defect reflects the structures developing when the damage occurs. The time during prenatal development when a genetic mutation or exposure to a teratogen can alter a specific structure is called its *critical period.* Clinical Application 20.1 discusses some teratogens.

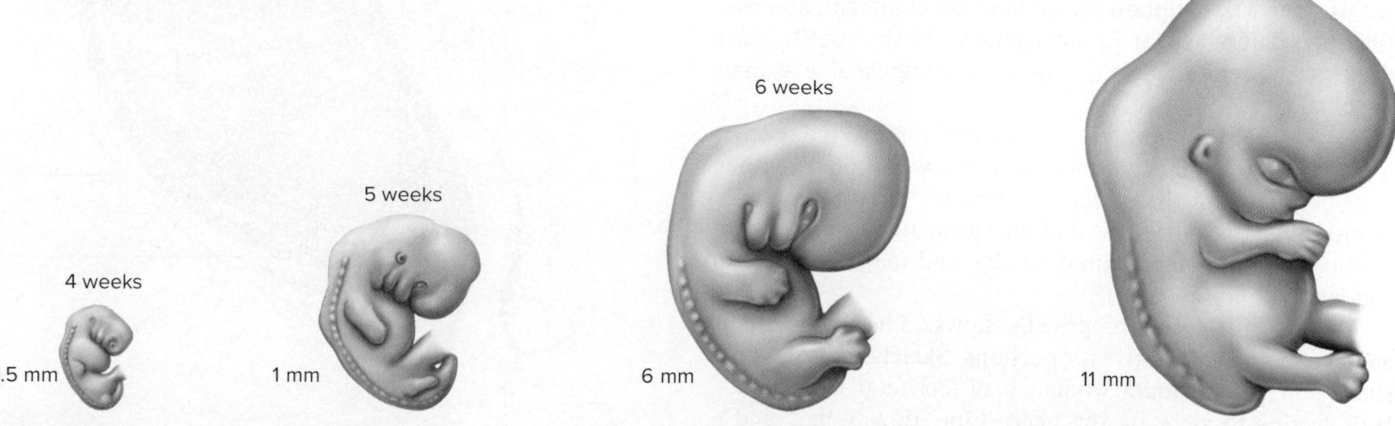

Figure 20.10 In the fifth through the seventh weeks of development, the embryonic body and face develop. Figures are not drawn to scale. Actual size indicated in millimeters (mm).

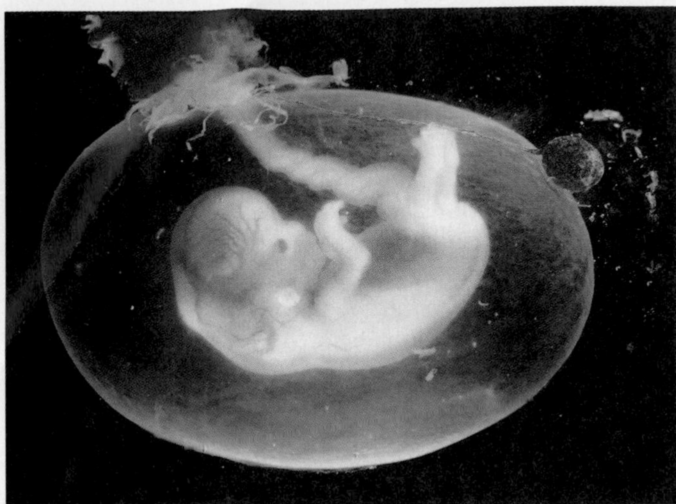

Figure 20.11 This human embryo is in its eighth week of development. Dr G. Moscoso/Science Source

PRACTICE 20.3

8. What is gastrulation?
9. Which tissues develop from the ectoderm? From the mesoderm? From the endoderm?
10. What is a teratogen?

Fetal Stage

The **fetal stage** begins at the end of the eighth week of development and lasts until birth. During this period, the **fetus** (fē′tus) grows rapidly and body proportions change considerably. At the beginning of the fetal stage, the head is disproportionately large and the lower limbs are short. Gradually, proportions come to more closely resemble those of a child.

During the third month, body lengthening accelerates but head growth slows. The upper limbs achieve the relative length they will maintain throughout development, and ossification centers appear in most bones. By the twelfth week the external reproductive organs are distinguishable as male or female.

In the fourth month, the body grows rapidly and reaches a length of up to 20 centimeters. The lower limbs lengthen considerably, and the skeleton continues to ossify. A four-month-old fetus will startle and turn away from a bright light flashed on a pregnant woman's belly, and may also react to sudden loud noises.

In the fifth month, growth slows. The lower limbs achieve their final relative proportions. Skeletal muscles contract, and the pregnant woman may feel fetal movements. Hair begins to grow on the head. Fine, downy hair and a cheesy mixture of dead epidermal cells and sebum from the sebaceous glands coat the skin.

During the sixth month, the fetus gains substantial weight. Eyebrows and eyelashes grow. The skin is wrinkled and translucent. Blood vessels in the skin cause a reddish appearance.

In the seventh month, the skin becomes smoother as fat is deposited in subcutaneous tissues. The eyelids, which fused during the third month, reopen. At the end of this month, the fetus is about 40 centimeters long.

In the final trimester, fetal brain cells rapidly form networks, as organs specialize and grow. A layer of fat is laid down beneath the skin. In the male, the testes descend from regions near the developing kidneys, through the inguinal canal, and into the scrotum. The digestive and respiratory systems mature last, which is why premature infants may have difficulty digesting milk and breathing. Approximately 266 days after a single sperm fertilized an egg, a baby is ready to be born. It is *full-term.* It is about 50 centimeters long and weighs 2.7–4.5 kilograms. The skin has lost its downy hair, but sebum and dead epidermal cells still coat it. Hair usually covers the scalp. The fingers and toes have well-developed nails. The skull bones are largely ossified. The fetus is usually positioned upside down, with its head toward the cervix, as shown in **figure 20.12**.

PRACTICE 20.3

11. What major changes occur during the fetal stage of development?
12. How is a fetus usually positioned in the uterus as birth nears?

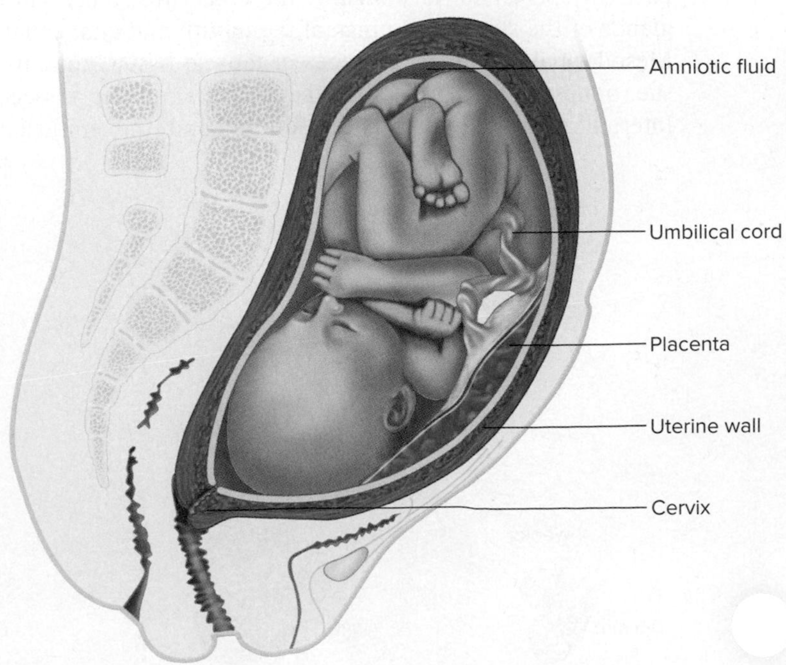

Figure 20.12 A full-term fetus is usually positioned with its head near the cervix.

Fetal Blood and Circulation

Throughout fetal development, the maternal blood supplies oxygen and nutrients and carries away wastes. These substances diffuse between maternal and fetal blood through the placental membrane, and the umbilical blood vessels carry them to and from the fetus.

The fetal blood and cardiovascular system are adapted to the intrauterine environment. The concentration of oxygen-carrying hemoglobin in fetal red blood cells is greater than in maternal red blood cells, so fetal red blood cells can bind more oxygen. Also, fetal hemoglobin has a greater attraction for oxygen than does adult hemoglobin, so fetal blood picks up oxygen preferentially to maternal blood in the placenta. At a particular oxygen partial pressure, fetal hemoglobin can carry 20% to 30% more oxygen than adult hemoglobin. Different genes encode the protein subunits of hemoglobin in embryos, fetuses, and individuals after birth. The different subunits have different attractions for oxygen.

Figure 20.13 shows the path of blood in the fetal cardiovascular system. The umbilical vein transports blood rich in oxygen and nutrients from the placenta to the fetal body. This vein enters the body and continues along the anterior abdominal wall to the liver. About half the blood it carries passes into the liver, and the rest enters a vessel called the **ductus venosus** (duk′tus vē′nō-sus), which bypasses the liver.

The ductus venosus extends a short distance and joins the inferior vena cava. There, oxygen-rich blood from the

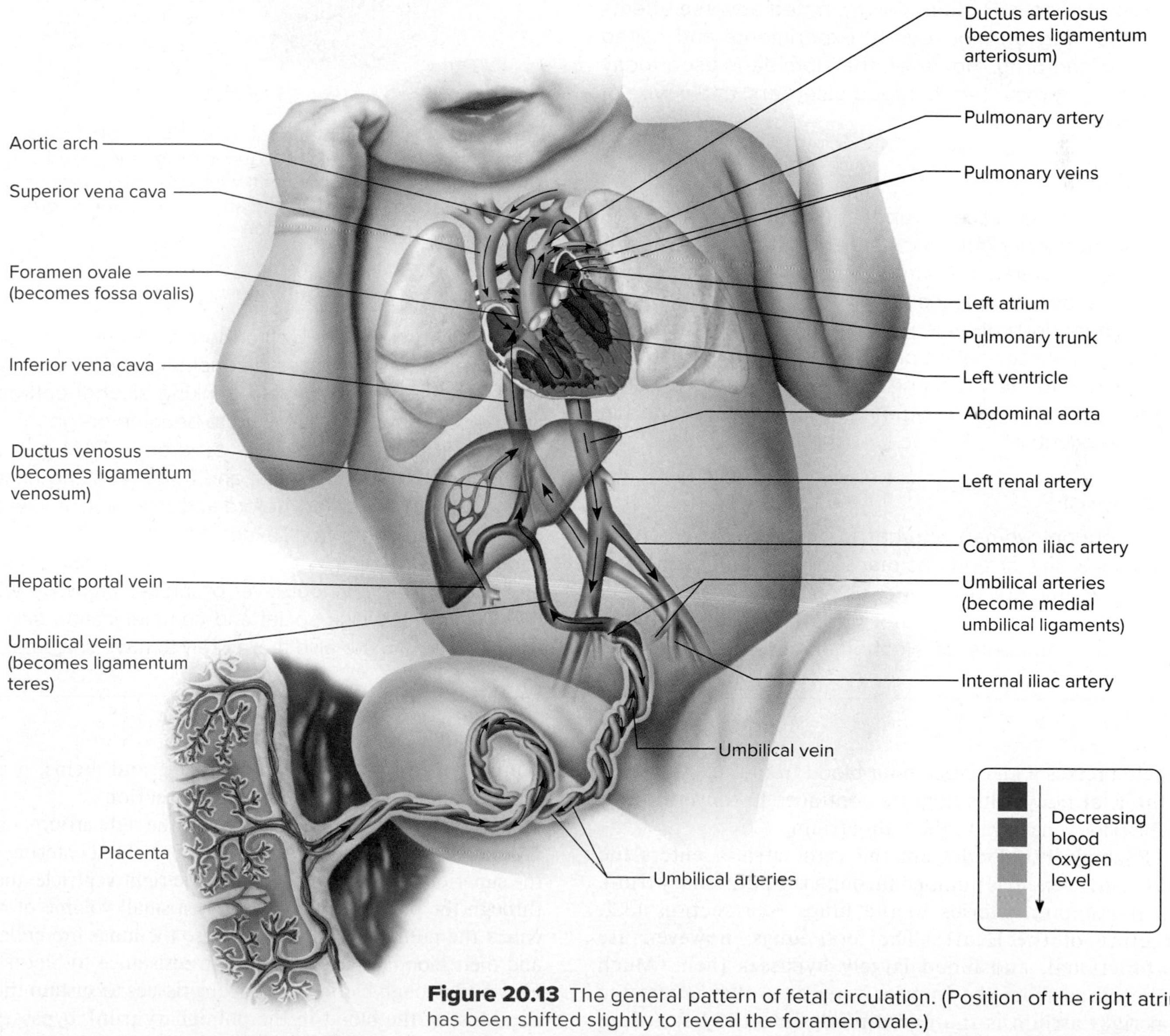

Ductus arteriosus (becomes ligamentum arteriosum)
Pulmonary artery
Pulmonary veins
Aortic arch
Superior vena cava
Foramen ovale (becomes fossa ovalis)
Left atrium
Pulmonary trunk
Inferior vena cava
Left ventricle
Abdominal aorta
Ductus venosus (becomes ligamentum venosum)
Left renal artery
Common iliac artery
Hepatic portal vein
Umbilical arteries (become medial umbilical ligaments)
Umbilical vein (becomes ligamentum teres)
Internal iliac artery
Umbilical vein
Placenta
Umbilical arteries
Decreasing blood oxygen level

Figure 20.13 The general pattern of fetal circulation. (Position of the right atrium has been shifted slightly to reveal the foramen ovale.)

PRACTICE FIGURE 20.13

Which structures are unique to the fetal circulation?

Answer can be found in Appendix E.

CLINICAL APPLICATION 20.1
Some Causes of Birth Defects

Thalidomide

The idea that the placenta protects the embryo and fetus from harmful substances was disproven between 1957 and 1961, when 10,000 children in Europe were born with malformed, stunted limbs due to exposure to a mild tranquilizer drug, *thalidomide,* during the time of limb formation. Although some women in the United States did use thalidomide and had affected children, the country was spared a larger number of affected children because an astute government physician, Frances Oldham Kelsey, noted adverse effects of the drug on monkeys in experiments and halted use of the drug. However, thalidomide is used today to treat leprosy, certain blood disorders, and a type of severe nosebleed.

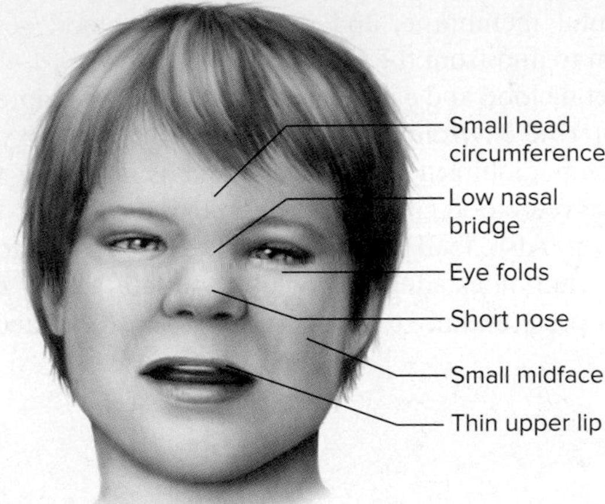

Figure 20A Fetal alcohol syndrome (FAS). Some children whose mothers drank alcohol during pregnancy have characteristic flat faces. Women who drink while pregnant have a chance of having a child affected to some degree by prenatal exposure to alcohol.

Rubella

The virus that causes rubella (German measles) is a powerful teratogen. Exposure in the first trimester leads to cataracts, deafness, and heart defects, and later exposure causes learning disabilities, speech and hearing problems, and type 1 diabetes mellitus. Successful vaccination programs provide for maternal immunity to the virus, and have since greatly lowered the incidence of "congenital rubella syndrome" in many countries.

Alcohol

A pregnant woman who has just one or two alcoholic drinks a day, or perhaps many drinks at a crucial time in prenatal development, risks *fetal alcohol spectrum disorder (FASD)* in her unborn child. The effects of small amounts of alcohol at different stages of pregnancy are not yet well understood, and because each woman metabolizes alcohol slightly differently, women are advised to avoid drinking alcohol entirely when pregnant or when trying to become pregnant.

A child with fetal alcohol syndrome (FAS) has a small head, misshapen eyes, and a flat face and nose (fig. 20A). Growth is slow before and after birth. Intellect is impaired, ranging from minor learning disabilities to intellectual disability. Many individuals with FAS remain at an early grade-school level of intellectual development, and many lack social and communication skills. People with FAS are also more likely to have seizures.

placenta mixes with oxygen-poor blood from the lower parts of the fetal body. This mixture continues to move through the inferior vena cava to the right atrium.

Postnatally, blood from the right atrium enters the right ventricle and is pumped through the pulmonary trunk and pulmonary arteries to the lungs (see section 13.2, Structure of the Heart). The fetal lungs, however, are nonfunctional, and blood largely bypasses them. Much of the blood from the inferior vena cava that enters the fetal right atrium is shunted directly into the left atrium through an opening in the atrial septum called the **foramen ovale** (fo-ra′man ōva′lē). Blood passes through the foramen ovale because the blood pressure is greater in the right atrium than in the left atrium. Furthermore, a small valvelike structure on the left side of the atrial septum overlaps the foramen ovale and helps prevent blood from moving in the reverse direction.

The rest of the fetal blood entering the right atrium, including a large proportion of the oxygen-poor blood entering from the superior vena cava, passes into the right ventricle and out through the pulmonary trunk. Only a small volume of blood enters the pulmonary circuit because the lungs are collapsed and their blood vessels have a high resistance to blood flow. However, enough blood reaches lung tissues to sustain them.

Most of the blood in the pulmonary trunk bypasses the lungs by entering a fetal vessel called the **ductus arteriosus** (duk′tus ar-te″rē-ō′sus), which connects the pulmonary trunk to the descending portion of the aortic arch. As a result of this connection, blood with a relatively low oxygen concentration, returning to the heart through the superior vena

In the United States today, FAS is the third most common cause of intellectual disability in newborns. Each year in the United States, about 5,000 children are born with FAS, and many more are born with milder "alcohol-related effects." Statistics on the prevalence of fetal alcohol spectrum disorders are unreliable because many cases are likely unrecognized, according to the Centers for Disease Control and Prevention.

Cigarettes

Chemicals in cigarette smoke stress a fetus. Carbon monoxide crosses the placenta and binds to fetal hemoglobin in a way that prevents the hemoglobin from delivering oxygen to the fetal tissues. Other chemicals in smoke prevent nutrients from reaching the fetus. Smoke-exposed placentas lack important growth factors. The result of these assaults is poor growth before and after birth. Cigarette smoking during pregnancy is linked to spontaneous abortion, stillbirth, prematurity, and low birth weight.

Nutrients and Malnutrition

Certain nutrients in large amounts, particularly vitamins, act in the body as drugs. A vitamin A–based drug used to treat psoriasis, as well as excesses of vitamin A itself, can cause harm because some forms of the vitamin remain in the body for up to three years after ingestion.

Malnutrition during pregnancy causes intrauterine growth retardation (IUGR). Malnutrition before birth also causes shifts in metabolism to make the most of calories from food. This protective action, however, sets the stage for developing obesity and associated disorders, such as type 2 diabetes and cardiovascular disease, in adulthood.

Occupational Hazards

The workplace can be a source of teratogens. Women who work with textile dyes, lead, certain photographic chemicals, semiconductor materials, mercury, and cadmium have increased rates of spontaneous abortion and delivering children with birth defects. Men whose jobs expose them to sustained heat, such as smelter workers, glass manufacturers, and bakers, may produce sperm that can fertilize a secondary oocyte, possibly leading to spontaneous abortion or a birth defect. Another way that a man can cause a birth defect is if a virus or a toxic chemical is carried in semen.

Zika Virus

Zika virus is the first mosquito-borne virus identified that causes birth defects. If a pregnant woman is exposed during the first trimester or early second trimester, the baby may be born with an extremely small head with diminished brain matter, termed *microcephaly.* Zika virus disease in adults causes only fever, rash, joint pain, and red eyes. The virus is carried in semen. We still do not know the effects of exposure later in pregnancy. The virus was discovered in Uganda in 1947, but came to the world's attention in 2015, after it was associated with a dramatic increase in the incidence of microcephaly in newborns in Brazil.

---

cava, bypasses the lungs. (This might seem odd, but remember that fetal blood becomes oxygenated in the placenta, not in the fetal lungs.) At the same time, the low-oxygen blood is prevented from entering the portion of the aorta that branches to the heart and brain.

The more highly oxygenated blood that enters the left atrium through the foramen ovale mixes with a small amount of oxygen-poor blood returning from the pulmonary veins. This mixture moves into the left ventricle and is pumped into the aorta. Some of it reaches the myocardium through the coronary arteries, and some reaches the brain tissues through the carotid arteries.

Blood carried by the descending aorta includes the less oxygenated blood from the ductus arteriosus. Some of the blood is carried into the branches of the aorta that lead to the lower regions of the body. The rest passes into the umbilical arteries, which branch from the internal iliac arteries and lead to the placenta. There, the blood is reoxygenated (fig. 20.13).

Table 20.3 summarizes the major features of fetal circulation. At the time of birth, important adjustments must occur in the fetal cardiovascular system when the placenta ceases to function and the newborn begins to breathe.

PRACTICE 20.3

13. Which umbilical vessel carries oxygen-rich blood to the fetus?
14. What is the function of the ductus venosus?
15. How does fetal circulation allow blood to bypass the lungs?

TABLE 20.3 Fetal Cardiovascular Adaptations

Adaptation	Function
Fetal blood	Fetal hemoglobin has greater attraction for oxygen than adult hemoglobin
Umbilical vein	Carries nutrient-rich, oxygen-rich blood from placenta to fetus
Ductus venosus	Conducts about half the blood from the umbilical vein directly to the inferior vena cava, bypassing the liver
Foramen ovale	Conveys a large proportion of the blood entering the right atrium from the inferior vena cava, through the atrial septum, and into the left atrium, bypassing the lungs
Ductus arteriosus	Conducts some blood from the pulmonary trunk to the aorta, bypassing the lungs
Umbilical arteries	Carry oxygen-poor blood containing carbon dioxide and other wastes from the internal iliac arteries to the placenta

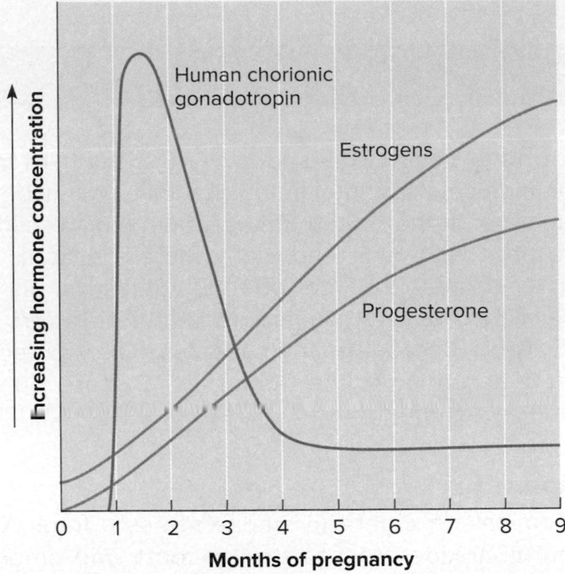

Figure 20.14 The relative concentrations of three hormones in maternal blood change during pregnancy.

Maternal Changes During Pregnancy

During a typical menstrual cycle, the corpus luteum degenerates about two weeks after ovulation. Consequently, concentrations of estrogens and progesterone decline rapidly, the uterine lining breaks down, and the endometrium sloughs away as menstrual flow. If this occurs following implantation, the embryo is lost in a spontaneous abortion.

The hormone hCG normally helps prevent spontaneous abortion. It functions similarly to luteinizing hormone (LH), and it maintains the corpus luteum, which continues secreting estrogens and progesterone. Thus, the uterine wall continues to grow and develop. At the same time, hCG inhibits the anterior pituitary gland's release of follicle-stimulating hormone (FSH) and LH, halting the normal menstrual cycles.

Secretion of hCG continues at a high level for about two months and then declines by the end of four months. Detecting this hormone in urine or blood is the basis of pregnancy tests. Although the corpus luteum persists throughout pregnancy, its function as a hormone source becomes less important after the first three months (first trimester), when the placenta secretes sufficient estrogens and progesterone (fig. 20.14).

For the remainder of the pregnancy, *placental estrogens* and *placental progesterone* maintain the uterine wall. The placenta also secretes a hormone called **placental lactogen** that, with placental estrogens and progesterone, stimulates breast development and prepares the mammary glands to secrete milk. Placental progesterone and a polypeptide hormone called **relaxin** from the corpus luteum inhibit the smooth muscle in the myometrium, suppressing uterine contractions until the birth process begins.

The high concentration of placental estrogens during pregnancy enlarges the vagina and external reproductive organs. Also, relaxin relaxes the connective tissue of the pubic symphysis and sacroiliac joints during the last week of pregnancy, allowing greater movement at these joints and aiding the passage of the fetus through the birth canal.

Other hormonal changes of pregnancy include increased adrenal secretion of aldosterone, which promotes renal reabsorption of sodium and leads to fluid retention. The parathyroid glands secrete parathyroid hormone, which helps maintain a high concentration of maternal blood calcium (see section 11.6, Parathyroid Glands). Table 20.4 summarizes the hormonal changes of pregnancy.

PRACTICE 20.3

16. What are the sources of the hormones that sustain the uterine wall during pregnancy?
17. What other hormonal changes occur during pregnancy?

Birth Process

Pregnancy usually continues for thirty-eight weeks from conception and ends with the *birth process.* During pregnancy, progesterone suppresses uterine contractions. As the placenta ages, the progesterone concentration in the uterus declines, which stimulates synthesis of a prostaglandin that promotes uterine contractions. At the same time, the cervix thins and then opens. Changes in the cervix may begin a week or two before other signs of labor appear.

Stretching of the uterine and vaginal tissues late in pregnancy also stimulates the birth process. This action initiates impulses to the hypothalamus, which in turn signals the posterior pituitary gland to release the hormone **oxytocin** (see section 11.4, Pituitary Gland). Oxytocin stimulates powerful uterine contractions. Combined with the

TABLE 20.4	Hormonal Changes During Pregnancy
1.	Following implantation, cells of the trophoblast begin to secrete human chorionic gonadotropin (hGC).
2.	Human chorionic gonadotropin maintains the corpus luteum, which continues to secrete estrogens and progesterone.
3.	As the placenta develops, it secretes abundant estrogens and progesterone.
4.	Placental estrogens and progesterone: a. stimulate the uterine lining to continue development. b. maintain the uterine lining. c. inhibit secretion of follicle-stimulating hormone (FSH) and luteinizing hormone (LH) from the anterior pituitary gland. d. stimulate development of mammary glands. e. inhibit uterine contractions (progesterone). f. enlarge the reproductive organs (estrogens).
5.	Relaxin from the corpus luteum also inhibits uterine contractions and relaxes the pelvic ligaments.
6.	The placenta secretes placental lactogen that stimulates breast development.
7.	Aldosterone from the adrenal cortex promotes renal reabsorption of sodium.
8.	Parathyroid hormone from the parathyroid glands helps maintain a high concentration of maternal blood calcium.

greater excitability of the myometrium due to the decline in progesterone secretion, stimulation by oxytocin aids *labor* in its later stages.

During labor, rhythmic muscular contractions begin at the top of the uterus and travel down its length. Because the fetus is usually positioned head downward, labor contractions force the head against the cervix (fig. 20.15). This action stretches the cervix, which elicits a reflex that stimulates stronger labor contractions. Thus, a **positive feedback system** operates in which uterine contractions produce more-intense uterine contractions until effort is maximal. At the same time, dilation of the cervix reflexively stimulates the posterior pituitary to increase oxytocin release. As labor continues, positive feedback stimulates abdominal wall muscles to contract, helping to propel the fetus through the birth canal (cervix, vagina, and vulva) to the outside.

An infant passing through the birth canal can stretch and tear the perineal tissues between the vulva and anus. To avoid a ragged tear, a physician may make an *episiotomy*, which is an incision along the midline of the perineal tissues from the vestibule to within 1.5 centimeters of the anus.

Following birth of the fetus, the placenta separates from the uterine wall and is pushed by uterine contractions through the birth canal. This expelled placenta, called the *afterbirth,* is accompanied by bleeding, because vascular tissues are damaged in the process. However, oxytocin stimulates continued uterine contraction, which compresses the bleeding vessels and minimizes blood loss. Breastfeeding also contributes to returning the uterus to its original, prepregnancy size, because suckling by the newborn stimulates the mother's posterior pituitary gland to release oxytocin.

PRACTICE 20.3

18. Describe the role of progesterone in initiating labor.
19. Explain how dilation of the cervix affects labor.
20. Explain how bleeding is naturally controlled after the placenta is expelled.

Milk Production and Secretion

During pregnancy, placental estrogens and progesterone stimulate further development of the mammary glands. Estrogens cause the ductile systems to grow and branch and deposit abundant fat around them. Progesterone stimulates the development of the alveolar glands at the ends of the ducts. Placental lactogen also promotes these changes.

The breasts may double in size during pregnancy because of hormonal action. Mammary glands become capable of secreting milk. However, milk secretion does not begin until after birth, because placental progesterone inhibits milk production and placental lactogen blocks the action of **prolactin** (see section 11.4, Pituitary Gland). Prolactin is synthesized from early pregnancy throughout gestation, peaking at birth.

Following childbirth and the expulsion of the placenta, maternal blood concentrations of placental hormones decline rapidly. The action of prolactin is no longer inhibited. Prolactin stimulates the mammary glands to secrete milk. This hormonal effect does not occur until two or three days following birth. In the meantime, the glands secrete a thin, watery fluid called *colostrum* that has more protein, but less carbohydrate and fat, than milk. Colostrum contains antibodies from the mother's immune system that protect the newborn from certain infections.

Milk ejection requires contraction of specialized *myoepithelial cells* surrounding the alveolar glands (fig. 20.16). Suckling or mechanical stimulation of sensory receptors in the nipple or areola elicits the reflex action that controls this process. Impulses from these receptors go to the hypothalamus, which signals the posterior pituitary gland to release oxytocin. Oxytocin travels in the bloodstream to the breasts and stimulates the myoepithelial cells to contract. Within about thirty seconds, milk squirts into a suckling infant's mouth.

As long as milk is removed from the breasts, release of prolactin and oxytocin continues, and the mammary glands produce milk. If milk is not removed regularly, the hypothalamus inhibits prolactin secretion, and within about one week the mammary glands stop producing milk.

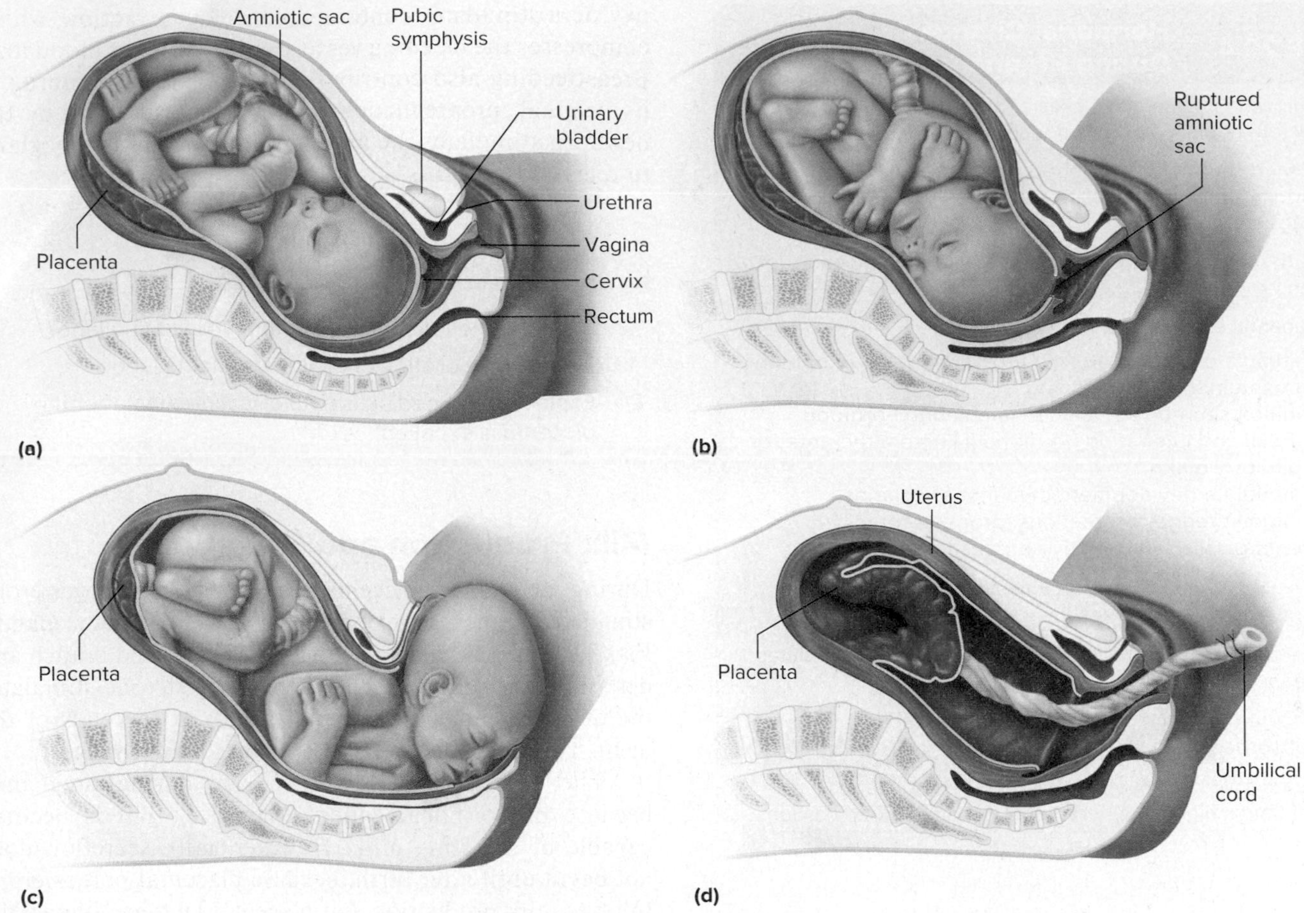

Figure 20.15 Stages in birth. (**a**) Fetal position before labor, (**b**) dilation of the cervix, (**c**) expulsion of the fetus, (**d**) expulsion of the placenta.

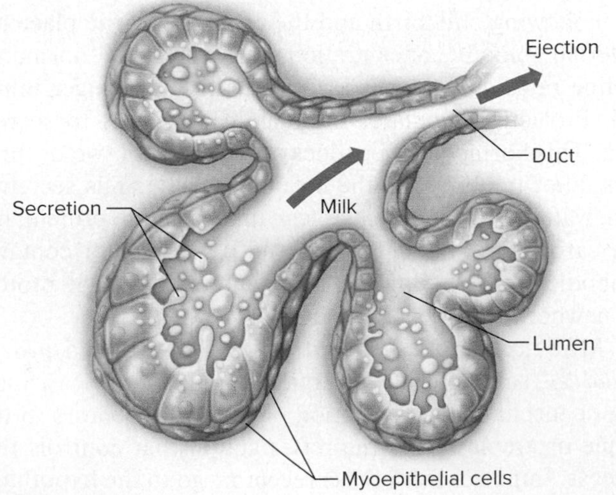

Figure 20.16 Myoepthelial cells contract, releasing milk from an alveolar gland.

PRACTICE 20.3

21. How does pregnancy affect the mammary glands?
22. What stimulates the mammary glands to produce milk?
23. What causes milk ejection?

OF INTEREST Human milk is the best possible food for human babies. The milk of other mammals contains different proportions of nutrients.

Human milk is 4.5% fat and 1.1% protein, which is suitable for supporting synthesis of myelin, the lipid electrical insulation that enables neurons in the developing brain to communicate effectively. In contrast, cow's milk has about 3.5% fat and about 3.1% protein.

20.4 | Aging: The Human Life Span

LEARN

1. Describe the major cardiovascular and other physiological adjustments in the newborn.
2. Describe the stages of postnatal development.

Aging begins before birth, as certain cells die as part of the developmental program encoded in the genes. This process of programmed cell death, called **apoptosis** (ap-uh-tō′sis), occurs regularly in the embryo, degrading certain structures to pave the way for new ones. The number of neurons in the fetal brain, for example, is reduced by nearly half. Those that survive make certain synaptic connections important for proper neurological development. Throughout life, apoptosis and mitosis, opposite yet complementary processes, enable organs to maintain their characteristic shapes.

Overall, the aging process is difficult to analyze because of the intricate interactions of the body's organ systems. Breakdown of one structure ultimately affects the functioning of others. Table 20.5 outlines aging-related changes. Following birth, the newborn experiences physiological and structural changes. The postnatal period of development and aging lasts from birth to death.

Neonatal Period

The **neonatal** (nē″ō-nā′tal) **period** begins abruptly at birth and extends to the end of the first four weeks. At birth, the newborn must make quick physiological adjustments to become self-reliant. It must respire, obtain and digest nutrients, excrete wastes, and regulate body temperature.

A newborn's most immediate need is to obtain oxygen and excrete carbon dioxide. The first breath must be particularly forceful, because the newborn's lungs are collapsed and the airways are small, offering considerable resistance to air movement. Also, surface tension holds the moist membranes of the lungs together. However, the lungs of a full-term fetus continuously secrete *surfactant* (see section 16.3, Mechanics of Breathing), which reduces surface tension. After the first powerful inhalation inflates the lungs, breathing eases.

The newborn has a high metabolic rate. To supplement the liver's supply of glucose to support metabolism, the newborn typically utilizes stored fat for energy.

A newborn's kidneys are usually unable to concentrate urine, so the baby excretes dilute urine. This may cause dehydration and a water and electrolyte imbalance. Also, certain homeostatic control mechanisms may not function adequately. For example, during the first few days of life, body temperature may respond to slight stimuli by fluctuating above or below normal levels.

When the placenta ceases to function and breathing begins, the newborn's cardiovascular system changes. Following birth, the umbilical vessels constrict. The umbilical

TABLE 20.5 Aging-Related Changes

Organ System	Aging-Related Changes
Integumentary system	Degenerative loss of collagen and elastic fibers in dermis; decreased production of pigment in hair follicles, which eventually causes hair to turn white; reduced activity of sweat and sebaceous glands; thinning, wrinkling, and drying of skin
Skeletal system	Degenerative loss of bone matrix; thinner, less dense, bones more likely to fracture; possible shortened stature due to compression of intervertebral discs and vertebrae
Muscular system	Loss of skeletal muscle fibers; degenerative changes in neuromuscular junctions; loss of muscular strength
Nervous system	Degenerative changes in neurons; loss of dendrites and synaptic connections; accumulation of lipofuscin in neurons; decreases in sensation; decreasing efficiency in processing and recalling information; decreasing ability to communicate; diminished sense of smell and taste; loss of elasticity of lenses and consequent loss of ability to accommodate for close vision
Endocrine system	Reduced hormonal secretions; decreased metabolic rate; reduced ability to cope with stress; reduced ability to maintain homeostasis
Cardiovascular system	Degenerative changes in cardiac muscle; decrease in lumen diameters of arteries and arterioles; decreased cardiac output; increased resistance to blood flow; increased blood pressure
Lymphatic system	Decrease in efficiency of immune system; increased incidence of infections and neoplastic diseases; increased incidence of autoimmune diseases
Digestive system	Decreased motility in gastrointestinal tract; reduced secretion of digestive juices; reduced efficiency of digestion
Respiratory system	Degenerative loss of elastic tissue in lungs; fewer alveoli; reduced vital capacity; increase in dead air space; reduced ability to clear airways by coughing
Urinary system	Degenerative changes in kidneys; fewer functional nephrons; reductions in filtration rate, tubular secretion, and tubular reabsorption
Reproductive systems	
Male	Reduced secretion of sex hormones; enlargement of prostate gland; decrease in sexual energy
Female	Degenerative changes in ovaries; decrease in secretion of sex hormones; menopause; regression of secondary sex characteristics

arteries close first, and if the umbilical cord is not clamped or severed for a minute or so, blood continues to flow from the placenta to the newborn through the umbilical vein, adding to the newborn's blood volume. Similarly, the ductus venosus constricts shortly after birth and appears in the adult as a fibrous cord (ligamentum venosum) superficially embedded in the wall of the liver.

The foramen ovale closes as a result of blood pressure changes in the right and left atria. As blood ceases to flow from the umbilical vein into the inferior vena cava, the blood pressure in the right atrium falls. Also, as the lungs expand with the first breathing movements, resistance to blood flow through the pulmonary circuit decreases, more blood enters the left atrium through the pulmonary veins, and blood pressure in the left atrium increases. As the blood pressure in the left atrium rises and that in the right atrium falls, the tissue flaps on the left side of the atrial septum close the foramen ovale. In most individuals, the tissue flaps gradually fuse with the tissues along the margin of the foramen. In an adult, a depression called the *fossa ovalis* marks the site of the past opening.

The ductus arteriosus, like the other fetal vessels, constricts after birth. After this, blood can no longer bypass the lungs by moving from the pulmonary trunk directly into the aorta. In an adult, a cord called the *ligamentum arteriosum* forms from the remnants of the ductus arteriosus.

In patent ductus arteriosus (PDA), the ductus arteriosus does not close completely. After birth, the metabolic rate and oxygen consumption in neonatal tissues increase. If the ductus arteriosus remains open, bypassing the pulmonary circuit, the neonate's blood oxygen concentration may be too low to adequately supply tissues, including the myocardium. If PDA is not corrected surgically, the heart may fail, even though the myocardium is normal.

Changes in the newborn's cardiovascular system are gradual. Although constriction of the ductus arteriosus may be functionally complete within fifteen minutes, the permanent closure of the foramen ovale may take up to a year. Figure 20.17 illustrates cardiovascular changes in the newborn.

Fetal hemoglobin production falls after birth, and by the time an infant is four months old, most of the circulating hemoglobin is the adult type.

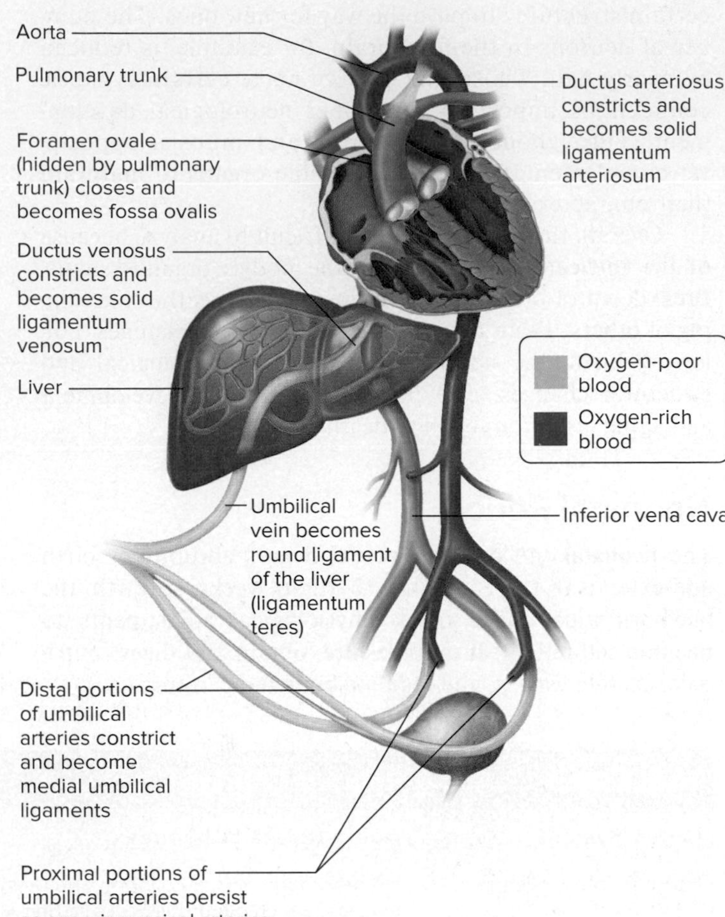

Figure 20.17 Major changes in the newborn's cardiovascular system.

PRACTICE 20.4

1. Why must a newborn's first breath be particularly forceful?
2. What does a newborn use for energy during its first few days?
3. How do the kidneys of a newborn differ from those of an adult?
4. What changes occur in the newborn's cardiovascular system?

Infancy

The period of continual development extending from the end of the first four weeks to one year is called **infancy.** During this time, the infant grows rapidly and may triple its birth weight. Teeth begin to erupt through the gums, and the muscular and nervous systems mature so that movements become increasingly coordinated. The infant is soon able to visually follow objects, reach for and grasp objects, sit, crawl, and stand.

Infancy also brings the beginning of the ability to communicate. The infant learns to smile, laugh, and respond to some sounds. By the end of the first year, the infant may be able to say two or three words. The infant and young child have particular nutritional requirements to fuel rapid growth. In addition to an energy source, the body requires proteins to provide the amino acids necessary to form new tissues, calcium and vitamin D to promote the development and ossification of skeletal structures (see section 7.4, Bone Development, Growth, and Repair), iron to support blood cell formation, and vitamin C for production of structural tissues such as cartilage and bone.

Childhood

Childhood begins at the end of the first year and ends at puberty. During this period, growth continues at a rapid rate. The primary teeth erupt, and then secondary teeth replace them (see section 15.3, Mouth). The child develops voluntary muscular control and learns to walk, run, and climb. Bladder and bowel controls are established. The child learns to communicate effectively by speaking, and later learns to read, write, and think. At the same time, the child is maturing emotionally.

Adolescence

Adolescence is the period of development between puberty and adulthood when anatomical and physiological changes result in reproductively functional individuals (see section 19.4, Hormonal Control of Male Reproductive Functions; and section 19.6, Oogenesis and the Ovarian Cycle). Most of these changes are hormonally controlled, and they include the appearance of secondary sex characteristics, as well as growth spurts in the muscular and skeletal systems.

Females usually experience the changes of adolescence at younger ages than males. Early in adolescence, females may be taller and stronger than their male peers. However, females attain full growth at earlier ages, and in late adolescence the average male is taller and stronger than the average female. The periods of rapid growth in adolescence, which usually begin between the ages of eleven and thirteen in females and between thirteen and fifteen in males, increase demands for certain nutrients. It is not uncommon for a teenager to consume a huge plate of food, go back for more—and still remain thin. In addition to energy sources, foods must provide ample amounts of proteins, vitamins, and minerals to support the growth of new tissues. Adolescence also brings increasingly refined motor skills, intellectual ability, and emotional maturity.

PRACTICE 20.4

5. What is the period of time called *infancy?*
6. What are some motor skills that develop during *childhood?*
7. Describe the changes that occur during *adolescence.*

Adulthood

Adulthood (maturity) extends from adolescence to old age. As we age, we become gradually aware of certain declining functions—yet other abilities remain adequate. It is interesting to note the peaks of particular structures or functions throughout an average human life.

By age eighteen, the human male is producing the most testosterone he will ever have, and as a result his sex drive is strong. In the twenties, muscle strength peaks in both sexes. Hair is at its fullest, each hair its thickest. By the end of the third decade of life, signs of aging may appear as a loss in the elasticity of facial skin, producing small wrinkles around the mouth and eyes. Height is already starting to decrease, but not yet at a detectable level.

The age of thirty is a developmental turning point. After this, hearing often becomes less acute. Heart muscle begins to thicken. The elasticity of the ligaments between the small bones in the back lessens, setting the stage for the slumping posture that becomes apparent in later years. Researchers estimate that beginning roughly at age thirty, the human body becomes functionally less efficient by about 0.8% every year.

During their forties, many people weigh 10 to 20 pounds (4.5 to 9 kilograms) more than they did at the age of twenty, thanks to a slowing of metabolism and a decrease in activity level. They may be 0.125 inch (0.3 centimeter) shorter. Hair may be graying as melanin production wanes, and some hair may fall out. Vision may become farsighted. The immune system is less efficient, making the body more prone to infection and cancer. Skeletal muscles lose strength as connective tissue appears within them; the cardiovascular system is strained as the lumens of arterioles and arteries narrow with fatty deposits; skin loosens and wrinkles as elastic fibers in the dermis break down.

The early fifties bring further declines. Nail growth slows, taste buds die, and the skin continues to lose elasticity. For most people, the ability to see close objects becomes impaired, but for the nearsighted, vision improves. Women stop menstruating, although interest in sex continues (see section 19.6, Oogenesis and the Ovarian Cycle). Delayed or reduced insulin release by the pancreas, in response to a glucose load, may lead to diabetes. By the decade's end, muscle mass and weight begin to decrease. A male produces less semen but is still sexually active. His voice may become higher as his vocal cords degenerate. A man has half the strength in his upper limb muscles and half the lung function as he did at age twenty-five. He is about 0.75 inch (2 centimeters) shorter.

The sixty-year-old may experience minor memory losses. A few million of the person's billions of brain cells have been lost over his or her lifetime, but for the most part, intellect remains sharp. By age seventy, height decreases a full inch (2.5 centimeters). Sagging skin and loss of connective tissue, combined with continued growth of cartilage, make the nose, ears, and eyes more prominent.

Senescence

Senescence (se-nes'ens) is the process of growing old. It is a continuation of the degenerative changes that begin during adulthood. The body becomes less able to cope with the demands placed on it by the individual and by the environment.

Senescence is a result of the normal wear and tear of body parts over many years. For example, the cartilage covering the ends of bones at joints may wear away, leaving the joints stiff and painful. Other degenerative changes are caused by disease processes that interfere with vital functions, such as gas exchanges or blood circulation. Metabolic rate and distribution of body fluids may change. The rate of division of certain cell types declines, and immune responses weaken. The person becomes less able to repair damaged tissue and more susceptible to disease.

TABLE 20.6 Stages in Development: From Birth to Death

Stage	Time Period	Major Events
Neonatal period	Birth to end of fourth week	Newborn begins to respire, eat, digest nutrients, excrete wastes, regulate body temperature, and make cardiovascular adjustments
Infancy	End of fourth week to one year	Growth rate is high; teeth begin to erupt; muscular and nervous systems mature so that coordinated activities are possible; communication begins
Childhood	One year to puberty	Growth rate is high; primary teeth erupt and are then replaced by secondary teeth; high degree of muscular control is achieved; bladder and bowel controls are established; intellectual abilities mature
Adolescence	Puberty to adulthood	Person becomes reproductively functional and emotionally more mature; growth spurts occur in skeletal and muscular systems; high levels of motor skills are developed; intellectual abilities increase
Adulthood	Adolescence to old age	Person remains relatively unchanged anatomically and physiologically; degenerative changes begin to occur
Senescence	Old age to death	Degenerative changes continue; body becomes less able to cope with demands; death usually results from mechanical disturbances in the cardiovascular system or from diseases that affect vital organs

Decreasing efficiency of the central nervous system accompanies senescence. The person may lose some intellectual functions. Also, the physiological coordinating capacity of the nervous system may decrease, and homeostatic mechanisms may fail to operate effectively. Sensory functions decline with age also. Table 20.6 summarizes the major events in the stages of postnatal development.

PRACTICE 20.4

8. At what age is there a developmental turning point?
9. How is aging a "process"?

20.5 | Genetics

LEARN

1. Distinguish among the modes of inheritance.
2. Describe the extensions of Mendelian inheritance.

The newborn enters the world, and the elated parents look for family resemblances. Does she have her father's nose, or her grandmother's curly hair? Or they may be concerned about inherited health conditions rather than appearances.

As the child grows, a unique mix of traits emerges. Inherited traits are determined by DNA sequences that comprise genes, which instruct cells to synthesize particular proteins, as discussed in section 4.6, Protein Synthesis. When a gene's DNA sequence changes, or *mutates,* illness may result (see Clinical Application 4.1). The environment influences how most genes are expressed. For example, inherited gene variants that confer susceptibility to lung cancer might affect health only if a person smokes or is exposed to air pollution for many years.

The field of **genetics** (jĕ-net′iks) investigates how genes confer specific characteristics that affect health or contribute to our natural variation, and how genes are passed from generation to generation. Now that it is possible to very quickly and accurately determine the sequence of DNA bases that constitute a human genome, the field of genomics has arisen to consider many genes and their variants simultaneously. Clinical Application 4.1, for example, describes clinical applications of determining the sequence of the protein-encoding portion of the genome. However, single genes and chromosome sets are still checked for clues to health. As discussed in Clinical Application 20.2, fetal chromosome checks provide clues to an individual's future health.

Chromosomes and Genes Are Paired

Chromosome charts called karyotypes display the 23 chromosome pairs (homologous pairs) of a human somatic cell in size order (fig. 20.18). Pairs 1 through 22 are **autosomes** (aw′to-sōmz), which do not carry genes that determine sex. The other two chromosomes, the X and the Y, include genes that determine sex and are called **sex chromosomes.** A female has two X chromosomes, and a male has one X and one Y.

Each chromosome except the tiny Y includes hundreds of genes. Somatic cells have two copies of each autosome, and therefore two copies of each gene. Gene copies can be identical or slightly different in DNA sequence. Such variant forms of a gene are called **alleles** (ah-lēls′). An individual who has two identical alleles of a gene is **homozygous** (hō″mo-zī′gus) for that gene. A person with two different alleles is **heterozygous** (het′er-o-zī′gus) for it.

The combination of alleles, for one gene or many, constitutes a person's **genotype** (jē′no-tīp). The appearance, health condition, or other characteristics associated with a particular genotype is the **phenotype** (fē′no-tīp). Alleles often come in only two varieties: **dominant** and **recessive.** A dominant allele masks expression of a recessive allele. That is, it "dominates" during genetic expression.

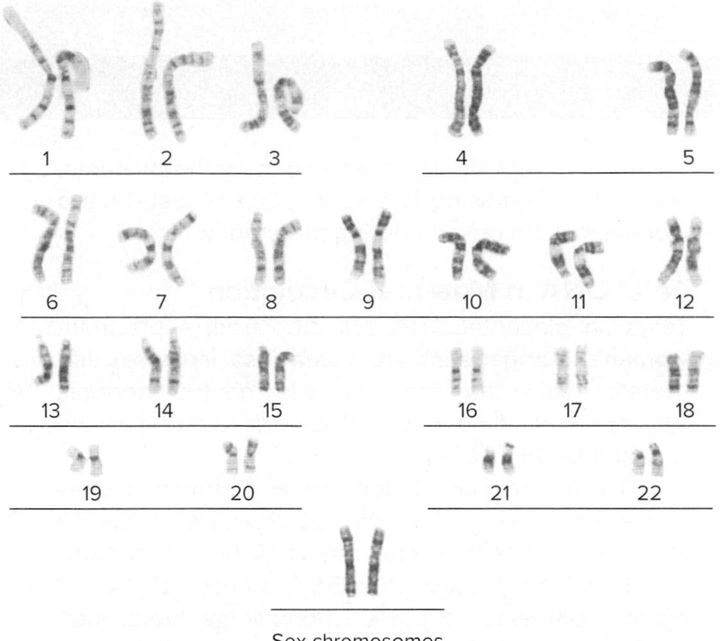

Figure 20.18 A normal human karyotype. Each somatic cell has 22 pairs of autosomes and a pair of sex chromosomes. The two X chromosomes indicate that this size-order chromosome chart (karyotype) is from a female. Courtesy Genzyme Corporation

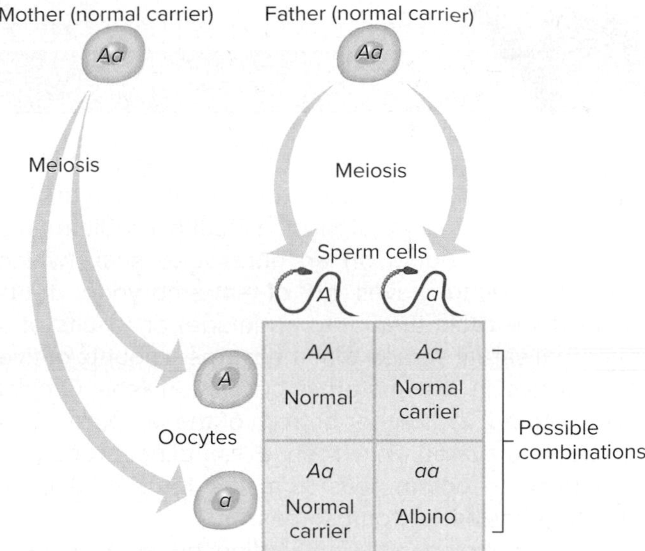

Figure 20.19 "A" represents the dominant allele for normal pigmentation, and "a" represents the recessive allele for albinism. The Punnett square reveals the genotypes and phenotypes from a mating between two heterozygote carriers with normal pigmentation.

PRACTICE 20.5

1. Distinguish between autosomes and sex chromosomes.
2. How does a homozygote differ from a heterozygote?
3. Distinguish between genotype and phenotype.

Mendelian Inheritance

We can predict the probability that a certain inherited trait will occur in the offspring of two individuals by considering how genes and chromosomes are distributed in meiosis, and the combinations in which they can unite at fertilization. Patterns in which genes are transmitted in families are called *modes of inheritance.*

Discovered by Gregor Mendel in 1866, the existence of dominant and recessive alleles is one of the core, fundamental concepts in the field of genetics. Therefore, when traits are controlled in a purely dominant/recessive fashion, it is known as Mendelian Inheritance.

A **Punnett square** (fig. 20.19) symbolizes the logic used to deduce the probabilities of inheriting particular genotypes in offspring. Each box records a possible allele combination at fertilization, and the entire square is used to exhibit the ratio of all possible combinations. Dominant alleles are usually indicated with a capital letter, whereas lowercase letters are used to represent recessive alleles. Figure 20.19 illustrates this for a cross between a male and a female, both of whom are heterozygous for expression of an enzyme necessary for the production of the pigment melanin. Melanin is responsible for skin, hair, and eye color. They each have one dominant allele for "normal" pigmentation and one for the recessive allele for a condition known as "albinism," or lack of pigmentation. Both parents have normal pigmentation, but are considered "carriers" of the abnormal allele. The square reveals that 25% of their offspring will be homozygous dominant "AA," with normal pigmentation, and 50% will be heterozygous "Aa" (carriers) with normal pigmentation. The only way to express the trait of albinism is to be homozygous recessive "aa." Therefore, only 25% of the offspring are affected with albinism.

Cystic fibrosis is a disease that is inherited in the same fashion, where the abnormal allele is recessive. Affected individuals are homozygous recessive, giving rise to poorly functioning chloride channels in the pancreas, respiratory tract, intestines, and testes.

Huntington disease is the result of a dominant allele. Symptoms, including loss of coordination and behavioral changes, do not occur until the late thirties or early forties. This dominant allele remains in human populations, because most often the affected individuals do not realize they have it until after they have passed it on to their children.

Extensions of Mendelian Inheritance

Modes of inheritance not strictly governed by one gene with only two allele versions (dominant and recessive) are known as Mendelian extensions. These include codominance, incomplete dominance, polygenic inheritance, multiple allele inheritance, and pleiotropy.

CLINICAL APPLICATION 20.2

Fetal Chromosome Checks

Visualizing chromosomes and analyzing DNA from the placenta can provide clues to fetal health. For many years, the first sign of a health problem was an abnormal finding on an ultrasound scan (which bounces sound waves off of an embryo or fetus and reconstructs them into an image) or results of a maternal serum screen that measures amounts of five biomarkers in the maternal circulation (see Clinical Application 2.2). If either or both of these approaches revealed elevated risk, then a sampling procedure was used to obtain cells from the fetus to directly check genes and chromosomes.

Today, ultrasound is still routine, but some women are having tests performed on pieces of DNA from the placenta that are present in the maternal circulation. These DNA pieces are presumed to represent parts of the fetal genome. Results from such "cell-free DNA" indicating higher risk are followed up with the more invasive and direct sampling procedures, which confirm whether the fetus, and not only the placenta, is affected. Following is a description of tests that a woman may be offered during pregnancy.

Fetal DNA in Maternal Circulation

Tests on placental DNA collected from a pregnant woman's bloodstream are much less invasive, and therefore safer, than sampling cells from the chorionic villi or amniotic fluid. Such a test is called non-invasive prenatal testing (NIPT).

The maternal circulation normally contains pieces of placental DNA, which can be separated because they are shorter than maternal DNA pieces. A chromosome type present in a 50% excess number of pieces compared to other chromosome types indicates an extra chromosome—such as the trisomy 21 that causes Down syndrome.

Entire fetal genomes have been sequenced from DNA pieces in maternal circulation (fig. 20Bc). This

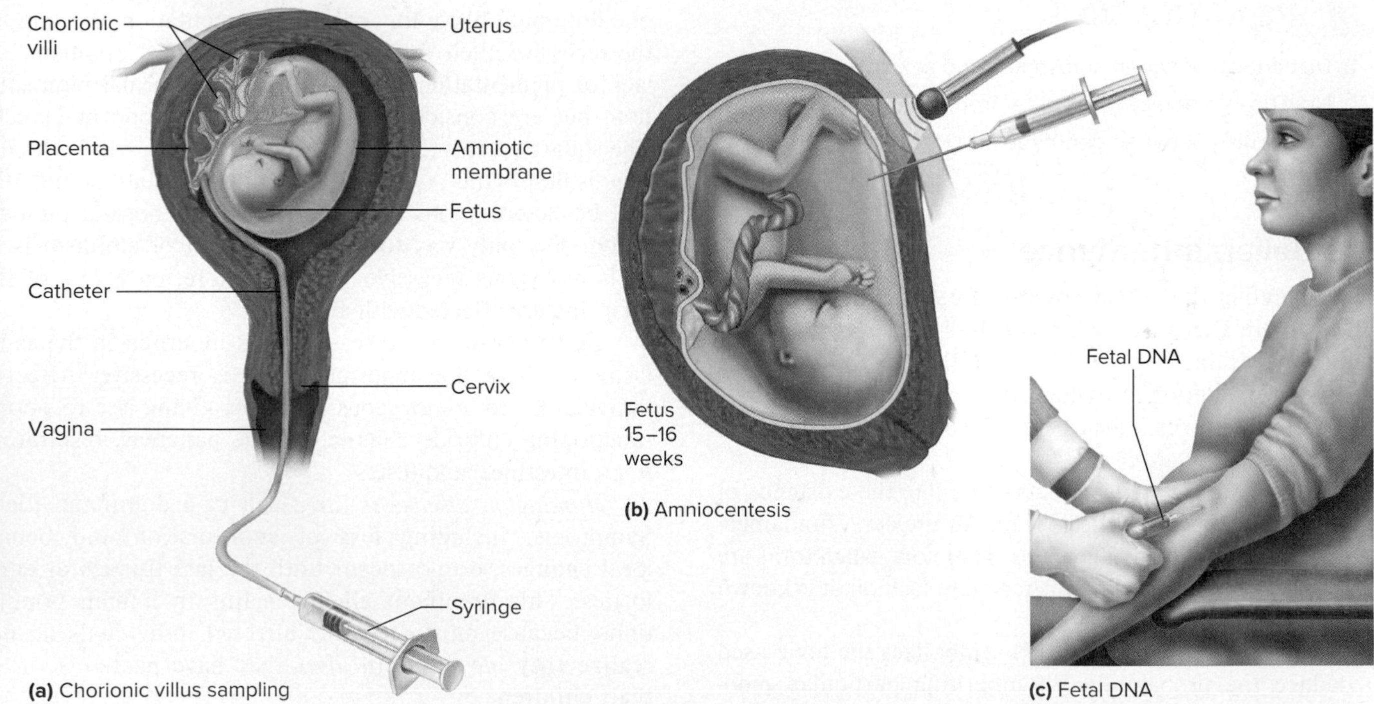

Figure 20B Three ways to check a fetus's chromosomes. **(a)** Chorionic villus sampling (CVS) removes cells of the chorionic villi, whose chromosomes match those of the fetus. **(b)** Amniocentesis withdraws amniotic fluid, which contains fetal cells. **(c)** Some chromosomal conditions can be inferred by analyzing fetal DNA in the maternal circulation. For CVS and amniocentesis, fetal chromosomes are stained and examined. Additional tests are required to detect mutations that cause specific genetic disorders.

technology may become a routine way to detect genetic disease before birth. It is likely that in the coming years, seeking clues to fetal health in the maternal bloodstream may replace the more invasive, older techniques.

Chorionic Villus Sampling

Chorionic villus sampling (CVS) (fig. 20B*a*) examines the chromosomes in chorionic villus cells, which are genetically identical to fetal cells because they are derived from the same fertilized egg. On rare occasions, the test causes spontaneous abortion. Due to this risk, women are advised to have the procedure only if maternal blood screening indicates increased risk of an extra chromosome, or if the couple has had a child with a detectable chromosome abnormality. CVS is performed at the tenth week of gestation.

Amniocentesis

Amniocentesis is performed after the fourteenth week of gestation. A physician uses ultrasound to guide a needle into the amniotic sac and withdraws about 5 milliliters of fluid (fig. 20B*b*). Fetal fibroblasts in the sample are cultured and their chromosomes are checked. It takes about a week to grow these cells, but a test using fluorescent dyes is used to detect the most common extra-chromosome conditions within 48 hours.

Until recently, amniocentesis carried a risk of about 0.5% of being followed by spontaneous abortion. But the procedure has become much safer. Previously, amniocentesis was offered only to women over age thirty-five, when the risk of conceiving a fetus with abnormal chromosomes is about 0.5% (risk increases with age), and to women who had already had a child with a detectable chromosomal abnormality. Some medical practices offer the test earlier in pregnancy because of the improvement in safety.

Preimplantation Genetic Diagnosis (PGD)

If a couple has a family history of a chromosomal or single-gene condition that could affect their offspring, a procedure called preimplantation genetic diagnosis (PGD) offers the ability to select embryos that have not inherited the condition (fig. 20C). After secondary oocytes are fertilized *in vitro* and allowed to divide to the eight-celled stage, one cell from each of several embryos is removed and tested for the disease-causing mutation or chromosome abnormality. If the genes or chromosomes are unaffected, a tested seven-celled embryo is allowed to continue development *in vitro* briefly, then it is introduced into the woman.

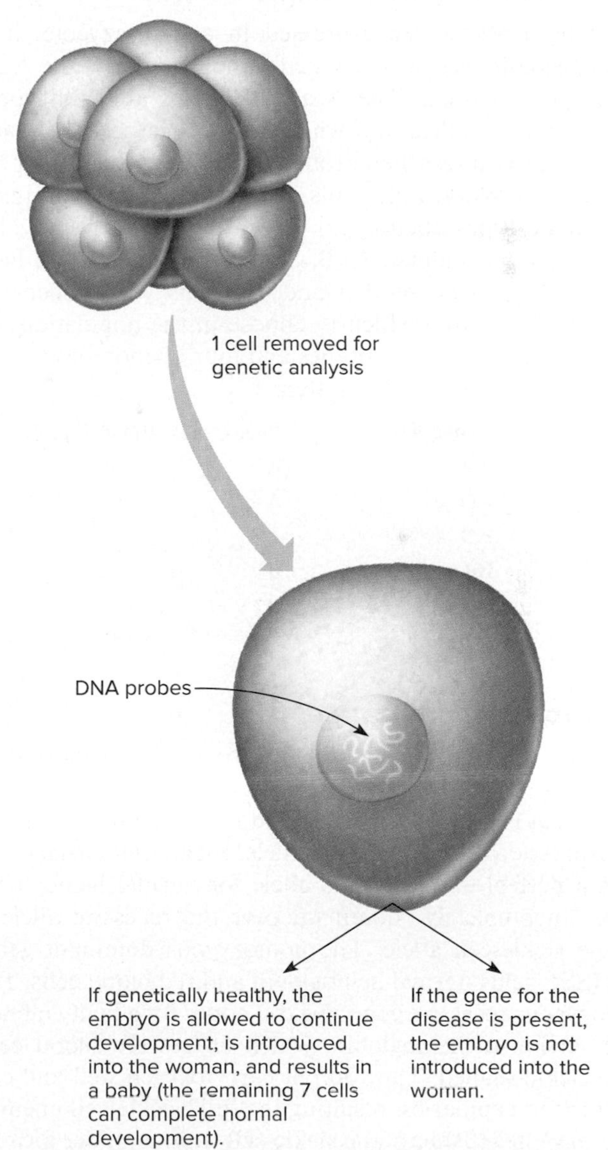
1 cell removed for genetic analysis
DNA probes
If genetically healthy, the embryo is allowed to continue development, is introduced into the woman, and results in a baby (the remaining 7 cells can complete normal development).
If the gene for the disease is present, the embryo is not introduced into the woman.

Figure 20C Preimplantation genetic diagnosis probes disease-causing genes in an eight-celled embryo.

Figure 20.20 Height is a polygenic trait. These students in a genetics class lined up, with shorter individuals on the left of the photo and taller individuals on the right, to display a classic bell-shaped distribution. David Hyde/Wayne Falda/McGraw-Hill Education

Codominance and Multiple Alleles

When both alleles are expressed in a heterozygote, it is called **codominance.** The classic example is that of the ABO blood typing system. The A and B alleles are both dominant over the O allele. But when paired together, the A and B alleles both express their protein, or antigen (see chapter 12) found on red blood cells. This gives rise to the codominant "AB" blood type (phenotype). In addition, this system also exhibits **multiple alleles:** A, B, and O. You only get to have two—one from your mother and one from your father—but there are three from which to choose in the population. As a result, there are six genotypes and four phenotypes (blood types) associated with this system:

Genotypes	Phenotypes (Blood Types)
AA	A
AO	A
AB	AB
BB	B
BO	B
OO	O

Incomplete Dominance

In **incomplete dominance,** the heterozygous phenotype is somewhat of a hybrid between the two homozygous phenotypes. That is, the dominant allele does not completely mask the expression of the recessive allele. **Sickle-cell anemia** provides a perfect example. The allele for normal hemoglobin (S) is "incompletely" dominant over the recessive allele(s) for the sickle-cell allele. The homozygous dominant genotype (SS) yields normal hemoglobin and red blood cells. The homozygous recessive genotype (ss) codes for a faulty (abnormal) version of hemoglobin, which causes red blood cells to be sickle-shaped. They do not carry oxygen well and can get stuck in capillaries, resulting in "full" sickle-cell anemia. Heterozygotes (Ss) have the **sickle-cell trait,** a milder form of sickle-cell anemia. Some of the hemoglobin is normal and not all of the red blood cells are sickle-shaped. Interestingly, this milder form of the disease gives affected individuals a heightened resistance to malaria. The sickle-cell allele remains in the population in a particular geographic region in Africa where the mosquitoes that carry malaria are common.

Polygenic Inheritance and Pleiotropy

Traits controlled by more than one gene are called **polygenic** (many genes) traits, and show a spectrum of variation, also known as continuous variation. Height (fig. 20.20), eye color, and skin color are examples of polygenic traits. When individuals with a polygenic trait are categorized into classes and then the frequencies of the classes are plotted as a bar graph, a bell-shaped curve emerges. The curves are strikingly similar for different polygenic traits. Figure 20.20 shows the bell curve for height of students in a genetics class.

If one gene contributes to several traits or has multiple effects, it exhibits **pleiotropy.** Pleiotropy is seen in genetic diseases that affect a single protein found in different parts of the body. This is the case for Marfan syndrome, a defect in an elastic connective tissue protein called fibrillin. The protein's abundance in the lens of the eye; in the bones of the limbs, fingers, and ribs; and in the aorta explains the symptoms of lens dislocation, long limbs, spindly fingers, and a caved-in chest. The most serious symptom is weakening in the aorta wall, which can burst the vessel. If the weakening is detected early, a synthetic graft can be used to patch that part of the vessel wall, saving the person's life.

Chromosomal Inheritance and Sex-Linked Traits

As opposed to following only alleles, inheritance of whole chromosomes can be tracked with a Punnett square. This is especially so for the sex chromosomes X and Y. Females (XX) have two X chromosomes and males (XY) have an X and a Y chromosome. During meiosis and gamete formation, females make egg cells of only one type, those with an X chromosome. Sperm cells will contain either an X or a Y chromosome. Figure 20.21 shows the possible combinations of egg and sperm cells during fertilization. Note that the probability of having a female or male offspring is 50%.

The other 22 pairs of chromosomes, the autosomes, can also be tracked. Of particular interest is chromosome number 21. If an individual is born with a third copy of chromosome number 21, it is known as *trisomy 21* (fig. 20.22). This is caused by nondisjunction, the failure of the chromosomes to separate properly during meiosis. Trisomy 21 results in a condition called Down syndrome.

A person with Down syndrome is short and has straight, sparse hair and a tongue protruding through thick lips. The eyes slant and have upward skin folds in the inner corners. Ears are abnormally shaped. The hands have an unusual pattern of creases, the joints are loose, and reflexes and muscle tone are poor. Reaching developmental milestones (such as sitting, standing, and walking) is slow, and toilet training may take several years. Intelligence varies greatly, from profound intellectual disability to being able to follow directions, understand conversations, respond appropriately (such as laughing), read, and use a computer. Down syndrome is also associated with many physical problems, including heart and kidney defects, susceptibility to infections, and blockages in the digestive system.

Sex-linked traits are coded for by genes found on the sex chromosomes, most notably the X chromosome. Because males have only one X chromosome, they are affected by recessive X-linked traits with more frequency than females. Red-green color blindness is a classic example. Using "C" as the dominant full color vision allele, and "c" as the recessive color blindness allele, table 20.7 shows the possible genotypes and phenotypes of females and males regarding full color vision and color blindness. Females with full color vision are either homozygous dominant or are heterozygous carriers of the color-blindness allele. If they are color blind, they are homozygous recessive. Because males only receive one X chromosome from their mother, the one allele they get dictates their phenotype. Males are never carriers; they either have full color vision or are color blind.

Hemophilia A and Duchenne muscular dystrophy are also inherited as X-linked recessive traits.

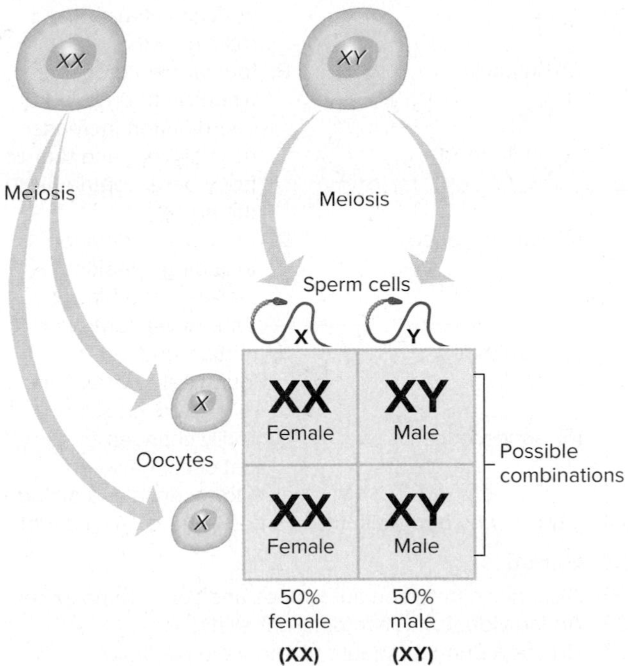

Figure 20.21 Inheritance of sex. The female produces oocytes containing one X chromosome, whereas the male produces sperm cells with either an X or a Y chromosome. There are four possible combinations of an oocyte with a sperm cell, half of which produce females and half of which produce males.

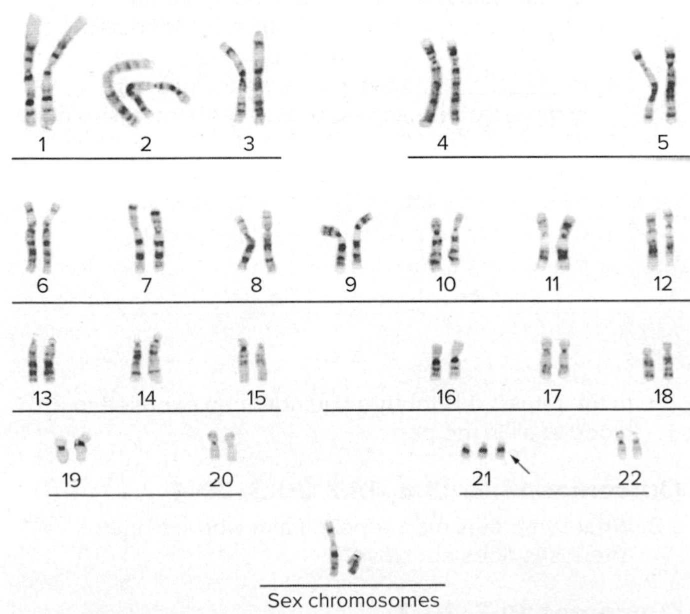

Figure 20.22 Down syndrome karyotype. A karyotype of a male with Down syndrome caused by trisomy 21. Source: National Human Genome Research Institute

PRACTICE 20.5

4. Distinguish between dominant and recessive alleles.
5. How do Punnett squares depict gene transmission?
6. How does the ABO blood typing system exhibit codominance and multiple alleles?
7. How is sickle-cell anemia an example of incomplete dominance?
8. What is a polygenic trait?
9. Explain nondisjunction of chromosome number 21.
10. Why are males never carriers of X-linked traits?

TABLE 20.7 Possible Genotypes and Phenotypes for Red-Green Color Blindness, an X-Linked Trait

	Genotype	Phenotype
Females	$X^C X^C$	Full color vision
	$X^C X^c$	Full color vision carrier
	$X^c X^c$	Color-blind
Males	$X^C Y$	Full color vision
	$X^c Y$	Color-blind

C = Allele for full color vision; *c* = Allele for color blindness.

ASSESS

CHAPTER ASSESSMENTS

20.1 Introduction

1. __________ is an increase in the size of the individual, whereas __________ is the continuous process by which an individual changes from one life phase to another.
2. __________ is the period of development from fertilization to birth, whereas __________ is the period of development from birth to death.

20.2 Fertilization

3. Describe how sperm cells are assisted in their movement through the female reproductive tract.
4. Summarize the events occurring after the sperm cell head enters the egg's cytoplasm.

20.3 Pregnancy and the Prenatal Period

5. Define *pregnancy*.
6. Describe the process of cleavage.
7. Distinguish between a morula and a blastocyst.
8. List the function(s) of the placenta.
9. Distinguish between the chorion and the amnion.
10. Explain the function of the amniotic fluid.
11. Describe the formation of the umbilical cord.
12. Describe the structure of the placenta.
13. Ectodermal cells of the developing embryo give rise to __________.
    a. bone tissue
    b. the kidneys
    c. the lining of the urethra
    d. the epidermis
14. List the major changes that occur during fetal development.
15. Describe a full-term fetus.
16. Trace the pathway of blood from the placenta to the fetus and back to the placenta.
17. List hormones associated with pregnancy, and describe the function of each.
18. Explain positive feedback and the role of hormones in the expulsion of the fetus and the afterbirth.
19. Describe milk production and secretion.

20.4 Aging: The Human Life Span

20. Explain why a newborn's first breath must be particularly forceful.
21. Discuss the difficulties of the fetus in maintaining water/electrolyte and body temperature homeostasis.
22. Describe the changes in the newborn's cardiovascular system.
23. Match the stage of postnatal development that best fit its description.

(1) neonatal period	A. appearance of secondary sex characteristics and rapid growth
(2) infancy	B. four weeks to one year when teeth erupt and coordination increases
(3) childhood	C. normal wear and tear of body parts contribute to the aging
(4) adolescence	D. motor skills develop, including speaking, walking, and bladder and bowel control
(5) adulthood	E. birth to end of first four weeks when self-reliance occurs
(6) senescence	F. bodily changes that occur between adolescence and old age

24. Explain why people in their forties tend to gain weight.

20.5 Genetics

25. Distinguish between autosomes and sex chromosomes.
26. An individual with two different alleles is ______.
27. Create a Punnett square to show the possible genotypes in offspring of a mother who has cystic fibrosis and a father who is heterozygous (a carrier) for cystic fibrosis. What is the probability (in percent) that an offspring will have this disease?
28. Match the mode of inheritance with its description.

(1) autosomal recessive	A. inherited from one affected parent
(2) autosomal dominant	B. inherited from two carrier (unaffected) parents

29. In __________, the dominant allele does not completely mask the expression of the recessive allele.
30. Why are males affected by recessive X-linked traits with more frequency than females?

ASSESS

INTEGRATIVE ASSESSMENTS/CRITICAL THINKING

Outcomes 11.4, 19.7, 20.3

1. Kari, a new mother, was surprised when she had milk dripping from her nipple when she heard another baby cry or when she saw another woman nursing her infant. Sometimes, it even occurred when Kari was just thinking about nursing her baby. Why was Kari experiencing lactation when her baby was not suckling on her breast?

Outcomes 12.5, 20.3

2. A woman with Rh^- blood type is pregnant with a Rh^+ fetus. Explain why there would be no adverse effects to the fetus if this mother had not been exposed to Rh^+ blood at all in the past.

Outcomes 13.2, 13.6, 13.7, 20.3, 20.4

3. What symptoms may appear if a newborn's ductus arteriosus does not close?

Outcomes 14.7, 20.2

4. Why can twins resulting from a single fertilized egg (monozygotic) exchange blood or receive organ transplants from each other without rejection, whereas twins resulting from two fertilized eggs (dizygotic) sometimes cannot?

Outcomes 15.10, 20.5

5. Janet was fifty years old when she had her first colonoscopy screening. She had a couple polyps removed that turned out to be adenocarcinomas, or cancerous, after a biopsy was performed. Fortunately, the cancer was caught early. Given her "young age" as the gastroenterologist commented, Janet was referred to a genetic counselor. The counselor discussed with Janet reasons to get genetically tested. What benefit(s) do you see in Janet getting genetic testing?

Outcomes 19.3, 20.3

6. What kinds of studies and information are required to determine whether a man's exposure to a potential teratogen can cause birth defects years later?

Outcome 20.2

7. Explain how premature hardening of the zona pellucida would affect a woman's fertility?

Outcomes 20.3, 20.4

8. What technology would enable a fetus born in the fourth month to survive in a laboratory setting? (This is not yet possible.)

Outcome 20.5

9. Bob and Joan know from a blood test that they are each heterozygous (carriers) for the autosomal recessive gene that causes sickle-cell anemia. If their first three children are healthy, what is the probability that their fourth child will have the disease?

Chapter Summary

20.1 Introduction

***Growth** is an increase in size. **Development** is the process of changing from one life phase to another.*

20.2 Fertilization

***Fertilization** occurs with the union of a secondary oocyte and a sperm cell.*

1. Transport of sex cells
   a. A male deposits semen in the vagina during sexual intercourse.
   b. A sperm cell lashes its tail to move and is aided by muscular contractions in the female reproductive tract.
2. Sperm cell joins egg cell
   a. With the aid of enzymes, a sperm cell penetrates the corona radiata and zona pellucida.
   b. When a sperm cell head fuses with the egg cell membrane, changes in the membrane and the zona pellucida prevent entry of additional sperm cells.
   c. Completion of meiosis forms the second polar body.
   d. Fusion of the pronuclei from the two sex cells completes fertilization.
   e. The product of fertilization is a **zygote** with 46 chromosomes.

20.3 Pregnancy and the Prenatal Period

***Pregnancy** is the presence of a developing offspring in the uterus. The prenatal period consists of the period of cleavage, the embryonic stage, and the fetal stage.*

1. **Embryonic stage**
   a. Cells undergo mitosis, giving rise to smaller and smaller cells during **cleavage.**
      (1) The developing offspring moves down the uterine tube to the uterus, where it implants in the endometrium.
      (2) The inner cell mass gives rise to the **embryo proper.**
      (3) Eventually, embryonic and maternal cells form a **placenta.**
   b. Extraembryonic membrane formation and placentation occur after implantation.
      (1) The trophoblast and its lining layer of cells form the **chorion.**
      (2) **Chorionic villi** develop and are surrounded by spaces filled with maternal blood.
      (3) A fluid-filled **amnion** develops around the embryo.
      (4) The **umbilical** cord forms as the amnion envelops the tissues attached to the underside of the embryo.
      (5) The **yolk sac** forms on the underside of the embryonic disc. It helps form the digestive tube, and gives rise to blood cells and cells that later become sex cells.
      (6) The **allantois** extends from the yolk sac into the connecting stalk. It forms blood cells and gives rise to the umbilical vessels.
      (7) The placenta consists of an embryonic portion and a maternal portion.
      (8) The **placental membrane** consists of the epithelium of the chorionic villi and the endothelium of the capillaries inside the villi.
         (a) Oxygen and nutrients diffuse from maternal blood across the placental membrane and into fetal blood.
         (b) Carbon dioxide and other wastes diffuse from fetal blood across the placental membrane and into maternal blood.
   c. **Gastrulation** is the movement of cells within the embryonic disc to form layers, and **organogenesis** is the process by which primitive tissues in the germ layers form organs.

2. **Fetal stage** (extends from the end of the eighth week of development until birth)
   a. Existing structures grow and mature.
   b. The **fetus** is full-term at approximately 266 days from conception.
   c. Fetal blood and circulation promote reception of oxygen and nutrients from maternal blood and wastes being carried away by maternal blood.
      (1) Umbilical vessels carry blood between the placenta and the fetus.
      (2) Fetal red blood cells carry more oxygen than do maternal red blood cells because the concentration of oxygen-carrying hemoglobin is greater in fetal red blood cells, and fetal hemoglobin has greater attraction for oxygen.
      (3) Blood enters the fetus through the umbilical vein and partially bypasses the liver by means of the **ductus venosus.**
      (4) Blood enters the right atrium and partially bypasses the lungs by means of the **foramen ovale.**
      (5) Blood entering the pulmonary trunk partially bypasses the lungs by means of the **ductus arteriosus.**
      (6) Blood enters the umbilical arteries from the internal iliac arteries.
3. Maternal changes during pregnancy
   a. Embryonic cells produce human chorionic gonadotropin (hCG), which maintains the corpus luteum.
   b. Placental tissue produces high concentrations of estrogens and progesterone.
      (1) Estrogens and progesterone maintain the uterine wall and inhibit secretion of follicle-stimulating hormone (FSH) and luteinizing hormone (LH).
      (2) Progesterone and relaxin inhibit contraction of uterine smooth muscle.
      (3) Estrogens enlarge the vagina.
      (4) Relaxin helps relax the connective tissue of the pelvic joints.
   c. **Placental lactogen** stimulates development of the breasts and mammary glands.
   d. During pregnancy, increased aldosterone secretion promotes retention of sodium and body fluid. Increased secretion of parathyroid hormone helps maintain a high concentration of maternal blood calcium.
4. Birth process
   a. A variety of factors promote birth.
      (1) A decreasing progesterone concentration and the release of prostaglandins may initiate the birth process.
      (2) The posterior pituitary gland releases **oxytocin.**
      (3) Oxytocin stimulates uterine smooth muscle to contract, and labor begins.
   b. Following birth, placental tissues are expelled.
5. Milk production and secretion
   a. Following childbirth, concentrations of placental hormones decline, the action of prolactin is no longer blocked, and the mammary glands begin to secrete milk.
   b. A reflex response to mechanical stimulation of the nipple stimulates the posterior pituitary gland to release oxytocin, which causes the alveolar ducts to eject milk.

20.4 Aging: The Human Life Span

1. Neonatal period
   a. The **neonatal period** extends from birth to the end of the first four weeks.
   b. The newborn must begin to respire, obtain nutrients, excrete wastes, and regulate body temperature.
   c. The first breath must be powerful to expand the lungs.
   d. The liver is immature and unable to supply sufficient glucose, so the newborn depends primarily on stored fat for energy.
   e. A newborn's immature kidneys cannot concentrate urine well.
   f. A newborn's homeostatic mechanisms may function imperfectly, and body temperature may be unstable.
   g. The cardiovascular system changes when placental circulation ceases.
      (1) Umbilical vessels constrict.
      (2) The ductus venosus constricts.
      (3) Tissue flaps close the foramen ovale as blood pressure in the right atrium falls and pressure in the left atrium rises.
      (4) The ductus arteriosus constricts.
2. The neonatal period is followed by infancy, childhood, adolescence, adulthood, and senescence.

20.5 Genetics

1. Chromosomes and genes are paired
   a. Karyotypes are charts that display the two copies of each of the 22 **autosomes,** which do not carry genes that determine sex, and the sex chromosomes (X and Y), which do.
   b. A person with two identical variants, or **alleles,** for a gene is **homozygous.** A person with two different alleles is **heterozygous.**
   c. The combination of alleles is the **genotype;** their expression as a trait is the **phenotype.**
2. Modes of inheritance
   a. A dominant allele masks the expression of a recessive allele.
   b. Punnett squares are used to predict genotypic and phenotypic ratios for offspring.
3. Extensions to Mendelian inheritance
   a. Some alleles are codominant, and some traits are controlled by multiple alleles.
   b. Some alleles are incompletely dominant.
   c. The continuously varying nature of polygenic traits can be depicted in a bell-shaped curve. When a gene controls many traits, it is called pleiotropy.
   d. Chromosomes can be tracked in a Punnet square.
   e. Sex-linked traits are most often controlled by genes on the X chromosome.

APPENDIX A

Aids to Understanding Words

acetabul-, vinegar cup: *acetabul*um
adip-, fat: *adip*ose tissue
agglutin-, to glue together: *agglutin*ation
aliment-, food: *aliment*ary canal
allant-, sausage: *allant*ois
alveol-, small cavity: *alveol*us
an-, without: *an*aerobic respiration
ana-, up: *ana*bolism
andr-, man: *andr*ogens
append-, to hang something: *append*icular
ax-, axis: *ax*ial skeleton, *ax*on
bil-, bile: *bil*irubin
-blast, bud: osteo*blast*
brady-, slow: *brady*cardia
bronch-, windpipe: *bronch*us
calat-, something inserted: inter*calat*ed disc
calyc-, small cup: major *calyc*es
cardi-, heart: peri*cardi*um
carp-, wrist: *carp*als
cata-, down: *cata*bolism
chondr-, cartilage: *chondr*ocyte
chorio-, skin: *chorio*n
choroid, skinlike: *choroid* plexus
chym-, juice: *chym*e
-clast, break: osteo*clast*
cleav-, to divide: *cleav*age
cochlea, snail: *cochlea*
condyl-, knob: *condyl*e
corac-, a crow's beak: *corac*oid process
cort-, covering: renal *cort*ex
cran-, helmet: *cran*ial
cribr-, sieve: *cribr*iform plate
cric-, ring: *cric*oid cartilage
-crin, to secrete: endo*crin*e
crist-, crest: *crist*a galli
cut-, skin: sub*cut*aneous
cyt-, cell: *cyt*oplasm, osteo*cyt*e
de-, separation from: *de*hydration
decidu-, falling off: *decidu*ous teeth
dendr-, tree: *dendr*ite
derm-, skin: *derm*is
detrus-, to force away: *detrus*or muscle
di-, two: *di*saccharide
diastol-, dilation: *diastol*ic pressure
diure-, to pass urine: *diure*tic
dors-, back: *dors*al
ejacul-, to shoot forth: *ejacul*ation
embol-, stopper: *embol*us
endo-, within: *endo*plasmic reticulum, *endo*crine gland
epi-, upon: *epi*thelial tissue, *epi*dermis, *epi*glottis
erg-, work: syn*erg*ist
erythr-, red: *erythr*ocyte
exo-, outside: *exo*crine gland
extra-, outside: *extra*cellular fluid
fimb-, fringe: *fimb*riae
follic-, small bag: hair *follic*le, ovarian *follic*le
fov-, pit: *fov*ea capitis
funi-, small cord or fiber: *funi*culus
gangli-, a swelling: *gangli*on
gastr-, stomach: *gastr*ic gland
-gen, to be produced: aller*gen*
-genesis, origin: spermato*genesis*
glen-, joint socket: *glen*oid cavity
-glia, glue: neuro*glia*
glom-, little ball: *glom*erulus
glyc-, sweet: *glyc*ogen
-gram, something written: electrocardio*gram*
hema-, blood: *hema*tocrit
hemo-, blood: *hemo*globin
hepat-, liver: *hepat*ic duct
hetero-, other, different: *hetero*zygous
hom-, same, common: *hom*ozygous
homeo-, same: *homeo*stasis
humor-, fluid: *humor*al immunity
hyper-, above, more, over: *hyper*tonic, *hyper*trophy, *hyper*thyroidism
hypo-, below: *hypo*tonic, *hypo*thyroidism
im-, not: *im*balance
immun-, free: *immun*ity
inflamm-, set on fire: *inflamm*ation
inter-, among, between: *inter*phase, *inter*calated disc, *inter*vertebral disc
intra-, inside, within: *intra*membranous bone, *intra*cellular fluid
iris, rainbow: *iris*
iso-, equal: *iso*tonic
kerat-, horn: *kerat*in
labi-, lip: *labi*a minora
labyrinth, maze: *labyrinth*
lacri-, tears: *lacri*mal gland
lacun-, pool: *lacun*a
laten-, hidden: *laten*t period
-lemm, rind or peel: neuri*lemm*a
leuko-, white: *leuko*cyte
lingu-, tongue: *lingu*al tonsil
lip-, fat: *lip*ids
-logy, study of: physio*logy*
-lyt, dissolvable: electro*lyt*e
macr-, large: *macr*ophage
macula, spot: *macula* lutea
meat-, passage: auditory *meat*us
melan-, black: *melan*in
mening-, membrane: *mening*es
mens-, month: *mens*es
meta-, change: *meta*bolism
mict-, to pass urine: *mict*urition
mit-, thread: *mit*osis
mono-, one: *mono*saccharide
mons-, mountain: *mons* pubis
morul-, mulberry: *morul*a
moto-, moving: *mot*or neuron
mut-, change: *mut*ation
myo-, muscle: *myo*fibril
nat-, to be born: pre*nat*al
nephr-, pertaining to the kidney: *nephr*on
neutr-, neither one nor the other: *neutr*al
nod-, knot: *nod*ule
nutri-, nourish: *nutri*ent
odont-, tooth: *odont*oid process
olfact-, to smell: *olfact*ory
os-, bone: *os*seous tissue
-osis, abnormal condition: leukocyt*osis*
papill-, nipple: *papill*ary muscle, renal *papill*ae
para-, beside: *para*thyroid glands
pariet-, wall: *pariet*al membrane
patho-, disease: *patho*gen
pelv-, basin: *pelv*ic cavity
peri-, around: *peri*cardial membrane, *peri*pheral nervous system, *peri*stalsis
phag-, to eat: *phag*ocytosis
pino-, to drink: *pino*cytosis
pleur-, rib: *pleur*al membrane
plex-, interweaving: choroid *plex*us
-poie, make, produce: hemato*poie*sis, erythro*poie*tin
poly-, many: *poly*unsaturated
pseud-, false: *pseud*ostratified epithelium
puber-, adult: *puber*ty
pyl-, gatekeeper: *pyl*oric sphincter
sacchar-, sugar: mono*sacchar*ide
sarco-, flesh: *sarco*plasm
scler-, hard: *scler*a
seb-, grease: *seb*aceous gland
sens-, feeling: *sens*ory neuron
-som, body: ribo*som*e
squam-, scale: *squam*ous epithelium
-stasis, standing still, halt: homeo*stasis,* hemo*stasis*
strat-, layer: *strat*ified
sudor-, sweat: *sudor*iferous gland
syn-, together: *syn*thesis, *syn*ergist, *syn*apse, *syn*cytium
systol-, contraction: *systol*ic pressure
tachy-, rapid: *tachy*cardia
tetan-, stiff: *tetan*ic contraction
thromb-, clot: *thromb*ocyte
toc-, birth: oxy*toc*in
-tomy, cutting: ana*tomy*
trigon-, triangle: *trigon*e
-troph-, well fed: muscular hyper*troph*y, *troph*oblast
-tropic, influencing: adrenocortico*tropic*
tympan-, drum: *tympan*ic membrane
umbil-, navel: *umbil*ical cord
ventr-, belly or stomach: *ventr*icle
vill-, hairy: *vill*i
vitre-, glass: *vitre*ous humor
-zym, ferment: en*zym*e

APPENDIX B

Scientific Method

Our knowledge of how the human body works is based on centuries of careful observation and experimentation. A way of thinking and organizing information, called the **scientific method,** guides the acquisition and interpretation of new information about the human body. The scientific method is used to either support or challenge a scientific theory, which is a systematically organized body of knowledge that applies to a variety of situations. It is a key part of a general process known as *scientific inquiry,* which includes other ways of investigating the natural world. The scientific method is a framework in which to consider ideas and evidence that involves use of familiar skills of observing, questioning, reasoning, predicting, testing, interpreting, and concluding.

The scientific method provides a sequence of logical steps that address a specific question. A proposed answer to the question is stated as a *hypothesis,* or "educated guess." More than a hunch, a hypothesis is based on existing knowledge and is often phrased in the form "If this is true, then that must follow." It is specific and testable, and should examine only one changeable factor, or *variable.* Even though the hypothesis may seem attractive and may make sense, it is not accepted until sufficient experimental evidence supports it. Indeed, it may be proven wrong. Experimental findings that go against the hypothesis are just as important as those that support it.

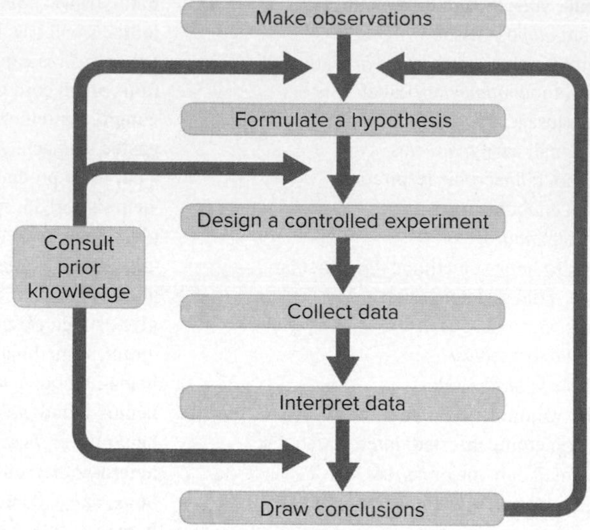

The process of drug development illustrates the use of the scientific method in anatomy and physiology. Consider a hypothesis to test a new cancer treatment:

If a drug blocks the receptor proteins to which a growth factor binds on cancer cells, then the cells may stop dividing and halt progression of the cancer.

This hypothesis is based on past experiments that identified the specific receptors and growth factors that fuel a particular type of cancer and described how they interact, and observations that the cells of certain cancers have excess growth factor receptors compared to healthy cells. Using that knowledge and the hypothesis, researchers design an experiment (a clinical trial in this example), which is a test that yields information:

One group of patients receives the experimental treatment (new drug) and a second group (the control) receives an existing treatment.

Experiments are carefully planned to make the results as meaningful as possible. In drug development, the control group provides a point of comparison for the new treatment group, because the participants are selected to be as alike as possible. In this way, differences due to the drug can stand out. Studies should also be "double blinded" whenever possible, which means that neither participants nor researchers know which patients receive which treatment until the end of the study. The results that will determine whether the hypothesis is accepted or rejected must be measurable and clearly identified. At the end of six months, if significantly more of the patients taking the experimental drug show measurable signs of improvement compared to the control group, then further study is warranted. It is essential that experiments be repeated to confirm experimental results. Before a new drug is prescribed for patients, it is tested on thousands of individuals in multiple clinical trials.

The scientific method is not a linear process, but a cycle, because results and conclusions often suggest new questions to explore. Something about the new cancer drug, for example, may suggest how it could be "repurposed" to treat a different condition. Testing begins anew.

In scientific investigations, so-called "negative" results are just as important as results that support the hypothesis. If the drug being tested doesn't work, researchers can either modify the hypothesis (perhaps the drug works only on certain subtypes of the disease) or develop more effective versions of the drug. Medical journals provide diverse examples of how the scientific method lies behind the drugs and procedures that have vastly improved the lives of many of us.

APPENDIX C

Metric Measurement System and Conversions

Measurement	Unit & Abbreviation	Metric Equivalent	Conversion Factor Metric to English (approximate)	Conversion Factor English to Metric (approximate)
Length	1 kilometer (km)	1,000 (10^3) m	1 km = 0.62 mile	1 mile = 1.61 km
	1 meter (m)	100 (10^2) cm 1,000 (10^3) mm	1 m = 1.1 yards = 3.3 feet = 39.4 inches	1 yard = 0.9 m 1 foot = 0.3 m
	1 decimeter (dm)	0.1 (10^{-1}) m	1 dm = 3.94 inches	1 inch = 0.25 dm
	1 centimeter (cm)	0.01 (10^{-2}) m	1 cm = 0.4 inch	1 foot = 30.5 cm 1 inch = 2.54 cm
	1 millimeter (mm)	0.001 (10^{-3}) m 0.1 (10^{-1}) cm	1 mm = 0.04 inch	
	1 micrometer (µm)	0.000001 (10^{-6}) m 0.001 (10^{-3}) mm		
Mass	1 metric ton (t)	1,000 (10^3) kg	1 t = 1.1 tons	1 ton = 0.91 t
	1 kilogram (kg)	1,000 (10^3) g	1 kg = 2.2 pounds	1 pound = 0.45 kg
	1 gram (g)	1,000 (10^3) mg	1 g = 0.04 ounce	1 pound = 454 g 1 ounce = 28.35 g
	1 milligram (mg)	0.001 (10^{-3}) g		
	1 microgram (µg)	0.000001 (10^{-6}) g		
Volume (liquids and gases)	1 liter (L)	1,000 (10^3) mL	1 L = 1.06 quarts	1 gallon = 3.78 L 1 quart = 0.95 L
	1 milliliter (mL)	0.001 (10^{-3}) L 1 cubic centimeter (cc or cm^3)	1 mL = 0.03 fluid ounce 1 mL = 1/5 teaspoon 1 mL = 15–16 drops	1 quart = 946 mL 1 fluid ounce = 29.6 mL 1 teaspoon = 5 mL
Time	1 second (s)	1/60 minute	same	same
	1 millisecond (ms)	0.001 (10^{-3}) s	same	same
Temperature	Degrees Celsius (°C)		°F = 9/5°C + 32	°C = 5/9 (°F – 32)

APPENDIX D

Periodic Table of the Elements

9 **F** Fluorine 19.00	Atomic number (9) Atomic mass (19.00)

1 1A	2 2A	3 3B	4 4B	5 5B	6 6B	7 7B	8 8B	9 8B	10 8B	11 1B	12 2B	13 3A	14 4A	15 5A	16 6A	17 7A	18 8A
1 **H** Hydrogen 1.008																	2 **He** Helium 4.003
3 **Li** Lithium 6.941	4 **Be** Beryllium 9.012											5 **B** Boron 10.81	6 **C** Carbon 12.01	7 **N** Nitrogen 14.01	8 **O** Oxygen 16.00	9 **F** Fluorine 19.00	10 **Ne** Neon 20.18
11 **Na** Sodium 22.99	12 **Mg** Magnesium 24.31											13 **Al** Aluminum 26.98	14 **Si** Silicon 28.09	15 **P** Phosphorus 30.97	16 **S** Sulfur 32.07	17 **Cl** Chlorine 35.45	18 **Ar** Argon 39.95
19 **K** Potassium 39.10	20 **Ca** Calcium 40.08	21 **Sc** Scandium 44.96	22 **Ti** Titanium 47.88	23 **V** Vanadium 50.94	24 **Cr** Chromium 52.00	25 **Mn** Manganese 54.94	26 **Fe** Iron 55.85	27 **Co** Cobalt 58.93	28 **Ni** Nickel 58.69	29 **Cu** Copper 63.55	30 **Zn** Zinc 65.39	31 **Ga** Gallium 69.72	32 **Ge** Germanium 72.59	33 **As** Arsenic 74.92	34 **Se** Selenium 78.96	35 **Br** Bromine 79.90	36 **Kr** Krypton 83.80
37 **Rb** Rubidium 85.47	38 **Sr** Strontium 87.62	39 **Y** Yttrium 88.91	40 **Zr** Zirconium 91.22	41 **Nb** Niobium 92.91	42 **Mo** Molybdenum 95.94	43 **Tc** Technetium (98)	44 **Ru** Ruthenium 101.1	45 **Rh** Rhodium 102.9	46 **Pd** Palladium 106.4	47 **Ag** Silver 107.9	48 **Cd** Cadmium 112.4	49 **In** Indium 114.8	50 **Sn** Tin 118.7	51 **Sb** Antimony 121.8	52 **Te** Tellurium 127.6	53 **I** Iodine 126.9	54 **Xe** Xenon 131.3
55 **Cs** Cesium 132.9	56 **Ba** Barium 137.3	57 **La** Lanthanum 138.9	72 **Hf** Hafnium 178.5	73 **Ta** Tantalum 180.9	74 **W** Tungsten 183.9	75 **Re** Rhenium 186.2	76 **Os** Osmium 190.2	77 **Ir** Iridium 192.2	78 **Pt** Platinum 195.1	79 **Au** Gold 197.0	80 **Hg** Mercury 200.6	81 **Tl** Thallium 204.4	82 **Pb** Lead 207.2	83 **Bi** Bismuth 209.0	84 **Po** Polonium (210)	85 **At** Astatine (210)	86 **Rn** Radon (222)
87 **Fr** Francium (223)	88 **Ra** Radium (226)	89 **Ac** Actinium (227)	104 **Rf** Rutherfordium (257)	105 **Db** Dubnium (260)	106 **Sg** Seaborgium (263)	107 **Bh** Bohrium (262)	108 **Hs** Hassium (265)	109 **Mt** Meitnerium (266)	110 **Ds** Darmstadtium (269)	111 **Rg** Roentgenium (272)	112 **Cn** Copernicium (285)	113 **Nh** Nihonium (286)	114 **Fl** Flerovium (289)	115 **Mc** Moscovium (289)	116 **Lv** Livermorium (293)	117 **Ts** Tennessine (294)	118 **Og** Oganesson (294)

58 **Ce** Cerium 140.1	59 **Pr** Praseodymium 140.9	60 **Nd** Neodymium 144.2	61 **Pm** Promethium (147)	62 **Sm** Samarium 150.4	63 **Eu** Europium 152.0	64 **Gd** Gadolinium 157.3	65 **Tb** Terbium 158.9	66 **Dy** Dysprosium 162.5	67 **Ho** Holmium 164.9	68 **Er** Erbium 167.3	69 **Tm** Thulium 168.9	70 **Yb** Ytterbium 173.0	71 **Lu** Lutetium 175.0
90 **Th** Thorium 232.0	91 **Pa** Protactinium (231)	92 **U** Uranium 238.0	93 **Np** Neptunium (237)	94 **Pu** Plutonium (242)	95 **Am** Americium (243)	96 **Cm** Curium (247)	97 **Bk** Berkelium (247)	98 **Cf** Californium (249)	99 **Es** Einsteinium (254)	100 **Fm** Fermium (253)	101 **Md** Mendelevium (256)	102 **No** Nobelium (254)	103 **Lr** Lawrencium (257)

Metals

Metalloids

Nonmetals

The 1–18 group designation has been recommended by the International Union of Pure and Applied Chemistry (IUPAC).

Elements up to 122 have been reported. Of these, some of the higher-numbered elements have been observed only under laboratory conditions. Four such elements, 113, 115, 117, and 118, were officially confirmed in December 2015 and have been named: nihonium, moscovium, tennessine, and organesson, respectively.

APPENDIX E

Figure Question Answers

Chapter 1

Figure 1.6: The heat would come on until the room temperature reached the new set point.
Figure 1.13: The hand is more lateral than the hip (remember that anatomical position is the reference).

Chapter 2

Figure 2.11: The solution at pH 6.4 has 100 times the hydrogen ion concentration of a solution at pH 8.4.
Figure 2.14: Fats are a subgroup of lipids. Other lipids include the steroids and the phospholipids.

Chapter 3

Figure 3.10: chromatin and nucleolus
Figure 3.18: A carrier protein and cellular energy (ATP) are required for active transport. Synthesis of ATP stops after death, and Na^+–K^+ pumps require ATP to function, while osmosis does not.

Chapter 4

Figure 4.10: in the cytosol
Figure 4.14: messenger RNA

Chapter 5

Figure 5.8: cross section, although the lumen is not shown completely and has an irregular shape
Figure 5.21: fluid

Chapter 6

Figure 6.2: palms of hands, soles of feet
Figure 6.4: lunula

Chapter 7

Figure 7.4: receptors (detect changes in blood Ca^{2+} levels), control center (compares changes to set point), effectors (glands secrete hormones to bring about responses)
Figure 7.28: Female obturator foramen is more triangular; male is more oval. Female pubic arch is a wider (greater) angle than the male pubic arch. Female ilia are more flared than those on the male hip bones.

Chapter 8

Figure 8.5: Neurotransmitters cross the synaptic cleft by diffusion.
Figure 8.8: Their lengths stay the same.

Chapter 9

Figure 9.12: The action potentials of myelinated axons occur only at the nodes of Ranvier.
Figure 9.33: inferior

Chapter 10

Figure 10.6: temporal bone
Figure 10.16: Yes, they are skeletal muscles. Look up at the ceiling without moving your head if you have a question about this!

Chapter 11

Figure 11.2: specific receptors for the chemicals to which they respond
Figure 11.15: exocrine

Chapter 12

Figure 12.2: With the loss of water, hematocrit will increase. The actual number of RBCs would be unaffected.
Figure 12.6: RBC production and oxygen-carrying capacity will be decreased if the kidneys are unable to efficiently produce erythropoietin.
Figure 12.15: The conversion of soluble fibrinogen into insoluble fibrin is the major event in forming a blood clot. A person with hypocalcemia may have difficulty forming a blood clot if injured because Ca^{2+} is required for blood coagulation.

Chapter 13

Figure 13.12: ventricular depolarization and atrial repolarization
Figure 13.21: plasma proteins
Figure 13.30: radial artery and ulnar artery
Figure 13.33: axillary vein

Chapter 14

Figure 14.4: thoracic duct
Figure 14.17: secondary immune response

Chapter 15

Figure 15.2: 19.7 inches (based on the measurements in the figure)
Figure 15.19: The pancreas is stimulated to secrete digestive enzymes.
Figure 15.30: The major secretion from these cells in the large intestinal wall is mucus to protect against the abrasiveness of material flowing through the tube. The large intestine needs more lubrication because its contents are less liquid than those of the small intestine.

Chapter 16

Figure 16.8: The pulmonary artery carries oxygen-poor blood.
Figure 16.15: upward–refer to the vertical axis on the left side of the figure

Chapter 17

Figure 17.5: The afferent arteriole has a larger diameter to ensure adequate blood flow into the glomerulus.
Figure 17.11: afferent arteriole
Figure 17.13: high blood pressure (hypertension)
Figure 17.14: tubular reabsorption
Figure 17.18: reduce urine flow

Chapter 18

Figure 18.2: potassium ion
Figure 18.8: Because of the bicarbonate-buffer reaction ($CO_2 + H_2O \leftrightarrow H_2CO_3 \leftrightarrow HCO_3^- + H^+$), when pH is lower than 7.35, exhaling more CO_2 lowers the hydrogen ion concentration in the plasma (as the reaction proceeds to the left), reestablishing the normal internal environment pH range of 7.35–7.45.

Chapter 19

Figure 19.3: It is important for the sperm and egg to each possess 23 chromosomes, so that upon fertilization of the egg by the sperm, the developing zygote will have the full complement of 46 chromosomes (no more, no less) found in human body cells.
Figure 19.11: corona radiata
Figure 19.15: oral contraceptives

Chapter 20

Figure 20.2: 23 chromosomes, each with two chromatids
Figure 20.13: umbilical vein, umbilical arteries, ductus venosus, ductus arteriosus, foramen ovale

GLOSSARY

A phonetic guide to pronunciation follows each glossary word. Any unmarked vowel that ends a syllable or stands alone as a syllable has the long sound. Thus, the word *play* is phonetically spelled *pla*. Any unmarked vowel followed by a consonant has the short sound. The word *tough*, for instance, is phonetically spelled *tuf*. If a long vowel appears in the middle of a syllable (followed by a consonant), it is marked with a macron (¯), the sign for a long vowel. Thus, the word *plate* is phonetically spelled *plāt*. Similarly, if a vowel stands alone or ends a syllable, but has a short sound, it is marked with a breve (˘).

A

abdominal (ab-dom'ĭ-nal) The region between the thorax and pelvis.

abdominal cavity (ab-dom'ĭ-nal kav'ĭ-te) Space between the diaphragm and the pelvic cavity that contains the abdominal viscera.

abdominopelvic cavity (ab-dom"ĭ-no-pel'vik kav'ĭ-te) Space between the diaphragm and the pelvic outlet that contains the abdominal and pelvic organs.

abduction (ab-duk'shun) Movement of a body part away from the midline.

accessory organ (ak-ses'o-re or'gan) Organ that supplements the functions of other organs.

accommodation (ah-kom"o-da'shun) Adjustment of the lens of the eye for close or distant vision.

acetylcholine (as"ĕ-til-ko'lēn) **(ACh)** Type of neurotransmitter, which is a chemical secreted into the synaptic clefts at the axon ends of neurons.

acetylcholinesterase (as"ĕ-til-ko"lin-es'ter-ās) Enzyme that catalyzes breakdown of acetylcholine.

acid (as'id) Substance that ionizes in water to release hydrogen ions.

acidic (as id'ik) A solution with pH less than 7.0.

acidosis (as"ĭ-do'sis) Decrease in the pH of body fluids below pH 7.35.

acromial (ah-kro'me-al) The point of the shoulder.

ACTH Adrenocorticotropic hormone.

actin (ak'tin) Protein in a muscle fiber that forms the thin filaments that slide between thick filaments of the protein myosin, contracting muscle fibers.

action potential (ak'shun po-ten'shal) Sequence of electrical changes in part of a nerve cell or muscle cell exposed to a stimulus of threshold or greater; an impulse.

active site (ak'tiv sīt) Part of an enzyme molecule that temporarily binds a substrate.

active transport (ak'tiv trans'port) Process that requires energy and a carrier molecule to move a substance across a cell membrane against the concentration gradient.

adaptation (ad"ap-ta-'shun) Ability of the nervous system to become less responsive to a sustained stimulus.

adaptive defense (a-dap'tiv dē-fenc) Specific defenses carried out by T and B cells.

adduction (ah-duk'shun) Movement of a body part toward the midline.

adenosine triphosphate (ah-den'o-sēn tri-fos'fāt) **(ATP)** Organic molecule that transfers energy, used in cellular processes.

ADH (an"tı˘-di"u-ret'ik hor'mōn) Antidiuretic hormone; also known as *vasopressin*.

adipose tissue (ad'ĭ-pōs tish'u) Fat-storing tissue.

adrenal cortex (ah-dre'nal kor'teks) Outer part of the adrenal gland.

adrenal gland (ah-dre'nal gland) Endocrine gland on the superior portion of each kidney. Adrenal hormones play roles in maintaining blood sodium levels and responding to stress. They also include certain sex hormones.

adrenal medulla (ah-dre'nal me-dul'ah) Inner part of the adrenal gland.

adrenergic fiber (ad"ren-er'jik fi'ber) Axon that secretes the neurotransmitter norepinephrine at its terminal.

adrenocorticotropic hormone (ad-re"no-kor"te-ko-trōp'ik hor'mōn) **(ACTH)** Hormone secreted by the anterior pituitary that stimulates activity in the adrenal cortex.

aerobic respiration (a"er-o'bik res"pĭ-ra'shun) Complete, energy-releasing, breakdown of glucose to carbon dioxide and water in the presence of oxygen.

afferent arteriole (af'er-ent ar-te're-ōl) Vessel that transports blood to the glomerulus of each kidney nephron.

agglutination (ah-gloo"tĭ-na'shun) Clumping of blood cells in response to a reaction between an antibody and an antigen.

agonist (ago'-nist) A muscle that causes a particular movement.

agranulocyte (a-gran'u-lo-sīt) Nongranular leukocyte.

albumin (al-bu'min) Plasma protein that contributes to the colloid osmotic pressure of plasma.

aldosterone (al-dos'ter-ōn") Adrenal cortical hormone that regulates plasma sodium and potassium ion concentrations and fluid volume.

alimentary canal (al"i-men'tar-e kah-nal') Tubular part of the digestive tract from the mouth to the anus.

alkaline (al'kah-līn) Basic; a solution with a pH greater than 7.0.

alkalosis (al"kah-lo'sis) Increase in the pH of body fluids above pH 7.45.

allantois (ah-lan'to-is) Structure in the embryo from which the umbilical cord blood vessels develop.

allele (ah-le'el) One of two or more forms of a gene.

allergen (al'er-jen) Chemical that triggers an allergic reaction.

alveolar duct (al-ve'o-lar dukt) Fine tube that conducts air to and from an alveolar sac of the lungs.

alveolar sac (al-ve'o-lar sak) A small outpouching at the end of an alveolar duct that has alveoli branching off of it.

alveolus (al-ve'o-lus) (plural, *alveoli*) Any of the microscopic air sacs of the lungs; a hollow grapelike structure.

amino acid (ah-me'no as'id) Organic compound that includes an amino group ($-NH_2$) and a carboxyl group ($-COOH$); the structural unit of a protein molecule.

amnion (am'ne-on) Extraembryonic membrane that encircles a fetus and amniotic fluid.

amniotic fluid (am"ne-ot'ik floo'id) Fluid in the amniotic cavity that surrounds the developing fetus.

ampulla (am-pul'ah) Expansion at the end of each semicircular duct that houses a crista ampullaris, the sensory organ of dynamic equilibrium.

anabolism (ah-nab'o-lizm) Synthesis of larger molecules from smaller ones; anabolic metabolism.

anaerobic respiration (an″er-o′bik res″pī-ra′shun) Energy-releasing reactions that can occur in the absence of molecular oxygen.

anaphase (an′ah-fāz) Stage in mitosis when replicated chromosomes separate and move to opposite poles of the cell.

anatomical position (an″ah-tom′e-kal po-zish′un) A standard position to which anatomical terminology refers, in which the body is standing erect, face forward, upper limbs at the sides, and palms forward.

anatomy (ah-nat′o-me) Branch of science dealing with the form and structure of body parts.

androgen (an′dro-jen) Male sex hormone such as testosterone.

antagonist (an-tag′o-nist) Muscle that opposes a particular movement.

antebrachial (an″te-bra′ke-al) Pertaining to the forearm.

antecubital (an″te-ku′bī-tal) Region anterior to the elbow joint.

anterior (an-te′re-or) Pertaining to the front; the opposite of *posterior.*

anterior pituitary (an-te′re-or pī-tu′ĭ-tār″e) Anterior lobe of the pituitary gland. It secretes growth hormone (GH), thyroid-stimulating hormone (TSH), adrenocorticotropic hormone (ACTH), prolactin (PRL), luteinizing hormone (LH), and follicle-stimulating hormone (FSH).

antibody (an′tĭ-bod″e) Protein that B cells of the immune system produce in response to a nonself antigen that then reacts with the antigen; also called *immunoglobulin*.

anticodon (an″tī-ko′don) Three nucleotides of a transfer RNA molecule that are complementary to a specific mRNA codon.

antidiuretic hormone (an″tī-di″u-ret′ik hor′mōn) **(ADH)** Hormone of the posterior pituitary gland that enhances water conservation in the kidneys. It is also called *vasopressin* because of its ability to cause vasoconstriction at high concentrations.

antigen (an′tĭ-jen) Chemical that triggers an immune response.

aorta (a-or′tah) Major systemic artery that receives blood directly from the left ventricle.

aortic sinus (a-or′tik si′nus) Swelling in the aortic wall, behind each cusp of the semilunar valve.

aortic valve (a-or′tik valv) Flaplike structure in the wall of the aorta near its origin that prevents blood from returning to the left ventricle of the heart; also called the *aortic semilunar valve.*

apocrine gland (ap′o-krin gland) Type of gland whose secretions contain parts of secretory cells.

aponeurosis (ap″o-nu-ro′sis) Sheet of connective tissue that attaches muscles to bone, skin, or to the coverings of adjacent muscles.

apoptosis (ap-o-toe′sis) Programmed cell death.

appendicular (ap″en-dik′u-lar) Pertaining to the upper or lower limbs.

aqueous humor (a′kwe-us hu′mor) Watery fluid that fills the anterior cavity of the eye.

arachnoid mater (ah-rak′noid ma′ter) Delicate, weblike middle layer of the meninges.

areolar tissue (ah-re′o-lar tish′u) Connective tissue composed mainly of protein fibers.

arrector pili muscle (ah-rek′tor pil′i mus′l) Smooth muscle in the skin associated with a hair follicle.

arteriole (ar-te′re-ōl) Small branch of an artery that communicates with a capillary network.

artery (ar′ter-e) Vessel that transports blood from the heart.

articular cartilage (ar-tik′u-lar kar′tĭ-lij) Hyaline cartilage that covers the ends of bones in synovial joints.

ascending tract (ah-send′ing trakt) Any of a number of bundles of axons in the spinal cord that conduct sensory impulses up to the brain.

association area (ah-so″se-a′shun a′re-ah) Any of the areas of the brain controlling memory, reasoning, judgment, and emotions.

astrocyte (as′tro-sīt) Type of neuroglia that connects neurons to blood vessels in the CNS.

atmospheric pressure (at″mos-fēr′ik presh′ur) Pressure exerted by the weight of air surrounding the earth, equivalent at sea level to the weight of a column of mercury 760 millimeters high.

atom (at′om) Smallest particle of an element that has the properties of that element.

atomic number (ah-tom′ik num′ber) Number of protons in the nucleus of an atom.

atomic weight (ah-tom′ik wāt) The combined number of protons and neutrons in the nucleus of an atom.

ATP Adenosine triphosphate.

ATPase Enzyme that transfers the energy stored in the terminal phosphate bonds of ATP molecules so it can be used by the cell.

atrioventricular bundle (a″tre-o-ven-trik′u-lar bun′dl) **(AV bundle)** Group of specialized muscle fibers that conducts impulses from the atrioventricular node to the ventricular muscle of the heart.

atrioventricular node (a″tre-o-ven-trik′u-lar nōd) **(AV node)** Specialized mass of cardiac muscle fibers in the interatrial septum of the heart that transmits cardiac impulses from the sinoatrial node to the AV bundle.

atrium (a′tre-um) (plural, *atria*) Either of the two upper chambers of the heart.

auditory ossicle (aw′di-to″re os′i-kl) Any of the three bones of the middle ear: the malleus, the incus, and the stapes.

auditory tube (aw′di-to″re tūb) Tube that connects the middle ear cavity to the pharynx; also called the *eustachian tube.*

auricle (aw′ri-kl) The external ear; also an outpouching of the wall of an atrium of the heart.

autocrine (aw″to krin) A secretion that affects only the cell secreting the substance.

autonomic nervous system (aw″to-nom′ik ner′vus sis′tem) **(ANS)** Motor portion of the peripheral nervous system that controls the viscera and operates without conscious effort.

autosome (aw′to-sōm) Any one of the 44 chromosomes that does not include a gene that determines sex.

axial (ak′se-al) Pertaining to the head, neck, and trunk.

axillary (ak′sĭ-ler″e) Pertaining to the armpit.

axon (ak′son) The process that conducts an impulse away from the neuron cell body. also called a *nerve fiber.*

B

B cell (b sel) Type of white blood cell that produces and secretes antibodies that bind and destroy nonself molecules; also called a *B lymphocyte*.

baroreceptor (bar″o-re-sep′tor) Sensory receptor in a blood vessel wall stimulated by changes in blood pressure.

basal nuclei (bas′al nu′kle-i) Masses of gray matter deep within each cerebral hemisphere that facilitate voluntary movement.

base (bās) Substance that ionizes in water, releasing hydroxide ions (OH^-) or other ions that combine with hydrogen ions.

basement membrane (bās′ment mem′brān) Layer of nonliving material that anchors epithelial tissue to underlying connective tissue.

basic (bās′ik) A solution with a pH greater than 7.0; alkaline

basophil (ba′so-fil) White blood cell containing cytoplasmic granules that react with basic stain.

beta oxidation (ba′tah ok″sĭ-da′shun) Chemical process that breaks fatty acids down to form acetyl coenzyme A, which can enter the citric acid cycle.

bilateral (bi″-lat′er-al) On both sides of the body.

bile (bīl) Fluid secreted by the liver and stored in the gallbladder that acts to emulsify fats.

bile duct (bīl dukt) Tube that transports bile from the cystic duct and common hepatic duct to the duodenum.

bilirubin (bil″ĭ-roo′bin) A bile pigment produced from hemoglobin breakdown.

biliverdin (bil″ĭ-ver′din) A bile pigment produced from hemoglobin breakdown.

blastocyst (blas′to-sist) Early stage of prenatal development when the embryo is a hollow ball of cells.

blood (blud) A connective tissue consisting of formed elements in a liquid matrix called plasma that circulates through the heart and vessels carrying substances throughout the body.

bone (bōn) Any of the 206 individual parts of the skeleton composed of cells and a matrix containing inorganic mineral salts and organic protein fibers; also a type of connective tissue and a type of cell.

brachial (bra′ke-al) Pertaining to the arm.

brainstem (brānstem) Part of the brain that includes the midbrain, pons, and medulla oblongata.

bronchial tree (brong′ke-al trē) The bronchi and their branches that carry air between the trachea and the alveoli of the lungs.

bronchiole (brong′ke-ōl) Small branches of bronchi in the lung.

bronchus (brong′kus) (plural, *bronchi*) Tube branching from the trachea that leads to a lung.

buccal (buk′al) Pertaining to the mouth and inner lining of the cheeks.

buffer (buf′er) Substance that can combine with hydrogen ions when they are in excess, or release them when they are in short supply, thus resisting a change in pH. Alternatively, substance that can react with a strong acid or base to form a weaker acid or base and thus resist a change in pH.

bulbourethral gland (bul″bo-u-re′thral gland) Gland that secretes viscous fluid into the male urethra during sexual excitement; also called *Cowper's gland*.

bursa (ber′sah) (plural, *bursae*) Saclike, synovial fluid-filled cushioning structure, lined with synovial membrane, near a joint.

C

calcaneal (cal-ca-ne-al) Pertaining to the heel.

calcitonin (kal″sĭ-to′nin) Hormone secreted by the thyroid gland that helps regulate the blood calcium and phosphate concentrations.

calorie (kal′o-re) Unit that measures heat energy and the energy content of foods.

cAMP cyclic adenosine monophosphate.

canaliculus (kan″ah-lik′u-lus) Microscopic canal that connects lacunae of bone tissue.

capacitation (kah-pas″i-ta′shun) Activation of a sperm cell to fertilize an egg cell.

capillary (kap′ĭ-lar″e) Small blood vessel that connects an arteriole and a venule and allows diffusional exchange of nutrients and waste.

carbaminohemoglobin (kar-bam″ĭ-no-he″mo-glo′bin) Carbon dioxide bound to the amino groups of hemoglobin.

carbohydrate (kar″bo-hi′drāt) Organic compound consisting of carbon, hydrogen, and oxygen, in a 1:2:1 ratio.

carbonic anhydrase (kar-bon′ik an-hi′drās) Enzyme that catalyzes the reversible reaction between carbon dioxide and water to form carbonic acid.

carboxypeptidase (kar-bok″se-pep′tĭ-dās) Protein-splitting enzyme in pancreatic juice.

cardiac conduction system (kar′de-ak kon-duk′shun sis′tem) System of specialized cardiac muscle fibers that conducts cardiac impulses from the SA node into the myocardium.

cardiac cycle (kar′de-ak si′kl) Sequence of myocardial contraction and relaxation that constitutes a complete heartbeat.

cardiac muscle tissue (kar′de-ak mus′l tish′u) Specialized muscle tissue found only in the heart.

cardiac output (kar′de-ak owt′poot) The volume of blood per minute that the heart pumps (stroke volume in milliliters multiplied by heart rate in beats per minute).

cardiovascular (kahr″de-o-vas′ku-lur) Pertaining to the heart and blood vessels.

cardiovascular system (kahr″de-o-vas′ku-lur sis′tem) The organ system that includes the heart and the blood vessels.

carpal (kar′pal) Pertaining to the wrist or any of the individual wrist bones.

cartilage (kar′ti-lij) Type of connective tissue in which cells are in lacunae separated by a semisolid extracellular matrix.

cartilaginous joint (kar-tĭ-laj′ĭ-nus joint) Two or more bones joined by cartilage.

catabolism (kă-tab′o-lizm) Breakdown of large molecules; also called *catabolic metabolism*.

catalyst (kat′ah-list) Chemical that increases the rate of a chemical reaction, but is not permanently altered by the reaction.

celiac (se′le-ak) Pertaining to the abdomen.

cell (sel) The structural and functional unit of an organism.

cell body (sel bod′e) Part of a neuron that includes cytoplasm and a nucleus, and from which nerve processes extend.

cell cycle (sel si′kl) The series of events that a cell undergoes, from when it forms until when it divides.

cell membrane (sel mem′brān) The selectively permeable outer boundary of a cell consisting of a phospholipid bilayer embedded with proteins; also called a *plasma membrane*.

cellular immune response (sel′u-lar i-mūn′ ri-spons′) The body's attack by T cells and their secreted products on nonself antigens; also called *cell-mediated immunity*.

cellular respiration (sel′u-lar res″pĭ-ra′shun) A biochemical pathway that releases energy from organic compounds.

cellulose (sel′u-lōs) Polysaccharide abundant in plant tissues that human digestive enzymes cannot break down.

cementum (se-men′tum) Bonelike material that surrounds the root of a tooth.

central canal (sen′tral kah-nal′) Tiny channel in bone tissue that houses blood vessels and nerve fibers; tube in the spinal cord continuous with brain ventricles and containing cerebrospinal fluid.

central nervous system (sen′tral ner′vus sis′tem) **(CNS)** The brain and spinal cord.

centriole (sen′tre-ōl) Cellular structure built of microtubules that organizes the mitotic spindle.

centromere (sen′tro-mēr) Region of chromosome where spindle fibers attach during cell division; region where chromatids are attached.

centrosome (sen′tro-sōm) Cellular organelle, consisting of two centrioles, that separates chromosome sets into two cells from one during cell division.

cephalic (sĕ-fal′ik) Pertaining to the head.

cerebellar cortex (ser″ĕ-bel′ar kor′teks) Outer, gray matter layer of the cerebellum.

cerebellum (ser″ĕ-bel′um) Part of the brain that coordinates skeletal muscle movement.

cerebral cortex (ser′ĕ-bral kor′teks) Outer, gray-matter layer of the cerebrum.

cerebral hemisphere (ser′ĕ-bral hem′ĭ-sfēr) Either of the large, paired structures that constitute the cerebrum.

cerebrospinal fluid (ser″ĕ-bro-spi′nal floo′id) **(CSF)** Fluid in the ventricles of the brain, subarachnoid space of the meninges, and the central canal of the spinal cord.

cerebrum (ser′ĕ-brum) Part of the brain in the upper part of the cranial cavity that provides higher mental functions.

cervical (ser′vĭ-kal) Pertaining to the neck.

cervix (ser′viks) Narrow, inferior end of the uterus that leads into the vagina.

chemical bond (kem′ikel bond) Attractive force holding atoms together.

chemistry (kem′is-trē) The study of the composition of matter.

chemoreceptor (ke″mo-re-sep′tor) Receptor stimulated by the binding of certain chemicals.

chief cell (chēf sel) Cell type in a gastric gland that secretes digestive enzymes.

cholecystokinin (ko″le-sis″to-ki′nin) Hormone the small intestine secretes that stimulates release of pancreatic juice from the pancreas and bile from the gallbladder.

cholesterol (ko-les′ter-ol) A lipid produced in the body and acquired from food that cells use to synthesize steroid hormones.

cholinergic fiber (ko″lin-er′jik fi′ber) Axon that secretes the neurotransmitter acetylcholine at its terminal.

chondrocyte (kon′dro-sīt) A cartilage cell.

chordae tendineae (kor′de ten′dĭ-ne) Fibrous strings attached to the cusps of the tricuspid and mitral valves in the heart.

chorion (ko′re-on) Extraembryonic membrane that forms the outermost covering around a fetus and contributes to formation of the placenta.

chorionic villus (ko″re-on′ik vil′us) Projection that extends from the outer surface of the chorion and helps attach an embryo to the uterine wall.

choroid coat (ko′roid kōt) Vascular, pigmented middle layer of the wall of the eye.

choroid plexus (ko′roid plek′sus) Any of several masses containing specialized capillaries that secretes cerebrospinal fluid into a brain ventricle.

chromatid (kro′mah-tid) One longitudinal half of a replicated chromosome.

chromatin (kro′mah-tin) DNA and complexed proteins that condense to form chromosomes during mitosis.

chromatophilic substance (kro″mah-to-fil′ik sub′stans) **(Nissl bodies)** Membranous sacs in the cytoplasm of neurons that have ribosomes attached to their surfaces.

chromosome (kro′mo-sōm) Threadlike structure built of DNA and protein that condenses and becomes visible in a cell's nucleus during cell division.

chylomicron (ki″lo-mi′kron) Microscopic droplet of fat encased in protein that forms during fat absorption.

chyme (kīm) Semifluid mass of partially digested food that passes from the stomach to the small intestine.

chymotrypsin (ki″mo-trip′sin) Protein-splitting enzyme in pancreatic juice.

cilia (sil′e-ah) Microscopic, hairlike extensions on the exposed surfaces of certain epithelial cells.

ciliary body (sil′e-er″e bod′e) Structure associated with the choroid layer of the eye that secretes aqueous humor and houses the ciliary muscle.

circadian rhythm (ser″kah-de′an rithm) Pattern of repeated behavior associated with cycles of night and day.

cisterna (sis-ter′nah) (plural, *cisternae*) Enlarged portion of the sarcoplasmic reticulum near the actin and myosin filaments of a striated muscle fiber.

citric acid cycle (sit′rik as′id si′kl) Series of chemical reactions that oxidizes certain molecules, releasing energy; Krebs cycle.

cleavage (klēv′ij) Early successive divisions of the zygote into a ball of progressively smaller cells.

clitoris (klit′o-ris) Small, erectile organ in the anterior vulva of the female, corresponding to the penis in the male.

clone (klōn) Group of cells that descend from a single cell and are therefore genetically identical to it.

CNS Central nervous system.

coagulation (ko-ag″u-la′shun) Blood clotting.

cochlea (kok′le-ah) Portion of inner ear that has hearing receptors.

codon (ko′don) Set of three contiguous nucleotides of a messenger RNA molecule that specifies a particular amino acid.

coenzyme (ko-en′zīm) Nonprotein organic molecule required for the activity of a particular enzyme.

cofactor (ko′fak-tor) Small molecule or ion that must combine with an enzyme for the enzyme to be active.

collagen fiber (kol′ah-jen fi′ber) White fiber consisting of the protein collagen, common in connective tissues, including bone matrix.

common hepatic duct (kom′mon hepat′ik dukt) Tube that transports bile from the liver to the bile duct.

compact bone (kom′-pakt bōn) Dense bone tissue in which cells are organized in osteons without apparent spaces; cortical bone.

complement (kom′ple-ment) Group of proteins activated when an antibody binds an antigen; enhances reaction against nonself substances.

complete protein (kom-plēt′ pro′te-in) Protein that contains adequate amounts of the essential amino acids to maintain body tissues and to promote normal growth and development.

complete tetanic contraction (tĕ-tan′ik kon-trak′shun) Continuous, forceful muscular contraction without relaxation.

compound (kom′pownd) Substance composed of two or more chemically bonded atoms of different elements.

cone (kōn) Type of light receptor that provides color vision.

conformation (kon″for-ma″shen) Three-dimensional form of a protein.

conjunctiva (kon″junk-ti′vah) Membranous covering lining the eyelids and on the anterior surface of the sclera of the eye covered by the eyelids.

connective tissue (kŏ-nek′tiv tish′u) Basic tissue type that consists of cells within an extracellular matrix, including bone, cartilage, blood, and loose and dense connective tissues.

contralateral (kon″trah-lat′er-al) On the opposite side of the body.

convergence (kon-ver′jens) A coming together. Two or more presynaptic neurons forming synapses with the same postsynaptic neuron.

cornea (kor′ne-ah) Transparent anterior portion of the outer layer of the wall of the eye.

coronary artery (kor′o-na″re ar′ter-e) An artery that supplies blood to the wall of the heart.

coronary sinus (kor′o-na″re si′nus) Large vessel on the posterior surface of the heart into which cardiac veins drain.

corpus callosum (kor′pus kah-lo′sum) Mass of white matter in the brain composed of nerve fibers connecting the right and left cerebral hemispheres.

corpus luteum (kor′pus loot′e-um) Structure that forms from the tissues of a ruptured ovarian follicle and secretes progesterone and estrogens.

cortisol (kor′tĭ-sol) Glucocorticoid hormone secreted by the adrenal cortex.

costal (kos′tal) Pertaining to the ribs.

covalent bond (ko′va-lent bond) Chemical bond formed by electron sharing between atoms.

coxal (kok′sal) The hip. Pertaining to the hip bone.

cranial cavity (kra′ne-al kav′i-te) Space in the skull containing the brain.

cranial nerve (kra′ne-al nerv) Any of the nerves that arise from the brain or brainstem.

creatine phosphate (kre′ah-tin fos′fāt) Molecule in muscle that stores energy.

crista ampullaris (kris′tah am-pul′ar-is) Sensory organ in a semicircular canal that functions in the sense of dynamic equilibrium.

crural (kroor′al) The leg. Pertaining to the leg.

cubital (ku′bi-tal) Pertaining to the elbow.

cutaneous (ku-ta′ne-us) Pertaining to the skin.

cyclic AMP (si′klik a-em-pē) **(cAMP)** A second messenger molecule in a signal transduction pathway.

cystic duct (sis′tik dukt) Tube that connects the gallbladder to the bile duct.

cytocrine secretion (si′to-krin se-kre′shun) Transfer of melanin granules from melanocytes into epithelial cells.

cytokinesis (si″to-kī-ne′sis) Division of the cytoplasm during cell division.

cytoplasm (si′to-plazm) The contents of a cell including the gel-like cytosol and organelles, excluding the nucleus, enclosed by the cell membrane.

cytoskeleton (si″to-skel′e-ten) A cell's framework of protein filaments and tubules.

D

deamination (de-am″ĭ-na′shun) Removing amino groups ($-NH_2$) from amino acids.

decomposition (de″-kom-po-zish′un) Breakdown of molecules.

deep (dēp) More internal, not near the surface.

dehydration synthesis (de″hi-dra′shun sin′thĕ-sis) Anabolic process that joins small molecules by releasing the equivalent of a water molecule; also called *synthesis*.

dendrite (den′drīt) Process of a neuron that receives input from other neurons.

dense connective tissue (dens kō-nek′tiv tish′u) A connective tissue with many collagenous fibers, a fine network of elastin fibers, and sparse fibroblasts.

dentin (den′tin) Bonelike substance that forms the bulk of a tooth beneath the enamel.

deoxyhemoglobin (de-ok″se-he″mo-glo′bin) Hemoglobin that has not bound oxygen.

deoxyribonucleic acid (de-ok′si-ri″bo-nu-kle″ik as′id) **(DNA)** The genetic material; a double-stranded polymer of nucleotides, each containing a phosphate group, a nitrogenous base (adenine, thymine, guanine, or cytosine), and the sugar deoxyribose.

depolarized (de-po″lar-īzd) Condition in which the voltage difference across a cell membrane becomes less negative on the inside (more positive) than the resting potential.

dermis (der′mis) The thick layer of the skin beneath the epidermis.

descending tract (de-send′ing trakt) Any of a number of bundles of axons that conduct motor impulses from the brain down through the spinal cord.

detrusor muscle (de-trūz′or mus′l) Muscular layer of the wall of the urinary bladder.

diapedesis (di″ah-pĕ-de′sis) Movement of leukocytes between the cells of blood vessel walls.

diaphragm (di′ah-fram) (1) A sheetlike structure largely composed of skeletal muscle and connective tissue that separates the thoracic and abdominal cavities. (2) A contraceptive device inserted in the vagina.

diaphysis (di-af′ĭ-sis) Shaft of a long bone.

diastole (di-as′to-le) Phase of the cardiac cycle when a heart chamber wall relaxes.

diastolic pressure (di-a-stol′ik presh′ur) Lowest arterial blood pressure during the cardiac cycle; occurs during diastole.

diencephalon (di″en-sef′ah-lon) Part of the brain in the region of the third ventricle that includes the thalamus, hypothalamus, pineal gland, and other structures.

differentiation (dif″er-en″she-a′shun) Cell specialization.

diffusion (dĭ-fu′zhun) Random movement of molecules from a region of higher concentration toward one of lower concentration.

digestion (di-jest′yun) Breaking down of large nutrient molecules into molecules small enough to be absorbed; occurs by *hydrolysis*.

digestive system (di-jest′iv sis′tem) The organ system that includes the mouth, tongue, teeth, salivary glands, pharynx, esophagus, stomach, liver, gallbladder, pancreas, small intestine, and large intestine.

digital (dij′ĭ-tal) Pertaining to the finger or toe.

dipeptide (di-pep′tīd) Molecule composed of two joined amino acids.

disaccharide (di-sak′ah-rīd) Sugar produced by the union of two monosaccharides.

distal (dis′tal) Further from a point of attachment; opposite of *proximal*.

divergence (di-ver′jens) Spreading apart. A single presynaptic neuron forming synapses with two or more postsynaptic neurons.

DNA Deoxyribonucleic acid; the genetic material.

dominant allele (dom′eh-nant ah-lēl) An allele that masks another allele.

dorsal (dors′al) Pertaining to the back surface of a body part.

dorsal root (dor′sal root) Sensory branch of a spinal nerve by which it joins the spinal cord.

ductus arteriosus (duk′tus ar-te″re-o′sus) Blood vessel that connects the pulmonary artery and the aorta in a fetus.

ductus deferens (duk′tus def″er-ens) (plural, *ductus deferentia*) Tube that leads from the epididymis to the urethra of the male reproductive tract.

ductus venosus (duk′tus ven-o′sus) Blood vessel that connects the umbilical vein and the inferior vena cava in a fetus.

dura mater (du′rah ma′ter) Tough outer layer of the meninges.

dynamic equilibrium (di-nam′ik e″kwī-lib′re-um) Maintenance of balance when the head and body are suddenly rotated or otherwise moved.

E

eardrum (er′drum) The tympanic membrane, a thin membrane that is part of the external ear. It separates the external ear from the middle ear.

eccrine gland (ek′rin gland) Sweat gland that maintains body temperature.

ECG Electrocardiogram.

ectoderm (ek′to-derm) Outermost primary germ layer of the embryo.

edema (ĕ-de′mah) Fluid accumulation in tissue spaces.

effector (e-fek′tor) A muscle or gland, either of which can effect change in the body.

efferent arteriole (ef′er-ent ar-te′re-ōl) The vessel that transports blood away from the glomerulus of each nephron.

egg (eg) The oocyte and its surrounding layers at any stage before feritilization.

ejaculation (e-jak″u-la′shun) Discharge of semen from the male urethra.

elastic cartilage (ē-las-tik kar′-ti-lij) Opaque, flexible connective tissue with many elastic fibers.

elastic fiber (e-las′tic fi′ber) Stretchy, yellow connective tissue fiber consisting of the protein elastin.

electrocardiogram (e-lek″tro-kar′de-o-gram″) **(ECG)** Recording of the electrical activity associated with the cardiac cycle.

electrolyte (e-lek′tro-līt) Substance that dissociates to release ions in water.

electrolyte balance (e-lek′tro-līt bal′ans) Condition when the quantities of electrolytes entering the body equal those leaving it.

electron (e-lek′tron) Small, negatively charged particle that encircles the nucleus of an atom.

electron transport chain (e-lek′tron tranz′port chān) Series of metabolic reactions that takes high-energy electrons from glycolysis and the citric acid cycle to form ATP, water, CO_2, and heat.

element (el′ĕ-ment) Any of the fundamental chemical substances, each characterized by a distinct type of atom.

embolus (em′bo-lus) Blood clot, gas bubble, or other object carried in circulation that may obstruct a blood vessel.

embryo (em′bre-o) A prenatal stage of development from fertilization through the eighth week of development, when the rudiments of all organs are present.

embryonic disc (em″brē-on′ik disk) Flattened area in the developing embryo from which the germ layers arise.

emission (e-mish′un) Movement of sperm cells and prostate and seminal vesicle secretions into the urethra.

emulsification (e-mul″sĭ-fĭ-ka′shun) Breaking up of fat globules into smaller droplets by the action of bile salts.

enamel (e-nam′el) Hard covering on the exposed surface of a tooth.

endocardium (en″do-kar′de-um) Inner lining of the heart chambers.

endochondral bone (en″do-kon′dral bōn) Bone that begins as hyaline cartilage that is subsequently replaced by bone tissue.

endocrine gland (en′do-krin gland) Gland that secretes hormones into the bloodstream; hormone-secreting gland.

endocrine system (en′do-krin sis′tem) The organ system that includes the glands that secrete hormones into the blood.

endocytosis (en″do-si-to′sis) Process by which a cell membrane envelops a substance and draws it into the cell in a vesicle.

endoderm (en′do-derm) The innermost primary germ layer in the embryo.

endolymph (en′do-limf) Fluid in the membranous labyrinth of the inner ear.

endometrium (en″do-me′tre-um) Inner lining of the uterus.

endomysium (en″do-mis′e-um) Sheath of connective tissue surrounding each skeletal muscle fiber.

endoplasmic reticulum (en′do-plaz-mik rĕ-tik′u-lum) Organelle composed of a network of connected membranous tubules and vesicles.

endosteum (en-dos′te-um) Tissue lining the medullary cavity in a bone.

endothelium (en″do-the′le-um) Layer of epithelial cells that forms the inner lining of blood vessels and heart chambers.

energy (en′er-je) An ability to move something and thus do work.

enzyme (en′zīm) Protein that catalyzes (speeds) a specific biochemical reaction.

eosinophil (e″o-sin′o-fil) White blood cell containing cytoplasmic granules that turn deep red with acidic stain.

ependymal cells (ĕ-pen′dĭ-mahl selz) Neuroglia that line the ventricles of the brain and the central canal of the spinal cord.

epicardium (ep″ĭ-kar′de-um) Visceral part of the pericardium on the surface of the heart.

epidermis (ep″i-der′mis) Outer, epithelial layer of the skin.

epididymis (ep″ĭ-did′ĭ-mis) (plural, *epididymides*) Highly coiled tube that leads from the seminiferous tubules of the testis to the ductus deferens.

epidural space (ep″ĭ-du′ral spās) Space between the dural sheath of the spinal cord and the bone of the vertebral canal.

epigastric region (ep″ĭ-gas′trik re′jun) Upper middle part of the abdomen.

epiglottis (ep″ĭ-glot′is) Flaplike structure of elastic cartilage and mucous membrane at the back of the tongue that covers the opening to the trachea during swallowing.

epimysium (ep″i-mis′e-um) Sheath of connective tissue, deep to the fascia, surrounding a skeletal muscle.

epinephrine (ep″ĭ-nef′rin) Hormone the adrenal medulla secretes during times of stress.

epiphyseal plate (ep″ĭ-fiz′e-al plāt) Cartilaginous layer between the epiphysis and diaphysis of a long bone that grows, lengthening the bone.

epiphysis (ĕ-pif′ĭ-sis) (plural, *epiphyses*) Either end of a long bone.

epithelial tissue (ep″ĭ-the′le-al tish′u) One of the basic types of tissue; it covers all free body surfaces. Varieties are classified by cell shape (squamous, cuboidal, or columnar) and number of layers (simple, stratified, or pseudostratified).

erythrocyte (ĕ-rith′ro-sīt) Red blood cell.

erythropoietin (ĕ-rith″ro-poi′ĕ-tin) **(EPO)** Hormone secreted by kidney and liver cells that promotes red blood cell formation.

esophagus (ĕ-sof′ah-gus) Tubular part of the digestive tract connecting the pharynx to the stomach.

essential amino acid (ĕ-sen′shal ah-me′no as′id) Amino acid required for health that body cells cannot synthesize in adequate amounts; must be obtained in the diet.

essential fatty acid (ĕ-sen′shal fat′e as′id) Fatty acid required for health that body cells cannot synthesize in adequate amounts; must be obtained in the diet.

estrogens (es′tro-jenz) Group of hormones (including estradiol, estrone, and estriol) that stimulates the development of female secondary sex characteristics and produces an environment suitable for fertilization, implantation, and growth of an embryo.

eumelanin (u-mel′ah-nin) Brownish-black pigment.

eversion (e-ver′zhun) Turning the plantar surface of the foot outward, away from the midline.

exchange reaction (eks-chānj re-ak′shun) Chemical reaction in which parts of two kinds of molecules trade positions.

exocrine gland (ek′so-krin gland) Gland that secretes its products into a duct or onto an outside body surface.

exocytosis (ek″so-si-tō′sis) Transport of substances out of a cell in membrane-bounded vesicles.

exome The part of the genome (about 1.5%) that encodes protein.

expiration (ek″spĭ-ra′shun) Breathing out; also called *exhalation*.

extension (ek-sten′shun) Movement increasing the angle between bones at a joint.

extracellular fluid (eks″trah-sel′u-lar floo′id) Body fluid outside cells.

extracellular matrix (eks″trah-sel′u-lar ma′triks) Fibers and ground substance in spaces between cells, especially between connective tissue cells.

extrapyramidal tract (ek″strah-pĭ-ram′i-dal trakt) Nerve tracts, other than the corticospinal tracts, that transmit impulses from the cerebral cortex to the spinal cord.

F

facilitated diffusion (fah-sil″ĭ-tāt′ed dĭ-fu′zhun) Diffusion in which a carrier molecule transports a substance across a cell membrane from a region of higher concentration to a region of lower concentration.

facilitation (fah-sil″ĭ-ta′shun) Repeated impulses on an excitatory presynaptic neuron may cause that neuron to release more neurotransmitter in response to a single impulse, making it more likely to bring the postsynaptic cell to threshold.

fascia (fash′e-ah) Sheet of dense connective tissue that separates individual muscles and helps hold them in position.

fascicle fascicle (fas′ĭ-k′l) (plural, *fascicles, fasciculi*) Small bundle of skeletal muscle fibers; also called a *fasciculus*

fat (fat) Adipose tissue; an organic molecule that includes glycerol and fatty acids.

fatty acid (fat′e as′id) Building block of a triglyceride (fat) molecule.

feces (fe′sēz) Material expelled from the digestive tract during defecation.

femoral (fem′or-al) Pertaining to the thigh.

fertilization (fer″tĭ-lĭ-za′shun) Union of a secondary oocyte and a sperm cell.

fetus (fe′tus) Prenatal human after eight weeks of development.

fibrin (fi′brin) Insoluble, fibrous protein formed from fibrinogen during blood coagulation.

fibrinogen (fi-brin′o-jen) Plasma protein converted into fibrin during blood coagulation.

fibroblast (fi′bro-blast) Cell that produces protein fibers in connective tissues.

fibrocartilage (fi′bro-kar′tĭ-lij) Strongest and most durable cartilage; made up of cartilage cells and many collagen fibers.

fibrous joint (fi′brus joint) Two or more bones joined by dense connective tissue.

filtration (fil-tra′shun) Movement of material through a membrane as a result of hydrostatic pressure.

fissure (fish′ur) Narrow cleft separating parts, such as the lobes of the cerebrum.

flagellum (fla-jel′um) (plural, *flagella*) Relatively long, motile process that extends from the surface of a sperm cell.

flexion (flek′shun) Movement decreasing the angle between bones at a joint.

follicle-stimulating hormone (fol′ĭ-kl stim′u-la″ting hor′mōn) **(FSH)** Hormone secreted by the anterior pituitary that stimulates development of an ovarian follicle in a female, or sperm cell production in a male.

follicular cell (fŏ-lik′u-lar sel) Ovarian cell that surrounds a developing oocyte and secretes female sex hormones. Thyroid gland cell that produces the thyroid hormones triiodothyronine (T3) and tetraiodothyronine (T4).

fontanel (fon″tah-nel′) Membranous region between certain developing cranial bones in the skull of a fetus or infant.

foramen magnum (fo-ra′men mag′num) Large opening in the occipital bone of the skull through which the spinal cord passes.

foramen ovale (fo-ra′men o-val′e) Opening in the interatrial septum of the fetal heart.

fovea centralis (fo′ve-ah sen-tral′is) Depressed region of the retina at the center of the macula lutea, consisting of densely packed cones. It provides the sharpest color vision.

free nerve endings (frē nerv end-ingz) Branched dendrites of sensory neurons abundant in epithelium, associated with sensing itching, temperature, and pain.

frontal (frun′tal) Plane that divides a structure into anterior and posterior portions; pertaining to the forehead.

FSH Follicle-stimulating hormone.

functional syncytium (funk′shun-al sin-sish′e-um) A mass of cells performing as a unit; those of the heart are joined electrically.

G

gallbladder (gawl′blad-er) Saclike organ associated with the liver that stores and concentrates bile.

ganglion (gang′glē-on) (plural, *ganglia*) Mass of neuron cell bodies outside the central nervous system.

gastric gland (gas′trik gland) Any of the glands in the stomach lining that secretes gastric juice.

gastric juice (gas′trik joos) Secretion of the gastric glands in the stomach containing mucus, digestive enzymes, and hydrochloric acid.

gastrin (gas′trin) Hormone secreted by the stomach that stimulates gastric juice secretion.

gene (jēn) Part of a DNA molecule that encodes information to synthesize a protein, a control sequence, or tRNA or rRNA; the unit of inheritance.

genetic code (jĕ-net′ik kōd) Information for synthesizing proteins encoded in the nucleotide sequence of DNA molecules.

genital (jen′ĭ-tal) Pertaining to the external reproductive organs.

genome (je′nōm) Complete set of genetic instructions for an organism.

genotype (jē-no-tīp) The alleles (gene variants) of a particular gene in an individual.

GH Growth hormone.

globulin (glob′u-lin) Type of protein in blood plasma.

glomerular capsule (glo-mer′u-lar kap′sūl) Double-walled enclosure of the glomerulus of a nephron; also called *Bowman's capsule.*

glomerular filtrate (glo-mer′u-lar fil′trāt) Water and solutes filtered out of the glomerular capillaries in the kidney.

glomerular filtration (glo-mer′u-lar fil-tra′shun) Filtration of plasma by the glomerular capillaries of the nephron.

glomerulus (glo-mer′u-lus) Filtering capillary tuft in the renal corpuscle of a nephron.

glottis (glot′is) The true vocal folds (or vocal cords) and the opening between them.

glucagon (gloo′kah-gon) Hormone secreted by the pancreatic islets that raises the blood sugar concentration by stimulating the liver to break down glycogen and convert certain noncarbohydrates, such as amino acids, into glucose.

glucocorticoid (gloo″ko-kor′tĭ-koid) Any of several hormones that the adrenal cortex secretes that affects carbohydrate, fat, and protein metabolism.

glucose (gloo′kōs) Monosaccharide in the blood that is the primary source of cellular energy.

gluteal (gloo′te-al) Pertaining to the buttocks.

glycerol (glis′er-ol) Organic compound that is a building block for triglyceride (fat) molecules.

glycogen (gli′ko-jen) Polysaccharide that stores glucose primarily in the liver and muscles.

glycolysis (gli-kol′ĭ-sis) The energy-releasing breakdown of glucose to produce two pyruvic acid molecules and a net of two ATP.

goblet cell (gob′let sel) Epithelial cell specialized to secrete mucus.

Golgi apparatus (gol′je ap″ah-ra′tus) Organelle that prepares and modifies proteins and glycoproteins for secretion.

gonadotropin (go-nad″o-trōp′in) Hormone that stimulates activity in the gonads (testes and ovaries).

granulocyte (gran′u-lo-sīt) Leukocyte with granules in the cytoplasm.

growth hormone (grōth hor′mōn) **(GH)** Hormone secreted by the anterior pituitary that promotes growth of the organism; also called *somatotropin*.

gyrus (ji′rus) (pl. *gyri*) Elevation on the brain's surface; convolution.

H

hair follicle (hār fol′ĭ-kl) Tubelike depression in the skin where a hair develops.

haploid (hap′loyd) Having half the normal number of chromosomes, in humans 23.

hapten (hap′ten) Small molecule that combines with a larger one, forming an antigen.

hematocrit (he-mat′o-krit) **(HCT)** The percentage by volume of red blood cells in a sample of whole blood; packed cell volume (PCV).

hematopoiesis (he″-mă-to-poi-e′sis) Production of blood cells from dividing stem and progenitor cells.

heme (hēm) Iron-containing part of a hemoglobin molecule.

hemoglobin (he″mo-glo′bin) Oxygen-carrying protein in red blood cells.

hemostasis (he″mo-sta′sis) Stoppage of bleeding.

hepatic lobule (hĕ-pat′ik lob′ul) Functional unit of the liver.

heterozygous (het′er-o-zi′gus) Different alleles in a gene pair.

holocrine gland (ho′lo-krin gland) Gland whose secretion contains entire secretory cells.

homeostasis (ho″me-o-sta′sis) Dynamic state in which the body's internal environment is maintained in the normal range.

homeostatic mechanism (ho″me-o-stat′ik mek′ah-nizm) Any of the control systems that help maintain a normal internal environment in the body.

homozygous (ho′mo-zi′gus) Identical alleles in a gene pair.

hormone (hor′mōn) Chemical messenger secreted by an endocrine gland, and transported in the blood, that acts on target cells.

human chorionic gonadotropin (hu′man ko″re-on′ik gon″ah-do-tro′pin) **(hCG)** Hormone, secreted by the embryo, that helps support pregnancy.

humoral immune response (hu′mor-al i-mūn′ ri-spons′) Circulating antibodies′ destruction of pathogens bearing nonself antigens; also called *antibody mediated immunity.*

hyaline cartilage (hī-ah-lin kar′ti-lij) Semitransparent, flexible connective tissue with ultrafine collagen fibers.

hydrogen bond (hi′dro-jen bond) Weak bond between a hydrogen atom and an atom of oxygen or nitrogen, between molecules or between different regions of a very large molecule.

hydrolysis (hi-drol′ĭ-sis) Enzymatically adding a water molecule to split a molecule.

hydrostatic pressure (hy″dro-stat′ik presh′ur) Pressure exerted by a fluid in response to a force, such as gravity or pumping of the heart. Blood pressure is an example of hydrostatic pressure.

hymen (hi′men) Membranous fold of tissue that partially covers the vaginal opening.

hypertonic (hi″per-ton′ik) Solution with a greater osmotic pressure than the solution (usually body fluids) to which it is compared.

hyperventilation (hi″per-ven″tĭ-la′shun) Deep and rapid breathing that lowers the blood CO_2 levels.

hypochondriac regions (hi″po-kon′dre-ak re′junz) Portions of the abdomen on either side of the epigastric region.

hypogastric region (hi″po-gas′trik re′jun) Lower middle portion of the abdomen.

hypothalamus (hi″po-thal′ah-mus) Part of the brain below the thalamus and forming the floor of the third ventricle.

hypotonic (hi″po-ton′ik) Solution with a lower osmotic pressure than the solution (usually body fluids) to which it is compared.

hypoxia (hi-pok′se-ah) Deficiency of oxygen reaching the body′s tissues.

I

iliac region (il′e-ak re′jun) Portion of the abdomen on either side of the lower middle or pubic (hypogastric) region.

ilium (il′e-um) One of the bones making up the hip bone.

immunity (ĭ-mu′nĭ-te) Resistance to the effects of specific disease-causing agents.

immunoglobulin (im″u-no-glob′u-lin) Globular plasma protein that functions as an antibody.

impulse (im′puls) An action potential.

incomplete protein (in″kom-plēt′ pro′te-in) Protein with inadequate amounts of essential amino acids.

inert (in-ert′) Elements that do not react with other elements.

inferior (in-fer′e-or) Situated below something else; pertaining to the lower surface of a part.

inflammation (in″flah-ma′shun) Tissue response to stress that includes pain, warmth, redness, and swelling.

inguinal regions (ing′gwĭ-nal re′junz) Pertaining to the groin, the depressed area of the abdominal wall near the thigh on either side of the pubic region; iliac region.

innate defense (in′ate dē-fens) Inborn, nonspecific defense that blocks entry of or destroys pathogens.

inorganic (in″or-gan′ik) Chemical substances that do not include both carbon and hydrogen atoms.

insertion (in-ser′shun) End of a muscle attached to a movable part.

inspiration (in″spĭ-ra′shun) Breathing in; also called *inhalation.*

insula (in′su-lah) Cerebral lobe deep within the lateral sulcus.

insulin (in′su-lin) Hormone the pancreatic islets secrete that lowers the blood glucose concentration by stimulating cells to take up glucose.

integumentary system (in-teg-u-men′tar-e sis′tem) The organ system that includes the skin and its accessory structures.

intercalated disc (in-ter″kah-lāt′ed disk) Connection between cardiac muscle cells.

internal environment (in-ter′nĕl en-vi-ruh-ment) Conditions inside of the body, surrounding the cells. The environment the body′s cells live in.

interneuron (in″ter-nu′ron) Neuron within the CNS, often connecting a sensory neuron and a motor neuron; also called *internuncial* or *association neuron.*

interphase (in′ter-fāz) Period between cell divisions when a cell metabolizes and may prepare to divide.

interstitial cell (in″ter-stish′al sel) Hormone-secreting cell between seminiferous tubules of the testis; also called *cell of Leydig.*

intervertebral disc (in″ter-ver′tĕ-bral disk) Fibrocartilage structure between the bodies of adjacent vertebrae.

intestinal gland (in-tes′tĭ-nal gland) Tubular gland at the base of a villus in the intestinal wall.

intestinal villus (in-tes′tĭ-nal vil′us) (plural, *intestinal villi*) Tiny, fingerlike projection that extends from the inner lining of the small intestine into the lumen.

intracellular fluid (in″trah-sel′u-lar floo′id) Fluid inside cells.

intramembranous bone (in″trah-mem′brah-nus bōn) Bone that forms from membranelike layers of primitive connective tissue.

intrinsic factor (in-trin′sik fak′tor) Substance that gastric glands produce that promotes intestinal absorption of vitamin B_{12}.

inversion (in-ver′zhun) Turning the plantar surface of the foot inward, toward the midline.

ion (i′on) Particle that results when an atom or molecule becomes electrically charged.

ionic bond (i-on′ik bond) Chemical bond that results from the attraction of two oppositely charged ions.

ipsilateral (ip″si-lat′er-al) On the same side of the body.

iris (i′ris) Colored, muscular part of the eye around the pupil that regulates its size.

isotonic (i″so-ton′ik) Solution with the same osmotic pressure as the solution (usually body fluids) to which it is compared.

isotope (i′so-tōp) Atom that has the same number of protons as other atoms of the same element but a different number of neutrons in its nucleus. Thus, an atom of an element with a different atomic weight than other atoms of that element.

J

joint (joynt) Union of two or more bones; also called an *articulation.*

juxtaglomerular apparatus (juks″tah-glo-mer′u-lar ap″ah-ra′tus) A group of cells in the wall of the afferent arteriole of the nephron that plays a role in the control of renin secretion by the kidney. Specialized cells monitor conditions of the tubular fluid before it enters the distal convoluted tubule.

K

keratin (ker′ah-tin) Intracellular protein in epidermis, hair, and nails.

keratinization (ker″ah-tin′i-za′shun) Process by which cells form fibrils of keratin and harden.

ketone body (ke′tōn bod′e) Compound produced during fat catabolism, including acetone, acetoacetic acid, and betahydroxybutyric acid.

Kupffer cell (koop′fer sel) Large, fixed macrophage in the liver that removes bacterial cells from the blood by phagocytosis.

L

labyrinth (lab′ĭ-rinth) System of connecting tubes in the inner ear, including the cochlea, vestibule, and semicircular canals.

lacrimal gland (lak′rĭ-mal gland) Tear-secreting gland.

lactase (lak′tās) Enzyme that catalyzes the breakdown of lactose into glucose and galactose.

lacteal (lak′te-al) Lymphatic capillary associated with a villus of the small intestine.

lactic acid (lak′tik as′id) Organic compound formed from pyruvic acid in the anaerobic pathway of cellular respiration.

lacuna (lah-ku′nah) Small chamber or cavity.

lamella (lah-mel′ah) Layer of matrix surrounding the central canal of an osteon.

laryngopharynx (lah-ring″go-far′ingks) Lower part of the pharynx, posterior to the larynx, that leads to the esophagus.

larynx (lar′inks) Structure inferior to the laryngopharynx and superior to the trachea that houses the vocal cords.

latent period (la′tent pe′re-od) Time between application of a stimulus and the beginning of a response in a muscle fiber.

lateral (lat′er-al) Pertaining to the side, away from midline.

lateral regions (lat′er-al re′junz) Parts of the abdomen on either side of the umbilical region; lumbar regions.

leukocyte (lu′ko-sīt) White blood cell.

lever (lev′er) (pl. *levers*) Simple mechanical device consisting of a rod, fulcrum, resistance, and a force that is applied to some point on the rod.

LH Luteinizing hormone.

ligament (lig′ah-ment) Cord or sheet of connective tissue binding two or more bones at a joint.

limbic system (lim′bik sis′tem) Connected structures in the brain that produce emotions.

lingual frenulum (ling′gwahl fren′u-lum) Fold of tissue that anchors the tongue to the floor of the mouth.

lipase (li′pās) Fat-digesting enzyme.

lipid (lip′id) Group of organic compounds that includes triglycerides (fats), steroids, and phospholipids.

lumbar (lum′bar) Pertaining to the region of the lower back.

lumen (lu′men) Hollow part of a tubular structure such as a blood vessel or intestine.

luteinizing hormone (lu′te-in-iz″ing hor′mōn) **(LH)** A hormone that the anterior pituitary secretes that controls formation of the corpus luteum in females and testosterone secretion in males.

lymph (limf) Fluid, derived from interstitial fluid, that the lymphatic vessels carry.

lymph node (limf nōd) Mass of lymphoid tissue located along the course of a lymphatic vessel.

lymphatic pathway (lim-fat′ik path′wā) Connected vessels that transport lymph.

lymphatic system (lim-fat′ik sis′tem) The organ system that includes lymphatic vessels, lymph nodes, the thymus, the spleen, and a fluid called lymph. It plays a role in the body's defense against infection.

lymphocyte (lim′fo-sīt) Type of white blood cell that provides immunity; B cell or T cell.

lysosome (li′so-sōm) Organelle that contains enzymes that break down worn-out cell parts and debris.

M

macromolecule (mak-rō-mol′ĕ-kūl) Very large molecule, such as protein, starch, or nucleic acid.

macronutrient (mak-rō-nu′tree-ent) Nutrient (carbohydrate, lipid, or protein) required in a large amount.

macrophage (mak′ro-fāj) Large phagocytic cell.

macula lutea (mak′u-lah lu′te-ah) Yellowish patch in the retina associated with sharp color vision.

malnutrition (mal″nu-trish′un) Symptoms resulting from lack of specific nutrients.

maltase (mawl′tās) Enzyme that catalyzes breakdown of maltose into glucose.

mammary (mam′er-e) Pertaining to the breast.

marrow (mar′o) Connective tissue in spaces in bones that includes blood-forming stem cells.

mast cell (mast sel) Cell to which antibodies formed in response to allergens attach, causing the cell to release allergy mediators, which cause symptoms.

matter (mat′er) Anything that has weight and occupies space.

mechanoreceptor (mek″ah-no-re-sep′tor) Sensory receptor sensitive to mechanical stimulation, such as changes in pressure or tension.

medial (me′de-al) Toward or near midline.

mediastinum (me″de-as-ti′num) The compartment in the thoracic cavity between the lungs.

medulla oblongata (mĕ-dul′ah ob″long-gah′tah) Part of the brainstem between the pons and the spinal cord.

medullary cavity (med′u-lār″e kav′ĭ-te) Cavity containing red or yellow marrow within the diaphysis of a long bone.

megakaryocyte (meg″ah-kar′e-o-sīt) Large cell in red bone marrow that breaks apart to yield blood platelets.

meiosis (mi-o′sis) Cell division that halves the chromosome number, producing egg or sperm cells (gametes).

melanin (mel′ah-nin) Dark pigment generally found in skin and hair.

melanocyte (mel′ah-no-sīt) Melanin-producing cell.

melatonin (mel″ah-to′nin) Hormone that the pineal gland secretes.

memory cell (mem′o-re sel) T lymphocyte or B lymphocyte produced in a primary immune response that can be activated rapidly if the same antigen is encountered again.

menarche (mĕ-nar′ke) A female's first reproductive cycle.

meninges (mĕ-nin′jēz) (singular, *meninx*) The three layers of membrane that cover the brain and spinal cord.

meniscus (mĕ-nis′kus) (plural, *menisci*) Fibrocartilage that separates the articulating surfaces of bones in the knee.

menopause (men′o-pawz) Cessation of the reproductive cycles.

mental (men′tal) The region associated with the chin; relating to the mind.

merocrine gland (mer′o-krin gland) A structure whose cells remain intact while secreting; a type of sweat gland.

mesentery (mes′en-ter″e) Fold of peritoneal membrane that attaches abdominal organs to the posterior abdominal wall.

mesoderm (mez′o-derm) Middle primary germ layer of the embryo.

messenger RNA (mes′in-jer) **(mRNA)** RNA that transmits information for a protein's amino acid sequence from the nucleus to the cytoplasm.

metabolic pathway (mĕ-tab′o-lik path-wa) Series of linked, enzymatically controlled chemical reactions.

metabolic water (mĕ-tab′o-lik wot′er) Water produced as a by-product of aerobic metabolism.

metabolism (mĕ-tab′o-lizm) In cells, the combined chemical reactions of anabolism and catabolism that use or release energy.

metacarpal (met″ah-kar′pal) Any of the five bones of the hand between the wrist bones and finger bones.

metaphase (met′ah-fāz) Stage in mitosis when chromosomes align in the middle of the cell.

metatarsal (met″ah-tar′sal) Any of the five bones of the foot between the ankle bones and the toe bones.

microfilament (mi″kro-fil′ah-ment) Rod of actin protein in the cytoplasm that provides structural support or movement. Part of the cytoskeleton.

microglia (mi-krogl′e-ah) Neuroglia of the CNS that support neurons and phagocytize bacterial cells and cellular debris.

micronutrient (mi-kro-nu′tree-ent) Nutrient (vitamin or mineral) required in small amount.

microtubule (mi″kro-tu′būl) Hollow rod constructed of many molecules of the protein tubulin. Part of the cytoskeleton.

microvillus (mi″kro-vil′us) Any of the cylindrical processes that extends from some epithelial cell membranes, increasing membrane surface area.

micturition (mik″tu-rish′un) Urination.

midbrain (mid′brān) Small region of the brainstem between the diencephalon and the pons.

mineral (min′er-al) Inorganic element essential in human metabolism.

mineralocorticoid (min″er-al-o-kor′tĭ-koid) Hormone the adrenal cortex secretes that affects electrolyte concentrations in body fluids.

mitochondrion (mi″to-kon′dre-on) (plural, *mitochondria*) Organelle housing enzymes that catalyze the aerobic reactions of cellular respiration.

mitosis (mi-to′sis) Division of a somatic cell, forming two genetically identical somatic cells.

mitral valve (mi′trul valv) Heart valve between the left atrium and the left ventricle; also called the *bicuspid valve* or *left atrioventricular (AV) valve.*

mixed nerve (mikst nerv) A nerve composed of axons (fibers) of both motor and sensory neurons.

molecular formula (mo-lek′u-lar fōr′mu-lah) Abbreviation for the number of atoms of each element in a compound.

molecule (mol′ĕ-kūl) Particle composed of two or more bonded atoms.

monocyte (mon′o-sīt) Type of white blood cell that can leave the bloodstream and become a macrophage.

monosaccharide (mon″o-sak′ah-rīd) Simple sugar, such as glucose or fructose.

motor area (mo′tor a′re-ah) Any of the areas of the brain that send impulses controlling skeletal muscles.

motor end plate (mo′tor end plāt) Specialized part of a muscle fiber membrane at a neuromuscular junction.

motor nerve (mo′tor nerv) A nerve composed of axons (nerve fibers) of motor neurons.

motor neuron (mo′tor nu′ron) Neuron that conducts impulses from the central nervous system to an effector.

motor speech area (mo′tor spēch ār′e-ah) Region of the frontal lobe that coordinates complex muscular actions of mouth, tongue, and larynx, making speech possible; also known as *Broca's area.*

motor unit (mo′tor u′nit) A motor neuron and the muscle fibers that it controls.

mucosa (mu-ko′sah) Innermost layer of the alimentary canal.

mucous cell (mu′kus sel) Glandular cell that secretes mucus.

mucous membrane (mu′kus mem-brān) Type of membrane that lines tubes and body cavities that open to the outside of the body.

mucus (mu′kus) Fluid secretion of the mucous cells.

muscle tissue (mus′el tish′u) Contractile tissue consisting of filaments of actin and myosin, which slide past each other, shortening cells.

muscle tone (mus′el tōn) Ongoing low-level contraction of some fibers in otherwise resting skeletal muscle.

muscular system (mus′ku-lar sis′tem) The organ system that includes the skeletal muscles.

muscularis (mus″ku-lar′is) Smooth muscle layers of the alimentary canal.

myelin (mi′ĕ-lin) Lipid material that forms a sheathlike covering around some axons. It provides electrical insulation.

myelin sheath (mi′ĕ-lin shēth) Lipid-rich layer formed from certain neuroglia that wraps around an axon, providing insulation.

myocardium (mi″o-kar′de-um) Muscle layer of the heart.

myofibril (mi″o-fi′bril) Contractile fiber in striated muscle cells.

myoglobin (mi″o-glo′bin) Oxygen-storing protein in muscle tissue.

myometrium (mi″o-me′tre-um) Layer of smooth muscle tissue in the uterine wall.

myosin (mi′o-sin) Protein in a muscle fiber that forms the thick filaments that pull on the thin filaments of the protein actin, contracting muscle fibers.

N

nail (nāl) Protective plate at the distal end of a finger or toe.

nasal (na′zal) Pertaining to the nose.

nasal cavity (na′zal kav′ĭ-te) Space posterior to the nose.

nasal concha (na′zal kong′kah) Any of the shelf-like bones or bony processes extending medially from the wall of the nasal cavity; also called a *turbinate bone.*

nasal septum (na′zal sep′tum) Midline wall of bone and cartilage that separates the nasal cavity into right and left parts.

nasopharynx (na″zo-far′inks) Part of the pharynx posterior to the nasal cavity.

negative feedback (neg′ah-tiv fēd′bak) A mechanism that returns the level of a chemical or other substance or condition in the internal environment to its set point level.

neonatal (ne″o-na′tal) The first four weeks after birth.

nephron (nef′ron) Functional unit of a kidney, consisting of renal corpuscle and renal tubule.

nerve (nerv) Bundle of axons in the peripheral nervous system.

nervous system (ner′vus sis′tem) The organ system that includes the brain, the spinal cord, nerves, and sense organs.

nervous tissue (ner′vus tish′u) Neurons and neuroglia composing the brain, spinal cord, and nerves.

net filtration pressure (net fil-tra′shun presh′ur) The driving force for glomerular filtration in the kidneys.

neurilemma (nu″rĭ-lem′ah) Outer layer formed from Schwann cells on the exterior of some axons, outside of the myelin sheath.

neurofilaments (nu″ro-fil′ah-ments) Fine, cytoplasmic threads that extend from the cell body into the axon for its entire length.

neuroglia (nu-rog′le-ah) Specialized cells of the nervous system that, depending on the type of neuroglia, produce myelin, maintain the ionic environment, provide growth factors that support neurons, provide structural support, and play a role in cell-to-cell communication.

neuromuscular junction (nu″ro-mus′ku-lar jungk′shun) Synapse between a motor neuron and a skeletal muscle fiber.

neuron (nu′ron) Nerve cell.

neurotransmitter (nu″ro-trans′mit-er) Chemical that an axon secretes at a synapse that stimulates or inhibits an effector (muscle or gland) or other neuron.

neutral (nu′tral) Neither acidic nor alkaline; pH 7.0.

neutron (nu′tron) Electrically neutral particle in an atomic nucleus.

neutrophil (nu′tro-fil) Type of phagocytic white blood cell containing cytoplasmic granules that react with neutral pH stain.

nodes of Ranvier (nōdz uv ron′vee-ay) The many gaps in the myelin sheath along axons of myelinated neurons of the peripheral nervous system.

nonelectrolyte (non″e-lek′tro-līt) Substance that does not dissociate into ions when dissolved in water.

nonprotein nitrogenous substance (non-pro′te-in ni-troj′ĕ-nus sub′stans) A nitrogen-containing molecule that is not a protein.

norepinephrine (nor″ep-ĭ-nef′rin) Type of neurotransmitter, which is a chemical secreted into the synaptic cleft at axon ends of neurons. Also secreted as a hormone by the adrenal medulla during times of stress.

nuclear envelope (nu′kle-er ahn-veh-lop) Double membrane surrounding the cell nucleus that separates it from the cytoplasm.

nucleic acid (nu-kle′ik as′id) A molecule formed of a chain of nucleotides; RNA or DNA.

nucleolus (nu-kle′o-lus) (plural, *nucleoli*) Small structure in the cell nucleus that contains RNA and proteins.

nucleotide (nu′kle-o-tīd″) Building block of a nucleic acid molecule, consisting of a sugar, a nitrogenous base, and a phosphate group.

nucleus (nu′kle-us) (plural, *nuclei*) (1) The dense core of an atom, composed of protons and usually neutrons. (2) Cellular organelle enclosed by double-layered, porous membrane and containing DNA. (3) Masses of interneuron cell bodies in the central nervous system.

nutrient (nu′tre-ent) Chemical that the body requires from the environment.

O

occipital (ok-sip′ĭ-tal) Pertaining to the lower, back part of the head.

olfactory (ol-fak′to-re) Pertaining to the sense of smell.

olfactory nerves (ol-fak′to-re nervz) The first pair of cranial nerves, which conduct impulses associated with the sense of smell.

oligodendrocyte (ol″ĭ-go-den′dro-sīt) Type of neuroglia that produces myelin in the CNS.

oocyte (o′o-sīt) Cell formed by oogenesis; egg cell.

oogenesis (o″o-jen′ĕ-sis) Formation of an oocyte.

optic chiasma (op′tik ki-az′mah) X-shaped structure on the underside of the brain formed by optic nerve fibers (axons) that partially cross over to the visual cortex on the opposite side.

optic disc (op′tik disk) Region in the retina where nerve fibers (axons) exit to form the optic nerve.

oral (o′ral) Pertaining to the mouth.

orbital (or′bi-tal) Pertaining to the bony socket of the eye.

organ (or′gan) Structure consisting of two or more tissues that performs a specialized function.

organ system (or′gan sis′tem) Group of organs coordinated to carry on a specialized function.

organelle (or″gah-nel′) A structure in a cell that has a specialized function.

organic (or-gan′ik) Chemicals that contain both carbon and hydrogen.

organism (or′gah-nizm) An individual living thing.

orgasm (or′gazm) An intense sensation that is the culmination of sexual stimulation.

origin (or′ĭ-jin) End of a muscle that attaches to a relatively immovable part.

oropharynx (o″ro-far′inks) Part of the pharynx posterior to the oral cavity.

osmosis (oz-mo′sis) Movement of water through a semipermeable membrane toward a concentration of an impermeant solute.

osmotic pressure (oz-mot′ik presh′ur) Pressure needed to stop osmosis; a solution's potential pressure caused by impermeant solute particles in the solution.

ossification (os″ĭ-fĭ-ka′shun) Formation of bone tissue.

osteoblast (os′te-o-blast″) Bone-forming cell.

osteoclast (os′te-o-klast″) Cell that breaks down bone matrix.

osteocyte (os′te-o-sīt) Mature bone cell.

osteon (os′te-on) Cylinder-shaped unit containing bone cells and matrix lamellae that surround a central canal; Haversian system.

OT Oxytocin.

oval window (o′val win′do) Opening in the inner ear to which the stapes attaches.

ovary (o′var-e) Primary female reproductive organ; egg cell-producing organ.

ovulation (o″vu-la′shun) Release of an egg cell from a mature ovarian follicle.

oxidation (ok″sĭ-da′shun) Process by which oxygen combines with another chemical; removal of hydrogen or the loss of electrons; opposite of reduction.

oxygen debt (ok′sĭ-jen det) The amount of oxygen that liver cells require, after anaerobic exercise, to convert the accumulated lactate into glucose, plus the amount muscle cells require to restore ATP and creatine phosphate to their original concentrations and to return blood and tissue oxygen levels to normal.

oxyhemoglobin (ok″sĭ-he″mo-glo′bin) Compound formed when oxygen binds hemoglobin.

oxytocin (ok″sĭ-to′sin) **(OT)** Posterior pituitary hormone that contracts smooth muscle in the uterus and mammary gland myoepithelial cells.

P

pacemaker (pās′māk-er) Mass of specialized cardiac muscle tissue that controls the rhythm of the heartbeat; also called *the sinoatrial (SA) node*.

pain receptor (pān re″sep′tor) Sensory receptor that transmits impulses interpreted as pain.

palate (pal′at) Roof of the mouth.

palatine (pal′ah-tīn) Pertaining to the palate.

palmar (pahl′mar) Pertaining to the palm of the hand.

pancreas (pan′kre-as) Glandular organ in the abdominal cavity that secretes the hormones insulin and glucagon into the bloodstream, and a variety of digestive enzymes into the small intestine.

pancreatic juice (pan″kre-at′ik joos) Digestive secretions of the pancreas.

papilla (pah-pil′ah) Tiny, nipplelike projection.

papillary muscle (pap′ĭ-ler″e mus′l) Muscle that extends inward from the ventricular wall of the heart and to which the chordae tendineae attach.

paracrine (par″ah krin) A secretion that affects only neighboring cells.

paranasal sinus (par″ah-na′zal si-nus) Any of the several air-filled cavities in a cranial or facial bone lined with mucous membrane and connected to the nasal cavity.

parasympathetic division (par″ah-sim″pah-thet′ik de-vijh′in) Part of the autonomic nervous system that arises from the brain and sacral region of the spinal cord; parasympathetic nervous system.

parathyroid gland (par″ah-thi′roid gland) One of four small endocrine glands embedded in the posterior part of the thyroid gland. The hormone they secrete helps control blood calcium and phosphate levels.

parathyroid hormone (par″ah-thi′roid hor′mōn) **(PTH)** Hormone secreted by the parathyroid glands that helps regulate the levels of blood calcium and phosphate ions.

parietal (pah-ri′ĕ-tal) Pertaining to the wall of a cavity.

parietal cell (pah-ri′ĕ-tal sel) Cell of a gastric gland that secretes hydrochloric acid and intrinsic factor.

parietal pericardium (pah-ri′ĕ-tal per″ĭ-kar′de-um) Membrane that forms the outer wall of the pericardial cavity.

parietal peritoneum (pah-ri′ĕ-tal per″ĭ-to-ne′-um) Membrane that forms the outer wall of the peritoneal cavity.

parietal pleura (pah-ri′ĕ-tal ploo′rah) Serous membrane that covers the inner surface of the thoracic cavity wall.

parotid gland (pah-rot'id gland) Large salivary gland on the side of the face just in front of and below the ear.

partial pressure (par'shal presh'ur) The pressure that one gas produces in a mixture of gases.

patellar (pah-tel'ar) Pertaining to the front of the knee.

pathogen (path'o-jen) Disease-causing agent.

pectoral (pek'tor-al) Pertaining to the anterior chest.

pectoral girdle (pek'tor-al ger'dl) Part of the skeleton that supports and attaches the upper limbs.

pedal (ped'al) Pertaining to the foot.

pedigree (ped'eh-gree) Chart that displays relationships among family members and their inherited traits and disorders.

pelvic (pel'vik) Pertaining to the pelvis.

pelvic cavity (pel'vik kav'ĭ-te) Space within the ring formed by the sacrum and hip bones that encloses the terminal part of the large intestine, the urinary bladder, and the internal reproductive organs.

pelvic girdle (pel'vik ger'dl) Part of the skeleton to which the lower limbs attach.

pelvis (pel'vis) Basin-shaped structure formed by the sacrum and hip bones.

penis (pe'nis) Male external reproductive organ through which the urethra passes.

pepsin (pep'sin) Protein-splitting enzyme that the gastric glands secrete.

pepsinogen (pep-sin'o-jen) Inactive form of pepsin.

perception (per-sep'shun) Mental interpretation of sensory stimulation.

pericardial cavity (per″ĭ-kar'de-al kav'ĭ-te) Potential space between the visceral and parietal pericardial membranes.

pericardium (per″ĭ-kar'de-um) Serous membrane that surrounds the heart.

perichondrium (per″ĭ-kon'dre-um) Layer of dense connective tissue that encloses cartilaginous structures.

perilymph (per'ĭ-limf) Fluid in the space between the membranous and bony (osseous) labyrinths of the inner ear.

perimetrium (per-ĭ-me'tre-um) Outer serosal layer of the uterine wall.

perimysium (per″i-mis'e-um) Sheath of connective tissue that encloses a bundle of skeletal muscle fibers (encloses a fascicle).

perineal (per″ĭ-ne'al) Pertaining to the perineum, the inferior-most region of the trunk between the buttocks and the thighs.

periodontal ligament (per″e-o-don'tal lig'ah-ment) Dense connective tissue that surrounds the root of a tooth and attaches the tooth to the jawbone.

periosteum (per″e-os'te-um) Dense connective tissue covering the surface of a bone.

peripheral nervous system (pĕ-rif'er-al ner'vus sis'tem) **(PNS)** Parts of the nervous system outside the brain and spinal cord.

peripheral resistance (pĕ-rif'er-al re-zis'tans) Resistance to blood flow due to friction between the blood and the blood vessel walls.

peristalsis (per″ĭ-stal'sis) Rhythmic waves of muscular contraction in the walls of certain tubular organs.

peritoneal cavity (per″ĭ-to-ne'al kav'ĭ-te) Potential space between the visceral and parietal peritoneal membranes.

peritubular capillary (per″ĭ-tu'bu-lar kap'ĭ-ler″e) Capillary that surrounds a renal tubule and functions in tubular reabsorption and tubular secretion during urine formation.

peroxisome (pĕ-roks'ĭ-sōm) Membranous sac abundant in kidney and liver cells that contains enzymes that catalyze reactions that decompose hydrogen peroxide.

pH (pH) A shorthand system that indicates the acidic or basic (alkaline) condition of a solution; values range from 0 to 14. The lower the pH number, the more acidic the solution.

phagocytosis (fag″o-si-to'sis) Process by which a cell engulfs solids from its surroundings.

phalanx (fa'langks) (plural, *phalanges*) Bone of a finger or toe.

pharynx (far'inks) The space posterior to the nasal cavity, the oral cavity, and the larynx.

phenotype (fe'no-tīp) A trait or health condition caused by the expression of a gene or genes.

pheomelanin (fe″o-mel'ah-nin) A reddish-yellow pigment.

phospholipid (fos″fo-lip'id) Molecule consisting of two fatty acid molecules and a phosphate group bound to a glycerol molecule.

photoreceptor (fo″to-re-sep'tor) Sensory receptor sensitive to light; rods and cones of the eyes.

physiology (fiz″e-ol'o-je) Branch of science concerned with the study of body functions.

pia mater (pi'ah ma'ter) Inner layer of meninges that is in direct contact with the brain and spinal cord.

pigment (pig-ment) A substance that imparts color to a cell or tissue.

pineal gland (pin'e-al gland) A small gland in the brain that secretes the hormone melatonin, which controls certain biological rhythms.

pinocytosis (pi″no-si-to'sis) Process by which a cell engulfs droplets of fluid from its surroundings.

pituitary gland (pĭ-tu'ĭ-tār″e gland) Endocrine gland attached to the base of the brain consisting of anterior and posterior lobes.

placenta (plah-sen'tah) Structure that attaches the fetus to the uterine wall, delivering nutrients to and removing wastes from the fetus. The placenta secretes several hormones associated with pregnancy.

plantar (plan'tar) Pertaining to the sole of the foot.

plasma (plaz'mah) Fluid portion of the blood.

plasma cell (plaz'mah sel) Type of antibody-producing cell that forms when activated B cells proliferate.

plasma protein (plaz'mah pro'te-in) Protein dissolved in blood plasma.

platelet (plāt'let) Cellular fragment found in the blood that helps blood clot.

pleural cavity (ploo'ral kav'ĭ-te) Potential space between the visceral and parietal pleural membranes.

pleural membrane (ploo'ral mem'brān) Serous membrane that encloses the lungs and lines the chest wall.

plexus (plek'sus) Network of interlaced nerves or blood vessels.

PNS Peripheral nervous system.

polar (po'lar) A molecule with equal numbers of protons and electrons, yet having a slightly positive region and a slightly negative region due to uneven distribution of those charged particles.

polar body (pō'lar bod'e) Small, nonfunctional cell that is a product of meiosis in the female.

polarized (po″lar-īzd) Condition in which there is a voltage difference across a cell membrane due to an unequal distribution of positive and negative ions inside compared to outside of the membrane.

polysaccharide (pol″e-sak'ah-rīd) Carbohydrate composed of many joined monosaccharides.

polyunsaturated fatty acid (pol″e-un-sach'ĕ-ra-ted fat'e as'id) Fatty acid with many double bonds between carbon atoms.

pons (ponz) Part of the brainstem above the medulla oblongata and below the midbrain.

popliteal (pop″lĭ-te'al) Pertaining to the region behind the knee.

positive feedback system (poz'ĭ-tiv fēd'bak sis'tem) Process by which changes cause additional similar changes, producing unstable conditions.

posterior (pos-tēr′e-or) Toward the back; the opposite of *anterior*.

posterior pituitary (pos-tēr′e-or pĭ-tu′ĭ-tār″e) Rear (posterior) lobe of the pituitary gland. It secretes oxytocin and antidiuretic hormone.

postganglionic fiber (pōst″gang-gle-on′ik fi′ber) Axon of a postsynaptic neuron reaching from an autonomic ganglion to an effector.

postnatal (pōst-na′tal) After birth.

preganglionic fiber (pre″gang-gle-on′ik fi′ber) Axon of a presynaptic neuron reaching an autonomic ganglion.

pregnancy (preg′nan-se) Condition in which a female has a developing offspring in her uterus.

prenatal (pre-na′tal) Before birth.

primary germ layers (pri′mar-e jerm la′erz) Three layers of cells in the embryo that divide and differentiate into specific tissues and organs; ectoderm, mesoderm, and endoderm.

primary sex organs (pri′ma-re seks or′ganz) Organs that produce sex cells; testes in males and ovaries in females.

prime mover (prīm moov′er) Muscle that provides most of a particular body movement; also called an *agonist*.

primordial follicle (pri-mor′de-al fol′ĭ-kl) Oocyte enclosed by a single layer of cells in the ovary.

PRL Prolactin.

progenitor cell (pro-jen′ĭ-tor sel) Daughter cell of a stem cell whose own daughter cells are restricted to follow specific lineages.

progesterone (pro-jes′tĭ-rōn) Female hormone secreted by the corpus luteum of the ovary and by the placenta.

projection (pro-jek′shun) Process by which the brain causes a sensation to seem to come from the region of the body being stimulated.

prolactin (pro-lak′tin) **(PRL)** Hormone secreted by the anterior pituitary that stimulates milk production in the mammary glands.

pronation (pro-na′shun) Downward or backward rotation of the palm.

prophase (pro′fāz) Stage of mitosis when chromosomes become visible in the nucleus when stained and viewed under a microscope.

prostaglandins (pros″tah-glan′dins) Group of compounds that have powerful, hormonelike effects.

prostate gland (pros′tāt gland) Gland surrounding the male urethra below the urinary bladder that secretes a fluid into semen prior to ejaculation.

protein (pro′tēn) Nitrogen-containing organic compound composed of a chain of many bonded amino acid molecules.

prothrombin (pro-throm′bin) Plasma protein that functions in blood clotting.

proton (pro′ton) Positively charged particle in an atomic nucleus.

protraction (pro-trak′shun) Forward movement of a body part.

proximal (prok′sĭ-mal) Closer to the point of attachment; opposite of *distal*.

pseudostratified columnar epithelium (soo″do-strat′ĭ-fīd co-lum′nar ep″ĭ-the′lē-um) Single layer of cells appearing as more than one layer because the nuclei occupy different positions in the cells.

PTH Parathyroid hormone.

puberty (pu′ber-te) Stage of development in which the reproductive organs become functional.

pubic region (pu′bik re′jun) Lower middle portion of the abdomen, between the left and right inguinal regions; hypogastric region.

pulmonary circuit (pul′mo-ner″e ser′kit) System of blood vessels that transports blood between the heart and the lungs.

pulmonary valve (pul′mo-ner″e valv) Valve leading from the right ventricle to the pulmonary trunk; also called the *pulmonary semilunar valve*.

pulse (puls) Surge of blood felt through the walls of arteries due to the contraction of the heart ventricles.

Punnett square (pun-it sqware) A grid diagram that displays possible genotypes in offspring based on parental gametes.

pupil (pu′pil) Opening in iris through which light enters the eye.

Purkinje fibers (pur-kin′je fi′berz) Specialized cardiac muscle fibers that conduct cardiac impulses from the AV bundle into the ventricular walls.

pyruvic acid (pi-roo′vik as′id) Three-carbon compound that is the breakdown product of the 6-carbon sugar glucose in glycolysis. One glucose molecule splits to yield two pyruvic acid molecules.

R

radioactive (ra″de-o-ak′tiv) Property of some atoms that releases energy or pieces of matter at a constant rate.

rate-limiting enzyme (rāt lim′ĭ-ting en′zīm) Enzyme, usually present in small amounts, that controls the rate of a metabolic pathway by regulating one step.

receptor (re-sep′tor) Specialized cell or organ that provides information about the environment. Also, cell membrane protein that binds specific molecules, called ligands, thereby sending a signal inside the cell.

receptor-mediated endocytosis (re-sep′tor-me-de-ay-ted en″do-si-to′sis) A type of endocytosis (transport into a cell in a vesicle from the cell membrane) that is specific because the substance being transported binds to a protein receptor it fits on the cell surface.

recessive allele (re-sess′iv ah-lēl) Form of a gene not expressed if the dominant form is also present.

recruitment (re-kroot′ment) Increase in the number of motor units taking part in a muscle contraction.

red marrow (red mar′o) Blood-cell-forming tissue in spaces within bones.

referred pain (re-ferd′ pān) Pain that feels as if it is originating from a part other than the site being stimulated.

reflex (re′fleks) A rapid, automatic (involuntary) response to a stimulus.

reflex arc (re′fleks ark) Nerve pathway, consisting of a sensory receptor, a sensory neuron, interneuron, motor neuron, and an effector, that forms the structural and functional bases for a reflex.

refraction (re-frak′shun) Bending of light as it passes between substances of different densities.

relaxin (re-lak′sin) Hormone from the corpus luteum that inhibits uterine contractions during pregnancy.

releasing hormone (re-le′-sing hor′mōn) Any of a group of hormones from the hypothalamus, each of which stimulates release of a specific anterior pituitary hormone.

renal corpuscle (re′nal kor′pusl) Part of a nephron that consists of a glomerulus and a glomerular capsule.

renal cortex (re′nal kor′teks) Outer part of a kidney.

renal medulla (re′nal mĕ-dul′ah) Inner part of a kidney.

renal pelvis (re′nal pel′vis) Funnel-shaped cavity in a kidney that channels urine to the ureter.

renal tubule (re′nal tu′būl) Part of a nephron that extends from the renal corpuscle to the collecting duct.

renin (re′nin) Enzyme that kidneys release that maintains blood pressure and blood volume.

replication (rep″lĭ-ka′shun) Copying of a DNA molecule.

reproductive systems (re″pro-duk′tiv sis′tems) The organ systems in the male and female that work together to produce offspring. The male reproductive system includes the scrotum, testes, epididymides, ductus deferentia, seminal vesicles, prostate gland, bulbourethral glands, penis, and urethra. The female reproductive system includes the ovaries, uterine tubes, uterus, vagina, clitoris, and vulva.

respiration (res″pĭ-ra′shun) The entire process of exchanging gases between the atmosphere and the body cells, including the utilization of oxygen and the production of carbon dioxide in the process of cellular respiration.

respiratory capacity (re-spi′rah-to″re kah-pas′ĭ-te) The sum of any two or more respiratory volumes.

respiratory cycle (re-spi′rah-to″re si′kl) An inspiration followed by an expiration.

respiratory membrane (re-spi′rah-to″re mem′brān) Layers including a capillary wall, an alveolar wall, and their basement membranes through which blood and inspired air exchange gases.

respiratory system (re-spi′rah-to″re sis′-tem) The organ system that obtains oxygen for the body cells and removes carbon dioxide. The nasal cavity, pharynx, larynx, trachea, bronchi, and lungs are parts of this system.

respiratory volume (re-spi′rah-to″re vol′ūm) Any one of several distinct volumes of air that can be moved into or out of the lungs.

resting potential (res′ting po-ten′shal) Difference in electrical charge between the intracellular side and the extracellular side of an undisturbed nerve cell membrane.

resting tidal volume (res′ting tīd′al vol′um) Volume of air moved in, then out, of the lungs in a respiratory cycle at rest.

reticular fiber (rĕ-tik′u-lar fi′ber) Thin protein fiber.

reticular formation (rĕ-tik′u-lar for-ma′shun) Complex network of nerve fibers in the brainstem that arouses the cerebrum to a wakeful state.

retina (ret′ĭ-nah) Inner layer of the wall of the eye that includes the visual receptors.

retraction (rĕ-trak′shun) Movement of a part toward the back.

retroperitoneal (ret″ro-per″i-to-ne′al) Behind the peritoneum.

reversible reaction (re-ver′sĭ-bl re-ak′shun) Chemical reaction in which the products can react, reforming the reactants.

rhodopsin (ro-dop′sin) Light-sensitive pigment in the rods of the retina; visual purple.

ribonucleic acid (ri″bo-nu-kle′ik as′id) **(RNA)** Single-stranded polymer of nucleotides in which each nucleotide includes the sugar ribose, a phosphate group, and a nitrogenous base (adenine, uracil, guanine, or cytosine).

ribosome (ri′bo-sōm) Organelle composed of RNA and protein that is a structural support for protein synthesis and provides enzyme activity to help join amino acids.

RNA Ribonucleic acid.

rod (rod) Type of light receptor that provides colorless (grayscale) vision.

rotation (ro-ta′shun) Movement turning a body part on its longitudinal axis.

round window (rownd win′do) Membrane-covered opening between the inner ear and the middle ear.

S

SA node (nōd) Sinoatrial node.

saccule (sak′ūl) An enlarged part of the membranous labyrinth of the inner ear. It contains some of the receptors involved in static equilibrium.

sacral (sa′kral) Pertaining to the posterior region between the hip bones.

sagittal (saj′ĭ-tal) Plane or section that divides a structure into right and left portions.

salivary amylase (sal′i-ver-e am′ĭ-lās) Enzyme that hydrolyzes (digests) starch in the mouth.

salivary gland (sal′i-ver-e gland) Any of the glands, associated with the mouth, that secrete saliva.

salt (salt) Compound composed of oppositely charged ions; compound produced by the reaction of an acid and a base.

sarcomere (sar′ko-mēr) Structural unit of a myofibril and the functional unit of muscle contraction.

sarcoplasmic reticulum (sar″ko-plaz′mik rĕ-tik′u-lum) Membranous network of channels and tubules within a muscle fiber, corresponding to the endoplasmic reticulum of other cells.

saturated fatty acid (sat′u-rāt″ed fat′e as′id) Fatty acid molecule that includes maximal hydrogens and therefore has no double-bonded carbon atoms.

Schwann cell (shwahn sel) Type of neuroglia that surrounds an axon of a peripheral neuron, forming the neurilemma and myelin sheath.

sclera (skle′rah) White, fibrous outer layer of the wall of the eye.

scrotum (skro′tum) Pouch of skin in males that encloses the testes.

sebaceous gland (se-ba′shus gland) Skin gland that secretes sebum.

sebum (se′bum) Oily secretion of the sebaceous glands.

secretin (se-kre′tin) Hormone from the small intestine that stimulates the pancreas to release pancreatic juice rich in bicarbonate ion.

selectively permeable (se-lek′tiv-le per′me-ah-bl) Membrane that allows some types of molecules through but not others; semipermeable.

semen (se′men) Fluid containing sperm cells and secretions discharged from the male reproductive tract at ejaculation.

semicircular canal (sem″ĭ-ser′ku-lar kah-nal′) Bony, tubular structure in the inner ear that houses receptors providing the sense of dynamic equilibrium.

seminiferous tubule (sem″ĭ-nif′er-us tu′būl) Tubule in the testes where sperm cells form.

sensation (sen-sa′shun) An awareness that impulses associated with a sensory event have reached the brain.

sensory adaptation (sen′so-re ad″ap-ta-′shun) Ability of the nervous system to become less responsive to a maintained stimulus.

sensory area (sen′so-re a′re-ah) Any of the areas of the brain that receive and interpret sensory impulses.

sensory nerve (sen′so-re nerv) A nerve composed of axons (fibers) of sensory neurons.

sensory neuron (sen′so-re nu′ron) Neuron that conducts impulses from sensory receptors to the central nervous system.

sensory receptor (sen′so-re re″sep′tor) Specialized structure associated with the peripheral end of a sensory neuron specific to detecting a particular stimulus and triggering an impulse in response.

sensory speech area (sen′so-re spēch ār′e-ah) Region of the parietal lobe and the temporal lobe just posterior to the lateral sulcus that is necessary for understanding written and spoken language; also referred to as *Wernicke's area*.

serosa (sĕ′ro-sah) Outer covering of the alimentary canal.

serotonin (se″ro-to′nin) Vasoconstrictor that blood platelets release when blood vessels break, controlling bleeding. Also a neurotransmitter.

serous cell (ser′us sel) Glandular cell that secretes a watery lubricating fluid (serous fluid).

serous membrane (ser′us mem′brān) Type of membrane that lines a cavity without an opening to the outside of the body.

serum (ser′um) Fluid portion of coagulated blood.

set point (set point) Target value of a physiological condition maintained in the body by homeostasis. For example, normal body temperature.

sex chromosome (seks crō-mo-some) Chromosome that carries genes responsible for the development of characteristics associated with femaleness or maleness; an X or Y chromosome.

signal transduction (sig′nahl trans-duk′shun) Series of chemical reactions that allows cells to respond to signals reaching the outside of the cell membrane.

simple sugar (sim′pl shoog′ar) Monosaccharide.

sinoatrial node (si″no-a′tre-al nōd) **(SA node)** Specialized tissue in the wall of the right atrium that initiates cardiac cycles; also called the *pacemaker.*

skeletal muscle tissue (skel′ĭ-tal mus′l tish′u) Type of voluntary muscle tissue in muscles attached to bones.

skeletal system (skel′ĭ-tal sis′tem) The organ system that includes the bones, and the ligaments and cartilages that hold them together.

smooth muscle tissue (smooth mus′l tish′u) Type of involuntary muscle tissue; does not have striations.

solute (sol′ūt) Chemical dissolved in a solution.

solvent (sol′vent) Liquid portion of a solution in which a solute is dissolved.

somatic nervous system (so-mat′ik ner′vus sis′tem) Motor pathways of the peripheral nervous system that lead to skeletal muscles.

special senses (spesh′al sen′ses) Senses that stem from receptors associated with specialized sensory organs in the head, such as the eyes and ears.

spermatid (sper′mah-tid) Intermediate stage in sperm cell formation.

spermatogenesis (sper″mah-to-jen′ĕ-sis) Sperm cell production.

spermatogonium (sper″mah-to-go′ne-um) Undifferentiated spermatogenic cell in the outer part of a seminiferous tubule.

spinal cord (spi′nal kord) Part of the central nervous system extending from the brainstem below the foramen magnum through the vertebral canal.

spinal nerve (spi′nal nerv) Nerve that arises from the spinal cord and gives rise to peripheral nerves.

spleen (splēn) Large organ in the upper left region of the abdomen that processes old red blood cells.

spongy bone (spun′jē bōn) Bone that consists of bars and plates separated by irregular spaces; also called *cancellous bone.*

static equilibrium (stat′ik e″kwĭ-lib′re-um) Maintenance of balance when the head and body are motionless.

stem cell (stem sel) Undifferentiated cell that can divide to yield two daughter stem cells, or a stem cell and a progenitor cell.

sternal (ster′nal) Pertaining to the region in the middle of the thorax, anteriorly; pertaining to the sternum.

steroid (ste′roid) Type of lipid formed of complex rings of carbon and associated hydrogen and oxygen atoms. Cholesterol is an example.

stomach (stum′ak) Digestive organ between the esophagus and small intestine.

stratum basale (strat′tum ba′sal-e) Deepest layer of the epidermis, where cells divide; also called the *stratum germinativum.*

stratum corneum (stra′tum kor′ne-um) Outer, horny layer of the epidermis.

stress (stres) Response to factors perceived as life-threatening.

stressor (stres′or) Factor capable of stimulating a stress response.

stroke volume (strōk vol′ūm) Volume of blood the ventricle discharges with each heartbeat.

structural formula (struk′cher-al for′mu-lah) Representation of the way atoms bond to form a molecule, using symbols for each element and lines to indicate chemical bonds.

subarachnoid space (sub″ah-rak′noid spās) Space in the meninges between the arachnoid mater and the pia mater. It contains cerebrospinal fluid (CSF).

subcutaneous layer (sub″ku-ta′ne-us la′yer) Loose connective tissue layer beneath the skin; also called the *hypodermis.*

sublingual (sub-ling′gwal) Beneath the tongue.

submucosa (sub″mu-ko′sah) Layer of the alimentary canal beneath the mucosa.

substrate (sub′strāt) Target of enzyme action.

sucrase (su′krās) Digestive enzyme that catalyzes the breakdown of sucrose.

sugar (shoog′ar) Sweet-tasting carbohydrate.

sulcus (sul′kus) (plural, *sulci*) Shallow groove, such as that between gyri on the brain surface.

summation (sum-ma′shun) Increased force of contraction by a skeletal muscle fiber when a twitch occurs before the previous twitch relaxes.

superficial (soo′per-fish′al) Near the surface of the body or a specified body structure.

superior (soo-pe′re-or) Situated above something else; pertaining to the upper surface of a part.

supination (soo″pĭ-na′shun) Upward or forward rotation of the palm.

sural (su′ral) Pertaining to the calf of the leg.

surface tension (sur′fis ten′shun) Force due to the attraction of water molecules that makes it difficult to inflate the alveoli of the lungs.

surfactant (ser-fak′tant) Substance produced by the lungs that reduces the surface tension in alveoli.

sweat gland (swet gland) Exocrine gland in skin that secretes a mixture of water, salt, and wastes such as urea.

sympathetic division (sim″pah-thet′ik de-vijh′in) Part of the autonomic nervous system that arises from the thoracic and lumbar regions of the spinal cord; also called the *sympathetic nervous system.*

sympathetic nervous system (sim″pah-thet′ik ner′vus sis′tem) Part of the autonomic nervous system that arises from the thoracic and lumbar regions of the spinal cord.

synapse (sin′aps) Functional connection between the axon terminal of a neuron and the dendrite or cell body of another neuron or the membrane of another cell type.

synaptic cleft (sĭ-nap′tik kleft) A narrow extracellular space between two cells at a synapse.

synaptic knob (sĭ-nap′tik nob) Tiny enlargement at the end of an axon that secretes a neurotransmitter.

synergist (sin′er-jist) Muscle that assists the action of an agonist.

synovial joint (sĭ-no′ve-al joint) Freely movable joint.

synovial membrane (sĭ-no′ve-al mem′brān) Membrane that forms the inner lining of the capsule of a freely movable joint.

synthesis (sin′thĕ-sis) Building large molecules from smaller ones.

systemic circuit (sis-tem′ik ser′kit) Vessels that transport blood between the heart and all body tissues except the lungs.

systole (sis′to-le) Phase of the cardiac cycle when a heart chamber wall contracts.

systolic pressure (sis-tol′ik presh′ur) Highest arterial blood pressure during the cardiac cycle; occurs during systole.

T

T cell (tee sel) Type of white blood cell that interacts with antigen-bearing cells and particles, and secretes cytokines, contributing to the cellular immune response; also called the *T lymphocyte.*

target cell (tar′get sel) Cell on which a hormone exerts its effect, with specific receptors for that hormone.

tarsal (tahr'sul) Pertaining to the ankle or any of the individual ankle bones.

taste bud (tāst bud) Organ containing receptors associated with the sense of taste.

telophase (tel'o-fāz) Stage in mitosis when the two chromosome sets complete movement toward the centrioles and elongate and unwind. Nuclear envelopes appear around each set of chromosomes, and nucleoli form within the new nuclei.

tendon (ten'don) Cordlike or bandlike mass of dense connective tissue that connects a muscle to a bone.

testis (tes'tis) (plural, *testes*) Primary male reproductive organ; sperm cell-producing organ.

testosterone (tes-tos'tĕ-rōn) Male sex hormone secreted by the interstitial cells of the testes.

thalamus (thal'ah-mus) Mass of gray matter at the base of the cerebrum in the wall of the third ventricle. It is a processing and relay center for almost all sensory information.

thermoreceptor (ther″mo-re-sep'tor) Sensory receptor sensitive to temperature changes; warm and cold receptors.

thoracic cavity (tho-ras'ik kav'ĭ-te) Hollow space inside the chest containing the thoracic organs.

threshold potential (thresh'old po-ten'shul) Stimulation level that must be achieved to elicit an action potential.

thrombus (throm'bus) Blood clot that remains where it forms in a blood vessel.

thymosins (thi'mo-sinz) Group of peptides the thymus secretes that increases production of certain types of white blood cells.

thymus (thī'mus) Gland in the mediastinum, superior to the heart. It secretes hormones involved in development of T lymphocytes.

thyroid gland (thi'roid gland) Endocrine gland consisting of two connected lobes located in the anterior neck, just below the larynx and in front and to the side of the trachea. Hormones from the thyroid gland help control how many calories the body consumes and play a role in bone growth and maintenance of blood calcium and phosphate levels.

thyroid-stimulating hormone (thi-roid stim-ū-lay-ting hor-mone) **(TSH)** Hormone secreted from the anterior pituitary that stimulates secretion of thyroid hormones from the thyroid gland.

thyroxine (thi-rok'sin) One of the thyroid hormones; also called *T4*. It plays a role in controlling the basal metabolic rate.

tidal volume (tīd'al vol'um) Volume of air entering and leaving the lungs in a respiratory cycle.

tissue (tish'u) Assembled group of similar cells that performs a specialized function.

trabecula (trah-bek'u-lah) Branching bony plate that separates irregular spaces within spongy bone.

trachea (tra'ke-ah) Tubular organ that leads from the larynx to the bronchi.

tract (trakt) A bundle of axons within the central nervous system (CNS).

transcellular fluid (trans″sel'u-lar floo'id) Part of the extracellular fluid, including the fluid within special body cavities.

transcription (trans-krip'shun) Manufacturing a complementary RNA from DNA.

transfer RNA (trans'fer) **(tRNA)** RNA molecule that carries an amino acid to a ribosome in protein synthesis.

translation (trans-la'shun) Assembly of an amino acid chain according to the sequence of base triplets in an mRNA molecule.

transverse (tranz-vers') Plane that divides a structure into superior and inferior portions.

transverse tubule (tranz-vers' tu'būl) Any of several membranous channels that extend deep into the cell from a muscle fiber membrane.

tricuspid valve (tri-kus'pid valv) Heart valve between the right atrium and the right ventricle also called the *right atrioventricular (AV) valve.*

triglyceride (tri-glis'er-īd) Lipid composed of three fatty acids and a glycerol molecule; also called *fat.*

triiodothyronine (tri″i-o″do-thi'ro-nēn) One of the thyroid hormones; also called *T3.* It plays a role in controlling the basal metabolic rate.

trypsin (trip'sin) Enzyme in pancreatic juice that breaks down protein molecules.

TSH Thyroid stimulating hormone.

tubular reabsorption (tu'bu-lar re-absorp'shun) Movement of substances out of the renal tubule into peritubular capillaries. (Some substances are reabsorbed by the collecting duct and returned to the bloodstream.)

tubular secretion (tu'bu-lar se-kre'shun) Movement of substances out of the peritubular capillaries into the renal tubule for excretion in the urine.

twitch (twich) Single contraction of a muscle fiber followed by relaxation.

U

umbilical cord (um-bil'ĭ-kal kord) Cordlike structure that connects the fetus to the placenta.

umbilical region (um-bil'ĭ-kal re'jun) Central portion of the abdomen surrounding the navel.

unsaturated fatty acid (un-sat'u-rāt″ed fat'e as'id) Fatty acid molecule with one or more double bonds between carbon atoms.

urea (u-re'ah) Substance resulting from amino acid catabolism.

ureter (u-re'ter) Tube that carries urine from the kidney to the urinary bladder.

urethra (u-re'thrah) Tube leading from the urinary bladder to the outside of the body.

uric acid (u'rik as'id) Substance resulting from nucleic acid catabolism.

urinary system (u'rĭ-ner″e sis'tem) The organ system that includes the kidneys, ureters, urinary bladder, and urethra.

urine (u'rin) Wastes and excess water and electrolytes removed from the blood and excreted by the kidneys into the ureters, to the urinary bladder, and out of the body through the urethra.

uterine tube (u'ter-in tūb) Tube that extends from the uterus on each side toward an ovary; also called the *fallopian tube* or *oviduct.*

uterus (u'ter-us) Hollow, muscular organ in the female pelvis where a fetus develops.

utricle (u'trĭ-kl) An enlarged part of the membranous labyrinth of the inner ear. It contains some of the receptors involved in static equilibrium.

uvula (u'vu-lah) Fleshy part of the soft palate that hangs down above the root of the tongue.

V

vaccine (vak'sēn) Preparation that includes antigens that stimulate an immune response to prevent an infectious disease.

vagina (vah-ji'nah) Tubular organ that leads from the uterus to the vestibule of the female reproductive tract.

vasoconstriction (vas″o-kon-strik'shun) Decrease in the diameter of a blood vessel.

vasodilation (vas″o-di-la'shun) Increase in the diameter of a blood vessel.

vein (vān) Vessel that transports blood toward the heart.

vena cava (ve'nah kav'ah) One of two large veins (superior and inferior) that convey oxygen-poor blood to the right atrium of the heart.

ventral root (ven'tral root) Motor branch of a spinal nerve by which it connects with the spinal cord.

ventricle (ven'trĭ-kl) Cavity, such as a brain ventricle that contains cerebrospinal fluid, or a heart ventricle that contains blood.

venule (ven'ūl) Vessel that transports blood from capillaries to a vein.

vertebral (ver'te-bral) Pertaining to the bones of the spinal column.

vertebral canal (ver'te-bral kah-nal') Canal formed by hollow areas in the vertebrae that contains the spinal cord.

vesicle (ves'ĭ-kal) Membranous cytoplasmic sac formed by an infolding of the cell membrane or pinching off of membranes within the cell.

viscera (vis'er-ah) Organs in the thoracic and abdominopelvic cavities.

visceral (vis'er-al) Pertaining to the organs within a body cavity.

visceral pericardium (vis'er-al per"ĭ-kar'de-um) Membrane that covers the surface of the heart.

visceral peritoneum (vis'er-al per"ĭ-to-ne'-um) Membrane that covers organ surfaces in the abdominal cavity.

visceral pleura (vis'er-al ploo'rah) Serous membrane that covers the surface of each lung.

viscosity (vis-kos'ĭ-te) Tendency for a fluid to resist flowing due to the internal friction of its molecules; thickness.

vitamin (vi'tah-min) Organic nutrient other than a carbohydrate, lipid, or protein needed for normal metabolism that the body cannot synthesize in adequate amounts and must therefore be obtained in the diet.

vitreous humor (vit're-us hu'mor) Fluid between the lens and the retina of the eye.

vocal cords (vo'kal kordz) A pair of folds of tissue of the larynx that produce sound when air movement causes them to vibrate.

vulva (vul'vah) External female reproductive parts that surround the vaginal opening.

W

water balance (wot'er bal'ans) When the volume of water entering and produced by the body is equal to the volume leaving it.

Y

yellow marrow (yel'o mar'o) Fat storage tissue in the medullary cavities of certain bones.

Z

zygote (zi'gōt) Cell produced by the fusion of nuclei from an egg and a sperm.

zymogen granule (zi-mo'jen gran'ūl) Cellular structure that stores inactive forms of protein-splitting enzymes in a pancreatic cell.

APPLICATION INDEX

Boxed Readings

Career Corner

Clinical Applications

Tables

SUBJECT INDEX

Page numbers in **bold** indicate definitions. An italic *f* indicates figures, *t* indicates tables, and *p* indicates color plates.

A

B

D

E

F

G

H

I

M

N

O

P

Q

R

S

T

U

V

W

Y

Z